History of Agriculture in India

Publications in PHISPC—CONSSAVY Series

HISTORY OF SCIENCE, PHILOSOPHY AND CULTURE IN INDIAN CIVILIZATION

General Editor & Project Director: D.P. Chattopadhyaya

Conceptual Volumes

* Part 1	Science, Philosophy and Culture: Multi-disciplinary Explorations	D.P. Chattopadhyaya & Ravinder Kumar (eds.)
* Part 2	Science, Philosophy and Culture: Multi-disciplinary Explorations	D.P. Chattopadhyaya & Ravinder Kumar (eds.)

VOLUME I The Dawn and Development of Indian Civilization

* Part 1	The Dawn of Indian Civilization (up to c. 600 BC)	G.C. Pande (ed.)
* Part 2	Life, Thought and Culture in India (from c. 600 BC to c. AD 300)	G.C. Pande (ed.)
* Part 3	India's Interaction with Southeast Asia	G.C. Pande (ed.)
* Part 4	A Golden Chain of Civilizations: Indian, Iranian, Semitic and Hellenic (from c. 600 BC to c. AD 600)	G.C. Pande (ed.)
† Part 5	A Golden Chain of Civilizations: Indic, Iranic, Semitic and Hellenic (upto c. 600 BC)	G.C. Pande (ed.)
† Part 6	Purāṇas, History and Itihāsa	Vidya Niwas Misra & N.S.S. Raman (eds.)

VOLUME II Life, Thought and Culture in India (AD 300–1100)

* Part 1	Life, Thought and Culture in India (AD 300–1000)	K. Satchidananda Murty (ed.)
* Part 2	Advaita Vedānta	R. Balasubramanian (ed.)
* Part 3	Theistic Vedānta	R. Balasubramanian (ed.)
* Part 4	Origin and Development of the Vaiśeṣika System	Anantalal Thakur
† Part 5	A Social History of Early India	B.D. Chattopadhyaya (ed.)
* Part 6	Pūrvamīmāṁsā from an Interdisciplinary Point of View	K.T. Pandurangi (ed.)

VOLUME III Development of Philosophy, Science and Technology in India and Neighbouring Civilizations

* Part 1	History of Indian Science, Technology and Culture (AD 1000–1800)	A. Rahman (ed.)
* Part 2	India's Interaction with China, Central and West Asia	A. Rahman (ed.)
* Part 3	Development of Nyāya Philosophy and its Social Context	Sibajiban Bhattacharyya (ed.)
* Part 4	Philosophical Concepts Relevant to Science in Indian Tradition	Pranab Kumar Sen (ed.)
† Part 5	Philosophical Concepts Relevant to Science in Indian Tradition	Pranab Kumar Sen (ed.)
* Part 6	India and China: Twenty Centuries of Civilizational Interaction and Vibrations	Tan Chung & Geng Yinzeng (eds.)
† Part 7	The Trading World of the Indian Ocean, AD 1500-1800	Om Prakash (ed.)

VOLUME IV Fundamental Indian Ideas of Physics, Chemistry, Life Sciences and Medicine

* Part 1	Chemistry and Chemical Techniques in India	B.V. Subbarayappa (ed.)
* Part 2	Medicine and Life Sciences in India	B.V. Subbarayappa (ed.)
* Part 3	Indian Perspectives on the Physical World	B.V. Subbarayappa
† Part 4	The Tradition of Astronomy in India: Jyotiḥśāstra	B.V. Subbarayappa

VOLUME V Agriculture in India

* Part 1	A History of Agriculture in India	Lallanji Gopal & V.C. Srivastava (eds.)
† Part 2	A History of Agriculture in India	Lallanji Gopal & V.C. Srivastava (eds.)

VOLUME VI Culture, Language, Literature and Arts

† Part 1	Aesthetic Theories and Forms in Indian Tradition	S.S. Barlingay, D.P. Chattopadhyaya & Kapila Vatsyayan (eds.)
† Part 2	Architecture in India	M.A. Dhaky (ed.)
* Part 3	Indian Art: Forms, Concerns and Development in Historical Perspective	B.N. Goswamy (ed.)
† Part 4	Language, Grammar and Linguistics in Indian Tradition	V.N. Jha (ed.)
‡ Parts 5-10	PHISPC in Regional Languages and Literatures	

VOLUME VII The Rise of New Polity and Life in Villages and Towns

* Part 1	The State and Society in Medieval India	J.S. Grewal (ed.)
* Part 2	Religious Movements and Institutions in Medieval India	J.S. Grewal (ed.)
† Parts 3, 4 & 5	Religious Systems of India	S.R. Saha, N.S.S. Raman, M. Rafique and others (eds.)

VOLUME VIII Economic History of India

† Part 1	Economic History of India from Thirteenth to Seventeenth Century	Irfan Habib (ed.)
† Part 2	Economic History of India from Eighteenth to Twentieth Century	B.B. Chaudhuri
* Part 3	Economic History of India from Eighteenth to Twentieth Century	B.B. Chaudhuri (ed.)

VOLUME IX Colonial Period

* Part 1	Medicine in India: Modern Period	O.P. Jaggi
† Part 2	Women in India: Ancient and Medieval Period	Bhuvan Chandel (ed.)
* Part 3	Women of India: Colonial and Post-colonial Periods	Bharati Ray (ed.)

VOLUME X Towards Independence

* Part 1	Development of Indian Philosophy from Eighteenth Century Onwards	Daya Krishna
† Part 2	Colonial Development, Education and Social Awareness up to 2000	S. Gopal, Ravinder Kumar & S. Bhattacharya (eds.)
* Part 3	Historical Perspectives of Warfare in India: Some *Morale* and *Material Determinants*	S.N. Prasad (ed.)

* Already Published
† In the Process of Publication
‡ Under Plan

History of Science, Philosophy and Culture in Indian Civilization

General Editor: *D.P. Chattopadhyaya*

Volume V Part I

History of Agriculture in India (up to c.1200 AD)

LALLANJI GOPAL *AND* V.C. SRIVASTAVA

An imprint of **Concept Publishing Company**
Project of History of Indian Science, Philosophy and Culture

CENTRE FOR STUDIES IN CIVILIZATIONS

Cataloging in Publication Data–DK
Courtesy: D.K. Agencies (P) Ltd. <docinfo@dkagencies.com>

History of agriculture in India, up to c. 1200 AD / edited by Lallanji Gopal and V.C. Srivastava.
p. cm. – (History of science, philosophy, and culture in Indian civilization ; v. 5 pt. 1)
Contributed articles.
Includes bibliographical references.
Includes index.
ISBN 13: 9788180695216
ISBN 10: 8180695212

1. Agriculture, Ancient–India. 2. Agriculture–India-History. 3. India–Civilization–To 1200. I. Gopal, Lallanji, 1934-1999. II. Srivastava, V. C. (Vinod Chandra), 1938- III. Centre for Studies in Civilizations (Delhi, India) IV. Series: History of science, philosophy, and culture in Indian civilization ; v. 5 pt. 1.

DDC 630.934 22

Publication of this Volume and much of the research it represents have been made possible by continuing grants with full financial assistance from the Department of Secondary and Higher Education, Ministry of Human Resource Development, Government of India, which has supported multidisciplinary exploration of the Project of History of Indian Science, Philosophy and Culture.

Jointly published by Professor Bhuvan Chandel, Member Secretary, DD-24, Kalkaji, New Delhi 110019 and Concept Publishing Company, A/15-16, Commercial Block, Mohan Garden, New Delhi-110059 (INDIA), *Phones*: 25351460, 25351794, *Fax*: 091-11-25357103, *Email*: publishing@conceptpub.com for the Project of History of Indian Science, Philosophy and Culture

ISBN -13: 978-81-8069-521-6
ISBN -10: 81-8069-521-2

First Published 2008

Printed at
Concept Publishing Company, New Delhi

Typeset in Nalandabas, 11.5/14 pts.
Typeset by Mindways Design

Centre for Studies in Civilizations

Governing Board

Contents

SECTION ONE
ORIGIN AND BEGINNINGS OF AGRICULTURE

SECTION TWO
HARAPPAN AND VEDIC FACETS OF AGRICULTURE

SECTION THREE
AGRICULTURE FROM THE CHALCOLITHIC PHASE TO C. 600 BC

SECTION FOUR
AGRICULTURE IN EARLY ANCIENT INDIA (FROM C. 600 BC TO C. 600 AD)

SECTION FIVE
AGRICULTURE IN EARLY MEDIEVAL INDIA (FROM C. 600 AD TO C. 1200 AD)

General Introduction

I

It is understandable that man, shaped by Nature, would like to know Nature. The human ways of knowing Nature are evidently diverse, theoretical and practical, scientific and technological, artistic and spiritual. This diversity has, on scrutiny, been found to be neither exhaustive nor exclusive. The complexity of physical nature, life-world and, particularly, human mind is so enormous that it is futile to follow a single method for comprehending all the aspects of the world in which we are situated.

One need not feel bewildered by the variety and complexity of the worldly phenomena. After all, both from traditional wisdom and our daily experience, we know that our own nature is not quite alien to the structure of the world. Positively speaking, the elements and forces that are out there in the world are also present in our body-mind complex, enabling us to adjust ourselves to our environment. Not only the natural conditions but also the social conditions of life have instructive similarities between them. This is not to underrate in anyway the difference between the human ways of life all over the world. It is partly due to the variation in climatic conditions and partly due to the distinctness of production-related tradition, history and culture.

Three broad approaches are discernible in the works on historiography of civilization, comprising science and technology, art and architecture, social sciences and institutions. Firstly, some writers are primarily interested in discovering the general laws which govern all civilizations spread over different continents. They tend to underplay what they call the noisy local events of the external world and peculiarities of different languages, literatures and histories. Their accent is on the unity of Nature, the unity of science and the unity of mankind. The second group of writers, unlike the generalist or transcendentalist ones, attach primary importance to the distinctiveness of every culture. To these writers human freedom and creativity are extremely important and basic in character. Social institutions and the cultural articulations of human consciousness, they argue, are bound to be expressive of the concerned people's consciousness. By implication they tend to reject concepts like archetypal consciousness, universal mind and providential history. There is a third group of writers who offer a composite picture of civilizations, drawing elements both from their local as well as common characteristics. Every culture has its local roots and peculiarities. At the same time, it is pointed out that due to demographic migration and immigration over the centuries an element of compositeness emerges almost in every culture. When, due to a natural calamity or political exigencies people move from one part of the world to another, they carry with them, among other things, their language, cultural inheritance and their ways of living.

In the light of the above facts, it is not at all surprising that comparative anthropologists and philologists are intrigued by the striking similarity between different language families and the rites, rituals and myths of different peoples. Speculative philosophers of history, heavily relying on the findings of epigraphy, ethnography, archaeology and theology, try to show in very general terms that the particulars and universals of culture are 'essentially' or 'secretly' interrelated. The spiritual aspects of culture like dance and music, beliefs pertaining to life, death and duties, on analysis, are found to be mediated by the material forms of life like weather forecasting, food production, urbanization and invention of script. The transition from the oral culture to the written one was made possible because of the mastery of symbols and rules of measurement. Speech precedes grammar, poetry, prosody. All these show how the 'matters' and 'forms' of life are so subtly interwoven.

II

The PHISPC publications on History of Science, Philosophy and Culture in Indian Civilization, in spite of their unitary look, do recognize the differences between the areas of material civilization and those of ideational culture. It is not a work of a single author. Nor is it being executed by a group of thinkers and writers who are methodologically uniform or ideologically identical in their commitments. In conceiving the Project we have interacted with, and been influenced by, the writings and views of many Indian and non-Indian thinkers.

The attempted unity of this Project lies in its aim and inspiration. We have in India many scholarly works written by Indians on different aspects of our civilization and culture. Right from the pre-Christian era to our own time, India has drawn the attention of various countries of Asia, Europe and Africa. Some of these writings are objective and informative and many others are based on insufficient information and hearsay, and therefore not quite reliable, but they have their own value. Quality and view-points keep on changing not only because of the adequacy and inadequacy of evidence but also, and perhaps more so, because of the bias and prejudice, religious and political conviction, of the writers.

Besides, it is to be remembered that history, like Nature, is not an open book to be read alike by all. The past is mainly enclosed and only partially disclosed. History is, therefore, partly objective or 'real' and largely a matter of construction. This is one of the reasons why some historians themselves think that it is a form of literature or art. However, it does not mean that historical construction is 'anarchic' and arbitrary. Certainly, imagination plays an important role in it.

But its character is basically dependent upon the *questions* which the historian raises and wants to understand or answer in terms of the ideas and actions of human beings in the past ages. In a way, history, somewhat like the natural sciences, is engaged in answering questions and in exploring relationships of cause and effect between events and developments across time. While in the natural sciences, the scientist poses questions about nature in the form of hypotheses, expecting to elicit authoritative answers to such questions, the historian studies the past, partly for the sake of understanding it for its own sake and partly also for the light which the past throws upon the present, and the possibilities which it opens up for moulding the future. But the difference between the two approaches must not be lost sight of. The scientist is primarily interested in discovering laws and framing theories, in terms of which, different events and processes can be connected and anticipated. His interest in the conditions or circumstances attending the concerned events is secondary. Therefore, scientific laws turn out to be basically abstract and easily expressible in terms of mathematical language. In contrast, the historian's main interest centres round the *specific* events, human ideas and actions, not *general* laws. So, the historian, unlike the scientist, is obliged to pay primary attention to the circumstances of the events he wants to study. Consequently, history, like most other humanistic disciplines, is concrete and particularist. This is not to deny the obvious truth that historical events and processes consisting of human ideas and actions show some trend or other and weave some pattern or other. If these trends and patterns were not there at all in history, the study of history as a branch of knowledge would not have been profitable or instructive. But one must recognize that historical trends and patterns, unlike scientific laws and theories, are not general or purported to be universal in their scope.

III

The aim of this Project is to discover the main aspects of Indian culture and present them in an interrelated way. Since our culture has influenced, and has been influenced by, the neighbouring cultures of West Asia, Central Asia, East Asia and South-East Asia, attempts have been made here to trace and study these influences in their mutuality. It is well known that during the last three centuries, European presence in India, both political and cultural, has been very widespread. In many volumes of the Project considerable attention has been paid to Europe and through Europe to other parts of the world. For the purpose of a comprehensive cultural study of India, the existing political boundaries of the South Asia of today are more of a hindrance than help. Cultures, like languages, often transcend the bounds of changing political territories.

If the inconstant political geography is not a reliable help to the understanding of the layered structure and spread of culture, a somewhat comparable problem is encountered in the area of historical periodization. Periodization or segmenting time is a very tricky affair. When exactly one period ends and another begins is not precisely ascertainable. The periods of history designated as ancient, medieval and modern are purely conventional and merely heuristic in character. The varying scopes of history, local, national and continental or universal, somewhat like the periods of history, are unavoidably fuzzy and shifting. Amidst all these difficulties, the volume-wise details have been planned and worked out by the editors in consultation with the Project Director and the General Editor. I believe that the editors of different volumes have also profited from the reactions and suggestions of the contributors of individual chapters in planning the volumes.

Another aspect of Indian history which the volume-editors and contributors of the Project have carefully dealt with is the distinction and relation between civilization and culture. The material conditions which substantially shaped Indian civilization have been discussed in detail. From agriculture and industry to metallurgy and technology, from physics and chemical practices to the life sciences and different systems of medicines—all the branches of knowledge and skill which directly affect human life—form the heart of this Project. Since the periods covered by the PHISPC are extensive—prehistory, proto-history, early history, medieval history and modern history of India—we do not claim to have gone into all the relevant material conditions of human life. We had to be selective. Therefore, one should not be surprised if one finds that only some material aspects of Indian civilization have received our pointed attention, while the rest have been dealt with in principle or only alluded to.

One of the main aims of the Project has been to spell out the first principles of the philosophy of different schools, both pro-Vedic and anti-Vedic. The basic ideas of Buddhism, Jainism and Islam have been given their due importance. The special position accorded to philosophy is to be understood partly in terms of its proclaimed unifying character and partly to be explained in terms of the fact that different philosophical systems represent alternative world-views, cultural perspectives, their conflict and mutual assimilation.

Most of the volume-editors and at their instance the concerned contributors have followed a middle path between the extremes of narrativism and theoreticism. The underlying idea has been this: if, in the process of working out a comprehensive Project like this, every contributor attempts to narrate all those interesting things that he has in the back of his mind, the enterprise is likely to prove unmanageable. If, on the other hand, particular details are consciously forced into a fixed mould or pre-supposed theoretical structure, the details lose their particularity and interesting character. Therefore, depending on the nature of the problem of discourse, most of the writers have tried to reconcile in their presentation, the specificity of narrativism and the generality of theoretical orientation. This is a conscious editorial decision. Because, in the absence of a theory, however inarticulate it may be, the factual details tend to fall apart. Spiritual network or theoretical orientation makes historical details not only meaningful but also interesting and enjoyable.

Another editorial decision which deserves spelling out is the necessity or avoidability of duplication of the same theme in different volumes or even in the same volume. Certainly, this Project is not an assortment of several volumes. Nor is any volume intended to be a miscellany. This Project has been designed with a definite end in view and has a structure of its own. The character of the structure has admittedly been influenced by the variety of the themes accommodated within it. Again it must be understood that the complexity of structure is rooted in the aimed integrality of the Project itself.

IV

Long and in-depth editorial discussion has led us to several unanimous conclusions. Firstly, our Project is going to be *unique,* unrivalled and discursive in its attempt to integrate different forms of science, technology, philosophy and culture. Its comprehensive scope, continuous character and accent on culture distinguish it from the works of such

Indian authors as P.C. Ray, B.N. Seal, Binoy Kumar Sarkar and S.N. Sen and also from such Euro-American writers as Lynn Thorndike, George Sarton and Joseph Needham. Indeed, it would be no exaggeration to suggest that it is for the first time that an endeavour of so comprehensive a character, in its exploration of the social, philosophical and cultural characteristics of a distinctive world civilization—that of India—has been attempted in the domain of scholarship.

Secondly, we try to show the linkages between different branches of learning as different modes of experience in an *organic* manner and without resorting to a kind of reductionism, materialistic or spiritualistic. The internal dialectics of organicism without reductionism allows fuzziness, discontinuity and discreteness within limits.

Thirdly, positively speaking, different modes of human experience—scientific, artistic, etc., have their own individuality, not necessarily autonomy. Since all these modes are modification and articulation of *human* experience, these are bound to have between them some finely graded commonness. At the same time, it has been recognized that reflection on different areas of experience and investigation brings to light new insights and findings. Growth of knowledge requires humans, in general, and scholars, in particular, to identify the distinctness of different branches of learning.

Fourthly, to follow simultaneously the twin principles of: (a) individuality of human experience as a whole, and (b) individuality of diverse disciplines, is not at all an easy task. Overlapping of themes and duplication of the terms of discourse become unavoidable at times. For example, in the context of *Dharmaśāstra*, the writer is bound to discuss the concept of value. The same concept also figures in economic discourse and also occurs in a discussion on fine arts. The conscious editorial decision has been that, while duplication should be kept to its minimum, for the sake of intended clarity of the themes under discussion, their reiteration must not be avoided at high intellectual cost.

Fifthly, the scholars working on the Project are drawn from widely different disciplines. They have brought to our notice an important fact that has clear relevance to our work. Many of our contemporary disciplines like economics and sociology did not exist, at least not in their present form, just two centuries or so ago. For example, before the middle of nineteenth century, sociology as a distinct branch of knowledge was unknown. The term is said to have been coined first by the French philosopher Auguste Comte in 1838. Obviously, this does not mean that the issues discussed in sociology were not there. Similarly, Adam Smith's (1723–90) famous work *The Wealth of Nations* is often referred to as the first authoritative statement of the principles of (what we now call) economics. Interestingly enough, the author was equally interested in ethics and jurisprudence. It is clear from history that over time the nature and scope of different disciplines undergo change, at times very radically. For example, in India the term '*Arthaśāstra*' does not mean the science of economics as understood today. Besides the principles of economics, the *Arthaśāstra* of ancient India discusses at length those of governance, diplomacy and military science.

Sixthly, this brings us to the next editorial policy followed in the Project. We have tried to remain very conscious of what may be called indeterminacy or inexactness of translation. When a word or expression of one language is translated into another, some loss of meaning or exactitude seems to be unavoidable. This is true not only in the bilingual relations like Sanskrit-English and Sanskrit-Arabic, but also in those of Hindi-Tamil and Hindi-Bengali. In recognition of the importance of language-bound and context-relative character of meaning we have solicited from many learned scholars, contributions, written in vernacular languages. In order to minimize the miseffect of semantic inexactitude we have solicited translational help from those bilingual scholars who know both English and the concerned vernacular language, Hindi, Tamil, Telugu, Bengali or Marathi.

Seventhly and finally, perhaps the place of technology as a branch of knowledge in the composite universe of science and art merits some elucidation. Technology has been conceived in very many ways, e.g. as autonomous, as 'standing reserve', as liberating or enlargemental, and alienative or estrangemental force. The studies undertaken by the Project show that, in spite of its much emphasized mechanical and alienative characteristics, technology embodies a very useful mode of knowledge that is peculiar to man. The Greek root words of technology are *techne* (art) and *logos* (science). This is the basic justification of recognizing technology as closely related to both epistemology, the discipline of valid knowledge, and axiology, the discipline of freedom and values. It is in this context that we are reminded of the definition of man as *homo technikos.* In Sanskrit, the word closest to *techne* is *kalā* which means any practical art, any mechanical or fine art. In the Indian tradition, in *Śaivatantra,* for example, among the arts (*kalā*) are counted dance, drama, music, architecture, metallurgy, knowledge of dictionary, encyclopaedia and prosody. The closeness of the relation between arts and sciences, technology and other forms of knowledge are evident from these examples and was known to the ancient people. The human quest for knowledge involves the use of both head and hand. Without mind, the body is a corpse and the disembodied mind is a bare abstraction. Even for our appreciation of what is beautiful and the creation of what is valuable, we are required to exercise both our intellectual competence and physical capacity. In a manner of speaking, one might rightly affirm that our psychosomatic structure is a functional connector between what we are and what we could be, between the physical and the beyond. To suppose that there is a clear-cut distinction between the physical world and the psychosomatic one amounts to denial of the possible emergence of higher logico-mathematical, musical and other capacities. The very availability of aesthetic experience and creation proves that the supposed distinction is somehow overcome by what may be called the bodily self or embodied mind.

V

The ways of classification of arts and sciences are neither universal nor permanent. In the Indian tradition, in the *Ṛgveda,* for example, *vidyās* (or sciences) are said to be four in number: (i) *Trayī,* the triple Veda; (ii) *Ānvīkṣikī,* logic and metaphysics; (iii) *Daṇḍa-nīti,* science of governance; (iv) *Vārtta,* practical arts such as agriculture, commerce, medicine, etc. Manu speaks of a fifth *vidyā,* viz., *Ātma-vidyā,* knowledge of self or of spiritual truth. According to many others, *vidyā* has fourteen divisions, viz., the four Vedas, the six Vedāngas, the Purāṇas, the Mīmāṁsā, Nyāya, and Dharma or law. At times, the four *Upavedas* are also recognized by some as *vidyā. Kalās* are said to be 33 or even 64.

In the classical tradition of India, the word *śāstra* has at times been used as synonym of *vidyā. Vidyā* denotes instrument of teaching, manual or compendium of rules, religious or scientific treatise. The word *śāstra* is usually found after the word referring to the subject of the book, e.g. *Dharma-śāstra, Artha-śāstra, Alaṁkāra-śāstra* and *Mokṣa-śāstra.* Two other words which have been frequently used to denote different branches of knowledge are *jñāna* and *vijñāna.* While *jñāna* means knowing, knowledge, especially the higher form of it, *vijñāna* stands for the act of distinguishing or discerning, understanding, comprehending and recognizing. It means worldly or profane knowledge as distinguished from *jñāna,* knowledge of the divine.

It must be said here that the division of knowledge is partly conventional and partly administrative or practical. It keeps on changing from culture to culture, from age to age. It is difficult to claim that the distinction between *jñāna* and *vijñāna* or that between science and art is universal. It is true that even before the advent of modern age, both in the East and the West, two basic aspects of sciences started gaining recognition. One is the *specialized character* of what we call scientific knowledge. The other is the concept of *trained skill* which was brought close to scientific knowledge. In the medieval Europe, the expression 'the seven liberal sciences' has so often been used simultaneously with 'the seven liberal arts', meaning thereby, the group of studies by the *Trivium* (Grammar, Logic and Rhetoric) and *Quadrivium* (Arithmetic, Music, Geometry and Astronomy).

It may be observed here, as has already been alluded to earlier, that the division between different branches of knowledge, between theory and practice, was not pushed to an extreme extent in the early ages. *Praxis,* for example, was recognized as the prime *techne.* The Greek word, *technologia* stood for systematic treatment, for example, of Grammar. *Praxis* is not the mere application of *theoria,* unified vision or integral outlook, but it also stands for the active impetus and base of knowledge. In India, one often uses the terms *Prayukti-vidyā* and *Prodyogiki-vidyā* to emphasize the practical or applicative character of knowledge. *Prayoga* or application is both the test and base of knowledge. Doing is the best way of knowing and learning.

That one and the same word may mean different 'things' or concepts in different cultures and thus create confusion has already been stated before. Two such words which in the context of this Project under discussion deserve special mention are *dharma* and *itihāsa.* Ordinarily, *dharma* in Sanskrit-rooted languages is taken to be conceptual equivalent of the English word *religion.* But, while the meaning of religion is primarily theological, that of *dharma* seems to be manifold. Literally, *dharma* stands for that which is established or that which holds people steadfastly together. Its other meanings are law, rule, usage, practice, custom, ordinance and statute. Spiritual or moral merit, virtue, righteousness and good works are also denoted by it. Further, *dharma* stands for natural qualities like burning (of fire), liquidity (of water) and fragility (of glass). Thus one finds that meanings of *dharma* are of many types—legal, social, moral, religious or spiritual, and even ontological or physical. All these meanings of *dharma* have received due attention of the writers in the relevant contexts of different volumes.

This Project, being primarily historical as it is, has naturally paid serious attention to the different concepts of history—epic-mythic, artistic-narrative, scientific-causal, theoretical and ideological. Perhaps the point that must be mentioned first about history is that it is not a correct translation of the Sanskrit word *itihāsa.* Etymologically, it means what really happened (*iti-ha-āsa*). But, as we know, in the Indian tradition *purāṇa* (legend, myth, tale, etc.), *gāthā* (ballad), *itivṛtta* (description of past occurrence, event, etc.), *ākhyāyikā* (short narrative) and *vaṁśa-carita* (genealogy) have been consciously accorded a very important place. Things started changing with the passage of time and particularly after the effective presence of Islamic culture in India. Islamic historians, because of their own cultural moorings and the influence of the Semitic and Graeco-Roman cultures on them, were more particular about their facts, figures and dates than their Indian predecessors. Their aim to bring history close to statecraft, social conditions and the lives and teachings of the religious leaders imparted a mundane character to this branch of learning. The Europeans whose political appearance on the Indian scene became quite perceptible only towards the end of the eighteenth century brought in with them their own view of historiography in their cultural baggage. The impact of the Newtonian Revolution in the field of history was very faithfully worked out, among others, by David Hume (1711–76) in *History of Great Britain from the Invasion of Julius Caesar to the Revolution of 1688* (6 Vol. 1754–62) and Edward Gibbon (1737–94) in *The History of the Decline and Fall of the Roman Empire* (6 Vol., 1776–88). Their emphasis on the principles of causality, datability and continuity/linearity of historical events introduced the spirit of scientific revolution in European historiography. The introduction of English education in India and the exposure of the elites of the country to it largely account for the decline of the traditional concept of *itihāsa* and the rise of the post-Newtonian scientific historiography. Gradually, Indian writers of our own history and cultural heritage started using more and more European concepts and categories. This is not to suggest that the impact of the European historiography on Indian historians was entirely negative. On the contrary, it imparted an analytical and critical temper which motivated many Indian historians of the nineteenth century to try to discover and represent our heritage in a new way.

VI

The principles which have been followed for organizing the subjects of different volumes under this Project may be stated in this way. We have kept in view the main structures which are discernible in the decomposible composition of the world. The first structure may be described as physical and chemical. The second structure consists, broadly speaking, of biology, psychology and epistemology. The highest and the most abstract structure nests many substructures within it, for example, logic, mathematics and musical notes. It is well known that the substructures within each structure are interactive, i.e. not isolable. The more important point to be noted in this connection is that the basic three structures of the world, viz., (a) physico-chemical, (b) bio-psychological, and (c) logico-mathematical are all simultaneously open to upward and downward causation. In other words, while the physico-chemical structure can causally influence the bio-psychological one and the latter can causally influence the most abstract logico-mathematical, the reverse process of causation is also operative in the world. In spite of its relative abstractness and durability, the logico-mathematical world has its downward causal impact on our bio-psychological and epistemological processes and products. And the latter can also bring about change in the structures of the physical world and its chemical composition. Applied physics and bio-technology make the last point abundantly clear.

Many philosophers, life-scientists, and social scientists highlight the point that nature loves hierarchies. Herbert Simon, the economist and the management scientist, speaks of four steps of partial ordering of our world, namely: (i) chemical substances, (ii) living organisms, tissues and organs, (iii) genes, chromosomes and DNA, and (iv) human beings, the social organizations, programmes and information process. All these views are in accord with the anti-reductionist character of our Project. Many biologists defend this approach by pointing out that certain characteristics of biological phenomena and process like unpredictability, randomness, uniqueness, magnitude of stochastic perturbations, complexity and emergence cannot be reduced without recourse to physical laws.

The main subjects dealt with in different volumes of the Project are connected not only conceptually and synchronically but also historically or diachronically. For pressing practical reasons, however, we did not aim at presenting the pre-historical, proto-historical and historical past of India in a continuous or chronological manner. Besides, it has been shown in the presentation of the PHISPC that the process of history is non-linear. And this process is to be understood in terms of human praxis and an absence of general laws in history. Another point which deserves special mention is that the editorial advisors have taken a conscious decision not to make this historical Project primarily political. We felt that this area of history has always been receiving extensive attention. Therefore, the customary discussion of dynastic rule and succession will not be found in a prominent way in this series. Instead, as said before, most of the available space has been given to social, scientific, philosophical and other cultural aspects of Indian civilization.

Having stated this, it must be admitted that our departure from conventional style of writing Indian history is not total. We have followed an inarticulate framework of time in organizing and presenting the results of our studies. The first volume, together with its parts, deals with the pre-historical period to AD 300. The next two volumes, together with their parts, deal with, among other things, the development of social and political institutions and philosophical and scientific ideas from AD 300 to the beginning of the eleventh century AD. The next period with which this Project is concerned spans from the twelfth century to the early part of the eighteenth century. The last three centuries constitute the fourth period covered by this Project. But, as said before, the definition of all these periods by their very nature are inexact and merely indicative.

Two other points must be mentioned before I conclude this General Introduction to the series. The history of some of the subjects like religion, language and literature, philosophy, science and technology cannot for obvious reason be squeezed within the cramped space of the periodic moulds. Attempts to do so result in thematic distortion. Therefore, the reader will often see the overflow of some ideas from one period to another. I have already drawn attention to this tricky and fuzzy as also the misleading aspects of the periodization of history, if pressed beyond a point.

Secondly, strictly speaking, history knows no end. Every age rewrites its history. Every generation, beset with new issues, problems and questions, looks back to its history and reinterprets and renews its past. This shows why history is not only contemporaneous but also futural. Human life actually knows no separative wall between its past, present and future. Its cognitive enterprises, moral endeavours and practical activities are informed of the past, oriented by the present and addressed to the future. This process persists, consciously or unconsciously, wittingly or unwittingly. In the narrative of this Project, we have tried to represent this complex and fascinating story of Indian civilization.

Centre for Studies in Civilizations
New Delhi

D.P. Chattopadhyaya
General Editor

Editors

D.P. CHATTOPADHYAYA has studied, researched on law, philosophy and history, and has taught at various universities in India, Asia, Europe and the USA from 1954 to 1994. Founder-Chairman of Indian Council of Philosophical Research (1981-90) and President-cum-Chairman of Indian Institute of Advanced Study, Shimla (1984-91), Chattopadhyaya is currently the Project Director of the multidisciplinary 96-volume PHISPC and Chairman of CSC. Among his 35 books, of which he has authored 18 and edited 17, are *Individuals and Societies; Individuals and Worlds; Sri Aurobindo and Karl Marx; Anthropology and Historiography of Science; Induction, Probability and Skepticism* and *Cultures and Ideologies*. He has also held high public offices, namely, of Union cabinet minister and state governor. He is a Life Member of Russian Academy of Sciences and a Member of International Institute of Philosophy, Paris.

LALLANJI GOPAL was Head, Department of Indian Philosophy and Religion, B.H.U., 1970–71. Head, Department of Philosophy, B.H.U., 1971–73, Director, Centre of Advanced Study in Philosophy, B.H.U., 1971–73. Professor of A.I.H.C. & Archaeology, B.H.U., 1973–94. Head Department of A.I.H.C. & Archaeology, B.H.U., 1966–67, 1973–77, 1979–81, 1983–85. Dean, Faculty of Arts, Banaras Hindu University, 1975–77, 1983–85, 1992–94. Rector and Acting Vice-Chancellor, B.H.U., December 1993–February, 1994. He has published very extensively on Indian Culture and Ancient Indian History.

V.C. SRIVASTAVA, formerly Director of Indian Institute of Advanced Study, Shimla and Manindra Chandra Nandi Professor and Head of the Department of Ancient Indian History, Culture and Archaeology, Banaras Hindu University, Varanasi, was also a Deputy Coordinator of its UGC Special Assistance Programme. His publications include *Hellenistic Sabhyata; Sun-Worship in Ancient India; The Samba Purana* (Tr. In Hindi); *Prehistoric Afghanistan: A Source Book; Protohistoric Afghanistan: A Source Book; Historical Probings in Afghanistan; Revision in Puranic Sun-cult*. He is the editor of prestigious Indological journal viz. Bharati. At present he is UGC Professor Emeritus at University of Allahabad and also a member of Sashtri Indo-Canandian Institute. He was a member of General Assembly, Indian Council of Cultural Relations, New Delhi. He has been Visiting Professor at Kabul University and member of UNSCO sponsored Centre of Kusan Studies at Kabul and a Visiting Fellow at several European, Asian Universities.

Contributors

UDAI PRAKASH ARORA is currently Professor and Head of the Department of Ancient History and Culture, M.J.P. Rohilkhand University, Bareilly. He was Post-Doctoral Fellow in Athens and Thessaloniki Universities. His published books include *Motifs in Indian Mythology: Their Greek and Other Parallels* (1981), *Graeco-Indica* (1991) and *Greeks on India* (1996). He is the editor of Graeco-Roman Journal *Yavanka.* He wrote several articles and better cultural relationship between India and Greece earned him the Golden Cross Order of Honour from the President of Greece.

SIBESH CHANDRA BHATTACHARYA is the former Professor and Head of the Department of Ancient History, Culture and Archaeology, University of Allahabad. He had been Coordinator of its UGC Special Assistance Programme. He was British Council Scholar in England (1972–73). He had been Fellow of Indian Institute of Advanced Study, Shimla (1999–2001), and National Fellow of the same Institute from 2001 to 2004. He has contributed several articles on social history of ancient India. His published works include *Some Aspects of Indian Society from 2nd Century BC to 4th Century AD.*

RITA CHAUDHURI is Reader, Department of Ancient Indian History and Culture, University of Calcutta. She has contributed a number of research articles on social and cultural history of India in reputed journals.

H.N. DUBEY is at present Professor in the Department of Ancient History, Culture and Archaeology, University of Allahabad. His field of specialization is cultural history of ancient India specially Puranic research. His published books include *Purāṇa Vimarș, Purāṇa Samīkṣa and Vāman Purāṇa*, 1992, *Bhartiya Sanskriti Eavam Kala*, 1999 (in Hindi). He has contributed several research papers on Purāṇas and cultural history of ancient India in journals.

V.K. DUBEY is currently working in National Council of Development Communication. He was the Professor & Head Department of Extension Education, Institute of Agriculture Sciences, Banaras Hindu University, Varanasi. His field of specialization is Ancient & Traditional wisdom of Indian Rural Society (Agriculture & Health). His major published work is *Prachin Bharat Me Krishi Gyan.* Also he has published books and articles on various aspects of agriculture in India.

D.K. GANGULY is former Professor and Head of the Department of Ancient History, Culture and Archaeology, Vishwa Bharati University, Santiniketan. His specialization is political administration in ancient India. He has contributed several research papers on social and economic history of ancient India.

M.C. Gupta is an Archaeologist from the Department of Ancient History, Culture and Archaeology, University of Allahabad. He has published four books and fifty research papers on various facets of Indian Archaeology. His major published work is *The Study of Beads of Kausambi*, 2002.

Rasesh Jamindar is former Professor and Head of the Department of History and Culture, Gujarat Vidyapith, Ahmedabad. He had been Senior Fellow of Indian Council of Historical Research, New Delhi. His field of specialization is cultural history of Gujarat. He is a prolific writer in Gujarati, Hindi, and English. He has published many books and research papers in reputed journals. He was elected as the President of Indian History and Culture Society, session 1997.

N.M. Kansara is a scholar of ancient Indian history and culture. His special field of study is cultural history of India specially animal husbandry in the Vedas. His major work is *Agriculture and Animal Husbandry in the Vedas*, 1995. He specializes in Vedānta, Saṁkhya-Yoga, Swaminarayana, Poetics/Sanskrit literature.

Anup Ranjan Mishra is an archaeologist and is at present working as Technical/ Archaeological Officer in the Department of Ancient History and Culture, M.J.P. Rohilkhand University Bareilly. His (unpublished) Ph.D. thesis from University of Pune, is entitled *Chalcolithic Ceramics of Balthal, District Udaipur*, Rajasthan. He has contributed many articles on Chalcolithic culture and farming in reputed journals.

V.D. Misra is a former Professor and Head of the Department of Ancient History, Culture and Archaeology, University of Allahabad and Coordinator of its UGC Special Assistance Programme. His major work is *Some Aspects of Indian Archaeology*, 1975. He, along with his colleagues, has edited several archaeological reports of the excavations and seminars of the Department. Along with Professor G.R. Sharma, he published a significant monograph entitled *Beginnings of Agriculture*, (1980). At present he is engaged in ICHR's project on Excavations at Jhusi (Allahabad). He is currently a member of Indian Council of Historical Research, New Delhi.

A.V. Narasimhamurthy is former Professor and Head of the Department of Ancient History and Archaeology, Mysore University. He had been UGC's Emeritus Fellow. His specialization is cultural history of ancient India, Numismatics and Epigraphy. He has contributed several research papers on ancient Indian religion and culture in reputed journals.

K.P. Nautiyal was Professor and later Vice-Chancellor H.N. Bahuguna Garhwal University Srinagar. Also Vice-Chancellor Dr. R.M. Lohia Avadh University, Faizabad. Visiting Fellow University of California, Berkeley. Fellow IIAS Shimla and Emeritus Fellow UGC. His specialization is Archaeology (Protohistory). Has published four books and a hundred articles.

VINOD NAUTIYAL specializes in Analytical Archaeology at H.N. Bahuguna Garhwal University. He has published 35 papers.

PUSPA NIYOGI had been Professor of Ancient Indian History and Culture, Calcutta University. Her field of specialization is economic history of India. Her published books include contributions to the *Economic History of Northern India from 10th to 12th Century* AD. She has contributed several significant research articles on socio-economic history of ancient India in national and international journals.

A.P. OJHA is a Reader in the Centre of Advanced Study, Department of Ancient History, Culture and Archaeology, University of Allahabad. His field of specialization is Social History of India. His published work is *Prachin Bharat Men Samajika Starikarana,* (in Hindi) (1992). He has published more than two dozen research papers in reputed journals of India.

J.N. PAL is at present Professor in the Department of Ancient History, Culture and Archaeology, University of Allahabad. His field of specialization is Archaeology. His major published work is *Ceramic Industries in Northern India,* 1986. He is Co-editor of several reports of excavations of his department. He has published more than four dozen articles on archaeological aspects of ancient Indian history and culture.

SUSMITA PANDE is a teacher of Ancient History Culture & Archaeology, for over 25 years. Dr. Pande has many books and research articles to her credit. Her major published works are '*Birth of Bhakti in Indian Religion and Art*' and '*Medieval Bhakti Movement*'. She has been awarded D.Litt. for the study of Vaiṣṇava Purāṇas with reference to the development of Bhakti and its impact on Society, Religion and Art of Northern India. She is currently working as Reader, Madhya Pradesh Bhoj Open University and is actively involved in the development of Multi Media based lessons on Indian Culture & Civilization, and production of video films on the rich heritage of India.

ANIL KUMAR PANDEY is Dy. Superintendent of Police, U.P. Police Service. His field of specialization is Ancient Agriculture wisdom. He has published books and articles. He was a co-author in the book *Prachin Bharat Me Krishi Gyan.*

J.N. PANDEY is at present Professor in the Department of Ancient History, Culture and Archaeology, University of Allahabad. His specialization is Archaeology, Paleography, Art and Architecture. His published works include *Mesolithic in the Vindhyan Plateau* and the *Gangetic Plains, Puratatva Vimarsh* (Hindi) and *Bhartiya Kala* (Hindi).

NILAM PANDEY is a research scholar in the faculty of agriculture in Banaras Hindu University, Varanasi.

OM PRAKASH is Vice-Chancellor, M.J.P. Rohilkhand University, Bareilly. He was former Professor and Head of the Department of Ancient History, Culture and Archaeology, University of Allahabad and had been Director of its Centre of Advanced Study of U.G.C. His published works include *Political Ideas in the Purāṇas,* 1977, *Early Indian Land Grants* and *State Economy,* 1988. *Conceptualization and History,* 1992. He has published more than four dozen significant research papers in the fields of socio-economic history, political ideas and institutions and historiography. He had been Fellow of Indian Institute of Advanced Study, Shimla.

G.K. RAI is at present Professor and former Head of the Department of Ancient History, Culture and Archaeology, University of Allahabad and Director of its UGC Centre of Advanced Study. He is Coordinator of many projects of interdisciplinary nature and of Institute of Professional Studies. His field of specialization is socio-economic history. His major published works include *Involuntary Labour in Ancient India*, 1981, *Those in Between: Essays in Socio-economic History and Historiography of Ancient, India, 2002* an Approach paper and *Status Report on Mathematical Techniques in Archaeology*, 2003.

K.V. RAMAN is former Professor and Head of the Department of Ancient History and Archaeology, University of Madras. His published works include *Excavation at Uraivur and Brahmi Inscriptions of Tamil Nadu.* He has published several research papers on south Indian culture in National and International journals.

VIJAYA RAMASWAMY is presently Professor, School of History, Jawaharlal Nehru University, New Delhi. She is an alumnus of the School of Oriental and African Studies, London, was Senior Fulbright scholar at Department of History, University of California, Berkeley. She was a fellow at Indian Institute of Advanced Study, Shimla (1992–1995). She was also an awardee of Shastri Indo-Canadian Fellowship at York University, Toronto. Her major books are *Textiles and Weavers in Medieval South India*, 1985, *Divinity and Deviance, Women in Virasaivism*, 1996, and *Walking Naked; Women, Society, Spirituality in South India* (IIAS Shimla, 1997). Her fields of specialization are economic and religious history and women studies.

S.R. RAO is a well-recognized archaeologist and scholar. He had held high positions in Archaeological Survey of India. He has been associated with marine archaeology also. His major published works include *Lothal and Indus Civilization*, 1973, *Lothal Harappan Fort Town*, 2 volumes, ASI Memoir No. 78, Delhi 1979–85, *Frontiers of Archaeology*, 1996, *Dawn and Devolution of the Indus Civilization*, 1991.

BRAJDEO PRASAD ROY is former Professor and Head of the Department of Ancient Indian History and Archaeology, Patna University. His major published works are *Political Ideas and Institutions*, 1975, *Later Vedic Economy*, 1994 and *Dimensions of Indian History and Culture*, 1997.

PRADEEP M. SAKLANI specializes in Ethno Archaeology at the Department of Archaeology, HNB Garhwal University. He has published one book and twenty papers.

B.D. SHARMA is an agricultural scientist who worked at N.B.P.G.K. Regional Station, Shimla. He has contributed a number of articles on different aspects of Horticulture and Agriculture in reputed journals of India. He is currently Head, Research and Development, Divya Yog Mandir (Trust), Kankhal.

HARI SHANKER SHUKLA is presently Professor in the Department of Pali and Buddhist Studies, Banaras Hindu University. His field of specialization is Theravada Buddhism–Abhidhamma Philosophy. He has published books on *Pali Buddhism and its Philosophy.* He has contributed more than three dozen articles in reputed journals.

ABHAY KUMAR SINGH is presently Professor of History, NEHU, Shillong. His field of specialization is cultural history of north-west India. He has published works on Northwest India and allied issues.

PURUSHOTTAM SINGH is former Professor and Head of the Department of Ancient Indian History, Culture and Archaeology, Banaras Hindu University. He had been coordinator of Special Assistance Programme of the Department. He was a Fellow at Indian Institute of Advanced Study, Shimla. His field of specialization is archaeology. His major published works are *Neolithic Culture of the Western Asia,* 1974, *Neolithic Origins,* 1991, *Excavation at Narhan,* 1999, *Agiabir,* IIAS Shimla, 2004.

RANJIT PRATAP SINGH had been Post-Doctoral Fellow of U.G.C. at the Department of Ancient Indian History, Culture and Archaeology, Banaras Hindu University. His field of specialization is proto-historic archaeology. His major published work is *Agriculture in Proto-Historic India,* 1990.

A.K. SINHA is currently Professor in the Department of Ancient History, Culture and Archaeology, M.J.P. Rohilkhand University, Bareilly. His specialization is History of Ideas, Philosophy of History and Historiography, India and the Greek World. Has authored six books and edited another six. Has published more than sixty research papers.

O.P. SRIVASTAVA is presently, Professor in the Department of Ancient History, Culture and Archaeology, University of Allahabad. His field of specialization in socio-economic history. His major published work is *Commercial Taxation in India c.* AD *600–1200,* 1999.

PUSHPA TIWARI is presently Reader in the Department of Ancient History, Culture and Archaeology, University of Allahabad. Her field of specialization is history of Indian art. Her major published work is *Prachin Bhartiya Abhusana,* 1992. She has contributed several research papers on history of Indian art in reputed journals.

R.P. TRIPATHI is Professor and Head, Department of Ancient History, Culture and Archaeology, University of Allahabad. He has published extensively on socio-economic history, polity and epigraphs. His specialization is socio-economic history, Ancient Indian Polity.

VIBHA TRIPATHI is currently Professor and Head of the Department of Ancient Indian History, Culture and Archaeology, Banaras Hindu University. Her field of specialization is Proto-historic and early historic Archaeology, Archaeo-metallurgy specially Iron Age Culture. Her major published works are *Painted Grey Ware and Iron Age Culture of Northern India,* 1976, (ed.) *Archaeo-Metallurgy in India,* 1997, *The Age of Iron in South Asia, Legacy and Tradition,* 2001. She has an Interdisciplinary project on Metallurgy and Ancient Indian History. She has contributed more than three dozen papers in National and International journals.

RADHA KANT VARMA is Chairman, Allahabad Museum Society, Allahabad. He is a former Professor and Head of the Department of Ancient Indian History, Culture and Archaeology, A.P.S. University, Rewa (M.P.). His field of specialization is prehistoric archaeology. His major published works are *The Mesolithic Age in Mirzapur*, 1986, (ed.) *Art and Archaeology of the Vindhyan Region and Aryans—A Myth or Reality*, 1994. He has produced a number of standard books on archaeology of India in Hindi and contributed more than three dozen articles on prehistoric archaeology and rock-art.

A.L. YADAV is presently Professor in the Department of Ancient Indian History, Culture and Archaeology, Banaras Hindu University. His field of specialization is socio-economic history. His major published work is *Prachin Bharat Men Krisi*, Varanasi 1980. He has contributed more than two dozen articles on socio-economic history of ancient India in reputed journals.

MEENASHRI YADAV is presently, Senior Lecturer in Ancient History, Culture and Archaeology at Jagat Taran Degree College, University of Allahabad. Her field of specialization is socio-economic history. Her Ph.D. thesis is on *Agriculture and Peasants in Northern India from c.600–1200 AD*, Allahabad, 1982.

B.N.S. YADAVA is former Professor and Head of the Department of Ancient History, Culture and Archaeology, University of Allahabad, and Coordinator of its UGC Special Assistance Programme. His national eminence was recognized by his election as General President, Indian History Congress, 1993. He has published well-known works on Ancient Indian and early medieval social history. He is lifetime Honorary Fellow, Indian Institute of Advanced Study, Shimla. His major published work is *Society and Culture in Northern India in the Twelfth Century*, 1973. He has contributed more than three dozen articles in reputed journals of India.

Introduction

Food is one of the essential needs of human beings (as of all creatures) and agriculture is the prime source of food of all living beings. Man has been responsible for the origin of agriculture very early in his evolution. Human beings continued to practice, evolve and develop this science of producing food in its manifold aspects in all ages and regions of the world. The history of agriculture is practically co-terminus with the history of civilization. The story of agriculture from the earliest stage of incipient farming along with gathering in the Mesolithic period to biotechnological and genetic farming of the present century is interesting, engrossing and instructive. It is linked with the evolution of man on one hand and associated with the latest scientific and technological culture of recent times on the other.

The history of Indian agriculture is a very complex, complicated and controversial discipline with a plethora of problems, unresolved issues, dark corners and extended gaps. The complex nature of the history of agriculture in India stems from a variety of circumstances. There is practically an absence of a literary text on agriculture for the early period. The literary sources available to us in respect of agriculture belong to later-Gupta and post-Gupta and even early medieval eras.

Besides its appearance, the literary sources available to us are mostly religious texts and there are scarce material in them on economy. Moreover, chronology of the early texts are uncertain. Another limitation of literary sources on agriculture is the official character. The earliest source on agriculture in India is the *Arthaśāstra* of Kauṭilya which discusses agriculture and herding under the heading of revenue administration. So is the case with innumerable copper and stone inscriptions of the first millennium of common era. This predominance of state revenue has overshadowed the real face of farmer's culture. The history of agriculture is the history of peoples' attachment to land but this has been submerged in the flood of state, revenue and official orientation. The history of agriculture in India is again circumscribed by its amorphous nature. It is intertwined with so many rather, all the aspects of human life and civilization—state, society, trade, commerce, transport, communication and other aspects of *Vārtā* (economy). Above all, agriculture is a way of life, a philosophy and a culture of its own. Naturally, such an amorphous and wide/expansive discipline may be researched from a variety of sources—folklore, music, rituals, magic, religion, technology, science, etc., and by multiple experts in an interdisciplinary manner as this volume proposes to do in a modest way.

Another difficulty in studying the history of agriculture in India lies in the fact that in spite of many common features, Indian agriculture has regional variations and differentiations. Moreover, it has intimate linkages with other forms of economy like, pastoralism, tribal economy and forestry. In view of paucity of early literary works on agriculture, along with the predominance of official orientation of literary and epigraphical evidences and amorphous nature and regional variations, it is a very onerous task to present the history of agriculture in India.

As compared to other countries, India has not given due attention to her history of agriculture. Before the middle of the twentieth century, history of agriculture in ancient India formed a part of the chapter on economic history in the general history of India, with focus on revenue administration and dynastic achievements. With the introduction of Marxist interpretation of history, Indian agriculture came to be studied in the light of Marxist ideology. Agriculture became an appendage to feudalism of early medieval India. However, some scholars presented certain aspects of history of agriculture in ancient India on the basis of indigenous sources against the cultural backdrop of Indian idealist tradition. However, a systematic factual history of agriculture in India was presented for the first time by Dr. M.S. Randhawa during 1980–86 in four volumes. Its first part covers the ancient period upto 1200 AD. The remaining volumes deal with agriculture in medieval and modern periods upto 1980. In spite of Dr. Randhawa's encyclopaedic, laudable factual attempt, there remain many academic issues unanswered. Moreover, much archaeological evidences have accumulated on the origin, antiquity, diffusion and development of agriculture in India, after Randhawa. In addition to this, some technical texts of agriculture have also been made available now. Some popular literary sources have also been ransacked. In addition to these advantages, there are many critical researches on different aspects of agriculture of India. This has inspired us to reconstruct the history of agriculture in India in two parts. The first part covers ancient India upto 1200 AD, while the second part is concerned with medieval and modern periods upto the present. The present volume is a synthesis and summation of existing knowledge on the history of agriculture in ancient India on the combined bases of archaeological and literary sources against the backdrop of Asian history in general. Besides summing up the existing knowledge, it opens new vistas for further research on many debated issues in the history of agriculture in ancient India.

The origin, antiquity and sources of agriculture in India are very vexed, hazy and controversial questions. As these issues are associated with prehistoric culture, literary evidences are of no use. Fortunately, archaeological materials have recently been retrieved to throw welcome light on these problems. These materials have been reported from sites of the Vindhya-Ganga region—Koldihwa, Mahagara, Kunjhun, Panchoh, Chopani Mando, Tokwa, Jhusi, Mahadaha, Damdama, Sarai Nahar Rai, Lahuradeva, Sohgaura, Imlikhurd, Narhan, Chirand, Senuwar, etc. Plant remains, agricultural tools and pots, dental pathology, settlement remains, etc., have been utilized to throw light on the origin and antiquity of cultivation in India. It appears that there was a smooth, gradual transformation from hunting-gathering in the late Mesolithic to domestication of animals and cultivation of plants in the Neolithic phase in the Ganga-Vindhya regions as well as in the northwest of the Indian subcontinent. The antiquity of cultivation of plants in the Indian subcontinent may be traced back to the seventh-sixth millennium BC, if not earlier, in our present state of knowledge. There is every likelihood that it may be pushed back earlier in future. From available evidences, it may be suggested that the Indian subcontinent had two regional centres for farming of cereals—northwest with barley, wheat complex (Mehrgarh) and Vindhya-Ganga region with rice cultivation. The Vindhya region appears to be earlier than the Gangetic complex being of sixth-fifth millennium BC but it is also quite probable that both developed the practice of cultivation of rice almost simultaneously. No doubt, the chronological horizon of rice cultivation in China (12000+120, 12500+120 years BP) is much earlier than that of India but that does not give us authority to assume that India learned the practice of rice cultivation via diffusionary medavism from China. The problem of a wide spatial gap and other circumstantial evidences from early transition from Mesolithic to Neolithic in India (harvested rice at Chopani Mando) (17th–7th millennium BC) inspire most of us to discard the theory of diffusion and propose a case for indigenous development of rice cultivation in the Vindhya-Gangetic region. It is a common view that wheat, barley, lintel and chickpea are of the Mediterranean origin. We have evidence of cultivation of these cereals in the northwestern part of the Indian subcontinent (Mehrgarh) in around seventh millennium BC. Bannu has *Triticum* sp. in third millennium BC or even earlier, introduced by the Harappans. Harappan crops have been documented from OCP phase at Atranjikhera (C. 3000–1500 BC) and all other sites of Middle Ganga plain.

Domesticated rice is found at Atranjikhera (2000–1500 BC) and Lal Qila. Imlikhurd, Narhan, Chirand, Senuwar, etc., have provided us evidence of rice-wheat-barley legume agriculture by the end of the third millennium BC and two crops a year practice has started. It had a settled farming community. It is significant to point out that the mid-Ganga plain seems to be responsible for the diffusion of cultivated rice to northwest in competition to wheat-barley in around 2700–2800 BC, as we have evidence of cultivated rice in Haryana around this time.

Millets are of African origin. They were introduced in the third millennium BC, as we have found them associated with Mature Harappan Culture (2500–2000 BC). Probably they came to middle Ganga plain around 1800 BC because we have evidence of them at the earliest level of Imlikhurd and Narhan (1800 BC). Route of diffusion is yet to be researched but it is certain that once the complex was introduced, the phenomenon had a major impact on the subsistence system of the subcontinent. In the border land Afghanistan, there is a strong case for the domestication of animals and plants in the Neolithic phase and its full beginning in the protohistoric period (Mundigak, Said Qala, Deh Narasi, Shortugai). The most remarkable feature of agriculture in Afghanistan appears to be the symbiotic development of sedentary agriculture and pastoral nomadism—a feature which is quite common in the mountainous regions. These issues and problems have been critically analysed in first seven chapters. The problems of individual cereals and plants such as millet, barley, wheat, grain and potato have been analysed in the context of origin, and development in next five chapters (8–12).

Harappan and Vedic facets of agriculture present many problems on account of nature of evidences (archaeological only for the former and literary only for the latter) available and the prevalence of several hypotheses. Ethnographic perspective of agriculture in Garhwal Himalayas reveal that agriculture in this ecological framework is well related to pastoral management of the hill community—a hypothesis which is substantiated by symbolic development of agriculture and pastoralism in Afghanistan. The Indus civilization, in spite of its urban-industrial character, appears to be well provided by the development of agriculture and animal husbandry. Some interesting areas of research in the far and wide Indus civilization are the hypotheses such as the presence of higher precipitation, existence of some sort of irrigation, cultivation of barley at Kalibangan, wheat at Banawali and rice at Atranjikhera (late Harappan/OCP), use of chemical fertilizers (Gypsum calcium sulphate) (Kalibangan and Lothal), minor role of grains in Lothal economy, and continuation of the barter system. The history of agriculture in early and later Vedic tradition is beset with various problem areas—Nomadic pastoralism *versus* sedentary agriculture, the peasant proprietorship, type of produces—presence of barley and *dhanya*, absence of wheat and rice, plough-cultivation, irrigation with mechanical devices, nature of manure, association of agriculture with rituals and deities in early Vedic period along with other issues like occupation (agriculture, cattle rearing and handicrafts) are interesting and controversial issues. The later Vedic age has better records for the development of agriculture—dangers to crops, variety of crops, integrated view of life in which economy and spirituality marched together, use of more than two oxen, the familiarity with two harvests a year, rotation of crops, storage mechanism of food grains, ignorance of rice cultivation and use of iron/wooden tools. Moreland's concept of agrarian system may not be applicable to ancient India. On the basis of scanty and hazy data, the elements of agrarian system of the Harappan and Vedic traditions are to be researched in which caste and class might or might not have played any crucial role. Animal husbandry, an integral part of agriculture in ancient India with special reference to the Vedic literature, has interesting accounts. All these facets of agriculture in the Harappan and Vedic times have been enlightened in chapters 13–19.

The so-called Dark Period (from the fall of the Indus civilization to the emergence of the Mauryan empire) has been enlightened now with the discovery of the chalcolithic cultures of Central India and the Deccan—the Kayatha, the Svalda, the Ahar/Banas, the Malwa and Jorwe cultures. They cultivated a variety of crops, the principal cereal being barley and, to a lesser extent, wheat. These cultures appear to represent peasant settlements. In some Maharashtra sites, we have adjustments with different agricultural systems. Further works in the area may provide us with an understanding the way of such adjustments. The development of agriculture in the first millennium BC with problems of technology adaptation in early agricultural system is also a fascinating area of research as it involves the questions of role of territorial state *versus* lineage in this process. All these aspects of agriculture in the chalcolithic historic India upto 1st millennium BC have been addressed in chapters 20–23.

The historic period from c. sixth century BC to c. 600 AD has a variety of sources for the study of agriculture—Pali, Sangam, Sanskrit (Kauṭilya's *Arthaśastra* and the *Dharmaśastras*), and the Graeco-Roman literatures. All these sources have been utilized in chapter 24–30, while agricultural taxes which formed an important subject of agrarian study has been delineated in chapter 31. The classification of lands, processes of cultivation, irrigation, different produces and export items, taxes, etc., have been discussed in this section. Foreign sources indicate that some Indian medicinal and aromatic plants were experimented in Arabia and Indian sugar was an export item. Enhancement of rate of agricultural taxes in early medieval India is a point to be discussed in connection with feudalism.

The early medieval period in India (c. 600 AD–c. 1200 AD) is the most controversial period in the socio-economic sphere. Regional history becomes prominent. Some sort of agrarian structure appears to be prevalent. New types of tools and technology appear to have been operative. Some literary texts on agriculture came to be produced. All these aspects have been discussed in the last section of the volume—chapters 32–47. While chapters 34–37 tell the story of agriculture as depicted in the *Lokopakāra, Kṛiṣi Parāśara, Khana's vacanas* and *Vrikṣāyurveda.* The chapters 41, 42 and 43 deal with regional agriculture in South India, Bengal and Gujarat respectively. Types, techniques and processes of agriculture have been addressed in chapters 32, 33, 38, 39, 40 and 47. These chapters deal with the prevalence of share cropping throughout the early medieval period, technique and management of irrigation in south India, irrigation works, ownership of irrigation water, forecasting of weather, manuring, seed and sowing, agricultural implements and agricultural processes. The problem of *arhaṭṭa* with gearing mechanism has been discussed in the light of art evidences (Mandor and Jogeshwar). The question of forced labour has been discussed with the conclusion that it was neither essential nor specific to the socio-economic formations of early medieval India. Peasant differentiation precludes us to suggest that there was a high degree of class consciousness and peasantry cannot be called to be a class by itself at the time, though local oppressions may have organized them locally for a short time for a some specific purpose.

Art is a very effective source for reconstructing the history of agriculture. Chapter 45 has been devoted to this aspect, throwing light on agriculture, horticulture, fishing, animal husbandry, etc. The *Purāṇas* provide valuable source of information on medicinal and nutritional properties of agricultural produces. Chapter 46 has been devoted to this aspect.

Ancient history cannot be treated only as an object of museum. The ancient knowledge system of agriculture has relevance for the present also. Chapter 44 is devoted to this aspect. It is suggested that traditional knowledge of integrated approach of Indians to farming, animal husbandry and forestry is still relevant. It may be responsible—combined with modern science and technology—for sustainable development for which we are so much concerned today because therein lies the prosperity and destiny of India.

I am conscious of the limitations of this work. There may be many gaps to be filled, many dark corners to be lighted, many problems to be resolved, many questions and answers to be revised. However, if this ancient history of agriculture is able to provide us with some incentive for further research and give some direction to the development of agriculture in India in the right path, I will feel amply satisfied and rewarded. I crave the indulgence of learned readers for errors and omissions in the volume.

At the last but not the least, I take this opportunity to heartily thank all the learned authors who have taken pains to contribute in this volume in spite of their busy schedules and other commitments. Without their cooperation and timely action the volume could not have been completed. I am deeply beholden to Professor D.P. Chattopadhyaya, a philosopher and extraordinary scholar and Director of the Project of History of Indian

Science, Philosophy and Culture, who has kindly entrusted the editorial responsibility to me in 1999 after the sad and sudden demise of my teacher Professor Lallanji Gopal. I had been conscious of my limitations but accepted the responsibility with the sacred desire to complete the unfinished work of my teacher. Words simply fail to express my deep sense of gratitude to my teacher—Prof. G.C. Pande, a historian, philosopher, original thinker and a poet who had helped me in various ways in this respect. He has always been a source of inspiration to me. I am thankful to Prof. Bhuvan Chandel, Project Coordinator, for help and cooperation from time to time in completing this work. I acknowledge with thanks the fact that Mrs. Gopal has permitted me to include an important article of Prof. L. Gopal entitled, 'Technique and Process of Agriculture in Early Medieval India' in a revised form as Chapter No. 47 in the present volume which was originally published in *University of Allahabad Studies* (Ancient History Section), 1963–64. My wife Madhuri Srivastava gave me unstinted support and exemplary cooperation in this work and provided me necessary leisure and freedom from domestic responsibilities for which I am sincerely grateful to her. Thanks are due to entire staff of the project—especially Prof. A.K. Anand and Mr. S. Sreekumaran for helping me in many ways, especially in the organization of two seminars. I am thankful to Mr. T.K. Majumdar, P.S. to Director, Indian Institute of Advanced Study, Shimla, who provided me during my Directorship (2000–2003) all sorts of help and cooperation in my discharge of project work purely as love and commitment to me. Dr. Jeewan Rai, Research Associate (1999–2000) with me and Mr S.C. Kesarwani, Project Computer Operator, deserve thanks for their help and hard work.

V.C. Srivastava

SECTION ONE

ORIGIN AND BEGINNINGS OF AGRICULTURE

CHAPTER 1

Origin of Agriculture in the Middle Ganga Plain

Purushottam Singh

GEOGRAPHICAL BACKGROUND

The plain of northern India, stretching over 3200 km from the Arabian Sea to the Bay of Bengal, is largely the creation of the mighty Ganga and its tributaries. Sir Dudley Stamp was greatly impressed by the flatness of the plain and its gentle seaward slope. He wrote, "One is the dead flatness of the plain—not a hill, scarcely a mound to break the monotony of the level surface. So gentle is the seaward slope that it is imperceptible to the eye. Nearly a thousand miles from its mouth, the Ganga is only 900 feet above sea level. Another feature is the sudden rise of the Himalayas from the level plain.... But otherwise the uniformity is amazing. Not a rock, not even a pebble is present, to alter the uniform character of the alluvium" (Stamp, 1967: 198–99).

The entire Ganga plain is divided into three units: (i) the upper Ganga plain; (ii) the middle Ganga plain; and (iii) the lower Ganga plain.

The middle Ganga plain stretches for about 300 km from the Himalayan foothills in the north and the Vindhyan ranges in the south. This landmass attained its present form during the post-Tertiary period, when this deep trough was filled with fine alluvium brought down by the Himalayas in the north with an average thickness of 1300–1400 metres. In prehistoric times, the main river Ganga, watered by several tributaries, flowed sluggishly in a meandering fashion forming ox-bow lakes, some of which were even perennial. These lakes were rich in aquatic fauna and the lands around them were covered with wild grasses, many with edible grains. With the onset of milder climate of the Holocene, the marshy land gradually turned into good grassland, which attracted small animals.

The Ganga plain is a peripheral foreland basin which came into existence in the early Miocene, expanded in the middle Miocene, and attained its present form in the late Quaternary. The stage-wise evolution of the Ganga plain occurred in response to the thrust-fold belt loading events in the Himalayas and consequent southward expansion of the basin (Singh, 1996: 133). The basin expanded in the middle Miocene and attained its present configuration in the late Quaternary. Presently, the Ganga plain foreland basin is in a mature stage of evolution, with oversupply of sediment and undefiled condition, building to 300–50 m above sea level, exclusively by fluvial processes (Singh, 1996: 99). The main rivers of the Ganga plain, the Ganga, the Yamuna, the Gomati, the Ghaghara, etc., possess very wide river valleys in which the present-day active channel and its flood plain is intrenched. The regions are areas of sand deposition in the Ganga plain. The residence time of these main rivers in their valleys is of the order of 10^3–10^5 years, during which several climatic and base-level changes have also transformed the character of the river channels, along with the nature of sediment (Singh, 1996: 129).

The middle Ganga plain is bounded by the Ganga-Yamuna confluence in the west and the West Bengal and Bihar border in the east, the Himalayas in the north and the Vindhyas in the south. The area includes modern eastern Uttar Pradesh and parts of Bihar. On the basis of river systems, the middle Ganga plain is further subdivided into: (1) the Ganga plain north; and (2) the Ganga plain south—the former further subdivisible into: (i) the Ganga–Ghaghara Doab; (ii) the Ghaghara–Gandak interfluve; and (iii) the Gandak–Kosi interfluve and (iv) the the Kosi–Mahananda interfluve. The Ganga plain south is also further subdivisible into: (a) the region to the west of Karmnasa; (b) Karmnasa east interfluve; (c) lower Son valley; and (d) Magadha–Anga plain. Needless to say, the Ganga remains the lifeline of the middle Ganga plain (Singh, 1971: 190).

The middle Ganga plain, particularly its western section, is characterized by horseshoe lakes. Obviously, these lakes were formed by meandering rivers. It is not unlikely that the river might have been the Ganga shifting southward. In fact, it has been claimed that the Ganga shifted its course some 55 km southward during terminal Pleistocene time (Kennedy, 2000: 202). At some places, these horseshoe lake formations have been studied (Gupta, 1976). The available evidence suggests that in the beginning, there was very arid climatic condition followed by arid, semi-arid and semi-humid conditions (Misra and Gupta, 1995: 22).

THE DRAINAGE SYSTEM

The drainage system of the middle Ganga plain is governed by the general life of the land which slopes from the northwest to the southeast. The entire region is drained by the Ganga system. Most of the left-sided tributaries of the Ganga have their sources in the Himalayas and are, thus, perennial. The right-sided tributaries rise in the plateau and are seasonal. Again, most of these have only a narrow ribbon of water straggling through vast span of sands in summer and winter.

THE FLORA

The annual average rainfall of the area is about 1000 mm. The natural vegetation cover now been completely obliterated on account of increasing population pressure and consequent human manipulation. However, remains of typical tropical dry deciduous woodland can be seen in heavily degraded and isolated patches in the area. As indicated by pollen analysis (Gupta, 1976, Pant and Pant, 1980), in the past, the area was covered with grassy vegetation. The small shrubs and trees, which still survive in the bad land ravines, include dhak (*Butea menosperma*), sheesham (*Dalbergia sissoo*), siras (*Albizzia lebbek*), banyan (*Ficus glomerata*), pakar (*Ficus infectorisa*), pipal (*Ficus religiosa*), mahua (*Basia latifolia*), tamarind (*Tamarindus indica*), neem (*Azadirachta indica*), babul (*Acacia arabica*), kaitha (*Feronia elephantum*), chilbil (*Ulmus integrifolicus*), lisora (*Cordia myxa*), etc. (Pal, 2002: 61).

(i) *Palynological Studies*

In order to reconstruct the ancient climatic conditions and to discover the earliest traces of agricultural activity, if any, the pollen sequence from an ox-bow lake in Pratapgarh district was collected by H.P. Gupta (1976). This lake is situated about 15 km southwest of Pratapgarh near the Mesolithic site of Sarai Nahar Rai. The biostratigraphy of the lake basin was grouped into four phases (I–IV). This study showed that in the last 8,000 years, there have been four phases of vegetational development reflecting directly on four brief phases of climate such as very arid, arid, semi-arid and semi-humid. The abundance of grasses throughout the pollen diagram has revealed the existence of an open savannah forest. The arboreal vegetation is poor and represented by a few scattered trees and shrubs.

It may be emphasized that the commencement of arable and pastoral economy, chiefly based on cereal pollen, has been recognized during *circa* 4500 BP, which prospered in the succeeding phases.

Further, palynological studies of lake-filled deposits in central Ganga plain currently under progress—by the Birbal Sahni Institute of Palaeobotany—and the University of Lucknow reveals that the Ganga plain was mainly grassland at least for the last 15,000 years, with local forest thickets. Significantly, cultural pollens are present throughout the lakes' depositional history of last 15,000 years, suggesting anthropogenic activity in the region throughout this period. The cultural pollens in these profiles exhibit changes with the changing climate history. For example, there is a decrease in cultural pollens around 11,500–10,500 years BP. An increase in the cultural pollen in the last 5000 years indicates advanced anthropogenic activity (Sharma, *et al.*, 2000, 2003; Tewari, *et al.*, 2002; Tewari, *et al.*, 2003).

(ii) *Mesolithic and Proto-Neolithic Sites*

Intensive explorations conducted by the Universities of Allahabad and Benaras during the sixties and seventies of the previous century resulted into the discovery of more than 500 Mesolithic sites, mainly in the eastern Vindhyas and the middle Ganga valley. In the Vindhyas, the discovered Mesolithic sites lie in the districts of Sidhi, Satna, Rewa in Madhya Pradesh and Banda, Allahabad, Mirzapur, Sonbhadra and hilly parts of Chandauli in eastern Uttar Pradesh. In the middle Ganga valley, the Mesolithic sites were located in parts of Allahabad, Varanasi, Pratapgarh, Jaunpur and Sultanpur districts of Uttar Pradesh (Misra, 1997: 32). Of the 500 sites listed above, nearly 200 sites are located in the middle Ganga valley. This increase in the number of Mesolithic sites is an indication of a spurt in the population. Evidence regarding harvesting of wild cereals (rice) comes from Chopani Mando. Hence, this evidence has been examined in detail in the following pages.

Chopani Mando *24° 55′ N; 82° 5′ E*

This site is located about 77 km southeast of Allahabad on the middle terrace of the Belan river. This site, spread over an area of 15000 square metres, was subjected to excavation in 1967 and again in 1977–78 by Professor G.R. Sharma. The excavation brought to light a total habitational deposit of 1.55 m in thickness, comprising 10 layers divisible in the following three phases:

1. Epipalaeolithic (comprising tools of upper Palaeolithic as well as early Mesolithic culture).
2. Early Mesolithic (comprising non-geometric microliths in sub-phase A unassociated with pottery. In sub-phase B, geometric microliths are present).
3. Advanced Mesolithic or Proto-Neolithic: Plans of 13 circular or oval huts and four hearths have been exposed. This phase is characterized by the emergence of handmade pottery with bowls and vases as the main types.

Remains of charred or carbonized rice were found embedded in lumps of burnt clay from the Proto-Neolithic levels of this site. These have been dated to the 9th-8th millennium BP (Agrawal and Kharakwal, 2002: 85) and represent a wild variety of rice. However, the samples are still under examination. This is a very important find, particularly in view of the occurrence of domesticated rice and handmade pottery from the Neolithic levels at Koldihwa and Mahgara in the same region.

In the absence of any C-14 dates from Chopani Mando, it is difficult to assess the chronology of this site. However, keeping in view the date of Epipalaeolithic of the Belan river (17765±40 BC), together with the radiocarbon date from Sarai Nahar Rai (8395±10 BC) and the date from Koldihwa (5440±240 BC), the entire occupational deposit of Chopani Mando has been tentatively placed within a timespan of 17000 to 7000 BC. (Misra, 1989: 108).

(iii) Neolithic Sites

The succeeding stage of human evolution is termed as Neolithic. This stage was marked by the breeding of wild animals and cultivation of selected wild grasses. The first animals to be domesticated were dog, cattle, sheep and goat and the first plants to be cultivated were rice, wheat and barley. This change occurred around 10,000 years ago or even earlier in the vast stretch of mountainous and hilly environment, extending from the east coast of the Mediterranean to the eastern edge of Baluchistan plateau, as also in the Yangtze valley in China (Yasuda, 2002). In India, evidence of Neolithic culture have been found in northwestern India, Kashmir, the Vindhyan region, middle Ganga valley, southeastern India, northeastern India and southern India. The Neolithic cultures of these regions are characterized by sedentary settlements, cultivation of cereal plants, domestication of selected animals, ground stone industries including a microlithic component, handmade pottery, etc. In the middle Ganga valley, evidence of Neolithic settlements comes from Lahuradeva (Sant Kabir Nagar), Sohgaura and Imlidih Khurd in Gorakhpur district of Uttar Pradesh and Chirand, Chechar-Kutubpur, Taradih and Senuwar in Bihar. The excavations, though conducted on a limited scale, have brought to light the evidence of wattle-and-daub structures. Traces of post-holes have been noticed. Screen walls of huts were made of split bamboo reeds, the impression of which have been found in the burnt clay lumps. The hutments appear to be round or oval on plan. At Chirand, evidence of pit dwelling has also been reported.

The excavations have yielded rich data regarding the ceramics used by the Neolithic inhabitants of the Ganga valley. In the early stages, they used to make handmade pottery but the use of wheel-turned pot has also been reported from some of the Neolithic sites (Senuwar) from the upper strata of this culture. Corded ware appears to be a characteristic trait of the Neolithic culture, both of the Vindhyas and of the middle Ganga valley. Other ceramic industries comprise ordinary red ware, burnished ware, rusticated ware, black-and-red ware, lustrous ware and lustrous red ware. The principal types are bowls with varying profiles, vases, squatted vessels and miniature jars. The clay used for pottery contains gritty rice husk and chaff as degraissant. A limited number of painted sherds have also been reported from Chirand and Senuwar. These post-firing paintings confined to the rim have been executed in red ochre. At Chirand, the painted motifs include a linear design of criss-cross lines and concentric circles. This diversity of ceramic industries and types strongly suggests that the potters had acquired sufficient technical skill in ceramic tradition.

The stone objects of Neolithic culture comprise flakes, blades, scrapers, arrowheads, points, lunates and borers made on cilicious stones such as chert, chalcedony, agate, jasper, quartz, etc. Rounded celts of basalt and granite have been found at both Chirand and Senuwar. Bone objects constitute another trait of the Neolithic culture of the middle Ganga valley. They include celts, scrapers, stylie and arrowheads. Excavations at Chirand have yielded a vast amount of such objects. In this context, it may be noted that with the exception of the Neolithic sites of the Kashmir valley, Chirand is the only Neolithic site in India to produce bone objects, both in quality and variety (Misra and Gupta, 1995: 26). Terracotta objects from the Neolithic sites in the middle Ganga valley are not plenteous but Chirand had yielded terracotta wheels, beads, bangles, cakes, birds, serpents, etc.

As stated above, the domestication of animals and cultivation of cereal plants are the hallmark of the Neolithic culture. Thus, the Neolithic people of the middle Ganga valley also cultivated wheat, barley, rice, field pea, lentil, *masur, moong,* etc. Evidence from Senuwar and Chirand is copious. Rice appears to be the staple food of the Neolithic people. Evidence of cultivated rice, along with its wild progenitor, has been recently found (2002) embedded in the core of a number of pot sherds from Lahuradeva from Period IA. The two radiocarbon dates found from this level have been dated to 5298 ±160 and 4196±90 BC (BS 1966 and BS 1951). Thus, the cultivation of rice now goes back to the late sixth and the fifth millennium BC in the Sarayupar area of the middle Ganga valley. Bones of cattle, buffalo, elephant, rhinoceros, stag and deer have been found from these sites. Of these, cattle and buffalo were domesticated while stag and deer were obviously hunted. As rhinoceros has never been domesticated, during the Neolithic phase also it would have roamed wild in the marshy area in the middle Ganga valley (Misra and Gupta, 1995: 27).

To fix the chronology of Neolithic culture of the middle Ganga valley, the dates of Chirand and Senuwar may be taken into consideration. Two radiocarbon dates reading 1760±150 BC and 1680±135 BC push back the beginning the Neolithic culture around 2000 BC at Chirand. The overlap phase of the Neolithic-Chalcolithic levels at Senuwar has provided four radiocarbon dates. All these dates, taken together, push back the beginning of the Neolithic culture earlier than 2000 BC. The newly obtained dates from Lahuradeva place this culture around 5000 BC but this hypothesis needs to be supported by more data.

Koldihwa *(24° 54′ N; 82° 2′ E)*

Koldihwa is the key-site for understanding the beginning of agriculture in the middle Ganga plain, as it provides the earliest evidence of domesticated rice in India. Situated on the left bank of river Belan in the Meja Tehsil of district Allahabad, this site was excavated by the University of Allahabad from 1972–73 to 1975–76, which yielded remains of a Neolithic culture comprising corded ware, incized ware, and ill-fired crude black-and-red ware—all handmade, together with Mesolithic tools made of chert and chalcedony. The Neolithic of Koldihwa was followed by a chalcolithic phase and, finally, by Iron Age deposits.

Remains of rice from the Neolithic deposits is one of the most significant discoveries of this excavation. These domesticated rice remains comprise seed without pericarp, impression of caryopsis, stalk, glume and probably also spikelets. They have been found in loose form as well as embedded in pot sherds. The imprints of caryopsis, stalk, glume, etc., have been found only on pot sherds, specially on the cord-impressed and rusticated wares. It is obvious that they have been used to temper the clay so as to manufacture the parts. A preliminary palaeobotanical analysis of the rice husk by Vishnu-Mittre of BSIP, Lucknow and Te-Tzu Chang of the International Rice Research Institute, Manila, Philippines, reveals that they belong to the domesticated *Oryza sativa* group. The radiocarbon dates from Koldihwa give a bracket of *Circa* 9000–8000 years BP. A C-14 date from the pre-Neolithic Mahagara is 8080±115 BC (Sharma *et al.*, 1980; Pal, 1990; Agrawal, 2002: 85).

Imlidih Khurd *26° 30′ N; 83° 12′ E.*

Imlidih Khurd is an inconspicuous, featureless mound located on the left bank of the Kuwana, a tributary of the Ghaghara river. It was excavated during 1992, 1993 and 1995 by the present author (Singh, 1991–92; 1992–93). The culture sequence has been divided into three periods, viz., Neolithic, Narhan Culture and early historical. The Neolithic culture is represented by cord-impressed pottery as its diagnostic trait. The Neolithic man lived in wattle-and-daub huts represented by reed marks found in large numbers. Several floors made of mud, ovens and hearths marked the other structural activities. The small finds comprise several micro beads of steatite, other beads of terracotta, agate and faience, bone points and pottery discs. A tentative taxonomic classification of animal bones shows that cattle, sheep/goat and pig had already been domesticated in Period I. The earliest settlers cultivated two crops a year. Cereals comprised rice, barley, wheat, jowar millet, and bajra (pearl millet). Pulses included lentil, field pea, grass pea (*Khesari)* and green gram or *mung*. The oilseeds comprise field brassica and sesame or *til*. Jubube, awla and grapes were the fruits eaten by the inhabitants.

The crop remains, as reported by Dr. K.S. Saraswat, are given in the following table:

Crop Remains From Imlidih-Khurd

Imlidih-Khurd, Gorakhpur: Crop Remains	**Pre-Narhan Pre-1800–1400** BC	**Overlap Phase 1400–1300** BC	**Narhan Phase 1300–800** BC
Rice (*Oryza Sativa*)	■	■	■
Barley (*Hordeum vulgare*)	■	■	■
Dwarf wheat (*Triticum sphaerococcum*)	■	■	■
Bread wheat (*Triticum aestivum*)	■	■	■
Ragi millet (*Eleusine coracana*)			■
Jowar millet (*Sorghum bicolor*)	■		■
Kodan millet (*Paspalum* cf. *scrobiculatum*)			■
Lentil (*Lens culinaris*)	■	■	■
Field pea (*Pisum arvense*)	■	■	■
Grass pea (*Lathyrus sativus*)	■		■
Green gram (*Vigna radiata*)	■	■	■
Chickpea (*Cicer arietinum*)			■
Mothbean (*V. aconitifolia*)			■
Horse gram (*Dolichos biflorus*)	■	■	■
Sem/Hyacinth bean (*Lablab purpureus*)			■
Cotton (*Gossypium arboreum/ herbaceum*)			■
Field brassica (*Brassica juncea*)	■		■
Linseed (*Linum usitatissimum*)	■		■
Safflower (*Carthamus tinctorius*)			■
Til (*Sesame indicum*)	■		■
Melon/Phut/Kheera (Some var. of *Cucumis melo*)			■
Watermelon (*Citrullus lanatus*)			■

Senuwar *24° 56´ N; 83° 56´ E*

The ancient settlement of Senuwar is located 7 km south of Sasaram in district Rohtas (Bihar). This site was under excavation by the Banaras Hindu University for two field seasons (1986–87; 89–90). This excavation revealed a thick deposit divisible into four periods, the earliest being Neolithic-Chalcolithic. A large collection of archaeobotanical data was made by K.S. Saraswat, who reports that agriculture was practised on a large scale during the first two periods. Here the earliest occupation, subdivisible into two subperiods, is similar to that of Chirand and has been dated from 2000 to 1400 BC. In Senuwar, the earliest subperiod (IA), designated pure Neolithic, has given grains of rice (*Oryza sativa*), barley (*Hordeum vulgare*), field pea (*Pisum sativum* var. *arvense*), lentil (*Lens culinaris*) and some millets. In the upper layers of the same subperiod, dwarf wheat (*T. Sphaerococcum*), grass pea (*Lathyrus sativus*) and Kodon (*Paspalum scrobiculatum*) have been identified.

In the subperiod IB (Neolithic-Chalcolithic), the carbonized grains comprised rice, barley, two forms of wheat (dwarf wheat and bread wheat), kodon, lentil, field pea (*Pisum sativum var arvense*), grass pea, chickpea (*Cicer arietnum*) and green gram or *mung* (*Vigna radiata*). The radiocarbon date for this level is 1770±120 BC. A detailed report on these food grains prepared by Dr. Saraswat is in press and its publication will go a long way in understanding the beginnings of agriculture in the middle Ganga plain.

Chirand 25° 45′ N; 84° 50′ E

This ancient site (also spelt as Cherand or Cheran) is situated on the bank of the Ghaghara, about 11 km east of Chapra town. This site was known to archaeologists since 1877 as it was visited by Hunter, A.C.L. Carlleyle and N.L. Dey. However, the true nature of the habitational remains could be known only after seven seasons' work done during 1962–63 to 1971–72 by the Directorate of Archaeology and Museum, Bihar. These excavations revealed five broad cultural periods, viz., Period I (Neolithic), Period II (Chalcolithic), Period III (NBPW culture), Period IV (first century BC to third century AD) and Period V (late Historical) (Roy, 1989: 103–105).

Period I, covering a 3.5 m thick deposit has revealed a full-fledged Neolithic culture represented mainly by bone tools, handmade pottery, microliths and Neolithic celts, hammers, pestles, querns, etc., made of quartzite, basalt and granite.

The Neolithic Chirand is well developed and almost on the verge of a Chalcolithic stage. The lithic component is very meagre, only four diminutive celts and some pestles and querns along with a limited number of microliths have been reported. Neolithic Chirand has a very developed bone and antler industry, comprising needles, bodkins, points, borers, pins, etc., and personal ornaments like pendants, earrings and combs, beads and bangles. The pottery comprises red, grey and black-and-red wares, occasionally decorated with linear designs painted in red ochre. Structural remains are not many but pit dwellings with thatched roofs are suggested. There is plenty of evidence to suggest wattle-and-daub constructions and the discovery of a series of hearths in a semi-circular form suggests communal cooking. The Neolithic economy was based on the cultivation of wheat, six-rowed barley (both naked and hulled), lentil and green gram. Paddy husk impressions on pieces of burnt clay and the discovery of charred grains of rice suggest its use in the economy of the inhabitants. There are eight radiocarbon dates from the Neolithic strata of Chirand (Agrawal and Kusumgar, 1974: 64) which give a time bracket of circa 1800–1400 BC (discounting certain incompatible dates). The picture obtained from Chirand has also been substantiated by Chechar but this site remains largely unpublished.

(iv) Chalcolithic Sites

The Neolithic period was succeeded by the Chalcolithic phase. The general pattern of life did not change during this period, barring some significant developments. The introduction of metals (copper/bronze)—though limited in quantity—had a profound effect on the quality of life. However, as the supply of metal was very limited, stone tools continued to be used side by side and hence this stage of human evolution has been aptly termed as Chalcolithic. This stage witnesses a marked increase in the number of settlements, introduction of copper-bronze for the manufacture of tools, weapons and ornaments, improvement in architecture, introduction of wheel-made pottery, diversification of wares and profuse decoration of vessels by painted, incized and applique designs (Misra, 2001: 512). A number of Chalcolithic sites have been found in different parts of country and the middle Ganga plain is no exception to it. Some of the important Chalcolithic sites of this region are listed below:

Pure Deojani, Pelkhawar and Bhelwani in Pratapgarh, Akahua in Jaunpur, Rajghat, Prahladpur, Sarai-Mohana and Kamauli in Varanasi; Masondih in Ghazipur, Nahusa Raja Ka Tila in Mau; Banwarighat, Gularihwaghat, Susipar, Ramnagarghat, Baragaon and Gerar in Basti; Lahuradeva in Sant Kabir Nagar; Khairadih, Waina and Bhunadih in Ballia; Sohgaura, Narhan, Dhuriapar and Imlidih Khurd in Gorakhpur; Purainadih in Maharajganj; Malhar, Raja Nala Ka Tila in Sonbhadra; Jhusi in Allahabad districts of Uttar Pradesh and Chirand and Manjhi in Saran; Maner in Patna; Oriup and Champa in Bhagalpur; Chechar-Kutubpur in Vaisali; Sonpur and Taradih in Gaya and Senuwar in Rohtas district in Bihar. Out of these, Rajghat, Prahladpur, Sarai-Mohana, Kamauli, Masondih, Sohgaura, Narhan, Dhuriapar, Imlidih Khurd, Waina, Bhunadih, Khairadih, Lahuradeva, Jhusi, Malhar, Raja Nala Ka Tila, Chirand, Manjhi, Maner, Oriup, Champa, Chechar-Kutubpur, Sonpur, Taradih and Senuwar have been excavated. Most of these sites are located on river banks, big or small, or on horseshoe lakes. The size of settlements ranges from small to medium. In the absence of large-scale excavations at any of these sites, no definite statement can be made regarding the planning of the settlement. However, the occurrence of burnt clay lumps with split bamboo or reed impressions at a number of sites shows that the Chalcolithic people used to construct walls with this material. The walls were plastered with mud on both sides (Misra and Gupta, 1995: 27). Evidence of post-holes comes from a number of sites including Imlidih Khurd which indicates that the construction of huts was circular or oval on plan. Plastered floors and various types of hearths have been excavated at a number of sites including Narhan, Imlidih Khurd and Waina.

For the chronology of the Chalcolithic culture of this region, the dates obtained from Sohgaura, Narhan, Khairadih, Chirand, and Senuwar may be taken into account. There are five radiocarbon dates each from Senuwar and Chirand while Khairadih has yielded three dates. Sohgaura and Narhan have given two radiocarbon dates each. The combined testimony of all these dates places the Chalcolithic culture of the middle

Ganga valley in a time bracket of 1500 BC–900 BC. However, the recent radiocarbon dates from Malhar, Raja Nala Ka Tila and Lahuradeva put the beginning of the Chalcolithic culture around 2000 BC.

The economy of the people as obtained from several sites of the middle Ganga valley was based on plant cultivation, animal domestication, hunting and gathering. Cultivated plants include wheat, barley, rice, jowar, *mung*, gram, kodon (a coarse millet grown only in tropical countries), lentil, *til* (sesame), linseed and pea.

Narhan *26° 19′N; 83° 24′E*

Narhan is situated on the left bank of the Ghaghara river in tehsil Gola of district Gorakhpur. Five field seasons' investigations at this place have revealed the existence of continuous human settlement from about thirteenth century BC to the end of the Gupta period. This long story of human civilization has been divided into five periods on the basis of various diagnostic traits of each period (Singh, 1994).

The highlights of Narhan excavations are the discovery of a culture loosely termed as Chalcolithic by earlier scholars but more aptly termed Narhan Culture by the author. This culture is characterized by wattle-and-daub houses, white painted black-and-red ware, bone tools and an extremely meagre quantity of copper but with no lithic component. Two charcoal samples from Narhan Culture levels have been dated by the radiocarbon method to 1380±110 BC and 1390±110 BC, respectively.

A good quantity of archaeo-botanical remains obtained by flotation from Narhan give a fair idea of the agricultural practices of the Chalcolithic people of the Sarayupar plain. Grains of hulled and six-row barley (*Hordeum vulgare*), club wheat (*T. compactum*), bread wheat (*T. aestivum*), dwarf wheat (*T. sphaeorococcum*) and cultivated rice (*Oryza sativa*) have been identified. The pulses include pea (*Pisum sativum*), green gram (*moong Vigna radiata*), gram/chickpea (*Cicer arietinum*), *Khesari* (*Lathyrus sativus*), besides mustard oilseeds and flax/linseed. Seeds of jackfruit (*Artocarpos integrifolious*) found from these levels is an important discovery. Jackfruit grows wild in the western ghats from where it was taken to cultivation in north India. The present evidence suggests that the Narhan inhabitants cultivated it for its fruit (Saraswat *et al.*, 1994). Commenting upon the state of preservation of archaeo-botanical samples from Narhan, Saraswat (1994: 256) observes:

"The majority of plant remains recovered from Narhan turn out to be a mixture of carbonized seeds and fruits of cultivated and wild plants along with a bulk of wood charcoal pieces. These remains could survive in the carbonized state by being exposed to the heat of fire. The carbonized remains inevitably provide an incomplete picture of the spectrum of plant resources utilized by the ancient settlers at the site. A good number of plant material are unlikely to have come in contact with fire and become destroyed by decomposition. Carbonized remains, thus, represent an adventitiously scorched portion of a larger amount. It is difficult to explain as to how these plant remains were dispersed into the deposits. They are likely to have resulted from specific human activities. Wholly or partially, the occurrence of some material may be incidental as well. The samples are of mixed content. The possibility that the composition of samples of plant remains might vary according to the context of activity, could not be investigated.

The seeds and fruits in quite a good proportion are found deformed and altered due to burning and also damaged by puffing. The degree of puffing varies greatly even in the same sample of seeds and fruits. Quite a few seeds are found in fragile and poor state of preservation, due to which the characteristic details to determine their specific identity are not discernible.

"Apart from chance discovery of sizeable catches of carbonized material, archaeobotanical material consists of impressions of grains and spikelets in pot sherds. The rice husk seems to have been mixed with clay prior to turning it into pots. A few rice grains have also been noticed within the matrix of pottery, but in a peculiarly shrunken state and not always determinable. Wet clay particles formed a fine coating all over the surface of husk and grains during the course of kneading. When the vessel was fired, the organic content of husk and spikelets burnt away, leaving cavities in the hardened clay, lined with fine-grain material in which minute morphological details are found to be represented. A complete impression is less frequently met with. Firing of pottery also brought further shrinkage of the impression, resulting in a reduction in the size of spikelet. When broken open, quite a few sherds have revealed the copious impressions."

DISCUSSION AND CONCLUSIONS

The study of plant material from Imlidih Khurd, Narhan, Chirand and Senuwar demonstrates that rice-wheat-barley-legume agriculture was firmly established in the middle Ganga plain by the end of the third millennium BC and the inhabitants had already commenced cultivating two crops a year. The finds depict the traditional composition of vegetable food available to the people. The main ingredients were cereals and legumes, flvoured with fruits. The culmination of a settled farming community of Narhan culture has already been established. The cultures of Imlidih Khurd and Narhan, both in same ecological zone, shared the same kind of crop economy and enjoyed mutual contacts.

It is commonly held that wheat, barley, lentil and chickpea are crops of Mediterranean origin. These cereals were grown in the northwestern part of this subcontinent in and around the seventh millennium BC. Several varieties of wheat such as the einkorn, emmer and bread wheat have been reported from the Aceramic Period I at Mehrgarh. *Triticum* sp. occurs in the Early Harappan Kot Diji context at Tarakai Qila in the Bannu area in the third millennium BC or possibly a little earlier (Chakrabarti, 1988: 89). These were introduced by the Harappans in the northwestern part of India, during the mid-third millennium BC. Cultivation of Harappan crops have been documented from the OCP phase at Atranjikhera (*ca* 2000–1500 BC) and at all the sites in the areas of our study. The facts suggest the spread of Harappan cultural products within a short span of time over the vast area of the Gangetic plain. It is also clear that grain trade acquired dietary preferences of diverse cultural groups as early as the beginning of the second millennium BC. The grain finds from the earlier phase of cord-impressed ware culture at Imlidih Khurd for the first time pushed back the introduction of crops of the Harappan nutritional trait in the middle Ganga plain as early as the beginning of second millennium BC.

Rice

Rice, an annual grass belonging to the same family as wheat, barley, oats, and rye is the only major grain crop grown almost exclusively as human food. Hence, it was one of earliest plants to be cultivated or domesticated in both tropical and warm temperate region. T.T. Chang (1976a, 1976b, 1989) has shown that the original habitat of *Oryza* was the Gondwana subcontinent. When Gondwana broke up and became Africa, Antarctica, Australia, Malagasay, South America and southeast Asia, *Oryza* species drifted into distinct geographical habitats (Swaminathan, 1984). It is interesting to note that wild rice continues to be gathered by the tribal populations of the Jeypore tract of Orissa state, India and the Batticaloa district of Sri Lanka even today (Chang, 1989: 409).

As regards the origin of rice cultivation, middle Ganga plains provide important evidence. The fact that this cereal was harvested (though not cultivated as yet) in the Proto-Neolithic phase at Chopani Mando has been documented in the preceding pages. The tentative timespan of seventeenth-seventh millennium BC suggested by the excavators for this site, has also been supported by other scholars (Chakrabarti, 1988: 91). The presence of cultivated rice at Koldihwa, dating from 9,000 to 7,000 years BP provides the earliest record of rice cultivation in this subcontinent. Therefore, Wenming (2002: 152) is right in concluding that rice agriculture originated in India no later than 9,000 years BP. The presence of cultivated rice at Lahuradeva (district Sant Kabir Nagar, UP) in the Neolithic context and dated by C-14 method to 5300 BC provides confirmatory data to the chronological framework given to Koldihwa and Mahagara (Tewari *et al.*, 2002). Domesticated rice is present in the lowest levels of Imlidih Khurd, Narhan, Senuwar and Chirand. Other early finds of cultivated rice beyond the middle Ganga plain comprise Atranjikhera (2000–1500 BC) and Lal Qila (1880 BC), but far more important are the evidences of rice cultivation as early as 2700–2800 BC in the region of Haryana. These Early and Mature Harappans in the northwest India were primarily growers of wheat and barley and the rice moved to this region in competition to those crops at such an early stage. Commenting on the diffusion of rice from the mid-Ganga plains to the northwestern part of subcontinent, K.S. Saraswat (in press) writes, "The dissemination of rice cultivation would have been a slow-moving process from the primary homestead or clan in the Gangetic plain to the northwest. The cultivation and early domestication could have taken place concurrently and independently in many localities and moved in northwestern direction through zigzag routes, combined forces of natural and human selection, diverse climates and soils and varied cultural practices must have led to the great ecological diversity in this crop."

It is interesting to note that the origin and antiquity of rice cultivation in China has undergone a substantial change during the last three decades. By the late seventies and early eighties of the previous century, many rice remains dated from 6,000–5,500 years BP have been found in Hemudu, Zhejiang Province, and some rice remains from the early period were also found at Majiabang and Daxi sites. Therefore, it was believed that the lower reaches of the Yangtze Valley might be one of the centres of origin. Soon after, rice remains dated from even earlier time—9,000–7,000 years BP—were found at Pengtoushan and Chengbeixi sites located in the middle Yangtze Valley. Since these

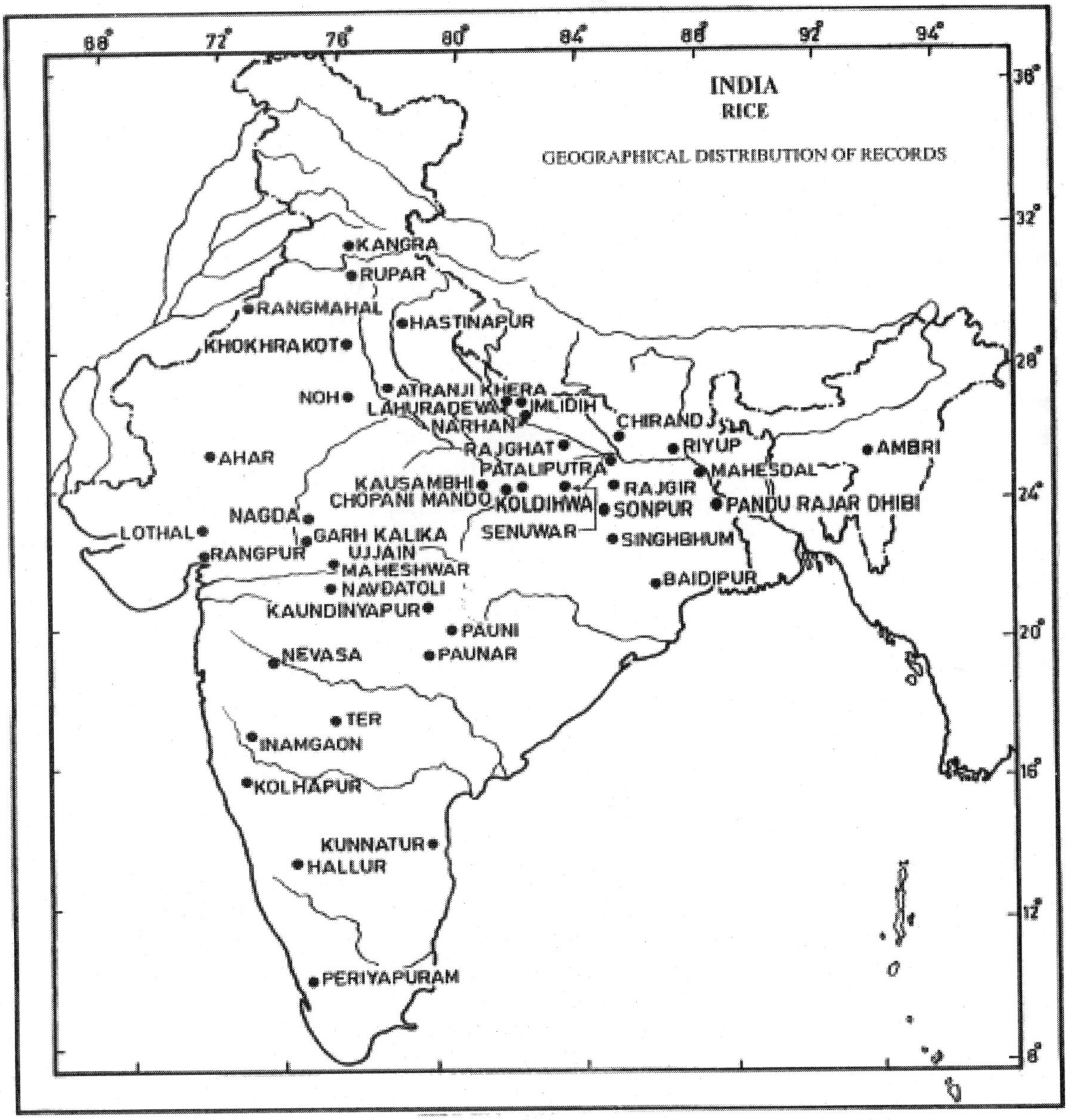

Fig. 1: Geographical distribution of records of rice

represented the earliest remains of rice agriculture known at that time, the theory that it originated in the Yangtze Valley gradually became axiomatic. Recent discoveries from Yuchanyan, Hunan Province and Xianrendong and Diaotonghuan, Jiangxi Province, dated the origin of rice agriculture to more than 10,000 years ago. Using C-14 dating, the charcoal and pieces of pottery collected from Yuchanyan can be dated to 12,060 ±120 and 12,320±120 years BP, respectively, and samples collected from the two sites in Jiangxi Province were nearly contemporary, dating to much earlier than the origins of domesticated wheat and barley in West Asia (Wenming, 2002: 151–52).

Millets

The presence of Jowar millet (*Sorghum bicolor*) in the earliest levels of Imlidih Khurd and the occurrence of Ragi or finger millet in the Narhan culture levels at the same site raise interesting questions. Both these millets, together with pearl millet or Bajra (*Pennisetum typhoides*) are considered to be plants of African origin. They are associated with the Mature Harappa culture datable to 2500–2000 BC. At present, their introduction is believed to have taken place some time in the third millennium BC and once introduced, these millets had a major impact on the subsistence system of subcontinent (Possehl, 1999: 240). Presently, these cereals of African origin account for over 13 per cent of the total cereals produced in India (Possehl, 1999: 240). The presence of Jowar millet at Imlidih Khurd in the pre-Narhan phase, datable to about 1800 BC, raises interesting questions about the date and possible route of diffusion of this cereal from Africa to the middle Ganga plain. It must be confessed that it is difficult to pinpoint the time bracket and route of diffusion of this cereal in the middle Ganga plain in the present state of research because there are still significant gaps in our knowledge.

REFERENCES

Agrawal, D.P. and J.S. Kharakwal, 2002. *South Asian Prehistory*, New Delhi, Aryan Books International.

Agrawal, D.P. and Sheela Kusumgar, 1974. *Prehistoric Chronology and Radiocarbon Dating in India*, Delhi, Munshiram Manoharlal.

Chakrabarti, D.K., 1988. *Theoretical Issues in Indian Archaeology*, Delhi, Munshiram Manoharlal.

Chang, T.T., 1976a. The Rice Cultures, *Philosophical Transactions of the Royal Society of London* B 275, 143–157.

Chang, T.T., 1976b. The Origin, Evolution, Cultivation, Dissemination and Diversification of Asian and African Rice. *Euphytica* 25: 425–41.

Chang, T.T., 1989. Domestication and spread of the Cultivated Rices. *Foraging and Farming* (eds) David R. Harris and Gordon C. Hillman, London, Unwin Hyman, 408–17.

Costantini, L., 1984. The Beginning of Agriculture in the Kachi Plain: The Evidence of Mehrgarh, *South Asian Archaeology 1981*, (ed.) B. Allchin, Cambridge, 29–33.

Gupta, H.P., 1976. Holocene Palynology from Meander Lake in the Ganga Valley, District Pratapgarh, UP, *The Palaeobotanist*, 25: 109–119.

Jarrige, J.F., 1985. Continuity and Change in the North Kachi Plain (Baluchistan, Pakistan) *South Asian Archaeology 1983*, (eds) J. Schotsmans and M. Taddei, Naples, 35–68.

Kennedy, K. A.R., 2000. *God-Apes and Fossil Men: Paleoanthropology in South Asia*, Michigan, The University of Michigan.

Misra, B.B., 1989. Chopani Mando, *An Encyclopaedia of Indian Archaeology*, (ed.) A. Ghosh, New Delhi, Munshiram Manoharlal.

Misra, V.D. and M.C. Gupta, 1995. Pre-NBP Cultures in the Mid-Ganga Valley, *Prof. Agam Prasad Mathur Felicitation Volume*, Vol. III, Agra, Huzuri Bhavan, Peepal Mandi: 21–34.

Misra, V.D., 1997a. The Mesolithic Cultures of the Vindhyas and the Ganga Valley with Special Reference to Environment and Archaeology, *Him-Kanti* (Prof. K.P. Nautiyal Felicitation Volume) (eds) B.M. Khanduri and Vinod Nautiyal, Delhi : Book India Publishing Co.: 41–42.

Misra, V.D., 1997b. *Some Aspects of Indian Archaeology*, Allahabad, Prabhat Prakashan.

Misra, V.D., 1999. A Brief Note on Indian Stone Age Archaeology, *The Dawn of Indian Civilisation* (upto *c. 600 BC*) (ed.) G.C. Pande, Vol. I, Part I: 129–138.

Misra, V.D, 1999. Agriculture, Domestication of Animals and Ceramic and Other Industries in Prehistoric India: Mesolithic and Neolithic, *The Dawn of Indian Civilisation (upto c. 600 BC)* (ed.) G.C. Pande, Vol. I, Part I: 233–266.

Misra, V.N., 2001. Prehistoric Human Colonisation of India, *Journal of Biosciences, 26* (4), Supplement: 491–531.

Misra, V.N., 2002. Keynote Address *Mesolithic India* (eds) V.D. Misra and J.N. Pal, Allahabad, University of Allahabad, 1–66.

Pal, J.N., 1986. *Archaeology of Southern Uttar Pradesh*, Allahabad, Swabha Prakashan.

Pal, J.N., 2002, Mesolithic Gangetic Plain, in *Mesolithic India*, V.D. Misra and J.N. Pal (eds), Allahabad, University of Allahabad, 289–305.

Pant, D.D. and Rekha Pant, 1980, Preliminary Observations on Pollen Flora of Chopani Mando (Vindhyas) and Mahadaha (Ganga Valley) in *Beginnings of Agriculture*, G.R. Sharma and Others, Allahabad, Abinash Prakashan, 229–230.

Poesshl, G.L., 1999. *Indus Age: The Beginnings*, Philadelphia, University of Pennsylvania Press.

Roy, S.R., 1989, Chirand, *An Encyclopaedia of Indian Archaeology*, (ed.) A. Ghosh, New Delhi, Munshiram Manoharlal.

Saraswat, K.S., N.K. Sharma and D.C. Saini, 1994, Plant Economy at Ancient Narhan in *Excavations at Narhan 1984–89*, P. Singh (ed.), New Delhi, B.R. Publishing Corporation, 255–342.

Sharma, G.R., V.D. Misra, D. Mandal, B.B. Misa and J.N. Pal, 1980. From Hunting and Food-Gathering to Domestication of Plants and Animals—Epi-Palaeolithic to Neolithic—Excavation at Chopani Mando, Mahadaha and Mahagara, *History and Archaeology*, I, Allahabad University.

Sharma, G.R., V.D. Misra, D. Mandal and J.N. Pal, 1980. *Beginnings of Agriculture*, Allahabad: Abinash Prakashan,

Sharma, S., M. Joachimski, M. Sharma, H.J. Tobschall, I.B. Singh, C. Sharma, M.S. Chauhan and G. Morgenroth, 2000. Holocene Palaeoclimate Reconstruction Using Multiproxy data from Sanai lake, Central Ganga Plain, India, *Quaternary Science Review* (in press).

Sharma, S., M. Joachimski, M. Sharma, H.J. Tobschall, I.B. Singh, C. Sharma, M.S. Chauhan and G. Morgenroth, 2003. Monsoon Variability, Vegetation Change and Human Inhabitation over 15,000 years from multi proxy record of a lake deposit in Ganga plain, India, Terra Nova (in press).

Singh, B.P., 1989–90, The Chalcolithic Culture of Southern Bihar, *Puratattva*, 20, 83–92.

Singh, Indra Bir, 1996. Geological Evolution of Ganga plain—An Overview, *Journal of Palaeontological Society of India*, 41, 99–137.

Singh, P., 1974. *Neolithic Cultures of Western Asia*, London-New York, Seminar Press.

Singh, P., 1991. *The Neolithic Origins*, Delhi, Agam Kala Prakashan.

Singh, P., A.K. Singh and I.J. Singh, 1991–92, Excavations in Imlidih Khurd, *Puratattva, No.* 22, 120–122.

Singh, P., 1992–93, Archaeological Excavations in Imlidih Khurd 1992, *Pragdhara, No.* 3, 21–36.

Singh, P., 1994, *Excavations at Narhan 1984–89*, New Delhi, B.R. Publishing Corporation.

Singh, R.L., 1971. *India: A Regional Geography*, Varanasi, Banaras Hindu University; National Geographical Society of India.

Stamp, Sir Dudley, 1967, *Asia: A Regional and Economic Geography*, London, Methuen & Co. Twelfth Edition.

Swaminathan, M.S., 1984. Rice, *Scientific American*, 250 (1): January 1984, 80–93.

Thomas, P.K., P.P. Joglekar, V.D. Misra, J.N. Pandey and J.N. Pal, 1995. Faunal Evidence for the Mesolithic Food Economy of the Gangetic plain with Special Reference to Damdama, *Bio Archaeology of Mesolithic India: An Integrated Approach:* 322–28.

Thomas, P.K., P.P. Joglekar, V.D. Misra, J.N. Pande and J.N. Pal, 1996. Faunal Evidence for the Mesolithic Food Economy of the Gangetic plain with special Reference to Damdama, in *Bioarchaeology of Mesolithic India: An Integrated Approach*, Colloquium XXXIII of the International Union of Prehistoric and Protohistoric Sciences (eds) G.D. Afanas'ev, S. Cleuziou, J.R. Lukacs and M. Tosi, 255–66. Forli: ABACO Edizioni.

Tewari R., R.K. Srivastava, K.K. Singh, 2002. Excavation at Lahuradewa, District Sant Kabir Nagar, U.P., *Purātattva*, 32: 54–62.

Tewari, R., P.C. Pant, I.B. Singh, S. Sharma, M. Sharma, P. Srivastava, A.K. Singhvi, P.K. Mira and H.J. Tobschall, 2002. Middle Palaeolithic Human Activity and Palaeoclimate at Kalpi in Yamuna Valley, Ganga Plain, *Man & Environment*, 27, No. 2: 1–13.

Tewari, R., R.K. Srivastava, K.K. Singh, K.S. Saraswat, and I.B. Singh 2003. 'Preliminary Report of the Excavation at Lahuradewa, District Sant Kabir Nagar, U.P. 2001–2002: Wider Archaeological Implications', *Prāgdhārā*, 13 (in press).

Varma, R.K., 1999. The Emergence of Culture in the Vindhya-Ganga Region with Special Reference to Old Stone Age, *The Dawn of Indian Civilization (upto c. 600 BC)* (ed.) G.C. Pande, Vol. I, Part I: 175–204.

Wheeler, R.E.M., 1946. Archaeological Fieldwork in India: Planning Ahead, *Ancient India*, 5: 4–11.

Wenming, Yan. 2002. The Origins of Rice Agriculture, Pottery and Cities, *The Origins of Pottery and Agriculture* (ed.) Yasuda, Y. New Delhi, Roli Books, 151–56.

Yasuda, Yoshinori (ed.), 2002. *The Origins of Pottery and Agriculture*, New Delhi, Roli Books.

CHAPTER 2

Beginnings of Agriculture in the Vindhyas (North–Central India)

V.D. Misra

INTRODUCTION

The Vindhyas loom large in the geology, geography and history of India. This mountain range is a dividing line between the north India and the Deccan.[1] The economic potential of the area is well known. However, the historical and cultural significance of the Vindhyas can hardly be overlooked. The prehistoric rock art of India is heavily concentrated in this region. It is worth mentioning that recently, Bhimbetka has been declared a world heritage site. The Vindhyas also shelter a number of tribes which, with their ethnic and cultural diversity, constitute a living museum. By studying them from close quarters, one can get some glimpses about the functioning of the past. These tribes provide, needless to say, an insight for understanding and appreciating the archaeological objects discovered in explorations and excavations.

The geologico-archaeological investigations were conducted by the Department of Ancient History, Culture and Archaeology, University of Allahabad, independently or occasionally in collaboration of Geological Survey of India or the Department of Anthropology, University of California, Berkeley, USA. These studies concentrated on the northern Vindhays covering parts of Allahabad, Mirzapur and Sonbhadra districts of Uttar Pradesh and Rewa, Satna, and Sidhi districts of Madhya Pradesh in the river valleys of the Belan, a tributary of the Tons and in that of the Son, a tributary of the Ganga. These surveys have added a new chapter in the studies of the Stone Age cultures of India. The explored area measures 5400 sq km in the Belan Valley and 13601 sq km in the Son Valley. A continuous story of man from the mid-Pleistocene to early Holocene has been unfolded. In this tract of land, the Pleistocene formations are found exposed over a considerable area. In fact, an 18-m thick section of the Belan[2] and 30-m thick section of the Son[3] present a vertical record of evolution of human culture right from the Lower Palaeolithic to the Mesolithic through the Middle Palaeolithic and Upper Palaeolithic. The exposed geological units not only furnish information about the evolving trends in the lithic tool-typology and technology, but also throw a welcome light on the climatic fluctuations that the region underwent in different times. H.D. Sankalia, the doyen of Indian Archaeologists, was so overwhelmed by seeing the Belan section and its archaeological contents that he called it a 'textbook section'.[4] From the geological formations of the Belan and Son, thousands of animal fossils have been obtained. The species include elephas, bos, equus, gavialis, hippopotamus, etc. In this connection, it may be pointed out that these investigations placed the Vindhyan region on the fossil map of India for the first time. Prior to these investigations, this honour was given to the Siwalik in the north and the Narmada Valley in central India. These archaeological investigations in the Belan and Son Valleys established the stratigraphic position of the Upper Palaeolithic in India. Prior to this discovery, there was a good deal of debate regarding the status of the Upper Palaeolithic culture in India. According to one group of scholars, the Palaeolithic cultures in India evolved in the same fashion as in Western Asia and Europe, particularly in the Western Europe, the order of succession being Lower Palaeolithic, Middle Palaeolithic and Upper Palaeolithic[5], the last one leading to the Mesolithic. However, another group of scholars opined that in India, the Mesolithic succeeded the Middle Palaeolithic. There was no Upper Palaeolithic at all. Before these scholars, there was African Model.[6] The investigations in the Belan and Son Valleys sealed the controversy and proved the independent status of the Upper Palaeolithic. The Mesolithic sites such as Chopani-Mando[7] in the Belan Valley, Baghor II[8], Medhauli[9], Banki[10] and Kunjhun I[11] in the Son Valley and Lekhahia on the Kaimur[12] have furnished valuable information not only of the inner evolution of the Mesolithic culture but also its linkage with the Upper Palaeolithic on the one hand and Neolithic on the other. In the Palaeolithic and early Mesolithic cultures, man practiced hunting and gathering. But in the late phase of the Mesolithic, there are records of a quasi-sedentary settlement. In this phase, as the available evidence from Bagor[13] in Rajasthan and Adamgarh[14] in Madhya Pradesh indicate domestication of some animals originated.

The emergence of Neolithic culture in human history marks a distinct phase. Domestication of animals and cultivation of plants are supposed to be the characteristic features of the emerging Neolithic cultures. In the initial stage of the studies, scholars attributed five salient features: sedentary settlement; domestication of animals; cultivation of plants; occurrence of polished celts; and introduction of handmade pottery[15]. However, further researches brought to light such sites in a wide range of area covering Greece[16], Turkey[17], Western Asia[18], Iran[19], Afghanistan[20] and North-Western Indian subcontinent,[21] where a preceramic stage of Neolithic culture has also been recorded. Besides these traces, such Neolithic sites were also brought to light where there were either no polished celts at all or they were a few and far between, thereby indicating that these implements did not play any significant role in the life of the said Neolithic community. Gordon Childe[22] marks this as the emergence of the Neolithic phase heralding a revolution in human history. For him, this was the first revolution. But it was contended by equally eminent scholars that in history, revolution is a rare phenomena. We are generally confronted with evolution. One form evolves into another form with the passage of time. Today is born out of the womb of yesterday. Every tomorrow is genetically related with today. Hence, it was stated that the emergence of the Neolithic phase would not have been a sudden phenomenon in the history of mankind. Its roots should be looked into the preceding cultures. Over a wide stretch of area, the symptoms of the genesis of the Neolithic culture started coming as archaeological investigations gathered momentum not only in western Asia but in India also. The forthcoming evidence displayed a gradual evolution from Mesolithic to Neolithic. In western Asia, people living in the Mesolithic stage had already started collecting wild grains.[23] Such discoveries, needless to say, added a new dimension to the problems related to the genesis of Neolithic cultures. It was a complex phenomena and a result of long-drawn experiments which man had been carrying on across the centuries.

The discovery of Neolithic implements in archaeological context in the Vindhyan region, covering Allahabad, Mirzapur and Sonbhadra districts in Uttar Pradesh and Rewa and Sidhi districts in Madhya Pradesh, is a recent event in Indian archaeology. In this connection, it may be pointed out that Neolithic celts fashioned on basalt had been collected from the hilly tracts of Lalitpur, Hamirpur, Banda, Allahabad and Mirzapur districts from time to time[24]. Triangular celts, conforming to the characteristic tools type of south Indian Neolithic group, was the main tool type discovered in the area. Besides, a few rounded celts, supposed to be the characteristic tool type of the Neolithic cultures of northeast India and further afield of southeast Asia, were also found. As these two distinct tool types, i.e. triangular celts and rounded celts were surface collection and were found together, regarding their respective chronological position, three possibilities were raised:

(1) Two distinct tool traditions represented by triangular celts and rounded celts coming from south India and northeast India, respectively, met together in the hilly tract of the Vindhyas. And as such these should be treated as contemporaries.

(2) The tradition of triangular celts coming from south India entered the Vindhyan area prior to the rounded celt tradition.

(3) The rounded celt tradition constituted the base over which the triangular celt tradition was superimposed.

Before the excavation at Koldihwa[25], the consensus of the archaeologists was tilted in favour of the second possibility. But as the round celts have been obtained in the course of excavations in a definite archaeological context not only from Koldihwa, but also from Mahagara[26] and Panchoch[27] in the Belan Valley in Allahabad district, from Indari[28] and Tokwa[29] in the Adwa valley in Mirzapur district in Uttar Pradesh and from Kunjhun[30] on the Son in Sidhi district of Madhya Pradesh, its high antiquity is suggested both by its archaeological context and C^{14} dates. Till now, at no site the triangular celt is found in a proper archaeological context. The available evidence would thus suggest a greater antiquity to the rounded celts in comparison to the triangular celts at least in the Vindhyan area.

As indicated earlier, the Neolithic culture of the northern Vindhyas has a district personality. After the excavation at Koldihwa, persistent and planned archaeological investigations were carried out in the Vindhayan area over decades by the Department of Ancient History, Culture and Archaeology, University of Allahabad, aimed at discovering more and more Neolithic sites. The explored area ranges between 24° and 28° 30′ north parallel and the 80° and 84° east meridian, roughly bounded in the north by the Ganga and in the south by the Son. The explored sites are situated in the northern Vindhyan zone of Uttar Pradesh and the adjoining plateau region of Madhya Pradesh. These Neolithic sites are found located in the valleys of the Belan, Adwa and Son. Of the important sites, mention may be made of Koldihwa, Mahagara, Panchoha, Kukarahata[31], Deoghat[32], Koilariha[33], Futaha[34], Chauridih-Kotia[35], Bansghat[36], Daiya[37] and Patehari[38] in the Belan Valley in Allahabad district; Magha[39], Indari, Berauncha[40], Sinduria Mahalpur[41], and Tokwa in the Adwa Valley in Mirzapur district; and Kunjhun, Lalnahia[42] and Dhodauhi[43] in Sidhi district of Madhya Pradesh. Of these, Koldihwa, Mahagara, Panchoha, Indari, Tokwa, and Kunjhun have been excavated. These excavations, needless to say, have placed the Neolithic culture of the Northern Vindhyan region in a proper archaeological context. The following pages present a brief report on these excavated sites.

KOLDIHWA

The archaeological site Koldihwa (Lat 24° 54′ 30″ N; Long. 82° 2′ E) is situated on the left bank of the Belan at distance of 85 km southeast of Allahabad city in Koraon subdivision, Allahabad district. The site, roughly rectangular in outline and flat in elevation, is fairly extensive (500 m × 200 m), the longer axis being west-east. The mound is eroded partly by river and nullah and partly by badland topography, as a result of which, it is broken into a number of small mounds of which two are situated along the river and the other one lies at the southeast corner. Excavated in 1972–73, 1973–74, 1974–75 and 1975–76, Koldihwa yielded threefold cultural sequence—Neolithic, Chalcolithic and early Iron Age. The Neolithic deposit (0.45 m) rests on compact natural soil. The salient feature of the Neolithic culture here included handmade pottery, ground stone implements, rounded celts furnished on basalt, Neolithic blades and microliths.

MAHAGARA

The site of Mahagara (Lat. 24° 54′ 50″N; Long 82° 3′ 20″ E), lying in the hilly tract of the southern part of Koraon subdivision of Allahabad, is situated on the right bank of the river Belan to the west of the confluence of the old and new channels of the river at a distance of about 85 km southeast of Allahabad city. Mahagara is just apposite Koldihwa. The mound, irregular oval in form, is bounded on the southeast by the old and on the southwest by the new channel of the Belan, measuring about 800 sq m. It was excavated in three sessions, 1975–76, 1976–77 and finally 1977–78. It is a single culture site. The Index trench brought to light 2.60-m thick occupational deposit. The excavation at the site exposed as many as 18 hutments, circular or oval on plan. These were found littered with potsherds, animal bones, fragments of querns, mullers, hammer stones, microliths, rounded polished celts, etc. Postholes were detected on the periphery of the floors of these hutments.

A cattle pen measuring 12.5 × 7.5 m surrounded by eight floors (four house units) was exposed at Mahagara. The pen, irregular rectangular on plan, was enclosed by

28 post holes with three openings. According to the excavator, the absence of any post hole inside the cattle pen indicated that it would have been open to the sky. Hoof-prints of cattle of different age groups were detected in the pen.

PANCHOHA

The Neolithic site of Panchoha (Lat. 24° 55′ 45″ N; 82° 2′ 30″ E) is located at a distance of 2.5 km northwest of Koldihwa on the right bank of the Belan. The site was found disturbed on account of constant cultivation. A small trench laid out in 1975–76 brought to light a 0.60-m thick deposit. Traces of circular or oval floors were detected at the site. Archaeological objects identical to those of Neolithic Koldihwa ware obtained.

INDARI

Indari is located on the left bank of the Kahenjua Nala, a seasonal tributary of the Adwa in the revenue village of Manigara in Mirzapur district. It has been largely disturbed on account of tilling. As Neolithic objects were obtained in course of exploration, a small trench was laid out at the site in 1981. A 0.75-m thick occupational deposit was exposed. There is reason to believe that, when intact, the thickness of the Neolithic strata would have been more. The site yielded handmade pottery, a few ground stone tools and microliths. Traces of hutments were also exposed.

TOKWA

The site (Lat. 24° 54′ 50″ N; Long 82° 16′ 45″ E) of Tokwa is situated on the influence of the Belan and Adwa in Lalganj subdivision of Mirzapur district at a distance of 68 km from Mirzapur city in southeastern direction and 8 km from the east of Baraundha. It is by far the most extensive and comparatively well-preserved Neolithic site discovered till date in the Vindhyas. Measuring about 27579 sq m, the site has been excavated by the Department of Ancient History, Culture and Archaeology, University of Allahabad in two seasons, 2000 and 2003. A total cultural deposit of 4 m was exposed, of which 2.47 m was Neolithic. A threefold culture sequence—Neolithic, Chalcolithic and Iron age—was recognized. The objects obtained included handmade pottery, bone arrowheads, beads of terracotta and semi-precious stones, and fragments of foodprocessing equipments like a quern, muller, hammer stones, rounded celts and microliths. A large number of animal bones in comparatively well-preserved condition were also obtained. Cereals were also found in flotation. Traces of oval or round hutments with post holes on the periphery were also brought to light. Wattle and daub fragments were also obtained.

KUNJHUN

The site (Lat 24° 30′ 30″ N; Long. 82° 12′ 32″ E) is situated on the right bank of the Son in Sidhi district of Madhya Pradesh. Handmade pottery, microliths and a few celts of the rounded variety were obtained from the Khetaunhi formation, the last formation of the Son. With a view to ascertaining the character of Neolithic culture at the site, a step trench was laid out in 1982. In all, 14 sq m, was exposed. Potsherds and animal bones were found scattered, starting from a depth 3.7 m of the bottom.

The ceramic assemblage consisted of corded ware, red ware, coarse black-and-red ware and gritty ware. A sizable number of sherds was found burnished. A few painted sherds having affinity with Malwa ware were also found from the last phase of the deposit. The presence of rice and rice-husk in pottery was also noticed. Of the other finds, mention may be made of animal bones of cattle, deer, antelope and dog. A few microliths including cores, flakes and blades were also obtained.

Though no evidence of settlement was obtained in the excavated trench, the absence of rolling on potsherds suggested that these had not travelled much before being deposited here. The ceramic industry, in comparison to that of Koldihwa and Mahagara in the Belan Valley, exhibited evolved traits both in firing and finish. The occurrence of a few sherds of the Malwa ware, as indicated above, in the assemblage may suggest a contact between the two cultures—Neolithic and the Malwa.

From the aforesaid, it would be evident that the Vindhyan Neolithic culture is characterized by sedentary settlement, handmade pottery comprising corded, rusticated, burnished red, burnished black and occasionally crude black-and-red wares, ground stone celts of rounded variety with rectangular or oval cross-section, food-processing equipments like quern, muller, hammer stone, microliths and domestication of animals and cultivation of plants. Single-tanged bone arrowheads, earthen discs with central perforation generally made of broken pottery, spherical terracotta beads and shell-pendants—are some other objects found at these sites. Of the animal bones obtained from Koldihwa, Mahagara, Tokwa and Kunjhun, mention may be made of bones of cattle, sheep, goat, deer, antelope, etc. Remains of aquatic creatures like turtle, fish, etc., were also recorded along with the bones of birds. Cattle, sheep and goat appear to have been domesticated.[44]

Both burnt rice and rice-husk have been found embedded in the pottery. In some cases, rice-chaff has also been discerned. The available evidence indicates that rice husk was used as degraissant. Palaeobotanical studies have revealed that the inhabitants here were using both wild and cultivated rice (*Oryza sativa*). Recent studies by Emma Harvey and Dorian Fuller of the botanical material of Koldihwa and Mahagara have thrown welcome light on the range of cereals cultivated by the Vindhyan Neolithic people.[45] Crops identified at Koldihwa and Mahagara included rice, barley, wheat, pulses, sesame and small millets (*Bracharia ramosa, panicum sunatrensa and Setaria verticillata*). The available evidence suggests that both at Koldihwa and Mahagara, wild and cultivated varieties of rice and small millets were used from the beginning of the Neolithic settlement. Barley, wheat, pulses and sesame were introduced at these sites subsequently. At Tokwa, Dr Saraswat has tentatively come to a similar conclusion. He has identified rice, wheat, barley, green gram, lentil and small millets. According to his preliminary observations, the Neolithic people were cultivating rice and small millets in the beginning. Subsequently, they also started cultivating barley, wheat, green gram, lentil, etc.[46] Supportive evidence has also been obtained from Senuwar[47] in Rohtas district of Bihar, where the Neolithic period is divided into two subgroups IA and IB. From the lowest part to the middle of the period

IA, only grains of cultivated rice (*Oryza sativa*) were found. But the people were also using wild plants like job's tear, foxtail/bandra, wild rice, *Jharberi*, *Chaulai* and *Jangali Palak*. But in the later phase of IA, there was a major shift in the economy. New cereals were added including barley, wheat, *jowar* millet, lentil, field pea, finger millet/ *ragi* and *khesari*. The introduction of the double crop pattern clearly indicates that agriculture had started playing a major role by the later phase of the Neolithic culture in the Vindhyas.

The range of cereals cultivated during the Vindhyan Neolithic raises some interesting issues. While rice, small millet, green gram, etc., constituted kharif crops to be raised during the rainy season, barley, wheat, lintel etc., the winter crops, were raised during the winter season. In this connection, it would be pertinent to point out that as rice needs sufficient water, its cultivation would have been initiated in an area receiving enough rainfall. But for winter crops like barley, wheat, etc., that amount of water was not needed. Cultivated cereals—whether falling in the category of Kharif or Rabi (winter) crops—were wild grasses. Stone Age man started gathering seeds of these grasses for consumption. This would indicate that the first cultivation of any cereal would have taken place only in that area where its wild prototypes were available. Viewed against this background, one finds the distribution of wild rice over a wide area covering eastern India, southeast Asia and southwestern China.[48] The distribution of wild barley, wheat, lentil, etc., on the other hand, is recorded from the border of Pakistan-Afghanistan to Iran, Levant, Turkey, etc[49]. This would suggest that rice and possibly small millets were available in the wild form in the Vindhyas before these were cultivated. But wheat, barley, etc., would have been introduced in this area, i.e. Vindhyan area only when people of the Vindhyas came in contact with the people of north and northwestern India. The absence of rice not only in the Neolithic sites of northwestern India like Mehrgarh but also at the early Harappan and mature Harappan sites provides significant readings. This would suggest that by that time northwestern India had not acquired knowledge of rice. Northwestern and north India acquiring knowledge of rice cultivation and central India acquiring expertise of barley, wheat, lentil, etc., cultivation would have been possible only through north-south dialogue. Ideas and informations could have exchanged hands in that dim distant past.

Who were the authors of the Neolithic culture of the Vindhyas? Had they come from outside with readymade ingredients of a Neolithic culture, or had their moorings in the pre-Neolithic cultures of the Vindhyas? In the present state of our knowledge, the Vindhyan Neolithic represents its independent personality. In this connection, the excavations at Chopanimando on the old Belan in Koraon subdivision of Allahabad deserve our attention. This site is situated at a distance of only 3 km from Koldihwa and Mahagara. A total habitational deposit measuring 1.55 m. was exposed at the site. The entire deposit has been divided into three cultural phases—Phase I: Epipalaeolithic; Phase II: Mesolithic; and Phase III: Advanced Mesolithic or Proto-Neolithic. The last phase, i.e., Proto-Neolithic is characterized by handmade pottery, some new tool types such as the isosceles triangle and tranchets on the one hand and plans of hutments, hearths along with food-processing equipments like querns and mullers, anvils, hammer stones, ring stones, etc., on the other. Burnt clay lumps with reed impression suggest wattle-and-daub structure. Remains of wild rice were also found embedded in the burnt clay lumps with reed impression that suggested wattle and daub structure. The presence of wild rice, however, when viewed in the context of the presence of food-processing equipments and traits of semi-sedentary settlement, would suggest the collection of wild rice during this phase. The appearance of handmade pottery in the Proto-Neolithic stage, at Chopanimando and all the Vindhyan Neolithic sites excavated so far is another linkage between the two. The first occurrence of tranchets during the Proto-Neolithic phase at Chopanimando vis-à-vis its presence from Koldihwa, Mahagara, Panchoha, etc., would also suggest continuity between the two, i.e. Proto-Neolithic and Neolithic. This would indicate transition from the stage of gathering wild rice to that of its cultivation. In this connection, it may also be pointed out that the presence of wild sheep, goat, and cattle has been recovered from gravel III and IV and from Chopanimando in the late Mesolithic/Proto-Neolithic context. At Mahagara, these animals are found to have been domesticated. This evidence would suggest that these animals were domesticated from their wild prototypes available in the area during the earlier period and were not introduced in the area as readymade breeds from outside. The microlithic component of the Vindhyan Neolithic culture also constitutes a running theme from Mesolithic to Neolithic without any breack. The evolution of the Vindhyan Neolithic from the underlying Mesolithic is, thus, clearly indicated.

As far the agricultural practices prevalent in the Vindhyan Neolithic culture were concerned nothing definite can be said whether the people had started tilling land with plough or they were simply digging with hoe-like implements to sow seeds. It is not unlikely that for reaping, they might have been using a sickle-like implement prepared by inserting microlithic blades in the frame of wood or bone for which analogies are available from western Asia. The agricultural practice, needless to say, might have been in an experimental stage ascribable to the advanced Mesolithic/Proto-Neolithic phase. Besides cereals, flesh of animals, aquatic creatures and birds were also consumed. Hunting and catching continued along with agriculture and domestication of animals.

To understand the antiquity of agriculture in the Vindhyas, one has to know the antiquity of the Vindhyan Neolithic culture; the two issues are intertwined. To compute the chronology of the Neolithic culture of the Vindhyas, both archaeological data and C^{14} dates are available. Even then, one has to admit that the question is still in a melting pot. In this connection, the following points deserve attention:

(1) In the Vindhyas, transition from the late Mesolithic to Neolithic is documented. We start getting evidence of quasi-sedentary settlement during the late Mesolithic period. It is demonstrated by the presence of hutments, food-processing equipments like quern, muller and handmade pottery. Collection of wild rice is also indicated.

(2) There is linkage between the late Mesolithic and Neolithic on the construction of hutments, use of food-processing equipments, handmade pottery and microliths.

(3) It is not unlikely that one may, in furture, gather evidence of domestication of animals at some Late Mesolithic sites in the Vindhyan area, as such traces have been already obtained from Bagor and Adamgarh.

(4) As per available evidence in the Late Mesolithic phase, people had started collecting wild rice available in the area but in the Neolithic period, along

with collecting such rice, the domestication of rice (*Oraza sativa*) was also resorted to.

(5) All the excavated Vindhyan Neolithic sites are characterized by handmade pottery. These sites are also devoid of metal. Bone tools are very very limited. This would indicate that the inhabitants heavily depended on stone tools, whether microliths or heavy duty foots. These points demonstrate the primitive character of this culture.

(6) C^{14} dates have been obtained from three Vindhyan Neolithic sites, viz., Koldihwa,[50] Mahagara[51] and Kunjhun[52]. From Koldihwa, three dates reading 4530±185 BC, 5440±40 BC and 6570±210 have been obtained. Mahagara has yielded six dates, of which two are thermoluminiscent (TL) dates and the rest C^{14} dates. The two TL dates read 2265 and 1616 BC. The four C^{14} dates read 1400±150 BC, 1330±120 BC, 1440±100 BC and 1480±110 BC. Three calibrated dates from Kunjhun River-face read 1565–1265; 2675–2515 and 3530–3335 BC. Possehl and Rissman have proposed a date for the Vindhyan Neolithic at between 4000 to 1200 BC.[53] But if the dates of Koldihwa are taken into consideration, the antiquity of the Vindhyan Neolithic might be pushed back to 7th millennium BC. At one stage, doubts were expressed about the high antiquity of the dates obtained from Koldihwa. At that time, the dates obtained from Mahagara were treated as model. But the dates obtained from Kunjhum tilted the balance in favour of Koldihwa dates. Supportive evidence has recently come from Lahuradeva[54], Sant Kabir Nagar in the Ganga Valley. Two C^{14} date read 5320±90 BP (Cal. 4220, 4196 and 4161 BC); 6290±160 BP (Cal. 5298 BC). In the light of these dates, the antiquity of the Neolithic culture of Lahuradeava may be pushed back to the middle of 6th millennium BC. If the Vindhyan area served as springboard for the Neolithic Culture of the middle Ganga Valley, the greater antiquity for the Vindhyan Neolithic would be only a logical hypothesis.

In the light of the above data, the antiquity of the farming in the Vindhyas area may be pushed back to 7th millennium BC. There appears to be a gradual transformation from hunting-gathering to domestication of animals and cultivation of plants. This Neolithic phase with rice cultivation appears to be indigenous, Confirming the thesis of the famous Russian plant geneticist Vavilov who observed that rice, sugarcane, a large number of legumes and many tropical fruit plants including the mango, etc., originated in India.[55] In the light of the available data, it may be pointed out that in the Indian subcontinent, two regional centres for the cultivation of cereals were developing in 7th millennium BC. In the northwest, barley, wheat, etc., were being cultivated, while in the Vindhyas, the cultivation of rice was initiated.

NOTES AND REFERENCES

1. Wadia, D.N., 1973. The Vindhayan System, *Geology of India*, New Delhi, Tata McGraw-Hill Publishing Company, pp. 121–131.

2. *Indian Archaeology: A Review,* 1966–97 (Abbreviated as *I.A.R.*), pp. 35–38, see also V.D. Misra, 1977, Prehistory in Uttar Pradesh in *Some Aspects of Indian Archaeology,* Allahabad, Prabhat Prakashan, pp. 30–31.
3. Williams, M.A.J. and K. Royce, 1983, Alluvial History of the Middle Son Valley North Central India, in G.R. Sharma and J.D. Clark (eds), *Palaeo-environments and Prehistory in the Middle Son Valley,* Allahabad, Department of Ancient History, Culture and Archaeology, University of Allahabad, pp. 9–21.
4. Sankalia, H.D., 1974, *Prehistory and Protohistory of India and Pakistan,* Deccan College, Postgraduate and Research Institute, Pune, p. 172.
5. Allchin, B. and A. Guddi, 1971, Dunes, Aridity and Early Man in Gujarat, Western India, *Man* Vol. VI, pp. 248–65, see also V.N. Misra, 1962, Problems of Terminology in Indian Prehistory, *Eastern Anthropologist,* Vol. XV, No. 2, pp. 113–124.
6. Subbarao, B., 1958, *Personality of India,* Second edition, M.S. University of Baroda, Baroda, p. 39.
7. Sharma, G.R., V.D. Misra, D. Mandal, B.B. Misra and J.N. Pal, 1980, Excavations at Chopani-Mando, *Beginnings of Agriculture,* Allahabad, Abinash Prakashan, pp. 33–76.
8. Sussman, C., R. Blumenschine, J.D. Clark and B.B. Misra, 1983, Preliminary Report on Excavations at the Mesolithic Occupation Site at Baghor II Locality *Palaeo-environments and Prehistory in the Middle-Son Valley,* Sharma, G.R. and J.D. Clark (eds) pp. 161–196.
9. *Indian Archaeology: A Review,* 1982–83.
10. Misra, V.D., B.B. Misra, U.C. Chattopadhyaya, D.K. Shukla and P. Sinha, 1992–93, Excavation at the Primary Context Mesolithic Occupation site of Banki, District Sidhi, M.P.: A Preliminary Report, *Prāgdhārā* No. 3, pp. 5–10.
11. *I.A.R.* 1981–82, pp. 43–44.
12. *I.A.R.*, 1963–64, pp. 51–53, See also, Sharma, G.R. 1965, Excavation at Lekhahia. *Indian Prehistory,* 1964, Poona, Deccan College Postgraduate and Research Institute, Poona, pp. 76–77.

    See also V.D. Misra, 1977, Prehistory in Uttar Pradesh, *Some Aspects of Indian Archaeology,* Allahabad, Prabhat Prakashan, p. 53.
13. Thomas, P.K., **Yr?** Role of Animals in the Food Economy of the Mesolithic Culture of Western and Central India: *A Source Book of Indian Archaeology,* Vol. I (eds) F.R. Allchin and D.K. Chakravarty, Delhi, Munshiram Manoharlal, pp. 341–47.
14. R.V. Joshi and M.D. Khare, 1966, Microliths Bearing Deposit of Adamgarh Rock-shelter in D. Sen and A.K. Ghosh (eds) *Studies in Prehistory: Robert Bruce Foot Memorial Volume,* Calcutta, Firma K.L. Mukhopadhyay, pp. 90–95; R.V. Joshi, 1978, *Stone Age Cultures of Central India,* Pune; see also G.L. Possehl and Paul C. Rissman, 1992, in *Chronologies in Old World Archaeology,* (3rd edition), Chicago and London, University of Chicago Press, pp. 475–76.
15. George, A. Christopoulos (ed.), 1974, *Prehistory and Protohistory*. Theocharis, D.R. *History of the Hellenic World,* Athens, Ekpotike Athenon S.A., pp. 56–57.
16. *Ibid.*
17. Singh, P., 1991, *The Neolithic Origins,* New Delhi, Agam Kala Prakashan, pp. 39–58.
18. *Ibid.*, pp. 59–92.

19. *Ibid.,* pp. 93–118.
20. *Ibid.,* p. 121, see also Srivastava, V.C. *The Prehistoric Afghanistan: A Source Book,* 1982, (Allahabad), Indological Publications, pp. 45–46, 212–216.
21. Shashi Asthana, 1985, *Pre-Harrapan Cultures of India and the Border Land,* Delhi, Books and Books, pp. 110–20; see also V.D. Misra 'Agriculture, Domestication of Animals and Ceramic and other Industries in Prehistoric India: Mesolithic and Neolithic, pp. 240–41. *The Dawn of Indian Civilization,* Vol. I, Part 1 (ed.) G.C. Pande, Delhi, Project of History of Indian Science, Philosophy and Culture, (1999).
22. Childe, Gordon, *What Happened in History,* 1957, Harmondsworth: Pelican, p. 49.
23. *Ibid.,* p. 48; see also Moore, A.M.T. A Pre-Neolithic Farmer's Village on Euphrates, *Scientific American* 241(2) Anjust 1979 pp. 62–70; E. Kislev-Emergence of Wheat Agriculture, *Paleorient,* Vol. 1012–1984 pp. 61–70; Moore, AMT, Agricultural origins in the Near East: A model for 1980s *World Archaeology,* 14(2), Oct. 1982 pp. 225–36. Henri'de Contenson, 'Early Agriculture in Western Asia'; *The Hilly Flanks and Beyond* (eds) T.C. Young, P.E.L. Smith and P. Mortensern. The University of Chicago, the Oriental Institute Studies in Ancient Oriental Civilization No. 36, pp. 57–67.
24. Allchin, B. and F.R. Allchin, 1968, *Birth of Indian Civilization,* Harmondsworth: Pelican, p. 172.
25. IAR 1970–71 pp. 36–37; IAR 1971–72, p. 44; IAR 1973–74, pp. 26–27; IAR 1975–76, pp. 45–46; Misra, V.D. 'Excavations at Koldihwa: *Some Aspects of Indian Archaeology,'* 1977, pp. 107–19, see also Sharma, G.R. et al., 1980, Excavation at Mahagara, *Beginnings of Agriculture,* 1980, Allahabad Avinash Prakashan, pp. 135–116.
26. Sharma, G.R. 1980, et al. *Beginnings of Agriculture,* Allahabad, Avinash Prakashan, 133–200.
27. *Ibid.,* p. 136; I.A.R. 1975–76, p. 47 see also Misra, V.D. *Some Aspects of Indian Archaeology,* 1977, Allahabad, p. 57.
28. I.A.R., 1980–81, pp. 72.
29. V.D. Misra, B.B. Misra, J.N. Pal and M.C. Gupta 'Explorations at Tokwa: A Neolithic-Chalcolithic settlement on the Confluence of the Belan and Adwa Rivers *Peeping Through the Past'* (eds) S.C. Bhattacharya, V.D. Misra, J.N. Pandey and J.N. Pal, 2000, Allahabad, Department of Ancient History, Culture and Archaeology, University of Allahabad, pp. 45–57; V.D. Misra, J.N. Pal and M.C. Gupta 'Excavation at Tokwa: A Neolithic-Chalcolithic Settlement,' pp. 59–72 *Prāgdhārā* No. 11.
30. *IAR* 1981–82, pp. 43–44.
31. Explored by the Department of Ancient History, Culture and Archaeology, University of Allahabad see also J.N. Pandey, 'Northern Vindhyan Neolithic Settlement' in *Rural Life and Folk Culture in Ancient India,* (eds) U.N. Roy, V.D. Misra and J.N. Pandey, Allahabad, Department of Ancient History, Culture and Archaeology, University of Allahabad, 1988, p. 13.
32. *Ibid.*
33. *Ibid.*
34. *Ibid.*
35. *Ibid.*
36. *Ibid.*

37. *Ibid.*
38. *Ibid.*
39. *Ibid.*
40. Explored by the Department of Ancient History, University of Allahabad.
41. I.A.R. 1981–82, p. 47.
42. Explored by the Department of Ancient History, Culture and Archaeology, University of Allahabad in 1983.
43. I.A.R. 1982–83, p. 58.
44. Alur, K.R. 'Faunal Remains from Vindhyas and the Ganga Valley' *Beginnings of Agriculture* (eds) G.R. Sharma, V.D. Misra, D. Mandal, B.B. Misra and J.N. Pal, 1980, Allahabad, Abinash Prakashan, pp. 220–26.
45. Harvey Emma, Dorian Fuller and J.N. Pal 2003. Early Agriculture of Neolithic Koldihwa and Mahagara submitted to the Department of Ancient History, Culture and Archaeology, University of Allahabad, Allahabad.
46. Dr. Saraswat participated in the excavations at Tokwa in 2003. A lot of cereals were collected. Dr. Saraswat has tentatively identified the presence of small millet, barley, wheat, green gram, lentil, etc. Currently, he is working on the material.
47. Singh, B.P., Stages of Cultural Development in the Middle Ganga Plains—A Case Study of Senuwar, *Prāgdhārā* No. 11 pp. 109–118.
48. Chang, T.T. Domestication and spread of the Cultivated Rices, *Foraging and Farming* (eds) David R. Harris and Gordon C., Hillman 1989, Unwin Hyman, London, pp. 408–17; Chang, T.T., 1976, The Rice Cultures *Philosophical Transaction of Royal Society*, Oxford University Press, London. B. 275, pp. 143–157; see also Singh, P., op. cit., pp. 149–152.
49. Moore, A.M.T. 1982, Agricultural Origins in the Near East: A Model for the 1980s, *World Archaeology* 14 (2), pp. 225–36; Moore A.M.T. 1989, The Transition From Foraging to Farming in South West Asia: Present Problems and Future Directions, *Foraging and Farming* (eds) David R. Harris and Gordon C. Hillman, Unwin Hyman, London, pp. 620–639; See also Kislev, M.E. Emergence of Wheat Agriculture-*Paleorient* Vol. 10/2, 1984, pp. 61–70; Peake Harold and Fleure, H.J. *Peasant and Potters*, 1927, London, Oxford University Press, see also, Singh, P. *The Neolithic Origins*, Agam Kalan Prakashan, Delhi (1991), pp. 1–16.
50. Possehl, G.L. and Paul C. Rissman, 1992. *Chronologies in Old World Archaeology* Volume I, Chicago and London, University of Chicago Press, p. 461.
51. *Ibid.*
52. *Ibid.* See also pp. 473–474 for Kunjhan, p. 474 for Mahagara and p. 474 for Koldihwa.
53. Possehl, G.L. and Paul C. Rissman, 1992, Chicago and London, University of Chicago Press, pp. 473–474.
54. Tiwari Rakesh, R.K. Srivastava and K.K. Singh, Excavation at Lahuradeva, District Sant Kabir Nagar, Uttar Pradesh, *Purātattva* No. 32, pp. 55–56.
55. Vavilov, N.I., *Phytogeographic Basis of Plant Breeding*, 1951, The Origin, Variation, Immunity and Breeding of Cultivated Plants, New York: Ronald Press, p. 29.

CHAPTER 3

Beginnings of Agriculture in the Vindhya-Ganga Region

Radha Kant Varma

INTRODUCTION

The beginning of animal domestication and agriculture are the most important processes in the evolution and development of human culture. Domestication of anima and plan set in motion the processes of such cultural changes, which enabled man to create an artificial environment for himself. It brought in changes in the hunter-gatherer pattern of life and effected the natural ecosystem. Cultivation brought in large-scale replacement of the wild biota by domesticated and semi-domesticated organisms. In other words, it transformed a natural ecosystem into a largely artificial one created and maintained by man. It provided man with an assured supply of plant and animal food. This transformation resulted in a large number of chain reactions that ultimately ushered in an era of all-round development.

The beginning of agriculture is shrouded in obscurity. How, when and where it started are questions to be investigated. Before we attempt to take up these issues, it would be worthwhile to examine the general concepts regarding the evolution of economic sequence—hunting, intensified hunting-gathering and early farming. Each of these activities is associated with the Palaeolithic, Mesolithic and Neolithic stages of man's cultural evolution. It is mostly understood that during the Mesolithic period, man still was nomadic, supplemented his diet with gathering and thus gradually became acquainted with grain. This period was followed by an era of primary village farming efficiency associated with domestication of plants and animals. These stages reflect, in a very broad sense, the cultural stages of development. Culture change or modification is too complex a process to be simplified thus. For example, it is not correct to postulate that permanent settlements grew up as a result of agriculture, though there is hardly a doubt that it helped in creating permanent settlements. There are a number of pre-agricultural societies belonging to the Mesolithic period, which indicate that man had started making permanent or semi-permanent bases. Instances may be quoted of agricultural communities living in seasonal camps till recently. There are a number of agricultural communities in the hilly regions that maintain two sets of habitations—an upper village for the summer period and a lower one for the winter.

Domestication of flora and fauna was a gradual process and must have taken a sufficiently long period of time. The available records show that the man became aware of seeds and wild grains most likely in the period of transition from advanced hunting to the hunting and gathering phase, which is known as the Mesolithic period. It has been rightly suggested (Harris, 1969: 7–9) that cultivation probably originated with the manipulation of relatively complex ecosystems in which a wide variety of wild plants and animals were used for subsistence by forager bands localized within circumscribed territories. Gradually, such bands of gatherers, fishermen, and hunters developed an intimate knowledge of the plants and animals in the limited area they inhabit. Close involvement with certain of those item in a seasonally scheduled pattern of wild food procurement (Rannery, 1968), or as organism of utilitarian or ritual significance, gradually led to proto-cultivation.

Archaeology has it own limitations. We must appreciate the fact that incipient cultivation or domestication had only a minor subsistence role, and in no case does it seem to have had any marked effect on cultural change. The new economy was established so gradually and with such little disruption to established patterns that we can expect it to be scarcely visible in the archaeological records. The hint of something new can be gathered mainly by the reaping and processing tools which the grain eaters used and left. These tools are the main indicators of culture change. In order to investigate and understand the beginnings of agriculture in the Vindhya-Ganga region, it is essential to examine the cultural components of the Mesolithic settlements where evidence for the use of vegetal food processing is available. Microlithic blades and bladelets, which are used in sickles for harvesting plants and edible seeds, attain a particular type of gloss on their cutting edges. This gloss can be detected by microscopic examination on a microscope. On such evidence, it may be presumed that vegetal food formed an essential part of their subsistence economy. Likewise, quern and mullers are supposed to be the tools used for processing edible grains. It is really unfortunate that only a few such Mesolithic sites have been excavated so far and at most of the places food grains have not been retrieved through floatation or other techniques. The other indicator for the use of processed vegetal food grains is the study of dental pathology conducted on the tooth of skeletal remains. It has been observed that the societies that subsist more on vegetal food show a greater degree of teeth decay because the grinding of vegetable foods inevitably produces a certain amount of grit from the quern mixed with the floor: excess consumption of such food wears down the teeth. These are only circumstantial evidence on the basis of which presumption for the use of vegetal food can be made.

It is true that transition from gathering to producing economy may or may not have taken place on a Mesolithic settlement but there is hardly any doubt that grounds for the transition must have been prepared on such settlements. The close observations made by the foraging populations regarding the seasonality of the crops—its sprouting and ripening time and the type of environment needed for its propagation—must have helped the beginning of cultivation. From a wide range of wild grasses, some must have been selected for repeated harvesting. From among them, a few plants would have been chosen for deliberate planting.

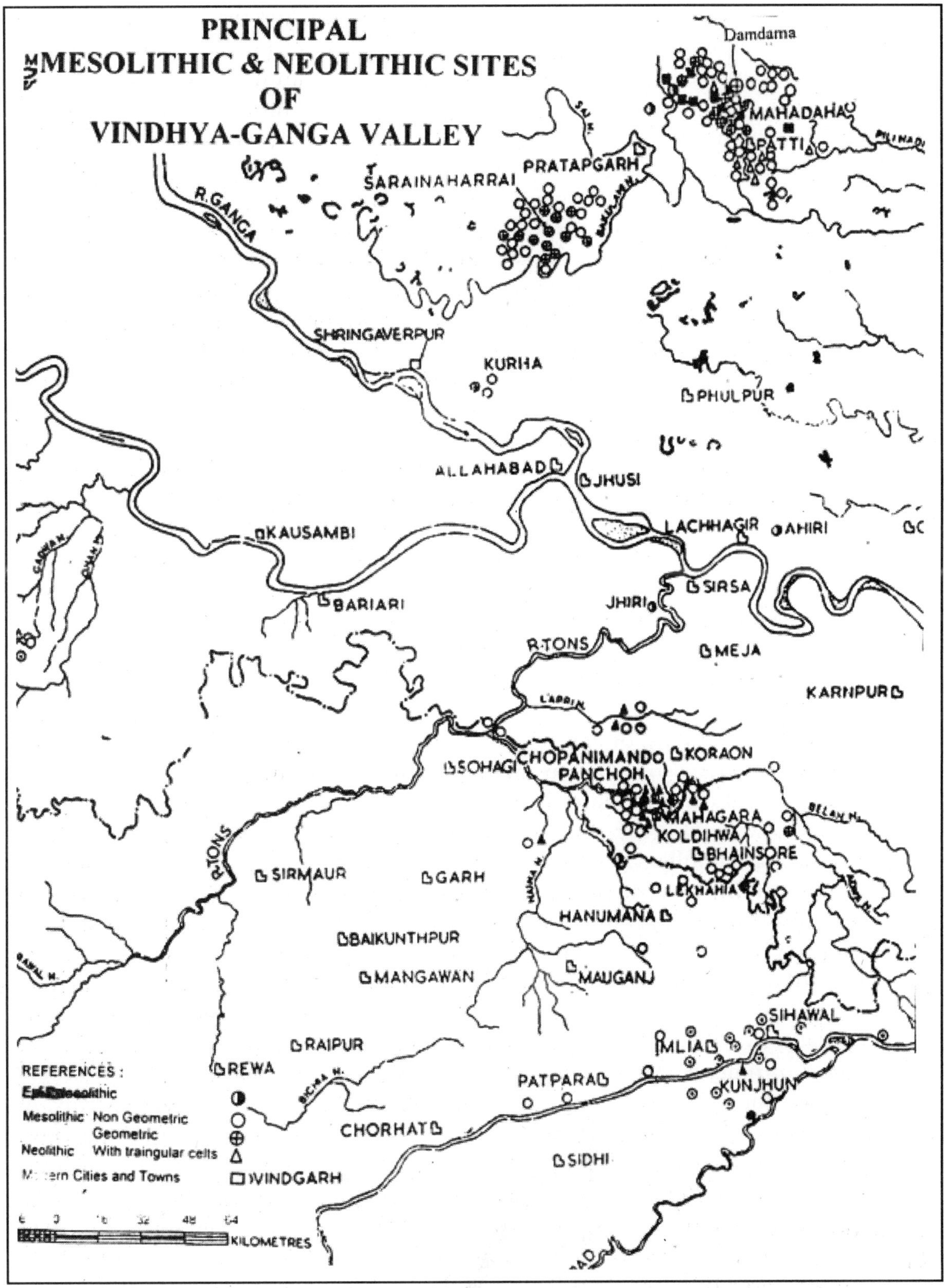

Fig. 1: Principal Mesolithic and Neolithic Sites of Vindhya-Ganga-Valley

Explorations and excavations in the Vindhyan region and the Ganga Valley have given a complete sequence of culture right from the earliest times upto the beginning of the historical period. They are especially important for understanding the process of transition from advanced hunting to the beginning of agriculture. From the viewpoint of subsistence economy, the Mesolithic settlements may be broadly divided into two broad categories:

1. In the first category have been included such sites where no evidence for the use of vegetal food processing is available. In other words, the subsistence economy at such sites depended basically on hunting. These sites are devoid of stone objects such as querns and mullers that are used for processing cereals or seeds. In this category are included the excavated sites of Morahana Pahar, Baghaikhor and Lekhahia in the Vindhyan region and Sarai Nahar Rai in the Ganga Valley. All the Vindhyan sites are located in close vicinity of each other on the Great Deccan road near the village Bhainsore, which is nearly 40 km from Mirzapur and 2.5 km from Hanumana, in Rewa district (M.P.). Excavations at these sites reveal that the Mesolithic industries locally evolved from the Epi-Palaeolithic stage. Within the Mesolithic, four stages (Varma, 1965; 1986) have been marked:

    i. Non-geometric stage
    ii. Geometric without pottery
    iii. Geometric with pottery
    iv. Diminutive microlithic stage

    In the absence of querns and mullers from such sites, it is presumed that their subsistence economy was mainly based on hunting.

2. In the second category have been placed sites that reveal clear-cut evidence of the use of vegetal food also. On such sites, besides microliths, a large number of querns and mullers have been noted. Among such sites, special mention may be made of Chopani Mando in the Belan Valley under the shadow of the Kaimur hills of the Vindhyan ranges and the sites of Mahadaha and Damdama in the Ganga Valley. Of the three excavated sites in the Ganga Valley, only Sarai Nahar Rai is devoid of heavy stone tools such as quern and muller. The accumulation of habitation deposit also at Sarai Nahar Rai site is of only one period. But the habitation at the sites of Mahadaha and Damdama continued for a fairly long period of time. At Mahadaha, a 60-cm habitation deposit has been recorded, divisible into four layers corresponding to the four periods of burial activity. At Damdama, a habitation deposit of 1.5 metres was observed which has been divided into ten layers and nine periods of burial activity. Each period is represented by a floor level (Pl. 1) also. Among all the so far excavated Mesolithic sites of India, the site of Damdama is unique for its thick habitation deposit and continuous habitation. Five seasons of excavations have not revealed any evidence that the site was ever deserted. The people migrated to the Ganga Valley during the Epi-Palaeolithic period when the river was carving out its older terrace, Khadar,

and stabilizing its channel by breaking old meanders. The people who came from the Vindhyan region settled on the banks of the left out meanders that looked like a lake. In the beginning, the migrations were probably seasonal. But soon they became stable bases and the people who settled on these sites continued to live for generations together till the site was finally abandoned.

Here it is worthwhile to remember that the Ganga Valley is completely devoid of stones and the nearest source of raw material for the manufacture of tools is the Vindhyan ranges from where the original inhabitants of Damdama had migrated. It has also been observed that the number of microliths decreased in the later levels. It was probably because fresh inputs of raw material were not being received. It is also important to mention here that the use of bone tools increased with the decrease in the number of microliths. It seems that the dwellers were locally trying to solve the problem of paucity of raw material by using bones and antlers as an alternative to stones.

Significantly, a very large number of querns (Pl. 2) and mullers have been obtained from both Mahadaha and Damdama. The importance of querns and mullers can be easily understood if we consider the fact that stones for manufacturing querns and mullers were not available locally and they had to be imported from the Vindhyan region. Furthermore,

Plate 1: Floor levels of Damdama

Plate 2: Rice husk (Uncultivated) in pot sherd-Chopanimando

instead of importing raw material for the manufacture of microliths, these people were getting querns and mullers. It seems that in their subsistence economy, querns and mullers played a more important role. Under the circumstances, it may be presumed that vegetal food played an important role in their economy. Kennedy (1986) and Lukacs (Lukacs and Pal, 1993) have made a detailed study of the Dental Anthropology of these people. They are of the view that the tooth crown size is large and is interpreted as well adapted to a coarse diet that contains grit derived from grindstones (Lukacs, 1993: 762). A preliminary report on plant remains from Damdama reveals evidence of several wild plant species (Kajale, 1990). Three taxa have been identified to species; wild grasses (*Heteropogon contortus* and several of indeterminate type), goosefoot *(Chenopodium album)* and purslane (*Partulaca oleracea)*. Seeds belonging to three additional plant groups have been identified to family: buck wheat (*Polygonaceae)* nightshade *(Solanaceae)* and mint (Labiatae). Today, *C. album* is found in the wild as well as a weed in cultivated fields. The young shoots are eaten as greens, and boiled seeds are mixed with or substituted for grains (Weber, 1991). Purdane is a succulent, yellow flowered herb. In addition, there is evidence of millet-like grains and the Indian jujube (Zizyphus). They are commonly harvested in February and March and consumed fresh, dried, candied, stewed, or smoked. But the excavations have so far not yielded any evidence that cultivation of any type whatsoever had commenced. Since there is no evidence of any habitation overlaying the Mesolithic deposits, it is presumed that the site was deserted before the beginning of cultivation. What actually was the cause of the desertion? It is difficult to say, but probably the rise in population and the depletion of the natural resources within the catchment area contributed to the abandonment of the Mesolithic settlements of the Ganga Valley.

Where did the Mesolithic folk of the Ganga Valley shift or migrate is an another important question. In the Ganga Valley, where Mesolithic sites are located, till date no Neolithic settlement has been reported. It seems that these people once again returned to the Vindhyan region from where they had come. Since the Ganga Valley sites are completely devoid of pottery, it is reasonable to presume that the Mesolithic dwellers of the Ganga Valley shifted back to the Vindhyan region before the introduction of pottery. The people from the Ganga Valley, on their return, introduced quern and mullers in the Vindhyan region with which they had become familiar while living in the Ganga Valley. Pottery had already emerged in the Vindhyan region. There is yet another question to be settled. Where from did the pottery come? Did it evolve locally or was it the result of borrowing? This question is yet to be solved satisfactorily though there are scholars who think that it was a local development (Pal, 1986) because handmade fragmentary pottery was found during excavation from the third phase of the Vindhyan Mesolithic, which so far, is the earliest evidence of pottery, at least for this part of the country.

Among all the sites excavated so far in the Belan Valley, Chopanimando is of special significance. Chopanimando (Lat. 24° 30″ N and Long. 82° 4′ 45″ E) is situated on terrace T2 on the old channel of River Belan at a distance of 77 km from Allahabad in east and southeasterly direction in Koraon subdivision of Allahabad (Sharma, et al., 1980). The site is spread on an area of roughly 15,000 sq m. An area of nearly 680 sq m was excavated to varying depths from 30–35 cm to 80 cm. The excavations brought to light 10 layers divisible into three uninterrupted cultural phases: Epi-Palaeolithic, Mesolithic, Advanced Mesolithic or Proto-Neolithic.

The lowermost layer (10) yielded the tools of Epi-Palaeolithic, which represents a period of transition from the upper Palaeolithic to Mesolithic.

Layers 9 to 4 belong to the Early Mesolithic phase. It has been further divided into two sub-periods IIA, Early Mesolithic with non-geometric microliths. It comes from layers (9) and (8). Geometric microliths were obtained from layers (7) to (4). It has been designated as IIB, Early Mesolithic with geometric microliths.

The Advanced Mesolithic or Proto-Neolithic phase has been represented in the layers (3) to (1). Fragile handmade pottery makes its appearance at this stage. The ground plan of 13 huts and four hearths were also unearthed. Other significant finds associated with this stage are anvils and hammers, fragments of quern and mullers, ring stones, burnt clay lumps with reed marks, animal bones, etc.

The Proto-Neolithic phase at Chopanimando is important from a number of viewpoints:

1. This phase brings out a change in the living pattern within the Mesolithic. The appearance of quern and mullers indicates that these people substituted their subsistence economy with vegetal food. It further finds support from the discovery of wild rice in the lumps of burnt clay and the use of rice husk and straw as

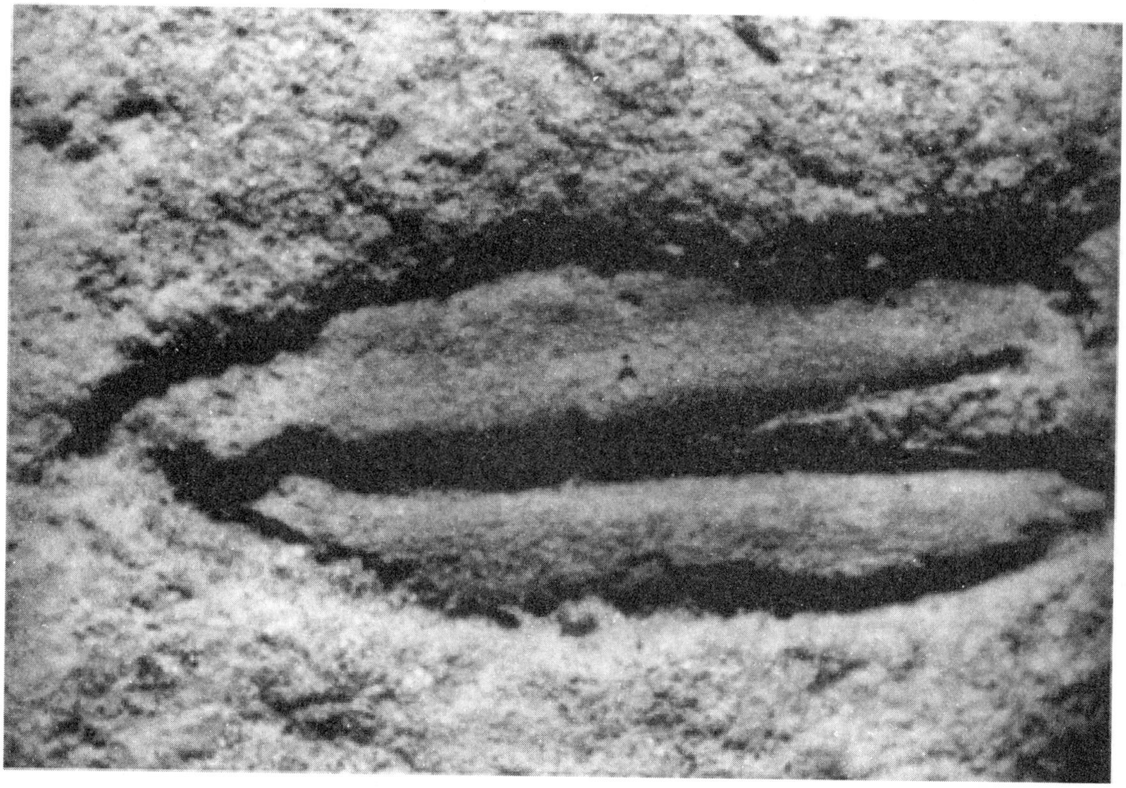

Plate 3: Microphoto of rice husk(*Oryza Sativa*) – Mahagara

Plate 4: Rice and rice impressions on a sherd of corded ware – Mahagara

degraissants. In the Vindhya-Ganga region, wild rice was discovered for the first time on a Mesolithic settlement.

2. It is presumed that the return of the people of the Ganga Valley gave a new impetus to the Mesolithic folk of the Vindhyas. These people probably introduced the exploitation of vegetal food and its processing. Out of all the wild cereals and seeds, rice was chosen as the favourite item. Rice seems to have been available in plenty in wild form in river valleys of the Belan and its tributaries. We find that in Phase III at Chopanimando, wild rice is available in a burnt lump of clay. Querns and mullers are also introduced at this stage. We have seen that purely handmade pottery is also introduced at this juncture of time. Each piece of pottery differs from the other. "It is evident that each pottery piece was an individual creation" (Sharma, et al., 1980: 15). Clearly, it was a completely recent innovation at the site. Here it would not be out of place to mention that excavations carried out in the Bhainsore group of rock-shelters consisting of Morahana Pahar, Baghaikhor and Lekhahia offer evidence of pottery in the third stage, i.e. after the introduction of geometric shapes. The pottery collected from these places are extremely fragile. Some pieces are impressed and some resemblance may be observed with the cord impressed ware of Koldihwa and Mahagara. But no quern and muller was found associated with them. It seems that pottery appears before the introduction of querns and mullers.

3. Also, the site of Chopanimando was probably deserted before the introduction of cultivation. Though cultivation did not start here, but the stage was set for a take off. It is a well-known phenomenon that cultural development is a cumulative process in which inventions and innovations take place only when the necessary antecedents already exist. Here, everything was at a breaking point but unfortunately, before the introduction of agriculture or cultivation, the site was deserted and the focus now shifts to Koldihwa and Mahagara, which are at a distance of not more than three kilometres southwest as the crow flies from the site of Chopanimando in the Belan Valley.

Koldihwa (Lat. 24° 54′ 30″ N, 82° 2′ E) is situated on the left bank of the River Belan near Deoghat and the site of Mahagara (Lat. 24° 54′ N, Long. 82° 3′ 20″ E) is on the right bank to the west of the confluence of the old and new channels of the river. Besides these, a few more Neolithic sites were located in the close vicinity (Misra, V.D. 1999), out of which, a few have been excavated. The Neolithic site of Panchoh (Lat. 55° 45′ N, Long. 82° 2′ 30″ E) is at a distance of 2.5 km northwest of Koldihwa on the right bank of the Belan. Another site, Kunjhun (*IAR*, 1976–76:27), is on the right bank of River Son at a distance of 36 km from Sidhi towards the northeast. It is the southernmost site of the Vindhyan Neolithic. The surface finds from Kunjhun closely correspond to the assemblage recovered from the sites in the Belan Valley. The easternmost site is Tokwa (Lat. 24° 54′ 20″ N, Long. 82° 16′ 46″ E) on the Belan river itself. It is being excavated at present (Misra, et al., 2001:59) by the Department of Ancient History, Culture and Archaeology, University of Allahabad. There is hardly a doubt that there must have been many more sites but due to bad land topography, they are completely destroyed.

Excavation at Koldihwa has revealed evidence of three cultural phases: Neolithic, Chalcolithic and Iron Age. The Neolithic deposit that rested on gravel IV was of 45 cm thickness only. Excavations at Mahagara revealed a Neolithic deposit of 2.60 m. The excavations at Koldihwa and Mahagara present conclusive evidence of the cultivation of rice and its use as the staple food of the people inhabiting these places. The querns recovered from these places show greater concavity due to constant use than those collected from the Mesolithic sites. The querns recovered from the Mesolithic sites are shallower (Sharma, et al., 1980: 20).

Plentiful evidence of rice has been found from the excavations of Koldihwa and Mahagara in loose form as well as embedded in potsherds. The imprints of caryopsis, stalk, glume, etc., have been found only on potsherds, especially of the cord-impressed and rusticated wares.

The material collected from the potsherds of cord impressed ware, excavated at Koldihwa was examined by Vishnu Mittre of the Birbal Sahani Institute of Palaeobotany, Lucknow. He identified it as a domesticated variety of Oryza Sativa (Vishnu Mittre, 1977: 141). Later, Te-Tzu-chang of the International Rice Research Institute, Philippines, also made SUMP study of five samples. He confirmed "From the texture of the glume surface, we may say that they represented the cultivated form" (Sharma, et al., 1980:182).

There are a number of C14 dates based on charcoal samples from Koldihwa which are given as under:

PRL 224 6570±210 BC
PRL 100 5440±240 BC
PRL 101 4530±185 BC

Besides, there was one more date PRL 223 reading 1440±120 BC. But this date is obtained from a sample which was collected from two consecutive layers—one Neolithic layer and the other of the Chalcolithic period (Sharma, et al., 1980 : 198). It is worthwhile to mention here that besides the above dates, there are six more carbon 14 dates from Mahagara. Out of these, two are TL dates. All the dates range from 2265 BC to 1330±120 BC. These dates are not consistent with the dates from Koldihwa. The excavators are of the view that the Mahagara dates need not be considered because they are inconsistent (Sharma, et al., 1980:198) within themselves. Even the samples from one layer have given extremely inconsistent dates; hence they seem to be contaminated.

Besides the evidence of the domestication of rice, we have evidence of the domestication of cattle, sheep, goat, and horse. A cattle pen measuring 12.5 × 7.5 m surrounded by huts on tour sides was recognized at the site of Mahagara. It was enclosed by 28 post holes with three openings. Hoofprints of cattle of different age groups were detected in the pen. It clearly shows that domestication of animals had reached an advanced stage. It may be mentioned that the bones of all these animals in their wild form have been found from pre-Neolithic levels. "The evidence thus suggests that these animals were domesticated from their wild prototypes available in the area" (Misra, 1999: 244).

On the basis of the then evidence (Nov. 1980), Prof. Sharma opined that "As the evidence stands, the possibilities of incipient cultivation of rice at that stage may not be ruled out. The evidence of domestication of rice (*Oryza sativa*), in view of the radio-carbon dates from the early Neolithic levels of Koldihwa and its earlier history in the Advanced Mesolithic, it is the earliest in the world." Prof. Sharma (Sharma, et al., 1980, op. cit.) does not think that rice cultivation in the region was introduced as a result of cultural diffusion. "The transition was achieved in a small compact area of 12 sq km in the valley of the Belan and was an indigenous development." The theory of diffusion from an original culture in Western Asia is ruled out on account of its absence there. Prof. Sharma goes on to say, "equally untenable is the view of diffusion from Southeastern Asian sources, as the evidence for cultivation of rice in all these areas is much later and the lithic tools and other components of the Vindhyan Neolithic are almost absent. He concludes "till more coherent and conclusive evidence of greater antiquity is available from some other centre, this area of India, the Belan Valley in the Vindhyas, will remain an original, primary and nuclear centre of Neolithic transformation, for the beginning of agriculture, of rice and of domestication of animals."

There is nothing final in archaeology. Its last page always remains open to accommodate new discoveries. In recent years, Japanese and Chinese archaeologists have added much to our knowledge in the field of early agriculture and ceramics. They have pushed back the dates of earliest rice technology to 12000 BP and beyond (Agrawal, 2001). Recent progress in archaeological studies has pointed to rice domestication in the middle reaches of the Yangtze River as early as 10,000 years BC. At the Bashidang archaeological site, located about 20 km northeast of the Chengtoushan site, rice husks date back more than 10,000 years. From this evidence, the Liyang plain is considered to be one of the earliest places where rice cultivation originated (Agrawal, 2001:17).

In view of the early dates of rice cultivation in Japan and China, Dr. Agrawal feels "there seem to be a distinct possibility that early rice technology may have entered the Ganga Valley, through the South Asian route through Mekong delta, Brahmaputra, and central Himalayas from south China" (Agrawal, 2001:17). He further adds, 'there is every reason to expect that in the Ganga Valley, Vindhyas and Rajasthan, the beginning of agriculture may go back to early Holocene'.

There is hardly a doubt that cultural evolutions have mostly taken place on account of cultural diffusion of ideas and technology. Man has always been more prone to borrow than to invent but there is hardly a doubt that all cultural advancement is not the result of diffusion. Parallel evolution of ideas and technology cannot be ruled out. In the present case, on account of early date of rice in Japan and China and similarity of cord impressed pottery, the possibility of 'rice technology' coming to the Vindhya-Ganga region from the Yangtze Basin (China) cannot be ruled out, but more evidence is required. On the contrary, there is circumstantial evidence, which points to an independent origin of rice technology in the region, as suggested by Sharma.

1. It is known that the geological formations in the Belan Valley have preserved the evidence of gradual evolution of man's culture right from the Lower Palaeolithic period onwards. The evolution from the Upper Palaeolithic to Epi-Palaeolithic and Mesolithic is well documented through the excavations in and around the rock-shelters in the Vindhyan region, Belan Valley and the Ganga Valley. The

evolution of a distinct pottery in the advanced Mesolithic phase and the use of wild rice in Chopanimando have no parallel elsewhere.

2. The Neolithic settlements of Koldihwa and Mahagara are located directly on Cemented Gravel IV from which non-geometric microliths have been obtained. Gravel IV geologically correlates with the last terrace formation in the Belan Valley on which the site of Chopanimando is situated. The excavation at Chopanimando has given an unbroken sequence of evolution from the Epi-Palaeolithic to advanced Mesolithic or Proto-Neolithic, where the use of wild rice is undisputed. From the cemented gravel IV, we have now four dates from two different laboratories which are given as under:

12,190±390 BC (PRL 603)
9,350±130 BC (PRL 602)
8,395±110 BC (SUA)
8080±115 BC (SUA)

Previously, only one date, i.e. 8080±115 (BC) was available.

Similarly, the new dates have pushed back the antiquity of the Vindhyan Mesolithic also. From the Lekhahia R.S. we had two dates—2410±115 BC and 1,710±110 BC. Now, from the same place, we have two dates more, which are being given as under:

8370±75 (GX–20983–AMS) BP
8000±75 (GX–20984–AMS) BP

From the excavations in the Ganga Valley-Sarai Nahar Rai and Damdama, we have matching dates. Previously, there was only one date from Sarai Nahar Rai, i.e. 8395±110 BC. Now, from Damdama, there are two additional dates:

8,640±65 (GX–20829–AMS) BP
8,865±65 (GX–20827–AMS) BP

The above survey shows that the antiquity of Mesolithic is being pushed back further with new discoveries in the Vindhya-Ganga region. There are matching dates from other sites such as Paisara and Adamgarh. From Paisara, we have a date from the Mesolithic settlement which reads 7420±110 BC and from Adamgarh, we have a date which reads (7450±130) BP (TF 120). Similarly, the date of corded ware Neolithic culture of NE India, which were previously dated to c.2500 and 1500 BC have now been pushed back to 4460±120 BC (Agrawal, 2001 p. 18). Unfortunately, neither any new site in the Belan Valley has been excavated in recent years nor is any fresh date from the earlier excavated sites available. Misra, Pal and Manik Chand Gupta are excavating a new Neolithic site Tokwa, where 2.47-m thick Neolithic deposits has been marked. The Neolithic assemblage consist of cord impressed pottery, rusticated ware, bone arrowheads, beads of terracotta and semi-precious stones, microliths, quern and mullers, etc. Besides, rice grains include barley, *til* and *moong*. Bones of both wild and domesticated animals have been excavated.

So far, no absolute date is available from Tokwa. We hope that the dates from there will be able to solve some of the problems. But looking at the importance of the problem, it is necessary to re-excavate Mahagara and Koldihwa.

Recent excavations carried out between November 2001 and March 2002 at the site of Lahuradeva, district Sant Nagar (Uttar Pradesh) by State Department of Archaeology, U.P. have brought important evidence for the antiquity of rice in the trans-Saryu region. This region is formed by an alluvial deposit with an average depth of nearly 85 metres. The alluvial formation is divided into Khader and Bhanger. Except for some sand-mounds (*Dhus*) and horseshoe lakes, the entire region is featureless and almost plain. Towards its north are the Himalayan ranges, gradually rising higher and higher. Saryu and Rapti are the principal rivers that drain the water of this land finally into the Ganga. Gorakhpur is the most important modern city. The ancient site of Lahuradeva (26° 40′ N; 82° 57′ E) is located at a distance of about 5 km south to the Bhujaini crossing which is situated on Basti Gorakhpur Road (NH28) under village Jagdishpur on the bank of an ancient lake (Fig. 2). The excess water of the lake during the rainy season is drained through a small River Katrahia, a tributory of Kuwano river. The credit of the discovery of this site besides other sites of trans-Saryu region goes to Krishnananda Tripathi of the Department of Ancient Indian History Culture and Archaeology, University of Gorakhpur. Shailnath Chaturvedi excavated the site of Sohagaura which highlighted the importance of these sites (Chaturvedi, 1980: 339–40; 1985: 105). The excavations of Sohagaura gave a cultural

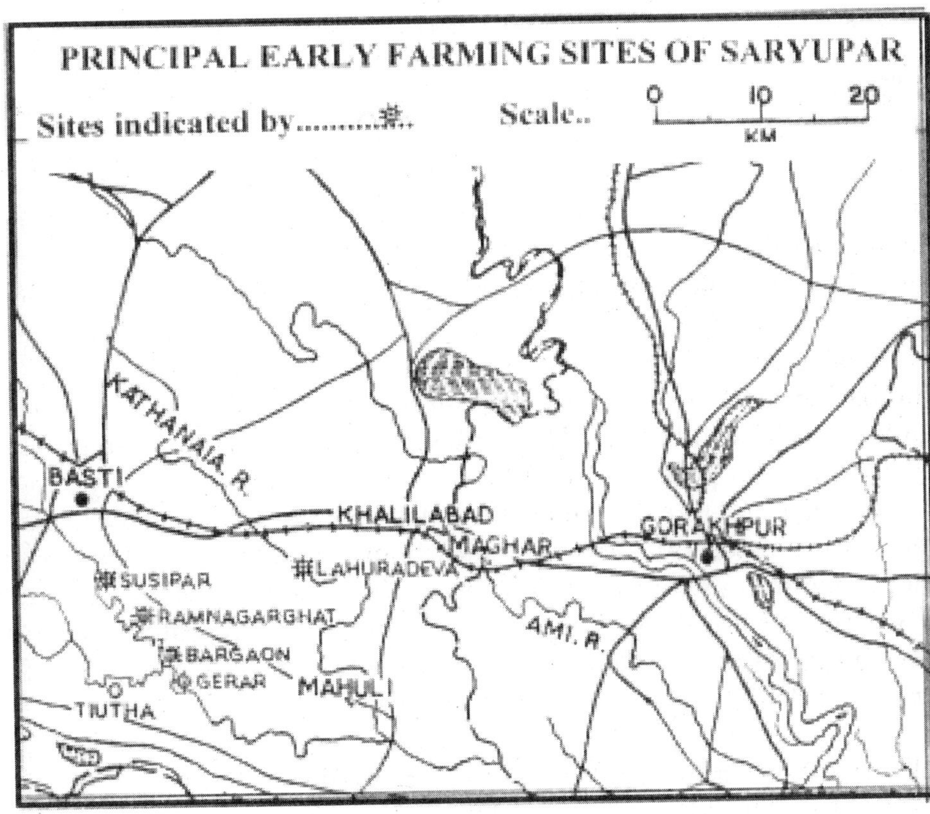

Fig. 2: Principal Early Farming sites of Saryupar

sequence right from the Neolithic to the Medieval period. It has been subdivided into VI periods. The earliest phase—Pd.-I (Neolithic) had a deposit of 70 cms. Though excavated in a very limited area, it brought to light the cord impressed ware of gritty clay mixed with rice husk, straw, etc. Besides, microlithic flakes and tools made of chert and chalcedony were also associated with this phase, just as we find in the Vindhyan Neolithic assemblage. It should be noted that the lithic material was not locally available; hence it was imported in all probability from the Vindhyan region. This phase had a fairly wide distribution in the Saryupar region. Corded ware has been found at Susipar, Ramnagar ghat, Gerer, Lahuradeva and a number of other sites recently excavated by Purushottam Singh and his colleagues of Department of Ancient Indian History Culture and Archaeology Benaras Hindu University. Rice played an important role in the subsistence economy of all these early farming communities. Unfortunately, Chaturvedi could not get any C14 date for Pd.-I and Pd.-II of Sohagaura. On the then evidence, he opined that "Neolithic of the Vindhya reached the trans-Ghaghara region at its advanced stage" (Chaturvedi, 1985:106). No neolithic tool was reported from that region till then. In November 2002, Krishnanand Tripathi found a neolithic Celt near the Lahuradeva mound as a result of boring in a field along with some pieces of corded ware. From a nearby field, he also found a broken mace head. (I was informed by Krishnanand Tripathi about these discoveries in June 2003 through personal communication.) In mid-eighties, Tripathi had reported some microlithic sites also. Some of the material is in the museum of the Department of Ancient Indian History Culture and Archaeology, University of Gorakhpur. All the discoveries taken together make a strong case to suppose that the entire Saryupar region was inhabited during the early Neolithic phase, if not even earlier. I do not agree with Chaturvedi when he opines that the people occupied this region during the advanced Neolithic phase. The horseshoe lakes formed by the Saryu and its tributaries provided fertile semi-permanent or permanent bases to the Mesolithic/Neolithic newcomers from the Vindhyan region, where edible seeds from wild grasses were available. Uncultivated rice *oryza rufipogen* and wild grass such as *Foxtail grass Sataria cf Glauca* have been reported from the excavation at Lahuradeva. The new settlers must have exploited these fertile tracts. The scenerio must have been quite similar to the Ganga Valley where men shifted from the Vindhyan region during the Epi-Palaeolithic/Mesolithic periods in search of food and fodder.

In the background of the above discussion, the recent excavation and the early C14 dates gather special significance. The earliest phase, Pd.-I (Early farming phase) has been subdivided into two periods: Pd.-IA and Pd.-IB. The deposits of IA have yielded, besides red ware, black-and-red ware (mostly hand made), corded ware, burnt clay chunks and nodules bearing reed and straw marks, carbonized material containing grains of cultivated rice along with uncultivated rice *oryza rufipogen* and a few wild grasses such as *Foxtail grass Sataria cf Glauca.* According to K.S. Saraswat of the Birbal Sahani Institute of Palaeobotany, the grains of cultivated rice are *Oryaza sativa* (Tewari, et al., 2001–2002:55). The two carbon 14 dates for the earliest period of Lahuradeva are:

1. B.S.1951; BP 5320±90 (Cal. BC 4220, 4196, 4161)
2. B.S.1966; BP 6290 ±160 (Cal. BC 5298)

These dates are very significant because they not only support the early carbon dates obtained from Koldihwa but also support the thesis that this (Vindhya-Ganga-Ghaghara) region must have been one of the nuclear zones for the origin of rice. On the basis of these dates, the origin of rice in this region can be easily pushed back to 6th–7th millennium BC if not still earlier. The discovery of the Neolithic Celt and the mace head indicate that there must be a still earlier Neolithic phase in this region which has not yet been obtained in the excavations. It is reasonable to believe that the Saryupar region must have been occupied almost at the same time when the Ganga Valley was inhabited. Chaturvedi has rightly observed that "The advance of the Vindhya–Ganga people to the Saryupar region was a logical consequence of their movement" (Chaturvedi, 1985:108). The discovery of a Neolithic phase in the excavation at Jhusi, Allahabad is also very significant because now the beginning of culture in the Ganga Valley can be traced uninterruptedly right from the Mesolithic phase. A large number of rice samples have been collected from the Neolithic level along with corded ware and associated pottery. It indicates that rice was a staple food during the Neolithic period throughout the Ganga-Ghaghara region. It also supports the views of Prof. Sharma regarding the origin of rice in the Vindhya-Ganga valley. Unfortunately, no C14 date has been obtained from this level so far, yet but there is sufficient ground to suppose that this part of Uttar Pradesh and adjoining Bihar was inhabited by the Neolithic rice-growing farmers in 7th–6th millennium BC.

The Vindhyan Neolithic has a distinct character of its own and is not comparable in totality with the Neolithic culture of the Northeast India and Southeast Asia. The common factors between the two are: firstly, the corded ware and secondly, rounded celts. The corded ware recovered from these places differ considerably from each other in their colour as well as in the range of decoration patterns. The rounded celts at the northeastern sites are associated with shouldered celts also, which are absent on the Vindhyan Neolithic sites. The association of shouldered axes show an advanced stage of cultural development. The third component or an integral part of the Vindhyan Neolithic is the stone blade industry which seems to be a continuation of the Advanced Mesolithic or Proto-Neolithic phase of Chopanimando. It suggests that the Vindhyan Neolithic was a culmination of Advanced Mesolithic or Proto-Neolithic of Chopanimando. This continuous cultural development makes a strong case for the indigenous development of cultivation in the region. The traces of the Neolithic in the Saryupar region is nearer to the Vindhyan Neolithic than to any other. The pre-Neolithic stage is absent in the culture complex of northeast. Regarding the question of diffusion from the northeast, more evidence is required. So far, no earlier date for Mesolithic or Neolithic is available from elsewhere in India. The space gap between the northeast and the Vindhya-Ganga region has to be filled in and the chronological discrepancy also needs proper explanation. Under the circumstances, we are not in a position at the present stage of our knowledge to brush aside the hypothesis of 'indigenous development' earlier proposed by Sharma. Only because the date of rice in China and Japan is earlier than those in India do not make a case for borrowing of rice technology from China or Japan. On the other hand, the very fact that in the Belan Valley we are getting the evolutionary stages from hunting-gathering to cultivation make a strong case for indigenous development of cultivation in the Vindhya-Ganga region. We can reasonably hope that the antiquity of cultivated rice will also be pushed back, as is the case in respect of Mesolithic industries of the region. Dr. Agrawal (2001:18) is of the opinion that the new discoveries in China and Japan, mentioned above, push back the antiquity of rice and ceramic technology further and make it possible to expect early Holocene dates of earliest rice technology in the Ganga Valley and the Vindhyas.

REFERENCES

Agrawal, D.P., 1999, A Note on the Earliest Agriculture and Pottery in South Asia. In Newsletter of the Grant-in-aid Program for COE Research Foundation, Science, Sports and Culture in Japan. March 1999, Vol. 2, Number 1 ISSN 0917–1045.

Agrawal, D.P., 2001, 'The idea of India and its Heritage: The Millenial Challenges', *Man and Environment.* Vol. XXVI, No.1. 2001.

Chaturvedi, S.N., 1980, 'Advance of Vindhyan Neolithic and Chalcolithic Cultures to the Himalayan Terai: Excavation and Exploration in Saryupar region of Uttar Pradesh', *Man and Environment,* IX: pp. 101–108.

Flennery, K.V., 1968, 'Archaeological systems theory and early Mesoamerica', In Meggers, B.J. (ed.) *Anthropological Archaeology in the Americas,* 67–87 Washington, D.C., Anthropological Society of Washington.

Harris David R., 1968, *The Prehistory of Tropical Agriculture: An Ethno-ecological Model,* 1969, 'Agricultural systems, ecosystems and the origin of agriculture', In Ucko, P.J. and Dimbleby, G.W. (eds), *The Domestication and Exploitation of Plants and Animals,* 3–15 London, Edward Arnold.

Kennedy, K.A.R., N.C. Lovell and C.B. Burrow, 1986, Mesolithic Human Remains from the Gangetic Plain: Sarai Nahar Rai, Cornell Report University South Asia Program Occasional Papers and Thesis No. 10.

Kajale, M.D., 1990, 'Some Initial Observations on Palaeo-botanical evidence for Mesolithic Plant Economy from Excavations at Damdama, Pratapgarh, Uttar Pradesh', in *Adaptation and Other Essays.* N.C. Ghosh and S. Chakrabarti, (eds) pp. 98–102. Shantiniketan: Visva Bharati.

Kajale, M.D., 1996, Archaeology and Domestication of Crops in the Indian Subcontinent. Diversity, Vol. 12. No. 3. pp. 23–24. 1996.

Lukacs, J.R. and J.N. Pal, 1992, Dental Anthropology of the Mesolithic Hunter Gatherers: A Preliminary Report on the Mahadaha and Sarai Nahar Rai Dentition, *Man and Environment* Vol. XVII No. 2, 1992.

Lukacs, J.R. and J.N. Pal, 1993, Mesolithic Subsistence in North India: Inference from Dental Attributes. *Current Anthropology,* Vol. 34 No. 5 Dec. 1993 pp. 745–765.

Misra, V.D., 1999, Agriculture, Domestication of Animals and Ceramic and other Industries in Prehistoric India: Mesolithic and Neolithic, in *The Dawn of Indian Civilization upto 600 BC* Edited by Pande, G.C. New Delhi, PHISPC.

Misra, V.D., J.N. Pal and M.C. Gupta, 2000, Further Excavation at Jhusi, *Prāgdhārā* 10: 23–30.

Misra, V.D., J.N. Pal and M.C. Gupta, 2001, Excavation at Tokwa: A Neolithic Chalcolithic Settlement. *Prāgdhārā,* No. 1. *Journal of the U.P. State Archaeological Department.*

Pal, J.N., 1986, *Ceramic Industries of Northern India,* Allahabad, Swabha Prakashan.

Sharma, G.R., 1973, Mesolithic Lake Cultures of the Ganga Valley, India, *Proceedings of the Prehistoric Society* 39: 129–46.

Sharma, G.R., et al., 1980, *Beginnings of Agriculture* Allahabad, Abinash Prakashan.

Tewari, Rakesh, R.K. Srivastava, and K.K. Singh, 2002, Excavations at Lahuradeva, District Sant Nagar, Uttar Pradesh, *Purātattva* No. 32.

____, 2003, Preliminary Report of the Excavation at Lahuradeva District Sant Kabir Nagar, U.P. 2001–2002: Wider Archaeological Simplifications. *Prāgdhārā* 13 (in press).

Varma, R.K., 1981–83, The Mesolithic Cultures of India, *Purātattva* 13–14, pp. 27–36.

Varma, R.K., 1986, *The Mesolithic Age in Mirzapur,* Allahabad, Paramjyoti Prakashan.

Varma, R.K., V.D. Misra, J.N. Pandey and J.N. Pal, 1985, A Preliminary report on the excavation at Damdama (1982–1984), *Man and Environment* 9: 45–65.

Vishnu Mittre, 1977, Discussion on 'India: Local and Introduced Crops' by J. Hutchinson in the *Early History of Agriculture,* a joint Symposium of the Royal Society and the British Academy, 1977, p. 141.

Waber, S.A., 1991, Plants and Harappan Subsistence: An Example of Stability and Change From Rojdi. New Delhi, Oxford and IBH Publishing.

CHAPTER 4

Recent Excavations at Tokwa: Fresh Light on the Early Farming Culture of the Vindhyas

J.N. Pal

INTRODUCTION

The Vindhyan Plateau, one of the archaeologically richest regions of the country lies in northern India south of the Gangetic Plain. Northern Vindhyan plateau is the northern projection of the peninsular block broadly composed of Bundelkhand plateau in the west and Baghelkhand plateau in the east.[1] This plateau is characterized by step formations gradually rising from the northeast to the southwest. This hilly track, marked by diverse and rugged topography, is made of granite gneiss, sandstone, shale and serrated ridges of quartz reefs. The undulating plateau gradually merges at many places with the Ganga alluvium in the north.

The Vindhyan plateau has a unique place in the history of development of human cultures in India, where the longest Quaternary geological and archaeological sequence, uninterrupted human habitation from the Lower Palaeolithic to early historic period, and coming into existence of the domesticating/cultivating Neolithic culture transformed from the foraging-gathering Mesolithic culture have been brought to light.

The first Neolithic implement in the area was found as early as 1860 when Le Mesurier[2] collected celts or hatchets from the Tons valley. Subsequently, Theobald,[3] Cockburn[4] and Rivett Carnac[5] collected a good number of Neolithic celts from Banda district. Cockburn[6] had also found human celvarea with 'glazed earthen ware', stone hammers, flakes and spalls from a 'Neolithic burial' and a number of polished celts of dolerite, a black basalt celt of lanceolate type, ring stones and hammer stones in the Kon ravines of Mirzapur district.

Systematic investigation of the area was planned in 1950s by G.R. Sharma of Allahabad University and numerous triangular celts were collected during initial surface surveys from Banda district.[7] In subsequent years, similar celts and hammer stones were collected from the river valleys of Bundelkhand and Baghelkhand region.[8] Neolithic celts fashioned on basalt have been collected from the hilly tracts of Lalitpur, Hamirpur, Banda, Allahabad and Mirzapur districts.

Most of the Neolithic celts were collected from the fields or from the platforms below the trees, apparently not in primary context. A morphological study of the celts shows that only in the Bundelkhand region the triangular variety of celts marked with curved medial-ground edges, corresponding to the principal type of Southern Neolithic celt, were found while in the Baghelkhand, the most frequent type was the rounded variety of celt marked with rectangular or ovaloid cross-section, fully ground and polished and corresponding to the Eastern Neolithic celts. Being surface collections in secondary context, unassociated with any other cultural material, nothing was known about their primary archaeological context. The rounded celts were found for the first time in definite stratigraphical context in 1972 during the excavations at Koldihwa. But the triangular celts have not been found till now in the primary context in the Vindhyan plateau; therefore, their archaeological context is still shrouded in obscurity.

Some of the habitation sites associated with rounded celts were put to excavation and the results obtained have shed new light on the different facets of the Neolithic culture of the area. The excavated sites (Fig. 1) included Koldihwa, Mahagara and Pachoha in the Belan Valley, Indari and Tokwa in the Adwa Valley and Kunjhun in the Son Valley.[9] In this chapter, an attempt has been made to present the salient features of the Neolithic culture of the Vindhyas with special reference to the results revealed from recent excavations at Tokwa.

EXCAVATIONS AT TOKWA

Tokwa (Lat. 24° 54′ 20" N., Long. 82° 16′ 45″ E.), an important archaeological site, is situated at an elevation of 146 m MSL on the confluence of the Belan and Adwa in Lalganj subdivision of Mirzapur District of Uttar Pradesh, at a distance of 68 km from Mirzapur city in southeastern direction and at a distance of 8 km east of Baraundha (Fig. 2). The major portion of the site is undisturbed but on the northern and southern sides, it has been cut vertically by the rivers Belan and Adwa, showing cultural deposits in the section (Plate I).

Excavations conducted at the site during 2000[10] and 2003 have revealed cultural relics of Neolithic, Chalcolithic and early Iron Age periods.

The northern margin of the site is flanked by the Belan River, while the southern is facing the Adwa. The western margin of the site looks like the peak of a triangle. This site is quite extensive, presently covering an area of about 27579 square m. As one moves eastward, the width of the site widens. The eastern limit of the site is upto the village Tokwa. For the shake of convenience, the entire site is divided into three subparts: Tokwa 1, Tokwa 2 and Tokwa 3 from west to east (Fig. 3).

Though the site was discovered long back in seventies, it was revisited in 1999 and extensive explorations were done.[11] The explorations at Tokwa resulted in the discovery of handmade pottery comprising cord impressed, rusticated, burnished red, burnished black, black-and-red and ordinary red wares, on the one hand and black slipped ware, fine black-and-red ware and red ware on the other. The second assemblage of the pottery is wheel thrown. Surface exploration also resulted in the discovery of Kotia type of pottery and iron objects as well. On piecing the scattered evidence together,

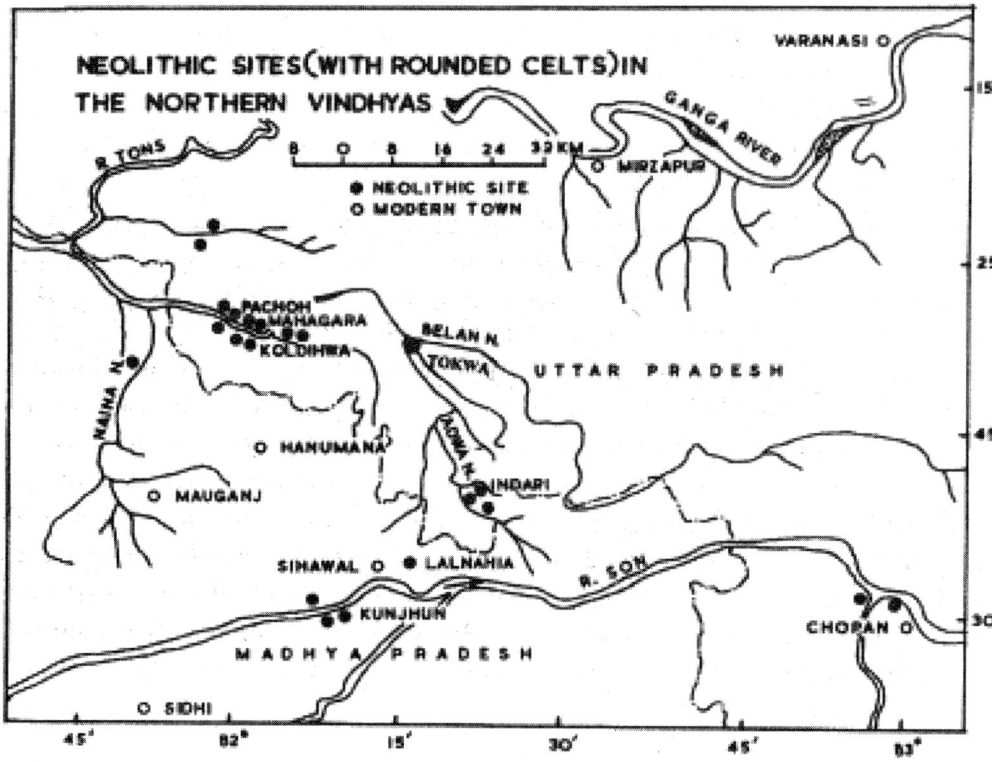

Fig. 1

Plate I: Tokwa, General view of the Western Part of the Mound

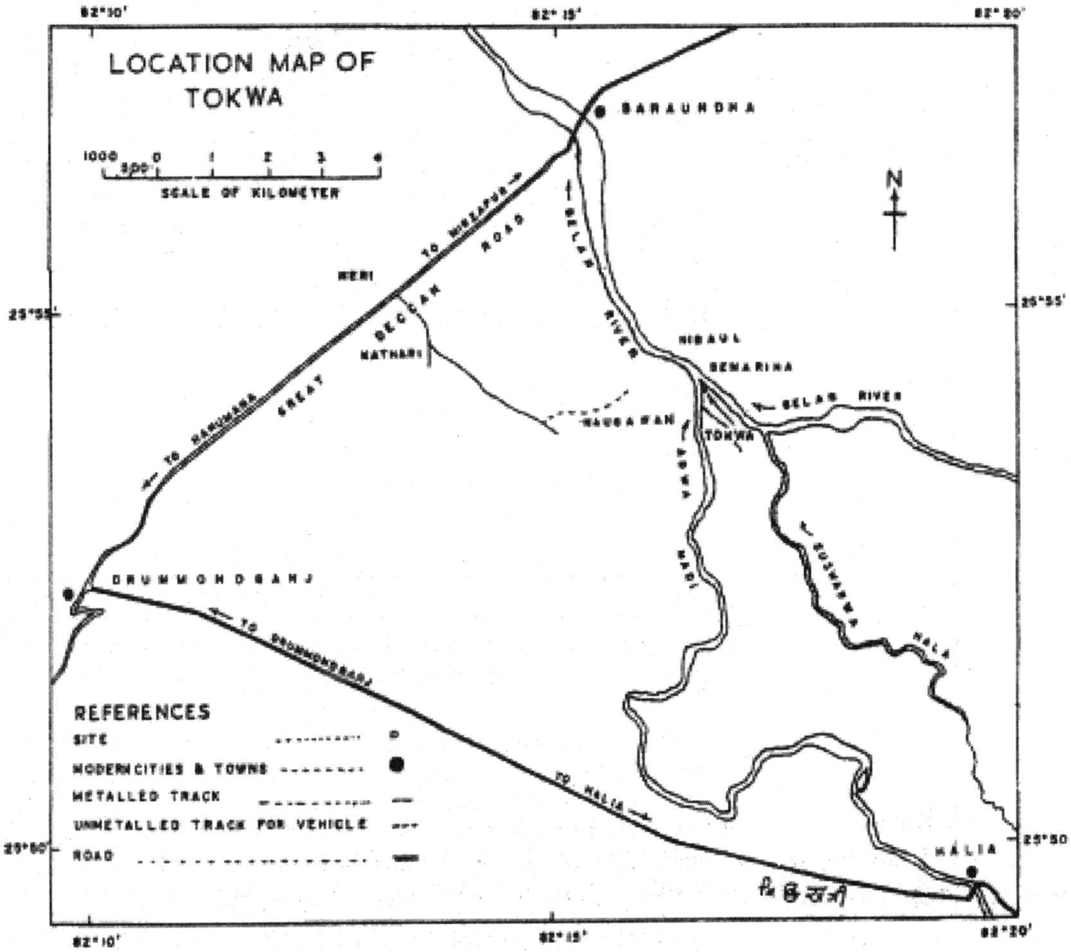

Fig. 2

it was evident that site had been occupied during the Neolithic period. In this way, it is a midway station in the Vindhyan Neolithic culture complex between Koldihwa, Mahagara, Pachoh on the one hand, all lying in the Belan Valley in the Koraon subdivision of Allahabad district and Kunjhun in the Son Valley in Sidhi District (MP). The Department of Ancient History, Culture and Archaeology Allahabad University if had not only located but also conducted a small-scale excavation in 1981 at a Neolithic site of Indari in Mirzapur District in the Adwa Valley. The excavation yielded cord impressed pottery, rusticated pottery, some burnished red and burnished black sherds along with postholes. The available evidence indicated that it was an integral part of the Vindhyan Neolithic culture complex. The artefacts collected from the site of Tokwa indicate that the site was under occupation in the Chalcolithic and early Iron Age as well. The site was excavated in 2000 and 2003.

At Tokwa I two trenches, H-8 and H-9 in 2000 and two trenches, I-8 and I-9 in 2003, each measuring 5 square metres were laid out (Plate II). The general slope of

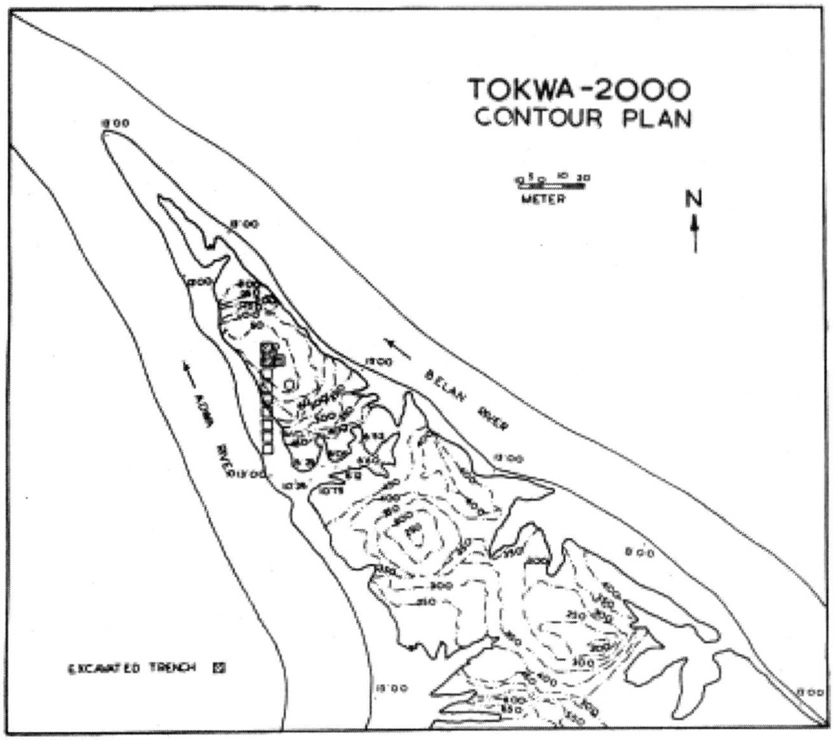

Fig. 3

the area is from east to west and north to south. To the west of trench, H-8, a control pit measuring 1 square m was also laid out in 2000. Trench C-9 was also selected for excavation. As it is located just on the section cut by river Adwa, the area was selected for excavation to learn the nature of deposit towards the bank of the Adwa. In this trench, only one square (E-4) of 1 × 1 metre area was excavated.

The combined testimony of the excavations brought to light the archaeological evidence of three cultures: Neolithic, Chalcolithic and Iron Age. The occupational strata, divisible into as many as 16 layers, measured 4.00 m (Plate III, Fig. 4).

Layers 1 and 2 measuring 58 cm, slightly compact and yellowish deposit, yielded sherds of NBPW, black slipped, black-and-red and red wares, the last one further subdivisible into a number of subgroups. In this connection, it may be pointed out that before conducting excavation at the site, one had no inkling of getting the NBP ware at the site. On the other hand, on the basis of the occurrence of Kotia group of potsherds, it was suspected that the people using this type of pottery might have settled at the site. However, the occurrence of NBPW sherds opened a new vista. Some of the dishes have stamp designs in the centre, as is case with several NBPW sites in the Ganga Valley. Here, it is pertinent to note that dishes—though occurring in the black slipped ware as well—don't bear this type of stamp design. Along with the ceramic assemblage noted above these two layers yielded a good number of bone arrowheads, animal bones, beads fashioned on semi-precious stones, terracotta beads, glass bangles, copper and iron objects, fragments of querns and mullers, sandstone fragments, etc. A container made

Plate II: Tokwa, General view of the Excavated Trenches

Plate III: General view of Hut Floor in Trench H-8, Top of Layer 9

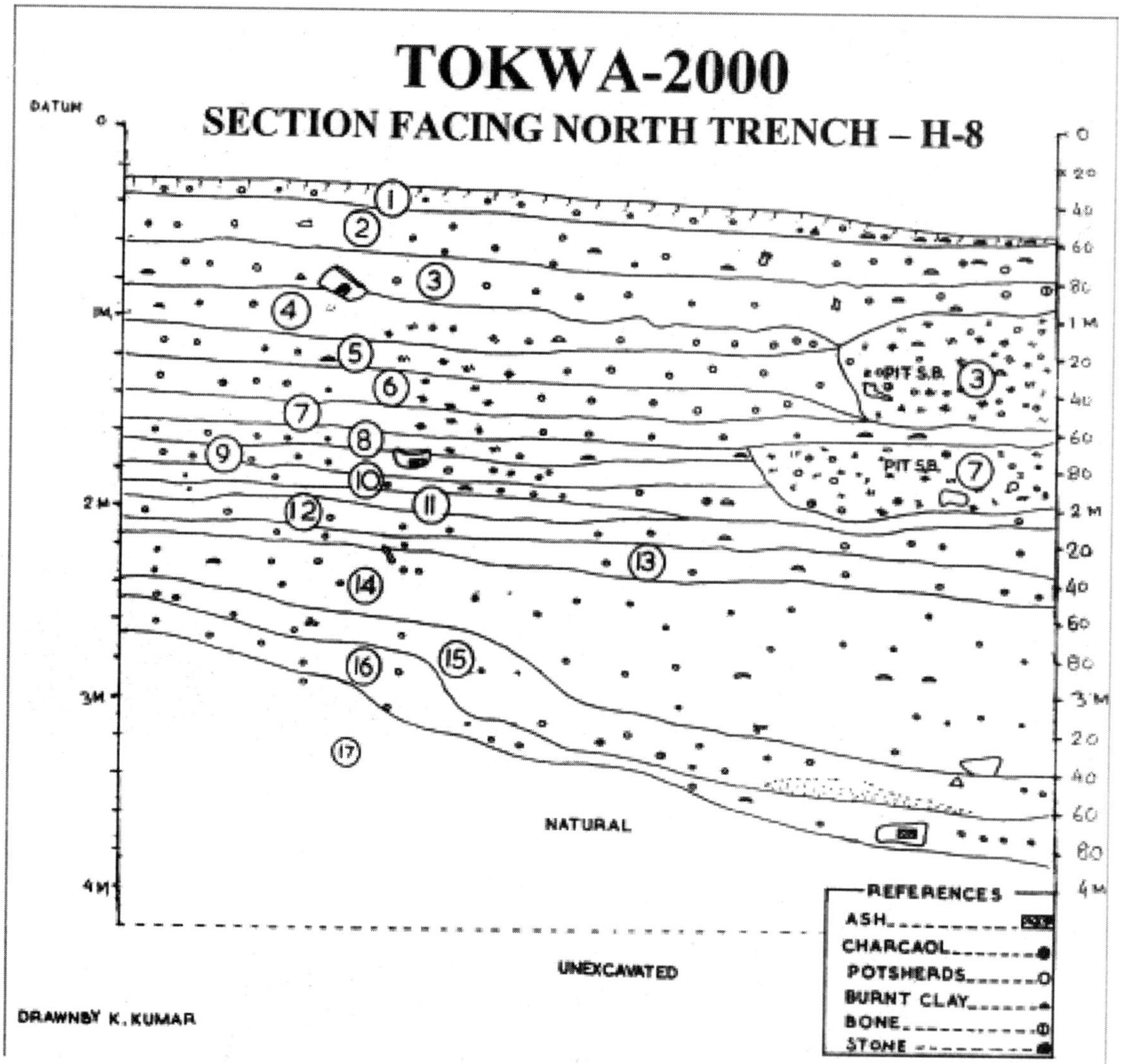

Fig. 4

of a cylindrical hollow bone having a cover on both the sides is an interesting find of this phase. Structural activities—pit hearths and post holes—are associated with these two layers. Burnt clay lumps with reed impression suggest the wattle and daub construction. In some of the pits, Kotia type of pottery was also obtained, suggesting thereby that even though periodically, the site was also used by Kotia group of people.

Layers 3 and 4, measuring 46 cm, constitute the pre-NBPW horizon associated with iron. These layers have yielded sherds of black-and-red ware, black slipped ware and various types of red ware, bone arrowheads, a few copper objects, beads of semi-precious stones and terracotta, fragments of querns, mullers, hammer stones, animal bones and carbonized cereals in good number. These layers are also associated with pit hearths and structural hearths. Occurrence of post holes and clay lumps with reed impressions are suggestive of wattle and daub structure. Probably, they used to build round huts. In this connection, it may be pointed out that ceramic assemblage of this phase is a continuum of the ceramic tradition associated with underlying Chalcolithic deposit of the site. However, the introduction of iron at this stage marks a new chapter. Here, it may also be pointed out that this stratigraphic position at Tokwa tallies well with the evidence obtained from a number of sites like Chechar Kutubpur,[12] Chirand,[13] Taradih[14] and Senuar[15] in Bihar on the one hand and Narhan,[16] Khairadih,[17] Masondih,[18] Sohgaura,[19] Rajghat,[20] Prahladpur[21] in Uttar Pradesh on the other. This stratigraphic status of iron has been confirmed by the excavations at Jhusi,[22] Agiabir,[23] Raja Nal Ka Tila[24] and Malhar[25] also. At all these sites, iron antedates the NBPW and is found with a ceramic tradition and in archaeological context which indicates the continuity of the Chalcolithic tradition in respect of pottery, bone arrowheads, etc. This indicates that the Chalcolithic culture of the Vindhyas and the Middle Ganga Valley acquired knowledge of iron technology at the fag end of its career.

Layers 5 to 7, light, yellowish, loose and ashy in character, measured 50 cm in thickness and represent the Chalcolithic culture of the site. These layers have yielded sherds of black-and-red ware, black slipped ware and red ware, the last one further divisible into number of subgroups. The characteristic pottery types include—different types of bowls, dishes, basins, footed and perforated basins, medium-sized vases, storage jars, pedestalled bowls, etc. Occasionally, painting is also met with though it is not a characteristic feature of the culture at the site. The clay used in manufacturing these pots is moderately levigated. Pots are well fired. Ceramic industries and pottery types associated with this phase compare well with its counterparts at Chechar Kutubpur, Chirand, Senuar, Taradih in Bihar and Sohgaura, Narhan, Khairadih, Jhusi, Agiabir, Raja Nal Ka Tila, Malhar, etc. in up. Besides the ceramic assemblage, these layers have also yielded bone arrowheads, a few copper objects, beads of semi-precious stones and terracotta, flakes and flake/ fragments, blades and blade fragments, fragments of querns, mullers, hammer stones, a good number of animal bones and carbonized cereals as well. Burnt clay lumps, suggesting wattle and daub structure, have also been obtained. A number of hearths, both structural and pit hearths, were also exposed. Occurrence of post holes suggested the existence of hut-like structures. Another significant point associated with this culture is digging of various pits. As a result of this, the underlying Neolithic horizon has been disturbed.

EVIDENCE OF EARLY FARMING CULTURE AT NEOLITHIC TOKWA

Layers 8 to 16, measuring 2.47 m, represent the Neolithic horizon at the site. These layers have yielded cord impressed pottery, rusticated ware and burnished red and burnished black ware sherds as well. Other antiquities of the period include bone arrowheads, beads fashioned on semi-precious stones and terracotta, fragments of querns, mullers, hammer stones, etc. Microliths fashioned on semi-precious stones include flakes, flake fragments, blades, blade fragments, scrapers, triangles, etc. Animal bones obtained form these deposits are significant, both from the point of view of the number as well as from state of preservation. A preliminary observation to identify the grains obtained by floatation includes rice, barley, *til, moong*.[26] Some fruits and beans were also found. A sampling by floatation for macrobotanical remains and collection of phytolith samples was carried out at sites of the Belan Valley by Dorian Q. Fuller and Emma Harvey of

Institute of Archaeology, University College, London and the preliminary result is that barley, wheat and pulses were introduced after rice.[27] Similar observation has been made by K.S. Saraswat of Birbal Sahani Institute of Palaeobotany, Lucknow at Lahuradewa in the Gangetic Plain.[28] The analysis and identification of animal species is under process. The bones are from both wild and domesticated animals.[29]

Oval or round hut floors encircled by post holes were also exposed from this phase (Plate IV; Figs. 5, 6). Burnt clay lumps with reed impression indicate wattle and daub structure. Ceramic industry of this phase is handmade. Rice husk is used as degraissant, which is found both from the surface as well as in the core of potsherds. These potsherds are generally thick in fabric even though pots in medium and fine fabric are also met with. The pots are not well fired and the clay used in the manufacture of these pots is not lavigated. As a result, small granules are discernible in the core of the pots. The black core indicates ill firing. It appears that people at this stage did not possess the technical knowhow of getting their pots well fired. Shapes in their ceramic tradition are also limited in comparison to the functional types of overlying Chalcolithic culture. The functional types associated with the Neolithic phase comprise bowls (shallow and deep), basins (shallow), medium-sized vases, etc. The prevalence of spouted vases in the assemblage is also reported.

The Neolithic horizon at the site witnessed disturbance in subsequent periods, as deep pits have been cut into these deposits mostly by the Chalcolithic people. The excavation has also yielded cereal remains, the identification of which is under process. Tentatively, it appears that the inhabitants had been cultivating wheat, rice, green gram

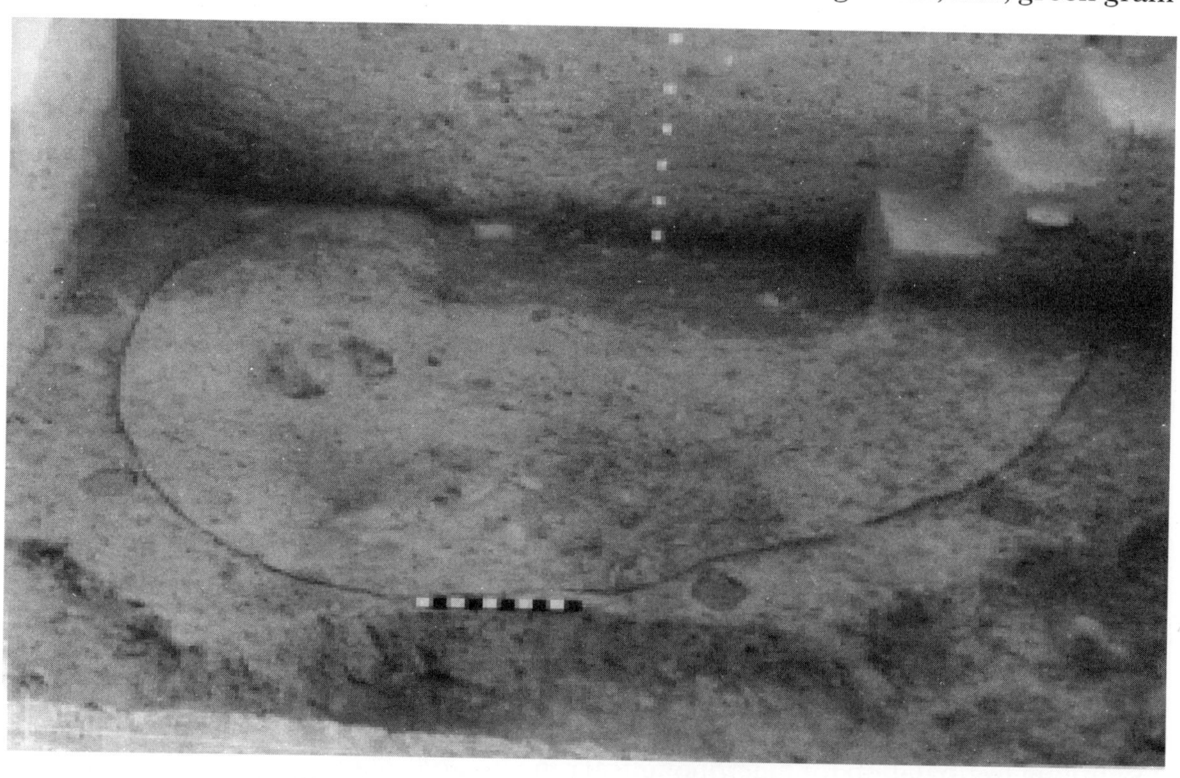

Plate IV: Tokwa, General view of the Section

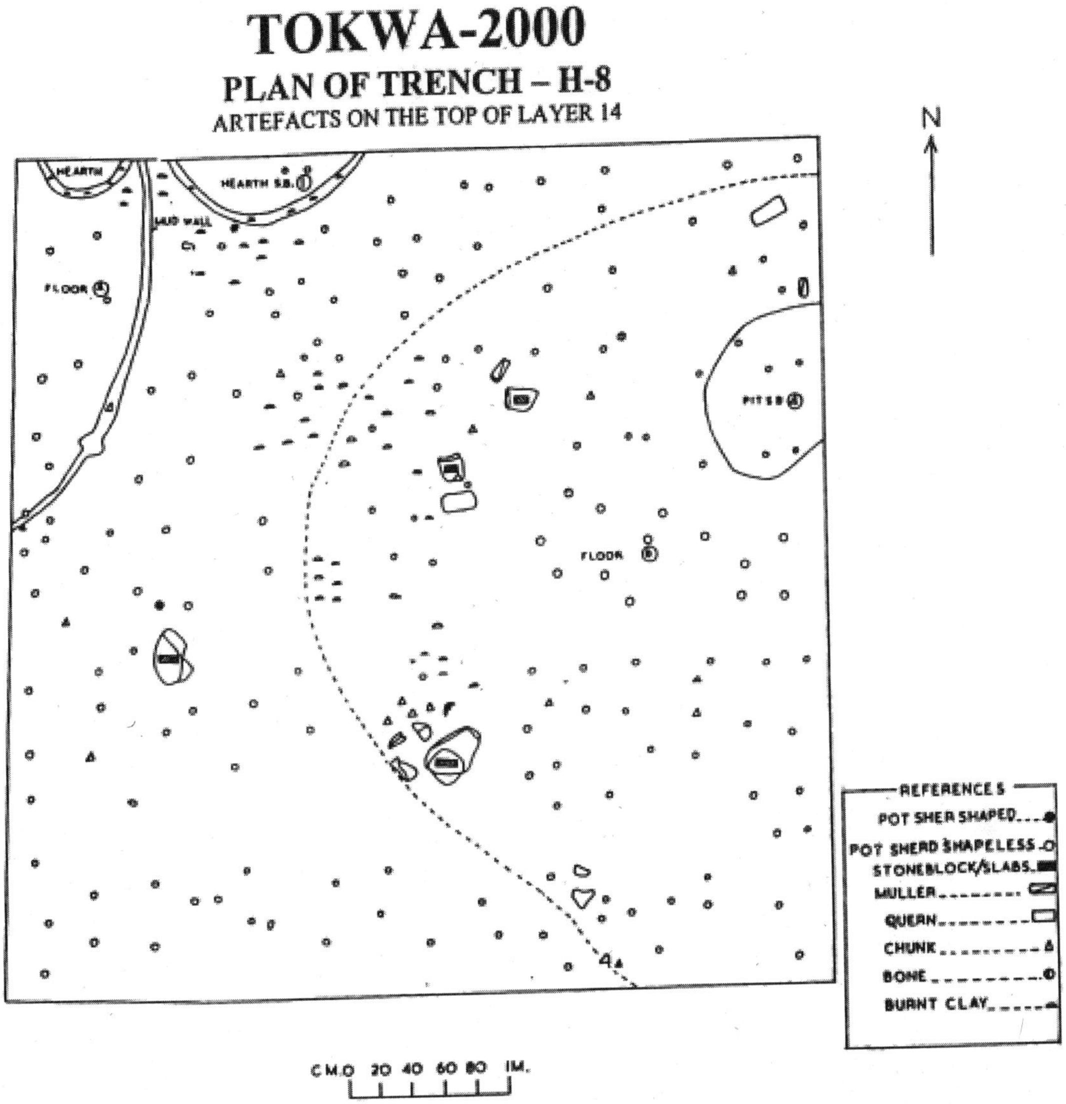

Fig. 5

(*moong*) and mustard. Fragments of querns, mullers, hammer stones, etc., have also been obtained from this horizon, as was charcoal. Burnt clay lumps with reed impressions are also met with, thereby suggesting wattle and daub structures. The charcoal samples have been sent to the Physical Research Laboratory, Ahmedabad and the International Research Centre for Japanese Studies, Kyoto, Japan for C-14 dating. These dates, when received, may throw welcome light on the chronology of the Neolithic culture of area.[30]

On the basis of the combined testimony of the excavation, the following tentative observations may be made about the Neolithic phase at Tokwa:

Habitation at the site was started by the Neolithic people. The traits of this culture unearthed at the site associate this culture with the Vindhyan Neolithic culture. In this

TOKWA-2000

PLAN OF TRENCH – H-9

ARTEFACTS ON THE TOP OF LAYER 15

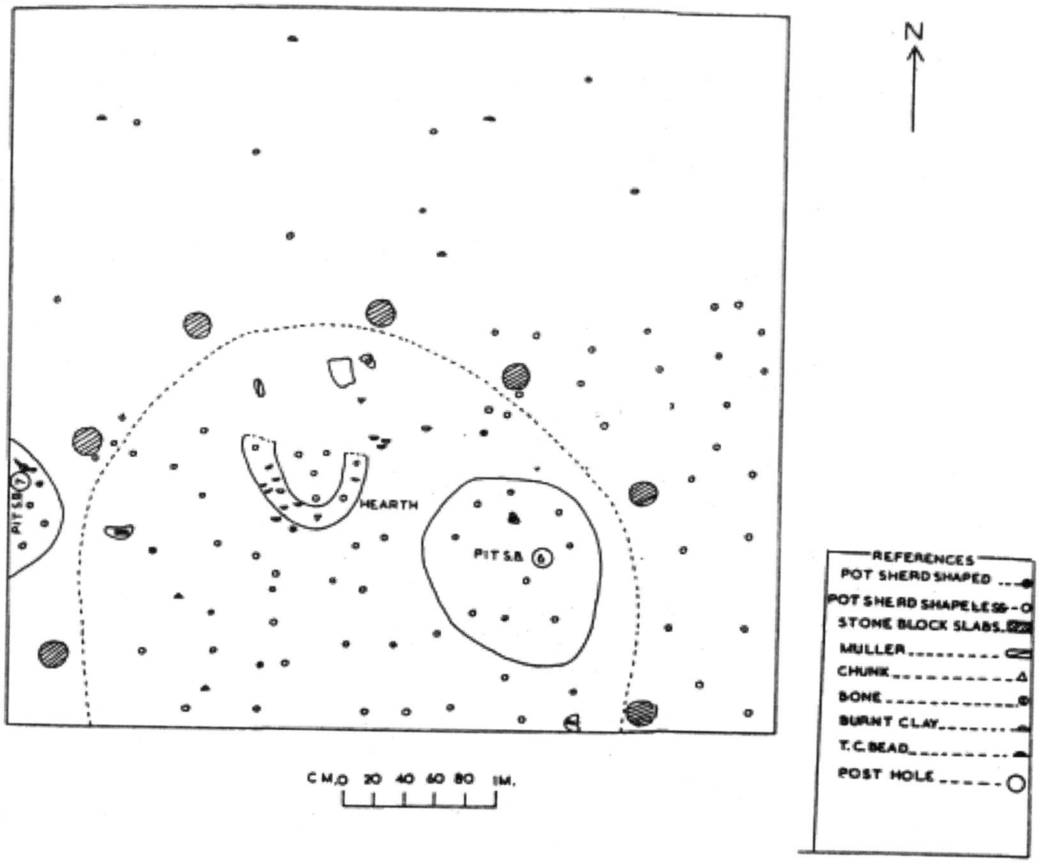

Fig. 6

connection, it may be pointed out that of the sites associated with Vindhyan Neolithic excavated/explored so far, Tokwa appears to be most extensive and well preserved. Sustained excavation at the site over a number of years may unfold different aspects of the Vindhyan Neolithic culture. It is not unlikely that besides confirming data obtained from other excavated sites, it may also throw new light on some other aspects. Here, it may be pointed out that animal bones obtained from the Neolithic horizon of Tokwa are most plentiful, well preserved and well distributed. After scientific examination, these animal bone carpus may help us in identifying not only the animal species domesticated but also animal species hunted by these Neolithic people. A scientific analysis of the data would furnish clues to understand the nature and extent of hunting activities practiced by these people. A scientific study of the cereal grains obtained from this site may throw a welcome light on the nature of cultivation and range of cereals of domesticated plants.

From excavation at Tokwa, it appears that there was a gap between the disappearance of the Neolithic culture and arrival of the Chalcolithic culture, as these two cultures exhibit different traits in ceramic tradition. While Neolithic pottery is handmade, the Chalcolithic is wheel thrown. There is also difference in preparation of clay in manufacturing of the pots. The characteristic ceramic tradition of the Neolithic is not represented in the Chalcolithic horizon. Moreover, the uppermost Neolithic horizon has yielded small potsherds, thereby indicating that after the desertion of the site by the Neolithic people, it remained unoccupied for a considerable length of time before the arrival of Chalcolithic people. However, this point needs closer observation.

OTHER EXCAVATED SITES OF THE NEIGHBOURING AREA

A brief review of the sites of the neighbouring area excavated earlier is given below to strengthen the points regarding early farming culture as revealed from the excavations of Tokwa.

KOLDIHWA

Koldihwa is a small village situated at a distance of about 80 km to the southeast of Allahabad on the left bank of River Belan of the Koraon subdivision of Allahabad district. To the northeast of the village, there is an extensive (500 × 200 m) mound (Lat. 24° 54′ 30″ N, Long. 82° 2′ E). This mound is roughly rectangular in form and has now been divided into small mounds—mainly eastern, western and southern—by nalas.

Under the supervision of G.R. Sharma, R.K. Varma and V.D. Misra, the site was excavated by B.B. Misra and R. Singh in 1972–73 and 1973–74. Subsequently, it was excavated by D. Mandal and J.N. Pal during 1974–75 and 1975–76. The excavations revealed that on the western mound, occupation started in the Chalcolithic period, while on the eastern and southern mounds, evidence goes back to the Neolithic period. The habitational debris of 1.90 m, as revealed at the eastern mound, contained the relics of three cultures: Neolithic, Chalcolithic and Iron Age.

The Neolithic deposit at Koldihwa, resting on the hard compact natural soil, is of 45 cm thickness, represented by layers 4 and 3. Of the tools and weapons of the Neolithic people of Koldihwa, mention may be made of rounded celts with flat sides and rectangular cross section made on basalt and granite, ring stones and microliths. The microlithic assemblage fashioned on chert, chalcedony, agate, carnelian, jasper and quartz include cores, flakes and tools. The tools consist of retouched blades, backed blades and scrapers. No piece evincing the employment of crested ridged technique has been found. The bone tools, though found in appreciable number in the Chalcolithic phase at the site, are conspicuous by their absence.

The Neolithic people at Koldihwa appear to have practiced domestication and cultivation of plants and animals. On account of the limited area exposed in the Neolithic level, only a few fragmentary animal bones have been found. Remains of rice from the Neolithic deposit constitute one of the most significant discoveries of this excavation. A preliminary palaeobotanical analysis of the rice husks by Vishnu Mittre of Birbal Sahani Institute of Palaeobotany, Lucknow and Te-Tzu Chang of the International Rice Research Institute, Manila, Philippines reveals that these belong to the domesticated variety. However, the wild grain also was part of their diet, as is suggested by finding of millet. Querns and mullers also inferentially indicate that the people had commenced cultivation. There was a surplus economy, as may be inferred from the occurrence of big storage jars.

The evidence of the structure at Neolithic Koldihwa is very poor due to the limited area of excavation and disturbance of Neolithic strata by later pit activities. However, burnt clay lumps with wattle and daub impression, obtained in profusion, indicate that the dwellers used to live in thatched huts made of bamboo and wooden post with mud-plastered screen walls. The ceramic assemblage includes cord impressed, rusticated, burnished red and burnished black wares.

MAHAGARA

The Neolithic site of Mahagara (Lat. 24° 54′ 30″ N, Long. 82° 3′ 20″ E) is situated to the west of the revenue village of Kotia on the confluence of the old and new Belans just opposite Koldihwa, 80 km southeast of Allahabad in Allahabad district. The site was explored in 1976 and excavated during 1976–79 by D. Mandal and J.N. Pal under the supervision of G.R. Sharma, R.K. Varma and V.D. Misra[31]. The surface exploration at the site had revealed that cord impressed ware along with other wares, celts and microliths were confined to an oval area measuring about 8,000 sq m surrounded by natural ridges on three sides, i.e. north, east and south. The ridge is composed of geological formations representing, in succession, reddish clay, yellow silt, brown red clay and cemented gravel IV. The cemented gravel III, ancient humus (palaeosol) and the succeeding two layers, present elsewhere in the river section, are replaced here by the cemented gravel IV. The top of the ridge at Mahagara is marked by gravel IV of the Belan section. The ridge surrounding the site has a steep slope on the outer side but the inner slope of the ridge is gentle and centrally directed, which forms a basin-shaped morphology of the site. A trial excavation of gravel IV on the ridge yielded Epi-Palaeolithic blade implements along with shell and animal bones.

Rich archaeological potentiality and good state of preservation of the site at Mahagara initiated to plan a horizontal excavation. To ascertain the nature of habitation deposit of the site, one of the trenches, i.e. the Index trench, revealed a habitation deposit of about 2.60 m comprising as many as 17 layers. Mahagara is a single-culture site containing only Neolithic material. The Neolithic deposit is divisible into six structural periods demarcated mainly by floors and post holes of huts, and pits. The excavation brought to light three successive floors and three separate pits in the Index trench.

In other squares, digging was restricted only to the uppermost phase of the Neolithic deposit. The total area excavated so far measures about 1,622 sq m. As many as eighteen hutments have been brought to light. The floors of the huts—circular or oval on plan—vary from 6.70 × 6.25 m to 5.0 × 3.50 m. At the periphery of these floors, post holes varying from 6 to 9 in number have also been exposed. The floors of the huts have not been rammed or paved. The occurrence of burnt clay lumps with wattle and daub impression indicates the use of mud plaster on the screen walls of these huts. The floors are well represented by occupational debris containing potsherds, household implements, bones of animals, etc.

One of the most remarkable discoveries was the exposure of a cattle pen in the southeast sector of the site. It measure 12.5 × 7.5 m and is irregular rectangular in shape, marked by 28 post holes of varying dimensions and surrounded by huts. Hoof impressions of cattle of different age groups have been found in cluster inside the pen.

Of the tools and weapons found in excavations, mention may be made of microliths, consisting retouched blades, backed blades, lunates, scrapers, borers, triangles, trapezes and tranchets; rounded celts and bone arrowheads. The other excavated objects comprise querns, mullers, hammers, sling-balls, sharpeners, ring stones, perforated shell objects (ornament), clay dabber, terracotta beads and perforated pottery discs.

The pottery recovered from excavations is grouped into: (a) cord impressed; (b) rusticated; (c) burnished red; and (d) burnished black wares.

Rice husk and rice grain have been found either embedded in pottery as degraissant or separately in a charred condition. Similar evidence of rice from Koldihwa are ascribed to the domesticated variety by Vishnu Mittre and Chang. Some of the Mahagara species may also belong to the same category. The occurrence of the wild variety side by side the domesticated one may not be ruled out.

Animal bones have also been found in profusion. The assemblage consists of the skeletal remains of both domesticated and wild animals. Of the identified species, mention may be made of cattle (*Bos indicus-Bovidae*), sheep/goat (*Ovivignei-Capridae*), deer (*Cerevidae*), wild boar (*Sus scrofa*), horse (*Equidae*), tortoise (*Chelonia*) and fish.[32]

PACHOHA

The Neolithic site of Pachoha (Lat. 24° 55′ 45″ N, 82° 2′ 30″ E) is also in the Koraon subdivision of Allahabad district, only 2.5 km northwest of Koldihwa near the right bank of the Belan. The site at present does not make any mound and is under cultivation. Microliths, celts, pottery, etc., were found from the surface. The site was also marked by black circular or oval patches, indicating hut floors. The site was excavated in 1975–76 session by B.B. Misra under the supervision of G.R. Sharma.[33] The total thickness of cultural deposit—though major portion was washed away and disturbed by cultivation—is 60 cm. Besides the pottery, excavation brought to light burnt clay lumps, ring stones, rounded celts and microliths consisting of fluted cores, retouched blades, backed blades, scrapers and points.[34]

INDARI

The Neolithic site of Indari (Lat. 24° 37′ 38″ N, Long. 82° 20′ E) is located on the left bank of a nala of the Kahenjua, a tributary of the Adwa in the revenue village of Manigara, Mirzapur district.[35] The site is flanked by the Adhesar hillock in the west and northwest and the Kaimur in the southeast. The site, measuring about three acres, is on the eastern and southeastern margin of a cup-shaped depression. A trail trench measuring 3 × 3 m revealed a 75 cm thick Neolithic deposit.

The excavation yielded pottery, microliths and traces of hut floors as that of other excavated sites of the region.

SETTLEMENT PATTERN

The Neolithic settlements have been located on the banks of rivers or nalas, generally above the flood plain. Neighbouring flood plains provided fertile fields for agriculture by annual inundations, which could be cultivated without tilling and irrigation. The situation of settlements in a shallow basin-shaped bluff surrounded by natural ridges, is significant in view of safety and security against flood, cold winds, and wild animals. The ridge played the role of a defence wall. As the evidence shows, the Neolithic people were not dependent on agriculture alone for their subsistence. They domesticated animals and exploited wild plants and animals in the nearby forests, which were rich in woodland and grassland vegetation. The rivers and nalas not only provided suitable land for agriculture, they also provided food in the form of aquatic animals like fish, tortoise, snail, etc.

STRUCTURES

The evidence of structures has been found in the form of hut floors and cattle pen from the excavation at Mahagara. On the other excavated sites, the structural evidence is very poor mainly due to limited area of excavation and disturbance of Neolithic strata by later activities. However, the burnt clay lumps with wattle and daub impressions obtained in profusion from all the habitation sites indicate that thatched huts were prepared by bamboo and wooden posts and mud-plastered screen walls. Traces of a hut floor were also observed at Indari during excavation. As is evident from the excavations from Mahagara, there appears a pattern in the layout of hutment. All the 18 hut floors exposed at the site are circular or oval on plan. The area covered by huts varies from 6.70 × 6.25 m to 5.0 × 3.50 m. The average living area in a hut is 15.74 sq m. The floors are surrounded by post holes, varying from 6 to 9 in number, on their periphery. The posts supported the upper roof as well as the side screen wall.

The cattle pen exposed at Mahagara is one of the most remarkable discoveries of structures. This cattle pen, located in the southeast sector of the site, is irregular rectangular on plan and measures 12.5 m in length and 7.5 m. in width. It was surrounded by four houses on its four corners so that it could be constantly watched against attack by wild animals. Having three openings, two to the east and one to the west, the cattle pen was enclosed by thatched screen walls, as is evident by the discovery of 28 post holes around the pen. As there is no evidence of post holes inside the pen, it may be suggested that the pen was open to sky and the cattle were put inside it untied. At least 23 pairs of hoof imprints of cattle of different age groups have been located within the pen.

SUBSISTENCE

The subsistence of the Neolithic people was based on the exploitation of plants and animals, by domestication as well as by hunting and gathering. The evidence of cultivation of plants has been found in the form of rice husk, used as degraissant in the pottery as well as charred grains of rice of domesticated variety. The presence of a wild variety of rice along with a domesticated variety is significant, which presents the practice of collecting wild rice along with cultivation of rice. The cultivated variety of rice has been identified as *Oryza sativa* and wild variety are annual and perennial *Oryza nivara* and *Oryza rufigona.*[36] The wild variety of rice was also present in the last phase of the Mesolithic culture at Chopani Mando. Imprints of *Ischaemum rugosum* also have been identified from Mahagara, which is a common weed grown in marshy paddy fields. That wild grain was a part of the diet of the Neolithic people is also attested by the finding of millet-like grains from Koldihwa. Other botanical remains recovered from excavations include charred seeds of jujube and charred bamboo fragments.

A large number of animal bones have been collected from the excavation at Mahagara, Koldihwa and Kunjhun, including both the domesticated as well as wild species. The domesticated animals have been identified by Alur[37] as cattle (*Bos indicus*), sheep/goat, and wild ones such as deer (*Cervidae*), boar (*Sus scrofa*) and horse (*Equidae*). The animal bones recovered from Kunjhun include cattle, deer, antelope and dog. A good number of bones of tortoise, fish and bird also have been found from almost all the excavated sites; these supplemented the diet of Neolithic people.

MATERIAL CULTURE

The artefacts recovered from the habitation sites give ample evidence of the life pattern of the first farming communities of the Vindhyas. Recording of the artefacts distributed on the hut floors indicates a degree of specialized craftsmanship connected with some huts. The artefacts connected with food producing, food processing and hunting-gathering include celts, querns, mullers, bored stones, microliths, bone tools, pottery, etc.

Nearly 44 Neolithic tools, fully ground and polished, were recovered from the habitation sites. The main types are celt, adze and chisel. The celts predominate in the assemblage. The celts and adzes are of a small rounded variety with rectangular or ovaloid cross section. Many of the beautiful small celts and adzes have use–marks on their working edges. Nearly 75 per cent of the neoliths are made of basalt, though some are also made of granite (20.45%) and quartzite (4.54%). The source of igneous rocks, used as raw material for manufacturing the neoliths, could not be located in the Vindhyas. But to the south of the Son, there are granite formations in Semari series of lower Vindhyan formation. As there is no evidence of waste material of neoliths at these sites, it appears that these were manufactured at the source of raw material and were brought to the settlements in their finished forms.

The querns, mullers, sharpeners, rubber stones, hammers, anvils, stone discs, sling balls and bored stones, most of which were food-processing equipment, are made of sandstone or quartzite. Morphologically, the querns are divisible into two groups: (i) basin-shaped; and (ii) flat. Basin-shaped querns are heavily used. There are ample circular marks on the slope of the basin-shaped querns, indicating that the upper grinding stone was rotated on the quern. The mullers are marked with a pitted surface and also have an unifaced or multifaced working surface. Most of the stone objects have been fashioned by pecking and grinding. Some of the hammers, anvils and sling balls also have evidence of pitted surface on their working edges. The sharpeners and rubber stones have heavily rubbed use marks. The bored stones with hourglass perforation have also been made by pecking and grinding.

The microlithic industry is an integral part of the Neolithic cultural assemblage. An analysis of microliths recovered from the excavations of Mahagara shows that chalcedony (56.45%), chert (21.19%), agate (19.88%), carnelian (1.41%) and quartz and crystal (1.82%) comprised the raw material for manufacturing microliths. The microlithic artefacts may be broadly divided into two groups: (i) waste material; and (ii) tools. The waste material (86.63%) includes cores, flakes and chips. The tools (13.32%) on typo-technological considerations have been grouped into retouched blades (32%), backed blades (6.19%), backed and truncated blades (2.17%), serrated blades (1.45%), points (3.62%), awls (0.17%), trapezes (1.45%), lunates (1.09%) and tranchets (0.72%). Some of the blades have a shining gloss on their working edges, presumably used for cutting green vegetation. There are a good number of large-sized flakes made of grey chert, some of which have use marks on their edges. With a sharp working edge, these flakes possibly played the role of heavy duty tools and were used for cutting and scraping.

Bone arrowheads marked with a pointed end and occasionally with a single tang are the tools other than microliths used in hunting by the Neolithic people. But the evidence of bone tools is available only from Mahagara.

Ceramic industry is one of the most diagnostic traits of the Neolithic culture of the northern Vindhyas. Statistically also, it is the richest among the cultural assemblages. On the basis of surface treatment, the ceramic industry has been divided into four major groups: (i) Cord impressed ware; (ii) Rusticated ware; (iii) Burnished red ware; and (iv) Burnished black ware.[38]

The clay used for shaping the pottery is not well levigated and contains calcium granules and small iron nodules. Husk of rice and millet, chopped straw and leaves and cowdung have been mixed as degraissant in the clay. The pottery is handmade, as is evident by palm and finger impressions on both sides of the uneven pot surfaces. The fabric varies from coarse to medium. The surface colour of first two groups is generally mat red and those of third and fourth groups are bright red and black, respectively. The core colour of the pots is smoky grey to blackish due to the organic temper mixed in the clay and ill firing.

The functional pottery types are very simple but, to some extent, standardized. These include convex, straight or tapering-sided deep and shallow bowls; tubular spouted bowls; straight, concave or carinated necked jars; basins, handis and platters. The carinated neck of jars and the tubular spouts were luted to the jars and bowls, respectively, made separately. The decoration of pottery is confined only to applique and incized designs.

The corded ware, one of the characteristic features of the Neolithic culture of the northern Vindhyas, appears to have been the dominant ware at these sites.[39] The ware is represented by 56.64% at Koldihwa, 17.54% at Mahagara, 30% at Pachoha and 80% at Kunjhun. The pots were provided a cord-impressed decorative pattern when they became leather hard. An experiment on the manufacturing technique of the cord-impressed ware shows that the cord-impressed pattern can be obtained by using tortoise bones as dabber; thus, the ware may be termed as tortoise bone impressed ware.[40]

However, the employment of cord-wrapped paddle to impart the cording pattern on some of the pots cannot be totally ruled out. The cord-impressed pattern varies from thick to thin and deep to indistinct. The cording strokes are vertical, horizontal, oblique, slanting and occasionally multidirectional.

The rusticated ware is marked by roughened external surface of the pots. The rustication has been done by clay solution mixed with husk, chopped leaves and straw, grits and calcium granules. Possibly, the rustication was done to make the pots more sturdy and hardened. Sometimes, pots of cord impressed ware and burnished red ware have also been rusticated. The ware is represented by 22.63% at Koldihwa, 62.96% at Mahagara, 40% at Pachoh and 9% at Kunjhun.

The burnished red ware, represented by 17.63% at Koldihwa, 16.41% at Mahagara, 22% at Pachoha and 9% at Kunjhun, is characterized by a bright red ochrous slip and burnishing. The bowls have been slipped and burnished on both the surfaces, while the jars have this treatment only on the outer surface.

The burnished black ware marked by black slip and burnishing has the same features, techno-typologically, as those of the burnished red ware. It is represented by 3.78% at Koldihwa, 3.09% at Mahagara, 8% at Pachoh and 1.4% at Kunjhun.

Among the ceramic industries, the cord-impressed ware has immense archaeological importance as it denotes the cultural contact of Vindhyas with other regions. The cord-impressed ware has been found in Neolithic context at Chirand, Chechar Kutubpur, Taradih, Sohagaura and Lahuradewa in the middle Gangetic plain that has techno-typological similarity with that of the Vindhyas. The evidence suggests that the Vindhyan Neolithic pottery has considerably influenced the Neolithic pottery of the middle Gangetic plain.[41] The cord-impressed pottery is also considered as a distinctive ware of the eastern Asiatic Neolithic complex.[42] It has been found at Daojali Hading[43] and Sarutaru[44] in Assam, but it differs in colour as also in the range of decorative patterns from that of the Vindhyas.

The other artefacts associated with the Neolithic culture of the Vindhyas include perforated pottery disc (possibly used as whorl for spinning), terracotta dabber, spherical beads of terracotta and a shell pendant.[45]

ORIGIN AND ANTIQUITY

Though there is no direct stratigraphic evidence to demonstrate the transformation of farming culture from the hunting-gathering culture, but it is clear from the comparative study of the cultural contents of both the cultures of the northern Vindhyas that several features of the Neolithic culture have their source in the Mesolithic culture of the area. The microlithic industry; bone arrowheads; food-processing equipment like querns and mullers; tool fabricators like hammers, anvils; and curvilinear hut floors may be treated as the Mesolithic remnants in the Neolithic culture. The wild rice found in the form of husk embedded in burnt clay lump at Chopani Mando in the Belan Valley indicates the presence of rice in wild form in the area which was cultivated in the Neolithic period. Similarly, some of the wild animals present in the area were domesticated in the Neolithic period. The ceramic tradition in the Vindhyas dates back to pre-Neolithic times. Some of the pottery traditions, especially the cord-impressed ware recovered in the Mesolithic rock shelters, may be the result of contact with Neolithic people. But some specific features of the Mesolithic pottery of the Vindhyas, such as engraved beater-impressed floral designs on brownish grey or dull red ware, the simplicity of form, the initial stage of technological perfection and surface decoration by impressing natural objects amply demonstrate the beginning of pottery tradition in the Mesolithic period. All these points present a picture of smooth transition of cultural traditions of Mesolithic to the Neolithic in the region and tend to support the indigenous origin theory of the Vindhyan Neolithic.

The problem of the antiquity of the Neolithic culture of the Vindhyas is still not finally settled. Considering three of C-14 dates reading 4530±185 BC (PRL 101), 5440±240 BC (PRL 100) and 6570±210 BC (PRL 224) obtained from Koldihwa as dependable, the culture was dated to the 7th–6th millennium BC.[46] But as these dates were not in correct stratigraphic context, doubts were raised about such early antiquity.[47] The C-14 date belonging to the transitional phase of the Neolithic to Chalcolithic is 1440±120 BC (PRL 223). The absolute dates obtained from Mahagara also indicate a late date to the culture, though these dates have the possibility of contamination of samples. Two TL dates 2265 BC and 1616 BC and four C-14 dates 1440±150 BC (PRL 409), 1330±120 BC (PRL 408), 1440±100 BC (PRL 407) and 1480±110 BC (BSI) have been obtained from the samples from Mahagara. These dates are not consistent with the stratigraphy possibly due to contamination of samples. In the light of C-14 dates obtained from Kunjhun, reading 4600±80 BC, the revised chronology of the Vindhyan Neolithic culture has been proposed to 4th millennium BC.[48] Some C-14 dates have come to light from recent excavations at Lahuradewa in the middle Gangetic plain, which read as 5320±90 BP (BS 1951) (Cal BC 4220, 4196, 4161) and 6290±160 BP (BS 1966) (Cal BC 5298).[49] The antiquity of the early farming culture of the Gangetic plain on the basis of these dates may be pushed back to 6th-5th millennium BC. If the Gangetic Neolithic owes to that of the Vindhyas for its origin, the antiquity of the Vindhyan Neolithic now may be fixed in the 6th millennium BC.

NOTES AND REFERENCES

1. Spate, O.H.K. and A.T.A. Learmonth 1967, *India and Pakistan: A General Regional Geography*, London, Methuen & Co. Ltd., p. 14.
2. Le Mesurier, H.P. 1861, in *Proceedings of the Asiatic Society of Bengal*, pp. 81–85.
3. Theobald, W. 1862, Celts found in Bundelkhand, *Proceedings of the Asiatic Society of Bengal.*
4. Cockburn, J. 1879, Notes on Stone Age Implements from Khasi Hills and the Banda and Vellore Districts, *Journal of the Asiatic Society of Bengal*, XLVIII, pt. 2.
5. Rivett-Carnac, H. 1882, Stone Implements Found in Bundelkhand, *Proceedings of the Asiatic Society of Bengal.*
6. Cockburn, J. 1894, On Flint Implements from the Kon Ravines of South Mirzapur, 20 Miles Southwest of Benaras, *Journal of the Asiatic Society of Bengal*, LXIII.

7. *I.A.R.* 1959–60:48; *I.A.R.* 1961–62:56; *I.A.R.* 1969–70:37; *I.A.R.* 1970–71:36; *I.A.R.* 1973–74:25; *I.A.R.* 1975–76:44; *I.A.R.* 1975–76:27; *I.A.R.* 1980–81:72.
8. Pal, J.N. 1990, The Early Farming Culture of the Northern Vindhyas, *Bulletin of the Deccan College Post-Graduate Research Institute*, Volume 49, pp. 297–310.

   Misra, V.D. 1975, *Some Aspects of Indian Archaeology*, Prabhat Prakashan, Allahabad.
9. Pal, J.N. 1990, op. cit.
10. Misra, V.D., J.N. Pal and M.C. Gupta, 2001, Neolithic Culture of the Northern Vindhyas with Special Reference to Tokwa, *Bharti*, Bulletin of the Department of Ancient Indian History, Culture and Archaeology, B.H.U. Varanasi 25 (I and II) 1998–99: 211–233.

    Misra, V.D., J.N. Pal and M.C. Gupta, 2000–2001, Excavation at Tokwa: A Neolithic-Chalcolithic Settlement, *Prāgdhārā* 11, pp. 59–72.
11. Misra, V.D., B.B. Misra, J.N. Pal and M.C. Gupta, 2000, Exploration at Tokwa: A Neolithic-Chalcolithic Settlement on the Confluence of the Belan and Adwa Rivers, in S.C. Bhattacharya, V.D. Misra, J.N. Pandey, J.N. Pal (Eds) *Peeping through the Past: Prof. G.R. Sharma Memorial Volume*, Department of Ancient History, Culture & Archaeology, University of Allahabad, pp. 45–57.
12. *I.A.R. 1977–78*, pp. 17–18.
13. *I.A.R. 1963–64*, pp. 5–6; *I.A.R. 1968–69* to *1971–72* and *I.A.R. 1981–82*, pp. 13, 14.
14. *I.A.R. 1981–82*, pp. 10–12; *1982–83*, pp. 16,25, *1983–84*, pp. 12–13; *1984–85*, pp. 12–13, and Prasad, A.K. 1980–81. Excavation at Taradih, *Purātattva* 12: 138–139.
15. Singh, B.P. 1989–90. The Chalcolithic Culture of Southern Bihar as Revealed by the Explorations and Excavations in District Rohtas, *Purātattva* 20: 83–92.
16. Singh, P. and M. Lal 1985. Narhan 1983–85: A Preliminary Report, *Bharati, New Series* 3:120 ff.; also Singh, P. 1994. *Excavations at Narhan (1984–89)*, Benaras Hindu University and B.R. Publishing Corporation, New Delhi.
17. *I.A.R. 1981–82*, pp. 67–70, also Singh, B.P. 1987–88. Khairadih—A Chalcolithic Settlement, *Purātattva* 18: 28–37.
18. *I.A.R. 1963–64* pp. 57–58.
19. *I.A.R. 1961–62*, p. 56; *I.A.R. 1974–75*, pp. 46–47, also Chaturvedi, S.N. 1985. Advance of Vindhyan Neolithic and Chalcolithic Cultures to the Himalayan Tarai: Excavations and Explorations in the Saryupar Region of Uttar Pradesh, *Man and Environment* IX: 120 ff.
20. Narain, A.K. and T.N. Roy 1977. *Excavations at Rajghat*, Benaras Hindu University, Varanasi.
21. Narain, A.K. and T.N. Roy 1968. *Excavations at Prahladpur*, Benaras Hindu University, Varanasi.
22. Misra, V.D., B.B. Misra, J.N. Pandey and J.N. Pal 1995–96. A Preliminary Report of Excavation at Jhusi, 1995, *Prāgdhārā* 6: 63–66.

    Misra, V.D., J.N. Pal and M.C. Gupta 1998–89, Further Excavation at Jhusi, District Allahabad (Uttar Pradesh): 1998. *Prāgdhārā* 9, pp. 43–49.

    Misra, V.D., J.N. Pal and M.C. Gupta 1999–2000, Further Excavation at Jhusi, District Allahabad (Uttar Pradesh): 1999. *Prāgdhārā* 10: 23–30.

23. Singh, P. and Ashok Kumar Singh, 1999–2000. Excavations at Agiabir, District Mirzapur (U.P.), *Prāgdhārā* 10, pp. 31–55.
24. Tewari, R. and R.K. Srivastava 1996–97. Excavations at Raja Nal Ka Tila, District Sonbhadra, U.P.: Preliminary Observation, *Prāgdhārā* 7: 77–96, and also Tewari, R. and R.K. Srivastava 1997–98. Excavations at Raja Nal Ka Tila (1996–97), District Sonbhadra, U.P.: Preliminary Observation, *Prāgdhārā* 6: 99–106.
25. Tewari, R., R.K. Srivastava, K.S. Saraswat and K.K. Singh 1999–2000, Excavations at Malhar, District Chandauli (U.P.) 1999: A Preliminary Report, *Prāgdhārā* 10: 69–98.
26. The author is thankful to Dr. Dorian Q. Fuller of University College, London for tentative identification of grains recovered from the excavation at Tokwa.
27. Harvey, Emma, Dorian Fuller and J.N. Pal. 2003. Early Agriculture of the Neolithic Vindhyas (North-Central India), Poster displayed in the South Asian Archaeology Conference in Bonn, 2003.
28. Tewari, Rakesh, R.K. Srivastava, K.K. Singh, K.S. Saraswat and I.B. Singh, 2002–2003, Preliminary Report of the Excavation at Lahuradewa, District Sant Kabir Nagar, U.P. —2001–2002: Wider Archaeological Implications, *Prāgdhārā* 13: 37–68.
29. The study of faunal remains from Tokwa is being done by Dr. P.P. Joglekar of Deccan College, Pune. The author is extremely thankful to Dr. Joglekar and the Deccan College for undertaking this study.
30. The author is also thankful to Dr. R. Ramesh of Physical Research Laboratory, Ahmedabad and Dr. Y. Yasuda and Dr. V. Shinde of International Research Centre for Japanese Studies, Kyoto, Japan, for processing the samples for C-14 dating.
31. Sharma, G.R. and D. Mandal 1980, *Excavations at Mahagara*, Department of Ancient History, Culture and Archaeology, University of Allahabad, Allahabad.
32. Alur, K.R. 1980. Faunal Remains from the Vindhyas and the Ganga Valley. In G.R. Sharma, V.D. Misra, D. Mandal, B.B. Misra and J.N. Pal, *Beginnings of Agriculture*, Allahabad, Abinash Prakashan.
33. *I.A.R.* 1975–76: 47.
34. Sharma, G.R. and D. Mandal 1980. *ibid.*
35. *I.A.R.* 1980–81: 72.
36. Vishnu Mittre and Aruna Sharma, n.d. Neolithic-Chalcolithic Food Economy of Eastern Uttar Pradesh.
37. Alur, K.R. 1980. Faunal Remains from the Vindhyas and the Ganga Valley. In G.R. Sarma, V.D. Misra, D. Mandal, B.B. Misra and J.N. Pal, *Beginnings of Agriculture* Allahabad, Abinash Prakashan.
38. Pal, J.N. 1986, *Archaeology of Southern Uttar Pradesh: Ceramic Industries of Northern Vindhyas,* Allahabad, Swabha Prakashan.
39. *Op. cit.*
40. Pal, J.N., 1987, Neolithic Cord-Impressed Ware of the Vindhyas, *Man and Environment* XI, 61–65.
41. Pal, J.N. 1983, Bases of the Neolithic Culture of the Middle Ganga Valley, *Baldev Upadhyaya Felicitation Volume (Journal of the G.N. Jha Kendriya Sanskrit Vidyapeeth),* xxxvii: 205–16.

42. Sharma, T.C., 1997, The Neolithic Pattern of Eastern India. In V.D. Misra and J.N. Pal (Eds), *Indian Prehistory: 1980.* Department of Ancient History, Culture and Archaeology, University of Allahabad, Allahabad, pp. 145.
43. Sharma, T.C. 1966, Prehistoric Archaeology of Assam, Unpublished Ph.D. Thesis London, London University.
44. Rao, S.N. 1977, Excavation at Sarutaru, *Man and Environment,* 1, 39–43.
45. Mandal, D. 1997, Neolithic Culture of the Vindhyas : Excavations at Mahagara in the Belan Valley. In V.D. Misra and J.N. Pal (Eds) *Indian Prehistory: 1980.* Department of Ancient History, Culture and Archaeology, University of Allahabad, pp. 163–67.
46. Sharma, G.R. and D. Mandal. 1980 *Ibid.*
47. Pal, J.N. 1986. *Ibid.*
48. Clark, J. Desmond and Gurucharan S. Khanna, 1989, The Site of Kunjhun II, Middle Son Valley, and its Relevance for the Neolithic of Central India, in J.M. Kenoyer (Ed) *Old Problems and New Perspectives in the Archaeology of South Asia.* Wisconsin Archaeological Reports, Vol. 2, Department of Anthropology, University of Wisconsin, Madison: 29–46.

    Possehl, G.L. and Paul C. Rissman, 1992, The Chronology of Prehistoric India: from Earliest Times to the Iron Age, in Robert W. Ehrich (Ed) *Chronologies in Old World Archaeology,* Volume I, India, Chicago and London, The University of Chicago Press, p. 461.
49. Tewari, Rakesh, R.K. Srivastava, K.K. Singh, K.S. Saraswat and I.B. Singh. 2002–2003. *Op. cit.*

    Tewari, Rakesh, R.K. Srivastava and K.K. Singh. 2001–2002. Excavation at Lahuradewa, District Sant Kabir Nagar, Uttar Pradesh. *Purātattva* No. 32: 54–62.

CHAPTER 5

The Settlement and Subsistence Pattern of the Early Farming Cultures in the Middle Ganga Plain

(With Special Reference to Lahuradeva, Jhusi and other Recent Sites)

J.N. Pandey and *R.P. Tripathi*

INTRODUCTION

The study of ecological and environmental setup of the Middle Ganga Plain is essential for the study of the early farming culture of the region. The first basic ingredient of an ecological setting is the underlying land formation. The region of Middle Ganga Valley under study is located between the Ganga–Yamuna in the south and Ghaghra or Gogra in the north. Geologically speaking, the Middle Ganga Plain is of a very recent age and its surface has been built up by the silting action of its streams during the Pleistocene. In the Middle Ganga Plain, River Ganga is the master stream and is the recipient of all other water lines. The Ganga Plain is an area of active sedimentation and the age of the surface is mostly Holocene (12000 BP). For the proper understanding of the microgeomorphic features, landscape evolution, the past climate can help in understanding of the human settlements in the Ganga plain. These studies are also important to investigate the changes in the landscape due to anthropogenic activity of the early settlers. It would help us to differentiate between changes in the landscape due to natural causes and the those from the anthropogenic activities (Singh, 2004 n.d.). There is great development of dead arms, deferred junctions, *jhils* (lakes) and natural *tals* (ponds) in the Middle Ganga Plain (Spate, 1967: 547). The general slope of the region is towards the southeast. The drainage pattern is dendritic in nature. The tributary streams of Ganga have occasional bad-land and ravines. On the basis of aspect and hydrography, the area under study may be divided into two regions: (1) Upper Ganga Valley; and (2) Middle Ganga Valley (Spate, 1967; Singh, 1971; Deshpande, 1992). The Gangetic alluvium is divisible into *bhangar* and *khader* (Wadia, 1957: 394–95). *Bhangar* is the older alluvium and covers upland tracts of the plain beyond the annual flood limits; it is under the process of denudation and contains patches of barren land (*usar*). The formation of *kankar* (Calcium carbonate) through capillary action is also noteworthy. At places, there may be as much as 30% of calcareous matter in the *bhangar*. *Bhangar* is richer in lime and *kankar* content than *khadar*. It is sticky and is usually not well drained. *Bhanger* grades from sandy clay loam but is often heavier with higher clay factor in depressions. The *bhangar* formation may, tentatively, be attributed to the terminal Pleistocene and the warm and wet early Holocene. Mesolithic artefacts are also reported to occur in the *bhangar* formation (Sharma, 1973: 130).

Khadar covers the flood plains in the vicinity of rivers, including the lower reaches of smaller rivers and also the old beds. It is replenished annually by new deposits. *Khadar* mostly consists of fine silt but is sandy at places. It is newer in age than *bhanger* (Wadia 1957: 95).

The enormous interest generated in theoretical questions over the latter half of twentieth century in Europe (Renfrew, 1968; 1973; Clarke, 1968, 1972; Coles, 1973) and in the United States of America (Binford, 1966, 1968, 1972; Mueller 1975) has brought about a change in the aims and methods of archaeological studies. The 'New Archaeology' is preoccupied with methodological and theoretical formulations. It reflects passion for conferring scientific respectability to a rather *indisciplined emperical discipline* (Clarke, 1968: XIII). From one point of view, it indicates an aspect of the professionalization of prehistoric archaeology. From another point of view, it is only talking about old concepts in a new guise. Archaeological science is also being applied to what are obviously scientific investigations, as for instance, in the environmental field. It must, however, be the responsibility of the archaeologists to control the application of methodological and theoretical formulations to their problems (Hodder, 1991, 1992). Jacquetta Hawkes' (1968) nightmare of a *Frankenstein monster* that theatens to take over archaeology is not the product of the 'New Archaeology' or processual archaeology *per se*. It is true that the theoretical archaeology contains many examples of methodology treated out of all proportion to the result they produce. To some extent, it is inevitable in a developing field, but balance has to be maintained. The prevalent approaches to the study of Neolithic cultures in the Vindhyas and adjacent parts of the Middle Gangetic plain have been empirical, typological and functional (Sharma *et al.*, 1980). The question is not whether empirical approach is good or bad, but how this can be used effectively. The present study aims at delineating the patterns of Neolithic human settlement against the ecological background of the Middle Ganga plains and the adjoining Vindhyan plateau on the basis of the available evidence. In many respects, prehistoric archaeology depends upon environmental factors and justifies the importance placed on the landscape (Butzer, 1992). The spatial study includes various aspects of settlement patterns comprising size and distribution of settlement and spacing of settlements, development of site hierarchies, and population and site catchment areas. Settlement patterns have a close relationship to the makers of subsistence and environment. Mesolithic sites have been discovered between the Ganga in the south and Gomati in the north, covering the northern part of Allahabad district, Pratapgarh, Sultanpur, Jaunpur, Bhadohi and Varanasi districts of Uttar Pradesh (Pandey, 1985: 86–198). In the Ganga Valley, 204 Mesolithic sites have been discovered so far. A total of 177 sites are recorded in Pratapgarh district as compared to 6 in the northern part of Allahabad, 5 in Sultanpur, 14 in Jaunpur and one each in Bhadohi and Varanasi districts, respectively. Thus Mesolithic sites have a fairly dense but discontinuous distribution in Pratapgarh district and sporadic occurrences in northern Allahabad, Sultanpur, Jaunpur, Bhadohi and Varanasi districts. This distribution disparity is probably the result of uneven exploration but could also be due to microlevel environmental differences among the district concerned. This latter possibility is yet to be investigated (Pandey, 1985:130).

The Mesolithic settlement in the Gangetic plain is the earliest evidence of human colonization of the virgin Ganga plains. These colonizers were evidently migrants from the hilly and forested Vindhyan region which have been inhabited by Stone Age hunter-gatherers right from the Lower Paleolithic times (Misra, 1977: 22–75; Sharma and Clark, 1983: 25–150). All the explored Mesolithic sites in the Middle Ganga Valley till 1985 do not exhaust the number of sites; they are sufficient in number and are among those particular sites that have great relevance for the present study. Information from published and unpublished sources has been used wherever available but more often than not, it has been found to be insufficient specific as to the location of sites and dimensions of many surface sites. The published accounts in most of the cases are in the form of brief summaries. The publication of an excavation report of Sarai Nahar Rai is pending since 1973 and that of Damdama since 1987 (Pandey, 1990: 311–316). The only available detailed reports are those of Chopanimando (Sharma *et al.*, 1986), Mahadaha (Sharma *et al.*, 1980), Morhana Pahar, Baghai Khor (Varma, 1980), Baghor II (Sharma and Clark, 1983), Ghaghara rock shetler (Sharma and Clark, 1983), Medhauli (Indian Archaeology. A Review, 1982–83) and Banki, (Misra *et al.*, 1992–93), (Mishra, 2002).

The Mesolithic settlement pattern in the Gangetic plain has a somewhat different character to the ones in the south. Associated with old lakes—now in many cases dry —which appear to be cut off meanders of former smaller streams such as Sai and Gomti, most of Mesolithic settlements in the Gangetic Plain are small in dimension and seem to have been temporary camping places, measuring between 5m^2 and 4.5m^2 in area (Pandey, 1990: 312). However, a few larger sites like Sarai Nahar Rai: 1800m^2 (Sharma, 1973:130), Mahadaha, 8000m^2 (Sharma *et al.*, 1980) and Damdama: 8750 m^2 (Varma, *et al.*, 1985:45) appear to have been regularly inhabited (Pandey, 1990: 312). The excavated sites in the Gangetic Plain had occupational areas with hearths in large oval pits. A pattern of differential density was discerned at Sarai Nahar Rai. According to late G.R. Sharma, the excavator of the site, "it is significant that though the area is compact and there is no intervening encroachment between the living area and burial area, the dead were buried on the habitation site itself, but care was taken to separate the burial ground from the area in which hearths were carved out (Sharma, 1973:130). One hut floor was made of rammed burnt clay. The maximum length and width of the floor measured 5.64.02 metres (Sharma, 1973:140). It has four post-holes on the oblong-shaped hut which was built on wooden posts, four corners suggesting that vertical posts supported the walls and roof. (Pandey, 1985:138). It may be pointed out in this connection that it was inadvertently identified as 'community hearth' (Sharma, 1973:140). According to the excavator, 'the floor has not been fully excavated; a trench across it revealed the existence of five hearths scooped in the floor'. Mesolithic people at Sarai Nahar Rai would have different structures for different seasons. A winter hut could have included hearths or fire-pits inside the floor, possibly towards the centre. A summer hut might have essentially the same features as that of the winter hut floor, although the hearths or fire pits would have been outside the hut. The question arises whether the floor at Sarai Nahar Rai was enclosed by walls or not. Because of the four post-holes on the four corners, obviously the hut has wattle and daub superstructure, which was supported by the four posts. The hut was most probably built during the winter season because the fire pits (heaths) have been found inside the hut. The site of Sarai Nahar Rai was considered by the excavator to be a small open-air settlement of short duration (Sharma, 1973:135).

The excavations at Chopanimando, a Mesolithic site on the Belan river, have laid bare the ground plans of 20 circular or oval huts. Of these huts, 2 belong to phase II A (Early Mesolithic), 5 belong to phase II B, and 13 belong to Phase III (Advanced Mesolithic/Proto-Neolithic (Mishra, 1980, 37:40). One of the huts of Phase IIA has 12 post holes and its diameter was 3.80 metres. One of the huts of Phase II B is 3 m in diameter. Of the 13 huts of Phase III, 6 are round and the remaining 7 are oval in shape. The excavations at Mahadaha between 1977–78 and 1978–79 were carried out as a salvage operation in the scorching heat of May-June. They have, nevertheless, produced interesting finds in respect of hitherto unknown aspects of Mesolithic culture (Sharma, *et al.*, 1980: 1–2). The excavations at Mahadaha brought to light three distinct complexes: (1) habitation-cum-burial complex; (2) the so-called butchering complex, which was most probably the dumping ground; and (3) the lake area (Sharma *et al.*, 1980: 83–101). The habitation-cum-burial site constituted the main area within the settlement. It is mainly confined to the west of the Jaunpur (Minor) Canal and comprises 28 burials containing 30 human skeletons, the two graves contained the skeletal remains of two individuals each, and 35 hearths or fire pits (Sharma *et al.*, 1980: 83–98; Pal, 1988: 28–37). The area lying between habitation-cum-burial complex in the west and the western margin of the horseshoe lake in the east had been designated as the butchering complex (Sharma *et al.*, 1980:9). The site of Mahadala was situated on the western margins of a horseshoe lake and much of the discarded animal bones and other organic material found their way into the water-logged deposit. This led to the survival of an unusually large number of animal bone fragments and other objects made of animal bones and antlers. Animal bone fragments consist of horn core, antlers, skulls, mandibles, teeth, scapulae, patellae, vertebrae, ribs, metacarpals and metatarsals. No complete animal bone was found in this complex. The available evidence strongly suggests that it was most probably a dumping ground of discarded animal bones or a part of the lake itself. Butchering/kill sites are generally not situated within the habitation site itself (Pandey, 1985:140). Excavations in the 'lake area' situated on the eastern side of the site clearly indicated the lacustrine nature of the deposit (Pandey, 1985:141). A large number of animal bones had been recovered from the upper layers of the lake, bones of wild *gaur*, pig, deer and rhino had been found in the aforesaid area (Alur, 1980). No evidence relating to the structure had been found at Mahadaha. The excavator had paid little attention to the process of decay and disturbance and, as a result, had little success in distinguishing the post holes and other patterns indicative of structure/huts at Mahadaha (Pandey, 1985:141). Damdama was excavated by the faculty members of Allahabad University between 1982 and 1987 (Pandey, 1990:314). It revealed a 1.50-metre thick habitation deposit, divisible into 10 layers. Excavations at the site had thrown new light on the Mesolithic culture of the Ganga basin (Varma *et al.*, 1985: 45–65). Fire pits, living floors and human burials were discovered at the site. In all, 41 graves had been excavated at the site. While five graves contained skeletal remains of two individuals each, one grave had the skeletal remains of three individuals. Of the remaining thirty-five graves, each contained skeletal remains of a single individual (Pal, 1988: 115–22). Besides the burials, microliths, bone artefacts, querns, mullers, hammer stones and bones of wild animals were other noteworthy finds from the site. To sum up the results, the sites excavated had occupation areas with hearths, fire pits in the shape of large oval pits, traces of hut/living floors, and burials within the living areas. There was no suitable stone for tool making available in the Gangetic plains and microlithic blade tools were made at the sites from the nodules of chert, chalcedony, agate and carnelian and other semi-precious stones originating in various parts of the Vindhyas, and presumably carried across by their makers or obtained through contact or exchange. A number of sources supplied stone artefacts to the settlements in the Gangetic plains. The areas supplied can be identified by the colour of the stone which varied from source to source, showing that there were regular ongoing relationships along the established routes between certain localities in the hills and Mesolithic communities in specific areas in the plains from then onward (Sharma *et al.*, 1980). All the stone artefacts are extremely small and cores used to the maximum, presumably because of the need for economy. A preliminary study has been carried out of the Mesolithic assemblages from 100 surface sites from Pratapgarh district, selected at random (Pandey, 1985: 164). The total number of assemblages is 2051, of which the numbers of shaped artefacts is 392 (19.11%) and simple artefacts are 1659 (80.89%). The smaller sites are poor in shaped artefacts, which are completely missing at 25 sites. Eighteen sites have one shaped tool each, and only ten have shaped artefacts numbering between 10 and 18. Most of the retouched bladelets are broken either at the proximal or at the distal end. Retouching is very frequently found on the dorsal surface of bladelets, while a few pieces show retouch on ventral surface. Unmodified and modified broken bladelets predominate in the assemblages. Chips (798)—very small irregular flake pieces, less than 10 mm in length and without retouch—occur in great frequency. A large number of bone tools and ornaments have been found at Sarai Nahar Rai (Sharma, 1973: 142–43), and Mahadaha (Sharma *et al.*, 1985: 64). Bone tools comprised arrowheads, points, blades, scrapers, a chisel and a saw. Among the bone ornaments, mention may be made of circular bone rings, pendants and beads. Heavy-duty stone artefacts are also present, but largely in the form of broken fragments, including pieces of querns and some complete hammers which are closely similar to those from Chopanimando and other sites south of the Ganga (Sharma *et al.*, 1980: 61–64).

Numerous animal bones have been found at the stratified sites at Sarai Nahar Rai, Mahadaha and Damdama, situated on the margins of lakes which are now dry. Much of the animal materials found their way into the water-logged deposit. This had led to the survival of an unusually large number of animal bones and other objects made of organic material. Faunal remains had undergone considerable mineralization. Most of the animal bones are heavier than the recently macerated bones. In addition to high mineral components in the compact and concellous bone tissues, there are varying degrees of calcareous concretion deposited over the external surfaces of many animal bones, helping in the preservation of bones. A large percentage of animal bones is fragmentary, broken and charred. A preliminary study of animal bones from Sarai Nahar Rai and Mahadaha was done by K.R. Alur (Sharma *et al.*, 1980: 89–115). He had indentified cattle, sheep and goats. These animal skeletal remains were subsequently analyzed by P.K. Thomas and P.P. Joglekar of the Deccan College, Post-graduate and Research Institute, (1995, 1996). The most common animals identified include Indian humped cattle (*Bos indicus*), gaur (*Bos gaurus*), sambhar (*Cervus unicolor*), chital (*Axis axis*), nilgai (*Boselephas Tragocamelus*), chinkara (*Gazella gazella*), antelope and wild boar (*Sus scrofa*). Considerable dependence on aquatic food resources is indicated by fish bones and tortoise shells. Aquatic resources would have been non-seasonal, predictable, relatively non-fluctuating in availability than terrestrial resources and were capable of supporting localized human population. In the absence of quantitative data, it is not possible to determine whether there were any significant temporal changes in the pattern of animal resource procurement at Sarai Nahar Rai, Mahadaha and Damdama. However, the available evidence is suggestive of broad spectrum hunting rather than specialization on any particular game (Pandey, 1985: 186). A recent study of the faunal remains of the Mesolithic sites of the Gangetic plain by P.K. Thomas *et al.*, of Deccan College, Pune indicates that cattle, sheep and goats are totally absent in assemblage. This evidence contributes substantially to our knowledge of Mesolithic economy and animal environment. No signs of domestication appear to be present in the animal bones from Mesolithic levels. In many ways, these sites have the character of seasonal hunting camps and one feels that their occupants spent part of their time elsewhere. The question is, where?

In India, Mesolithic burials are circumscribed in some parts of Uttar Pradesh, Madhya Pradesh, Gujarat and Rajasthan. The burials at Sarai Nahar Rai, Mahadaha and Damdama were extended and mostly west-east oriented like those in the Vindhyan Rock shelters but the skeletons were those of considerably larger, more robust people (Kennedy *et al.*, 1986, 1992). These studies have confirmed a lesser degree of sex-differentiation in height and bone structure in the skeletons from the Ganga Valley sites than those from the Vindhyas (Lukacs, 2002: 261–288). This further raises an interesting question as to the relationship of two communities to one another. Were the Mesolithic inhabitants of the Gangetic plain part of population whose habitat extended into the Vindhyas and who perhaps moved in a yearly cycle between the Vindhyan hills and the Gangetic plains possibly following seasonal movement of the animals they hunted? Was there at that time a cultural and perhaps an ethnic interface between those whose roots were in the south and those who had connection in the north? If so, a certain amount of contact and exchange of goods evidently took place between these two groups, which perhaps had differing ethic and or cultural histories. The regional pattern that we recognize today within India are important as a basis of understanding the past, but further careful research is required as to how they have been modified in order to understand the surroundings in which the past cultures existed, and the complex inter-relationship of environmental and cultural changes.

A slight increase in average rainfall and in its distribution throughout the year or a drop in average temperature which reduces the amount of evaporation, may cause the margins of the Vindhyas to have a little more vegetation and support more grazing wild animals. Large trees may extend into the areas of grassland and thorn forest beyond their former limits. Such changes can be crucial to the survival of a hunting-gathering communities. They may necessitate a change in the pattern of life which would be registered in the archaeological record in a variety of ways—changes in the tools and/ or in the nature of and distribution of Mesolithic sites. We may be able to see a number of changes as we follow the story of cultural development during the Mesolithic phase. Minor climatic changes had a more noticeable effect in a marginal region of low rainfall than in an area of high rainfall, i.e. the Middle Ganga Plains, where a few centimetres of rainfall more or less would not make much difference. In the dry Vindhyan region, over-exploitation of the resources could have a similar effect to that of natural decrease in rainfall. Earlier, during the Pleistocene age, more profound changes in climate of India, particularly of the Vindhyas, took place (Williams and Royce, 1983: 9–21). Climatic changes that had taken place at the onset of Holocene some 12,000 years ago have been of relative minor kind. During the early Holocene, the climate of the Vindhyas was warmer and wetter (Williams and Royce, 1983:17). It is sometimes not clear whether the changes have been brought about by natural causes or by human activities, or by a combination of both. The late Quaternary is a time of repeated events of global climatic changes affecting the sea levels, atmospheric circulation system, rainfall and temperature. The fluvial system responded to these changes by the process of channel migration and abandonment, changes in sediments and water discharge which are manifested in the alluvial landforms. The Ganga Plain exhibits a large variety of alluvial landforms which were formed under different climatic conditions with strong interaction of base-level and tectonic events during last about 100000 years. Systematic geomorphic studies, geochemical-minerological investigations, chronometry using luminescence method, radiocarbon dating and palynological studies have helped the archaeologists to reconstruct the events based on base-level change, tectonic activity and climate change during the late Quaternary, which is often intermixed and difficult to separate (Singh, 2004 n.d.).

An important interference of these studies is that the Ganga Plains, at least for the last 45000 years, was a grassland with few thickets and patches of forest. The landscape had some high grounds, represented by alluvial ridges and natural *levees*. A large number of small and large water bodies (ponds and lakes) and minor channels were prominent features of the landscape. There is a continuous record of palynological evidence of grassland for the last about 15000 years. During the Ist Glacial Maxima (LGM), the landforms supported a dense network of minor tributaries, many of which were ephemeral in character. These channels became very prominent in around 13000 years with increased monsoonal rainfall. A tectonic event between 8–5000 years transformed this landscape of dense river channels into one of ponds and lakes. High monsoon rainfall during 13–6000 years supported large lakes. Five to four thousand years was a dry climate phase, causing shrinkages and siltation of the lakes. The study of bovid teeth enamel suggest century-scale monsoonal rainfall change in the last 4000 years.

The Ganga Plains, with their high grounds for settlement, plenteous of water and rich fauna and flora in the grassland attracted humans from Paleolithic times. The finds of Middle Palaeolithic archaeological site at Kalpi in the Yamuna Valley, dated around 45000 years, with rich faunal remains and a large variety of bone tools, suggests early occupation of the Ganga plains by humans. The upwarp and ponds developed during early Holocene provided ideal sites for the Mesolithic people (8–4000 years) in Pratapgarh district and the adjoining areas. There is also evidence of Mesolithic occupation at Kalpi. Large-scale human settlements in the Ganga Plains took place around 3500 to 3000 years on the high grounds close to the lakes and small rivers. Later, during the last 3–2000 years, sites close to large rivers were occupied.

It seems reasonable to assume that the Ganga Plains were occupied by human since Middle Palaeolithic times (Tiwari *et al.*, 2002). There is evidence of agricultural activities in the form of cultural pollen since the last 15000 years. The Ganga Plains is an area of active sedimentation and the age of the surface sediments is mostly late Holocene. Evidence of early human settlements are most probably buried in the subsurface and can be studied in cliff-sections, deeper excavations in appropriate sites and areas of upliftment undergoing erosion, for example, in Pratapgarh district of Uttar Pradesh. Proper understanding of the microgeomorphic features, landscape evolution, past climate can help in learning about the human settlements in the Ganga Plains. These studies are also important in investigating the changes in the landscape due to anthropogenic activity of early settlers. It would help us to differentiate between changes due to natural causes and those—less massive but still very important changes—that took place during the last 12,000 years will be further discussed in the following paragraphs, where they are relevant to the study of the domestication of animals and cultivation of plants. The excavations at Chopanimandos (Mishra, 1980: 69–76) have yielded remains of paddy. The remains of charred or carbonized paddy were found embedded in lumps of burnt clay. These remains of rice from advanced Mesolithic/Proto-Neolithic level represent the wild variety (Mishra, 1980:69). The last phase, advanced Mesolithic/Proto-Neolithic of the occupation at Chopanimando, is characterized by the emergence of handmade and very fragile pottery, which is generally found in weathered and worn out condition and is insufficiently backed (Misra, 1980:65). The clay is not well levigated and contains sand and sometimes straw and husk as *degraissants.* It varies in fabric from medium to coarse and is represented by simple shapes like medium- and small-sized bowls and vases. However, it may be observed that shapes are yet not standardized. No evidence of surface treatment like slip, wash and burnishing is discernible. Pottery makes its first appearance in layer 3 at Chopanimando (Mishra, 1980:65), where its number is very limited. However, the frequency increased gradually in subsequent layers—2A, 2 and so on. The entire ceramic assemblage is divisible into Red Ware and brownish Grey Ware (Mishra, 1980:65). In this context, it may be noted that pottery and ring stones which are functionally associated generally with agriculture, make their appearance for the first time during phase III (advanced Mesolithic/Proto-Neolithic).

Archaeologists are familiar with the impact of food-producing economies on the environment and landscape but it is necessary to emphasize the possibility of change at pre-agriculture stage also. Usually, it is taken for granted that the Mesolithic people, with their primitive culture, could have but little effect on the ecology. There had been a suggestion that Mesolithic man was modifying the forest-cover partly in course of burning of forests but also to improve the plant cover for animal and human consumption (Mellars, 1976: 15–45). The use of fire on a large scale is attested at Sarai Nahar Rai (Sharma, 1973), Mahadaha (Sharma *et al.*, 1980) and Damdama (Varma *et al.*, 1985) by a number of fire pits. Whether the Mesolithic people of the Middle Ganga Valley were burning forest or not, we have no direct evidence (Pandey, 1985:187).

Archaeological evidence for the study of subsistence patterns of the Mesolithic culture of the middle Ganga Valley may be divided into three categories: (1) artefactual evidence; (2) faunal evidence; and (3) floral and palynological evidence. The available microlithic bladelets are biased in favour of hunting. Besides, a large number of bone arrowheads, fragments of querns and rubbers (mullers) have also been found in excavations at Sarai Nahar Rai (Sharma, 1973), Mahadaha (Sharma *et al.*, 1980) and Damdama (Varma *et al.* 1985). Animal bones make up the bulk of organic remains. Butchering marks and charring of bones indicate that these were used for food (Thomas and Joglekar *et al.*, 1995, 1996). Floral and palynological evidence has been found at Mahadaha (Pant and Pant, 1980) and Damdama (Kajale, 1990), respectively. The attractions which drew Mesolithic people to parts of the middle Ganga Valley must have been primarily those of more abundant terrestrial, aquatic and avian food resources (Thomas and Joglekar *et al.*, 1995, 1996). Besides, in the Vindhyas, there was vigorous artistic activity during the Mesolithic period (Varma, 1986, Mathpal, 1984). Mesolithic people have left a vivid record of themselves, their way of life and contemporary diverse natural fauna (Pandey, 1985:84). Antelopes, deer, boar, cattle/rhinoceros were chief animals hunted by Mesolithic people. Hunters are sometimes shown wearing masks. Bows, arrows and spearheads were the main weapons of hunt. In rock paintings of the Vindhyas, besides predominant animal hunting scenes, there are scenes showing collecting fruits and honey, catching fish and tortoise and hunting/trapping birds. It is not unlikely that Mesolithic people of the Middle Gangetic Plains were gatherers rather than hunters by choice. Gathering would have been supplemented by fishing, fowling and hunting (Thomas and Joglekar, 1995, 1996). For vegetal food gathered by Mesolithic people of the Middle Ganga Valley, one has to rely upon inferences drawn from artefacts used to procure and process foods (Pandey, 1985:179) and upon ethnographic analogy. Besides, the faunal and floral data have prompted attempts to model Mesolithic diet from ecological sources. Some, such as Malti Nagar's (1978) derive estimates on resource abundance, availability and distribution, almost entirely, from contemporary studies of present-day tribals of central India. According to Nagar (1978), quoted by Sharma and Clark (1983: 272–73), as many as 67 wild plant species are still used as food by the tribals of Bhimbetka area in Raisen district of Madhya Pradesh and Bastar district of Chhattisgarh. These comprise 8 leaves, 7 flowers, 30 fruits, 4 seeds and 18 tubers, shoots and roots species. There is hardly any month of the year when a combination of wild plant resources is not available for exploitation. During the monsoon, some 19 species, 3 leaves, one flower, 8 fruits and 7 tubers are available; during the winter there are 22—one leaf, one flower, 13 fruits and 7 tubers; and during the summer the number is 23—two leaves, four flowers, 11 fruits, 2 seeds, 3 tubers and one gum.

In the Middle Ganga Plains, there are bushes of *jharber* (*Ziziphus Jujuba*) in and around Damdama. *Ber* or *jharber* form dense masses of thorn bushes. Fruits ripen in October–November. The ripe fruits were eaten either off the bush or collected and stored after drying and preserved for future consumption during the summer and monsoon. Flowers of mahua (*Bassia latifolia*) and fruits of *bel* (*Aegle marmelos corr*), kathajamun (*Egenia jambolna wild*), *umar* or *gular* (*Ficus glomerate*), *peepal* (*Ficus religiosa*), *bargada* (*Ficus bengalensis*) and *khajur* (*Phoenix sylvestris*), were available during the seasons in the summer in the Middle Ganga Valley. *Khair (Acacia catechu)* produces gum during summer, which is collected and consumed. Flowers of the *mahua* tree are one of the most important edible produces of the summer. Wild and planted *mahua* trees grow in great luxuriance throughout the region. The trees lose their leaves when the flowers appear at the end of March or April. *Mahua* flowers are eaten in a variety of ways. The fresh flowers may be chewed raw or cooked in water and then consumed. They are generally dried, beaten and winnowed to remove their inner husk *jeera (rind)* and then stored. The dried flowers are roasted and crushed with or without *til* (*sesame*) and eaten as a delicacy. The flowers are also cooked mixed with gram and *dal*. The ripen *Mahua* fruits, which are sweet in taste, are also consumed. Liquor is also distilled from *Mahua*. Oil extracted from *Mahua* seeds is also used in earthen lamps and occasionally for cooking. The importance of *Mahua* flowers and fruits as a staple food is significant in primitive tribal economy. Near Damdama, there are a number of wild *Chhiul* or *dhak* (*Butea frandosa*) trees. Tender roots of *Chhiul* saplings are plucked and consumed occasionally in the monsoon. Besides *ber*, the other main wild fruit which is available during the winter at Damdama is *makuiya* (*Caesalpinia sepiaria roxb*). So far, no floral evidence has been found in the excavated Mesolithic sites of the middle Ganga Valley. Carbonized remains of wild paddy (*Oryza nivara*) have been found from the late Phase (Phase III) of Chopanimando in the Belan Valley (Sharma *et al.*, 1980:75). That India appears to be native home of paddy is born out by the presence of a number of wild paddy species, as well as common paddy, growing wild, as a weed, and possessing a character common to wild grasses, namely, shedding of the grains at maturity, which ensure self-sowing (Sharma *et al.*, 1980:23). Here are also found intermediate forms connecting wild and cultivated paddy. The varietal diversity of the cultivated paddy of India is the richest in the world, the coarse-grained primitive varieties being especially typical.

The Neolithic period is easy to define as a way of life, as it has been generally taken to involve a settled life, cultivation of plants and domestication of animals. New discoveries continue to widen and deepen our knowledge about the Neolithic phase in the Ganga Valley. Recent archaeological researches have, once again, pushed back the frontiers of our knowledge of Indian prehistory and have brought to light a number of Neolithic sites hitherto unknown. The present study is devoted to Neolithic settlement patterns in the Middle Ganga Valley. The sites—some on the banks of major rivers including the Ganga—have been reported over the last one or two decades. The well-known excavated Neolithic sites in the Middle Ganga Valley in Bihar are Chirand in Saran district; Checher-Kutubpur, Vaishali district; Taradih, Gaya District; and Maner, Patna district. Chirand (Lat. 25°45′ N; Long. 84°45′ E) is a small village on the vast alluvial plain of north Bihar near the confluence of the rivers Ghaghara and Ganga about 11 km east of Chapra, the headquarters of Saran district.

The site of Chirand was excavated by the State Directorate of Archaeology, Bihar from 1962–63 to 1969–70 (Verma, 1970–71). The most important discovery of the Neolithic assemblage came to light in two trenches, namely *CRD XI* (15 × 10m) and CRD XIII (10 × 10m). The Neolithic deposit here is 3.50 metres in thickness.

The average rainfall in the area is 112.50 cm which occurs in June and July, when the snow melts and monsoon sets in; the Ganga gets water much beyond its capacity to contain. The excess water inundates the adjoining area and spreads alluvium every year.

The Neolithic pottery from Chirand has been divided into two categories: the pottery of phase II, which is refined and superior in comparison to those of Phase I. The pottery is fine. More than 25,000 potsherds have been recovered from excavation. The pottery is by and large handmade. A few examples of vessels made on turntables or by dabbing methods are known. Mica appears to have been mixed in the levigated clay.

The bulk of pottery at Chirand is Red Ware (50%) and Black and Red ware. The main shapes are that of vases, vases with spouts, miniature vases, vases with a pointed base, *handi* with blunt carination on the shoulder, deep bowls, hemispherical and perforated bowls, lipped bowls, bowls on stands, footed bowls, oval-shaped bowls with broad lip and knobbed pottery. B.S. Verma (1970–71) suggested that the pottery form and types are entirely different from the Chalcolithic black and red ware pots and pans.

The settled way of life needed tools and weapons which could be effective in field clearance and in meeting many other needs of daily life. The tool kit of the Chirand Neolithic Community is very distinct. It is significant to note that Chirand has yielded only four Celts (Verma, 1970–71). Neolithic stone axes are rectangular in outline, and have flat faces with a cutting edge towards one corner. Most probably, these heavy-duty tools were imported from elsewhere. The stones used were quartzite, basalt and granite. A large number of microliths have been found grouped under nine types such as parallel-sided bladelets, scrapers, lunates, points, serrated points, notched bladlets, borers, arrowheads and simple cores. Waste flakes and some in manufacturing stage indicate a flourishing local microlithic industry as an integral part of the Neolithic phase at Chirand. The raw material used for microlithic bladelets is chert, chalcedony, agate and jasper. Most likely, raw material was collected in the form of nodules from the dry bed of the Son River, flowing not far from Chirand. Such material can still be seen in that river bed. Stone beads of agate, jasper and chalcedony are excellent in finish and quality. Unfinished beads and nodules in the form of raw material indicate the local nature of the industry. The most common shapes of beads are the long tubular disc, long barrel, short barrel, cylindrical and triangular varieties.

The Neolithic community of Chirand exploited the environment and untilized bones of animals and antlers as raw material on a larger scale for the manufacture of bone tools, which is a special feature of the Neolithic culture of middle Ganga Valley. The antler artefacts at Chirand number about 400. Lala Aditya Narain (1970) has published an analysis of about 150 artefacts from the site and, on the basis of their supposed functional uses, he has classified them into thiry varieties, viz. chisels, bar celts, hammers, burnishers, wedge scrapers, shaft-straightners, punching tools, javelin heads, tongs, borers, awls, needles, tanged arrowheads, spear points, drills, sockets, discs, besides ornaments like bangles, combs and pendants. The material used in their manufacture was the antlers, long bones of deer and cattle. In some cases, for example, in the manufacture of ornaments, tortoise shell and ivory were also used. This tool typology appears to novel to Neolithic culture of the middle Ganga Valley and is unique to Chirand.

Chirand had a mixed economy based on early farming and domestication of animals supplemented by hunting and fishing game. Subsistence largely rested on the cultivation of cereal crops. It is apparent that paddy was known to the people at Chirand, as paddy husk impressions have been noticed on some burnt clay pieces. It has further been confirmed by the discovery of a few examples of carbonized rice along with wheat, barley, *moong* and *masur.* Remains of carbonized cereals suggest that the inhabitants at Chirand knew of raising summer and winter crops. Vishnu Mittre has classified the paddy remains from Chirand into cultivated and wild variety. No evidence of ploughing was available. The antler's pick has been identified as an agriculture tool by archaeologists.

A study of animal bones from Chirand revealed that the majority of the remains are domesticated animals; the identified bones represented *Bos indicus, Bubalis bubalis linn, Ovis aries lichure durest, Cepra hirascus acqaqrus exable, Sus-cristatus wagher, Canies familiarie linn.* Apart from the domesticated animals mentioned above, wild animals such as *Rhinoceros Unicer, Elephas matnum Carvices duvancellu cuv,* antelope, deer, *sambhar, stag, cheetal* (*Axis axis*), dog and bear occurred at Chirand.

Checker-Kutubpur in Vaishali district of Bihar is situated on the northern bank of the vast alluvial tract of the Ganga on the road between Hajipur and Mohnar. The site was excavated by R.S. Bisht on a small scale in 1977–78 (IAR) on being brought to his notice by Sri Rampukar Singh, a local villager. The Neolithic assemblage has been classified into three subphases A, B and C on the basis of pottery. Polished stone celts are collected in good number. The tools made of (on) bone and antlers have also been found. An interesting bone tool is a double-forked pick-axe on an antler, having a meticulously carved socket-hole for hafting. Period IB yielded ordinary bone tools and points (Sant, 1990:120); however, sophisticated bone and antler implements are absent.

Taradih is situated southwest of the famous Mahabodhi Temple at Gaya in Bihar. A trial excavation was taken up in the year 1981–82 (the work is still in progress) by the Directorate of Archaeology and Museum, Government of Bihar. In 1984–85, a 60-cm thick Neolithic deposit was located at the site. It indicates the potentiality of site. The excavations revealed a sixfold cultural sequence: Neolithic followed by Chalcolithic, early Historical, the Kushana, Gupta down to the remains of the Pala period.

Senuwar (Lat. 25°, 56′ N; Long. 83°, 56′ E) is on the right bank of Kudra River, which flows about 1 km away from the mound, in Rohtas district of Bihar. The site was excavated by late K.K. Sinha and B.P. Singh of the Department of Ancient Indian History, Culture and Archaeology, B.H.U. Varanasi in the year 1986–87. The ancient mound at Senuwar is 360 × 300 metres from north to south, with a maximum height of 9 metres from the ground. Period I is Neolithic at Senuwar.

Maner is in Patna district of Bihar. A.K. Singh, Professor and Head of the Department of Ancient Indian History, Culture and Archaeology, University of Patna, Patna is conducting excavations at the site. Period I is Neolithic in character.

Recently, Neolithic material has been reported from Jhusi (Lat. 25° 26′,10″ North and Long 81° 54′30″ East), the ancient *Pratisthana*, located on the left bank of the Ganga within a marked meander very close to the Ganga-Yamuna confluence (*Sangama*). The site is at a distance of about 7 km to the east of Allahabad city. To reach Jhusi, if one does not have one's own transport, one can take a bus from Allahabad bound for Varanasi or any other place further beyond Varanasi, and alight at Jhusi. The ancient site extends about 3 km along the Ganga from the railway bridge in the north. Its width from east to west measures about 1.5 km. Villages Jhusi-*Kohna*, Jhusi *Hawellia* and Chhatanaga, occupy the major portion of the ancient site at present. Several imposing mounds mark the site, the best preserved and highest being the Samudrakupa Mound.

The ancient name of Jhusi was *Pratisthana*, which was part of *Prayaga-Kshetra*. According to the *Ramayana* of Valmiki, *Pratisthana* was the capital of *Pururavasa* and other kings of the same dynasty. A mass of mythological narratives centre round *Pratisthana*. According to the *Mahabharata* and *Puranas*, it contains localities like *Hamsaprapatana* and *Urvasipulina* on its northern side. Kalidasa, the famous Sanskrit poet, lays the scene of his drama *Vikramorvasiyama* in Pratisthana.

The epigraphical reference indicates that Pratisthana had its political importance down to twelfth century AD. A copper plate discovered in 1830 of the times of Trilochana Pal, the Pratihara king, dated AD 1027 informs us that Trilochana Pal, while camping on the banks of the Ganga near *Prayaga*, donated certain villages to Brahamanas belonging to Pratisthana. Sri Pratisthana is referred to in a grant of King Govinda Chandra (1114–1154 AD) of Gahadawala dynasty.

The Department of Ancient History, Culture and Archaeology, University of Allahabad had conducted the excavation at the site for five seasons, 1995, 1998, 1999, 2001, 2002 (Misra *et al.*, 1998–99, Misra *et al.*, 1999–2000), exposing a number of trenches. The total thickness of cultural deposit is about 16.50 metres, divisible into 63 layers. The entire deposit is divisible into six cultures: (1) Neolithic period; (2) Chalcolithic period; (3) N.B.P.W. period; (4) Sunga-Kushana period; (5) Gupta period; (6) Early Medieval period. After the Neolithic period and also after the Gupta period, the localities excavated appear to have been deserted for some centuries.

The evidence of Neolithic culture is marked by handmade cord-impressed pottery, microliths, etc., and has been obtained in the lowest habitation deposit towards the southern fringe of Samudrakupa mound (about 50-cm thick deposit) during the excavation conducted in 2002.

Another newly discovered site is Lahuradeva in Sant Kabir Nagar (Basti) district of Uttar Pradesh. The mound of Lahuradeva (Lat. 26°46′N.; Long. 82° 57′ E.) is located at a distance of about 5 km south to Bhujaini Crossing, which is situated on Basti-Gorakhpur Road (NH28) under village Jagdishpur. Earlier, a lake surrounded it from three sides. Presently, however, most of the lake area is under cultivation, and only its western portion still retains water throughout the year. Its excess water spills over into a rivulet called Katnahia, which is tributary of the Kuwano and flows about 500 metres away to the west of the lake. During the rainy season, it almost merges with the lake. The eastern portion of the mound has been considerably levelled in order to convert it into an agriculture field. Courtesy Samai Ma-Ka Than, a local deity located on the western part of the mound—

that grace as well as fear of the deity could prevent the villagers to further extend their activity—the remaining western portion of the mound is in a good state of preservation. It extends over an area of about 220 metres from east to west, 140 metres from north to south and is about 4 metres elevated from the surrounding levels (Tiwari *et al.*, 2000–2002: 54).

The site was excavated by Rakesh Tiwari *et al.*, Directorate of Archaeology, Government of Uttar Pradesh between 1997 and 2005. (The work is still in progress.) The excavation revealed a fivefold cultural sequence: Neolithic followed Chalcolithic (Copper Age), early historical (Early Iron Age), the NBPW and early centuries BC/AD. On the basis of the nature of deposits, radiocarbon dates and cultural material, period I is divided into two subperiods—I.A. and I.B. Remains of period IA have been found in lower most Kankar mixed layers of trench Nos. YA2 Qdt I (layer No 16) and 4 (layer Nos. 12, 13 and 14); YA1 Qdt. I (layers 3 and 14). The total thickness of this deposit is about 45cm. The cultural remains of this deposit are represented by a considerable quantity of potsherds, a few charred/uncharred animal bones, scattered small pieces of charcoal, small burnt chunks of clay, a small piece of stone and tortoise-shell.

The main pottery types are red ware and black-and-red ware, which includes most hand made varieties. The proportion of black and red ware sherds is as high as about 50% of the pottery assemblage. The inner core of the pottery is generally black, while a small portion of the exterior core is red. Some of the black-and-red ware and red ware potsherds bear black and red ship, respectively, on their interior and exterior surfaces. Burnishing is also evident (Tiwari *et al.*, 2000–2002: 55).

The pottery is generally ill-fired, having coarse and porous surface and an uneven core. The main shapes include convex-sided bowls, pedestal bowls, knobbed vessels, miniature bowls and vases.

The inhabitants of period I lived in wattle and daub houses which is testified from the fragments of burnt clay-lumps with reed marks and patches of floors and post holes.

Carbonized material of Period I.A., collected by the flotation method, is very signficant because, according to K.S. Saraswat of Birbal Sahani Institute of Palaeobotany, Lucknow, it contains grains of cultivated rice (*Oryza sativa*) along with a few wild grasses. Besides, the husk impressions of rice is found embedded in a number of potsherds.

Although the first settlers at Lahuradeva practised agriculture, meat was an important component of their diet, as is evident from the presence of a large number of charred animal bones and antlers, some of them having cut-marks.

Recent archaeological investigations in the Middle Ganga Plain have significantly enriched our knowledge and understanding about archaeological sites, material culture and chronological sequence of the region. The changing features and landscape from pre- to post-Tertiary period filling the trough-shaped formation by fine flood plain alluvium brought down from the Himalayas in the north through Ganga and its tributaries flowing sluggishly and forming ox-bow lakes while meanderings have provided a habitat to Mesolithic man who settled in parts of the region having lakes and waterbodies with aquatic fauna and land around them, producing edible grasses, which were soon cultivated for producing rice (paddy) and other grains in atleast sixth-fifth millennium B.C.E. During the Holocene, new fauna and flora created new environmental conditions for man. We cannot at present illustrate the process of his response to this change. We only know the results; it is clear that the change was gradual. There may have long periods when man collected and consumed the seeds of wild rice along with seeds of other grasses, before satisfactory cultivated varieties were selected. The Middle Ganga Valley has produced definite material to show that food-producing communities developed more or less simultaneously in the Vindhyas and the Ganga Valley, as is evidenced by the sequences at Chirand, Jushi and Lahuradeva in the Gangetic plain (Pandey, 2004 n.d.).

Archaeologists and historians were undecided about two decades ago in taking back the antiquity of settlements in the region before the first millennium B.C.E. But the pioneering work at Sohagaura by a team of archaeologists from Gorakhpur University led by S.N. Chaturvedi in 1961–62 (1980: 339–40, 1985:105), followed by excavations for seven seasons commencing from 1962–63 at Chirand by the Directorate of Archaeology and Museums, Bihar, established the Neolithic-Chalcolithic antiquities of the region. Quite a large number of sites such as Chechar-Qutubpur, Taradih, Senuwar, Narhan (Singh, 1994), Imlidih Khurd, Dhuriapar, Khairadih, Waina, Bhunadih, Jhusi and Lahuradeva have confirmed the earlier findings and have further enriched our knowledge and understanding about the material culture, dating and chronological sequence of Pre-Northern Black polished ware cultures of the Middle Ganga Plain.

REFERENCES

Alur, K.R., 1980, Faunal Remains from the Vindhyas and Ganga Valley in G.R. Sharma *et al., Beginnings of Agriculture.* 221–227 Allahabad, Abinash Prakashan.

Binford, L.R., 1966, A Preliminary Analysis of Functional Variability in the Mousterian of Levallois Facies. *Recent Studies in Palaeoanthropology.* 238–295. American Anthropologists Special Pub. Volume 68, p. 2.

___, 1968, Methodological Considerations of Archaeological use of Enthnographic Data In R.B. Lee and Devore I. (eds), 1968. *Man, the Hunter* 268–273, Chicago.

___, 1972, *An Archaeological Perspective,* Seminar Press, New York, London.

Butzer, K.W., 1992, *Archaeology as Human Ecology,* Cambridge: Cambridge University Press.

Chaturvedi, S.N., 1980, Excavations at Sathiaon—Fajil Nagar District Deoria and Explorations in the Districts of Gorakhpur and Basti of Uttar Pradesh, *History and Archaeology* Volume I, Nos. 1–2, 333–340. Department of Ancient History, Culture and Archaeology, University of Allahabad, Allahabad.

___, 1985, Advance of Vindhyan Neolithic and Chalcolithic Cultures of the Himalayan Tarai: Excavation and Explorations in Sarayupar Region of Uttar Pradesh. *Man and Environment,* IX 101–108.

Clarke, D.L., 1968, *Analytical Archaeology,* London.

Clarke, D.L., 1972, (ed.) *Models in Archaeology,* London.

Coles, J., 1973, *Archaeology by Experiments,* New York, Scribner.

Goel, Anshu, 2003, *Architecture and Sculpture of Lower Ganga-Yamuna Doab (Circa* 600 BC. to *Circa* 600 AD) (Unpublished) D.Phil. Thesis: Allahabad University, Allahabad.

Hawkes, Jacquetta, 1968, The Proper study of Mankind, *Antiquity* 42: 255–62.

Hodder, Ian, 1991, *Archaeological Theory in Europe: The Last Three Decades* Routledge: London.

___, 1992, *Theory and Practice in Archaeology,* Routledge: London.

Kajale, M.D., 1990, 'Some Initial Observation on Palaeobotanical Evidence for Mesolithic Plant Economy from Excavations at Damdama, Pratapgarh Uttar Pradesh', in N.C. Ghosh and S. Chakrabarti (eds) *Adaption and other Essays,* Shantiniketan: Visva Bharati, pp.98–102.

Kennedy, K.A.R. *et al.*, 1992, *Human Skeletal Remains from Mahadaha: A Gangetic Mesolithic site:* South Asian Occasional Papers and Thesis No.11, New York: Cornell University.

___, 1986, *Mesolithic Human Remains from the Gangetic Plain: Sarai Nahar Rai* South Asian Occasional Papers and Thesis No.10. New York: Cornell University.

Lukacs, J.R. and V.D. Misra, 2002, 'Human Skeletons at Lekhahia' in, V.D. Misra and J.N. Pal. *Mesolithic India* (eds), Department of Ancient History, Culture and Archaeology, University of Allahabad, Allahabad.

Mathpal, Y., 1984, *Prehistoric Rock Painting of Bhimbetka, Central India,* New Delhi, Abhinava Publication.

Mellars, P., 1976, Fire, Ecology, Animal Populations and Man: A Study of Some Ecological Relationships in Prehistory. *Proceedings of Prehistoric Society* 42: 15–45.

Mishra, B.B., 2002, 'Mesolithic Culture of the Belan Valley' in V.D. Misra and J.N. Pal *Mesolithic India (*eds*)*, 26–36, Allahabad University, Allahabd.

Misra, V.D., 1977, *Some Aspects of Indian Archaeology,* Allahabad, Prabhat Prakashan.

Misra, V.D. *et al.*, 1995–96, A Preliminary Report on the Excavations at Jhusi, *Pragdhara* No. 6: 63–66.

___, 1998–99, Further Excavation at Jhusi, District Allahabad 1998, *Pragdhara* No. 9: 43–49.

___, 1999–2000, Further Excavations at Jhusi 1998–99 *Prāgdhārā* No.10: 23–30.

Mueller, J. (ed), 1975, *Sampling in Archaeology,* Tuscan: University of Arizona Press.

Nagar, Malti, 1978, The use of wild Plant Foods by tribal communities of Central India with Special Reference to Bastar and Bhimbetka Area. Paper presented at the Xth International Congress of Anthropological and Ethnological Sciences, Pune, India.

Narain, Lala Aditya, 1970, The Neolithic Settlement at Chirand JBRS LVI. 1–5.

Pal, J.N., 1988, Mesolithic Double Burials from Recent Excavations at Damdama, *Man and Environment* XII: 115–22.

Pandey, J.N., 1985, *Settlement Patterns and Life in the Mesolithic Period in Uttar Pradesh.* D.Phil Thesis: Allahabad University, Allahabad.

___, 1990, Mesolithic in the Middle Ganga Valley, *Bulletin of the Deccan College Post-graduate and Research Institute,* 49: 311–16.

___, 2004, Settlement Pattern in the Mesolithic and the Neolithic of the Middle Ganga Plains, Paper presented in the Workshop on 'The Archaeology of the Middle Ganga Plain' held at Lucknow, March 13–14.

Pant, D.D. and Pant, Rekha, 1980, Preliminary Observations on Pollen Flora of Chopani-mando (Vindhyas) and Mahadaha (Ganga Valley) in Sharma, G.R. (*et al.*), *Beginnings of Agriculture* 229–230. Allahabad, Abinash Prakashan.

Renfrew, C., 1968, Models in Prehistory. *Antiquity* 42: 132–134.

___ (ed.), 1973, *The Explanation of Cultural Change: Models in Prehistory*. London, Duckworth.

Sant, Urmila, 1990, *Neolithic Settlement Pattern of Northeastern and Northern India,* Delhi, P.S. Negi for Sarita Book House.

Sharma, G.R., 1973, Mesolithic Lake Cultures in the Ganga Valley, India, *Proceedings of Prehistoric Society* 39: 129–146.

Sharma, G.R. *et al.*, 1980, *Beginnings in Agriculture,* Allahabad Abinash Prakashana.

Sharma, G.R. and Clark, J.D.(eds), 1983, *Palaeo-environment and Prehistory in the Middle Son Valley,* Allahabad, Abinash Prakashan.

Singh, P., 1994, *Excavations at Narhan,* Deptt. of Ancient Indian History, B.H.U. Varanasi.

Singh, R.L. (ed.), 1971, *India—A Regional Geography,* Varanasi, National Geographical Society.

Spate, O.H.K. *et al.*, 1967, *India and Pakistan (A General Regional Geography.* London, Methuen.

Thomas, P.K. and Joglekar, P.P. *et al.*, 1995, A Preliminary Report on the faunal Remains from Damdama, *Man and Environment* XX(1) 29–36.

Thomas, P.K.,1996, Joglekar Faunal Evidence far Mesolithic Food Economy of the Garngetic Plain with Special Reference to Damdama Colloquim xxxii Bio-archaeology of Mesolithic

Tiwari *et al.*, 2002, Preliminary Report on Excavations at Lahuradeva District Sant Kabir Nagar, U.P. 2001. *Prāgdhārā* No. 13: 54–83.

Varma, R.K. *et al.*, 1985, A Preliminary Report on the Excavations at Damdama (1982–84) *Man and Environment,* IX: 45–65.

Varma R.K., 1986, Mesolithic Age in Mirsapur Allahabad, Paramjoti Prakashan.

Verma, B.S., 1970–71, Excavations at Chirand *Purāttava* No.4: 18–22.

Wadia, D.N., 1957, *Geology of India,* London, Macmillan.

Williams, M.A.J. and Royce, K., 1983, Alluvial History of the Middle Son Valley, North-Central India. In Sharma, G.R. and Clark, J.D. (eds) *Palaeoenvironments and Prehistory in the Middle Son Valley* (Madhya Pradesh, North-Central India), 9–21 Allahabad, Abinash Prakashan.

CHAPTER 6

Beginnings of Agriculture in the Middle Ganga Plain with Special Reference to Recent Excavations at Jhusi

M.C. Gupta, J.N. Pal and *V.D. Misra*

INTRODUCTION

In the Indian context, the significance of two rivers, the Indus and the Ganga, cannot be overemphasized. Sir Mortimer Wheeler has correctly pointed out that while the Indus has given this country a name, the Ganga has given her faith.[1] The Ganga Plain has been the cradle of Indian civilization right from ancient times. The entire Ganga Plain is divided into three units: (i) Upper Ganga Plain; (ii) Middle Ganga Plain; and (iii) Lower Ganga Plain. The middle Ganga plain (Lat. 24° 30′ N–27° 50′ N and Long. 81° 47′ E– 87° 50′ E) is bounded by the Ganga-Yamuna confluence in the West Bengal and Bihar border in the east, the Himalayas in the north and the Vindhyas in the south. The area, measuring about 144,409 sq km, has been a distinct cultural and economic entity even though it does not clearly comprise a defined physical unit.[2] The middle Ganga plain includes modern eastern Uttar Pradesh and plains of Bihar. On the basis of river system, the middle Ganga plain is further subdivided into: (i) the Ganga Plain north, and (ii) the Ganga Plain south, the former further subdivided into (a) the Ganga-Ghaghara Doab, (b) the Ghaghara-Gandak interfluve, (c) the Gandak-Koshi interfluve and (d) the Koshi-Mahananada interfluve. The Ganga plain south is also further subdivided into (a) west of the Karmanasa, (b) the Karmanasa east interfluve, (c) the lower Son Valley and (d) the Magadha-Anga plain. As would be evident, the Ganga is the lifeline of the middle Ganga plain.

The region outlined above, has been the nerve centre of the Indian culture and civilization, a meeting ground of different races, cultures and creeds. People of this region were credited with elasticity, both in their thought and action. It was the home of the *Vrātyas*. Divine gospels, contained in some of the *Upaniṣadas*, were conceived and preached in this region. It had the privilege of hearing the philosophical discourses of Yagyavalkya. It is the land associated with the story of the *Rāmāyaṇa*. The deductive portions of the *Mahābhārata* are also supposed to have been written in this region. The middle Ganga plain not only gave birth to the Buddha and Mahavira, but also heard their teachings with rapt attention. Right from the sixth century BC onward, the middle Ganga plain played a crucial role in the political history of India. It witnessed the rise of the early Magadhan empire as also that of the Guptas. Viewed against such illustrious history, tradition and legends, the recorded archaeological relics of the region upto the fifties of the previous century seemed prosaic. The archaeological investigations over the last three decades under the auspices of the Department of Ancient History, Culture and Archaeology of the University of Allahabad, Benaras Hindu University, Sampurnananad Sanskrit Vishwavidyalaya, D.D.U. Gorakhpur University and the University of Patna on the one hand and the State Department of Archaeology and Museum, Government of Bihar, the State Department of Archaeology, Government of Uttar Pradesh and Archaeological Survey of India on the other, have brought to light the relics of human culture right from the late Upper Palaeolithic[3] to the early historic periods, through the Mesolithic, Neolithic and Chalcolithic ages.

The beginning of human settlement in the Gangetic plains commences from the Epi-palaeolithic period (transition from Upper Palaeolithic to Mesolithic). The archaeological evidence indicates that the first colonizers of the Gangetic plain came from the Vindhyas and it was a seasonal migration in the beginning, as is indicated by thin deposits of Epi-palaeolithic and early Mesolithic (non-geometric) sites.[4] The Mesolithic (geometric) sites are marked by thick occupational deposit, indicating semi-sedentary settlement. The excavation of three of the sites of this phase, viz., Sarai Nahar Rai, Mahadaha and Damdama have enabled us to reconstruct the history of Mesolithic period of the region. Man living on the bank of horseshoe lakes exploited the floral and faunal resources of the area. The human skeletons from the habitation area have helped in the study of bio-archaeology of the period. The next culture in the area belongs to the early farming (Neolithic) culture. In the following pages, an attempt has been made to trace the antiquity of agriculture in the middle Gangetic plain and delineate the salient features of the first farming culture of region with special reference to evidence received from recent excavations at Jhusi.

The excavated Neolithic sites of the middle Gangetic plain include Chirand[5] in Saran district, Chechar Kutubpur[6] in Vaishali district, Taradih[7] in Gaya district, Majhi[8] in Saran district and Senuar[9] in Palamu district in Bihar; Sohgaura[10] and Imlidih[11] in Gorakhpur district, Bhunadih[12] and Waina[13] in Ballia district, Lahuradeva[14] in Sant Kabir Nagar district and Jhusi in Allahabad district in Uttar Pradesh. Excavations of these sites have presented the evidence to reconstruct the history of the first farming culture of the area.

EXCAVATIONS AT JHUSI

Jhusi (Lat. 25° 26′10″N. Long. 81° 54′30″E.), the ancient Pratiṣṭhānpur,[15] is located on the left bank of the Ganga within a marked meander very close to the Ganga-Yamuna confluence at a distance of about 7 km to the east of Allahabad city. The site can be conveniently reached by metalled road from Allahabad. The ancient site extends about 3 km along the river from the railway bridge in the north to Chhatanaga locality in the south. Its width from west to east measures about 1.5 km. The major portion of the ancient site is occupied at present by villages—Jhusi-Kohana, Jhusi-Hawelia and Chhatanaga. Originally one compact mound, it has now been dissected into a number of smaller mounds by rain gullies and encroachment by peoples. The *Samudrakūpa* mound[16], however, is comparatively well preserved and has the maximum height of more than 16 metres from the surrounding ground.

The ancient name of Jhusi was Pratiṣṭhāna, a part of Prayāga-Kshetra. According to the *Rāmāyaṇa* of Vālmīki (VIII), Pratisṭhāna was founded by Ila, the legendary king of the Lunar dynasty and it was the capital of Pururavāsa and other kings of the Lunar dynasty. A mass of mythological narration centres around Pratiṣṭhā. According to the *Mahābhārata* and the Purāṇas, it contains localities like Hamsaprapātana and Uravashīpulina on its northern side. Kalidasa, the famous Sanskrit poet, lays the scene of his drama, *Vikramorvashīyama* (II.45), in Pratiṣṭhāna.

Epigraphical references trace the political importance of Pratiṣṭhāna down to twelfth century AD. The copper plate discovered in 1830 AD of Trilochanapāla, the Pratihāra King, dated AD 1027, informs that Trilochanapāla, while camping on the bank of the Ganga near Prayāga, donated certain villages to the Brahmaṇas belonging to Pratiṣṭhāna[17]. Śri Pratiṣṭhāna is referred to in a grant of King Govinda Chandra of Gāhadavāla Dynasty (AD 1126).

The Department of Ancient History, Culture and Archaeology, Allahabad University, has conducted excavations at the site on Samudrakūp mound (Plate I) for five seasons —1995[17], 1998[18], 1999[19], 2002[20] and 2003[21]—exposing a number of trenches.

Plate I: Jhusi: Samudrakūp mound under excavation

The total thickness of cultural deposit at the site is about 16.5 m, divisible into 6 cultures: 1. Neolithic Period, 2. Pre-NBPW Period (Chalcolithic and early Iron Age), 3. NBPW Period, 4. Sunga-Kushan Period, 5. Gupta Period and 6. Early Medieval Period. There is also evidence of the presence of Mesolithic culture on the top of geological formation underlying the archaeological deposit. Of the cultures outlined above, there is a continuous settlement from Pre-NBPW to Gupta period. However, after the Gupta period, the locality excavated appears to have been deserted for some centuries. It was again occupied in early medieval period.

It is noteworthy that the archaeological deposit at Jhusi rests on the natural formation measuring about 10 m in thickness. This geological formation is marked by several units featuring highly cemented gravel formations, hard concretionary clay formation and yellowish silty formations. The microliths found from the top of this formation at Jhusi and at Nibikalan and Jamunipur near Jhusi belong to the Mesolithic phase.

The evidence of the Neolithic culture, presenting for the first time the remains of first farming culture, were obtained in square SG-8, where the lower layers revealed handmade ill-fired pottery, marked by cord impression and rustication on the outer surface during the fag end of excavation in 2001–2002[22]. To confirm the status of the Neolithic horizon, further excavation was planned in this area in 2002–2003. In this connection, it may be mentioned that the sporadic occurrence of microliths both in the form of flake fragments and blade fragments on the one hand and finished tools on the other was noticed in the lowest level of occupational deposit. Four trenches—SF-7, SG-7, SF-8 and SG-8—each measuring 5 × 5 m, were excavated. Trench SF-8 was partly excavated in 2001–2002. The latest cultural deposit in this trench is of NBPW period (from layers 21 to 45). The NBPW period deposit is marked by a number of pit activities, as a result of which, the pre-NBPW horizon of this trench is found disturbed. Layers 46 and 47 constitute the Chalcolithic horizon in this trench, while layers 48, 49 and 50 represent the Neolithic phase. The trench SG-8 lying to the south of trench SF-8 also yielded a similar cultural sequence. The trench SF-7 is located east of SF-8. The excavation in this trench yielded the archaeological relics of NBPW period and Neolithic period. The Chalcolithic horizon in this trench was not encountered. There is stratigraphical evidence of a gap between the underlying Neolithic deposit and overlying NBPW period deposit. This barren deposit (layer 44) is very significant. The layer contains erosional material and its base is zigzag. It indicates that after the end of Neolithic phase and beginning of NBPW phase, there was a considerable time gap. The layers 45 to 56, measuring 1.5 m in thickness, belong to the Neolithic horizon. The layers associated with the Neolithic phase are compact and light yellowish in character.

FIRST FARMING CULTURE AT JHUSI

The Neolithic phase has yielded cord-impressed (Plate II), rusticated, burnished red and burnished black wares. There are also some crude black-and-red ware potsherds. These pots are handmade and ill fired. The clay used in their manufacture is not levigated. It contains rice husk and chaff as tempering material. The range of pots is very limited.

Plate II: Jhusi: Cord-impressed Pottery, Neolithic level

The occurrence of spouted pots is noteworthy. Some of the spouted pots bear appliqué decoration on their neck (Plate III).

Besides the pottery, the Neolithic phase also yielded animal bones and remains of cereals. Among the latter, the occurrence of rice, wheat, *moong* and *masoor* is noteworthy.

From the Neolithic horizon were obtained evidence of hutments (Plate IV) and pits. Three huts, circular in shape, were noticed. On the floors of these huts were found scattered animal bones, food-processing equipments made of sandstone or quartzite in the form of querns and mullers (Plate V), sling-balls (Plate VI) and potsherds. The querns, mullers and sling balls have pitted surface with heavy use mark. Of the other antiquities of Neolithic phase, mention may be made of bone arrowheads (Plate VII) and beads of semi-precious stone, steatite and fiance. Tiny pieces of microliths have also been found from this horizon. A few blades, triangles and trapezes (Plate VIII) along with flakes and flake fragments and blades and blade fragments were also found. The Neolithic phase at the site presents clear cut evidence of domestication and cultivation. The thick deposit of 1.5 m pertaining to this phase also suggests sedentary settlement.

Some very significant discoveries have been brought to light by these diggings, which may be listed as follows:

1. There are indications that during the Mesolithic period, man had started operating in and around Jhusi. Microliths have been obtained not only at the site but also at Nibikalan and Jamunipur near Jhusi. It appears that the Mesolithic people

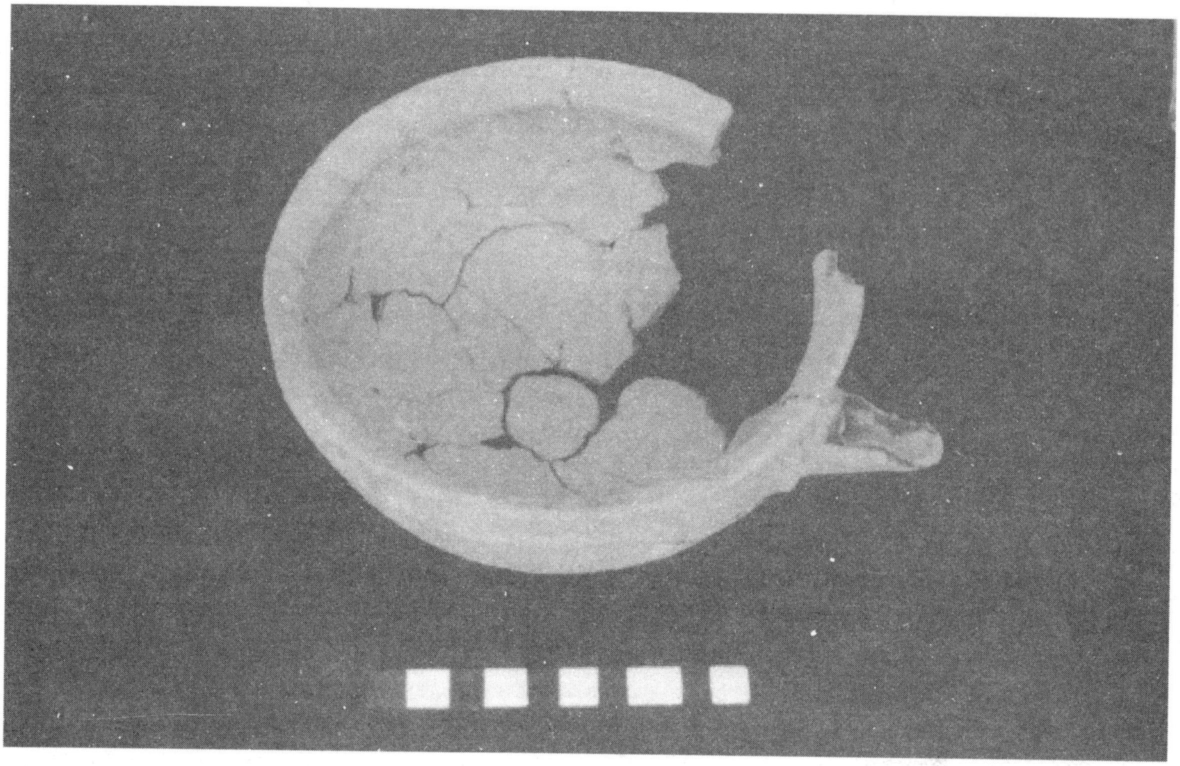

Plate III: Jhusi: Spouted Basin with appliqué decoration, Neolithic level

Plate IV: Jhusi: Circular hut floors on plan, Neolithic level

Plate V: Jhusi: Muller of sandstone with pitted surface, Neolithic level

Plate VI: Jhusi: Quern fragments with pitted surface and heavy use mark and sling ball, Neolithic level

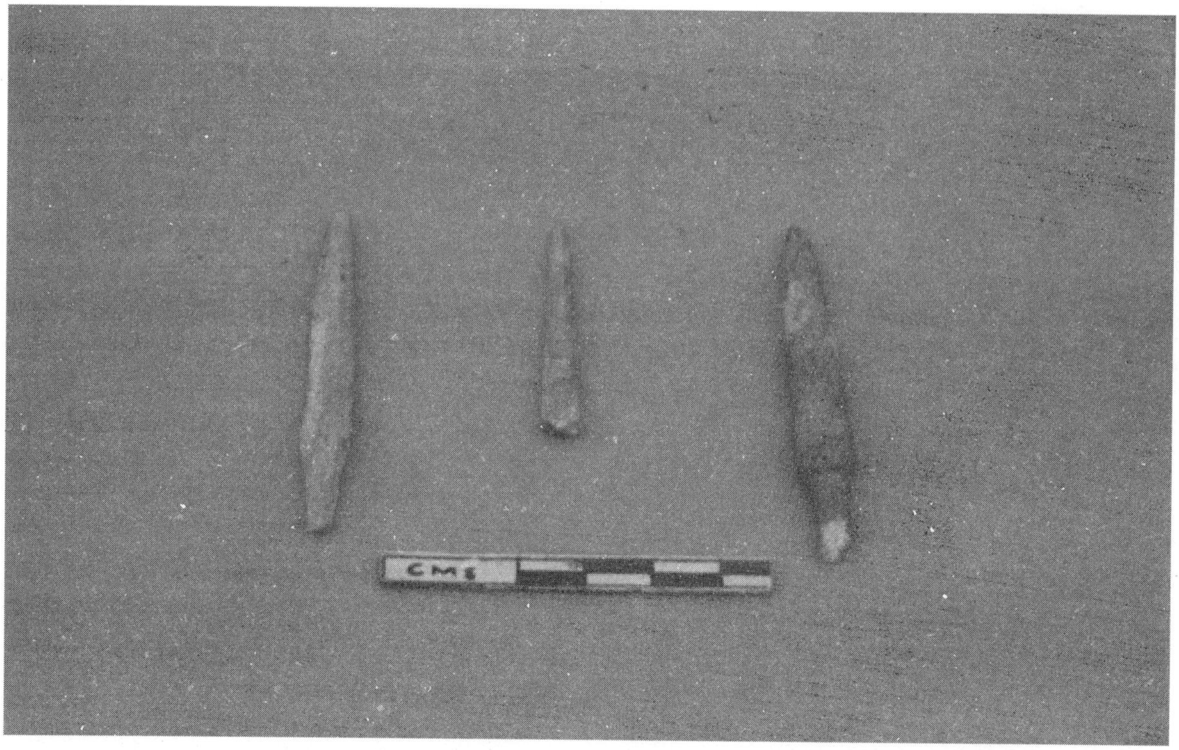

Plate VII: Jhusi: Bone arrowheads, Neolithic level

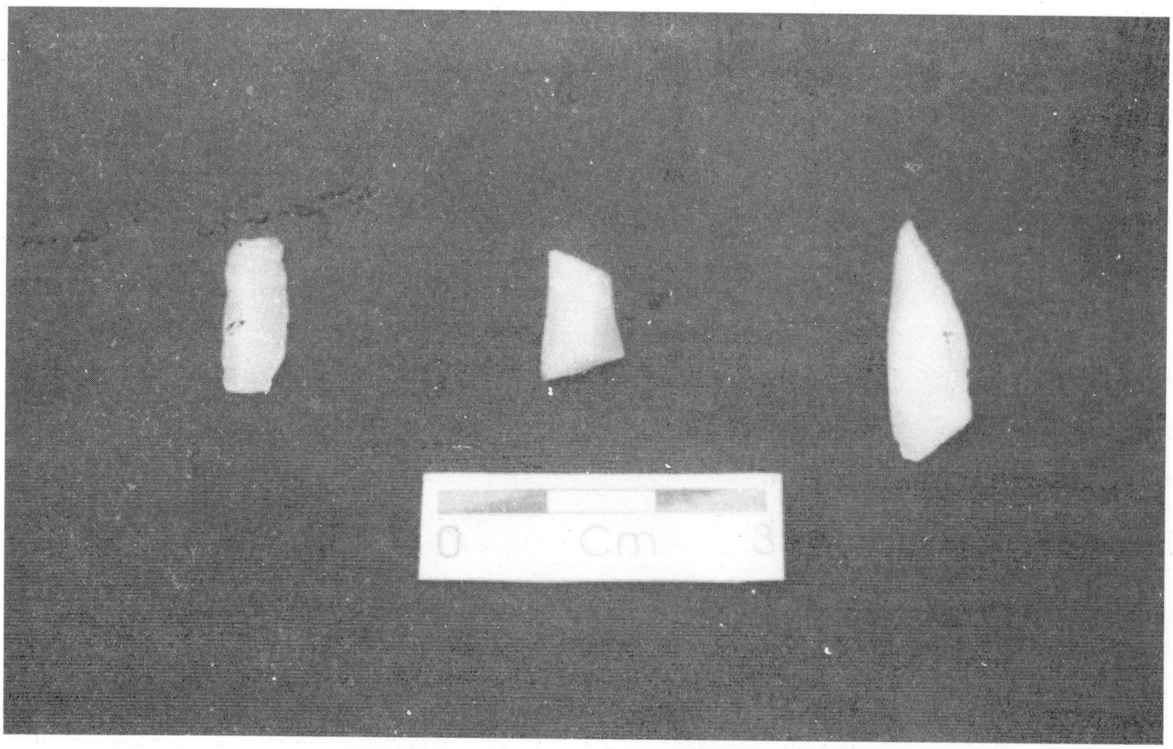

Plate VIII: Jhusi: Microlithic tools, Neolithic level

would have come in this area at a time when silty and sandy formation of Bhagar was taking place.

2. Many years ago, the discovery and excavations at the Mesolithic sites of Sarai Nahar Rai, Mahadaha and Damdama in the Gangetic Plain presented a surprise because the Stone Age culture of the Mesolithic phase was discovered in an area totally free of hills—the source of stone supply. Indications of the Mesolithic culture were available from more than 200 sites in the middle Ganga plain but Neolithic was conspicuous by its absence. It was, therefore, surmized that there was a cultural break after the Mesolithic age in this region and settlers of the Chalcolithic phase at Jhusi and other sites probably were not native to this region. But the current excavations at Jhusi have succeeded in discovering below the Chalcolithic phase a nearly 1.5-metre thick Neolithic deposit. This excavation is significant because it presents the proof of cultural continuity right from the Mesolithic through the Neolithic to the Chalcolithic, which again continued into the historical period down to the time of the Buddha and, further on, upto the early medieval period. In this way, it is an epoch-making discovery, as it has taken the antiquity of Allahabad in an unbroken manner to the Mesolithic times.

3. The discovery also highlights the fact that Jhusi has been a cradle of human civilization right from the Mesolithic age down to the blossoming of urban phase of the culture of this region. The caravan of the human succession journeyed through the hunting-gathering, foraging, farming and village life before striking upon the legendary urban theme of being the capital city of the Pururvā. It is born, bred and developed at its own epicentre, the middle Ganga plain, as exemplified by the present excavations at Jhusi. Outside influences, if any, were of an interactive rather than originary character.

4. The Neolithic phase at Jhusi is characterized by handmade pottery, which is commonly recognized as ill-fired cord-impressed, rusticated and burnished ware. Bone tools, bone arrowheads, fish and animal bones and stone tools, including querns, mullers and sling balls, constitute the other cultural material of the period. As an evidence of cooking may be mentioned a large-sized spouted basin with soot marks at the bottom. This is further substantiated by a big structure, which might have been used as community hearth-cum-pottery kiln. The environment of the Neolithic people at Jhusi appears to have been characterized by grassy land, frequented by occasional trees, which probably did not constitute a thick forest, though bamboo groves appear to have predominated. Marshy land and lakes also seem to have marked the landscape. The companions of the Neolithic man of Jhusi were cattle, sheep, goat, boar, *barasingha*, etc. Fish definitely constituted an important item of their diet. The food grain produced by them comprised rice, millet, *moong* and *masoor*.

5. The carbon samples, when processed, may yield definite information regarding the chronology of the Neolithic phase at the site. In this connection, it may be pointed out that at Lahuradeva in Sant Kabir Nagar district, the Neolithic phase is dated to 5000 BC. There is every possibility that the carriers of the Neolithic culture might have entered in the Ganga plain from the Vindhyas. Jhusi, nearer

to the Vindhyan area, and would have imbibed the Neolithic element earlier than those sites lying further east and north. In the light of the above discoveries, the antiquity of Neolithic Jhusi might be earlier that 5000 BC.

The settlements of the Neolithic people have left behind the evidence of circular huts known through the patterns of post-holes that have come to light. The walls of the huts appear to have been constituted by bamboo and reeds smeared with mud plaster, as is evident by the burnt clay lumps with wattle and daub impressions on them.

On the basis of combined testimony of excavations at the Neolithic sites of the Gangetic plain, it can be concluded that the excavations have brought to light evidence of wattle and daub structures. Traces of post-holes have been noticed practically at all the excavated sites. These hutments appear to be round or oval on plan. At Chirand, evidence of pit dwelling has also been reported.

The ceramic assemblage, generally handmade, includes red, burnished (red, grey or black), rusticated, corded, crude black-and-red and occasionally lustrous red wares. In this connection, it may be pointed out that corded ware appears to be a characteristic ware of the Neolithic cultures of the Vindhyas and that of the middle Ganga plain. The clay used for pottery is not levigated, containing grit, rice husks and chaff as degraissant. The principal pottery types comprise bowls (wide mouthed, channeled, etc.), vases, spouted vessels, miniature jars, etc. Post-firing painting is reported from Chirand and Senuwar, executed in red ocher, the motifs bearing linear designs of criss-cross lines and concentric circles. Wheel-thrown pots have been reported from the late Neolithic phase of Senuwar.

The excavated sites have yielded stone objects in an appreciable number. Bladelets, flakes, blades, scrapers, arrowheads, serrated points, lunates, borers, etc., fashioned on chert, chalcedony, agate, jasper, and quartz have been found from some of the sites. However, celts of basalt and granite have been obtained from Lahuradeva, Chirand and Senuwar. Other stone objects include fragments of querns, mullers, balls, hammer stones, etc., fashioned on sandstone or quartzite. Beads, finished and unfinished and fashioned on chalcedony, agate, etc., have been found at Jhusi, Chirand and Senuwar. Chirand has also yielded beads of steatite and faience.

Bone tools have been found at Jhusi, Senuwar and Chirand. The last site has yielded a corpus of bone tools and weapons, including celts, scrapers, chisels, hammers, needles, points, borers, awls, arrowheads, etc. Other bone objects comprise ornaments such as pendants, earrings, bangles, discs, combs, etc. In this connection, it may be pointed out that with the exception of the Neolithic sites of Kashmir-Burzahom and Gufkral-Chirand is the only Neolithic site in India to have produced bone objects both in quantity and variety.

Terracotta objects, including edge ground potsherds (triangular or rectangular in shape), spherical beads with central perforation were obtained from Senuwar. Chirand has yielded terracotta wheels, beads, bangles, cakes, birds, snakes, etc.

These people cultivated barley, wheat, rice, field pea, lentil, green gram, etc. However, the botanical evidence obtained from Senuwar and Jhusi in middle Ganga plain, read with its counterpart of Tokwa in the Vindhyas, suggests that in the early phase, only rice and some primitive millets were being cultivated. With the passage of time, other cereals were also added to the corpus of cultivated variety. From the IB (Late Neolithic) phase of Senuwar, besides rice, wheat, barley, *jowar* millet, lentil, field pea, finger millet (*Ragi*) and *Khesari* have also been obtained. The available evidence is demonstrative of the fact that by the late phase of Neolithic culture in the middle Ganga plain, double crop pattern had become an accomplished fact.

Domesticated animals included cattle, buffalo, sheep, goat and pig. Besides these, the bones of elephants, rhinoceros, stag, deer, etc., have also been found from some of the sites. Of aquatic fauna whose bones have been found, mention may be made of fish and turtle. Bones of birds have also been discovered. The available evidence, thus, indicates that besides agriculture and domestication, the Neolithic people of the middle Ganga plain also practiced hunting, catching and fishing.

NOTES AND REFERENCES

1. Wheeler, R.E.M. 1948, in *Ancient India,* No. 4, p. 2.
2. Singh, R.L. 1971. *India: A Regional Geography.* Varanasi, pp. 192–193; see also Singh, S.C. Delimination of the Middle Ganga Plain, *National Geographical Journal of India* XI (2 June, 1965), pp. 74–83.
3. Sharma, G.R., V.D. Misra, D. Mandal, B.B. Misra and J.N. Pal, 1980. *Beginnings of Agriculture,* Abinash Prakashan, Allahabad, pp. 117–122; see also Sharma, G.R. 1973, Mesolithic Lake Cultures in the Ganga Valley, India, *Proceedings of the Prehistoric Society* Vol. XXXIX, pp. 129–156; Sharma, G.R. 1975, Seasonal Migrations and Mesolithic Lake Cultures of the Ganga Valley, *K.C. Chattopadhyaya Memorial Volume,* Allahabad, pp. 1–20; Pal, J.N. 1988, The Antecedents of the Village and Folk Culture in the Gangetic Plain, *Rural Life and Folk Culture in Ancient India,* U.N. Roy, V.D. Misra and J.N. Pandey (eds), Department of Ancient History, Culture and Archaeology, University of Allahabad, Allahabad, pp. 117–127; Varma, B.S. 1971, Excavations at Chirand: New Light on the Indian Neolithic Culture Complex, *Purātattva* No. 4, pp. 19–22; Lala Aditya Narayan 1970, The Neolithic Settlements at Chirand, *Journal of the Bihar Research Society,* Vol. IV, Part I–IV, pp. 16–35.
4. Pal, J.N. 2002. The Mesolithic Phase in the Ganga Valley, *Recent Studies in Indian Archaeology,* Indian Council of Historical Research, New Delhi, Monograph Series VI, pp. 60–80; Pal, J.N. 2002. Mesolithic Gangetic Plain, in *Mesolithic India.* V.D. Misra and J.N. Pal (eds), pp. 289–305. Department of Ancient History, Culture and Archaeology, University of Allahabad, Allahabad, pp. 289–305.
5. *I.A.R.* (*Indian Archaeology: A Review*) 1962–63, p. 6; 1963–64, pp. 6–8; 1964–65, pp. 6–7; 1968–69, pp. 5–6; 1969–70, pp. 3–4; 1970–71, pp. 6–7; 1971–72, pp. 6–7; 1972–1973, pp. 7–8; 1980–81, pp. 9–10; 1981–82, pp. 13–14.
6. *I.A.R.* 1977–78, pp. 17–18.
7. *I.A.R.* 1981–82, pp. 10–12; 1982–83, pp. 16, 25; 1983–84, pp. 12–13; 1984–85, pp. 9–10.

8. *I.A.R.* 1983–84, pp. 15–16; 1984–85, pp. 12–13.
9. Singh, B.P. 1989–90. Chalcolithic Cultures of Southern Bihar as revealed by the explorations and excavations in District Rohtas, *Purātattva* No. 20, pp. 83–92; and see also Singh, B.P. 1995–96, Transformation of cultures in the Middle Ganga Plains: A case study of Senuwar, *Prāgdhārā* No. 6, pp. 75–93. Singh, B.P. 1988–89. Early Farming Communities of Kaimur Foot-hills, *Purātattva* No. 19, p. 18.
10. Chaturvedi, S.N. 1985. Advance of Vindhyan Neolithic and Chalcolithic Cultures to the Himalayan Tarai: Excavation and Exploration in the Saryupar Region of Uttar Pradesh, *Man and Environment*, Volume IX, pp. 120 ff.
11. Singh P. 1992–93. Archaeological Excavation at Imlidih Khurd, *Prāgdhārā* No. 3, pp. 21–35; see also Singh P. Ashok Kumar Singh and Indrajeet Singh, 1991–92 Excavation at Imlidih Khurd, *Purātattva* No. 22, pp. 120–122.
12. Singh, P. and A.K. Singh, 1997–98. Trial excavation at Bhunadih, District Ballia, U.P., *Prāgdhārā* No. 8 pp. 11–29.
13. Singh P. and A.K. Singh, 1995–96. Trial Excavation at Waina, District Ballia (U.P.), *Prāgdhārā* No. 6, pp. 41–42.
14. Tiwari R., R.K. Srivastava and K.K. Singh, 2001–2002 Excavation at Lahuradeva, District Sant Kabir Nagar, Uttar Pradesh, *Purātattva* No. 32, pp. 55–56.
15. Ancient Pratiṣṭhānpur, according to Purāṇas, was the capital of kings of Lunar Dynasty. Ilā, daughter of Manu was married to Budha, son of Som (Chandra). Their son Pururavā established the Lunar Dynasty at Pratiṣṭhānpur, Āyu, eldest son of Pururāva and then Nahuṣ, eldest son of Āyu became kings of Pratiṣṭhānpur. Nahuṣ was succeeded by his second son Yayāti and Yayāti was succeeded by his youngest son Puru. *See Matsya Purāṇa,* 24.33.35; *Vāyu Purāṇa,* 65.51; 92.1.2: *Viṣṇu Purāṇa,* 4.8.1; *Brhmāṇḍa Purāṇa,* 3.66.22; *Bhāgawata Purāṇa,* 9.15.1. The story of Yayāti is mentioned in the *Mahābhārat, Ādi Parva,* 81.90.
16. A big pucca well on the mound known as Samudrakūpa is traditionally associated with Samudra Gupta. Local tradition says that this well is connected with the sea. *See,* Srivastava, Saligram, 1937 *Prayāg Pradeep,* Hindustani Academy, Allahabad, p.24.
17. The copper plate is housed in the library of Asiatic Society of Bengal; see *Indian Antiquary,* Vol. 18.
18. Misra, V.D., B.B. Misra, J.N. Pandey and J.N. Pal, 1995–96. A preliminary report on the Excavations at Jhusi, 1995. *Prāgdhārā* No. 6, pp. 63–66.
19. Misra, V.D., J.N. Pal and M.C. Gupta, 1998–99. Further Excavation at Jhusi, District Allahabad (Uttar Pradesh): 1998. *Prāgdhārā* No. 9, pp. 43–49.
20. Misra V.D., J.N. Pal and M.C. Gupta, 1999–2000. Further Excavations at Jhusi (1998–99), *Prāgdhārā* No. 10, pp. 23–30.
21. Misra, V.D., J.N. Pal and M.C. Gupta, 2002–2003. Further Excavations at Jhusi: Evidence of Neolithic Culture, *Prāgdhārā,* No. 13, pp. 227–229.
22. The excavations at Jhusi are being carried by a team of archaeologists of the Department of Ancient History, Culture and Archaeology, University of Allahabad under the direction of Prof. V.D. Misra, Prof. J.N. Pal and Dr. M.C. Gupta. Dr. K.S. Saraswat and Dr. A.K. Pokharia of the Birbal Sahni Institute of Palaeobotany, Lucknow are collaborating in the field of archaeobotany.

23. Misra, V.D., J.N. Pal and M.C. Gupta, 2002. 'Jhusi: Archaeological and Historical Implications as Revealed from the Excavations at Ancient Pratishthānpur'. In *Allahabad: Aspects of Historical and Cultural Profile,* pp. 1–14, J.N. Pal, O.P. Srivastava, Anamika Roy, M.C. Gupta (eds), Department of Ancient History, Culture and Archaeology, University of Allahabad, Allahabad.

CHAPTER 7

Origin and Beginning of Agriculture in Indian Borderland: The Case of Afghanistan

V.C. Srivastava

INTRODUCTION

Vavilov and Buknin (1929) were the first scholars who have recognized the potential of Afghanistan as one of the areas which witnessed the domestication of plants and animals, identifying many varieties of wild wheat in Afghanistan. They opined that Afghanistan might have been one of the original hearth areas for domestication of plants and animals. It is unfortunate that even after a lapse of seventy-five years from that prophecy no systematic archaeological research could be undertaken to substantiate this statement. There are very few sites for the later prehistoric period in Afghanistan, which may throw light on these revolutionary socio-economic and cultural processes, i.e. domestication of plants and animals. From the available data, it may be safely said that later prehistoric period of Afghanistan witnessed two very important, almost revolutionary transformations in the socio-economic and cultural profile (Shaffer, 1978, p. 71):

(1) Domestication of plants and animals; and
(2) Development of stratified society.

AGRICULTURE IN NEOLITHIC AFGHANISTAN

There are three Neolithic sites in Afghanistan—Aq Kupruk I (Snake Cave or *Ghari-i-Mar*), Aq Kupruk II (House Cave or *Ghari-i-Asp*) and Darra-i-Kur. The reported Neolithic cultures from Harzar Sum and Gurziwan need further confirmation and, therefore, have not been taken into consideration here. While Aq Kupruk I and II represent the classic Neolithic phase, Darra-i-Kur has later Goat-cult Neolithic. Needless to point out that the Neolithic phase had been responsible for the beginning of cultivation as a subsistence pattern all over the world (Childe, 1942). Let us examine the Neolithic data from Afghanistan.

Aq Kupruk I has two phases: A and B of the Non-Ceramic Neolithic and Aq Kupruk II also has Non-Ceramic Neolithic. While at Aq Kupruk I and II (Dupree, 1967), ceramic Neolithic has been reported from a changed stratum. The industry consisted mostly of stone but bone artifacts were also reported such as polished points, plain and decorated needles, awls and spatulas. The lithic industry includes sickle blades with sheen on edges, one pressure-flaked unifacial leaf point, a bifacial point fragment, many notched flakes, hoes, ground stone querns and pounders, celts, bowl fragments, bladelets, drills, carved turtleshell fragments, etc., in both the ceramic and non-ceramic Neolithic phases but there are some differences also between non-ceramic A and B phases. Phase A did not have the stone hoes, querns, pounders and steatite bowl fragments which have been traced in phase B. These items are absent in non-ceramic Neolithic at Aq Kupruk II, as also from Ceramic Neolithic at Aq Kupruk I. In non-ceramic Neolithic phase B, more sickle-blades plus cores, microblades, and side-scrapers, points, burins, occasional backed blades, one pressure-flaked unifacial point have been found. There appears to be some sort of continuity from the Kuprukian Upper Palaeolithic because in both the areas, there were large blades, points and dihedral burin but the departure from the earlier tradition is evidenced by the presence of sickle blades. From the point of view of lithic industry, non-ceramic and ceramic Neolithic at Aq Kupruk I and II form one cultural assemblage. The difference was due to the presence of pottery. The ceramic ware has been described as a crude, soft, chaff, crushed limestone and crushed tempered ware with flat bases and simple rounded rims. Dupree calls it black ware but Shaffer has found it to be buff ware. Actually, due to differential firing, there were wares of many colours. The pottery appears to be handmade, and is coarse in manufacturing.

The chronological bracket for the classical Neolithic in Afghanistan may be fixed from 8000 BC to 4000 BC on the basis of the following C-14 dates available to us: 8566 BC and 6960 BC for non-ceramic Neolithic and 2685 BC, 5018 BC, 4549 BC and 5214 BC for ceramic Neolithic. This indicates that the Neolithic phase may be placed from 9th millennium BC to 5th–4th millennium BC.

Perkins (1972) has identified domesticated sheep (*Ovis* sp.), goat (*Capra hircus hircus* and *Capra hircus* spp.) and unidentified sheep-goat (*Ovis/Capra*). There is a controversy regarding the domestication of cattle from non-ceramic Neolithic at Aq Kupruk I.

There is no floral evidence for the domestication of plants in Afghanistan at this stage but the presence of sickle blades with sheen may be taken to supply archaeological evidence for this phenomenon in the non-ceramic Neolithic at Aq Kupruk. The domestication of animals and some sort of domestication of plants generally go together as they are interconnected. It has rightly been suggested that these developments—domestication of animals and plants—are processes rather than events. Therefore, the suggestion that the process must have begun some time between 14665 BC and 8566 BC in Afghanistan is not without foundation, especially in view of other circumstantial evidences collected by Shaffer (1978). Other Neolithic developments like the manufacturing of the pottery and the polish of the tools are well documented here. However, there is no evidence of architecture or of any artificial structure at Aq Kupruk I and II. On the other hand, natural caves were selected to be habitats. It appears that the climate of Afghanistan in the post-Pleistocene period with snowfall and cold winds prevented the Neolithic population to erect temporary shelters, particularly so when the area abounds in natural rock shelters and caves. Even today, there are evidences of human habitation in caves and rock shelters in Afghanistan. Even in historic times, the Afghans continued to occupy caves for homes, as has been shown at Shamshir Ghar. There is a controversy as to whether Neolithic peoples of Aq Kupruk I and II were specialized pastoral nomads or agriculturists. Arguments and counter-arguments have been put forth in favour of these interpretations. If we have to learn anything from the present-day condition in Afghanistan, we may say that the Neolithic culture witnessed a symbiotic development of pastoral nomadism and agricultural sedentatrism. The presence of steatite bowls in Afghanistan at this time shows the interaction between pastoral nomads and sedentary agriculturists. From this point of view, the Neolithic culture of Afghanistan is slightly different from the sedentary agricultural Neolithic of West Asia and even India.

The chronological position, as noted above, shows that Afghanistan was in the nuclear zone of food-producing economy. Rather, it was one of the many centres for agricultural beginnings along with other changes. However, we have to wait for a definite answer till the cultivation of plants is proved for this phase by the pollen analysis. Since the Darra-i-Kur Goat cult Neolithic is later in date, it has not been considered here for the origin of domestication of plants and animals.

AGRICULTURE IN THE PROTO-HISTORIC AFGHANISTAN

Unfortunately, we do not have sites like Mehrgarh (Pakistan) or Shahr-i-Sokhta (Iran) in Afghanistan which may throw sufficient light on the development of agriculture in protohistoric Afghanistan. However, we do have the following protohistoric sites in Afghanistan, which may throw light on agriculture and other subsistence patterns:

(1) Mundigak
(2) Said-Qala
(3) Deh-Morasi-Ghundai
(4) Shortugai
(5) Dashli
(6) Daulatabad
(7) Farukhabad

Agriculture, pastoralism, mining and trade were the main components of the subsistence system of protohistoric Afghanistan. Remains of plants from these sites form the main basis of agriculture for this period.

Plants remains include domesticated wheat (*Triticum campactum*) from Mundigak Pd. II and Said Qala may have wheat and barley, while Deh Morasi II has remains of a fodder grass (*Aegilopas tauchi* syn. *A. squarrosa*) related to wheat and domesticated six-row barley (*Hordeum vulgare* var. *afghana*). Animal remains at Mundigak included both domesticated (sheep, goat, cattle, ass, horse and dog) and wild (gazelle, ibex and lynx, bird of prey) animals, while at Said Qala and Deh Morasi, we have probable evidence of domesticated sheep, goat and bovid. These remains of plant and animals must have supplied food for human population such as wheat, barley and some grass, meat and milk of sheep/goat/cattle as well as food for animals (forage, fodder, crops). These must have supplied fibres, leather skin/hide for clothing as well as fuel (wood and dung) and building material (wood, twigs, grasses, dung) and other materials like bone and shell. The tool kit at these sites such as bronze sickles, blades, axes and adzes were suitable for agriculture (Srivastava, 1996).

The environmental condition of Mundigak region appears to have been similar to that of the present times (the theory of wetter phase, even if it is correct, is not very effective because it must have been marginally wetter). The Kandahar region is a desert with small villages maintained by scarce water from rivers and tributaries and there is not much scope for developed irrigation agriculture. Wheat and barley are winter crops. Mundigak has an eccentric location and its development as an urban centre cannot be accounted for by agricultural activities. The ethnographical data of Mundigak (Kandahar region) includes, at present, sedentary agriculturists as well as nomads and mixed varieties to tribes. In Kandahar, the tribes migrate in summer towards the central region and return for agriculture and livestock breeding. There is evidence of a canal system in the region of very ancient origin. Presently, Kandahar is one of the major areas of cultivation. The dry farmed area known as *Lalmi* as against the irrigated land (*abi*) support wheat and barley. It may be surmized that dry farming (rain-fed) was done specially in winters as wheat and barley as evidenced in archaeological record are winter and *lalmi* crops. Nomadic pastoralism was the main occupation during summer when these tribes used to move to better grazing areas in Central Afghanistan, just as they do today. This grain-livestock subsistence system appears to have been thoroughly integrated as it is even today in Afghanistan (Dupree, 1973).

In northern Afghanistan, the evidence from Shortughai (Francfort, 1989) may be taken as a model. There is evidence of artificial irrigation of a very sophisticated type (Gardin and Gentelle, 1976, 59–99) as well as of dry farming. The latter is confirmed by the discovery of a ploughed field covered with flax seeds in an area hardly suitable for artificial irrigation. There is also evidence of small irrigation canals. The irrigation technique appears to be a result of the local tradition. Domesticated plants included wheat (notably *Triticum aestivam*) (Central Asian variety), barley, millet, lentils, small peas and vetch. Seeds identified included grapes, almonds, oilseed (*Sesame*) and the following trees and bushes were identified—pistachio, willow, poplar, tamarisk, almond and lycium (Willcox, 1989, Tab. 65, in Francfort, 1989). Animal remains include sheep, goat, ox, zebu, buffalo. The wild animals included *ovis ammon, Gazella dorcas, cervus axis, equeus hemionus, capra aegagrus* (Desse, 1989, Table 67, in Francfort, 1989). Thus, there is definite evidence of settled agriculture in phase A and dry farming in phase B as a means of subsistence but it was accompanied by pastoralism, transhumance, and seasonal nomadism during both phase A and B (Srivastava, 1996).

It may be mentioned here that both at Mundigak and Shortughai, the subsistence system was not exhausted by cultivation and nomadic pastoralism. Other important components of the subsistence system were handicrafts, mining, and trade and exchange.

The details of these aspects may be seen in the work of Jarrige and Tosi, 1981. It is rightly said that protourbanism of Mundigak appears to be a case of holding a key position to organize the exploitation of various mountainous minerary bases. Seasonal exploitation of minerary resources by Shortughai and specially lapis and copper also formed a major determinant of protourbanism here in Pd I and II (Francfort, 1985). Besides Sar-i-sang (Afghanistan), there were other active sources of lapis lazuli (Delmas and Casonova, 1990, 504). The subsistence system of the early second millennium BC in Daulatabad oasis consisted of farming, cattle breeding supplemented by hunting wild boar and perhaps jarjans (Sarianidi, 1977).

To sum up, the subsistence system of protohistoric Afghanistan had four main components: agriculture, pastoralism, mining and trade. With the advent of metal technology, exploitation of minerary resources and protoubanism, there appears to be a compulsion that a political entity existed in these societies and this may be regarded as the beginning for a stratified society. The control of minerary exploitation over such an extensive area and trading could not have been possible without a controlling authority with the power to regulate (may be tribal chief) as well as trade unions/heads to manage such an extensive trade contacts. Cultivation involved ploughing of land, sowing, irrigating, harvesting threshing, etc., which also demanded various types of labour/specialization. Thus, Afghanistan must have developed a stratified society with classes and groups of people engaged in similar economic activity, though no archaeological evidence has come to light so far to confirm it. Keeping in view the close proximity in date between the archaeological culture of Afghanistan and the literary data in the *Ṛgveda* (2nd millennium BC), a *Varṇa* type social order appears to be in the making. The existence of Casal's 'Palace', 'Temple' and special function buildings in Mundigak IV and a monumental building for human sacrifice—something like Mesopotamian "Ziggurat" in Pd. V, along with common room and habitation structures as well as 'Temple' (round building) and 'Palace' at Dashli and an unfired brick-paved floored room at Shortughai—may indicate social stratification based upon functions such as political, religious and economic controls. Shaffer (1978) has tried to show on the basis of the presence of metal artifacts from AK-Kupruk, goat cult Neolithic at the Darra-i-Kur, Tash-Kurghan, Dashli, Mundigak, Deh Morasi Ghundai and Said Qala that there was development of stratified society in protohistoric Afghanistan.

Shaffer (in Allchin and Hammond, 1978, pp. 83–87) has utilized data from a few neighbouring sites such as Djetun, Kile Gul Mohammad and Gumla to create a picture of agricultural village in protohistoric Afghanistan. In brief, it may be said that Djeitun culture was characterized by the presence of domesticated barley and wheat as well as domesticated sheep and goat along with wild animals. It had mud brick rectangular or square houses with internal hearth, storage bins and pits, grain storage structures, handmade and wheel-made pottery and figurines. This culture demonstrates a prosperous agricultural development. Kile Gul Mohammad culture was characterized by mud and *pise* structures, presence of sheep, goat and cattle, handmade pottery, etc. Gumla gives us the impression of mobile population. It is relevant to quote Shaffer (1978, p. 87): "This brief summary of the Djeitun, KGM-I-II and Gumla I cultures provides a possible picture of early sedentary agriculturists in Afghanistan, although when contemporary cultural complexes are defined in Afghanistan, they will doubtless present differences from as well as similarities with those groups. The transition to a subsistence dependent upon utilization of domesticated plants and animals resulted in the development of specialized nomadic and agricultural groups. However, this transition stimulated additional processes of cultural, economic and ecological change, which resulted in even more complex ecological and cultural adjustments, which in turn, resulted in another equally important transition—the transition to a socially stratified way of life."

Sophia R. Bowlby (in Allchin and Hammond, 1978, pp. 25–30) has given us a picture of present-day agriculture in Afghanistan. He is of the opinion that the present farming system is not unlike that of the prehistoric times. It is desirable to quote him (pp. 29–30): "It seems pertinent at this point to consider what aspects of the farming system of the past could be similar to the modern situation. Certainly the tools used today are not more advanced than those that might have been used by an Iron Age farmer. Some of the crops grown today would have been unknown but many probably grew as wild plants in Afghanistan in the past, as they do today. Vavilov (1951) suggested that the area of Afghanistan and Pakistan form an independent cultural hearth for a variety of crops—notably the wheat, chick peas and beans that still form an important part of the farming system. Moreover, most Afghans still use fairly primitive forms of wheat. The simple riverine system of irrigation would have been feasible for most prehistoric farming communities but the *karez* system is technically more difficult."

To recapitulate and conclude, Afghanistan witnessed the origin of domestication of plants in Neolithic stage and its development may be traced in the protohistoric period. The most significant feature of agriculture in Afghanistan appears to be the symbiotic development of sedentary agriculture and nomadic pastoralism. Within these groups, there were many fine and overlapping functional differences. It appears that this feature has marked the entire history of Afghanistan upto the present times. Afghanistan had agricultural villages with structures of different purposes. The village was supported by industries such as pottery-making, tool-making, cattle-rearing, handicrafts, nearby trade on an exchange basis. The village had some sort of stratified society. It is more than probable that agriculture was an indigenous process. The old theory of diffusionism (Fairservis, 1971, p. 105) may be discarded in view of recent researches in Afghanistan. Thus, Afghanistan was one of the centres for the birth and growth of agriculture in ancient times.

NOTES AND REFERENCES

1. Allchin, F.R. and Hammond, N. (Ed.) 1978, *The Archaeology of Afghanistan,* Academic Press, London.
2. Ball, W. in collaboration with J.C. Gardin, 1982, *Archaeological Gazetteer of Afghanistan,* Two Volumes, Paris.
3. Casal, J.M. 1961, *Fouilles de Mundigak,* Two Volumes. MDAFA XVII, Klineksick.
4. Childe, V.G., 1942, *What Happened in History,* Penguin, Hammondsworth.
5. Delmas, A.B. and Casonova, M. 1990, The Lapis Lazuli Sources in the Ancient East, in *South Asian Archaeology,* (ed.) M. Taddei, Rome, pp. 493–506.

6. Dupree, L. 1963, Deh Morasi Ghundai: A Chalcolithic site in South Central Afghanistan, *Anthropological Papers of American Museum of Natural History*, 50, pp. 59–135.
7. Dupree, L. 1967, Prehistoric Period of Afghanistan, *Afghanistan,* 20(3), Ministry of Information & Culture, Kabul, pp. 8–27.
8. Dupree, L. 1973, *Afghanistan,* Princeton University Press, Princeton.
9. Fairservis, W.A. Jr, 1971, *The Roots of Ancient India,* George Allen and Unwin, London.
10. Francfort, H.P., 1989, *Fouilles De Shortughai: Recherches Sur L' Asie Centrale Protohistorique,* 2 Vols, Centre National de la Recherche Scientifique, Paris.
11. Jarrige, C. and Tosi, M. 1981, The Natural Resources, in Hartel, ed. *South Asian Archaeology,* (1979) Berlin, pp. 115–142.
12. Perkins, D. Jr., 1972, The Fauna of the Aq Kupuruk Caves: A Brief Note in *Prehistoric Research in Afghanistan* (1959–1966), (ed.) L. Dupree, in Transactions of the American Philosophical Society, Vol. 62(4), Philadelphia.
13. Sarianidi, V.I., 1977, *Drevnie zemledelcy Afganistana,* Moscow.
14. Shaffer, J.G., 1971, Neolithic-Preliminary Field Report on Excavations at Said Qala, Tepe, *Afghanistan,* Vol. XXIV, Nos. 2 & 3, pp. 89–127.
15. Srivastava, V.C., 1979, Economy and Society in Prehistoric Afghanistan, *Afghanistan,* 32(2), 54–59, reprinted in *Eastern Anthropologist,* 34(2), 1981.
16. Srivastava, V.C., 1982, *The Prehistoric Afghanistan: A Source Book,* Indological Publications, Allahabad.
17. Srivastava, V.C., 1996, *The Protohistoric Afghanistan: A Source Book,* Tara Book Agency, Varanasi.
18. Vavilov, N.J., 1951, The Origin, Variation, Immunity and Breeding of Cultivated Plants. Selected writings translated from the Russian, translated by K.S. Chester, *Chronica Botanica,* Vol. 13, i–vi, Waltham, Mass.
19. Vogelsang, W., 2002, *The Afghans,* Blackwell Publishers Ltd., Oxford.

CHAPTER 8

History of Millet Cultivation in India

Purushottam Singh

INTRODUCTION

Millets are annual, warm weather grasses which comprise a number of cereals belonging to different genera and species. They are mostly grown on dry lands as rain-fed crop. Millets are grown primarily for grain and their straw makes valuable cattle fodder. Millets, among the world's most important food crop for the inhabitants of the semi-arid tropics, are the main source of protein and energy. Millets form a major portion of subsistence system in India today and constitute 45% of all land planted to food plants (Weber, 1991: 171). Today, various kinds of millets are grown in southern Uttar Pradesh, central Madhya Pradesh, Rajasthan, Haryana, western Andhra Pradesh, western Tamil Nadu, eastern Maharashtra and parts of Karnataka. The commonest varieties are *jowar (sorghum)*, *bajra* (pearl millet), ragi (finger millet), and *kodon* millet. In the following pages, an attempt has been made to present the data pertaining to the cultivation of millet as are available in the archaeological record till date and trace the history of their cultivation through the ages.

Archaeological studies of plant and grain remains obtained from the excavated sites in India during the past three decades have resulted in an almost continuous history of millet cultivation, at least from 3rd millennium BC. Various types of millets have been reported from Neolithic levels in Karnataka (Hallur); from pre-Harappan and Harappan sites in Punjab; Haryana and Gujarat (Rohira, Sanghol, Banawali, Rangpur, Surkotada, Rojdi, Oriyo Timbo); and from Chalcolithic levels of a number of sites in the Deccan and from southeastern Rajasthan. The Satavahana and Kushana layers of a number of sites have provided further evidence of millet cultivation. We discuss below the history of five major varieties of millets as found in the archaeological record in the following pages.

JOWAR MILLET (*SORGHUM BICOLOR*)

The *jowar* millet, along with several varieties of wheat, barley, field pea, etc., has been reported by K.S. Saraswat from pre-Harappan culture (*circa* 2750–2500 BC) as also from the mature Harappan phase (2500–2000 BC) at Banawali (district Hissar) (*IAR*, 1994–95: 96); from Sanghol (district Ludhiana) from Baran/Late Harappan levels (1900–1400 BC) (*IAR*, 1992–93: 123); and the late Harappan context (1700–1000 BC) at Hulas (district Saharanpur) (*IAR*, 1986–87: 132). *Jowar* millet comes along with wheat, barley, gram from the pre-Harappan levels of Period IA (2300–2000 BC) from Rohira (district Sangrur, Punjab) (*IAR*, 1983–84: 177). *Sorghum* has also been reported from Period I at Pirak (2200–1500 BC) (Chakrabarti, 1988: 90). *Sorghum* is absent in the early levels of occupation at Rojdi but it is recovered in the upper levels, implying its introduction into Rojdi after 2000 BC. Since it is almost certainly of African origin and yet the Rojdi record is found in levels after the occurrence of *Eleusine*, sorghum may have entered south India independently of *Eleusine*, implying either continuous or multiple periods of interaction with people in Africa or with people who had contact with African populations (Weber, 1991: 174). *Jowar* millet, along with wheat, barley and lentil, has also been reported from Neolithic levels (2200–1800 BC) at Senuwar, situated on the Kudra river in district Rohtas (Bihar) (Saraswat, 2004). It also continued to be cultivated in the succeeding Neolithic–Chalcolithic (1800–1200 BC) and Chalcolithic periods at this site.

Among the Chalcolithic cultures in the Deccan, *Jowar* was introduced at Inamgaon from the end of early Jorwe period (*IAR*, 1972–73: 68). *Jowar* was found at Daimabad in the Savalda phase (Period I, 2200–2000 BC) and later, at Inamgaon (ca. 1400–1000 BC) (Dhavalikar, 1995: 107). The occurrence of sorghum in Savalda culture level at Daimabad shows that in Maharashtra, this variety goes back to the third millennium BC (Chakravarti, 1988: 90). *Jowar*, along with food grains of other 18 species of cultivated and wild plants, was recovered from Tuljapur Garhi (district Amaravati) but its exact cultural context is not known (*IAR*, 1986–87: 129). *Jowar* was also found from early historical levels at Hulaskhera (district Lucknow) (*IAR*, 1987–88: 152). *Jowar* millet has been found along with wheat, barley and rice from the Kushana levels (1st–3rd centuries AD) at Sanghol (district Ludhiana) (*IAR*, 1986–87: 130; *IAR*, 1993–94: 144). This grain also comes from Satavahana levels at Bhokardan (district Aurangabad) (*IAR*, 1976–77: 91), and from the Satavahana site of Kausan and Mungi (district Aurangabad) on the right bank of Godavari (*IAR*, 1976–77: 91).

Evidence for the cultivation of *jowar* also comes from Bhatkuli (district Amaravati (5th century BC) and Paunar (3rd century BC).

At present, there are about 18 million hectares under *jowar* with a production of 9 millions of grains (Randhawa, 1980: 245). The largest area under *jowar* is in Maharashtra as it is the favourite cereal of the inhabitants of the state.

BAJRA (PEARL MILLET) (*PENNISETUM TYPHOIDES*)

Pearl millet, bulrush, or spiked millet is the most important of all millets. It is a quick-growing summer cereal grass and has a higher level of heat tolerance than *sorghum* and maize. It thrives on light-textured and well-drained soils, but does not tolerate waterlogging and flooding as well as *sorghum* (Randhawa, 1980: 318). *Bajra* is one of the principal millets grown in India, next only to *jowar* among the coarse grains. It is presently grown in Rajasthan, Gujarat, Maharashtra, Andhra Pradesh and Haryana.

In archaeological record, the earliest evidence for pearl millet cultivation comes from Hallur in Karnataka, where it was cultivated during the Neolithic times.

Charred remains of *bajra* were found along with rice husk in Period III at Rangpur (Chakrabarti, 1988: 90). This grain was also reported from Rojdi (Possehl, 1986: 195–236). *Bajra* was reported from Period I at Ahar, the type site of Banas culture. Pearl millet comes from early historical levels at Mungi (district Aurangabad) situated on the right bank of the Godavari along with rice and wheat (*IAR*, 1976–77: 91). *Bajra* was also grown at Narhan (district Gorakhpur) in the Narhan culture levels (1300–800 BC) (Saraswat *et al.*, 1994: 272).

RAGI/FINGER MILLET (*ELEUSINE CORACANA*)

Ragi or finger millet is the hardiest crop suited for dry farming. Its grain has great nutritive value and can be stored for many years. Presently, Karnataka is the largest area under *ragi* but this variety is also grown in the adjoining Andhra Pradesh, Tamil Nadu and Maharashtra.

Perhaps the earliest report of *ragi* comes from Hallur in Karnataka, dating to approximately 2300 BC (Vishnu-Mittre, 1971). It comes from Harappan levels (2500–2200 BC) from Shikarpur (district Kutch) (*IAR*, 1990–91: 103). *Ragi* has also been reported along with wheat and pulses from the late Harappan levels at Hulas (district Saharanpur) (*IAR*, 1982–83: 149; 1986–87: 132). At Surkotada, two lumps of charred masses yielded as many as 574 carbonized seeds, an overwhelming majority of which were of wild plant species. These lumps contained charred remains of *ragi* and *Setaria italica* (*IAR*, 1974–75: 78). *Ragi* was most abundant millet species at Rojdi, as also at Kuntasi. While Rojdi is an extensive Harappan settlement (7 hectares) located in the heart of Saurashtra, Kuntasi is a 2-hectare small settlement of the Harappa culture in district Rajkot, Saurashtra. This site is just 4 km from the Arabian Sea and was excavated by M.K. Dhavalikar in 1987–90. Dhavalikar (1995: 210) believes that *ragi* may have been the major ingredient of the Harappan diet in Saurashtra. The presence of *Eleusine* from the earliest levels of Rojdi suggests that if these plants came from Africa, there was direct or indirect contact with African populations prior to 2600 BC (Weber, 1991: 173). Evidence of *ragi* cultivation comes from Daimabad from Malwa and Jorwe cultures (*IAR*, 1977–78: 91; 1979–80: 113; 1980–81: 108–109; Vishnu Mittre *et al.*, 1986: 592 ff). In all, 123 grains of this cereal were represented from Malwa culture levels and 29 from Jorwe culture at this site. *Ragi* has also been reported from Inamgaon (*IAR*, 1977–78: 92; 1981–82: 106). Seeds of finger millet were reported from Surkotada (district Kutch) but it is not clear whether it belongs to the wild species or had already been taken for cultivation (*IAR*, 1975–76: 86). It was found from the early historical levels at Hulaskhera (district Lucknow) (*IAR*, 1987–88: 152) and comes from Manjhi in red ware levels (250 BC–250 AD) (*IAR*, 1992–93: 123). This grain also comes from Satavahana levels at Bhokardan (district Auragabad) (*IAR*, 1976–77: 91).

Recently, *ragi* millet has been reported from the Neolithic levels of Senuwar (district Rohtas, Bihar). Besides, there are some traces of the distorted seeds in an utterly fragile state giving the appearance of *ragi* millet in the succeeding Neolithic-Chalcolithic phase at this site, but their identification remains doubtful (Saraswat, 2004).

KODON MILLET (*PASPALUM SCROBICULATUM* L.)

Kodon millet has been reported from Daimabad from Malwa and Jorwe levels (Vishnu Mittre, *et al.*, 1986: 588–623; *IAR*, 1980–81: 108–109). Imprints of Kodon were discovered on the burnt clay lumps from Magha in the Adwa Valley, district Mirzapur from Chalcolithic levels (*IAR*, 1981–82: 106). Besides, six complete spikelets and some broken pieces of Kodon millet have been reported from Senuwar (district Rohtas, Bihar) from the Neolithic–Chalcolithic levels (1800–1200 BC) along with barley, dwarf wheat, *jowar* millet, lentil, field pea and *khesari* (Saraswat, 2004). Kodon millet, along with barley, has been reported from the Megalithic phase (*c.* 1000–300 BC) at Veerapuram (district Kurnool, Andhra Pradesh) (*IAR*, 1980–81: 106); Naikund along with rice, field pea and lentils (*IAR*, 1979–80: 113); from red ware level (250 BC–AD 250) at Manjhi (district Saran) (*IAR*, 1993–94: 143); Satavahana levels of Paithan (district Aurangabad) situated on the left bank of the Godavari (*IAR*, 1976–77: 91); and Nevasa (district Ahmadnagar) from early historical levels (*IAR*, 1976–77: 91). Kodon millet was also cultivated at Narhan in the Sarayupar plain of the middle Ganga valley in the Narhan culture levels (1300–800 BC), as also in the succeeding NBPW levels (Saraswat *et. al.*, 1994: 309). This cereal has been mentioned as *kodrava* in the Arthashastra, which records that it was grown along with *Sali, vrihi* and *tila* during the Mauryan Age (Randhawa, 1980: 360).

FOXTAIL MILLET (*SETARIA ITALICA*)

As many as 40 carbonized grains of *setaria* were discovered from Surkotada. However, it is not clear whether it was cultivated at this site. Vishnu Mittre believes that the grains of *setaria* was collected from wild species (*IAR*, 1978–79: 108). The discovery of Italian millet is of considerable interest as it has been discovered for the first time in India (*IAR*, 1974–75: 78). It occurs in Harappan levels (2500–2200 BC) at Shikarpur (district Kutch) (*IAR*, 1990–91: 103). This grain also comes from Sanghol from late Harappan levels (1900–1400 BC) (*IAR*, 1992–93: 123). *Setaria Italica* has been recovered from the earliest strata of Rojdi and it can be placed within 400 years of the oldest finds of domesticated *Setaria* anywhere else in the world, including those in the highlands of central China from the Yang-Shao culture some 5000 years ago (Rao *et al.*, 1987). As regards the earliest centre of domestication of *Setaria Italica,* the consensus seems to be leaning towards east Asia (Weber, 1991: 174). Three carbonized seeds of foxtail millet were reported from Jorwe culture levels and Daimabad located on the left bank of the Pravara river, a tributary of the Godavari in Ahmednagar district of Maharashtra (Vishnu Mittre, 1986: 600). This millet was also grown at Manjhi (district Saran, Bihar) in the red ware levels (250 BC–250 AD) (*IAR*, 1993–94: 143).

DISCUSSION

The fact that the mighty Harappan culture flourished on a sound agricultural base has been amply proved by the discovery of a ploughed field from pre-Harappan levels at Kalibangan, the terracotta model of a plough from Banawali and actual recovery of archaeo-botanical remains of cereals, legumes and other related objects from several sites. The discovery of granaries at Harappa and Mohenjo-daro suggest that cereals were produced in such a large quantity that not only were all the immediate needs of the people duly met with, but there was also a surplus to face any future emergency (Lal, 1997: 159). Until recently, it was believed that the staple crops of the Harappans were wheat and barley, while rice was unknown to the mature Harappans, at least in the Indus Valley. However, the occurrence of rice and millets in the pre-Harappan and the Harappan levels at Rohira and from mature Harappan levels at Banawali has forced us to change this assumption. Similarly, Harappan agriculture in Gujarat presents an altogether different picture. The occurrence of rice on several sites in Gujarat (Lothal, Rangpur, Rojdi) shows that rice and sorghum were cultivated as monsoon crops in Gujarat at the beginning of the Harappan phase around 2600 BC (Mark Kenoyer, 1998: 163). Dhavalikar (1995: 98) believes that there was a distinct climatic change in Saurashtra around 2000 BC and the region became more and more arid. It was at this time that new cultigens such as coarse millets were introduced in the Harappan subsistence system. Towards the close of the mature Harappan phase in Gujarat, two new cultivars, viz., *bajra* and *jowar* appeared (Dhavalikar, 1997: 46). Rangpur and Rojdi have provided ample proof of *jowar* and *bajra* cultivation and *ragi* was a predominant crop at Kuntasi. Besides, *Setaria* and *ragi* are also present at Surkotada (Vishnu Mittre and R. Savitri, 1982: 215).

The stratigraphic position of millets in the Ahar culture is rather confusing. Impressions of spikelets of *jowar* were found on the pottery of this site from the top layers of phase IC but these deposits are believed to be disturbed (Dhavalikar, 1997: 93). Among the Chalcolithic culture of central India, the economy of the Kayatha culture has not been worked out but the excavations at Navdatoli give us a fair idea of the economy of the Malwa people. Wheat, rice, lentil, black gram and green gram were the staple food of the these people but there is no record of millet cultivation (Sankalia, *et al.* 1971: 418). However, *jowar* was grown by the Malwa people in Maharashtra (Dhavalikar, 1997: 137). Excavations at Kaothe (district Dhule, Maharashtra) show that the first farmers of Maharashtra (Savalda culture) cultivated *bajra* in ca 2300–2000 BC (Dhavalikar, 1971: 161). The Savalda culture inhabitants of Daimabad cultivated *jowar* along with barley and lentil (Vishnu-Mittre, 1982: 215).

Among the Chalcolithic culture of the Deccan, the Jorwe farmers practised crop rotation of *Kharif* and *Rabi*. Inamgaon has provided evidence of the cultivation of *jowar* and *ragi* along with wheat, barley and rice. Similar evidence comes from Daimabad also (Dhavalikar, 1997: 191). *Ragi* was the staple crop in the Neolithic culture of Karnataka (Hallur, Tekkalkota). Archaeological record shows that once millets came into the dietary system of the Indians, they continued to be grown by the megalith builders of south India, and the authors of early historic culture in the Vidarbha region. Evidence of *sorghum* comes from Periods II and III (Satavahana and Vakataka-Vishnukundin times) at Paunar (ancient Pravarapura, the capital of the Vakatakas). Spikelets of rice and millet were found mixed with clay for making terracotta figures, tiles and ringwells. These samples were analyzed by Vishnu-Mittre and H.P. Gupta (1968: 128–130), who observe that during the early Satavahana times, *sorghum* was abundantly used over rice. This position remained more or less the same during the late Satavahana times (Ist century

BC to 3rd century AD), but there was a change in the plant economy during this period and the use of *jowar* became rare. Millets remained the staple food during the times of the Mauryas, Satavahana and Kushanas. P.K. Gode has shown that the term *Yāwaāngla,* mentioned in the Charaka Samhita, (dated to AD 100–200) stands for *jowar.* We are told that the Romans sought an Indian millet grown in Pliny's days (first century AD) which had a high yield capacity; one grain produced approximately 1.65 kilograms (Sidebotham, 1986: 21, Singh, 1992: 17). Ibn Batuta, a native of Tangiers in Morocco who visited India in the reign of Muhammad-bin-Tughlak (AD 1325–1351), records that *kudhru* (*Paspalum Scrobiculatum*) was the commonest of all the grains (Randhawa, 1982: 67). Writing about the Vijayanagar empire, the Portuguese traveller, Fernao Nuniz (AD 1535) observes that wheat, gram, rice and millet were the main crops and the last named grain was consumed the most in the land (Randhawa, 1982: 113). The Ain-i-Akbari records that millets which included *jowar, bajra, kodon, sawan* and *manduva* and formed the *Kharif* crops, were cultivated in Malwa, Gujarat, Ajmer, Khandesh, Delhi, Lahore, Agra, Allahabad, Awadh and Multan (Randhawa, 1982: 211). Francisco Pelsaert, a Senior Factor of the Dutch East India Company at Agra (1621–1627) writes that *jowar, bajra* and *kangni* were the food grains eaten by the poor in the seventeenth century (Randhawa, 1982: 282) and this practice is prevalent in many parts of the country even today. At present, there are about 18 million hectare under *jowar* with a production of 9 million tonnes of grain (ICAR, 1980: 815).

DIFFUSION OF MILLET CULTIVATION

It is commonly believed that millets are indigenous to equatorial Africa. The area of probable domestication of *ragi* was in the highlands of Ethiopia to Uganda; of *jowar* in a wide zone in the broad-leaved Savanna belt that stretches from Lake Chad to eastern central Sudan and of pearl millet in the dry Savanna from Sudan to Senegal (Harlan, 1971: 471). In this context, mention may be made of the excavations conducted during 1975–77 and again in 1990–91 at site E-75–6 at Nabta Playa near the Egyptian-Sudanese border in southernmost Egypt. Here, cultural deposits belonged to two phases of early Neolithic dated to 8800–8500 BP. These excavations yielded hundreds of carbonized seeds of *sorghum* and millets, with consistent radiocarbon dates of 8000 years BP, thus providing the earliest evidence for the use of these plants. They are morphologically wild, but the lipid fraction of the *sorghum* grains shows a close relationship to domesticated than the wild varieties (Wendorf *et al.,* 1992, 721). Research carried out in the Atbara and Khartoum regions of central Sudan shows that cultivation of *sorghum* was practised from the 6th millennium BP (Haaland, 1992: 43).

It is now reasonably certain that the first cultivation of *sorghum* and finger millet was in the regions of Ethiopia and that the southern Sudan–Chad region represents the early belt of wild-growing *sorghum.* In this context, mention may be made of the discoveries made at Kadero, located 18 km north of the junction of the Blue and White Niles and 6.5 km east of the main Nile channel. Potsherds obtained from this site carried impressions of cultivated type of grains of *sorghum* and finger millet. The calibrated dates of this site belong to the end of the fifth millennium BC. It is believed that sorghum and finger

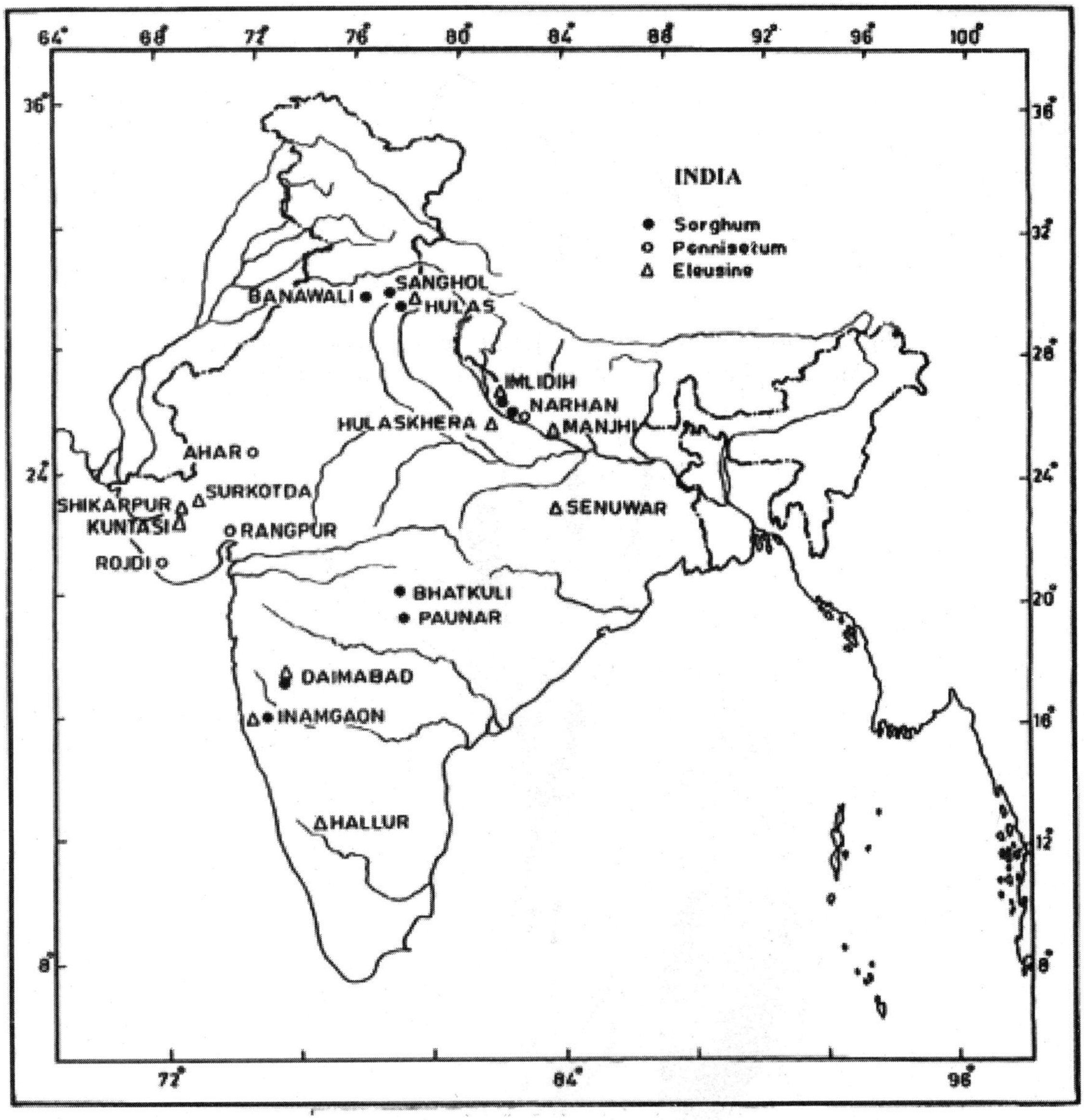

Fig. 1: Geographical distribution of records of *Sorghum*, *Pennisetum* and *Eleusine*

Fig. 2: Sorghum bicolour or Jowar (after Reddy, 1994)
Mature plants are just over two metres in height

Fig. 3: Panicle of Sorghum (after Randhawa, 1980)

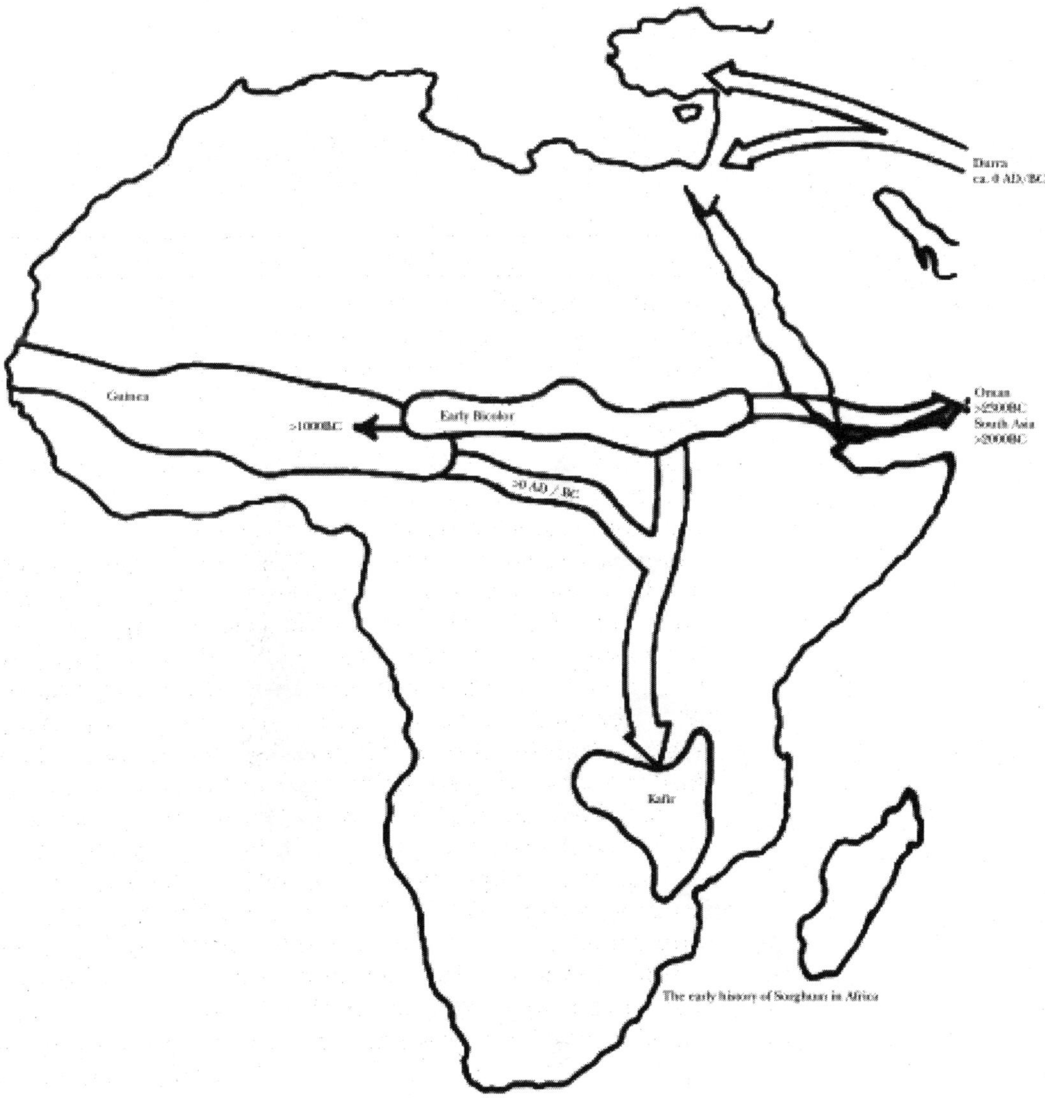

Fig. 4: The evolution and diffusion of Sorghum (after Poesshl, 1999)

Fig. 5: The distribution of pearl millets. Circles indicate the location of wild millet, Pennisetum violaccum, dark shading is the northern pearl millet belt; light shading, areas (after Harlan, 1994)

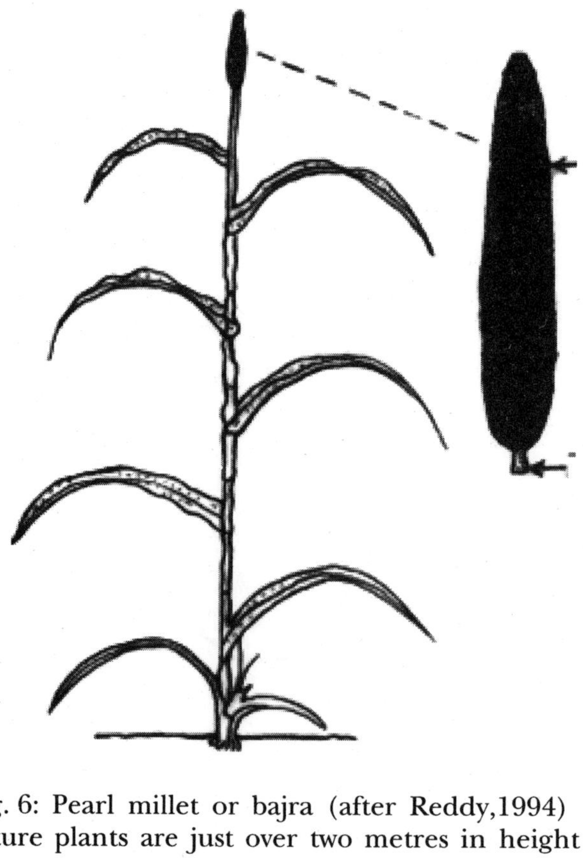

Fig. 6: Pearl millet or bajra (after Reddy,1994)
Mature plants are just over two metres in height

millet reached India from this region around 1500 BC (Doggett, 1989: 45) and trade with Ethiopia and Egypt moved by caravan through Somalia and across the straits to Aden. Besides, there is evidence of sea trade between Kulli culture of south Baluchistan and Early Dynastic Sumer soon after 2800 BC.

Pearl millet is another African cereal to have been brought to India. It was developed in the African savanna to the south of the Sahara. Only race typhoides spread to India in early times (Doggett, 1989: 39). In this context, it is worthwhile to mention the early occurrence of at least two types of millets—foxtail millet (*Setaria italica*) and broomcorn millet (*Panicum miliaceum*)—in the Neolithic culture of the Yellow River in China (An Zhimin, 1989: 645). Foxtail millet has been found at more than 20 sites, often stored in pits or pottery vessels and it was the main cereal food along with rice. An Zhimin observes that at least 8000 years ago, foxtail millet began to be widely cultivated in the Peiligang and Cishan cultures and it was the main crop in North China from prehistoric times to the historical period (An Zhimin, 1989: 646). Broomcorn millet has been recorded from two sites belonging to the Dadiwan and Yangshao cultures, respectively. Its first appearance and geographical distribution corresponds to that of foxtail millet, and it was cultivated in parallel with the latter (An Zhimin, 1989: 647). All these evidences prove that China is one of the centres of origin of agriculture in the world. However, at present, we do not have any evidence to show any connection with Chinese Neolithic and the Indian cultures of the corresponding period.

Based on limited archaeological data, as well as linguistic information and comparisons of morphological and genetic features of modern samples of millets, a number of hypotheses have been developed to explain where these millets came from, the time at which they appeared, and their influence on the cultural development of South Asian populations. Three possible centres have been suggested (Weber, 1991: 171–72):

1. African centre of origin: *Sorghum*, finger millet (*ragi*) and pearl millet.
2. East Asian centre of origin *Penicum, Setaria.*
3. South Asian centre of origin: *Sorghum, ragi,* kodon millet and *Penicum.*

From the above account, it is clear that there is serious divergence of opinion among scholars on this issue. However, there seems to be a consensus that *sorghum*, finger millet and pearl millet are all of African origin and were probably brought to India by the third millennium BC. It has been suggested that these crops were moving within an interactional sphere involving the Indus Valley and the Persian Gulf, and that the arrival of African millets coincided with an increase in sedentism and agriculture in peninsular India (Possehl, 1986: 237–250).

REFERENCES

An Zhimin, 1989, Prehistoric Agriculture in China, in *Foraging and Farming* (ed.) David R. Harris and Gordon C. Hillman, London, Unwin Hyman, 643–49.

Chakrabarti, D.K., 1988, *Theoretical Issues in Indian Archaeology*, Delhi, Munshiram Manoharlal.

Dhavalikar, M.K., 1995, *Cultural Imperialism: Indus Civilization in Western India*, New Delhi, Books & Books.

Dhavalikar, M.K., 1997, *Indian Protohistory*, New Delhi, Books & Books.

Dogett, Hugh, 1989, A Suggested History of the Crops Common to Ethiopia and India, *Late Prehistory of the Nile Basin and the Sahara* (eds) Lech Krzyzaniak and Michail Kobusiewicz, Poznan, 27–47.

Gode, P.K. 1961, Studies in the History of Indian Plants: Antiquity of *Jowar* or Jondhala, *Studies in Indian Cultural History*, Vol. I, Vishveshvaranand Vedic Research Institute: 266–282.

Haaland, Randi, 1992, Fish, Pots and Grains: Early and Mid-Holocene Adaptations in the Central Sudan, *The African Review*, 10: 43–64.

Harlan, Jack, R. 1971, Agricultural Origins: Centres and Noncentres, *Science*, 174 No, 4008: 468–474.

Harlan, Jack, R. 1989, The Tropical African Cereals, in *Foraging and Farming* (eds) David R. Harris and Gordon C. Hillman, London, Unwin Hyman: 335–343.

IAR, *Indian Archaeology—A Review*, Archaeological Survey of India, New Delhi.

ICAR, 1980, *A Handbook of Indian Agriculture*, New Delhi.

Lal, B.B. 1997, *The Earliest Civilization of South Asia*, New Delhi, Aryan Books International.

Mark Kenoyer, J. 1998, *Ancient Cities of the Indus Valley Civilization*, Karachi, Oxford University Press.

Magid, Anwar A. 1989 Exploitation of plants in the eastern Sahel (Sudan), 5000–2000 BC, *Late Prehistory of the Nile Basin and Sahara* (eds) Lech Krzyzaniak and Michael Kobusiewicz, 459–468.

Possehl, G.L., 1986. African Millets in South Asian Prehistory. In: J. Jacobson (ed.) *Studies in the Archaeology of India and Pakistan*, New Delhi, Oxford & IBH Publishing Co. 237–256.

Randhawa, M.S. 1980, *A History of Agriculture in India*, Vol. I, New Delhi, Indian Council of Agricultural Research.

Randhawa, M.S. 1982, *A History of Agriculture in India*, Vol. II, New Delhi, Indian Council of Agricultural Research.

Rao, K.E., J.M.J. de Wet, B.E. Brink and M.H. Mengesha. 1987, Infraspecific Variation and Systematics of Cultivated *Setaria italica*, Foxtail Millet. *Economic Botany* 41(1): 108-116.

Rissman, Paul, C. and Y.M. Chitalwala, 1990, *Harappan Civilization and Oriyo Timbo*, New Delhi, Oxford & IBH Publishing Co.

Sankalia, H.D., S.B. Deo and Z.D. Ansari, 1971, *Chalcolithic Navdatoli*, Pune-Baroda.

Saraswat, K.S., N.K. Sharma and D.C. Saini, 1994, Plant Economy at Ancient Narhan, in Singh P., *Excavations at Narhan 1984–89*, New Delhi, B.R. Publishing Corporation, 255–346.

Saraswat, K.S., 2004, 'Plant Economy of Early Farming Communities at Senuwar' in B.P. Singh (ed), *Early Farming Communities of the Kaimur (Excavations at Senuwar)*, Jaipur, Publication Scheme.

Sidebotham, Steven, E. 1986, *Roman Economic Policy in the Erytrha Thalossa,* Leiden, E.J. Brill.

Singh, P., 1992, The Origin and Dispersal of Millet Cultivation in India, Paper read in the Conference on Inter Regional Contacts in Later Prehistory of North Eastern Africa, held at Poznan (Poland).

Ucko, Peter J. 1989, Foreword in *Foraging and Farming* (eds) David R. Harris and Gordon C. Hillman, London, Unwin Hyman IX–XVIII.

Vishnu Mittre and H.P. Gupta, 1968, Ancient Plant Economy at Paunar, Maharashtra, in S.B. Deo and M.K. Dhavalikar, *Paunar Excavations 1967,* Nagpur, University Press: 128–130.

Vishnu Mittre, 1971, 'Ancient Plant Economy at Halhar' in *Protohistoric Cultures of the Tungabhadra Valley* (ed) M.S. Nagaraja Rao, Dharwar, Karnataka University, 125–133.

Vishnu Mittre, 1988, Forty Years of Archaeobotanical Research in South Asia, Presidential Address read at the XVIth Annual Conference of the Indian Archaeological Society held at Visvabharati, Santiniketan, West Bengal, December 1–3, 1988.

Vishnu Mittre and R. Savitri, 1982, Food Economy of the Harappans, in *Harappan Civilization,* G.L. Possehl (ed), New Delhi, Oxford and India Book House, 205–229.

Vishnu Mittre, Aruna Sharma and Chanchala, 1986, Ancient Plant Economy at Daimabad, in *Daimabad 1976–79* by S.A. Sali (ed.), New Delhi, Archaeological Survey of India: 588–627.

Weber, Steven A. 1991. *Plants and Harappan Subsistence: An Example of Stability and Change from Rojdi,* New Delhi, Oxford & IBH Publishing Co.

Wendorf, Fred, Angela E. Close *et. al.* 1992, Saharan Exploitation of Plants 8000 years B.P. *Nature* 359, 22 October 1992: 721–723.

CHAPTER 9

The Origin and History of Barley in India

B.D. Sharma

Barley is one of ancient crops in India. It was widely consumed as human food and horsefeed in early times, particularly by Aryans. Even today, barley is used in religious programmes such as in *havan samagri* as an important ingredient. In higher elevations (above 3000 m), this cereal is still used as a food crop, especially in Kinnaur, Lahual-Spiti, Uttaranchal, and Ladakh region. It is a staple food crop in Tibet and higher elevations of Nepal and Bhutan. Barley thrives well in temperate regions and can be grown on soils unsuited for wheat.

The barley, *Hordeum vulgare* (Linn.) Emend, is a cultivated species and is either six rowed or four rowed. The other cultivated species are *H. distichon, H. intermedium* and *H. deficiens.* The barley may be hulled hull (naked or pearl barley). Barley has chromosome number 2n=14 and genome formula VV. The wild parent of barley is *Hordeum spontaneum* C. Koch (2n=14) (Zeven & Zhukovsky, 1975). It occurs in eastern Mediterranean region, Turkmenia and Afghanistan. *H. spontaneum* is two rowed and has a brittle rachis. In 1886, De Candolle suggested that evidence for the origins of cultivated plants should be sought in tradition, religion, language, archaeology and the present distribution of related wild species. Many ancient religions taught that barley was the first crop and that is held a special place in the Eleusinian mysteries. It was the cereal gift of the Goddess Ceres or to the Egyptians, the gift of Isis. The Egyptians believe that the germination of barley is symbolic to the resurrection of Osiris. Barley was sown and germinated in Nile mud held in pottery trays shaped in the form of the god. The germinating grain symbolized 'Osiris' return to life.

The wild barley, *Hordeum spontaneum* Koch, is concentrated in an arc in the Middle East with scattered stands on a wide area spread between Tunisia to Afghanistan. In the arc, it occurs wild in undisturbed as well as disturbed habitats. It also occurs as a common weed in the crops of wheat and barley. Outside this main area (Map 1) it occurs only in vegetal habitats and as a weed in crops. There is high probability that this 'wild' barley occupied at least the 'high density' zone at the time of beginning of agriculture about 10,000 BC. *H. spontaneum* is two rowed and has brittle rachis. Its domestication probably took place in the 'Fertile Crescent' resulting in two-rowed type (ssp, *distichon* syn. *H. distichon* L.). The start of its domestication might have been about 7000 BC. Helbeck (1966) suggested that the barley found in Beidha dating C. 7000 BC was a cultivated species (Zeven & Zhukovsky, 1975). Braid wood *et al.* (1967) found domesticated barley at Cayonu, Trukey (dated back C. 7000 BC). Domesticated barley has a tough rachis and the change of brittle to tough rachis may have occurred between 7000 to 6000 BC. The tough rachis is controlled by two closely linked recessive alleles bt1 and bt2. Hull less or naked barley was first identified at Beidha, Aceramic Hacilar and Ali Kosh C. 7000 BC.

Archaeological evidence indicates that the wild barley and wild *emmer einkorn* wheat's were growing in the ancient past in the 'fertile crescent'. About 9000 BC, the 'Natufians', who had a Mesolithic culture, gathered and used wild grains. Agriculture in which planting and harvesting of grains took place, was probably occurring in and around 8000 BC, since by 7000 BC small villages were in existence throughout the 'fertile crescent' from southwest Iran, Taurus and Zagros mountains of Iraq, Turkey, Anatolia and Palestine.

About 5000 BC remains of excavations in Egypt show the presence of a mixture of two-rowed distichon and deficiens forms, irregulare, and six-rowed lax and dense-eared, husked and naked forms. In Mesopotamia, by 5000 BC, six-rowed, lax-eared husked barley was being grown and, by 2000 BC, this comprised some 90% of the cereal crop. According to Briggs (1978), six-rowed barley which was being grown in the Indus Valley at Mohenjo-daro and Harappa by about 2500 BC, was almost certainly introduced from Mesopotamia. The crop was introduced in China by about 2000 BC to 1300 BC From China, it spread to Japan. In India, two-rowed barleys were not grown until 1874 or Japan until 1868.

By 3000 BC, the lake dwellers of Switzerland were growing husked and naked six-rowed barleys and these had spread to North Europe by 2500 BC. Two-rowed barley was not introduced to Northern Europe until Medieval times.

There are two barley, 'Centres of diversity', namely: (i) the area including Abyssinia and Eritrea; and (ii) the far east from Nepal, through China to Japan. In Euthopia, highly varied mixtures are found growing, that contain two-rowed deficiens and irregular forms, six-rowed husked and naked types, grains having range of colours, including black, etc. In Bhutan three rowed barley had also been reported to occur.

Briggs (1978) opined that the diversity in the far eastern centre in six-rowed varieties, hooded forms with naked grains and intermedium forms may have arisen after an early introduction from Mesopotamia and subsequent genetic isolation and a gradual spread into a range of environments from the Indus Valley into the high Himalayan mountains, the Tibetan plateau and into China and Japan. There, these were subjected to stringent selection by the climate and farmers, combined with the use of a major portion of the grains for human food.

The species diversity in genus Hordeum may be divided into two major groups: (i) cultivated; and (ii) wild. These are further divisible into: (a) two-rowed; and (b) six-rowed forms. The two-rowed species include *H. zeocriton* (dense eared), *H. erectum* (medium dense), *H. distichum* (lax) and six-rowed species *H. hexastichon* (dense), *H. parrallelum* (medium dense), *H. vulgare* (lax), *H. intermedium, H. irregulare.* The wild barley species include *H. spontaneum, H. agriocrithon, H. pusillum, H. paradoxon, H. lagunculiforme, H. murinum, H. bulbosum, H. jubatum, H, nodosum.* The maximum use of barley grains at present is in malt industries, and in the form of feed.

EARLY HISTORY OF BARLEY CROP IN INDIA

As has been seen in earlier paragraphs, the barley crop has not originated in India but has been introduced from Mesopotamia or elsewhere in the antiquity. Mittre (1974) has argued that barley was used throughout the Harappan period from 2300 BC to 1750 BC.

The earliest available data of Mehargarh site (northern kachhi plains) dated back to 5000 BC to 4000 BC in period II and III provide evidence of barley cultivation and barley remains belong to two-row and six-row types (Asthana, 1985). The material from Harappa is referred to *H. vulgare* var. *hexastichum* (Vats, 1940). The Neolithic, Chalcolithic and Harappan cultures make up a pattern that is basic to south Asian prehistory. There had been distinct groups of people in a common region possibly interacting, yet maintaining their different socio-economic living such as farmers, hunter-gatherers and pastoralists. The remains of fully domesticated wheat and barley from the early Neolithic levels of Mehrgarh are believed to be contemporary with early west Asian Neolithic sites in the Deh Luran plains of Iran and in Syria (Meklean, 1983: 77). These finds the earliest of wheat and barley in the subcontinent to date imply that cultivation of these crops occurred at least as early as the 7000 BC. The cultivation of wheat and barley had been widely spread in village farming communities of Gujarat by the mature (Harappan) phase between 2500 and 2000 BC, (Possehl and Raval, 1989) as quoted by Weber in his book 'Plants and Harappan Subsistence'. He had further stated that on all the Harappan and Harappan-like sites, six crops such as wheat, barley, oats, date, grape and jujube were commonly cultivated in the early Harappan phase before 2600 BC. The barley taxa were identified by Weber from the Kojdi site as *Hordeum distichum* and *H. vulgare.* A sizeable quantity of barley grains was found at Kalibangan and is dated to 2090 to 2075 BC (Vishnu Mittre and Savithri, 1982; Agarwal and Kussumgar, 1968 a, b). There is reason to believe that barley was used throughout the Harappan period from 2300 BC to 1750 BC and belonged to *Hordeum vulgare* L. and *Hordeum spontaeum* Koch. (Vishnu Mittre 1974; Vishnu Mittre and Savithri, 1982).

The *Susruta Samhita* has made mention of *yava* (Su, 14, 29; 16, 18; 17, 31). Lal (1976) has mentioned in his book *India: Cradle of Cultures* that the relics of a pre-Harappan culture have been found at Rangpur and Prabhas—Somnath as well as all over north and central Gujarat. Still earlier, there were far numerous nucleated village settlements of the late Stone Age scattered all over Gujarat, dating back in antiquity to about 6000 BC. The *yava* (barley) was widely grown during the period of Rishi Patanjali (Agnihotri, 1963). During the time *Mahābhārata* war took place, barley was cultivated throughout the country and it is dated to 1400 BC (*Mahābhārat Mimānsā,* 1990). The remains of barley have been discovered from the Gangetic plain, from Atranjikhera in U.P. dated to 2000–1500 BC (Choudhary, Saraswat, Hassan and Gaur, 1971) and from the Neolithic of Chirand, Bihar estimated to date from 2500–1800 BC or earlier (Vishnu

Mittre, 1971c, 1972). Varahamihira (2449 BC), in his *Brhatasamhita* in Kusumlata Adhayaya 29 pp. 188-190, states that heavy flowers on banyan trees indicate a heavy crop of barley in that year.

Chalcolithic, Iron Age and early historical records are known from Atranjikhera dated 1200–600 BC. (Buth and Chowdhary, 1971) and findings at Ter, Osmanabad, Maharashtra date to 155 BC to 160 AD (Vishnu Mittre *et al.*, 1972). In south India, Apegaon excavations in district Aurangabad (Maharashtra) at the confluence of a small rivulet Virbhadra with Godavari river, Chalcolithic habitations existed in 1600–1000 BC and were cultivating *Hordeum vulgare* L. along with other crops (Deo, 1976). Dhavalikar (1999) reported remains of barley from U.P., M.P. date to 600 BC and 400–500 AD Barley was an important crop in agriculture date to 200 BC to 700 AD (Auboyer, 1961). In 1985, Auboyer had given an account of the early settlement and subsistence pattern in the Deccan and had mentioned that the first farmers of Deccan cultivated a variety of grains including barley, wheat, *sorghum,* rice and bajra. The first settlers of Malwa culture (C. 1600–1400 BC) grew only barley as a cereal crop. The early Jorwe agriculture involved the cultivation of three cereals wheat, barley and *sorghum.* The jain canouns (400 BC–500 AD) had mentioned field crops, which, among others, included barley as one of the important crops.

THE BARLEY CROP IN INDIAN AGRICULTURE

Barley is widely grown in India even today. In area, it comes next to wheat as a winter cereal and occupies about 2.4 million hectares producing nearly 3.0 million tonnes of grain annually. Barley is the best suited for areas where winter rains are deficient or where soils are alkaline. The main barley growing regions is in Bihar and Madhya Pradesh. In plains and low hills, only the hulled barley is grown commercially. At higher altitudes, especially in cold arid region, hull less six-rowed and two-rowed varieties are grown.

The bulk of barley grain in India is used as human food. It is consumed either as flour mixed with wheat flour for making chapatti or as parched grain ground to make the favourite item, '*Jau Sattu*'. This was very important prevalent in the days of *Mahābhārata* wartime and even today, it has a preferred food article in tribal and villages in certain areas. A small portion of produce is utilized in industry for malting and brewing. If a farmer has some surplus grain, it is used for livestock feeding.

As was the practice in Harappan time, barley replaces wheat only in areas which are either too dry or too saline to grow a satisfactory wheat crop. Hence barley breeding has not received much incentive in India and even barley-growing areas has not been supported with public distribution through which supply is only for the main cereals (wheat, rice). This is causing change in food habits of people who used to be traditionally barley consumers. There is another reason for this change, i.e. the barley replacing crop in these areas is remunerative and profit making. For example, in Lahaul-Spiti and Kinnaur (NAKO) barley hulled and hull-less was the main cereal food and was grown under irrigated conditions but with the introduction of pea (green pods), potato, etc., barley has been replaced. In general, in India, the barley crop is still confined to poor marginal lands and hence it is still a low-yielding crop. Due to these reasons, barley crop improvement work is still continuing and a series of varieties were released in Uttar Pradesh, Punjab, Bihar, Rajasthan and at the I.A.R.I., Pusa. The phase of improvement involved intervarietial hybridization mainly between the existing indigenous varieties. These varieties though had yield advantage over the local varieties but had a narrow range of adaptability and susceptibility to diseases and lodging.

BARLEY BREEDING IN INDIA

The modern crop improvement work was designed to meet the requirement of crop growing in four agro-climatic regions, the northern hills, the north western plains, the north eastern plains and the central plains. The All India Co-ordinated Barley Improvement Project is in operation, it tests the new strains in the four zones and if any strain shows its superiority in trials at different co-ordinating centres in a particular zone, it is released as a variety for commercial growing. Most important barley varieties developed by 1990 were Ranjit, DL 85, DL 88, DL 171, P103, RD 118, DL 165, P267, Sonu (HBL 87), Kailash, Himani, RD 103, RD 137, Rajkiran, DL 260 and K 257 (Gulati & Verma, 1988). The major emphasis on barley breeding in the country, is on the following aspects:

Lodging Resistance: This is a major drawback in the Indian varieties and a serious limiting factor in increasing barley production. Development of dwarf or semi-dwarf barley varieties with good straw strength, responsive to higher fertility soils is essential to increase barley productivity. This type of barley varieties are also in cultivation in other parts of the world. Mutations causing dwarfing have been induced in India and Sweden. In the Indian breeding programme, five dwarf-induced mutants are in use. Among these, two mutants BM 10 and BM 12 (plant height 71 to 74 cm against normal above one metre) appear to be the most promising. During 1969–70 eighty promising barley varieties from Australia, Canada, USA, Mexico, Sweden, Hungary, Denmark and Japan were introduced. The most promising among these are being employed in breeding work.

Disease Resistance: The important diseases of barley in India are stripe rust, Heliminthosporium and smuts. Aphids are a serious pest. Twelve races of stripe rust (*Puccinia striformis* West) have been recorded in India. Of these races, viz. G, A, 57, 31 and 24 do much damage to the crop. Resistance sources are EB 410, EB 145, EB 1626 (West Germany), EB 1556 (Ethiopia) and Heitpas (USA), the last two are six-row varieties.

A two-row variety from Taiwan, EB 921, is immune to aphids (Rhopalosiphon maidis Fitch.) and EB 2434, EB 2436, EB 2457 are highly resistant introductions from Balkans.

Salt Tolerance: Ratna and R 56 varieties have a high degree of tolerance to saline and alkaline conditions. Ratna was among the top-yielding varieties.

Nutritional and malting qualities: Two-row barley are usually preferred for malting, whereas the food barley varieties in India are all six-rowed. For malting, grain with high starch and low protein is required, whereas for food, high protein content in barley is needed. The quality germplasm from countries like Australia may be instrumental to the improving Indian barley.

REFERENCES

Agnihotri, P. 1963. *Pātanjalikāleen Bhārat,* pp. 207–231, Bihar Rashtrabhasha Parishad, Patna.

Agnipurana (600–900 AD) Presentation by Rakesh Sastri, 2002. Sadhana Pocket Books, Delhi.

Allchin, F.R. 1995. *The Archaeology of Early Historic South Asia,* pp. 75–97. Cambridge University Press, New York.

Asthana, Shashi. 1985. *Pre-Harappan Cultures of India and Borderlands.* Books & Books, New Delhi.

Auboyer, J. 1961. *Daily Life in Ancient India,* pp. 62–113. Asia Publishing House, London.

Briggs, D.E. 1978. *Barley.* pp. 76–88. Chapmans Hall, John Wiley & Sons, New York.

Choudhary, S.L., G.S. Sharma and Y.L. Nene. 2000. *Ancient and Medieval History of Indian Agriculture,* Rajasthan College of Agriculture, Udaipur.

Das, A.C. 1925. *Ṛgvedic Culture.* R. Cambray & Co., Calcutta.

Dhavalikar, M.K. 1999. *Historical Archaeology of India.* Books & Books, New Delhi.

Dwivedi, B. and Dwivedi, K. 2001. *Vedon Mein Ayurveda,* Viswabharati Anusandhan Parishad, Gyanpur.

Gulati, S.C. and N.S. Verma. 1988. 'Genetic resources in barley, their diversity and utilization' In R.S. Paroda, R.K. Arora and K.P.S. Chandel (eds.) *Plant Genetic Resources—Indian Perspective,* pp. 134-149. National Bureau of Plant Genetic Resources, New Delhi.

Hutchinson, J. (ed.). 1974. *Evolutionary Studies in World Crops,* pp. 18–19 & 47–52 Cambridge University Press, London.

Jha, Achutanand. 1997. Varaha Mihir *Virachita Brhata Samhita,* pp. 188–190. Chaukhamba Vidya Bhavan, Varanasi.

Mishra, V.N. and Peter Bellwood (eds.) 1985. *Recent Advances in Indo-Pacific Prehistory.* Oxford & IBH Publishing Co. Pvt. Ltd., New Delhi.

Matsya Purana (200–400 AD) 1985. Part I & II Adhayaya 217, pp. 35–38. Gita Press. Gorakhpur.

Randhawa, M.S. 1986. *History of Agriculture in India.* Vol. I. ICAR, New Delhi.

Vaidya, C.V. 1990. *Mahabharata Mimansa,* pp. 258–259. Haryana Sahitya Academy, Chandigarh.

Raychaudhuri, S.P. 1953. *Agricultural Practices in Ancient India.* Indian Council of Agricultural Research, New Delhi.

Vishnu Mittre and R. Savithri. 1982. 'Food Economy of the Harappans' In G.L. Possehl (ed.) *Harappan Civilization,* pp. 205-221. Oxford & IBH Publishing Co. Pvt. Ltd., New Delhi.

Weber, Steven. 1991. *Plant and Harappan Subsistence—An Example of Stability and Change from Rojdi.* Oxford & IBH Publ. Co. Pvt. Ltd., New Delhi and American Institute of Indian Studies.

Zeven, A.C. and P.M. Zhukovsky. 1975. *Dictionary of Cultivated Plants and their Centres of Diversity.* Centre of Agricultural Publishing and Documentation, Wagemingen.

CHAPTER 10

The Origin and History of Wheat in Indian Agriculture

B.D. Sharma

It is remarkable that the word *Godhuma* (wheat) is not mentioned in the *Ṛgveda.* Probably the grain was not known nor cultivated in *Ṛgvedic* times, and was indigenous to the Punjab (*sapta sindhu*). It must have been introduced from elsewhere in a later age, as it is mentioned in the *Yajurveda samhita* (*Maitriya samhita* 1, 2, 8; *Vaj.samhita* 18, 12, 19, 22, 89, etc.) and the Brahmanas (sat bra 12, 7, 1, 2, 2, 9 etc.). The absence of any mention of wheat (*godhuma*) in the *Ṛgveda* has historical significance. "The centre of diffusion of wheat," says H.G. Wells in his *Outline of History* was somewhere in the eastern Mediterranean region. A wild form is still found in the neighbourhood of Mount Hermon. When the lake dwellers sowed their little patches of wheat in Switzerland they were already following an immemorial practice of mankind. The seed must have been brought age by age from that distant centre of diffusion. In the ancestral lands of the southeast, man had already been sowing wheat perhaps for thousands of years. The lake dwellings were probably occupied continuously from 5000 or 4000 BC down almost to historic times (pp. 56–57). He further elaborates "De Candole asserts that it is only in the Euphrates-Tigris district that wheat has ever been found growing wild".

It may be that from Mesopotamia as a centre, the cultivation of wheat spread over entire eastern hemisphere. It is also possible that the growing of wheat had spread from the Atlantic to the Pacific with the distribution of the Neolithic culture by perhaps 10,000 to 9000 BC before the beginning of civilization. It is most likely that Indian merchants brought the knowledge of the cultivation of wheat to India from Mesopotamia about 10,000 BC or even later, which undoubtedly corresponded to the later *Ṛgvedic* or post *Ṛgvedic* age, as the grain has been mentioned in the *Yajurveda Samhitas* and the Brahmanas. Chaman Lal (1976), in his "India: Cradle of Cultures" revealed the pre-Harappan cultures have been found at Rangpur and Prabhas Somnath as well as all north and central Gujarat. Still earlier, there were far numerous nucleated village settlements of the late middle stone age scattered all over Gujarat dating back in antiquity to about 6000 BC.

Ever since man's first success some 10,000 years ago to producing food in Southwestern Asia, the history of cultivated wheat and that of human civilization have been closely interwoven. In the course of domestication the wheat plant has lost the ability to disseminate its seeds effectively and is now completely dependent on man for dispersal. But man has fostered this cereal to such an extent that it is now the world's famous crop plant (Feldman, 1976). Wheat has a complex history of its origin. The modern bread wheat (*Triticum aestivum var. aestivum*) is *hexaploid* and its varieties are widely cultivated all the world over. It is highly valuable for bread making because of the high gluten content necessary for dough making. Durum wheat, *Triticum turjidum* var. durum is a modern tetraploid type and is mainly grown in rainfed drier regions. Today, there are no economically important diploid wheats, Schulz (1913). A German had classified the known wheat on morphological Grounds alone, into three main groups (Einkorn, Emmer and Dinkel). These groups were as follows:

I.	Einkorn series	*T. aegilopoides (T. boeoticum)*
		T. monococcum
II.	Emmer series	*T. dicoccoides*
		T. dicoccum
		T. durum
		T. turgidum
		T. polonicum
III.	Dinkel series	*T. Spelta*
		T. compactum
		T vulgare

T. capitatum (a cross between *T. compactum* and *T. vulgare*) has been excluded and this grouping remains unchanged. Now it is known with certainty that the cultivated Einkorn (*T. boeoticum*) had descended directly from the wild Einkorn (*T. boeoticum*) and that the einkorns were ancestors of all present-day wheat species. The pioneering cytogenetic work of Sakamura, Sax, Kihara and others (Riley, 1965); Morris and Sears (1967) have shown that the different species of *Triticum* had evolved through the process of polyploidy, starting with a basic x=7 number and consisting of three different ploidy levels, diploids (2n=2x=14), tetraploids (2n=4x=28), and hexaploids (2n=6x=42).

The tribe Triticeae (=Hordeae) is economically the more important group of the family Poaceae. It has contributed food species such as wheat, barley and rye. The related genera are *Aegilips, Agropyron, Eremopyron* and Haynaldia. These may hybridize among themselves and, consequently, may have direct exchange of genetic material or *amphidiploidy*.

Origin of Bread and other Wheats: The bread or common wheat, *Triticum aestivum* (L) Thell. (2n=42; genome AABBDD) has its primary centre of origin in transcaucasia and adjacent areas. There natural cross pollination within the species, between species and even with related species like *Aegilops* and *Secale* is still taking place. In the Fertile Crescent the wild einkorn, *Triticum boeoticum* (Boiss) from which the

Table 1
Classification of cultivated wheat and closely related wild species

Species	*Genomes*	*Wild*	*Cultivated*	
		Hulled	*Hulled*	*Free threshing*
I. Diploid (2n=14) species				
T. boeoticum	AA	”	-	(Wild einkorn)
T. monococcum	AA	”	-	(cult einkorn)
Aegilops speltoides	BB	”	-	-
Aegilops bicornis	BB	”	-	-
Aegilops longissimum	BB	”	-	-
Aegilops caudata	CC	”	-	-
Aegilops squarrosa	DD	”	-	-
II. Tetraploid (2n=28) species				
T. dicoccoides (Wild Emmer)	AABB	-		
T. dicoccum (Emmer)	AABB	-		
T. durum (Durum)	AABB	-		
T. carthlicum (Persian wheat)	AABB	-		
T. polonicum (Polish wheat)	AABB	-		
T. turgidum (Solid stem wheat)	AABB	-		
T. timopheevi (Timopheevi)	AABB	Var araraticum	Var timopheevi	
Aegilops cylindrica (goat grass)	CCDD			
III. Hexaploid (2n=42) species				
T. compactum (Club wheat)	AABBDD			
T. Spelta (Spelt wheat)	AABBDD			
T. aestivum (Common wheat)	AABBDD			
T. zhukovski (Zanduri)	AABBDD			

cultivated einkorn *T. monococcum* had evolved and was domesticated dating c 6500 BC and had a tough rachis. Einkorn spread over Europe, North Africa, Asia minor, Caucasus, Iraq and Iran and is still cultivated in some areas as a fodder crop.

The cultivated emmer wheat have 2n=28 and genome formula AABB. These tetraploid wheats can be subdivided as follows (MacKey, 1966).

Wild emmer ssp *dicoccum* (Schrank) Thell. Syn, *T. dicoccum* Schull, English wheat ssp *turgidum.* Syn, *T. durum* Desf (cultivated), Durum hard wheat ssp *durum* (derf) Mackey Syn *T. poloniculm,* Persian ssp *carthlicum* (novske) Mac Kay Syn. *T. carthlicum* Novske.

These species have originated as a result of hybridization and chromosome doubling. The hybridization may have involved *T. boeoticum* having *genome* AA and a diploid wild grass, probably Aegilops spetloides. This gave rise to a new wild species having *genomes* AABB, which we know today as *T. dicoccoides.* This natural synthesis of

T. dicoccoides probably took place a long time before man began to domesticate the wild wheats. The domestication of *T. dicoccoides* took place in southwestern Asia about the beginning of the seventh millennium BC if not earlier through spontaneous mutation in *T. dicoccoides*, natural crossing of the diverse mutant forms, and natural selection under the pressure of farming methods or purposeful selection by the farmers, the cultivated species now known as *T. dicoccum* was evolved. The cultivated emmer became widespread and there arose far greater opportunities for mutation, natural crossing, and natural or manmade selection to take place then it was possible for the wild (uncultivated) *T. dicoccoides.* The remaining *tetroploides* may, therefore, have evolved from this singe species (*T. dicoccum*), since their morphological differences are only due to a single major gene and possibly with some minor ones. In this group of tetraploid wheats, durum wheat is cultivated over a large area in the Mediterranean coastal region, Ethiopia (a secondary centre), areas north of Black Sea, India, SSR, Syria, Turkey, and Iraq but rarely seen in Iran and Afghanistan. It is the second most important wheat in the world.

The most important cultivated bread wheat is hexaploid (2n=42) with the genomic constitution AABBDD. The scientific investigations made so far indicate that their 'A' genome was probably contributed by *T. boceticum,* the 'B' genome probably derived from a related grass Aegilops speltoides and the 'D' *genome* probably contributed by a wild grass, *Aegilops squarrosa* (=*T. tauschii*) (Peterson, 1965). A single gene mutation gave rise to *T. aestivum* from *T. spelta* and both *T. compactum* and *T. sphaerococcum* arose from *T. aestivum.* There are the far known facts of origin of wheat gathered from several investigations done mainly through the work of Kihara and his associates in Japan (1944), Mcfedden and Sears in U.S.A. (1944–1946) and Riley (1965) in U.K. The Indian dwarf wheat (*T. aestivum* (L) Thell. ssp. *Sphaerococcum* (Perc) MK. syn. *T. Sphaerococcum* Perc.) is indigenous to northwest India and adjacent Afghanistan. It is characterized by short-statured non-lodging culms, erect leaves, globular grains and susceptibility to diseases. Club wheat (2n=42 and constitution genome AABBDD) *T. aestivum* (L) Thell. ssp. Compactum (Host) MK. syn. *T. compactum* Host. has developed in the mountains of the Hindukush. Vavilov—a Russian explorer, regarded Ethiopia as an independent centre of origin of many of the world's cultivated plants, including sub-species of *T. durum, T. turgidum, T. dicoccum* and *T. polonicum.* However other studies have indicated Ethiopia to be a secondary centre for wheat.

Geographical Distribution of Wild wheat species and their role in Evolution of Bread Wheat: The cytotaxonomic work on wheat and its wild related species by Sakamura, Sax, Kihara, Riley, Morris and Sears have shown that wheat had originated through hybridization in nature and subsequent polyploidization. Each polyploid species can be identified as a product of hybridization followed by chromosome doubling. Since the different genomes are closely related (Morris and Sears, 1967), polyploid wheats are segmentel allopolyploid rather than a true genomic *allopolyploid,* as was speculated earlier. The diploid like behaviour of polyploid wheat is due to suppression of pairing of homoeologus chromosomes (i.e. related *chromosomes* of different genomes) by a specific gene. In hexaploid, *T. aestivum,* this gene is located on the long arm of chromosome 5 of genome B and known as the 5BL (Riley, 1965). To confirm the wheat evolutionary trend, scientists have attempted artificial synthetic polyploid wheat which do not contain this gene, and are partially sterile.

In wheat evolution, three groups of polyploids are recognized (Zohary and Feldman, 1962). Species in each group have one *genome* in common and differ in their other genomes. Polyploids of group A-genome share the genome of diploid wheat, *T. monococcum*: those of group D share the genome of *T. taushii (=Ae. squarrosa)*; and those of group C share the *genome* of *T. umbellulatum* (=*Ae. umbellulata*). The cultivated polyploid wheats (Fig. 1) belong to group A derived from its wild diploid. *T. monococcum* var. *boeoticum* is involved in all the wild polyploids since it has a brittle ear, while the non-brittle ear of cultivated, *T. monococcum* var monococcum is round in all the cultivated polyploids. Group A polyploids comprise the tetraploids, *T. turgidum* (AABB), *T. timopheevi* (AAGG), and the hexaploids, *T. aestivum* (AABBDD).

Tremendous variability occurs in hexaploid wheat which probably is found as a result of recurrent numerous hybridizations involving different genotypes of tetraploid varieties and various races of *T. tauschii* (donor of D genome) which grows even today within and at the edges of wheat fields in Iran and Armenia. So far conclusive identification of donor of B genome has not been possible. However, recent cytogenetic data (Kimber, 1974) indicate that *T. speltoides* may have donated the G genome of *T. timopheevii* rather than the B of *T. turgidum.* The studies further have shown that the initial amphidiploids AASS could have exchanged *chromosome* segments with other *amphidiploids* or with *diploids* such as *T. longissimum* and *T. bicorne* or others giving rise to G and B *genome.* To provide further support to the theory of origin of wheats it will be of interest to examine the geographical distributions of wild species. *T. monococcum* var. *boeoticum* and *T. tauschii* (=*Ae. squarrosa auct.* Non 2.8 overlap and the distribution of *T. turgidum* var *dicoccoides* (=*T. dicoccoides*) overlaps the region common to the distributions of these progenitors. *T. Speltoides* (*genomes*) is in contact with wild *T. monococcum* var. *boeoticum* in eastern Turkey, northwestern Iraq and western Iran. In mixed populations of two species wild *T. timepheevii* and more sporadically that of wild *turgidum* are also found. However, the main distribution of wild *T. turgidum* var. *dicoccoides* is southwest of this area and occurs in northwestern Israel, northwestern Jordan and southern Syria where it does not form mixed populations with *T. speltoides* but does grow with *T. longissimum.* The *polyploidy* of wheat has conferred upon it the distinct advantage of wide range of genetic flexibility and adaptation to a great variety of environments.

The historical account of wheat in India will remain incomplete unless centers of diversity are discussed in this context because this knowledge may aid in solving historical riddles. Below are given the geographical distribution areas or centres of diversity for wheat and its wild related species:

1. *T. aestivum* (L) Thell var *aestivum* (=*T. vulgare* Vill.) soft wheat at present occupy a great area. The centre of origin is in southwestern Asia.
2. *T. aestivum* (L) Thell. ssp. *spelta* (L) Thell. (=*T. spelta* L) spelt wheat is found in Asturias, southern mountains of Germany, Tyrol and possibly Northwestern Turkey.
3. *T. aestivum* (L) Thell. ssp *vavilovi* (Tum) Sears (=*T. vavilovi* (Tum) Jakubz.): Endemic in Turkish Armenia.
4. *T. aestivum* (L) Thell. ssp. *Macha* (Dek & Men.) MK. syn. *T. macha* (Dek & Men) macha wheat is endemic in western Georgia (Russia).

Fig. 1. Evolutionary Relationships of the Wheat

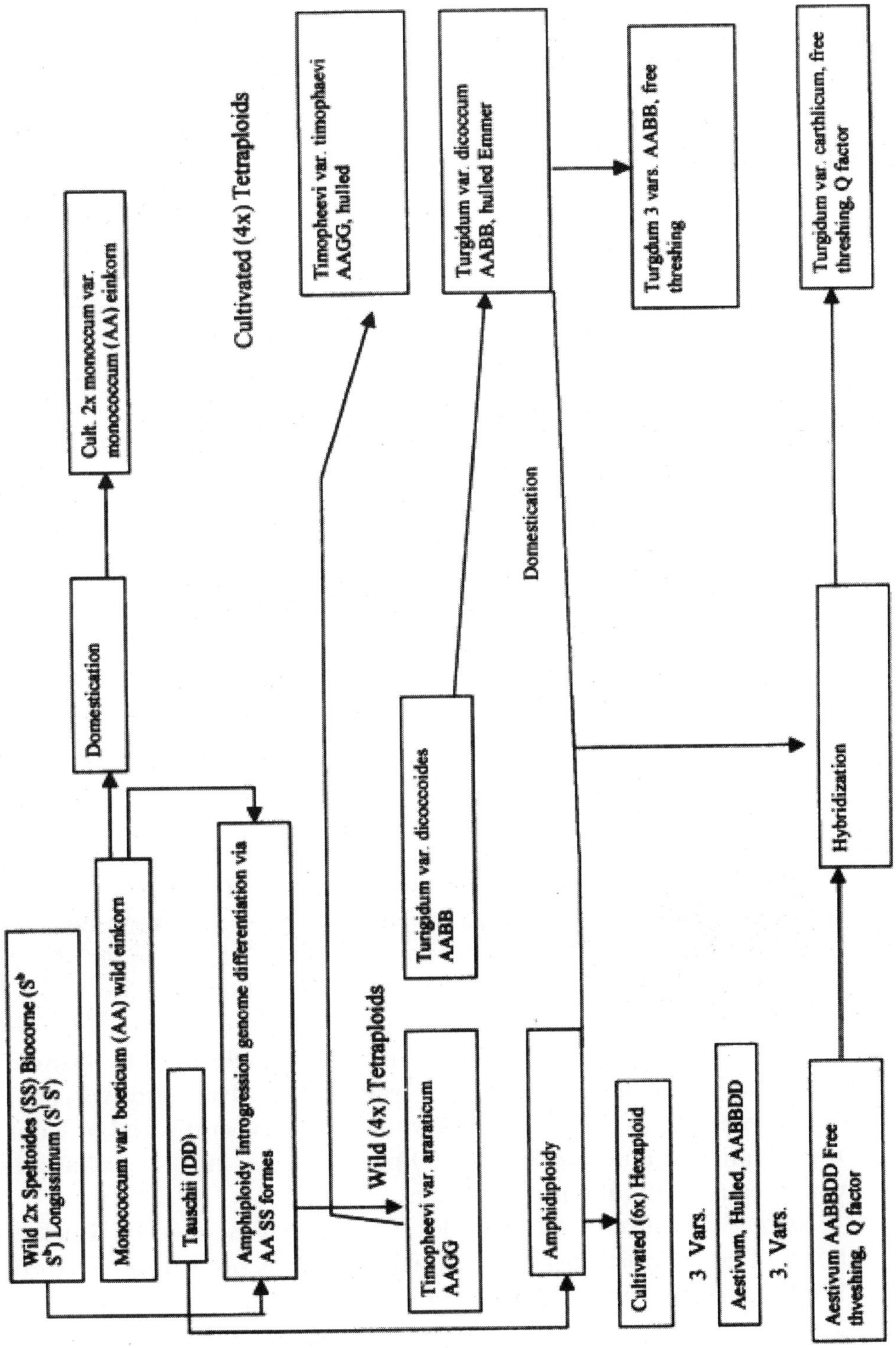

5. *T. aestivum* (L) Thell. ssp. Vulgare (Vill.) MK. (=*T. vulgare* Vill.) macha wheat or common wheat or bread wheats are especially cultivated in the Sahara Oases and Sudan zone of Africa.
6. *T. aestivum* (L.) Thell. ssp. *compactum* (Host) MK. (*T. compactum* Host) club wheat occurs in southwestern. Asia, specially in Afghanistan and Armenia (Hindukush) and probably in Switzerland, Austria.
7. *T. aestrivum* (L) Thell. ssp *sphaerococcum* (Perc) MK. (=*T. sphaerococcum* Perc.) Indian dwarf wheat originated in northwest India and adjacent regions.

(A) Tetroploid wheat species

8. *T. turgidum* (L) Thell. ssp *dicoccoides* (Korn.) wild *emmer* is derived from a natural amphidiploidization of an unknown diploid species and *T. monococcum* ssp. *Boeoticum.* The source of B genome may be *T. tauschii.* It occurs in southern America, northeastern Turkey, western Iran, Syria and northern Palestine.
9. *T. turgidum* (L.) Thell. ssp. *dicoccum* (Schrank) Thell. (=*T. dicoccum* Schubl.) cultivated emmer wheat is found distributed in Abyssinia, Yemen, India, mountainous region of western Europe, Georgia, Armenia, northern Iran and central Asia.
10. *T. turgidum* (L.) Thell. ssp *paleocolchicum* (Men) MK. (= *T. poleocolchicum* Men., *T. georgicum* Dek.) Georgian emmer Kolchic emmer wheat was formerly cultivated in a mixture with *T. aestivum* ssp. *Macha* in western Georgia (Russia).
11. *T. turgidum* (L.) Thell. var *turgidum* (=*T. turgidum* L.) poulard wheats, English wheat occur in Abyssinia, Eritrea, southern Europe (Portugal, Spain, Italy, Balkans) and in the regions along the shores of the Mediterranean Sea.
12. *T. turgidum* (L.) Thell. var. *durum* (desf.) MK. (*T. durum* Desf) durum wheat, hard wheat occurs in Abyssinia Eritrea, Yemen, Mediterranean region, southeast European part of USSR, western Siberia, northwestern region of USA, India and Asia Minor.
13. *T. turgidum* (L.) Thell. var. *polonicum* (L.) MK. (=*T. polonicum* L) Polish wheat is found in Abyssinia, and regions along the shores of the Mediterranean sea.
14. *T. turgidum* (L.) Thell. var. *carthlicum* (Nevski) MK. (=*T. carthlicum* nevski) Persian wheat was cultivated in Iraq, Iran, and Caucasia. This wheat carries Factor (present in hexaploid wheat for free threshing) and suggests its recent origin.
15. *T. timopheevii* Zunk. var *araraticum* and var. *timopheevii.* The farmers grows this variety in Transcaucasia, northern Iraq, western Iran and eastern Turkey. Var. *timopheevii* is apart of the zanduri wheat, which is cultivated in Georgia (SSSR). It is an ancient cultivated wheat (Diploid wheats, 2n=14). It has a brittle rachis which cause difficulty in threshing.
16. *T. speltoides* (Tausch.) Gren ex Richter (=*Aegilops speltoides* Tausch.) occurs in South Turkey, north Syria and north Iraq. It is often found together with the Wild. *T. monococcum* var. boeoticum.
17. *T. bicorne* Forsk (=*Aegilops bicornis* (Forsk.) Jaub & Sp. is found in xeric sandy soils of south Israel, lower Egypt and Cyrenacia.
18. *T. longissimum* is found in south Turkey, north Syria and north Iraq.

19. *T. tauschii* (Coss.) Schmalh. (=*Ae. squarrosa* Auct. non. L. *Ae tauschii* Cosson) occurs in eastern Turkey, Iraq, Crimea and Caucasus, Pakistan and Kashmir (India). Its primary centre is South Caspian area.
20. *T. monococcum* L. wild and cultivated einkorn. The wild var. boeoticum includes the types aegilopoides (*T. aegilopoides* (Link) Bal., *T. thaoudar* Reut., *T. spontaneum* Flaksb. and *T. urartu* Tuman.). It is spread over Greece, Turkey, Syria, north Iraq and Transcaucasus. var. *monococcum* has been domesticated in the Fertile Crescent from the wild einkorn and was spread over Europe, N. Africa, Asia minor, Caucasus, Iraq and Iran.

Early History in centres of Diversity: The earlier evidence of domesticated wheats C-14 date to approximately 7500 to 6500 BC, found in prehistoric sites of the 'Fertile Crescent' (comprises an arc of mountain chains flanking the plains of Mesopotamia and the Syrian Desert), and also in Anatolia and the Balkans. In ancient Mesopotamia, wheat and barley were the most common crops but lentil, flax and fig were also grown (Pollock, 1999). This was about 4000 BC and barley was still the chief cereal crop under cultivation (Pollock, 1999).

The Paleolithic Age: There are no archaeological remains evidences which indicate that wheat was growing somewhere in the old world during the Paleolithic (old stone) age. This failure to find remains of plants is due to the perishable nature of cereal or other food plants and even their mature grains. However, tools and implements for grinding food materials have been found in sites indicating an age of about 75000 years (Horder *et al.*, 1954).

The Mesolithic Age: The Mesolithic or middle Stone Age (the period of transition from the Paleolithic to the Neolithic Age) lasted roughly from 9000 to 7000 BC in southwestern Asia. During this period, man still lived as a hunter or gatherer of food rather than its producer. Excavations of the famous caves at Mount Carmel have yielded indirect evidences such as sickles made of flint, different blades, flint tools which have been used for cutting stems of cereals or other grasses besides hunting implements of stone and bone. There were also mortars and pestles for grinding grain as well as bone hoes. No actual cereal remains were found. It is significant, however, that wild wheat and wild barley grow in this region even now. The caves date back to somewhat earlier than 8000 BC. The evidence indicates that the cave dwellers (Natufians) were primarily hunters and food gatherers but were, nevertheless, beginning farmers.

The Neolithic, Bronze and Iron Ages: The archaeological findings dating back to about 7000 BC (Neolithic or stone Age) provide the clue of existence of agriculture, animal husbandry and making of polished stone axes. The evidence of pottery making by Neolithic people came later, about 5000 BC. The Bronze and Iron Age cultures also appear to have begun in this region between c.3000 BC and c.1000 BC. Most of the wheat grown in prehistoric times and had grains enclosed in thick, hard, tight glumes and were difficult to thresh. When heads of such a wheat are heated over fire, or on a hot stone surface, the glumes become brittle and can be readily be removed from the grain by rubbing. This has been the practice of grain thrashing in Neolithic times, as has been evidenced from the archaeological investigations in Asia, Africa and Europe.

Grains or spikelets stored for centuries or millenia in containers under perpetually dry conditions, as found in ancient Egyptian storage pits and tombs, even today, have the appearance of almost grains or spikelets. They are not carbonized, but have turned coppery or bronze.

The remains of ancient grains or other plants have been found in excavations, in a good state, buried in peat at sites of the lake dwellers of Switzerland and southeastern Europe as also in Scandinavia and Britain. Impressions of grains, spikelets, glumes, etc., have been frequently found in pieces of pottery excavated from sites of Neolithic or later periods. Such grains or other plant parts have often been accidentally or imbedded in soil mixed for preparation of pottery showing, thereby, impressions such evidence has also been found in India.

So far, the oldest remains of carbonized spikelets of wheat and barley have been found in excavations at the site of a Neolithic village community at Jarmo, between Mosul and Suleman in Iraq (Braidwood and Braidwood, 1950) and the cereal remains and imprints were identified by Helbaek (1959).The estimated date of this site is 6700 BC (Braidwood, 1958). The carbonized wheat were identified as that of *T. aegilopoides* (syn. *T. turgidum* var. *boeoticum*), *T. dicoccoides* (syn. *T. turgidum* var *dicoccoides*), and a type of wheat resembling *T. dicoccum* (syn. *T. tugridum* var. *dicoccum*). Imprints of grains and spikelets found in baked clay and sun-dried bricks were identified as belonging to *T. dicoccoides* (=*T. turgidum* var. *dicoccoides*) as shown in plate 22 and to *T. turgidum* var. *dicoccum* (cult. Emmer). The barley remains resembled the wild two-rowed *Hordeum spontaneus* but had a less brittle rachis in these samples.

Helbaek (1959) has given (Table 2) below information on the wheat in early times as revealed by archaeological investigations of plant remains.

From the above data, it is clear that man had already begun to domesticate the two wheats species *T. dicoccoides* and *T. boeoticum* in southwestern Asia by about 7800 BC. It seems equally clear that emmer became the main wheat and spread with human migrations into much of Asia, Africa and Europe and occupied a place of pride in agriculture till 300 BC.

The early agriculture based mainly on Emmer wheat and barley spread to the interior of Iron Age settlements and the Indus Valley or India in 5th millennium BC or even earlier and by the beginning of third millennium BC, bread wheat *T. aestivum* was being grown commonly in the Indus Valley or elsewhere in India.

(B) Historical chronology of wheat in India

The available carbon-14 dates indicate that the upper Paleolithic culture had flourished in India from about 20,000 to 10,000 years ago (Patne 1979–80 Asia). These were spread widely in the country from the Ganga basin in the north to Tamil Nadu in south and from Assam in the east to Gujarat, Rajasthan in the West. The available data indicate that the Mesolithic phase began some 10,000 years ago and continued till the Neolithic, some 4,500 years ago. There existed industries of tool manufacture about 4000 BC in Karnataka at over 30 Mesolithic sites (Sankalia, 1987). Data excavations from Chopani-mando (Allahabad district) carbon date about 17th to 7th millennium BC indicate that

Table 2
Wheat of Early Primary Agricultural Settlements Found in Archaeological Excavations

Period (Millennium BC)	*Site region or country*	*Wheat remains identified*
7th	Damascus basin	Emmer
6th	Matarrah, Iraq	*T. monococcum* var. boeticum (Wild einkorn), *T. turigdum* var. dicoccoides (wild emmer), *T. turgidum* var. dicoccum (Cult. Emmer)
5th	Matarrah, Iraq	Emmer
5th	Mehargarh, India	Einkorn Wheat, bread wheat
5th	Halafian communities of the upper Euphrates-Tigris, Iraq	Emmer, *T. monococcum* (Einkorn, few grains only)
5th	Alluvial plains of lower Iraq and Fayum, Egypt	Emmer
5th	Merimbole beni Salame, Egypt	Emmer having stray grains of club wheat
5th	Intermittent loess plains from the Danube delta to the mouth of the Rhine, Europe	Emmer, Einkorn
4th	El Omari, Egypt	Emmer (few grains of club wheat)
3rd	Rojdi (Neolithic Harappan India)	*T. aestivum, T. sphaerococcum, T. compactum*
3rd	Switzerland, France, Northern Italy, Spain, Britain, Central Europe, Scandinavia	Emmer, Einkorn (large deposits of club in Switzerland and imprints in Denmark)

the Neolithic made use of the available edible (cattle, sheep, goat, horse, deer, wild bear, tortoise, fish) for their food and rice was the main food. Evidence of cultivation of rice was found from Mesolithic sites (Sharma & Mishra, 1980).The above account clearly points out the antiquity of Indian (dravadian) culture.

The word Godhuma and Brihi is not mentioned in *Ṛgveda* indicating thereby that neither the grain was known nor cultivated in Ṛgvedic times but was indigenous to the Punjab (then known as *Sapta Sindhu*). But is mentioned in the *Yajurveda Samhitas* (1.2.8 Vaj. *Maitriya Samhita* 18.12; 19, 22, 89, etc).

Further discussion on early history of wheat in India is continue. As Pollock (1999) has pointed out, ancient Mesopotamia was cultivating wheat and barley, lentil, flax and fig and barely was the chief cereal crop around 4000 BC. On the other hand Ratnagar (2000) had indicated that the Harappan had trade with Mesopotamia around 2000 BC.

This clearly brings out the fact that wheat cultivation in India may be much older than that indicated by the available data from excavations. This view is further supported by Harlan (1961) when he says 'I only wish to point out that pre-Columbian transport by man is not necessary to account for distributions of wilds species but there is evidence that man may have transported a crop like sweet potato at that time.'

The chronology of early History of wheat in India based on the facts from available data, is given in the Table below:

Table 3
Chronology of early history of wheat in India

Period	*Reference/Site*
5000 BC	Mehargarh pre-ceramic Neolithic Site in N-region of Kachhi plains—a Pre-Harappan culture, grew einkorn emmer and bread wheat (14).
5000 BC–4000 BC	*Yajurveda Samhita.*
3000 BC	Early Harappan crop—*T. aestivum, T. sphaerococcum, T. campactum, T. vulgare* (Lal, 1976; Weber, 1991).
2500–1750 BC	Harappan period *T. aestivum, T. sphaerococcum.*
2040 BC	Kalibangan (15)
2000–1800 BC	Kayatha culture near Ujjain in M.P. produced wheat (Ansari & Dhavalikar, 1975).
1600–1400 BC	Diamabad (16) in Malwa wheat was grown.
2449 BC	Varahmihir Samhita mentions Godhuma (17).
2600 BC	*Susruta Samhita.*
1900 BC–1200 BC	Excavations at Ahar (18).
1600–1000 BC	Apeagaon in Aurangabad, Maharashtra *T. durum* (Deo, 1976).
1400 BC	*Mahabharata.*
1200 BC	Post-Harappan at Atranjikhera (19).
1340–1290 BC	Inamgaon in Maharashtra (20).
200 BC–700 AD	Wheat was in cultivation (21).

The antiquity of wheat may be even older than the above evidences. The crop was spread over large areas in pre-Harappan, Harappan, Mahabharata and later periods. The abundance of food grains was the only reason that foreign invaders were attracted to there cultures. However, under the rule of Islamic Kings and the British monarchy, the decline in Indian agriculture began. Bajaj and Srinivas (2001) wrote that the yields of lands in India continued to either decline or stagnate at a low level throughout the yield of wheat from 796 kg/ha. in 1901 declined to 646kg/ha. in 1947.

(C) History of wheat improvement work in India

The systematic work of wheat improvement was started in India towards the beginning of the twentieth century. The improvement work of wheat involved three phases of selection, viz.(1) selection of productive types by the farmers in their fields during harvesting and planting; (2) selection by the breeders in the mixed stands of the primitive farmer; and (3) scientifically planned breeding. Some early varieties such as NP 114 and NP 775 arose through natural crossing on Federation and NP 4 and Vijay from natural cross between Motia durum and khapli dicoccum. Howards (1909b) reported that Punjab local type 9 wheat is a cross between *T. durum* and *T. sphaerococcum.* The systematic work on wheat was started in 1905 at the Indian (then Imperial) Agricultural Research Institute, Pusa (Bihar) at Government Agricultural College, Lyallpur (now Faislabad, Pakistan) as well as in areas which fall in the states of U.P., M.P. and Maharashtra. Sir Albert Howard assisted by his wife Gabriele Howard and Abdur Rahman collected wheat variants from all parts of the country and made a selection of them. This work resulted in the development of Pusa 4, Pusa 6 and Pusa 12. Pusa 4 had higher grain yield and better grain quality and won a prize for grain quality at the international grain exhibition held in Canada in 1919. In Punjab, type 11(a selection) was recommended to the farmers in 1913. Mr. D. Milne selected Pb 8A in 1919 which replaced Type 11. Another selection Pb 9D was recommended for cultivation under rainfed conditions. At Kanpur, Uttar Pradesh pure line selection from 1928 to 1958 resulted in the development of K13, K46, K49, K53 and K54. Similarly, in Madhya Pradesh, at Power Kheda, pure line selection gave varieties such as AO-13, AO-85, AO-88 and AO-90. In the eastwhile Bombay state at Niphad, the breeders developed Bansi 202, Motia and Gulab by selection from local mixtures.

VARIETIES DEVELOPED THROUGH HYBRIDIZATION

In the early phase of hybridization of wheat followed by pure line selection with great success could be achieved as the enormous variability was present in land races grown by the farmers. As a result of this, through early hybridization work, several varieties such as NP52, NP80-5, NP-120, NP125, NP165; Pb518, Pb591, Pb281, Pb273, Niphad4, K65, K68, AO68, AO113, RS31-1, HY65, etc., were developed and were released for cultivation.

During this early phase of hybrid pureline selections, the grain yield increased but diseases, particularly rusts, caused heavy losses. Therefore to achieve stability in wheat yields, Prof. K.C. Mehta and Dr. B.P. Pal developed varieties possessing good quality and resistant to major disease of wheat production, since they had resistance to rusts and loose smut diseases. The varieties developed and released in this series were NP 710, NP 718, NP 720, NP 770, NP 798, NP 808, NP 809, NP 818, NP 824, NP 830, NP 846, etc. In some states, several good varieties were also evolved .These were Pb 228, Pb 250, Pb 253, Hy 23, Hy 38, C 286, C 285, C 281, C 273, C 303, C 306, etc. Though a commendable wheat breeding work was undertaken from 1947 to 1965 in improving grain quality, yield and disease resistance, etc., a realbreak through in yield could not be achieved. All these wheat varieties were tall in stature and under high fertility conditions, they used to lodge and break, thus resulting in loss of yield. Among early wheat breeders in India, mention may be made of T. Milne, Choudhary Ramdhan Singh in Punjab, H.M. Leake and Ram Prasad in Uttar Pradesh and Geoffrey Evans in Madhya Pradesh.

USE OF NORIN DWARFING GENES AND SEMI-DWARF WHEATS

Dr Vogel in USA developed dwarf winter wheat variety 'Gaines' by using norin dwarfing genes. Dr. N.E. Borlaug (Nobel Laureate) transferred norin dwarfing genes to spring wheat in Maxico. Dr Swaminathan, the Director, Indian Agricultural Research Institute (I.A.R.I.) at Pusa, Delhi, made efforts to introduce the Maxican dwarf spring wheat varieties Lerma rojo 64, Sonora 63, Sonora 64, and Mayo 64 in November 1963. The seeds of these varieties were distributed to wheat breeders located at P.A.U. Ludhiana, G.B. Pant Agricultural University; Pantnagar, Kanpur, Indore and Pusa. The initial trial reported in 1964 were highly encouraging as they had enormous yield potential. Bulk imports of two varieties in 1966 enabled the farmers to grow these in their fields as these had been found promising in the national demonstration trials. These varieties were Lerma hoja 64A and Sonora 64 and were fit for cultivation under irrigated conditions. In 1965-66, selection at I.A.R.I. New Delhi S 227(popularly known as Kalyan sona) and S 308 (Sonalika) and selections at P.A.U. Ludhiana PV 18 and Kalyan sona and Pantnagar Kalyan sona were released, which brought about a green revolution in wheat production in India. These varieties became very popular and spread all over the country within a short span of time and produced good yield under high fertility and irrigated soils. However, rusts and other fungal diseases continue to cause heavy toll in yield in some years.

ERA OF STABILIZING WHEAT PRODUCTION IN INDIA

The yields per unit area had increased tremendously as well as the production of wheat with the introduction of semi-dwarf and dwarf wheat, in the country. However, diseases started playing havoc in certain years and varieties under rain-fed conditions still did not show a marked progress. It was, therefore, realized that the vertical spectrum of wheat varieties will have to be diversified in order to prevent disease epidemic and a steady flow of new varieties, deriving resistance genes from diverse sources, will have to be insured to replace the old ones in order to stabilize wheat production. Accordingly efforts were intensified to develop wheat varieties possessing a high degree of diverse resistance to diseases, particularly rusts, by the wheat breeders. As a consequence of these concerted efforts, several wheat varieties such as Arjun (HD 2009), Nilgiri (HD 2135), HD 2189, HD 2204, Girija, Shailja, HW 517, Meghdoot, HD 1209, PBW 34, PBW 65, Raj 1555, HUW 206, WH 281, etc., were released in different regions in the country. Some of the most popular wheat varieties in the country are Arjun, HD 2135, HD 2189, HD 2204, Girija, HS86, HD 1209, Meghdoot, HW 517, RAJ 1555, K 7410, UP 310, MACS 1967, WL 410, PBW 34, DWL 5023, WH 331. In spite of all these efforts the scene in the country is that only two varieties are grown in the greater areas of the country PBW 343

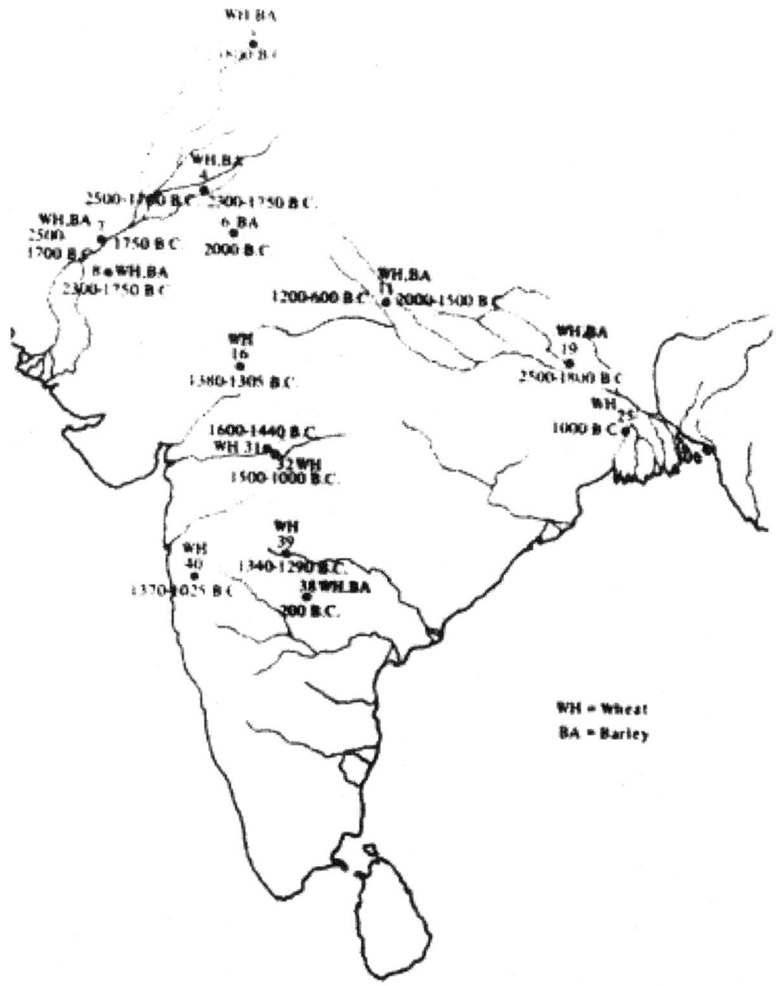

Map 1. Geographical distribution of records together with dates of wheat and barley in the Indian subcontinent.

Key to sites plotted on Maps 1–3

1 Burzahom
2 Kangra
3 Rupar
4 Harappa
5 Rangmahal
6 Kalibangan
7 Mohenjo-daro
8 Chanhu-daro
9 Khokhra Kot
10 Hastinapur
11 Atranji Khera
12 Noh
13 Ahar
14 Nagda
15 Garh Kalika, Ujjain
16 Kayatha
17 Kausambhi
18 Rajghat
19 Chirand
20 Rajgir
21 Patliputra
22 Oriyup
23 Sonpur
24 Mahesdal
25 Pandu Rajar Dhibi
26 Singhbhum

27	Baidipur	36	Paunar
28	Ambri	37	Nevasa
29	Lothal	38	Ter
30	Rangpur	39	Sonegaon
31	Maheshwar	40	Inamgaon
32	Navdatoli	41	Koihapur
33	Bhatkuli	42	Hallur
34	Kaundinyapur	43	Kunnatur
35	Pauni	44	Periyapuram

Key to symbols for crops on Maps 1-3 and Fig. 1.

BA	Barley, Hordeum	PE	Peas, Pisum
BM	Bajra, Pennisetum	PH	Phyllanthus
CA	Castor, Ricinus	PS	Paspalum scrobiculatum
CO	Cotton, Gossy plum	RA	Ragi, Eleusine
DP	Date palm, Phoenix	RI	Rice, Oryza
HG	Horse gram, Dolicos	SE	Sesame, Sesame
JO	Jowar, Sorghum	WH	Wheat, Triticum
LE	Lentils, Lens	WD	Weeds
LI	Linseed, Linurn	ZI	Zizyphus
MA	Maize, Zea	MS	Melon seeds, Cucumis
MU	Mung, Phaseolus		

from Punjab and Lok 1 from Gujarat, PB 343 is being grown in Punjab, Haryana, Rajasthan, U.P. and Bihar and Lok 1 is popular in Gujarat, Madhya Pradesh and Maharashtra.

India today ranks fourth among the wheat-growing countries in the world. Presently wheat is grown over nearly 26 million hectares. The wheat productions after 1966–67 has crossed 50 million tonnes in 1983–84 and during 1994–95, the production stands at 65.47 million tonnes. In the last 30 years (1970 to 2000), the country's wheat production has increased 4.5 times with an annual growth rate of 5.6490. The increase in production is a cumulative effect of expansion of area, development of irrigation infrastructure, fertilizer use, plant protection and breeding of high yielding disease-resistant varieties.

NOTES AND REFERENCES

1. Agrawala, V.S. 1946. *Paninikaleen Bharatvarsa.* Chaukhamba Vidya Bhavan, Varanasi.
2. Allchin, F.R. 1995. *The Archaeology of Early Historic South Asia,* Cambridge University Press, New York.
3. Anandmurti, 1993. *Discourses on Tantra,* Vol. I. Anand Marga Publications, New Delhi.

4. Ansari, Z.A. and M.K. Dhavlikar, 1975. *Excavations at Kayatha,* Deccan College, Pune.
5. Bajaj, J. and M.D. Srinivas, 2001. *Restoring the Abundance.* Indian Institute of Advanced Study, Shimla.
6. Braidwood, R.J. and L. Braidwood, 1950. *Jarmo: A Village of Early Farmers in Iraq. Antiquity,* 24: 189–195.
7. Choudhury, K.A., K.S. Saraswat, S.N. Hassan and R.C. Gaur 1971. 4000–3500 years old barley, rice and pulses from Atranjikhera. *Science and Culture* 37(11): 531–532.
8. Chaudhary, S.L., G.S. Sharma and Y.L. Nene. 2000. *Ancient and Medieval History of Agriculture.* Rajasthan College of Agriculture, Udaipur.
9. Das, A.C., 1925 *Ṛgvedic Culture.* R. Cambrays Co., Calcutta.
10. Deo, S.B. 1976. *Apegaon Excavations.* Deccan College Post-Graduate & Research Institute, Pune.
11. Dhavlikar, M.K. 1999. *Historical Archaeology of India,* Books and Books, New Delhi.
12. Dwividi, K. and B. Dwivedi, 2001. *Vedon Mein Ayurved,* Viswabharti Anusandhan Parishad, Gianpur.
13. Feldman, M. 1976. Wheat in Smarti, J. and N.W. Simmonds (ed.) *Evolution of Crop Plants,* pp. 120–128, Longman, Essex, UK.
14. Harlan, J.R. 1961. Geographic origin of plants useful in agriculture. In R.E. Hodgson (ed) *Germplasm Resources,* pp. 3-19, American Association for the Advancement of Science, Washington DC.
15. Hutchinson, J.L. (ed) 1974. *Evolutionary Studies in World Crops,* Cambridge University Press, London.
16. Jha. A. (commentators) *Varahmihir Virachit Brihat Samhita.* Chaukhamba Vidya Bhavan, Varanasi.
17. Lal, Chaman 1976. *India: Cradle of Cultures.* Chaman Lal C/o Modern School, New Delhi.
18. Mehra, K.L. 2000. History of crop cultivation in pre-historic India. In S.L. Chaudhary, G.S. Sharma and Y.L. Nene (eds.) *Ancient and Medieval History of Indian Agriculture,* pp. 11-15, Rajasthan College of Agriculture, Udaipur.
19. Mode, H. 1959. *The Harappa Culture and The West.* Principal, Sanskrit College, Calcutta.
20. Peterson, R.F. 1965. Wheat: Botany, Cultivation and Utilization, Inter-Science Publishers Inc., New York.
21. Ratnagar, S. 2000. *The End of the Great Harappan Tradition,* Manohar Publishers and Distributors, New Delhi.
22. Randhawa, M.S. 1982. *A History of Agriculture in India,* Vol. II, pp. 211–213. ICAR, New Delhi.
23. Sankalia, H.D., S.B. Deo and Z.D. Ansari, 1969. *Excavations at Ahar* (Tambavati) Deccan College and Post-Graduate Research Institute, Pune.
24. Sankalia, H.D. 1987. *Prehistoric and Historic Archaeology of Gujarat.* Munshiram Manoharlal Publishers Pvt. Ltd., Delhi.

25. Sastry, D.L.N., M. Kasturi and G.V. Ramkrishna Rao, 1999. *Studies in Prehistoric Culture of Andhra Pradesh.* Bhartiya Kala Prakashan, Delhi.
26. Sharma, G.L. 1980. *History to Prehistory.* Department of Ancient History, Culture and Archaeology, University of Allahabad, Allahabad.
27. Sharma, G.R. and B.B. Mishra 1980. *Excavations at Chopani-mando* (Belan Valley) 1977–78 *Epipalaeolithic to Proto/Neolithic.* Department of Ancient History, Culture and Archaeology, University of Allahabad, Allahabad.
28. Sharma, G.R., V.D. Mishra and J.N. Pal. 1980. *Excavations at Mahādaha* (1977–78): A Mesolithic Settlement in the Ganga Valley. Department of Ancient History, Culture and Archaeology, Allahabad University, Allahabad.
29. Sharma, R. 1985, *Vedic Sampatti.* Dayanand Sansthan Ved Mandir, New Delhi.
30. Singh, M. and M.K. Upadhyay, 1986. Wheat Improvement in India. *Indian J. Genet.*, 46 (Suppl.): 187–197.
31. Vaidya, C.V. 1990. *Mahābhārat Mimānsā.* Haryana Sahitya Academy. Chandigarh.
32. Vishnu-Mittre and R. Savitri 1982. 'Food Economy of the Harappans', In G.L. Possehl (ed.) *Harāppān Civilization,* pp. 205–221. Oxford & IBH Publishing Co. Pvt. Ltd., New Delhi.
33. Watt, G., 1893. *A Dictionary of the Economic Products of India.* Vol. VI. Part IV. Reprint 1971. Cosmo Publications, Delhi.
34. Weber, Steven A. 1991. *Plants and Harappan Subsistence.* Oxford & IBH Publishing Co. Pvt. Ltd., New Delhi and American Institute of Indian Studies.
35. Zeven, A.C. and P.M. Zhukovsky. 1975. *Dictionary of Cultivated Plants and Their Centres of Diversity.* Centre for Agricultural Publishing and Documentation, Wageningen.

CHAPTER 11

The Origin and History of Gram in India

B.D. Sharma

Gram (*Cicer arietinum* Linn.) also known as chickpea or *Chanak* in Sanskrit has chromosome 2n-14 & 16 and is believed to have originated from an unknown wild species probably in western Asia with a secondary centre in India, Afghanistan and Mediterranean region (Zeven & Zhukovsky, 1975). Large-seeded type, Kabuli, must have been evolved in Mediterranean region and Afghanistan. Kabuli gram have been introduced from Afghanistan in about 1700 AD (Van der Maesen, 1972). The genus Cicer has only few species. These are:

(i) *C. microphyllum* Royle (Syn. *C. songaricum* Staph., *C. jaquemontii* Jaub. & Spach) is widely distributed in the west Himalayas in India (Lahaul-Spiti), Tibet and Afghanistan.

(ii) *C. pinnatifidum* Jaub. & Spach (2n=16) is found distributed in Anatolia, Armenia, Syria, North Iraq and Cyprus.

(iii) *C. echinospermum* P.H. Davis is found in eastern Anatolia.

(iv) *C. bijugum* K.H. Rech. (2n=16) occurs in southeast Anatolia, north Syria and north Iraq.

Some of the above mentioned *Cicer* species indigenous to Anatolia may have played a role in its ancestory (Van der Maesen, 1972). *Cicer arietinum* has four recognized races, as mentioned below:

1. Race orientale Pop. is characterized by its very small seeds (1000 seeds weight 100 to 120g). It is common in Ethiopia, Sudan, Egypt, India, Pamir, Tadzhikistan and Iran. Those from Ethiopia are black seeded.
2. Race asiaticum Pop. has some what bigger, but still small seeds (1000 seeds weight 140–200g). It occurs in central Asia, Afghanistan, western China, Iran and eastern Turkey.
3. Race eurasiaticum has medium large seeds (1000 seeds weight 200–300g). It is cultivated in the near East, Armenia, Azerbaydzan, Ukraine upto Central SSR and near large cities in the eastern part of the area of cultivation. The seeds are white.

4. Race mediterraneum Pop. has the largest seeds (1000 seeds weigh over 350g). It is found in Spain, Italy, Morocco, Algeria, Tunisia and western Turkey. The seeds are white. In Turkey near Burdur, earliest deposits were dated about 5450 BC (Helbaeck, 1970) and its cultivation in Iraq, was c. 3300 BC during the Bronze Age (Randhawa, 1986).

The cultivation of gram has been recorded from Angola, Abyssinia, Ethiopia, Italy, Greece, Russia, Rumania, Egypt, North Africa, Spain, Central and South America, Australia, the Nile Delta, Iran, Moracco, Paksitan, India, western China and Myanmar. However, India is the biggest producer and the crop is grown over the largest acreage here.

EARLY HISTORY OF GRAM IN INDIA

The Harappan civilization has been divided into (i) Early (before (2600 BC); (ii) Mature (2500–200 BC); (iii) Late (2200–1700 BC). According to Weber (1991), the Harappan people were growing a variety of crop plants and that a sophisticated cultivation strategy was in existence at the beginning of the mature phase (2500 BC onwards) which continued till the Late Harappan Phase (2200–1700 BC). Both summer and winter cultivation were probably being practised in most regions from 2600 BC onward. Although indigenous plants were grown, the focus of farming appears to have been on both indigenous and non-indigenous species, with an increased reliance in later periods on indigenous taxa. Though there seems to be a greater emphasis on plants with origins in Africa or western Asia, there is ample evidence that plants have been introduced from east Asia and South-east Asia. A number of new taxa had appeared in Harappan agriculture by about 2000 BC.

Grams' introduction may be attributed to the Harappan, who had been growing it during the Mature Phase (2500 BC). It seems probable that gram might have been introduced in India by about 2600 BC or even before. As there is no reference of gram in the *Ṛgveda* but the *Yajurveda* and *Atharvaveda* have made mention of *Chanak* (gram) under the name *Khalava* according to Dwivedi (2001). However, Majumdar (1945) disagrees and points out that *khalava* indicates (Mung, *Vigna radiata*). It is just possible that Aryans got aquainted with wheat and gram when they had come in contact with Harappan people and were settling in Sapta Sindhu (Punjab). Ratnagar (2000) had pointed out that Harappan people had trade with Mesopotamia around 2000 BC, which indicates that crop of that or nearby regions must have got the attention of Harappan traders and there is every reason to believe their introduction during this period or even earlier had taken place.

The remains of gram (only three seeds) had been found from Kalibangan date back to 3000–2000 BC—(Vishnu Mittre & Savithri, 1982). Mehra (2000) has reported finds of gram from Bihar (c. 2000–1200 BC), Gangetic plains (c. 2200–800 BC), Rajasthan (c. 2200–800 BC), Maharashtra (c. 2000–800 BC) and Punjab (c. 2300–1400 BC). The remains of gram from Atranjikhera in U.P. date to 2000 BC (thermoluminescent dating) (Choudhary *et al.*, 1971). There is no mention of gram in the Vedic period.

Indian ancient literature also provides evidences of the cultivation and consumption of gram. The *Kama Sutra* of Shri Vatsayana Muni (commentator Devadutta Sastri),

(1982) mention that gram was used for food. It is believed that this treatise was written in about 300 AD. Jain canons have depicted life in ancient India and there while mentioning field crops, gram (*Chanaka*) has been listed as one of the pulse crops during the period between 400 BC to 500 AD (Chandra, 1947).

Gode (1961) gives a detalied account of gram and some references are being produced here under. The laxicon *Amarakosa* mentions two types of *Chanaka* (gram), i.e. *Chanaka* and *Harimanthak*. It may be that latter name is given to the green-coloured grain variety. *Manasollasa* of the king Somesvara also confirms it be so, as both these names are also mentioned in *Charaka Samhitā*. Dallahana (c. 1100 AD) and Cakrapanidatta (c. 1060 AD) has given a reference that the *chanaka* is well known crop.

The *Markandeya Purana* mentioned chickpeas (Canto xv, 8, pp. 84–85). The *Matsya Purana* among other crop mentions *chanaka*. The crop of gram was being grown during the period (600–900 AD) when the *Agni Purana* was composed (p. 129). In the *Brhatasamhita* of Varahamihira (c. 537 AD) Chap. 15 and Verse 14 and chapter 16 verse 34 gives a narration of *chanaka*. The *Rajnighantu* of Narahari (c 1450 AD) and *Madana Vinoda* (1374 AD) also defines *chanaka* as a legume. Nakula, in his *Asvachikitsā* (Bib. Indica, Calcutta, 1887, p. 39), prescribes *chanaka* moistened with water for horses in the absence of *yava* (barley). The practice of feeding horse on *chanaka* moistened in water referred to by Jayadatta is corroborated by Travernier (1641–1668 AD), who observes in his Travels (Vol. I, pp. 102–103) as follows:

> "They receive a measure of chickpeas which the groom has crushed between two stones and steeped in water. It is these which take the place of hay and oats."

The Vijaynagar horses in the 16th century were fed on *chanaka*. Barbosa (1500 AD) writes:

> "The food is rice boiled with chickpeas and other pulse. And each man comes to draw the ration for his horse or elephant". (Vide pp. 130–131 of Third Dynasty by Venkataramanayya, Madras). The *Brhdagargiya Samhita* (MS No. 542 of 1895–1902 written 1825 AD) contains a chapter on Tulakosha in which there is reference to *Chanaka* and other grains. In the *Arthashastra* of Kautulya (300 BC) there is no mention of gram being used as ration for horses whereas Manasollas (1130 AD) has given details of it. So it shows that gram appears to have been adopted as horse food much later than the time of *Arthashastra*. In *Asvayurveda*, probably written earlier than 1000 AD there is clear mention of use of barley (*yava*) and *chanaka*, which tells us that *yava* was the principle horse food in the region between the Himalayas and the Vindhya mountains, while to the south of Vindhyas *chanaka* was favoured as horse food and in the western India, it was *Makustaka* (*Motha*) as the principle horse food. This is also supported by *Sarsamuchaya* of 1866–68 written by Kalhana. The date of this *Sarsamuchaya* is not known but it may be later than 1000 AD. After about 1000 AD *yava* fell into background and *chanaka* took its place as food for horses. However there are now archaeological evidences of occurrence of this crop in Rajasthan, Punjab, U.P. and Bihar by about 2000 BC or earlier and it must

have been used as a food pulse for human consumption. Hence, the conclusion of Gode (1961) that *chanaka* was used as food for men along with wheat and other grains only after 1000 AD and has a history of 2000 years in India is not true and lacks scientific credibility to the subject. The *yava* was known to Aryans since *Ṛgvedic* times and was used both as man's and horse food and when Aryans could be successful in migration to the southern parts beyond the Vindhyas, *yava* had to fight with *chanaka*, its rival that must have been then cultivated in large quantities. The gram replaced barley as horse feed. The barley-wheat partnership was dissolved more than 1000 years ago and gram-wheat alliance has governed Indian kitchens without a break till 1960.

RECENT HISTORY OF GRAM IN INDIA

India has more than 80 percent of the world's chickpea (gram) area (10.5 million ha.); about 10 percent of the world's area under the crop is in other Asian countries (Pakistan, Myanmar and Bangladesh). Most of the remaining area of chickpea is in Ethiopia, Mexico, Spain, Morocco, Turkey and Iran. In India, chickpea ranks fifth in area and fourth in production among food grains. Among the grain legumes grown in India, chickpea ranks first both in area and production (Gill, 1980).

The major area under chickpea in India is covered by cultivars with small to medium seeds (12–20 g/100 seeds) which are brown and wrinkled popularly known as '*Kala Chana*' or simply '*desi chana*'. These cultivars are adapted to poor marginal lands. Earlier, say before 1960 gram was grown often mixed with wheat and these were harvested and threshed together and ground into flour to make '*chapaties*'. The pure gram crop was usually grown for use as '*dal*'. The average yields of gram, however, have fluctuated between 600 to 700 kg/ha. and national average of 683 kg/ha. which is almost half what is harvested in Turkey (1478 kg/ha.) and Morocco, Mexico, central America harvest higher grain yield. The kabuli types are also grown over a sizeable area and require fertile soil and irrigated conditions. After independence, the breeding work on gram has yielded good results and in recent years, yield of gram has risen to 1500 to 2000 kg/ha. even under rainfed conditions but more efforts are needed to stabilise the yield and production in the country.

The varietal improvement of gram was taken up in the different states of India in the early part of this century with the establishment of Agriculture Department in the states. The first approach was to collect local land races from within the country and these were evaluated for their yield potential, resistance to diseases and pests etc. As a result of selection, good-yielding strains, such as Pb.7, G 24, S 26 in the Punjab, TI, T2, T3, T87 in Uttar Pradesh and many other promising selections were released for cultivation in other states namely Bijapur 3, Dohad 1597, var. 1543, T-15, NP 58, NP 28, NP 100, C 49 (Kabuli) and C-24.

Considerable work on gram was done in Punjab by selection and hybridization for early maturity, and resistance to draught and wilt. Hybridization was taken up in Punjab in the late thirties and the first variety released in Punjab in 1964 was C 1234 (pb 7 x f 8) suitable for cultivation in blight-prone submontaneous region. Other varieties released were C235 (released in 1960), G 543 in 1977. The varieties developed for

irrigated area were G 130, kabuli type C 104, L 550 for Punjab and Harechhole 1, C 214 and G 24 were released on all India bases.

With the initiation of All India Coordinated Pulse Improvement Project in mid-sixties, followed by the establishment of Directorate of Pulses Research (ICAR) at Kanpur, the work has been further strengthened. Variety C 235 tolerant to blight and wilt, is a popular variety. Variety Gaurav is also resistant to Ascochyta blight, a highly damaging disease of gram crop and associated with thunderbolt for heavy incidence of disease on the crop. Other good varieties are BG 261, BG 244, BG 256, Mahamaya-1 (B 108), B 110, B 124, Mahamaya 2 (B 115). In Maharashtra, a remarkable bold-seeded variety Phule G 5 was developed. Among the released varieties with very bold seed and kabuli grain type for culinary purposes C 104 is outstanding. From the nutritional quality angle, NP 44 has protein 24.4 percent and NP 17 has the highest amount of iron (10.3 mg/ 100 g). The GN 1 of Rajasthan has the highest calcium. Variety NP 100 has deep green seeds at maturity and yields 30 q/ha.

Thus, within less than a decade, the gram improvement work of All India Co-ordinated Project led to the development of several genotypes which are higher yielding, disease resistant and more widely adaptable. In gram, there are three seed shapes, viz., Desi wrinkled, round kabuli and pea type and the colour of testa though shows great variation, but three main types are grown by the farmers are black (*kala chana*), white kabuli type and green type. These three types have also reference in ancient literature as Harimantha, Harimanthka and Hirimantha. In recent years, the use of chemical fertilisers such as urea (high in nitrogen) in wheat crops has caused a severe setback to gram cultivation as it fails to grow in such fields. This is due to the reason that the plant bears nodules in its roots which contain nitrogen-fixing bacteria. These symbiotic bacteria are killed in acidic high salt conditions created by the application of fertilizers, resulting in failure of plant to survive. This factor has caused great reduction in production of gram in many traditional gram-growing areas and, as a consequence, the price of gram has gone very high (Rs. 40 to 50/kg). The demand of gram as *dal, besan,* etc., is growing in recent years. It is an excellent source of protein in the diet of predominantly vegetarian population of India. The recent breeding programmes of gram are also paying equal attention to the improvement nutritional quality along with grain yield, disease and pest resistance and better adaptability or stability in yield and production.

REFERENCES

Agricultural Year Book—New Vistas in Crop Yields. 1970. Indian Council of Agricultural Research, New Delhi.

Agni Purana (600–900 AD) presentation by Rakesh Sastri. 2002. Sadhana Pocket Books, Delhi.

Gill, K.S. 1980. 'Objectives, Breeding Approaches and Achievements in Chickpea', In K.S. Gill (ed.) *Breeding Methods for the Improvement of Pulse Crops,* pp. 243–259. Punjab Agricultural University, Ludhiana.

Gode, P.K. 1961. *Studies in Indian Cultural History.* Visvesvaranand Vedic Research Institute, Hoshiarpur.

Hutchinson, J. 1974. *Evolutionary Studies in World Crops,* pp. 23–24. Cambridge University Press, London.

Mehra, K.L. 2000. 'History of Crop Cultivation in Prehistoric India', In S.L. Choudhary, G.S. Sharma and Y.L. Nene (eds.) *Ancient and Medieval History of Indian Agriculture,* pp. 11–15. Rajasthan College of Agriculture, Udaipur.

Mishra, V.N. and Peter Bellwood (eds.). 1985. *Recent Advances in Indo-Pacific Prehistory.* Oxford & IBH Publishing Co. Pvt. Ltd., New Delhi.

Randhawa, M.S. 1986. *History of Agriculture in India.* Vol. I. I.C.A.R. New Delhi.

Vishnu Mittre and R. Savithri. 1982. 'Food Economy of the Harappans', In G.L. Possehl (ed.) *Harappan Civilization* pp. 205–221. Oxford & IBH Publishing Co. Pvt. Ltd., New Delhi.

Weber, S. 1991. *Plants and Harappan Subsistence.* Oxford & IBH Publishing Co. Pvt. Ltd., New Delhi & American Institute of Indian Studies, pp. 23–24.

Zeven, A.C. and P.M. Zhukovsky. 1975. *Dictionary of Cultivated Plants and their Centres of Diversity.* Centre for Agricultural Publishing and Documentation, Wageningen.

CHAPTER 12

The Origin and History of Potato in India

B.D. Sharma

In India, potato is so popular today that it is difficult to think of a diet without potato either alone or in combination with other vegetables. However, a few people give a thaught to the interesting history of this vegetable. Yet 400 years ago, it was quite unknown to most of the Indians.

Early domesticated potatoes were found in central Peru which has been dated R-14 to 7000 B.P. It is likely that the potato might have been domesticated in the lake Titicaca to lake Poopo in region of north Bolivia and may have originated from the wild diploid species *Solanum leptophyes* some 10,000 to 7000 years ago and that the first domesticated potato became *Solanum stenotomum.* Evidences are available to show that three wild species *Solanum sparsipilum, Solanum acaule* and *Solanum megistacrolobum* had been involved in the initial domestication step.

The more important event in the evolution of cultivated potato was the hybridization between *Solanum stenotomum* and *S. sparsipilum* to form amphidiploid *Solanum tuberosum* (potato of today). The sweet potato (*Ipomea l-atata*) was the earliest to be cultivated in the Caribbean, where it was known as 'camata' or 'batata'. As batata or patata, it became widely grown in parts of Spain in the 16th century and tubers were exported to other parts of Europe. From the word 'patata' in England, it was known as 'potato' (*Gerard,* 1597). In Europe, the potato was introduced during the period between 1535 and 1585 into Spain from where its cultivation spread into Portugal, Italy, France, Belgium and Germany. It was introduced into Ireland in 1585 or 1586 and in England in 1590 AD (Hawkes, 1967). From its Spanish introduction, potato diffused throughout Europe and Asia (Yadava, 1993). The cultivated potato varieties mainly belong to either *Solanum tuberosum* ssp *Tuberosum* (long-day) or ssp andigena (short day).The cultivated potato is tetraploid with chromosome number equal to 48 but the tuber bearing *solanum* may be diploid (2x=24), Triploid (3x=36), Pentaploid (5x=60) and hexaploid (6x=72).

In South America, farmers grow a great variety of potato stocks that differ in colour, shape and texture or flesh colour (plate photo). Many varieties have red or purple skins possess deep eyes and also have flesh tinted red, purple or deep yellow. A variety known as 'chapina' is jet black all through and is used for dyeing cloth.

In his Phytopinax' published in 1596, the herbalist Caspur Bauhin gave the potato its scientific name *solanum tuberosum* though in his later work, he added the adjective *esculentum.* Several contemporary references indicate that the potato was commonly grown in Italy and has been brought there from Spain by the carmelite friars. Dr R.N. Salaman a physician turned potato expert, had gathered references to the potato having been grown in Spain in about 1570 to 1573, and had concluded that it was introduced into that country in 1570. The second introduction of the potato into Europe was mentioned by John Gerard in his Herball, 597. Drake had been sailing along the northern coast of South America, it was likely that in his raids on the Columbian coastal towns, he obtained potatoes as store or loot. Potatoes were certainly brought to Cartagena to be stored aboard a Spanish vessel that returned to Spain. Dr. R.N. Salaman had mentioned in his paper in the *Journal of the Royal Horticultural Society* that potato was a market crop in Ireland by 1623. Towards the end of seventeenth century, it was widely grown in Lancashire area of England where it was introduced from Ireland. In Scotland it was not cultivated untill 1683. In other countries of Europe, the spread was slow unlike countries as Ireland where the standard of life was so low as to accept potato in their diet for bringing improvement.

The first mention of potato in Indian history is that of Edward Terry's voyage account of the banquet at Ajmer given by Asaph Khan to Sir Thomos Roe, the British Ambassador in 1675. Fryer's travel record (1672–81) described the gardens of Surat and in Karnataka in1675 as containing, among other vegetables, brinjals (*Solanum melongena*) and potatoes (Watt, 1908). In India, potato was introduced in the early seventeenth century by the Portuguese who started its cultivation along the western coast, where it was called Batata. Hawkes (1990, 1992) suggested that potato may have been brought to India by British missionaries in the late seventeenth century. The British traders also introduced potato into Bengal where it was known as *Alu* (a root crop). Johnson (quoted by Stuart, 1928) saw good potatoes produced in 1842 in the immediate vicinity of Kolkata and still finer ones in the hills of Cherrapunjee (presently in Meghalaya). Capt. David Scott (1802–1831), the agent of the Governor General of British India in the northeast frontier introduced potatoes in the North-East region of India during 1830 at Nongkhlaw village (msl 1000 m) in the Khasi hills. During almost the same period, an army officer, Major Young, introduced potatoes in Dehradun hills. Captain Mundi, an army officer has mentioned that in 1828, many families were already cultivating small strips of potatoes in Shimla hills. The potato cultivation generally spread widely throughout the northwestern hill region because the trade between the hills and plains developed by 1839. The potato cultivation had become well established as a gainful source of income in the hill areas.It would appear that potato cultivation received a great boost during the time of Governor General Warren Hastings (1772–1785). He also attempted to open up trade with China and Tibet. Apparently, potato was introduced into Tibet in the early part of the nineteenth century through trade routes originated from India. Until 1935, potato improvement work did not start for developing suitable varieties for Indian conditions. Foreign commercial varieties were imported from time to time to raise production.

Although the process of introduction and selection of suitable varieties continued for nearly a century, little success could be achieved in maintaining the productive potential, of these varieties. Despite various unsuccessful attempts to achieve high production potential a large number of varieties were introduced and no breeding work was initiated. All efforts to raise potato production in the country remained ineffective. In October 1934, Dr. F.J.F. Shaw the Director of the Imperial (now Indian) Institute of Agricultural Research conceived a scheme on potato breeding for northern India, which on 1st April 1935, was sanctioned by the government. Dr. B.P. Pal the then economic botanist and the officer-in-charge of the scheme, provided the necessary direction. True seeds of two varieties, viz., Phulwa and Darjeeling Red Round (DRR); which flowers under Indian plain conditions, but only Phulwa produced seed because DRR was male sterile, were collected and sown in November 1939. However none of the seedling population was superior to the parent. The next few years were spent in collection of potato germplasm from within the country and introduction of commercial varieties from abroad. In 1945, more than 500 commercial varieties were imported mostly from England. These varieties were tested in yield evaluation trials but none of these and the selection from old germplasm proved a success. A disease-free clone of variety 'Up-to-Date' was the only exception. In 1935, the potato breeding station at Shimla started functioning under the control of Indian (then Imperial) Agricultural Research Institute as a scheme sponsored by Indian (then Imperial) Council of Agricultural Research (ICAR). Dr. S. Ramanujan was appointed as Officer on Special Duty in November 1946 in order to establish a full-fledged institute for research on all aspects of potato. Later, he became the first Director of the Central Potato Research Institute (CPRI), which was started at Patna in August 1949. In the same year, three substations of Institute were also established at Shimla, Kufri and Bhowali. The Institute conducted varietal trials in order to identify superior varieties in collaboration with the states from 1950 on wards. These trials ultimately led to the selection of O.N 1645, O.N. 2236, P.S. 194, P.S. 196 and P.S. 555 for northern plains and Hybrid 8 for Kumaon hills and Hybrid 9 for Himachal hills. The period from 1949 to 1956 was mainly devoted to study the problems of potato production in different states and for the development of new concepts and solution to the problem. In 1952 the coordinated scheme for applied research and development of potatoes at 26 centres (16 in plain and 10 in hills) came in operation.

In 1956, the headquarters of Central Potato Research Institute was shifted from Patna to Shimla and Dr. S. Pushkarnath became the Director. Between 1957 to 1960, six stations were established at Jalandhar, Babugarh, Ootacamund, Rajgurunagar, Shillong and Mukteshwar. By this time, the problems of potato production in India had been clarified. These were: (1)varied climate (subtropical in plains to temperate in hills), over 90 per cent of the area under potatoes lie in the subtropical plains, which require short duration varieties;(2) varieties for hills require long day adaptation, late blight resistance; (3) degeneration of potato seed stocks due to viral infection; and (4) dormancy of varieties for which seed stocks procured from hills for planting immediately in the plains but, due to dormancy, it was not possible.

In 1952, six potato varieties were released by Central Potato Research Institute, Shimla. These were Kufri kisan, Kufri kuber, Kufri kumar, Kufri kundan, Kufri red, Kufri

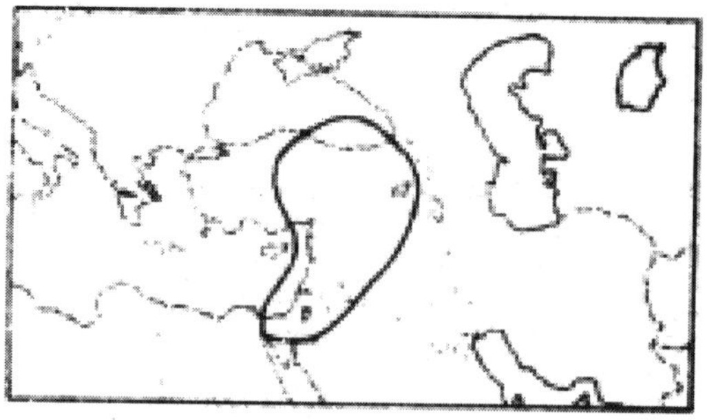

Aegilops speltoides

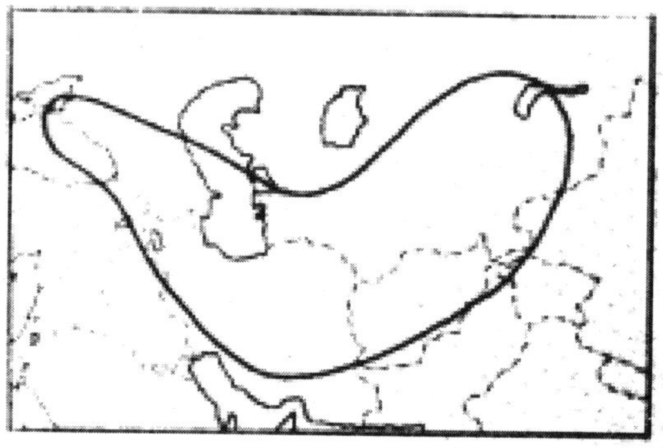

Aegilops squarrosa

Potato tubers variation

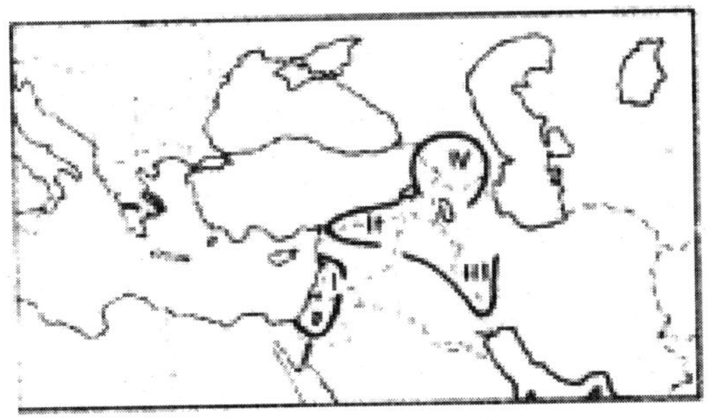

Wild tetraploid Triticum species: Triticum tugidum ssp. dicocoides (1,2,3) ssp. dicoccoides var. nudiglum (3) and T. timopheevi ssp. araticum (4) (Johnson, 1972)

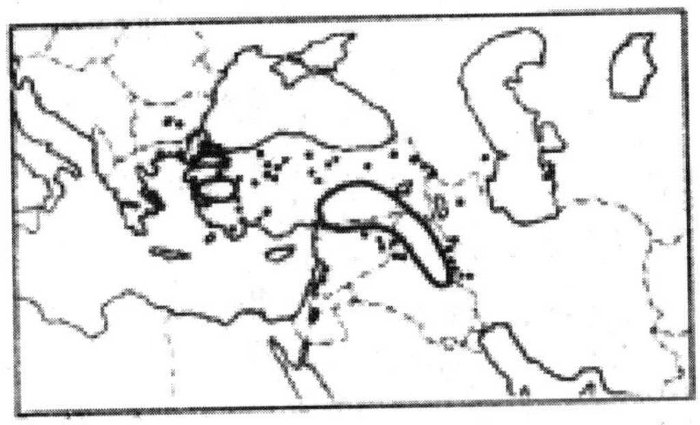

Triticum boeticum (Harlan & Zohhary, 1969)

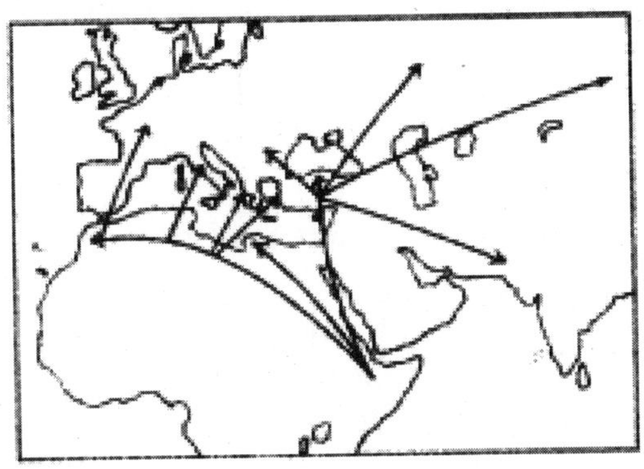

Spread of drum wheat in the old World (Ciferri, 1939)

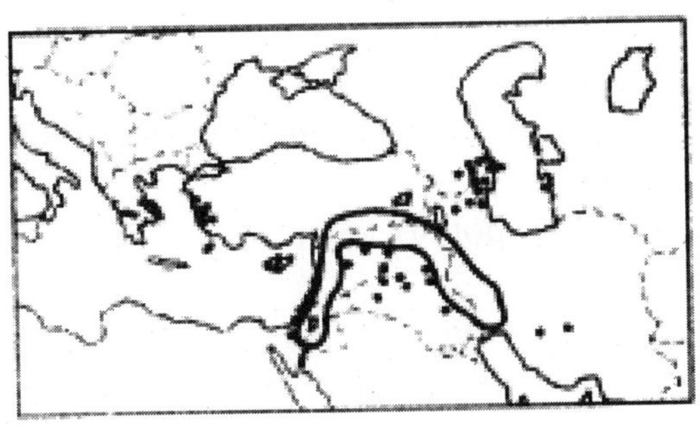

Hordeum spontaneum (Harlan & Zohary, 1969)

safed. All these varieties were released in1967–68. So far, 35 potato varieties have been released by CPRI. Potato production has been revolutionised in the country by: (1) development of high-yielding, disease-resistant varieties, (2) divizing a seed plot technique to facilitate healthy seed production in plains, and (3) standardization of agronomic practices. As a consequence of these measures, potato production in the country rose from 1.54 million tonnes in 1949–50 to 25.00 million tonnes in 2003. The per hactare yield increased from 6.58 t/ha. in 1949–50 to 17.58 t/ha. in 1998-99. The annual compounded growth rate (ACGR) of production, area and average yield of potato in India were 5.94, 3.48 and 2.37, respectively. The ACGR for yield of potato was next to wheat but was higher than others. The nutrient production per hectare/day for potato was also found to be higher than other food crops. Today, the per capita production of potato is 21 kg/year. Potato produces more food per unit area and per unit time. Out of 35 released varieties, only 14 potato varieties, viz., Kufri jyoti, Kufri lalima, Kufri badshah, Kufri bahar, Kufri chandramukhi, Kufri anand, Kufri ashoka, Kufri chipsona 1 & 2, Kufri giriraj, Kufri jawahar, Kufri sindhuri, Kufri sutlej, and Kufri pukhraj are the most important in India in recent years.

India ranked fifth in area and production of potato (Bist and Sharma, 1997) potato-based institutions are also on the rise in India. The biological value of potato protein is high. Nutritionally, potato is a complete food lacking only in two essential amino acids and milk supplement can make up the deficiency. One hundred grams of potato provides 90 to 100 calories (Shekhawat, Grewal & Verma, 1992). Thus, potato is credited to be a common man's food in India as also elsewhere.

REFERENCES

1. Bist, B.S. and H.C. Sharma, 1997. Potato Statistics. *Tech. Bulletin*, No. 40, C.P.R.I., Shimla.
2. Gaur, P.O., P.S. Naik; S.K. Kaushik and S.K. Chakarbarti, 1999. Indian potato varieties; *Tech. Bulletin*, No. 51, C.P.R.I., Shimla.
3. Hawkes, J.G. 1945. *The Story of the Potato. Discovery*, February, pp. 38–46.
4. Hawkes, J.G. 1990.*The Potato: Evolution, Biodiversity and Genetic Resources,* pp. 5761, Belhaven Press, London.
5. Hawkes, J.G. 1992., 'History of the Potato' In P.M. Harris (ed.) *The Potato Crop: The Scientific Basis for Improvement*, pp. 1–13.
6. Pal, B.P. and Pushkarnath. 1951. *India Potato Varieties, Bulletin* No. 82. I.C.A.R. New Delhi.
7. Pushkarnath, 1956. Breeding potato varieties in South Asia. *Indian J. Genet. Pl. Breed.*, 197–211.
8. Pushkarnath, 1969. *Potato in India: Varieties.* I.C.A.R., New Delhi, p. 493.
9. Pushkarnath, 1970. *Problems of Potato Varieties in India. Expl. Agric.*, 6:181-190.
10. Pushkarnath, 1976. *Potato in Subtropics*, Orient Longman, New Delhi, pp. 289.
11. Sharma, B.D.; H.S. Gupta and K.R. Dhiman, 1984. *Performance of Advanced Hybrids and Released Potato Varieties in Northeastern states.*

12. Shekhawat, G.S., J.S. Grewal and S.C. Sharma, 1992. Potato in India. *Tech. Bulletin* No. 1. C.P.R.I., Shimla.
13. *Souvenir,* 25 years of potato research in India 1949–74. pp. 16–18, C.P.R.I., Shimla, 1974.
14. Upadhyaya, M.D. 1974. Potato. *In Evolutionary Studies in World Crops* (ed.) Sir Joseph Hutchinson, pp. 139–148, Cambridge University Press, London.
15. Watt, G. 1908. *Dictionary of the Economic Products of India.*
16. Wealth of India, Vol. IX. 1972. pp. 396–397 C.S.I.R., New Delhi.
17. Yadava, S.R. (ed.) 1993. Potato Research in India (1901–1991). A Bibliography, Vol. I, C.P.R.I., Shimla.

SECTION TWO

HARAPPAN AND VEDIC FACETS OF AGRICULTURE

CHAPTER 13

Agriculture in the Garhwal Himalayas: An Ethnographic Perspective

K.P. Nautiyal, Pradeep M. Saklani and *Vinod Nautiyal*

INTRODUCTION

It has been estimated that about seventy percent of the population living in the Himalayan villages draws its sustenance from agriculture and livestock rearing. The Himalayas are known to be the storehouse of many traditional agricultural crops (Maikhuri, *et al.*, 1994). As far as Garhwal Himalayan region is concerned, it has been found that the villagers mainly depend upon rain-fed terraced agro-ecosystem which consists of mono as well as mixed crops, including *Dhān*, or *Sathi* (*Oryza sativa*), *Gehūṇ* (*Triticumaestivvum, T. durum*), *Jau* (*Hordeum Vulgare*), *Rāgī*, *Maṇḍuā* (*Eleusive corncana*), *Jhangorā* (*Echinochola crusgali*), *Kauṇī* (*Setavia italica*), *Mārsā* (*hypochondriacus*), *Phāphar* (*F. tataricum*) and Bhaṭ *(Glycine Max)* (Dutta, 1993). Singh, in general, has studied the history of some of these crops (Singh, 1984). It has been found that the cropping patterns upto 1800m AMSL are built around two major seasons, i.e. Kharif (rainy season, April-October) and Rabi (winter season, October-March) (Maikhuri,. *et al.*, 1996).

HISTORICAL SURVEY

The early history of agriculture in the Himalayas in general has not been investigated extensively. However, the retrieval of the carbonized seeds of *Lithospermum arvanse, Medicago denticulata, Lotus corniculatus,* etc., from the Neolithic sites of Burzahom (Agarwal, 1982) and the seeds of wild wheat and barley from the aceramic and ceramic levels belonging to pd I and II A and II B from Gufkral (Sharma, 1982), suggest that the Neolithic settlers in Kashmir valley cultivated both wild and domesticated varieties of plants. Thus, it may be rightly said that agriculture began with the Neolithic culture in the Himalayas. Nevertheless, moving towards the north in the Uttarakhand, we do not find sufficient evidences on the antiquity of agriculture, or for that matter, the development of agro-pastoralism. The major handicap in studying the early history of agriculture is the non-retrieval of any palaeobotanical remains in the archaeological deposits of the excavated sites in Garhwal. Except for the early historical and Painted Grey Ware sites of Ranihat and Purola in the Alaknanda and the Yamuna valleys, most of the other excavated sites in Garhwal Himalayas belong to some kind of pastoral settlements and, therefore, the reconstruction of the history of agriculture in this region which is a storehouse of some of the endangered plant species and traditional crops has always been a challenge to archaeologists.

In view of this the authors have tried to re-examine afresh the historical and archaeological data. The textual references related to the movement of the ancient tribes, coupled with an ethnographic background of the Himalayas, bring out an insight into the age-old agro-pastoral practices, thereby indicating a possible development of agriculture in this part of the Himalayas. The present study has been divided into two parts, i.e.,

(I) Archaeological background.
(II) *Kharak-Chhān* practice associated with agro-pastoralism.

ARCHAEOLOGICAL BACKGROUND

Early Settlement in the River Valleys of Garhwal

The excavation of the PGW site at Thapli in Alaknanda valley in 1980 not only established the northernmost extension of the Painted Grey Ware culture in Garhwal Himalayas, but the nature of the site also suggests that the P.G.W. settlers moved from the Gangetic plain into the river valleys of Garhwal around the beginning of the first millennium BC (Nautiyal, *et al.*, 1981). The excavation of a large number of faunal remains of domestic animals of such as *Bos indicus* (Indian humped cattle), *Sus scrofa cristatus* (domestic pig), *Equus caballus* (domestic horse), confirms the fact that these animals must have supplemented for the dietary needs. The analysis of the faunal remains also shows that contrary to the preference for younger animals, the P.G.W. settlers at Thapli butchered matured animals for their proteins (Badam, 1991). Based on the carbon isotope analysis of the faunal remains of Thapli, it has been shown that the domesticated animals at Thapli were fed with a C_4– rich diet (Nautiyal, *et al.*, 1995). This further indicates that the inhabitants at Thapli around the first millennium BC might have possibly cultivated some of the crops like maize, which is essentially a C_4 plant. However, in the absence of any paleobotanical remains from Thapli and Purola (P.G.W. sites in the Alakananda and the Yamuna valleys) and from an early historical sites at Ranihat datable from *circa* 600 BC to 200 AD, it is difficult to identify the crops cultivated by the early settlers of the Himalayan river valleys. As far as an indirect archaeological evidence is concerned, it is limited to the discovery of a sickle from an excavation of the Vedika at Purola in the Yamuna valley, datable to the second century AD (Nautiyal and Khanduri, 1988–89); and also by the presence of a few iron implements, including a plough-share from the late levels at Moradhwaj, an early historical site, datable between fourth century BC to fourth century AD (Nautiyal and Khanduri, 1986). These evidences indirectly suggest that the inhabitants in the river valleys as well as in the foothills must have engaged themselves in some sort of an agricultural activity and applied such implements in harvesting the crops.

From Pastoralism to Agriculture

It has been generally assumed that pastoralism in the Himalayas is primarily associated with a purely transhumance groups, and not with the sedentary agricultural community occupying a large part of the Himalayas with a limited livestock husbandry (Berremen, 1963; Philmore, 1989). Philmore (1989), on the basis of his observations, has tried to differentiate between the sedentary agriculturist and the pastoralist. For the former, according to him, the herding is a secondary activity, whereas for the pastoralist, the herding with long-distance movements is a primary activity. However, based on our own study in the western part of the Garhwal Himalayas, adjoining the Himachal Pradesh, it is observed that pastoralism became an important and an integral part of the sedentary agricultural farming and formed the very basis of the survival of these communities inhabiting the valleys in the mountainous environment. Various scholars, in order to understand their distinctive practices (Sherring, 1905; Pant, 1935; Dabral, 1964; Berreman, 1963; Philmore, 1989), have investigated the transhumance pastoralist communities of the central Himalayas.

From the Present to the Past

To understand the past, we have attempted to draw an account of the adaptability of the practice of transhumance by the contemporary pastoralist communities in the mountains, indicating that theirs must have been a continuing practice of combining the two important aspects of agriculture from the distant ancient past, exclusively, at least in the Himalayan region. This practice involves the seasonal movement of livestock from low altitude to the high altitude by various nomadic and pastoralist groups. Hence, it is referred to by various names as nomadism or transhumance pastoralism, as the case may be. This practice in the Himalayas is generally associated with the nomadic groups, particularly in the Gujjar, Bakarwal of the Himachal and the buffalo transhumance herding community of Kashmir.

In the case of the Garhwal Himalayas, attempts have been made to discover the traits of a pastoral community within a historical perspective (Dabral, 1964). Recently, an excavation of the remains of a horse from a megalithic site at Malari at an altitude of 3500 m in Garhwal provides sufficient archaeological evidences to prove that the early pastoral communities like the Khasas and those migrating from the Central Asian steppes came along with a sizable livestock herd around the beginning of second millennium BC (Bhatt and Nautiyal, 1987–88). This finding suggests that the transhumance pastoralism might have evolved in the high altitude region of Garhwal around the second millennium BC.

The widespread long-distance movement of these transhumance pastoralists in the inner valleys of Garhwal Himalayas triggered a kind of nomadic movement. This process continued for a longer span of time until these groups encountered the sedentary agricultural communities or the aborigines of the area, known as the Kol-Kirātas (Naithani, 1994) settled in the middle region of Uttarakhand. As such, they are taken to be the first initiator to transfer their subsistence on sedentary agriculture. Thus an interaction and a mutual inter-dependence of shifting cultivation along with a small-scale animal husbandry must have evolved a mixed strategy for a viable livelihood in these Himalayan terrains. In other words, this must have naturally and firmly established the sedentary agricultural pastoralism. The cattle sheds, designated as *Kharaks* and the *Chhāns* in the local dialect, must have been further reinforced in response to such an adaptive strategy.

It may not be out of place to mention here that similar to the process of development of pastoralism evolving in the Ganga-Yamuna Doab (Thapar, 1984), it also developed in the Himalayas as a result of the interaction of the Khaśas and the other non-Aryans aborigines. One of the hymns in the *Ṛgveda* refers to the conflict of the Ahīs (the aborigines) and the Devatās (the Aryans as mentioned in *Ṛgveda)*. 'That Dasyurāja Shambara headed a combined army of Ahīs and Dānavas against the Devatās, who were led by Devar raja (Indra)' (RV.2.12.11).

Yh Shambara Parvateshu Kshīwantam Chatvārishyam Saradhyanvāvindatah
Ojayamanām yo Ahīṁ Jaghāna Dānum Shayanaṁ Sa janāsa Indraḥ.

The Ṛgvedic reference of a combined force of the Ahī and the Dānavā is significant in this respect and has been taken by some scholars to indicate that by the time of the forty-year war, the serpent propitiating Austric people designated as the Nishādas and the Dravidians, identified variously as the Daṣyus, Dānavās and Dāsas in the Vedic tradition, had established friendly contacts among themselves after having resolved their initial differences and offered a combined resistance to the army led by Divodās (Handa, 1994).

It may be noted that the aborigines (Ahīs) carried out the agrarian activities and the Khaśas developed the Rudra-based cult in the Himalayas. The interaction of Khaśas and the Ahīs in the mountains may have followed the same process as has been recorded for the Ganga-Yamuna Doab. Thus mutual dependence of both the tribes led to the development of sedentary agriculture with pastoralism (Handa, 1994). Ethnographically, the serpent cult subsequently gained more popularity and covered the entire Himalayan belt including Himachal Pradesh and Uttarakhand in its fold (Handa, 1994).

With the above background, the ethnography of the present-day practice *of Kharak* and *Chhān* in the Garhwal Himalayas deserves a detailed discussion.

Study Area

The Yamuna valley (Long 78° 74′ to 78° 31′-Lat. E 30° 31′ to 31°), covering an area of 8289 sq km, forms the westernmost flank of the Garhwal Himalayas (Fig. 1).

Administratively, the entire region of Yamuna valley comprises the three districts of Uttarkashi, Tehri and Dehradun. The three distinct ethnic groups described geographically through different names inhabit this valley right from the high-altitude region to the foothills. Accordingly, the inhabitants of the area of Ranwāin are called

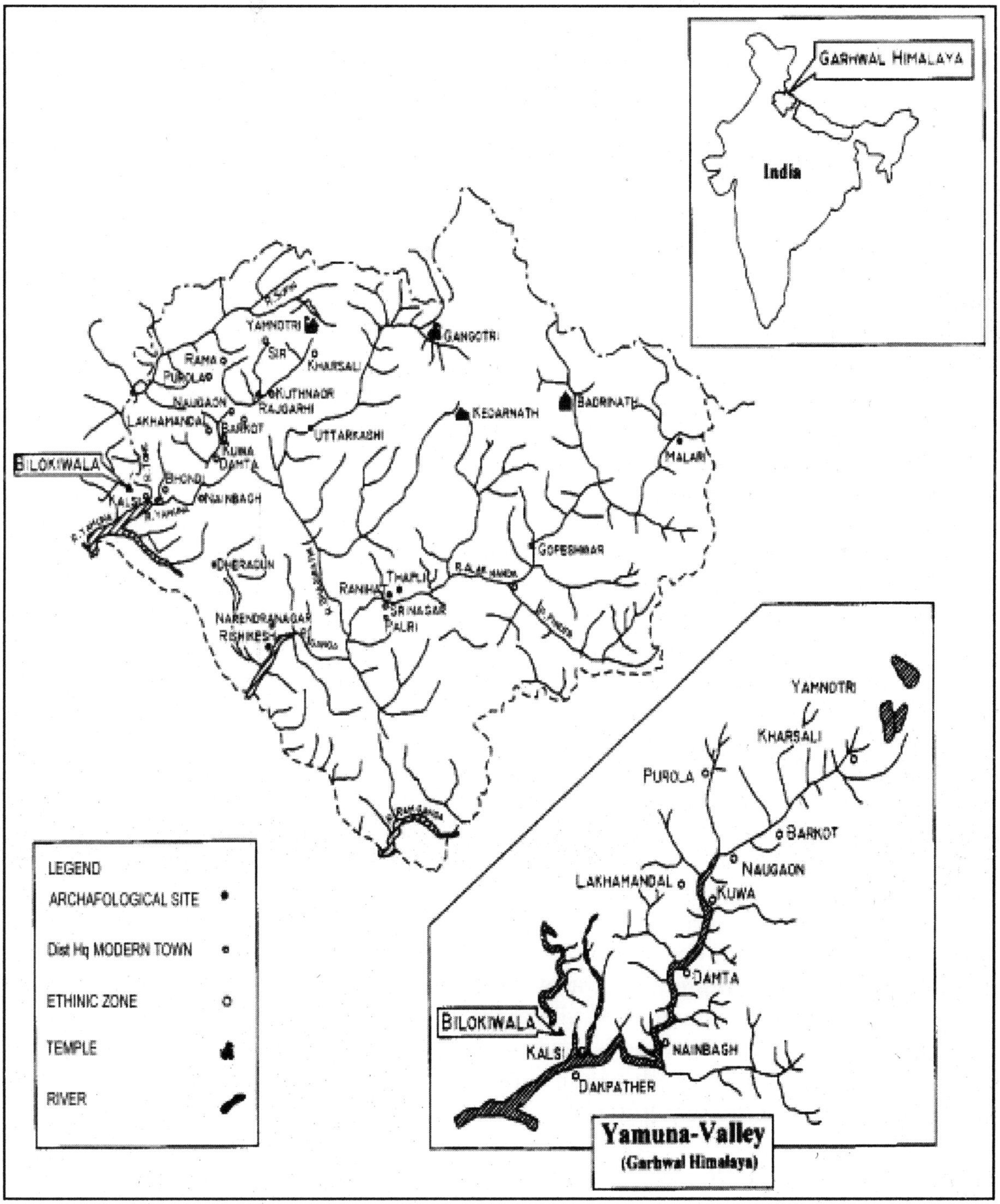

Fig. 1: Principal Archaeological Sites with Ethnic Zone in Yamuna Valley

the Raṅwāltā; those of Jaunsār-Bāwar the Jaunsārīes; while those from the Jaunpur are called the Jaunpuriās, respectively. From the anthropological and sociological point of view, this entire region—on account of its unique social system of marriage—has been considered as a distinct polyandrous zone (Berreman, 1963).

KHARAK-CHHĀN: PRESENT-DAY STRUCTURES ASSOCIATED WITH AGRO-PASTORALISM

The present-day practice of transhumance pastoralism in the Yamuna valley also owes its existence to the peculiar characteristic of the area and has evolved out of a gradual interaction between the aborigines and the Khaśas. The transhumance pastoralism in the Himalayas can be taken as a cyclic and rhythmic process, synchronized with the changing annual climatic cycles. In the recent years, due to a massive environmental degradation, the traditional practices have also been badly affected to a larger extent; nevertheless, there are certain pockets in the inner Himalayas where such practices associated with transhumance pastoralism are still intact and preserved.

In some remote villages of the Yamuna valley, particularly in the upper Jaunsār and Rawāin areas, the herd boys (called *Aṅwals* in local dialect) move away from the area along with their cattle in the month of March and return again in October. This practice may be called a long-duration transhumance, since the animals are not required for intensive agricultural work as compared to the river valley agriculture. Another very interesting observation made during the field studies is that the *Aṅwals* (herd boys) use whistles made out of the trachea of eagle to control their cattle from going astray. Interestingly, in this context, it may be added that far off in Maharashtra, a large number of skeletal remains of phalanges with a hole on the surface of each one of them, have been excavated from a Chalcolithic site at Prakash, which have been identified as a kind of whistle used by the Chalcolithic settlers of Prakash, dating back to 2500 BC (Thomas, 1988).

During their movement high up in the Alpine meadows, the principal work of the herdsmen is to build temporary sheds called *Kharaks* to reside in after reaching the high pastures of *Bugyāls*. Each group of villages in the same valley has a common pasture or *khātā* assigned to it over which it exerts community ownership (Pant, 1935). In the pasture, a spot is selected, depending upon the availability of fodder, water along with safety and other facilities. After selecting the spot, these herdsmen build the sheds of huts of a very temporary nature. The frame of the shed consists of wooden poles. The poles are tightened together and interlaced with the branches of a tree. Sometimes, the plank wall is raised upto a height of 10 to 12 feet and the door is kept very small as a precaution against the entry of wild animals. In the middle of the floor, a hole is made for the fire where the food is cooked and the milk is boiled. The herdsmen dig an underground pit lined with wooden plank and stone slab for keeping the milk products and their daily requirement.

Chhān—A Residential Unit

One of the distinctive adaptive features evolved by the Himalayan inhabitants in the mountainous landscape is the construction of a simple temporary single–storied structure or enclosure for keeping the livestock or rarely as a residential unit, called the *Chhāns*. In local dialect, the word *Chhān* represents a thatched covered structure built for a shelter (Andress, 1966). These structures are made for a temporary occupancy and shelter rather than for housing purpose. These temporary or semi-permanent types of cattle sheds are seen in the grassland zone of the high altitude and also in the terraced fields in the river valleys. As the name implies, the *Chhāns* in the high altitude grazing land are not occupied continuously, but only for a part of the year. These structures are primarily constructed to keep the cattle and other livestock safe, particularly during the night. They are also provided a small space for human occupation. These *Chhāns* a form very important components of the agro-pastoral, life raised by the inhabitants of the Garhwal Himalayas and particularly that of the Yamuna valley. The significance of these *Chhāns* has largely remained unnoticed. The only commendable work on such structures from anthropological perspective has been done by Andress (1966). It has also been observed in this regard that the *Chhāns* are primarily constructed because the cultivable land is not sufficient near the villages to support a large number of families. Secondly, the fodder is also not enough to support the livestock for the whole of the year. In this way, the construction of *Chhāns* not only enables the inhabitants to move their livestock for intensive grazing from the low-altitude *Chhāns* to the high-altitude *Chhāns* between the summer and the autumn months. Simultaneously, it also helps the farmers to save their time in collecting the manure in the form of cowdung at one place, very near to their ploughing fields.

The *Chhān* sometimes proved to be a multi-purpose enclosure. Villagers also occupied them during the agricultural or herding period. It has been observed that all the people affiliated with a particular village live in the village settlement permanently and continuously (Andress, 1966), but go to the *Chhāns* at the time of harvesting and planting. The people living quasi-permanently in the *Chhāns* visit the village occasionally. Therefore, these *Chhāns* also function as multiple houses in the Himalayan landscape. It has been observed that the site for a *Chhān* construction is selected on the basis of economic and agricultural viabilities. If cultivation is a major activity, the quality and quantity of the soil, steepness of the ground and the irrigation possibilities are given special consideration. If the *Chhān* is to be used primarily as a livestock shelter or milk station close to the town, an ample fodder must be locally available (Berreman, 1963). *Chhāns* are normally constructed near the wild vegetation for supplying fuel and enough stock of fodder, depending upon the number of animals that can be kept in them. Agricultural *Chhāns* are not established in large deforested areas already under cultivation, but at a place wherefrom the cowdung could be easily collected as manure.

The construction of *Chhāns*, as stated above, is a novel strategy evolved not only to obtain a maximum yield from the crops but also to manage the livestock productivity for farming purposes. Rather, this dual strategy forms a basis of the entire agro-based economy of the villages in the mountainous habitat.

Types of Chhāns

The *Chhān* in the Garhwal Himalayas, particularly in the Yamuna valley, can be classified into two types based upon the construction and the manner of their use (Fig. 2): *Puccā* (permanent) *Chhāns* and *Kacchā* (semi-permanent) *Chhāns*.

Fig. 2: *Chhān* in Yamuna Valley

A *puccā Chhān* is built of stone and strong timbers with the roof normally made of locally available thin sheets of slate stone. The walls are entirely masonary, having beam set-up without any mortar. The exterior surface is plastered with mud only once. The dimension of the *Chhān,* though varying, is generally 40 feet in length, 14 feet in width and 9 feet in height from the ground level to the roof. The *Chhān* works both for the human abode as also that for the livestock.

The architecture of the *puccā* (permanent) *Chhān* is characteristic to the needs of the user, studded with paved-stone courtyard, living quarters in the form of a circular threshing floor, etc. The intention to build such a *Chhān* was primarily to provide all the comforts in one unit for a longer period of time.

In a few instances, the *puccā Chhāns* are improved to such a degree that they differ little in appearance from the normal village dwellings. This occurs only when the site of *Chhān* is economically viable with all the needs available at one place. Andress (1966) considers it as an extension of the household economy of the village.

On the other hand, the *Kacchā Chhān* is the simplest kind of a structure to which the word *Chhappar* (hut, hovel) has been used (Andress, 1966). It is about 50 feet long, 16 feet wide and 10 feet in height with a simply thatched roof resting on bamboo pieces. The floor is packed with earth and is plastered with dung. There is no provision for the outlet of smoke caused by the cooking fire. It is furnished very simply, with a *chullhā* (cooking hearth), a wooden chest for clothes and food staples and a few cooking utensils, baskets, brass water pots, etc. The occupants sleep on the floor space. In such a *Chhān,* people and animals live together on the same floor and essentially in the same room. It is primarily a shelter for livestock and only secondarily for men. A *kacchā Chhān* can be used occasionally either during some period of the year or it may be occupied only once and then left abandoned.

Morphogenesis of Chhāns—A Conceptual Framework

The tradition of *Chhān* or semi-permanent enclosures—what we see today in all likelihood—might have evolved through various stages of development. The most formative and early stage of its evolution can be traced back to the early Neolithic pastoralist societies. The archaeological remains of similar structures, identified as enclosures, have been excavated at Mahagara, a pre-Neolithic site in district Allahabad, dated to 6000 BC. They can also be correlated with the cattle pens associated with the ash mounds of southern Neolithic culture (Allchin, 1963).

However, the present-day ethnographic evidence of *Chhāns* in the Himalayas sheds a new light on its concept and gradual development from a simple practice of the *Kharaks* and *Chhān* transforming through various stages into the houses and subsequently into the entire village.

The concept behind the evolution of the traditional village morphology from *Chhān* has been explained from the socio-economic criterion propounded by Berreman (1963) and Andress (1966). Therefore, based upon the anthropological study of the ethnic communities and their associated adaptive strategies, it has been found that due to the polyandrous and polygamous system of the marriage institution in the region of Jaunpur, Jaunsār-Bāwar and Rawāīn region of Yamuna valley, a large family could not be supported under a single roof. Hence, this enlarged family was shifted or technically extended into a different house, called the *Chhān*, which served the multifunctional requirement of the family. The temporary or permanent *Chhāns* served a dual purpose of an extended house and also due to their closeness to the agricultural fields, it enabled the villagers not only to look after their agricultural fields more effectively but also saved their time and energy in procuring the manure.

The *Chhān* serves simultaneously as an organic link between the livestock management and the agricultural practice adopted by the users. This sensitive relationship between these two components of human ecology remained preserved over centuries in the Garhwal (mid-Central) Himalayan region. Therefore, it may be conclusively stated that the present practice of such an age-old tradition in the Yamuna valley provides a good case study of continuity and the change in terms of agro-pastoral activity in Garhwal. In the light of this trend of development of houses from the various stages of *Chhān* and *Kharak* in deeper Himalayas, we have tried to trace such practices in ancient literature. Interestingly, a close parallel may be drawn between the present-day practices and those mentioned in the ancient literature (Fig. 3). In Vedic literature, there are references regarding the development and emergence of villages during the *Ṛgvedic* period. The Vedic text refers to the three stages of human settlement (Agarwal, 1953).

(I) *Guhā* and *Parvat Niwās*
(II) *Araṇya Niwās*
(III) *Kṣhetra Niwās—Shaya Se Dharvasthiti Tak*

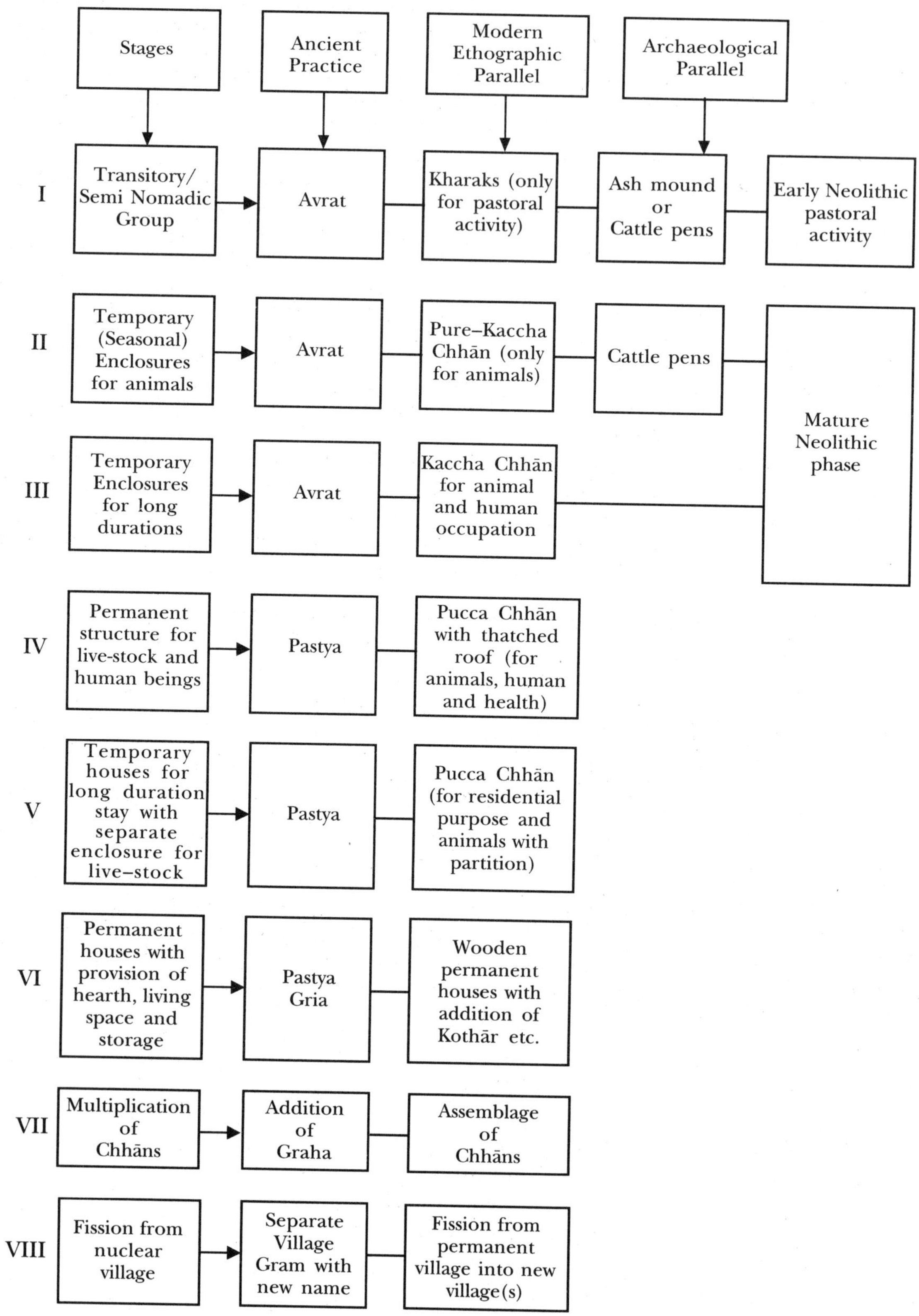

Fig. 3: Hypoth model showing different stages of development of *Kharak, Chhāns* and houses

The first two stages towards the development of village were semi-permanent ones, without the houses. However, the third stage is interesting in context to the present study which has been referred to as *Dharva Stithi* or temporary settlement. Its development is seen in two direction—first the herdsmen or a group of herdsmen started establishing a kind of enclosures for the animals in an area where water and pasture were in plenty. The herdsmen for temporary living also used this kind of temporary structure. This may be correlated with the present-day Kharaks. These temporary structures have been referred to as *Pastya, Vraja* or *Varjana* or *Pastyamiti Grihaṇām,* as can be seen from the following passage of (*Ṛgveda* 10.96.11).

Gawam-Haryantam Sprihnīyam Pastyam
Griham Gorudakasyokta Gunakam Sthānam.

The word *Pastya* denotes a semi-permanent enclosure (Mishra, 1984). Based on this reference, it suggests that in the beginning, *Pastya* was used in the context to the animals and was distinctively similar to the kachhā *Chhāns.* This subsequently became a semi-permanent structure for both animals and herdsmen (Puccā *Chhān*). Similarly, it has been mentioned that just like the terms *pastya, Vrajon* or *Vrajan,* the herdsmen were referred to in terms such as *Gau chāraṇa* or *Gau chārak.* Accordingly, the *Vrajana* denotes a place, which was made only for keeping the animals but later on the herdsmen also started to use these enclosure as residences. With the gradual development of agriculture, the inhabitants started living permanently in one place.

In conclusion, it may be pointed out that in the trans-Himalayan ecological framework, the growth and development of agriculture in the Himalayan region is very well related to the pastoral management of the hill community.

NOTES AND REFERENCES

Agrawal, D.P., 1982, *The Archaeology of India,* Curzon Press, London.

Agarwal, V.S., 1953, *India as known to Panini,* Chaukhamba Sanskrit Series Press, Benaras.

Allchin, F.R., 1963, *Neolithic Cattle-Keepers of South India,* Cambridge University Press, Cambridge.

Andress, J.M., 1966, *Culture and Habitat in the Central Himalaya,* Unpublished Thesis, University of California, Berkeley.

Badam, G.L., 1991, *Faunal Remains from Thapli,* in K.P. Nautiyal and B.M. Khanduri, (eds) *Emergence of Early Culture in Garhwal, Central Himalaya,* Srinagar (Garhwal).

Berreman, G.D., 1963, *Hindus of the Himalayas: Ethnography and Change,* University of California Press, Berkeley.

Bhatt, R.C. and Nautiyal, K.P., 1987-88, Trans-Himalayan Burials, vis–à–vis Malari: An Assessment. *Journal of Himalayan Studies and Regional Development,* Vol.11 and 12:95–101.

Dabral, S.P., 1964, *Uttarakhand Ke Pasucharak,* Veergatha Prakashan, Dugadda, Pauri Garhwal.

Dutta, M., 1993, Conservation of Genetic Resources: *The Himalaya Imperative, Eco-Crisis in the Himalaya,* Ed. by Vir Singh; IBD, Dehradun.

Handa, O.C., 1994, *Buddhist Art and Antiquities of Himachal Pradesh,* Indus Publication Company, New Delhi.

Maikhuri, R.K., Semwal, R.L., Singh, A. and Nautiyal, M.C., 1994, Wild fruits as a contribution of sustainable rural development: A case study from the Garhwal Himalaya, *International Journal of Sustainable Development and World Ecology,* 1: 56–68.

Maikhuri, R.K., Saxena, K.G. and Rao, K.S., 1996, Agricultural Crop Diversity in Central Himalaya: Conservation Problems and Prospects. In *Conservation and Management of Biological Resources in Himalaya,* P.S. Ramakrishnan, A.N. Purohit, K.G. Saxena, K.S. Rao and R.K. Maikhuri (eds), Oxford and IBH Publishing Co. Pvt. Ltd., New Delhi 443–53.

Mishra, S.N., 1984, *Prachhin Bharat Me Gram Evam Gramya Jeevan,* Purva Sansthan, Gorakhpur.

Naithani, S.P., 1994, *Uttarakhand Ke Pracheen Tirth Evam Mandir,* Pavetri Prakashan, Srinagar.

Nautiyal, Vinod and Farswan, Y.S., and Uniyal, A.K., 1995, Stable Isotope Analysis of Archaeological, Bone from Garhwal Himalaya, *Science and Culture,* 61 (7–9): 139–140.

Nautiyal, K.P. and Khanduri, B.M., 1986, *New Cultural Dimension in the Central Himalayan Region of Uttarakhand—An Archaeological Assessment,* ANNALI, Vol. 46, 77–100.

Nautiyal, K.P. and Khanduri, B.M., 1988–89, Excavations of a Vedic Brick Altar at Purola, District Uttarkashi, Central Himalaya: *Purātattva,* 19. New Delhi: 68–69.

Nautiyal, K.P., Khanduri, B.M., Nautiyal, Vinod and Rajput, D.L., 1981, Thapli: A PGW Site in Garhwal, *Purātattva,* 10. New Delhi: 94.

Pant, S.D., 1935, *The Social Economy of the Himalayans,* George Allen and Unwin Ltd., London.

Phillimore, Peter, 1989, 'Folk Forests and land in Himachal Pradesh', in *Conservation of the Indian Heritage,* B. Allchin, F.R. Allchin and B.K. Thapar (eds), New Delhi, Cosmo Publications, 56–66.

Ṛgveda, 2.12.11.

Ṛgveda, 10.96.11

Sharma, A.K. 1982, Excavation at Gufkral, *Purātattva,* II. New Delhi: 9.25.

Sheering, C.A., 1905, *Western Tibet and British Borderland,* Edward Arnold Publisher, London.

Singh, C.D., 1984, *Modern Techniques of Raising Field Crops,* IBH Publishing Company, New Delhi.

Thapar, R., 1984, *From Lineage to State,* Oxford University Press, Bombay.

Thomas, P.K., 1988, *Faunal Assemblage, in Excavation at Inamgaon,* Vol.I, Part-11, 823–862, Deccan College Post-Graduate and Research Institute, Pune.

CHAPTER 14

Agriculture in the Indus Civilization

S.R. Rao

INTRODUCTION

The announcement by John Marshall, Director General of the Archaeological Survey of India in an article in the *Illustrated London News* (September 20, 1924) stating that a civilization as old as and as great as the Mesopotamian flourished in the Indus valley took the historians and archaeologists all over the world by surprise. As a result of the preliminary excavations at Harappa in Punjab and Mohenjo-daro in Sindh, large brick buildings and hundreds of antiquities including seals with writing on them were found. Excavations continued in the thirties and reports of Mohenjo-daro's excavation by J. Marshall (1931), J.H. Mackay (1938) and M.S. Vats' report on Harappa excavation (1940) were all published. In the meanwhile, Aureil Stein surveyed Baluchistan and discovered a few sites of Indus civilization. Before the partition of India, Rangpur in Saurashtra, excavated by M.S. Vats, was considered as an Harappan outpost, but doubt still lingered about its Harappa affinity, especially when the Deccan College, Poona resumed excavations in 1947. However, the present author's excavation at Rangpur in 1953–55 confirmed it to be a mature Harappan settlement which continued to be occupied for a few centuries more after the decline of the metropolitan centres of Harappa and Mohenjo-daro. Further survey of the Sabarmati-Bhogavo river estuaries in Gujarat in 1954 brought to light a full-fledged mature Harappan port at Lothal, where excavation revealed a unique tidal dock, thereby adding a Maritime province to the Indus civilization south of the Indus valley (Rao, S.R., 1973 and 1991). The excavation at Kalibangan in Bikaner division of Rajasthan by B.B. Lal and B.K. Thapar highlighted a Pre-Harappan phase of the Indus civilization developing into a mature Harappan phase. Other important sites of Indus civilization recently discovered and excavated are Kotada-Dholavira and Surkotada in Kutchh, Banawali, Kunal and Bhagwanpura in Haryana, Prabhas, Kanewal, Zekda and Bhagatrav in Gujarat, Dadheri in Punjab and Daimabad in Maharashtra.

The Indus civilization, also known as the Harappa civilization, is no longer confined to the Indus valley. It extends from the border of Iran on the west, Turkmenia and

Kashmir in the north to the Godavari valley in the south and beyond Delhi in the east, covering an area of more than 1.5 million square kilometres. In view of the fact that more than 300 sites of this civilization are found in the ancient Sarasvati valley in Punjab, Haryana and Rajasthan (Fig. 1), it is deemed fit to redesignate the Indus civilization as Indus Sarasvati or Sindhu-Sarasvati civilization. Since the term Indus civilization is popular, it is retained here. Three distinct phases of the Indus (Harappa) civilization distinguishable on the basis of material equipment found are the early Harappa or Pre-Harappa culture, the Harappa (mature phase) culture and late (declining phase) Harappa culture. Sir John Marshall, under whose supervision Harappa and Mohenjo-daro (now both in Pakistan) were excavated in the twenties and early thirties of the previous, century, assigned the seven building periods of Mohenjo-daro to 3250–2750 BC with a rider that a thousand years must have taken for the Indus civilization to attain maturity by 3250 BC. D.P. Agrawal on the basis of C 14 dates, suggested a time bracket of 2300–1750 BC which, however, Wheeler rightly discarded on the grounds that dates based on dendrochronology were more accurate and futher, the archaeological finds of Mesopotamian origin found in Indus sites and *viceversa* are reliable for cross-dating. In the meanwhile, the C 14 dates earlier than 1200 BC were found to be much younger and corrections known as the MASCA correction have been accepted. The callibrated C 14 dates for the pre-Harappa, Harappa (mature) and late Harappa culture sites are given below:

Pre-Harappa Culture

Mehrgarh III-IV (Baluchisthan) Kotdiji I (Sindh)	3900–2800 BC
Sothi (Rajasthan) Kalibangan I (Rajasthan)	3100–2800 BC
Kotada-dholavira (Kutch)	2800–2500 BC
Kunal (Haryana)	3000–2000 BC
Harappa Culture	
Mohenjo-daro (Sindh)	3100–1900 BC
Lothal Period A	3000–1900 BC
Kalibangan I	2900–1800 BC
Mehrgarh V	3200–2500 BC
Kotada-Dholavira	2500–1900 BC
Banawali (Haryana)	2250–1700 BC
Kunal (Haryana, Dist. Kurukshetra)	2100–1900 BC
Manda IA (Kashmir)	2350–1900 BC
Late-Harappa Cultures	
Lothal B	1900–1600 BC
Rangpur IIB – IIC	1900–1600 BC
Manda IB	1900–1750 BC
Prabhas II (Saurashtra)	1800–1500 BC
Bet Dwarka (Saurashtra) (TL Date)	1700–1500 BC
Bhagwanpura IB (Haryana Dist. Kurukshetra) (TL Date)	1700–1500 BC

Note: If pre-Harappa ceramic/Neolithic phase of Mehrgarh is considered the origin of Indus civilization, its beginning could be pushed back to 5000 BC in Baluchistan and Ganga valley (Koldihwa), but the absence of Indus type seals and script does not make it Early Harappa Culture.

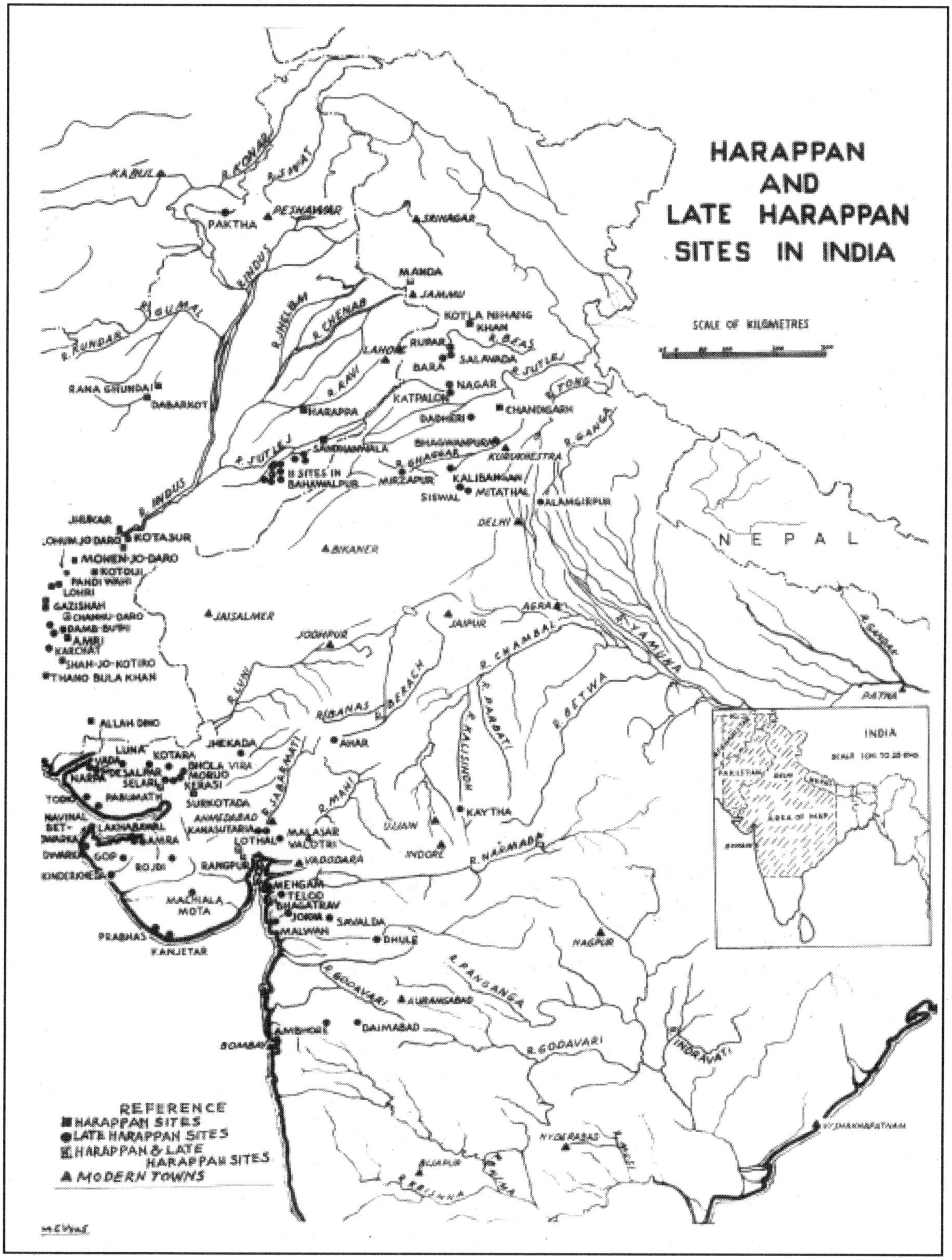

Fig. 1: Harappan Culture (map)

PROVINCES OF THE INDUS EMPIRE

Within the geographical limits of the Indus empire, four provinces are recognizable: Baluchistan on the west, the Indus valley (Sindh and Punjab) in the centre, Sarasvati-Yamuna valleys in the east and Gujarat, including Kutch and Saurashtra, in the south.

ACHIEVEMENTS OF THE INDUS CIVILIZATION

The most outstanding achievement of the Indus civilization is the cultural integration of different groups of people—urban and rural folk, pastoral people and forest dwellers—by peaceful means. Common economic interests brought them together. With competent leaders at the centre and in the provinces, the people enjoyed not only material prosperity but also the freedom to observe one's own faith. For instance, the worship of the mother goddess, popular in the Indus valley, was hardly known in Kalibangan and Lothal. On the other hand, fire worship and offering of sacrifices were practiced in Saurashtra, Rajasthan and Haryana. Yet the cultural intergration of the forest dwellers venerating the tiger and elephant deities with the fire (Agni) worshippers is clearly indicated in the so-called Pasupati seal of Mohenjo-daro.The great discipline of the people in every walk of life resulting from the practice of yoga is an important contribution of the Indus civilization. The ruler brought about discipline in civic administration, ensured proper planning of towns and standardized weights and measures, which facilitated internal and international trade. The Harappans were the first in the world to build a tidal dock at Lothal after a careful study of the waves, tides and currents. Their achievement in the field of Hydraulic engineering is remarkable for the age. As metallurgists, they produced the purest copper which was in demand in Bahrain and Mesopotamia. They manufactured on a mass scale the finest beads of semi-precious and precious gems for export to Bahrain, Mesopotamia and Africa. The decimal graduation of their weights and linear scale is a pointer to the early realization of its advantages.

The development of arts, crafts, science and technology of the Harappans was possible because there was enough surplus food available to sustain the non-agricultural population. If so, the question, that arise are: 'Did they apply scientific knowledge to agricultural operation?' and 'Did they have a balanced diet?' Before answering these questions, it is necessary to examine the food crop that they cultivated, which itself demands a knowledge of the environment, rainfall, soil, method of cultivation, agricultural tools, use of fertilizers and method of storing grains. It also involves a study of animals domesticated for food, transport and other purposes.

ENVIRONMENT

The environment, which includes the area's climate, is an important factor that determined the growth or decline of the Indus civilization in general and agriculture in particular. The region which the Harappans occupied 6000–5000 years ago vary from the present-day barren hilly tracts of Baluchistan and the desert and semi-desert condition of Sindh, Rajasthan and north Gujarat to fertile plains of the Ganga-Yamuna valleys and heavy rainfall area of south Gujarat. According to one school, the climate during the pre-Harappan and Harappan times was more favourable and wetter than what it is today in Sindh and Rajasthan. There was a thick forest cover in the hilly regions of Saurashtra, north Gujarat and Baluchistan, so that enough timber for building houses and boats and fuel for baking millions of bricks for construction work could be had. Another view is that not much fuel is needed for baking bricks. It must, however, be acknowledged that the archaeological evidence for frequent flooding of Indus settlements, cultivation of food crops such as rice and wheat and the presence of rhinoceros and elephants indicates the existence of swamps, grasslands and better vegetation cover which, in turn, implies wetter conditions. Even in 1500 BC, the Dwarka region was fertile and had a forest cover.

It must be remembered that not only climate but all surroundings that affect an organism is its environment. There are also biotic (inter and interaspecific competition) and abiotic (soil, atmospheres, light, temperature, precipitation, etc.) components of the environment (Vishnu-Mittre, 1993 : 34). As far as man is concerned, his adjustability and level of tolerance also have to be taken into account. Even today, the Indian desert is said to be not as bad as elsewhere. There has been a rainfall of 10–25 cm more volume during the last 10 years.

The river Sarasvati, which provided abundant water and rendered fertile its vast flanking plains with alluvium during the pre-Harappa and mature Harappa periods lost its two main tributaries, the Yamuna and Satluj, resulting in gradual drying up of the river, further aggravated by less rainfall was greatly responsible for the decline of the Indus civilization. However, hundreds of small, de-urbanized rural settlements sprang up in Gujarat, Haryana, Punjab and Sindh in the late Harappa period. It is interesting to find that the late Harappans not only knew the presence of groundwater even in the sandstone and basalt beds of Saurashtra and other regions, but also exploited them. There is a reference in the *Mahābhārata* to Sri Krishna shooting an arrow and getting potable water to quench the thirst of Rukmini near Dwarka. Even now, there is plenty of groundwater in Okhamandal, where Dwarka is situated. The epic Mahabharata describes the plants and animals in the forests around Dwarka (Rao, S.R., 1987).

The satellite imagery shows that there were several habitation sites in the little and great Rann in ancient times. As late as seventeenth century AD, tamarind fruit was exported from Saurashtra. The leguminous plants Indigo and lathyrus odoratus, the roots of which help in the natural fertilization of the soil, were cultivated even two centuries earlier in western India. But the British stopped their cultivation (Vishnu-Mittre, 1993:35). Apart from floods and fluctuating climatic condition, earthquakes too seem to have contributed to the decline of Harappan prosperity. In conclusion, it can be said that varied environmental context must have resulted in the decline and devolution of the Indus civilization in the second millennium BC all over the Indus and Sarasvati valleys as also in Gujarat. During this period, the less fertile black soil covered the more fertile alluvium of the earlier phases.

SOIL

One of the late Harappan sites excavated in Dholka district of Gujarat is Kanasutaria, 40 km northeast of Lothal (Rao, S.R., 1963: 188), where recent wind-blown sand has completely covered the protohistoric mound. In the habitation deposit of 2 m thickness, post-Harappan pottery of Rangpur periods IIC-III types was discovered. Mudbricks, sandstone weights and terracotta sling balls are other finds here. What is, however, significant is the stratigraphic evidence for ecological change. At the bottom is a 0.6 m thick deposit of coarse river sand and pebbles overlain by fine-grained river sand in which geometric microliths and flake blades of chalcedony unassociated with pottery occur. This layer is succeeded by buffish *kankar* (*calcareous nodules*). It is succeeded by alluvium and fine sand locally known as *muram*. The subsequent overlying deposit of black clay is about 1m thick, over which the Chalcolithic (post-Harappan) dwellers settled using lustrous red ware and black and red ware.

In this connection, it is necessary to mention the trial trenches dug at several places within a radius of 20 km of Lothal. Here sand overlies coarse gravel. The river bed on the western flank of Lothal town revealed a coarse gravel bed with microliths and a fine gravel bed underlying successive deposits of loam brought in from annual floods interspersed with 0.3 to 0.5 m thick deposit of loam during abnormal floods, which indicate a wetter condition. This loam is similar in composition to the *muram* found underlying the black clay at Kanasutria. In the habitation area of Lothal period A, buffish *muram* underlying the early structures was used for making pots and bricks. In Lothal period B, however, black clay overlying buffish *muram* was used for the same purpose. This sequence of buffish loan underlying black clay is noticeable in Rangpur periods I to IIc and also at Koth and other mature Harappa sites.

Prior to the advent of the Harappans, Lothal was a small village. The main occupation of the residents was stockraising, fishing and agriculture, besides bead making. From the fine levigated buffish loam pottery of superior fabric, christened 'micaceous red ware' could be produced. The Harappans, too, used the same clay, but five centuries later, this fertile loam was not easily available as it was covered by black clay. This change indicates a climatic and environmental variation.

In the low-lying Bhal region around Lothal, where the soil is fertile and rainfall is 75 to 100 cms, even now rice and wheat are cultivated. It is, therefore, no wonder if both cereals were cultivated is the Harappan period too. The present cultivators resort to contour bunding (Rao, 1973, 50) for locking up rainwater in the fields during the monsoon. This measure was meant to leach out saltpetre and retain the moisture necessary for cultivation of cotton and wheat in winter. The canals dug near the dock for sluicing small boats suggest that a few irrigation canals too might have been dug for cultivating rice in Harappan times.

Malvan and two other late Harappan sites, Jokha and Dhatva in Tapti (Payoshni) valley, are situated on black cotton soil, while Mehgam in Narmada estuary is on buffish *muram*. The flood-stricken population of the Indus estuary rushed to safer places in north Gujarat and Saurashtra and temporarily settled on the coast. Thus came into existence clusters of late Harappan villages at Lakhabawal, Amra and Vasai in Jamnagar district and at Chachana, Samadhiala, Devaliyo and Bhimpatal around Rangpur. Most of these settlements are on black cotton soil and a few on *kankary* soil. These inhabitants depended on small streams for water supply.

In Haryana, Punjab, Sindh and Rajasthan, however, the late Harappans continued to settle on fertile alluvium on the plains flanking the fast drying up Sarasvati and

Drishadvati rivers and their feeders. Earlier, the pre-Harappans and Harappans had enjoyed living on the banks of these snowfed and rainfed rivers, which brought down alluvium to render the land fertile. Most of the fifty late Harappan sites in the upper plans in the Ganga-Yamuna doab in Saharanpur and Muzaffarpur districts are situated at a distance of 3 to 12 km from one another, with an occupation deposit of 1 m or even less. It appears that an emotional attachment to their ancestral home led some of the late Harappans to Lothal, Rangpur, Dwarka and a few other places in Saurashtra and to the upper Ganga-Yamuna doab in north India.

When a virile civilization like the Harappan declines owing to causes other than political and the people move to new areas or resettle in abandoned sites, the settlement pattern differs from the original. The urban character is lost and there is a conscious effort to live in clusters of villages in order to readjust themselves to new hydrological and socio-economic conditions, which is reflected in their occupations such as agriculture, cattle rearing, pottery making and shell working. Villages like Vaniavadar and Bhimpatal in Saurashtra specialized in pottery making, while Zekda in north Gujarat became a farming village, like many others.

PRE-INDUS PEASANT CULTURES OF BALUCHISTAN AND GANGA VALLEY

For a proper understanding of the Indus civilization, it is essential to study the peasant culture as revealed in the excavations at Mehrgarh in Baluchistan (J.F. Jarrige and R.H. Meadow) and Koldihwa in Uttar Pradesh (G.R. Sharma). Both Baluchistan and Ganga valley should be considered as primary centres for two important food crops, namely, wheat and rice, respectively.

Mehrgarh: Seven cultural periods have been distinguished at Mehrgarh, which is situated on the river Bolan. The earliest (period I 6000 BC) is an Aceramic Neolithic culture while the latest (period VII 4000 BC) represents the Chalcolithic culture.

The early farmers of Mehrgarh cultivated emmer, einkorn and two-row barley, thereby indicating that the northwestern boundaries of the Indus civilization were occupied by agriculturists as early as the sixth millennium BC. Slightly later and perhaps beyond the eastern boundary of the Indus civilization, rice was being cultivated. Both these regions were successful in selecting superior cereals. The Mehrgarh cultivators selected free-threshing wheat and six-row barley. Here, one can see a clear transition from hunting to herding (Meadow, 1981). In the lower levels of the Neolithic culture, the faunal remains are characterized by bones of hunted animals such as wild cattle, onagers, gazelle, wild goat and even elephant, a discovery which suggests a savanna type of vegetation in Baluchistan 8000 years ago.

FROM HUNTING TO HERDING

By the end of the Neolithic phase, the faunal remains included mostly domesticated cattle. In Mehrgarh, the domesticated cattle, zebra and water buffalo became prominent after 5000 BC. In Iranian Neolithic culture, goats and the cattle were dependent on the same soil sources as man. He had to open more fields to meet the needs of the cattle.

It is in the riverine plains of the Indus and Sarasvati and not in the hilly tracts that he found enormous opportunity to open more field and increase his cattle wealth in the sixth millennium BC. The Vedic people considered cattle as the real wealth. Farming also increased with the selection of cereals suitable for the local condition of soil, climate and water supply. The compact and short variety of wheat *triticum sphaerococcum*, which was well adopted to flood irrigation, was cultivated in Mehrgarh as early as 4000 BC. At the same time or even earlier, the farmer in Koldihwa (Gangetic plain) cultivated rice. The spread of farming from the plateau to the plains caused disturbance in natural vegetation and habitat of several species of wild animals, which stimulated the process of domestication, thereby reducing hunting for diet. By the fifth millennium BC, farming had spread greatly in the alluvial lowlands.

The Mesolithic culture of Adamgarh in Madhya Pradesh is datable to 5500 BC. In the later phase of Microlithic culture at Langhnaj and Mitli (near Lothal) in Gujarat, pottery was already being created and even copper was known to the inhabitants.

DIVERSIFICATION OF AGRICULTURE IN NEOLITHIC CULTURE

Some more particulars of the diversification of agriculture in Baluchistan in Aceramic to Ceramic Neolithic phases of Mehrgarh Period I may be noted here for a proper understanding of the diversification of agriculture in Indus valley civilization (Jarrige and Lachevallier, 1979). In the 5-metre thick deposit of early Neolithic phase (seventh Millennium BC), sickle, grinding stones and cereal impressions in lumps of clay have been reported. Impressions of two-row hulled barley (*Hordeum distichum*), six-row barley (*H. Vulgare* and *H. Vulgare, varnudm*), einkorn wheat (*Triticum monococcum*) and bread wheat (*T. durum/aestivum*) indicate that agriculture in Baluchistan was not only well developed but also diversified by the end of the sixth millennium BC. As noted earlier, the domesticated sheep (*Ovis orientales*), goats and cattle dominated over wild animals such as gazelle, wild sheep and swamp deer. Here, traces have been found of the water buffalo, the earliest of its species in south Asia. Large rectangular buildings with symmetrical rooms were exposed in the Neolithic phase of Mehrgarh. Interspaced among these buildings are large cemeteries, in some of which baskets coated with bitumen were found along with imported ornaments of local stones as also lapis lazuli and turquoise. A copper bead and traces of red ochre-coloured textile on polished stone axes are among other noteworthy finds. In the later phase potsherds were also found.

CHALCOLITHIC PHASE

South of the Neolithic habitation of Mehrgarh is a fifth millennium BC habitation, wherein is a large rectangular building divided into 10 narrow compartments symmetrically disposed: on each side of a narrow corridor there are no doors. A large number of imprints of wheat and barley are visible in the compartments. Moreover, two sickles with stone bladelets hafted slantingly in bitumen were also found in a compartment. The fireplace outside the wall contained charred grains of several kinds of wheat such as *Triticum aestivum–compactum* and barley, besides some seeds of the cultivated variety of cotton (*Gossypium* sp.) used either for oil or its fibrous properties. The cut marks on animal bones indicate butchering activity. Pottery consisted of wheel-turned burnished pear-shaped vessels.

The granary of Mehrgarh anticipates the Harappan granaries of the Indus valley. Outside the granary of Mehrgarh, terracotta human figures in sitting position comparable to Harappan figures were noticed. In Mehrgarh, the deposits of the fifth-sixth millennium BC reveal the antiquity of a farming economy, of craft specialization and long distance contact with Baluchistan, all of which definitely indicate the cultural process leading to urbanization of fourth millennium in the Indus valley. This cultural process is deeply rooted in the Neolithic culture. At the beginning of the fourth millennium BC, Mehrgarh (Period III) becomes a mass production centre of pottery, and micro drill bits of stone, which the Harappans had also produced. Period III is one of technological innovation and diversification of agriculture. The earliest appearance of oats (*avence* sp.) is remarkable here. *Triticum sphaerococcum* was added to other varieties of wheat. The Neolithic cultures of Mehrgarh (Baluchistan), Koldihwa (Ganga basin) and Kunal (Sarasvati valley) can be treated as the precursor of the Indus–Sarasvati urban civilization. Kalibangan I (Pre-Harappa) witnessed the planning of the town, use of kiln-fired bricks and devicing antiflood measures. The Neolithic-Chalcolithic phase of Kunal I anticipated Harappan religious rites and produced bone and copper tools, Harappan pottery types and terracotta figurines.

MEHRGARH PERIOD IV

In this period polychrome pottery, large semi-conical vessels on pedestals, goblets, stamp seals and female terracotta figures with pendulous breasts anticipate Harappan figures.

MEHRGARH PERIOD V

Between 3200 and 2500 BC Mehrgarh (V) shows contacts with Indus cities on the one hand and eastern Iran (Sahr-i-Sokhta), Afghanistan (Mundigak III) (Jarrige, 1979:85–87) and Damb Sədaat II in Quetta Valley on the other.

In brief, Mehrgarh unfolds the history of man's adaptation to the Indus valley, just as Kunal, and Koldhihwa reveal man's adaptation to Sarasvati-Yamuna Valleys. The large variety of Terracotta human figures suggest a sophisticated society in Mehrgarh V, just as the structure, utensils and ritual objects of Kunal suggest a sophistication in religious rites and beliefs similar to those of the Indus and Vedic people.

Chirand on the river Ghaghra in District Chapra (Bihar) is a very important Neolithic-Chalcolithic site which had direct or indirect contact with the late Harappans. Its earlier phase is datable to 4000 BC. The earliest Neolithic sites are on hilltops or foothills but Chirand, a pure Neolithic Complex, is situated in low-lying plain with a monsoon climate. It was excavated by B.S. Verma as early as 1960. Its Chalcolithic phase came to light in subsequent excavations. The alluvial soil of the Ganga and the Ghaghra was suitable for agriculture. The most significant evidence of cultivation at Chirand is the presence of carbonized remains of grains like rice, wheat and barley and lentils such as *moong, masoor* and peas. The discovery of microliths in the Neolithic strata of Chirand indicates their use as sickle teeth (A.K. Verma, 1988: 28–29). The animal and bird bones from Neolithic level include those of domesticated humped cattle, buffalo, sheep, goat, pig and dog. The bones of wild animals belong to one-horned rhinoceros, swamp deer and *chital.* Bone and stone arrowheads, spearheads and tanged arrowheads indicate hunting activity, while bone needles and terracotta spindle whorls are indicative of weaving. Stone tools such as, axes, pestles, querns, hammers and celts with butt end, burnished stone wedge, chisels and hammers were used for exploiting wood and bamboo. Terracotta human and animal figures occur in Neolithic Chirand, and cotton occurs in Alamgirpur in the Harappan context (Sankalia,1962:54). Vishnu-Mittre dates rice to 6000–7000 BC (*Indian Express* Madras Nov. 19, 1975 Science Diary). The Neolithic horizon of Koldihwa in Allahabad District of U.P. in which rice (perhaps cultivated) was found, yielded rounded celts, microliths and pottery with cord impressions (G.R. Sharma, 1974). The dates of middle and late Neolithic levels of Chirard range from 1755±155 BC to 1580±100 BC. The earlier level may go to 2000 BC. Sankalia has proposed 4000 to 3000 BC for Neolithic Chirand on the basis of the rate of sedimentation (Sankalia, 1962).

In conclusion, it can be said that cotton, lentils and cereals including rice and wheat, were cultivated by the Neolithic people by 4000 BC and continued to be cultivated by the Harappans at Lothal, Rangpur and Alamgirpur.

PREPARATION OF THE FIELD

The Neolithic food producers did not plough the field. They used to dig pits with a pointed wooden stick, at the top of which a stone ring for weight was fixed. This method continued for some time even after metal came into use.

Lambrick (1967) observes that 'wheat and barley, the principal food grains on the flood plains are cultivated without ploughing, manuring or producing additional water'. This may be the manner by which the Harappans produced crops. But there is positive evidence from Kalibangan I to prove that during the pre-Harrappan phase itself, the field was ploughed. According to B.K. Thapar (1975), an outstanding discovery of the excavation, however, was a ploughed field situated to the southeast of the settlement, outside the town wall. This is perhaps the earliest ploughed field excavated so far (Fig. 2), showing a grid of furrows with one set more closely spaced (about 30 cm apart) running east-west and the other, widely spaced (about 1.90 m apart) running north-south these. The former seemed to have been ploughed first. This pattern bears a remarkable resemblance to modern ploughing in the neighbourhood, wherein two types of crops (pulse in one direction and mustard in the other) are grown in the same field, the combination being dependent upon the size and growth behaviour of plants. At Banawali, R.S. Bisht found two terracotta models of ploughs, one of which is intact (Fig. 3).

A partly broken terracotta plough, mistaken for a horn, was found at Lothal also. It is, therefore, obvious that the plough was used by the Harappans. It is assumed by the excavators that since cultivation during this period (Kalibangan I) depended on flood irrigation supplemented by seasonal precipitation, only the winter crop (*rabi*) was

Fig. 2: Kalibangan: Ploughed Field

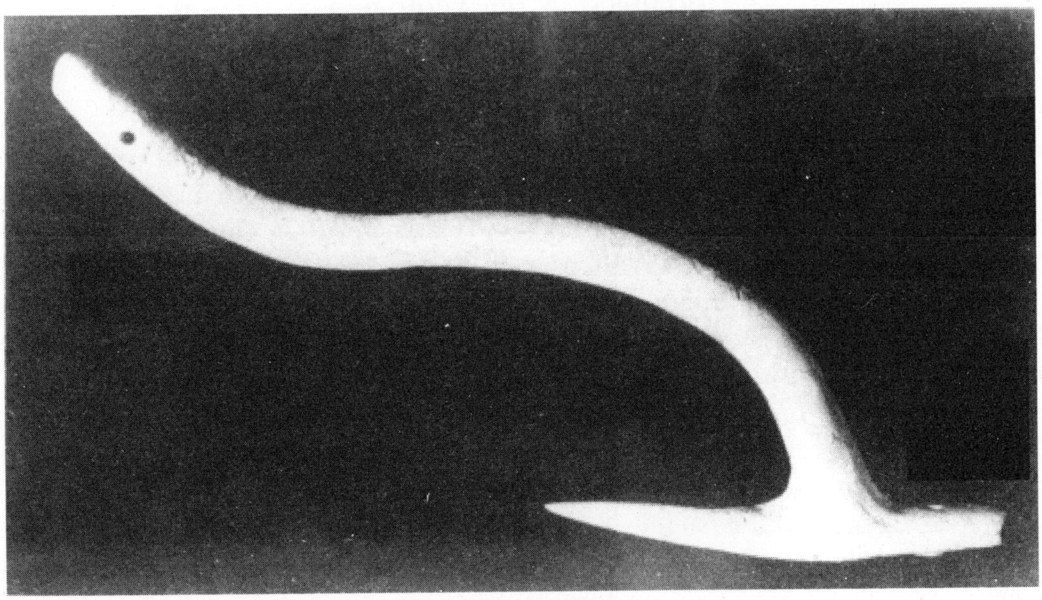

Fig. 3: Banawali: Terracotta model of a plough

grown and that too in autumn after the river flood had subsided. Although no cereals were found in the course of excavation, cereal type pollen has been attested to in good numbers in the deposits of Period I (Singh *et al.*, 1974). It appears that the people were capable of dealing with the demands of the river. Fields may have been surrounded by earth embankments in the Harappan period as is done even now in Saurashtra. The natural fertility of the alluvium was exploited together with the annual inundation.

As a result of the research and experiments conducted by the Punjab Agricultural University, Ludhiana, to achieve a minimum tilling, a bullock-drawn till planter has been developed to shear off the soil crust and stubbles, to open a furrow, sow the seeds and apply fertilizer. A seed drill with two hooks similar to the one depicted on a terracotta seal from Lothal (Fig. 4) was used by the farmers. It could serve the purpose of a tiller as well as sower in a single operation.

Fig. 4: Lothal: A seed drill motif

Harappa is situated on a minor geomorphological anomaly—within the entrenchment of the river Ravi, a feature of high ground juts into the valley near the alluvium. Some Harappan settlements like Dholavira could have been made close to cultivable land and more or less safe from floods, but this is not true of late Harappan sites.

In Rajasthan and Haryana, there are a number of early Harappan settlements, namely Kalibangan I, Siswal A, Balu I, Kunal I and Banawali I, the rest being either mature or late Harappan sites. A couple of early and mature Harappan sites must have been engaged in exploiting copper from Ganeswar (Agrawal and Kumar 1993), but others depended on agriculture in the flood plains. Most of these sites are located in the region occupied by the Vedic people (Fig. 5).

SOIL AND WATER CONSERVATION

For purposes of conserving water and preventing soil erosion, contour bunding was done in the Harappan period. Huges Buller was the first to write on the Gabarbunds of Baluchistan (1903–04). G.L. Possehl also refers to Gabarbunds in Gujarat (Possehl 1975). There are several forms of bunds, which play an important role in the adaptation of agriculture to factors like climate and rainfall in semi-desert regions. The bunds in Saurashtra have a long history, dating back to the period of Asoka (272–232 BC), as attested by the inscription of Kshatrapa Rudradaman mentioning the improvement made by Asoka in the bund of Sudarsan Lake at Girnar. It is obvious that large-scale irrigation facilities were constructed in Saurashtra by third century BC Gabarbunds in Baluchistan, according to Raikes (1965), date from pre-Harappan and Harappan times. Recently, R.S. Bisht has also suggested that across the two rivulets close to Kotada Dholavira, the pre-Harappans and Harappans in Kutchh must have built bunds of rubble for water supply. There is also an underground water storage tank of Harappan times in Dholavira town. In Saurashtra and north Gujarat, the late Harappan farming

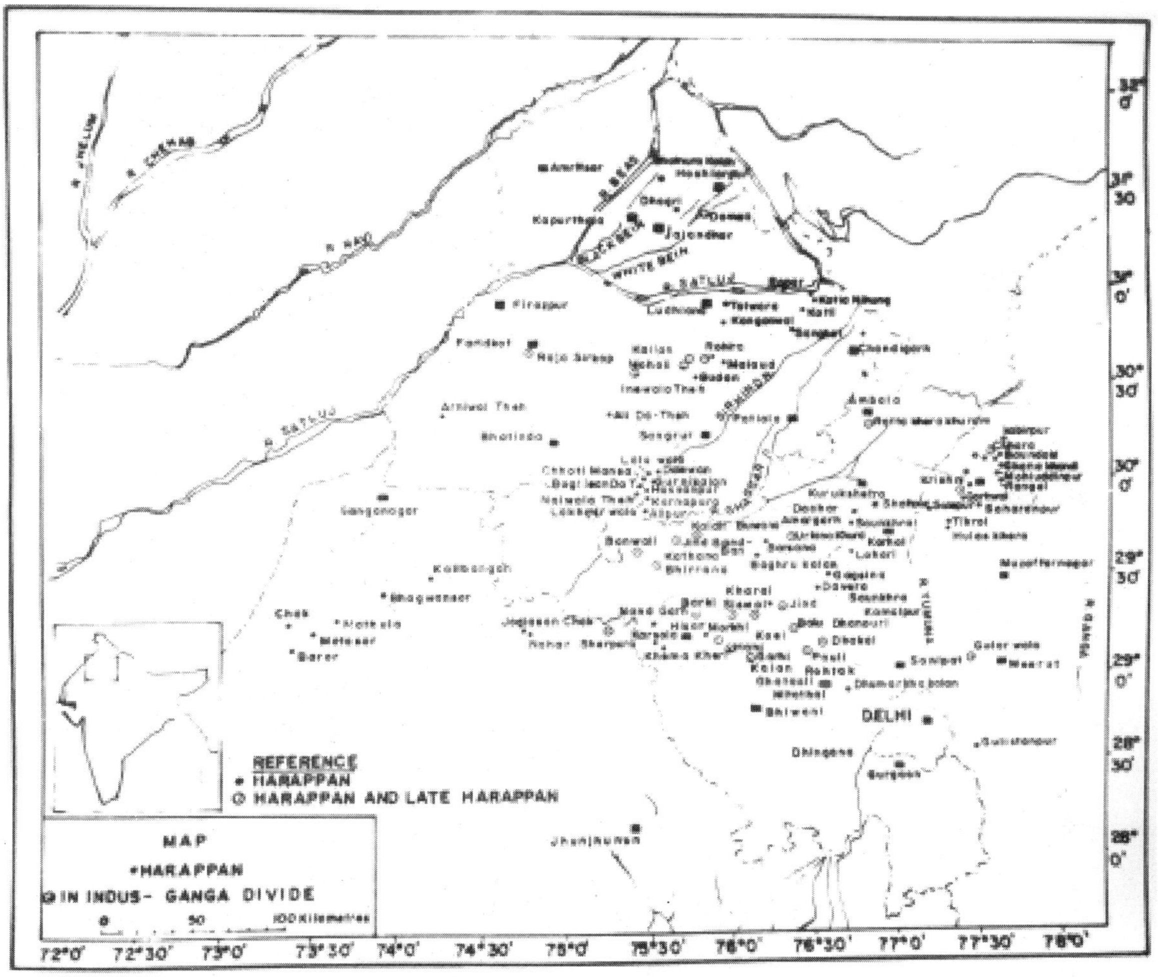

Fig. 5: Movement of the late Harappans and the Vedic people (map)

communities living in villages on the banks of the fast-flowing monsoon rivulets must have built Gabarbunds, some traces of which were visible at the late Harappan site of Vaniavadar near Amreli. The Harappans at Khirasar (Kshirasaras) built one such bund in Kutchh. A careful survey of the Harappan sites in Mehsana district of north Gujarat is likely to reveal bunds across the Rupen and Sarasvati rivers and their tributaries. The credit for building very large irrigation tanks such as the Sahasralinga tank at Patan and the Dholka tank goes to the Solanki rulers of Gujarat (eleventh century AD). The Sahasralinga tank was fed by a canal dug from the river Sarasvati, according to the ancient text *Sarasvati Purana.*

Around Lothal and Rangpur, contour bunding operations must have been undertaken, and perhaps irrigation canals must have been dug from the small feeders of major rivers like, Bhogavo, Bhadar and Sabarmati in order to supply water to the fields in winter, but this assumption needs confirmation by exploration based on satellite imageries. One such imagery suggests fields and canals near Lothal.

FORECASTING RAIN

Forecasting rain involves the study of winds, temperature and seasons. The Vedic calendar was based on Harappan system of counting of days in a year. In order to keep track of the seasons and the number of days in each season, the Harappans devised the method of performing sacrifices of long duration, lasting a whole year and adjusting the lunar calendar to solar calendar. It is they who divided the Zodiac into *nakshatras* (stellar constellations).

It was mentioned earlier that the Harappan religion was not different from that of the Vedic Aryans. They spoke an old Indo-Aryan language and practised yoga. They had full knowledge and use of rice and horse. It is, therefore, reasonable to consider the Vedic age as the product of Harappan intellectual activity. A close examination of the agricultural practices mentioned in the Vedic texts reveals that they are extremely similar to the Harappan agricultural operation. The implements used and the crops grown in particular seasons are common to both.

SEASONS AND ASTRONOMY

The beginning and end of the different seasons were determined on the basis of the heliacal rising of the stars which could help the farmers in fixing the date of annual events such as sowing and harvesting. The Egyptians and, later on, the Greeks also took advantage of the heliacal rising of stars. The Harappans devised a unique instrument to track the heliacal rising of a fixed star for comnencing the sacrifice (*sattra*) of long duration and determining the seasons. It appears that 2000 years before the Greek had thought of an 8-fold or 12-fold division of the sky, the Harappans had already achieved it and devised an instrument of a hollow shell ring with 8 or 12 slits cut into the upper and lower margins. The shell ring from Dholavira has 6 slits, in the upper margin and 6 more in the lower margin. The ring from Lothal has 4 slits in each margin (Fig. 6). When the lines passing through opposite slits on both margins are drawn on a plain surface, they cut at an angle of 30° in the Dholavira ring and at 45° in the Lothal ring. The purpose of having slits on both margins was not for measuring merely the line, points and degrees on a plane surface but to measure 8 or 12 entire sections of the horizon by

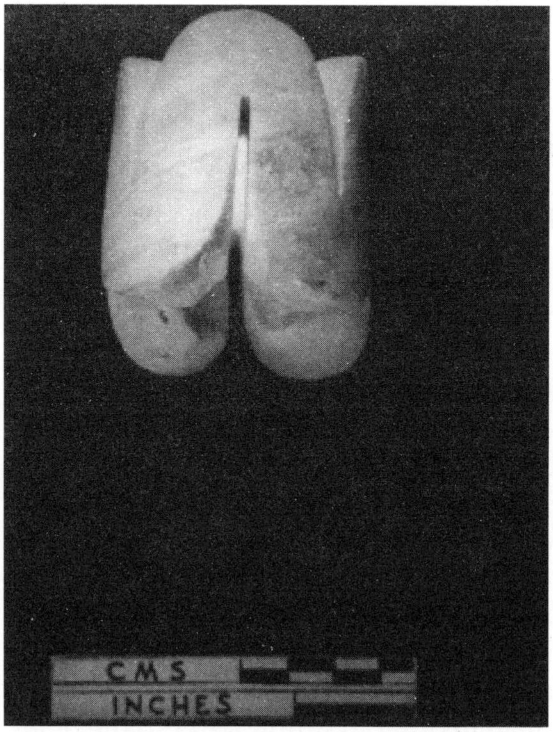

Fig. 6: Lothal: Shell ring used as a sextant

viewing the object simultaneously through opposite slits of the lower and upper margins when the instrument is suspended by a thread or wire passing through the central hole in the circular disc of metal or terracotta supporting the ring. It acts exactly like a sextant (Rao, S.R., 1996, 36–37). That the Harappans, like the Vedic people, had studied the star groups is evident from a Harappan seal in which a reference to the 7-star group Pleiades (*krittika*) is made. The inscription reads *haptaśasā,* which conveys the sense of seven shining stars) (Rao, S.R. 1996, 38–42).

SACRIFICES FOR KEEPING COUNT OF DAYS

The stars are nature's guide to navigators as well as agriculturists. The Indus priests were guided by the Heliacal rising of the *krittika* (*Pleiades*) for the commencement of sacrifices. *Krittika* was the first star group in the list of *nakshatras* of the *Yajurveda* and of the fire-worshipping Harappans. For *krittika,* the presiding deity was Agni. This explains the depiction of the fire god (Agni) in a loop of *asvattha* (pipal tree) in the Harappan Seal no. 677. A row of seven holy men (*saptarishis*) symbolizing the 7-star group of *krittika* is depicted in the lower register of the seal. In the upper register, a devotee is seen making an offering to the fire god. Behind the devotee is a goat. Above the Agni motif is the inscription *tridha sasaar,* conveying the sense 'triply shining praiseworthy', which is a reference to Agni referred to as *tridha* in the *Ṛgveda.* The tricephalic-horned deity surrounded by animals in the Mohenjo-daro seal no. 420 is also Agni and not Siva or Rudra. One of the eight epithets of Agni is Pasupati as he is surrounded by animals. He is their sustainer. For the Greeks, the rising of the Pleiades in May proclaimed that the season for harvesting the grain (grown in winter in the Mediterranean countries) had arrived. For the Harappans too, Pleiades, the 7-star group (*krittika*) was important for commencement of sacrifices and agricultural operations. The Indus seals refer to various kinds of sacrifices, namely, *ekaha, haptaha, aśvasatra, ashtaka,* etc. Actual sacrificial altars of brick and remains of bones of animals sacrificed have been found along with ritual objects like terracotta cakes in the altars at Kalibangan and Lothal (Rao, S.R., 1991). On a terracotta cake from Kalibangan a horned deity is engraved along with motif of a sacrificial animal being led by a man. The gold disc found in the Lothal altar is similar to the Vedic *rukma* shown on the forehead and arm of the statue of a priest from Mohenjo-daro. The sacrifices performed at Lothal and Kalibangan correspond more or less to the *gavāmayana* sacrifice of the Vedic period. This ritual symbolized the course of the sun (Rao, S.R., 1970). The *sattra* of this period must have lasted for a whole year, as did the *sattra* of the Vedas. The word *gavām* stands for the solar gait in the year. 'By holding the session of *gavām* ayana they also hold the walk of the Adityas. According to N. Mahadevan, the equinox occurred in the middle of the *sattra.* Atiratra (RV VII 103–7) is the ceremonial commencement of the *gavamayana* in the Vedic period (Dange, S.S., 1987). The cylindrical perforated jar of the Harappan and late Harappan period seem to have been used in *avabhrita snāna* (sacred bath) which needed a sieve-like vessel through which sacred water was poured over the bride and bridegroom, symbolizing the waters of hundreds of rivers (N. Mahadevan, 1931). This practice of pouring water through a sieve is observed even today in some Hindu marriages. The mention of *Magha* in Indus seal inscriptions, according to Mahadevan, seems to be a reference to the suspension of the cow's gait or the precession of days in *Maghas* and resumption in the *Phalgunis*, for the *Aitareya Brahamana* says, *Aghāsuhanyante gāvo Arjunyo Paryubate.* The Harappan *gavamayana* might have been a symbolic way of keeping count of the days and months and adjusting the lunar calender of 350 days to the solar calendar of 365¼ days (Rao, S.R., 1996: 40).

VEDIC SOURCE FOR INFORMATION ON INDUS AGRICULTURE

In the absence of Harappan records relating to land survey, grains stored in granaries, wages paid and seasons in which sowing and harvesting had to be done, scholars have to rely on whatever information can be gathered from the *Ṛgveda* and other texts. Most Harappans themselves being Aryans, it is reasonable to conclude that the Vedic texts reflect the climate, agricultural operations, tools, crops, etc., of the Indus civilization. S.A. Dange observes that the *Ṛgveda* is full of references to the rainy seasons (Dange, 1977:3–7). 'Parjanya, the god of rain, is said to be the seeder who plants the seed in the herbs, the mares, the cows and human females' (*RV.* VII 102.2). The Vedic people distinguished between various seasons, namely *Sarad* (autumn), *Hemanta* (winter) and *Vasanta* (spring). The *Ṛgveda* vioces it (R.V. X 161.4). The *Atharvaveda* mentions six seasons, *Grishma, Hemanta, Sisira, Vasanta, Sarad* and *Varsha.* The reference in *Ṛgveda* to *Purah Saradih* is misconstructed as a reference to the autumnal forts of the Harappans (*Dasas*) by the Aryan invaders (Wheeler, 1968). Dange has clearly explained that the reference to *purah saradih* (RV I 131.4, 174.2 and VI 20.10) is not to forts built to defend the city against invaders, but to autumnal rains brought by smashing the rain-bearing clouds (Dange, 1977:3–5). If we remember that the months *Asvina* and *Kartika* form the autumn and that the Asvins are associated with divine honey (rain), which is important for crops wheat and barley, it becomes obvious that the '*Ṛgvedic* seer is using a figurative expression for the gain of the much desired rain' (Dange, 1977:4). The autumnal forts were rain clouds and not actual forts. The *Ṛgveda* mentions Vrtra and Sushna as the demons of drought; but they indicate a spell of rainless period and not necessarily severe summer. It is only in later times (RV X) that acute cold was experienced in *Hemanta.* The *Śatapatha Brāhmaṇa* (1.5.4.5) says, '*Hemanta* verily controls all the creatures. Hence, in the *Hemanta,* herbs fade away; the sprout of the trees falls off; the birds turn as it were, man becomes wretched'.

The *Ṛgveda* says that agriculture is the best occupation and recomends that people should not play with dice but take to agriculture (*RV X* 34–13). The Asvins associated with autumnal rain were important for agriculture. The cutting of crops was done with a sickle (*datra*). *Ṛgvedic* references in this connection are significant—'may the sickles go close to the crop to cut it' (*RV X* 101.3) *nediva it srnyah pakvam a iyat.* The crop thus cut was tied in bundles. *Ṛgveda* refers to barley harvesters who cut shafts of *yava* by selective binding (*RV X* 131.2).

The same text mentions pounding the dried up crop in the field in a special spot called *khala* (*RV X* 48.7); the corn was then separated from the chaff by using a winnowing basket *surpa,* prepared from bamboo. This is referred to in the *Atharvaveda.*

The grain was thus cleaned and then measured and stored in a reservoir. The reservoirs or granaries of Harappa (Fig. 7) and Mohenjo-daro were immense. At Harappa, seven dehusking platforms (Fig. 8) were found in a row near the granary. Reference has been made earlier to the storage of grains in Mehrgarh, Surkotada and Banawali. The *Ṛgveda* refers to the reapers of corn called *dhanya krt* and to the men with measures as *sthivimantah.* Lothal has yielded beakers like earthen vessel of various sizes for measuring grain. They are in the ratio of 1:½:¼ (Fig. 9). Storage silos are found in most Harappan sites. Lothal has a warehouse (Fig. 10). The *Atharvaveda* mentions *methi,* which Dange thinks is a reference to the post in the threshing floor (Griffith: *Hymns of the Atharvaveda.* Benares, 1917, 167 note to the Verse *AV.*XIV 1. 40) The *Śatapatha Brāhmaṇa* introduces a legend about seasons which demanded a share of grains from the gods. On being refused by the gods, they went to Asuras. 'some of the Asuras went on tilling; others sowing the seed; cutting did others: thrashing at others; for them the herbs (corn-shoots)

Fig. 7: Harappa: Granary

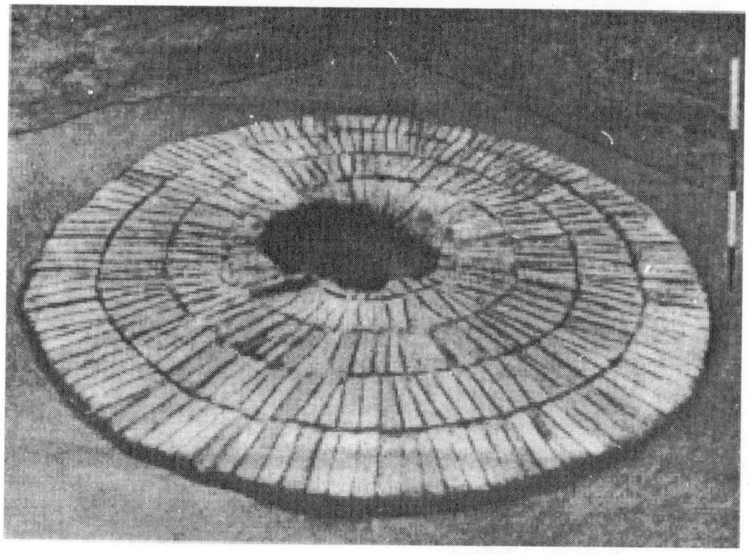

Fig. 8: Harappa: Dehusking platform

Fig. 9: Lothal: Terracotta beaker-like vessels for measuring grain

Fig. 10: Lothal: Warehouse

ripened without tilling (as it were *S.Br.* I.6.1.3). The *Ṛgveda* mentions *vrka* and *langala*, the two types of ploughs, the latter being associated with bullocks. *Vrka* may be a crude hoe-type instrument used for digging, but the Harappans used the *langala* type of plough, as attested by the terracotta models from Banawali. The *Atharvaveda* mentions that the *sira* was to be yoked with 6 to 8 (and double the number of bullocks) (*AV* VI. 91.1). Obivously, the *sira* must have been large and heavy. On the other hand, *langala* had only one share. It was light and easy to handle (Dange, 1977:9). According to the *Ṛgveda* ploughing meant digging, when in a symbolic way, it alludes to the sex act (*RV* X 101–12).

AGRICULTURAL PRODUCE IN THE VEDIC PERIOD

Yava (barley) was very popular in the Vedic period. The *saktu dhānah* (parched grain) is mentioned in the *Ṛgveda* (*RV* X 71–2). This crop was cleansed with the sieve, cooked and roasted. *Bhrstayavah* is parched *yava*, according to Sayana.

The *Vajasaneyi Samhita* mentions the following agricultural produce (*Vaj. Sam.* XVIII.12) *vrihi* (paddy), barley, black gram, sesame, green gram, *khalvas* (gram?). *Priyangus* (*Panicum italicum*), *anus* (*Panicum miliaceum*), *syamaka* (*Panicum ferumentaceum*), *nivaras* (?) *godhuma* (wheat) and *masuras*. Other *Samhitas* and the *Atharvaveda* also mention these agricultural products. *Symaka* is not mentioned in *RV* and *AV*, nor does *vrihi* find mention in *RV*. *Tandula* and *Vrihi* are mentioned in *AV*. The most significant part of information is the mention of season wise crop. Juice in *Vasanta* (Spring), *yava* (barley) in *Grishma* (Summer), herbs in *Varsha* (rainy season), *vrihi* (rice) in autumn *māsa*, and *tila* in *Hemanta* and *Śiśira* (four winter months). *Māsas* were not considered to be sacrificeworthy (*Tait. Sam.* V 1.2–1). The *tilas* were cultivated and also obtained without cultivation. The *jartilas* were both *gramya* and *arānya* when grown on untilled land. *Tila* was cooked with rice for *caru* and *odana* and the food thus prepared was called *tilaudana*. A kind of grass was cooked with rice for preparing *caru*, perhaps to give flavour to the preparation (Dange, 1977:10).

TYPE OF LAND

The *Ṛgveda* mentions two types of land: (1) *urvara* (fertile land), and (2) *artana* (unfertile land). Agricultural land was divided into suitable plots of field, and families could own this land (*urvara*).

Manure: Dried dung (*śaka/karisha*) was used as a natural fertiliser. The *Atharvaveda* refers to cows staying in pasture full of *karisha* (*asmin goshthe karishinih A.V.* III 14.3).

IRRIGATION

Whether irrigation was by canals is not known, although Indra is said to have released the rivers *sira*, which are off-currents diverted from the mainstream. Well irrigation too was in vogue. The well was provided with a big stone wheel, over which water was drawn in large buckets. There is mention of *kucakra* water wheel with circular jars fitted to the well. This was an important irrigation device for supplying water to the field

(RV X 102.1). Water hose for carrying water to a particular spot finds mention in the *Ṛgveda* (VIII 69–12). Although the normal size of the field is not mentioned, the use of *sira* type of plough with 24 bullocks suggests that the field was large. Between two fields, there was a strip of wasteland called *khila,* kept for grazing and temporary habitation during the reaping season (*RV* VI 28.2).

ANIMALS

In reference to animals, the *RV* mentions 3 categories (*vayavya* flying?), *Aranya* (wild beasts) and *grāmya* (domesticated). Among wild animals, lion and wolf are mentioned in *RV* and the tiger in later texts. The chariot of the Asvins was yoked with the bull and the *sisumāra,* a crocodile-like animal (*RV I.* 116–18). The propitiation of the frog for rains is referred to in the *Ṛgveda* (VII 103) and *Atharvaveda* (IV 15.13). *Nakra* is another type of crocodile mentioned in AV.

DOMESTICATED ANIMALS OF THE VEDIC PERIOD

Animals such as cow bull/bullock, horse, sheep and dog were popular beasts. The stud bull was left to roam among the cows for procreation. The zoomorphism of Indra becomes evident from the hymn (*RV,* VI. 28.8) which says 'may this sprinkling be sprinkled in the cows. That which is in the semen of the bull. O Indra, be it the same is in thy manly fluid'. Perhaps the Indus seal depicting the bull as symbolic of Indra.

THE BULL AND PLOUGH RITUALS

Dange remarks that the stud bull was also used for the symbolic fructification of the field by making the beast walk and urinate in the field. The plough was sprinkled with *soma* juice and the ploughshare was symbolically sharpened by stone axes (*RV* X 101.10–11). This is a clear indication of Neolithic antecedents of the Vedic cultures, as was the case with Harappan culture. Bullocks and horses were used for dragging the cart and as beasts of burden.

Agni is compared to the horse, both being called *akra, atya, aśva,* etc. (*atya* is derived from root at 'to advance' *asva* from the root as or *asin*: as to pervade). The words *akra* and *atya* went out of use later, although *asva* continued to be used. Perhaps this explains the use of the words *at* and *akra* in Indus seals for Agni and horse besides the more common word *asva.* The ass was the best favourite of burden and was not inauspicious as it was yoked to the chariot of the Asvins (*RV* I 34.9).

In the ritual of the horse sacrifice, the goat was sacrificed before the horse. The ram was also sacrificed and eaten. In one of the Indus seals, the sacrificial animal, a ram is shown standing behind a *hotr* or *yajamana,* who is making an offering to the fire God. In the lower register of the seal is a row of 7 holy men, *rishis,* symbolizing the *Saptarishi maṇḍala*—the seven star group (*pleiades–krittika*). The castrated ram was called *petva* and the hornless one was known as *basta* (Dange, 1977:17). *Vajasaneyi Samhita* (XXX.11) mentions *hastipa* for the cemetery, *asvapa* for speed, *gopala* for nourishment and *avipala* for strength. The elephant, horse, goat and cow were all taken care of by the respective protectors. There is a reference to *asvapa* in an Indus seal inscription. Another word for *hasti* (elephant) was *ibha.*

The habitat of the elephant is said to be the Himalayas (*Vaj. Sam.* XXIV. 30). Camel and elephant were well known to the Vedic people and so also to the Harappans. The elephant is said to be wild (*mrga*) but the office of the *hastipa* suggests that the beast was tamed and protected. Elephant tusk was used for making ivory combs, gamesmen and seals in Indus civilization. In the Acropolis at Lothal, there is an ivory workshop, and a seal made of ivory was found during excavations. It is observed that the words *kimpurusha* and *vrshakapi* and the identification of this beast with the man shows that the beast thus caught must have been very large and similar to the *yati,* famous in modern times. The hawk was known as *syena* and *suparna,* the destroyer of snake poision.

AGRICULTURE PRODUCE IN THE INDUS CIVILIZATION

PRE-HARAPPAN

(1) Neolithic Culture at Mehrgah (Period I: 4th–7th Millennium)

From the upper layers of this period, the findings include a sickle, a few grinding stones, cereal impressions of six-row barley (H.Vulgare and H.Vulgare var *Nudum*), two-row barley (*Hordeum distichum*), einkorn wheat (*Triticum monococcum*), *emmer* wheat (*T. dicoccum*) and bread wheat (*T. durum/aestivum*). Agriculture in Baluchistan was, thus, diversified (Jarrige and Lechevallier, 1979).

(2) Mehrgarh II

Among a series of agricultural settlements of 6000 BC on the Bolan River in Baluchistan, Mehrgarh has made a unique contribution to the history of agriculture. Seeds of cotton (*gossypium*) in Mehrgarh Period II (5000 BC) are found in association with seeds of other cultivated plants near the granary. Apparently, cotton too must have been cultivated for its fibre or oil rich seeds. Other cultivated crops of the sixth millennium BC are emmer, einkorn and two-row barley. It is obvious that on the border of Indus civilization, the Neolithic Chalcolithic folk were successful agriculturists. They could also select superior cereals such as free-threshing wheat and six-row barley. The ancient wheat in Indus civilization, namely *T. sphaerococcum,* a hexaploid is even today basically adapted to northwestern India (V. Mittre, 1974:8).

This compact and short variety of wheat well adapted to field irrigation was present in Mehrgarh in 4000 BC. The fast aggradation of the surface resulted in burying the Neolithic deposit in 9 m thick alluvium. In fact, if the pre-Harappan habitation deposits in waterlogged area of Mohenjo-daro are curefully excavated, more remains of cultivated crop in the early fourth millennium BC may be found.

(3) *Mehrgarh III*

A large building of the fifth millennium BC, divided into ten narrow compartments symmetrically disposed is exposed south of the Neolithic settlement. It has no doors. In one compartment, a sickle of wooden frame hafted with stone bladelets was recovered. Outside the granary, charred grains of wheat, barley and cotton seeds were found. Since wheat, rice and barley were all cultivated in Indus civilization, it can be said that the agriculture produce and methods of the Indus civilization are partly rooted in Baluchistan and partly in the Ganga-Yamuna valley wherefrom earliest rice reported is dated 5th millennium BC.

Technological innovations were made in the 4th millennium BC alongside diversification of agriculture. For instance, mass production of pottery and beads of semiprecious stones has been witnessed side by side the introduction of oats (*avenia* sp) and *triticum sphaerococcum* in the group of wheats. Man's adaption to the Indus valley begins with the farming people of Mehrgarh on the one hand and to the Ganga-Yamuna at Koldihawa on the other. Further progress in farming was made during the fourth and third millennium BC all over the Indus-Sarasvati-Yamuna Valleys and in the Saurashtra-Kutchh region.

(4) *Mehrgarh IV (3200–2500 BC)*

Major elements of the Pre-Harappa culture such as massive brick platforms, suggesting monumental architecture, terracotta triangular cakes, earthen jars and long parallel-sided blades of silicious stone are found in Period IV.

Although Mehrgarh was abandoned before 2500 BC, Nowshera, 8 km south of Mehrgarh, was occupied by mature Harappans. This fact confirms that the Kutchh plain was part of the Harappan territory (J.F. Jarrige and M. Lechevellier, 1979). Earlier we have noticed the domesticated and undomesticated animals flourishing in periods I–IV.

FOOD ECONOMY OF THE PRE-HARAPPAN PERIOD

(Ref. M. Casal, 1961, Allchin and Allchin, 1968: cited by V. Mittre and R. Savithri, 1993: 206, 207). The following food crops were grown in Pre-Harappan period at Mundigak: (1) Wheat *triticum* (compactum) and (2) *Jujube* (*zyzyphus jujube*). Kalibangan I. Sample No. 4 jujube (?) from Pre-Harappan level may not be real *ber* but not truly wild either. J. *jujuba lam non hill* is said to be the real *ber*. It is frequently cultivated and self sown wild.

THE HARAPPAN FOOD ECONOMY

At Harappa (M.S. Vats, 1941: 406–407; Mittre, 1974a) wheat (*Triticum compactum* and *T. sphaerococcum*): barley (*hordeum vulgare.* a small-seeded six-rowed variety); peas (*pisum arvense L*): sesame (*sesame indicum*) have been found.

From Mohenjo-daro and Chanhudaro (Pakistan) the wheat collected from late levels of the site was dated recently by C 14 method, which gives the date as 1650 BC. The sample consists of 56 entire caryopses of wheat with several broken fragments as well as fragments of grain barley. The Chanhudaro sample consists of several hundred caryopses of wheat and barley. Stapf (1931) found the wheat grains from Mohenjo-daro to be of variable size and shape and referred most of them to *Triticum compactum* and the plumpest ones to *T.sphaerococcum.* Luthra referred to them as *T.aestivum* subsp *vulgare* and *T. aestivum* subsp *compactum* and stated that the *T. compactum* of Mohenjo-daro was very similar to that from Harappa (Vishnu Mittre and R. Savithri, 1993, 207). After comparing modern wheat grains with those of Chanhudaro and Mohenjo-daro, Vishnu Mittre concludes that 'the Harappans at Mohenjo-daro and Chanhudaro were growing and consuming grains of both the, *T. sphaerococcum* and *T. aestivum/ T. compactum* groups'.

Mohenjo-daro produced only one broken hulled grain of barley while Chanhudaro yielded two. V. Mittre cites a report by Andrus and Mohammed (1958) on the occurrence of rice at Harappan sites in Sind and Punjab (Mittre, *Ibid.*, 208). Pea (*pisum arvevse*) and *Brassica Juncea* were also found in Mohenjo-daro.

Lothal and Rangpur (India): Rice imprints of husks and spikelets were recognized by Ramesh Rao and Krishna Lal (Rao, S.R., 1985: 279). The possibility of rice cultivation in Bhal region, where Lothal and Rangpur are situated, is very high. Even now, rice is grown in the flood plains of Sabarmati and Bhogavo estuaries. Ramesh Rao and Krishna Lal (1985) observe in their report that the plant remains excavated at Lothal are fairly rich in variety and number, comprising several kinds of dicotyledonous wood and charcoal, husk and epidermis of rice, leaf and stem impression of a monocot and some seeds of dicot monocot. Apart from botanical interest, their identification is of considerable importance as it gives some idea of the vegetation, climate and culture that prevailed at that time. Among five varieties of hard wood species teak (*tectona grandis*) and *haldu* (*adina cardifolia*), an unidentified wood belonging to the family of lauranceae and two other species, namely, *acacia* sp *albazzia* sp are noteworthy forest species furnishing valuable timber in mature Harappan Phase II (2350 BC) at Lothal. *Acacia* sp, *albizzia* sp *adina cordifolia* and *soymida febifuga* (*rohini*) occur in Phase III, whereas in Phase IV (Late Harappa Culture) the luxuriant forest of 3rd millennium BC in Saurashtra was fast dwindling by mid-second millennium. At present, only scrub to forest can be seen except in the Gir forest where teak is grown.

Rangpur rice husks have been identified from among the collection of cultivated plants during excavations at Rangpur (Rao, S.R., 1963).

Kalibangan sample No: 4, pea is from Pre-Harappan levels and samples 1, 2, 3, 5 and 6 are from Harappan levels. They are identified as follows:

Barley (*hordem* sp): The grains are 3 to 5 mm smaller than the modern grains of barley. The carbonized barley kernels comprise the bulk of the plant remains in the Kalibangan samples.

VARIETIES IN KALIBANGAN SAMPLE

The varieties or barley cultivated at Kalibangan are hulled barley grains. Naked barley-producing varieties were less common (Mittre and Savithri, 1993: 213). The hulled barley grains were produced by a six-rowed variety.

Wheat (*triticum sphaerococcum*): The small number of these grains occurring along with hundreds of barley grains possess, characterstics intermediate between wheat and barley. They may be either aberrant forms or the result of a few wheat plants that grew in the barley fields.

Gram/chickpeas (*cicer arietinum*): The three seeds of this species are squat-shaped, angular and pointed at one end. The seed coat is well preserved.

Pea (*pisum arvense linn*): This single seed is spherical with an ovate hilum.

LATE HARAPPAN FOOD ECONOMY

I. Surkotada (Kutchh)

The charred lumps of carbonized seeds in the earthen pot belong to Period III Layer 5 (1660 BC). The Period III ranges in date from 1970 to 1660 BC (V. Mittre and R. Savithri 1993, 214). Among several charred lumps two yielded 574 seeds, a majority of which were from wild plants. Seven per cent were cereals. The wild ones comprise the following:

1. *Setaria* Spp: It appears that grains of Italian Millet (*S. Italica*)—the cultivated species and the two wild species—the green millet (*S. verticillata*) are also present.
2. *Eleusine coracana* (*ragi*), comparing well with finger millet.
3. *Wild grasses*: There are 257 obovoid, ovoid, oblong and grooved seeds of grasses, some comparing with those of *phragmites karka*, a reed grass. The rest may belong to wild grasses such as andropogon, bracharia, panicum, etc.
4. Sedges comparing well with seeds of *scirpus supinus*. Some glumeless nuts compare with those of *carex* spp. and *eriophorum* spp.
5. *Cheno-amranthus*: A few among them compare with *amaranthus* spp. and some others with those of *atriplex* spp.
6. *Polygonum* spp.: There are 200 seeds of this genus.
7. *Eupharbia* spp. compare with those of *Eupharbia pyenostagia*.

II. Daimabad (Maharashtra)

The Savalda culture people had a food economy consisting of the following plants (Mittre and Savithri 1993)—jowar (*sorghum* sp), barley (*hordeum* sp), lentil (*lens culinaria*).

In period II, only barley was found, while period III had barley (*hordeum* sp), pea (*pissum* sp), horse gram, lentil (*lens culinaris*), mung (*vigna radiatus*) and Indian jujube (*zyzyphus* sp).

III. Daulatpur (Kurukshetra, Haryana)

The only charred grains found here are identified by V. Mittre as *vigna mungo* (L) H Hepper (*phaseolus mungo L*). The most significant result of recent carbon 14 and thermoluminiscence dating of Neolithic Chirand (1900–1700 BC) and OCP ware cultures in western Uttar Pradesh, Punjab and Haryana is that these cultures are found to have co-existed with late Harappa culture. In fact, OCP is a late Harappan ware.

FOOD ECONOMY OF CONTEMPORARY CULTURES

Contemporary with Harappa Culture are the Neolithic cultures of Burzahom in Kashmir, the socalled Ochre-coloured pottery (OCP) culture which itself is the late Harappa culture in Rajasthan and Ganga Yamuna valley and the Copper hoard culture of Uttar Pradesh, Punjab and Haryana. The OCP (late Harappa) culture people cultivated rice, barley and gram at Atranjikhera in UP, while the Neolithic folk of Chirand in Bihar used rice, wheat, barley and pisum. Rice, wheat and barley crops were cultivated earlier in the mature Harappan phase. Rice was perhaps not known in the Pre-Harappan period. The cytogeneticists (Darlington and Janaki Ammal) have included northwestern India in the area where wheat and barley were domesticated. With the discovery of rice in Neolithic Koldihwa as early as 6000–7000 BC, the Gangetic valley may be included in the area where rice was first domesticated and later went to Indus valley and Saurashtra (V. Mittre 1993, 219).

LAND ECONOMY OF VEDIC PEOPLE AND HARAPPANS

The ownership of land in Harappan times is not known. According to S.A. Dange (1977: 79), agricultural land was measured into tillable plots and owned by individuals. Perhaps this was the position in the Harrapan times also. There is a reference to *gavya* pasture land in one of the Indus seal inscriptions (Rao, S.R., 1982). The term *adripataka*, occurring in another seal, refers to the ruler of hill region. The term *kṣhetra*, used in seals, must have meant field used for agricultural purpose. The main question is whether land was owned by the king. In *Ṛgveda* I.110.5, there is a reference to measurements of fields. For example, RV. says 'they measured the one cup as field with a bamboo' (*kṣetramiva mamus* to *janena*) and 'this fertile land of ours' (*RV* VIII-91.6). Perhaps such fields owned by the individuals could be given in gift.

In view of the fact that the economy, religion, and location of settlements of Indus civilization correspond with those of Vedic people, textual reference to agriculture in the *Vedas* may be considered for extracting information on Indus agricultural economy.

In the early stage, the land appears to have been owned by individuals or groups of individuals who could clean up forest land, turn it into agricultural land which could be divided among individual families. The *Ṛgveda* (VI.47.20) hints at a classification of land into non-pastoral (forest) land and pastoral. The *urvara* was a tillable fertile land and *artana* was arid land. The saline land was known as *ushara* or *usha*. The land used for grazing was called *khilya*, which lay close to *urvara*. In pastureland (*khilya*), dwellings too were permitted (RV. VI. 2S. 2). The *khala* land used for thrashing corn was in the *urvara* or near it. Dange observes that as the individual or group was the owner of the land, there was no question of the king taking any tax for allowing cultivation of the land, but the location of the granaries in or near the citadel at Harappa and Mohenjo-daro suggests that the king might have collected tax in kind and also paid wages in kind. The term *bali*, according to Dange, may refer to contributory tax paid to the king for protection (Dange, 1977:80). We know from Indus seal incriptions the king was a protector and there was a hierarchy of such protectors. The reference to *bali* in the *Ṛgveda* is taken to mean war tribute, for it is said 'Ajas, Sigrus and the Yaksus offered bali in the form of the heads of horses' (*R V.* VII 18–19). In the later period, the king could gift his own land just as an individual or group could gift one's land. When King Janasruti Pautrayana gave certain villages to the recluse Raivaka (*Chhandogya Upanishad* IV 4.5), it meant taxes from the land and not actual land. According to *Jaimaniya Sutra* (VI 7.3), 'the land must not be given as it belongs to all and not any one in particular'.

Barter being the order of the day, *Ṛgveda* mentions *vasna,* which meant price. The *panis* were the bankers who lent money. At the Harappan port town of Lothal, small gold discs weighing 50 mg, 100 mg, 200 mg, etc., upto 3250 mg stood for *nishka* which might have served as currency or for exchange price of valuable land and goods. However, this is only a surmise.

Measures

For measurement of grains, the Harappans used beaker-like earthern wares which were in the ratio of 1:1/2:2:4 and 5.

FAUNA IN INDUS CIVILIZATION

Domesticated and Wild animals (Pre-Harappan Period)

Mention has been made of the animals in Mehrgarh Periods I to IV. A resume will be useful to understand whether any new species were domesticated in the mature Harappan Period.

Richard Meadow (1993: 295–295) has received various suggestions about animal domestication in the Middle East with special reference to the eastern margin, taking into account the evidence from Mehrgarh periods I-IV. Meadow observes that to evaluate claims for animal domestication, one must bear in mind that there is a conceptual difference between ascertaining the presence of domestic animals at a site and determining the phenomena that a process of local domestication took place. According to him, 'Animal domestication in the case of food and species can be defined as being a selective diachronic process of change in human animal relationships involving, at the very least, a shift of focus from the dead to the living, and more particularly from obtaining and distributing the products of the dead animal to securing and selectively maintaining the most important product of the animal, the progeny.' This process manifests itself in the structural trasformation in the socio-economic dimensions of the human societies which interact with the living animals and in the behaviour morphology and physiology of the animals.

To the Vedic Aryans, both the bull and the cow were of great economic importance and were, therefore, considered to be sacred. For the Harappan, the bull appears to be more sacred, as can be inferred from the seals on which the long-humped bull and the short-horned humpless bulls are represented.

Hundreds of terracotta/figures of cattle of both humped and humpless variety have been found in all major Harappan settlements, including Harappa, Mohenjo-daro,

Lothal, Rana Gundai, Kalibangan and Dholavira. At Lothal, there is a solitary terracotta figure of a cow with its udders (Fig. 11) prominently shown, but no such figure occurs in other Harappan sites. A large variety of domesticated dogs occur at Lothal.

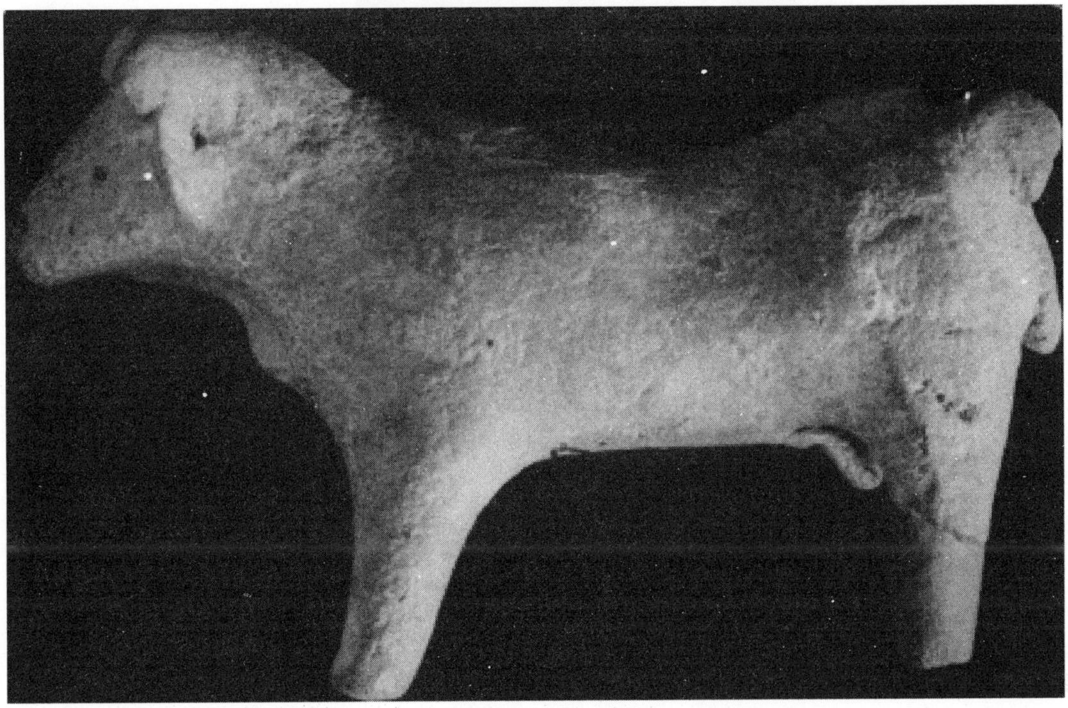

Fig. 11: Lothal: Terrcotta figure of a cow

Mountain goats were also considered to be sacred by the Harappans. They are beautifully carved on Indus seals. That the goat was offered in sacrifices is obvious from an engraving of a goat-like animal being led by a man to a deity. Both the animal and deity are depicted on a terracotta triangular cake of Kalibangan (Fig. 12). The presence of bovid bones in the sacrifical altars at Lothal and Kalibangan and among the grave goods of burials at both the places indicate that the cattle were offered to gods in sacrifies as also to the humans after death.

MIGRATORY HERDERS

Mehrgarh is situated in the Kutchhi plain at the tip of the vast Indus alluvial belt. This plain is considered 'the bread basket of Baluchistan'. Though affected by intense heat and occasional rains in summer, it is pleasant in winter when snow falls in mountains and the plain witnesses heavy rains. The sheep and goat herders come from the mountains to Kutchhi plains in winter and go back in summer so as to avoid the heat. It is interesting to find that the cattle and sheep herders of Kutchh and Saurashtra migrate to south Gujarat in search of fodder in winter and summer and return to their home during the monsoon. This community of herders is known as the Bharwads. The Bharwad

Fig. 12: Kalibangan: Terracotta cake graved motif of a scarificial animal being led by a man

herder from particular villages in Kutch and Saurashtra camp with their cattle and sheep in particular villages both on their way to south Gujarat and their return journey. The Bharwads from other villages are not permitted to camp in the villages, which by tradition, are allotted to a particular group. The local farmers either pay in cash or kind (grains) for the dung and milk products of the cattle, sheep and camel.

In Acermaic cultural period I A of Mehrgarh, bones of gazelle, blackbuck, sheep and goat are found in large numbers. A few bones of rodents, fish, birds, river turtle, onegar and boar also occur. In the later levels, goat, sheep and cattle dominate. In period II, the cattle bones are most numerous. In period III, numerous limbs of segments, *eqqus of hemionus* (wild horse) found here suggests they were used for making tools. During the Acermaic Neolothic period, there was greater dependence on animal keeping and herding. By the sixth millennium BC, (period III), Meadows has noted that the size dimunition of animals had advanced (Meadows, 1993: 31a), and they were as short and light as in later periods. This decreasing size in mammals is often attributed to domestication.

As far as the popularity of cattle (*bos indicus*) in mature Harappan and late Harappan periods is concerned, this species is well adapted to life in hot, semi-arid environment. Incidentally, Mehrgarh and Koldihwa prove beyond doubt that there were Neolothic farmers in Baluchistan and Ganga valley long before any influence from the Middle East ('Fertile Cresent') came and urbanism was a natural cosequence of the fast-growing industries, crafts and trade supported by agricultural surplus and cheap water transport.

LOTHAL ANIMALS (REF: BHOLA NATH AND G.V. SREENIVAS RAO)

The vertebrates of Lothal identified by Bhola Nath and C.V. Srinivasa Rao from the bones are quite significant:

1. Carp fish
2. *Rita rita* (freshwater fish)
3. *Lissemys punctala* forma *typica* (common soft-shelled box turtle)
4. *Chitra indica* (river turtle)
5. *Canis familiaris linnaeus* (domestic dog) the pariah dog
6. *Canisaureus linnaeus* (the Indian jackal)
7. *Herpestes aeropunctatus nodgson* (the small Indian mongoose)
8. *Elephas maximus linnarus* (the Indian elephant)
9. *Elephas caballus linnarus* (the horse) (Bhola Nath observes that the remains of horses have also been recorded from Mohenjo-daro, Harappa, Taxila, Ropar and Ujjain (Bhola Nath *Ibid.*, p. 642)
10. *Rhinoceros unicornis linnaeus* (the great one-horned rhinoceros)
11. *Sus scrofa cristatus wagner* (the Indian domestic pig)
12. *Muntacus mumtzak zimmermann* (the barking deer)
13. *Axis axis erxleben* (the spotted deer)
14. *Cervus unicolour kerr* (the sambhar)
15. *Cervus duvanceli cuvier* (the barasingha)
16. *Baselaphus tragocamelus pallas* (the nilgai)
17. *Bos gaurus* H.smith (the Indian bison)
18. *Bubalus bubalis linnaeus* (the Indian buffalo)
19. *Bos indicus linnaeus* (the *zebu* or the domestic humped cattle of India)
20. *Antelope cervicapra linnaeus* (the black buck)
21. *Capra hircus asgarus erxleben* (the Indian domestic goat)
22. *Ovis orientalis vignie blyth* (the Indian domestic sheep)
23. *Lepus nigricollis cuvier* (the Indian hare)
24. *Rattus rathus linnaeus* (the house rat)

The Lothal collection consists of 4860 identified animal remains represented by 31 species of animals, including 6 invertebrates and 23 vertebrates. These findings resemble animal remains from Harappa, Mohenjo-daro, Ropar and Sialk.

CONCLUSION

The environmental analysis of Indus civilization, when compared with the present environment in northwestern, western and northern parts of the Indian subcontinent suggests that water, being a great limiting factor, some sort of irrigation must have existed unless the proximity of river alluvium—as in the Indus-Sarasvati and Yamuna valleys and to a limited extent in the Sabarmati estuary—could offset low precipitation. Natural irrigation in flood plains was taken advantage of Vishnu-Mittre does not agree with

Singh's view that the precipitation was higher than the present in Harappan times in western and northwestern India and in other areas it must be the same as is at present (Mittre and Savithri, 1993: 218–19) but the frequent floods at Mohenjo-daro, Chanhudaro, Lothal and Kalibangan indicate higher precipitation in mature Harappan times which, was however, reduced by the beginning of the second millennium BC. The fact that only barley was found at Kalibangan is said to suggest that the environment was harsher but it must be noted that wheat was grown in Banawali and rice in Atranji Khera in late Harappan/OCP period. Hence, the environment could not have been harsher in the third millennium BC in the Sarasvati valley, and it is most likely that the late Harappan and post Harappan periods in the 2nd millennium witnessed a harsher environment as is indicated by the crops of Rojadi, Daulatpur and Daimabad.

CHEMICAL FERTILISERS

The use of gypsum calcium sulphate ($CaSo_4 Zh_{20}$) as a chemical fertilizer in Kalibangan and Lothal is suggested by the storing of the chemical. At present, gypsum is used to reclaim saline land and make it fertile. Ground gypsum is used in agriculture as a surface plaster for conserving moisture in the soil in order to aid nitrogen absorption from manure to reduce salinity and alkanity, as in the case of Rajasthan and parts of Sabarmati estuary.

BARTER SYSTEM

The Mehrgarh structures in periods II A and III served as a central storage facility around which the population would have concentrated. Harappa and Mohenjo-daro had such facilities in or near the citadel where the ruler lived and controlled commercial transaction. In Lothal, however, there is no granary but there is a much larger structure of 64 cubical blocks which served the commercial purpose of storing and examining goods. It is likely that the barter system was declining, as is indicated by the gold discs of Lothal which may stand for a monetary system. If, however, the conical, cubical and spherical terracottas objects served as symbols for exchangeable goat, sheep and cattle as in the Elamite economy, barter might have continued in Harappan times, but the absence of a granary and in its place the presence of a warehouse suggests that grains did not play a prominent role in the economy of Lothal.

REFERENCES

1. Agrawal, R.C. and Vijay Kumar. 1993. 'Ganeshwer-Jodhpura Culture: New Traits, in Indian Archaeology, 125–134.
2. Agrawal, D.P and R.K. Sood. 1993. 'Ecological Factors and the Harappan Civilization' in Possehl, G.L., 223–231.
3. Allchin, B. and Allchin, F.R. 1968. *The Birth of Indian Civilization*, Hammondsworth, London.

4. Bhola Nath and Srinivasa Rao, G.V. 1985. 'Animal Remains from Lothal' in Rao, S.R., *A.S.I. Memoir* 78 Vol. II.
5. Bisht, R.S. 'Excavation at Banavali', 1974–77 in *Possehl*, 1993, 113–124.
6. Casal, J.M. 1961. *Excavations at Mundigak.*
7. Casal, Jean Marie. 1961. *Fouilles De Mundigak*, 2 Vols Paris.
8. Dange, S.S. 1987. *Sacrifice in the Ancient India—Concept.*
9. Griffith, 1917. *Hymns of the Atharvaveda*, Benaras.
10. Jarrige, J.F. 1979. 'Exavation at Mehrgarh' in *South Asian Archaeology*, M. Toddel (Ed.), Naples.
11. Jarrige, J.F. and Lachevillier. 1979. 'Excavation at Mehrgarh' in *South Asian Archaeology*, 463–536, Naples.
12. Jarrige, J.F. 1993. 'Excavations at Mehrgarh: Their Significance for Understanding the Background of the Harappan Civilization' in G.L., Possehl, 1992, 79–85.
13. Lambrick, 1967. 'The Indus Flood Plain and the Indus Civilization,' *Geographical Journal*, 133: 483–94.
14. Mackay, E.J.H. 1938. *Further Exacavations at Mohenjo-daro*, Vol. I–II (A.S.I.).
15. Marshall, J. 1931. *Mohenjo-daro and the Indus Civilization*–3 Vols., Arthur Probsthain, London.
16. Meadow, R.H. 1981. 'Early Animal Domestication in South Asia', a First Report of the Faunal Remains from Mehrgarh, Pakistan in *South Asian Archaeology* (H. Hartal Ed.) Berlin, Dietrih Reimer Verlag, 143–179, Berlin.
17. Meadow, 1993. Animal Domestication in the Middle East: A Revised View from the Eastern Margin in Possehl G.L., 295–320.
18. Mehta, R.N. 1993. 'Some Rural Harappan Settlements in Gujarat' in Possehl, 167–174.
19. Possehl, G.L. 1993. *Harappan Civilization* (2nd Ed.) Oxford & IBH, New Delhi.
20. Rao, S.R. 1963. 'Exacavation at Rangpur and other Explorations in Gujarat' in Ghosh, A. (Ed.) *Ancient India* No. 18–19, 5–207, Archaeological Survey of India, Delhi.
21. Rao, S.R. 1973. *Lothal and the Indus Civilization*, Asia Publishers House, Mumbai.
22. Rao, S.R. 1979–1985. *Lothal Harappan Port Town*, 2 Vol., *ASI Memoir* 78, Delhi.
23. Rao, S.R. 1996. *New Frontiers of Archaeology*, Popular Prakashan, Mumbai.
24. Rao, S.R. 1991. *Dawn and Devolution of the Indus Civilization*, Aditya Prakashan, Delhi.
25. Ramesh Rao Krishna Lal. 1985. 'Wood Specimens from Lothal' in Rao, S.R., *ASI Memoir* 78, Vol. II.
26. Sharma, G.R. 1974. *Excavations at Koldihwa: Some Observations, Some Aspects of Archaeology* in IAR, 1973–74, 26.
27. Sankalia, H.D. 1962. *Pre-History and Proto-History of India and Pakistan*, Deccan College, Pune.
28. Stap, F. 1931. in J. Marshall, *Mohenjo-daro and Indus Civilization*, London Arthur Probethain, Vol. II.
29. Thapar, B.K. 1975. Kalibangan, A Harappan Metropolis Beyond the Indus Valley, *Expedition*, Vol. 17 no. 2, Pennsylvania University Museum, 19–32.

30. Vats, M.S. 1941. *Excavation at Harappa,* 2 Vols., Delhi, GOI.
31. Vishnu-Mittre and R. Savithri. 1993. Food Economy of the Harappans in G.L. Possehl *Harappan Civilization* (2nd revised edition) 205–222, Delhi, Oxford & IBH.
32. Vishnu-Mittre. 1974a. 'Palaeobotanical Evidence' in *Evolutionary Studies in World Crops Diversity and Change in the Indian Subcontinent,* Cambridge, Oxford Universiιy Press.
33. Vishnu-Mittre. 1974b. 'Neolithic Plant Economy of Chirand, Bihar', *Palaeobotanist* 22 (1) 18–22.
34. Vishnu-Mittre. 1978. 'Origin and History of Agriculture in the Indian Subcontinent', *Journal of Human Evolution* 7, 31–36.
35. Vishnu-Mittre and R. Savithri. 1975. Supposed Remains of Rice (*Oryza saliva*) in Terracotta Cakes and Pai at 'Kalibangan, Rajasthan', *Palaeobotanist* 22 (2), 124–26.
36. Verma, A.K. 1988. *Neolithic Culture of Eastern India.*
37. Vishnu-Mittre. 1993. 'The Harappan Civilization and the Need for a New Approach' in Possehl, G.L. 1993, pp. 31–39.
38. Wheeler, Sir Mortimer. 1968. *The Indus Civilization,* 3rd Ed., Cambridge.

CHAPTER 15

Agriculture in the Vedic Age: A Review

V.C. Srivastava

INTRODUCTION

Agriculture in the Vedic age has been the subject of study by most of socio-economic historians in the past.[1] In spite of much ingenuity shown by these historians, many dark corners remain to be lighted, many gaps to be filled, many questions to be answered and many problems to be satisfactorily resolved. Such a situation is due to the fact that there is no early text on agriculture[2] which may throw sufficient light on these dark corners. The *Ṛgveda Saṁhitā* is the only source for this period and this source, being associated primarily with religion and nature, does not devote itself to the material aspects of life of the period. We get only indirect references to the material aspects of life and this is the reason that the agriculture, *inter alia*, is known from myths and metaphors found in this oldest text of the world. The difficulty is further enhanced due to various assumptions and theories[3] in respect of the Aryans and the Harappans. The problem is further accentuated by the efforts of a group of modern historians to prove the Marxian philosophy of economic determinism with theories of class-struggle, changes in the mode of production, etc. It is a pity that no archaeological culture, including P.G.W., can be conclusively identified with the Vedic culture.[4] The information in the present versions of the *Ṛgveda* may be fragmentary and incomplete and, therefore, misleading, because there were many versions (*Śākhās*) of the *Ṛgveda* which are not available to us. It is also interesting to find that the *Vedic Index* of Keith and Macdonell, on which most of scholars depend, does not contain full information in many cases.[5] In any case, there is no option but to rely on the internal evidence of the *Ṛgveda* and later Vedic literature supplemented by archaeological sources.

Broadly speaking, the areas which need further probing are the developmental sequence of agriculture from the early to the later Vedic age, the place of agriculture *vis-à-vis* pastoralism, the process of agriculture, the tools and technology of agriculture, watering-management, nature of crops, ownership of agricultural land, system of manure, agricultural taxes, relationship between agriculture and philosophy of life, etc.

There are some assumptions which require further scrutiny. For example, it is a very common assumption that the Vedic society was wholly rural.[6] Similarly, it is assumed that early Vedic society was predominantly pastoral[7] and agriculture played a secondary role, though it developed as a primary occupation in the later Vedic age. It is to be noted that these assumptions do not stem from a careful study of the Vedic literature, which is the only source for knowing the life pattern for the Vedic people but from the hypothesis about Aryan migrations and discovery of the Indus culture. Since the Indus culture was urban and Aryans were the enemies and the destroyers of the Indus culture, the hypothesis presumes that the Vedic culture was totally rural and pastoral, at least in the earlier phase. The theory of the predominance of pastoralism is based upon the desire of the Marxist historians to locate the changes in economic life in different periods of Indian history with a view to illustrating the historical development of Indian history in accordance with Marxian prophesy. But this ideology has been questioned by many historians[8] and as such cannot be the basis of a theory such as predominance of pastoralism in the early Vedic age. It may not be out of place to quote an eminent historian, G.C. Pande[9] in this regard,

> '...Some historians' are...assiduously seeking to reconstruct changes in economic life in the different epochs of Indian history and on that basis seeking to illustrate the historical development of India in accordance with Marxian prophesy. This would be, in part at least, a significant historical enterprise but the limits of its success are somewhat narrow. And that is because the advancement of production technology since the Iron Age was only of marginal significance till the Industrial Revolution. Population fluctuated at a high stationary level. Cities remained few and far between, the villages numerous, isolated and largely self-sufficient. Agriculture continued on traditional lines, transport was slow and risky. Trade was confined largely to luxury items except for the needs of the bigger cities. Crafts depended on traditional skill and remained with guilds or hereditary subcastes. Different stages of economic development peacefully coexisted. Tribal settlements or *Pallis,* villages and towns represented such different stages. No one thought of a general economic improvement of the whole society as a realistic goal. The economic order appeared to be perennial, subsisting by its own little-understood laws. No one thought it possible to transform it and tradition preserved no memory of any revolutionary changes in it in the past except that of fall from a primeval golden age when man did not have to labour to produce his food. Nor was wealth regarded as the *summum bonum.*'

Besides this fundamental nature of Indian economy, other points also exist for questioning the Marxian theory and other assumptions in respect of Indian history. The assumption that the Indus culture was urban and, therefore, the Vedic culture was rural is also based on indefinite and uncertain basis. The relationship of the Indus culture with the Vedic culture is not definitely settled and various theories have been propounded in this respect. There is, *inter alia,* one view which regards the Indus and the Vedic cultures as two sides of a coin. In view of these uncertainties, the assumption stands discredited. The theory that the Aryans were the destroyers of the Harappa culture also stands rejected nowadays.[10]

It is necessary to take note of the fact that while taking the *Ṛgveda* as the source for economy and other material aspects of culture, the *Ṛgveda saṁhitā* is concerned more with natural and ritual phenomena than with human habitations and material objects. In view of this, it is natural that we do not find many references to human habitations and material objects such as agricultural operations and processes in this primarily religio-spiritual text.

It is well known that the Vedic seers (*ṛṣis*) used to live in hermitages outside the village. Needless to say that the Vedic seers neither traded nor farmed. They lived on the herds of cattles tended by their disciples (*Brahmacāris*) and the produce of the forest. The epics clearly mention that the seers lived in hermitages. The Sūtras confirm that the Vedas should be studied only outside the village. There are ample references in the Vedic literature to the effect that the *Brahmacāris* attached to seers tended their cattles and lived on uncultivated produce of the forest. As will be shown later on in this chapter, there used to be big unending forests beyond the pastures and village producing a variety of fruits and wild cereals (uncultivated). G.C. Pande[11] has very aptly remarked that 'despite outward similarities, life in the hermitages ought not to be confused with the life of pastoralists in *ghoṣas* or of tribals in *pallis*'. There are profuse references to cattles in the *Ṛgveda* which has been responsible for the confusion among historians who thought that the *Ṛgveda* represents the predominance of pastoralism.

It is also to be noted that there are some references to human settlements[12] also in the *Ṛgveda saṁhitā*. The human settlements referred to in this text are termed as *grāma* and *pur. Grāma* occurs nine times in the *Ṛgveda,* while *Grāmya* occurs only once. *Pur* occurs eighty-five times in this text. There are three references to *Kṛṣ*/ *Kṛṣi* and two to *Vaṇij*. In view of the above statistics, it is wrong assumption that the Vedic culture was of a wholly rural background. It is further argued that while *grāma* means village, *pur* does not mean a city. *Pur* may be a temporary place of refuse or natural stronghold or mountain fastness. It is definitely known that in later times, *pur* meant a city. It is difficult to imagine as to why a different meaning for *pur* may be given for the earlier period. I am conscious of the view that *pur* has been interpreted as a fortified town of Harappans but again, this view has been questioned. Similarly, there is no valid ground for supporting the hypothesis of racial conflict between Aryans and the Dāsas and regarding the *Paṇīs* as non-Aryans. To quote G.C. Pande,[13] 'if we exclude unproved hypotheses, we are left with the references in the *Ṛksaṁhitā* to *araṇya, grāma* and *pur.* The *pur* appears to have been a fortified place with defensive ramparts and a surrounding ditch'. It is to be remembered that there is no qualitative difference in village and town. Ghurye[14] has also argued in favour of the existence of towns in the Vedic literature. His main argument is that the Ṛgvedic society was aware of the use of two or more horse-drawn chariots in *pur* which indicates the existence of big roads and towns. The hypothesis of rural-urban dichotomy should not be pressed too far in the Vedic context.

The above background note will give us a correct perspective for the study of agriculture in the Vedic period. In brief, it may be said that the Ṛgvedic society was a settled society with villages and towns and the main subsistence pattern was agriculture and animal husbandry along with handicrafts. G.C. Pande[15] has summed up the position when he says that 'agriculture, cattle rearing and handicrafts constituted the principal occupation of the people in the village'. V.M. Apte[16] has also made a similar assessment: 'Agriculture made real headway during the Ṛgvedic age, although the practice of ploughing can be traced to Indo-Iranian times'. Likewise, Ghurye[17] has also assessed the position as follows: 'The principal economic activity of the Ṛgvedic people would appear to have been agriculture or plough-culture, which, students of Indo-European culture assures us, was already practised by early Indo-Europeans'. From these assessments, it may be safely said that the Ṛgvedic Aryans had an agricultural background from the Indo-Europeans through the Indo-Iranians. It may also be mentioned that the Ṛgvedic Aryans had the background of Harappan agriculture in India. Some scholars have remarked that the theory of predominant pastoralism for early Vedic people will make the Ṛgvedic agriculture as a retrograde step. But, as may be discussed in the following pages, Ṛgvedic agriculture was not a retrograde step.[18] Rather, it was an improvement. S.P. Raychaudhari[19] remarks that 'RV. I. 117.21 refers to ploughshares which appears to be an improvement upon the toothed arrow or the ploughshare of the Indus valley civilization'. However, Lal's discovery of a ploughed field at Kalibangan[20] takes out the weight of the theory of retrograde step. Actually, there is no data with us to compare the Ṛgvedic agriculture with the Harappan agriculture. In both cultures, there is paucity of material for such comparison.

There are references to at least twenty occupations in the *Ṛgveda*[21] as follows (1) rower (*aritr*), (2) boatman, (3) corn-grinder or grinder on a mill; (4) priest; (5) *ṛṣi*, hymner, poet; (6) smith; (7) smelter and welder (*draviṭri, dhamaṭri*); (8) ploughman; (9) herdsman; (10) village headman; (11) carpenter; (12) chariot-builder; (13) charioteer; (14) barber; (15) merchant; (16) weaver (*vaya*); (17) vintner; (18) *bhiṣaj*, physician; (19) *kāru*, singer-poet; (20) *brahmaṇ* (priest). Two out of these occupations are related to agriculture. These are ploughman and herdsman who are to be associated with the agriculture and animal husbandry. Agriculture had been a serious social concern, as is evident from many references to its importance in the *Ṛgveda*. For instance, agriculture has been contrasted with such wasteful activities as gambling. Naturally, agriculture was the basis of livelihood in the Ṛgvedic society and later times. The *Ṛgveda* (*R.V.* X.34. (*Tatragavah kitya tatra jaya*) asserts that agriculture would mean wealth (*vitta*) and a proper home with cattle and a wife. The importance of agriculture in this society may be realized from the fact that the cultivator has been assigned divine dignity and related to cosmic mythology. Agriculture is compared with sacrifice (*R.V.* X. 101. 3–6), which had been the most important activity of the Ṛgvedic society. Cultivation is related to Vedic rituals also. Some examples of these aspects may be referred to. The *Ṛgveda* mentions that agriculture was taught by Aśvins/Nāsatya, who in turn, taught it to Manu, the progenitor of the Indo-Aryan peoples, who then taught it to the Aryans. Another hymn of the *Ṛgveda* credits Aśvins with ploughing the fields and sowing barley for Aryans. Similarly, there is a reference in the *Ṛgveda* (*R.V.* VIII.22.6) that Trasadasyu was enabled to win fields and plough lands for the Aryans by Mitra-Varuṇa with the gift of the fabled horse, Dadhikrā. At another place (*R.V.* II. 13.6), Indra, the national god of the Aryans, is requested in a prayer 'to dwell as a guard to our tilled fields in our home steeds.' It has rightly been pointed out by G.C. Pande[22] that, 'they represent the divine archetype of the cultivator'. Sobharī Kaṇva likens his new hymn designed to propitiate Maruts to the one or ones which a cultivator sings to his steers (*R.V.* VIII. 20.19). It indicates that 'ploughing with bullocks was a long-established practice'. The significance of agriculture may be realized from the fact that it has been compared with the sacrifice *Yajna*. It is the principal form of worship in the Ṛgvedic society. The *Ṛgveda* compares the process of agriculture—ploughing, sowing and reaping—with the sacrifice. This tradition is continued in the later Vedic age, as is evident from the *Śatpath a Brāhmaṇa* (I.6.1.3).

Agriculture was associated with rituals in the *Ṛgveda* and later texts. For example, the Ṛgvedic hymn, IV. 57 was recited at the commencement of the ploughing. It mentions many agrarian deities—Kṣetrasyapati/Kṣetrapati—the lord of fields, *Sunā, Sīra, Sītā* along with Pūṣan and Parjanya. The *Ṛgveda,* IV.57 is devoted to an account of agriculture. Out of eight stanzas, the first four are related to an appeal to Kṣetrapati which ends as follows: 'happily work our steers and men, may the plough (*lāṅgala*) furrow happily. Happily be the traces bound (to secure the bullocks to the yoke of the plough); happily may the goad be plied'. Ghurye[23] has rightly pointed out that the term—*lāṅgal* for plough continues to stay in many Indian languages. The three deities Sūnā, Śira and Sītā referred to in this hymn are personifications of some agricultural implements or some agricultural operations. The other gods, Indra, Pūṣan and Parjanya are associated with the field, seed and rain, respectively. It is interesting to note that at least once, Indra is called *Urvarāpati,* lord of plough-land or lord of corn-land. Sunā and Sīra are referred to by Pāṇini but they do not figure in later ritual literature. Sītā is closely connected with plough-field in the *Yajurveda,* the *Rāmāyaṇa, Pāṇini,* the *Arthaśāstra,* and later literature. She has a long continuity. The formula in her respect is 'may the divine Sītā prosper in my seeds and my grains'. Kṣetrapati is invoked in the Gṛihyasūtras in the beginning of the agricultural operation. Kṣetrapati, later on, was merged in Rūdra-Śiva/Bhairava in south India. This Ṛgvedic hymn informs us that Sunāsīrau are a pair of deities personifying the plough. Sītā is the furrow goddess. *Sīra* has been connected with Sītā or furrow by Grassman. It means scratching lines or furrows. Kṛṣi, it may be noted, has the etymological meaning of scratching the earth with a stick. G.C. Pande[24] suggests that the *lāngala* was probably connected with the later word for stick, i.e. *laguḍa.* The meaning of the word *lāngala* is not certain but the probability, as indicated above, may be accepted in view of the practice of the scratching by a stick in case of cultivation in early times. Later on, the stick was replaced by share tip or point. It is also to be noted that D.D. Kosambi's opinion that Harappan agriculture was carried on with a hoe has been disproved by B.B. Lal's discovery of a ploughed field at Kalibangan. It appears that the Ṛgvedic *lāngala* was provided with a metal share in later Vedic times.

The reference to agricultural deities in the family *manḍala* (*R.V.* IV. 57) disproves the theory that agriculture was little known in the earlier phase of the *Ṛgveda* because the occurrences of references to ploughing are mainly found in the first and the tenth *maṅdalas,* which are later than that of the family *manḍalas.*

In the opinion of Prof. Pande[25], the *Ṛgveda* has in all about thirty-six references to agricultural fields (*Kṣetra*) and arable land (*Urvarā*) occurring separately or in compounds with *pati, vid, zith,* etc. *Kṣetra* or *Urvarā* is contrasted with *Khilya* (wasteland) as much as with *gavyūti* (pasture) as with *araṇya* (forest). There is some controversy about meaning of *Khilya.* Some take this term to mean an infertile land or waste land, while others believe that it may refer to fallow land. The *Ṛgveda* (VI.28.2) probably indicates that the *Khilya* also had the sense of land in between cultivated fields.

Broadly speaking, there are three types of theories regarding the ownership of agricultural[26] land in ancient India—communal theory, king theory and peasant theory. There are indications in the *Ṛgveda* to the effect that peasant ownership of the agricultural land was in practice. Most of the references to agricultural fields and arable lands in the *Ṛgveda* relate to the winning or gaining of *Kṣetra.* Naturally, such acquisitions makes one the master of the field. In another reference (*R.V.* I. 105), it is said that the fields (*Kṣetra*) were carefully measured with a rod. It reads: 'The Ribhus (artificier gods) with a rod measured, as it were a field, the single sacrificial chalice wide of mouth'. There is another reference in the *Ṛgveda* (*R.V.* IV.41.6; VI. 25.4) relevant to this point. A lady named Apālā valued her father's field as a personal possession on the same level as the hair on his head and her own body. The arable land is treated at par with one's possessions like cattle or children. 'It is clearly employed as personal and private possession.' The terms *Kṣetrapati* and *Urvarāpati* also indicate the same conclusion. There is another reference in the *Ṛgveda* in support of this view. This reference is as follows: 'With strength a limb the hero slays the hero... when two opposing hordes contend in battle for seed and offspring, waters, kine or corn-lands.' The view is further supported by the description of Agni as our well-loved friend, the god who prospereth our fields by Kaṇva sage (*R.V.*, VI. 25.4). Thus, Indra and Agni, the two chief gods of the *Ṛgveda,* are concerned with procuring and protecting the fields. The *Vedic index* rules out the possibility of communal property, community ownership and communal cultivation[27]. G.C. Pande rightly opines that there is no question of ownership of land by the king or the village community. Lallanji Gopal[28] has also suggested that peasant ownership of agricultural land is indicated in the Ṛgvedic society. It is not clear as to how far the personal ownership of land was individual and how far it was vested in joint family. Property belonging to the head of the family was partible to some extent. Whether land was also partible is not certain, though on later analogy, it was very likely.

Agriculture was widely practiced by Ṛgvedic people. With increased agricultural activity, the role of gods also changed. For example, Indra the war god becomes associated with agriculture and cattles. The *Ṛgveda,* II. 13.6 associates Indra with wet trunks and dried produce of agriculture. He is said to provide ample food to the *Yajmāna.* In the *Ṛgveda,* II. 13.7 he is referred to as guarding the fields. In the *Ṛgveda* II.14.11, it is prayed that Indra should be made full of Soma drink as the granaries are made full with barley (*Yava*). There is another reference to granaries in the *Ṛgveda,* X. 68.3, where it is said that the *yava* comes out of the granaries as cows come out running after they are released. Here the term used for granary is *Isthavi,* which probably used to be opened in the downward side. There are other references also to agriculture in the family *maṇḍalas.* The *Ṛgveda,* V. 83.7 invites Maruts to give rain along with speedy wind. Uneven lands are said to be made even in this reference. Obviously, the reference is to making the field even after due ploughing. The *Ṛgveda* V. 53.13 probably refers to the preservation of seeds which shows that agriculture was a regular occupation from year to year.

On the basis of historical study of family *maṇḍalas*[29], it has been said that there are only 21 references to agriculture in this portion of the *Ṛgveda*. But this statistical data has been questioned. Taking into consideration of references to all items connected with agriculture such as agricultural tools, agricultural products, agricultural labours, irrigation management, etc., it may be said that there are more than 100 references to the agriculture in the *Ṛgveda* out of which most of them are in the family *maṇḍalas*.[30] As such, it is not fair to say that early Vedic society was predominantly pastoral and agriculture was a minor affair in the Ṛgvedic age because references to agriculture are to be found mostly in the later *maṇḍalas*, i.e. the Xth and the 1st *maṇḍalas*.

Among the agricultural tools, we may refer to *lāngala, phala, yuga, sīra*. All these terms represent different portions of the plough. At the same time, all these terms are synonyms of the plough. The *Ṛgveda*, 117.21 refers to a rather peculiar name for the plough. This is *vṛka*, meaning wolf. This name was due to the cutting nature of the plough like the teeth of the wolf ("यवं व केन वपन्त" cf. VIII.22.6 यवं व केन कषर्थः"). The *Ṛgveda* V. 57.6–7 refers twice to *Sītā* and twice to *Yoni* (*R.V.* X.101.3). Bullocks with ploughs have been referred to many times (*R.V.* IV. 57.8; X. 101.3). *Kināsa* is the term for the cultivator. *Sṛnī* and *Dātra*, as agricultural implements, are referred to many times (*R.V.* X. 101.3; VIII. 78.10). The protector of the agricultural fields (*Kṣetrarakṣaka*) is also referred to (*R.V.* X. 68.1). Words derived from *Kṛṣ* and *Vap* have been used in the *Ṛgveda* several times. The *Ṛgveda* 128.5 and X. 71.2, respectively, refer to two agricultural implements *Ūlūkhalaka* (modern *Ūkhala*) and *Titaunā* (strainer). These implements are connected with two important agricultural operations. References to *Yava* in about twenty-six mantras and to *Dhānya* in three mantras and its derivative *Dhāna* (popped corn) for thirteen times enlighten us about the agricultural background of the Ṛgvedic peoples. An agriculturally rich person used to be termed as *Yavamata* or *Yavamanta*. It may not be out of place to mention the food items made out of agricultural produce. One such item is *puroḍāsa* (something like *parāṭhā* of modern times). The other item is *Dhāna* (something like modern popcorn). The Ṛgvedic peoples use to sacrifice thrice a day and *puroḍāsa* was to be used in the morning, the noon and the evening sacrifice with the addition of *Dhāna* for the noon sacrifice. There are other terms for other eatables—*apūpa, karambha* and *śaktu* (*R.V.* III. 52.4–6).

Some scholars[31] have expressed the view that there is no reference to the winning of or forcibly taking away of fertile fields in the *Ṛgveda*. This view may be questioned[32] because there are direct references to fertile fields in the *Ṛgveda* and many of them are in the family *maṇḍalas*. The *Vedic Index* itself contains references to *urvarā* but it is a pity that scholars, in order to prove their Marxian hypothesis, have ignored them. To illustrate the point, we may refer to references to fertile fields for which there used to be fights. The word *urvarā* in the *Ṛgveda* is referred to several times, denoting the meaning of fertile fields or *Kṣetra*. It has been[33] estimated that there are more than a dozen references to *urvarā* and its derivatives (*R.V.* II. 21.1; IV. 38.1; IV. 41.6; V. 33.4; VI. 20.1; VI.25.4; VIII. 91.5–6; VIII. 21.3; I. 27.6; X. 142.3; IV. 57.7–8). The number of references to word, *Kṣetra*, and its derivatives are not insignificant (*R.V.* I. 100.18; I. 110.5;

VII 35.10; IV 57.1–3; III. 31.15; IX. 85.4; VII. 19.3; X. 68.1). There are references to the acquisition or winning of fertile fields by fights (*R.V.* VI. 25.4; VII. 19.3). This indicates that the main source of wealth was agriculture in the Ṛgvedic society. Indra forcibly acquired the lands of Dāsas and distributed them among the Aryans (*R.V.* I. 100.18). The story of Trasadasyu getting *Kṣetra* from the gods is in support of the agricultural basis of the Ṛgvedic society (*R.V.* IV. 38.1). Words like *Urvarājita* (*RV* II. 21.1), *Urvarāsama* (*R.V.* VI. 20.1; IV. 38.1) and *Kṣetrasāma* (*R.V.* IV. 38.1) have been used in family and other *maḍalas* of the *Ṛgveda*. There is another important reference in the family *manḍala* (*R.V.* VI 20.1(सहस्रभरम् उर्वरासाम्) which refers to Indra as the conqueror of fertile field capable of feeding thousand persons. Indra has been termed as *Urvarājita* (*R.V.* II. 20.1) and, in the same *manḍala,* he has been referred to as the guard of the cultivated fields (*R.V.* II. 13.6). In this very *manḍala,* we get reference to rich granaries which have been compared to Soma drink of Indra. Indra is *Urvarāpati,* i.e. lord of fertile field (*R.V.* VIII. 21.3). The *Ṛgveda,* I. 127.6 uses the word *Apnasvatī* (अप्नस्वती) as an adjective to fertile field. Apnas means wealth or property. Naturally, the wealth-producing fertile field had been the basis of subsistence in the Ṛgvedic society. The *Apālā-sūkta* (*R.V.* VIII. 91.5–6) gives us the idea that in case of infertile fields (known *Ārtanā*) (आर्तना), the *Ṛgvedic* people used to lament and weep because fertile field and cultivation had been the main source of wealth. There are many references to the protection of cultivated fields from birds, fire and pest (*R.V.* X. 68.1; X. 142.3). The references to Kṣetrapati, Savitā, Parjanya, etc., (*R.V.* VII. 35.10) are indicative of agriculture. There are many prayers in the family *manḍalas* as well as other *manḍalas* for fertile fields along with sons, grandsons, cattles and water-resources (*R.V.* IV. 41.6; V. 33.4; IV. 57.7–8).

Labour-resources for agricultural operations are also an important ingredient in farming culture. We have referred to fertile lands capable of feeding thousands persons. Naturally, such a field must have been quite extensive and must have involved large labour force. This large labour force must have included the family men, close relations, close kinsmen and non-relations also along with the cattle force. There is a very important reference in this respect in the *Ṛgveda* (*R.V.* VII. 1.12; VI. 49.15). For non-relations, the words used are *Prajā, Nṛ, Carsaṇī.* There is a very significant word *Asinvan* (असिन्वन्) *R.V.* II. 13.4). *Asinvan* means a person who is not busy in making boundary in the cultivated land. It has been surmized that this refers to interval for food for agricultural labour. In this very *mantra,* it has been said that these labourers ate food given by the family men. It is not definitely known whether these labourers were voluntary, paid labourers or were forced, unpaid labourers. The sources for the labour force had been the *Dāsas* made captive in war or received in donation from the king. The *Ṛgveda* (*RV.* V. 34.6) refers to a *Dāsa* being brought under control by an Arya (वंश नयति दासमार्यः). There is another reference in the *Ṛgveda* (*R.V.* II. 12.4) to the effect that such *Dāsas* were given a lower land among the Aryans (दासं वर्ण अधरं गुहाकः). There is a reference (*R.V.* XIII.56.3) where it is said that a king gave one hundred donkeys, one hundred sheep and one hundred Dāsas to a *ṛṣi* in charity. Even if we take the number of hundred as traditional and figurative, there is no doubt that many Dāsas must have been given to the *ṛṣi* This source must have been utilized for cultivating the fields. The term *Dāsa* does not mean here slave but they might have been forced labour (*Viṣṭi*). But it is certain that it was this labour force which was responsible for most of the agricultural operations such as ploughing, levelling, sowing, watering, threshing, collecting, loading on cattles the produce and carrying them to the owner's house. There are references to the distribution of the produce (*Puṣṭi*) (पुष्टि) among the labour force known as *Prajā.*

Agriculture in the Ṛgvedic society was dependent largely on rainwater (*R.V.* V.33.4; IV. 57.7; VII 35.10) but it appears that they knew and practiced artificial irrigation also. We get references to *Avata.* The *Vedic index* makes the following observations on its note on *Avata*: (it) 'denotes a well artificially made in contrast to natural springs (*Utsa*), though the latter expression is also applied to an artificial well'. The well or *Avata* is described as *Iṣkṛtāhāvam avatam suvaratram* and again as *droṇāhavam avatam aṁsacakram aṁsatrakośam siṅcata nṛpānam.* Griffith translates'...from the well with pails prepared and goodly straps...' and 'pour forth the well with stone wheel, wooden buckets, the drink of heroes, with the trough for armour'. The *Vedic Index* understands this to mean as follows: 'The water was raised by a wheel (*cakra*) of stone, to which was fastened a strap (*varatra*), with a pail (*kosa*) attached to it when raised it was poured (*sinc*) into buckets (*ahava*) of wood.' Ghurye[34] thinks that the *Vedic Index* has missed the point when it says that the water was being raised by one pail. In his opinion, there must have been a number of pails. He advances the following arguments in favour of his opinion:

(1) If the water was being raised by one pail, it could not be so large in quantity that it could be poured out into a number of buckets; and if the water was poured out into a bucket, how could it be used to carry it off to the plants, unless somebody poured out the bucket again?

(2) The *Ṛgveda,* (*R.V.* X. 101.5) refers to a number of pails arranged on a strap or bell.

(3) In the Ṛgvedic society, there was currency of water channels, as is clear from the word *Kulya.* We have *Kulyavapa* from *Kulya* in the *Arthaśāstra* of Kauṭilya. From this later meaning of water channels, it may be presumed that *Kulya* denoted water channels in earlier society also.

(4) The *Vedic Index* accepts this interpretation of the word *Kulya* in the *Ṛgveda* (*R.V.* VII. 49.2) and concludes that 'Intensive agriculture by means of irrigation is clearly referred to in the *Ṛgveda* as also the *Atharvaveda.*

(5) Another passage from the *Ṛgveda* (*R.V.* IV 58.5) has been cited in support of underground artificial channels. This hymn has *Ghṛta* for its deity. It reads: 'From inmost reservoir (*Hṛidyāt Samudrāt*) in countless channels (*Śatvraj*) flow down these rivers which the foe beholds not'.

(6) The *Atharvaveda* (III. 13) definitely evidences for canals or channels for carrying water to fields because this *mantra* was to be recited on the occasion of cutting a channel for irrigation or to avert the flood. This may be true for the Ṛgvedic society also.

The *Ṛgveda* (*R.V.* VIII 49.2) refers to *Khaniṭramā,* which probably refers to small canals. It also refers to underground canals/drains by words such as *surmī, susīra.* Probably we may find similar practice in northwest India. The practice of canals or channels for irrigation purposes was quite popular in Afghanistan right from the protohistoric times onwards. They were known as *Karez or Qanat or Jui*. Archaeological works in Afghanistan have proved the existence of this practice[35]. It may be remembered that the Ṛgvedic people inhabited Afghanistan and northwest India before they spread to the Gangetic valley.

According to Ghurye[36], the above reference presents before us a picture of a large wheel to be moved by a bullock going round and round and the row of pots placed in a belt over the wheel coming up full and pouring out their water contents into the channels or tub on the edge of the well. S.A. Dange[37] speaks of buckets but does not believe that they were joined to form a chain or a circle. He takes it as an ordinary type of well. Ghurye sees in these passages a reference to the *rahat*. Some scholars are inclined to view it as the so called Persian wheel[38] or *rahat*. Another reference (*R.V.* I 16.9) speaks of Nāsatya lifting up the well and setting the base on high to open downwards stream. Lallanji Gopal thinks that the descriptions are not clear enough regarding its functioning. He says[39]: 'In the absence of any reference to the bulls operating the device and to the gearing system, it is not safe to infer the existence of Persian wheel 'in the Ṛgvedic society'. A very pertinent question arises as to why this device is not referred to in the later Vedic literature. The conclusion appears to be that there was some simple device to uplift the water from the well and flow them to agricultural fields through channels. The question of nomenclature for this device in the Ṛgvedic society should be left as open question.

The processes of agriculture have been compared with those of the sacrifice in the *Ṛgveda* (*R.V.* X. 101.3–6). These are ploughing, sowing and reaping which have been mentioned here graphically. These processes have been further elaborated by the *Śatapatha Brāhmaṇa* (I. 6.1–3) by the addition of threshing. The ripe grain was cut with a sickle (*Dātra sṛṇi*) (*R.V.* X 131. 2). It was then bound into bundles (*parśa*) and threshed at the floor of the granary (*Khala*) (*R.V.* X. 48.7). The grain was then separated from the straw and refined either by a sieve (*Titau*) or a winnower fan (*Śūrpa*). The winnower was called *dhānyakṛt* and grain was measured in a vessel called *urdara* (*Vedic Index,* 1, p. 182).

Ghurye[40] suggests that the Ṛgvedic Aryans borrowed the knowledge of seasons and the system of two crops in a year from the Asuras (non-aryans; probably indigenous peoples). The basis of this hypothesis is a story in the *Śatapatha Brāhmaṇa* (I. 6.1.1–3) about a bickering of seasons with the gods and their going to Asuras and return to gods with the mediation of Agni, assuring them a share in the sacrifice. In this story, it is said that the Asuras thrived with the presence of seasons with them. While one section of them was engaged in reaping of the harvest, the other section had begun ploughing and sowing. Agni brought back the seasons, promising them a share in the sacrifice. From the above reference, it may said that the Vedic Aryans learned the proper appreciation of the seasons with the help of which they also started cultivating two crops in a year. This may also be relevant for later Vedic period.

S.P. Raychaudhuri[41] has drawn our attention to the *Ṛgveda,* X.27.8 which is in praise of land, bullocks, seeds and peasants. It indicates the importance attached to crop husbandry with different types of field grasses for food and fodder.

The Ṛgvedic peoples cultivated *yava, tokman* and *dhānya. Dhānya* is the generic term denoting corn or grain in the *Ṛgveda. Dhān* is found in the *Ṛgveda,* always in the plural, denoting grains of corn. *Dhān* in the *Ṛgveda* (I. 16.2) meant popped corn. It refers that Indra was invited to partake of *Dhānā*: Oozing ghee may refer either to popped corn turned into ghee or rice cooked in ghee or eaten with ghee. It may be noted that the presence of rice has been questioned. *Yava* is the most often commonly mentioned cultivated corn in the *Ṛgveda.* Most of the Vedic scholars identify *yava* with barley. It has rightly been pointed out that the term *yava* may be a generic term for any sort of grain, not merely barley in the *Ṛgveda* (*Vedic Index,* Vol. II, p. 187). It was in the later Vedic period that *yava* becomes barley. Most of the scholars believe that the rice (*vṛhi*) was not cultivated by the Ṛgvedic society as the term *vṛhi* is absent from the *Ṛgveda.* It is supposed that the rice came to be grown from the later Vedic time onwards when the Aryans penetrated further east. Prof. Ghurye[42] has suggested that probably the Ṛgvedic people knew and ate rice for which they used the word *odana,* not *vṛhi.* The following arguments have been advanced by Prof. Ghurye in favour of his hypothesis:

(1) *Vedic Index* adduces a few references from the *Ṛgveda* (VIII. 58.69.14; VIII. 66. 77.10) for odana of different varieties.

(2) Griffith notes on *odana* that it is general word for the mess or massed rice. In later Vedic literature, we have references to *odana* as a food preparation. In Pāṇini, it means boiled rice.

It is true that the antiquity of rice in India is very high, as is proved by archaeology but it is difficult to say that the Ṛgvedic peoples cultivated rice and used another word, i.e. *odana* in place of *vṛhi.* The hypothesis of Prof. Ghurye is based on insecure grounds and may be questioned. *Godhūma* (wheat) is not mentioned in the *Ṛgveda.* It is surprising that the Indo-Aryans did not cultivate wheat in earlier phase because cultivation of wheat in Afghanistan is well attested by archaeology from protohistoric times. Barley was a fair season crop and ripened in the summer while the rice was an autumnal crop.

Agriculture depended upon natural fertility of the soil. The use of manure appears to have been known in the Vedic society, at least in the later phase. For the Ṛgvedic period, we do not have direct evidence of the use of manure, though cowdung (*kariṣa, śākrt, śakan, saka* and *saka*) as such is mentioned in the *Ṛgveda* which has been taken by scholars as indicative of the use of manure. The *Ṛgveda* (I .161.10) refers to Ṛbhus separating *sakrt* from other parts. Lallanji Gopal[43] says that 'there is nothing in these references to prove the use of cowdung as manure' in the Ṛgvedic society. We get positive references to manure in the *Atharvaveda* (II 2.4.5, 8.3; VI. 91.1; VIII. 2.19; X. 6.12). The *Atharvaveda* refers to *śāriśāka* (cowdung) as a useful manure for the *śāli* variety of rice.

There were significant developments in agriculture in age of later Samhitas.[44] The *Yajurveda* mentions twelve grains:

1. *Vṛhi* (rice)
2. *Nīvāra* (a form of wild rice)

3. *Yava* (barley)
4. *Godhūma* (wheat)
5. *Priyaṅgu* (a kind of small millet)
6. *Aṇu* (another kind of millet)
7. *Śyāmaka* (syama grass)
8. *Māsa* (pulse)
9. *Mudga* (Lentil)
10. *Masūr* (Lentil)
11. *Khalva* (a wild grain)
12. *Tila* (sesame)

Ikshu (sugarcane) is mentioned in the *Taittiriya Saṁhitā* (VII. 3.76) and the *Atharvaveda* (I. 34.5) but it is not certain whether sugarcane was cultivated. The *Vajasaneyi Saṁhitā* (18.22) refers to the above 12 grains in *rudraaṣtadhyāyī*[45] *mantras.*

Six, eight, twelve or twenty-four oxen were used to drag the plough which must have been very heavy in the later Vedic age. The oxen were yoked and harnessed with traces and guided by the goad of the ploughman. The furrow (*Sītā*) is often mentioned in the later Vedic literature. The plough is described as having a smooth handle, well lying and lance pointed (*paviravat*). *Kināsa* is a special name for ploughman. The *Atharvaveda* credits introducing the art of ploughing to Prithi Vainya but ploughing is constantly referred to in the text of this period as practiced by the Vedic Aryans while the *Vrātyas* (non-Aryans) were non-cultivators.

The later Vedic text refer to *yava* (barley) sown in winter and harvested in summer. Rice was sown during the rains and ripened in autumn. Beans and sesame were planted in the summer rain and ripened in winter. *Godhūma* (wheat) came to be cultivated in later Vedic age and is distinguished from rice and barley. Groats (*Śaktvāh*) made of wheat is mentioned according to the *Vedic Index.* It is interesting to note that in later Vedic literature, only the *Bṛihadāraṇyaka Upaniṣad* (VI.3.13) mentions wheat as one of twelve cereals. Barley seems to have overshadowed wheat in this period.

The later Vedic literature informs us that agriculture suffered from usual pests/ moles that destroyed seeds. Other creatures harmed the tender plants and shoots. To prevent these evils, the cultivators employed ritual charms as mentioned in the *Atharvaveda.* It may be noted that in this remote age, the knowledge of pesticides was not known or practiced. The *Atharvaveda* (VI. 50.1–2; VII. 11) refers to crop problems such as insects, animals, excessive rains, drought. For such eventuality, the *Atharvaveda* recommends charms to prevent them. Needless to point out, medicine and magic were mixed up in this age.

Agriculture became a very popular intensive activity in the age of the Upaniṣads and the Sūtras. It has been rightly[46] mentioned that the village agriculturist was the true representative of ancient India as we have a domestic rite for each stage[47] in the agricultural life of the people and practically none for the urban life. There are rites for putting the bulls and oxen to the plough, for sowing, for honoring *Sītā*—the goddess of agriculture. There is a sacrifice to Kṣetrapati, the lord of the field. The *Āgrāyana* sacrifice is the ritual for partaking of the first fruits of the harvest. There are many other sacrifices symbolizing various agricultural operations such as threshing, reaping, etc. A few examples of the charms for agricultural protection and abundance may be cited here:

For protection against fire:

Fires that are in the waters
in Vṛtra, in man, and in stones,
Who have entered the plants and the trees
to those fires be this oblation (*AV.* III. 21)

For abundance of grain:

As a mountain of a hundred jets
of a thousand jets, unexhausted.
Collect O with hundred hands,
pile O with thousand hands;
Bring about the amplitude here
of what was and will be done (*A.V.* III. 24)

In spite of intensive agriculture, there appear to be extensive uncultivated lands and forests. This is the rationale for the injunction of the Gṛihyasūtras that the *snātaka* has to live on uncultivated corn. Wild rice and grains used to grow on untilled lands.

The measurement of cultivated fields was done on the basis of the quantity of seeds liable to be sown in it. This is clear from the term *Prasthika* in Pāṇini applied to a field that accommodated *Prastha*, full of seeds. It is relevant to mention that the Vedic period had advanced knowledge of Botany, which must have contributed[48] to the development of agriculture along scientific lines.

To recapitulate, early Vedic society was a rural society with agriculture, cattle-rearing and handicrafts as the main occupations. The theory of the predominance of nomadic pastoralism is not borne out by the internal evidence of the *Ṛgveda*. Cultivated fields formed the nucleus of the village, beyond which were pastures and forests. Early Vedic society points towards the peasant's proprietorship. Barley and *Dhānya* were common products of the cultivation. Rice and wheat were not cultivated in early Vedic age. Plough with deferent parts and other implements of agriculture were known. Some device for artificial irrigation in the rudimentary form (something like *rahat*) was developed as early as during the *Ṛgveda*. Agriculture was raised to cosmic dignity. Agriculture in the early Vedic age came to be associated with agricultural deities such as Kshetrapati, Sītā, Sunā and Śīra. Besides, other gods like Indra, Parjanya, Aśvīns and Savitr came to be associated with the prosperity of agriculture,[49] a feature which continued in the later Vedic age. The later Vedic age witnessed significant developments in various fields of agriculture. Rice and wheat came to be cultivated in this age when the Aryans moved to the Gangetic valley. Pulses of different varieties and other products were developed. Agricultural operations came to be associated with domestic rituals in the later Vedic age. The use of manure, which was probably not known in the Ṛgvedic age, came to be developed in the later Vedic age. The dangers to crops from pests, flood, drought, etc., were well known and efforts were made to avert them in the later Vedic age.

It may be noted that agriculture in the Vedic times was an integral part of the philosophy of life. Agriculture was just a means of subsistence for the realization of the fourfold aims of life (*Puruṣārthas*)—*Dharma, Artha, Kāma* and *Mokṣa.* Economic activity (*Artha*) of which agriculture was an essential ingredient was in harmony with nature, ecology and spirituality. To use a modern terminology, agriculture in the Vedic age was in perfect harmony with sustainable development because it was inspired by human considerations.[50]

NOTES AND REFERENCES

1. Some prominent works in this respect are S.C. Sarkar; 1928, 1934, *Aspects of Ancient Indian Economic Thought,* Benaras, 1955, *Economic History of Ancient India,* Calcutta, K.V.R. Aiyangar, J.N. Samaddar, *Some Aspects of the Earliest Social History of India,* G.L. Adhya, 1966, London, *Early Indian Economics,* Bombay, *Social and Rural Economy of Northern India,* Two volumes 1961, 1945 A.N. Bose; *Vedic India,* Bombay, 1979 by G.S. Ghurye; *Aspects of History of Agriculture in Ancient India,* Varanasi, 1980 and *The Economic Life of Northern India,* Delhi, 1965 by L. Gopal; *Economic Condition in Ancient India,* Allahabad, 1980 (Rep.), Pran Nath; *Ancient Foundations of Economics in India,* Bombay, 1954, by K.T. Shah; *Society in Culture in Northern India in the Twelfth Century,* Allahabad 1973 by B.N.S. Yadava; *Material Culture and Social Formations in Ancient India,* Delhi, 1983 and *Perspectives in Social and Economic History of Early India,* New Delhi, 1983, by R.S. Sharma; *The Culture and Civilization of Ancient India,* London, 1965 by D.D. Kosambi; *Ancient Indian Social History: Some Interpretations,* New Delhi 1978 by Romila Thapar; *Veda Men Krsi Vidhya,* by S.D. Satavlekar; *Prachin Bharat Men Krsi* by A.L. Yadav. For detailed bibliography see David Luden, *An Agrarian History of South Asia, The New Cambridge History of India,* Volume IV. 4, Cambridge, 1999.
2. L. Gopal, 1980, *Aspects of History of Agriculture in Ancient India,* Varanasi, p. 91 writes 'the available texts belong to the early medieval period'.
3. G.C. Pande, 1995, *Foundations of Indian Culture:* Volume II, *Dimensions of Ancient Indian Social History,* Delhi (Rep.), p. 70.
4. *Ibid.*
5. R.N. Nandi, *Ṛgvedic Samaj: Eka Punarvichara* (in Hindi) in *Itihasa* (Journal of Indian Council of Historical Research), Volume II, Jan.–Dec. 1993, New Delhi, pp. 71–72.
6. Pande, op. cit., 1995, p. 70.
7. R.S. Sharma, 1983, *Material Culture and Social Formations in Ancient India,* Delhi, pp. 28–39.
8. Pande, *op. cit.,* pp. 70–71.
9. *Ibid,* p. 58.
10. *Ibid,* p. 71.

11. *Ibid,* p. 71.
12. *Ibid,* p. 71, for details see R.N. Nandi, *op. cit.*, pp. 76–78.
13. Pande, *op.cit.*, p. 72.
14. G.S. Ghurye, 1979, *Vedic India,* Bombay, cf. Nandi, *op. cit.*, pp. 76–77.
15. G.C. Pande, *op.cit.*, p. 73.
16. V.M. Apte, 1971, in *The Vedic Age,* Bombay (Rep.), p. 398.
17. Ghurye, *op. cit.*, p. 224.
18. S.P. Raychaudhuri, Agriculture in Ancient and Medieval India, in *Cultural Heritage of India,* Vol. VI, Calcutta, p. 178.
19. *Ibid.*
20. D.D. Kosambi, *Ancient Indian Culture and Civilization,* and B.B. Lal, in *Purātattva.*
21. Ghurye, *op. cit.*, p. 224.
22. Pande, *op. cit.*, p. 76.
23. Ghurye, *op. cit.*, p. 225.
24. Pande, *op. cit.*, p. 74.
25. *Ibid.*
26. L. Gopal, 1980, Ownership of Agricultural Land in *Aspects of History of Agriculture in Ancient India,* Varanasi, pp. 42–43.
27. Pande, *op. cit.*, p. 75.
28. Gopal, *op. cit.*, p. 43, cf. Schrader, *Prehistoric Antiquities,* p. 289; Macdonell and Kzith, *Vedic Index,* Vol. I, p. 210; Bandopadhyaya, *Economic Life and Progress in Ancient India,* p. 114; U.N. Ghoshal, *Hindu Public Life,* Part I, p. 60.
29. R.S. Sharma, *op. cit.*, p. 39.
30. Nandi, *op. cit.*, p. 80
31. R.S. Sharma, 1958, *Aspects of Political Ideas and Institutions in Ancient India,* (Sec. ed.) 1968, p. 115.
32. Nandi, *op. cit.*, p. 78.
33. *Ibid,* pp. 78–80.
34. Ghurye, *op. cit.*, p. 226f.
35. V.C. Srivastava, 1996, *The Protohistoric Afghanistan,* Varanasi, pp. 8, 195. Cf L. Dupree, 1973, *Afghanistan,* pp. 40, 42, 132.
36. Ghurye, *op. cit.*, p. 226.
37. Quoted by L. Gopal, in Araghatta, *op. cit.*, p. 154 fn. – 2.
38. Sushila Malti Devi, 1961, Irrigation in northern India from earliest times to 1200 AD, *Patna University Journal,* Vol. XVI.
39. *Op. cit.*, p. 154.
40. *Op. cit.*, p. 228.
41. *Op. cit.*, p. 179.
42. *Op. cit.*, p. 230
43. *Op. cit.*, p. 92.
44. Quoted by Raychaudhuri, *op. cit.*, p. 179. *V.S.*, 18.22 Cotton was not cultivated which is surprising since the Harappans knew of it.
45. G.C. Pande, 2001, *Vadika Sanskriti,* (in Hindi) Allahabad, p. 263.
46. V.M. Apte, *op. cit.*, p. 529.

47. *Ibid.*
48. S.C. Datta, in *Cultural Heritage of India,* Vol. VI, pp. 286–304.
49. V.C. Srivastava, 1972, *Sun-worship in Ancient India.* Allahabad, associated with the Sun god and fertility so essential for the agriculture.
50. G.C. Pande, *op. cit.*, 1995, pp. 59–60.

CHAPTER 16

The Spirituality and the Development of Agriculture in the Vedic Age

A.K. Sinha

INTRODUCTION

It is generally admitted that the identity and individuality of Indian culture in the crowd of the cultures of the world is established on account of a high sense of spirituality that dominates all other aspects of life and culture and makes itself a defining factor for Indian way of life and its multi-dimensional manifestations.[1] Having related intimately and inseparably with the basic religiousity of human mind expressing itself in the form or notions of some god and godly affairs, spirituality is often thought to have developed a sense of aversion and indifference towards the world and worldly life[2] and Indian culture is often cited as an example of it.[3] But looking at the spirituality in a worldwide historical context, this view does not seem to be entirely correct, for there are spiritual cultures in the ancient world which have been successful in proving their might in the worldly planes too. Therefore, spirituality is not an essentially world-negating attitude, rather, it becomes an inspiring force for worldly objects and attainments when god or gods are considered to be the best 'giver' of all things provided they, anyhow, are pleased. Spirituality in the Indian context is often confused with the idea of transcendentalism which is a philosophy developed in the centrality of the idea of transcending this or that world and, hence, of final liberation from all type of bondages, including the cycles of birth and death. This is called *mokṣa* and is considered as the *summum bonnum* of life, according to the Indian philosophy. Spirituality is an otherworldly and divine affair, which is totally different to transcendentalism.[4] It is taken to have held its influence and control over the business of the earthly life through a number of 'mediums' and 'performances' with variations of time and space.

It would be well, at the outset, to have a clear understanding about the general notion of spiritualism before touching the point of interrelation between Indian spirituality and agricultural development, which is the central theme of the present study. Spiritualism is the philosophy of spirituality; hence, it deserves to be grasped first. Spiritualism (from *spiro* meaning breath coming from the high)[5] is generally considered as a philosophy opposite to materialism (a philosophy developed in the centrality of matter). It is often interchangeable with idealism, although the two differ in the strict sense of their meaning. If 'idealism renders all reality relative to and dependent upon mind, spiritualism regards it as consisting of "spirits" which are not wholly or chiefly minds, and so does not make the cognitive process essentially to reality'.[6] Spiritualism, again, stands for "the belief or doctrine that the spirits of the dead survive after mortal life and communicate with the living through the intercession of a medium; the practice or phenomena associated with this belief; the doctrine that reality is spiritual rather than material; in philosophy, the type of idealism which defines the ultimate reality as spiritual rather than material; in philosophy, the type of idealism which defines the ultimate reality as spirit and the world as a realm of thought and ideas."[7] According to the Dictionary of Philosophy, Spiritualism (1) is the doctrine that the ultimate reality in the universe is spirit (*Pneuma, Nous, Reason, Logos*) over mind, akin to human spirit but pervading the entire universe as its ground and rational explanation. It is opposed to materialism. Spiritualism (2) is sometimes used to denote the idealistic view that nothing but an absolute and finite spirit exists. The world of sense in this view is a realm of ideas. Spiritualism (3) is used in religious terminology to emphasize the direct influence of the holy spirit in the sphere of religion and to especially indicate the teaching of St. John's Gospel that God is Spirit and that worship is the direct correspondence of Spirit with Spirit. Spiritualism (4) means the faith that spirit of the dead communicate with the living through persons who are 'mediums' and through other forms of manifestation'.[8] Spiritualism, thus, seems to be inseparably related with the philosophy of otherworldliness and is closely expressed in the theories regarding gods, demons, heaven and hell, etc., which form a very important part of almost every religion of the world that establishes the essential religiosity of the former.[9] With all these traits of spiritualism, spirituality develops as an important characteristic of a culture or society and, by and by, comes to become a distinct feature of the same that exercises its influence on the other features.[10]

The essential 'otherworldliness' of the spirituality is not necessarily opposed to the 'worldly', as is often thought. It is, rather, an image of the worldly at its level of perfection[11] in both the positive or negative forms. If the positive form makes the basis for the development of the concepts such as gods, angles, heavens, etc., all of which have been conceptualized as the highest totality of the earthly goodness, pleasure and attainments, the negative form is represented in the notions of demons, hell, etc.[12] It is in this manner that the 'otherworld' exercises its influence on a worldly human being with a view to inspiring him to promote the goodness and avoid the evils of life on the 'reward and punishment' pattern, which has been very common in almost every society. Spirituality, thus, creates a god who, through a number of demigods and angels, etc., controls the affairs of this world and offers rewards and punishments to the human beings in this world and the other world for their good and evil deeds. This is the basic idea of spirituality revealed through the various scriptures in a specific way and manner in case of a specific religious society. Indian spirituality and society has also been developed on the same line but what makes Indian spirituality distinct and different is the next stage at which spirituality crosses the boundaries of this and that world and develops a sense of transcendentalism aimed at the final liberation.[13] And then begins the journey of Indian spiritualism from the 'spirit' to a higher level of 'self' which reaches its final destination only after the 'self' is 'realized'. In this realization, everything other than the 'self' is considered to be unreal or non-existent. The world, worldly objects, God, heaven, and all divine attainments and even a desire for anything does not remain any longer and a state of desirelessness actualizes the blissful state of final liberation, called *mokṣa*. This is the highest stage of Indian spirituality, which is altogether different from the spirituality of other, especially Semetic cultures on the grounds of the latter being a spirituality of 'realization', resulting into the development for a homogenous religious plurality with a sense of 'Ekaṃ sad viprāḥ bahudhā vadantī', the 'revealed' spirituality has firm belief in the 'onliness' of the truth revealed and which it expects to be recognized as the final truth by others also.[14] This caused a sense' of religious superiority, leading to the feelings of intolerance resulting into the confrontation between different religious groups of which history is full of examples.[15]

Having, thus, a general view of spirituality, particularly Indian spirituality, one would agree that spirituality is a higher state of mind that develops from a consciousness of a union with the divine forces, enabling one to rise higher continuously. It is an integrated and uninterrupted journey from the body to the spirit and from the spirit to the self. Thus, there seem to be three stages or dimensions of Indian spirituality and all the three stages are linked together, playing important roles in shaping and reshaping the mood and the mind of Indian culture and society as a whole. The first stage of Indian spirituality begins with a belief in an other world with a number of gods and goddesses whose support is always required for every good of life, mainly and exclusively worldly. Every success of life and the world is thought to depend on the pleasure of gods which should be sought first through prayers and then by performing rituals and sacrifices. The second way, viz., performance of rituals and sacrifices, developed a mechanized from in the later days, giving rise to the second dimension of Indian spirituality in which rituals became more important than the god in whose honour the ritual was performed. The third dimension of Indian spirituality could appear in the days of Āraṇyakas and later developed in the Upaniṣadic days when a sense of transcendentalism developed from the *jñānamārga* came to replace ritual complexity of the earlier age. This dimension adopted its functional and behavioural form in the moral or ethical code of human behaviour on the social plane that paved the way for the development of some heterodox religions, especially Buddhism and Jainism which were originally the social movements of an ethical nature. Thus, in the *devavāda, yajñavāda* and *ātmavāda* may be categorized the three dimensions of Indian spirituality and the process of the development of an agrarian culture was in no ways uninfluenced by any of them.

Spirituality, thus, is inseparably related with the religiosity of human nature and is expressed through the acts and conduct of man as prescribed in the scriptures of one's religion. It may have some dogmatic, ritualistic or mystic form but in no way is it unconcerned with the world of human beings for the betterment of whom divine powers are invoked.[16] In the case of Indian culture, spirituality appears in the concept of *dharma* that makes in its broad ethical religiosity a bridge between the two worlds—physical and divine,[17]—being an earthly manifestation of *ṛta,* the controlling factor of all universal, natural and divine affairs.[18] *Dharma* presents a code of conduct for an individual and also for a society, following which all the worldly and otherworldly objects and attainments could be achieved. Having associated with some specific cult or sect, *dharma* may adopt its religious and dogmatic character but in a general sense, it denotes all ethical or moral virtue to only that which inspires people for promoting the moral and general goodness in the society. The true nature of *dharma* is clearly manifested in the ideological framework of Indian value-system, viz., *Puruṣārtha-Catuṣṭaya,* in which *Dharma* is simultaneously a value to be achieved and a means to the highest value, *mokṣa* besides being a check on the unbalanced growth of the physical and sensuous worldliness.[20] Like spirituality, *dharma* also has three dimensions in the total cultural perspective—moral, ritual and spiritual, including the transcendental. It is in all the three dimensions that the *dharma* becomes inseparably and intimately related with Indian spirituality and its different manifestations, as has been already indicated.

Without going into details regarding the material and the spiritual aspects of the Harappan people, one may very safely conclude that a kind of fertility cult bearing, in some form, divine influence on the process of agriculture was in existence in those days as is evident from a seal discovered from the Harappan phase which portrays an agricultural harvest scene. On one side of this seal, the figure of a nude female has been depicted upside down whose legs are apart and a plant is issuing from her womb.[21] There is another seal where a horned female figure stands between two branches of a tree with her hair falling down in coils.[22] These figures, of course, have been associated with the concept of the Mother earth as the goddess of fertility and vegetation, but they may also be associated with the idea of the otherworldly influence on the development of agriculture because the fertility and the vegetation cannot be dissociated from the agriculture; they are rather closely and intimately related with it. It may again be pointed out in this connection that a well-developed urban civilization, like that of the Harappans, would certainly have a solid foundations of surplus production for which, apart of the hard labour and technological advancement etc., they would also have sought by any means the help and support of divine forces. This, to a religions mind, was essential for any success and the religiousity of the Harappans, is an established fact on the basis of the archaeological remains found from various sites of this civilization.

The spiritual character of the Vedic culture may be clearly understood by the faith and belief of the Vedic people in a number of gods and goddesses for whom a number of hymns had been composed that forms the main content of the *Ṛgveda* and other Vedic texts of the later period. A study of the earliest passages of the *Ṛgveda* clearly explains about the life and culture of the earliest Aryans who had a highly positive and world-affirming attitude with much interest in the various means of worldly pleasure and enjoyment.[23] This includes a healthy long life, wealth, cows, sons, victory over enemies and many more things which could make their lives happy and pleasurable.[24] To achieve these objectives, the Aryans sought help from gods by pleasing them through recitation of mantras as prayers composed in their honour and some simple rituals often performed by the head of the family, for a priestly system had yet to be developed.[25] Belief in the number of deities, as may be seen in the Vedic pantheon with their supernatural powers and heavenly abode, brings before us the first stage of Vedic spirituality, viz., *devavāda,* in which it was believed that the gods, if pleased, were the best 'giver' of everything desired.

A content-wise analysis of the Ṛgvedic passages would reveal that the earliest Vedic society comprised the various mobile *janas* which, in their eastward expansion, were most of time engaged in war with the non-Aryan and sometimes even with the Aryan *janas* in want of the material and other possessions on which depended their own existence. Victory in war was considered to be the most desirable thing for it guaranteed the existence of the *jana*.[26] Importance of land had yet to be developed and consequently, the practice of agriculture too had not started, indicating no possibility of any settled life in those earliest days.[27] All social, political, economic and religious activities were centralized to war and the victory depended not only on the physical or military strength and valour but mainly on the divine help and support, thus, giving this affair a spiritual touch.[28] This is clearly indicated by the fact that in the *Ṛgveda*, the highest importance has been given to Indra who was considered to be a warrior god and whose victorious deeds have, repeatedly, been referred to with an aim to inspire the fighting warrior of the *jana* for defeating the enemies.[29] For this purpose, Indra's support was sought by every means to please him. References to some other gods are also found in these passages of the *Ṛgveda* but their importance was based mostly on their role in the victory of the respective *jana* over the enemy in the war.

The victory of King Sudās in the famous ten-king war fought on the bank of Paruṣṇi river marks a significant change in the life and culture of the people. If the reason of this war as believed later was to have a control over the waters of river,[30] perhaps, for irrigation purposes, the possibility of the beginning of agriculture before this event becomes strengthened and then the meanings of some early passages indicating agricultural activities become more clear.[31] This would, certainly, have initiated the process of settled life in the Vedic Aryans of those days developing the importance of land that led a shift from the tribal to the territorial form of state, identifying one's territory as an important and the defining factor of the state, later termed as *janapada*.[32] A good number of such janapadas, e.g. Kuru, Pañcāla, Kāśī had been referred to in the later Vedic texts. Land now was taken to be the most important source of prosperity[33] and then developed an ambition in every *janapada* to possess more and more area of land which could be cultivated to a greater extent and as a result, the *janapada* could be made more and more prosperous and develop into a *Mahājanapada* in the days to come. Thus, agricultural land could be procured.[34] The development of agriculture was taking place slowly, as may be seen in the relevant passages of the *Ṛgveda*. One finds mentions of the gradual process of agricultural operations in the *Ṛgveda*.[35] At first, the soil was prepared by ploughing the land, followed by sowing the seed. Then the plants were irrigated and proper care was taken for plant protection. When the crop was ripe, it was harvested and brought to the threshing floor where the grain was separated from the stalks and stored in houses. This process had been summed up in a Sūkta of the *Ṛgveda* dedicated to the *kśetrapati*.[36] Various references to the *kṛṣti*[37], *carṣaṇa*[38], *kṣetra*[39], *kṣetrapati*[40], *kināśa*[41], *lāñgala, sīra*[42], *sītā*[43], *dātra*[44] *Sṛṇi*[45] (sickle) with ox-driven plough[46] and a number of natural and artificial means of irrigation[47] (which all belong to the process of agriculture) make us believe of an early society of Aryans[48] where agriculture was growing in importance so much so that as Pande writes, "it was implied that agriculture would mean wealth (*vitta*) and a proper home with cattle and wife"[49]. In the *Akṣasūkta* of the *Ṛgveda,* a gambler is advised to give up his dice and to cultivate his corn field,[50] which again, is an indication of the growing importance of agriculture in the later half of the Ṛgvedic society.

The development of agriculture, like all other developments of the Ṛgvedic society, was considered to be an unparalleled gift from the gods, who, when pleased, gave a spiritual touch to the process of agriculture. In one passage of the *Ṛgveda,* it is said that the twin Aśvins taught men the art of agriculture. The mantra reads, "wishing to teach men, O Aśvins, you first cultivated grains by a plough in heaven."[51] The Aśvins, thus, represent the divine archetype of the cultivator. Plough in this connection is called Vṛka and the act of ploughing, as is indicated by a scholar, 'was raised to a cosmic dignity and thus rendered fit for the ritual ceremonies.'[52] It may be pointed out here that the idea of the union between Ṛgvedic father *Dwaus* and mother *Pṛthivi* termed Dwāvāpṛthivī,[53] would have, indirectly, influenced the process of developing a spiritual sense to agriculture that later came to be regarded as a divine affair involving the roles of a number of gods and goddesses, without whose blessing and support, it would not have been developed as an important means of livelihood and prosperity. The roles of gods in various operations of agriculture can be well understood through the fact that almost every important god was, in one or the other way, related with the process and the development of agriculture. Indra was the warrior god whose pleasure was sought for the victory in war for making possession over the land to be cultivated.[54] Men prayed to Indra for the 'release of water' which he had made possible by killing asuras like *Vṛtra, Śambara* and others and destroying their forts (*puras*), etc.[55] 'This release of water' could, in later days, be interpret as 'rains, thus, transforming Indra from warrior god to rain god and making an indication towards the transformation of a pastoral society into an agricultural society, as Kosambi thinks.'[56] But in all capacities, Indra remained important to the farmers and the agriculturists and his pleasure was sought for the development of agriculture. Varuṇa seems to have no direct connection with agricultural affairs but as the Guardian of Ṛta[57], he had full control over the rotatory changes of the weathers or seasons.[58] We know that the knowledge of the seasons was a very important factor that promoted agriculture. We find five seasons mentioned in the *Ṛgveda.*[59] Apart from this, Varuṇa also controlled the flow and the course of rivers, etc.,[60] which had their irrigational role in agriculture, thus adding greater importance on the part of Varuṇa. Agni as 'Purohita' was the 'medium' between gods and man.[61] He also helped the Aryans in burning down the forests and thus making land fit for cultivation, as is known by the story of Videgh Mathava in the *Śatapath Brāhmaṇa.*[62] This story, though, belongs to a later date, yet the idea of clearing forests by burning them for cultivation may be taken to be a continuation of an early tradition.[63] Similarly, Viṣṇu, an unimportant god of the earlier days, gained importance in the later tradition, probably on account of his association with agriculture. By his *trivikrama* he procured land from the *Asuras* in the heaven and on the earth[64] and thus, emerged as the protector of foetus in the womb and consequently, the crop to be grown in later tradition.[65] Pūṣan was the god to guide the travellers but sometimes he was also considered as the god of fertility[66] and hence, important for the farmers. Parjanya as the rain god was also associated with the agriculture. Besides these, we find in the *Ṛgveda*[67] mention of several deities which personify agricultural operations.

Kṣetrapati was the lord of the field, sunāsīrau a pair of deities personifying the plough. Sītā was the furrow goddess.[68] Thus, spirituality in its first dimension, viz., *devavāda* seems to have held its influence over the development of agriculture since the very beginning of the latter and this tradition continued for a long time, though there seems some changes in the exteriors of the two.

From the simple rituals performed in the honour of gods to seek their pleasure for fulfilling one's desires could grow a well-defined mechanism of sacrifices as the second dimension of spirituality that dominated each and every aspect of life in the first half of the later Vedic age, including the development of agriculture also. The *Ṛgveda* refers to such rituals performed on the first ploughing of the season when elaborate invocations and supplications were offered to the gods presiding over agriculture operations[69]. In such a ritual, curds, rice, fried grains and other things were offered to the gods and the bullocks were fed with honey and ghee.[70] Similarly, sowing of seed was also considered as an 'auspious' occasion on which 'prayers were offered to the god such as *Kṣetrapati, Soma, Pūṣan* and *Indra* for letting out the tender shoots of plants, resulting in increased yield of grain'.[71] *Sita* was the goddess of furrow who was invoked for grant of blessing and rich crop.[72] These rituals were named as *lāngalyajña* and *Sītāyajña* in the later traditions,[73] clearly indicating their sacrificial nature and developing a background for the inseparable relation between the sacrifices and the agricultural affairs which had already come in existence though in a rudimentary form in the *Ṛgveda,*[74] wherein, Pande sees 'the process of sacrifices in compared with that of agriculture.'[75]

The development of sacrificial culture in the later Vedic period dominated ritualism over the other aspects of life and culture of the people of that age. Even the gods in whose honour these sacrifices were performed, were, gradually, made subservient to the flawless performance of the sacrifices, resulting into the development of a group of priests as the expert and also a systematized—later mechanized—form of the sacrifices. *Yajña,* in this culture, was brought to the highest level of both the thought and action when it came to be considered as the source of everything. It was the centre of the universe (*bhuvanasya nābhiḥ*),[76] the source of Ṛta (*ṛtasya yoniḥ*)[77] the formation of good deeds, acts piety (*sukṛtasya yoniḥ*)[78] and 'the unvanquished person of the gods.[79] It was the first *Yajña* performed by the *Puruṣa,* the primal being, that the universe sprang from. Thus *Yajña* came to be recognized as the pivot of all social, political, religious and economic activities of the day and nothing could be thought of beyond its fold. If on the social plane, it had its influence on the value-system and social-structure defining the stages of their development,[80] on the political horizon, it greatly accelerated the process of transformation of the earlier (mobile) tribal states into the settled territorial states with their continuously changing geographical boundaries by inspiring the kings to actualize the political ideals like *Ekrāṭa, Adhirāja* and *Samrāṭa* etc., through the performance of the great sacrifices such as *Rājasūya, Aśvamedha* and *Vājapeya,*[81] etc., that practically meant nothing but establishing one's control over more area of land by defeating one's enemies.[82] With this was connected the development of agriculture because the first condition to it was to acquire a piece of land for cultivation. One wonders why the development of ritualistic activities and the agricultural operations took place simultaneously in those days, which indicate an interrelation between the two.

The later Vedic age witnessed rapid development in agriculture as we find a number of references to in the later Vedic texts. One important fact in this regard is that in the *Athravaveda*,[83] credit for starting agriculture has been given to *Pṛthu Vainya* indicating a departure from the Ṛgvedic tradition of choosing god Aśvins for the same.[84] This may be taken to indicate towards the lessening importance of gods in the growing importance of sacrifices, especially when the latter was raised to the 'dignity and status of the cosmic process'. The accelerated growth of agriculture may again be seen in the fact that we find in the later Vedic texts names of a number of crops[85] cultivated as compared to the only name *Yava* in the *Ṛgveda*, although in one opinion, it might have been used as a common name for many grains.[86] Also, there are references to ploughing the field by yoking 4, 6, 8, 12 and 24 oxen[87] to the plough. Though it has been argued that 'such number of oxen are symbolical'[88] and not real in nature,[89] this reference, if taken within a total perspective, may be accepted as an evidence for a speedy development of agriculture in that period. A *Sūkta* of the *Atharvaveda*, called as the Vedic farmer's song, very clearly describes the gradual process of agricultural operations.[90] Similarly, a passage in the *Śatapath Brāhamaṇa* mentions the different processes of agriculture as 'ploughing and sowing, reaping and threshing, tilling the plants', etc.[91] In one passage of the same text is mentioned a person who was going to sow seeds in his field.[92] 'Agriculture is verily food' says *Śatapath Brāhamaṇa*,[93] which indicates that then, the society was depending mainly on agricultural product of its livelihood and that the agriculture till then had developed as a very important part for the economy. The importance of agriculture is again proved when we find that in the *Rājasūya* sacrifice, the priests proclaimed the monarchy of the king saying 'O king! this state is given to thee for agriculture (*kṛṣyai*), for the common good (*kṣemāya*) and for prosperity (*poṣāya*)'.[94] This means that the progress in agriculture was considered to be one of the prime duties of the king.[95] A good number of references may be gathered from the later Vedic texts to show that agriculture was developing very rapidly in those days. This may be corraborated with the fact that iron was known to the later Vedic people and though its use in agriculture is debatable,[96]one may say that if not directly, it would have indirectly affected the agricultural development of those days along with some other factors like fertility of recently occupied land between the rivers,[97] use of cowdung,[98] knowledge of weather,[99] refinement in the quality of agricultural tools,[100] a good network of the means of irrigation, etc.[101] But during this development of agriculture, the role of sacrifices was never ignored. It rather came to be not only intimately connected with the agriculture, but was sometimes made to equalize with the agricultural operations. We have a number of references to the later Vedic texts wherein different agricultural operations have been given a sacrificial touch.[102] This intimacy and inseparability between the agricultural and sacrificial operations led to the development of the belief that, as we have already indicated, agriculture itself is a sacrifice and so is everything related to it. Then ox which is yoked in the plough is sacrifice; *Saṁvatsara* which gives knowledge of season is sacrifice; field is the *vedi* or alter of the sacrifices from which everything may be achieved.[103] *Tāṇḍya Brāhmaṇa* says that the cultivated land is *vedi* that produces everything.[104] *Yajña*, thus, came to appear as a favourable divine means for the development of agriculture which later developed as a mechanism to intensify the entire process of the latter, making an increase in the production[105] that created a 'surplus' to be used in the development of an urban culture in the days of the Mahājanapadas of the next phase of Indian history.

The overemphasis on the ritualism in the first half of the later Vedic period resulted into a sharp reaction against it that led the performance of sacrifices from the physical to the mental level giving rise to the notions of *antaryajña* and *manoyajña* in the days of the *Āraṇyakas.*[106] The *Āraṇyaka* texts were composed by the forest-dwelling sages who had attempted to trace the meanings and purposes of the sacrifices which were lost in their highly complex and over-mechanized form of physical performances. The *Āraṇyaka* texts reflect such a culture in which natural attachment towards vegetation and cultivation can be very easily inferred. The insistence on the mental performance of the sacrifices did not, in any way, negate the importance of the same but it certainly raised doubts regarding the hollowness or shallowness of its physical performance and declared it unnecessary.[107] It provided a symbolical meaning to *yajña* and advocated for its performance on a mental plane. This was the beginning of an 'inwardness' of thought that, gradually culminated in the *Ātman-Brāhmaṇ* philosophy of the Upaniṣads that declared the union of the two as the ultimate end of life and named it as *mokṣa.*[108] Self-realization was the only means for this the highest attainment that could provide one with a sight to look beyond the world and the otherworld too. This presented the third stage of spirituality, especially and specifically Indian, which, as is generally thought, initiated a discussion on the futility and the meaninglessness of the sensuous world and the worldliness and developed finally into the idea of transcendentalism rejecting both this and that worlds. This caused the rise of *nivṛttimārga,* promoting a sense of *vairāgya* and *sannyāsa,* allegedly, the dominating factor in Indian way of life that led people towards a philosophy of inaction.[109] But this picture represents only half of the truth. The whole truth is that the *Upaniṣads,* undoubtedly, had developed a philosophy of spiritualism and transcendentalism in the centrality of *mokṣa* but it had never been in opposition of the reality of life. Seeing the elements of *nivṛttimārga* as the seed of the philosophy of inaction in the name of *vairāgya* and *sannyāysa* in the Upaniṣadic thought is also a contextual fallacy. The *Upaniṣads,* in reality, advocated the philosophy of '*Ātmavāda*' in which world and worldly achievements were, certainly, not given any importance but it was all with a special reference to the Upaniṣadic spirituality and not as an independent and isolated development of thought. The philosophy of '*Ātmavāda*' was a culmination of the philosophy of *jñānamārga* and not *nivṛttimārga* and the two differed as the positive and negative approaches towards the world and worldly life. Upaniṣadic philosophy, thus, is not a philosophy of denial and negation but a philosophy of admission and affirmation, of course, of the highest truth, as also the high and the higher truths. Similarly, *mokṣa* does not necessarily, deny *artha* and *kāma* but it provides a *prajñā* to realize their temporary and perishable nature and warns against the 'lust' for these. It also presents a sense of continuous progression in the human being from his lowest to the highest level of consciousness as is very clearly manifested in the *Pañcakośa* theory propounded in the *Taittirīya Upaniṣad.*[110] This may be confirmed by the example of the Upaniṣadic sage *Yājñavalkya* and others who were great champions of *Brahmavidyā* but had shown, at the same time, no less interest in the wealth and the other means of worldly prosperity.[111] They had big establishments as their *āśramas* with almost everything required for leading a satisfactory and good life.[112] The *Bṛhadāraṇyka Upaniṣad* mentions Yājñavalkya, showing his interest in one thousand cows with golden horns which King Janaka had offered for the *Brahmajñānī* and Yājñavalkya claimed for that.[113] The same text again mentions Yājñavalkya who desired to divide his property between his two wives before proceedings for *pravrajyā*.[114] There are a number of references to in the Upaniṣads wherein sages demanded or accepted, when offered, gifts in the form of movable or immovable property for imparting knowledge of Brahmavidyā to king and others.[115] This presents a clear evidence of the positive or affirmative attitude of the Upaniṣadic tradition which has often been ignored in the crowds of the spiritualistic ideas contained therein. The transcendental spirituality of the Upaniṣads can be clearly understood only with a reference to the theories of the āśramas and Puruṣārthas, which in the Upaniṣads, are in a rudimentary form[116] but could develop later on as the defining characteristic of Indian culture. Similarly, the philosophy of the Upaniṣads finds the fullest meaning in a wider and holistic context of the general wisdom of ancient Indian people and their philosophy of life, including the value-system prevailing therein.

On the basis of the above account, it would be evident that although we find very rare references to the processes and the operation of agriculture in the Upaniṣads on account of their being, primarily and ultimately, the philosophical texts, the absence of references cannot, necessarily, be taken as an aversion or opposition to it. In such a situation, one has to bank upon the indirect and circumstantial evidences to discover the truth of history.[117] We have already seen that the transcendental spirituality is not necessarily opposed to the worldly prosperity and in Indian philosophy of life manifested through the philosophy of the *puruśārath catuṣṭaya*, a balance between the two has been attempted. If the historical evidences present for us a picture of the continuous progress in the field of agriculture on account of various factors indicated earlier during the whole of the later Vedic period, there remains nothing to doubt regarding the positive influence of third dimension of Indian spirituality over the progress of agriculture. It has been argued that the check on the slaughtering of animals in the sacrifices had been an important factor in the progress of agriculture[118] and it is an accepted fact that the Upaniṣads in teaching the principles of *ātmavat sarvabhūtānāma* and *sarvaṃ khalvidaṃ brahma*, had raised, for the first time, their voice against animal slaughtering in the sacrifices and advocated for the vegetation sacrifices. The Upaniṣadic philosophy of *ātmavāda* could lead on the level of human behaviour and conduct, an atmosphere for the development of the moral virtues of which *Satya* and *Ahiṃsā* were the most important. We know that in spirit the two are one and the same, carrying a sense of 'making no injury to all living beings'.[119] This too had prepared a ground for the check over even the 'Vedikīhiṃsā' that again made the cause of agriculture promote. At various places in the Upaniṣadic literature, *anna* has been equated with or identified as Brahma,[120] the final reality, and there seems no doubt to think that *anna* means none the else than the 'cultivated' grain which again is an indication of the people's interest in agriculture. One may also point out here that the Upaniṣads belong to the tradition of the philosophy and the philosophers, while agriculture is related to common man. Therefore, absence or rare reference to agriculture in the Upaniṣadic literature is not very unnatural and surprising and it might not be taken on the face value. But if, as we have already said, on the basis of other historical evidences we find no interruption in the agricultural activities during the period to which the Upaniṣads are thought to have composed, one can safely say that the spirituality even in its third dimension had never been a negative force to the progress of agriculture.

The foregoing accounts make it very clear the despite its spiritual nature, Indian culture had no aversion for worldly well-being and a keen interest in agriculture as the best source of worldly prosperity. Spirituality also had never ignored this world but had presented the other world as the 'best' picture of the world and its belongings, including even the philosophical ideas and hence, had, through and through, played its important role in the progress and the development of agriculture which, like *dharma*, 'sustains and maintains' the world. Spirituality inspires one to get away from sin and as the first step in this direction, it promoted agriculture to feed everyone, because the first source of sin lies in hunger *bubhukṣitaḥ kiṃ na karoti pāpaṃ*'. It then further leads to divinity and ultimately to REALITY but with a consistent growth of agriculture and they two make a rhythm to sing jointly '*annaṃ vai Brahma*' and thus spirituality and agriculture march together.

NOTES AND REFERENCES

1. Murty, K. Satchidanand, 1965, *The Indian Spirit*, Andhra University Press, Waltair, pp. 1–2; Mukhopadhyay, P.K. 'History of Science and the Two Metamorphoses' in *Language, Logic and Science*, (ed.) D.P. Chattopadhyay, Project of History of Science, Philosophy and Culture, Delhi, 1995, p. 22.
2. cf. Hick, J., 1979, *Philosophy of Religion*, Prentice Hall of India Pvt. Ltd., Delhi, pp. 101–102; also Pande, G.C., 1995, *Foundations of Indian Culture*, Vol. I, (Reprint), Motilal Banarasidass Publishers Pvt. Ltd., Delhi, pp. 5–6.
3. Mukhopadhyay, P.K., *op cit.*, p. 22; Murty, K.S. *op cit.* p. 24.
4. Sinha, A.K., 'Scientific Temper of Upaniṣadic Spiritualism' paper presented to World Philosophical Congress on 'Spirituality, Science and Technology' organised by Indian Philosophical Congress, The Platinum Jubilee Session Dec. 28, 2001–Jan.1, 2002) J.N.U., New Delhi (To be published).
5. Tripathi, B.N., 1987, *Indian View of Spiritual Bondage*, Aradhana Prakashan, Varanasi, 1987, p. 3.
6. Schiller, F.C.S., (1981), 'Spiritualism' in Encyclopaedia of Religion and Ethics (ed.) J. Hastings, Vol. II, 1981, p. 808.
7. *Webster's Encyclopaedic Unabridged Dictionary of the English Language*, (1994), Gramercy Books, New York, p. 1372.
8. Runes, D.D., *Dictionary of Philosophy*, p. 300.
9. Hick, J., *op. cit.*, p. 102; See also, Pande, G.C., 1985, *An Approach to Indian Culture and Civilization*, Department of Ancient History, Culture and Archaeology, Benaras Hindu University, Varanasi, p. 22.
10. Indian culture is one of the best examples in this regard.

11. cf. Pande, G.C., 1981, *Bharatiyā Parampara Ke Mūla Svara,* National Publishing House, Delhi, pp. 47–48; Tripathi B.N., *op. cit.* p. 18.
12. Hick., J. *op cit.*, pp. 100–101.
13. Pande, G.C. *Bharatiya Paramparā Ke Mūla Svara,* pp. 30–31.
14. *Ibid.*
15. Crusades and Jehads may be mentioned as examples in this connection.
16. Sinha, A.K., 2000, *Readings in Early Indian Socio-Cultural History,* Anamika Publishers Pvt. Ltd., Delhi, pp. 62–63.
17. Sinha, A.K., 1981, *Prācīna Bhārtīya Vaiyaktika evam Sāmājik Mūlyabodh,* Uma Prakashan, Allahabad, Ch. V.
18. Hiriyanna, M. 1975, *Indian Conception of Values,* Kavyalaya Publishers, Mysore, p. 156; Bhagwan Das, 1910, *The Science of Social Organization,* Theosophical Publishing Society, Madras, p. 123.
19. *Yato'bhyudaya nisśreyassiddhi so dharmaḥ, Vaiḥesika Sūtra,* 1.1.3.
20. Pande, G.C. *Bhārtīya Paramparā Ke Mūla Svara,* pp. 71–72.
21. Vats, M.S., 1940, *Excavations at Harappa,* Archaeological Survey of India, Delhi, p. 42.
22. Mackay, E.J., *Further Excavations at Mohanjodaro,* pl. XCIV-430, quoted by Roy, B.P., 1984, *Later Vedic Economy,* Janaki Prakashan, Patna, p. 157, fn.15.
23. Raju, P.T., 1966, 'The Concept of Man in Indian Thought' in *The Concept of Man* (ed.) S. Radhakrishnan and P.T. Raju, Motilal Banarasidass, Delhi, p. 233. Kunhanraja, C., 1957, *The Vedas,* Andhra University Press, Walta, p. 186.
24. *Ṛgveda,* 1.79.4; 89.1, 92.13, 157.2 III. I.19; IV.8.7; V 10.1, 23.2; VI. 10.5, 16.20; VII. 41.3, 72.1; Kulkarni, Chidambara, 1973, *Vedic Foundation of Indian Culture,* Shri Dvaipayan Trust Bombay, p. 223.
25. Kosambi, D.D., 1977, *The Culture and Civilization of Ancient India in Historical Outlines,* (Hindi Tr.) Raj Kamal Prakashan, Delhi, p. 105.; Sinha, A.K., 'Value and Social Structure in Ancient India.' *Journal of Oriental Institute,* Vol. XL., Nos. 1–2; 1990, Baroda.
26. Sinha, A.K., *op. cit.*
27. ".....agriculture was associated with settlement because in a pastoral economy agriculture operations could not be practised on a permanent basis...." Roy, B.P., 1984, *Later Vedic Economy,* Patna, p. 118.
28. Sinha, A.K., *Readings in Early Indian Socio-Cultural History,* 24, p. 18.
29. Barth, A., 1990, *The Religions of India,* (Reprint) Low Price Publication, Delhi, 1990, p. 12; Williams, M. 1978, *Religious Thought and Life in India,* (Reprint) K.P. Bagchi & Company, Calcutta, p. 15; also Chattopadhyay, K., *Studies in the Vedic and Indo-Iranian Religion and Literature,* Vol. 1, Bhartiya Vidya Prakashan, Delhi, pp. 101–102.
30. Kosambi, D.D., *op. cit.*, p. 104.
31. See, Mazumdar, G.P., 1951, 'Origin and Development of Agriculture in Ancient India' in *Proceedings of All India Oriental Conference,* 1985, pt. III., pp. 113–116; Ganguli, R., 'Cultivation in Ancient India' in *Indian Historical Quarterly,* Vol. VI (Reprint), Delhi, pp. 737ff.

32. Pande, G.C. 2001, *Vedic Samskriti,* Lokbharti Prakashan, Allahabad, pp. 54, 55.
33. The Pṛthivi Sūkta of the *Atharvaveda* is the best example of it.
    Atharvaveda, 14.1.1.66; 4.39.2; 18.3.39; 18.4–6; *Taittirīya Saṃhīta,* 1.72; *Vājasneyi Saṃhitā,* 25.17.
34. "...... *The Ṛgveda Saṃhitā* has in all thirty-six reference to 'agricultural field' (Kśetra) and 'arable land' (Urvarā)........Most of the references relate to the winning or gaining of Kṣetra'.
    Pande, G.C., *Foundation of Indian Culture,* Vol.II., p. 74.
35. *Ṛgveda,* 1.23.15; 10.34.13; 10.117.7.
36. *Ṛgveda,* 4.57.1–8; Roy, B.P., op. cit., p. 129; also Ṛgveda, X. 101 quoted by Chidambara Kulkarni 1973, in *Vedic Foundation of Indian Culture,* Shri Dvaipayan Trust Bombay, pp. 160–61.
37. *Ṛgveda,* 1.4.6.
38. *Ibid.* 1.3.7.
39. *Ibid.* 5.62.7; 9.85.4; 91.6.
40. *Ibid.* 4.57.1.
41. *Ibid.* According to G.C. Pande, farmers using the plough were called Kināśa, a word of uncertain etymology.
    Pande, G.C., *Foundation of Indian Culture,* Vol. II, p. 74.
42. *Ṛgveda,* 4.57.8.
43. *Ibid.* 4.57.6.
44. *Ibid.* 8.78.10.
45. *Ibid.* 1.58.4; 4.5.
46. *Ibid.* 8.6.48; 10.10.1.
47. Rain and floods were natural means while there were also the artifical means of irrigation known as *Avata* (Ṛv. 1.105.19.), *Utsa* (Ṛv.8.49.6), *Kulya* (Ṛv. 1.43.7). A device named as asmacakra (10.110.6) is also referred to for details see, Raghvan, D., *Agriculture in Ancient India,* Delhi, 1964, p. 13; Pande, G.C., *Vedic Samskrti,* pp. 48, 52; Kulkarni, Chidambara op. cit., p. 161; Roy, BP *op. cit.*, pp. 150–153.
48. Throwing light on the natural interest of the Vedic Aryans in agriculture it has been pointed out that the "very name *Arya* is said to have originated from a root (*Krish*) which means to cultivate., "Dutt, R.C., *History of Civilization in Ancient India,* p. 35, quoted by Das, S.K., 1987, *Economic History of Ancient India,* (Reprint), Vohra Publishers and Distributors, Allahabad, p. 10.
    Das mentions a reference to this meaning of *Arya* in the *Ṛgveda,* 1.117.21; Das, *Ibid.* p. 10. According to R.S. Sharma, the word *Arya* is originated from the root '*ar*' which named to cultivate. Sharma, R.S., 1992, *Material Progress and Social Structures in Ancient India* (Hindi Tr.) Rajkamal Prakashan, Delhi, p. 55.
49. *Pande,* G.C.; *Foundation of Indian Culture,* p. 73.
50. *Ṛgveda,* 10.34.13; Pandey, R.B., 1965, *Bhārtīya Niti Kā Vikāsa,* Bihar Rashtrabhasha Parishad, Patna, Pant, p. 1.
51. *Ṛgveda,* 8.2.26.
52. Pande, G.C., *Foundation of Indian Culture,* p. 76.

53. cf. Williams, M., 1978, *Indian Wisdom,* (Reprint), Cosmo Publications, Delhi, p.12; Macdonell, A.A., 1974, *Vedic Mythology,* (Reprint), Motilal Banarasidass Publishers Pvt. Ltd, Delhi, 1974, p. 22.
54. Pande, G.C.; *Vedic Samskrti,* p. 79, Pandey, R.B., *op. cit.,* p. 36.
55. *Ṛgveda,* 1.32.11; 1.57.6; 103.2; 2.11.2, 4.19.2; 5.30.6; 5.33.1; Macdonell, A.A., *Vedic Mythology* p. 59.; Pande, G.C., *Vedic Samskriti,* p.79, Kosambi, D.D., *op. cit.,* p. 100.
56. Kosambi, D.D., *op. cit.,* p. 101.
57. *Ṛgveda,* 7.86.56, 88.3–5 Pandey, R.B., *op. cit.,* p. 34.
58. *Ṛgveda,* 1.25.7–8.
59. Sharma, R.S., 'Economic Life and Organisation in Ancient India in *Studies in the Cultural History of India* (ed.) (1965), Metraux, G.S. and Crozet, F. Shivlal Agrawal & Co. (P) Ltd., p. 30. Also Roy B.P., *op. cit.,* pp. 129–130.
60. *Ṛgveda,* 1.24.6; 151.9; Macdonell, A.A., *op. cit.,* p. 26.
61. *Ṛgveda,* 1.14; 3.5.4; 4.9.4; Macdonell, A.A., *op. cit.,* pp. 96–97, Pande, G.C., *Vedic Samskriti,* p. 78.
62. *Śatapath Brāhmaṇa,* 1.4.1. 15–16.
63. The Taittirīya Samhitā speaks of the earth being afraid of excessive burning which indicates of the practice of burning forest on a very large scale. *Taittirīya Saṃhitā,* 5.2.10.3, Sharma, R.S., 1966, *Light on Early Indian Economy and Society* (Hindi Tr.) Motilal Banarasidass Publishers Pvt. Ltd, Delhi, p. 45.
64. R.S. Sharma refers to one mantra of the Ṛgveda,(7.100.4) which says that *Viṣṇu* made the land fit for agriculture and then the he gave it to people.
    Sharma, R.S., *Materials Progress and Social Structures* (Hindi Tr.), p. 77.
65. For details, see, Singh, B., 1997, *Harappa Sabhyata aur Vedic Sahitya,* Radhakrishna Prakashan, Delhi, pp. 381–382.
66. Roy, B.P., *op. cit.,* p. 136.
67. *Ṛgveda,* 4.5.7.
68. Pande, G.C.; *Foundation of Indian Culture,* p. 74; Das, S.K., *op. cit.,* p. 13.
69. *Ṛgveda,* 4.12.1–8; Roy, B.P., *op. cit.,* p. 133 fn 155.
70. Pāraskara Gṛhyasūtra, 2.13; Roy. B.P., op. *cit.*
71. *Ṛgveda,* 10:101.1–4; 2.21.1; 4.57.7; 8.21.3.
72. Roy, B.P., *op. cit.,* p. 136.
73. Pāraskara Gṛhyasūtra, 2.13; 4.49, Armstrong, A.E., 'The Rituals of the Plough', *Folklore,* Vol. 54, London, pp. 250–259.
74. *Ṛgveda,* 10.101. 3–6.
75. Pande, G.C., *Foundation of Indian Culture,* p. 76.
76. *Taittirīya Brāhmaṇa,* 3.9.5.5.
77. *Satapath Brāhmaṇa,* 1.3.4.16.
78. *Aitareya Brāhmaṇa,* 1.5.2.
79. *'Etad vai Bevānāmaparājita māyatanam yad yajñah.*
    Taittirīya Brāhmaṇa, 3.3.7.7.
80. See Sinha, A.K., 'Value and Social-Structure in Ancient India', *Journal of The Oriental Studies,* Baroda, Vol. 40, No. 1–2, pp. 31–49.

81. Basu, J. 1969, *India of the Age of the Brahmanas,* Sanskrit Pustak Bhandar, Calcutta, p. 91.
82. Sinha, A.K., 1983, 'Dharma, Rājadharma and the Process of State-formation in Ancient India.' *Bhārti* (NS. No. 1), Department of Ancient Indian History, Culture and Archaeology, Benaras Hindu University, Varanasi, p. 54.
83. *Atharvaveda,* 8.22.6.
84. *Ṛgveda,* 8.22.6 also, 1.117.21.
85. Vājasneyi Saṃhitā for example, refers to Vṛhi, Māsa, Tila, Mudga, Khala, Priyāngu, Anu, Shyāmlaka, Nivāra, Godhuma and Masura, etc. *Vājasneyi Samhita,* 18.22, also *Atharvaveda,* 4.140.2.
86. Macdonell, A.A., and Keith, A.B., 1967, *Vedic Index of Names and Subjects.* Vol. II, Delhi, p.187; Sharma, R.S., *Light on Early Indian Economy and Society,* p. 45.
87. Macdonell, A.A. and Keith, A.B., *op. cit.,* p. 451; also Mazumdar, R.C., and Pusalkar, A.D., 1971, *The Vedic Age,* Bhartiya Vidya Bhavan, Bombay, p. 460.
88. The six Oxen represent six seasons and twelve signify twelve months of year. Likewise, twenty-four oxen represent twenty-four oxen of a year. *Taittirṃya Saṃhitā,* 5.2.5.2.
89. Roy, B.P., *op. cit.,* p. 134.
90. *Ibid.,* p. 129; fn 133.
91. *Satapath Brāhmana,* 1.6.13; Basu, J., *India of the Age of the Brahmanas,* Calcutta, 1969, p. 67.
92. *Satapath Brāhmaṇa,* 3.3.3.17.
93. '*annam vai kṛsih*', *Satapath Brāhmaṇa,* 7.2.2.6.
94. Basu, J., *op. cit.,* p. 68.
95. Mishra, G.S.P., *Prācīna Bhārtīya Samāja evam Arthavyavasthā,* Rajasthan Hindi Granth Academy, Jaipur, 1983, p. 118.
96. Sharma, R.S., *Light on Early Indian Society and Economy,* p. 46; also Material Progress and Social Structures, p. 113; Pande, G.C., *Foundation of Indian Culture,* p. 77.
97. See Sharma, R.S., *Material Progress and Social Structures,* pp. 109–111.
98. Śatapath Brāhmaṇa says 'cowdung surcharges the earth with sap; hence cowdung is collected (for cultivation) *Satapath Brāhmaṇa,* 2.1.1.7.

    Atharvaveda (3.14.3.–4) refers to the value of natural mannure of animals. Basu, J., *op. cit.,* p. 68.
99. Das, S.K., *op. cit.,* pp. 40–41; Sharma, R.S., *Light on Early Indian Society and Economy,* p. 45.
100. Roy, B.P., *op. cit.,* pp. 130–131.
101. *Ibid.,* pp. 151 ff.
102. References to the lāngalyajña and *Sītāyajña* in the later tradition clearly confirm this point.
103. Singh, B., *Harappā Sabhyatā aur Vedic Sāhitya,* Delhi, 1987, pp. 272–273.
104. Tāṇḍya Brāhmaṇa, 16.13.6.
105. Singh, B., *op. cit.,* p. 272; Chattopadhyay, D.P., *Lakayata* (Hindi Tr.) Macmillan India Ltd., Delhi, 1982, p. 483.

106. Sinha, A.K., *Readings in Early Indian Socio-Cultural History,* pp. 23–24.
107. Ranade, R.D., *A Constructive Survey of Upaniṣadic Philosophy* (Hindi Tr.) Rajasthan Hindi Granth Academy, Jaipur, 1983, pp. 5–6; Basham, A.L., 1987, *The Wonder That was India,* Rupa & Co. Calcutta, p. 248.
108. Pandey, R.B., *op cit.,* p. 52.
109. Pande, G.C., (ed.) *Itihāsa Svarūp evaṃ Sidhānta,* Rajasthan Hindi Granth Academy, Jaipur, 1973, pp. 47–48; 56–57.1.
110. *Taittirya Upaniṣad,* 2.3.
111. Murty, K.S., *op. cit.,* p. 10; also, Pande, G.C., 1984, *Bhārtīya Paramparā Ke Mūla Svara,* Delhi, p. 46.
112. Murty, K.S., *op. cit.,* p. 11–12.
113. *Bṛhadāraṇyaka Upaniṣad,* 3.12.
114. *Ibid.,* 4.5.2.
115. Yājñavalkya and Gārgya in the Bṛhadāraṇyaka and Raikva in the *Chāndogya Upanisad* may be referred to as examples in this regard *Bṛhadāraṇyaka Upaniṣāda,* 2.1; 1.3. *Chāndogya Upaniṣada,* 4.2.3.
116. See, Kane, P.V., 1980, *History of Dharmashastra* (Hindi Tr.), Pt I, Hindi Samiti, Uttar Pradesh, Lucknow, p. 265.
117. A number of things connected with the agriculture and agricultural operations have been mentioned in the attempts to explain the Brahma and its relation to worldly affairs which all can be taken as example in this regard. Beside, we find references to agricultural activities in the Āśramas of the sages. The well-known story of Āruni who lay down on the border of his teacher's field to stop the flow of soil with the rain waters is also an indirect reference to the agricultural development in the Upanisadic age.
118. Kosambi, D.D., *op. cit.,* pp. 129–130; Sharma, R.S., *Material Progress and Social Structures in Ancient India,* p. 159.
119. See, Sinha, A.K., 1994, 'Ahimsa and National Integration', *Bhārti,* Vol. 21, Pt. II, Department of Ancient Indian History, Culture and Archaeology, Benaras Hindu University, Varanasi, pp. 82–83.
120. *Taittirīya Upaniṣada,* II-2; III 7–9; also *Bṛhadāraṇyaka Upaniṣada,* IV, 12; *Maitri Upaniṣada,* VI 11.2; Chattopadhyay, D.P., *Lokayata,* (Hindi) Macmillan Ltd., Delhi, pp. 523–525.

CHAPTER 17

Some Materials for the Study of Agriculture in Vedic India: Problems and Perspectives

A.L. Yadav

INTRODUCTION

It is now generally agreed upon that the Vedas are the oldest existing literature in the world, although the exact date of their composition is still in controversy. Scholars suggest different dates regarding the composition of the Vedas right from 10,000 BC to 1300 BC. Whatever be the exact period of the Vedas, it is definite that they are the oldest Sanskrit texts available and subsequent Hindu scriptures are directly or indirectly based on their ideas and expositions.

Like other spheres of Indian culture, we find a lot of information regarding the economic life of the people of the period in the Vedas but, comparatively speaking, they provide lesser information about the agriculture of the day. The information that we have about Vedic agricultural practices needs an extremely careful and critical observation by the scholars; otherwise, it may create confusion, giving rise to several controversies as well. Hence, the practical and traditional knowledge of agriculture along with the proper understanding of the 'village life' among the scholars is one of the fundamental requirements to tackle and interpret the Vedic references regarding the agriculture during that period. The proverbial conservative character of Indian agriculturists is also one of the major hurdles in tracing the correct position and different stages of the evolution of Indian agriculture. At the same time, the uncertain date of ancient literary texts, particularly the Vedic literature, also invites difficulty with the result that the changes in society and economy cannot be placed in any perspective. For example, the epics contain several strata; hence, it is erroneous to ascribe the entire epic material to Vedic times. Similarly, certain Vedic traditions can be traced in the epics but they were compiled in the early centuries of the Christian era. The history of agriculture in Vedic times can be written only on the basis of Vedic literature. Of course, the archaeological material from Western U.P. and neighboring areas belonging to the later Vedic period can be used in tracing the history of agriculture of 1000–1500 BC (Sharma, 1995, 9.8).

The *Ṛgveda*, is the oldest Vedic text. Like many other things about social organizations, the *Ṛgveda* also provides vital information about farming in its tenth book, especially the *puruṣaśukta*. We many note, says R.S. Sharma, that it was a common practice to interpolate material in ancient texts and thus make it acceptable to the common people. Interpolation could be made easily either in the beginning or at the end of the text. Therefore, says R.B. Sharma, unless we take account of the different strata found in Vedic and other texts, it will not be possible to construct solid history.[1] Hence, keeping in mind the statement of Sharma, we may be careful enough to handle the references of the Vedic literature pertaining to the agriculture of that period. The literary traditions preserve legends about the introduction of agriculture. We have to determine the manner in which nomadic society slowly moved to settled cultivation of slash and burn and of the first use of agricultural implements and cattle. For later Vedic period, the archaeologists give us agricultural implements made of various metals, including iron. By being more careful, he may unearth the plant and *grain remains,* which need scientific analyses in order to identify them.

It is obvious from the different nature of the problems indicated above that the study of the agriculture in Vedic India is a difficult task and needs more attention and sufficient care in handling the sources of information as well as the material for study. Keeping in view these precautions, an attempt has been made in the following pages to analyze the materials—literary as well as archaeological—so far available for the study of agriculture in Vedic India.

Vedic India was essentially agricultural India. After conquering the fertile land of India, the Aryans settled down and commenced cultivation and animal husbandry on the Indo-Gangetic plains of northern India. The very name *Arya,* by which Aryan conquerors of India have distinguished themselves from the aborigines, is said to be derived from a root, which means 'to cultivate'. Agriculture was the most important industry and occupation of the people as also the biggest source of income of the society and state.

It has been admitted that cultivation of soil was known to the Indians before they separated from the Iranians, as is indicated by the identity of expressions *Yavaṁ Kṛṣ* and *Śaṣya* in the *Ṛgveda* with *Yao karesh* and *hahya* in the *Avestā* referring to ploughing of the seed and the grain which resulted[2].

The fourth book of the *Ṛgveda* (IV. 57.1–8) contains a few hymns which conveyed prayers to several important deities related to different agricultural activities which echo even today in the hearts of every Indian cultivator as it did in days of yore. In these hymns, various agricultural personifications have been addressed, the deity of the first three stanzas being *kṣetrapati,* of the fourth, *Suna* of the fifth and eighth, *sunasīra,* of the sixth and seventh, *sītā, suna* and *sīra* being the plough and the ploughman.[3] The view of R.S. Sharma[4] is perhaps not acceptable when he says that 'consolidated references to ploughing are found at one place in the fourth book of the *Ṛgveda,* but it is thought that these are later interpolations'. Commenting of this particular hymn on the *Ṛgveda,* R.C. Dutt has well observed. 'In those two remarkable verses the furrow, *sītā,* is addressed as a female and asked to yield copious harvests. In the *Yajurveda* also, the furrow is similarly worshipped. When the Aryans gradually conquered the whole of India and primeval jungles and wastelands were marked with the furrow, the furrow, *sītā* assumed a more definite human character and became the heroine of the epic, which describes the Aryan conquest of southern India.'[5]

We find references to ploughing and sowing barley (*yava*) and milking out food for men in the *Ṛgveda* (I. 117.21). On the basis of this reference, Sharma says that in spite of their practice of ploughing, the Aryans are known to have had one variety of grain known as *yava*, which may be taken either as generic name for the various kinds of grain or in its literal meaning, barley.[6] But Samaddar opines that rice cultivation hardly seems to be knowledge in the *Ṛgveda*. Rice grain is mentioned in the *Atharvaveda*.[7]

The verse eight of hymn 28 of the tenth book of the *Ṛgveda* states, 'Burnt the grass up where they found it growing'. Ludwig considers this as a reference to the beginning of agriculture. Apart from it, the *Ṛgveda* (I. 101) gives us a glowing picture of the agricultural habits of the Aryans and the implements used by them where we read 'Fasten the ploughs, spread out the yokes and sow the seed on this field which has been prepared. Through song may we find hearing fraught with plenty; near to the ripened grain approach the sickle. The ploughs have been fastened; the labourers have spread the yokes; the wise men are uttering prayers to gods. Prepare the trough for the drinking of the animals. Fasten the leather string, and let us take out water from this deep and goodly well which never dries up. Take out water from the well. Refresh the horses, take up the corn stacked in the field and make a cart which will convey it easily.' Ploughing is distinctly referred to in the *Ṛgveda* I.23; VIII. 20; 22 and X. 166.

The *Ṛgveda* of takes note of other processes relating to agriculture also, e.g. showing, threshing (X.48) and winnowing (X. 27). It also contains many references regarding irrigation, wells (X. 25), water for irrigation (X. 93), irrigation of field by means of canals (X.99), cultivators irrigating their fields (X. 68) and digging the canal for the forward course (X. 75).

Plough was the main implement of agriculture during the Vedic period. It was known as *sīra* (*RV*, IV. 58.8), *sītā* (*Kapisthal Saṁhitā* 28.8 V.I. 1451) and *tangala* (*RV*., IV. 57. 4; X.101.4; *AV*. V.III.173) in Vedic literature. The fourth verse of hymn 101 of the tenth book of the *Ṛgveda* describes ploughing by the wise men. In one of the other hymns of the same text (I.117.2), the ...svins are spoken of as concerned with the sowing of grain by means of the plough. The *Atharvaveda* regards *Prithivanya* (VIII. 20.24) as the inventor of the plough. The ploughs during the Vedic period were made of wood and were generally drawn by a team of two bullocks. But we get references of drawing the ploughs by a team of six (*A.V.* VI. 91.1), eight (Ath. VIII. 9.16), twelve (*T.S.*, I. 8.7.1; V. 2.5.2; *Mait. Saṃ*. II. 6.2) or even twenty-four oxen (*Kathak Saṃ*. XV. 2) as also is clear from *Khadyoga, dvādaśayoga, khadagava, dvādaśgava* used in the Vedic literature. Even the *Ṛgveda* (VI. 91.1) mentions the use of a team of six to twelve oxen in a plough.

There is a great controversy among the scholars regarding the reason for the use of more than two oxen in a plough during the Vedic period. Few scholars like Macdonell and Keith,[8] Damodar Satavalekar[9] and V.B. Rao[10] are of the opinion that the long shape and heavy weight of the Vedic plough was the main reason behind the use of more than two oxen in a plough. Aiyar[11] 'and Banerjee[12] also support the view and state that the use of horses in the plough during Vedic period cannot be denied on this ground. But

R.S. Sharma,[13] condemning the views of Aiyar and Banerjee, says, 'Kine and horses were not yoked to a plough, although they may have been used as pack animals for carrying grain and fodder'.

Randhawa has given another argument based on the feeling of a *ahiṃsā* prevailing in the Vedic society. He opines that plough oxen may be employed for the whole day when they are eight strong per plough. If the number of oxen yoked to a plough is six, they should not be engaged for more than three-quarters of the day; if the number is four, not more than half of day; and if the number is two, not more than one quarter of the day. In support of his view, he has quoted Atri, Parāśara and...Apastamba and others who say that one who yokes eight oxen to a plough is a pious man. One who yokes six is just a businessman. Cruel are those who employ four and those who employ two are but beef eaters.

It is notable here that the justification given by Randhawa for yoking more than two oxen in a single plough is limited upto eight bullocks only, while we get references in Vedic literature for the use of twelve, sixteen and even twenty-four oxen in a plough, which does not come within the limit of a day fixed by Randhawa. Secondly, may it not be more practical to use more than one plough, i.e. three or four, as the case may be instead of employing a team of two oxen separately in a single plough for the entire day in parts? It was also not impossible for a peasant to arrange ploughs and ploughmen according to the needs of the oxen of his own capacity in order to save his time and man labour as well. It is needless to mention that if Randhawa is so careful about the capacity and pain of the oxen yoked in the plough he would also have been equally careful about the same of the ploughman.

Randhawa has offered another argument for employing more than two oxen in a single plough. He argues that employment of a large team of bullocks indicates that the bullocks were weaker than the present-day breeds. Besides, the ploughs were made of wood and they were not provided with iron ploughshares and to plough hard clay soil, more bullock power was needed. This argument also cannot be accepted, as it is clear from the biological factors that the breeds of the bullocks of those days were not weaker at all in comparison to the present-day breeds. The reason also stands defective in the sense that it is different to agree that the soil was hard in general.

Lallanji Gopal has been of the opinion that applying more than two oxen for ploughing were not motivated by a desire for deep ploughing, but resulted from a compassionate concern for the animal which can ultimately be traced to the principle of non-violence. At a time I was also of the opinion that since the agriculture was prescribed for all the *varṇas* of the society, the number of bullocks recommended for use in a plough was decided by the position of *varṇas* in the society, i.e. the highest number to the Brāhmaṇas and the lowest number for the Sūdras. But now I take it in an other way.

On my agricultural background in general and practical experience and experiments on this particular subject, I have reached to the conclusion that the use of more than two oxen in a single plough at a time is impossible. It is not practical to use them team by team also. The words used in the Vedic literature testifying the use of more than two oxen do not stand only for the literal meaning, i.e. for the yoking of oxen in a plough but they stand for a hidden sense as well. Actually, they are used in the sense of glorifying the 'symbol status' of the agriculturists during the period. In other words, it can be said that the number of oxen mentioned in the vedic literature by using the words like *Dvādaśgava, Dvādaṣāyoga, ṣaḍayoga, ṣadayava* etc.[14] shows the arable land lying in possession of a particular farmer as we see it even today.

Later on, during the time of *Pāṇini Halya* and *Sītya* became two popular words for this Vedic expression. The land which could be cultivated by one plough (i.e. by two oxen) was called *Halya* or *Sitya* and that which could be cultivated by two ploughs or three ploughs (i.e. by 4 or 6 oxen) was known as *Dwihalya and Trihalya* and so on.[15]

The shape of the Vedic plough was perhaps similar to the present-day one, something like English alphabet 'V'. One of the important parts of the plough used in its bottom was known as phāla, phāra or stega in the Vedic literature.[16] It was lance-pointed pavīrvat pavīrvam *lāngala* and *Sīra* were synonyms in the *Ṛgvedic* period, while in later Vedic period, they had different meanings. The *lāngala* was lance pointed, having a smooth handle, whereas the *sīra* was a large and heavy plough. The handle of the plough was known as *tsru*,[17] which is described as soft and attractive. Oxen were yoked in the plough with the help of *vatā*[18] (thick rope). This word appears to be the ancestor of the present-day word *varta/vatta*, popularly used in the villages of Uttar Pradesh.

Spade, known as *Khanitra* (*RV.* I. 179.6) was yet, another agriculture implement used to dig land. Sickle was the harvesting tool in those days as is today. It is called *Dātra* or *Sriṇi* in the *Ṛgveda* (VIII.78.10; X. 101.3) and *Parś* or *Paraśu* in the *Atharvaveda* (XII.3.31; also *Tat. Saṃ.* 3.2.4.1). After harvesting, the crops were brought to the *Khalya* for threshing. There were two methods adopted by the Vedic peasants for the threshing of their crops (1) *Sūrpa* or *Chhāja* (*RV.* X. 71.2), and (2) *Titau* (*RV:* Ibid) or *Chhananī*. *Titau* (Sieve) was in used to separate the grains from the straw and *sūrpa* (winnowing) was resorted to. *Urdara* (*RV* II. 14.21) was the vessel used for measuring grain and *Sthivi* (*R RV* X. 68.3) was used to store grain. It was something like today's granary.

Different processes of agriculture have also been mentioned in the Vedic literature. In the *Satapatha Brāhmaṇa*,[19] the main agriculture operations have been described as ploughing (*karṣaṇa*), sowing (*Vepa*), reaping (*Lava*) and thresting (*Niṣpāva*).

The word *Kṛṣi* originally meant ploughing and this meaning of *Kṛṣi* seems continue till the epic period. In the *Mahābhārata*,[20] the word *Kṛṣi* is used in the sense of agriculture and tilling of field as well. Later on, this word covered other operations of agriculture also. *Kṛṣival* (*RV.* IV. 58.80) was the term used for the ploughman in the Vedic period.

The system of cultivating the soil was more or less similar to the one prevailing in India at present. The ploughing of the field commenced on an auspicious day with sufficient rainfall. The first ploughing of the season was inaugurated amidst much rituals. Offerings were made to the agricultural deities on the occasion. The *Yajurveda* (189.13–20) mentions a prayer to ploughs and ploughing, 'Our auspicious ploughs with their ornamental handles and their sharp-pointed shares. Cleave the ground to the happiness of cows, sheep and well-grown maidens. May the ploughmen plough round and round, happy with prospect of heavy crops—may the rain god grant us plenty with milk and honey', etc.

After ploughing the fields, crops were sown. Seeds were sown into the fields when the soil was fully prepared. Terms referring to ploughing twice or more indicate the realization of the need for deep ploughing. *Vapan* (*RV.* I. 117.21) was the popular word in the Vedic literature used for sowing. Sowing of the seed was known as *Vapa.* The *Ṛgveda* (X. 9.2.3) gives a very clear reference of seed and sowing as under, 'Harness the ploughs, fit on the yokes, now that the womb of the earth is ready to sow the seeds therein; and our praise to Indra, may there be abundant food, may the grains fall ripe towards the sickle'.

Though presently we are not in a position to understand the method adopted by the Vedic farmers to sow the seeds of different crops in the fields, on the basis of their sufficient—rather surplus—production, we may assume that they were well acquainted with the skill of sowing. Our contention may also be supported by the fact that the Asvina are spoken of in the *Ṛgveda* as concerned with the sowing of grain. Many rituals were also being performed at the time of sowing of the seed. They have been fully described in the Gṛhyasūtras (*Kaus. G.S.* 24.1–2; *Kath. G.S.* 71.8).

When the crops were ripe, they were reaped and the plants were stocked in a corner of a field. It was called *Lavan* or harvesting, which began with a prayer. The harvesting tool was the sickle (*Dātra, Lavitra).* 'I take the sickle also in my hand with a prayer to thee' (*RV.* VIII. 8.9–10). 'May the crop swell at my prayers; let the sickles cut down the heavy crop of grain' (*RV.* V. 6.12, 9,10). 'May there be abundant food, may the grain fall ripe towards the sickle' (*RV.,* X. 9.2–3).

Both the popular methods of present-day's harvesting were in practice in those days also—one by cutting down the crop at the level of the ground and the other by cutting the ear heads. The harvesting of the crops like *Vrīhi, Māṣa, Mudga,* etc., was done by the first methods, while yava Jwara, Bajara (millet) were harvested by second method, as is clear from the reference in the *Yajurveda* (122.1)., 'As barley is harvested by separating the ear heads from the stalk'. When the harvesting was done, bundles (*Parṣa, RV.* X. 10.1–3) were prepared and plants were stocked in a corner of a field from where it was brought to the *Khalya* either with the help of men or by bullock carts.

Sometimes, many farmers stored their harvest at a common place, which they called *Khalya.* Finally, there was threshing of crop in order to separate the grain from the hay. The word *Mardana* was used in the *Śatapatha Brāhmana* (I. 6. 20.3) for this purpose. Threshing was done by beating the crop repeatedly on the stone. Sometimes, it was also done with the help of bullocks. Both the methods can be seen even today in some of the villages of northern India. To separate the grain from the straw, *atitau* (Sieve) or *Sūrpa* (Winnowing) was in use. The word *Pavan* is used in the *Atharvaveda* (IV. 34.2) for this agricultural operation. After measuring the grain by a vessel named *Urdara* (*RV.* II. 14.11), the total production of a farmer of that particular season was stored in the granary called *Stīvi* in the *Ṛgveda* (X. 68.3). When all the agricultural processes were completed smoothly, rituals were performed and offerings were made to several deities. At the time of the partaking of the first fruits of crops, *Āgrayaṇa Navaprāsana* offering was made.

Classification of land for various purposes seems to have been a distinctive feature in ancient India. In the Vedic period, land was generally divided into cultivated areas, pastures or grazing grounds and forests. A large area was used for the cattle ranches. A village included both cultivated and uncultivated land. *Urvara, Kṣetra, Pharavara,* etc., were the words used for arable and un-arable land in the *Ṛgveda.* On the basis of the fertility, the land in the Vedic period was divided into two categories, i.e. (1) *Urvara* (fertile), and (2) *unuravara* (unfertile). Further the *Uravara* or fertile land was divided into three categories: (1) *Ārtanā,* (2) *Apnasavtī,* and (3) *Urvarā* or *Kṣetra.*

Ārtanā was a barren or unfertile land, while *Apnasvatī* was a fertile tract. *Urvarā,* as is clear from its name, was perhaps the most useful for the agricultural point of view. Maximum production in minimum labour and cost was the special feature of this kind of land.

Unurvarā or unfertile land was also divided into three categories: (1) *Usar,* (2) *Banjara* (barren land), and (3) *Jangala* or waste (forest). *Usar* was absolutely useless land. *Banjara* or barren land, known as *Khila* or *Khilya* in the Vedic literature (*A.V.* VIII. 115.4; *S.B.*, 8.3.4.1; *RV.* 1.142.3; VI. 28.2), was used for grazing puposes of the village cattle. Forest was the centre of *vānaprasthis* and *sanyāsins.* They also supplied woods, grasses, leaves and even uncultivated (अक ष्टच्य) corn like *Nivāra,* Tinnī, etc.

Agriculture had to depend on irrigation as is the practise even today. In fact, in a country like India, water storage was and is absolutely necessary. The Vedas contain many references to irrigation. On the basis of the references available in the Vedas, it is clear that there were two means of irrigation during the Vedic period—first, natural and second artificial. Rain, rivers, falls, likes, etc., come under the first category while wells, canals, tanks and other reservoirs come under the second category.

We can read certain references in the *Ṛgveda* of wells (X.25), water for irrigation (X. 93), irrigation of fields by means of canals (X. 99), cultivators irrigating their fields (X. 68) and digging the canal for the forward course (X. 75). In the *Atharvaveda* also, we come across such references. This text (III. 13) states that the canal has been cut and the water of the river is to be admitted. Here, this Veda designates the former (canal) as a calf and the latter (river) as cow. We may also release the importance of such channels because the *Atharvaveda* (VII. 45) mentions the punishment to be meted out to the undiscovered thief, viz., his conduit was to be bound. The word *kulyā* is used for a canal in the *Ṛgveda* (III. 45.3) as well as in the *Yajurveda.* The *Kauśika Sūtra* (40.I.10) describe the ritual to be performed at the time of directing the course of a river; and in fact, this seems to suggest the opening ceremony of letting water into a canal.

The cultivation of the soil was undoubtedly dependent on rains to a great extent. The seasonal rainfall primarily served the purpose of irrigation. The Vedic period witnessed two crops a year. We find prayers in the Vedas invoking deities like Indra, Varuṇa, etc., to induce rains. The Śrauta and Gṛhya Sūtras frequently refer to rains and describe the ritual of procuring rains.

However, cultivation was not entirely dependent on rains, Apart from rains, rivers constituted another natural source of irrigation in the Vedic India, as is the case even today. Frequent references to rivers in the Vedic literature indicate that the people were mostly living in the vicinity of rivers. There is reference of craftsmen (*ribhus*) who led forth the rivers (*RV.* IV. 4.1.7). The reference is to irrigation by channels taken from the rivers. There is also reference to soil erosion by rivers, 'Rivers, the corroders of their banks, like armies destructive of their foes' (*RV.*, X. 9.2.5).

Indian agriculture is still regarded a gamble because of uncertain and irregularity of the monsoon rains. As a precaution against drought and famine, the significance of artificial irrigation facilities was well recognized by the people as early as the Vedic period. Hence, the system of artificial irrigation was quite prevalent in those days. Wells were in use for supplying drinking water to men as well as to animals, and also for irrigation. 'Tie the rope tight to the water pots, let us draw water from this unfailing well. Set up the cattle trough, bind the straps to it, let us pour out the water from the well which is not easily exhausted' (*RV.*, X. 9.2.5). These references in the *Ṛgveda* suggest some devices which were used by the Vedic agriculturists for lifting water from wells, viz., *Asma-Chakra, Yuga-Varatrrā, Charma-Rajju,* etc.

Water and manure are two primary elements, necessary for the survival and growth of crops. Though the ancient Indian agriculturists knew the proper methods of manuring their fields and crops, but on the basis of the references available in the Vedic literature, it is not clear as to whether they were manuring their crops or not. Some scholars like Macdonell and Kieth,[21] R.B. Pandey,[22] Damodar Satavalekar[23] and V.B. Rao[24] have taken the Vedic words such as *Šakrit, Karīṣa, Šaka* (*AV.*, XII. 4.9; XIX 31.3; XIV. 3.4 and *RV.* 1.161.10), etc., in the sense of manure and expressed their views very strongly that the Vedic farmers were well acquainted with the knowledge of manuring their crops. Munsi Ram Sharma[25] opines that the adjective *Karīṣiṇī* used for *Pṛithvī* in the *Atharvaveda* denotes the same sense. But actually, these words were not used in the sense of manure but they were really used in the sense of dung or dry cattle-dung as interpreted by *Sāyaṇa* also in his commentaries of the *Ṛgveda* and the *Atharvaveda.*

On the basis of the above discussions, a few main characteristics of the Vedic agriculture can be pointed out as under:

(1) Generally, in the Vedic period, two harvests a year were gathered. The characteristics and requirements of different types of soil were realized. There were terms signifying the suitability of fields for certain crops and their seed requirements.

(2) The Vedic farmers knew the method of improving the fertility of soil by using the method of rotation. The practice of keeping lands fallow is evident. What is generally taken to imply the principle of rotation of crops refers to the double crops system.

(3) In the Vedic texts, we find views of seers regarding the methods of storage of food gains. *Kausṭhi* (Kothi) was a huge container for preserving the food grains (S.B., I. 12.7). *Kumbha* (Ibid.) was another huge pitcher used for the same purpose.

(4) Rice is not mentioned in the early Vedic literature. It is highly probable, says Randhawa,[26] that the Aryans were not aware of rice in the early stages of their spread in India and learnt its cultivation from the aborigines of this country when they reached the areas now described as Uttar Pradesh and Bihar (including present Uttaranchal in UP and Jharkhand in Bihar). Rice does not find a place

in the basic rituals of the Aryans; however, it has a place in secondary rituals which were later elaborations.

(5) The mastery over iron technology is undoubtedly an outstanding achievement of mankind. It almost led to a 'Green Revolution', which was sparked off as a result of the invention of two most important tools, viz., the socketed iron axe and the iron ploughshare in the later Vedic period. In due course of time, all small tools for agriculture, viz. sickles, hoes and spades, etc., were also manufactured from iron.

From the modern viewpoint, 'the Aryans were very superstitious and they observed many rituals, totems, some of which were nothing but short of black magic.'[27] In the *Atharvaveda,* we find a detailed description of such rituals. These superstitions made them of conservative character, which stands as an obstacle to trace the different stages of the evolution of Indian agriculture. Hence, in determining the nature of the techniques and implements in cultivation, one has to avoid the tendency of glorifying the past and to view recent scientific advancements reflected in the past. It need not be emphasized that whatever scientific principles or practices existed in those times were not the result of any particular experiment but they were developed as a result of observation and experiments spread over a long period.

NOTES AND REFERENCES

1. Sharma, R.S., 1995, *Perspectives in Social and Economic History of Early India,* p. 9, Munshiram Manoharlal, New Delhi.
2. *Vedic Index* I, 181 (In the *Ṛgveda* the Āsvins are spoken of as concerned with the sowing of grain but means of the plough).
3. Samaddar, J.N., 1922, *Lectures on the Economic Condition of Ancient India,* p. 13, Kolkata (Calcutta).
4. *Op. cit.,* p. 151.
5. Dutta, R.C., 1893, *History of Civilization in Ancient India,* E.N., p. 39, London.
6. Sharma, R.S., *op. cit.,* p. 151.
7. Samaddar, J.N.: *op. cit.,* p. 16 f.n. (cf. *Vedic Index* I. 398).
8. *Vedic Index,* II. 451.
9. *Veda Mein Krisi Vidya,* pp. 23–24.
10. *Uttar Vaidika Samaj Evam Samskriti,* p. 5.
11. Aiyar, A.K. Yajna Narain, 1949: *Agriculture and Allied Arts in Vedic India,* pp. 15–16, Bangalore.
12. Banerjee, S.C.: 1972, *Aspects of Ancient Indian Life,* p. 2.
13. Sharma, *op. cit.,* pp. 151–52.
14. Kāthak Saṁ "Śīraṃ vā Dvadaśāyogaṃ" Tait. Saṃ "Śāḍgaven Kṛṣati" 1.8.7.1 "Dvādaśa gavam Śīraṃ" *S.B.* 13.8.2.3 "Śīraṃ Yunākti Ṣadgavaṃbhavaṃ."
15. *Aṣṭādhyāyī* 4.4.91, 97.

16. *RV.*, X. 117.7; IV 58.8; *AV.* X. 6.33; XVIII. 1.39 and *RV.* X. 31.9.
17. *RV.*, VII. 50.1; *AV.* III. 17.3.
18. RV., IV. 57.4; X. 102.8.
19. I. 6.2.3. "Kṛṣanto ha smaiva purvē vapanto,
yauti lunants Aparē Mṛṇantah."
20. *MBH* Śanti 186.20 Paṇyanām Śobhanaṃ paṇyam Kṛiṣēnāṃ Vādhate Kṛṣi Vahukāraṃ Ch Sasyānāṃ Vāhane Vāhanaṃ tathā gavāṃ.
21. *Vedic Index,* I. 182.
22. Bhārati, Vol. III, 1959–60, p. 149.
23. *Veda Mein Kṛṣi Vidyā,* pp. 19–20.
24. *Op. cit.*
25. Sharma, Munshi Ram, 1987, *Vedic Sanskṛiti Evam Sabhyatā,* pp. 255–56, Prachārak Publishers, Varanasi.
26. *A History of Agriculture in India,* p. 300, 1964, ICAR, New Delhi.
27. Sharma, B.L., 1987, *Economic Ideas in Ancient India Before Kautilya,* p. 126, Ramanand Vidya Bhavan, New Delhi.

CHAPTER 18

Agrarian System in Ancient India: Harappan and Vedic

Om Prakash

INTRODUCTION

Agrarian system has been conceived by Moreland, a pioneer in the field of its historical study in India, as an institution of intermediaries assessing and collecting the state share of the peasant's produce in return for their share in collection and authority as determined by royal policy, customary law and historical circumstances from time to time[1]. This model, worked out in the specific context of Mughal India, had set the trend of institutional history for historians of ancient India as well. U.N. Ghoshal[2], D.D. Kosambi[3], Lallanji Gopal[4], B.N.S. Yadava[5], R.S. Sharma[6] and numerous other scholars have dug up diverse sources of ancient Indian history to outline the structure and operations of the ancient Indian agrarian system basically understood in the aforesaid sense. Moreland, however, tried to counter the nationalist projections blaming Muslim rulers for the legacy of pre-colonial oppression in contrast to the glorious past of the traditional modes of agriculture reflected in ancient sources. Without denying this legacy left behind by Muslim administrators, he held the unfavourable and political social environment responsible for the barren struggle of the revenue farmers and the assignees to translate the idea of agricultural development into practice by applying the regimes of Delhi Sultanate and Mughal Empire.[7] Ghoshal attempted the reconstruction of the ancient Indian agrarian system on the basis of original sources, which might have facilitated the nationalist agenda to some extent but, in all possibility, without being harboured as a preconceived notion on the part of the historian. Credit goes to D.D. Kosambi and R.S. Sharma for inducting into this institutional historiography the theoretical component and jargon of Marxist ideology, and to Lallanji Gopal[8] and Randhava[9] for working out in detail the perspective of technological, industrial and commercial advancements in the field of ancient Indian agriculture. The expanding horizons, precisions and refinements of this basic concept as achieved by these pioneers of the historiography of ancient Indian agrarian system have brought the whole thing to bear upon a central debate: whether Indian feudalism, perceived as a major change in the ancient Indian agrarian system, constitutes the total explanation of the epochal change called medievalism or is only a part or attribute thereof. This issue implicitly present in the works of B.N.S. Yadava and particularly in the reviews of his book *Society and Culture in India in the Twelfth Century* appears to represent the finale of this approach.[10] This approach, nevertheless, triggered the trend of intensive area studies based on local epigraphy, literary sources and traditions with a view to determining the social formations around class interests, conflicts and coordination, leading to environmental, anthropological, sociological and other model-based interventions. Entitlement and deprivation[11], servitude or bondage of agricultural labour, status of peasantry, production and natural and human hazards, credit and patronage, commercialization, feudal appropriation, etc., have come up as recent issues related to agriculture.[12] Subaltern studies also represent a fallout of the Marxist intervention, seeking to view the agrarian system from the two vantage points of state and economy, commonly called political economy.[13]

To my mind, an agrarian system is more than the web of production relations between the peasants and the state consisting of a number intermediaries. It is a dynamic complex, which makes social as well as political forces and formations to interact with itself. The interacting partners determine and get determined, in turn, under the impact of each other with a persistent element of traditionality during the course of their historical development. The historical studies of ancient Indian agrarian system have emphasized the institutional aspect of this dynamic process and produced more or less static images of it in their normative-cum-factual settings in the first phase; on the basis of which was determined the class structure of haves and have-nots in the rural segment of pre-modern political economy of India, in the second phase. What has been constantly ignored in the second phase is the element of stubborn traditionality, first as the die-hard tribalism and then as religio-social and ethno-occupational phenomenon of caste. The trouble with these notions is that they do not easily evaporate as notional mist and work at the grassroots level to unsettle the socio-economic generalizations on class lines.[14] It is not enough to point out that classes cut across caste divisions and manage to determine social and historical change in spite of the persistence of caste. Even sociologists are now constrained to conclude:

> Thus caste is more a historical reality and infrastructure rather than merely an ideological or ritualistic institution. Caste-based movements and mobilizations have been more effective and intense compared to the ones involving inter-caste involvement and class-based participation.

And further:

> The class question, however, remains central in several studies of agrarian structure and change. Thus, it is a *mixed situation* characterized by forces of caste and feudalism on the one hand and capitalism on the other.

In particularly ancient Indian context, transition from tribal to caste society has not even been adequately explored with the idea of a persistent traditionality in view. In this context, moreover, we have very few formulated agrarian systems but there was hardly any age in early Indian history—except the pre-production ones—which did not have an agrarian system of their own. The system pervasively present has, therefore, to be exhumed from the invisible depths of the historical phenomenon of the age. I am fully conscious of the inadequacies and pitfalls of such a venture. But I am also confident of the fact that the research cannot advance if it is made to proceed wholly on the beaten tracks alone. After all, all investigations and formulations advancing the frontiers of knowledge are essentially provisional. Our endeavour in the present chapter will be to present the ancient Indian agrarian system on these rather unconventional lines, making full use of the knowledge and skills attained by conventional methods and approaches.

QUESTION OF HARAPPAN AGRARIAN SYSTEM

Harappan urbanism has not only been over-emphasized, but has also been studied in isolation.[15] Neither the spade of the archaeologists nor the theories and hypotheses of those who attempt a synthesis of the archaeo-ethnological and biological evidence of the Harappan civilization have lodged it in its agrarian context. I am reminded of the apt remark of Heather Margaret and Louise Miller that 'in discussion of technologies and apprenticeships, archaeologists often overlook the fact that farming too is a craft, involving specialized knowledge passed on through "apprenticeship" and a variety of technologies.'[16] To extend this remark to synthesizers of evidence, it may be said that while correlating the results of the technological study with different activity areas within the site and at the macro-level, agrarian activity areas are often ignored by the keenness to locate non-agrarian activity areas diagnostic of the urban character of the site. It is conveniently forgotten that non-agrarian activity areas in any specific site or sites are unsustainable in the absence of vast agrarian activity areas either at that or those sites or other sites clustered at a manageable distance from such site/sites. It is unthinkable that non-agrarian activity centres of Harappan civilization were deriving their food supplies from the nearby hunters/gatherers or agriculturists of distant Mesopotamia, Iran or Turkmenia in exchange for the products of the crafts. The inference of the presence of a well-organized agrarian system in Harappan civilization is, therefore, not exclusively contingent upon the identification of workmen's quarters, granaries or central organizational control over production of craft-goods. The hallmarks of Harappan civilization, viz., urban centralization, craft specialization and diversified subsistence economy do imply the existence of a Harappan agrarian system. The argument put forward by Brian E. Hamphill, John R. Lukacs and K.A.R. Kennedy that continued utilization of non-agricultural and non-domesticated foodstuffs may call into question earlier models of the development of Harappan civilization which presumed an overwhelming reliance on intensive agricultural exploitation does not make much difference[17] unless, however, it can be shown that 'evidence of increased dependence on agricultural foodstuffs is abundant from Indus civilization sites of Harappa, Kalibangan and Mohenjo-daro'.[18] After all, one should not lose sight of the main issue that the Harappan urbanism based on production of craft goods and mercantilism and covering a geographical area almost as vast as that of France can sustain on such casual supplies as non-agricultural and non-domesticated foodstuffs can ensure. The argument of meager evidence of agricultural implements too is not valid in view of the question as to why should agricultural implements be present in the centres of non-agrarian activities. The existence of some kind of Harappan agrarian system is, therefore, not in doubt.

The Indus urbanism surely did not drop clean from the skies; nor did it emerge directly from the primitive economy of hunting and gathering people, bypassing the agrarian stage.[19] This cannot happen simply in spite of any amount of foreign inspiration. Why should one resort to wild speculation if Indus urbanism is clearly preceded by early/pre-Harappan cultures already practicing plough cultivation, double cropping and living in well-laid out settlements (sometimes fortified and apparently planned), standardized crafts and long-distance trade for procuring raw material for their craftbased industries?[20] This order of development clearly indicates the fact that the Indus urbanism is a consequence of the transformation of a predominantly agrarian economy into craft-based mercantilism.

The Harappan agrarian system belonged to a pre-monetary age with manual and animal force-driven devices of land and water transport and primitive machines and tools of agricultural and craft-based production. Iron had yet to enter the scene and copper and bronze, though predominant, could not fully drive out the use of stone, wood and bone as tools and implements fashioning material. Technologically, the Harappan level of achievement was not too far ahead of the early Harappan level. At least, it does not give the look of a technological revolution *vis-à-vis* the early Harappan, as did Industrial Revolution as compared to Middle Age technologies of Europe. The feat of Harappan civilization was, therefore, more organizational than technological in nature. Whatever technological skills were added by the Harappans of the mature phase were surely obtained in response to the organizational challenges to be overcome. This organizational revolution represented by Harappan civilization of the mature phase may be identified in terms of the parameters of context, record-keeping and complementarily coordinated distribution of labour.[21] Rita P. Wright has already subjected the non-agricultural production to investigation under these parameters but her conclusions suffer from inevitable distortions because she was doing it without a reference to agrarian production understood through the same parameters. I propose to work out the main features of the Harappan agrarian system under these parameters and with reference to non-agrarian production system as understood by Mark Keynoir and Wright.

The context of Harappan agrarian system has, of necessity, not to be the context of individual sites (as very few agrarian sites have been excavated), but of the settlement pattern in different domains of Harappan civilization identified as eco-zones by Possehl. This context conforms more or less to the eco-zones[22] within the area occupied by the Harappans with particular reference to their agricultural resources and in complementarity with the settlement pattern structured with a view to non-agricultural production and trade on the one hand and procurement of the raw material for these activities, on the other.

Harappan penetration in the Kutchh and Saurashtra is chronologically a later event. The process started in the case of Lothal in 2400 BC and in case of Dholavira and Surkotda in 2300 BC. Ranging of the settlements from smaller size to no settlement and traces of temporary hutment-type structure show pastoral, fishing, hunting and gathering rather than agricultural activity by autochthones interacting with Harappans and gradually getting acculturated.[23] The location of Harappan centres in a non-cultivable area like the Rann of Kutchh also indicates that this extension of Harappan civilization was not with a view to exploiting the agricultural resources of the area. No agrarian system, except some incipient beginnings, can, therefore, be looked for in that area. But colonizing such non-agricultural eco-zones most certainly presupposes its correlation with agricultural eco-zones in the Harappan civilization, for Harappan settlements in this non-agricultural eco-zone cannot entirely subsist on non-agrarian supplies of foodstuffs alone. At the most, Harappans of the non-agrarian eco-zones would have supplemented their diet by the non-agrarian foodstuffs available locally.

The domain of Gedrosia too appears to be less agrarian and more non-agrarian, pastoral and commercial by virtue of its ecology and topography. Despite the altitude of over 2000 metres in the Baluchistan plateau, the rainfall is scanty and affords excellent grazing grounds only to sheep, goats and wild ibex. The mountain range of the Kirther and Sulaiman form a piedmont zone between the western highland and the eastern alluvium with slopes gently descending into plains such as Kachi on the mouth of Bolan Pass constituted by the alluvium of the Bolan river. Another alluvial tract is the plain of Las, stretching northward from the sea coast to a distance of about 90 km and formed by the deposits brought by Porali, Hab and Malir rivers.[24] The third agrarian resource in the domain of Gedrosia may be identified around the freshwater lake, Manchhar, which is annually fed by the flood channels of Indus besides the western Nara and the drainage of the hill country. On the Kachi plain is the earliest evidence of agrarian activity at Mehergarh, several thousand years before the beginnings of Harappan civilization.

How rich is the Harappan domain in agrarian resources is a question which can be answered on the basis of more negative than positive evidence available at the present state of our knowledge. The reliability of the answer will always rest on the final position of our understanding of the domain, subject to future investigations. At present, however, two considerations alone appear to sway the verdict in favour of the view that it was a non-agrarian zone. The first consideration is that there are few mature Harappan sites with the exception of Chak Purbane Syal and Vainiwal.[25] The nearest cluster of 175 Harappan sites falls in the dried up bed of Hakra in the Cholistan area which, judging from the smaller size of the ruins of the settlement, appears to be an agrarian tract occupied by villages some of which are industrial in character. This tract is some 300 miles long and 10 to 15 miles wide.[26]

The second consideration is the character of activity of Indus and its tributaries and the geomorphic composition of the terrain in its Panchnad and Sindh sections. The geomorphic composition of the Sindh section is alluvial with only one rocky straight between Rohri and Sukkur hills, whereas the Panchnad section is characterized by the fingers of rocky high ground between the courses of the rivers. Due to their degradational activity, the rivers in the Panchnad are entrenched and flow in a bed below the general level of their flood plains available for agrarian use, which are significantly narrower than the extensive alluvium of the Indus in Sindh. The aggradational activity of the Indus in Sindh makes it to flow on a ridge higher than the surrounding plain. This renders the Panchnad region largely unfit for agrarian production on any appreciable scale. It has been pointed out by Ratnagar, on the basis of District Gazetteer of 1884 of Montgomary (now Sahiwal) district, that vast tracts of land annually flooded by Ravi and Sutlej remained uncultivated till the nineteenth century and the village communities occupying the arid bars of southern Panjab did not come up until the introduction of canal irrigation during the British rule.[27] According to Bharadvaj, nomadic herding was the dominant occupation in the harsh environments of the bars.[28]

This understanding of the Harappan domain and the location of Harappa *vis-à-vis* agrarian resources based mainly on the writings of B.K. Thapar and Shireen Ratnagar is contradicted by that of Marcia Frentress, who is of the opinion that 'prime agricultural land around Harappa is found immediately northeast of this site between an old bed of the Ravi and the present Ravi.' She further thinks that the area is low-lying land and could have been irrigated by flow irrigation through old *bandhs* of the river. The Kamalia plain across Ravi too—she presumes—has 'sources of irrigation in the form of hill *nai* flowing down from high ground of the Rechna Doab and low-lying *jhils* or *dhands* which retain water much of the year.'[29] But Harappa occupied the bank of Ravi when it was flowing in its old bed. The prime agricultural land, of which Frentress is talking, must have been the outcome of the river action when it was shifting its bed in post-Harappan times. The *jhils* or *dhandhs* too are products of river action and might not have been present in the days of Harappan civilization. The simple presence of hill *nai* flowing from high ground to the Kamalia plain cannot ensure a rich agrarian base unless there is evidence to show that it was exploited for irrigation. The absence of smaller settlements indicating agrarian villages in the Kamalia plain as well as in the vicinity of Harappa, however, demonstrates to the contrary. Therefore, the conclusion arrived at by Frentress that these two areas (between the old and new beds of Ravi and the Kamalia plain), taken together, provide almost 6600 square kilometres of immediately available land for Harappa precluding any necessity for dependence on a wide ranging network linking with it the agricultural villages for its food needs, cannot be accepted. Such a wide-ranging network was, in fact, a necessity for the sustenance of a city like Harappa, so far removed situationally from all the agrarian bases of the civilization.[30]

Contrary to the rivers of the Panchnad, the Indus is an aggrading river. After its confluence with rivers carrying the drainage of Hindukush and Panchnad, its course forms a deep S curve adding more land to its fertility effect. Its bed is higher than its flood plain and it is subject to two annual floodings: one as the result of monsoon and the other when the snow at its source melts. Vast Indus flood plains, the delta and the shifting of channels to a distance of 160 km to the west from the Rann of Kutchh (into which it originally emptied) thus constituted the greatest agrarian resource of Harappan civilization. It also contributed to the fertility of the agrarian resources of Gedrosia, as already noted.[31] The city of Mohenjo-daro did not occupy the centre of this domain. It was strategically located at the northern fringes. This further shows that the Harappans planned the location of their major cities with a view to industrial and commercial advantages rather than closer proximity to agrarian resources. The reason for this kind of locational planning appears to be their greater transportation capacity and industrial-cum-commercial orientation. Large-scale river-faring and sea-faring activity must have made all the difference between early Harappans and Harappans. The wide-ranging network sought to be explained away by Marcia Frentress, thus, appears to represent the organizational feat which ushered in the Harappan revolution. Undoubtedly it was made possible by the achievement of their greater transportation and record-keeping capability which the Early Harappans did not possess. Without this the agrarian resource areas could not have been organizationally made the launching pad of major Harappan cities strategically located in non-agrarian areas or on the peripheral points commanding strategic trade and industrial resource inflow routes rather than in the centres of the agrarian resource areas.

Of the three largest cities—Mohenjo-daro, Ganweriwala and Harappa—Ganweriwala occupying an extent of 81.5 hectares is located in the agricultural zone of Cholistan (Bahawalpur). Mohenjo-daro (83 hectares) too is similarly located in the agricultural zone of Sindh. The smallest of the three, Harappa alone is far removed from the proximity of any agricultural zone, as is Dholavira. Each of the six domains of Harappan civilization of which the main centres were Kalibangan, Harappa, Ganweriwala, Mohenjo-daro, Lothal, including Gedrosia—which has no specific centre—was marked by the diversity of topography, vegetation and natural resources, which unless exploited in a well-coordinated and complementary manner, would not have supported a highly developed civilization. Cultural and ethnic diversity is intruding fast and prominently upon the assumed monotonous homogeneity and stagnation of the Harappan civilization. Regional variation in belief-systems marked by the use and non-use of fire altars and figurines of mother goddess, sacred structures for ceremonial public worship, area-specific custom of double burial amply demonstrate civil integration of Harappan society and polity on liberal and perhaps non-military lines. This integration holding all diversities[32] into a complementarily organized whole, appears to have been achieved through economic and political activities of which the latter is less in evidence in the excavated material of the civilization. The economic activities criss-crossing not only all the Harappan domains but also some of the neighbouring and contemporary civilizations is very much in evidence. The urban, commercial and overseas aspects of economic activities of the Harappan civilization constituting a glorious attainment is well known but the agrarian counterpart of it silently sustaining the pinnacle of this glory has always been a victim of academic imbalance.

The very fact that such a vast civilization sustaining huge townships, like those of Harappa and Dholavira, quite removed from its agricultural bases must have inevitably contained not only an efficient agrarian system but also an equally strong water and land transport system managing commodity exchange through different eco-zones spread over a vast area within and other contemporary civilizations overseas and overland. This undoubtedly makes the Harappan urbanism and industries widespread, 'dispersed and embedded in agriculture' to use a phrase by Ludden. The latter system would have been responsible for collecting the surplus from primary producers and making it available in abundant quantities to the consumers of cities and townships away from agricultural areas of the civilization. The complex of granaries, servant quarters and pounding platforms at Harappa clearly shows adequate arrangements for receiving, processing and storing the surplus thus obtained. Was this elaborate arrangement for sustaining the inhabitants of Harappa alone or for catering to the needs of redistribution of surplus to different adjacent towns is, however, a moot question which cannot be conclusively answered at the present state of our knowledge. The mode of collection of surplus of the agrarian production too can only be tentatively inferred from the evidence of its impact and the possible beneficiaries of its appropriation, if any. The most blatant utilization of agrarian surplus in Harappan civilization does not appear to be royal, military or religious—as in Sumerian and Egyptian civilizations—because large-scale monumental buildings, palaces and religious monuments such as Pyramids and Ziggurats are conspicuous by their absence in Harappan civilization, except, however, the two enclosures of fire altars inside and outside the citadel area at Kalibangan and the Great Bath adjacent to the Pillared Hall at Mohenjo-daro. The beneficiary of the appropriated agrarian surplus thus does not appear to be limited to a small circle of rulers with a king and his personal retinue at the top. Lack of evidence of military activity rules out utilization of agrarian surplus in army building and warfare too. What is most evident in the Harappan context is the industrial, mercantile and commercial activity. To facilitate it on a large scale, elaborate networks of the supply of raw material from afar had been developed and the efforts had gone to the extent of colonizing areas of Kathiawar, Saurastra and Kutchh in Gujarat; Shortughai in Afghanistan; and Bahrain, Failka and Oman in the Persian Gulf. Land, river and overseas transport had been developed and huge cities had been planted at strategic points with a view to making them function as collection centres of the supplies of the raw material and finished products to be transported across the length and breadth of the civilization and to overseas destinations.[33] Agrarian surplus might not have gone to constitute the capital of this large-scale industrial, commercial and transport activity, in the absence of the developed fluidity of coined money and banking facility. Also, it cannot be believed that Harappans were obtaining the supplies of their raw material in exchange for agricultural production, for then each of their cities should have been equipped with elaborate food grain collecting complex comparable to that of Harappa. At the most, the agrarian supplies must have been utilized in providing subsistence to the large cities planted at strategic points far removed from the areas of agricultural activity[34] for sustaining traffic of raw material and finished product of Harappan craftsmanship from and to distant destinations.

Another index of working of Harappan agrarian system is its impact on social statuses that may be inferred from the material remains of the civilization. These social statuses can be broadly identified urban nulcei as those of citadel dwellers, dwellers of the middle and downtowns, and population living outside the cities. As far as the complexion of the rural Harappan is concerned, the mapping of settlement pattern in Cholistan by M.R. Mughal may provide an indication. Out of 414 sites mapped ranging from early Harappan to Medieval times, mature Harappan number 174. Of the rest, 30 sites are smaller than 5 hectares and twenty-five per cent range between 5 to 10 hectares. About 44 per cent Harappan sites are industrial and the rest are agricultural.[35] The lower Sarasvati valley and lower Sindh appear to be the most densely populated areas while transhumance seem to have been practiced to the northwest of Harappa.[36] There is some evidence of pastoral camps of the Harappans in Gujarat.[37] Outside Dholavira might have been living non-agrarian autochthones, depending mostly on fishing and food collection. Dwellers of servant quarters outside Harappa and worshippers at the public fire altars outside Kalibangan might have denoted the lowest social status in the urban context. No evidence of a residential discrimination among the dwellers of smaller towns and settlements has, however, been recorded in the densely populated agrarian belts. Harappan society was, thus, obviously stratified residentially as well as socially and religiously. There is no data to infer the status of Harappan peasantry but judging from the smaller size of agrarian settlements, the life of Harappan farmers appear to be too moderate as compared to that of city citadel dwellers. A significant proportion of the inhabitants of cities like Mohenjo-daro and Harappa were engaged in activities other than food production. The presence of 44 per cent industrial settlements in the mature Harappan phase in Cholistan, moreover, indicates considerable industrial activity even in rural areas. At least they do not appear to have been entirely dependent for their essential industrial products on the urban areas. There appears to have been no neat division of labour. It is possible to infer from this data that the agrarian surplus was not obtained wholly in exchange for industrial goods supplied from the cities. But partly it was, for example, in exchange for chert blades used probably for harvesting and mass produced at Rohri and Sukkur hills of Sindh alone. Their supply to rural destinations must have been arranged through the urban distributional network.

The Harappan agrarian system thus appears to have contained an element of authority because of which the farmers must have parted with their surplus production. The vast extent of the civilization, coupled with the lack of evidence of elaborate military and bureaucratic network, indicates either the presence of local intermediaries or the exercise of collecting authority by the mercantile network of the civilization. The evidence at our disposal does not permit us to infer anything about the modes of collection, rewards of collectors and coordination of collection at the top. The state of peasantry too cannot be clearly worked out apart from the fact that the growth of industrial centres in the rural areas may lead to the inference that intermediaries who could invest their gains and official influence in organizing the local industrial activity worked through the artisans. This again indicates the possibility of intermediaries being traders connected with the wider network of commerce-cum-authority in the civilization.

The question of the ownership of land in Harappan civilization is indeed a moot point. The view stressing the possibility of the entire land being the property of and directly administered by the great temple or its priesthood[38] can no longer be tenable. The attempt to project Harappan civilization as a unitary imperial government[39] too is doubtful. The only clear evidence suggesting the possibility of a theocratic control is available at Kalibangan, where institutionalized public worship of fire enshrined inside and outside the citadel area with its priests probably occupying the other half of the citadel area prominently figures through the material remains of the site.[40] But fire worship was only a regional phenomenon in the Harappan civilization. Priesthood, managing the affairs of the Great Bath and the Pillared Hall at Mohenjo-daro, also must be having a limited religious jurisdiction confined to mother goddess worship in that and some other areas alone.[41] Establishment of a theocratic control over diverse belief-systems with freedom and equality-loving structures of traditionalism among tribes preceding its emergence and, through it, over the means of production had been achieved in most of the civilizations through warfare carried out in the name of the god of the aggressor. This is perhaps why theocracies seldom existed in the ancient world without aggressive militarism of which there is hardly any trace in the Harappan civilization. The theocratic control system is more coercive than the authority, as in Harappan civilization, emerging from an attempt to organize need-based coordination among diverse peoples with their diverse resources. These resources were further augmented by the gradual agro-commercial expansion adjusting complimentarity also with the forest dwellers and nomads. The autocratic and coercive methods of theocratic regimes created the institution of slavery—the binary opposite of freedom and equality-loving traditionalism of the tribes from which it emerges.

The experiment of Harappan civilization had to be more tolerant and coordinating in nature in the absence of a strong military tradition in it. It is erroneous to say that the surplus of the thriving and expanding Harappan civilization remained static throughout its long life and the same contributed to its decline and ultimate fall. There appears to be greater room for communal or even private ownership of land in Harappan civilization than state ownership of it. The existence of a large-scale state or private slavery in Harappan civilization too is doubtful, particularly in its agrarian areas. At the base it would have been a civilization of peasant proprietors or independent co-sharers and cultivators of land distributed on family or clan lines. The farmers were obliged to part with their surplus production for a share of power and material benefit in the wider regime of mercantile exchange-cum-control at the top that developed hand in hand with the emerging and expanding urbanism.

THE VEDIC AGE AND ITS AGRARIAN SYSTEM

Conceptualizing the character of the Vedic age, particularly that of the early Vedic, is one of the knottiest problems of ancient Indian history. The locale of early Vedic India overlapped that of the Harappan civilization, though it was shorter in extent than that of the Harappan civilization. The early Vedic age is supposed to have succeeded the Harappan civilization without learning anything from it but with the continuity of some elements of the surviving traditions of its skills and attainments. The early Vedic Indra, once held guilty of the destruction of the Harappan civilization,[42] is now generally absolved of charge[43] so unequivocally levelled against him on the basis of rather tenuous circumstantial evidence. The material remains of the early Vedic age defy identification. However, J.P. Joshi and R.S. Sharma are of the opinion that the pre-iron PGW deposits of Manda, Bhagwanpura, Dadheri, Nagar and Katpalon, located in the areas of Jammu, Kurukshetra, Ludhiana and Jullundur, respectively, constitute the material remains of the early Ṛgvedic age, i.e. the earlier part of the early Vedic age of which the main or rather the only source of knowledge is the *Ṛgveda*.[44] These supposedly material remains of the early Ṛgvedic age are shown to have overlapped with those of late Harappans which indicates that the Harappans and the early Ṛgvedic people came into contact with each other, but again, their contact in terms of cultural and institutional exchange cannot be established on the basis of the literary data available in the *Ṛgveda*, except on the lame racial supposition that *dasas* and *dasyus*, the sworn enemies of the Ṛgvedic Aryans, necessarily referred to the Harappans, which is no longer held tenable.[45] What is most astonishing in the *Ṛgveda* is the absence of any name (proper noun) for human settlements, although common nouns for what later denoted such settlements, e.g., *grama* (village), *pura* or *puri* (city or town) frequently occur. It is all the more striking in view of the background of Harappan civilization full of rural and urban settlements, which would have had names surviving in social memory even after their supposedly wholesale destruction and the alleged contact between Harappans and the Ṛgvedic Aryans. Even as nomads and stark pastoralists the Aryans, wandering through the long trek from their controversial original homeland, must have come across or at least have heard about Harappan civilization full of settlements having names. At least some of them should have found reflection in their lore as recorded in the *Ṛgveda*. Nomadism and pastoralism, therefore, can hardly be swallowed as the reason for the absence the names of human settlements in the *Ṛgveda*.

Conscious of the fact that the records other than the family books of the *Ṛgveda* contain later material which cannot be used in composing the image of the Ṛgvedic society, authors of the *Vedic Index* conceived it as pastoral-cum-agricultural, having clusters of houses called *grama* which were associated with different *viśas* or cantons, many of which ultimately formed a *jana* or tribe.[46] The age was given to inter-tribal warfare of which the biggest event was the celebrated battle of ten kings or chiefs of the ten tribes confederated against the king or chief of the Bharata tribe, Sudas.[47] They also show the emergence of the *varṇas* in the Ṛgvedic society as the result of the growth of militarism which led to class formation first of the warrior class or *rājanya* and then of the priest or *Brāhmaṇa* and next of the *viśas* or commoner consisting the people engaged in crafts, agriculture, and cattle rearing, who could not be included in the warrior or the priest classes. The fourth *varṇa*, however, did not appear until the composition of the *Puruṣasūkta*. The appropriation of the surplus, according to the authors of the *Vedic Index*, started as the tribute imposed by victorious kings and their claims on the defeated kings and their tribes. It also included the voluntary offerings or *bali* tendered by the *viśas* of the victorious king's own tribe, but made compulsory later.[48]

As against this picture of the Ṛgvedic society, R.S. Sharma has tried to conjure up a case of early Ṛgvedic society drawn exclusively on the data of the family hymns alone, as if the above view was based on a mix of early and late material, which is not the case. He projects early Ṛgvedic society as pastoral and only marginally agricultural. Agriculture in it is supposed to have been practiced on collectively held fields naturally irrigated by floods and tilled by ploughs equipped with wooden ploughshares. He tries to show that only a few references to terms like *cariṣṇī* and *kṛiṣṭī*, supposedly denoting agriculture, occur in the family hymns and seeks to explain them away by laboriously bringing them to bear different connotations, and where it is not possible, as in *Ṛgveda* IV. 57, by showing that the IV book is the latest in the group of family hymns and, therefore, likely to contain later material. The Ṛgvedic society, according to him, was a pre-agrarian phenomenon in which the inter-tribal warfare for booty consisting primarily of cattle wealth was rampant. It gave birth to the ordinary and ceremonial (*potlach*) exchange systems and social differentiation. He further thinks that this war-and-booty generated differentiation could not give rise to class structure unless economic surplus, the would be outcome of agricultural economy, was available. The differentiation created was limited to the emerging patterns of ritual ranking. To his mind, the *varṇa* structure denoting the class formation was a later Vedic phenomenon.[49]

It is significant to note that militarism appears as the crucial factor in both the views and that it was at its lowest ebb in the Harappan civilization. One thing that the onset of the Ṛgvedic age may be unanimously held to have ushered in was militarism but not against the Harappans who were probably already a thing of the past, unless the Ṛgvedic epoch is supposed to be prior to that of the Harappan. The supposition making the Harappan and the Ṛgvedic civilizations one and the same or co-existent in the same time and clime is precluded by the predominant tradition of militarism in the latter and conspicuous absence of it in the former, not to speak of the high possibility of the unmistakable presence of place names in the former and their surprising absence in the latter. In the circumstances, it can be assumed that the Ṛgvedic warfare was directed mostly against their own tribes or *janas*. The story of the Aryan struggle against the non-Aryans is again a tissue of racial bias of the Orientalists and may be easily disregarded. The Semitic type of Chalcolithic theocracies and the religious functions of the Greek and Roman kings show that militarism usually springs from under the wings of religion which, if not challenged by the might of the ruling class symbolized by the king, becomes a theocracy where the roles of the king and the priest are performed by one and the same person. The militarism of the Bharatas of the *Ṛgveda* too was initiated under the tutelage of the priest Viśvāmitra as per the tradition recorded in the family books.[50] The great battle of ten kings was fought and successfully won by Sudāsa when Viśvāmitra, replaced by Vaśiṣṭha, decided to avenge his honour by organizing a confederation of ten tribes and their kings against him.[51] This was perhaps the first assertion of independence of the military power against the absoluteness of the divine dispensation controlled by the priest. It is, therefore, the turning point in the universal traditional pattern obtaining in the early Ṛgvedic age prior to the battle of ten kings. It ended the possibility of the rise of theocracy coupled by the rise of chattel slavery and slave-based agricultural and industrial production. The origin of *varṇa* order and *varṇa*-based agricultural-cum-industrial production and agrarian system too may be traced to it.

Sudāsa's success, however, does not appear to be an unmixed blessing. It could neither generate imperial structures like those of the Chalcolithic Middle East and Egypt nor create the coercive institution of chattel slavery for working the productive apparatus. Instead of annexing the opposing ten tribes to his kingdom, the victory of Bharatas led to their amalgamation. The Purus, for example, are shown as amalgamating with the Bharatas after their defeat and the two together paving the way for the rise of Kurus in the later Vedic period.[52] Similarly, the challenge thrown to his priest Viṣvāmitra was no doubt successful in terms of military achievements but it could not bust the hold of emerging priesthood and ritual which expanded beyond all proportions in the later Vedic age and came up with a large number of elaborate religious ceremonies for the kings and conquerors.[53] However, in spite of this tendency of ritualistic spurt and expansion, one significant outcome of the battle of *dāśrājña* was that no Viśvāmitra could in future throw a military challenge to the might of a king, at least in the later Vedic period. The relationship between religion and politics ceased to be confrontational and became, to a large extent, a cooperative venture between the military might of the king and the gradually institutionalizing power of religious ministrations of the priest. But this did not happen all at once. On the social and economic front, the defeated people appear to have been not so completely defeated as to be coercively pushed into chattel slavery and harnessed in agricultural and industrial production. The tendency of amalgamation of the *janas* might have led to their three-fold differentiation into *brāhmaṇa, rājanya* and *viśaḥ* categories the last division of which became subservient to the king. Perhaps in the wake of having a section of people subservient to priestly as well as the warrior classes, the fourth class of the *śūdras* emerged on account of the disabilities coercively imposed on some helpless section of the people jointly by the subsequently ascendant warrior and priestly classes.[54]

Some indications of developments on these lines is afforded by the evidence contained in the family books of the *Ṛgveda* and certain indications of the later Vedic period referring to ancient traditions. References to *pañca janāḥ, pañca kṛṣṭayaḥ, pañca carṣaṇyaḥ* and *pañca kṣitayaḥ* in the family books of the *Ṛgveda* appear to indicate the emergence of a new social and economic differentiation among the five tribes of Yadus, Purus, Anus, Druhyus and Turvasas after their defeat in the battle of *dāśrājña*, rather than informing us about the pastoral or contra-agrarian character of the early age, as shown by R.S. Sharma.[55] Their amalgamation with the victors might have been effected on the condition of their abandoning warlike activities and turning themselves as *viśaḥ* without being reduced to slavery. It is in this light that the expressions like *viśas* of Tritsus[56] (royal family of which Sudāsa was a member) *viśas* of Triniskand[57] (a king), the first one occurring in the family books of the *Ṛgveda* becomes meaningful. That is to say a sort of lordship of the king and the ruling class was established over the *viśas,* as is indicated by the occurrence of the term *viśapati* or *viśāmpati* in the family books of the *Ṛgveda.*[58] This lordship was different from the ownership of *helots* held in state slavery, as in Sparta and it was also different from the ownership of land over which the *viśas* worked as serfs, as in feudal formation of Europe. It was, instead, a lordship structurally compatible with the emerging *varṇa* structure which was also, simultaneously, a production system different from the slave system and the feudal systems of Europe. Failure to recognize this aspect of the *varṇa* system and to make it fall in line with the pre-conceived notions of pre-production relation phenomenon in the early Ṛgvedic age and slave or feudal modes of production in the Mauryan and early medieval times, respectively, has led to a number of laboured distortions of ancient Indian economic and social history. Transformation of the five famous tribes and also probably of other less famous ones of early Ṛgvedic period into *viśa* component of the emerging *varṇa* formation is also hinted at by the answer to the question as to what is the meaning of the *pañca janāḥ* by a predecessor of Yaska, Aupamānavya. Quoted with disapproval in the *Nirukta*, the answer is that the expression meant the four *varṇas* and the Niṣādas.[59] The *varṇa* mode of production emerging in the early Ṛgvedic period after and as the result of the battle of the ten kings was undoubtedly agrarian-cum-artisanal in character.

The age of booty-based distribution and differentiation, of the concept of R.S. Sharma, might have been prior to or even faintly concurrent with the mainstream of the emergence and consideration of the *varṇa* mode of production but it was already on its way out.

In spite of clear cut unmistakable references to *kṣetrasya pati,*[60] lord or owner of the field, *vāhaḥ*[61] oxen in harness, *kṛṣatu lāṅgalam,*[62] plough in action, *varatrābdhyāntam,*[63] tying (the oxen) with strap, *sītā,*[64] furrow, *phālam vikṛṣantu bhūmim,*[65] ploughshare tearing the earth, *kīnāśaḥ,*[66] ploughman and the metaphor of milking the earth by cultivation as one milks a cow in a single hymn in one of the family books of the *Ṛgveda* hardly leaves any doubt about the agrarian character of the emergent *varṇa* system of production even in the early Ṛgvedic period. Stating that the fourth book is the latest of the family books and, therefore, does not contain material characteristic of the early Ṛgvedic period is arguing in a circle, for the lateness of the fourth book is determined on basis of the very material which is *a priori* considered late. The battle of ten kings figures most prominently in the family hymns and also the activities of as many as 23 tribes out of the 66 listed by the authors of the *Vedic Index.* An indication of the late character of any portion of the *Ṛgveda* may have references to rest of the tribes and their activities, which do not figure in the IV book. The importance of ploughshare and the material of which it was made has been unfortunately blown out of all proportions. When agriculture of the region is known to have sustained as vast a civilization as the Harappan and agriculture dates back to circa 6000 BC at Mehergarh,[67] it is futile to argue that agriculture of the early Ṛgvedic period was of no consequence because of the absence of iron ploughshare. Such an argument, I am afraid, is nothing but an example of Cleopatra's nose theory of history. In view of the clear cut evidence above, it is by no means too much to ascribe individual or group/community ownership of land to Ṛgvedic *viśas.* But it should always be kept in mind that this individual ownership of land is to be seen and understood with a compulsory reference to the emergent *varṇa* system of production and not in the sense of slave or feudal modes of production. The *varṇa* mode does not consist of absolute ownerships of a man either on another man or land; it is always a case of only concurrent ownerships of different degrees. Being a lord is not also being an owner in the western sense of the term. The notion of lordship is linked with the ability to protect and exploit without harming the source. The *pañca janāḥ,* after being defeated in the battle of ten kings by the Bharata King Sudāsa, for example, appear to have been amalgamated partly among the warriors of the Bharata *jana* but by and large among the *viśa* portion of that *jana.* Their right to protect themselves as a tribe must have ceased, as their protection became the responsibility not of their king but of Sudāsa, though warriors or *rājanyas* who formerly belonged to the respective tribes of the *pañca janaḥ* also had to help him. The references to *pañca kṣitayaḥ, pañca kṣiṭayaḥ* and *pañca carṣanyaḥ* only show that the people of the erstwhile five tribes, may be of the rest also, were now by and large obliged to confine themselves to economic pursuits alone, abandoning their warlike functions for self-protection or the protection of their tribes to Sudāsa and the amalgamated *varṇa* of *rājanyas.* Their erstwhile identity too particularly in case of the *rājanyas,* appears to have survived this transformation in terms of *vaṃśas* or lineages as even after the rise of the later Vedic *janapadas* lineages referring back to some of these tribal names are very much in evidence and they continued down to Puranic genealogies and even later. But among the *viśas,* these identities tended to get diluted and mostly lost.

The juxtaposition of *pañca,* the numerical epithet of the five tribes, used as symbolic of them, with words like *kṣitayaḥ, kṛṣṭayah* and *carṣaṇyaḥ* denoting earth and agriculture shows that with the beginning of the *varṇa* system of production agriculture came to occupy the centre-stage of economic activities. This must have led to the reformulation of the relationship between the king and his amalgamated people, *viśas.* Till such time as booty capture and distribution were the order of the day, the defeated tribes and kings used to pay tribute to the victorious king and his own tribe's *bali,* the voluntary gifts, but after the defeated got amalgamated with the victorious tribe, the distinction between tribute and *bali* must have tended to disappear. It is at this juncture, perhaps, that the way should have been paved for the theory of sharing the agrarian produce worked out by *viśa varṇa* by the king as the cost or wages of protection given to them and the kingdom should have tended to acquire a territorial implication. These developments appear to have been at the root of the origin of the ancient Indian agrarian system and its basic notions. Needless to say, this agrarian system and the *varṇa* mode of production were the two sides of the same coin; they originated simultaneously and continued simultaneously through history with all sorts of changes and extraneous influences interacting with this basic structure. Though there are some indications of the practice of not alienating any land without the consent of the *viśa,* there is also some evidence even in the family books of the *Ṛgveda* showing that the priests expected grant of *kṣetra* or agricultural field and *araṇyāni* or forest land from generous rulers such as Divodāsa Vadhryaśva. They were obliged to do so by deities like Sarasvati and Gayatri.[68] At another place,[69] the generous donor Paijavana is said to have distributed *rodisī* or earth among the leading priests. Needless to say, such gifts would have tended to turn some real cultivators into tenants in due course of time and the king as the donor of land could not have remained a mere protector. Without some kind of overlordship on land, such gifts were not possible. Religion, thus, started impinging on the socio-economic transformations taking place as the result of the battle of ten kings and the rise of the *varṇa* mode of production. Later Vedic developments appear to have been largely connected with this trend.

It is difficult to find a pre-*varṇa* stratum in the *Ṛgveda.* Therefore, the economy of booty capture and distribution unmixed with *varṇa* mode of production should have been on its way out even in the early Ṛgvedic period. But the *varṇa* system was not formed in a day. One may visualize that it could not receive the fourth *varṇa* till the period of the composition of the tenth book of the *Ṛgveda* which is supposed to be the latest in the text and marks the transition to the later Vedic period. Early Ṛgvedic period too appears to have been a transition from the primitive economy of booty distribution to that of *varṇa* system. The most important development in the emergent *varṇa* mode through the passage of time appears to be its consolidation from early Ṛgvedic period to later Vedic period, starting from the origin of the Śūdra *varṇa,* the mythical theory of the origin of which is enshrined in the Puruṣasūkta. Śūdras appear to have been in the beginning either a section of the *viśas* differentiated functionally with the emergence of the religious/orthodox trend, degrading the artisans and workers or the tendency of the conquering kings to offer stricter terms to the rank and file of tribes conquered during the later Ṛgvedic period. It might also have been a mixed achievement of both these trends. Obviously, the power of the king or priest at this juncture was not so decisive as to give birth to a full-fledged institution of slavery by reducing the erstwhile *viśas* as well as the newly conquered people to the status of slaves. Technological constraints of the age (lack of iron implements such as ploughshare, axe, spade, sickle, etc.), as pointed out by some scholars, do not appear to have been a genuine hindrance in the emergence of slave-based agrarian system at this or subsequent later Vedic stage, for even Chalcolithic civilizations are known to have developed slave-based agrarian systems long before the appearance of iron smelting technology, both in India and abroad. Power constraints among the iron-age Spartans, pitched against a ten times stronger slave force, obliged them to dedicate their entire life and civilization to the needs of stark militarism. The same power constraint of emerging territorial and priest-ridden monarchies, faced with the prospect of ever-swelling number of the agrarian and artisanal *viśas*, appear to have resorted to the policy of divide and rule by politically supported religious game of status differentiation heralded by the progressive imposition of disabilities on the *varṇa* of the Śūdras. The same exercise was repeated at a similar juncture subsequently, when women were discriminated against and the fifth *varṇa* of untouchables was added to the system. The historical development of ancient Indian agrarian system as encapsulated in the *varṇa* mode of production thus appears to have been propelled by the central motive of not allowing emergence of the economic power of the *viśas* or Vaiśyas as a counterpoint to the power of the priest-supported monarchies and king-supported priesthood. Their constant effort was to break any combination and consolidation of agriculturists, artisans and workers into an economic force through the instrumentality of a politically backed religious discourse silently working out a hierarchical order of divisions and subdivisions based on restrictions on inter-*varṇa* marriages and status-lowering disabilities of various kinds.

The coin of the politically backed religious discourse of the later Vedic period, however, had an ideal as well as material side. The ideal side, examined elsewhere,[70] shows the Vedic order as an undifferentiated whole, and yet it was not without differentiation. As an undifferentiated whole, it was conceived as the order of *ṛta, satya* or *yajña*. But on the differentiated plane it consisted of the orders of gods, manes, men, animals, vegetation, space, landforms, skies and waters—all held together by a non-exploitative interrelationship both within and without their respective orders. Sacrifice or *yajña* symbolized the idea of living for others and not for oneself. Conceived on this principle, the *varṇa* system could not have been deemed to be of an exploitative character. It may very well be projected as representing a social order based on the sublime principle of self-sacrifice of all the *varṇas* for the whole order and for each other, always reaping the reward of their self-sacrifice either in this or in any of the subsequent births as ensured by the theory of *karma* or rebirth. Reduction of social and economic discrimination in the material life to the principle of purity and pollution is not possible in the absence of any principle of dichotomous opposite of purity on the ideal plane. Impurity, exploitation or living for the self alone can only be a contingent and not an absolute category in the Vedic idea of order, ontologically speaking. The ideal side of the discourse was thus brought to ideological perfection in its own way, concealing, under its penumbra, the fact of exploitation inherent in the emerging socio-economic and political orders as an inevitable deviation from the ideal. Instead of rationalizing things at the level of empirical reality, the ideal side preferred to telescope it as a mere contingency unfit even to figure in the learned spiritual and metaphysical debates of the age. The crudity of this contingent material life is best illustrated by the physical acts of sacrifice condemned as unsteady boats wholly unreliable in the pursuit of *śreya*—the higher aim of human life.

The pursuit of *preya*, the lower aim of human life, is intimately linked with sacrifice and the exploitative nexus of state, society, and individual as integrated through the *varṇa* mode of production on the material plane. A sort of ambivalence towards the Śūdras prevails throughout the later Vedic period. There are certain passages in the Brahmaṇas and the Śūtras, which categorize them with the dogs and crows, debar them from sacrifice, and declare them unfit even for the leavings of the sacrificial food. Their touch defiles like that of a dog, a Candāla, a crow, a *patita*, a donkey and a woman in her courses. A rite has been prescribed by which one may ward off the possibility of being born as a Śūdra.[71] This shows that except for the accident of birth in another *varṇa*, there was no escape from the unwelcome Śūdrahood and this was probably true of other *varṇas* too, though birth in them might have been welcome. It would have been possible for people outside the *varṇa* order to fall into different *varṇas* at the time of their admission to the order, but once included in a particular *varṇa*, exit from it and entry into another *varṇa* appears to have already become nearly impossible in the later Vedic age, except notionally through a rebirth into a *varṇa* other than that of the undesirable Śūdra for which a rite was there. The son of a Śūdra or any of the three other *varṇas* was thus bound to belong to his respective *varṇa*. *Varṇa* status had become practically hereditary. It gives the semblance of openness because entry into a particular *varṇa* could be secured, in case of the people outside the *varṇa* order without being born into it. But this would have been by no means a free choice of the individual and must have been decided by the forces obliging them to fall in line with the *varṇas*.

On the other hand, there are some passages in the Saṃhitās and the Sūtras which appear to sound concern and regard for the Śūdras. For example, one such passage in the *Taittirīya Saṃhitā*[72] lays down a ritual correction (*kratvartha prāyaścitta*) for a ritual fault, viz., touching of the spilt *pṛṣadājya* by a crow. This fault if left uncorrected is said to lead to the destruction of the Śūdras. Sacred fire is allowed to be set up even by a Śūdra, and any one desirous of prosperity. By setting up a sacred fire, even a Śūdra was permitted to procure fire from a Vaiśya or Śūdra, rich like an *asura*.[73] Milk drawn by Śūdra was allowed to be used for *agnihotra* after putting it on the *gārhpatya* fire[74] and in Soma sacrifice, *soma*, after being taken from his nest and given a seat, is washed by a Śūdra following its consecration by a Brāhmaṇa.[75] A passage in the *Māitrayaṇī Saṃhitā* prays for the prosperity and glory of the Śūdras.[76] On the material plane, therefore, Śūdras were a useful and valuable part of the *varṇa* order and the prosperity and glory acquired by some of them was not only tolerated but also respected and ritually recognized. This is what gives the semblance of the efficacy of *karma* in the practically birth-based *varṇa* order of the later Vedic times and vouchsafes for their active participation in the *varṇa* mode of production, as it actually worked. The orthodox opinion of the time, honoured more in breach than in observance, was for their outright degradation and imposition of all sorts of disabilities on them with a view to turning them into a *persona non grata* by virtue of their total enslavement and subjugation.[77] But, presumably for power constraints, the emerging territorial monarchy of the later Vedic age was not in a position to translate it into action. Consequently, agriculture of the time could not become a wholly tenant and/or slave-based enterprise. The Śūdras too could not be turned into a *persona non grata,* as a wergeld of 10 cows per head is fixed for them.[78] But the Śūdras were discriminated as the least important and unequal citizens of the *varṇa* commonwealth even in the eyes of the law, for their wergeld was the lowest of all. The *varṇas* of Vaiśyas and Śūdras could not be driven wholesale into tenancy and slavery, respectively, but a complicated schematic pattern of agrarian system appears to have emerged from this situation in which, like the transit shades of colours on the spectrum, tenures ranging from peasant proprietorship to near slavery and manned by individuals and groups from among the Vaiśyas and Śūdras circumstantially reduced to different degrees of subjection and helplessness propped up. It was this social economic destiny that the orthodox discourse referred to above gradually drove the Śūdras to.

To be or not to be a victim of this power backed discourse depended on how poor or rich the material condition of the individual or group of Vaiśyas and Śūdras to be subjected to it was. This selective discrimination, initially within these two *varṇas,* led to a further status differentiation, which must have paved the way for the emergence of caste among them. The same process appears to have simultaneously replicated itself among the Brāhmaṇas and Kṣatriyas too, obviously for the entitlement to the shares in the agrarian surplus. The basis of the *varṇa* mode of production was thus neither the principle of open coercion nor that of a formalized and recorded contract and yet it had elements of both, being based on the contingent factors of *ad-hocism* and the inevitability of the need to divide in order to dominate. The discourse about Vedic order, as consolidated in the later half of the later Vedic period, philosophically dismissed and disregarded material life as a contingent phenomenon. An attitude based on contingency, opportunism and *ad-hocism,* instead of one prizing careful rational approach to material existence and its orientation to a sense of justice and higher values thus took care of the worldly life, particularly on the social and agrarian planes. This first led to the blurring of the distinction between *varṇas* and classes and prevented the *varṇa* mode of production from turning into or even approximating to the class-based mode of production. Then appeared the unique phenomenon of rigidly closed endogamous groups within what would otherwise have been classes (the *varṇas*), killing once for all, the potential to shake off disabilities and subjection by recourse to conflict for which the class formation is known. These group proliferated to an enormous number with the passage of time. *Varṇas* with these segmentations are often mistakenly equated with classes, which actually they are not. Fragmented into caste groups or even prior to such fragmentation, *varṇas* were never more than pseudo-classes and always obstructed the emergence of what would have been a neat class formation leading to a possible class struggle in the Marxist sense of the term. Relations of production in later Vedic times thus tended to produce a fractured and rigid formation known as *varṇa-jati* system, the schematic units of which, big or small, are always divided against themselves and, therefore, an ineffective vehicle of social change on rational and judicious lines. The *varṇa* formation has a dynamism of its own, which is self-sustaining and continuous, absorbing all opposition into itself, unlike self-liquidating and successive class formations of Marxist conception, which succumb to inherent self contradiction. To my mind, it was this inhibitive phenomenon preventing free economic growth rather than the lack of the iron ploughshare and other agricultural implements that was responsible for the limited availability of the limited agrarian surplus in later Vedic age. It appears to have been also at the root of the meagre development of technology and craftsmanship in the period, as the emerging *varṇa-jati* order can be demonstrated to have worked for the social and economic degradation of the artisans who alone, if left with a little surplus and freedom to innovate, would have achieved adequate development of technology as and when required.

In a nutshell, the main features of later Vedic agrarian system, as reflected in the sources, consisted of a component of Vaiśyas who were obliged to contribute *bali* to others (*anyasya balikṛta*)[79] and *bhāga*, a share of the agriculture produce to the king. Others obviously were the Kṣatriyas or Rājanyas and the Brāhmaṇas by whom he could be oppressed at will (*yathākāmajyeyaḥ*)[80], directly by the former and indirectly, i.e. through the power of the king and ritual by the latter. The Vaiśyas appear to have had greater affinity with the Śūdras, for to their sustenance they were not forced to provide but it might have been arranged through exchange generated by the complementarity of their needs. But the fractured class structure generated by the power constraint originating from the limited availability of agrarian surplus, the situation of the emergence of *varṇa*-based monarchy being hemmed in on all sides by too numerous tribals not belonging to *varṇa* system and the inability of their conquerors to impose slavery on them, complicated this simple picture of the later Vedic agrarian system. The fractured class structure, as pointed out above, replicated itself even on the ruling and the priestly classes known as Kṣatriya and Brāhmaṇa in the *varṇa* system on account of the survival of tribal identities in the form of lineages and *gotras,* respectively, among them, which could not have been effaced because of discrimination in entitlements to shares in the surplus. Older identities might have been retained in order to justify this discrimination. This is demonstrated by the categories of *Rājanyabandhu*[81] and *Brahmbandhu*[82] besides those of Rājanyas and Kṣatriyas and the frequent references to lineages like Paurava, Ānava, Sṛnjaya, Bharata, etc., and of the *gotras.* All the Kṣatriyas and Brāhmaṇas, therefore, may not have been men of consequence in the later Vedic times. There would have been many whose material condition would have approximated to that of the Vaiśyas and Śūdras and who were not in a position to oppress them. The converse would also have happened among the Vaiśyas and Śūdras, some of whom must have been rich like the traditional *asuras* as one of the sources referred to above indicates and whose oppression even by the Kṣatriyas and Brāhamaṇas would have been simply out of the question. Had it all been confined to the limits of exception in the later Vedic society, it would have contributed to easy social mobility and class dynamism but unfortunately it tended to casteify, perhaps because of the need to divide in order to dominate and oppress in the restricted areas of class or *varṇa* division by subjecting them to force, indoctrination and socio-economic compulsions/discrimination, as and when possible. There is enough evidence in the texts of the later Vedic period to show that the priests competed with other priests for netting in a royal and generous patron and always tried to impress them of the superiority of their knowledge of the ritualistic details of the sacrifice. A similar scramble, though not adequately reflected in the texts of the time given to religious objectives, appears to have characterized the Kṣatriyas out for grabbing a greater share of *bali* from the Vaiśyas, while at the same time, keeping the king pleased. Replication of this scramble for land and cattle and favour of the powers that had been, would have created a multiplicity of tenures and economic statuses among the Vaiśyas and Śūdras over and above the norms recorded in the Dharmasūtras which were rather recommendatory in nature. Multiformity of institutions and customs was inherent in the situation out of which the Dharmasūtras were trying to bring about a semblance of uniformity through recommendatory norms in consonance of the Vedic tradition but such an effort was always open to compromise and accommodation of actually prevailing institutions and customs. The agrarian system thus appears to be at the root of the melting pot of the later Vedic norms, practices and institutions characterized by all sorts of multiformity and multiplicity. That is why one can gather indications of communal, private, royal and even concurrent tenures of land in the texts of this period while yet noticing the fact that pastoralism persisted side by side with agrarian economy. Emergence of the fractured pseudo-class structure in the form of the *varṇas* already travelled a good deal into casteification[83] coupled with power constraint referred to above which led to the growing of agrarian potential by degraded landless artisans to be used both as skilled and unskilled labour. It might have created a setback in the normal pace of proliferation of metallurgy and other technology. As to landownership, there appear to have been pulls and counter pulls for and against its donation as sacrificial fee. This is reflected in the story of Vaiśvakarman Bhauvana[84] besides the anticipations of the oft-repeated precept that land belongs to him who recovers it from the forest or fallow land for agriculture. The king's share was still due on such land because of the protection offered.

The foregoing account of the agrarian system, as reflected in the sources of the later Vedic period, shows that the age was passing virtually through the throes of state formation and the agrarian system was an essential concomitant of it. The hallmark of this phenomenon was the emergence of the Kṣatriyas as the protectors as well as the *eaters* of the Vaiśyas (*viśamattā*) who probably constituted the peasantry of the age. A comparable case of state formation and the accompanying social formation, again in the framework of *varṇa* mode of production but in a different ecological, temporal and spatial situation, is afforded by south India in the sixth-seventh century AD. Significantly enough in this case, the traditional warrior tribes occupying the hills and allied with the pastoralists, hunters and men of the dry land, forest and coastal areas, emerged as adversaries of peasants settled in the fertile deltas and riverine plains of the regions and their Brāhmaṇa allies. This situation, characterized by frequent raids from times immemorial by the warriors and marauders—the erstwhile chiefs and their kin—on the peaceful peasant villages, drove a permanent wedge between them and precluded the possibility of the rise of a Kṣatriya *varṇa* in south India. It further compelled the peasantry to be their own soldiers and confide with the Brāhmaṇas who not only promoted agriculture, irrigation and reduction of the population of hunters, pastoralists, and even fishers into Śūdras in the emerging *varṇa* order sans Kṣatriyas and Vaiśyas but also assigned to the peasants, also (conceptually) put among the Śūdras, a higher ritual status as compared to the formers. Both Brāhmaṇas and peasants were naturally allied to the king, the former as enjoying his patronage and the latter supporting him as well as the Brāhmaṇas by their productive and military endeavours. The state that was born here too was like the later Vedic one—'custodial' rather than 'managerial' or 'patrimonial' in character. The emergence of Kṣatriyas and Vaiśyas in the later Vedic order, when viewed in the light of this comparison, amply demonstrates the differentiation as agriculturists and pastoralists of the rank and file of the Vedic tribes opposing the Bharatas in the battle of ten kings as *viśas* and taking away from them the worries of defending themselves by bearing arms. The victors must have undertaken their protection in lieu of a part of their agrarian production, a thing which could not take place in the case of a south Indian state formation. The warlike gentry of the amalgamated tribes, however, appears to have joined the rank of Kṣatriyas as supporters of the king of the commonwealth. Śūdras should have been the outcome of the second grade amalgamation following the take off stage of the Vedic Commonwealth. It is comparable with the entry into the South Indian Commonwealth of the erstwhile allies of the traditional warriors and marauders. A similar position of the Vedic Śūdras as lesser allies of the main *janas* too cannot be ruled out.

Expansion of Vedic civilization to the Ganga valley, as indicated in the story of Videgh Mathava, is actually a narration of the ritual fulfilment of a king's ambition for territorial expansion.[85] Historians first took the story as symbolic of military conquest of the region by Aryans, following which, the aborigines were amalgamated in the Vedic civilization causing necessary changes and adjustment in the latter.[86] Other scholars have referred to it as indicative of colonization[87] of the region by the Vedic Aryans or their migration[88] to middle Ganga valley along the base of the foothills of the Himalayas and reaching as far as Vindhyan outcrops. In either case, it is supposed to lead to the enormous augmentation of resources causing the emergence of a full-fledged agricultural society[89] and marking the transition from lineage society to state.[90]

The movement of the Vedic people has been conceived in two phases. In the first phase, it was confined to Sarasvati valley and the region of Indo-Gangetic Divide, which, including the northernmost portion of the Bari doab, covered the vast area of 35000 sq. miles.[91] Archaeologically, Harappan and Painted Grey Ware (PGW) sites mark this region. In the second phase, it embraced the transition zone between the upper Ganga plains and deltaic region of Bengal, which covers an area of 62000 sq. miles.[92] Northern Black Polished Ware (NBPW) sites constitute the archaeological profile of this region. The PGW culture has been attributed to later Vedic people and NBPW culture to the age of Dharmasūtras and the Buddha on quite cogent grounds. But the story of Videgh Mathava, coming from a text of the first phase and reporting a movement that embraces the region covered the second phase too and constitutes an element of anachronism in this scheme of chronology. Introducing a finer distinction, R.S. Sharma divides the first phase into two sub-phases on the basis of archaeological material. The first of these is taken to be constituted by layers of Grey Ware deposit without iron and is confined to Haryana sites alone and for which the dating (from Bhagwanpura) ranges between 1500 BC and 1000 BC. This sub-phase he assigns to the age of the *Ṛgveda*. The second sub-phase is taken to represent PGW deposits with iron produced by primitive technology for fashioning artifacts for non-agrarian use alone. It has been assigned to 1000 BC to 600 BC.[93] The cultural profile of these phases, including sub-phases, has been respectively characterized as pastoral, proto-urban and urban. 'All these,' writes R.S. Sharma, 'presuppose a full-fledged agrarian society not typical of the Ṛgvedic phase.[94] These match, according to him, with the journey of the Vedic culture from the Ṛgvedic pastoralism to the beginning of the territorial state formation, advent of social stratification and emergence of administrative machinery reflected in the later Vedic texts. If one excludes the second phase too, which is fully agricultural and urban, one gets the Later Vedic society, which is agricultural but not yet full-fledged, nor urban. Romila Thapar spells out the reason for this underdeveloped state of agriculture and state formation (urbanization is not the focus of her study, but the statement applies on that too) in the following words: 'In the western Ganga valley, the resources were neither sufficient to finance the institutions required for the establishment of a state nor were they directed towards the creation of such institutions. Archaeological evidence from Painted Grey Ware culture points to the size of these communities (although larger and more numerous than previous settlements), being smaller than those of subsequent period, that of the Northern Black Polished Ware. Territory was not seen merely as an area over which a *jana* had political control, for the territorial dimensions of marriage alliances were far wider, particularly for *kṣatriya* caste. Like lineage, *varṇa* was a mechanism for assimilation but reflected a stratified society.'[95]

It is significant to note that the role of lineages *vis-à-vis varṇa* order and agrarian production has not been taken into account in this statement and, therefore, it suppresses more than what it reveals so far as its bearing on emerging agrarian system of the age is concerned. This overall and generalized summation of the later Vedic situation at the root of its upcoming agrarian system by the pioneers in the field, though of much use, is problematic and a question begging proposition. For example, it may be asked as to how, without the beginning of some sort of agrarian activity and social stratification, *varṇas* started emerging already in the early Vedic period? Why did early Vedic society tend to become divided along lines of lineage-fractured *varṇas* of Brāhmaṇas and Rājanyas, reducing the subdued lineages to the status of *viśas* held in agrarian and occupational subjection different from slavery, as noted above? How did disability-distinguished Śūdras emerge just at the advent of later Vedic age as a social category different from Spartan helots? Can we say that the early Vedic age was absolutely without a notion of territoriality, particularly when such a territorial civilization as Harappan either preceded or succeeded or else coincided with it and battles like *dāśarājña*, which were not of the nature of cattle raids, were fought among *janas*? How can the place (proper) names be totally blotted out of the entire text of the *Ṛgveda* despite several Harappan settlements which could not have been without names containing Harappan and Grey Ware sherds, supposedly characteristic of the presence of early Vedic culture in the same layers? Why assume a dramatic switch over to the notion of territorial state, social stratification and emergence of administrative machinery if the advancing antecedents of all these were already there in the preceding epoch? If there is an unbroken continuity between the early Vedic and later Vedic texts, why should one imagine sharp breaks and absolutely new advents in the agrarian, social and political developments of the later Vedic period? Did social stratification bring out class-based social formation or only contributed to the transformation of the received *varṇa* order? The following observations emerging from this paper may be offered as a focused response to these questions.

The first question highlights the fact that domestication of animals and cattle tending signify a mode of production and that some kind of social formation starts the moment cattle is acquired by force of arms or religious obligation and not just by tending and breeding. Use of force of arms or religion for acquisition of cattle perhaps starts social differentiation based on organized aggression or coordinated ministration of ritual needs made compulsory by custom and tradition. From the attempt to organize an aggression emerges what is called military ranking and from the effort at organized ritual performance appears the system of ritual ranking. When agriculture started moving from the periphery to the centre of economy, these modes of social ranking became transformed already in the early Vedic culture into what has been described as *varṇa* mode of production. This emergent mode of production, as noted above, was essentially different from the class-based mode of production in that the dominating section of society created by the two modes of ranking gave birth to more or less rigid blocs of lineages and *gotras*, one formed on the principle of ever-widening circle of inter-marriage alliances among arms-wielding kin groups aimed at consolidating power and the other on that of the embargo on intra-kin marriages aimed at securing the purity of lines of priests and teachers believed to have started by the *gotra*-and-*pravara*-founding *ṛṣis* or patriarchs. These divisions within divisions, called *varṇas*, made the latter consist of aggregates of the former kin-groupings of two diverse kinds rather than individuals living in small nuclear families. This imparted the first two emerging *varṇas*, viz., Brāhmaṇa (*brahma*) and Rājanya a character different from that of a social class. Lineage identity in the *varṇa* formation was perhaps more basic than *varṇa* identity. This prevented the absorption of the identities of the former into the latter and led to relations of mutual conflict and subservience among them and also less frequently between the two *varṇas* as well. *Varṇa* formation is, therefore, not the same as class formation. Because of the aforesaid basic difference, the Brāhmaṇa and Rājanya *varṇas* could not dispense with their separate identity in spite of their common function of dominating the emerging *varṇa* of *viśas*. Developing relations of subservience between the lineage group of Bharatas with the *janas* of the confederacy defeated by them in the battle of *dāśarājña* and other such earlier conflicts that might have gone unrecorded in the text of the *Ṛgveda* appear to have given rise to *viśa varṇa*. The dominated *viśas* too had lineage groupings but agrarian and occupational subjection imposed upon them confined their activities to economic pursuits alone and subdued their lineage groupings to the extent of preventing an armed rebellion, but not totally, which kept them divided against themselves and prevented them from becoming consolidated into a class of individuals. The tribute collected from the third *varṇa* freed the dominating *varṇas* largely from economic pursuits and allowed them to devote more time to the pursuits of arms and ritual, leading to unprecedented expansion of these pursuits in the later Vedic age. A sort of commonwealth was thus created with the emergence of *varṇa* mode of production or power relations. A fourth *varṇa* of Śūdras was added to this *varṇa*-based commonwealth last of all at the threshold of transition to the later Vedic period on terms of servility, as they were not admitted to the commonwealth at par with the *viśas* as bearers of tribute and/or *bali*. But again, in the interest of security against a possible rebellion, terms of servility imposed upon them did not allow them to lapse into chattel slavery or land-bound serfs. Social and ritual disabilities and the initial trend of placing in this *varṇa* alone all the *varṇasaṃskaras* produced by inter-*varṇa* marriages and nearly all new entrants in the Vedic commonwealth, as it expanded to the Ganga Valley, tended to subdivide the last *varṇa* into *jatis* or castes which later crept into all the *varṇas* from different channels made available to them from time to time.

Unless one is prepared to rule out the perpetuation of any form of Harappan contribution to early Vedic civilization despite controversies about its positioning *vis-à-vis* Harappan civilization, the notion of territoriality in the early Vedic period can hardly be absolutely denied. Even the nomads and cattle traders have their own notion of territoriality centring around the pasture land in their possession, to defend which they frequently take recourse to armed struggle against all aggressors. The only thing is that their notion of territoriality is essentially temporary and lasts either till the occupied land is useful for grazing and breeding or is snatched away by an adversary. It tends to become permanent as agriculture in their economy moves to the center and agrarian fields are sculpted out of the wild and held in continuous cultivation. So long pastoralism was containing marginal agriculture in its fold; now agrarian environment centred on permanent and continuous occupation of arable land, holds cattle tending as its integral part. It also contextualizes and promotes the pastoral mode of economy, not falling in line to the borders of the agrarian domain beyond which lay hills and forests. Agrarian people not only dress up the natural landscape to their needs and taste but also impart it a personality distinguished by specific 'emotions, gods, poetry, ritual, architecture, outsiders, frontiers, myths, borderlands, landmarks and families which give farms meaning and purpose.'[96] Pastoral people too impart to the land of their temporary occupation such cultural imprints and marks of their activities as survive not so much in the written records but in the folklore and archaeological remains left behind by them. Pastoralism and agriculture are not exclusive of each other. They frequently stay together, complement each other and collide too if agriculture starts encroaching upon the land given to pastorialism or pastoral people start raiding and looting the agriculturists to get over scarcity conditions.

The absence of place names in the *Ṛgveda* is intriguing and might also have been deliberately ruled out by their composers and redactors to make it more inclined to the culture of divinity than that of humanity. Administrative machinery is presupposed by any attempt at organization and cannot be totally ruled out in the early Vedic age, which furnishes clear references to *rājan, purohita, grāmaṇī, spaśaḥ, sabhā, samiti,* etc. As, unlike class formations—which are self-liquidating—*varṇa* formations are self-perpetuating, dialectical relations consisting of sharp breaks, struggles and disjunctions between the early Vedic and later Vedic institutions and developments cannot be deemed necessary. Given the kind of cultural elements and presuppositions informing the Vedic culture, continuity and transformation rather than dramatic change and revolution appear to be inherent in the process of historical change operating there.

Romila Thapar prefers to view the later Vedic society through the model of *potlatch* economy constructed by anthropologists and seeks to conjoin this conception with that of proto-polity of later Vedic age. Both these conceptions have been grounded on what has been called ritual prestation, tribute and burning of surplus through *yajña*, the great religious project of gifting to gods and Brāhmaṇas for glory rather than booty and *bali* (voluntary gift). Its objectives in the early Vedic period transform into compulsory revenue collection in the form of *bhāga* and *bali* (now a compulsory surcharge) in the post-Vedic period.[97] Without taking into account the emergence of the *varṇa* mode of production, Thapar tries to hem in all the developments of the later Vedic age in her theory of lineage society based on *potlatch* type economy. But a number of facts refuse to fall in line and militate against this construction. Firstly, if tribute continued to be the backbone of emergent organization in the later Vedic age besides booty in war, what was the need to invent a new one, viz., *bhāga,* meaning share besides *bali,* which might have very well come to denote the tribute obtained from subordinated lineages? Secondly, the designation *bhāgdugha* for one of the *ratnins* suggests the gradual and direct drawing of the share of agrarian produce like the share of milk from a cow rather indirectly through subordinate lineages exacting it from their *viśas.* Lineages as persisting kin groups perpetuated the tribal element of the past of the later of Vedic age and served to fracture and complicate *varṇa* orders, as noted above. They also paved the way for the later rise of birth-based castes with which *varṇas* finally became interchangeable. It appears that with the maturing of the *varṇa* order in the later Vedic age and its complication by the birth-based lineage groupings finally culminating into innumerable castes, the die of ancient Indian agrarian system was already cast with its inherent and potential tendency to ascriptive hierarchy, birth-based social and professional differentiation and creation and maintenance of those higher in hierarchy by entitlement to surplus by virtue of their high status in the society rather than by producing it. The vibrancy of this agrarian system is vouchsafed by its expansion, self-perpetuation and ability to resolve conflicts between different modes of economy and people of diverse background. Its inhibitive character too is at the same time evident from the factors that made *varṇa* formation different from class formation and ultimately making them rigid and caste-ridden.

NOTES AND REFERENCES

1. Moreland, William H., 1968, *The Agrarian System of Moslem India,* Delhi, Oriental Book Reprint Corporation, Book Publishers, pp. 6–10.
2. Ghoshal, U.N., 1930, *The Agrarian System in Ancient India,* Kolkata, University of Calcutta; 1929, *Contributions to History of Hindu Revenue System,* (Moreland's book first appeared in 1929).
3. Kosambi, D.D., 1956, *An Introduction to the Study of Indian History,* Bombay, Popular Book Depot.

4. Gopal, L., 1965, *Economic Life of Northern India, Varanasi.* Motilal Banarasidass.
5. Yadava, B.N.S., 1973, *Society and Culture in Northern India in the Twelfth Century,* Allahabad, Central Book Depot; Also his presidential addresses to Ancient India Section of Indian History Congress, 1980 and to the inaugural session of the same body in 1993.
6. Sharma, R.S., 1965, *Indian Feudalism,* Kolkata, University of Calcutta.
7. Moreland, *op. cit.,* pp. 205–6. Irfan Habib's work, 1963, *The Agrarian System of Mughal India* (1556–1707) Bombay, Asia Publishing House and that of his father, Mohammad Habib, 1981, *Politics and Society during Early Medieval India,* New Delhi, (Reprint first published by Aligarh Muslim University) focusing largely on Sultanate period seek to rule out the nationalist theory of the legacy of loss left behind by Muslim rulers conceded by Moreland. Irfan Habib later rejected this nationalist projection as communal.
8. Gopal, Lallanji, 1980, *Aspects of Agricultre in Ancient India,* Varanasi, Bharati Prakashan, and the work of his student, A.L. Yadav, 1980, *Prachin Bharat men Krishi,* Varanasi, Siddharth Prakashan, in Hindi as also that of B.N.S. Yadava noted above apply the tradition of Rankean historiography in economic and social history. G.C. Pande initiated this trend at the Allahabad University at a time when political history was ruling the roost. Gopal and Yadava were research students of G.C. Pande.
9. Randhava, M.S., 1980–86, *A History of Indian Agriculture,* in 4 Volumes, New Delhi, Indian Council of Agricultural Research.
10. Om Prakash, 'B.N.S. Yadava as a Pioneer of Allahabad School of History', *Purātattva,* No. 1, 2000–2001.
11. Sen, Amartya K., 1981, *Poverty and Famine: Entitlement and Deprivation,* New York, Oxford University Press.
12. A number of studies appeared on each of these topics appeared from time to time. One can view them at the website http://www.sas.upenn.edu/~dludden/aha-bib2.htnl.
13. The 10 volumes of Subaltern Studies edited by Ranjit Guha. For a more exhaustive bibliography of these studies, visit the website mentioned in the footnote 12.
14. Om Prakash, 2000, 'Class as the Starting Point of Initial Enquiry in Ancient Indian Social Historiography', *Social History and Social Theory,* Department of Ancient History, University of Allahabad.
15. Piggot, Stuart, 1980, *Prehistoric India,* London, Harmondsworth, Penguine Books and Wheeler, R.E.M., 1968, *The Indus Civilization,* Cambridge, Cambridge University Press (Reprint) depict Indus urbanism as a product of foreign inspiration, which appeared all of a sudden and in a fully developed form. Its disappearance too was supposed to be as sudden and complete as its origin. These authors further present Harappan urbanism as an isolated phenomenon, not needing a rural background.
16. Meadow, R.H., 1986–1990, *Harappan Excavations,* Madison, Prehistory Press, p. 121.
17. *Ibid.,* 138.
18. *Ibid.*

19. Dyson, Robert H. Jr., 'Paradigm of Changes in the Study of Indus Civilization', in Possehl, G.L. (ed.), *Harappan Civilization: A Contemporary Perspective,* Walnut Creek, Altamira Press, p. 572.
20. Lal, B.B., 1997, *The Earliest Civilization of South Asia,* New Delhi, Aryan Books, Chapter IV.
21. Wright, Rita P., 1984, 'Technologies Style and Craft Specialists: Spheres of Interaction on the Indo-Iranian Borderland', (Harvard University Ph.D. thesis).
22. *Ibid.*, pp. 236ff.
23. Dhavlikar, M.K., 'Harappans in Saurashtra; The Mercantile Enterprise as Seen from Recent Excavations at Kuntasi', Possehl, *op. cit.,* pp. 555–567.
24. Thapar, B.K., 'The Harappan Civilization: Some reflections on its Environments and Resources and their Exploitation' in Possehl, *op. cit.,* pp. 3–13.
25. Ratnagar, Shireen, 'Location of Harappa', in Possehl, *op. cit.,* p. 261.
26. Mughal, M.R., 'The Protohistoric Settlement Patterns in the Cholistan Desert' in Taddei, M. 1990, (ed.), *South Asian Archaeology-1987,* Rome, ISMEO, pp. 143–156.
27. Ratnagar, *op. cit.,* p. 201.
28. Bharadvaj, O.P., 'The Arid Zone of India and Pakistan' in Stamp, L. Dudley (ed.), *Year A History of Land Use in Arid Regions,* Paris, UNESCO, p. 157.
29. Frentress, M., 'From Jhelum to Yamuna: Cities and Settlements in the Second and Third Millennium BC' in Possehl, *op. cit.,* pp. 245–260.
30. Thapar, B.K., *loc. cit.,* p. 248.
31. *Ibid.*, pp. 248ff.
32. The source of this diversity, as pointed out by David Ludden is the kind of 'cuisine, ritual, folklore and aesthetics' entailed by natural settings and physical landscapes of the forest dwellers, fisher folk, nomads and agriculturists. *The New Cambridge History of India,* IV. 4, *An Agrarian History of South Asia,* CUP, 1999, pp.144f. Agrarian intrusions in environments other than its own generates the character of territoriality and patriarchy (pp. 60–112) which is probably the first step towards cultural unity in the midst of prevailing diversities with or without coercively dealing with them. The Harappan approach, however, is to be distinguished from the Pre-Harappan one in that latter was essentially agrarian, whereas the former depended more on exchange oriented commercial activities across different echo-zones within Harappan and non Harappan civilizations.
33. Lal, B.B., *loc. cit.,* pp. 181–202.
34. Ratnagar, *op. cit.,* pp. 3–13.
35. Mughal, M.R., 'Recent Archaeological Research in the Cholistan Desert' in Possehl, *op. cit.,* pp. 85–95.
36. Ratnagar, S., *op. cit.,* 201.
37. Rissman, Paul C. and Chitwala, Y.M., 1990 *Harappan Civilization* of *Oriyo Timbo,* New Delhi, American Institute of Indian Studies/Oxford-IBH Publishing. Another site is Zedka, Report in Possehl, *op. cit.*
38. Jacobson, J., 1986, 'The Harappan Civilization: An Early State' in Jacobson (ed.), *Studies in the Archaeology of India and Pakistan,* New Delhi, American Institute of American Studies/Oxford–IBH Publishing, pp. 137–173.

39. Ratnagar, S., 1991, *Enquiries into the Political Organization of the Harappan Society,* Pune, Ravish Publishers.
40. Lal, B.B., *op. cit.,* pp. 123–125.
41. Lal, B.B. (*op. cit.,* p. 232) is of the view that 'the authority holding the Acropolis may have derived its strength partly, if not wholly, from a religious background'. But the religious background of the Harappans is far from being as homogenous as their civilization appears to be. This is not to say that religion exercises have no homogenizing effect on the diversities of culture. They have, but for that it has to become dominant and overcome religious diversities, which is not so much in evidence in Harappan culture. What is in greater evidence, however, is the economically induced standardization promoted by commercial, industrial and urban needs. The triumph of agrarian territoriality too brought new ritual, folklore, cuisine, aesthetics and culture in its train and started taming and humanizing the wilderness of different terrains and landscapes at first differently and then homogenously. Pre-Harappan cultures probably represent the first stage of this process which led to the final stage of Harappan civilization and its subdued-diversity reflecting uniformity. The character of Harappan political authority, therefore, must have been agro-commercial rather than religio-militaristic. Gods here did not have to take recourse to armed aggression through human aggregates led by their priest-kings to establish their supremacy over other gods and pave the way for the rise of a religious empire.
42. Wheeler, R.E.M., 'Harappa 1946: The Defences and Cemetery R37', *Ancient India,* No. 3, p. 82.
43. Dales, G.F., 'The Mythical Massacre at Mohenjo-daro', *Expedition,* 6 (3), pp. 63–43.
44. Joshi, J.P., 'Interlocking of Late Harappan Culture and the Painted Grey Ware Culture in Recent Excavations', *Man and Environment,* 2, pp. 98–101, Joshi, J.P., and Madhubala, 'Life During the Period of Overlap of Late Harappa and PGW Cultures', *Journal of the Indian Society of Oriental Art,* NS, ix, 1977–78, pp. 20–29, Sharma, R.S., 1983, *Material Culture and Social Formations in Ancient India,* Delhi, McMillan India Ltd., p. 23.
45. Sharma, R.S., "Booty Capture, Distribution and Differentiation in Ṛgvedic Society", *Ibid.,* p. 36.
46. *Vedic Index,* I, 269–271.
47. *Ibid.,* pp. 355f., 315f., II, p. 454.
48. *Ibid.,* pp. 247–271.
49. Sharma, R.S., 'Forms of Property and Subsistence in the Early Portions of the *Ṛgveda*', *op. cit.,* pp. 23–35 and also the next chapter.
50. *Ṛgveda* (RV), III, 33; III. 53. 9–11.
51. *Vedic Index,* II, pp. 275f.
52. Sudden disappearance of the name of Purus, Turvasas and Kṛvis shows their amalgamation with the victorious Bharatas. Together, they are supposed to have created the later Vedic people and their habitat (*janapada*) known as Kur-Panchala. *Ibid.,* I, p. 12.

53. The institution of soma sacrifices requiring many priests and rites of political importance such as *aśvamedha, rājasuya, vājapeya,* etc., including those necessitating animal sacrifice as per *Āprīsūkta* amply illustrates the point.
54. The subservience of the *viśa* and the Śūdras effected earlier finds a mention in the *Aitareya Brāhmaṇa* (AB), VII. 35. 3. 27.
55. Sharma, R.S., *op. cit.,* p. 25.
56. *RV,* IV. 33. 6.
57. *RV,* I. 172. 3.
58. *RV,* III, 13. 5; VII. 55. 6; VII, 39. 2.
59. *Nirukta,* I. 1; II. 6. 11, etc.
60. *RV,* IV. 37. 1. 2; VII. 35. 10; X. 66 Authors of the *Vedic Index* explain this term as denoting 'lord of the field' to be understood as the god presiding over each field and not as the owner of the field (*RV,* III. 31. 15; V. 62. 7), which was carefully measured (*RV,* I.110.5). They also take notice of the expression *kṣetra-jeśa* (*RV,* I. 33. 15) and *kṣetra-sa* (*RV,* IV. 38. 1) meaning 'acquisition of land' and 'gaining land', respectively. The reluctance to see *kṣetrasyapati* as the owner of the land is there in spite of all these references and interpretations because the authors of the *Vedic Index* do not want to subscribe to the view of Baden Powell (*Indian Village Commuities,* p. 190) that the Vedic society was a landholding aristocracy, superimposed upon agricultural aboriginal stock. Conceding that vaiśyas were agriculturists (owning land), they are not prepared to grant the same position to the early Vedic *viśa,* for they were subject to removal at will by the *rājanyas.* Still, they appear to hold with Hopkins that *viśas* were not without property, though they were also not irrevocable owners of land either. *Vedic Index,* II, p. 333. Needless to say, there is extraordinary effort at too guarded a statement.
61. *RV,* IV. 57. 4. 8.
62. *RV,* IV. 57. 4.
63. *RV,* IV. 57. 4; X. 102. 8.
64. *RV,* IV. 57. 6. 7.
65. *RV,* IV. 157. 8; X. 117. 7.
66. *RV,* IV. 57. 8.
67. Jarridge, J.F., 1981, 'Economy and Society in Early Chalcolithic/Bronze Age of Baluchistan: New Pespectives from Recent Excavations at Mehergarh' in Haertel (ed.), *South Asian Archaeology-1979,* Berlin, Reimer Varlag, pp. 93–114.
68. *RV,* VI. 61. 14.
69. *RV,* VII. 18. 24.
70. Om Prakash, 1992, 'Nature of Vedic Order and Taboos' in his book *Conceptualisation and History,* Allahabad, Excellence Publishers, pp. 79–105.
71. *Ibid.,* pp. 91–96.
72. *Taittirīya Saṃhitā,* III. 2. 6 cf. *Baudhāyana Śrauta Sūtra* (*BSS*), XIV. 9.
73. *Asvalayana Srauta Sutra* (ASS), V. 11–18
74. *ASS,* VI. 1–29.
75. *BSS,* VI, 16–18.

76. *Maitrāyaṇī Saṁhitā*, IV. 2. 7. 10; III. 4. 8. 14.
77. Om Prakash, *op. cit.*, p. 96.
78. *Vedic Index*, p. 331.
79. AB, VII. 35. 29.
80. *Ibid.*
81. *Śatapatha Brāhmaṇa* (*SP*), XI. 6. 2. 5; VIII. 1. 4.10; *Brihadāraṇyaka Upaniṣad*, VI. 1.5. Confusion on account of fracture within the *varṇas* on individual or group lines is illustrated by as passage of *SP*, I. 1. 4. 12, which, in an enumeration of *varṇas* places Vaiśyas prior to Rājanyabandhu in clear violation of the traditional order of ritual ranking.
82. *AB*, VII. 27; *Chāndogya Upaniṣad*, VI. 1.1 also *Lāṭayana Śrauta Sūtra*, VIII. 6.28; *Kātyāyana Śrauta Sūtra*, XXII. 4.22; *Śankhāyana Śrauta Sūtra*, XVI. 29.9.
83. See discussion under headings of 'caste in the *Ṛgveda*', 'caste in the later Vedic Saṃhitās and Brāhmaṇas', 'the names of castes', 'restriction on intermarriage', 'occupation and caste', etc., under the entry Varṇa in the *Vedic Index*, II to have a direct feel of how the fractured structure of the *varṇas* was already tending to casteification.
84. *SP*, XXIII. 7. 15.
85. *SP*, I. 4. 1. 10 et. seq.
86. Wheeler, R.E.M., 1966, *Civilizations of Indus Valley and Beyond*, London, Thames & Hudson, Cambridge, p. 102. This is perhaps the latest reiteration of this oft-repeated position taken as a settled fact of history in almost all the textbooks and other relevant research publications on the subject.
87. Sharma, R.S., *op. cit.*, 90.
88. Thapar, Romila, 1992, *From Lineage to State Social Formations in the mid-First Millennium BC in the Ganga Valley, New Delhi*, OUP, p. 70.
89. Sharma, R.S., *op. cit.*, 89–116.
90. Thapar, Romila, *op. cit.*, 70–115.
91. Sharma, R.S., *op. cit.*, p. 56.
92. *Ibid.*, p. 89.
93. *Ibid.*, p. 59f.
94. *Ibid.*, p. 64.
95. Thapar, Romila, *op. cit.*, pp. 68f.
96. Ludden, David, *op. cit.*, p. 60.
97. Thapar, Romila *op. cit.*, pp. 38, 40–42; 'Dāna and Dakṣiṇā and Forms of Exchange', *Ancient Indian Social History*, Delhi, 1978, pp. 105–121.

CHAPTER 19

Animal Husbandry in the Vedas

N.M. Kansara

AGRICULTURE IN LIVESTOCK

Agriculture and animal husbandry was developed in India from pre-Vedic times, for in the *Ṛgveda* itself we find references to hundreds and thousands of cows, to horses yoked to chariots, to sheep and goats offered to gods as sacrificial oblations, and to the use of wool for clothing. The famous cow-hymn of the *Ṛgveda* (6.28) indicates that cow had already become the very basis of rural economy. In another hymn (8,101), cow is designated as mother of Vasus, of Rudras and of Ādityas, as also the pivot of immortality. The Vedic people appear to have had large forests at their disposal for securing timber, and a vast amount of green fodder to animals. The farmer's vocation was held in high regard, and both agriculture and animal husbandry developed in India to a degree of skill rarely known in other parts of the world.[1] The ancient Vedic *ṛṣis* invented such beautiful sciences like those of agriculture and cattle breeding, both interdependent on one another, and both rooted in the practical spiritual outlook of the one god residing in all beings and becomings, in everything from the galaxies to the nucleus of the sub-sub-atom, animate and inanimate. Agricultural, animal husbandry, handspun cloth, *pañcāmṛta* and *pañcagavya* were their most perfect scientific inventions interlinked with each other, and in tune with the unified field of life, of man in relation to its environment both sentient and insentient.

Since very ancient times in India, the Vedas have been traditionally regarded as a veritable repository of all knowledge conducive to both mundane and spiritual growth. By Vedas, we here understand not merely the Vedic texts, known by the name of *Saṃhitas* of the four Vedas, but also their *Brāhmaṇas, Āraṇyakas, Upaniṣads,* ancient *Vedāṅgas,* their *Upavedas,* the *Sūtras,* the ancient *Smṛtis* like the *Manusmṛti,* as also the celebrated great epics like the *Rāmāyaṇa* and the *Mahābhārata,* along with some ancient *Purāṇas* like the *Vāyupurāṇa,* all of which cover a vast period of time from several thousand years before the commencement of the Christian era to about few continues from that of the Vikraṃa Saṃwat. This vast literature contains stray oblique references to these sciences, which when culled and arranged systematically, afford us a glimpse of the developments in these fields in very ancient times. Agriculture and animal husbandry have thus been developed and practiced in India for at least last six or seven thousand years since the pre-Vedic times, the practical knowledge being handed down from generation to generation since then in the families of farmers and tillers of land, and developed by their practical experience in their field of activity and vocation. It was rarely codified in the form of specialized treatises, and that too was done much later during the subsequent ages as it grew in extent and the need to record and preserve it for prosperity was felt.

In India, it was almost impossible to think of practicing agriculture without having good cattle. Holdings were small; agricultural practices were old-fashioned and marketing of produce continued to be done with bullock carts. They provided the required motive power for various agricultural operations including irrigation and rural transport, and also manure for the fields. Again, milk and milk products, which were the only source of animal proteins in the diet of the predominantly vegetarian population of the country, came from them. The farmer also supplemented his meagre resources through milk and milk products, especially during the wide gaps that occurred between sowing and the reaping of agricultural crops. Thus, development of cattle, both in regard to their milk yield and draught capacity, as also for daily cost-free ready availability of organic manure at the doorstep, was very important for the farmer. Cattle was, therefore, indispensable for the rural cost effective economy of the country[2].

Cattle rearing being a part and parcel of husbandry, animal husbandry deals with the production and preservation of domestic animals or livestock, and it embraces all the phases of their breeding, feeding, wellbeing and management. The cattle are so intimately associated with the science and practice of agriculture, that in ancient literature, they have been referred to simultaneously. This is more so in the case of cows and bulls. In ancient India, cattle tending was one of the items of solemn religious *vow* (*vrata*), and it was entrusted to a certain section of the people who thoroughly understood and mastered the business. Agriculture, cattle rearing, trade and commerce constituted the fourfold pursuit suitable for making fortune. Agriculture and animal husbandry flourished hand in hand. There are records in ancient literature showing that livestock and animal farming were not the business of a particular group or section of people. The kings, being the warrior class (*kṣatriya*), owned and managed cattle rearing and cattle wealth, as it was the mainstay of their household finances. Besides horses and elephants, which were mostly used for transport both civil and military, the kings maintained cows, buffaloes, sheep, goats, camels, asses, mules, swine and dogs for a variety of purposes directly or indirectly connected with agriculture. Animal husbandry was the occupation, solely or partly, of most of the people. For the classes of people practicing the occupation of animal husbandry, their economy was mainly dependent upon agriculture. Some of them were really very rich and possessed extensive farms. For those who practiced it partly, it was the major part of their economic affairs, while they busied themselves ruling people as feudal chieftains or owners of land or landlords. The village generally maintained —either on pay or on a share of the produce—full time shepherds who were entrusted with the work of taking the animals to the pasture grounds early in the morning and bring them back in the evening. The herds were supposed to be kept in forests in spite of great inconvenience and constant threat from wild beasts and cattle lifters. Villages happened to be surrounded by vast arable land, and beyond was situated the pasture land interspersed with wild tracts of dense forest infested with wild animals. But, along with this, every villager also used to keep a few animals for draught purposes with him at his house to meet the supply of his own household[3].

The Vedic society was pastoral in its earliest phase, and people were always in search of new pastures.[4] The Ṛgvedic seer Mathita Yāmayāna prays to Indra and Agni, his favourite deities, to undertake the task of a herdsman entrusted with taking the cattle to the pastures and forests for grazing and bringing them back in the evening safely.[5] The seer Maitrāvaruni Vasiṣṭha takes Agni as a deity who looks upon the people of the world as a herdsman watching over his cattle.[6] The seer Bārhaspatya Bharadvāja addresses Pūsan in five hymns of the *Ṛgveda* as a deity who holds a tuft of leather strap as the controller of cattle (*paśu-sādhanī*) playing the role of a herdsman.[7] The seer Kaṇva Ghaura prays to Pūṣan for the favour of wide pastures.[8] The seer Gāthina Viśvāmitra praises Indra as an expert herdsman who nourishes the cows and feeds them with grass and guards them.[9] Similarly, prayers were offered to Indra to lead and protect in the battle, as does a herdsman lead his cattle to pastures and protects them.[10] Cattle grazed in wide pasture grounds and their calves were tied to ropes.[11] As is vouchsafed in the *Ṛgveda*, the Vedic people showed anxious solicitude for welfare of their cattle, cows and horses, as is found in the case of the seer who prays thus: 'May Pūṣā follow near our kine; may he keep our horses safe; may Pūṣā gather gear for us; may he follow the kine of his sacrificer who pours libations out and worships; let none of the cattle be lost, none injured, none sink in a pit and break a limb; may he return with them safe and sound.'[12] Another seer speaks of the herdsman going to the grazing ground with kine along various paths and returning home with them to the delight of all.[13] The seer Sabara Kākṣivata entreats Rudra endow the blowing wind with healing powers for the kine so that they may eat herbage full of vigorous juice and drink waters rich in life and fatness.[14] The seer Budha Saumya shows how cattle were given pure drinking water poured into wooden troughs and how they were bound with straps so that they could be carried easily from the wells to the cow-stalls.[15]

Cattle were housed in stalls. The Atharvavedic seer Brahmā has devoted a whole hymn to the cow-stall (*goṣṭha*).[16] This hymn, along with another one,[17] is used by the Kauśika Sūtra[18] as useful in a ceremony for the prosperity of cattle. Here the seer prays thus: 'with a comfortable stall, with wealth, with wellbeing, with that, which is the name of the dayborn one, do we unite you. Let Aryaman unite you, let Pūṣan, let Indra, who is conqueror of riches; in my possession gain by what is good having come together, un-affrighted, rich in manure, in this stall, bearing the sweet of Soma, come ye hither, free from disease. Come ye just here, O king, and flourish here like Śaka; also multiply just here; let your complaisance be towards me. Let your stall be propitious; flourish ye like Śāriśāka; also multiply just here; with me we unite you. Attach yourselves, O king, to me as lord of king; this your stall here (be) flourishing; to you, becoming numerous with abundance of wealth, to you living, may we living be near.'[19] The Atharvavedic seer Cātana seeks to obviate sterility in cattle by the following charm; 'We drive you out of the table, out of the axle (of the wagon) and the body of the wagon; we chase you,

O ye daughters of Magundi, from the house.'[20] A whole hymn has been devoted to the safety and increase of cattle, wherein the Atharvavedic seer Savitā prays for the prosperity of cattle, thus: 'Here shall come the cattle which have strayed, to a distance, whose companionship Vāyu (the wind) enjoys! (The cattle) whose structure of form *Tvastṛ* knows, Savitā shall hold in place this stable. To this stable the cattle shall flow together, Bṛhaspati skilfully shall conduct them hither. Sīnivālī shall conduct hither their van: 'do thou, O Anumati, hold them in place after they have arrived: May the cattle, may the horses, and may the domestics flow together; may the increase of the grain flow together sacrifice with an oblation that causes them to flow together. I pour together the milk of the cows; I pour together strength and sap with the ghee. Poured together shall be our heroes, constant shall be the cows with me the owner of the cows. I bring hither the milk of the cows; I have brought hither the sap of the grain brought hither are our heroes; brought hither to this house are our wives.'[21] The seer Bhāgali says, 'The kine have sat down in the stall; the wild beasts have gone to rest; the waves of the streams, the unseen ones, have disappeared.'[22] And the seer Brahmā recommends; having given a bull to brahmans, one makes his mind wider; he beholds prosperity of the inviolable (kine) in his own stall.'[23] In another prayer for sovereign power addressed to the god Rohita and the goddess Rohiṇī, he prays: 'O! Vācaspati, good cheer and spirit, cattle in our stable, children in our wombs beget thou. Right here life's breath shall be to our friend thee, O Parameṣṭhin, I envelop in life and luster.'[24] Cattle was won as the prize of victory during the tribal wars and Indra was always prayed for recovering the cattle of his patron chief from his enemies, and the recovered cattle became the common property of the entire tribe as they were distributed among the members of the tribe by the tribal chief; it became also the common property of a family.

In the later Vedic period, agriculture flourished as a means of occupation of the Vedic people, but cattle remained their principle wealth.[25] The *Kauṣītaki Brāhmaṇa* and the *Aitareya Brāhmaṇa* also testify to the fact that cattle had been the companions and collaborators of Indians in the field of economic development that they were the solid foundation of their economic management and, consequently, they were considered as the real wealth.[26] It was from the cattle that they obtained milk or food and leather for means of daily use. Tilling could be done only with the help of bullocks. Cattle were also a medium of exchange.[27] Animals like horses, elephants, and perhaps dogs too, played an important part in battles and racing games.[28] For all these reasons, cattle rearing received much care and were held as highly important.[29]

The *Rāmāyaṇa* of Vālmīki also shows that cattle rearing and agriculture were sister occupations in rural areas. Pastoral villages, known as *ghoṣas*, were probably in close proximity to the *grāmas* or agricultural villages. *Grāma-ghoṣa* is another term frequently met within the epic. But cattle rearing were not confined to rural areas only. Even cities like Ayodhyā had abundant cows; and the day Rāma departed for the forest, cows in Ayodhyā were so much in grief that they refused to suckle their calves.[30] That the cattle wealth of the country was substantial is evident from the references to the numerous occasions when myriads of cows were given away as gifts to supplicants.[31] The king himself was a large owner and breeder of cattle and, likewise, the breeding of cattle, no less than their rearing, engaged equal attention of the agriculturists. Cattle breeding was done on an extensive scale throughout the kingdom. At the same time, cattle lifting, which was common practice of Kṣatriyas during the Vedic times, seems to have continued during the post-Vedic age of the *Rāmāyaṇa* too, as is evidenced by the legend of Visvamitra's attempt at taking away the cattle of Vasistha's *āśrama* by force.[32] Similarly, in the *Mahābhārata* too, an incident of cattle lifting by the Kauravas as a clever strategy to expose the Pāṇdavas from their stay incognito with King Viraṭa during their stipulated thirteenth year, is well known.[33]

The *Śatapatha Brāhmaṇa* refers to the food and fodder of the cattle, particularly *gavedhuka, śaspa, jai* and chaff, while *narakula* is mentioned as the grass for horses.[34] The cattle were tethered and fed in their stalls or stables[35] and cattle feed was prepared by adding pinches of salt, since salt-mixing played an important role in the feed and drinking water of cattle.[36]

The twofold classification of animals, viz., domestic (*grāmya*) and wild (*āraṇya*) is found in the *Śatapatha Brāhmaṇa*, but these seven varieties—each one of these two classes—are not enumerated. It is interesting to note here that the *Vāyupurāṇa–na* agrees closely with the classification of the *Śatapatha Brāhmaṇa* and further mentions their names.[37]

A draught-animal is referred to by the peculiar term *patra* from the root 'pat', to move, in the *Aṣṭādhyāyī of Pāṇini* (3.1.121; 4.3.122-123). An animal fit to be yoked is called *yugya* (3.1.127). In the *Taddhītā* section (4.4.76-81), *Pāṇini* classifies animals on the basis of their being yoked to different vehicles and their capacity to draw various loads, as

(1) *rathya,* bullocks to draw chariots (4.4.76);
(2) *śākaṭa,* bullocks for carts (4.4.80);
(3) *hālika*; and
(4) *sairika,* bullocks for ploughs (4.4.81). These distinctive terms were used in connection with the care and quantity of rations prescribed for the animals of each class.[38]

COWS

Cow enjoyed a very important status and prominent place in the life of the people in Vedic times. Among the innumerable boons prayed for by the Vedic seers, the cow may be deemed one of the most cherished one. Milk, butter, ghee and curd, the dairy produce from cows were not only highly esteemed foods but they were also some of the main offerings to the gods. Cows supplied the other needs of agriculture and of life in general, providing plough bullocks, the pack bullocks and cart bullocks. Their leather provided many articles in general use or required for special purposes, such as garments, curiasses, sandals, straps and thongs, whips, bellows, bottles and large receptacles, etc. They were kept in herds and probably no man or homestead was so poor as not to own a few. Wealth consisted as much in the number of cows owned as in the number and size of full granaries. Herds were common targets of cattle raids, and pilfering as also expeditions both to raid and to rescue stolen cattle were common features in their life.

Thus, supplications for aid, victory and the praise of successful exploits are a subject of numerous hymns. Cows were carefully guarded inside walled enclosures. Watchdogs were posted both to warn the inmates and to attack raiders. Cows of outstanding merit tempted covetous individuals or tribes, so much so that even sages like Viśvāmitra succumbed to the temptation; as witness the famous episode of the stealing of the 'wonder' cow *Kāmadhenu* of Vasiṣṭha. The milking of cows, the making of curds, of butter and ghee were domestic duties of everyday importance: besides being offered to the gods, these milk products comprised the daily food, both by themselves and by being put up in many forms in mixture with other foods. Ghee was required in large quantities both for the sacred fire in the house and in sacrificial halls, and this was one of its main utility. Cows were tended with great care. Much store was set by green grass and grazing for which vast pastures were prayed for. Cows were also fed on grains as concentrated feed. Ample supplies of pure water were secured for them either from clear streams and pulls, or if necessary, by lifting from wells.[39]

Cows were treated against injuries and diseases, for which many herbal and other specifics were known. Barren cows and cows of susceptible abortion were dreaded and avoided, since the points or factors which distinguished good cows from bad ones were known.[40]

The lore of the cow forms a frequent feature in the imagery of the Vedic hymns and similes as well as illustrations referring to their nature and ways abound, in as much as cow was a symbol of wealth and prosperity while milk, butter and ghee were the symbols of fertility and abundance. Gifts of charity, reward or religious merit, were made in the form of cows, whose number varied with the wealth or status of the donor and their disposition, liberal or otherwise. Lavish praises are bestowed upon liberal donors, while wealthy but miserly traders and others were comended to the wrath of the gods.[41]

The Ṛgvedic seer Kuśika Aiṣīrathi says: 'wide fields, vast treasures and spacious pastures have Indra bestowed upon his friends'.[42] The seer Vāmadeva Gautama prays: 'we seek to bring down from thee (Indra) thousands and hundreds of cattle. May riches come to us from thee'.[43] The seer Kṛṣṇa Āngirasa says: 'may we escape poverty by means of cattle'[44]; 'grant us, Indra, wealth of barley and cattle'[45]. The Yajurvedic seer solicits the god of fire, saying 'O Agni, may I have milking cows in thousands and tens of thousands.'[46] The Ṛgvedic seer Madhuchanda Viśvāmitra entreats Indra thus: 'shedder of rain, set open these clouds, set open the cow pastures, send us liberal kine liberally'.[47] The seer Prājāpatya Saṁvaraṇa is confident that Indra 'places the devout man in a pasturage stocked with cattle'.[48] The Atharvavedic seer Brahmā has devoted a full hymn of seven mantras to the extollation of cows, and therein he says thus: 'the king have come, and have done what is excellent; let them stay in the stall; let them take pleasure with us; may they be rich in progeny here, many-formed, milking for Indra many dawns. To the sacrificer and singer, to the helpful one, Indra verily gives further, steals not what is his; increasing more and more the wealth to him, he sets the godly man in an undivided domain. They shall not be lost; no thief shall harm (them); no hostile (person) shall dare attack their track; with whom he both sacrifices to the gods and gives, long verily with him does the kind lord go in company. No dust-raising horseman reaches them; not unto the slaughterhouse do that go; those kine of that sacrificing mortal roam over wide-going fearlessness. 'the kine (are) Bharga; Indra has seemed to me the kine are the draught of first Soma; these kine, that, O people, (is) Indra; with whatever heart (and) mind I seek Indra. Ye, O kine, fatten whoever is lean; the unlovely one ye make of good aspect; ye make the house excellent, O ye of excellent voice; great is your vigour called in the assemblies. Rich in progeny, shining in good pasture, drinking clean waters at a good watering-place let not the thief master you, nor the evil-plotter; let Rudra's weapon avoid you.'[49] The seer Kakṣāvīn Dairghatamasa Auśija prays to the Aśvins; 'direct us to food associated with kine.'[50] The seer Śainyu Bārhaspatya says; 'the liberal donation of thousands of cattle has been given to me.'[51] The seer Bhauma Atri praises the gods Viśvedevas as 'givers of horses, cows and garments.'[52] The Ṛgvedic seer Vaśa Aśvya invites Indra and prays; 'Lord of wealth, visit us as of old to give us cows, horses and chariots'[53]. He further says: 'I have received sixty thousand horses, and tens of thousands—a score of camels, a thousand brown mares, and ten times ten thousand cows with red patches.'[54] The seer *Bārhaspatya Bharadvāja* prays: 'Let not the cow be lost, let no thief carry them away; let no hostile weapon fall upon them; may the master of cattle be long possessed of those which he sacrifices and presents to the gods.'[55]

The Vedic people prayed for cows which gave abundant milk and unfailingly, and which can be milked with great ease. Thus, the Yajurvedic seer refers to cows milking with ease and yielding abundant milk and ghee.[56] The seer Priyamedha Āngirasa points out to these white kine giving milk like wells[57]. The seer Dirghatama Aucathya prays to Viśvedevas: 'I invoke the milk cow that is easily milked, that the handy may milk her'[58]. The quality of some cows which let down their milk merely at the sight of their calf has been well observed and referred to by the seer Hiraṇyastupa Āngirasa thus: 'as a cow having a copious stream of milk yields it coming into the presence of the calf.'[59] On the contrary; cows which are troublesome to milk and which need control at the time of milking, mostly by the well-known method of shackling their hind legs with a rope as is done at present, have also been taken note of. Thus, the seer Kṛṣṇa Āngirasa says: 'control your friend like a cow for milking'[60]. The barren cow and the cow prone to abort are dreaded and avoided, and the restoration of barren cows to fecundity and lactation greatly appreciated. Thus, the seer Kakṣivān Daighatamasa Auśija praises Aśvin: 'you filled the milk less barren and emaciated cows of Śayu with milk'[61]. The seeress Kākṣivaì Ghoṣā further corroborates this, thus: 'you, Aśvins, renovated for Śayu the barren cow.'[62] Reference is obviously some medicinal treatment by which barrenness has been cured, for the Asvins are said to be great physicians.[63]

The *Brāhmaṇas* too contain references, which show that care was taken for both the maintenance and protection of the cattle.[64] The *Aitareya Brāhmaṇa* states that cows possess vitality, and hence people love them.[65] In the *Rāmāyaṇa*, the cow-stall was known as *pratyāgāra* (2.40.43). Herds of cows were called *go-kula* (246.17) *go-yutha* (2.49.10) or *go-vraja* (2.32.38). The cowherd or keeper of cows was called *go-pāla* (2.67.29); and the meadow was known as *śādvala* (3.16.20). At Bharadvāja's hermitage, the soldiers saw plots of tender grass all around, coloured like blue lapses, to serve as pastures for the cattle (2.91.79).[66] In the *Mahābhārata* too, we find that most of the people seemed to possess cows. Vaṣṭstha's cow, known as Kāmadhenu, has been extolled at length and cow was regarded to be most useful for the society in comparison with other domestic animals.

Sahadeva, the fourth of the five Pāṇḍavas, was an expert in the lore of cows.[67] Also, it was specially recommended that one should take care of cows oneself and should not be left to others.[68] In a dialogue with Indra in the *Anuśāsana Parva* of the *Mahābhārata,* Bhīṣhma propounds the importance of cow thus: 'cow is the principal means of sacrifice; since it cannot be accomplished without cow. Milk and ghee are the main food of man and farming is possible only because of the cows; cows are the principal source of all things of sacrificial oblations. Cow is, therefore, the best among other species of animals. Cow is verily the mother of all in this world. One who aspires to prosper should devote himself to the service of cow by all means.'[69] It was a popular belief in these days that if a domestic cow was not taken care of properly, the concerned householder would come to trouble. A sight of a cow was capable of destroying sins.[70] There are passages in praise of cow in the 50th and 80th *Adhyāyas* of the *Anuśāsana Parva.* There is a reference to *gavāhnika-dāna,* which denotes that it was one's duty to feed a cow just after one's daily *sandhya* rite.[71] Tawny or reddish (*kapila*) cow was considered to be the best.[72] 'Gift or donation of cows was valued most in comparison with that of other things'. The four *Adhyayas* from 71 to 74 treat the importance of this topic, and are known as *dāna-prakaraṇa.* 'Cowdung and cow-urine were regarded as purifying agents and were utilized respectably for plastering the floor of houses and for purifying one's body by drinking or bathing.'[73] A story in the 82nd *Adhyāya* of the *Anuśāsana Parva* is interesting in this connection: Once, the goddess Śri, adorned with excellent dress and ornaments, came to the cows. Cows asked her about the reason of her arrival. The goddess replied that since Indra, Viṣṇu and other gods were prosperous and powerful due to her grace, she wished that the cows might also have such a benefit due to her favour. The cows said that they did not need her since they were prosperous and pious by themselves. At that, the goddess was disappointed and solicited them to let her reside in their limbs, so that she would not be looked down upon in the eyes of the people in this world. The cows conferred among themselves and condescended to permit her to reside in their urine and dung, since both these were extremely pious. The goddess was highly pleased at this offer and disappeared. Then both the cow-urine and cowdung to be looked upon as a veritable seat of the goddess Lakṣmī. This story seems to emphasize the fact that both the urine and the dung of cows is calculated to serve as excellent organic fertilizer which brings profusion in farming and yields heavy crop. In religious ceremony, the back and the tail of a cow are considered very pious and people touch them and feel purified by it.[74] A vow was prescribed for the welfare of cows, was called *go-puṣṭi,* in which one was required to take bath with the dung and urine of cows, and offerings of cow-ghee was made in fire.[75] A number of cow-hymns are known as *gomati-vidya* or *go-upanisads,* and numerous benefits accruing by their recitation are enumerated.[76] Killing of cows and eating of cow-flesh was totally prohibited.[77] Looking after the cows and their herds, Sahadeva, the younger among the Pandavas, exhibits his knowledge of the cow-lore by his remark about bulls whose mere urine even is capable of making a barren cow conceive and yield progeny.[78] Although rearing of cows was considered to be the duty of the *vaisyas* in those days almost each family possessed at least one cow for the purpose of daily sacrifice and the Vaisyas took to cow rearing as a profession.[79]

According to the *Vāyu-Purāṇa,* the cow was born from the belly of the creator.[80] The cow that dropped from Maheṣvara's mouth when Brahmā was engaged in meditation, was but the Gāyatrī.[81] The cow figures considerably in ritual and gift of cows in a sacrifice is a highly commended act. In the description of the end of the Kali Age, we find a reference indicating that the slaughter of cows is a sin. We also find cows serving as means of exchange and *gavyūti* is a unit of measurement of distance.[82] In the *Kūrma-Purāṇa,* too, the cow is held in very high esteem. Though expiations are provided for the crime of killing a cow unintentionally, intentional killing of a cow, according to the *Kūrma-Purāṇa,* has no prescribed expiation, thus implying it to be a highly heinous crime.[83] Eating its flesh is prohibited.[84] The gift of cow is very efficacious.[85] Cow's urine (*go-mūtra*), cowdung (*gomaya or go-śakṛt*) and five cow-products (*pañca-gavya*) are held to be purifiers. Cowdung was used in smearing the ground in order to purify it for the purposes of *śraddhā,* etc., and it was essential for a brahmin going for his mid-day bath, as and for purificatory and expiratory purposes.[86]

Pāṇini, in his *Aṣṭādhyāyī,* refers to the cattle craving for salt as *lavaṇasyati* (7.1. 51), since salt was to be given to cattle as part of their food. Prosperity in cows and calves was blessed by the expression *svasti bhavate sa-gave sa-vatsāya* (Varttika on 6.3.83). A cowpen was *Vraja,* and cow-stall *gośalā* (4.3.35) and *goṣṭha* (8.3.97). *Goṣpada* was the place for the cows to roam (6.1.145). Dense forests impenetrable to cows were *Agoṣpada* (6.1.145). *Gotra,* in Panini, is an assemblage of cows (4.2.51), and its two new synonyms are *gavya* and *adhenava* (4.2.47). The cowherds were called *gopāla;* special officers in charge of royal cattle were called *tantipāla* (6.2.78). The son of a cowherd attaining the age when he was fit to take the cows out for grazing was called *anugavina.*[87]

The different stages in the lifecycle of a cow were expressed by suitable terms. The female calf attaining puberty was termed *upasaryā* (3.1.104), and after her first mating *upasarā* (3.3.71), a cow attains full youth at the age of three years. If she miscarried, she was called *vehat* (2.1. 65). On the eve of delivery, she was called *adyaśvīnā,* calving today or tomorrow (5.2.13), and after calving, *gṛṣṭi* (2.1.65). Panini also refers to *mahāgṛṣṭi* (6.2.38), a better cow whose milking period continues up to the next calving. *Dhenu* was a cow in milk (2.1.65), also called *asti-kṣīrā* by Kātyāyana (2.2.24). After about six months of calving, a cow became *baṣkayaṇī* (2.1.65). A cow calving every year was *strivatsa* (2.4.13) was one who gave birth to a heifer was marked out from the rest a *samāmsamīnā* (5.2.12). A cow pledged to the creditor to pay off the debt from her milk was called *Dhenuṣyā* (4.4.89).[88]

According to the *Śatapatha Brāhmaṇa,* cows were milked thrice a day.[89] The *Maitrāyaṇī Samhitā* names these three milkings, viz., the morning milking as *Pratar-dohā,* the pre-midday one as *saṁgava* and the evening one as *sāyam-dohā.*[90] Cows were taken out thrice for grazing in pastures. Calves were kept in the stall and in the evening, they were tied with their mothers. Bulls were also kept for mating with the cows and this ensured regular progeny of calves.[91] Livestock breeding was an important occupation of the Sūtra period. People kept big herds of cows as also of horses, goats and sheep. Herds of cows were so big that they were given away in hundreds and thousands as sacrificial fees. According to the *Kātyāyana Śrauta Sūtra* (15.4.43), a lac of cows was given away as sacrificial fees at the *rājasūya.*[92]

The Upaniṣads show the cow as a measure of value, a very precious commodity and an invaluable possession. The king gave away thousands of cows as prize-money or tutor's fees.[93] The students tended their teacher's cows.[94] The number of cattle he kept determined a man's position in society. In the order of merit, cattle stood second only to offspring. In order to dissuade Naciketas from his third boon, Yama offers him sons and grandsons who shall live a hundred years, herds of cattle, elephants, gold and horses in that order.[95] The *Chāndogya Upaniṣad* also promises long life, greatness with children and cattle and great fame to one who knows the Gāyatrī chant.[96] The *Kauṣītaki Upaniṣad* proclaims that meditation makes a man full of offspring, cattle, fame, and reach the full term of life.[97] The *Chāndogya* again prays to secure immortality for the gods, offsprings for the manes, hope for men and grass and water for cattle.[98]

The *Śrauta Sūtras* describe a number of ceremonies connected with cattle welfare.[99] Thus, the *sulagava* sacrifice was performed every year. The ceremony known as *vṛṣotsarga* was performed at that time when a bull was let loose for the purpose of breeding. The *Gṛhya Sūtras* mentions many other ceremonies for the welfare of cattle. It contains a ritual to be performed for restoring affection between a cow and her calf.[100] At another place, the same *Sūtra* describes rituals for the protection of cows and for the prosperity of young calves.[101] According to the same *Sūtra,* cows suffering from a disease, should be given saline water to drink.[102]

The outward characteristics in which cows may differ from one another were closely observed, and each one is said to be agreeable to a particular deity. The kinds described are too many to be listed with advantage. However, a few may be given as examples. Whether these outward differences were deemed to be related to any difference in performance or other quality of the animals, is an interesting speculation. Among the kinds mentioned are cows with many different colours, entire, patchy, piebald or brindled, with colours on one side only, colour on the face, on the ears, or on the tips of ears only, spotted, silky skinned, coarse skinned, with long tufted tail or a stumpy one, horned and hornless, with white patches on legs, feet, forehead or along the back and so on.[103] Thus, the *Śukla Yajurveda Saṃhitā*[104] mentions the following kinds: *śilpā, vaiśvadevī, rohiṇī, tryavi, avijñātā sarupā, vatsatarī, kṛṇṣagrīvā, śitibhru, rohitā, svetā, avarokin, nabhorupā, śukarūpā, vājinā, kalmaṣā, śyāmā, dvirūpā, vāmanā, anaḍuhī, vaśā, eṇi, kṛṣṇagrīvā, babhrū, saumyā, kṛṣṇā, dhumrā, divyā, śabalā, sidhmā, trivatsa, pṛsnī, bahurūpā, praśṛṅgā, babhrunīkāśā* and *dhumranīkāśā.* It is difficult to identify these breeds with certainty.

Broadly speaking, the Indian breeds of cows are classified under three heads: milch breeds, general utility breeds and draught breeds.[105] The cows of the milch breed are high yielders; the bullocks are of moderate type or of poor draught quality. They are generally ponderous in build, with pendulous dewlap and sheath, and having lateral and often curled horns. The gir, sindhi, sahiwal and deoni are some of the outstanding breeds of this group. The cows of the general utility breeds are fairly good milkers and bullocks are good for draught purpose. There are two types of cattle in this group, viz.,

(i) short-horned, white or light grey cows with long coffin-shaped skull and face, slightly convex in profile, e.g. hariana, ongole, krishna valley, etc.;

(ii) lyre-horned, gray cows, deep-bodied with a white forehead, prominent orbital arches, flat or dished-in profile, and good draught capacity, e.g. tharparkar and kankrej.

The cows of the draught breeds are poor milkers but the bullocks are excellent draught animals. In this group, there are four types of cattle, viz.,

(i) short-horned white and light gray cattle with long coffin-shaped skull and face, slightly convex in profile, e.g., nagori and bachaur;
(ii) lyre-horned gray cattle with a wide forehead, prominent orbital arches, flat or dished-in profile, deep body and powerful draught capacity, e.g. kenkaṭhā, mālvi and kherigarh;
(iii) 'mysore type', characterized by prominent forehead with long pointed horns which rise close together. They are, with a few exceptions, poor milkers, e.g., halikar, amritmahal, kangayam, khillari;
(iv) small black, red or dun cattle, often with large patches of white markings, found in the rugged mountainous areas of the Himalayan region or at the foot of the hills; they have tight sheaths and are either short-horned or slightly lyre horned; they are active, and are useful for light ploughing and miscellaneous work, but their milk yield is poor; they are able to thrive where large animals cannot survive; the ponwar and siri are notable breeds of this group.

Whatever the group or the breed might have been in the Vedic times, but the seers are clear that cows should have a compact body and a capacious udder. Thus, the seer Gṛtsamada Angirasa Śaunahotra prays for 'well-nourished compact-bodied cows',[106] and the seer Dirghatama Acathya prays: 'may the milk kine be possessed of well-filled udders'.[107] Milk has been the chief nourisher of the Indian people since the Vedic times, and milk as well as milk products is prayed for in innumerous hymns. Thus, the Ṛgvedic seer Parāśara Sāktya says: 'Saramā discovered the abundant milk of kine, with which man, the progeny of Manu, is still nourished'.[108] The Yajurvedic seer prays: 'may our riches in ghee and curds increase'.[109] Curd was churned with the churning rod and a rope as it has been done till the recent past in the villages. Thus, the seer Ājigarti Śunaḥśepa refers to the process of churning in which the churning staff is bound with a cord, like reins to restrain a horse.[110] Milk, in its progress from the initial bodily secretion as the food of the embryo to the form of ghee at the end, is traced and mentioned thus in the *Yajurveda*: 'He is the seed (of the calf), the embryo, the navel cord, and calf, the first milk, the colostrum, the fresh milk, the curdled milk and ghee.'[111]

Butter and ghee made the richest viands. Thus, the Ṛgvedic seer Sadhvaṃsa Kāṇva prays to the Asvins; 'give us food of many kinds, dripping with ghee'[112]. Gradual stages of curds, butter and ghee are also mentioned in the *Taittiriya Saṃhita*[113]. Milk, butter and ghee were the symbols of plenty and fertility.[114] The *Yajurveda* refers to 'the land wetted with ghee'[115], and prays that 'may the land so ploughed and rich with milk, strength and wet with honey and ghee come back to us'.[116] Large quantities of ghee were used up in the sacrificial fire. Thus, the seer Vāmadeva Gautama says: 'these streams of ghee descend upon the fire, like deer fleeing from a hunter; the streams of ghee fall copious and rapid as the waters of a river; the streams of ghee incline to Agni, as devoted wives to a husband'.[117]

Cows were fed on grains in addition to the grass on pastures.[118] Thus, the Ṛgvedic seer Aindra Vasukra says: 'the assembled cows feed upon barley'.[119] Good water of pools was also easily available for cattle. Thus, the seer Bārhaspatya Bharadvāja prays; 'may your cows have many calves, grazing upon good pastures and drinking pure water at accessible pools'[120]. And the seer Buddha Saumya urges: 'set up the cattle troughs, bind the straps to it, let us bale out the water from the well, which is not easily exhausted'[121]; thus vouchsafing that the cattle were also watered at wells.

Similes based on the nature of the cow, particularly about her love for her calf, which makes her run after it and allow the milk to flow spontaneously from her udder and her aggressive posture on sensing danger to it, are numerous among which the following are typical. Thus, the seer Hiranyastupa Arigirasa, while describing the waters released by Indra, says: 'Flowing waters rushed to the ocean, like cows to their calves'.[122] Another seer compares his rescue along with his wives by Indra, with a cow grazing in a meadow leading her calf out of danger.[123]

These numerous and varied references to the many aspects of the great importance of cows, milk and milk products make it abundantly clear that breeding of cow as a milk animal must have received great attention and reached a high state of perfection. Wonder cows were probably not mere myths or poetic exaggerations, but real.[124]

As is clear from the fact that a full hymn is devoted to the cow stable, cattle were protected behind walled enclosures and watch dogs stationed to guard them. Nevertheless, thieves caught by dogs could be silenced by appropriate incantations, and cattle could be stolen. *Yajurveda* mentions such an instance thus: 'like a cattle thief, getting over all obstructions, comes in'.[125] And in the famous Ṛgvedic hymn (7.55) known as the *Prasvāpinī Upaniṣad*, the seer Maitrāvaruṇi Vaśiṣṭha actually spells out the mantras meant to be the incantations for lulling the watch dogs and the inmates of the house to sleep; one of the mantras reads thus: 'may the mother sleep, let the father sleep, let the dog sleep, let the son-in-law sleep, let the kindred sleep, let the people stationed around sleep.'[126]

Cows often met with accidents and were open to attacks by wild beasts. Prayers were, therefore, offered for their protection, particularly to Pūṣan and Indra. Thus, the seer Bārhaspatya Bharadvāja prays to Pūṣan: 'let not, Pūṣan, our cattle perish, let then not be injured; let them not be hurt by falling into a well'.[127] And, he hopes, with Indra's grace, to protect his cattle thus: 'may no thief be your master; no beast prey assail you, may the fatal weapon of Rudra avoid you'.[128]

Cattle were raided and carried off by enemies, and expeditions for recovering them were undertaken with Indra's aid. Thus, the Ṛgvedic seer Garga Bharadvāja invokes Indra, saying: 'Recover thou our cattle, Indra; bring them back; the drum sounds repeatedly as a signal; our leaders mounted on their steeds, assemble; may our warriors be victorious'.[129] The famous cattle raid of the Kauravas against the cattle of King Virāṭ, to compel the Pāṇḍavas to expose themselves from their stay incognito during the thirteenth year of forest residence is well known. It is interesting, says A.K. Yegna Narayan Aiyer, to recall that cattle raids and organized thefts and counter-raids appear to have been rather a common feature of village life even in south India in the olden days.[130] We may also recall in this connection that during the pre-independence days in the old states of Kathiawar, now combined as Saurashtra with Gujarat after independence, the countryside is studded with stone memorials (*pāḷiā*) for warrior heroes to commemorate a departed warrior killed in such forays while fighting the raiders.

Cows were also treated against diseases and much herblore of this kind seems to have existed, as the references to them show. The Yajurvedic seer prays: 'give me medicinal herbs, suitable for cows, horses, men, goats and sheep'.[131] Medicinal herbs to which animals suffering from particular ailments naturally and instinctively resorted to, and ate, were carefully observed and noted, and were adopted as successful remedies for such ailments. The habits of the boar, mongoose, eagle, snake, swan, various birds, cows and sheep were studied in this connection and made use of by veterinary physicians.[132] According to the *Kṛṣiparaśara*,[133] a shed measuring five by five steps is good for the healthy growth of cattle. Washing of rice, hot scum of boiled rice, fish broth, cotton seeds and husk, if kept in the cowshed, prove baneful to the cattle. Incidentally, this gives an idea of the fodder of the cattle at that time. 'To safeguard against the breaking out of diseases, the shed should occasionally be fumigated with vapours of *devadaru* (pinus deodara), *vaca* (oris root), *mamsi* (pulp of nardostachys), *guggulu* (a fragrant gum resin), *hinga* (asafoetida) and mustard seeds, all mixed together,' according to the *Agni Purāṇa*.[134] The *Viṣṇudharmottara Purāṇa*[135] describes certain veterinary medical practices of treating diseased cattle for curing their affected horns, ears, eyes, teeth, tongue, throat, bladder, etc. The text says that oil in which the pounded mass of ginger, *bata* and *jaṭāmāṃsī* has been cooked, and rock salt and honey added, should be applied to the root of the horns. The powder of the roots of wood-apple tree, *apāmārga, dhātaki, paṭala and kuṭaja*, when rubbed into the gums, removes toothache. Ginger, turmeric and three myrobalans are indicated in the cure of sore throat and a certain prepared collyrium for the eyes. For the reunion of fractures, *priyaṅgu* mixed with salt is recommended. The billious disorders are stated to be cured by the administration of cow's ghee in which liquorice has been cooked. Ailing calves should be made to drink *pātha* stirred in the buttermilk or turmeric added in milk for a alleviation of their suffering. Oilcake in general was considered to be an elixir for the cattle. Salt is to be given to the cattle once in fifteen days to prevent constipation, colic diseases and loss of appetite. On the whole, animal husbandry was of sufficiently high order in the ancient period.[136]

Vasudev Sharan Agrawal had noticed[137] that Pāṇinī gives interesting information about branding certain marks (*lakṣaṇa*) on the body—generally ears—of cattle to distinguish ownership (6.2.115) and incidentally gives a list of some marks used to indicate different owners, e.g. *viṣṭa, aṣṭa, pañca, maṇi, bhinna, chinna, chidra, sruva* and *svastika* (6.3.115). The branding of cows was known in the Vedic period. The *Atharvaveda* refers to it as *lakṣṃa* and mentions the *mithuna* mark.[138] The *Maitrāyaṇi Saṃhitā* (4.2.9), the *Mānava Śrauta Sūtra* (9.5.1–3), and the *Gonāmika Pariśiṣṭa* of the *Vārāha Śrauta Sūtra* give details of this ancient cattle rite and a few more marks.[139] The *Māhābhārata* also refers to a census of the royal cattle (*samaraṇa*) by branding them (*anika, lakṣa*).[140] To the nine marks mentioned by Pāṇinī, other names may be added from the *Maitrāyaṇi Saṃhitā, the Ṛk-tantra and Kāśikā, Viṣṭa-karṇī.* A mark in Pāṇini's list is in the *Maitrāyaṇi Saṃhitā* as a mark of the cows of Agastya, those of *Jamadagni* having a lute and of Vasistha a stake[141]. The *aṣṭakarṇī* of Pāṇinī occurs in the *Ṛgveda*,[142] where Grassmann translates having the sign for (the number) marked on the ear. Some of these marks (*lakṣaṇa*) can be identified amongst the symbols stamped on puñch-marked coins, e.g., *sruva, svastika, ankuśā, kuṇḍalā, plihā, bana, mithunā.*[143]

BULLS AND BULLOCKS

Bulls, and particularly bullocks, were widely employed as draught animals. Their usefulness is distinctly portrayed in the *Bṛhat-parāśara Saṃhitā*, cited in the *Vācaspatya-kośa*, thus: 'they grow all kinds of food grains but themselves eat only grass and sustain the whole world. The gift of one of them is equal to that of ten cows and in him is embodied the very image of religious virtue (*dharma*) on this earth. They should be carefully tended and yoked in a proper manner. One who takes toil from them without tending them, goes to hell full of horrors. A bull with an extra or deficient limb, or with spots on the body, or otherwise suffering from a bodily injury should not be yoked; if one does so, one would be ruined. The reward that accrues to one rearing the oxen is ten times the merit that the wise have declared to accrue in protecting the cows. As the entire animate and inanimate world is supported by oxen, hence an ox should always be protected and well-fed. Brahma, the creator, intent on the welfare of the world, has made an ox the very image of virtue (*dharma*) on this earth capable of producing food grains and sustaining the three worlds. They roam about in remote and far-off places and eat with relish wild grasses which are unfit for (human) consumption. They make the cultivation of crops possible, they thrash the harvest and they carry it to far-off places. They draw and bring loads from afar by means of the strength of their shoulders; they do not demand food grains for their food, and they preserve the lives of others with their own lives, nourish them and make them flourish. The gift of one ox means giving away of ten cows and is equal to the gift of the whole earth. Therefore, no other animal than an ox deserves veneration. Having been instrumental in growing the food grains, they (oxen) themselves eat grass and then, carrying the entire harvest (to its destination), they do not complain even if they are fatigued. All the living creatures are sustained by the oxen. The nostrils of an ox should not be pierced for three or four days after it has become strong-limbed and should never be pierced before that time as long as he is weak.'[144]

The *Kṛṣiparāśara* by *Parāśara* contains some useful hints about the attention to be paid to draught cattle: 'One should do cultivation in such a way that the bullocks are not wearied. Crop raised by the fatigue of the draught animals are condemned by the gods and the manes. Cultivation done by over-working the animals, though yielding a fourfold harvest to the cultivator, is to be avoided. At the commencement of tilling the land, one should take care to select the bull of this kind and smear the sides of the mouth with butter and ghee. A bull that is rough or whose horns or hoofs are broken or whose horns are either too short or too long should, as far as possible, be avoided in ploughing. In the course of ploughing, the cultivator should avoid striking and hurting the tail and ears of the tired oxen and also refrain from beating them too often.'[145]

The bull figures in the *Vāyu Purāṇa* mostly as the vehicle of Siva[146] who is also mentioned as having the bull standard (*vṛṣabha-dhvaja*); Siva himself is addressed as bull by Daksa in his prayer to the god; and the bull is the 'consecrated lord' of the quadrupeds.[147] The *Rāmāyaṇa* testifies to the use of bullocks for ploughing, riding and drawing the country carts[148] and, the *Mahābhārata* mentions that bullocks were utilized for ploughing.[149]

We know from Pāṇini[150] that the wooden club hanging from the neck of an out calf was called *prasaṇga*, and the calf so restrained while grazing was called *prāsaṅgya* (4.4.76). A calf of two years was called *dityavāh* (7.3.1). A calf above the ordinary and selected to grow as a stud bull was called *ārṣabhya*, i.e. 'good for becoming a bull' (51.14). As such, he was termed *jātokṣa*: 'growing up as a bull' (5.4.77), and was not castrated. Calves intended to grow as stud bulls are given special food and care. A young (*taruṇa*) bull was called *ukṣa* and more developed one *ukṣatara* (5391), while he was called *mahokṣa* when fully grown up (5.4.77) and *vṛddhokṣa* or *ṛṣabhatara* (5.3.91) when declining in age. At the age of two and a half years, the young bull gets his first pair of permanent teeth (*dvidan*): at the age of three, he becomes one with four teeth (*caturdan*). Similarly, a draught bull was *vatsa* in the first stage, *damya* when broken, and *balivarda* as a bullock (Patanjali on 1.11, 1.42). After he was three years of age, the bull was given a nose-string (3.2.25), and was broken (*danya*) and castrated (4.4.76,77,80,81), The draught bulls were classified according to their work—drawing a chariot (*rathya*), carrying a yoke (*yugya, dhurya and dhaureya*), drawing a cart or cartload (*śakaṭa*), yoked to a plough (*hālika* or *sairika*). An ox accustomed to be yoked both on right and left of the yoke was called *sarva-dhuriṇa*, and the one accustomed to one side only was called *ekadhuriṇa* (4.4078–79), the latter being of less worth. Pāṇinī mentions the famous *sālvaka* breed of bulls reared in the Sālva country (4.2.136), which was a large confederacy of six member-states (4.1.173), three among them being Ajamiḍha, Ajākranda and Bodha (4.1.170) and the *Janapada* comprised the vast territory extending from Alwar to Bikaner. The sālvaka breed of Pāṇinī seems to be the same as the celebrated kāgaurī bulls reared in the jungle-covered tracts of Nagaur in the Jodhpur and Bikaner regions. Other breeds of bulls are vāhika, kāccha and rānkva (4.2.134: 4.2.100).

BUFFALOES

Apart from the cow and her progeny, there are several references to the male and female buffaloes which are mentioned by the term *mahiṣa* and *mahiṣī*[151], respectively. It seems the female buffaloes were used for milk, while the males were used only for load carrying. Buffaloes were also reared for their dung, which was useful as manure, and the dry cakes whereof were used as good fuel for cooking and other purposes too[152]. But it is difficult to identify the breeds of the buffaloes of the Vedic period, as it is being done now, viz., *murrah, bhadawari, jāffarābādī, surtī, mehsaṇā, nāgpuri, nīlī*, according to their peculiar physiological characteristics, utility and habitats.[153]

Curiously enough, buffalo (*mahiṣa*) is mentioned as one of the progenies of Mṛgamaṇḍa, the wife of Pulaha, one of the eight mental-sons (*mānasa-putra*) of Brahmā, the creator of the universe.[154] The *Kūrma-Purāṇa* refers to buffalo in connection with the *śrāddha* ceremony of the ancestors (*pitṛ*) who are said to be gratified if served the meat of a buffalo.[155] At the same time, its milk was forbidden for the *srāddha* purposes.[156]

HORSES

The horse, too, is mentioned in several Vedic hymns and *mantras* as the servant of man in capacity of pack horse, riding horse, chariot horse, war horse, race horse and even ploughing horse, though the sacrificial horse finds a rare mention. Description of horses are always apt and often beautiful, showing a keen appreciation of its performance, its importance, a great love for the animal, good judgement of its points, and close observation of its nature and even its anatomy. The gods are said to go about mounted on fleet steeds, or drive in chariots drawn by majestic horses; horsemen or cavalry of some kind forms an adjunct to the chariot force in warfare expeditions and forays are carried out with men mounted on horseback. Messengers and travellers use the horse both for riding and for drawing carts and wagons; and horse racing was a great pastime, race courses being mentioned. Horses and mares figure among the gifts made to the priests—generally along with the chariots, by the kings and other wealthy donors. Trappings of many kinds—some highly ornamental and rich in the case of those belonging to the wealthy—were in use; spurs and heel goads were used by riders; and traces, saddles, reins, bridles, heel ropes, halters are all specifically mentioned. The horse was also sacrificed and offered to the gods, perhaps as a rare and extraordinary event of *Asvamedha*[157] by an emperor after he successfully carried out his expedition of subduing all the others rulers in the vast continent of Asia in general or in India in particular.

The horse was called *vāha*, *aśva* or *arvat*, while the mare was known as *asva* or *aravatī*, and the mule was called *aśvatara*.[158] Among the numerous references to the horses, the following few may be interesting; thus, the Ṛgvedic seer Budha Saumya recommends to 'satisfy the (... 28) horses, accomplish the good work of ploughing'.[159] The seer Dīrghātamā Aucathya says: 'thou hast the wings of the falcon and limbs of the deer' (said of the horse).[160] The seer Bārhaspat Bharadvāja prays: 'may your equally spirited steeds bear you hither'.[161] The Yajurvedic seer employs the simile of 'horses breaking through fences.'[162] The seer Garga Bharadvāja entreats: 'recover thou our cattle, Indra? bring them back? the drum sounds rapeatedly as a signal; our leaders assemble mounted on their steeds; may our warriors be victorious'.[163] The seer Pāyu Bharadvāja informs: 'whip with which the skilled warriors lash their thighs and scourge their flanks, urge the horses in battle'.[164] The Ṛgvedic seer Medhatithi Kaṇva prays: 'Asvins, you stir up the sacrifice whip that is wet with the foam of your horses, and lashing loudly'.[165] The seer Gāthina Viśvāmitra describes the waters of the rivers which 'rush like drivers of their chariots to the goal (in a race).'[166] The seer Ajigarta Śunaḥṣepa *alias* Vaisvāmitra Devarāta says: 'we soothe thy mind by praises for our good, as a charioteer his weary steed'.[167] The seer Garga Bharadvāja acknowledges: 'I have received ten horses, ten purses, clothes, and ample food; and ten lumps of gold; thou has given ten chariots, with their horses and one hundred cows to the priests'.[168] The seer Viśvamana Vaiyāśva also says in the same strain: 'We have received a well-going chariot of silver, yoked with a pair of horses'.[169] The Atharvavedic seer Cātana says: I have surrounded their places like a race course for horses'[170]. The seer Gṛtsamada Bhārgava Śaunaka prays to Śakunta (Indra), thus: 'bring good fortune to us from all sides, like a horse neighing when approaching a mare'[171]. The seer Śyavāśva Atreya says: 'They have harnessed their horses fleet as the wind'[172]. He further asks the Maruts thus: 'Where are your horses? Where are your reins? How do you move? Where are you going? The saddle is on the back of the steeds, the bridle in their nostrils. The goad is applied to their flanks, the drivers force them spread their thighs apart like women in bringing forth children'[173]. The horse for the *Aśvamedha* sacrifice is described in the *Ṛgveda* hymns (1.162-163) in all its gruesome details and not without a pathetic element[174].

The *Taittirīya Brāhmaṇa* refers to forty-eight types of horses,[175] which testifies to the care taken in rearing of this animal. If the cow predominated the homefront, then the horse held the pride of place in warfare. It was yoked to the chariot and was the swiftest mode of transport. The Upaniṣads mention good *saindhava* horses, bad horses, fresh horses, impatient horses tearing up their tethers, and so on.[176] The *Bṛhadāraṇyaka Upaniṣad* describes the anatomy of a horse as the universe.[177] The horse was highly venerated at the *Aśvamedha* sacrifice. Mules were also yoked to chariots and carriages.[178] The horse was also a common domestic animal used for riding purpose and the *Gobhila Gṛhya Sūtra* describes a ceremony which was to be performed by that person who wishes that his stock of horses and elephants may increase.[179] The *dhruvāśva* ceremony of the *Mānava Gṛhya Sūtra* is intended for the welfare of horses.[180]

Next to cow and ox, horse was the most important domestic animal and played a vital part in the country's economy during the age. Breeding of good horses was a matter deserving of special care. The steeds of Rāma were fleet horses of famous breeds[181]. Rāma had spotted from a distance the grateful, well-bred horses of his father[182] on seeing Bharata and his party advancing towards him. Special breeds of horses must have been bred for specific purposes. War horses used for cavalry had to be specifically trained[183]. Horse-breaking was an art with which princes like Rama were conversant.[184] Horses for war-chariots (*sāngrāmika-haya*)[185] had to be trained to make them withstand the din and clatter of the battlefield and execute desired movements. Of special note is Indrajit's war-chariot whose horses with intelligent movements (*vidheyāśvaḥ*)[186] in the battlefield after the death of their driver evoked admiration even from the enemy.[187] Rāvaṇa's chariot was drawn by well-broken horses.[188] Horses drawing war-chariots, were well decked with gold-trappings provided with armoury.[189] Horses of pleasure-car, the royal state chariot and ordinary riding chariots[190] belonged to different breeds and were trained differently in view of their different natures. Hill-bred horses (*parvatīya-turaṇgama*),[191] which were fast and sure-footed, were used for transport on uneven hill tracks. These were used by Bharata when hurrying home from the Kekaya region on Daśaratha's death. Speed was the most coveted quality in horses.[192] Horses used by couriers for summoning Bharata from the Kekaya province must have been extraordinary (*sammata*).[193] Horse-breeding must have been carried out in certain provinces (*desajataulaja*) as the most famous breeds of the day, viz., the tartary breed (*kāmboja*), the Bactria breed (*bāhlika*), the soinde breed (*nadīja*) and Arab breed (*vanayu*) indicate[194] that these breeds must have been probably imported originally from neighbouring countries and then broken locally.[195] Mule was the beast of burden in times both of peace and war.[196]

In the *Mahābhārata*, we find King Nala to be well-versed in the lore of horses, and was an expert connoisseur of horses as also extraordinarily endowed with the skill of riding; he learned the skill of dice playing (*akṣa-hṛdya-vidyā*) from King Ṛtuparṇa in exchange of his own *aśva* vidya[197]. Nakula too was an expert in horse-lore, and while introducing himself to King Virāṭa, during the thirteenth year of staying incognito on the part of the Pāṇḍavas, he said that he was looking after the horses of King *Yudhiṣthira*, and that he is conversant with the nature of horses, their training, the ways and means of curing their defects, bringing to book the indisciplined horses, and with the cures and treatment of sick horses.[198] The knowledge of *Aśvasūtra*, a work on horse-diseases, was considered essential for rulers.[199]

In the *Purāṇas*, the horse is said to have been born of the creator's feet, and is included in the list of domesticated animals.[200] There are many references in the *Vayu Purāṇa* to stables for horses and to horse riding.[201] Horses were good gifts for brahmins in *srāddha* and were regarded useful in sacrifices. A horse was one of the select jewels (*atiśaya-ratnāni*) of a Cakravartin in the Tretā age.[202] The horses of the Gandhara country were said to be the best amongst the breeds and the *Aśvinī-nakṣatra* was said to be favourable for acquisition of horses.[203] The *Kūrma Purāṇa*, too, provides similar references and prescribes expiations for killing them.[204]

Pāṇini terms horse and mare together as *aṣva-vadavā* (2.4.27) and mentions a special breed of mares from across the Indus (*pārevāḍavā*).[205]

As in the case of cows, much attention seems to have been bestowed upon the breeding of horses and more than one type evolved, each suited to one or other of the several uses to which horses were put. It is possible that horses were imported from central and southwest Asia, as a part of a regular import trade across the mountains. Horses, as we have seen, were being used largely and for such a number of purposes and were presented by kings in such large numbers that it is more reasonable to presume that they were local bred stock and not purchases from outside.[206]

ELEPHANTS

While we hardly find any reference to elephant in the *Ṛgveda*, there are a few in the *Yajurveda* and the *Atharvaveda*. Thus, Yajurvedic seer recommends that 'for the service of the king, valiant soldiers and elephants should be secured' and that 'an elephant should be known for the snowy mountains'.[207] The Atharvavedic seer Vaśiṣṭha admires the splendour of the elephant thus: 'With what splendor the elephant came into being, with what the king among men ... with that splendor do thou, O Agni, now make me splendid... How great splendour there is of the Sun, and of the asura-like elephant ... so great splendor let the Aśvins assign to me.' The elephant has become the superior of the 'comfortable wild beasts'.[208] The seer *Cātana* refers to the anger of an elephant thus: 'They two anger me, making a noise, as flies an elephant, them I think ill off like mites on man'.[209] The seer Atharvan prays for the brilliancy and odour of the elephant thus: 'What briliancy is in elephant (let her come to us in union with splendour)'; and 'What odor of thine is in human beings, what in wild animals and in elephants? What splendor, O Earth, in a maiden... with that do thou unite us also; let no one so ever hate us'[210]. The seer Kāṁkāyana observes that 'the elephant strains foot with foot of the she-elephant; as of a lustful man'[211]... The seer Bhṛgvāngiras refers to the steadfast footing of an elephant thus : 'built on the earth thou standest, like a she-elephant having feet'.[212] And the seer *Pratyangiras* refers to the might of an elephant thus: 'I quit an ill-natured magic made by the witchcraft-makers, as an elephant the difficult haze'.[213]

A couple of oblique references to elephant in the *Ṛgveda* indicate that wild elephants were domesticated and considered fit for riding by kings[214] and there is a mention of the hook (*ankuśa*) which was used for controlling the elephants.[215]

The *Śatapatha Brāhmaṇa* testifies to the domestication of elephants and their use for riding.[216] The *Gobhila Gṛhya Sūtra* describes a ceremony to be performed by the person who wishes his stocks of elephants to increase.[217,218]

The Valmiki *Rāmāyaṇa* shows that elephants were largely requisitioned both for civil and military purposes, and were trained on considerations of majesty, beauty and pedigree. Kings kept huge forest reserves for breeding elephants. Rāma asked Bharata specifically to protect these reserves.[219] Special tricks were employed for catching wild elephants and taming them. They were caught with trained elephants as decoy, and broken and tamed by those already tamed and trained.[220] They were also caught in pits covered with grass into which they were driven with torches and firebrands.[221] They were also harassed by *goads* and *tomaras*.[222] The elephant drivers were called *mahāmātras*.[223] Captured elephants were then tied with huge waistbands.[224] Sītā in the *Aśokavāṭikā* is compared to a captured she-elephant, separated from her mate, tied to a post, sighing deeply and in great distress.[225] The slopes of the Himalaya and the Vindhya mountains were reputed for the finest breeds of elephants.[226] The breeds from Airavata and Indrasira Hills were also noted for their beauty.[227] The famous breeds of the day were *airāvatakula, mahāpadmakula, anjanakula* and *vāmanakula*; and elephants of various species could be seen in Rāvaṇa's palace.[228] Elephants of the age of sixty years were considered enhanced in their vintage value.[229]

Regarding the use of elephants in war, the *Mahābhārata* informs us that the army consisted of four wings, viz., those of *rathī, gajārohī, asvārohī,* and *padati* and due to these four limbs (*aṅga*), the army was called *catur-aṅga*. Each *rathī* was accompanied by ten elephants.[230]

In the *Vāyu-Purāṇa* also, we find that although the elephant is included in the list of wild beasts, there are numerous references to suggest that they were tamed, and there were stables for elephants.[231] The words by which the elephants were known are *nāga, gaja, dantin, dvirada, kuñjara, karin, dvīpa, mātaṅga,* etc. The famous mythological elephant of the gods, known by the name Airāvata, was said to be the son of Irāvatī, the wife of Pulaha. The word Sāmaja for elephant connects the animal with the *Sāmaveda.* In the account of cosmology, we find them described as being born of the creator's feet. It appears that the term *gaja* for elephant signified the class of the species born of domesticated parents.[232] Elephants are also recommended as good gifts to brahmins in *śrāddha* and it is also considered as a favourite animal of Prājāpati in a sacrifice. Giving gifts under the shadow of an elephant was considered meritorious.[233] The Kūrma Purāṇa too gives similar information.[234]

Pāṇinī also refers to elephants by the terms *hastin, nāga, kuñjara, śunḍāra,* etc.[235]; the last term was applied to a trumpeting elephant with a prominent trunk. A herd of elephants was called *hāstika.*[236] The height of an elephant constituted a measure, and the expressions *dvi-hasti, tri-hasti*[237] denoted the height as high as two or three elephants one upon another. These words were used with reference to the depth of a moat or the height of a rampart. The tusk of an elephant was called *danta,*[238] used also as ivory. Strength to kill or shoot an elephant was a mark of valour, expressed by the term *Hastighna.*[239] A tusker was called *dantāvala,* and a goad was called *totra.*[240]

SHEEP AND GOATS

Sheep and goats are mentioned in the *Ṛgveda* only once, as *aja* and *avi,* respectively, in the famous *Puruṣa-sūkta,* and this same hymn is found in the *Yajurveda,* too.[241] They are also mentioned in one more *mantra* where cows, sheep and goats—and the essence of food are invoked—for welfare, peace and happiness.[242] The *Taittiriya Samhitā* mentions the origin of human beings followed by that of animals, among which the sheep and goats are referred to.[243] The *Kāṭhaka Saṃhitā* refers to sheep and goats as meant for sacrificing to Prajapati,[244] but it is not specified that they were killed for what purposes. The *Atharvaveda* notes that these animals were scared at the sight of a wolf,[245] and that the grass and the plants eaten by sheep and goat grow again and give peace,[246] thereby indicating the usefulness of their dung as manure for enhancing the fertility of the field. Rudra is said to have classified animals in five species, viz., cows, horses, men, sheep and goats.[247] The *Taittirīya Saṃhitā* recommends the rearing of sheep and goats for prosperity.[248] The *Śatapatha Brāhmaṇa* mentions that sheep and goats procreate thrice in a year and each time they give birth, the milk is about two or three litres.[249] They were also reared for the milk as also the wool from which warm clothes were made. The *Bṛhadāraṇyaka Upaniṣad* mentions that sheep and goats were reared for wool, meat and milk.[250] A goat (*chāga*) was sacrificed and its meat was cooked on the sacrificial fire and was partaken of by the priests and the *yajamāna.*[251] Large flocks of goats and sheep were kept by the people.

According to the *Vāyu-Purāṇa,* goat (*aja*) is born of the creator's mouth, and is included in the list of domesticated animals (*grāmya-paśu*), which is useful in sacrifice (as a victim) along with cow and other animals.[252] It is said that at the end of the Kali age, people will be keeping *aja, eḍaka, khara,* and *uṣtra* as their pets.[253] At the time of the birth of Skanda Kārtikeya. Vāyu gave him *mayūra* and *kukkūṭa.* Lord Svayambhu Śiva gave him *meṣa* as presents.[254] But, the milk of sheep and goats was a taboo in *śrāddha.*[255] The *Kūrma Purāṇa* too gives similar information.[256] Sheep is said to have been born of the breast of Brahma, but its meat is said to have been used for *srāddha* purposes.[257] In the days of Pāṇinī, a herd of goats was called ājakā.[258] Goats and sheep together were called *ajāvi* and *ajaida. Jābāla* denoted a goatherd, *mahājābāla* was one who owned a big sheep-run.[259] Sheep was termed both *avi* and *avika,* and a flock of rams is referred to as *aurabhrakā.*[260]

CAMELS

Yajurveda refers to camel as the diverse coloured animal belonging to the Āditya (months).[261] There is an indirect reference to taming of camel in the *Yajurveda,* when the seer prays : 'let the wild, uncontrollable camel be put to grief by thee.' [262] The Ṛgvedic seer Brahmātithi Kāṇva says 'Become apprised, Aśvins, of my recent gifts, how that Kaśu, the son of Cedi, has presented me with a hundred camels and ten thousand cows'[263]... The seer Yatsa Kaṇva acknowledges... The exalted prince ... has given camels laden with four (loads of gold).[264] And, the seer Vaśa Aśvya, too, confesses : 'I have received... tens of thousands; a score of hundreds of camels[265]... The *Atharvaveda* refers to Nārāśaṃsa as having a caravan of twenty camels with their females.[266] The Ṛgvedic seer Parucchesa Daivodāsi refers to the use of camels in war, when he says: '... I exalt you, Pūṣan, that you may bear us across the combat, like a camel'.[267]

The Vālmiki *Rāmāyaṇa* shows that among the several divisions of the army, camels too comprised one. Rāvaṇa is said to have had one regiment of camels.[268] Camels were employed for other purposes also, such as being yoked to chariots, and ridden for pleasure and trade,[269] though there is no reference to the habitat of camels. According to both the Vāyu Purāṇa and the Kūrma Purāṇ, camels are said to be born of the Creator's feet.[270] The milk of the camel was a taboo in *srāddha.*[271] Pāṇinī refers to camel crops by the term *auṣṭraka,* to a young camel restrained by a chain during infancy, to camel riders and the army units employed for quick transport.[272] He also mentions *auṣṭraka* as the name of articles made from the body parts of dead camels, such as large and small sacks made of camel hair, and leather jars of large and small size, made of hides and intestinal integuments of camels.[273]

ASSES

The ass has been referred to in the *Ṛgveda, Yajurveda* and *Atharvaveda* by the word *gardabha.* Thus, the Ṛgvedic seer Kṛṣṇa Āngirasa urges to yoke the assess to their firmly built chariot.[274] The Yajurvedic seer refers to an ass as 'the long-eared animal'.[275] The Ṛgvedic seer Pṛsadhra Kāṇva acknowledges that Pūtakratu made a gift of a hundred asses to him.[276] But the Ṛgvedic seer Gathina Vaiśvāmitra considers an ass as a representative of a foolish person when he says: 'the wise condescend not to turn the foolish into ridicule; they do not head an ass before the horse.[277] But the seers of both the *Ṛgveda* and *Atharvaveda* dislike the noise of the ass. Thus, the Ṛgvedic seer Ajigarti Śunaḥśepa prays to Indra to destroy the ass (i.e. his adversary), praising with such discordant speech.[278] The Atharvavedic seer Mātṛnamā prays: 'They who dance around the dwellings, making donkey-noise,... these, O herb, with thy *Atharva*-smell do thou make to disappear scattered'.[279] The Athavavedic seer Bhutarnsa Kasyapa orders the evil spell to 'step away, making much noise like an unfastened she-ass, and attain thy makers, pushed hence by an energetic spell'.[280] On the other hand, the seer Śakra retaliates the evil spell directed against his ass and, the Atharvavedic seer Atharvāngiras prays to his deity *Śepa Arka* thus: 'As much as a limb as is that of the Parasvant, that of the elephant and that of the ass,[281] as great as of the vigorous horse, so great let thy member grow.[282] This is a spell meant for regaining virility, for rejuvenation.

The *Pāraskara Gṛhya Sūtra* refers to the ass as a domestic animal used for riding and as a beast of burden.[283] The *Gṛhya Sūtras* as well as the *Dharma Sūtras* lay down that the student who violated his vow of chastity should sacrifice as ass at the crossing of roads.[284] It was called *avakirṇi prāyaścittam.* The Vālmīki *Rāmāyaṇa* informs us that chariots were drawn by asses as well.[285] Asses, protected by breast-plates, drove the chariot of Rāvaṇa.[286] Fast-footed asses seem to have been bred in Kekaya country.[287] The *Vāyu Purāṇa* regards the ass as a rural domestic animal (*grāmya-paśu*) and the *Kūrma Purāṇa* says that its flesh is forbidden for food.[288] Pāṇini mentions asses in connection with stables for asses called *khara-śalā* (4.3.35).

DOGS

In the *Yamasūktas* of the *Ṛgveda,* we are introduced to the famous dogs of Yama, who guide the departed souls to his domains. Thus, the Ṛgvedic seer Vaivasvata Yama urges the souls of the departed ones thus: 'Pass by a secure path beyond the two spotted four-eyed dogs, the progeny of Saramā, and join the wise *pitrs* who rejoice joyfully with Yama.' He prays to Yama thus: 'Entrust him, O King, to your two dogs, which are your protectors, Yama, the four-eyed guardians of the road, renowned by men'.[289] The Atharvavedic seer Atharva praised the greatness of the heavenly dog thus: 'What the greatness of the heavenly dog with that oblation would we pay worship to them'.[290] The Ṛgvedic seer Indra presents the dog as the enemy of the boar thus: 'may the dog which chases the boar (seize) him by the ear (and) devour him'.[291] The Atharvavedic seer Atharva alludes to the fondness of the dogs for flesh and bones thus: 'Not as it were stuck in the flesh, not in the fat, not as it were in the arrows, let the spotted sling after birth come down, for the dog to act, let the afterbirth descend'.[292] But they are afraid of lions, as is observed by the Atharvavedic seer Cātana, 'like dogs on seeing a lion, they do not find a hiding place'.[293] The seer Atharva alludes to the practice of throwing victims to dogs, thus: 'Whoever shall curse us not cursing, and whoever shall curse us cursing, him withered, I cast forth for death, as a bone for dogs'.[294] But the same seer expresses his dislike for the practice, thus: 'make ye not bodies to the dog...away let the dogs go... To Rudra's howl making, unhymned, swallowing, great-mouthed dogs I have paid this homage.[295] The dogs were trained to guard the houses and their residents against thieves. Thus, the Atharvavedic seer Brahmā in his famous Prasvāpinī hymn says: 'The wind bloweth not over the earth; no one so ever seeth over (it); both all women and dogs do thou make to sleep, going with Indra as companion...let the mother sleep, the father sleep, the dog sleep, the housemaster sleep; let the relatives of her sleep; let the folk round about sleep'.[296] But the Ṛgvedic seer Agastya Maitrāvaruṇi disliked the barking of the dogs; he prays to Aśvins thus: 'annihilate the dogs who bark at us, slay them warring (against us), you know their (means of harm)...'[297] The Ṛgvedic seer Andhigu Śyāvāśvi urges: 'do you friends destroy the long-tongued dog'; 'drive off the dog that sacrifices not as the Bhṛgus drove off Makha.'[298] The Yajurvedic seer has a very low estimate about the character of dogs, when he says: 'He, low in character like a dog, is an enemy and must be checked by thee.'[299] One is surprised to find the seer Atharva saying: 'The gods, confounded, both sacrificed with a dog and sacrificed variously with limbs of a cow...'[300] The seer Brahmā warns: 'Let not the dark and the brindled one, sent forth, (seize) thee, that are Yama's dogs, red, defenders.'[301] The seer Atharva too prays: 'Who have defences, who have no defences, and the enemies who have defences—all those , O Arbuda, being slain, let dogs eat on the ground.'[302] But Yama's dogs are always friendly and protective;

thus, this same seer says: 'Run thou past the four-eyed, brindled dogs of Saramā, by happy road. 'What two defending dogs thou hast, O Yama, four-eyed, sitting by the road, man-watching, with them, O King, do thou surround him.'[303] 'Strangely, one is further surprised when one finds the Ṛgvedic seer Vāmadeva Gautama declare: In extreme destitution I have cooked the entrails of a dog; I have not found a comforter among the gods...'[304] The *Śatapatha Brāhmaṇa* mentions that dogs were domesticated for they scared away the thieves and other creatures.[305] A reference to the dogs occurs in a seemingly satirical passage of the *Chāndogya Upaniṣad* (1.12) but the Upaniṣadic people must have found them of great help for guarding their sheep, cattle and fields.[306] In the *Kalpasūtras* too, we find that the dog was, of course, the faithful servant of the people.[307]

The *Rāmāyaṇa* affords us valuable information that special breeds of ferocious dogs were bred in the Kekaya country. They were brought up in the inner apartments, large of body, tigers-like in prowess, with teeth as weapons.[308] Considered unclean and unfit to enter into houses and sacrificial grounds,[309] they were put to use only in hunting.[310] The incident of the dog that accompanied Yudhiṣṭhira to heaven in the *Mahābhārata* is well known. The *Purāṇas* consider the dog as a taboo in *śrāddha.* It is not to be allowed anywhere in the vicinity on that occasion. Food smelt by a dog is not to be eaten, and expiations are to be undertaken for killing a dog and eating its flesh.[311]

Pāṇini mentions the dog as one of the carnivorous animals (4.4.11). Domestic dogs bred in royal kennels were called *kauleyaka* (4.2.96).

PIGS

A pig is called *varāha* in the Vedic texts. The *Ṛgveda* mentions the pig as strong, having a bright family and purifying,[312] thus indicating the fact that female pigs roam with noise and in company of the young ones and keep the surroundings clean by eating up the filth. The *Taittiriya Saṃhitā* equates the pig with Viṣṇu who keeps Lakṣmi.[313] It also mentions the myth that Prajāpati assumed the form of a pig and took away the water, which he transformed into the earth in capacity of Viṣvakarmā.[314] The *Ṛgveda* refers to the incident of Viṣṇu shooting a pig with an arrow,[315] indicating that it was usual to kill the pigs by hunting. The *Śatapatha Brāhmaṇa* refers to the fat of the swine and sandals made of its skin.[316]

VEDIC SOLUTIONS TO THE PROBLEMS OF ANIMAL HUSBANDRY IN INDIA

According to the Vedic system of social management and structure, agriculture and animal husbandry were never an industry. Being a means to sustain life, agriculture and animal hunbandry have been a means to produce foodgrain so that both the human beings and other domestic creatures can be fed and can sustain their life. Thus, the ancient concept of cow breeding never viewed cow or cow milk from the commercial angle. The purpose of cow breeding was to rear it as a family member and to obtain from it the necessities of the day-to-day life such as milk, pure ghee and dung-cake fuel. The same concept has led to formulation of various social and religious customs and views to regulate cow breeding consistent with the geographical condition of India. But fresh milk and pure ghee started vanishing from the country after 1940, as a result of slaughter of cows in large number for export of meat to obtain foreign exchange for various modern national developments and plan projects. All our problems pertaining to agriculture and animal husbandry are rooted in our modern concepts of progress, and the Vedic solutions to them have been discussed in detail by the present author elsewhere.[317]

NOTES AND REFERENCES

1. K.M. Munshi, 1964, Foreword to *Agriculture in Ancient India,* Indian Council of Agricultural Research, New Delhi.
2. M.S. Randhava, 1958, *Agriculture and Animal Husbandary in India,* ICAR, New Delhi, p. 127.
3. D. Saghavan, 1964, *Agriculture in Ancient India,* ICAR, New Delhi, pp. 116–117.
4. Haripada, Chakraborty, *Socio-Economic Life of India in the Vedic Period,* Sanskrit Pustak Bhandar, Calcutta, 1986, p. 199.
5. *Ṛgveda (RV)*, 10.19.3–5.
6. *RV.*, 713.3.
7. *RV.*, 6.53–56, 58; 6.53.9.
8. *RV.*, 1.42.8.
9. *RV.*, 3.45.3.
10. *RV.*, 6.19.3.
11. *RV.*, 6.244.
12. *RV.*, 6.54.5–7.
13. *RV.*, 10.19.5.
14. *RV.*, 10.1691
15. *RV,* 10.105.5–7.
16. *Atharvaveda (AV)*, 3.14.
17. *AV.*, 2.26.
18. *Kauśika Sūtra,* 1914.
19. *RV.*, 3.14.1–6.
20. *AV.*, 2.14.2.
21. *AV.*, 2.26.1–5.
22. *AV.*, 6.52.2.
23. *AV.*, 9.4.19.
24. *AV.*, 13.1.19.
25. Haripada Chakraborty, *op. cit.*, pp. 200–201.
26. Dr. Balavir Acharya, 1991, *Ṛgvediya Brāhmaṇo kā Sāṃskṛtika Adhyāyana (Hin),* Vidyanidhi Prakashan, Delhi, p.122? Sat. Br., 1.8.1.38.
27. *Kau. Br.*, 1.1.1.5; 1.1.5.2; *Ait. Br.* 39.8–9.
28. *Kaus. Br.*, 18.1; *Ait. Br.*, 3.3; 17.3; 29.2; 39.8–9.

29. *AV.*, 1.16.4.
30. *Vālmīki Rāmāyaṇa (VR)*, Gita Press, Gorakhpur, V.S2017, 2.41.9.
31. *VR.*, 1.14.50; 18.20; 72.21.
32. *VR.*, 1.54.
33. *Mahābhārata (MBh)*, edited by Pt. Paneanan Tarkaratna, 1826, Calcutta, S.S. vir.pv.
34. *Śatapatha Brāhmaṇa* (*Sat. Br.*) edited by A.Weber, London, 1885, Eng. Tr. by J Eggling, S.B.E., Vol.12, 26, 41, 43, 44, Oxford, 1882–1900, 14.1.2.19; 12.7.2.9; 12.5.2.3; 1.6.4.15; 13.2,2.19.
35. *Sat Br.*, 1.2.4.16; 2.3.4.26.
36. Dhirendrakumar Sinha, *Brāhmaṇa-Granthoṅ Meṅ Pratibimbita Samāja Evam Saṃskṛti (Hin.)* (B&PSS) Penman Publishers, Delhi, 1990, p.198. rc Hvp.
37. Devendrakumar Rajaram Patil, 1973, *Cultural History From the Vāyu Purāṇa*, Motilal Banarsidass, Delhi, pp. 237–238.
38. V.S Agrawala, 1953, *India as Known to Pāṇinī*, University of Lucknow, p. 153.
39. A.K. Yegna Narayan Aiyer, 1940, *Agriculture and Allied Arts in Vedic India*, (AAAVI), Bangalore City, pp. 40–41.
40. *Ibid.*, p. 41.
41. *Ibid.*
42. *RV.*, 331.15.
43. *RV.*, 4.32.18.
44. *RV.*, 10.44.10.
45. *RV.*, 10.42.7.
46. *YV.*, 13.490
47. *RV.*, 1.10.7.
48. *RV.*, 5.34.5.
49. *AV.*, 4.21.1–7.
50. *RV.*, 1,120.9.
51. *RV.*, 6.48.15.
52. *RV.*, 5.42.8.
53. *RV.*, 8.40.10.
54. *RV.*, 8.46.22.
55. *RV.*, 6.28.3.
56. *YV.*, 19.86.
57. *RV.*, 8.69.3.
58. *RV.*, 1.164.26.
59. *RV*, 9.69.1.
60. *RV.*, 10 42.2.
61. *RV.*, 1.20.
62. *RV.*, 10.39.13.
63. *RV.*, 8.18.8
64. *Aitareya Brāhmaṇa (Ait. Br)*, 1952, Edited by Satyavrat Samashrami, Calcutta, 18.3.
65. *Ibid.*, 322–4 RAO.
66. Dr S.N. Vyas, 1967, *India in the Rāmāyaṇa Age*, Atma Ram & Sons, Delhi, p. 241.

67. Sukhamay Bhattacharya, 1966, *Mahābhārat Kalina Samaja* (*Hin*) (Mks) Lokabharati Prakashan, Allahabad, p. 165.
68. *MBh. Udyo. Pv.*, 38.12.
69. *MBh. Anuśā. Pv.* 85.17–22; 69.7–8; 126.29; 95.117; 94.52.
70. *MBh. Van. Pv.* 2.57; *Anusa.* Pv. 78.16; 162.42.
71. *BEBh Anuśā. Pv.* 155; *Udyo.* Pv. 58.12.
72. *MBh., Anuśā. Pv.* 71.51; 75.42.
73. *MBh., Anuśā._Pv.* 69.11; 146.48; 79.5; 78.19; 81.55; 128.9.
74. Sukhamay Bhattacharya, *MKS*, p. 168; *MBh., Anusa.Pv.*125.50; *San. Pv.*195.18.
75. *MBh., Anuśā, Pv,* 78 19–?1
76. *MBh., Anuṣā, Pv.* 78.5–8.
77. *MBh, Anusa, Pv.* 78. 17? 74.4.
78. *MBh., Vir, Pv.* 10e5 10.14.
79. *MBh., Bhi. Pv.* 42.44.
80. D.R. Patil, p. 116; *VR.*, 9.42–44.
81. *VR.*, 23.5–12.
82. D.R., Patil, CHVP, p. 116
83. G.K. Pai, *History from the Kūrma Purāṇa CHKP,* Sukrtindra Oriental Research Institute, Cochin, 1975, T 305 Kur.Pu. 2.33.39
84. *Kūrma Purāṇam*: (*Kur.Pu*) edited by Dr Ramashankar Bhattacharya, Indological Book House, Varanasi, 1968, 2.34.29.
85. *Kur. Pu.*, 2. 26.45?4–1.2–24.
86. *Kur. Pu,* 2.22.1; 22.14; 22.49; 2,18,56; 2.27.1; 32.; 34.4,16,23,35.
87. V.S. Agrawala, *IKP,* pp. 222–223.
88. *Ibid.*, pp. 223–224.
89. *Sat. Br.*, 2.2.39.
90. *Matrāyaṇī Saṃhitā (Mait, Sam)*, ed., Pandit Shripad Damodara Satavalekar, Svadhyay Mandal, Pardi, 1983; 7.5.3.1.
91. *Sat. Br.*, 12.7.2.3; 12.71.14.
92. Ram Gopal: *India of Vedic Kalpasūtras* (IVKS), National Publishing House, Delhi, 1959, p. 135.
93. *Bṛhadāraṇyaka Upaniṣad,* (*Br. Up.*), 3.1.1.2; 4.1.2.7.
94. *Chāndogya Upaniṣad* (*Ch. Up.*), 445.
95. *Kaṭha Upaniṣad (Kath. Up.*), 1.1.23.
96. *Ch.Up.*, 2,11.
97. *Kauṣītaki Upaniṣad (Katu. Up.*), 4.8.
98. *Ch. Up.* 2.22.2.
99. Ram Gopal, IVKS, p. 135.
100. *Kauśīka Sūtra (Kau.Su.*), 4118–20.
101. *Kau. Su.*, 51.1–6.
102. *Kau. Su.* 19. 1–3: *Kathaka Grhya Sutra (Kath. G. Su.*), 58.3.
103. Aiyer, *op. cit.*, p. 43.
104. *YV.* 24. 5–19.

105. Randhava, *op. cit.*, pp. 288–289.
106. *RV.*, 2.32.3.
107. *RV.*, 1.152.6.
108. *RV.*, 1.72.8.
109. *RV*, 18.65.
110. *RV.*1.28.4.
111. *YV.*, 10.8.
112. *RV.*, 8.8.15.
113. *Tattiriya Saṃhita* (*Tait. Sam*), ed., Pandit Shripad Damodar Satavalekar, Svadhyay Mandal, Pardi, 1983, 2.5.3.5; 2.3.10.1.
114. Aiyer, *op. cit.*, p. 44.
115. *YV.*, 5.28.
116. *YV.*, 12.70.
117. *RV.*, 4.58.6.
118. Aiyer, *op. cit.*, pp. 44–45.
119. *RV.*, 10.27.8.
120. *RV.*, 6.28.7.
121. *RV.*, 10.101.5.
122. *RV.*, 1.32.2.
123. *RV*, 2.16.8.
124. Aiyer, *op. cit*, p. 45.
125. *YV.*, 12,84.
126. *RV.*, 755.5.
127. *RV.*, 6,54.7.
128. *RV.*, 6.28.7.
129. *RV.*, 64751.
130. Aiyer, *op. cit.*, p. 46.
131. *YV.*, 3.59.
132. *AV.*, 8.723–25.
133. *Krsi-Parasara (KP)*, 89-92.
134. *Agni-Purana*, 292–33.
135. *Viṣṇu-dharmaottara-Purāṇa*, 2.31.1–27.
136. S.P., *et al.*, Raychaudhari, *A Concise History of Science in India*, I.N.S. Academy, New Delhi, Chap., 6, Agriculture, pp. 363–364.
137. *V.S.*, Agrawala, IKP, pp. 226–227.
138. *AV* 6.41.2–3; 12.4.6.
139. *Journal of Vedic Studies*, Lahore, 1934, p. 166.
140. *Mbh. Van. Pv Ghyoasatra* 239.41 240.4.
141. *Vedic Index*, 5.1.46.
142. *RV.*, 10.62.6.
143. Allen, *Index of Punch-marked Symbols*, Index IV.
144. D. Raghavan, *AAI*, pp. 124–125.
145. *Ibid.*, pp. 125–126.

146. *Vayu-Purāṇa (Vay. Pu.)*, 30.261; 54.107–108; etc. (Ananda Rama Sanskrit Series, 1905).
147. D.R. Patil, *CHVP*, pp. 116–117.
148. *VR.*, 2.74.23; 2.70.29; 2.83.16 pv.
149. *Mbh., Anuśā RV* 83. 18.
150. Agrawala, *IKP*, pp. 224–225.
151. *RV* 1.121.2 5.29.8; 5.37.3; *YV*, 12.105; *Tait. Sam*, 3.4.11.
152. *RV*, 1.16110; *Jaiminiya Brahmaṇa*, 2.267; *YV*, 22,8.
153. Randhava, *AAHI*, pp. 298–301.
154. Vay. Pu. 69.207; Citrava, M.M. Siddhesvara Shastri, 1964, *Prācina Caritra Kosa, Bharatiya* Caritra Kosa Mandala, Poona, P.438, Col. 2.
155. *Kur. Pu.* 11. 20.42.
156. *Ibid.*, II. 20.47.
157. Aiyer, *op. cit.*, p. 47.
158. *RV.*, 1.38.12; 1.73.9; 9.107.8; 105.2; 10.101.7; *AV*, 10.4.21; *Tait. Sam.* 71.1.2; *AV.*, 4.4.8.
159. *RV.*, 10.101.7.
160. *RV.*, 1.163.1.
161. *RV.*, 6.69.4.
162. *YV*, 29.44.
163. *RV.*, 6.47.31.
164. *RV.*, 6.75.13.
165. *RV.*, 1.22.3.
166. *RV.*, 3.36.6.
167. *RV.*, 1.25.3.
168. *RV.*, 6.47.23–24.
169. *RV.*, 8.25.22.
170. *AV.*, 2.14.6.
171. *RV.*, 2.43.2.
172. *RV.*, 5.58.7.
173. *RV.*, 3.61.2–3.
174. Aiyer, *op., cit.*, p. 48.
175. Balavir Acharya, *RBSA*, p. 124; *Tait. Br.* 3.8.1; *Tait. Sam.* 7.3.17.
176. Shubhra Shanna, 1985, *Life in the Upanishada* (LIU) Abhinav Publications, New Delhi, p. 151; *Br. Up.* 6.2.13; *Kath. Up.* 1.3.5; *Sv. Up.* 2.9; *Ch. Up.*, 5.1.12.
177. *Br. Up.* 1.1.
178. *Ch.Up.*, 4.2.3–4.
179. *Kaus. Su.*, 17.22; *Par. G.Su* 315.1–4; *Hir. G.Su* 1.12.1; *Ap.G.Su.* 8.22.16–17; *Gobho.G.Su.* 4.5.20.
180. Ram Gopal, *IVK*, p. 107.
181. *VR.*, 2.45.14
182. *VR.*, 2,97.24.
183. *VR.*, 2.83.5.

184. *VR.*, 2.1.28.
185. *VR.*, 6.43.3.
186. *VR.*, 6.8.2.2.
187. *VR.*, 6.90.28–29.
188. *VR.*, 6.11.30.
189. *VR.*, 3.22.22.
190. *VR.*, 5.6.4–5; 2.26.15; 2.97.24; 2.39.10.
191. *VR.*, 2.71.14.
192. *VR.*, 2.45.140.
193. *VR.*, 2.68.10.
194. *VR.*, 1.53.19; 1.6.22.
195. S.N., Vyas, *IRA.*, pp. 241–242.
196. *VR.*, 343.41.
197. *MBh., Van. Pv.* 72.28; *Vir. Pv* 12.6–7.
198. Sukhamay Bhattacharya, *MKS,* p. 165.
199. *MBh. Sabh.* Pv, 5.120.
200. *Vay, Pu.*, 9.42–44; 9.46–47.
201. DR., Patil, *CHW,* p. 115.
202. *Vay. Pu.* 80.15; 57.68.
203. *Vay. Pu,* 99.10; 82.14.
204. *Kur. Pu.*, 2.26.45; 2.33.31.
205. Agrawala, V.S., ICT., p. 219.
206. Aiyer, *op. cit.*, p. 48.
207. *YV* 29.30.
208. *AV.*J 3.22.3–4,6.
209. *AV.*, 4.36.9.
210. *AV.*, 6.38.2; 12.1.25.
211. *AV,* 6.70 2.
212. *AV.*, 9.3.17.
213. *AV.*, 10.1.32.
214. *RV.*, 4.4.1;1.138.2.
215. *RV.*, 10.44.9.
216. *Sat. Br.*, 1.2.3.9.
217. *Gobh.Gr. Su.* 4.5.29.
218. *VR.*, 5.6.32.
219. *VR.*, 2.100.50.
220. *VR.*, 356.31; 6.16.6–8.
221. *VR.*, 5.47.20; 2.21.54; 6.13.19; 6.24.38.
222. *VR.*, 2 74 .35.
223. *VR.*, 5.6.5.
224. *VR.*, 5.1.64.
225. *VR.*, 5.19.18.
226. *VR.*, 1.6.23.

227. *VR.*, 270.23000.
228. *VR.*, 1.6.24.
229. *VR.*, 2.67.20.
230. Sukhamay Bhattacharya, *MKS*, p. 482.
231. *Vay. Pu.* 10.83.
232. *Vay. Pu.* 69.211; 231–234; 214.
233. *Vay. Pu.* 5916; 80.15,45.
234. G.K., Pai, *CHKP.*, p. 304.
235. Pāṇinī's *Aṣṭadhayāyī* 5.2.133; 2.1.62; 5.3.88.
236. *Ibid.*, 4.247.
237. *Ibid.*, 52.38.
238. *Ibid.*, 5.2.113.
239. *Ibid.*, 3.2.54.
240. *Ibid.*, 32182.
241. *RV*; 10.9001; YV., 31.8.
242. *YV.*, 3.43.
243. *Tait. Sam.*, 6.5.10.1.
244. *Kath. Sam.*, 29.10.97.
245. *AV*, 5.21.5.
246. *AV.*, 8.7.75.
247. *AV.*, 11.2.9.
248. *Tait. Br.*, 3.9.8.3.
249. *Sat. Br.*, 4.5.5.6.
250. *Br. Up.*, 1.4.4.
251. *Sat. Br.*, 3.8.2–4.
252. *Vay. Pu.* 9.43,46; 59.16.
253. *Vay. Pu.* 90.406.
254. *Vay. Pu.* 72.46.
255. *Vay. Pu.* 78017.
256. *Kur. Pu.*, 1.7.55; 2.10.46.
257. *Kur. Pu.*, 1.7.54; 2.20.40.
258. *Pa. Ast.*, 4.1.4; 4.2.39.
259. *Ibid.*, 21.16; 6.2.38.
260. *Ibid.*, 5.1 8; 5.4.28; 4.2.39.
261. *YV.*, 24.39.
262. *YV.*, 13.50.
263. *RV.*, 8.5,37.
264. *RV.*, 8.6.48.
265. *RV.*, 8.46.22.
266. *AV.*, 20.127.2.
267. *RV.*, 1.138.2.
268. *VR.*, 6.53.5.
269. *VR.*, 2.70.29; 2.82.32.

270. *Vay. Pu.*, 9. 42–44; *Kur. Pu.*, 2.34.12.
271. *Vay. Pu.*, 78.17; *Kur. Pu.*, 1.756.
272. *Pa. Ast.*, 4.239; 5.2.79; 6.2.40.
273. *Ibid.*, 4.3.157; 5.3.90, 89.
274. *RV.*, 885.7.
275. *YV.*, 24.40.
276. *RV.*, 8.56.3.
277. *RV.*, 3.53.23.
278. *RV.*, 1.29.50.
279. *AV.*, 8.6.10.
280. *AV.*, 10.114.
281. *AV.*, 5.31.3.
282. *AV.*, 6.72.3.
283. *Par.G.Su.* 5.15.6.
284. *Ibid.*, 5,112.
285. *VR.*, 5.31.5; 6.51.26; 3.55.6.
286. *VR.*, 364. 46.
287. *VR.*, 2.70.25.
288. D.R., Patil, *CHVP.*, p. 115; *Kur. Pu.* 2.17.53.
289. *RV.*, 1014.10–11.
290. *AV.*, 6.80.1,5.
291. *RV.*, 10,86.4.
292. *AV.*, 1.11.4.
293. *AV.*, 4.36.6.
294. *AV.*, 6.57,3.
295. *AV.*, 11.2.2,11,50.
296. *AV.*, 4.5.2,6.
297. *RV.*, 21.182.4.
298. *RV.*, 9.101.1,13.
299. *YV.*, 22.5.
300. *AV.*, 7.5.5.
301. *AV.*, 8.1.9.
302. *AV.*, 11.10.2.
303. *AV.*, 18.2.11,22.
304. *R.V.*, 4.18.13.
305. *Sat.*, *Br.*, 11.1.51.
306. Shubhra, Sharma, *LIU,* p. 142.
307. Ram Gopal, *IVK.* p. 107.
308. *VR.*, 2.70. 20.
309. *VR.*, 7.18.6.
310. *VR.*, 3.55.5.
311. *Vay. Pu.* 78.38–40; *Kur. Pu.* 2.22.34; 2.17.33; 2.33.29–30; 2.34.8–18.
312. *RV.*, 9.97.7.

313. *Tait. Sam.*, 6.24.2.
314. *Ibid,* 7.5.1.1.
315. *RV.*, 1.61.7.
316. *Sat. Br.*, 4.3.19.
317. N.M. Kansara, 1995, *Agriculture and Animal Husbandry in the Vedas,* Dharam Hinduja International Centre of Indic Research, pp. 248–274.

SECTION THREE

AGRICULTURE FROM THE CHALCOLITHIC PHASE TO c. 600 BC

CHAPTER 20

Agricultural Background of the Chalcolithic Cultures of Central India and the Deccan

Ranjit Pratap Singh

The cultural gap or the 'Dark Period' after the fall of the Indus Civilization around 1700 BC and the emergence of the Mauryan empire has been filled by extensive explorations and excavations in different parts of central India and Maharashtra. This fieldwork has resulted into the discovery of at least five regional cultures with their characteristic ceramic industries. These are:

1. The Kayatha culture in the central Malwa region of the Chambal valley (*circa* 2100–1800 BC).
2. The Savalda culture in the Tapi and Pravara valleys (*circa* 2000–1700 BC uncalibrated; 2300–2000 BC calibrated).
3. The Ahar or Banas culture in south-eastern Rajasthan (*circa* 2100–1400 BC uncalibrated; *circa* 2600–1500 BC calibrated).
4. The Malwa culture in Central and Western India (*circa* 1700–1500 BC).
5. The Jorwe culture in the northern Deccan (*circa* 1500–900 BC).

Except the Jorwe culture, all these cultures shared a common level of subsistence economy and technology, but were distinguished from each other by a distinctive painted pottery. Their equipment comprised blades and microliths of siliceous material. Copper was extremely scarce except at Ahar. These people raised a variety of crops, the principal cereal being barley and, to a lesser extent, wheat. These cultures were represented essentially by peasant agricultural settlements without any pretensions to urbanity (Thapar 1985: 84). A summary of the essential features of each feature and its agricultural background is presented below:

THE KAYATHA CULTURE
(*Circa* 2100–1800 BC)

The type-site of this culture is Kayatha, situated on the right bank of the Choti Kali Sind, a tributary of the Chambal, about 25 km east of Ujjain. Excavations conducted by

Wakankar in 1965–66 and by Dhavalikar and Ansari in 1968 revealed 12-metre thick occupational strata, divided into five cultural periods labelled from bottom upwards. These are Period I–Kayatha culture; Period II–Ahar culture; Period III–Malwa culture; Period IV–Early Historical culture and Period IV–Sunga-Kushan-Gupta culture.

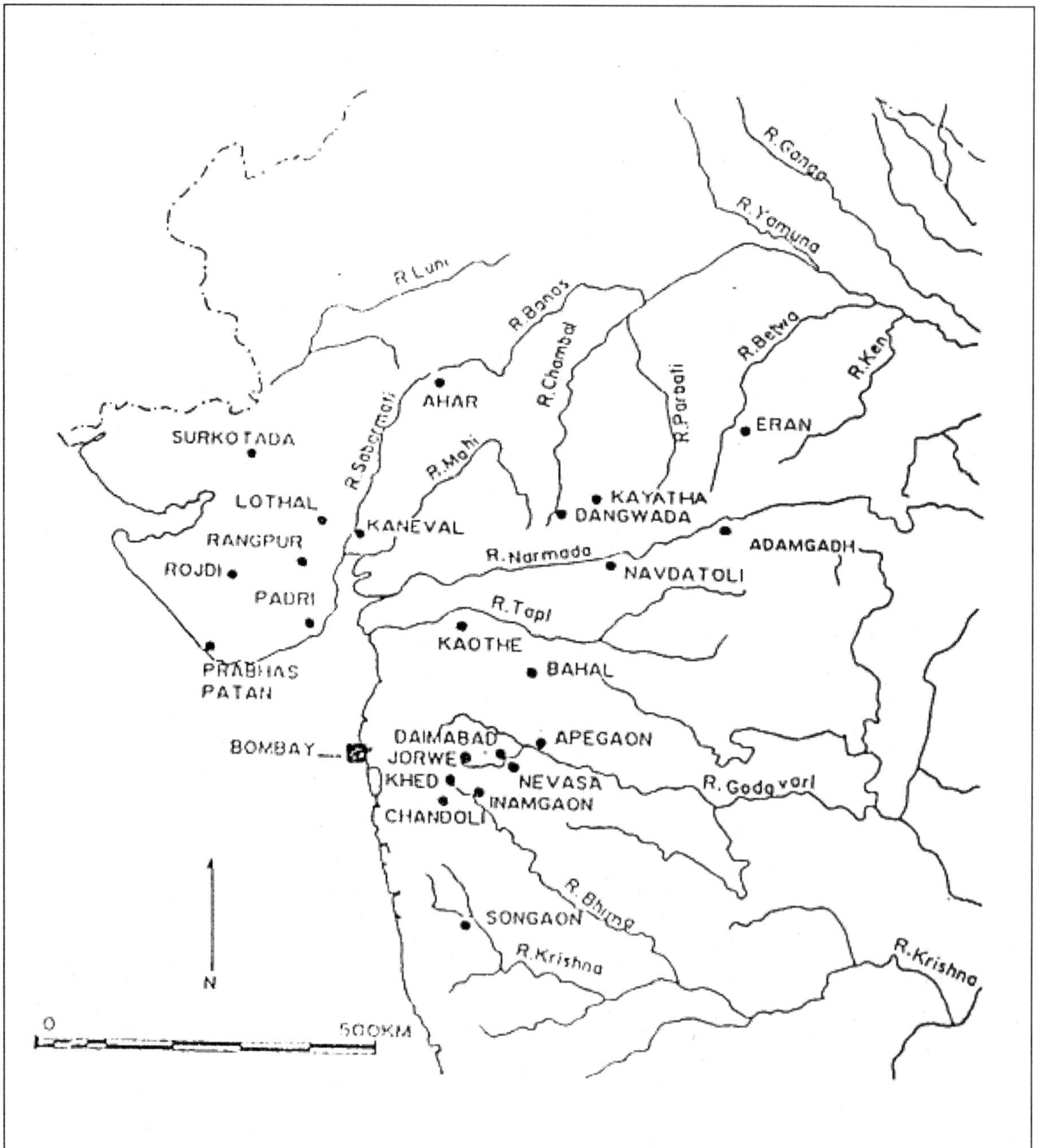

Fig. 1: Map showing Chalcolithic Sites of Central India and the Deccan (after Dhavalikar, 1997).

The Kayatha culture is characterised by three principal ceramic industries:

(i) Sturdy violet-painted pinkish red ware;
(ii) Red painted buff ware; and
(iii) Combed ware.

Other associated finds of this culture include: copper objects such as axes, bangles and chisels; a specialized blade industry of siliceous stones, stone macehead; blades of semi-precious stones and micro blades of steatite. The people lived in houses of mud and wattle-and-daub with floors made of hard compact, yellowish silt. Explorations conducted in north-western parts of Madhya Pradesh have brought to light over 40 sites of Kayatha culture, located mostly in the Chambal valley (Thapar, 1985: 89). Radiocarbon dates from this site range between twenty-first and fourteenth century BC (Deo in Rao, 1991: 347). Stratigraphically, this culture antedates the Banas culture which in turn, is earlier than the Malwa culture. This stratigraphic evidence has also been confirmed by a series of radiocarbon dates (Agrawal and Kusumgar, 1974: 112–116). The food habits of the Kayatha people are partially known as only wheat and Indian *Jujuba* have been reported from this site (Kajale, 1975: 164). The recent investigations of carbonised material in Sample TF 402 of the site brought out two kinds of wheat (*Triticum sphaerococcum* and *T. compactum*) and seeds of *Dolichos biflorus* Linn. horse gram (*IAR*, 1980–81: 108 & *IAR*, 1982–83: 148). Besides, they domesticated cattle, as is evidenced by bones of bovine species. They also included tortoise in their diet (Dhavalikar, 1979: 232). Surprisingly enough, they also had domesticated horse (Alur, 1975: 157–163). All these evidences bear ample testimony to the existence of a stable society dependent on agriculture and stock raising.

THE SAVALDA CULTURE

(*Circa* 2000–1700 BC; calibrated date 2300–2000 BC)

This culture has been named after the type-site of Savalda, on the southern bank of the Tapi river in Taluka Nandurbar, district Dhule, Maharashtra. This culture is characterized by a wheel-made painted pottery of medium to coarse fabric, copper-bronze rings, beads of carnelian and agate, microliths, tanged bone arrowheads, and stone mullers and querns. Remains of this culture have also been found in the earliest period (Period I) at Diamabad situated on the Pravara, a tributary of the Godavari in district Ahmednagar. Savalda was excavated in 1958 by the Archaeolgical Survey of India but the chronological position of the culture was placed on a secure footing in the course of excavations at Diamabad where the Savalda culture underlies the late Harappan debris (Sali 1982). Therefore, it has been dated to 2000–1800 BC.

A small-scale vertical excavation revealed a 3.60-m thick occupational strata. From the comparable deposits at Diamabad, evidence of some mud-walled structures with courtyards and hearths and floors made of rammed clay and silt, were discovered. Some querns and mullers provided the domestic equipments of this culture (Thapar, 1985: 87). At the present state of research, it is not possible to dwell on the origin of this culture. Suffice it to mention that being an indigenous regional culture, it seems to have flourished during the time of Indus civilization, the level of inter-relationship being at present, inadequately understood (Thapar 1985: 87). Here, it is pertinent to note that only a few Savalda sites (25 in number) have been discovered so far. All of them with solitary exception of Diamabad, are located in the central Tapi basin. The late Harappans, who entered the Khandesh region through the Tapi valley, do not seem to have driven out these people. What happened to them is anybody's guess (Dhavalikar and Shinde, 1989: 280; Dhavalikar, 1997: 161).

Carbonized grains from this site were collected by using flotation technique and their identification was carried out by Kajale (1977: 818–819; *IAR*, 1977–78; 9 and Vishnu-Mittre *et al.* 1986: 588–627) by noting their external morphological features under low power binocular microscope. This study has brought to light 16 wild and cultivated species of plants, some of them being the oldest records in the Indian subcontinent as well as in Africa. The Savalda culture (*Circa* 2300–2000 BC) people cultivated wheat, barley, lentil, common pea, grass pea, black gram/green gram horse gram and hyacinth bean (Sali, 1984: 235–242). In the later Harappan phase, lentil is absent. From the succeeding buff-and-cream (Diamabad) ware phase, six species, namely, wheat, barley, grass pea, horse gram, hyacinth bean, lentil and black gram/green gram were reported (Sali, 1986: 28).The Malwa culture has yielded seven species of plant remains while the Jorwe culture (*Circa* 1000 BC) has yielded remains of all the sixteen species. The carbonized food grains from Malwa culture have been identified as of wheat (*Triticum sphaerococcumn, T. compactum*), barley (*Hordeum* sp.), ragi (*Eleusine coracana*) and six kinds of pulses and beans (*Phaseolus* sp., *Vigna* sp., *Lens esculenta*), etc. Besides, fruit stones of *Zizyphus* also occur (*IAR*, 1980–81: 108). The carbonised seeds of Ragi (*Eleusine coracana*) from Malwa and Jorwe cultures have been found to show variability in size (*IAR* 1981–82: 106). Similar grains have been identified from the Jorwe phase but grains of *Paspalum scrobiculatum, Setaria* sp., *Pisum arvense* and *P.sativa* are the additional food grains in the Jorwe phase. These are not observed in the samples from the Malwa phase. The grains of wheat (165–173) outnumbered those of barley (35–36) in both the Malwa and Jorwe phases (*IAR*, 1980–81: 108). It has been noted that comparatively, the Jorwe samples are richer in kind and number of grains than those of the Malwa samples. However, the food economy in both was largely alike but for some references wheat, barley, *Eleusine*, peas, pulses, *zizyphus*, etc., are common ones (*IAR*, 1979–80: 113). It is noteworthy that wheat occurs only from Malwa and Jorwe levels while this cereal is significantly absent in the earlier three phases. Barley was the staple food in this region as it is present in all the five phases at Diamabad. Another important contribution of Diamabad plant economy is the finding of Sorghum (Jowar) which forms the earliest record of its charred grains in India. This evidence supports Dogett's hypothesis (Doggett, 1965) that Sorghum could have been introduced into India from Africa by the end of second millennium BC.

Evidence gathered from comparative morphology, cytogenetic distribution, history, linguistics etc. suggest that finger millet (*Ragi*) originated in Africa from where it was introduced into India. The present record of *Ragi* is, therefore, important from the point of view of its time and place of domestication in India and Africa. The grains of gram (chickpea, *Cicer arietinum* Linn.) and safflower also predate all other archaeo-botanical findings of the same species in India. The present findings are all the more important because they illustrate the sequential history of the earliest agriculture in Maharashtra (Kajale, 1977).

KAOTHE

Another site of Savalda culture excavated in recent years is Kaothe, situated in Sakri Taluka of Dhule district in Maharashtra. It is located 4 km west of Sakri on the Sakri-Navapur highway, on the left bank of the Kan, a tributary of the Panjhra, which, in turn, is a tributary of the Tapi. This site was excavated by Vasant Shinde (1985: 173–177; Shinde, 1998: 54), who believed that it was a single culture site of the Late Harappan period, having pottery similar to Rangpur IIB. However, after excavation, it was found to be a site of Savalda culture (Dhavalikar and Shinde, 1989: 277–280; Shinde, 1998: 18–19).

The ancient site of Kaothe is perhaps the most extensive early farming settlement in Maharashtra as it is spread over an area of 30 hactares (Dhavilkar and Shinde, 1989: 277). Unfortunately, the entire site is at present under intensive cultivation and as a consequence much of the cultural material has been lost. What remains now is only a thin deposit which varies from 50 cm to 1 metre. But since it is a single culture site, which was not occupied after it was deserted by the first farmers, there is no danger of the admixture of later material.

The excavation covered a large area of habitational deposit. A large number of dwelling pits, the storage pits and pits for keeping pottery, etc., were exposed in the excavations. The dwelling pits measured 5.60 m in length and 3.65 m in width. Its average depth was 80 cm (Dhavalikar and Shinde, 1989: 277; Dhavalikar 1997: 157). The pottery comprised a sturdy red ware and the distinct Savalda ware, named after the type-site.The single radio-carbon date from Kaothe 1920±80 BC (uncalibrated) also enables us to date the habitation to 2300–2000 BC (calibrated) (Dhavalikar, 1997: 160).

The excavations yielded evidence for the cultivation of *bajra*, pearl millet (*Pennisetum typhoideum*) (Dhavalikar, 1997: 160). This is a noteworthy feature because the early farming communities in central India and the Deccan produced mostly barley (*Hordeum vulgare*), while the Kaothe people cultivated *bajra* which was not grown by the succeeding Chalcolithic people. It needs to be noted that *bajra* was introduced in Saurashtra during the Harappan phase (CA, 2500–2000 BC) from the Gulf region. Here it needs to be emphasized that the occurrence of *bajra* at Kaothe is the earliest so far in the subcontinent as it can be dated to C 2000–1800 BC. In this connection it may be stated that the earliest *bajra* in the world is reported from South Africa where it is dated to 2400 BC and it is generally supposed that this millet was introduced into India from outside (Dhavalikar and Shinde 1989: 280). In the region around Kaothe today the main crop is *bajra* and not *jowar* (*Sorghum bicolour*) as in other parts of Maharashtra. Therefore, Dhavalikar (1997: 160) believes that millet was introduced into India from the Gulf region where it occurs from CA 4000 BC.

The excavations at Kaothe have thrown light on the first farmers of Maharashtra who were junior contemporaries of the Harappans in the adjoining Gujarat. However, the society does not appear to have been purely agricultural for hunting and fishing seem to have played a very important role in their subsistence pattern. It is suggested that they may have lived at the site during the monsoon when they raised the pearl millet whereas in the remaining parts of the year they may have been on the move, hunting, fishing and collecting plant foods (Dhavalikar, 1997: 161).

AHAR OR BANAS CULTURE
(2100–1400 BC, calibrated dates, 2600–1500 BC)

The Ahar culture, also known as the Banas culture, is among the earliest Chalcolithic cultures in India as the calibrated dates would show. It has been known after the type-site Ahar (Sanskrit *Aghatpura*) is located on the bank of the Banas river about 3 km to the east of Udaipur town. This site was discovered and excavated by R.C. Agrawala, then of the Rajasthan State Department of Archaeology, in 1954–55. Subsequently, large-scale excavations were undertaken by the late Professor H.D. Sankalia of the Deccan College, Pune, in 1961–62. This excavation revealed many interesting features of this culture.

The ancient mound, locally known as Dhulkot, literally, the mound of dust, is situated on the Ahar river. The site (500 × 275 m) is located in the southeastern corner of Rajasthan in the Banas valley and is adjacent to the ferrite tract of Malwa, the western part of Madhya Pradesh.

The excavations at Ahar revealed a two-fold sequence of cultures of which the first (Period I) is protohistoric and the other (Period II) belongs to the early historical period with a wide hiatus of over one thousand years in between them. Period I, according to the calibrated dates can be assigned to about 2600–1500 BC and Period II CA 500 BC–300 AD (Dhavalikar 1997: 80).

Period I is characterized by the distinctive black-and-red ware bearing white paintings which occur predominantly in all the phases of Period I. A series of calibrated radio-carbon dates allows us to date the different phases as follows:

Phase I A CA. 2600 – 2150 BC
Phase I B CA. 2150 – 1950 BC
Phase I C CA. 1950 – 1500 BC

The calibrated dates would thus make the Ahar culture contemporaneous with the Indus Civilization (Dhavalikar, 1997: 80).

The subsistence pattern of Banas culture comprised rice and millets together with meat of domesticated and hunted animals. The Aharians practised mixed economy based on agriculture and animal herding coupled with hunting and fishing. *Jowar* (a millet) is known from the pottery matrix of Period II. There are also impressions of long grained rice. No other grains, charred or otherwise, have been found. Animal bones, however, are in plenty: turtles, fish, goat, sheep, deer, pig and cattle were eaten. Bovines dominated the animal remains (Agrawal, 1982: 213).

The archaeo-botanical material from Ahar was examined by Vishnu-Mittre (1969: 229–235). It consists of impressions of grains and spikelets on the pottery sherds which have a black core and paintings in red. The impressions are of two kinds: the oblong ones which can hardly be mistaken for anything else than rice and the small round to ovid cavities on bodies referred to millet.

With the exception of a few sherds in which practically no impressions have been seen, impressions of rice husk have been noted in almost all the other sherds. The spikelets in all the sherds are oblong, flattened along the sides and strongly ribbed. From the size of spikelets and grains, it appears that the impressions belong to the long seeded strain of *Oryza sativa* Lin. Var. *Sativa* Bor.

Some potsherds have been found to bear fixiform impressions in the form of cavities which are mostly ovid in shape, subrotund to orbicular or elliptic oblong and slightly compressed more or less round, all round but slightly pointed at one of the ends. That these impressions belong to millets is largely borne out by the following considerations:

1. These impressions belong to the cultivated plants because of the presence of impressions of cultivated rice.
2. Among the cultivated plants they seem to belong to the cereals.
3. In their morphological characters they compare with the spikelets and grains of millets and preclude the possibility of their belonging to any wild grasses.

A comparison of the size and shape of the impressions with those of the living reveals that a large majority of the impressions appear to belong to Sorghum (Great millet, *Jowar*), while a few of them might belong to *Pennisetum typhoideum* (Pearly millet, *Bajri*) (Vishnu-Mittre, 1969: 231). Thus, Vishnu-Mittre rightly concludes that Ahar material has brought forth for the first time the factual history of millets in India.

Sorghum was present in the final stage of Period I at this site. There is also a possibility that *bajra* or the bulrush millet was cultivated here. Along with the similar evidence from Rangpur, this is the earliest known occurrence of this crop. This picture presented suggests a continuation of the earlier crop patterns like those of the Harappan province but augmented by millet (Allchin and Allchin, 1983: 264).

The origin and probable source of different varieties of millets cultivated in India has been a matter of speculation among scholars. Sir Joseph Hutchinson (1976: 129–141) believes that sorghum, pearl millet and finger millet were borrowed from Africa. However, this hypothesis is based on the botanical evidence alone and needs to be confirmed by archaeological data. It is surprising that a whole group of crops, the millets, would have been an African origin and introduced into Peninsular India during the second millennium BC. It is for archaeologists to investigate the medium through which the introduction took place (Allchin, 1976: 129–141). Regarding Sorghum at Ahar, Sankalia (1974) has drawn our attention to the disturbed condition on that part of the site from where the specimens came, and suggested that until further proof is forthcoming, these should be treated as belonging to the subsequent period datable to the last centuries BC. Nonetheless, as Sorghum occurs at Inamgaon in the Chalcolithic levels from the overlap phase between early Jorwe and late Jorwe culture dated to 1400–1000 BC its occurrence at Ahar in similar milieu should not be surprising at all.

Sorghum and other millets have also been reported from recent excavations at Diamabad.

After the excavations at Ahar much research has been carried out about the Ahar culture in recent years and nearly one hundred sites belonging to this culture have been located in the valleys of rivers Banas and its tributaries and sub-tributaries in Banswara, Udaipur, Chittorgarh, Bhilwara, Bundi and Ajmer districts. After the excavations of the type-site by Sankalia, three more sites, Gilund, Balathal and Ojiyana have been excavated. Balathal is located in Udaipur district, Gilund in Rajsamund district and Ojiyana is situated in district Bhilwara. Of these, Balathal is better known as it was excavated horizontally by V.N. Misra for seven seasons between 1994 and 2000 and it has provided an excellent picture of the culture of Ahar people (Misra *et al.* 1995: 57–80; Misra *et al.*, 1997, 35–60; Misra 2001: 491–531).

As stated above, Balathal is located about 40 km southeast of Udaipur city in the Vallabhanagar Tehsil of Udaipur district. The ancient mound covers an area of over five acres with a habitational deposit of about 7 metres. Excavations at this site have yielded a two-fold cultural sequence: Chalcolithic or Ahar culture (2600–1500 BC) and Iron age (fifth century BC – sixth century AD.). The chalcolithc period at Balathal is characterised by the presence of large multi-roomed rectangular or squarish mud-brick and stone-structures of varied functions, the profuse use of copper and the manufacture of a variety of fine and coarse ceramics. There are ten radiocarbon dates from the Chalcolithic levels at Balathal which ranged from 3800 BC to 1800 BC (Misra *et al.*, 1997: 57–58). Barring the first date, which is nearly 1200 years, older that the next oldest date, all other dates are reasonably consistent. Thus, excluding the first date, the duration of the Chalcolithic culture ranges between 2600 and 1800 BC. Commenting on these dates Misra writes; 'However, so far we have only two dates, including the deviant one, from the lowermost levels. Therefore, there is a strong possibility that when more dates become available from the lowermost levels, the antiquity of the settlement will go beyond 2600 BC. Even on the basis of presently available dates, Balathal would be the oldest known village outside the domain of the Indus (Harappan) civilization' (Misra *et al.*, 1997: 58).

The economy of the Ahar culture people at Balathal was based on cultivation, animal husbandry and hunting. They cultivated wheat (*Triticum* sp.), barley (*Hordeum vulgare*), lentil (*Lens esculenta,* Moench), common pea (*Pisum arvense* L.), finger millet (*Eleusine coracana* L.), Italian millet (*Setaria italica* Beauv) and penicum millet (*Panicum* sp.). The animals domesticated by them include cattle, sheep/goat, buffalo and pig. Wild animals hunted by them comprise gaur, nilgai, chausingha and black buck. Vast quantities of animal bones, almost all of them charred and broken, suggest that meat was an important component of diet and was consumed after roasting. There are also remains of fish, turtle and molluscs (Misra, 2001: 513).

The next excavated site of Ahar culture is Ojiyana, located on the hilly terrain of the Aravalis in district Bhilwara. This site was excavated horizontally by B.R. Meena and Alok Tripathi of the Archaeological Survey of India for two field seasons during 1999–2000 and January to June 2001. This excavation revealed a 7.5-metre thick single culture (Ahar culture) deposits divisible into three phases. The habitational period has been bracketed between 3000 BC and 1500 BC as per relative chronology offered by comparative study with excavated material from other sites of Ahar culture (Meena and Tripathi, 2000–2001: 76).

As regards the subsistence economy of Ojiyana, a variety of floral remains, charcoal and charred grains and seeds were collected by flotation. The main crops identified by

K.S. Saraswat of the Birbal Sahni Institute of Paleo-botany, Lucknow comprise barley, dwarf-wheat, bread wheat, rice, *ragi*-millet, Italian-millet, *jowar* millet, horse-gram, *kulthi*, lentil, field-pea, moth bean, gram/chickpea, *til* and sunflower/*kusum*. These crop remains are found in association with wide range of remains of weed and other wild taxa (Meena and Tripathi, 2001: 76).

Besides the archaeobotanical remains, indirect evidence of agricultural activity is provided by architectural remains. The Ahar culture people lived in single, double and multi-roomed rectangular, squarish and circular houses made of stone, mud-brick and mud. The houses were provided with overground and underground grain storage bins and stone saddle querns and rubbers for grinding cereals and pulses. At Balathal and Gilund a number of deep silos of various sizes lined with grass and plastered with lime have been found, meant for grain storage (Misra 2001: 513).

MALWA CULTURE
(*Circa* 1700–1500 BC)

The Malwa region of western Madhya Pradesh is drained by the Chambal, Narmada, Betwa and their tributaries. The Malwa culture flourished in the valleys of these rivers in the first half of the second millennium BC. More than a hundred settlements of this culture have been located by the archaeologists of the Deccan College, Pune under the leadership of late H.D. Sankalia and those of the Archaeological Survey of India and the University of Sagar. Among the excavated sites of this culture, mention may be made of Nagda, Kayatha (Period III), Maheshwar-Navdatoli and Eran. Of these, Navdatoli was excavated horizontally and has provided the best evidence of Malwa culture.

The Malwa region lies to the east of the Banas valley and Aravalli hills and it traversed by two great river systems, the Narmada and the Chambal. While the Narmada, the Tapi and the Mahi join the Arabian sea, the Chambal and the Betwa join the Yamuna. The climate is the typical monsoon type and the average rainfall is about 100 cm. The black soil which occupies the entire Malwa region, is much deeper in the Narmada valley. The earliest settlers of Malwa, the authors of the Chalcolithic culture, were attracted to this region because of its fertility.

The Malwa culture people lived in wattle-and-daub houses of rectangular and round shape as indicated by burnt wooden posts and clay plaster with bamboo and reed impressions. Round huts have diameters varying from 2.40 to 3.60 m and with walls from 30 to 60 cm thick. The rectangular structures were comparatively more spacious. A mud rampart made of mud and mud bricks has been reported at Nagda. It was probably constructed for protection against floods of the Chambal river (Misra, 2001: 515). A defence wall made of mud and having a width of 30 metres at the base and a height of 6.4 metres and with a moat running parallel to it has been reported from Eran. At Nagda, a drain built of mud-bricks and measuring 2.28 × 2.13 metres and with a height of about 1 metre has been reported. At Navdatoli, a squarish pit enclosed by mud walls and containing ash and burnt logs of wood has been identified as a sacrificial pit or *Yajña Kunda* (Sankalia *et al.*, 1971).

The origins of Malwa culture are still shrouded in mystery but it is reasonably clear that its origin took place in Central India because the beginning of the Malwa culture in Central India is slightly earlier than the Deccan (Shinde, 1998: 22). It appears that around 1600 BC the Malwa people started moving into the Deccan (in the Godavari-Pravara basins and still further into the Bhima valley) in which area, as the Inamgaon evidence demonstrates, they were the pioneer colonizers. Explorations carried out in the Deccan have brought to light over fifty Malwa sites, most of which were concentrated in the Tapi valley (Shinde, 1998: 23).

The Malwa people cultivated cereals, legumes, oilseeds and fruits. Cereals comprise bread wheat and rice. Among the pulses and legumes are lentil, black gram (*urd*), green gram and *khesari* (*Lathyrus* sativus). Oilseed is represented by linseed and fruit is represented by *ber* (*Zizyphus* jujuba). The Malwa people domesticated cattle, sheep, goat and pig. They also consumed the flesh of wild animals like *barasingha*, rat, fish, turtle and molluscs (Misra, 2001: 515–516).

The excavations at Navdatoli have provided excellent botanical evidence comprising the remains of cereals, legumes, oilseeds and fruits. Charred grains of wheat, barley and rice were found in phases II and IV. Rice was also found in phase III. Black gram, green gram, lentil, grass pea, linseed are other vegetable products used by them. The oilseeds are represented by *Linum usitatissimum* and the fruit remains by *Zizyphus jujuba* (*ber*) and *Phyllanthus emblica* (Awanla). It may be noted that linseed appears as well as two fruits, the Indian *Jujuba* are *ber* and the myrobalan (*Awanla*) (*Phyllanthus emblica*) at Mundigak (Allchin and Allchin, 1983: 267). It is interesting to note that almost all the crops are still grown in this region.

The plant remains obtained from the earlier excavations at Maheshwar-Navdatoli (1953–54) were studied by Vishnu-Mittre (1965: 13–32). It consisted of plant remains of cereals (wheat and rice), legumes with lentil, black gram (*urd*), green gram (*mung*), grass pea and *Lathyrus*, oilseeds of linseed and fruit remains of *Zizyphus jujuba* and *Phyllanthus emblica*. The re-examination (*IAR*, 1976–77: 91) of archaeo-botanical plant remains the site has yielded barley, in addition to the ones noted earlier by Vishnu-Mittre.

The entire Chalcolithic habitation was dated on the basis of radio-carbon dates to CA. 1600–1300 BC. However, the chronology has now been revised according to calibrated dates (Possehl and Rissman 1992: I, 486–87, II, Table 12, 465) as follows:

		Half Life 5568 b.p.	Half Life 5730 BC	Calibrated (Calib) BC
Phase I	TF 59	3380±105	1530	1875–1580
	P–200	3457±127	1610	1990–1660
	P–475	3455±70	1610	1905–1695
	P–201	3492±128	1645	2105–1670
Phase II	P–202	3503±128	1660	2120–1675
	P–476	4125±69	2300	2890–2640
Phase III	P–204	3449±130	1600	1980–1650
Phase IV	P–205	3294±125	1445	1780–1415

These dates are corroborated by those for the Malwa occupation at Kayatha and Diamabad. There are several dates from Eran, but there is a wide scatter from 1865 to 1020 BC. However, of the ten dates, a majority (seven) range between 1865–1365 BC. It will, therefore, be reasonable to assign CA. 1900–1400 BC. as the time bracket for the Malwa culture. The Malwa culture spread into Maharashtra later about 1700 BC. as the dates from Inamgaon and Diamabad would show.

Carbonised grains recovered from excavations at Navdatoli give us a fair idea of the agriculture of the Malwa people. There are cereals, legumes, oil seeds and fruits. Among cereals, there are wheat and rice; the former has been identified as bread wheat (*Triticum vulgare, compactum*) and the latter as *Oriza sativa* L. Malwa gets winter rains and even at present wheat is cultivated on a large scale but when the Malwa people went to Maharashtra in search of fresh posture and settled there, they adapted to the local environment. The evidence from Diamabad and Inamgaon shows that the Malwa people did not cultivate wheat there obviously because wheat is a winter crop, and there are no winter rains in Maharashtra. Hence, they cultivated barley (*Hordeum vulgare*) more in Maharashtra. Rice has been reported from Navdatoli from levels II–IV. It may have been grown there in small, carefully tended plots and as is done today.

Another cereal that the Malwa people cultivated, not in Malwa, but in Maharashtra, is jowar (*Sorghum vulgare*). Recent evidences from Saurashtra indicates that the millet was probably imported from the Persian Gulf region where it occurs much earlier. Jowar is a hardy crop which can be raised on rain fed lands, and was, therefore, adopted by the Harappans at the end of the third millennium when drastic change in climate was taking place and the climate was getting more and more arid. The Malwa people may have adopted it from the Late Harappans in the Tapi valley.

Among legumes, lentil (*Lens esculenta*) occurs in all the levels at Navdatoli. Black gram (*urd, vigna mungo*) and green gram (*Phaseolus mungo*) are also present. The latter is identified as *Phaseolus radiatus*. It is interesting that *Lathyrus sativus*, which is now known as *Gulabi chana*, was also known to the Malwa farmers. Mention should be also made of grass pea. (Sankalia *et al.*, 1971: 418 ff). Among oilseeds, only linseed (*Linum usitatissimum*) is reported. Ber (*Ziziphys jujuba*) is common at Chalcolithic sites.

Among the recently excavated Chalcolithic sites of Central India mention may be made of Dangwada located in district Ujjain. At this site, two season's work was carried out by V.S. Wakankar and M.D. Khare (*IAR*, 1978–79: 70–71 & *IAR*, 1979–80: 54–55) on the mound known as Boreshwar situated 1 km west of Dangwada village. The ancient settlement covers an area of 340 m in length, 50 m in width and 15 m in height. It revealed a fivefold cultural sequence ranging from Chalcolithic to early medieval times. Period I (Chalcolithic) is divisible into two sub-phases.

Period IA is characterized by Ahar pottery (plain and painted black-and-red ware) incised grey ware and lustrous red ware. The small antiquities include microlithic tools, copper, stone pesles, and terracotta bull figurines of archaic forms. No structural remains except for a piece of flooring made of small pebbles was encountered. An interesting evidence regarding the disposal of the dead is discovery of two urn burials, containing a few burnt bones, possibly of human. This type of evidence had not been discovered so far on the Ahar culture sites.

Period IB represents Malwa culture. The 4.3 m deposit of this sub-period has been divided into four phases (*IAR*, 1979–80: 54). Besides the characteristic Malwa pottery, microliths, stone objects like querns, pestles, balls and hammer stones were found. A noteworthy find is the discovery of structural remains of a shrine (1.2 m × 0.7 m) built of bricks and duly plastered with mud.

Period II belongs to pre-Mauryan and Mauryan times. This excavation has yielded important evidence of archaeo-botanical remains. In all, thirteen samples (*IAR* 1981–82: 105) of carbonised seeds ranging in age from the Chalcolithic to the early historical period were studied. Food grains identified from Chalcolithic (Malwa and Ahar) period are of *Oryza rufipogon, Oryza sativa, Triticum aestivum, Vigna mungo, Vigna radiatus, Lens esculentus* and *Zizyphus* sp. (*IAR*, 1981–82: 105).

However, it is not clear from the brief reports as to which grains belong to the Ahar culture and what is the economy of the succeeding Malwa culture.

Besides, various types of grains were found from the early historical (Sunga and Gupta periods) levels at the site (*IAR*, 1981–82: 105). These grains are lentil, rice, horse gram and Indian Jujube (*IAR*, 1980–81: 108). Besides, *Triticum compactum* and *Lathyrus sativus* were also found from these levels (*IAR*, 1981–82: 105).

JORWE CULTURE
(*Circa* 1500–900 BC)

This culture was identified at the type-site of Jorwe near Sangamner, a taluka headquarter in Ahmednagar district of Maharashtra in 1950. This culture was marked by its characteristic pottery which is wheel-made and well fired. The typical forms are spouted jar with a flaring mouth, carinated bowl and high-necked jar with globular profile. The pots were painted in black on red background with simple geometric motifs. Explorations carried out by several scholars have brought to light over 80 sites of this culture and excavations carried out at selected sites like Inamgaon, Chandoli and Walki in Pune district, Diamabad, Nevasa, Jorwe and Nasik in Ahmednagar district. Prakash in Dhule district and Bahal and Tekwada in Jalgaon district have thrown considerable light on many aspects of the material culture of this period. The systematic large scale excavations at Inamgaon is very significant not only because it has enabled us to reconstruct the socio-political organisation of the early farmers of the Deccan, but also because it has revealed a degenerate phase of the Jorwe culture termed as the 'Late Jorwe' which survived till c. 700 BC and thus has narrowed down the hiatus between the protohistoric and early historic periods (Shinde, 1998: 27). As regards the distribution of Jorwe culture sites, it may be noted that, of the 84 Jorwe phase sites, the highest number 52, are located in the Tapi basin, 22 in the Godavari basin and only ten in the Bhima basin. However, it must be recalled that compared to the Tapi basin, the Godavari and Bhima basins have not yet been thoroughly explored (Shinde, 1998: 30).

A characteristic feature of the Jorwe settlement pattern is the existence of large regional centres surrounded by smaller villages. Besides the regional centres, the Jorwe settlements can be classified into villages, which were in the majority, hamlets, farmsteads and camps. Most of the settlements are two hectares in size and their population many have been between 100 and 500 persons. However, larger villages like Bahal and Nevasa may have had populations between 500 and 1000 persons. Sites measuring only 1 hectare or less having a population of 50–100 persons may be classified as hamlets (Misra, 2001: 517). The largest Jorwe settlement is Diamabad which covers an area of 30 hectares and is one of the largest Chalcolithc settlements in the country. Even with a conservative estimate of 200 persons per hectare, the population of Diamabad would have been about 6000 individuals. Prakash in the Tapi valley, Diamabad in Godavari–Pravara valley and Inamgaon in the Bhima valley were three regional centres (Misra, 2001: 518).

As stated earlier, extensive excavations at Inamgaon have provided a very good picture of early farmers of the Deccan. These early farming communities subsisted on farming, stock-raising and hunting-fishing. They reared cattle, sheep/goat, buffalo and pig which were also slaughtered for food. Their subsistence agriculture was mainly dry farming since the entire Western India save the coastal strip is a semi-arid zone, with the annual rainfall ranging between 400–1000 mm. They also enjoyed the facility of artificial irrigation which is attested at Inamgaon (Dhavalikar, 1997: 190).

The Jorwe people cultivated barley, wheat, rice, *jowar*, kulthi, *ragi*, grass pea, lentil, green gram, black gram and hyacinth bean. The last named cultigen, hyacinth bean (*Doluichos biflourus*) is of special interest, as it requires a higher rainfall, over 800 mm and is grown today only in upper reaches of the Ghod and in the Konkan area (Dhavalikar, 1997: 190). Dhavalikar believes that although a single crop per year may have been the norm, the Jorwe farmers seem to have practised even crop rotation, Kharif and Rabi, as of today (Dhavalikar, 1997: 190).

As stated earlier, Inamgaon was excavated horizontally for 13 seasons between 1968 and 1983. It is the most heavily floated site in the country. This excavation has brought forth a sequence of cultures divisible into three periods with an applicable overlap between each. The chronology of these periods is as follows:

Period I:	Malwa culture (*c.* 1700–1400 BC)
Period IIA:	Early Jorwe culture (*c.* 1400–1000 BC)
Period IIB:	Late Jorwe culture (*c.* 1000 BC–700 BC)

The archaeo-botanical materials obtained during 1971 was studied by Vishnu-Mittre and R. Savithri (1975–76: 55–62 & *IAR*, 1968–69). This study showed that the rich and varied plant economy at Inamgaon, both during the Malwa and Jorwe cultures comprises barley, wheat, lentil, *Dolichos biflorus* (*Kulthi*), *Lablab*, *Pisum* sp., *Ziziphus nummularia* and *Buchanania.* The remains of *Phaseolus radiatus* and *Lathyrus sativus* have not been recovered from the late Jorwe phase. The stones of phoenix sylvestris are absent from Early and Late Jorwe, though present during the period of overlap between them. The authors have rightly concluded that these discontinuous records do not necessarily mean that there was a break in the consumption or use of these food plants. Perhaps their remains are not preserved (Vishnu-Mittre *et al.*, 1975–76: 62).

Subsequently, M.D. Kajale (1988b) published a detailed report on the data obtained from Inamgaon excavations on various season's materials excavated till 1982–83. Most of the materials recovered from this site were found in carbonized and semi-carbonized state. The grains were collected mostly by careful visual inspection of the habitational deposits during the course of excavations. Simple flotation techniques in the form of wet sieving was also undertaken for the representative soil samples from all the levels of the habitational deposits in order to recover small-sized grains which ultimately enabled strengthening the list of species recovered from the site. A list of the various cultivated and wild species is given below:

1.	Rice	*Oryza sativa* Linn.
2.	Wheat	*Triticum* spp., two morphotypes (*Triticum* cf. *sphaerococcum* Pers., *Triticum* cf. *aestivum* Linn.)
3.	Barley	*Hordeum vulgare* Linn. Var.polystichum
4.	Finger millet	*Eleusine coracana* (L.) Gaertn.
5.	Job's tears	*Jobi* Linn.
6.	Lentil	*Lens esculenta* Moench.
7.	Chick pea	*Cicer arietinum* Linn.
8.	Grass pea	*Lathyrus sativus* Linn.
9.	Common pea	*Pisum arvense* Linn.
10.	Horse gram	*Dolichos biflorus* Linn.
11.	Hyacinth bean	*Dolichos lablab* Linn.
12.	Black gram	*Phaseolus mungo* Linn. Syn. *Vigna mungo* Linn. Hepper
13.	Green gram	*Phaseolus radiatus* Linn. Syn. *Vigna radiata* Linn. Wilczek
14.	Indian jambos	*Eugenia jambolana* Linn. Syn. *Syzigium cumini* Lamk.
15.	Indian jujube	Lamk.
16.	Wild date	Phoenix sylvestris Roxb.
17.	Beleric myrabolan	*Terminalia belerica* Roxb.
18.	Emblic myrabolan	*Phyllanthus emblica* Linn.
19.	Cuddappah almond	*Buchnania lanzan* Spreng.
20.	Karanja (Marathi Name) Vent.	*Pongamia* cf. *pinnata* (L.) syn. *Pongamia* cf. *glabra*
21.	Indian mallow	*Abutilon indica* Lin.
22.	Indian cherry	*Cordia myxa* Linn.
23.	Common sedge	*Cyperus rotundus* Linn. Type
24.	Italian millet	*Setaria* cf. *italica* (L.) Beauv. (Foxtail millet)
25.	Wild grasses	*S. etaria* sp./*Panicum* sp Morphotypes
26.	Purslane	*Portulaca* cf. *oleracea* Linn.
27.	Polygonaceae type	*Polygonum* sp. Type
28.	Fat hen (Goose foot)	*Chenopodium* cf. *album* Linn.
29.	Solanaceae type	*Solanum* cf. *melongea* Linn. Morphotype
30.	Babul (Indian Gum Arabic)	*Acacia* cf. *arabica* Wild. Syn. *Acacia* cf. *nilotica* (L.) Del.
31.	Cucurbitaceae (Cucumber type)	*Cucumis* cf. *sativa* Linn. Type

32.	Tendu (Marathi Name)	*Diospyros* cf. *melanoxylon* Linn. Type
33.	Cyperaceae type (Common sedge type)	*Cyperus* cf. *rotundus* Linn.

Among the cereals noted above, the most problematic is the presence of rice as only three grains were recovered from the material excavated during 1974–75 field season. Of these, one of the grains is broken. Commenting on this, Kajale (1988b: 735) rightly believes that this cereal might have been imported from the other Chalcolithic sites like Diamabad or Maheshwar-Navdatoli.

Quantity wise, wheat also is not well represented at Inamgaon as compared to other species of grains like barley, horse gram, etc. Only two wheat grains have been found from the earliest habitation levels of Malwa culture but it appears in an appreciable quantity in Early Jorwe phase strata (Kajale, 1988b: 736). However, it continues to occur in layers of Late Jorwe phase in much lesser proportions.

Barley grains are found right from the beginning of Malwa culture until the Late Jorwe culture (Kajale, 1988b: 739). Obviously, this was the staple food of the inhabitants of Inamgaon. Barley is a hardy crop, well suited to dry climatic conditions and saline soils, especially where the other crops like wheat and pea fail to grow well. Kajale writes that looking at the vast quantity of grains from Inamgaon, it appears that barley was the most important human food plant and probably it could also have been an animal feed (Kajale, 1988: 740).

Among the pulses lentil occurs throughout all the cultural levels at Inamgaon. It appears first in layer (16), goes on increasing and maximum concentration is observed in layers (8), (9), (6) and (4). Therefore, it must have been a staple pulse from the beginning to the end of Chalcolithic culture at Inamgaon (Kajale, 1988b: 740). Common pea (*Pisum arvense* Linn.) occurs in meagre quantity in Malwa and early phases of early Jorwe culture. It is observed in highest proportions in the layers (7) and (6) and then declines subsequently in the upper layers (Kajale,1988b: 743). Horse gram (*Dolichus biflorus* Linn.) is present in the Malwa culture layers. Subsequently, its cultivation increased and probably reached its zenith during the Early Jorwe culture. Its cultivation continued during the Late Jorwe phase, though much less intensively (Kajale, 1988: 750). Hyacinth bean (*Dolichos lablab* Linn.) was most intensively cultivated during the terminal phases of Early Jorwe culture. It continued in Late Jorwe in a slightly declined state. It is not clear whether the Chalcolithic inhabitants of Inamgaon exploited it as vegetable or as pulse seeds. In all probability, it might have fulfilled both the requirements (Kajale, 1988b: 753). A few grains of gram (*Cicer arietinum* Linn.) were found during 1982–83 season's excavation. The grains certainly belonged to the cultivated forms. Considering the slender evidence for this species at Inamgaon, it has been argued that gram might have been imported at the site during the late phase of the Late Jorwe occupation (Kajale, 1988b: 754).

Grass pea (*Lathyrus sativus* Linn.) seeds occur in limited quantity at Inamgaon. Its record is much slender in the earlier phases of occupation but it goes on increasing in subsequent layers. Hence, it has been suspected that it might have been essentially a Late Jorwe cultigen even though it was introduced at the site during earliest phases of occupation (Kajale, 1988b: 757). Green gram (*Vigna radiata*) occurs essentially in Early Jorwe phase.

Among the millets, mention may be made of finger millet which has been found from all the three cultural levels. Besides, stored grains of millets were recovered from House No. 134 in 1982–83 season's excavations. They have been tentatively identified as *Panicum* sp. and *Setaria* sp. Besides, a few carbonized grains of foxtail millet were also recovered. But in view of the small quantity, it remains to be firmly established whether this species was cultivated in the vicinity of the site or was imported or got mixed up with the harvest of the other domesticated species (Kajale, 1988b: 762).

From the above discussion, it is clear that except perhaps rice, all the above noted species of grains were actually cultivated by the Chalcolithic people in the vicinity of the site. The occurrence of monsoon (summer) crops like rice, black gram, green gram, hyacinth bean, horse gram and winter crops like wheat, barley, lentil, grass pea, common pea etc. suggests that the Chalcolithic farmers carried out the agricultural practices during both the cropping seasons and they obviously had knowledge of double cropping. Mixed cropping might have prevailed, if the consistent occurrence of cereals and pulses is to be taken as any indication. Mixed cropping enabled the farmer in overcoming the loss of one of the crops due either to vagaries of rains or various other causes such as fungal attack or insect/pest menace (Kajale, 1988b: 783). We cannot know whether similar considerations prevailed in the prehistoric times at Inamgaon but it does not seem to be unlikely (Kajale, 1988b: 783). As stated earlier, the excavation yielded only three grains of rice, so it is not clear whether it was definitely cultivated at the site.

DAIMABAD

Another site of Jorwe culture which has furnished data regarding agricultural practices of Chalcolithic cultures of Maharashtra, is Daimabad, situated in Taluka Srirampur in district Ahmednagar on the left bank of the Pravara, a major tributary of the Godavari. It is the most extensive Chalcolithic settlement in Maharashtra, spread over 30 hectares and is considerably undisturbed because the nearest village is about 3 km away from it. This site was discovered by B.P. Bopardikar of the Archaeological Survey of India in 1958. It was excavated by M.N. Deshpande in 1959. The work was again taken up in 1974 by S.R. Rao because of the discovery of a unique hoard of four bronzes (an exquisite bull chariot, an elephant, a rhino and a buffalo). After this, excavation continued under the direction of Shri S.A. Sali from 1976 to 1979. His work (Sali, 1986) proved most rewarding and has thrown considerable light about the first farmers of Maharashtra. The culture sequence at Daimabad as obtained by Sali is as follows (from botton upwards):

Savalda culture	*c.* 2200–2000 BC
Late Harappa culture	*c.* 2000–1800 BC
Daimabad culture	*c.* 1800–1600 BC
Malwa culture	*c.* 1600–1400 BC
Jorwe culture	*c.* 1400–1000 BC

The excavations at Daimabad have yielded a variety of charred seeds. Carbonised grains from the site were collected by using flotation technique and their identification was carried out by Kajale (1977a: 818–819, *IAR* 1977–78: and Vishnu-Mittre *et al.*, 1986: 588–627) by noting their external morphological features under low power binocular microscope. This study has brought to light 16 wild and cultivated species of plants, some of them being the oldest records in the Indian subcontinent as well as in Africa.

The Savalda culture (*Circa* 2200–2000 BC) people cultivated wheat, barley, lentil, common pea, grass pea, black gram/green gram, horse gram and hyacinth bean (Sali, 1984: 235–242). In the later Harappan phase, lentil is absent. From the succeeding buff-and-cream (Daimabad) ware phase six species, namely, wheat, barley, grass pea, horse gram, hyacinth bean, lentil and black gram/green gram were reported (Sali 1986: 28). Malwa culture has yielded seven species of plant remains, while Jorwe culture (*Circa* 1000 BC) has yielded the remains of all the 16 species. The carbonized food grains from Malwa culture have been identified as those of wheat (*Triticum sphaerococcum, T. compactum*), barley (*Hordeum* spp.), ragi (*Eleusine coracana*) and six kinds of pulses and beans (*Phaseolus* spp., *Vigna* spp., Lens esculenta), etc. Besides, fruit stones of *Zizyphus* also occur (*IAR*, 1980–81: 108). Carbonized seeds of Ragi (*Eleusine coracana*) from Malwa and Jorwe cultures have been found to show variability in size (*IAR*, 1981–82: 106). Similar grains have been identified from the Jorwe phase but grains of *Paspalum scrobiculatum, Setaria* Spp., *Pisum arvense* and *p. sativum* are the additional food grains in the Jorwe phase. These are not observed in the samples from the Malwa phase. The grains of wheat (165–173) outnumber those of barley (35–36) in both the Malwa and Jorwe phases (*IAR*, 1980–81: 108). It has been noted that comparatively, the Jorwe samples are richer in kind and number of grains than those of the Malwa samples. However, the food economy in both was largely alike but for some differences. Wheat, barley, *Eleusine*, peas, pulses, *Zizyphus*, etc., are common ones (*IAR*, 1979–80: 113). It is noteworthy that wheat occurs only from Malwa and Jorwe levels while this cereal is significantly absent in the earlier three phases. Barley was the staple food in this region as it is present in all the five phases at Daimabad. Another important contribution of Daimabad plant economy is the finding of Sorghum (*Jowar*) which forms the earliest record of its charred grains in India. This evidence supports Dogett's hypothesis (Dogett, 1965) that Sorghum could have been introduced into India from Africa by the end of second millennium BC.

Evidence gathered from comparative morphology, cytogenitic distribution, history, linguistics etc. suggest that finger millet (*Ragi*) originated in Africa from where it was introduced in India. The present record of *Ragi* is, therefore, important from the point of view of its time and place of domestication in India and Africa. The grains of gram (chickpea, *Cicer arietinum* Linn.) and safflower also predate all other archaeo-botanical finding of the same species in India. The present findings are all the more important because they illustrate sequential history of earliest agriculture in Maharashtra (Kajale, 1977a).

Another Chalcolithic site of Jorwe culture is Songaon located on the right bank of the confluence of rivers Neera and Kasha, about 100 km southeast of Pune. It was excavated by S.B. Deo in 1965 (Deo *et al.*, 1969). The archaeo-botanical remains coming from Early Jorwe levels (*Circa* 1200 BC) were studied by Kajale (1978b: 84–86). The sample of charred grains was recovered from the debris of a burnt wall, at a depth of about 2 m below the datum line. Examination of this sample revealed that it comprised 120 grains of barley (*Hordeum vulgare* Linn.) preserved in an excellent condition. The average shape and size of the grains suggests their cultivated nature. All the grains show transverse ripplings on both the dorsal and ventral surface, an indirect evidence of the husked nature of the grains. The discovery of barley from Songaon serves as addition to the record of barley grains from other Chalcolithic sites of the Deccan such as Navdatoli and Inamgaon. Besides, carbonized material from this site consisted of charred kernels of wheat identified as *Triticum sphaerococcum* (*IAR*, 1972–73: 60).

The archaeo-botanical remains from Nevasa are not many. However, the carbonized grains from this Chalcolithic site in packet TR-39 were identified as of *Oryza* species (a single broken fragment), *Lathyrus sativus, Pisum arvense, P. sativum, Vigna mungo* and nuts of *Zizyphus* (*IAR*, 1982–83: 148).

A 15-metre thick habitational deposit of Chalcolithic culture was unearthed in 1985–86 by B.P. Bhopardikar of the Archaeological Survey of India at Tuljapur Garhi in Amaraoti district of Maharashtra. This site has yielded wheel-made pottery of Chalcolithic nature comprising the well-known Malwa and Jorwe wares. This Chalcolithic culture has been divided into phases: A and B with sign of overlap between the two. Here, Phase A represents the Malwa culture and Phase B is dated to the Jorwe culture. This Chalcolithic culture has got four radio carbon dates. Calculated on the Libby half-life of 5570 ± 30 years, these dates read 1360 ± 90 years; 920 BC ± 100 years and 460 BC ± 100 years. The fourth date was found to be erratic and out of context. Thus, the Chalcolithic culture falls in the time bracket of later half of second millennium BC or the first half of the first millennium BC.

The plant remains of this site have been examined by M.D. Kajale (1988a), who has identified seventeen plant species of which 15 are botanically diagnosed and two species are probably wild. Of the 15 diagnosed species, four belong to the cultivated cereals, eight to the domesticated pulses and one each to oil-yielding plant, fibre-cum-vegetable plant and fodder plant species. The plant economy of Tuljapur Garhi comprises the following species:

The evidence of Tuljapur Garhi demonstrates culmination of double-cropping practices in the Vidarbha region. The Chalcolithic farmers cultivated winter crops comprising wheat, barley, lentil, grass pea and gram a well as the summer (monsoon) crops like rice, great millet, hyacinth bean, horse gram, black gram and green gram. The occurrence of grains of great millet (sorghum), and pigeon pea is particularly important in view of the fact that the first named crop is supposed to have been introduced into India from Africa during second millennium BC. Sorghum millet comes from the Harappan levels at Rohira, Jorwe levels at Diamabad and Savalda levels at the last-named site. Similarly, the cultivation of pigeon pea is an important addition to our knowledge. Its cultivation was practised by the megalith builders of Bhagimohari (800–400 BC) in the Nagpur district, in Indo-Roman levels at Nevasa and the post-Satavahana (200 BC–300 BC) times at Bhokardan in Aurangabad district. Thus the discovery of cultivated grains of this summer/winter pulse from Tuljapur Garhi is of considerable significance and lends support to the hypothesis that pigeon pea might have been a locally domesticated pulse crop in western India (Kajale, 1988a: 378).

Mention may be made here of the small-scale excavations at the Chalcolithic site of Walki situated on the confluence of rivers Bhima and the Mula. This site was discovered in 1969 and archaeological sampling was done in 1972. It was excavated for two seasons by M.K. Dhavalikar (1988). This has provided indirect evidence of agricultural practices of the Jorwe culture. The site was found to be a very small (one hectare) seasonal settlement of the inhabitants of the Jorwe culture. Remains of dwelling huts, bin platforms, threshing floors and huts for domestic animals were found. The absence of walls clearly suggested that the huts were not lived in during the rainy season. Since no agricultural activity was carried out during summer, it was obvious that the site was occupied only during winter. Evidence from other Chalcolithic sites has shown that the principal cereal cultivated by the Jorwe people was barley which was a winter crop. It was thus obvious that the people were coming here during winter to cultivate barley in the 300 hectare large tract of black cotton soil on the confluence of the two rivers noted above (Dhavalikar, 1988).

DISCUSSION

Perhaps no account of the agricultural practices of the Chalcolithic cultures of the Deccan will be complete unless we take into account the soil-type, the climatic conditions, the problem and method of irrigation, if any, and the use of agricultural implements, particularly that of plough. It is common knowledge that the Chalcolithic cultures flourished in the area of black cotton soil which seems to have been the most important single factor in the formation of these cultures. This hard, compact soil is known for its fertility and moisture-retaining capacity. It is capable of yielding a rich harvest if cultivated effectively. However, it has been argued that the Chalcolithic agriculture was generally confined to alluvial patches only (Agrawal, 1982: 218), and the black cotton soil could not have been cultivated in the absence of iron plough (Kosambi, 1963: 315–16). In this context, the researches carried out by Dhavalikar (1997: 191) are quite interesting and revealing. He has worked out the population estimates of such large centres as Daimabad and Inamgaon which work out to be 4000 and 1000 individuals respectively. The population at Inamgaon during the Malwa and early Jorwe periods would be about 1000 and 650 persons during the Late Jorwe. If a person requires about 2000 calories a day, which can be obtained from less than half a kg of grains, the daily requirement of 1000 persons would be 400 kg and for a year it would be 1,50,000 kg. According to Leshnik (1967), a person requires 2 acres of cultivable land or 1000 ha. was required from the early Jorwe settlement at Inamgaon. As there are no alluvial strips worth the name at Inamgaon, the inescapable conclusion is that the Chalcolithic people did cultivate the black cotton soil (Dhavalikar, 1997: 192).

As regards the problem of irrigation, Kajale (1988b: 792) has rightly remarked that Inamgaon does not lie in the zone of winter rainfall and the cultivation of at least two cereals (wheat and barley) and four pulses (horse gram, hyacinth bean, lentil, common pea) suggests that the Inamgaon farmers probably utilized the then available water resources carefully. It is difficult to raise certain winter crops (such as wheat, lentil and pea) only by depending on the water naturally conserved in the medium black cotton soil and with winter dew. The area around Inamgaon is semi-arid with hardly 450 mm of rainfall and is perpetually draught affected. Further, the central parts of Maharashtra do not receive winter rains and hence irrigation is necessary. In this context, the embankment (about 240 m long; 2.25 m wide and perhaps equally high) uncovered at Inamgaon is of crucial significance (Dhavalikar, 1988–89: 72). Therefore, the inhabitants of Inamgaon built a sort of guidebund or an embankment parallel to a channel (extant length 118 m; 3.50 deep, 4 m wide) and diverted the flood water of the river. This channel was deep in the middle (3.50 m) and thus served also as a narrow tank for storing water which was probably used for irrigation. The traces of the channel suggest that the water could have been diverted to the adjoining fields by gravity flow (Dhavalikar, 1997: 194). Because of this facility of artificial irrigation the Early Jorwe farmers could cultivate wheat and hyacinth bean. This irrigation channel fell into disuse after 1200 BC (Dhavalikar, 1997: 194).

The occurrence of rice at Inamgaon is a bit enigmatic. Inamgaon is the best floated site and a vast quantity of charred seeds were collected by flotation. In this vast quantity of seeds, rice is represented by only three grains, recovered from the material excavated during 1974–75 field season. Hence, it has been rightly concluded that rice might have been imported at the site from other Chalcolithic sites like Daimabad or Navdatoli-Maheswar (Kajale, 1988b: 735). However, Dhavalikar believes that rice was grown locally in carefully tended plots.

Related to agricultural operations is the question whether the Chalcolithic farmers used the plough for agricultural operations. Neither plough, save that from Walki; nor hoe has been attested at Chalcolithic sites, but perforated stone discs which were used as weights for digging sticks abound. The digging stick can be useful only in the slash-and-burn or zoom cultivation. Dhavalikar and Possehl (1974) have suggested that the black soil may have been cultivated by means of a wooden plough, as is done even today in Malwa. It is also highly likely that antlers may have been used as hand ploughs. Antlers have been found in the excavations at Inamgaon (Dhavalikar, 1997: 192). In this context, mention may be made of the discovery of an ard made of the shoulder bone of cattle from Walki, situated on the confluence of the rivers Bhima and the Mula. Walki is a small (one hectare) seasonal settlement which was excavated for two seasons by M.K. Dhavalikar (1988) and yielded the remains of Jorwe culture. The ard discovered at this site is roughly triangular in shape (16 cm × 10 cm) and has two perforations. It has no parallel so far in the country but similar specimens of wood have been reported from neolithic sites in northern Europe and in stone in Bronze Age Syria (Dhavalikar, 1997: 193).

Walki has also furnished evidence regarding threshing floors which again are unique and have no parallel elsewhere in the country. The threshing floors comprised hard round mud floors which are specially prepared by beating the packed clay (Dhavalikar, 1997: 89–90).

CONCLUDING REMARKS

A survey of the archaeo-botanical data obtained from excavations of Chalcolithic sites of central India and the Deccan demonstrates that in spite of a large number of archaeo-botanical samples collected in recent years, there are significant gaps in our knowledge. This is because of the fact that the new technique of data collection through flotation was adopted only during the later nineteen seventies and on sites excavated earlier (Jorwe, Ahar, Maheswar-Navdatoli, etc.), the sampling remained grossly inadequate. Flotation was first carried out in India at Daimabad, Naikund and Inamgaon and soon came to be applied in research programmes at the Neolithic sites of Kashmir and later at Narhan in the Ganga plains (Fuller, 2002: 257). The impact of flotation on the scholarly understanding of ancient Indian agriculture was quite dramatic as a greater number of species and previously attested species became more reliably recorded (Fuller, 2002: 261). But such questions as crop diffusion in South Asia are still under debate (for a detailed discussion see Fuller, 2002: 306–322; Misra and Kajale, 2003). This is particularly true for the Chalcolithic villages of Maharashtra. Evidence obtained from Inamgaon, Daimabad and Kaothe indicate that Near Eastern cereals and pulses were grown in the same regions as African millets, indigenous millet and tropical pulses. Keeping this complex situation in mind, Fuller (2002: 316) rightly remarks that understanding the ways in which different agricultural systems, including diverse crops and cropping systems were integrated in Maharashtra during the late third millennium and early second millennium BC, remains a challenge to be addressed in future archaeo-botanical work.

REFERENCES

Agrawal, D.P. 1971. *The Copper–Bronze Age in India,* Munshiram Manoharlal, New Delhi.

Agrawal, D.P. 1982. *The Archaeology of India,* Curzon Press, London.

Agrawal, D.P. and Kusumgar Sheela, 1974. *Prehistoric Chronology and Radiocarbon Dating in India,* New Delhi, 112–116.

Allchin, F.R. 1976. Comments on Sir Joseph Hutchinson's Paper, *Phil. Trans. Royal Society,* London, B. Vol. 275. 129–141.

Allchin, F.R. and Allchin, B. 1983. *The Rise of Civilization in India and Pakistan,* New Delhi.

Alur, K.R. 1975. Animal Remains from Kayatha, in Z.D. Ansari and M.K. Dhavalikar, *Excavations at Kayatha,* Pune, 157–163.

Ansari, Z.D. and Dhavalikar, M.K. 1975. *Excavations at Kayatha,* Pune.

Deo, S.B. and Mujumdar, G.G. 1969. *Songaon Excavations 1965,* Pune.

Deo, S.B. 1991. Chalcolithic Culture of central India and Maharashtra. Appendix I in S.R. Rao, *Dawn and Evolution of Indus Civilization,* Aditya Prakashan, New Delhi, 345–356.

Dhavalikar, M.K. 1975–76. Settlement Archaeology of Inamgaon, *Purātattva,* No. 8, 51.

Dhavalikar, M.K. 1977A. Inamgaon: The Pattern of Settlement, *Man & Environment,* Vol. 1. 46–51.

Dhavalikar, M.K. 1979. Early Farming Culturs of Central India, *Essays in Indian Protohistory* (eds) D.P. Agrawal and Dilip K. Chakrabarti, Delhi, pp. 229–245.

Dhavalikar, M.K. 1984. Towards an Ecological Model for Chalcolithic Cultures of Central and Western India, *Journal of Anthropological Archaeology,* No. 3.

Dhavalikar, M.K. 1985. Early Settlement and Subsistence Patterns in the Deccan, *Recent Advances in Indo-Pacific Prehistory* (eds) V.N. Misra and Peter Bellwood, New Delhi. 281–290.

Dhavalikar, M.K. 1988. Excavation: The Scientific Method and the Scientific Aids. Paper presented at the *International Seminar on New Archaeology*, Indian Council of Historical Research, New Delhi, 14–16 October.

Dhavalikar, M.K. 1997. *Indian Protohistory,* Books and Books, New Delhi.

Dhavalikar, M.K. 1988–89. Daimabad: A Chiefdom Society, *Bulletin of the Deccan College Post-Graduate and Research Institute,* Vol. 47–48. 67–77.

Dhavalikar, M.K. and Gregory L. Possehl. 1974. Subsistence Pattern of an Early Farming Community of Western India, *Purātattva,* No. 7.39–46.

Dhavalikar, M.K. and Vasant Shinde, 1989. Excavations at Kaothe in the Central Tapi Basin, *South Asian Archaeology 1985* (eds) K. Frifelt and Per Sorensen, Curzon Press, London, 277–295.

Fuller, Dorian Q. 2002. Fifty Years of Archaeo-botanical research in India: Laying a Solid Foundation, *Archaeology and Interactive Discipline,* Indian Archaeology in Retrospect, Volume III (eds) S. Settar and Ravi Korisettar, New Delhi, ICHR and Manohar Publishers, 247–363.

Hutchinon, Sir Joseph. 1976. India: Local and Introduced Crops, *Phil, Trans. Royal Society,* London, B. Vol. 275. 129–141.

Kajale, M.D. 1975. Plant Economy at Kayatha, in Z.D. Ansari and M.K. Dhavalikar, *Excavations at Kayatha,* Pune.

Kajale, M.D. 1977a. On the Botanical Findings from Excavations of Daimabad, *Current Science,* vol. 46 (No. 23). 818–819.

Kajale, M.D. 1977b. Plant Economy at Inamgaon, *Man & Environment,* vol. 1.54–56.

Kajale, M.D. 1978a. *Bio-Archaeology of the Ghod Valley-Maharashtra,* Ph.D. thesis, University of Pune.

Kajale, M.D. 1978b. Barley Grains from Chalcolithic Songaon, *Man & Environment,* Vol. II. 84–86.

Kajale, M.D. 1988a. Ancient plant Economy at Chalcolithic Tuljapur Garhi, District Amraoti, Maharashtra, *Current Science,* Vol. 57 (No.7). April 3, 1988, 377–379.

Kajale, M.D. 1988b. Plant Economy in *Excavations at Inamgaon,* Vol. I, Part II by Dhavalikar, M.K., Sankalia, H.D. and Z.D. Ansari, Deccan College, Pune, 727–820.

Kajale, M.D. 1991. Current status of Indian Palaeo-ethnobotany. In J. Renfrew (ed.) *New Light on Early Farming: Recent Developments in Palaeo-ethnobotany,* Edinburgh University Press, Edinburgh, 155–189.

Kosambi, D.D. 1963. The Beginning of Iron Age in India, *Journal of the Economic and Social History of the Orient,* Vol. 6, pp. 309–318.

Leshnik, L.S., 1967. The System of dry Farming in the West Nimar district of Central India, *Yearbook of the South Asia Institute, Hydelberg University, 1966,* Wiesbaden, 56–74.

Meena, B.R. and Alok Tripathi, 2000. Excavations at Ojiyana, *Puāatattva* No. 30.

Meena, B.R. and Alok Tripathi. 2001. Further Excavations at Ojiyana, *Purātattva,* No. 31 (2000–2001), 73–77.

Misra, V.N., Vasant Shinde, R.K. Mohanty, Kusum Dalal, Anup Mishra, Lalit Pandey and Jeevan Kharakwal, 1995. Excavations at Balathal: Their contribution to the Chalcolithic and Iron Age Cultures of Mewar, Rajasthan, *Man & Environment,* vol. xx (1) 1995, 57–80.

Misra, V.N, Vasant Shinde, R.K. Mohanty, Lalit Pandey and Jeevan Kharakwal, 1997. Excavations at Balathal, Udaipur district, Rajasthan (1995–97) with special reference to Chalcolithic architecture, *Man & Environment,* vol. xxii (2), 35–60.

Misra, V.N. and M.D. Kajale (eds) 2003. *Introduction of African Crops into South Asia,* Pune: Indian Society for Prehistoric and Quaternary Studies.

Misra, V.N. 2001. Prehistoric Human Colonization in India. *J. Bio. Sciences* vol. 26, No. 4, Suppl. November 2001, 491–531.

Prakash, U. and N.A. Awasthi, 1971. Some Plant Remains from Navdatoli–India, in Sankalia, *et al. Chalcolithic Navdatoli,* Pune-Baroda, 440–448.

Sali, S.A. 1977. The First Evidence of a hafted Blade found from Daimabad, *Current Science,* 46 (No. 23), December 5, 1977, 818.

Sali, S.A. 1982. The Harappans of Daimabad, *The Harappan Civilization* (ed) Gregory L. Possehl, 175–184.

Sali, S.A. 1984. Late Harappan Settlement at Daimabad, *Frontiers of the Indus Civilization* (eds) B.B. Lal and S.P. Gupta, Delhi, 235–242.

Sali, S.A. 1986. *Daimabad 1976–79.* Memoirs of the Archaeological Survey of India, No. 83, New Delhi.

Sankalia, H.D. 1974. *Prehistory and Protohistory of India and Pakistan,* Pune, Deccan College, Second Edition.

Sankalia, H.D., S.B. Deo, Z.D. Ansari and Sophie Ehrhardt, 1960. *From History to Prehistory at Nevasa 1954–56,* Pune, Deccan College.

Sankalia, H.D., S.B. Deo and Z.D. Ansari, 1969. *Excavations at Ahar (Tambavati),* Pune, Deccan College.

Sankalia, H.D., S.B. Deo and Z.D. Ansari. 1971. *Chalcolithic Navdatoli,* Pune-Baroda.

Shinde, Vasant. 1985. Kaothe—A Late Harappan Settlement in the Central Tapi Basin, *Bulletin of the Deccan College Post Graduate and Research Institute,* 44, 173–177.

Shinde, Vasant 1998. *Early Settlements in the Central Tapi Basin,* Munshiram Manoharlal, New Delhi.

Singh, Ranjit Pratap. 1990. *Agriculture in Protohistoric India,* Pratibha Prakashan, Delhi.

Thapar, B.K. 1985. *Recent Archaeological Discoveries in India.* Tokyo: The Centre for East Asian Cultural Studies.

Vishnu-Mittre. 1965. Plant Economy at Ancient Navdatoli-Maheswar, *Technical Reports on Archaelogical Remains,* Pune, 13–32.

Vishnu-Mittre, 1969, Remains of Rice and Millet, in Sankalia *et al., Excavations at Ahar,* Pune, 229–235.

Vishnu-Mittre and R. Savithri, 1975–76. Ancient Plant Economy of Inamgaon, *Puratattva,* No. 8, 55–62.

Vishnu Mittre, Aruna Sharma and Chanchala. 1966. Ancient Plant Economy of Daimabad, in S.A. Sali, *Daimabad 1976–79*, Memoirs of the Archaeological Survey of India, No. 83, New Delhi, 588–627.

Wakankar, V.S. 1967, Kayatha Excavations, *Journal of the Vikram University*, Ujjain.

CHAPTER 21

Agriculture in Chalcolithic and Early Iron Age of North-central India

Anup Mishra

INTRODUCTION

Agriculture is one of the major modes of subsistence that brought a landmark change in the socio-economic and cultural life of man. V.G. Childe terms this as a 'Revolution' that started during the Neolithic age.[1] The beginning of agriculture is a very complex process. The environment, drainage system, soil or topography of the region must have played a greater role in its commencement. Apart from these, the demographic pressure, carrying capacity of the area and agricultural methodology, such as knowledge of crops and agricultural implements would have also contributed towards the same. Agriculture encompasses the practical application of the knowledge of sowing, nurturing and harvesting of plants and the subject of cultivating the soil. It is believed that man first learnt how to harvest before learning to sow.

This chapter aims to highlight the beginning of agriculture in the north-central India during the Chalcolithic and Early Iron Age. While referring to north-central India, the regions of western Madhya Pradesh and the eastern Rajasthan are kept in the mind. So, by and large, the land is confined within two hill ranges, Aravalli and Vindhya, and the adjacent valley of Narmada.

This chapter here attempts to understand the commencement of agriculture in the aforesaid region with the help of palaeo-environmental study carried out in the adjacent regions, direct or indirect evidence of agriculture through palynological studies—evidence of forest clearance or alteration of flora by man, and the direct and indirect evidences of agriculture gathered from the archaeological excavations in the study region.

A brief discussion of the extant geomorphic setting, climate and vegetation of the region will not be out of place here.

GEOMORPHIC SETTING

Physiographically, the northeast, east and southeast parts of Rajasthan are together known as Eastern Plain that covers about 23.3% of Rajasthan. Mainly the pediments and flood plains of the major rivers, Chambal, Banas and Mahi, form this tract. The northern extension of this plain in Bharatpur and Morena districts joins with the Upper Gangetic plains. In its southeastern boundary lies the Vindhyan plateau. Hillocks and plateaus represent the major landforms of the Chambal valley. The Chambal basin is characterized by an undulating floodplain, gullies and ravines. The Banas basin, a broad plain having an altitude of 150 to 300 m above MSL with a slope towards the east, is known as the Mewar Plain. The thickness of the alluvial deposits of this plain decreases towards the west where the plain is higher and more irregular, while in the east, the thickness of the alluvium increases. The Mahi basin, formed by the Mahi River and its tributaries, which join the Arabian Sea through the Gulf of Cambay, is popularly known as Chappan Plains. The land in the south is deeply dissected, resulting in the formation of separate hillocks. The southeastern plateau is generally known as Hadauti plateau. It is confined to the southern and southeastern parts of Rajasthan. This plateau has an average altitude of 500 m above MSL, and is dotted with isolated hillocks. This plateau in Rajasthan occurs in the upper catchment of the Chambal River to the southeast of the Mewar Plains. It merges with the Malwa plateau in the east. Physiographically, it can be divided into Vindhyan scrap land and Deccan Lava (Malwa) plateau.[2]

The eastern pediplain, occurring between the Vindhyan plateau and the Aravalli hill range, contains a thin veneer of Quaternary sediments, reworked soil and river channel fills. At least two erosional surfaces can be recognized within the pediplain, which according to Heron, are the Tertiary age. The Vindhyan upland, the adjoining Chambal valley and the Indo-Gangetic alluvial tract (Older Alluvium) are of Pleistocene to Sub-recent age. Badland topography is a characteristic feature of the Chambal valley, whereas *kankar* has extensively developed in the older alluvium.[3]

The Great Indian Water Divide distributes the river water of this region into the Bay of Bengal and the Arabian Sea. The rivers flowing to the west and south of Aravalli range drain the water into the Arabian Sea, whereas others flowing to the east join the Bay of Bengal. The rivers draining into the Arabian Sea are Mahi, Som, Jokham and Sabarmati. The rivers flowing into the Bay of Bengal are Chambal and its tributaries—Banas, Berach, Kothari, Khari, Parvati, Kali Sindh and Banganga. The Chambal is the only perennial river that originates from the Mhow near Manpur in the northern parts of the Vindhyas. It has total course of 960 km, out of which it covers only 135 km in Rajasthan and rest in the Madhya Pradesh and Uttar Pradesh, where it debouches into the Yamuna. The Chambal ravines, well known for their depth, width and length, are formed due to gully erosions. The surface is undulating and the maximum height is about 267 m above MSL and the minimum height is about 167 m above MSL. Most of the areas are still barren.

The western part of Madhya Pradesh is physiographically divided into three regions—the Malwa Plateau, the Vindhyan Plain and the Northwest Plain (Indo-Gangetic Alluvial Plain). The Malwa Plateau has an altitude of 460 m to 600 m and is bound by the Gujarat Plains on the west, the Vindhya range on the south, the Madhya Bharat Plateau and Bundelkhand Upland on the north, and the Vindhya Range on the east. It covers the land between the south of Gwalior and the Vindhyan ranges. At the southwest of the Ratlam town, the plateau descends gradually to the valley of the River Mahi. Having a volcanic origin, it comprises rounded base hills with flat tops interspersed by plains. The erosion has carved the lava into isolated mesas found throughout the plateau, together with an occasional sandstone hill. The slopes of the hills are gentle, steep and wooded. The major rivers of this region are Chambal and its tributaries—Sipra, Chamla, Gambhir, Lakhunder, Khan, Bangeri, Chhoti Kali Sindh, Kedel, Teelar and Badi Kali Sindh. The biggest river, Narmada, flows between the Vindhya and the Satpura hill ranges. The name *Malwa* is derived from the Sanskrit term Malav and means 'part of the abode of Lakshmi', a goddess of wealth. Vegetation is of the savanna type with scattered teak and sal (*Shorea*) forests.

CLIMATE

The climate of this region is subhumid to semi-arid. The annual rainfall in the east Rajasthan ranges from 55 cm to 103 cm, while that in the Malwa region it is about 100 cm and in the Vindhyan region, about 87.5 cm. The winter rainfall resulting from western disturbances is erratic. The plateau region of Malwa is largely temperate, while the eastern part of Aravalli is hot. The northern part is also extremely hot.

VEGETATION

The Vindhyan region is an important forest region, which is essentially tropical dry deciduous forest of Central Indian Type. The subsidiary edaphic type of dry tropical forests is also seen in this region. The common plants of these forests are—teak (*Tectona grandis*), sal (*Terminalia tomentosa*), dhanoda or monyen (*Lannea gransdis*), tinsa (*Ougeinia dalbegioides*), tendu (*Diospyros melanoxylon*), achar (*Buchanania latifolia*), aonla (*Emblica officinalis*), anjan (*Hardwickia binata*), amaltās (*Cassia fistla*), babul (*Acacia arabica*), bahera (*Terminalia belarica*), bar (*Ficus bengalensis*), bel (*Aegle marmelos*), ber (*Zizyphus jujuba*), bija (*Pterocarpus marsupium*), chheola/palās (*Butea forndosa*), dhaman (*Grewia tillaefolia*), dhaora (*Anogeissus latifolia*), dhawri (*Woodfordia floribudna*), dikamali (*Gardenia lucida*), imli (*Tamarindus indica*), mahuā (*Mahuca latifolia*), pipal (*Ficus religiosa*), semal (*Salmalia malabaricum*), shisham (*Dalbergia latifolia*), bamboo (*Dadrocalamus stictus*).

The eastern side of Aravalli is phytogeographically distinct from that of the western side. It constitutes Indo-Malayan type of vegetation, also including the Indian type.

SOIL

The soil of Malwa region is black cotton type, which is very fertile. The pediplain region of the eastern Rajasthan is also fertile.

PALAEO-ENVIRONMENT

The quaternary pollen stratigraphic evidence of India, although incomplete, sheds welcome light on the vegetational and climatic change in different times.

Pollen grain analysis of lake sediments of Rajasthan indicates a progressively drying and warming climate with increasing aeolian activity and moderate and graded advance of sand dune activity climaxed during 5000–3000 years ago when the desert thorn forest increased under maximum dry and high summer temperatures and intense sand dune formation activity. The precipitation estimates indicate a graded pattern of 150–400 mm from west to east of the desert and the climate continued to be predominantly dry.[4]

In the dry/moist deciduous forest region of Gujarat, the vegetational development in early Holocene prior to 7000 years ago commenced with Chenopod-grassland stage, into which *Holoptelea,* one of the sevral elements for the deciduous forest invaded the Chenopod grassland by about 7000 years ago and constituted the riverain forest which declined by about 5000–4000 years ago. Thereafter, the overall predominance of Savannah continued.[5]

In the vicinity of Pratapgarh, U.P., the forest comprised *Tecomella* and *Anogeissus,* a dry deciduous forest constituted about 5000-4000 years ago. It declined about 4000–2000 years ago. The overall predominance of grasses characterizes the sequence significantly, which was preceded before about 4000 years ago by Chenopod-grassland stage.[6]

To summarize, one can say that the *Maytenus* dominated thorn forest *Artemisia*-Gramineae-Sedge Steppe in northern Rajasthan, Chenopod-Grassland Savannah in Gujarat and Ganga Plains and seral stages for the dry deciduous forest formed towards 5000 years ago. There was rise in the sea level by two to three metres.[7]

The palaeo-environmental sequence built up from the vegetational history deduced from pollen analysis has been grouped into six phases. The Phase III pollen zone B (*c.* 7500 BC to 3000 BC) suggests a slight lowering of rainfall around 7500 BC, which was, however, not severe enough to alter the overall ecological patterns already established during the earlier phase.

In the salt-lake profiles from western Rajasthan, one can witness an extraordinary rise in carbonized vegetable remains, i.e., wood fragments, in the sediments at about the beginning of Phase III (pollen zone B) at all the sites. While it cannot be explained through the natural causes, it can be strongly argued that the increase in the burnt remains resulted from the introduction of the practice of scrub burning at the hands of early man. The close correlation between the occurrence of Cerealia-type pollen on one hand and the evidence scrub burning on the other raises the obvious question as to whether some sort of primitive cereal agriculture was in practice in this area as early as 7500 BC. However, this has to be corroborated with the archaeological evidence from the excavations at the microlithic settlements scattered over the southeastern Rajasthan.

In the next Phase IV (pollen zone C), *c.* 3000 BC to 1000 BC, the climate seems to have been changed to wetter conditions, which, however, lasted only upto 1800 BC. Thereafter, the climate shows a small-scale oscillation to drier conditions between c. 1800 to

1500 BC followed by a slight reversal to a relatively weak wetter interval lasting upto about 1000 BC. These three changes in climate are correlated to the three-fold oscillation of climate during the mid-post-glacial time.[8]

The vegetational fluctuations in the dry deciduous *Anogeissus* forest in the vicinity of Ajmer, Rajasthan desert and believed to date from 4000 to 3000 years ago (i.e., *c.* 2000 BC to 1000 BC) show the occurrence of *Anogeissus* forest within predominant grassland stage.[9]

The palaeo-ecological evidence suggests a significant increase in rainfall at the beginning of the third millennium BC that could have played an important part in the sudden expansion of the Neolithic-Chalcolithic cultures in northwest India.

ARCHAEOLOGICAL EVIDENCE OF EARLY AGRICULTURE

In general, the beginning of agriculture can be put within the bracket of Chalcolithic period of the region, which comprises the cultures of Ahar, Kayatha, Malwa and Ochre-coloured pottery (OCP). Since our focus here goes upto the early Iron Age, the Painted Grey ware (PGW) culture is a subject of discussion. On the flip side, it is imperative to look back to the antecedent cultures or the lifestyle of the people who lived in this region before the beginning of Chalcolithic culture. Unlike the common ecozone, we notice an identical cultural milieu is evinced in all parts of this region prior to the beginning of Chalcolithic culture. The most important excavations at the site of Bagor and Adamgarh can be cited here.

The excavation at the site of Bagor[10] has proved that the people who lived in this region before the Chalcolithic period were Mesolithic hunter-gatherers with little knowledge of agriculture. There are about 72 sites of this kind so far been reported in the southeastern Rajasthan, particularly on the Berach basin in the districts of Udaipur and Chittaurgarh. Besides, these sites are also located in the districts of Bhilwara, Ajmer, Tonk, Kota and Jhalawar on the banks of Banas. The sites are also located away from the river basins on the stone elevations in the gneissic plateau of Mewar.[11] The people used microliths of geometric and non-geometric types. V.N. Misra excavated this site.[12]

The area is covered by open woodland of Khejri (*Prosopis spicigera*), *babul*, dhok (*Butea frondesa*) and *ber*. The rainfall of the region is about 60 to 75 cm/annum. The alluvial deposit is fertile for agriculture.

The 1.4 m of habitation deposit yielded an uninterrupted occupation for a period of five millennium before Christ, which comprised three cultural phases (I–III).

Phase I (*c.* 5000–2800 BC) is the earliest, which is purely microlithic in nature. The economy seems to be based on hunting-gathering and herding. Phase II (*c.* 2800–600 BC) shows a marked change in the economy with a decline in the quantity of microliths, animal bones and ring stones that are suggestive of a stable economy with possibility of introduction of primitive agriculture. The presence of saddle quern and rubber stones also consolidate the fact in the absence of any direct evidence of carbonized grains or impression of grains on the burnt clay or pottery. A spindle whorl was found from a burial, which is an important indicator of weaving.

The Vindhyan Hills have housed a large number of natural rock-shelters. The shelters formed by the natural process of differential weathering of the sedimentary rocks are valuable for the investigations of the Stone Age cultures. The location of the Adamgarh Hill is about two kilometres south of the Narmada River at Hoshangabad. R.V. Joshi of the Archaeological Survey of India carried out a systematic excavation of the rock–shelters and the open ground in front of the shelters in 1961. The topmost deposit of the hill is the black clay. This thick deposit produced by water action was the horizon of Mesolithic culture. A large number of microlithis and waste materials were recovered from the trenches. According to the excavator, a detailed study of tools obtained from successive levels from a deposit of one to three metres thick clearly indicates that the earlier microliths belong to the Mesolithic stage and the later ones exhibit Neolithic-Chalcolithic affinities. The objects like mace heads, potteries[13] and animal bones further consolidate the assumption. There is no evidence of agriculture or pastoralism. The presence of domestic animals suggests a sort of Neolithic tradition. Thus, the Mesolithic industry reveals a hunting-based economy and a nomadic life. The available palaeoclimatic study suggests that the climate was more humid during the Palaeolithic times but gradually became dry towards the Mesolithic time. The C^{14} date of shell associated with the early Mesolithic deposit is 7450 ± 130 BP and that of an uncharred bone from upper level is 2845 ± 105 BP.

Both the sites have showed advancement of economy during the late Mesolithic period. The use of ceramics and mace heads are an indirect evidence of a sort of rudimentary agricultural practices. The excavator of Bagor concludes that in phase II, the people had established contact with the contemporary Aharians through whom they could have procured the copper implements and other artefacts.

BEGINNING OF AGRICULTURE

In the absence of Neolithic culture, we now ostensibly look into the Chalcolithic cultures of the region to find the modes of agricultural practices. As has been already mentioned, we find the Ahar culture sites in the southeast Rajasthan, which later on spread into the Vindhyan plain and Malwa regions. Before the extension of Ahar culture, Kayatha culture existed in the Vindhyan region. In the Malwa region, we find the Malwa culture following the decadence of Ahar culture. Apart from, the Ochre-coloured pottery culture, Black-and-Red ware culture and Painted Grey ware culture sites are found in the northern part of the region. The last two cultures fall within the ambit of the Iron Age.

AHAR CULTURE

About 90 sites of this culture have been reported from the southeast Rajasthan on the banks of river Banas and its tributaries—Wagan, Berach, Kothari, Khari, etc.—and River Mahi. The excavations at Ahar, near the city of Udaipur, have provided a few dates bracketed within 2100 BC to 1300 BC. However, the subsequent excavations at Balathal[14] have provided a date of an early beginning of the culture before 2500 BC.[15] Besides, there are two other important excavated sites—Gilund and Ojiyana, located in the Chittaurgarh and Bhilwara districts, respectively. However, the botanical remains have been reported so far only from the site of Ahar and Balathal.

The earliest evidence of this culture comes from the site of Balathal. Although the detailed palaeo-botanical report exclusively of the early level has not been published, it can be quite confidently asserted that the people cultivated barley (*Hordeum vulgare* Linn.) and panicum millet (*Panicum* sp.) during this time. Besides, a large number of coix shells or Job's tears (*Coix Lachryma jobi* Linn.) were obtained, which might have been collected from the wild plants for making ornaments.[16] Elliptical lime-plastered silos were found from the corridors of the houses, which reveal that the agriculture was well established and the surplus was carefully stored in these silos.

The palaeo-botanical remains of Ahar[17] have been studied by Vishnu Mittre.[18] Impressions of husks and spikelets of rice have been reported on the pottery and burnt clay lumps. Vishnu-Mittre observes that the rice belongs to the long-seeded strain of *Oryza sativa* Linn. *vor. Sativa Bor.* Evidences of millets have also been reported from the impressions of spikelets and husks on the potsherds and burnt clay lumps. The comparative study indicates that most of the impressions belonged to sorghum/jowar (*Sorghum vulgare Pers.*) and some to *bajri* (*Pennisetum typhoides Rich*). However, the excavators are sceptical about the context of the Sorghum and opine that these could have come from a later level, i.e., from the early historic period.

The excavations at Balathal have brought to light important information about the early history of agriculture and food habits of the people of Ahar culture. M.D. Kajale has published an exclusive report of the palaeo-botanical remains of Balathal.[19] He reports a large number of carbonized as well as uncarbonized plant remains. The latter includes coix shells, fibres, textiles and impressions on the mud clods and pottery pieces. The evidence of carbonized grains foretells that the people practised farming of two crops, winter (*Rabi*) and summer (*Kharif*) crops. These crops can be categorized into cereals, legumes, oil seeds, vegetables, fruits and others. The cereals included wheat (*Triticum* cf *aestivum* Linn. and *Triticum* sp.), barley (*Hordeum vulgare* Linn.), panicum millet (*Panicum* sp.), indet. Millet (*Setaria* sp.) and Italian millet (*Setaria italica*). The legumes comprised green gram (*Vigna radiata* (L.) *Wilczek*), black gram (*Vigna mungo* (L.) *Hepper*), common pea (*Pisum arvense Linn.*) and gram/Bengal gram (*Cicer arietinum* Linn.). The oil seeds—Linseed/Safflower types, were poorly preserved and, therefore, are not very clear. The fruits were Tranthema/Partulaca (*Trianthem* sp./*Partulaca* sp.), Bor (*Zizyphus jujuba* Linn.), etc. Among the vegetables, cucurbitaceae types (*Cucumis* sp.) and wild lady's finger types (*Ablemeschus* sp.) were common. Besides, there were some weed seeds.

The agriculture, as indicated by the above evidences, was a full time and regular practice, which played a crucial role in the economy of the people. Though it is hard to assume the tilling methodologies, the presence of kohl sticks and the ring-stones or mace heads provide some ideas about them. The evidence of ploughing has come from Kalibangan.[20a] So, it can be presumed that the later contemporary 'Aharians' might have also known the technology of ploughing.

Sites	Botanical Remains	Stone objects	Metal objects	T.C. objects	Ceramics	Bone objects	Faunal Evidence	Others
Ahar	Rice, Millet	Ring-stone, quern, muller	Kohl-stick, axe (copper)	Spindle whorl, Bull figurines, wheels	Storage jars, dough plates, cooking vessels	Kohl-stick	Cattle and Buffalo	Silos, micro-liths Hearths
Balathal	Wheat, barley, panicum millet, indet. Millet, Italian millet, green gram, Black gram, common pea, Bengal gram, Linseed/ Safflower, cucurbitaceae, wild lady's finger types, Tranthema/ Partulaca, Bor	Ring-stone, quern, muller	—	Spindle whorl, Bull figurines	Storage jars, dough plates, cooking vessels	—	Cattle and Buffalo	Silos, Hearths
Gilund	—	Ring-stone, quern, muller	—	Spindle whorl, Bull figurines	Storage jars, dough plates, cooking vessels	—	Cattle and Buffalo	Silos, micro-liths Hearths
Ojiyana	—	Ring-stone, quern, muller	—	Bull figurines	Storage jars, dough plates, cooking vessels		Cattle and Buffalo	Silos, Hearths

There are a lot of supporting evidences seen at the sites, which indirectly indicate the agricultural practices, such as the storage pits or silos plastered with lime, storage jars, grinding stones or saddle querns and mullers, terracotta spindle whorls for weaving, etc. Besides, the hearths, cooking vessels and dining utensils also throw substantial light on the food habits of the people. The use of cart and the bull worship can also be related to the agricultural practices.

KAYATHA CULTURE

More than 40 sites of Kayatha culture have been discovered in the Chambal valley and on the banks of its tributaries. The culture has a distinct geographical distribution and has shown a peculiar type of ceramics and other cultural traits. The excavations at the site of Kayatha[21] has revealed five-fold cultural phases. Opening with the Kayatha culture (*c.* 2000–1800 BC), the Chalcolithic cultures continued through the Ahar culture (*c.*1700–1500 BC) and the Malwa culture (*c.* 1500–1200 BC).

M.D. Kajale studied the botanical remains collected from the excavations.[22] The remains were mostly of Indian jujube or *bor* (*Zizyphus jujuba Linn.*). However, the indirect evidences of agriculture have been found aplenty. The large storage jars decorated with appliqué designs would have been exclusively used for storing grains. The ring-stones or the mace heads have been widely accepted as the agricultural implements, which could have been used for turning the soil. The copper axes with sharp cutting edges might have helped in clearing the jungles. The faunal evidence adds a note of participation of cattle in agriculture. A special study of the 'Pedosis of the third phalanx' was undertaken to know by their angle of inclination and pointing of the toe whether the animals were actually used for agricultural operations.[23] The study revealed a positive result of their use in the agriculture. This is a significant discovery that established the role of cattle in the agricultural revolution during the Chalcolithic period. The stylized and naturalistic bull figurines found in large quantities are suggestive of the importance of animals in the agricultural community.

The farmers of the Chalcolithic Kayatha culture exploited the fertile black cotton soil of the Chambal valley in the second millennium BC. However, unlike their counterpart 'Aharians' in the southeast Rajasthan, these people confined themselves to the riverbanks and raised the crops on the alluvial patches.

Sites	Botanical Remains	Stone objects	Metal objects	T.C. objects	Ceramics	Faunal Evidence	Others
Kayatha	Bor	Ring-stone, quern, muller	Axe (copper)	Bull figurines	Storage jars, dough plates, cooking vessels	Cattle and Buffalo, Pedosis of the III phalanx indicates their use in the cultivation	Silos, Hearths

MALWA CULTURE

A large number of sites of this culture have been reported from the Malwa region, the Narmada valley. The important excavated sites are Maheshwar, Navdatoli, Nagda, Manoti, Chichali, etc. The sites of Maheshwar and Navdatoli[24] are the best-studied sites, which yielded seven culture sequences starting from Palaeolithic period to the eighteenth century. The period III belonged to the Chalcolithic Malwa culture that has been dated back to *c.* 1500 BC to 1200 BC. In the point of view of the present study, the excavations of Maheshwar-Navdatoli and Chichali[25] are important.

MAHESHWAR-NAVDATOLI

The Chalcolithic people had occupied tops of the terraces and perhaps cultivated the riverbanks. Both *kharif* and *rabi* crops were sown, indicating that agriculture was a full-time occupation of the people. The excavations have yielded a large quantity of carbonized grains categorized into cereals, legumes and oil seeds. The cereals comprised wheat (*Triticum vulgare-compactum*) and rice (*Orzya sativa* L.) and the legumes were lentil (*Lens culinais medikus* and *Lens esculanta Moench*), black gram (*Phaseolus mungo* L.), green gram (*Phaseolus radiatus*) and grass pea (*Lathyrus sativus* and *Pisum arvense*). The oil seeds included linseed or flax (*Linum usitatissimum*).

Among the subsidiary evidences, ring-stones, saddle querns and rubber stones are important. The former was probably used for turning the soil up and down. These stones are found in a number of shapes and sizes, such as oval, rectangular, round, discoid, oblong, etc. Two types of querns were noted in a large number (67)—saddle quern and legged querns. There were 400 rubber stones found in different shapes and sizes, viz., oblong, oval, rectangular and circular. Besides, the copper celts, terracotta spindle whorls, ceramics and bull figurines of various types represent an agricultural society.

A large number of microliths were recovered from the Chalcolithic level. These, according to the experts, could have been used as sickle teeth, particularly the lunates and trapezes.

According to the excavators, wheat was probably cultivated with a plough and rice was grown manually in the ponds and ponded water of the river. The lentils, peas, *urd* and flax were grown in silt-laden, low river terraces immediately after the monsoon. These crops could have been easily grown in the 'hoe-cum-plough' agriculture practised by the people and a small settlement like Navdatoli could easily have sustained with the harvest.

CHICHALI

The excavations at Chichali revealed four-fold culture sequences, out of which the periods I–III belonged to Ahar, Malwa and Jorwe cultures. The Ahar culture people cultivated wheat, green gram (*mung*) and lentil (*tuwar*). The supporting objects, used by the agricultural community, were mace heads, saddle querns, storage jars and dough plates and hearths. Microliths were obtained from this level. In the subsequent period, the Malwa people cultivated cereals and legumes. The former included wheat, barley and the latter consisted black gram and green gram. There were storage jars found embedded in the floor and hearths of single, double and triple types were used for cooking. Some copper implements and microliths were also recovered. In the Jorwe period, people used querns, mullers and animal figurines, which throw light on an economy based on agriculture in absence of the direct evidences.

Sites	Botanical Remains	Stone objects	Metal objects	T.C. objects	Ceramics	Faunal Evidence	Others
Navdatoli	Wheat, rice, lentil, green gram, black gram, grass pea, linseed	Ring-stone, quern, muller	Axe (copper)	Bull figurines, spindle whorl	Storage jars, cooking vessels, dough plates	Cattle and Buffalo	Silos, microliths, Hearths
Nagda	—	Ring-stone, quern, muller, pounder	—	Bull figurines, spindle whorl	Storage jars, cooking vessels	Cattle and Buffalo	Hearths
Chichali	Wheat, barley, black gram and green gram	Ring-stone, quern, muller	—	Bull figurines, spindle whorl	Storage jars, dough plates, cooking vessels	Cattle and Buffalo	Hearths

OCHRE-COLOURED POTTERY CULTURE (OCP): c. 2000–1500 BC

The Ochre-coloured Pottery culture sites are found in the districts of Jaipur, Jhunjunu, Sikar and Bharatpur. However, only the last mentioned district falls within the study region. Noh[26] is the only important excavated site situated in this district. The site produced a five-fold culture, the earliest period of which belong to the OCP culture. However, no botanical remains have been reported from this period. The other two famous sites, Ganeshwar (District Sikar) and Jodhpura (District Jaipur) also have not yielded any botanical remains. It could be, however, due to the small-scale excavations at these sites.

These sites fall within the sphere of the Indo-Gangetic plain, which is a very fertile area and, thus, has been the centre of attraction since the OCP period. A number of such sites have been excavated in this region. Most of them have yielded remains of charred grains comprising cereals, legumes, etc. The example of Atranjikhera,[27] Hastinapura[28] and Lal Qila[29] can be cited here. The botanical remains recovered from Atranjikhera, in particular, are extremely important. The people grew two crops a year, rice in the summer and barley in the winter along with legumes like gram and *khesari.* The other antiquities also signify to a farming community.

BLACK-AND-RED WARE CULTURE (c. 1500–1200 BC)

The succeeding culture of OCP is the Black-and-Red ware culture, which has been found from most of the OCP sites. At Noh, no carbonized grains are found from this level. However, at Atranjikhera, there are evidences of rice and barley.

BEGINNING OF IRON AGE

In the end of the Black-and-Red ware culture, fresh groups of people occupied the Ganga-Yamuna doab in the name of Painted Grey ware culture (*c.*1200–600 BC) and Northern Black Polished ware culture (*c.* 600–50 BC). These people introduced the technique of iron smelting and casting. In the Malwa, Vindhyan and east and southeast Rajasthan after the decadence of Chalcolithic culture, there existed a cultural vacuum for at least five to six centuries. The succeeding people were of the early Historic period (*c.* 500–200 BC) who had the knowledge of iron. The iron implements were introduced in the cultivation, which gave an impetus to the agricultural practice and production.

The plant remains of Early Historic Balathal represent the agricultural practice in the central India during the Iron Age. The people produced two crops a year, *kharif* and *rabi,* which included cereals, legumes, oil seeds, fruits, vegetables, fibre plants and others. These are wheat, barley, rice, kodo millet (*Paspalum scrobiculatum* Linn.), panicum millet, indet, millet, green gram, black gram, common pea, Bengal gram, grapevine, trianthema/portulaca, Indian jujube, argemone type, wild lady's finger type, textile fibres, Italian millet, coix shell and some weed seeds. This shows a remarkable progress of agriculture in this time.

In the northern part of the region, people cultivated rice, wheat and barley along with the other crops. The agricultural implements included ploughs, sickles, knives, spades, hoes, axes, adzes and diggers (NBPW period at Atranjikhera, early historic period Balathal and Ahar).

CONCLUSION

In the palynological study, we have seen a wet phase prevailing over the central India in the beginning of third millennium BC that continued uninterruptedly till the 1800 BC. The archaeological evidence indicates the progress of life of man during this time. We notice the emergence of Ahar and Kayatha culture and beginning of agriculture in the meantime. While the latter bites the dust exactly in 1800 BC, we find disintegration of bigger sites of the former into smaller sites.[30] This is the time perhaps when the Aharians entered into the Malwa and the Vindhya regions. Again, the time matches with the palaeo-environmental study, which shows a dry phase from 1800 BC to 1500 BC. The improvement of climate, relapse of wet phase, in the beginning of 1500 BC saw the development of Malwa culture in the central India. There is no doubt that the economy of these cultures was based primarily on agriculture. The associate materials obtained from all the sites are almost identical, which reflect an integrated story of progression of agriculture based rural cultures.

In the above discussion, it is clear that rice, wheat and barley were cultivated invariably right from the beginning of agriculture in the study region. Rice, being native to India, is found in the wild form in some parts of the country. In the central India, only Ahar, Navdatoli and Maheshwar have yielded evidence of rice in the Chalcolithic period. Wheat and barley were cultivated regularly. Wheat being a winter crop, requires a supply of irrigated water. The surplus of this crop envisages a good winter rainfall during the Chalcolithic period. *Rabi* crops like black gram and green gram would have been cultivated on the riverbanks. Perhaps, these crops were harvested along with the major crops, a sort of mixed agriculture, as is being practised at present. The agriculture methodology seems to be very simple as no implement other than the kohl stick and mace heads have been encountered. There is evidence of participation of cattle in the agriculture. However, it cannot be asserted emphatically that they were being used for ploughing the field; they could have been used for dragging carts. The importance of cattle in the society can be enunciated from the presence of bull figurines or prevalence of 'bull worship'.

The evidence of granary from Balathal throws welcome light on the surplus production of the major crops, such as wheat and barley. Besides the granary, almost every household had storage pits for storing crops. These pits were plastered with lime and chopped grass. The crops, produced in a small scale, were stored in the jars inside the houses. The multi-armed hearths, different types of cooking vessels and dining vessels indicate a variety of food prepared and served in solid and liquid forms.

The invention of iron implements and their introduction in the agriculture gave it a new dimension. Clearing jungles was easier and there was rapid expansion of cultivated lands. Apart from rainwater, the people started tapping the groundwater by digging wells and channelling the river water through canals.

So, from the above discussion, it can be concluded that agriculture in the north central India started in the first half of third millennium BC when the climate changed from a spell of dry phase to a conducive wet phase. The evidences obtained from the Mesolithic sites of Bagor and Adamgarh are suggestive of the beginning of a sort of primitive agriculture around *c.* 2800 BC.

NOTES AND REFERENCES

1. V.G. Childe, *What Happened in the History*, Hammondsworth, (1942).
2. S. Sinha-Roy, G. Malhotra and M. Mohanty, *Geology of Rajasthan*, Geological Society of India, Bangalore, (1998).
3. A.M. Heron, 'The Geology of Central Rajaputana', *Memoir of Geological Survey of India*, Vol. 79: 389, (1953).
4. Vishnu-Mittre, 'Palaeoecology of Rajasthan Desert during the last 10,000 years', *Palaeobotany* 25: 549–558, (1978).
5. Vishnu-Mittre, 'Environment of Early Man in Northwest India: Palaeobotanical Evidence, in SRK Chopra (ed.) *Early Man in NW India*, Allied Publishers, Delhi, (1979).
6. H.P. Gupta, 'Holocene Palynology from meander lake in the Ganga Valley, District Pratapgarh, U.P.', *Palaeobotany*, 29:109–119, 1978.
7. Vishnu-Mittre, *et al.*, 'Pollen Stratigraphy of India', In K.N. Dikshit (ed.) *Archaeological Perspectives of India Since Independence*, New Delhi: Books and Books, pp. 115–122, (1985).
8. Gurudip Singh, 'Indus Valley Culture', in Gregory L. Possehl Ed. *Ancient Cities of the Indus*, Vikas Publishing House, New Delhi, (1979).

9. *Op. cit.* Vishnu-Mittre, 1978, 1979.
10. The site of Bagor is situated on the left bank of river Kothari, a tributary of Banas, 25 km west of the town of Bhilwara. The site is located on a stabilised fossil sand dune locally known as *Mahasati.*
11. V.N. Misra, *Pre-and-Protohistory of Berach Basin, South Rajasthan,* Deccan College, Pune, (1967); H.D. Sankalia, *Pre and-Protohistory of India and Pakistan,* Poona, pp. 259–60, (1974).
12. *Ibid.*, 1967; *IAR* 1967–68: 41–42; V.N. Misra, 'Bagor—A Late Mesolithic Settlement in North India', *World Archaeology* 5(1): 92–110, (1972).
13. A. Ghosh (ed.), *An Encyclopaedia of Indian Archaeology,* Munshiram Manoharlal, New Delhi, pp. 1–2, (1989).
14. Balathal (24° 43' N: 73° 59' E): The ancient mound is located on the eastern fringe of the village Balathal in Vallabhanagar *tehsil* of Udaipur district, Rajasthan. The excavations were conducted at the site from 1994 to 2000 by the department of Archaeology, Deccan College, Pune, and the Institute of Rajasthan Studies, Rajasthan University, Udaipur under the direction of V.N. Misra. (see Misra *et al.*, 'The Excavations at Balathal: Their Contribution to the Chalcolithic and Iron Age Cultures of Mewar, Rajasthan', *Man and Environment* XX (1): 57–80, (1995); V.N. Misra, 'Balathal: A Chalcolithic Settlement in Mewar, Rajasthan, India: Results of First Three Seasons' Excavation', *South Asian Studies* 13: 251–273, (1997).
15. Anup Mishra, 'Chalcolithic Ceramics of Balathal, District Udaipur, Rajasthan', Unpublished Ph.D. Thesis, University of Pune, Pune, (2000). Anup Mishra, 'The First Farming Community of Southeast Rajasthan', in *Prāgdhārā* 13: 1–27, (2002–03).
16. Personal Communication with M. Kajale.
17. The site of Ahar (*Sank.* Aghatpur) (24° 35' N: 74° 44' E) is situated on the right bank of river Ahad about three kilometres to the east of Udaipur town. H.D. Sankalia excavated the site in 1961–62.
18. Vishnu-Mittre, 'Remains of Rice and Millet', in H.D. Sankalia *et al., Excavations at Ahar (Tambavati),* Deccan College, Pune, pp. 229–239, (1969).
19. M. Kajale, 'Palaeobotanical Investigations at Balathal: Preliminary Results', in *Man and Environment* Vol.XXI (1): 98–102, (1996).
20. Kalibangan is located in the District Ganganagar, north Rajasthan. It is an important pre-Harappan site which evolved to the Mature stage. Remains of a cultivated field have been unearthed here, which belongs to *c.* 2600 BC. B.B. Lal, 'Perhaps the earliest Ploughed Field so far excavated anywhere in the World', *Purātattva* 4: 1–3, (1970–71).

20a. Allchin, F.R., 'Early Cultivated Plants in India and Pakistan', in Gregory L. Possehl (ed.) *Ancient Cities of the Indus,* Vikas Publishing House, New Delhi, (1979).

21. Kayatha (23° 14' N: 76° 02' E) is situated 25 km east of Ujjain, in the district of Ujjain and on the bank, of Chhoti Kali Sindh, a tributary of Kali Sindh, which in turn, is a branch of Chambal River. It was excavated first by the Vikram University, Ujjain in 1965–67 (*IAR* 1967–68:24) and later on jointly with Deccan College, Pune in 1968; Z.H. Ansari and M.K. Dhavalikar, *Excavation at Kayatha,* Pune, (1975).

22. M.D. Kajale, in Z.H. Ansari and M.K. Dhavalikar, *Excavation at Kayatha*, Pune, (1975).
23. K.R. Alur, 'Reports on the Animal Remains from Kayatha' (Appendix-A), in Z.H. Ansari and M.K. Dhavalikar, *Excavation at Kayatha*, Pune, p. 159, (1975).
24. The sites of Maheshwar and Navdatoli (22° 11' N: 75° 36' E) are located in the district of west Nimar on the north and south banks of river Narmada, respectively. The Universities of Baroda, Bombay and Pune jointly excavated these sites under the direction of H.D. Sankalia in 1952–53 and further horizontal excavations were undertaken at Navdatoli in 1957–59. [see H.D. Sankalia, S.B. Deo and Z.D. Ansari, *Chalcolithic Navdatoli*, Pune, (1971)].
25. Chichali (22° 08' N: 750 22' E) is located on the left bank of river Narmada about seven kilometres west of village Nimrani on the Mumbai-Agra National Highway. The prehistory branch of Archaeological Survey of India excavated the site in 1998–2000 under the direction of S.K. Mittra. [S.K. Mittra and Shivananda V., 'Chalcolithic Settlements at Chichali', in *Puratattva* No. 30:45–49, (1999–2000)].
26. The site of Noh is located in the Bharatpur district, about 6.5 km west of Bharatpur on the Agra-Bharatpur road. The site was excavated by the Department of Archaeology, Rajasthan in collaboration with J. Le Roy Davidson of the University of California. [see A. Ghosh, *An Encyclopaedia of Archaeology*, Vol. 2, Munshiram Manoharlal, New Delhi, pp. 318–319, (1989)].
27. Atranjikhera (27° 42' N: 78° 44' E) is located on the western bank of the River Kalinadi, a tributary of river Ganga in the district of Etah, U.P. The site was excavated for seven seasons starting from 1964. R.C. Gaur, *Excavations at Aranjikhera*, Motilal Banarsidass, Delhi, (1983).
28. Hastinapura (29° 09' N: 78° 03' E) is located in the Mowana *tehsil*, District Meerut, U.P. B.B. Lal excavated the site from 1951–52. B.B. Lal, 'Excavations at Hastinapura and other Explorations', in *Ancient India* Nos. 10–11, (1954–55).
29. Lal Qila (28° 30' N: 78° 15') is situated on the left bank of the Kalinadi, adjacent to the village Narainpur in the Bulandshahr District. The site was excavated by the Aligarh Muslim University from 1969–70 to 1970–71 under the direction of R.C. Gaur.
30. R.K. Mohanty *et al.*, 'Purani Marmi: A late Ahar Culture site in Chittaurgarh District, Rajasthan', in *Purātattva* No. 30: 132–141, (1999–2000).

CHAPTER 22

Agriculture in the Gangetic Plains During the First Millennium BC

Vibha Tripathi

INTRODUCTION

Agriculture has been the backbone of economy right since the subsistence level of human history. A land blessed with fertile soil and hospitable climate has drawn settlers towards it far more than any other commodity. Such an area has been categorically labelled as the area of attraction by B. Subbarao (1956). While narrating the sequential development of urban centres in his centre place theory, Chrystaller (1933) attributes the highest priority to agriculturally productive area. In his assessment, these were the primary centres around which a network of second order and third order centres grew. One hardly needs, therefore, to emphasize the importance of agriculture as the most significant and fundamental factor in the growth of civilizations.

This chapter focuses on archaeological evidences of agriculture in the Gangetic plain. The study of the agriculture in this region may be done within the framework of subzones, which precondition and limit human behaviour. Broadly speaking, the Gangetic plains may be divided into three subzones:

i. Upper Gangetic plain;
ii. Middle Gangetic plain; and
iii. Lower Gangetic plain.

The discussion of agriculture in the above three divisions will be preambled by the ecological conditions prevailing over the Gangetic plains. A close look at the data indicates little changes worth the name in this region. Only exceptions to it are certain changes in the drainage pattern in the upper Gangetic plain, especially in its upper reaches. However, any major ecological upheavals are ruled out during the first millennium BC in the Gangetic plain.

THE ECOLOGY

A large corridor of plain is formed by the rises of Himalayas on the one hand and the Vindhyas on the other. It is watered by major perennial rivers like Ganga, Yamuna, Ghaghra, Gomti, Gandak, Son and several minor tributaries having a sedimental mantle of approximately 1800 m. It expands over a stretch of 1050 km. This long stretch of land is not one homogenous mass. The eco-system and the landform divides it into a threefold division of upper, middle and lower Ganga plains, as indicated above. We need to look for certain fine distinctive features of soil, climate, vegetation, rainfall, etc., in these regions. These conditions affect the vegetation and crop pattern sufficiently to warrant a serious consideration for a study such as the present one.

The Upper Gangetic plain has the maximum length of 550 km from the east to the west and is about 380 km wide. The Allahabad-Faizabad tract is the dividing line between the upper and the middle Ganga plains. Climatically, it is a subhumid zone (Singh, 1955:69). It is comparatively drier and more arid than its more easterly parts. Wheat, followed by rice is the main crop growing in this part of the plain. Parts of Punjab, Rajasthan, Haryana and Western Uttar Pradesh may be included in this region. The weather here is divisible into four seasons as in most of the plains of India. The climate tends to be relatively more prone to extremes, comparatively speaking. Winters could be frosty, especially in January with hailstorms upto February (which damages sugarcane and other crops of the season). The temperature at times drops down to 6–7 degrees centigrade and goes up to 45° centigrade during May–June in summers when rains start cooling the scorching heat of summers. The annual rainfall varies from 50 cm to 140 cm with an uneven spatial distribution. The area around Agra and Mathura has only 25 inches of rainfall, while the north and east gets approximately 50 inches of annual rainfall.

The soil is by and large homogenous throughout the upper Gangetic plains with *Ūsar* and *Bhābar* being the two types of soils, depending on the drainage pattern of the particular area. *Khādar* and *Bhāngar* are other soils having a variety of local names. Khadar is a soil with rich nutrients found in the flood plains of rivers generally having a pH value of 6–8. *Bhāngar* soil is more extensive in inter pluvial zones. The soluble salt is present in a smaller percentage. This is neutral to slightly acidic in reaction. The soil closer to the Ganga is loamy to sandy-loamy in texture while the Yamuna region has a more silty composition. The *Bhāngar* soil is poor in phosphoric acid, nitrogen and other organic matters.

The drainage system of the region has Ganga and Yamuna as the major rivers with tributaries like Ghaghra and Gomti. A host of streams rising out of Siwaliks irrigate the western part of the plain. On the eastern side, there are Kalinadi, Sukantha, Hindon, etc. Most of the rivers are perennial, providing a hospitable land for human occupation. It is because of such hospitable conditions that this is one of the most cultivated parts of the country. In ancient days, thick vegetal cover, especially with moist deciduous forests must have covered the land. The plains are manmade, coming up as a result of deforestation for centuries. According to ancient literary sources, the area between Saraswati and Sadanira was covered with thick forests like Basundha, Naimiṣaraṇya and Utpalaraṇya, each associated with anecdotes related to different mythical contexts. *Sal, Babool, Khair, Shisham,*

Semal, etc., are the trees growing there. Many of these, along with several other plant remains, have been testified from excavations at Atranjikhera, Lal Quila, Hastinapur, etc.

The Middle Gangetic plain has the maximum length of approximately 600 km from the east to the west and the width of about 330 km. Stamp (1959: 316–326) has drawn the boundaries between upper and middle Gangetic plains on the basis of crops. According to him, rice becomes more prominent in this region as compared to wheat and barley, the primary crops of the subhumid upper Ganga plain. There is a distinction in housing scheme too, according to Stamp. Instead of tiled roofs of the east, wattle and daub mud houses are more commonly found in the middle Ganga plain; that area is demarcated along 100 cm isohyet. Stamp's demarcation line has been acceptable to Spate (1967: 351–354) but this line of middle and upper Ganga plain is vaguely described as 'a line running roughly from Ganga-Yamuna confluence at Allahabad across N NW to S SE section of the Ganga'. The middle Ganga plain covers the states of Uttar Pradesh and Bihar lying on either side of the Ganga and the Saryu (Ghaghra) within the Himalayan and peninsular ramparts on the north and south, respectively. This region has been an area of intense socio-political activity throughout the first millennium BC primarily because of its hospitable nature. It has a favourable climatic condition with a good rainfall ranging from 170 cm to 90 cm per annum. The rain is heavier in the eastern side than the western. It also gets good rains during winters. It is because of the good rainfall that this area is more densely populated when compared to the lower rainfall zones of the plains. This region has a broad alluvial soil cover, divisible into *Khādar* and *Bhāngar*. The *Khādar* land is closer to flood plains of rivers that get replenished during the annual floods. Generally, it is silty. But rivers like Ghaghra, Gandak and Son have more sandy beds. *Bhāngar* is the older alluvium and covers the upland tracts beyond the flood limits. There are patches of *Usar* land in these areas. The soil grades from sandy to clayey loam, in parts too clayey and, therefore, impervious. It is richer in lime content and *Kankar* and more suitable for rice cultivation, as it holds water. The Tarai region, with a narrow tip, is a pebbly *Bhābar* zone, exceedingly good for the rice cultivation. This area was once covered with thick forests.

Through the centuries man has preferred these fertile tracts compared to areas with lower fertility. The fact is borne out in archaeological remains. Indepth settlement studies taken up in Kanpur district in U.P. and Haryana (Makkhan Lal, 1984, Brehmadutt, 1980, 166) reveal that alluvial (*Bhāngar*) land was densely populated. In Haryana, out of 258 sites of PGW, 155 are located on alluvial land. In Kanpur also, the Ganga-Rind Doab is far more extensively colonized than the Yamuna-Rind Doab.

The lower Gangetic plain is an extension of the middle Gangetic plain with a greater humidity level. The rainfall here is around 170 cm and above. The soil is rich *Bhāngar*, suitable for rice cultivation. Part of this plain has a coastal climate and vegetation.

Ecological conditions invariably transform human behaviour and the cultural environment harmonizes with it. The specificities of cultural norms are reflected in material remains found in each zone defined above. Culturally also, as will become apparent in course of the discussion below, the Gangetic plains may not be treated as one homogenous entity. Therefore, our inquiry will have to be focused within a framework of the following categories:

Stage I: The Neolithic-Chalcolithic Background
Stage II: The Early Iron Age Cultures
Stage III: The Early Urban Cultures

STAGE I: THE NEOLITHIC-CHALCOLITHIC BACKGROUND

The second millennium BC laid down the foundations of the cultural growth of the Gangetic plains by the early farming communities. Thus, an uninterrupted human occupation of the Gangetic plains began with Neolithic settlements. The hilly terrains of the Vindhyas and Kaimurs are dotted with a plethora of small settlements wherever the ecological conditions were favourable.

The beginning of agriculture in the region may be traced with the evidence of Mahagarha and Koldihwa (Dist. Allahabad) in the Belan Valley. Rice grains identified as *Orizya sativa*—cultivated variety of rice—have been found in a Neolithic context dated as early as 5440±240 and 4530±185 BC by radiocarbon (Sharma, 1980) and more recently term Lahuradeva (Tiwari) . This is the earliest evidence of rice cultivation found anywhere in the world baring a new evidence forthcoming from China in recent years. Although these dates may not be acceptable to many archaeologists, yet the context is indisputably early and is of great significance to the reconstruction of history of agriculture in India. This revolutionizes the earlier evidence about the age and place of the origin of the rice. It acquires greater significance, especially in view of the presence of the wild rice at Chopanimando (Dist. Allahabad, on Belan River) in an earlier strata of advanced Mesolithic level. It is an evidence of the transformation of this species from its wild to the domesticated variety. With this evidence, India gets the credit to demonstrate an evolution in cultivation of rice from wild to cultivated variety.

In the middle and lower Gangetic plains also, there are a large number of sites like Sohgaura, Narhan, Imlidih, Chirand, Senuwar, Mangalkot, Nal ka Tila, and now Lahuradeva (Saraswat, 2004:2005, etc., with strong Neolithic traits. Here, bone industry dominates over stone. There are pottery types ranging from medium to coarse variety, viz., rusticated ware, red coloured corded ware of medium and finer variety besides other decorated types. Painted and plain Black and Red Ware, Black Slipped ware or Black wares (with a black core) and other ceramics, both handmade as well as slow wheel made varieties.

Thatched roofed wattle-and-daub structures were present both at Senuwar (Rohtas) and Narhan (Gorakhpur). A good evidence of cereals is recovered, viz., (*triticum*) Kodon millet (*paspalum scrobiclatum*), pearl millet (*bajra – pennisetum typhoides*) and pulses like chick pea or gram (*cicer arietinum* L.), field pea (*pisum sativum*), green gram (*mung*), (*vigna radiata*), horse gram or kulthi (*dolichos* biflorus) / grass pear, khesari (*lathyrus sativus*), lentil (*lens culinaris*), moth bean, and aconite bean (*vigna acoritifilia*). Oil seeds like *sarson* (*brassica juncea*), sesame (*sesame indicum*), linseed or flax (*linum usitatissium*) are also found. Interesting evidence is the pip of grapes, dates, and jackfruits as also of watermelons. Senuwar has also yielded an equally important evidence from the earliest period (Neolithic, dated to 2000 BC). Grains like rice, barley, dwarf wheat, bread wheat, sorghum millet, chickpea, green gram (*mung*), field pea, lentil, horse gram, grass pea, sesame and linseed have been identified. In the opinion of Saraswat (1995–96), the evidence of wheat from Senuwar is of special significance as it is of identical species as the Harappan wheat, suggesting Harappan connections. There is one more important evidence—the availability of hemp or *bhāng* (*cannabis sativa*) from this period. Summarizing the evidence of agricultural remains from Senuwar, Singh feels that 'the cereal remains suggest the spread of Harappan agricultural products in the middle Ganga plains' (Singh, 1995: 1996, 91).

With the introduction of copper almost in a similar cultural context, a new era of chalcolithism is heralded with certain innovations dated generally between 13/1200–8/700 BC, especially in the middle and lower Ganga plains. The important sites of this phase, in addition to the others mentioned above, are Kausambi, Ayodhya (Dist. Faizabad), Rajghat (Varanasi), Prahladpur (Varanasi), Mason (Ghazipur), Tilaurakot, Khairadih (Ballia), Sonpur, Vaisali, Rajgir, Champa, Taradih (Gaya), Pandu Rajar Dhibi (Dist. Burdwan), Mangalkot (Dist. Burdwan), Hathikhera, Mahishdal (Dist. Bankura). Such settlements extend from eastern Uttar Pradesh to Bihar and Bengal. The latest in this series is Nal ka Tila near Mirzapur in U.P., where iron appears in 1300–1200 BC context (Dr. Rakesh Tiwari, Personal Communication). At most of these sites, the earliest culture is predominantly a Black and Red Ware one. Copper is generally present at these sites in a small amount. Animal husbandry, substantiated by food gathering and hunting, was the primary mode of livelihood, though agriculture was being practiced. A large number of bones—many with cut marks or charred—suggest a heavy reliance of cattle rearing and hunting. River shell and snails were also eaten. Bones of elephant, rhino, buffalo, ox, stag and deer have been found from Chirand (*IAR*, 1963–64: 60. 1964–65: 46).

Bone artifacts dominate the material remains. We come across objects like hammers, points, borers, awls, shaft strengtheners, arrowheads, needles, etc. Stone objects include celts, chisels, scrapers, pestle and quern, balls, as well as microliths. Parallel-sided blades, scrapers, arrowheads, points and lunates are also met with. Wherever possible, semi-precious stones like chalcedony, chert, agate and jasper, etc., were collected from riverbeds and shaped into beautiful beads. Steatite and faience beads are in use right from the Neolithic stage and continue till BRW phase, as is evidenced at Narhan, Senuwar and Khairadih. Steatite microbeads are significant from the point of view of the infiltration of Harappan traits into this region.

Cord-impressed pottery of red and grey colour, many of them burnished on the outside, are the norm. Many pots have a shining slip on the outside with beautiful incised decorations (Narhan, Imlidih, Senuwar, Nal ka Tila, etc.). There are vases, basins, dish or bowls-on-stand, footed and perforated bowls, spouted *lotas* and spouted vases with narrow necks, channel spout are the common shapes. Postholes, indicating circular huts—many with *chulhas*—are found. The vegetal remains during this period are the same as mentioned above; the only addition being safflower from Narhan in 800–600 BC context from what is described as Black Slipped ware phase (Saraswat, 1994). Besides the cereals mentioned earlier, rice remains the main crop at Chalcolithic period in the lower Ganga plains. Mangalkot has yielded rice from a hearth (Amita Ray, 1987). At Mahishdal, according to Ghosh (1984: 17), rice could be stored in a large quantity in silos having a capacity of about 9 quintals (in every likelihood used for paddy). The overall picture is that of a prosperous agricultural community living in small-to medium-sized villages (Singh, 1990, 83).

STAGE II: THE EARLY IRON AGE CULTURES

It is in this very cultural milieu that iron is introduced in the middle and lower Ganga plains. Iron generally appears with Black and Red ware in pre-NBP levels at sites like Kausambi, Narhan, Chirand, Sonpur, Mangalkot, Pandu Rajar Dhibi, Hathigra and several other sites of this region. The culture is roughly contemporaneous with the PGW culture of upper Ganga plains. We propose to examine this phase in some detail now.

Much before the advent of the iron, the upper Gangetic plain was colonized by the Ochre-Coloured Pottery culture. This is borne out by sporadic and flimsy settlements in western Uttar Pradesh and the adjascent areas of Punjab, Haryana and Rajasthan. A thick Red ware with a rolled appearance is the hallmark of this period. This phase is associated with some bone and stone objects, a few terracotta specimens of rather crude variety, presenting a picture of an impoverished village culture. Its association with copper hoards is gaining ground with excavations at Saipai in District Etawah, as two copper hoard implements were found from OCP deposits itself over here during excavations. The culture is dated anywhere between 2600–1100 BC (Agrawal, 1982: 203). It is generally believed that there is a gap between this phase and the subsequent Black and Red and PGW cultures. It is, however, significant to the present discussion that grains sticking into mud clods have been found from this phase. Rice, barley, gram and *khesari* were identified by Chowdhary (1983:458) from excavations at Atranjikhera (see Table 2).

The picture in the pre-PGW, Black and Red Ware phase (dated between 12/ 1100–1000/900 BC) at Noh, Jodhpura, Atranjikherra and Jakhera is not much different from the OCP levels. A thin deposit bearing bone and stone objects, along with a few copper objects, represents it. Writing on the agricultural remains of OCP and BRW phase at Atranjikhera, Chowdhary states that these people were 'mainly food gatherers and growing of agricultural crops was an experiment'.

The cereals reported from the pre-PGW levels are rice and barley from Lal Qila, Atranjikhera and Sringverpura. Only rice could be noticed at Noh and Hulas. Among the legumes, garden pea *(pisum sativum)* at Atranjikhera, horse gram *(dolicobiflorous)* from Noh, *khesari (lathyrus)* from Atranjikhera, gram *(cicer arietinum)* from Lal Qila and Sringverpur and til or sesame also from Sringverpur have been recorded. But the nature of evidence suggests a scanty yield.

It was only with PGW during the next phase that a more stable and sound economy takes roots. This is the period of the beginning of the process of urbanization, a phase of uninterrupted human occupation of crucially significant loci that is the Gangetic plains. It is also the phase that has chronological parity with the time frame of the present discussion.

The thick cultural deposit, the expanding size of the area of occupation and the total nature of deposit is such that even a cursory look at the cultural components of the PGW set it apart from the preceding periods. There is an apparent prosperity. It is this culture on which the edifice of the socio-cultural structure of the 'Ganges civilization' (to use Wheeler's term) rests. Economic prosperity is the foundation of this structure. It needs no elaboration to suggest that the twin props—agriculture, trade and commerce supported the economy. Agriculture was the main economic activity of the early human societies, including that of the opening centuries of the first millennium BC. That there was a relative economic prosperity at this time is reflected in the growing size of settlements (Table 1), a rising demographic chart, intensified structural activity and a score of other features discernible in the material culture of this period. The agricultural bias of the society restricts its growth primarily to a developed village culture. The structural activity during this period is confined to wattle-and-daub or mud bricks. The only exception to this is a thirteen-room baked brick stricture complex at Bhagwanpura (Joshi, 1993). But since such house remains have so far not been seen at other PGW sites, it is more likely to assume that the brick structure belongs to the late Harappans who cohabited with the PGW folk here at this stage. We choose to pay little attention to evidence like that of baked bricks at Jakhera (Sahi, 1994) in view of their ritualistic affiliations and unwieldy size. There are, however, some interesting evidences of cultural advancements like the water channel at the latter site. It is significant from the point of view of the growth of the agriculture too as such channels could have been used aslo for irrigational purposes at this juncture.

The PGW culture also witnesses technological innovations like iron and glass in this region for the first time. The ceramic technology itself is sufficient to impress us. It manifests a strict control of temperatures in the kiln and perhaps its manipulation of fire in the furnaces too that were used in metallurgical process. We have come across evidence of iron working from sites like Jodhpura, Noh and Atranjikhera any Jakhera. The latter has yielded beautiful gold objects, including a sheet of gold, a proof of economic effluence of the PGW society. With this brief introduction of the PGW culture, we proceed to examine the agricultural activity during this period.

In the realm of agriculture, wheat cultivation was introduced in the true sense in the upper Gangetic plains by the PGW culture. Wheat grains (*titicumcompactum Host*), without their husk, were collected from the floor of a house at Atranjikhera. In the opinion of Chowdhary, wheat was not only cultivated for the first time at his level in this region but was found scattered all over the floor. It may be interpreted as a sign of surplus production. A variety of Harappan wheat *(triticum* L.*)* has been discovered at Narhan (Dist. Gorakhpur) and Senuwar in Rohtas Dist. (Singh 1995, 1996), indicating Harappan movement of the people into that otherwise rice zone, as is asserted by Saraswat. In the upper Gangetic plains too, wheat *(triticum)* was found at Harappan levels at Hulas in U.P. and Mohrana in Punjab. It is quite likely that wheat enters the upper Gangetic plains through Harappan interactions, as is well proven by sites like Bhagwanpura in Haryana. However, it is worth mentioning that the wheat at PGW sites is *triticum compactum host*, not *triticum L.* that is generally found at Narhan, Senuwar and other eastern U.P. sites. Large heaps of wheat at PGW levels at Atranjikhera also proved a noticeable advancement in farming as wheat is a crop that requires frequent ploughing, tilling as well as irrigation. Its high yield in itself is a proof of progress in agriculture and thus an evidence of setting in of surplus economy.

The PGW people (Hastinapur, Noh and Atranjikhera) had cultivated rice and barley, the staple food crops of this region. It is presumed that all the legumes mentioned in the previous phase were being cultivated even during this phase. The literary evidence of the later Vedic period that is contemporary with PGW mentions several cereals like *yava* (cereals in general, also barley), *dhānya* (paddy) with its several breeds like *plāsuka* (fast-growing rice), *śāli, vrīhi śastikā* (*sathi*), etc., find mention. *Upvāka* (barley), *godhūm* (wheat), *śyamāka* (millet), *māṣa* (lentil), *Khalakula* (Kulthi), *Khalva,* and *mudga* (munga) and *kulmaṣa* (*urad* or black gram), sesame (*til*), *bajra* (millet) are the crops that find mention. Incidentally, many of these cereals have been identified physically in archaeological remains. *Urvaru* and *urvauka* (cucumbers), *garmut* (wild beans), *karakandhu, kavala* and *badara* (jujube) are some other edible species mentioned in these texts that may be identified. In addition, there are several other names which are difficult to identify. Fruits such as *ām, amlā, gavidhuta* or *gavedhukta, nīmba, priyangu, masusya* and *śasya are mentioned* (Sharma,' 1983).

What may be deducted from the list of crops, both domesticated and wild, is that a large variety of grains were being consumed. A majority of the cereals were being cultivated, with substantial yields. Many wild varieties like beans and millets were also in circulation, as is indicated both by the archaeological and the literary evidences. The latter may have been in the process of domestication.

Wheat appears to have entered the Gangetic plains from two different streams. Two distinct varieties are clearly present *triticum* L. (bread wheat), and (*triticum compactum*) club wheat (Saraswat, 1994: 346). The former variety is found at Senuwar and Narhan. Writing about the presence of wheat (*Triticum* L) at Senuwar, Saraswat (*op. cit.*) suggests a Harappan source for this wheat in an otherwise rice zone. The other breed *triticum compactum host* found at Atranjikhera (Chowdhary, 1983) in PGW cultural horizon is distinct and may have been introduced into the region from a different source. There is a need to analyze and examine the precise species from various wheat-yielding sites to throw more light and put forth a positive theory based on the data.

It does not need an expert to state that wheat cultivation requires a better agricultural skill compared to *yava* that was being grown from earlier times. The fields are to be prepared thoroughly by ploughing it several times. It also requires irrigation and sowing to procure a good yield. Ecologically also, the PGW culture is at home in a predominantly wheat-growing zone. Even today, the states of Punjab, Haryana, along with western U.P. have wheat as the main crop; Punjab and Haryana being the wheat bowls of India. Does the PGW culture manifest the skills needed for wheat cultivation? It's a question that has to be answered here.

First of all, we have to look for the evidence of ploughing. Ploughshare of iron has not been generally found at PGW sites, though Jakhera has yielded it, suggesting its knowledge at a later stage during the span of the cultural period. However, the wooden ploughshare might have been in vogue. The *Satapath Brahmaṇa* refers to the use of upto six, eight or twelve and even twenty-four oxen-driven ploughs. The date of this text falls well within this period of 700–800 BC. Whether iron share was used or not cannot be ascertained from this reference. However, according to Keith and Macdonnell (1958: ii, 451), *Sira* was a large and heavy plough, driven by a number of oxen.

The *Atharvaveda* (III 17.3) uses *paviravant* interpreted as a lance-like ploughshare, perhaps of metal. The archaeological datas of 1000 BC–800 BC is not very positive in this context. The relatively dry and sandy soil of the upper Ganga region, however, could be ploughed even with wooden ploughshares. The *Atharvaveda* refers to a plough of

Khadir wood (AV X. 6.23). The *Satapath Brahmaṇa* describes the wooden plough to be as hard as of bone and that it was being used considerably (XIII, 4.4.9). The *Gṛhya Sutras,* however, attest the use of the iron ploughshare (Ram Gopal, 1959:134). This leaves little doubt about the prevalence of plough cultivation during 1000–800 BC in the upper Gangetic plains.

Whether the technological innovations like iron played any role in the growth of agriculture needs to be evaluated here. PGW culture was doubtlessly an iron using culture with the iron technology being harmessed for a variety of purposes during this period. We come across a considerable number of hunting tools and weapons like spearheads, arrowheads, points, dagger or knife blades, also socketed tangs and axes that could serve the dual purpose of weapon as well as cutting tools. In addition, rods, pins, nails, etc., occur in a good number. However, there is dearth of agricultural implements barring a few ploughshares and sickles from a few sites. But the presence of the iron had an indirect bearing on the production. To quote Sahi, 'Iron tools were useful not only for cutting down forests and breaking the hard alluvium of the Gangetic valley, but also for procuring water for irrigation by tapping artificial sources of water such as wells, tanks and canals. Copper tools could not have served these purposes effectively' (Sahi 1985: 83). There is no cause to doubt that the iron came in handy in digging irrigation channels and, subsequently, wells. We come across a large number and variety of wells during the next phase. There are, however, not many examples of wells, etc., during the PGW period. This could possibly be due to the lack of extensive horizontal excavations.

It is not easy to locate irrigation channels during excavations but an evidence of this nature has been reported from Jakhera (Sahi, 1994). Literary sources, right from the earliest times, mention about the water channels (Rv. III, 45.3 *kulya,* VII 49, 2, *khanitrima apah).* There is, thus, every possibility that irrigation as a practice was not unknown during the PGW period. With the concentration of PGW sites on riverbanks, it may not have been difficult either. It was only under these circumstances that wheat cultivation could be undertaken to an advantage.s

Thus, with good and easy-to-manage soil condition, a favourable ecological condition and knowledge of superior technology, the PGW folk favorably exploited the land for a greater yield and a richer variety of agricultural product. The upper Ganga plains attained a superior economic status in the course of about three-four centuries of the PGW culture.

STAGE III: THE EARLY URBAN CULTURE

A fresh era is heralded in the Gangetic plains with the emergence of a beautiful, glossy surfaced, fine ceramic known as the Northern Black Polished Ware, a deluxe ware of its times. Though at the outset it may appear to be a mere incidence of a new ceramic tradition in the existing matrix of material milieu impinging on the Black and Red Ware (in Eastern UP, Bihar and Bengal) and the PGW cultural horizons (in the Ganga-Yamuna Doab), yet the introduction of NBP proved to be of far-reaching consequences. It may be recounted as the harbinger of innovations in almost every socio-cultural sphere. Coinage and script are introduced in India for the first time at this stage. The pyro-technologies like ceramic, bricks and metals attain new dimensions, leading to the beginning and the intensification of the architectural activity. There is a perceptible improvement in the settlement system. The economy, with the growth in the agriculture and trade and commerce, touched new heights. This gave rise to the emergence of first cities in the Gangetic plains. It is during this period that heterodox sects like Buddhism and Jainism appear. Political system takes shape, culminating into the rise of Magadhan imperialism.

Now we focus on cereal remains, crop pattern, artificial irrigation, tools and implements used in farming and to evaluate the impact of advancements in science and technology on agricultural produce.

A close look at the cereal remains of NBP period may be illustraive. It is only with the application of the floatation technique that some detailed information comes forth. Excavations at Atranjikhera have yielded wheat, barley, rice and *khesari.* A new crop of pulse (*phaseolus mungo* L.) appears in the upper Gangetic plain. At Radhan in District Kanpur (Lal, 1984: 76), wheat, barley, pea and an unidentified cotyledon of legume are found (Table 3).

At Narhan (Singh, 1994), agricultural remains were recovered through floatation by experts of Birbal Sahni Institute of Palaeobotany, Lucknow. Dr. K.S. Saraswat (in Singh, 1994) has contributed a section on the plant remains. A large number of grains have been recovered, giving a complete picture of plant remains at the site at different cultural levels between 1300–100 BC (see Table 2).

Rice and barley were being cultivated from the earliest levels (Neolithic level) in addition to wheat (*Triticum*). Other crops that are known to have been grown are kodomillet (*paspaleum serobiculatum*). Pearl millet or *Bajra* (*Pennisetum typhoides*) and pulses like chick pea or gram (*Cicer arbetinum* L.), field pea (*Pisum sativum*), green gram (*mung*) (*Vigna redes*), horse gram or Kulthi *(Dolichas bifloris*), grass pea or *khesari* (*Lathyrus sativus*), lentil (*Lens culimaris*), moth bean or aconite bean (*Vigna aconitifolia*) and oil seeds like *sarson* (*Brassica juncea*), sesame (*Sesame indicum*) linseed or flax (*Linum usitatissunum*). Also interesting is the presence of grape pips, dates and jackfruit at the site right from the BRW level. All these cereals continue to be grown at the site even before the NBP times. The only new additions are black gram (*Pisum sativum*) and field pea (*Lathyrus sativus*). Fruits like jujube, *anwala* (*Embilica officinalus*) are added to the list of plant remains. This list goes to prove that most of the cropss found in this region today have a hoary antiquity and are in use from the first millennium BC or even earlier. There are, however, certain changes, which must have had a positive impact on the agricultural growth.

A change in the settlement pattern with size expansion and growth in demographic chart, availability of better tools and implements with advancements in technological expertise, understanding of weather, climate, soil, manure and last but not the least, emergence of science like astrology and astronomy came in handy to the forward-looking society taking strides towards growth and expansion.

As pointed above, there was imminent rise in the population, leading to the expansion of the size of the settlements as well as the colonization of newer areas in the alluvial plains of the importasnt rivers and their tributaries. It is during this period that people ventured to settle way from the rivers, closer to other sources of water like tanks and lakes. Even in more secluded areas closer to hilly tracts, we come across small settlements perhaps for procuring minerals and other raw material needed for a variety of purposes (Tripathi and Tripathi, 1997). Even a quick look at the distribution map of BRW/PGW and NBP is sufficient to highlight the pattern of the expansion of these cultures (See Table 1). The area under human occupation at these cultural levels indicates the population growth and new lands coming under occupation. The spatial distribution during NBP period covers more than double the area combined together by Chalcolithic BRW and PGW cultures. NBP succeeds both the above mentioned culture in the entire Gangetic plains—upper, mid and lower. In addition newer settlements come up during the NBP times in this region and in many distant lands. It extends in the Nepalese *terai* in northeast, Taxila and Udegram in west to Brahmagiri-Amaravati-Ter in south India and across the country from Gujarat, Saurashtra (Prabhas Patan) to Chandraketugarh in Bengal. Such an extensive distribution of a ceramic tradition may be taken to be a sure sign of the cultural expansion through the length and breadth of India. It also speaks of forces behind this phenomena. There has to be a strong political, social and economic backing for the movement of the people as well as cultural traditions. It is common knowledge that the period of 600–300/200 BC was the age of the consolidation of the power in the Gangetic plains. It was the era of the rise of the *Mahajanapadas* from the level of *Janapadas.* It witnessed the rise of the Magadhan imperialism. This is fully borne out by the archaeological records as may be seen in the NBP cultural horizon with its early and late phases—the two representing a beginning and culmination of culture into sophisticated urban centres.

A large number and a rich variety of agricultural implements came to be used gradually (see Table 4), though at the initial stage of advent of the iron, we have observed primarily weapons and hunting tools besides a few other types. Stray objects like axes and sickles have been found from PGW level from sites like Alamgirpur, Noh, Atranjikhera and Jakhera. The latter also yields a ploughshare from the PGW period. A ploughshare is also reported from the pre-NBP phase at Ganwaria (700 BC) from the middle Ganga plains. However, agricultural implements of iron during the PGW levels/pre-NBP level remain few and far between. There is a quantitative change in the NBP level. Atranjikhera has yielded sickle, spud, ploughshare, hoe, digger *(khurpi)* (Gaur 1983, Fig. 1, 2, 3: 426–431). Quite heavy ploughshare is reported from Ropar, Alamgirpur and Kausambi from the NBPW phase. The list may grow to the level of becoming exhaustive if one enumerates agricultural implements from sites of the Gangetic plains. Without going into such details, it may be suggested that there was greater emphasis on agriculture, which must have resulted into increased production, ensuring a better and more prosperous life.

The rich literary tradition of the time strongly supports this contention. A variety of iron implements for agricultural purposes have been referred to in various texts. Panini makes a clear reference to the iron ploughshare, *ayovikara kusi* (Panini IV 1.42) and also *hal.* This is in connection with the sugarcane cultivation (Agarwal, 1969). Subsequently, the early Pali texts, generally attributable to pre-Mauryan times refer to ploughshare. *Ayonangal,* an iron ploughshare appears in the later Pali texts. The *Suttanipata* mentions *Phala* or share that has been quenched producing a hissing sound (Rhys Davids and William Stede, 1921, and the *Kapalika sutta,* respectively. *Kuddala* or *kuddalika* is a term used for hoe. The *Baudhayana Dharma* sutra states that a person equipped with implements like *Kuddalika* can make a living out of it, perhaps by tilling other people's fields. Thus, the concept of agricultural labour of some sort may be seen in this reference. Needless to emphasize, better tools and implements were employed and even a level of perfection is perceptible in agricultural practices by 500/400 BC.

The technique of transplantation evolved over a period. Panini mentions *Sali* (a winter crop), a variety of rice that was an improvized variety of the Vedic *Vrihi* that is sown by scattering the seeds during the rainy season. *Sali* was grown by transplantation. The importance of the technique results into composition of a Jataka as *Sali Kedar Jataka* (Sharma, 1983: 97–98). We come across *ropeti, nihae* to uproot and replant is the technique mentioned in the Jain text *Nayadharmmkahao* or *Jñātadharmakathā*, a text datable to the third century BC for paddy transplantation, especially in Bihar, Anga, Magadh, Mithila and Rajgriha.

Such references indicate a consistent growth of agricultural technique. Beginning of double crop pattern is easily derivable from the available sources. Barley and rice were the chief crops, wheat or *godhuma* was not very common in eastern region, according to the testimony of literary sources. For sacrificial or ritualistic purposes too, it was barley, rice and sesame that had some kind of a sanctity about them. The *Srautasutras* mention barley, rice, sesame, panic seed, millet, wheat, mustard and lentils like *masoor, mudga, kulatha,* etc. (Ram Gopal, 1959:134). The list of crops clearly indicates the practice of double crops right from the 1000 BC level. To quote Saraswat, the practice of rotation of crops was pursued. Rice, *bajra, kodon,* millet, horse gram, green gram, moth bean and black gram are usually grown in warm-rainy season, while wheat, barley, gram, khesari, lentil and field pea are the winter crops (Saraswat, 1994: 321).

It is difficult to state about the understanding of the soil condition on the testimony of archaeological data. However, a close look at the settlement pattern may provide some insight into it. We have already seen the concentration of sites in certain types of soils. Even at the risk of being repetitive, it may be pointed out once again that in Kanpur region, the Ganga-Rind Doab is more densely populated than the Yamuna-Rind Doab because of better soil condition of the former.

The Buddhist texts, however, are a better guide. The early Pali texts like *Sutta Nipāta* (CIV 314–17) classify the fields into three categories of good, medium and bad. The *Khetta Sutta* of the *Anguttar Nikāya* (IV 237) refers to eight types of fields that have a low yield and cultivating them is a wasteful exercise. These are: undulating, rocky and pebbly, saltish, without depth of tillage, without water outlet without inlet, with no water courses and without dykes. Similarly, there are eight types of good fields according to the *Arthaśāstra* (II, 35; III, 10).

Land management had acquired new dimensions by the third century BC, if the *Arthaśāstra* of Kautilya may be taken as a guide. There were several categories of land, classified as:

1. *Krisṭa* (cultivated)
2. *Akrisṭa* (uncultivated, waste or fallow land)
3. *Sthala* (high and dry ground)
4. *Kedāra* (fields with crops)

5. *Ārāma* (grove)
6. *Sanda* (orchards)
7. *Mūla vapa* (fields with roots like turmeric, ginger, etc.)
8. *Vāta* (sugarcane fields)
9. *Vaṅa* (forests)
10. *Vivita* (grazing grounds or pastures)
11. *Pathī* (parts covered by roads *(Kautiliya Arthaśāstra,* Shama Shastri. Book II Ch. 35 III, 10).

CONCLUSION

To recapitulate, a large variety of grains was cultivated in different parts of the Gangetic plains, right from Neolithic–Chalcolithic times. Two crops were being grown right from 800 BC or even earlier. We have no means of verifying archaeologically, whether the yield was rich enough right from the earliest stage. One has to look into the material milieu as unearthed from the excavations. The impoverished remains of stage I do not support large-scale cultivation with high yields. The economy was heavily substantiated by animal husbandry and hunting, as is indicated by the presence of a large number of bones charred and with cut marks. One does not come across heavy storage jars or bins in a great number, suggesting surplus production. Settlements were confined closer to the riverbanks. The inhabitants must have confined themselves to the flood plains only, generally avoiding the dense forests bordering such areas. This did not require heavy duty tools and implements nor any complex agricultural skill. This is also borne out by the crops cultivated at the earlier stages. For example, both literary and archaeological excavations confirm that a coarser variety of rice, *vrihi* or *Sasthi* are the varieties of rice mentioned in the Vedic texts. Panini, at a subsequent stage, mentions *Sali* (a winter crop of paddy). Rice transplantation, an evolved technique, finds a clear mention in the Pali texts, as suggested above. Similarly, wheat though found at pre-NBP levels in middle Ganga plains and with PGW in upper Ganga plains, does not become the important crop of the former region as hinted at above. Barley and rice were the chief crops there. To come to a reasonable level, wheat cultivation had to await till much later.

Wheat, as discussed above, requires a thorough preparation of the fields by frequent ploughing/tilling at least for 6–7 times. This could be attained even with wooden ploughs in the relatively arid zone of upper Gangetic plains. We find references to *phāla* and *stega* (Keith and Macdonell, 1958, II, 58 II 484). The latter term in the *Ṛgveda* stands for a goad or a spear but by the time of *Atharavaveda* and the *Yajurveda,* it comes to connote the ploughshare s(Keith and Madonell, 1958:509). In the mid-lower parts of the plains, good metal implements were imperative for preparation of fields. Thus, the growth of agriculture is interlinked with the growth of technology also.

Wheat cultivation is introduced during the PGW period in the upper Ganga plains. No remarkable headway could have been made in the agricultural sector despite the knowledge of the iron to the PGW folk as there is hardly any utilization of this newly acquired technological skill in this field. The social priority was waging and winning wars and acquiring newer territories. This is clearly manifested in the iron tool repertoire at the PGW level. One can count the agricultural implements on one's fingertips. The emphasis is on hunting and war weapons. The economy, by and large, remains at the subsistence level, though some improvement is perceptibly visible. True utilization of the iron technology begins only with NBP level with agricultural implements present in an impressive variety and sizeable number at most of the Gangetic valley sites. Many iron implements were useful as digging tools. The sudden emergence of wells in sizeable numbers in the Gangetic plains could be a pointer to well irrigation too. Crops like wheat and sugarcane needed artificial irrigation. During the late phase of NBP, such data is remarkably frequent.

Rice cultivation also picks up with these tools that must have been extremely useful in handling the hard sticky soil of the middle and lower Ganga plains. There is evidence to suggest a richer variety and higher yield in rice cultivation. Both an innovative resource management and better technology promoted growth and prosperity in the Gangetic plains, especially during the NBP phase (7/600 BC–200 BC).

Among the most conspicuous attributes of this period is technological innovation and its adaptation on a much higher scale for a variety of purpose. Iron implements, as stated above, come to be used in agricultural sector at a noticeable level for the first time at this stage. If we juxtapose the itinerary of tool typology, especially of iron on the rising level of agricultural production (that is difficult to measure directly but for its reflection in material remains like storage jars, silos and bins, etc.), one can notice an inter-connection between the two. The variety of cereal remains more or less static from the Chalcolithic Black-and-Red ware through the PGW and the early NBP level but an increased area comes under occupation later. There is increase in the population. Newer areas were being brought under habitation as well as exploited for the cultivation. The society was better organized. Thus, a better resource management, including human resource (by formation of occupational groups, and emergence of social stratification), channelization of expertise into desirable directions led to organized and prosperous social structure. Agriculture, no doubt, played a key role in bringing about material growth with surplus production. Before the circulation of coinage, cereals must have been a key commodity for the exchange and barter.

The cultural scenario of the Gangetic plains acquired new dimensions over the centuries with experience and experimentations in diverse fields. There is a consistent growth in the socio-economic milieu that acts as a catalyst for growth of socio-political organization. The changes inspire emergences of the heterodox systems of philosophy, a mass of literature, arts and crafts, sciences and technologies, all deriving their sustenance from the growing production level. The culmination of the multidimensional growth is reflected in the rise of the *Mahajanapadas* from the *Janapadas* by the sixth century BC. As a consequence, the Gangetic plains witness of the growth of the earliest cities by the third century BC after the decline of the Indus Valley Civilization.

Table 1: Size of settlement at different Periods (in sq. m.)

	BRW &	*PGW*	*NBP*	*POST NBP*
Bijapur	15000	15000	22000	22000
Galatha	26250	26250	30000	56000
Indyan	7500	7500	17500	30000
Jajmau	30000	30000	50000	70000
Kheora	9000	9000	18750	32000
Musanagar	30000	30000	50000	60000
Nawabganj	15000	15000	40000	75000
Nonha Narsigh	10000	10000	12800	52000
Radhan-A	40000	40000	87500	112500

Table 2: Distribution of cereal remains in the Middle Ganga Plains

Cereals	*Senuwar*	*Chirand*	*Narhan*	*Sonpur*	*Rajghat*	*Prahladpur*
Rice	•	•	•	•	•	• ■
Wheat	•	•	•			• ■
Barley	•	•	•			
Masoor	•	•				
Green Gram	•	•	•			
Green Pea	•		•			
Jowar	•		•			
Lentil	•					
Horse gram	•					
Til	•					
Linseed	•		•			
Sunflower			•			
Urad						
Blackgram						
Colyledoen of legume						

- • Pre-NBP
- ■ NBP

Table 3: Distribution of cereal remains in the Upper Ganga Plains

Cereals	*Lal Qila*	*Atranji-khera*	*Sring-verpur*	*Hulas*	*Noh*	*Hastinapur*	*Radhan*
Rice	*	• ▪	N.A.	N.A.	N.A.	•	▪
Wheat	*	• ▪					▪
Barley	*	• ▪					▪
Green Gram							
Green Pea							
Urad							
Khesari		*					
Horse gram		*					
Til							
Sarso							
Black gram							
Pea						▪	
Cotaloden of legume						▪	

* OCP
• PGW
▪ NBP
N.A. Not available

Table 4: Distribution of Agricultural implements of iron

	Atranji khera	*Jakhera*	*Rajghat*	*Gangawaria*
Axe	• ▪	• ▪		
Ploughshare	•	• ▪		•
Hoe	•	•		
Digger	•			
Sickle	• ▪	• ▪	•	
Spud	• ▪	•		

▪ Pre-NBP – BRW/PGW
• NBPW

REFERENCES

Agarwal, D.P., 1982, *Archaeology of India*, London Scandinavian Institute of Asian Studies Monograph, Curzon Press, No. 48.

Agrawal, V.S., 1969, *India As Known to Panini*, Varanasi, Vishwavidyalay Prakashan.

Brehmdutta, 1980, 'Settlements of PGW Culture in Haryana'. Unpublished Ph.D. Thesis of Kurukshetra University, Kurukshetra.

Chowdhary, K.A., 1983, Motilal Banarsidass, New Delhi, 'The Plant remains', in appendix in *Excavations at Atranjikhera*, New Delhi.

Crystaller, W. 1933, 'Central Places in Southern Germany' tr. by C.W. Bakin, New Jersey Single Wood Cliff.

Ghosh, N.C., 1984, 'The Chalcolithic Background of West Bengal', in *Essays Presented in Memory of Prof. N.R. Ray*, (ed.) Amita Ray, Eastern Publication, Delhi, pp. 15–19.

Joshi, J.P., 1993, Excavation at Bhagawanpura 1975–76 and other Explorations and Excavations 1975–81 in Haryana, Jammu & Kashmir and Punjab. Archaeological Survey of India, New Delhi.

Lal, Makkhan, 1984, *Settlement History and Rise of Civilization in Ganga Yamuna Doab*, New Delhi, B.R. Publishing Corporation, New Delhi.

Macdonell, A.A. and A.B. Keith, 1958, Vedic Index I & II, Delhi: Motilal Banarasidass.

Ram Gopal, 1959, *India of Vedic Kalpasutras*, Delhi, National Publishing House.

Ray, Amita, 1987, 'Urbanization in Bengal', Presidential Address, 48th Session of Indian History Congress.

Rhys Davids and W. Stede, 1921, *Pali English Dictionary*, P.T.S. London.

Sahi, M.D.N., 1985, 'Comments on Issues of the Indian Iron Age', of D.K. Chakrabarty in: *Recent Advances in Indian Archaeology*, (eds) S.B. Deo and K. Padayya, Deccan College, Pune, pp. 82–83.

Sahi, M.D.N., 1994, *Aspects of Indian Archaeology*, Publication Scheme, Jaipur.

Saraswat, K.S. 1994, 'Plant Economy of Ancient Narhan (Ca. 1300 BC–300–400 AD)' pp. 254–346 in *Excavations at Narhan 1984–89*. Appendix IV, Varanasi, Benaras Hindu University Press, p. 272.

Sharma, G.R., *History to Prehistory, Archaeology of the Ganga Valley and the Vindhyas*, Allahabad, 1980, pp. 103–110.

Sharma, R.S., 1983, *Material Culture and Social Formation in Ancient India*, Delhi, Macmillan India Limited.

Singh, Birendra Pratap, 1990, 'Chalcolithic Culture of Eastern Uttar Pradesh' in *Archaeological Perspective of Uttar Pradesh and Future Prospects*, Part 1. (ed) Rakesh Tewari, Lucknow, pp. 77–91, UP Puratattva Sangathan.

Singh, Birendra Pratap, 1996, 'Transformation of Cultures in the Middle Ganga Plains: A Case Study' (ed) Rakesh Tewari, in *Pragdham*, No. 6 Lucknow, UP Puratattva Sangathan, pp. 75–93.

Singh, P., 1994, *Excavations at Narhan*, Varanasi, Benaras Hindu University Press.

Singh, R.L., 1955, *India—A Regional Geography*, Varanasi.

Spate, O.H.K. and A.T.A. Learmouth, 1967, *India and Pakistan*, third edition, London, Mathews, National Geographical Society of India, Banaras Hindu University, Banaras.

Saraswat, K.S. 2004-05 'Agricultural Background of the Early Farming Communities in the Middle Ganga Plain' Pragelhara, 15: 145–177. Ed Rakash Tewari, *Jl. of UP State Arehaeology Department,* Lucknow.

Stamp, L.D., 1959, *Asia—A Regional and Economic Geography,* London, Methuen & Co. Ltd., London.

Subbarao, B., 1956, *The Personality of India,* Baroda, MS University, Baroda.

Tripathi, V. and Amit Tripathi, 1997, 'Mineral Resources: A Possible Archaeometallurgical Correlation', *Archaeometallurgy in India,* (ed.) Vibha Tripathi, New Delhi, Sharda Publishing House, pp. 191–208.

CHAPTER 23

The Early Indian Agrarian Society and Technology Adaptation

Vibha Tripathi

INTRODUCTION

India is a country with a vast expanse of fertile *Śasya Śyāmalā* plain. Cultivating these fertile lands has always been the chief occupation of the masses. Agriculture, therefore, has a hoary antiquity here. Despite several efforts to reconstruct the history of agriculture in India, with ever increasing mass of data, there is indeed a need to work on a comprehensive and multi-faceted history of agriculture. Any discussion on agriculture should necessarily incorporate, besides the crops under cultivation at different stages, the amount of production and the techniques employed, the tools and implements and their efficacy as they positively govern the degree of production. All these are interrelated and, therefore, essential for a holistic reconstruction of history of agriculture in India through the ages.

Agricultural activity in India can be traced back to the pre-Neolithic period. The process of domestication of plants and cereals takes a long time. We find evidence of cereals at Mehergarh and in the Vindhyas in 6000 BC. Wild varieties of rice have been found in the Vindhyan region in a Mesolithic context at Chopanimando in Meja Tehsil of Allahabad (Sharma[1], 1980: 110) that comes to be domesticated in the following millennia (Tripathi[2], 2001). Towards the final stages of Neolithism, the preference of humankind had shifted from the hills to the plains. The lakes and streams came to be occupied first during seasonal migrations, as is evident by the 'camp sites' near lakes and ponds, and then on a more permanent basis. This prepares the background for the subsequent occupation of river banks in the plains, primarily because of attractive arable lands. The gestation period of this process was, however, long and extended over thousands of years. Long experiments with crops suited for different regions necessitates a heavy reliance on animal husbandry, hunting and gathering. It may be worth identifying these stages. Additionally, it may also be worth taking a closer look at the changes in technology from stone, bone to metal—copper, bronze and iron—and their impact on agriculture. The priorities of a society decide the pattern of adaptation and ensuing culture change as visible in the material milieu. It is thus important to examine: (i) the cereals being grown at different stages of cultural development; (ii) the degree of success in agricultural production; (iii) the techniques being followed at different stages, like sowing pattern, irrigation, manuring, tilling/ploughing, etc., and (iv) tools and implements being used for different purposes. In the end, an attempt will be made to seek an interrelationship between the growing knowledge by way of better understanding of nature and development of technology, the agricultural yield and growth of culture. This may be attempted at different nodal points of historical phases.

STAGE I: BEGINNING OF AGRICULTURE

Agricultural activities take root in the hilly zones near the rivers flowing through these tracts, such as in the valley of Belan in the Vindhya-Kaimur region of Uttar Pradesh-Bihar region. A large number of settlements belonging to different cultural stages have been located and excavated in this region. Important among these are Mahagarha and Koldihwa (Dist. Allahabad) in Belan Valley. Rice, identified as *Oryza sativa* and dated to 5440±240 BC and 4530±185 BC, have been recovered from here. In the northwest (in Pakistan), Mehrgarh[3] has yielded an equally early date of the beginning of agriculture with the earliest evidence of wheat, barley, cotton and grapes. Interestingly, both domesticated and wild varieties of cereals found there indicate a local growth of agriculture at the site.

The cultivation of rice, one of the earliest in the world, begins in the Vindhyas in Mesolithic Chopanimando in the Belan Valley. A transition from wild to domesticated species takes place here. In the subsequent centuries, when the Gangetic Plains comes to be occupied on a more permanent basis, we come across the Ochre-Coloured Pottery (OCP, henceforth) culture dated between 2000–1100 BC. In the upper Ganga Plain, i.e. western Uttar Pradesh and adjacent parts of Rajasthan, Punjab and Haryana, this is generally the earliest culture. It is partially contemporaneous with the Harappans and its late Harappan phase. Its epicenter, however is Saharanpur area. The Copper Hoards of the Doab are attributed to this culture. The settlements are generally flimsy and disturbed because of natural calamity and come to an abrupt end. Sites like Lal Qila (Dist. Bulandshahr) and Atranjikhera, Noh, Jodhpura, etc., have yielded better evidence, including that on agricultural remains. Cereals like rice, barley, gram and *khesari* were found from Atranjikhera from mud-clods. However, Chowdhary[4], who analyzed the agricultural remains on seeing the nature of deposit, commented that people of the pre-PGW stage were mainly food-gatherers and cultivation of crops was an experiment for them.

In the middle and lower Ganga Plains, there are a number of sites belonging to Neolithic-Chalcolithic period that have yielded considerable data on agriculture. Special mention may be made here of Narhan (Dist. Gorakhpur, UP) and Senuwar (Dist. Rohtas, Bihar), Lahuradeva (Dist. Sant Kabir Nagar). A systematic floatation work was undertaken at these sites by Saraswat[5] of the Birbal Sahni Institute of Palaeobotany, Lucknow. It yielded encouraging results. From period I of Narhan, mention may be made of cereals like *Kodon* millet (*Paspalum Scrobic culatum*), pearl millet or *bajra* (*Pennisetum typhoides*) and pulses like chick pea or gram (*Cicer arietinum* L.), field pea (*Pisum sativum*), green gram or *mung* (*Vigna radiata*), horse gram or kulthi (*dolichos biflorus*), grass pea or *khesari* (*lathyrus sativus*), lentil (*Lens culinaris*), moth beam, aconite bean (*Vigna aconitifolia*). Oil seeds like *sarson* (*Brassica juncea*), sesame (*Sesame indicum*), linseed or flax (*Linum usitatissimum*) are also found. Interestingly, there is also evidence of pips of grape, date, jackfruit and watermelons. Equally interesting is the evidence from Senuwar right from the Neolithic period dating back to 2000/1800 BC grains like rice, barley, dwarf wheat, bread wheat, sorghum, chick pea, green gram, field pea, lentil, horse gram, grass pea, sesame and linseed have been identified (Singh[6]). Of special interest is the occurrence of the Harappan species of wheat (*Triticum compactum* L.) at this eastern site. It has been found in good amount and may suggest Harappan immigration into this eastern zone. Equally significant is the occurrences of *bhāng* or hemp (*Cannabis sativa*) from this period, perhaps the earliest evidence of hemp in the region so far.

From Chirand[7] in District Saran, rice has been found right from the Neolithic level. From the Chalcolithic levels of Ahar in District Udaipur, Rajasthan, Pandu Rajar Dhibi, Mahishdal and Mangalkot in Bengal, rice has been recovered.

The Neolithic-Megalithic cultural phase of Hallur in Tungabhadra basin yields millet, horse gram, legumes, date palm, *bajra* (*Pennisetum typhoides*). Paiyampalli yielded green gram. Thus, from a scanty beginning at the Neolithic stage, there is a gradual increase in the number and variety of crops that come to be cultivated by the Chalcolithic period. At a number of sites, there is a continuity between Neolithic-Chalcolithic stages as is evident from the description above. The OCP culture is almost invariably followed by a desertion of settlements.

STAGE II: THE EARLY IRON AGE

Iron is introduced in a Chalcolithic milieu in almost every part of the country, especially in the Gangetic Plains. The Painted Grey Ware (PGW, henceforth) in the OCP zone upto western Uttar Pradesh in the upper Ganga Plain, the Black-and-Red (BRW, henceforth) bearing cultures in the middle and lower Ganga Plain and the Megalithic cultures in Vidarbha and peninsular India came to use iron in the first millennium BC plus-minus two hundred years. These cultures have been dated to the later Vedic period. It is a beginning of continuous and uninterrupted cultural setting of the Ganga Plains except for few exceptions of local nature. Speaking of plant remains, most of the crops at home in this ecological setting were already introduced in the previous stage. However, then, production was supposedly on a limited scale, as has been suggested by Chowdhary (*op. cit.*).

A new breed of wheat (*Titicum compectum, Host*), distinct from the Harappan variety was introduced in the Ganga Plain, if evidence of Atranjikheda is a guide. To qoute Chowdhary (*op. cit.*), wheat 'was not only cultivated sfor the first time at this level in this region but was found scattered all over the floor. It may be interpreted as a sign of surplus production'. Thus, there is a considerable increase in the yield at this stage. The Kuru-Panchal region, it appears, not only experimented with new breads of important grains like wheat but took a lead in irrigational practices like digging of canals (Jakhera) and using iron implements like ploughshare and sickles, both found at Jakhera. Could application of better implements for productions account for its prosperous status?

The other important crop is rice, a staple food of the more easterly parts. There is a likelihood that more than one type of rice was being cultivated. Vedic literature mentions *Vrīhi* and *śasthi*. In the *Aitareya Brahmana, kusuka, śāli* and *Dhānya* are referred to. Rice has been found in excavations from Hastinapur, Atranjikhera, Noh. The other cereal of this period is barley or *yava*. Later Vedic texts mention *upvaka* (barley) *śyāmāka* (millet), *māsa* (lentil), *khalkula* (kulthi), *khalva* and *mudga* (mung), *kulmāṣa* (*urad* or black gram), sesame (til), *bājrā* (millet) and pulses (Sharma[8]). Several of these grains have been recovered from excavations. Thus, a large variety of crops came to be cultivated by this stage. Incidentally, many of these require a higher level of agricultural skill to produce them in good quantity, a subject that will be discussed a little later below.

STAGE III: THE EARLY URBAN CULTURE

The sixth century BC heralded a new era in the Gangetic Plain that culminated in the rise of *Mahajanapadas* and, subsequently, of Magadhan imperialism. In this section, it is proposed to restrict our discussion to crops alone. Atranjikhera has yielded, besides wheat, barley, rice and *khesari*, a new crop of pulse (*Phaseolus munga* L). Radhan also yielded an unidentifiable cotyledon of a legume. At Narhan, black gram (*Pisum sativum*) and field pea (*Lathyrus sativus*) have been added to the list of the previous period (Fig. 1A).

II. Agricultural Technology, Tools and Implements

A reconnaissance of crops as they have been mentioned in literature and substantiated by excavations leads us towards an enquiry of techniques being followed in cultivating of different crops and tools and implements in use in this sector.

Stage I is an incipient stage of agriculture. As Chowdhary construed on the basis of botanical remains of Atranjikhera excavations, a food gathering society was experimenting with cultivation. The *Ṛgveda* mentions *yava* or barley (as well as a generic term for cereals). The Brahmanas refer to wheat and rice (Ait. Br., VII 29). *Vrihi* is also mentioned here. *Ṣasṭikā* (Ait. Br. VIII, 22) is said to have been used in rituals. Other grains that find mention are millets like *Syāmāka* (*Ait. Br.* VIII, 22). We have discussed the large variety of crops that were being cultivated in pre-Iron age period, viz., at OCP, Neolithic and Chalcolithic cultural levels. However, at this stage, almost every excavated site yields a large number of bones, many with cut marks. Both domesticated and wild animals, including aquatic beings, seem to have been part of the diet, a practice that remains in vogue for centuries to come. This may be attributed to a limited scale of crop production which could not fully sustain the population. Hunting-fishing was apparently a sport and a necessity. The earlier the period, the heavier the reliance on animal husbandry and game, as fully corroborated by the tool repertoire (Tables 1, 2).

The small areas under occupation, the flimsy settlements of OCP/Pre-PGW cultures fully subscribe to this contention that it was indeed an incipient stage of agriculture in Gangetic Plains and in the adjacent Indo-Gangetic Divide. In south India, the Neolithic traits continued for a longer period. There is hardly a Chalcolithic stage of culture there. One comes across a Neolithic-Megalithic overlap in large parts there.

The digging tools of stone, bone or wood, axes and celts of pure copper (bronze, that is much stronger, is rarely found in OCP/BRW levels) could not have been effective enough in cultivation and farming activities. Nor is there a noticeable number of agricultural implements. Some Chalcolithic sites in Deccan do show signs of affluence such as the Malwa and Jorwe cultures, but these were localized to very restricted areas and are followed by large desertions. However, the presence of Harappan traits in the middle and lower Ganga plains, as indicated by steatite beads, some pottery forms and especially the Harappan breed of wheat, is indeed significant. Ploughing must have possibly been resorted to, albeit on a limited scale, perhaps with wooden ploughs. The *Ṛgveda* also suggests the use of hard woods like *Khadira* for the purpose. There is a mention of *sītā* and *sunu*—the furrows left behind by a plough. This could, however, be undertaken only in narrow river belts with silty-sandy soil. The hard clayey soil of mid and lower Gangetic Plains could not be broken easily and had to await the advent of iron and better quality tools on a good scale.

In the Painted Grewarey/BRW/Megalithic cultures, a change sets in the material mileu. There is an expansion in the size of settlements. A much larger number of sites come to be occupied, many for the first time. Ghosh's statement that the PGW people were in a habit of 'breaking new grounds' may be recalled here. Perhaps burning of forests and removing the stumps with iron axes was taking place. Literary references to this effect are not infrequent during the later Vedic period. The Aryans had set out to occupy new areas (*Satapath Br.*, Vedagdh Mathva anecdote) and were gradually consolidating their position. The several-metre thick deposits of the PGW period is a definite proof. In a nutshell, the cultural foci was shifting from the Sarasvati-Drisadvati valley to the Ganga-Yamuna Doab. The Kuru-Panchal region was gaining prominence. The important PGW sites in this region like Atranjikhera, Jakhera, etc., substantiate this.

In keeping with the requirements of this age, a large number of war and hunting tools of iron were being manufactured from every site. Besides weaponry other objects like axes, clamps, nails, pins, chisels and rarely a pair of tongs (Atranjikhera) of iron have been found from PGW levels. Copper arrowheads also continued, even though in a very restricted number (Hastinapur). Gradually, copper came to be used for toiletries, ornaments and sometimes utensils. Over the period, even within the PGW cultural phases, there is a change in utilization and technology adaptation pattern. In due course of time (over a period of three-four centuries), sickles and ploughshare of iron came to be used, as evidenced by Jakhera and Ganwaria, along with axes and tangs. Accidental carburization must have led to an understanding of this technique and production of better tools and implements to be employed in carpentry (chisels and nails in a good number are found) and, subsequently, in agriculture.

We have seen a new breed of wheat coming up in upper Ganga plain. Wheat cultivation, if attempted on a larger scale, needs repeated ploughing and irrigation of fields. That there were water channels in use is well testified from *Ṛgvedic* times (*Kulyā*, Rv, III, 45.8; *Khanitrimā āpah*, VII, 49.2). Jakhera (Sahi[9]) has yielded evidence of water channels being brought to the settlement. Paddy cultivation, likewise, requires watering.

Several varieties of rice have been mentioned, though the transplantation technique had to await further development. Rice could also be cultivated in southern parts of the country where water was available. The Chalcolithic community, such as at Inamgaon,[10] cultivated some rice. The Megalithic cultures could also cultivate rice, perhaps near water sources, though their main crops remain *rāgī,* horse gram and green gram, which were easy to grow in the rugged terrains occupied by them. Iron objects found in Megalithic regions manifest a richer tool repertoire. Lances, swords, tridents, spearheads, arrowheads, axes (with cross fasteners), adze, sickle and chisels (Fig. 3) are important tool types, indicating a higher level of skill in the fashioning of these objects. However, sickles and axes (the latter could be used both as weapon and clearance tool) are the only tools that could assist in agricultural activity. Thus, apparently, agriculture was not the priority of these social groups. Their economy was based more on animal husbandry than agriculture. It was also due to the fact that pastoralism was a way of life with them (Deo[11] 1987).

In the third stage, the pace of economic development quickens. It eventually leads the society towards urbanism. The Northern Black Polished Ware (NBPW) culture sets the pace and lays down foundations of the imperialism at Magadha. The middle and lower Gangetic plains are colonized in a big way (Fig. 4). A comparison of settlements shows an expansion in size as well as number of sites under occupation (Fig. 1). Almost in the entire subcontinent from Udegram and Taxila to Brahmagiri, Ter and Amaravati from north to south and Prabhas Patan to Bengal from east to west, NBP sherds have been found. Recently, they have been reported from Ceylon. To talk of agriculture at this stage is almost meaningless. Almost every crop known today was being cultivated. However, noteworthy is the rise in agricultural implements, especially of iron. Ploughshare, hoe, spade, digger (*Khurpi*), etc., are commonly found at almost every site of the Ganga plains. Pāninī refers to several types of ploughshares, perhaps suited for different soil conditions and types of soils. *Ayovikara, kusi* and *hal* are mentioned (Agrawal[10]). The Buddhist texts substantiate the list further. *Ayonangal, Phala* (ploughshare) being prepared by a smith and quench hardening of the same are mentioned,[12] in *Kalika sutta* of the *Suttanipata. Kuddālikā* or *Kuddālā,* a term still in vogue, has been used in the *Baudhāyana Dharma Sūttra* in the context of a professional tiller who makes a living out of this. This suggests a beginning of agricultural labour by 500 to 400 BC.

Transplantation of rice is learnt by fourth century BC *Ropeti, ukkhāya nihāe* clearly suggest it (Sharma[13]). The Jain text of third BC, *Nayadhammkahāo* or *Jñatadharmakathā* elaborate the transplantation technique prevalent in Anga-Magadha-Mithila region. This was not only a deviation from the simpler techniques of the earlier times but a higher yield is an imperative of this practice. A double crop pattern is apparent. It appears that rice and barley were the chief crops on the more easterly region. Even in sacrifices, these are the primary ingredients. Wheat was not so common on easterly region. Saraswat[14] suggests, 'The practice of rotation of crops was pursued. Rice, *Bajra* kodon—millet, horse gram, green gram, moth bean and black gram are usually grown in the warm rainy season while wheat, barley gram, *khesari,* lentil and field pea are the winter crops.' The Greeks who visited India were highly impressed by the expertise shown by the farmers as the statements of Megasthanes and Arrian indicate (Agrawal, *op. cit.,* 199). Thus, the agricultural practices show a constant innovation through the above three stages, culminating into an expertise of high order. In the end, it may be interesting to see whether there is an interrelationship between developing technological skill and agricultural output.

III. The Interface: Technology Adaptation vis-à-vis Agricultural Practices

We are faced with the question of whether there is any link between the growing technological expertise, especially that of iron metallurgy and economic affluence. It has frequently been debated as to why is it that even after centuries, iron fails to bring about any economic change. It has also been argued frequently that the Megalithic culture, despite its richness in iron objects, fails to achieve an economic prosperity. These issues need to be closely examined so as to answer these questions.

Iron is said to be a common man's metal, unlike the precious metals that were a prerogative of the selected few of the society. Once metallurgy was developed sufficiently, iron objects could be made commonly and cheaply available to all. Even copper and bronze was scarce in India for a variety of reasons that need not be discussed here in detail. However, it has been argued that after nearly thousand years after the introduction of iron, it fails to bring about urbanization. What is the possible reason for such a situation? It is not very difficult to answer this question.

Iron metallurgy, because of its complex nature, could be developed very slowly. The wrought iron produced by the bloomery process was hardly very handy as an asset in production mechanism. At best, it could be utilized for hunting tools. This explains the presence of only such objects at the first stage. With development of carburization, there was some improvement in the quality of tools. With this, chisels, clamps, etc., came into use as carpenter's implements (perhaps for making, carts, chariots, etc.). It is only with mastering the metallurgical skills by the NBP period that a regular utilization of iron could be made in various areas of activity. The accompanying Tables (1, 2) (Fig. 2) will bear this out. The technological input is closely related to productivity. As agriculture gains priority, the society seeks better tools and implements to use for higher agricultural produce. From 700 BC in Eastern U.P., we have found a ploughshare at Ganwaria in District Gorakhpur. As the pattern of adaptation changes, the agrarian base of the society expands, attaining higher standards. We come across silos and other larger storage jars in excavations. The emerging technological skill could be harnessed to its full potentials only at NBP stage when the know-how developed sufficiently enough. The plough cultivation manifests the expansion of agricultural activity and exploitation of larger areas. The population moves away from the narrow strips of river banks to deeper areas away from the flood zones using wells instead of relying on river water alone. In this process of expansion of settlements also, better and efficient iron tools could have played a positive role.

The Megalithic cultures, on the contrary, were located in difficult hilly terrains. Their ecological setting did not promote expansion of agriculture. sI have argued about this feature elsewhere in greater detail (Tripathi[15], 2001: 232–35). Their adaptation pattern is oriented towards war-hunting. Their economy seems to be supported by war and marauding, as argued by Gurukkal[16]. The socio-technical background suggests a chiefdom society with little emphasis on agricultural activity. Thus, the social priority of adaptation of technology appears to have played a key role in deciding the course of their fate.

CONCLUSION

There are certain basic preconditions for adaptation of innovations as manifestations of technological advancements made by ingenious people at any age. First and foremost is an easy accessibility of the end product to the users. Secondly, the right environment for those involved in production mechanism of the innovative items including availability of raw material and infrastructure, and thirdly, the social will to adopt and absorb new products. When there is an optimum interplay of these conditions not only there is a quickened pace in the 'technics', but the related growth rate in terms of material output is also maximized. If this paradigm is tested against the early agrarian society of India, we come across different situations at different points of time and place. It is proposed to situate our enquiry into the context of the Gangetic plains and the Megalithic burial culture of the southern India.

The society of the Vedic age was engaged in consolidating its position and establishing suzerainty on the unchartered territories, gearing itself up for unforeseen hazards that lay ahead of them. They needed superior hunting and war weapons to meet these challenges. It is for this reason that at the first stage of the newly emerging iron technology, we come across a tool typology helpful in this objective. Economic growth took a second position in order of priority. Agricultural implements, building masonry, household objects follow.

In the Megalithic zone, the economy was generally based on hunting, warfare and marauding because of the ecological setting in which these groups operated (Rajan Gurukkal[16] and Tripathi[15]). There was little scope for change, leading them towards a more settled life in this zone. The agricultural-pastoral life remained the norm with a socio-political structure entrusting faith in a chiefdom that enjoyed special status. Lack of extensive plains with arable lands capable of high yield forced a prolonged system of existence based on the same lifestyle (the megaliths can be dated from 1100/1000 BC—early centuries of Christian era). Their tool typology reflects this attitude clearly (See Fig. 3). Only one or two per cent of their iron tools may be classified as agricultural implements. It is because of this that despite a rich iron tool repertoire, little economic growth is manifested in this zone even after centuries of use of iron.

For growth, expansion and prosperity of a region, technology alone is not sufficient. The ecological setting like fertile land, coupled with a network of rivers sustaining and promoting communication, contacts, agricultural requirements, play a dominant role. Equally, if not more important is the social will to accommodate, incorporate and adapt to new innovations and changes. Crystaller's centre-place theory subscribes to several orders of the circle which go into making of an important centre (Fig. 5). Good agricultural land, if exploited to its full potential with all the required gadgetry, will go a long way in making of an affluent economic system. This was a situation that was created by the Gangetic plain social system, taking the region to a rich, growing and forward looking world by the latter half of the first millennium BC.

Sites	Arrowhead	Spearhead	Shaft	Tong	Clamp	Nail	Barfrod	Hook	Borer	Chisel	Needle	Axe	Knife	Bangle	Indeterminate	Ploughshare	Spears/Javelins	Hoe	Sickle	Dagger	Slag	Blade	Pin	Peg	Disc	Point	Ring	Tong	Adze	Digger	Dish	Laddle	Lamp	Nail parer	Lump	Slag	Ore	Crucible	
ATRANJIKHERA	●	●	●	●	●	●	●	●	●	●	●	●	●	●	●																								135
JAKHERA	●	●				●	●			●						●		●	●	●	●																		
KAUSAMBI																	●																						
GANAWARIA					●											●																							1116
RAJGHAT						●				●			●		●						●	●																	
CHIRAND (14)																																							13
SONKH												●																											
ALLAHPUR (IB)	●	●																				●																	
ALAMGIRPUR	●	●				●																	●																
HULAS																																							
AHICHCHHATRA																																							
SORON						●																																	
MANGALKOT	●	●				●	●		●										●		●			●															
PANDURAJARDHIBI		●																	●	●	●	●			●	●													
MAHISDAL	●	●								●											●																		
HATIGRA							●													●																			
DIHAR	●					●															●																		
TARADIH	●	●				●			●																														
HASTINAPUR	●	●																				●																●	
NARHAN 12	●	●				●	●			●					●																								
KOLDIHWA	●												●										●					●						x					
TAKALGHAT-KHAPA	●	●				●	●			●		●	●	●		●				●	●											●				●	●		
NAIKUND					●			●		●		●	●						●		●	●	●								●	●		●	●	●	●		
MAHURJHARI																																							
HALLUR	●	●											●																										
NOH	●	●										●																											

Table 1: Distribution of iron objects from Archaeological sites
STAGE-I (Early Iron age)
Artifacts

Sites	Arrowhead	Spearhead	Shaft	Sickle	Spud	Ploughshare	Hoe	Digger	Tong	Clamp	Ring fastener	Lance	Socketed clamp	Blade	Staple	Balt	Razor	Plumb-bob	Nail	Bar/rod	Hook	Borer	Chopping knife	Disc	Chopper	Pipe	Scraper	Chisel	Axe	Knife	Lid	Bangle	Lump	Indeterminate	Jevlins tips	Adze	Ring	Elephant goad	Hoe	Socket	Dagger	Sword	Point	Dish	Handle	Spade
ATRANJIKHERA	●	●	●	●	●	●	●	●	●	●	●	●	●		●	●		●	●	●	●	●	●	●	●	●	●	●	●	●	●	●	●	●												
JAKHERA	●	●		●																●										●										●						
KAUSAMBI	●	●																	●		●													●	●	●	●									
GANAWARIA				●		●													●											●																
SRAVASTI	●																		●									●										●	●	●						
SORON																			●										●																	
DHAR																			●																						●	●				
CHIRAND							●												●																						●					
CHAMPA		●																										●													●					
VAISALI																																									●					
MASAON	●	●																																									●			
TUMAIN	●	●																	●																●											
JODHPURA	●																		●																											
JAJMAU (overlap)	●	●		●															●		●																●									
RAJGIR																			●																											
DAULATPUR				●																																										
SOHGAURA	●																																													
NOH	●	●																																										●		
TARADIH	●	●		●													●		●									●				●	●				●									
AWRA		●																	●									●					●													
JADERUA	●	●		●																								●					●													
DHAR																			●																						●	●				
RAJGHAT	●	●		●																								●	●				●													
HASTINAPUR	●			●																								●						●												
KHAIRADIH	●	●		●		●		●						●					●	●			●					●	●	●			●	●						●						
NARHAN																			●									●																		
SONPUR	●	●										●		●					●									●	●												●					
MANER																																														
MANJHI	●																		●											●									●		●					
SENUWAR	●	●		●				●		●		●							●	●								●	●				●	●	●										●	
AHICHCHHATRA	●	●		●															●	●	●					●		●									●						●			●

Table 2: Distribution of iron objects from Archaeological sites
STAGE-II [Middle Iron age] (NBPW Level)
Artifacts

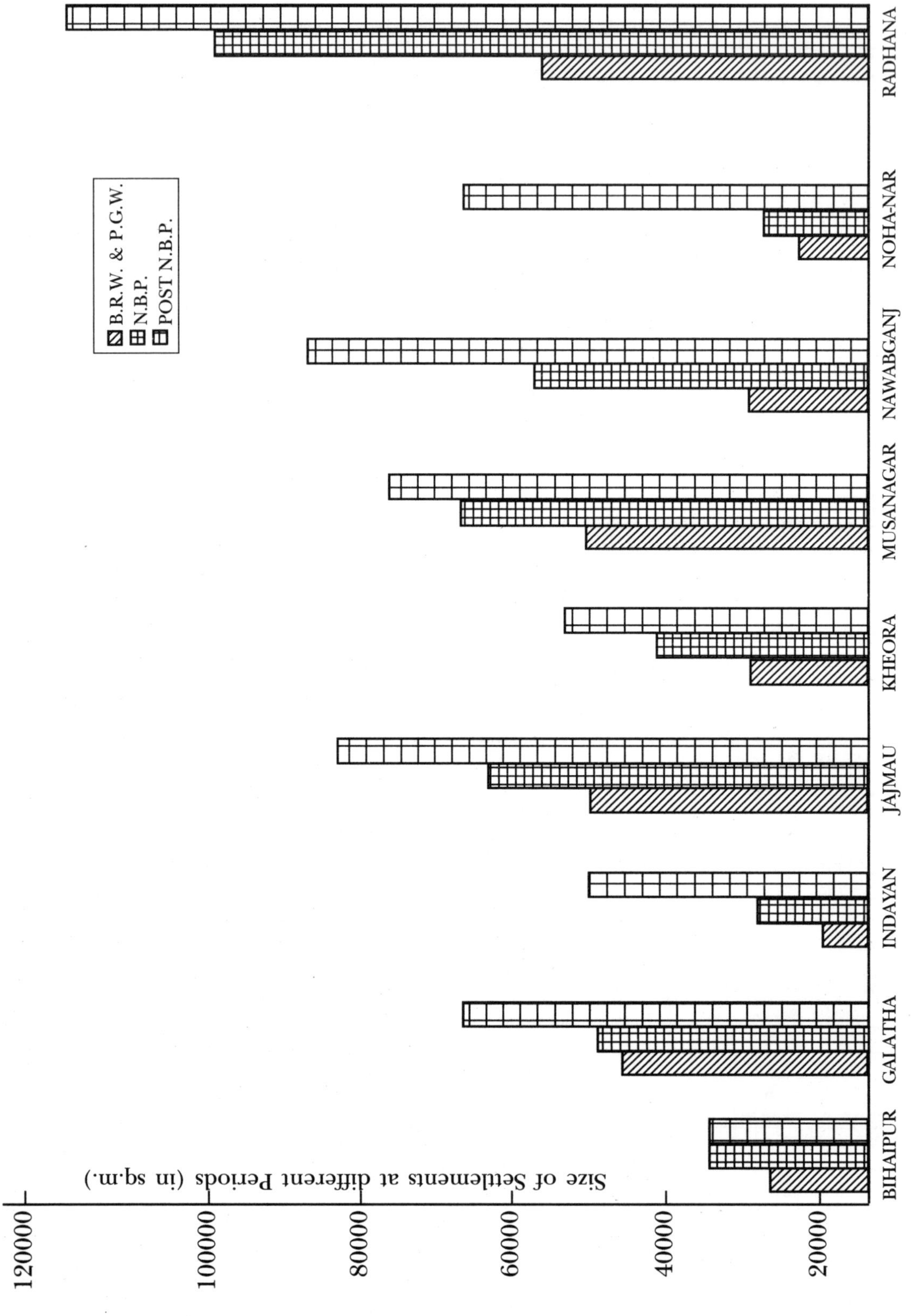
Size of Settlements at different Periods (in sq.m.)
120000
100000
80000
60000
40000
20000
BIHAIPUR
GALATHA
INDAYAN
JAJMAU
KHEORA
MUSANAGAR
NAWABGANJ
NOHA-NAR
RADHANA
B.R.W. & P.G.W.
N.B.P.
POST N.B.P.

Fig. 1: Graph

TOOL TYPE	NAME OF TOOLS	E.	M.	L.
HUNTING TOOLS	Spear heads	✪	✪	✪
	Arrow heads	✪	✪	✪
	Points	✪	◈	◈
	Socketed tangs	✪	◈	◈
	Blades	✪	✪	◈
	Spear lances	◈	✪	◈
	Dagger	◈	✪	✪
	Sword	◈	✪	✪
	Elephant goad	X	✪	✪
	Lances	X	X	✪
	Armour	X	X	✪
	Helmet	X	X	✪
	Horse bits	X	X	✪
	Caltrop	X	X	✪
AGRICULTURE TOOLS	Axes	✪	✪	✪
	Sickles	✪	✪	✪
	Spade	X	✪	◈
	Ploughshare	X	✪	◈
	Hoc	X	✪	✪
	Chisel	X	✪	✪
	Pick	X	◈	✪
HOUSEHOLD OBJECTS	Knives	✪	✪	✪
	Tangs	✪	◈	◈
	Discs	X	✪	◈
	Rings	X	✪	◈
	Spoons	X	✪	✪
	Sieve	X	X	✪
	Cauldron	X	X	✪
	Bowls	X	X	✪
	Disches	X	X	✪
BUILDING MATERIAL	Rods	✪	◈	◈
	Pins	◈	◈	◈
	Nails	✪	✪	✪
	Clamps	✪	✪	✪
	Pipes	X	✪	◈
	Sockets	X	✪	◈
	Plumb bob	X	✪	◈
	Chains	X	✪	✪
	Door hooks	X	✪	◈
	Hinges	X	X	✪
	Spikes	X	X	✪
	Tweezers	X	X	✪
	Anvils	X	X	✪
	Hammers	X	X	✪
	Scissors	X	X	✪
	Saw	X	X	✪

INDEX ✪ DEFINITE EXISTENCE ◈ CONFIRMED DATA NOT AVAILABLE X NON-EXISTENCE

Fig. 2: Distribution of Tool Typology

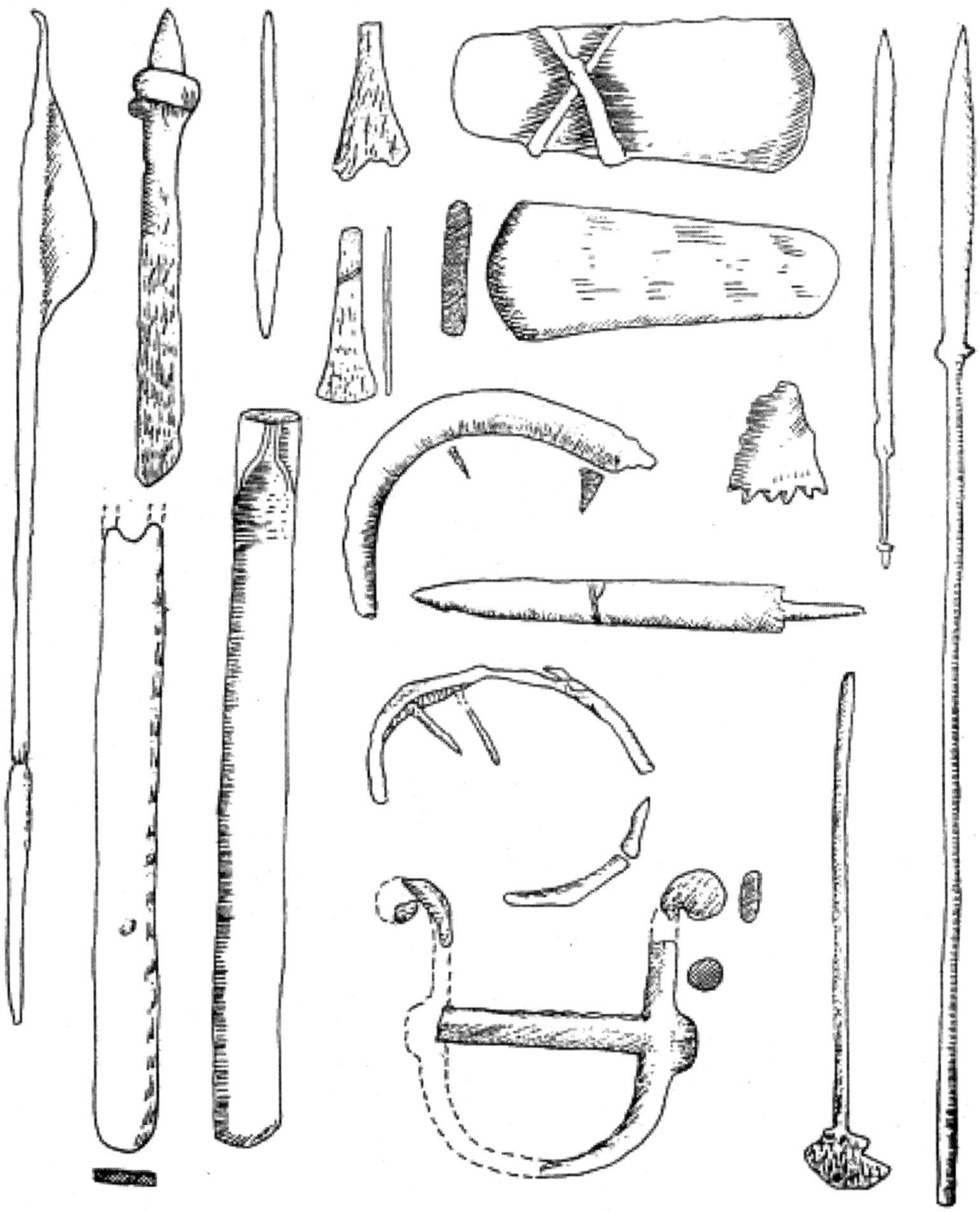

Fig. 3: Iron Objects (Megaliths, Vidarbh)

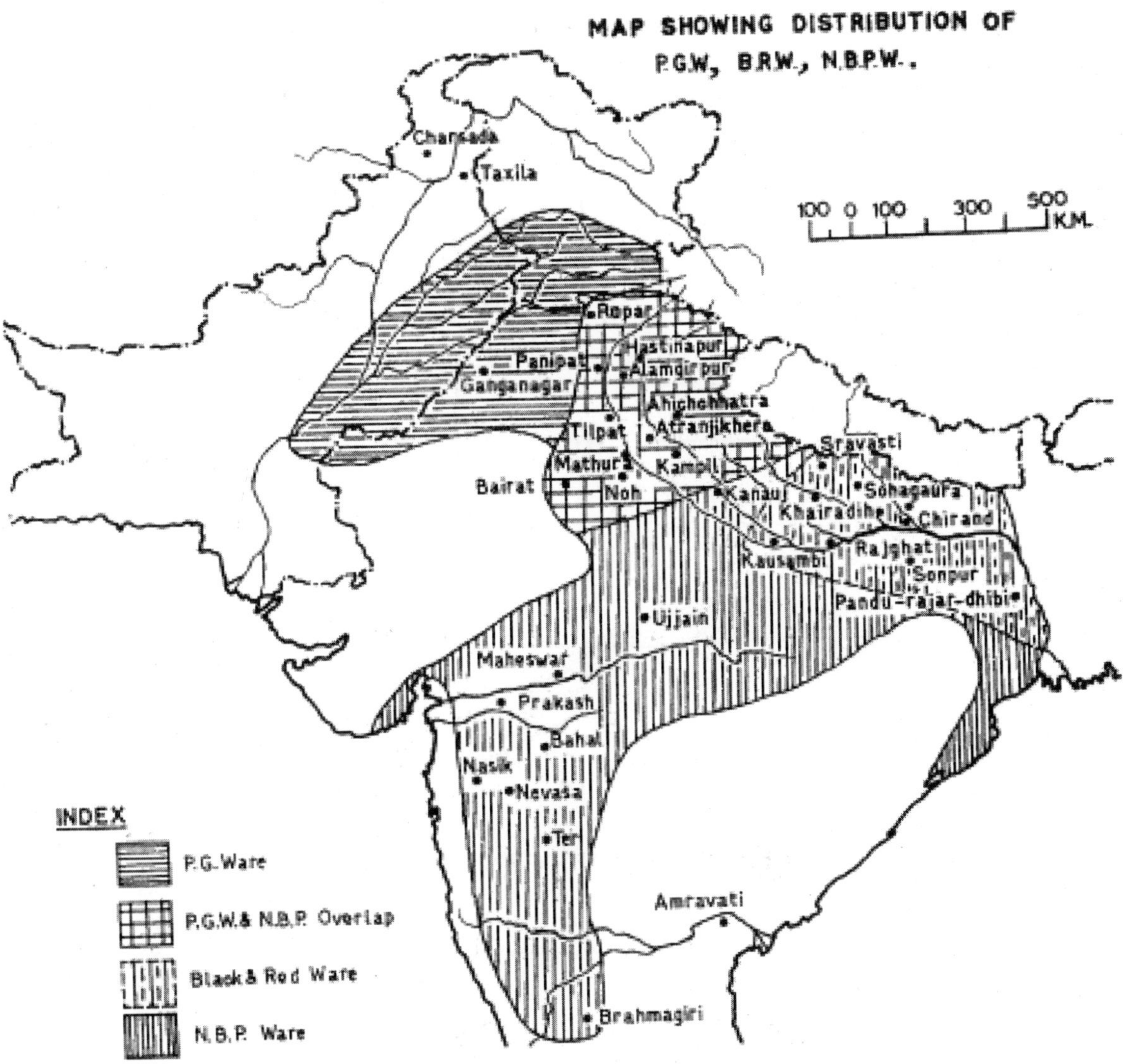
MAP SHOWING DISTRIBUTION OF
P.G.W., B.R.W., N.B.P.W..
100 0 100 300 500 K.M.
Charsada
Taxila
Ropar
Panipat
Ganganagar
Hastinapur
Alamgirpur
Ahichchhatra
Atranjikhera
Tilpat
Mathura
Noh
Kampil
Bairat
Sravasti
Sohagaura
Kanauj
Khairadih
Chirand
Kausambi
Rajghat
Sonpur
Pandu-rajar-dhibi
Ujjain
Maheswar
Prakash
Bahal
Nasik
Nevasa
Ter
Amravati
Brahmagiri
INDEX
P.G. Ware
P.G.W.& N.B.P. Overlap
Black & Red Ware
N.B.P. Ware

Fig. 4: Map

CHRYSTALLER'S CENTRE PLACE THEORY

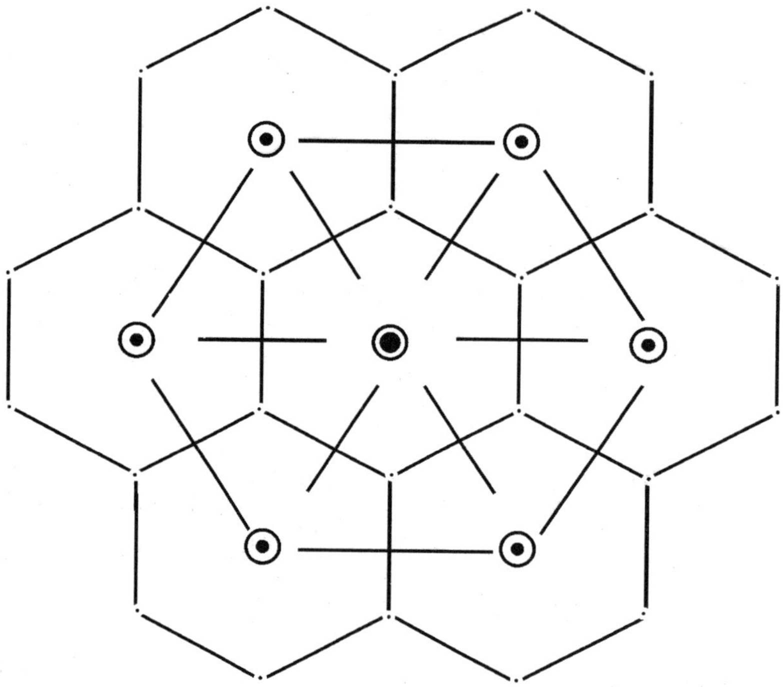

◉ FIRST ORDER CENTRE

⊙ SECOND ORDER CENTRE

● THIRD ORDER CENTRE

Fig. 5

NOTES AND REFERENCES

1. Sharma, G.R. 1980, *History to Prehistory,* Archaeology of the Ganga Valley and the Vindhyas, Allahabad.
2. Tripathi, Vibha 'From Hills to Plains: The Drainage Pattern and Growth of Culture in the Ganga Plain' paper presented at International workshop on Monsoon and Civilization Jan. 16–18, 2001, Pune.
3. Jarrige, J-F and Lechevallier 1977, 'Excavations at Mehargarh, Baluchistan: Their significance in the Prehistoric Context of Indo-Pakistan Border Lands' in M. Taddi (ed.), *South Asian Archaeology,* Naples Institute Universitario Orientale, 463–535.
4. Chowdhary, K.A. 1983, 'The Plant Remains' in *Excavations at Atranjikhera,* R.C. Gaur, New Delhi.
5. Saraswat, K.S. 1994, 'Plant Economy at Ancient Narhan' in Excavation at Narhan, P. Singh, Varanasi, pp. 255–347.
6. Singh, B.P. 1995–96, 'Transformation of Cultures in the Middle Ganga Plains: A Case Study', *Pragdhara* No. 6 ed. Rakesh Tiwari, U.P. State Organisation, Lucknow pp. 75–93.
7. Chirand, *Indian Archaeology, A Review* 1963–64, p. 60; 1964–65, p. 46.
8. Sharma, R.S. 1984, *Material Culture and Social Formation in Ancient India,* Macmillan, New Delhi.
9. Sahi, M.D.N. 1985, Comments on D.K. Chakravarti's paper, 'Issues of Indian Iron Age' in *Recent Advances in Indian Archaeology,* S.B. Deo and K. Paddayya (eds) pp. 87–88.
10. Dhavalikar, M.K., H.D. Sankalia and Z.D. Ansari, 1988, *Excavations at Inamgaon* Vol. I, II. Deccan College, Pune.
11. Deo, S.B. 1985, 'Megaliths: Their Culture, Ecology, Economy and Technology', *Recent Advances in Indian Archaeology,* S.B. Deo and K. Paddayya (eds.) Deccan College, Pune, pp. 89–99.
12. Agrawal, V.S. 1969, *India as known to Panini,* (in Hindi) Varanasi.
13. Rhys Davids and William Stede 1921, Suttanipata IV, 314–17 in Sharma (*op. cit.*).
14. Saraswat 1994, *op. cit.*
15. Tripathi, V. 2001, *The Age of Iron in South Asia, Legacy and Tradition,* Aryan Books International, New Delhi.
16. Gurukkal, R. 1987, 'Aspects of Early Iron Age Economy: Problems of Agricultural Expansion of Tamilakam. B. Chattopadhyay (ed.) *Essays in Ancient Indian Economy.* HC Series, Vol. II, pp. 46–57.

SECTION FOUR

AGRICULTURE IN EARLY ANCIENT INDIA (FROM c. 600 BC TO c. 600 AD)

CHAPTER 24

Agriculture as Revealed by the Pāli Literature

H.S. Shukla

INTRODUCTION

Pāli literature consists of *Tipiṭaka, Aṭṭhakathā, Ṭika, Vaṁsa* literature, *Sāragantha* (manual literature), books on grammar, prosody and rhetoric, etc. *Tipiṭaka* is the repository of the original teachings of the Buddha. It is so named as there are three *Piṭakas* in it namely, *Vinaya-piṭaka, Sutta-piṭaka and Abhidhamma-piṭaka,* dealing with monastic rules, popular discourses and the scholastic teachings, respectively.[1] The remaining types of literatures state the meanings of the words used in the *Tipiṭaka,* explanations of the ideas, and expository notes on the concepts.

After getting enlightenment under the Bodhi tree at Bodh Gaya, Lord Buddha preached the *Dhamma* for forty-five years for removal of the suffering and generation of harmony in life. In course of his preachings, he used the prevalent, popular, prominent and practical similies to simplify his speech and make clear the theme of the subject. These illustrations are so lively that they throw considerable light on social, religious, and economic conditions of the time. This chapter proposes to make a study of the agriculture as depicted in Pāli literature.

E.F. Schumacher is the first to relate the Buddha's teachings to economy. In his book *Small is Beautiful,* he looks forward to the Buddha's teachings of the noble eight-fold path for the purpose and affirms that the inclusion of the factor of Right Livelihood (*Sammā Ājivo*) is indicative of the necessity of Buddhist economy.[2]

India has been an agricultural country since ancient times. Agriculture was regarded as an honourable profession. Brāhmaṇas, Kṣatriyas, Sūdras, both rich and poor were engaged in agricultural operations.[3] This was not a profession of any particular class or caste. The people engaged in agricultural activities were known as farmers.

Farming was treated as a pious, religious and virtuous job at that time. The date and time of the first day of sowing was fixed by the king in consultation with his priest. That day was called the blessing day of sowing (*Vappa-maṅgala-divasa*). It was celebrated and inaugurated by the king himself along with his family members, ministers and other attendants. On this occasion, the whole city was decorated and the king offered new clothes and food to all the people connected with the function. After putting on new clothes and decorating themselves with garlands, they assembled at the palace. Thereafter, oxen (*balivaddāni*), string (*yottaṁ*), beams, ploughs (*Naṅgalāni*) were also decorated with silvers and the plough of the king was made of gold. There were one thousand such ploughs. The king started ploughing simply by going from this side to that side of the field with golden plough and others, like ministers and attendants too, started ploughing.[4]

It is evident from the text that the farmer Bhāradvāja *Brāhmaṇa* had five hundred ploughs. It was the time for sowing. The distribution of food was going on among the five hundred ploughmen. In the meantime, the Buddha reached there and was standing for alms. Having seen him, the Brāhamaṇa said "O Recluse! I plough and sow and having ploughed and sown, I eat. You should also plough and sow and then you eat." The Buddha replied "I too, O, Brāhamaṇa, plough and sow and having ploughed and sown, I eat." In replying to another question put up by the Brāhamaṇa regarding the appliance of agriculture, the Buddha said:

Saddhā bījaṃ tapo vuṭṭhi, Paññā me yuganaṅgalaṁ,
Hirī īsā mano yottaṁ, Sati me Phālapāchanaṁ.[5]

Faith is the seed and rain the discipline, Insight for me is plough fitted with yoke, My pole is consciousness and sense-mind the tie, and mindfulness is my ploughshare and my goad.[6]

The *Kasibhāradvājasutta* and its commentary proceed with such elaboration that we may infer the episode possessed great popularity in Buddhaghosa's day and even earlier. It is evident from the frequent allusions to the farmer and his activities in the *Tipiṭaka* that agriculture had already become a way of life.

FARMER (*KASSAKA*)

The farmer has been designated in Pāli literature as *Kassaka*[7] or *Khettapāla*[8], etc. It does not represent a caste but refers to the person who is engaged in the cultivation of land or agricultural operations. Lord Buddha says for a farmer:

Yo hi koci manussesu, gorakkhaṁ upajīvati,
Evaṁ Vāseṭṭha jānāhi, kassako so na Brāhamaṇo.[9]

Among persons, whoever lives by keeping cattles is known as farmer, not as a *Brāhmaṇa*.

A farmer has been characteristically described in the *Kassaka-sutta* of the *Saṅyuttanikāya* as a man with a plough on his shoulder holding an ox-goad in his hand, with dishevelled hair, raiment hempen and feet spattered with mud ...*kassakavaṇṇaṁ abhinimminitvā mahantaṁ naṅgalaṁ khandhe karitvā dīghapācanayaṭṭhiṁ gahetvā haṭahaṭakeso sāṇasāṭinivattho kaddamamakkhitehi pādehi yena Bhagavā tenupasaṇkami; upasaṇkamitvā bhagavantaṁ etadavoca – api, Samaṇa, valibadde addasā ti.*[10]

Thus, a farmer having these above mentioned characteristics, cultivates the land for growing crops, rearing animals, etc.

There were two types of farmers. Some were extremely rich and some were extremely poor[11]. They lived in villages, which were of different types. They were called *gāma* (an ordinary village), *gāmaka* (small but well-managed village), *nigamagāma* (a village having character of a town), *dvāra-gāma* (a village situated near the main gate of the city) and *paccanta-gāma* (a village situated on the border of the country).[12]

DUTIES OF A FARMER

The Buddha has prescribed the three main duties of a farmer. They are:

1. to get his field well ploughed and harrowed very quickly,
2. to sow the seeds promptly, and
3. to let the water in and then to let out very quickly.[13]

Other duties which we find in the text are to be prompt in reaping the crops, gathering them, harvesting, putting in stocks, treading them out, pulling off the stalks, winnowing away the chaff, collecting the rice, threshing it out and removing the husks. The Buddha says that the farmer's crops reach perfections in this way.[14] Further, the text mentions that the duty of the farmer is to look after the plants and remove the corrupt plants as they disturb the growth of the good plants.[15]

There are certain items essential for agriculture—land, plough, seeds, irrigation, etc. Some words in Pāli tell us about the condition of these items during the time of the Buddha.

LAND (BHŪMI)

Land is the first unit of production. We find not much difference in the conditions of land at present. During the Buddha's time, land was divided into small plots for the purpose of cultivation. They were divided and bifurcated by strips, embankments, etc., and allotted to each family by the ministers of a king. The entire land of the village was called *gāmakhetta,* on which collective right prevailed. A person had very limited right in respect of the disposition of land.

The size and the position of the then agricultural land can be seen in a passage of the *Mahāvagga* in the *Vinaya Piṭaka.* There, the Buddha notices the field of Magadha, laid out in strips, embankments and squires. On the basis of this, he instructed Ānanda to design robes for monks and he did this very successfully. This pattern became popular for making robes.[16]

This shows that the open fields were separated from one another by narrow and low earthen embankments. It does not mean that there were no big size plots at that time. According to the *Malalasekera,* the corn-fields of Magadha were rich and fertile and each plot was about one *gāvuta* in extent.[17] *Gāvuta* means a league, a measure a little less than two miles.[18] The *Suvaṇṇakakkaṭa Jātaka* and *Sālikedāra Jātaka* refer to a plot of one thousand *Karisas* in a *Brāhmaṇa* village named *Sālindiya* to the east of Rājagriha. According to *Malalasekar*, one thousand *karisas* is equivalent to about eight thousand acres.[19] Of course, this reference does not prove that this vast area is the size of one plot, but it shows a tendency towards cultivation on a large scale. It might be the total area of the field of a particular village. A.P. Buddhadatta Mahāthera is of the opinion that *karisa* is a square measure of land which may be nearly an acre.[20] Rhys Davids does not mention the area of a equivalence of one *karisa.* He simply says that *karisa* is a required measure of plot on which a *karisa* of seed may be sown.[21]

However, this does not prove that in size and shape, the Indian plots of those days were the same as of our own times. Accordingly, fields tended to be large rather than small. This characteristic is specific to the plots of Magadha. According, to the *Aṅguttara-nikāya,* an average plot was measured to two to three miles across.[22]

On one side of the field were village-houses and on the other side forests. The growing crops were always in danger from the village cattle and the herds of forest-deer. The farmers did not sow the crop for fear of these animals. The Jātaka story also frequently refers to the crop-destruction by these animals.

The following land-classification seems to be pertinent in the context of the period:

1. Fertile Land

It is the land on which seeds were sown and reaped perfectly. Such land was called *khetta* in Pāli literature. It was distributed among the farmers by the king. The border-line of a particular plot was fixed by a privy councillor appointed by the king[23]. According to the needs of the people, forest land was gradually converted into agricultural land by cutting and clearing the forest, *Sabbaṁ vanaṁ chinditvā khettāni karitvā kasikammaṁ karantu.*[24] Thus, the area of agricultural land was increasing day by day. It was really a very important aspect of the economy of those days because the large amount of revenue came from the farmers.[25] Therefore, it was regarded as the best land.

2. Less Fertile Land

It was a type of land which had less capacity to produce good crops. It was neither very fertile nor infertile but a middle group of land. The soil of such land was sandy. It required to be sufficiently irrigated. During the time of the Buddha, such land was regarded as less fertile due to the lack of sufficient irrigation facilities.

3. Infertile land

There was also a type of land which was of no use for agricultural purposes. It was neither fertile nor less fertile but totally infertile land. It was not at all cultivable. Forest, mountain and barren lands were included under this category. In this context, the text mentions that the plot (*khetta*) which possesses eight qualities is not very fruitful nor is it thought to be a flourishing plot. The eight qualities are: undulating, rocky and pebbly, saltish, without depth of tilth, without (water) outlet, without inlet, with no water-course, and without dyke.[26]

What type of land should be cultivated? Generally, it is supposed that the land or the field (*khetta*) which gives us a good yield of crops is regarded as cultivable land. It has also been pointed out in the text that the land on which the unbroken, unrotten, unspoiled and capable of sprouting seeds were sown, or planted and which came to growth abundantly, was regarded as the cultivable land.[27] The farmers sowed their seeds firstly on perfectly fertile land so that they got good crops.

In this way, according to the nature of land, plots were divided into two groups, namely: natural plots and artificial plots. Natural plots contained pure soil, pure clay with practically no stones or sands, etc., and artificial plots were full of pure stone, pure pebbles, pure sand with little soil, clay,[28] etc. Natural plots were regarded as most suitable for agriculture.

There were some types of plots, which were named after the plants which grew on them. They were: corn field (*yavakkhetta*)[29], rice-field (*Sālikkheta*)[30], sugarcane field (*Ucchukkhetta*)[31], grass field (*Tiṇakkhetta*),[32] etc. These plots were understood by the people by these names.

ESSENTIALS OF AGRICULTURE

Like today, in the Buddha's days too, there were some essentials for agricultural activities (*Kasi-kamma*) such as seed (*bīja*), yoke (*yuga*), plough (*naṅgala*), rain (*vuṭṭhi*), pole (*yugadanaḍa*), or ploughshares like *phāla* and beam of a plough (*naṅgalisā*) and oxen (*balivaddāni*).[33]

Seeds

Seeds were regarded as important and essential for the good production of crops. A combination of good quality of soil and seeds was considered to be the best to get good products of grains. The text says that the seeds sown in fertile land bring good crops as a great reward: *Etesu dinnāni mahapphalāni, bījāni vuttāni yathā sukhette.*[34]

There were five types of seeds prevalent in the time of the Buddha. They were: root germs (*mūla-bīja*), stem germs (*khandha-bīja*), node germs (*phala-bīja*), phumule germs (*agga-bīja*), and seed germs (*bīja-bīja*).[35]

Root Germs

It is a kind of seed produced from roots. There are many plants which grow by the sowing of their roots. Such plants are: turmeric, ginger, orris, white orris, garlic, black hellebore, khus khus, nut grass, etc.[36]

Stem Germs

It means the seed propagated from stems. Fig tree, banyan tree, Indian cedar-wood and others born from a stem are included in this category.[37]

Node Germs

It is a joint seed which produces a plant. In other words, it can be said that the joint part of some plants functions as a seed. Sugarcane, bamboo, reeds or whatever others are born from a joint, come in the category of node germs.[38]

Plumule Germs

It is one of the kinds of seeds propagated from cutting or budding or graftings. Basils, camel grass, a kind of andropogon or whatever others are born from buddings, belong to the group of plumule germs.[39]

Seed Germs

It is a special kind of grain selected for the use of sowing in the field for future production of grains. It refers to the seed which is born or propagated from seeds. Grains, pulses or whatever others of this group are born from a seed, come in the group of seed germs.[40]

The seed functions as a base or root of the plant under the ground and germinates on the ground.[41]

The above germinal order signifies the sprouts, shoots, trunks, branches, twigs, leaves, flowers and fruits, which spring from the seed. This explanation applies to all trees, creepers, shrubs, and grasses. This is neither made nor created by anyone. It is accomplished entirely by the natural order that we know as the germinal order. Taking into consideration the importance of seeds, Lord Buddha instructed the monks to abstain from the destruction of such seeds and vegetation, otherwise there would be an offence of expiation (*pācittiya*),[42] though such destruction is said to be no sin in the eyes of the laity.

Plough

A plough was supposed to be the most important tool for the cultivation of land during the time of the Buddha. It is called *Naṅgala* in Pāli literature. *Kasibhāradvāja-sutta,*[43] *Mahākanha-jātaka,*[44] *Therigātha*[45], *Naṅgala-kulathera*[46] and *Naṅgalisa-jātaka*[47] prove the fact that the plough was used for the cultivation of land at that time.

A plough was made of wood, iron, gold and silver. The ploughs made of gold and silver were used on the ceremonial day of sowing (*vappa-maṅgala-divasa*) by the king and his attendants, respectively, for cultivation of land.[48] A Jātaka story also makes mention of the iron ploughshares.[49] But the iron ploughshare could not become common in spite of its capacity of making deep ploughing in the time of the Buddha.

The plough in itself is nothing but a collection of parts or ploughshares such as beam (*Isā*), phāla, yoke (*yuga*), etc.[50]

Beam

A beam is pole of a plough made of wood. It is just like an elephant's trunk.[51] It bears the yoke and the plough.[52]

Phāla

It is a ploughshare or the part of a plough made of iron and fixed in the lower portion of the plough. It was used to tear or cultivate hard soil. A Jātaka story also says that sometimes, more than one phāla were fixed in one plough. Probably this system was prevalent to cultivate the hard or rocky land.

Yoke

A yoke is the part of a plough, but not fixed. Whenever the plough was used, the yoke was attached with the same and pressed on the neck of oxen. It was made of wood and tied with a leather rope to the plough. The text says that the yoke is the base for beam (*Isā*); it stands before it and remains tied with the beam (*Isā*). It is a helper of string (*yotti*) and keeps the oxen to move properly in one direction.[53]

String

A string is a tie of the yoke of a plough. It is a kind of rope which ties the yoke, beam and oxen together with the plough.

> *Yottaṁ ti rajjubandhanaṁ. Taṁ tividhaṁ isāya yugassa bandhanaṁ, yugena saha balivaddānaṁ bandhanaṁ sārathinā saha balivaddānaṁ bandhanaṁ ti.*[54]
>
> It controls the oxen to go in proper and desired direction. Therefore, Lord Buddha has compared it with mind *mano yottaṁ*[55] because mind is the forerunner of all activities[56] and it controls all physical and vocal deeds.

Oxen

It is evident from the text that oxen were used in the cultivation of land for agricultural purposes. In a Jātaka story, we are told that only two bullocks per plough were used then,[57] as is the mode now. Farmers bought the oxen, or loaned them from the king.[58] The selection depended upon the quality of an ox. The purchased oxen became the personal property of the farmer.

Other than ploughing, bullocks were also used for drawing carts to transport goods as well as for conveyance purposes.

In addition to the above mentioned essentials of agriculture, some other tools were also used for agricultural purposes. They are: spade (*Kuddāla*),[59] *phāvarā*, etc.

PROCESS OF CULTIVATION

It is believed that the agricultural process during the Buddha's period was the same as today—ploughing implements are changed in the modern times but the process is remained practically the same from tilling the land to collecting grains. The *Cullavagga* of the *Vinaya Piṭaka* gives us a more detailed account of the consecutive farming processes. Giving details about the agricultural process, Mahānāma Sākya says to his brother Anuruddha in the context of household life that first of all plots are ploughed; having them ploughed, they are sown; having sown them, water is led into them; after that the water is led away; weeds are dug up; the crop is ripe, it is reaped; then it is harvested, stocked and threshed; straw and chaff are winnowed. The stalks are then gathered into a heap and are treaded out. Thereafter, pure grains are stocked and has had to be done all over again in the next year in a never-ending continuity.[60]

Furthermore, the *Khetta-sutta* of the *Aṅguttara-nikāya* says that firstly the plots are ploughed and harrowed very quickly. Thereafter, the farmers sow their seeds, let the water in and turned it out very quickly. These chores were treated as urgent duties of the farmer.[61] In another Jātaka story, various agricultural operations are mentioned such as sowing, watering, fencing, weeding, reaping and treading.[62]

Thus, it is evident from the above that in the time of the Buddha, tilling or ploughing, making the field plane, sowing, watering in, letting away the water out, digging up weeds, watching the crops, reaping, harvesting, winnowing, etc., were familiar processes of agriculture.

How did these processes go on? The illustrations of the process have been so lively presented in the texts that they create a clear picture of the then agricultural economy of the country. In course of ploughing, the land was tilled deeply before sowing the seeds. The farmers ploughed not only once but twice or thrice in order to make soil soft. Thereafter, the plot was levelled with a planer and stones, pebbles and other hard matter were taken out.

After having been ploughed, the plot was ready for sowing. Farmers examined the quality of seeds before sowing. The good crop because depended upon the healthiness of seeds. Most of the farmers kept seeds from their former products and if the seeds were damaged or otherwise unsowable, they bought seeds from others. Poor farmers took the seeds on loan. In case of an emergency, the king provided seeds to the farmers.[63]

Good quality of seeds were always in demand because the yield depended upon the nature of seeds. In this context, it seems essential to quote a passage to highlight the importance of seed. The passage runs thus:

> The Buddha said: 'Suppose, monks, a nimb-seed or a seed of creeper or cucumber be planted in moist soil. Whatever essence it derives from earth or

water, all that conduces to its bitterness, its acridity, and its unpleasantness. What is the cause of that? The ill nature of the seeds is the root cause of that. Suppose, monks, a seed of sugarcane or paddy or grape be planted in moist soil. Whatever essence it derives from earth or water, all that conduces to its sweetness, pleasantness and delicious flavour. What is the cause of that? The happy or healthy nature of the seed is the root cause of that.'[64]

Just after sowing, the plot was irrigated and then the water was turned out from the plot very quickly. Generally, it is seen that after watering the plot, some unwanted grasses and weeds grow near the plants and harm their growth. Therefore, the Buddha emphasized the importance of clearing the weeds from plots. The enterprising peasant was thus expected to be ever vigilant and not to allow redundant grasses to grow near the plants. In the context of the negative effects of the association with evil persons, the Buddha pointed out the process of separating the chaff from the grain. He says, "Just when a great heap of corn is winnowed, the grain, which is sound, is piled in one place, but that of the poor quality, the more chaff the wind carries to one side. At once the farmers sweep it still further away". Why so? They say, "Let it not corrupt the corn of worth."[65]

From the above description regarding agricultural processes prevailing in the time of the Buddha, we come to know that they are exactly the same today even after more than 2500 years. Probably, these processes had been existing for many centuries since before Lord Buddha, but the agricultural technique had become matured at the time of the Buddha, so much so that the succeeding centuries simply carried them on. Plots were bifurcated from each other by low earthen dividers like today. Proper attention was paid to the suitability of the seed as well as the field. The peasants were industrious. They prepared the ground well and cleared the weeds constantly. Some peasant proprietors cultivated very big plots running to thousands of acres. Agriculture was not looked down upon as an occupation. In fact, it was regarded as an important and an honourable profession in India at that time.

As agriculture was regarded as the most honourable profession, so the Order (*Saṅgha*) in its early phase of development, having taken the seeds from other, cultivated its own land adjoining to monasteries and, at times, other's land with its own seeds. The one-tenth of the total production was given to the land owner. This was in practice in the time of the Buddha.[66]

IRRIGATION

Irrigation was supposed to be the most important and essential factor for the good production of crops. All the essentials of agriculture being well oganized, the good production of crops was not possible unless there would be an availability of sufficient irrigational sources. That is why the system of irrigation was fully developed during the period of the Buddha. The text mentions a farmer letting in and turning off the water very quickly after sowing the seeds in the field.[67] A *Jātaka* in a casual way gives us a glimpse of the irrigation system of *Mithilā*. It depicts flowing rivers, round going water under canals that were obviously taken out from the former and, as a consequence, green fields were lying about all over *Mithilā* region. It also refers to dug out tanks.[68]

Moreover, the reference to letting in and turning off the water very quickly shows that the people had made arrangements not only for supplying water to the plots, but also for regulating the volumes of the water supply according to the necessity. They knew that the excess of water was as harmful as was paucity of water. So, it can also be inferred that the farmers gave much attention to the drainage system for irrigation. In this way, irrigation was an established institution and common among the people.

SOURCES OF IRRIGATION

Sources of irrigation were the same as today, except that nowadays, with the development of science and technology, sources like tube well, pumping set etc., are increasing day by day. But, in spite of the new methods, even today vast areas do not have adequate irrigation facilities. In the Buddha's period too, irrigation facilities were not available everywhere. The farmers of such areas depended upon rainwater and they had to face many difficulties. They fought for survival at the time of famine.

There were two types of sources of irrigation available which have been indicated in Pāli literature. They are:

1. Natural Sources, and
2. Artificial Sources.

NATURAL SOURCES

Rain and rivers belong to this category. Farmers were fully dependent on rain for irrigation of plots. They were always worried about their crops because of the uncertainty of rain. Whenever it rained, people made dikes to store the water for the purpose of irrigation. The text mentions dikes in which water was collected: *Puriso mahato talākassa paṭikacceva aliṁ bandheyya yāvadeva udakassa analikkamanāya...*[69] Water was available in dikes throughout the year, even in the summer season, and the people took this water to the plot through canals. These dikes were gradually converted into wells. The farmers took the water from the well or dikes through a scoop-wheel (*Cakkavaṭṭaka*). It is a wheel revolving over the well with a string of earthen pots going down empty and coming up full after a dredger fashion. It is called *Rehaṭa* in Hindi.[70]

In such a situation, sometimes farmers had to face difficulty due to the shortage of rainfall, causing famines. We find some references that even in the lifetime of the Buddha, there happened a famine in the country of Vajjis.[71] *Kula-sutta* of the *Saṅyutta-nikāya* too refers to the famine of Nalanda, where many people died at that time.[72] *Viraka-jātaka, Vessantara-jātaka* and other texts refer to the oft-recurring famines. Droughts or famines lasting for three years have been mentioned in *Naḷinikā-jātaka.*[73] Due to drought or no rain at all, the crops were badly afflicted with mildew and grew to mere stubs. Under such a situation, it was hard to get adequate supply of food.[74]

The Buddha, throwing light on the causes of drought or famine, says thus, 'As kings are unrighteous, the ministers of kings also are unrighteous. When ministers are unrighteous, Brāhmaṇas and householders also are unrighteous. Thus, towns folk and villagers are unrighteous. This being so, moon and sun go wrong in their courses. This, being so, constellations and stars do likewise; days and nights, months and fortnights, seasons and years are out of joint; the winds blow wrong out of season. Thus, the *devas* (gods) are annoyed. This being so, the sky-deva bestows insufficient rain. Rains not falling seasonally, the crops ripen in the wrong season. When crops ripen in the wrong season, men who live on such crops are short-lived, ill-favoured, weak and sick.'[75]

Too much rain or no rain at all is equally harmful to crops because if the former washes away the standing crops, the latter dries them up completely. It was supposed that of the two extremes, the heavy rain was better than no rain at all, because in the case of heavy rain, some crops might have been produced on the higher land but no crop could be produced at the time of drought. So, the rain was considered as an indispensable factor of agriculture. The occurrence of severe famines due to the failure of rains underlines the importance of rainfall on agriculture.

Rivers had also been one of the main sources of irrigation, just like today. Canals were taken out from them. Water was available in them so long as the rivers were full, possibly throughout the year. We have a clear reference to this modern system in *Kuṇāla Jātaka*, which tells us that between the Sākya and Koliya countries, the river Rohiṇi was flowing. Canals were taken out from the river on both sides by these two tribes. The territories of the Sākyans and the Koliyans were adjacent and separated by the river Rohiṇi. Once a dispute started between these two tribes regarding the right to the water of the Rohiṇi through which the lands on both sides were irrigated. This dispute which could have led to bloodshed was, however, settled by the Buddha.[76]

ARTIFICIAL SOURCES

Artificial sources of irrigation were not well developed in the Buddha's time, as is the case today. Still, we find some references to them. Canals, dams, ponds, pool, tanks, etc., were included under this category.[77] Ponds were made in gardens and near the fields. Water was taken out from them to irrigate the land for agriculture purposes. Water channels had not been provided to each and every plot. But during the Buddha's period, most of the areas had the facilities for irrigation, because the cultivated area was much smaller than the forest area. The canal system was extended and came to play a greater part in agriculture than before Buddha's time.

CROPS

Generally, two types of crops were produced. They were rabi crops and kharif crops. During the Buddha's period, the system of rotation of sowing crops was developed. The rabi crops were followed by the kharif crops. As we know, this system of rotation is still continuing in the middle-north India.

Grains

In Pāli literature, the word *dhañña* stands for grains. It comprises seven types of grain, including pulses and seeds. The seven types of grain were: rice sorts (*Sāli, vīhi*), barley (*yava*), wheat (*godhum*), millet (*kaṅgu*), beans (*varaka*) and a kind of grain called kodo (*kudrusaka*). They were mainly divided into two groups such as natural (*pubbanna*) and prepared (*aparanna*). The seven types of grains mentioned above belong to the natural group and to the second group belong curry (*supeyya*).[78]

Sāli

Rice was called *Sāli,* which was the main food grain of the people at that time, chiefly in Magadh and northeastern range of Uttar Pradesh. So, it was specially produced in these areas. There were two types of rice: *Sāli* and *Vīhi. Sāli* was supposed to be superior to *Vīhi.* Rice were sown in the beginning of the rainy season and were ready in the month of December or January. The plot in which the *Sāli* was sown, was called *Sālikkhetta.*[79]

Vīhi[80]

Vīhi was also a kind of rice available at the time of the Buddha. It was treated as inferior to *Sāli.* It had been the main food of the people of poorer classes in the same areas mentioned above. It was totally different from *Sāli.* It was sown in the beginning of the rainy season and harvested at the end of the same season.

Taṇḍula[81]

Taṇḍula was also a kind of rice grain frequently mentioned in the *Pāli* text along with *Tila.* Probably, the rice ready for boiling or boiled was called *Taṇḍula* because it is mentioned in the context of offering alms to the monks.

Barley[82]

Barley was supposed to be a very important and healthy food grain in the time of the Buddha. In Pāli, it is called *yava,* which was understood as corn in general sense and as barley in particular. In the *Pācittiya* of the *Vinaya Piṭaka,* barley has been mentioned that the nuns used the corn field (*Yavakkhetta*) as a toilet and made the fields dirty. It was reported to the Buddha who forbade them against it. It was the main food of the people in general.[83]

Wheat

Wheat was treated as one of the rich food grains at that time. It was called *Godhuma* in Pāli. It had been the main food of the people living near the Ganga. Then, it was used with corn (*Yava*) as food because farmers did not sow it separately. They were sown in a mixed form in the areas near rivers. That is why, in the text, it is mentioned always with corn (*Yava*)—*Yavagodhumādibhedaṁ dhamiyaṁ ca.*[84]

Millet

In those times, millet was called *Kaṅgu,* which was used for rough food grains such as maize, etc. It was used as food by the poors. It was sown on the land which contained more sand than soil. So it was produced mainly on the banks of rivers or on deserted land as it did not require to be irrigated. The text mentions it as one of the seven kinds of grains.[85]

Beans

At that time, beans were called *varaka*[86] which belongs to the seven kinds of grain. It was used as vegetable and pulse.

Kudrusaka[87]

Kudrusaka is also a kind of grain used as a food mainly by the poor during the period of the Buddha. It was supposed to be inferior to others. It was produced near riverine areas. The straws of it were used in winter season for keeping warm.

In the *Cakkavatti-sutta* of the *Dīghanikāya,* the Buddha forecasted that a time would come, when there would be a life span of only ten years and marriageable age of maiden would be of five years. The taste of ghee, butter, oil, *tila,* sugar and salt would disappear and *Kudrusaka-grain* would be the highest kind of grain like rice and curry of today and treated as the highest kind of food.[88]

Nivāra[89]

Nivāra is a kind of food grain which grew on its own. At the time of the Buddha, it was not sown by anyone. It grew up automatically in a natural way mainly in the river areas or in low lands. So, it was not treated as a grain but as a fruit. People used to eat it in course of their fasting on special occasions. Maha Pandit Rahul Sanskrityayana called it *tinni*[90] and Monior Wiliams named it wild rice.[91]

Pulses

It is seen that the system of pulse eating was prevalent in the Buddha's period. It was a kind of liquid diet and was eaten with rice or any other solid or dry food. Generally, gram (caṇaka), beans (māsa)[92] and green peas (*mugga*)[93] were used for making pulses.

Oil

It appears from the texts that the people used oils in different forms, as a cooking medium and for purposes of massaging the body. Different types of oil were available, prepared from different things. The text mentions the following types of oils:

1. *Sesame*-oil (*Tila-tela*)
2. Mustard oil (*Sāsapa-tela*)
3. Honey or wine made from the blossom of *Basia Latifolia* (*Madhukatela*)
4. Castor oil (*Eraṇḍaka-tela*)
5. Melted fat of dead animals (*Vasā-tela*)[94]

These oils were used for food, anointing and burning purposes. Mustard seeds were sown after the rainy season with wheat and barley and taken out from the field in the month of February or March before the ripening of wheat and barley. This seed was of two kinds: black mustarad and yellowish brown mustarad. They were used for lighting lamps, cooking, massage and in the preparation of medicines.[95]

VEGETABLES

Vegetables were regarded as a rich food at the time of the Buddha. Eating of vegetables was common in the society. The people who were habituated to eat vegetables in a large quantity, were called *Sākabhakkhi*.[96]

Farmers, after cultivation and cleaning out the plots, sown the seeds of vegetables. They produced them in order to maintain their family. It had become the means of their livelihood. *Kuddāla Jātaka* gives an account of Kuddāla Paṇdita, who produced and sold vegetables for his livelihood.[97] In this way, the cultivation of vegetables had become one of the main occupations in the country. Vegetables like potatoes, peas, different kinds of gourd (*alābu*)[98], raddish, pumpkin (*tipusa*), garlic (*lasuṇa*)[99] and so on were available during the period of the Buddha. These were eaten with rice, meat, etc.[100]

FRUITS AND FLOWERS

We come to know through Pāli texts that the cultivation of various types of fruits and flowers was done in the Buddha's period. Even the kings of that time planted fruit and flower trees. A child was born with a medicinal plant in his hand in Mithila. In his childhood, he laid out a garden with the help of his companions.[101]

A number of trees and their fruits available in the Buddha's time have been enumerated in canonical and non-canonical literature. They are: jackfruit trees (*Panasa*)[102], coconut (*nālikāra*)[103], myrobalan (*Āmalaka*)[104], mango (*Amba*)[105], rose-apple (*Jambu*)[106], jujube (*Badarī*)[107], banana (*Kadaliphala*)[108], grape (*Muddikā phala*)[109], palm-tree fruit (*Tāla-pakkaṁ*)[110], timbaru-fruit, cucumber (*Kakkārika*)[111], sweet-lovi (*Phārusaka-fruit*)[112], citron (*Mātulungī*)[113], nimb-fruit[114], glomerous fig-tree (Udumbara)[115] and bread fruit-tree (*Labuja*)[116].

The *Pācittiya,* a text of the *Vinaya–Piṭaka* gives an account of plants in addition to the above mentioned trees in the following way:

Turmeric, ginger, orris root, white orris root, garlic, black hellebore, khus khus, nut grass, fig tree, banyan tree, the Indian Cedar wood, the wood apple, sugarcane, bamboo, reeds, basil, camel grass, a kind of andropogan,[117] etc.

A chapter on medicines in *Mahāvagga* of the same Piṭaka contains a list of fruits which were used as medicines in the time of the Buddha. Those fruits were: vilaṅga, pepper, black pepper, yellow myrobalam, beleric myrobalam, emblic myrobalam, goṭha-fruit, etc.[118] Different types of flowers like *Pātali, Kinsuka, Kalrṇikāra,* and *Jaisuoman*, etc., were planted and the people prepared garlands with these flowers to sell them for their livelihood.[119] The sellers of fruits and flowers were called *Paṇṇika* and *Mālākāra,* respectively.[120]

Further, it appears from the texts that the gardens were named on the basis of their trees or fruits. A lot of references to this regard are found such as *Ambalaṭṭhikā*[121], *Ambavana*[122], Sālavana[123] and so on.

RAW MATERIALS

Raw materials for cloth-making were found at that time. Among them, hemp (*Sāṇa*)[124] and cotton (*Kappāsa*)[125] were important. Cotton wool was made from trees, creeper and grass.[126] One of them was likely to be the cotton plant. The tree might be the Simbali tree.[127] The *Tunḷila-jātaka* and the *Mahājanaka-Jātaka*) mention that around Varanasi *Kappāsa* was cultivated on large scale and the women kept a watch over the fields of *Kappāsa.* Other raw materials were sugarcane,[128] *khus khus* (a kind of reed), bamboo, and a wild indigo. *Hirivera* was used for making perfumes as well as medicines. In this way, at the time of the Buddha, raw materials were also available for making various goods.

There is also a mention of the five stages through which a plant passes. They are: (1) primary germination (*ācaya*), (2) gradual growth or development (*upacaya*), (3) their standing in their fully developed state (*Santati*), (4) the decline of the plant (*jaratā*), and (5) the death of plant or the final disappearance of all its constituents. The commentator of the *Dhammasaṅgaṇi* in his book *Aṭṭhasālinī* explains this by an illustration of a well dug-out on the bank of a river. The first gushing out of water in the well is like the *Ācaya* or primary germination of the material phenomenon; the flushing up or the gradual increasing or the rising up of water to the full is like the *Upacaya*; and flooding is like the *Santati.*[129]

In a similar way, we may divide the life of a fruit-tree into five stages or periods such as the branches, the leaves, the buds, the flowers and the fruits. The first period is the appearance of branches, the second is of growth or development, the third is of standing, the fourth is of ripening and decaying and the fifth one is of falling from the stem or total destruction.[130]

CATTLE

Cattle were supposed to be an integral part of agriculture. So, cattle-rearing and agriculture or the cultivation of land were interdependent during the period of the Buddha. In *Kuṭadanta-sutta* of the *Dīghanikāya* and the *Esukārisutta* of the *Majjhimanikāya,* we find a reference that a king inspired the people to keep animals for agriculture purposes (*Kasigorakkha*).[131] At that time, cattle were treated as members of the family because they gave food, strength and happiness.[132] So, the people were advised not to kill them but, look after them with care and affection.

The word *Gopālaka*[133] has been used frequently in the canonical literature for the person who reared cattle and used them for economic purposes. In rearing or taming the cattle, there was no distinction of caste. People belonging to each and every caste could have cattle but most of Brāhmaṇas reared the cows. However, the cattle-rearing persons of any caste were called *Gopālaka.* Lord Buddha has said that the *Gopālaka* should possess eleven qualities to take care of the herd of cattle and make it prosperous. Those eleven qualities were supposed to be essential for being *Gopālaka.* They are:

1. The *Gopālaka* should be well versed in recognizing the cattle.
2. He should be skilled in distinguishing marks of character.
3. He should be capable in removing flies' eggs (*asatika*).
4. He should know how to dress sores.
5. He should be careful in making fumigation.
6. He should know what is a ford or a shallow place where a river or stream may be crossed by wading or in a boat.
7. He should know the watering placeds.
8. He should know the pathways.
9. He should be skilled in locating pastures or lands covered with grass and suitable for grazing animals, cattle or sheep.
10. He should not milk dry or the milch animals.
11. He should pay special respect to those bulls who are the leaders of the herd.[134]

Among the animals mentioned in the *Tipiṭaka,* the following are of special interest.

Cow

The cow was supposed to be the best one among all cattle at the time of the Buddha. A number of references that cows find mentions in the Tipiṭaka.[135] A headsman Dhaniya had thirty thousand oxen and twenty thousand cows. In a dialogue with the Buddha, he said that he had cows, calves, cows in calf and heifers and bull as lord over the cows.[136] The King Mohāsudassana had eighty-four thousand cows, with jute trappings and horns tipped with bronze.[137] Cattle were treated as mother, father, brother, friends and othe relatives. They were honoured and respected by the people in the society as they gave milk, curd, ghee, etc.,[138] and had been the source of being rich and poor. Probably, the status of a man in the society was decided on the basis of having a number of cattle.

Such a large number of cows brings us to the question of their management. Of course, cow-pens were provided to cowherds due to having a large number of cows in their care. They had to be very careful under pain of dire punishment, specially when they took out the cows for grazing during the day.

But, in fact, there were many people or majority of the people, who did not keep so many cows. They jointly hired a cowherd, who took the cows out in the morning for grazing and returned them to their respective owners in the evening counting the cows of several people.[139] He paid special care to those bulls, who were leaders of the herd.

As regards the grazing ground for cattle, the text makes mention that there was a field for grazing them in every village. The field was called *gocara-bhūmi,* a cattle-grazing ground.[140]

On the one side we see that cows were treated as mother, father, brother, etc., and on the other side was cow-killing, which comes across constantly in the *Tipiṭaka.* Cows were killed with other beasts such as pigs, goats, etc., in the great *yajñas,* in hundreds and thousands. The Buddha, however, as we find again and again, indicted these *yajñas;*[141] in fact, this was one of the planks of his revolutionary programmes. He did not appreciate all types of sacrifices, especially related to injury or killing of any living things. Addressing Ujjaya Brāhmaṇa, the Buddha says thus: 'In whatever sacrifice, Brāhmaṇa, cows are slaughtered, goats and sheep are slaughtered, poultry and pigs are slaughtered and divers living creatures come to destruction, such sacrifice, which involves butchery, I do not praise.[142]

Ox

The ox had been the most important part of agriculture at the time of the Buddha. The word *balivadda* stands for ox in Pāli literature. There are many references to ox in the *Tipiṭaka.* It is not possible to give all here but the *Kasibhāradvāja-sutta* of the *Suttanipāta*[143] and *Kasi-sutta* of the *Sañyuttanikāya* may be remembered. The texts mention that 'the oxen were used in drawing the ploughs for the cultivation of land. They were also used to drive the carts for transportation of goods and for conveyance to go here and there.'[144]

Horse

The horse was supposed to be a very important and valuable cattle at that time. Siddharth as a prince was also well acquainted with the horse as a part of his general knowledge.[145] Generally, horses were used for drawing chariots by the king and posed a valuable asset in battle. The science of breeding and training of horses was fully developed. In this context, the text mentions eight qualities of a horse by which a thoroughbreed becomes worthy of a king. The eight qualities are:

1. The horse should be of good breed on both sides, so that other good horses can be bred.
2. He should eat his food carefully without scattering it about.
3. He should not lie or sit in dung or urine due to feeling of abhorrence.
4. He should be easy to live with and should not cause other horses to stampede.
5. He should show his voices, tricks, faults or wiles to his driver so that he correct them.
6. He should pull his share of the weight correctly in case of harness.

7. While he is going, he should go straight.
8. He should be steadfast till the life ends in death.[146]

Furthermore, the Buddha defines the eight kinds of excitable horses and their vices. They are:

1. When an excitable horse is told to go on, being beaten and urged by the driver, he backs and twists the carriage round with his hind quarters.
2. ....being beaten and urged to go by the driver, the horse jumps back, batters against the carriage railing and breaks the triple bar.
3. ...he releases his hind quarters from the pole and tramples on it.
4. ...he takes the wrong road and makes the carriage go away.
5. ...he tosses high his breast and paws in the air.
6. ...heedless of the driver and the good, he champs the bit with his teeth and wanders at random.
7. ...when urged by the driver, he goes neither on nor back but halts and stands like a post.
8. ...being beaten and urged by the driver, he draws together his forelegs and hind legs and just sits down there on his four feet.[147]

Here, it seems essential to say that a horse is known by his speed, an ox by the amount of burden that it can carry and a cow by milking, etc.

Elephant

Though an elephant is supposed to be wild animal, yet was regarded as one of the most important beasts for kings. They used them for their conveyance as well as in the battle. A number of references to it are found in the respective literature. The *Rāhulovāda-sutta* of the *Majjhimanikāya* presents a full description of an elephant along with his activities. The Sutta enumerates that his tusks are as long as a plough-pole, his home is like a battlefield; when going forth to battle, he uses his forelegs, uses his hindlegs, uses the forepart of his body, uses his head, uses his ears, uses his tusks and tail, protecting only his trunk. Thereupon, it occurs to the mahout.[148] Jīvaka, an eminent physician during the time of the Buddha, had five hundred she-elephants. An elephant was treated as one of the seven treasures.[149] About its size, the text makes mention that an elephant is eight cubits in height and nine in girth and length.[150]

Camel

The camel was treated as a domestic animal.[151] We do not find any description about the function of this animal. But probably it was used for carrying loads from one place to another, specially in desert areas, just like today.

Goat

Goats were a very common animal and they were reared on a large scale, specially in the time of the Buddha. They were supposed to be very useful because both their flesh and skin were used, as is the case now. There was a group of people who used to eat their flesh and sold their skin, which was used as bedsheets.[152] Not only this, they were also considered as a fit object of sacrifice in yajñas.[153]

In the *Mahāvagga* of the *Vinaya–Piṭaka*, there is a reference to a banyan tree under the heading of *Ajapāla-nigrodha*. It was a shady tree and goatherds used to sit there. Probably, it was so named due to usual sitting of goatherds in the shade of the tree.[154]

Wild Goat

This animal[155] was also a victim of the sacrifice, but its more obvious utility, rather than meat and milk, is its wool. Its leather was also used in the form of bedsheets.

Pig

Pig was a very common and useful animal during the period of the Buddha. The people in general used its meat as food. So, they reared pigs.[156] The text makes mention that after eating the pig's meat and rice offered by Cunda, Lord Buddha felt ill and sharp pain came upon him even unto death.[157]

It does not seem essential to give details about each and every animal. So, I am quoting here only the names of animals available in Pāli literature. They are: dog (*sunakha*)[158], donkey (*gardabha*)[159], mare (*assa*)[160], cock (*kukkuṭa*)[161], hen (*kukkuṭi*)[162], squirrel (*kalandaka*)[163], panther (*dipikā*)[164], tortoise (*kumma*), crow (*vāyasa*)[165], monkey (*makkaṭa*)[166], cat (*vilāra*), mongoose (*nakula*), jackal (*singāla*), deer (*miga*), lion (*siha*), kakravaka bird (*cakkavāka*), penahika bird-house-pigeon (*ghara-kapota*), owl (*uluka*), crane (*satapatta*), leech (*Jalūkā*), serpent (*sappa*), rock-snake (*ajagara*), spider, peacock, fish, buffalo, bee, etc.[167]

All the cattle and animals mentioned above were well known to the people and played an important role in the economy of the times. Thus, the Pāli literature throws valuable light on different items, processes and role of agriculture in the economy of the Buddha's time.

NOTES AND REFERENCES

1. A.S. 18.
2. E.F. Schumacher, 1999, *Small is Beautiful.* Published by Hartley and Marks, London, 37
3. *K.N. I.* 281; S.N.I. 171–72.
4. *Tattha rājā suvaṇṇanaṅgalaṁ gaṇhāti. Āmacca ekaunaṭṭhasataṁ rajatanaṅgalāni, Kassaka sesanaṅgalāni. Te tāni gahetvā itocito ca kassanti, N.K.* 142–44; Sn. A.I. 167.

5. *K.N. I.* (Sn) 281; S.N.I. 171–72.
6. *Kindred Sayings*, I. 217.
7. *K.N.I.* 281, 364; S.N.I. 171; J.IV (Hindi) 376, 393.
8. *Thera.* A.I. 188.
9. *K.N.I.* (Sn) 364.
10. *S.N.I.* 114; *Kindred Saying* I, 144.
11. *J.* III (Hindi) 293.
12. *J.* Nọ. 232, 384, 396, 316; The Book of The Discpline, Vol. III, 16, 18, 78.
13. *Tīṇimāni, Bhikkhave, Kassakassa Gahapatissa accāyikāni karaṇiyāni, Katmāni tīṇi? Idha, Bhikkhave, Kassako gahapati sīghaṁ sīghaṁ khettaṁ sukaṭṭhaṁ karoti sumatikataṁ. Sīdghaṁ sīghaṁ khettaṁ sukaṭṭham karitvā sumatikataṁ sīghaṁ sīghaṁ bījāni patiṭṭhāpeti. Sīghaṁ sīghaṁ bījāni patiṭṭhāpetvā sīghaṁ sīghaṁ udakaṁ abhineti pi apaneti pi. A.N.I.*, 222.
14. *A.N.I.* I 224; *Gradual Sayings* I, 221.
15. *A.N.* III 285.
16. *Mv.* 303
17. *D.P.P.N.* II 403.
18. *C.P.E.D.* 97.
19. *D.P.P.N.* II 404.
20. *C.P.E.D.* 79.
21. *P.E.D.* 197.
22. *A.N.* II 379.
23. *J.* III (Hindi) 17.
24. *J.* II 385; **** A. (Roman), 136, 191.
25. *K.*N. II 319.
26. *Aṭṭhaṅgasamannāgate, Bhikkhave, Khette bījaṁ vuttaṁ na mahapphalaṁ hoti na mahassādaṁ na phātiseyyaṁ... Idha Khettaṁ unnāmaninnāmi ca hoti, pāsāṇa-sakkharikaṁ ca hoti.....na.....na mariyādasampannaṁ hoti. Evaṁ aṭṭhaṅgasamannāgate khette bījaṁ vuattaṁ na mahapphalaṁ hoti ...... A.N.* III 337 *Gradual sayings*, IV 161–62 .
27. S.N.V. 324; *Kindred Sayings* V 328-29.
28. *Pathavi nāma dve pathaviyo-Jātā ca pathavi Ajātā ca pathavi ....* Pac.53.
29. Pac. 71, 362.
30. Cv. 377.
31. *Ibid.*
32. Para 80.
33. *K.N.*I. (Sn.) 281; S.B.E. X (Sn.) 12-13.
34. *S.N.*I. 21.
35. *D.N.*I., 7; Pac. 55; Sn. A.I., 174; J.I. 176; S.N. III, 284.
36. Pac. 55; *S.A.V.* I., 99–100; *The Book of Discipline,* II, 228.
37. *Ibid.*
38. *Ibid.*
39. *Ibid.*
40. *Ibid.*
41. *Mūlabhūtaṁ bījaṁ dve kiccāni karoti-heṭṭha mūlena patiṭṭhāti, upari ankuraṁ uṭṭhāpeti,* Sn. A.I. 174.

42. *D.N.I.*, 7; S.V.I. 99-100; Pāca. 56; *The Book of Discipline* II, 227; *Dialogues of the Buddha* I, 6-7.
43. *K.N.*I. (Sn.), 180-81; Sn. A.I. 175.
44. *J.* No. 469.
45. *Theri*; 419
46. *Dh.* A. IV (P.T.S.) 115-17.
47. *J.* No. 123.
48. *N.K.* 142-43; Dh.A.I. (P.T.S.), 223.
49. *J.*I. 446–49.
50. *K.N.I.* (Sn.) 281; S.N.I. 172; Sn. A.I., 172-77.
51. *S.N.*I., 104.
52. *Isā ti yuganaṅgala-sandhārika-daruyatthi,* Sn. A.I., 176.
53. Sn. A.I. 175.
54. Sn. A.I., 176.
55. *K.N.*I. (Sn.), 281.
56. *K.N.*I. (Dha.), 17.
57. *J.* II, 165; J. No. 211.
58. *J.*I. (Hindi) 325.
59. *S.N.* II, 244.
60. *Pathamaṁ kettaṁ kasāpetabbaṁ kasāpetvā vapāpetabbaṁ, vapāpetvā udakaṁ abhinetabbaṁ, udakaṁ abhinetvā udakaṁ ninnetabbaṁ... atiharāpetvā āyatiṁ pi vassaṁ evameva kātabbaṁ, āyatiṁ pi vassaṁ evameva kātabbaṁ ti,* Cv. 279.
61. *Kassako gahapati paṭikacceva khettaṁ sukaṭṭhaṁ karoti sumatikataṁ... udakaṁ abhineti pi apaneti pi,* A.N.I., 213.
62. *J.*I., 125
63. *D.N.I.*, 166
64. *A.N.I.* 32-33; *Gradual Sayings* I, 28–29.
65. *A.N.* III, 285-86.
66. *Mv.* 263; *The Book of Discipline* IV, 347.
67. *Sīghaṁ sīghaṁ bījāni patiṭṭhāpetvā sīghaṁ sīghaṁ udakaṁ abhineti pi A.N.I.* 222.
68. *J.* IV, 358–59.
69. *A.N.* III, 373.
70. *Cakkavaṭṭakaṁ ti arahaṭaghaṭiyantaṁ,* S.P. III, 290; *Dictionary of Early Buddhist Monastic Terms,* 85.
71. Para, 17.
72. *S.N.* (Hindi) II, 585.
73. *J.* No. 526.
74. *A.N.*I., 147.
75. *A.N.* II, 79; *Gradual Sayings* II, 84–85.
76. *Kuṇāla Jātaka* (Jātaka No. 536).
77. *A.N.* IV, 219; D.N. II, 248.
78. *P.E.D.*, 334.
79. *M.N.*I., 79; D.N. II, 219; A.N.I., 224; Cv. 377.

80. *D.N.* II, 219.
81. *Cv.*, 377.
82. *S.N.* III, 179.
83. *Pac.* 71, 360, 36.
84. *Pac.* 360; Dn. A.I., 185; Sn. A. II, 135.
85. *Ibid.*
86. *J.* II, 75; Pac. 360
87. *D.N.* III, 57; Pac. 360
88. *D.N.* III, 57; *Dialogues of the Buddha* III, 70.
89. *A.N.*I. 223; M.N. II, 118.
90. *M.N.* (Hindi), 50.
91. *Sans. Eng. Dic.*, 567.
92. *M.N.*I., 79; *D.N.* II, 219.
93. *Mv.* 228-29; *M.N.* I., 113; *D.N.*II., 219.
94. *Pv.* 231; Para 357; *Pac.* 114, 354; 384, 336, 338.
95. *Mv.* 200, 223; *Pac. Ibid.*
96. *A.N.*I., 223, 275; M.N.II., 418.
97. *J.*I., 312–13.
98. *A.P.* II, 15–16.
99. *Cv.* 229–30; Pac, 352–53.
100. *J.II.*, 262–63
101. *Jātāka*, No. 546; DPPN II 465.
102. *J.*II., 160.
103. *A.P.* II, 88; *J.* II, 159.
104. *J.*I. 160; *AP* II 22; Mv. 220, 295; *Cv.* 242, *J.* II 159.
105. *J.*I. 160; J. II 159; *AP,* II. 22–33;. Cv. 198; etc.
106. *Ibid.*
107. *J.*I. 299, 303; S.N. II 346; Pac. 29.
108. *AP I* 372, 462; Cv. 258, 265; J I 111, V, 195.
109. *A.N.* I. 33; *A.N.* IV, 272; *Mv.* 260.
110. *A.N.*I. 167.
111. *D.N.* II 193; *API* 80, 399.
112. *A.P.*I. 371.
113. *A.P.* II. 86.
114. *Mv.* 220.
115. *API* 370; JI, 292; J. II., 356, 360; Pac. 55.
116. *J.*IV. 363; API. 369.
117. Pac. 55–56; *The Book of the Discipline* II 227–29.
118. *Mv.* 220; *Ibid* IV, 272–73.
119. *AP.* I. 145, 150–55; J. IV. (Roman) 285.
120. *D.N.I.* 44-45, 52; J. VI, 281.
121. *D.N.I.*, 3, 109; M.N.II., 92 etc.
122. *J.I.* 139, Mv 248; D.N. 41, 43 etc.

123. *AP* I 120, D.N.I. 104, 106, 108, 115, M.N. II, 420 etc.
124. *D.N.* II 260; Mv. 55, 100; Para, 363.
125. *Ibid.*
126. *Pac.* 225; *Cv.* 243.
127. *M.N.* II, 257, *J.* I., 202; *J.* II, 162.
128. *A.N.*I. 33; *A.N.* IV 272; Pac.55.
129. *A.S.* 263.
130. *Tha Manual of Buddhism,* (Mahātheraadi Sayadaw)17-18.
131. *D.N.*I. 116; M.N. II, 444; J.I., 227.
132. *Annadā baladā ceta, vaṇṇadā sukhadā tathā, K.N.* I. (Sn.) 312.
133. *D.N.* III 157; *M.N.* I 270; Ud. 107.
134. *A.N.* IV, 388, 397; *M.N.* I 270; *Middle Length Sayings,* I 274.
135. *D.N.*I, 121; *S.N.*I, 75; *K.N.* 1 (Sn.) 273, 312-13.
136. *Sn.* A.I. 33; *K.N.*I. (Sn.) 273.
137. D.N. II, 143; *Dialogues of the Buddha* II, 230.
138. *K.N.*I. (Sn.) 312-13.
139. *K.N.*I. (Dh.), 19.
140. *Pati.* 208, 210, 266; *M.N.* I, 270.
141. *D.N.* I. 109-10.
142. *A.N.* II, 44; *Gradual Sayings,* II 49.
143. *K.N.*I. (Sn.) 281.
144. *Mv.* 5-6.
145. *N.K.* 146-47, 154-55.
146. *A.N.* III, 300; *Gradual Sayings,* IV, 130.
147. *A.N.* III, 301-2; *Gradual Sayings,* IV, 132.
148. *Middle Length Sayings,* II, 88.
149. *D.N.*I. 77; *D.N.* II, 14, 16, 133; *D.N.* III, 60; *M.P.* 232.
150. *Q.M.* II, 128.
151. *Pac.*18, 25.
152. *Mv.* 214.
153. *D.N.*I. 109, 121.
154. *Mv.* 4-6; *S.N.I.* 103-4; 121, 137, 139.
155. *D.N.I.* 109, 121; *M.N.I.* 211; Pac. 18, 26.
156. *Sukara-posako.* D.N. II, 258.
157. *D.N. II* 99.
158. *Mv.* 236; J. II 128, 246; V. V. 219.
159. *A.N.I.* 212; J.II. 109, 110, V, 453; Pac. 12; Mv. 219.
160. *K.N.I.* (Dh.), 26.
161. *J.*IV 38,135; D.N. II, 74.
162. Cv. 33.
163. *M.P.* 259; Para 142, 215; *S.N.* II 18, 211, 228.
164. *M.P.* 260.
165. *M.P.* 262; *J. I.* 500; II 440; *D.N.*I. 10.

166. *M.P.* 263; *J.*I. 385; II 267; *Cu.* 256.
167. *M.P.* 363–95.

ABBREVIATIONS AND BOOKS CONSULTED

A.N.	*Aṅguttara-nikāya,* edited by Bhikkhu J. Kashyap, Nalanda, 1960.
A.P.	*Apadāna,* edited by Bhikkhu J. Kashyap, Nalanda, 1960.
A.S.	*Aṭṭhasālinī,* edited by V.P., Bapat, Pune, 1942.
C.P.E.D.	*Concise Pāli-English Dictionary,* A.P. Buddhadatta Mahathera, 1989.
Cv.	*Culavagga,* edited by Bhikkhu J. Kashyap, Nalanda, 1956.
D.N.	*Dīghanikāya,* edited by Bhikkhu J. Kashyap, Nalanda 1958.
D.P.P.N.	*Dictionary of Pāli Proper Names,* G.P. Malalasekara, 1995.
D.P.	*Dhammapada,* Nalanda, 1959.
Dn.A.	*Dīghanikāya-Aṭṭhakathā,* edited by Mahesh Tiwary, Nalanda, 1974.
J.	*Jātaka,* Edited by v. Farrsbold, P.T.S, London, 1963.
K.N.	*Khuddaka-nikāya,* Edited by Bhikkhu J. Kashyap, Nalanda, 1959.
K.N. (Sn.)	*Khuddaka-nikāya (Suttanipāta),* edited by Bhikkhu J. Kashyap, Nalanda, 1959.
M.N.	*Majjhima-nikāya,* edited by Bhikkhu J. Kashyap, Nalanda, 1958.
M.N. (Hindi)	*Majjhima-nikāya,* (Hindi), Rahul Sankrityayana, Mahabodhi, Sarnath, 1964.
M.P.	*Milindapañho,* edited by Shastri Dwarikadas, Bouddha-Bharati, Varanasi, 1979.
Mv.	*Mahāvagga,* edited by Bhikkhu J. Kashyap, Nalanda, 1956.
N.K.	*Nidāna-kathā,* Mahesh Tiwary, Chowkhambha, Varanasi, 1970.
Pati	*Paṭisambhidāmagga,* edited by Bhikkhu J. Kashyap, Nalanda, 1958.
Pac.	*Pācittiya,* edited by Bhikkhu J. Kashyap, Nalanda, 1958.
Para	*Pārājika,* edited by Bhikkhu J. Kashyap, Nalanda, 1958.
P.E.D.	*Pāli English Dictionary* (T.W. Rhys Davids), 1989.
Pv	*Parivāra,* edited by Bhikkhu J. Kashyap, Nalanda, 1958.
Q.M.	*Questions of Milinda* T.W. Rhys Davis/Motilal Banarasidass, Delhi, 1993.
S.B.E.	*Sacred Books of the East,* Vol. X, F. Maxmuller, Motilal Banarasidass, Delhi, 1980,
S.N.	*Saṅyuttanikāya,* edited by Bhikkhu J. Kashyap, Nalanda, 1959.

Sn.A.	*Suttanipāta-Aṭṭhakathā,* edited by Angraj Choudhary, Nalanda, 1975.
Sn.	*Suttanipāta,* Trans. V. Fausboll, Motilal Banarasidass, Delhi, 1980.
S.P.	*Samantapāsādikā,* edited by Birbal Sharma, Nalanda, 1967.
S.V.	*Sumaṅgala-vilāsinī,* edited by Mahesh Tiwary, Nalanda, 1974.
Thera	*Theragāthā,* edited by Bhikkhu J. Kashyap, Nalanda, 1959.
Thera A.	*Theragāthā-Aṭṭhakathā,* edited by Angraj Choudhary, Nalanda, 1976.
Theri	*Therigāthā,* edited by Bhikkhu J. Kashyap, Nalanda, 1959.
Ud.	*Udāna,* edited by Bhikkhu J. Kashyap, Nalanda, 1959.
V.P.	*Vinaya-Piṭaka,* (Hindi), Rahul Sankrityayan, Buddha Educational Foundation, Taipei, Taiwan, 2000.
V.V.	*Vimāna-vatthu,* edited by Bhikkhu, J. Kashyap, Nalanda, 1959.

I.B. Horner, 1971, *The Book of Discipline* (Trans. of *Mahāvagga, Cullavagga, Pārājika, Pācittiya and Parivāra),* Luzac and Company Ltd. London.

Prof. C.S. Upasak, 1975, *Dictionary of Early Buddhist Monastic Terms.*

Rhys David, 1971, *Dialogues of the Buddha* (Trans. of *Dīghanikāya*), P.T.S., London.

Rhys David, 1989, *The Book of Gradual Sayings* (Trans. of *Aṅguttaranikāya*), P.T.S., London.

Rhys David, 1989, *The Book of Kindred Sayings* (Trans. of *Saṅyuttanikāya*), P.T.S., London.

Ledi Sayadow, 1965, *The Manual of Buddhism,* Burma.

I.B. Horner, 1977, *Middle-Length Sayings* (Trans. of Majjhimanikāya), P.T.S., London.

E.F. Schumacher, 1973, *Small is Beautiful,* Firstly and later in 1999 by Hartley and Marks. Published by Blond and Briggs Ltd., London.

CHAPTER 25

Irrigation and Famine as Referred in Pāli Sources

H.S. Shukla

INTRODUCTION

Buddhism, being a psycho-ethical system of thought, enquires into the economic issues connected with the household life also. The basic needs of household life are: food, clothing, shelter and medicines. The fulfilment of these needs depends directly or indirectly, on the agricultural condition of the country. This chapter is intended to throw light on the topic 'Irrigation and Famine as Referred in Pāli sources', and may be treated as supplementary to previous one entitled 'Agriculture as Revealed by the Pāli Literature'.

So far as the process of cultivation is concerned, many references to the fact are available in the Pāli Canonical and Non-canonical literature. The Khetta-sutta of the *Aṅguttaranikāya* and Cullavagga of the *Vinayapiṭaka* give a detailed account of the agricultural process. According to this Sutta, firstly, plots were ploughed and harrowed very quickly. Then, the farmers would sow their seeds, let the water in and turned it out very quickly. These were treated as an urgent duties of the farmers[1]. However, the process was the same as today, but the ingredients to be used in the process have changed today due to the development in modern technology.

Agriculture is mainly based on irrigation. It was considered vital for agricultural purposes. All the essentials of agriculture being well organized, the good production of crops could not be possible unless there would be an availability of sufficient irrigational facilities. Besides rain, we find references of wells, canals, dams, tanks, ponds, lakes and rivers for watering the fields in Vedic literature, Jain literature, Buddhist literature, Kautilya's Arthasāstra, Hindu literature and inscriptions.

The Hathigumpha inscription of Khāravela gives an information of opening a canal during the period of Nanda dynasty. In the Junāgarh inscription, there is a reference of Sudarsan dam, which was initially constructed during the time of Chandragupta Maurya and due to the excessive rains, Sudarsan lake suddenly burst at the time of Rudradaman, which was repaired by his officer. Again, it was overflooded during the time of Skandagupta and was repaired by his officer (Junāgarh inscription of

Skandagupta). According to the Jātaka, there were engineers who constructed canals for irrigating the field.[2] The *Aṅguttara-nikāya* refers to dykes in which water was collected and the people took this water throughout the year to the plots through canal.[3] It seems that these dykes were converted into wells.

Another Jātaka, in a casual way, gives us a glimpse of the irrigation system of Mithila. It depicts the flowing rivers and round going water under canals that were obviously taken out from the rivers.

Good rainfall was supposed to be the most essential ingredient for good crops. It was not possible to provide irrigational facilities to each and every part of the country impartially and the people were also not be equally benefited by the artificial irrigational sources such as canals, dams, etc. Rain is the only source that can satisfy the people without any discrimination, though it also does not fall equally everywhere. Hence, it does not fulfil the necessities of the nation as a whole. Therefore, the development of irrigational sources was considered to be essential for agricultural operation. Even during the Buddha's period, the system of irrigation was not fully developed; it was developed to some extent in particular areas like Mithila, Sakyan and Koliyan states, etc., but not equally everywhere.

There are references to the devices to be used for lifting the water to the field for irrigation purposes. The *Cullavagga*[4] of the *Vinaya-piṭaka* refers three devices: *tulā, Karakaṭaka* and *Cakkavaṭṭaka. Tulā* is a pole to lift the water from the well like a balance. It is popularly known as *Iaṭhā* or *lāṭhā* or *ḍhekula* in the villages of Bihar and eastern U.P. This device is operated with the help of weight at one end of the pole and with vessel or water bag on the other end of pole.[5] *Karakaṭaka* is the device for taking out the water from the well, either by hand or by yoking bulls with the help of long straps or ropes.[6]

Earthen vessels were used with these devices in the primary stage, but they were broken again and again. This was reported to the Buddha. The Buddha allowed the monks to use vessels made of iron, wood and leather.[7]

From the above positive evidences and a reference to a farmer letting in and turning off water very quickly, it becomes clear that the practice of irrigation was quite common, popular and efficient during the period under study.[8]

Here, it would be pertinent to ask how far the development in the irrigation system would be registered in the Buddha's period. In this regard, we can simply say that a reference to irrigation by well water and canal is also available in the *Ṛgveda.* So, it might be said that by the time of the Buddha, canal irrigation system was extended and came to play a greater part in agriculture than before, though the Mauryan Empire was credited with a vast development in the irrigational network.

FAMINE

Famine is not a new phenomena in the history of Indian economy. We have references of famine in the *Ṛgveda,* Jain and early Buddhist literature and Mauryan sources. But here, we are only concerned with the early Buddhist literature.

After going through the description of irrigation system, contradiction arises in respect of the availability of the developed irrigation system *versus* the occurrence of famine from time to time. We come across with the accounts of harrowing visitations of famine, which seems to be a regular feature of the Indian economy throughout history, down to the time of the Buddha. Irrigation can only be a supplement, not supplant rainfall in our land. Occasionally, their local deficiency of water in case of a famine, can be combated to some extent by irrigational arrangements. Irrigation can bring about a regular increase in production. It does not seem adequate to remove famine at last.

Many references to famines give us a clear picture of hardship of our ancestors of those days. During the famine, crops are bad, food is hard to get and it is difficult to carry on. We find monks getting alms of small amounts of shrivelled grain. They threshed it in threshing pots and ate it.[9] In course of famine, some stricken place became white with man's bones and the crops grown to mere stubs. This situation is depicted with respect to Nālandā[10] and Vajji[11]. It means many people died due to the shortage of food in Nālandā and Vajji. A monk was found stealing a handful of rice from a shop during the famine in Sāvatthi.[12] The *Mahāvagga*, a book of the *Vinaya-Piṭaka*, refers to famine in Rājagaha. At that time, there was a shortage of food and the people brought salt, oil and rice to the monastery.[13] Due to the scarcity of food, the people ate elephant-flesh, horse-flesh, dog-flesh, snake-flesh, lion-flesh, etc., at the time of famine.[14] The Kula-sutta of *Sañyutta-nikāya* also gives an information about the famine-stricken Nālandā. Here, we are reminded of the much clear story of the great sage Visvāmitra stealing a dead dog's leg from a butcher's house during a famine. It is evident that people were in the habit of putting up with famines and they thought about it as a regular event.

The *Jātaka-aṭṭakathās* are also not lacking in references on the subject. The famine in Kāsi state is testified by the Jātaka, which happened due to washing away of seeds from the fields during too much rain. The villagers borrowed an old ox from the village-headman and ate its flesh, promising to return its value in two months.[15] Another Jātaka passage shows so severe famine in Kāsi that even crows, not getting food, had to leave the human inhabitation and to go into the jungale.[16] Further, it is perceived that there had been no rain for three years continuously in Kāsi. Agricultural activities were completely stopped. People went to the palace and blamed the king.[17] Famine in Kaliṅga and Kosala is also referred in Jātaka stories.[18] A terrible famine in the Magaha that lasted twelve years during the period of the great king Chandragupta Maurya has also been marked in Jain tradition (*History of Indian Literature*, Vol. II, 431).

As regards the causes of famine, it is said that the situation of famine arises not only due to the absence of rain but also due to the excess rains, and destruction of crops by locusts, earthquake, hailstorms and other natural calamities.

Buddhism believes in the reciprocal relationship between human morality and natural environment. There is a reference to the fact that when lust, greed and wrong values grip the heart of humanity, immorality becomes widespread in the society, timely rain does not fall, and the crops falls victim to pests and plant diseases.[19]

The *Cakkavattisihanāda-sutta* and its commentary go on to explain the process of mutual interaction in more detail. When humanity is detoriated or demoralized through greed, famine is the natural outcome; when moral degeneration is due to ignorance, epidemic is the inevitable result; when hatred is the demoralizing force, widespread violence is the ultimate outcome and so on.[20]

Such a relationship has been discussed systematically in the theory of five natural laws (*Pañca-niyāma*). These five natural laws are:

Season-law or physical laws (*Utuniyāma*), Seed-laws or biological laws (*Bīja-niyāma*) Psychological laws (*Citta-niyāma*), Action laws (*Kamma-niyāma*) and Universal laws (*Dhamma-niyāma*).

At last, the only question that remains for us to consider on the subject is whether the kings had developed any famine-code for preventing the subjects from regular visitations of famine. They felt quite helpless, as we find in *Kurudhamma-jātaka* and *Vessantara-jātaka* during the course of drought in Kaliṅga. However, Kautilya (later than the Buddha) seems to be feeling his way towards a famine relief policy.[21] He ask the king in such a calamity to show favour to his people by providing them with seeds and provisions, to construct buildings and roads, to distribute to the people either his own collections or to seek help from his friends among kings. And during the period of Chandragupta Maurya, measures were taken for protecting the people from the ravages of famine through giving work, free distribution of alms, etc.

It seems that the kings tried their best and gradually, by Kautilya's time, public works policy and other elements of famine relief policy were developing.

NOTES AND REFERENCES

1. *A.N.*I, 213; Cv. 279.
2. *J.* II, 25 (Nalanda); *J.*I. 25, 339.
3. *A.N.* III, 373.
4. Cv. 212.
5. Lallanji Gopal, *History of Agriculture in Ancient India,* 123.
6. *S.P* III, 1290.
7. Cv. 212; *The Book of Discipline* V, 168.
8. *A.N.*I, 222.
9. *Para* 9.
10. *S.N.* IV, 322 (R.).
11. *Para* 19, 109.
12. *Para* 9.
13. *Mv.* 229.
14. *Ibid.* 235-36.
15. *J.* I, 57.
16. *J.*II, 500; II, 440.
17. *J.*V, 193, J. No. 526.
18. *J.*II, No. 276, *J.* VI, 487.
19. *A.N.*I, 147.
20. *D.N.* III, 71.
21. *A.S.*, 220.

ABBREVIATIONS

A.S.	:	*Aṭṭhasālinī* edited by P.V. Bapat, Pune, 1942
A.N.	:	*Aṅguttara-nikāya,* Nalanda, 1960
Cv.	:	*Cullavagga,* Nalanda, 1956
D.N.	:	*Dīgha-nikāya,* Nalanda, 1958
J. Jataka (Trans.)	:	E.B. Cowel, London
K.N.	:	*Khuddaka-nikāya,* Nalanda, 1959
Mv.	:	*Mahāvagga,* Nalanda, 1956
N.K.	:	*Nidāna-Kathā,* Varanasi, 1970
Pac.	:	*Pācittiya,* Nalanda, 1958
Para	:	*Pārājika,* Nalanda, 1958
S.N.	:	*Sañyutta-nikāya,* Nalanda, 1959
S.P.	:	*Samantapāsādikā,* Nalanda, 1967
Therig.	:	*Therigāthā,* Nalanda, 1959

CHAPTER 26

Agriculture in the Age of Sangam

H.N. Dubey

INTRODUCTION

Scholars like Damodaram Pillai, Swaminath Aiyar, etc., published the texts of Sangam age at the close of the nineteenth century. These texts belonged to the earliest strata of Tamil literature, popularly known as Sangam literature. According to the tradition recorded in the commentary of Irayanar Ahapporil written in about 1200 AD, there existed three Tamil Sangam or Academies in which poets of that period used to assemble and in which literary works of that period were 'heard' and assessed. The first Sangam established at Madora, the capital city of Pandya kingdom, lasted for 4440 years. The second Sangam lasted for 3700 years and the third lasted for 1850 years. In this way, these three Sangam altogether lasted for 1990 years. Most of the scholars hold that the last phase of the third Sangam would have lasted with the beginning of the Christian era. Some of the scholars conjecture the date of first Sangam at about 1000 BC or so,[1] but such an early date for Sangam literature is not acceptable to most of the scholars. However, scholars like N.K. Shastri and others believe that the early Tamil Sangam literature can be dated from the first century AD to 300 AD.

Latest archaeological evidences reveal that the earliest cultivation in peninsular India would have started about the late phase of the new stone age, which can be dated in the first half of the second millennium BC. The people of the age started cultivation of the *ragi* and *bajara* and probably pulses like green gram and horse gram, etc. The new stone age people started making terraces on slopes of hills for cultivation and domestic settlements. This was the first phase of primitive agriculture in peninsular India. The second phase of development of agriculture started with the introduction of iron technology, characterized by plough agriculture. In this stage of development there was the spread of rice cultivation in south Indian river valleys. In the third phase of agricultural development in the peninsular India, the concentration of settlements in the river valleys grew with the result that harnessing of bullock to the plough and use of iron ploughshare gradually developed. This resulted in an increase in the area of cultivation and unprecedented growth in agricultural production. Increase in the agriculture resulted in the growth of population. During this phase, we find a remarkable change in the agrarian sector which witnessed the introduction of a non-cultivating groups. These groups were for better management and scientific knowledge of seasons and production of weather, etc. Also, during this period that we find the beginning of the practice of granting or donating cultivated land to religious beneficiaries like Brahmanas and Buddhist monks, development of agrarian settlements in Tamilham (Tamil Nadu) or land occupied by the Tamil people in the Sangam age.

In Sangam literature, we find several factors that determine the forms of subsistence of the people. These factors were geographical location of the land, position and nature of terrain, material culture and the development and level of tools and technology of the region, etc. Poets of Sangam age speak of *tinai* or five eco-zones of Tamil Nadu which are described as:

1. *Kurunji* (the hilly or forest region which were arid zones).
2. *Mullai* (the land for pasture with thin forest and grass).
3. *Marutam* (the plain fertile land in the river valleys more suitable for agriculture).
4. *Megtal* (the sea coast areas of remote peninsular India).
5. *Palai* (the barren or arid areas where storage of water was very low).

It is important to note that it was *marutam* tracts which was the most productive land and people of other *tinals* depended for livelihood on this area. Rice and sugarcane were the most important products of this zone, which were the main staple food of the people of the Sangam age. The *mullai* tract was the region occupied by the *ityas* or cowherds. Their main source of livelihood was cattle rearing, dairy products and sometimes production of pulses, lentils, millets, etc.

Archaeological researches related to iron age sites reveal that in entire peninsular India, we find a rapid increase in the spread of settlements, particularly from the upland areas to the fertile land of river valleys and sea coasts. This movement resulted in a great transformation of society. In Sangam literature, we find a transformation from partly cattle rearing and partly shifting cultivation to settled agricultural economy. Thus, in the river valleys, concentration of settlements gradually increased. The farmers of the Sangam age developed a certain level of craft specialization. During this age, we notice an extensive use of iron instruments/tools and new technology of iron ploughshare, etc., with the development of agrarian people of *marutam* zone developing minor irrigational facilities like construction of small tanks, wells and small canals, etc. With the development of new management of irrigational facilities, people changed the dry land crops to a more surplus yielding wet land, crop of paddy and sugarcane, etc., plenteous use of iron tools and iron ploughshare in this region started during the Megalithic-Iron period which lasted in Tamil Nadu in C. 300 AD. Iron tools like spades, sickles, hoes, etc., were produced by way to iron metallurgy. Archaeologists have obtained fumales and some pieces of iron slags from many iron sites. It is important to note that for better crops of rice and sugarcane, farmers needed deep ploughing with an iron-tipped plough. Buffaloes and bullocks were harnessed for ploughing in the fields. As K.A. Nilkantha Shastri and G. Srinivasachari observed regarding the state of agriculture in Tamilland in the Sangam age, 'cultivable land was abundant and the necessities of life plentiful. The fertility of the lands watered by the Cauvery, Palar and other rivers is the recurring theme of description of early Tamil poets. The natural forest produce of Pari's included bamboos, rice, jackfruit, the *valliroot* and honey, *Ragi*, sugarcane, pepper, turmeric, cotton, etc. Society in that period was organized along a caste system with habits and traditions of their own, but the population of the large cities and port towns tended to be cosmopolitan'.[2] It was during this period we notice for the first time about the cultivation of turmeric (*curcumalonga*) and *pepper* (*pipermigrum*). Scholars like Harlan observes that for the first time, the origin of turmeric took place in peninsular India and Malaysia.[3] Turmeric, popularly known as *Haldi* in Hindi, is a very important condiment used in every Indian home from very ancient times. Cultivation of turmeric was primarly confined to the coastal areas of Tamil Nadu, Andhra Pradesh and Orissa only, but nowadays, this crop is yielded in most of the northern Indian areas also. Turmeric, is popularly known in Tamil and Malayalam as *manijal*. It has a wild variety also, popularly known as *Curcuma Aromatica*, which possesses fragment stems. In Tamil Nadu, Salem, Coimbatore and Trichur districts are very famous for turmeric crops. Similarly, Krishna, Guntur and west and east Godavari districts are also the most important areas for the cultivation of turmeric. Like turmeric, the pepper plant is also a very important condiment. It is an indigenous plant cultivated in the forest areas of Kerala in the Sangam age. Most of the scholars say that its cultivation in Kerala is much earlier than the Sangam age. It is important to note that in Indonesia, extensive cultivation of pepper is done from very ancient times. It was exported to Rome and other Mediterranean countries because as a preservative of food and for flavouring meat, it was in great demand there.

The land in Tamil areas of Sangam age was very fertile. In it we find fertile river valleys of Kaveri, Pannar, Vaigai, Tamraparni, etc. The River Kaveri was worshipped like the Ganga in peninsular India because of its plenteous water throughout the year and fertile valleys. Sangam poets had written many poems in its praise. Every inch of land in its valley was a boon for the farmers because it was not only every fertile but also irrigated. In *Tirukulam*, a poet says that no state can prosper without the cooperation of farmers. Thus, the poet had admitted the importance of farmers as well as agriculture in the Sangam society.

So far as the question of the owner of land in the Sangam age is concerned, we do not find any specific reference of it in the Sangam literature. But unlike other societies of India, after the settlement of state authority in Tamil land, the ownership right of land was vested in the state and thus the kings and the rulers were the real owners of the land. The farmers were cultivating their fields on a rent fixed by their landlords. In the *Manimekalai*, we find a word *Bural* which means that cultivator of the land was empowered to take loan on his cultivated lands. In Sangam literature, we do not find any clear mention of the rate of interest fixed by the state. However, most of the scholars are of this view that the rate would have been 1/6th of production in general practice. The rent was collected either in the form of products of land or in the form of currency. According to V. Kanakasambhai, for giving irrigational facilities, states used to collect extra rent from the farmers. He says that in Sangam literature, we find some references where farmers had appealed to the kings for taking extra land tax by their officials. Besides, we find references of counselling the kings as how to store water, enrich the land and improve the conditions of the people. The dignity of labour was highly emphasized. The dignified cultivators or farmers of wetlands, popularly known in Sangam literature as *maratham*, were occupied by the people, referred as *vellalar* or *ulavar*. In some of the poems of Sangam age, *maratham* was also called *pannai*, which means fertile land or land having fertility and rich soil. For cultivable fields, we find a word *kalam*. The cultivators or farmers employed labours or tenants for their agriculture. In this way, they did not plough the land themselves. Ilango Adigal, the famous Sangam poet, states that *vallala* farmers were responsible for the triumphs of their kings. They were the supervisors or the managers of their cultivable land. We find several references to women labourers in the Sangam age, employed for transplanting and harvesting of paddy and digging of root crops. According to *Jivaga Chinthamani*, the land was ploughed with the aid of bullocks and male buffaloes.

In *Perumpanapruppadai*, which is a very famous early Tamil poem of the Sangam age, we find a vivid description of the different farming operations of that period like ploughing, levelling, transplanting the paddy, hoeing, weeding, harvesting, transporting the sheaves to the threshing floor, the threshing of crops by using cattle and winnowing, etc. The fertile land of Tamil Nadu was irrigated by the contructions of small dams on lakes and rivers like Cauvary, Vaigai, Palar, etc. The valleys of these rivers were most useful for transplanting of paddy seedlings. Fields were mostly irrigated from tanks, lakes and wells with the help of the bullocks; farmers drew water from the wells. The *Jivaga Chinthamani* describes the agricultural position and happiness of peasant communities of Sangam age. It says that when plentiful rains occur, the peasants feel great joy and happiness and hail the rainfall rivers or canals. The people are informed by the beating of drums loudly so that peasants rush to the spot and repair the breach with the cooperation of each other. The ladies of the peasant communities also rushed to the working places or fields or their husbands to encourage them to work untiringly by supplying liquor and cooperation.[4] In *Chillapathikaram* and *Pattinappalai*, we find description of crops in the Malaya mountain and western ghats. Both the texts describe the production of the sandal tree (*Santhamalbum*) in the Malaya hills. In *Chillapathikarm*, we find a vivid description of ships sailing from Southeast Asian countries to the Chola coast of pennisular India.

The main products of agriculture in the Sangam age were paddy, sugarcane, *ragi*, pulses like lab-lab, cajanas, etc. The cultivation was both by male and female peasants untiringly, with the result that food was produced in abundance in the Sangam age. We find the description in the *Jivaga Chinthamani* that the peasants used to help others by liberally supplying them grains. The surplus products of grain were sent for sale in the markets by bullock carts. We find descriptions of such carts carrying imported goods in the literature of the Sangam age. The peasants were living happily because of surplus production of grains and trades.[5] The *Chillpathikaram* gives a clear picture of agricultural operations of this period.

'Field labourers, their arms blackened by exposure to sun, came running with the farm owners. Their shouts could be heard from distance. The travellers could also hear the melodies of women, singing in drunken voices. Their broad shoulders and large breasts were soiled with mud. Having cast away the flowers from their hair they were sticking the tender sprouts of rice into the water-soaked ground. These graceful women looked like bronze statues sprung from the mire of the fields. Then the hymns sung by the ploughmen were heard. They walked behind their sharp ploughs, which ripped open the soil. From far the travellers could hear the farmers' threshing songs as their bullock tramped the harvest to separate the grain from the straw; and the cheers or these who were listening to mud-soiled drums played by vigorous young ministrels'.[6]

Regarding the modes of distribution of resources in the agrarian settlements in Tamil country of the Sangam age, we are informed by the poems of this period that gift was the most common mode of circulation of resources. For the services rendered by others, each producer gave a certain part of his product to them; a generally, a good meal and a piece of cloth was the simplest form of gift. Dancing women or poor singers, etc., were satisfied by getting a full meal and some cloth to cover their body. Soldiers and fighting heroes were honoured with feasts. Persons having higher position in the administration or in society were graced by providing them with silk clothes, fine imported wine and gold ornaments, etc., along with feast. In Sangam literature, we find some references to learned Brahmans or priests and warrior society. Brahman priests enjoyed great honour in the Sangam society. By providing them with lands or villages, rulers of the Sangam age tried to settle them in Tamilham. However, the acts of redistribution in the agrarian society of the Sangam age was mainly confined to the groups with wealth and power. In it first came the vender or crowned kings, the *Velir* or minor chieftains and *Vellalar* or rich agricultural householders. The gifts collected through plunder and pillage from agrarian areas were collected at the residence of these chiefs. In Sangam literature, we have several references to such acts of plunder and collection of resources like cattle and grains. In it is important to note that plunderers often set fire to the materials what they could not carry. Sometimes they set fire to the peasant settlements, killed cattle and devastated the gardens and harvesting fields of the enemies. For such heinous activities, *marva* fighters of the hilly and pasture lands were used by the warriors or chiefs of the Sangam age. The booty looted by *marva* plunderers were redistributed among the *marva* people, fighters and sometimes among the Brahman priests for getting rid of sinful acts of the chiefs. Though the poor defenceless peasants were used to be the sufferers, yet such acts of plunder and pillage were regarded as noble heroic acts of the chiefs of the Sangam age. The *pana* singers sang heroic songs in praise of such plunderers, fighters and chiefains. Not only this, even such wars were institutionalized, as we find some references of establishment of war memorial stone in honour of such warriors which, later on, became a cult object of worship in the Sangam age.[7]

So far as the society of Tamilham is concerned, it was basically tribal in character. In all eco-zones of Sangam age, totem worship, kinship organization, tribal customs cults and practices, etc., prevailed. Besides, we also find some references to developed agrarian society where social oraganization was becoming complex with the gradual breaking up of old kinship ties. In agricultural regions, gradual introduction of the Brahmanical *varna* or caste concept was taking place with the introduction of the *varna* system. Social stratification, with a broad distinction between 'high and low' among the social group, appeared. Brahman, priests, chiefs and the *velala* peasants and landed *vellalar* were enjoying a high status in the society. *Velala* were the basic producing groups in the agrarian society of the Sangam age. Craft organization was subsidiary to agricultural production and it was in rudimentary stage. *Techan* (carpenters), *Kollan* (blacksmiths) and weavers were the main groups in assistance of agrarian social groups.

With early village folk of Tamilham, old tribal rituals were prevalent. Their tribal religious worship and cult practices necessitated the formation of ritual groups like *Velan, Venttuvan,* etc. These groups looked after the supernatural elements and their ritual management. However, it is important to note that early agrarian society of Sangam age was not 'priest-dominated'. Brahaman priests were given importance in the late phase of this period. Trading groups were mainly dependent on agricultural products, since it was a considerable surplus for sales R.S. Sharma[8] observes that trading groups of Sangam age were not enjoying the status of the Vaisya groups of north India. According to *Tolkappium,* the famous grammar text of Sangam age, merchants of that period were accommodated in the caste system only perhaps in the late phase of Sangam age. V. Kanakasambhai[9] observes that due to the impact of the Brahmanic order of society, chiefly, groups who were originally tribal and agrarian started changing their descent, relating themselves either with *Suryavamsa* (Solar line) or with *Chandravamsa* (Lunar line), as the Kshatriya caste of northern India. Suvira Jaiswal[10] observes that due to the formation of a caste system in the Sangam society, the harmony prevailing in the Sangam society was disturbed. But N.K. Shastri[11] does not agree with this view. He observes that it is true that with the introduction of Aryan Brahmanic culture in the early Tamil society of Sangam age, their tribal social set up got a rapid change but the social groups formed due to caste system did not disturb the social harmony of the Sangam age.

NOTES AND REFERENCES

1. Shastri, K.A. Nilkanth (ed.), 1956, *A Comprehensive History of India.* Vol. II, Peoples Publishing House, New Delhi, p. 672.
2. Shastri, K.A. Nilkhanth and Srinivasachari, G., 1957, *Advanced History of India,* Madras University Press, Madras, p. 179.
3. Harlan, J.R., 1975, *Crops and Man,* Winsconsin, p. 75.
4. Stein, Burton, 1980, *State and Society in Medieval South India,* Oxford University Press, New Delhi, p. 27.
5. See, The translation provided by Dr. Rajmal Devadas, quoted in *A History of Agriculture in India,* Randhawa, M.S., 1980, Vol. I, Indian Council of Agricultural Research, New Delhi, p. 408.
6. Stein, Barton, 1997, *Essays on South India,* Munshiram Manoharlal, New Delhi, p. 16.
7. *Ibid;* 1980, *Peasant State and Society in Medieval South India,* Oxford University Press, New Delhi, p. 44.
8. Sharma, Ram Sharan, 1969, *Purva Madhya Kalin Bharat Me Samajik Parivartan,* Peoples Publishing House, New Delhi, p. 12.

9. *Ibid.* (ed.), 1977, *Indian and Society: Historical Probings*, Peoples Publishing House, New Delhi, p. 130.
10. *Ibid.* p. 151.
11. Shastri, K.A. Nilkhanth, 1955, *History of South India* (from Pre-historic times to the Fall of Vijayanagar), Oxford University Press, New Delhi, p. 64.

CHAPTER 27

The Graeco-Roman Accounts on Ancient Indian Agriculture and Agricultural Products

Udai Prakash Arora

INTRODUCTION

Ever since the Hellenic world became aware of India, it started taking a keen interest in this country. The Greek and Latin authors, on India, furnished varieties of material on Indian affairs—geographical background, flora and fauna, religion and philosophy, trade and commerce, customs and manners—in fact, everything connected with India and the Indians. Despite the fact that with an exception of the work *Periplus of the Erythraean* Sea, not a single work planned by an author who knew the country personally has come down to us, except in epitomes and some fragments that have survived the wreck of time, the Graeco-Roman accounts remain the best available literary source on ancient India. The development of their knowledge on India may be divided into the following main periods:

1. Early Greek Notices: Age of Achaemenian domination
2. Hellenistic Age Authors:
   (i) Accounts of Alexander's Companions
   (ii) Megasthenes and Eratosthenes
3. Roman-Age Accounts

EARLY GREEK NOTICES

The emergence of the Achaemenian empire may be stated as providing the first opportunity of interaction between India and the Greek world. But during this period of interaction, the Greek geographical knowledge of India was confined to the region of Indus only, i.e. the part of India included in the Persian empire. Beyond that corner of India, which Persians knew, it was thought that there was nothing towards the east but a continuous desert.[1] India was thus considered to be situated at the eastern end of the inhabited world. During this first phase of Indo-Hellenic contacts, our main source of information is the Ionian Greek authors. They are Skylax, Hekataios, Herodotos and Ktesias. Among these, only Skylax is said to have visited India, of the Persian empire. The book written by Skylax is lost and unfortunately, through fragments, survived we do not learn much about the author's Indian experience.

The early Greek notices on Indian agricultural products and industries occur mainly in the writings of Herodotos and Ktesias. We may firstly mention the cotton plant which was cultivated in India since the days of Harappans.[2] It was described by Herodotos as the wild tree that bore wool instead of fruit. The wool was reported to have surpassed that of a sheep in beauty and quality and was used by the Indians to make their clothings.[3] Herodotos' reference to wool-bearing tree is the first Greek mention of cotton, which in fact, grow on a shrub and not on a tree, as understood by him. The Indian mercenaries in Xerxes' army were mentioned by him as wearing uniforms of the same material.[4] In the work of Herodotos, there also occurs the word *sindon* for muslin, which appears to have been derives from Sindhu, i.e. the place of its origin.[5] The cotton clothes in Ktesias *Indika*[6] were mentioned as *Xylina Imatia* i.e. clothes prepared from some stuff obtained from the tree. But it might be flax or bark cloth (*valkala*) instead.[7]

The cotton fabrics from India appear to be in great demand in the Achaemenian empire. Herodotos' reference to types of dresses worn by soldiers of different nationalities[8] in the Persian empire may probably be made of Indian cotton.[9]

The above Greek references to cotton in northwestern India are in accordance with the Indian evidences. The area of Sindh was the home of cotton, long before the Greeks knew of that area. The discovery of several spindle whorls from Indus sites bear testimony of the practice of spinning as well as the existence of the textile industry. The dresses represented in terracotta human figurines also corroborate it. Curiously enough, the expressions used for the cotton of Sindh by the Asian king Sennacherib (704–681 BC) was also the Herodotean 'wool-bearing tree'.[10] Cotton in Sanskrit was called *Kāsāya* and it had three varieties, namely, the cotton produced on trees, on creepers and from the grass call *poṭaki.* The *Mahāvagga* and the *Sivi Jātaka* mention the high quality of the cloth of Sivi country (west of the river Chenab) situated in northwestern India.[11]

Our attention has been drawn towards the Indian reed, i.e. *Kalamos Indikos* in the accounts of both Herodotos and Ktesias. 'India produces reeds in abundance', remarked Ktesias[12]. In comparison to Herodotos[13], Ktesias exaggerated the size of reed, saying that it was two orguiai (about 12 feet) round the outside and that a couple of cargo boats could be made from a single joint of one of these reeds[14]. It was so thick that two men could scarcely encompass its stem and that it equalled in height of the mast of the largest ship. Ktesias mentioned the existence of male and female reeds and gave an accurate description of the differences between the two. According to him, the female reed had a pith. The male reeds had no pith but they were exceedingly strong and compact.[15] Despite the wide exaggeration of reed's size, Ktesis knew the plant to be dioecious.

The reed described above is generally taken as bamboo plant, but Ball opined that the bamboo plant is not in accordance with the account of the classical authors. He identified the plant with the palmyra tee.[16] It seems that these reeds were probably an exaggerated description of the large specimens of the Indian bamboo, which could attain a height of one hundred and twenty feet and was used commonly in all types of shipping.[17] They appear to be of the *Kaṇṭaka* or the *bhallūka* variety, mentioned as very big and long in the *Arthaśāstra*[18] of Kauṭilya. From Herodotos and Ktesias' notices of the Indian reed, it may be assumed that they were exported from India during their days. The exaggerated measures of bamboo plant in Ktesias' account are fully consistent with the ancient Greek idea of imagining large things in remote regions.[19] In Graeco-Roman accounts, it often becomes difficult to differentiate between the description of bamboo and sugarcane.

The Indian archers with their reed bows and iron-headed reed arrows, serving in the army of Xerxes, find mention in the work of Herodotos.[20] Their skill in archery was often admired in classical accounts. The Indian epic texts are full of episodes on the figthing skill of these archers, equipped with reed bows and arrows.

The fine black heart-wood of various trees, known as ebony, was a favourite trade item in ancient times. Herodotos refers to Kalantian Indians along with the Ethiopians, bringing among their gifts the logs of ebony for the Persian emperor Darius I.[21] During the ancient period, India was known for its quality of ebony in the western world. If the identification of the *Old Testament*'s[22] Dedan with the southern shore of the Persian Gulf is correct, the export of ebony from India may be then taken back prior to the emergence of Achaemenian empire. In fact, at this time in India, the houses were mainly of wood and the workers in wood and carpenters were greatly valued. In the *Jataka* [23] texts, there are numerous references to wood workers who made ships, boats, carts, chariots and furnitures.

Although the authors of this period had some idea of Indian fame in spices, medical plants and dye stuffs, they were not much aware of the plant itself. As most spices were produced only in south India, they remained unmentioned in the literature of this period and even those dealing with Alexander's campaigns.

The most well known among the Indian spices was certainly pepper, which was reported by Ktesias to have been collected from truncated plants by some primitive cave dwellers.[24] In Hippokrates' manual *Gynaikeia* on diseases of women, was prescribed an ingredient, which was the Indian medicine for the eyes, called *peperi* (Sansk, *pippalī*)[25]. The significant reference to "Indian medicine", called by the Persians pepper in Hippokrates' Corpus bears testimony to the fact that it was through Achaemenians that the Greeks had come to know about pepper. The pepper is frequently mentioned in the corpus of Hippokrates. It is also mentioned, before Alexander, by Antiphenes of Athens in the fourth century BC.[26]

Ktesias mentions the product *Karpion* in India [27], which may possibly be identified with cinnamon. The word in Tamil is *Karuva* and *Karppu* in Malayalam. The Sanskrit *Karpūra* (Camphor) is akin to the above words of south Indian languages.[28] Herodotos had erroneously pointed out that Arabia was the only place where frankincense, myrrh, cassia, cinnamon and the gum called ledanon were produced.[29] The error was repeated even by later authors. In fact, before the voyage of Eudoxos (late second century BC),[30] trade was carried through Arab middlemen. They were guarding their monopoly so zealously that no Indian was allowed inside Bad-el-Mandeb.

The Indian dye stuff appears to be in great demand in the Achaemenian world. The Indian purple dye was considered by the Persians to be exotically beautiful and far superior to their dye stuff. Even the famous Lydian purple at this time was regarded as inferior to the Indian variety. Ktesias mentions the two sources of purple in India, obtained from a kind of insect and from a specific flower growing near the source of the river marshes. The insects, according to Ktesias, were found on the trees producing amber. His description of the insect-producing purple dye coincides more with the lac-insect (*Tachardia lacca coccid*), the dye produced from which was well-known in ancient India (*Laktakah*).[31] The *varṇadhātu* in the *Arthaśāstra*, used for colouring purposes, appears to be the dye.[32] The tree-nurturing lac insect in Ktesias is difficult to identify. According to Ball,[33] it was definitely a shellac tree, but Watt[34] has made a list of as many as forty-three different trees which nurture lac insects. The northwestern India, i.e. India of the Persian empire, was the main area where lac was grown. The 'coloured lac' from this region continued to be exported during the period of Indo-Roman trade.[35] Herrmann's view of taking Ktesias' purple as well as insect dyes as indigo,[36] without stating any grounds, can hardly be accepted.

There is also a mention of another famous red dye exported from India, which was obtained from the plant cinnabar (Greek, *kinnåvari;* Sansk. *Vetasa*). It gave a fresh pink tint to a cloth. The plant was also used as a medicine for several ailments. Basing his information on Skylax, the Milesian Hekataios mentioned its cultivation in the region around Indus.[37] In fact, such a plant is cultivated in India mainly in the region of Deccan and Madhya Pradesh. It is viewed that the true cinnabar is a mineral product (red mercury sulphide, perhaps also red ochre) but in Graeco-Roman accounts, the name was thought to be a plant dye (*Dracaena* ssp. of South Arabia and East Africa and *Calamus draco* of India).[38]

Ktesias' reference to trees exuding amber[39] is difficult to explain. According to Lassen, it was probably some sort of gum, of which there are several varieties in India, especially in the east.[40] Ball identifies the elektron or amber of Ktesias with the shellac tree.[41] Amber is today unkown in India, but on the basis of Chinese[42] and classical sources, we may believe that it was once available in India.

Skylax[43] and Hekataios[44] mention the plant *kynåra,* i.e. artichoke grown in the region around Indus. The cultivation of artichoke at such an early date in the Indus region appears unlikely, for it is a western plant, rarely grown in the east even in modern times.[45] The *kynåra* cultivated in the Indus region appears to be some thorny bush, although it remains uncertain. Herzfeld's identification of the plant with the Persian *Chinar* (*Platanus Orientalis* L.) too appears unlikely.[46]

Among other botanical products, the fruits obtained from the *Siptakhora* tree, which were dried and packed up in plaited baskets for exporting to other parts of India, appear to be some kind of Indian fig.[47]

There are references to various types of Indian oils in Ktesias' work.[48] Among these, the oil made from sesame and nuts appear to be real, whereas the description of the rest cannot be explained. The oil obtained from milk[49] may be butter or clarified butter (*ghee*). That the edible oil of sesame and clarified butter were exported from Saurashtra in India in early centuries of Christian era may be attested by the *Periplus of the Erythrean Sea.* On the basis of Ktesias, their export from India may be postulated even in the fifth to fourth BC. Kauṭilya mentions a corporation in Saurashtra living on agriculture, trade and the wielding of weapons.[50] It is, thus, possible that as in the first and second century AD, similarly, in earlier period also, Saurashtra might have been trading in sesame oil and clarified butter along with many other items.[51] In fact, sesame seeds have been found at Harappa, and ever since the plant has had an important place in the lives of the Indians, both as food and as a part of religious ceremonies. Its oil was used as medicine both in India and the west. While the Greek doctors recommended it, for instance, in ear inflammations, eye diseases, burns and snake bites, the Āyurvedic authorities recognized it as being healthy for hair, skin, teeth and digestion, as a demulcent and tonic, conducive to general health.[52]

The scented oil of India was immensely favoured by the Persians. It has been pointed out that the use of perfume originated in India and Persians derived their knowledge of using scents and perfumes from India.[53] In his description of the Indian cinnamon flowers called *Karpion,* Ktesias reports, 'There oozes from them an oil in drops which are wiped off from the stem with wool from which they are afterwards wrung out and received into albaster boxes of stone. The oil is in the colour of a faint red and of a somewhat thick consistency. Its smell is the sweetest in all the world and is said to diffuse itself to a distance of five stadia around. The privilege of possessing this perfume belongs only to the king and the members of the royal family. A present of it was sent by the king of the Indians to the king of the Persians.' Ktesias claims himself to be an eyewitness. The perfume had such an exquisite fragrance that Ktesis found himself unable to describe it, for he knew of nothing comparable with it.[54] This scented oil may be identified with the oil known as *Karuṇṭa* in Sanskrit and *Karuppu* or *Karppu* in Tamil-Malayalam. The other forms of this word are *Karappu* and *Karuva.* The Sanskrit *Karpūra* (Camphor) is almost the same as the southern *Karuppa* and Ktesias *Kårpion.* Camphor and cinnamon are closely related. The camphor of commerce (*Camphora officinarum*) is from a cinnamon tree.[55]

Ktesias speaks about the cheese and wine of India, which he had himself tasted. He had found the Indian wine to be the sweetest of all.[56] The wine mentioned by Ktesias does not appear to be the product of grapes, for they were not much cultivated in India. This was probably prepared from the coconut palm juice which has a very old tradition in the Indus Delta. Ktesias' reference to large nuts, three times larger than the Babylonian variety of the cocoa palm or the date palm is a debatable issue.[57] Kauṭilya's *Śvetasurā,*[58] which was prepared in homes on special occasions, was probably the wine prepared from coconut palm.

So far as the cereal crops are concerned, the Greeks before Alexander probably had no knowledge of that. There is only one uncertain reference to rice in a fragment of writing of Sophocles.[59] Herodotos' description of some millet-sized grain of India[60] too remains unidentified. Whether it was a reference to rice or some leguminous plant or some fruit, remains unknown.

HELLENISTIC AGE: ACCOUNTS OF ALEXANDER'S COMPANIONS

Alexander had been a disciple of the great master of knowledge, and among the officers who had accompanied him into India, some were greatly distinguished for their scientific and literary achievements. They employed their pens in recording their emperor's warlike achievements and in describing the countries into which he had carried his army. The information on plants and agriculture, gathered during Alexander's campaigns, mainly come from Nearchos, Onesikritos and Aristoboulos.

Chares, Ephippos and Polykleitos, who have also been listed among the three historians of Alexander's companions as authorities on foreign plants, unfortunately, furnish no information on Indian plants in their survived fragments.

Theophrastos' treatise on Plant Researchers (*Peri Phytòn Historia*), followed by Plant Aetiology (*Peri Phytòn aitión*) furnish valuable accounts on Indian plants, including much otherwise unknown informations. Although he had not accompanied Alexander, he may well have oral eyewitness accounts in addition to written sources. His company with Aristotle had resulted in researches, especially in botany and perhaps also in political science.

Egypt, or Egypt along with Ethiopia, was the main yardstick of the Greeks through which they were trying to understand the Indian phenomena and affairs. This tradition was in fact, established by Alexander's companions, who were puzzled by the unfamiliar Indian physical phenomena. The Indian heat, rivers, plants, birds, animals, etc., all contributed to remind them of the land of the Nile, and comparisons were made between their climatic topographical and hydrographical conditions. Comparing the Indian flora and fauna with Egypt-Ethiopia, the Greek authors found many similarities.[61] Nearchos states that seeing Egyptian beans in the River Akesines (Chenab) and crocodiles in the Hydaspes (Jhelum), Alexander imagined that he had discovered the source of the river Nile.[62] They had heard of the production of cinnamon, spikenards and other aromatics in southern India, which reminded them of Arabia and Ethiopia[63]. The greater number, size and strength of Indian animals led them to believe in the superior fertility of the Indian land.[64] The Greeks noted several features of comparison between Nile and the Indian rivers. On seeing the rise of the Indian rivers through summer rains, Nearchos immediately jumped to the conclusion that the swelling of the river Nile was also caused by the rains.[65] Aristoboulos had the opinion that whereas the Indian rivers flooded from the rains of the north, the Nile flooded from the rains of the south.[66] Onesikritors' observation that in the Indus delta winds do not blow from the land and that this region is subject for the most part to winds that blow from the high sea,[67] reveals that he had at least understood the monsoon rains of Punjab and Indus Valley, which distinguish them from the valley of the Nile. Although he did not understand the significance of monsoon for navigating purposes, he was able to connect the summer rain and the fertility of the country with the southwest monsoon. It was further remarked that whereas the water of the Nile traversed in a straight course on a long and narrow tract of the country, undergoing many climatic and atmospheric changes, the Indian rivers poured their waters into plains of greater length and width and lingered in the same climate, which made the latter more nutritive than the Nile.[68]

In respect of the sun's rays, the temperature of India was regarded to be similar with Egypt and Ethiopia, but in the copiousness of its waters, it was said to have surpassed them, which made the atmosphere humid, nourishing and more productive.[69]

With the advent of Alexander and his armymen on Indian soil, the Greek world for the first time received first-hand knowledge of Indian agriculture. Agriculture was observed as the main occupation of Indians by Nearchos.[70] The soil was found to be exceedingly fertile. The historians reported that although the soil was sown when only half dried, yet the plant came to maturity and yielded excellent fruits after being furrowed by any sort of digging instrument.[71]

Alexander's companions had taken interest not only in various agricultural products, but they also tried to understand the system connected with the distribution of grain, cultivated on a common land by different groups. Nearchos reported that among certain Indian tribes, on the basis of kinship, the crops were cultivated in common and when the produce was collected, each carried off a loan sufficient for sustenance during the year, but burned the remainder in order to have work to do thereafter and not be idle.[72] The reason given here for burning the remainder is not altogether convincing; it appears an idealization of land in the Greek tradition of romanticizing the far-off regions.[73] However, the system of the common ownership and the distribution of produce, as indicated above, has in all its essential features, survied in India from the earliest times down to the present day.

Among the Indian crops mentioned in the accounts of Alexander's historians, the most important was certainly rice, which was hitherto unknown in the Greek world. An account of the cultivated rice, furnished by Aristoboulos the engineer, would be of more interest to the irrigation expert rather than a botanist. The rice, according to him, 'stands in water enclosures and is sown in beds, and the plant is four cubits in height, not only having many years but also yielding much grain; and the harvest is about the time of the setting of the Pleiades, and the grain is winnowed like barley'.[74] This is an accurate description of the rice plant (*Oryza sativa*), which is a tall grass to a height of about four feet and having one or two feet long leaves. Aristoboulos is here clearly describing the cultivation of lowland rice, most varieties of which require it only during the period of germination. Aristoboulos' report tallies with the Indian text *Milindapañho* (4), which informs us that the rice field was provided with canals for irrigation and embankments for storing water. A more accurate botanical description of rice was furnished by Theophrastos. He said that the rice grows in water and has no ears, but resembles millet.[75]

There were always many varieties of rice in India. The Greek word *Oryza* for rice appears to have been derived from Tamil *arisi.* Its cultivation in India goes to exceedingly high antiquity. It was cultivated in pre-historic times in northwest India, Indus sites, Gangetic Valley and other parts of India. In literary texts, it has been amply attested since days of the *Yajurveda.*[76]

Next to rice may be mentioned a crop called *Bosmoron* which according to Onesikritos, was smaller than wheat and grew in lands between rivers. In order to check the export of *Bosmoron's* seeds, they were immediately roasted after threshing.[77] As the botanical details of *Bosmoron* are absent in the account, this crop evades identification. McCrindle suggests that it may be wild barley or common millet. Schoff rejects McCrindle's view[78] and opines that it was probably *Panicum crus-galli* or 'Poor man's millet', which he says is extensively cultivated in India, China and Japan. The conservation of this plant and the desire to check its exportation, according to Schoff, is an indication of the native millet which was the grain most used for temple offerings and, therefore, thought to be especially pure.[79]

On account of being something very common, the Greek accounts on agriculture only occasionally mentioned wheat and barley. They are profusely mentioned in Indian literature and their cultivation in the northwest extends far back to pre-historic times. Nearchos found the cultivation of these two crops on the Gedrosian Coast.[80] Theophrastos briefly reported both. He spoke about the barley with side-shoots, a kind of 'Wild barley', which makes sweet bread and good porridge. The Macedonian horses had learnt to consume it. It appears to be some poor food mainly used as fodder, identified by Hort as *sorghum halepense.*[81] Some self-grown grain similar to wheat was noticed by Onesikritos in the land of Mousikanos, located somewhere in the lower Indus region.[82]

There are many varieties of pulses in India since early times. In western sources, we do not hear much of them. The chick pea, lentil and many other kinds which have been well-known in India but unknown to the Greeks, were mentioned by Theophrastos.[83] The 'Egyptian beans', however, seen by Alexander's companions[84] on the banks of the Chenab, were an entirely different plant. They have been identified as *Nelumbium speciosum,* now called *Nelumbo nucifera.*[85] It may be taken as a reference to Indian lotus also. Nearchos' description of the reeds yielding honey, although there are no bees,[86] certainly refers to sugarcane, a famous graminaceous plant of India of great economic significance. The sugarcane and the rice are the typical monsoon crops with which the Greeks had made their first acquaintance in India.

The sesame oil of India was known to Ktesias[87] but its importance in Indian agriculture was noted by Alexander's historians only. The best white seeds of sesame, according to Theophrastos, were available from India.[88] Aristoboulos refers to sesame oil and cakes of sesame used by the ascetics of Taxila. Aristoboulos refers, 'on every person whom they accost, they pour oil of sesame until it trickles down to their face; of honey, which is exposed for sale in great quantity, and of sesame they take enough wherewith to make cakes and their food costs them nothing.'[89]

The fruits referred to in the accounts of Alexander's historians are difficult to identify. Aristoboulos mentioned a small tree with bean-like pods, ten fingers in length, full of honey, but apparently deadly poisonous.[90] Ball identified this as *Cassia fistula,* the purging Cassia, which however, is not poisonous.[91] The reference to the poisonous effect of this fruit is not understandable. Combining this with Theophrastos' crooked fruits, Pearson identifies both as banana.[92] In fact, the Greeks never learned to eat bananas (or mangoes either) and so, consequently, all recollection of the fruit was lost.

Theophrastos draws our attention towards the varieties of fruits in India without naming them. Thus, he writes about some sweet, long and crooked fruit, which causes stomach problem and dysentery; a fruit resembling the carnelian cherry; some large-sized sweet fruit taken generally by the ascetics; and fruits grown on a tree of large, oblong-shaped leaves.[93] They have been identified by Horth as mango, jujube (*Zizypus jujuba*) fruit and banana tree,[94] respectively.

Nearchos furnishes several references to dates grown on the Gedrosian Coast.[95] Based probably on Nearchos' statement of wild dates of Gedrosia,[96] Theophrastos draws our attention towards the unripe dates of Gedrosia, which he considers dangerous for eating.[97] The wild date palm is grown in many parts of northern India and its literary antiquity may be taken back as early as the *Yajurveda*, but its cultivated variety appears to have been introduced in the medieval period only. The cultivated variety of date palm was never common in India. It appears to be an item of import since the Indus days.[98]

Like Herodotos, the companions of Alexander also mistook cotton as growing from tree, instead of shrub. Nearchos was uncertain on the point that the cotton produced from these trees was either of a brighter white colour than that of a cotton found elsewhere, or the darkness of the complexion of the Indians made the fabric white.[99] When Onesikritos claimed that the fibres were found in the flower around a stone, it may be explained that as he had never seen the plant with flowers, he erroneously interpreted the pods as flowers.[100] The Indian cotton plant was described by Theophrastos as similar to the rose plant with leaves like mulbery. About its cultivation, Theophrastos writes that it was planted in rows resembling vineyards.[101]

India as a country of spices and medicines was already mentioned by Ktesias. With Alexander and his armymen, this fact was established and has endured ever since. The companions of Alexander found that India produced many medicinal plants and roots, both of a salutary and a noxious quality and plants which yielded a great varieties of dyes.[102] Theophrastos knew that most of the aromatics were from India and Arabia.[103] One of the most famous Indian aromatic products known in ancient Greek world was certainly nard or spikenard, which seems to have become known at least in the time of Alexander. Theophrastos writes that spikenard has a biting quality as well as heat. He includes Indian spikenard among aromatics.[104] Onesikritos writes that the plant grows in the south[105] of India, like cinnamon and other aromatics. Herodotos' misconception that cinnamon and cassia were the products of Arabia[106] and not India, continued to persist in the time of Alexander also. Theophrastos had included them among the products of Arabia.[107] The Indian pepper was discussed by Theophrastos as not among Indian plants, but in his chapter on medical plants.[108] He described two kinds of pepper: a round reddish berry and the other elongated and black one. These have been identified as *piper nigrum* and *piper longum*. Among other such plants known to Alexander's companions, mention may be made of amomum,[109] bdellium[110], costus[111], silphium[112] and a kind of terebinth.[113]

Herodotos[114] and Ktesias[115] knew about the Indian reed or bamboo, but it was first seen by the companions of Alexander as growing by the river Chenab.[116] Theophrastos has furnished a more exact account of the plant. Another kind of reed, called *Kypeiron* was mentioned twice by Theophrastos in the lists of aromatics, might have come from India. The majority of aromatics, according to Theophrastos, came from India or Arabia. This species has been identified as *Cyperus rotundus* by Hort.[117] The skill of Indian archers with their reed bows and arrows was known to Herodotos[118] was often admired in literature by Alexander also.[119] The emperor had himself experienced their skill at least twice, once when he was wounded among the Assacenians and once again, when he was wounded more seriously among the Malli.[120]

Indian forests were often mentioned in classical sources. Forests of the country near to the Indus were known to Skylax.[121] Alexander's companions also noted the varieties of trees in Indian forests. Aristoboulos finds firs and pines being common in India but absent in Hyrcanian forests.[122] Theophrastos' mention of the plant resembling myrrh[123] may be identified with the balsam tree of India.[124]

The high quality of ebony, for which India was known even before the emergence of the Achaemenian empire, continued to attract the companions of Alexander. Theophrastos knew two kinds of Indian ebony, a rare variety but of high quality; a common, but inferior one. He knew the wood to be dark by nature, but had no idea of the tree.[125] The Greeks probably knew the timber only, not the tree[126]. The ebony tree was never described in classical sources.

The big banyan tree was a thing of great wonder for the army of Alexander. The aerial roots growing into supporting trunks were a completely new phenomenon. Onesikritos' description of banyan tree,[127] despite its errors, is a graphic one and seems to be based on personal observations. A more precisely detailed and botanically more correct description of banyan tree was reported by Theophrastos. He rightly pointed out that the tree did not send shoots or suckers into the ground to root there, but that those were really subsidiary roots.

Away from their homes, the companions of Alexander[128] were glad to see some familiar plants of their country in the distant Paropamisadae and northwest India. These included the vine, laurel, ivy and myrtle. Ivy was seen at Mount Nysa, sacred to Dionysos. Alexander had found that there was already a Greek community worshipping Dionysos at Nysa. He learnt that Dionysos had been there and the emperor was going farther than the God.[129]

HELLENISTIC AGE: MEGASTHENES AND ERATOSTHENES

Our main source of information on Indian agriculture after Alexander are Megasthenes and Eratosthenes. Whereas Alexander's companions were confined to the northwest and Punjab only, Megasthenes reached the heart of India, where the Indian caste system, rituals and Brahmanism with all its traits were widely prevalent. Dividing Indian society into seven classes, Megasthenes mentioned farmers as forming the bulk of India's population. They were in disposition most mild and gentle. They were exempted from all the public and military services and lived in the villages. Their whole business was to cultivate the soil and pay taxes, to the kings or to the free cities, as the case might be. They paid a land tribute to the king because all of India was the property of the king. They also paid into the royal treasury a fourth of the produce of the soil. The men of this group, according to Megasthenes, were protected from all injury and even the enemies during wars did not disturb them.[130]

There is no trace of Indian theory of the four-fold division in Megasthenes' division of classes, but his observations of craft exclusiveness and endogamy are undoubtedly a characteristic of the rigid caste system. Megasthenes' observation[131] that the farmers were left undisturbed during the wars may be corroborated by some Indian texts. In fact, there was a strong feeling in ancient India that the lives of non-combatants should be respected. An honourable Kṣatriya was never supposed to strike the unarmed, eunuch, fearful, the one who fled, and to that who supplicated. He was not allowed to strike that also, who was fighting with another foe.[132] The barbed, poisoned and the weapons in wood were not allowed to be used.[133]

However, in actual practice, the above ideal was frequently ignored. The *Arthaśāstra*, on the other hand, says nothing about fair play in battle and evidently looks in conquest of the demoniac variety as the most profitable and advisable.[134] Devastation of the crops in order to weaken enemy was said to be legitimate[135] and though the wholesale sacking of cities was not common, the townsmen were not always secure against the looting and exactions of enemy occupation.[136]

Megasthenes' observation of understanding the entire land of the country as king's own property does not comply with Indian evidences. The idea never existed in India. There are references which show king's authority over the land as his right of eminent domain but they didn't mean Megasthenes' beliefs.[137] We notice in the *Arthaśāstra* that without actually asserting the king's ownership of the soil, Kautilya advocated and doubtless introduced into the administration all the detailed supervision and control of agricultural and marketing operations that would have arisen from such an ownership.[138] It appears that Megasthenes, who saw such things with his Hellenistic background, thought that in India as in Ptolemaic Egypt, the king was the owner of the land and the cultivators were his tenants.[139]

Taxation on the land was made theoretically justified as a return for the protection granted by the king. 'No taxation without protection' is an Indian maxim. The fourth part of the land produce as a tribute to the king, as told by Megasthenes, was mentioned in the *Arthaśāstra*, which suggested one quarter or even one third for fertile land.[140] The figure generally given in various texts is one-sixth,[141] but in the time of king's distress, Manu recommended even the fourth part, if he protected his subject to the best of his ability.[142]

Alexander's companions knew only the monsoon crop of India. Megasthenes and Eratosthenes extended Greek knowledge of Indian agriculture by describing the two annual crops, the summer and the winter one.[143] The Greeks found it a great source of wonder that India was producing two crops a year. They knew that millet, bosmoron, rice, sesame and flax were cultivated during the rainy season; and wheat, barley and lentils in the winter.[144] Megasthenes knew that Indians made a beverage of rice, instead of barley, and that rice was the staple food in India. Speaking of the supper of the Indians, he said that at supper a table-like tripod was placed before each person and there was put upon that a golden bowl, into which they first put boiled rice and then they added many dainties prepared according to Indian recipes.[145] This must be a reference to the rich Indians only.

The sugarcane was well known to the companions of Alexander. Megasthenes made an attempt of scientifically explaining its sweet juice. He said that it was due to the water which it absorbed from the soil being so warmed by the Sun's heat that the plant was virtually cooked as it grew.[146] He also refers to sweet stones dug up in India which have been explained as candy sugar.[147] The wool-bearing tree (cotton) of Herodotos was mentioned by Eratosthenes as woolly blossoms which was included among products caused by heating.[148] The ebony, as an Indian product, was repeated by Megasthenes[149] from his predecessors. The Indian origin of cinnamon continued to remain unknown even after Alexander. The most well-informed man of his time, Eratosthenes, reported cinnamon as a product of eastern Arabia.[150]

Stressing the fertility of Indian soil, Megasthenes reported that there were numerous varieties of fruits produced in India.[151] The Indian soil yielded plenty of plants, useful for food, most of which grew spontaneously. He affirmed that famine had never visited India, and that there had never been a general scarcity in the supply of nourishing food.[152]

The staple crops of ancient India, as described by the Greeks, were the same as today, i.e. wheat and barley in the cool north and elsewhere as winter crops, rice in the irrigated plains, and millet in the dryer lands. Sugarcane was widely grown and numerous types of peas, beans and lentils were grown throughout the subcontinent. Cotton was at all times the major textile crop. Megasthenes' statement about the absence of famine in India cannot be taken literally, for various Indian literary works refer to famines and specially to one that occurred a few years after the Greeks' departure from India. The Jain tradition mentioned a famine at the end of the reign of Chandragupta Maurya. But it certainly shows that at that time Megasthenes wrote his *Indika*, there was plenty and prosperity, and famine was very uncommon; at least it did not occur within living memory. It may be said that rainfall was rather heavier than in recent years, and the pressure of population on the means of subsistence was not as great as it has become now. Hence, any outbreak of dearth and famine were probably less frequent in ancient days, but when they did occur, they caused even greater hardship and loss of life. To meet such situations, the *Arthaśāstra*[153] even suggested the shifting of entire population to another region, where crops could be grown. The state graneries would be opened in emergency, the charity of religious establishment and private persons was, no doubt, of some help, but with poor communications in those days, local famine might be even more severe than it is now.

ROMAN AGE ACCOUNTS

The imperial age authors differ from their predecessors in respect that their works are mainly the compilation of their predecessors. Barring only one doubtful exception, all of them wrote without personal knowledge of the country. So far as the knowledge of Indian agricultural products is concerned, new information reached the Graeco-Roman world through trade contacts. The texts refer to items of export and import being carried mainly between south India and the Graeco-Roman world. Thus, like Alexander's companions and Megasthenes, the authors of this period had, although not seen the plants, they knew about the products. Our main source of information are Plinius, the unknown author of the *Periplus of the Erythrean Sea,* Dioscorides and Ptolemaios, a geographer.

The authors of the imperial age continued to repeat the description of various Indian agriculture products from their predecessors. These include cereals like wheat,[154] barley,[155] bosmoron,[156] rice,[157] millet,[158] sesame,[159] and pulses;[160] various fruit trees,[161] pepper,[162] cinnamon,[163] and nard,[164] among the spices; cotton,[165] linen[166] and, flax;[167] ebony,[168] banyan,[169] reed plant,[170] and the sugarcane.[171]

'India enriches her sons with wealth in every form', remarked Dionysos Periegetes.[172] The author of *Periplus* found corn and rice in abundance in the region around the Gulf of Cambay.[173] In the fertile plains of the Ganges were reported the beans and millets of a very big size by Philostratos. The same writer mentioned the cultivation of vine and nuts in India.[174] Plutarch went forward erroneously to credit India even as the introducer of vine cultivation.[175] There are notices of various Indian fruits and vegetables in the writings of Roman age authors.[176]

The Indian sugar was pronounced as *Sakhron* by Dioscorides. He was perhaps the first to use the word from the Sanskrit *Śarkara* or Prakrit *Sakkhari* and described that as a kind of congealed honey.[177] In the *Periplus,* sugar was mentioned as an export item of India, and its quality was declared to be the best in the world.[178]

The Indian cotton was highly admired in the Roman empire. Loukianos commented that the 'Indian fabrics' (muslins) were brighter and softer than the Greeks.'[179] The *Periplus* rightly mentioned its production in Gujarat.[180] According to Philostratos, it was exported to Egypt for various religious purposes.[181]

Although India as a country of spices and medicines was known even before the emergence of Alexander, the full extent of this truth was realized only after the accelaration of trade contacts in the period of Roman domination. Theophrastos' work on plants appears to be well known among the imperial authors for the description of nard and spikenard, terebinth,[182] costus,[183] cardamon[184] and amomum[185] in imperial writings appear to have been derived from Theophrastos. There were also certainly some Indian botanical products which find their mention for the first time in the writings of the Roman period.

The famous 'Indian dye' which, in the Graeco-Roman world was simply called *Indikon* or *Indicum,* occurs for the first time in the works of Dioscorides[186] and Vitruvius[187] in the early imperical period. Plinius and the authors of *Periplus* write about their export from India.[188] This dye was considered as one of the few and expensive bright colours used by painters.[189] So far as the plant is concered, it remained unknown in the west for a long time. It was only mentioned for the first time by Marco Polo. There are many references to some Indian aromatic leaves known as *Malabathron* in Greek or *Malabathrum* in Latin in the writings of Plinius,[190] Dioscorides,[191] Ptolemaios[192] and the unknown author of the *Periplus.*[193] They became known to the western world only in the early Roman imperial period, but even now remain unidentified. A kind of barberry lycium was another Indian product which, according to the *Periplus,*[194] was an item of export from Barbarikon and Barygaza. Plinius knew it as the root of a spiny shrub from India.[195] It was used for cosmetic and medical purpose and its root also yielded a yellow dye. The aromatic macirbark from India,[196] introduced in the early imperial period, has been identified with the mace, the aromatic core of the nutmeg. The macir has been explained by Watt,[197] as *Holarrhena pubescens.* This plant grows all over India and is known as an ancient medicine against dysentery.

It appears that seeds of some Indian plants were also experimented with in Arabia, for Graeco-Roman authors refer to some plants as common in Arabia and India.

From Plinius we know that Seleucos attemped to cultivate Indian drugs and spices in Arabia.[198] Thus, according to Strabo, cinnamon is produced in India, Ethiopia and Arabia.[199] Discorides described the aloe leaf and claimed that it grows both in Arabia and India.[200] He also mentioned libanos and frankincense of Arabia and India.[201] According to Plinius, the mastich or *laina,* a kind of thorn-bush producing a gum resembling myrrh, is found both in Arabia and India.[202]

NOTES AND REFERENCES

'F' and 'No' refer to fragments and author's serial number, respectively in F. Jacoby's *Fragmente Griechischer Historiker* I-IID (Berlin, 1923–30); IIIa–IIIc (Leiden, 1954–58).

1. Herod., III, 98 (Loeb edn.).
2. R.E.M. Wheeler, 1981, *Indus Age,* Suppl. Volume to *Camb. Hist. of India,* p. 57 and 63; A.L. Basham, *The Wonder That Was India,* Paperback edn. Reprint, Delhi, pp. 18 and 26.
3. Herod., III, 106. It is interesting to note that the German word for cotton *baumwalle,* i.e. tree-wool coincides with Herodotus' description.
4. Herod., VII, 65.
5. Herod., I, 200; II, 95.
6. Ktes. Jac. No. 688, F.45(41).
7. Klaus Karttunen, 1997, *India and the Hellenistic World,* Helsinki, p. 135.
8. Herod., VII, 61–64; VII, 89.
9. R.N. Saletore, 1975, *Early Indian Economic History,* London, p. 55.
10. K.D. Sethna, 1981, *Karpasa, Pre-historic India,* New Delhi, p. 3.
11. *Mahāvagga,* VIII, 1; *Jataka* No. 499; *Cullavagga,* VI, 2, 6; IV, 44; VIII, 1, 3.
12. Ktes., F. 1b (Diod., II, 17).
13. Herod., III, 98.
14. F. 45C (Tzetz, *Chil.* 7, 738–41)., cf. Plin. *Nat. Hist.* XVI 36 & VII, 2; Theoph. *Hist. Plant,* IV, 11; Herod., III, 98; Strab., XV, 1, 21.
15. F. 45(14). cf. Theoph., *Hist. Plant.,* IV, 11, where he states that the male reed is solid and the female hollow. cf. Also Plinius, *Nat. Hist.* XVI, 36. The same reed is mentioned by Herodotos (III, 98).
16. McCrindle, 1901, *Ancient India as Described in Classical Literature,* Westminster, Repr. Amsterdam, 1971, p. 59, n.2.
17. See *Ency. Brit.* Vol.3, under 'Bamboo' (*Bambusa arundinacea* (1957 edn.), pp. 302–3.
18. *Arth.* Bk. II, Chap, 17, 5.
19. See Herod., II, 106.
20. Herod., III, 65.
21. Herod., III, 97.
22. Ezkiel, 27, 15.
23. *Jātaka* No. 315, 442, 465, 475, 506, etc.
24. Ktes., F.63 (loan. Lyd. *De mens* 4, 14).

25. The pepper as the Indian medicine is mentioned in Hippokrates *Morb Mul.* I, 81 and II, 185, 205. There are several other passages, where pepper is mentioned without being specified as an Indian product. The total number of occurrences of the world *peperi* in the Corpus is seventeen. See. J. Filliozat, 1949, *The Classical Doctrine of Indian Medicine,* Trans. By D.R. Channa from the French original, Reprint, Delhi, 1946, 253 ff.
26. K. Karttunen, *op cit.,* pp. 149–50.
27. F.45 (47).
28. McCrindle, *Ancient India as Described by Ktesias the Knidian* (Calcutta, 1982), pp. 29–30, n. 81.
29. Herod., III, 107.
30. The voyage was described by a Hellenistic author Posedonios, whose account has reached us through Strabo (Posed., Jac. No. 87, F.28=Strab, II, 34).
31. Ktes. F.45 (38–39); F.45p (Ael; *NA,* 4, 46). Also see McCrindle, op cit., pp. 22–23, 52.
32. *Arth. Bk.* II, Chap. 24 and Chap. 22, 6.
33. W. Ball, 'A Geologist's Contribution to the History of Ancient India', Paper presented to Royal Irish Academy, 1884.
34. G. Watt., 1908, *Dictionary of the Economic Products of India,* S.V. *Coccus Lacca,* London.
35. *Perip.,* 6. It writes that the coloured lac was exported from Ariaca (Roughly northwestern India) to the East African ports.
36. Abbert Herrmann, 1938, *Das Land der Seide und Tibet in Lichte der Antike,* Leipzig, p. 19.
37. Hekat., Jac. No.1, F.296 (Athen., II, 70B).
38. Karttunen, *op. cit.,* p. 164.
39. Ktes., F.45 (39); F.45 (36); F.450 (Plin. *N.H.,* 38, 39); F.45p. (Tzetz, *Chil.,* 7, 713); F.45p (Ael. *N.A.,* 4, 46).
40. cf. McCrindle, *op. cit.,* p. 70.
41. McCrindle, 1971, *Ancient India as Described in Classical Literature,* Westminster, 1901. Repr. Amsterdam, p. 129, n.3.
42. B. Laufer, 1919, *Sino-Iranica,* Field Museum of Natural History Pub. 201, Chicago, Repr. Taipei, p. 523.
43. Skylax, Jac. No. 709, F.3 (Athen., 2, 82, p.70, A-C); F.4 (Athen, 2, 82, p. 70 CD).
44. Hekat., Jac. No.1 F.296, (Athen., 2, 82, p. 79b).
45. W. Reese, 1914, *Die griechischen Nachrichten uber Indien bis Zum Feldzuge Alexanders des Grossen,* Leipzig, 47f.
46. E. Herzfeld, 1968, *The Persian Empire, Studies in Geography and Ethnography of the Ancient Near East,* edited from the posthumous papers by G. Walser, Wiesbaden, p. 286.
47. Ktes., F.45(36) and F.45o (Plin, *N.H.,* 37, 39). *Siptakhora* may be compared with the Persian *shiftekhor* 'agreeable to eat'. The apricot is called in Persian as *shifteh reng.* In Plinius (*Nat. Hist.,* XXXVII, 2), the word is disfigured and written as *Aphytacoras.* See McCrindle, *Ancient India as Described by Ktesias the Knidian,* p. 21, n. 60.
48. Ktes., F.45(25); F.45(46); F.45(47); F.45R (Ael., *N.A.,* 5,3).
49. F.45 (42).
50. *Arth.,* Bk XI, Chap. I, 4.

51. R.N. Saletore, 1973, *Early Indian Economic History,* Bombay, p. 201.
52. Karttunen, *op. cit.*, pp. 146–47.
53. R.N. Saletore, *op. cit.*, pp. 239.
54. F.45 (47); McCrindle, *op.cit.*, pp. 29–30.
55. McCrindle, *op. cit.*, pp. 29–30, n. 81.
56. F.45 (48).
57. F.45 (28). cf. Palladios, *De Brachm,* p. 4. See, K. Karttunen *op. cit.* pp. 137–138.
58. *Arth.* BK.II, Chap. 25.
59. Quoted by Karttunen, *op. cit.,* p. 143.
60. Herod., III, 100.
61. U.P. Arora, 1982, 'India vis-à-vis Egypt-Ethiopia in Classical Accounts', *Graeco-Arabica,* Vol. I., Athens, pp. 131–140.
62. Nearchos, Jacoby No. 133, F.20 (Strabon, XV, 1, 25).
63. Onesikritos, Jac. No. 134, F.22 (Strabin, XV, 1, 22).
64. Onesik, F.22 (Strabon, XV, 1, 22). See also F.114 (Strabon, XV, 1, 43) on the larger size of Indian elephants.
65. Nearchos, F.20 (Strabon, XV, 1, 25).
66. Aristob., Jac. No. 139, F.35 (Strabon, XVI, 19).
67. Onesik., F.8 (Strabon, XV, 1, 20).
68. Onesik., F.22 (Strabon, XV, 1, 23).
69. Onesik., F.22 (Strabon, XV, 1, 22).
70. Nearch., F.11 (Arr., *Ind.,* XVII, 5).
71. Nearch., F.18 and Aristob., F.35 (Strabon, XV, 1,18).
72. Nearch., F.23 (Strabon, XV, 1.66).
73. L. Pearson, 1960, *Lost Histories of Alexander the Great,* New York, p. 128.
74. Aristob., F.35 (Strabon, XV, 1.66).
75. Theoph., *Hist. Plan.*, IV, 4, 10.
76. Tuktuk Kumar, 1988, *History of Rice in India: Mythology, Culture and Agriculture,* Delhi, 9ff. 56ff. 58ff. 79ff. J.F. Hewitt in the *Journ. of Royal Asiatic Society* (Oct., 1980, p. 730. Quoted by McCrindle in *Ancient India as Described in Classical Literature,* Westminster 1901, Repr. Amsterdam, 1971, p. 24 n.1) writes, 'Rice was exported to Europe from the ancient sea port of Barygaza, the modern Broach, and Surparaka (Surat) and its export must date from a time when the people in the west of Bombay and at the mouths of the Indus spoke Dravidian tongues, and the Aryan Sanskrit and dialects derived from it were unknown to the country traders. But before a foreign trade began, numerous varieties must have been developed, and the development of these varieties, with the culture and agricultural skill necessary for their preservation, must have required a vast lapse of time, to be numbered by hundreds if not thousands of years.'
77. Onesik, F.15 (Strabon, XV, 1, 18).
78. McCrindle, *op.cit.*, p. 24 n. 2.
79. W.H. Schoff., *Periplus of Erythrean Sea* (New York, 1912, repr. New Delhi, 1974) pp. 178–79.

80. Nearchos, F.1 (Arrianos, *Ind.* 28,8).
81. Theoph., *Hist. Plant.* IV, 4, 9. See Karttunen, *op. cit.*, p. 145.
82. Onesik., F.22 (Strabon, XV, 1, 22).
83. Theoph, *Hist. Pl. IV,* 4,9 *In Hist. Pl.* IV, 4,10 he mentioned a kind of lentil, which was identified by Hort (Leob edn.) as *Phaseolus mungo.*
84. Nearchos, F. 20 (Strabon, XV, 1, 25). Also see Arrianos, *Anab.*, 6, 1, 2. It was described in connection with Egypt by Herodotos (II, 92) and Theophrastos (*Hist. Pl.* IV, 8).
85. Hugo Bretzl, 1903, *Botanische Forschungen des Alexanderzuges,* Leipzig, p. 203.
86. Nearch., F. 19 (Strabon, XV, 1, 20).
87. F.45 (25).
88. Theoph., *Hist. Pl.* VIII, 5. 1, In Suśruta, the white-seeded variety occupies the second place after the black seeded, but its medical significance was appreciated.
89. Aristob., F. 17a (Strabon, XV, 1, 62).
90. Aristob., F.37 (Strabon, XV, 1, 21).
91. Val. Ball, 1985, 'On the Identification of the Animals and Plants of India which were known to Early Greek Authors', *Indian Antiquary,* 14, p. 340.
92. L. Pearson (*op. cit.*, 174f) combines this with Theophrastos crooked fruits and identifies both as bananas.
93. Theoph., *Hist.Pl.*, IV, 4, 5.
94. Arthur Hort in Theophrastus *Enguiry into Plants* (Leob edn.) p. 484.
95. Nearch., F.1 (Arr., *Ind* , 26. 6; 27, 2; 28, 1).
96. *Ibid.* (Arr., *Ind.*, 29, 1).
97. Theoph., *Hist. Plant.*, IV, 4, 13. According to Strabon (XV, 2, 7). They are dangerous for beasts of burden.
98. K. Karttunen, *op. cit.*, p. 137.
99. Nearch., F.11 (Arr. *Ind.*, XVI, 1-2); Onesik., F.22 (Strabon, XV, 1, 21); Nearch., F.19 (Strabon, XV, 1, 20).
100. Onesik., F.22 (Strabon, XV, 1, 21).
101. Theoph., *Hist. Pl.*, IV, 4, 6.
102. Onesik., F.22 (Strabon, XV, 1, 22).
103. Theoph. *Hist. Pl.*, VII, 2, Also see IV, 4, 14.
104. Theoph., *De Odor.* 33; *Hist. Pl.*, IX, 7.
105. Onesik., F.22 (Strabon, XV, 1, 22).
106. Heord. III, 111.
107. Theoph., *Hist. Pl.*, IX, 4, 2; IX, 5, 1-3 and IX, 7, 2.
108. Theoph., *Hist. Pl.*, IX, 20.
109. Theoph., *Hist. Pl.*, IX, 7, 2. He mentioned together two different species, amomum and cardamonum, which according to some came from India and from Media according to others.
110. Aristob., F. 48a (Arr. *Anab.* VI, 22, 4); Theoph., *Hist., Pl.*, IX, 1, 2; IV, 4, 12; K. Karttunen, *op. cit.* p. 153.
111. Theop., *Hist. Pl.*, IX, 7, 3. He mentioned it in a list of aromatics. See G. Watt, *A Dictionary of the Economic Products of India,* I-VI: 4 (1889–93, and Index Vol. By Edger

Thurston, 1896; all repr. Delhi, 1972) s.v. *Saussurea lappa.* According to Watt, the plant grows in Kashmir and on the neighbouring mountains. Theophrastos came to know about the plant probably from Alexander's companions who had noticed it in India.

112. According to Aristoboulos (F.23= Arrianos, *Anabasis* III, 28, 6f. Also in Strabon XV, 2, 10), silphium and a kind of terebinth were the only plants commonly growing in the part of the Hindukush crossed by Alexander.
113. *Ibid.* and Theophras., *Hist. Pl.* IV, 4, 7. It is the tree *Pistacia terebinthus.*
114. Herod. III, 98.
115. Ktes. F 1b (17,5); F.45 (14) and F.45c.
116. Theoph., *Hist. Pl.* IV, 11, 13.
117. Theoph., *Hist. Pl.* IX, 7, 2, and *De Odoribus,* 33. Hort in Leob edn. of Theophrastus.
118. Herod., VII, 65.
119. Nearchos, F.11 (Arrianos, *Ind.* XVI); Plutarch, *Sayings of Kings,* 181, B.
120. Arrianos, *Anab. IV,* 26,4 and 6, 10f.
121. Skylax, Jac. No. F.4 (Athen., II, 82).
122. Aristob., F.19 (Strabon, XI, 72).
123. Theoph., *Hist. Pl.* IX, 1, 2; *Hist. Pl.* IV, 4, 12.
124. Arthur Hort's note in his Theophrastus (*Hist. Plant*) IX, 1, 2 (Leob edn.) text.
125. Theophrastus, *Hist. Pl.* IV, 4, 6. Ebony and its characteristics are further briefly mentioned in the *Hist. Pl.* I. 5, 4f; I, 6, 1; V3, 1f; V, 4, 2 and IX 20, 4.
126. Hugo Bretzl., *op. cit.,* p. 206.
127. Onesik., F.22 (Strabon, XV, 1, 21). Also see Aristob., F.36 (Strabon, XV, 1, 21); Nearchos, F.6 (Arrianos' *Indika* XI, 7).
128. Theoph., *Hist. Pl.,* 1, 7, 3 and IV, 4, 4. Also briefly referred to in the *Cause Pl.* II, 10, 2.
129. K. Karttunen, *op. cit.,* p. 129.
130. Meg. Jacoby, No. 715, F.4 (Diod. Sic., II, 40.4); F.19a (Arr., *Ind* XI, 9), F. 19b (Strabo, XV, 1, 40).
131. According to *Abhidharmakosavyākhyā,* kings while destroying the soldiers, was respecting the field labourer, who was common help of both the armies. See *Camb. Hist of India,* I, p.33; *Indian Historical Quarterly,* Vol. I, p. 369.
132. Manu, VII, 91–93; Gautama, X, 18; Baudhāyana, I, 18, 11; Ãpastamba, II, 10, 11; Yājñ., 1, 325.
133. Manu VII 90; Baudhāyana, 1, 18, 10; Yājña., 1, 323.
134. The Books 12, 13 and 14 of the *Arthaśāstra* are mainly devoted to the trickeries, stratagems, etc. All sorts of foul games for defeating the enemy were justified.
135. *Arthaśāstra,* XIII, 4, 6–7.
136. A.L. Basham, 1981, *The Wonder That Was India,* Delhi, pp. 126–28.
137. Manu (IX, 44) at one place says, "Land belongs to him who clears off the timber. In a *Jātaka* story a king tells his mistress that he cannot give her his kingdom, for he is not its onwer. When a legendary king, Viśvakarman Bhauvana, gave land to the priests, the goddess of earth rose up in person and rebuked him, saying that he had no right to give her away. A medieval commentator, probably basing his

statement on this old story, says that kings cannot give away land, because its is owned in common (Śabarsvāmin to Pūrvamīmāmsā Sūtras, VI 7, 3, quoted by Basham, *op. cit.,* p. 111). On the other hand, in the verse cited by Bhattasvamin in his commentary on *Arthaśāstra* (II, 24), it was declared that the king was lord of land and water, but that other things were the property of individual house holders. But even in this instance as viewed by N.K. Sastri (*Age of Nandas and Mauryas,* Delhi, 1967, p. 178) ''Pati' may indicate no more than the right of eminent domain, as is clearly the case with the expression 'bhusvāmi' in a closely argued passage of Katyāyana which has not been always correctly understood in spite of the excellent gloss attached to it.' The eminent domain of king over the land was also reported by Manu, who declared king as the ultimate lord (*adhipati*) of land (Manu, VIII, 39).

138. *Arthaśāstra,* I, 15. For discussion on the subject see U.N. Ghoshal, 1944, *Beginning of Indian Historiography and other Essays,* Calcutta, pp. 58–66.
139. N.K. Śāstri, *op. cit.,* 177-78. For Hellenistic view, see Avi-Yonah, *Hellenism and the East* (Monograph publishing on demand sponsor series, Ann Arbor; University Microfilms International, 1978), 70ff.
140. *Arth.,* V, 2,2.
141. *Arth.,* II, 15, 3; Manu VII, 131 and 308. In VII, 130 Manu says that eighth, sixth or twelfth part of the produce may be taken.
142. Manu, X, 118.
143. Meg., F.8 (Strabon, XV, 1,20). Basham (*op. cit.,* p. 196) comments that 'the Greek travellers were very impressed by the fertility of Indian soil and the energy and ability of its cultivators. The modern travellers impressions are diametrically opposite, but the Greeks judged Indian agriculture by standards lower than ours and the soil was less exhausted than it is now.'
144. Meg. F.4 (Diod. Sic. II, 36,4).
145. Meg. F.2 (Athen., IV, 39P. 1530).
146. Meg. F.8 (Strabon, XV, 1, 20).
147. Meg. F.21a (Strabon, XV, 1, 37). Accepted as sugar by V. Ball, *op. cit.,* p. 309 and E. Bevan, 1922, *Camb. Hist. of India,* Vol. 1., p. 363.
148. Eratosthenes, quoted by Strabon (XV, 1, 20).
149. Meg., F. 21a (Strabon, XV, 1, 37).
150. H. Berger, *De Geographischen Fragmenta Des Eratosthenes* (Leipzig, 1880), Frag. IIIB. 48, p. 291 (Strabon, XV, 4,4).
151. Meg. F.4 (Diod. Sic., II, 35, 3); F.8 (Strabon, XV, 1, 20).
152. Meg. F.4 (Diod. Sic., II, 36, 4).
153. *Arthaśāstra,* IV, 3, 17–20.
154. *Perip.* (14&32); Diod. Sic., II, 36, 3 & 4. Diodoros mentioned it as one of the winter crops.
155. Plinius' (*N.H., XVIII,* 13, 71) information about cultivated and wild barley in India seems to have been derived from Theophrastos.
156. Diod., II, 36, 3f. He twice lists 'bosporos', rice and millet as Indian summer crops (in the second passage also sesame).

157. *Ibid.;* Dioscurides (II, 95) included rice among medicines; Plin., *N.H.,* 13, 71; Horace, *Sat.2,* 3, 155; *Perip.* (14 & 13) writes its export from Barygaza; Ptolemaios, (*Geog.* VI, 4, 1) included rice among the products of Taprobane (Srī Lankā).
158. Diod., II, 36; Plin., *N.H.,* XVII, 10, 55 (black Indian millet).
159. Diod., II, 36,4; Plin., *N.H.,* XVIII, 22, 96; *Perip.,* 41.
160. Diod., II, 36.3; Strabon, XV, 1, 13.
161. Diod., II, 36; Plinius, (*N.H.,* XII, 12, 24) relies on Theophrastus for his account on fruits.
162. *Plin., N.H.,* VI, 26, 105; XII, 14, 26f.; XII, 15, 30f.; Athen., II, 66ff.; *Perip.* 49 and 56; Dioscurides, II, 159.cf. K. Karttunen, *op. cit.,* 149-151.
163. Strabon, XV, 1, 22; XVI, 4, 25; *Perip.,* 10; Plin., XII, 42, 85; XII, 44, 98.
164. Plin., *N.H.,* XII, 26, 42-46; XII, 20–35; XVI, 59, 135; Diosc., 1, 7, further 1, 99, 3; Horace, *Carm.,* IV, 12 & 2, 11, Ep. 13 cf. *K. Kartunen, op. cit.,* 161-162.
165. Plinius (*N.H.,* XII, 13, 25) bases his account on Theophrastus (*Hist. pl.,* IV, 4, 8,). Also see Plin. *N.H.* XII, 8, 17; XII, 13, 25; XII, 22, 36; XIII, 28, 90; XIX, 2, 15.
166. Curtius, IX, 8, 1; Plin. (*N.H.,* XII, 13, 25) mentioned *Vestes lineae* made of cotton.
167. Curtius VIII, 9, 15; Strabon in XV, 1, 13, quoting Erastosthenes writes that flax was one of those crops cultivated in the rainy season.
168. Vergil, *Georg.* 2; Plin., *N.H.,* XII, 8, 17–20 and XII, 10, 21; Diosc., 1, 98; *Perip.,* 36.
169. Diod., XVII, 90, 5; Curtius, IX, 1, 9f.; Plin., (*N.H.,* VII, 2, 21) dervies partly from Theophrastus (XII, II, 22f.)
170. Diosc. I, 5; 1, 18; Plin., *N.H.,* XII, 28; XXI, 70; XVI, 65, 159f (reed arrows); *Perip.* 24; Strabon, XVI, 4, 9; XVII, 3, 5.
171. Diod., II, 36, 5; Aelianos, *Nat. An.,* XV, 7; Seneca, *Ep.* 84.4; Diosc., II, 82, 5; Plin., *N.H.* XII, 17, 32; Aelian, *V.H.,* III, 39; *Perip.* 14.
172. Diog. Perieg., L. 1123.
173. *Perip.,* 41.
174. Philost., *Apoll. Tya.* III, 5.
175. Plut., *Moralia,* 957B
176. For various notices see E. Warmington, 1974, *Commerce between the Roman Empire and India,* London, revised and enlarged edn., 217ff.
177. Diosc., 11, 82.
178. *Perip.* 14.
179. Lucian, *Musc. Encom.* I, Cp, *Dialog. Merter.* V, 4 cf. Nearchos, Jac. No. 133, F.11 (*Arr. Ind.* XVI, 1).
180. *Perip.* 41. Also see *Perip* 6, 14, 31, 49, 51, 61, 63.
181. Philost., *Apoll. Tya.,* II, 20.
182. Plin., *N.H.,* XII, 13, 25.
183. Plin., *N.H.,* XII, 25, 41; Diosc., 1, 16; *Perip.,* 39. It is already mentioned in the *Atharvaveda,* VI, 102, 3.
184. Plin., *N.H.,* XII, 59, 50; Diosc., 1, 6.
185. Plin., *N.H.,* XII, 28, 48f; XVI, 59, 135.
186. Diosc., V, 92.

187. Vitruvius, VII, 9, 6 and VII, 14, 2.
188. Plin., XXXIII, 57, 163; *Perip.* 39.
189. Plin., XXXV, 12, 30.
190. Plin., *N.H.,* XII, 59, 129; XIII, 14 & 18; XIV, 108; XXIII, 48, 93.
191. Diosc., 1, 12, and 1, 63.
192. Ptolem., *Geog.,* VII, 2, 15.
193. *Perip.,* 56, 65.
194. *Perip.,* 39 and 49.
195. Plin., *N.H.,* XII, 15, 31.
196. See Plin., *N.H.,* XII, 16, 32; Diosc., 1, 82; K. Karttunen, *op. cit.,* 156–157.
197. Watt, *op. cit.,* s.v. *Holarrhena antidysenterica.*
198. Plin., *N.H.,* XVI, 59, 135.
199. Strabon, XV, 1, 22 and XVI, 4, 25.
200. Diosc., III, 22.
201. Diosc., I, 68.
202. Plin., *N.H.,* XII, 36.72.

CHAPTER 28

Agriculture in the Bactrian Economy

Abhay Kumar Singh

Be fruitful as a rice field, yea, be rich
In all good works! For that is the best field
Which yieldeth to the sower the goodliest crop.
Milindapāñho vii,7,9

INTRODUCTION

Riches, like possession of treasures and gold, have been synonymous with the prosperity of any country. Nevertheless, the role of agriculture in enriching the economy was understood and appreciated in the olden days, particularly for the wise Indians like Nāgasena, who believed that it was an essential feature for a mighty and prosperous city to possess filled up stores of food.[1]

The glitter of the Bactrian gold and silver often dazzled the eyes, preventing men from appreciating the agricultural achievements. Herodotus (III.94) wrote about 360 talents of gold dust that was sent as tribute to the Persian Emperor by the Indian satrapy. About 4680 talents of ants' gold was paid *per annum*. Darius had mentioned about the annual receipt of 300 talents of silver dust from Bactria and another 170 talents from Arachosia, Sattagydia and Gandhara as revenue.

However, some discreet Greek writers in the past also highlighted the agricultural prosperity of Bactria, that they believed to lie at the foundation of the Bactrian economy. Apollodorus of Artemita who had called the land *της συμπασης Αριανης προσχημα*—'the Jewel of all Ariana' had also noted that, 'it was on account of *the fertility of this country* that the Greeks who caused Bactria to revolt grew so powerful that they became masters not only of Ariana but also of India'[2] (Italics ours). The reference was for the Graeco-Bactrians who revolted under Diodotus and could set up an independent kingdom of Bactria on the strength of their self-sufficiency in food resources. Quintus Curtius has all praise for the fertility of Bactrian land since it could support a large population of both men and horses, including a cavalry of 30,000. Strabo commended Bactria as a first rate agricultural area which produced all kinds of crops except oil. There were orchards, vineyards and good grazing grounds too. The profile of the agriculture in Bactria is amply clear by these comments.

Obviously, the efforts in the agricultural development must have dated back to many centuries earlier. The growth in the annual yield of the crops sown must have been aimed at mustering up food stores for domestic and exchange purposes. The State's initiative also stepped into this endeavour because the land revenue accrual formed an important source to the economy. King Darius I introduced (*c*. 518 BC) the system where the land revenue was determined on the basis of the area of cultivated land and its fertility, as calculated through the mean annual yield.[3] The arrangements for irrigation suggest interest from the State in facilitating farming. The fact remains that agriculture in Bactria received at least the State's attention, even if it did not get patronage or encouragement.

BEGINNINGS OF AGRICULTURE IN AFGHANISTAN

Afghanistan possesses a variety of natural habitats though situated quite close together. Scholars like Humlum, Dupree and Bowlby have suggested several divisions of the country on climatic and physiographic criteria.[4] However, only the foothills on both sides of the Hindukush and the river valleys of Amu-darya and Helmand—Seistan could offer favourable environment for dry or irrigated farming. Bactria comprised the Amu-darya valley and slopes north of the Hindukush range.

In northern Bactria, the archaeological profile of the evolution of a farming culture can be observed from the middle or end of the third millennium BC. This culture took shape under the direct influence from Margiana, i.e. the ancient peasants of southern Turkmenistan, who moved into Bactria, bringing with them custom and practise connected with agriculture, stock-rearing, house-building, pottery making and metallurgy craft.

Kuzmina[5] has shown that farming started at the right bank of Amu-darya in the area of the town Kerki, where pottery of Namazga IV type was also found. Further development of farming culture could be traced through the finds at Sapal Tepe (west of Termez), where cereal plants and animal bones were discovered. The accompanying pottery finds were quite akin to that at Namazga VI. The Kuchuk Tepe shows that the farming culture developed throughout Namazga VI. An important fact that emerged from Kuzima's research is that Namazga VI pottery provided sure grounds for acknowledging the local Central Asiatic genesis of the complex and its diffusion was not connected with the Achaemenids, nor was it related to Persia.

In Bactria, Namazga VI traditions developed by local farming population and by the end of the first millennium BC, led to the formation of cultural complex of Kobdian–I that was feature of culture of north Bactria in Achaemenid Afghanistan.

DEVELOPMENT OF AGRICULTURE: CROPS, IMPLEMENTS AND FOLKLORE

The presence of settled agricultural communities in Bactria in 125 BC is attested in the eyewitness accounts of a Chinese envoy, Chang Kien. The *Annals of the Former Han Dynasty* or the *Chien Han Shu* of Pan Ku (d.92 AD) which relates to the conditions of Ki-Pin (Chi-Pin = Swat?) during 206 BC – 24 AD, mentioned that the people "cultivated the five grains (hemp, millet, rice, wheat and pulse) grapes and various fruits... In the low damp ground, they grow rice. In winter they eat raw vegetables..."[6]. The evidence highlights the knowledge about the variety of staple cereals that were cultivated; on the other hand, it suggests the acquaintance to different seasons and soils for growing rice and vegetables. Quite later in the seventh century AD, Huen Tsang during his visit in the lands, noticed the scented root of Yu-Kin (Skt. *Haridra* or *Curcuma Zedoaria*), saffron, cereals and a certain kind of 'upland rice' of Lan Po (= Laghman).[7]

On the side of the ancient Indian literature, the *Milindapãñho,* while describing a storehouse of the rich, mentioned that it comprised all kinds of wheat, rice, paddy, barley, dry grain, oil seeds, beans, peas and edible seeds in it.[8] The *Milindapãñho* calls the rice as 'chief of all grains'.[9] It points to many varieties of rice and contrasted the healthy *karumbhak* to the rusty red rice.[10] Interestingly, rice had a special significance in the Bactria, which is supposed to belong to the wheat-consuming region.

The advanced state of agriculture in Bactria, in regard to the techniques of cultivation, tools and implements and the cropping-cycle, can also be understood from the literary sources. Traditions inculcated by the Bactrians had them believe about their ancient connections with the practise of farming and their divine training in it. They were settled in the land by the Greek god Dionysus, who had initiated them to the arts of viticulture and farming with the oxen.[11] The use of plough in agriculture was another important characteristic of Bactrian farming. The reference to the 'task of ploughing in farming', oft-quoted[12] by Nāgasena during conversation with King Menander, depicts familiarity with the implement. Cultivation by ploughing and the harvesting of crops of barley and rice were specifically cited.[13]

The instrument also had a divine sanctity, as we shall note below. In contrast, among the Harappans, the hoe was the dominant tool, while the plough was rare.[14] But in Bactrian agriculture, the technique of cultivation by plough was well recognized. The plough was an attribute of Saṁkarṣana, the god whose connections with agriculture, serpentine antecedents, wine-drinking habit, and liking for wrestling and gourmet tastes—all match well with the Bactrian lifestyle. It is worthy of note that Saṃkarṣana-Balarāma, with his 'plough', is depicted on the *drachms* of Agathocles, an Indo-Greek king.[15]

Settled agricultural life raises peculiar deities and their myths. In Bactria, the estimation of the pastoralist and agricultural deities was enhanced. Dionysus remained a venerated god for long, as the coinage of Bactrian/Indo Greeks reveals.[16] Goddess Ardoxsho, with the 'cornucopia',[17] was also a relevant choice for the coin device, since that mythological 'horn of plenty' was the source that poured agricultural bounty and riches. The 'ears of grain' were also depicted as art-motifs in the Northwest.

IRRIGATION SYSTEMS: NURTURING RIVERS AND MARVELLOUS CANALS

The development of agriculture in Bactria owed much to the benevolence of its rivers. Recently, Paul Bernard, the architect of archaeology in Afghanistan (while referring to the Seleucids in Central Asia), summed up, 'Their wealth came from their oases, especially those in the valleys of the Oxus and Polytimetus (modern Amu-darya and

Zerafshan), which enjoyed an *agricultural surplus from the expansion of irrigated land,* and prosperity from the metals and precious gems found in the mountains of the region... '[18]

The value of the rivers was a well-recognized fact, even in the past. Originally, Anahita, the goddess of the land, was the personification of River Oxus, whose thousand arms represented a thousand canals that derive from the river. The River Zarafshan was spoken of as the Polytimetus as 'the most precious', the obvious reference being 'not to gold washing, but to its value for irrigation'.[19]

It may be noted here that agriculture in the Indus Valley civilization largely depended on the technique of flood irrigation. The mode of canal irrigation drawn from the river was not practised. The Bactrian irrigation and water management provided a contrast. In the words of Tarn, the archaeology in Bactria has shown what was 'a rich complex of water courses and husbandry that (*out of*) its river...most of the water was drawn off before it could reach the desert...'[20] The technique of irrigating the paddy fields through canals, by bounding the field by embankments to keep the water inside, and to bring the rice crop to maturity, also finds reference in the *Milindapãñho.*[21]

Archaeology of the ancient canal irrigation systems in Bactria caused some confusion for a while. Irrigation systems had begun in the pre-historic Kangui complex. But the most astounding discovery in 1975 pertained to a settlement situated south of Amu darya at Shortughai, which belonged to the last quarter of the third millennium BC at the earliest occupational level. The major characteristics, e.g. mud brick architecture, wheel-made pottery, bead making, and metallurgy connected it to the Indus Valley civilization, and it was declared an outpost of the same. The agriculture was extensively developed in the surrounding plains. A unique feature was an irrigational canal drawn from River Kokcha. This fact many scholars to believe that the irrigation construction was an imported system from Harappa. Gardin noted, '...the earlier irrigation system, in the Shortughai Plain (ca 2200 BC) was so sophisticated that we tended to regard it as the product of foreign craftsmanship brought full blown from the Indus Valley by Harappan settlers'.[22]

However, in 1978–79, Gardin and Lyonnet made another discovery of great significance. They found Chalcolithic sherds confined exclusively in the Taluqan region. The recent discoveries helped the scholars to opine that the existent settlements of the Chalcolitic culture had progressively expanded to reach the nearby Shortughai Plain, even before any Harappan colonization had done so. In consequence, the agricultural development in the Bactrian plain was indigenous and prior to the Harappan influx. As a result, the Shortughai irrigation system 'could be regarded as skilled version of the Taluqan system, developed a few centuries earlier some fifty miles south, across the Ambar Kuh.'[23]

The Shortughai system has been distinguished from the Taluqan system on certain characteristics. The Taluqan system consisted of 'a number of streams derived from the main course of the river. Several of them have cut deep ravines in the loess-covered plain.' In fact, this was long before the Chalcolithic period had become the 'natural arms' of the River Taluqan. These could be used as sources of artificial channels. In the other system—that of the Shortughai Plain—the land slope prevented the irrigation from Oxus, while the Kokcha flows 30 metres lower than the level of the terrace of the Plain. To bring the water to the top, a canal had to be drawn from a distant point upstream of River Kokcha, where the river was still at a higher level of ground. The task required a high degree of expertise, in which the ancient people had succeeded. Interestingly, the Harappan sherds have been found at the earliest levels of the number of superimposed ditches of the Bronze Age canal.

The irrigational systems in Bactria continued to develop further. After the decay of Shortughai, the focus shifted to the south of Qunduz region. The citadel of Bala Hissar was built and additional irrigational systems came up on Qunduz–Khanabad Plain. Again, south of Qunduz, the Shakh Tepe project of the ancient days drew water from a very distant river. In the pre-Achaemenid/Acahemenid period, the Rud-i-Shahrawan canal was dug up. During Hellenistic times, the Rud-i-Shahrawan was progressively extended from the Taluqan Plain to Amu-darya—a distance of about 50 kilometres—across the Ambar Kuh plateau, in order to provide irrigation to the Plain of Archi. Around the third century BC, a secondary foothill system was also prepared.[24] Polybius (x.28.3) testified that the Achaemenids had always paid great attention to the maintenance and development of irrigation and the system of underground channels, known to him as *υπονομοι* or *karezes,* illustrate the fact. Agriculture based on efficient irrigational arrangements had supported the Bactrian economy for a long period.

PHASE OF AGRICULTURAL DECLINE

The agriculture of Bactria suffered since the turbulent times of the nomadic invasions. The destruction of the water management system was considered as a tactic to subdue the enemy in any country where water was scarce. It seems that such damages to the canals must have adversely affected the region. The advent of the Kushanas also shifted the focus from canals to roads, and from agriculture to commerce. The abandonment of the irrigation canals since the Kushana period also point to the difficulties of the upkeep and maintenance of the systems.

The burst of urbanization during the Kushana Age could be another reason. It has been noted by Allchin that 'one of the basic factors of the rise of city life in the area was the draining of the river deltas and the shrinkage of the rivers towards the sources as more and more water was taken up for agricultural ends. Thus, appeared the balance between town and country, which in extreme cases, could starve the country to feed the city, and finally bring about the abandonment of the city. This process could be accelerated when war and devastation upset the balance and irrigation systems were destroyed.'[25] It may be added that the need of water for the city sacrificed the agricultural irrigation. Further, the Kushana economy was benefiting from trade, and trade routes facilitating commerce became the major concern of the rulers. The inflow of wealth from commerce of non-agricultural items reduced the estimation of agriculture and its contribution to the economy. It may be further added that the Kushana empire possessed a large territory and the agricultural yields from the Gangetic plains of India set off the loss of production in the Bactrian region. Thus, can be explained, the later neglect of the Bactrian agricultural development.

NOTES AND REFERENCES

1. *Milindapāñho,* V,4,330, 'The Questions of King Milinda', *Sacred Books of the East,* xxxvi, (Reprint, Motilal Banarassidass, Delhi, 1982) pp. 208–211.
2. U.P. Arora 'Dichotomy in Greek and Indian Sources: Bactria *vis-à-vis* Gangetic Valley', *Yavanika,* no. 9, 1999 (2004), pp. 35–53, esp. p. 37.
3. Dandamayev, M.A., 'Media and Achaemenid Iran' Chapter in Harmatta, J., Puri, B.N., Etemadi, G.F., 2002 (ed) *History of Civilizations of Central Asia,* (Vol. II) (Motilal Banarassidass, Delhi), p. 53.
4. Sophia B. Bowlby, 'The Geographical Background', Chapter in Allchin, F.R. and Hammond, Norman (ed). 1978, *The Archaeology of Afghanistan from earliest times to the Timurid period,* (Acamedic Press, London) p. 18ff.; Figures 1.6, and 1.7. Bowlby mentions the regions as: (i) the high mountains of Badakshan, Nuristan and Central mountains which were rich in mineral deposits and the mountainous forest zones provided food plants and hunting; (ii) the mountains and foothills on both sides of the Hindukush, offering favourable environment for dry or irrigated farming and abundant pasturage; (iii) the dry plains, e.g. Turkestan Plain and lowlands like the Herat-Farah lowland; (iv) the river valleys of Amudarya and Helmand–Seistan; and (v) the desert-like Dasht-i-Margo.
5. Kuz'mina, E.E., 'The "Bactrian Mirage" and the Archaeological Reality on the Problem of the Formation of North Bactrian Culture' in *East and West,* New Series, Vol. 26, Nos. 1–2 (March–June 1976), pp. 111–131.
6. Goswami, Jaya, 1979, *Cultural History of Ancient India,* (Agam Kala Prakashan, Delhi), p. 27.
7. Watters, T., 1973, *On Yuan Chwang's Travels in India,* Vol. I, (Reprint, Munshiram Manoharlal, Delhi) pp. 181-82. For a detailed account see, Sen Gupta, Nilima, 1984, *Cultural History of Kapisa and Gandhara,* (Sundeep Prakashan, Delhi), p. 106.
8. IV.1.27.
9. IV.3,32.
10. IV.6,37.
11. Dionysus settled people at Nysa. 'Indra-Dionysos had first brought knowledge of iron and metals, use of oxen for agriculture and the art of house building to India', notes Kosambi, D.D., 1992, *The Culture and Civilization of Ancient India in Historical Outline* (Reprint, Vikas Publishing House, Delhi), p. 117.
12. *Mil.* III,4,3; IV,3,7.
13. For the harvesting of barley (*Mil.* II,1,8; IV,2,16) and rice (*Mil.* VII,7,8).
14. Excavations at Banawali has brought to light an agricultural field which has visible plough marks in it.
15. For a discussion on the plough of Samkarshana, depicted on the Indo Greek coin of Agathocles see Guillaume, O., (edited and compiled) 1991, *Graeco-Bactrian and Indian Coins from Afghanistan,* (Oxford University Press, Delhi); Also see, Bopearachchi, O., 1990, *Monnaies-graeco bactriennes et indogrecques. Catalogue raisonne.* Bibliotheque Nationale, Paris.

16. Dionysus attributes have been represented in the coins of Indo-Greek kings Pantaleon and Agathocles.
17. Cornucopia was the horn held by Demeter and is shown on the coins of Philoxenus.
18. Dani, A.H. and Bernard, Paul, 'Alexander and his successors in Central Asia' in Harmatta, Puri and Etemadi, *op. cit.*, p. 91.
19. Tarn, W.W., 1980, *The Greeks in Bactria and India,* (Indian Reprint, Munshiram Manoharlal, Delhi) p. 102 (re. Zerafshan/Polytimetus) and p. 104 (re. Anahita/ Oxus).
20. *Ibid.*, p. 102.
21. In the *Milindapāñho,* information on irrigating of rice fields through canals (VII,7,7) and bounding the field by embankments to keep the water in to bring the rice crop to maturity (VII,7,8), finds reference.
22. Gardin, J.C., 'The Development of Eastern Bactria in Pre-Classical Times' in *Purātattva,* No. 10, 1978–79, p. 12.
23. Gardin, J.C., *op. cit.*, pp. 8–13.
24. *Ibid.*, For an update see also, Gardin, J.C. 'Canal Irrigation in Bronze Age Eastern Bactria' in B.B. Lal & S.P. Gupta (ed) 1984, *Frontiers of the Indus Civilization, Sir M. Wheeler Commemorative Volume,* (New Delhi, 1984), pp. 311–320.
25. Allchin, 'The Culture Sequence of Bactria', *Antiquity,* xxxi, 1957, p. 134, n. 5.

CHAPTER 29

Land and Land System and Some Allied Issues in the *Arthaśāstra*[1]

Sibesh Bhattacharya

Kauṭilya's *Arthaśāstra* is one of the richest sources of data relating to agriculture and agrarian life in early India. Although *Arthaśāstra* is not a work on agriculture, nevertheless, it is centrally concerned with land (*bhūmi*). According to Kauṭilya, the political power flows from 'land inhabited by men' (*manuṣyavatī bhūmi*).[2] In the conceptual universe of Kauṭilya, it is the man-land relationship or man in interaction with land that produces the political domain. A specifically agricultural text like *Kṛṣi Parāśāra* perhaps gives us more material on agricultural operation. But the agricultural data in the *Arthaśāstra*, in certain ways, are more interesting than those in the *Kṛṣi Parāśāra* because of the clearer political and economic contexts of the former.

Before we move on to the issues specifically dealing with the land system, we may take note of certain general background features of the agrarian setup.

In the Kauṭilya's scheme of administration and state finance, 'country' (*janapada, rāṣṭra*) and 'city' (*durga, pura*) seem to have been the two important and primary territorial units. The 'country' included areas of habitation as well as areas yielding various kinds of produce. The general pattern of landscape in the rural area, i.e. the territorial division technically called *janapada* or *rāṣṭra*, seemed to have remained practically the same from the Vedic period to the late medieval. On the basis of their produce, three kinds of broad areas appear to have been categorized. These were: agricultural land, mines and forests. Forests were again classified as material forest and elephant forests.[3] Some land was also set apart as pastures. The habitations also included public places like temples and parks. Then there were some game sanctuaries.[4]

Another feature of the socio-economic landscape that seems to have persisted over the entire span of Indian history since the Neolithic period is the simultaneous existence of hunting-gathering and agrarian economies and a running conflict and tension between the two. While some adopted the new mode of production—agriculture—and set on the path towards the growth of agriculture-based state, a few clung on to their old ways of hunting and gathering. In one of the accounts of the origin of state, it is said that after the enraged *ṛṣis* killed the power-drunk Veṇa, they proceeded to create a new king out of a sacrifice. The first to emerge from the sacrificial fire was the dark-skinned Niṣāda. He, however, was banished to the forest to become a hunter. Thereafter, the handsome and brilliant Pṛthu was created to become the first human ruler. And Pṛithu, for the first time, introduced agriculture and farming.[5]

This parting of ways symbolizes the beginning of a continuous competition between the two modes of production. The presence of hunting-gathering people on the fringes of the settled agrarian communities and a continuous tension between the two is a recurring motif in Indian history. Niṣāda, Śabara, Kirāṭa, Bhilla and numerous other such people, often lumped together with some other disparate groups under various terms such as *āṭavīka, mleccha, pañcama, vāhya, antya,* etc., figure continuously in Indian literature. Indian epics seem to suggest that the expansion of agrarian and market economy took place at the expense and in the teeth of opposition of hunting-gathering people. The Rākṣasas, especially in the *Rāmāyaṇa,* have been interpreted as symbolizing the receding hunting-gathering people fighting for the preservation of their way of life.[6] This confrontation was not perceived as just a competition between two modes of production and economy; it also assumed the proportions of a cultural and ethical contest.[7] We are not suggesting that the normal relation between the agrarian and hunting-gathering societies was one of aggressive confrontation. As a matter of fact, the normal relation was one of co-existence and non-interference. The mutual suspicion remained, but conscious efforts were made to keep the fallout of suspicion within control and not to allow it to develop into violent struggles. The political thinkers generally preferred tolerance of and non-interference in 'tribal' life and customs.[8] At the same time, however, generally, states appear to have pursued a vigorous policy of agrarian expansion by bringing virgin areas under plough. This process of the gradual extension of the domain of agriculture is reflected in Kauṭilya's *Arthaśāṣtra,* the *Manusmṛiti, Milindapañha,* etc.[9]

The city (*durga, pura*), expectedly, did not have much to do directly with agricultural operation, barring small-scale cultivation of corn, vegetable, fruits, etc., by some *kuṭumbikas.* The evidence suggests that these operations were not more than occasional maintenance of vegetable and fruit gardens by some city dwellers, as is the scene even now.[10] It may be mentioned, as an aside, that although there existed some kind of a rural urban divide from the cultural, economic and administrative viewpoints, the use of terms like 'urban revolution' in the context of pre-Industrial Revolution ages can be somewhat of an exaggeration.[11] From the fact that a variety of agricultural goods were brought to the city from which custom duties were collected, it is clear that the city populace formed an important group of customers for the agricultural produce of the neighbouring villages.[12] And then, from the reference to corn traders living in cities, it may be surmized that cities acted as trading marts for food grains and that the cities controlled the commerce in agricultural produce[13]. It needs no stressing that urban areas had practically no real role in agricultural production. In this context, it is noteworthy that in the list of the revenue items, the taxes on agriculture (*sītā* and *bhāga*) are mentioned only under the head *rāṣṭra* and do not figure under the *durga.*[14]

The forests and the mines too played important economic roles. But we are here concerned only with agriculture.

In a passage referring to the whole country between the Himalayas and the ocean, different kinds of land have been distinguished.[15] The distinctions seem to have been drawn in very general terms. The expressions that have been used for different kinds of land are: *āraṇya, grāmya, pārvata, audaka, bhauma, sama* and *viṣama*. *Bhauma* has been interpreted as dry land; the meaning of other terms are clear enough. This passage gives us a general idea about the topographical variations in the country. In addition, we are also provided with some information about the average annual rainfall in certain parts of the country like Aśmaka (upper Godavari region), Avantī (Malawa), Aparānta (Konkan), etc.[16] It is interesting that the rainfall mentioned by Kauṭilya accords quite well with the modern measurements of the rainfall of the concerned areas.[17] It has been stressed that not just the total rainfall alone, but a proper distribution of it over various periods of the year was important for agriculture. 'One-third of the (annual) rainfall in the first and the last months (together), two-third in the intervening two months' were considered to be ideal distribution.[18] The months most probably referred to are the monsoon months.

Land dependent on rainwater was known as *devamātṛka*. And the land with perennial water supply, say, fed by a river, was called *adevamātṛka* and was valued more because it yielded abundant crop.[19] The rainfed land was again divided into two categories: *jāṅgala* (dry) and *anupa* (wet), on the basis of rainfall difference; the land with sixteen *droṇas* of rainfall was called *jāṅgala*, and the land with twenty-four *droṇas* of rainfall was called *anupa*.[20] *Jāṅgala*, although translated as dry, however, did not mean arid land; the arid land was called 'desert like'.[21] On the basis of suitability to crops and humidity and water content, cultivable lands were again divided into the following types—*sthala* (dry), *kedāra* (marshy), *saṇḍa* (suitable for growing vegetable), *vāṭa* (flower and fruit garden).[22] Normally, *sthala* and *kedāra* are in contrast but at some places to each other, *sthala* has been contrasted with *anupa*.[23] *Anupa*, thus, seems to have been a common term used both for land with high rainfall as well as a marshy land.[24] *Sthala* has also been contrasted with *audaka*, which was considered much better than *sthala*; a small piece of the former was considered preferable to a large tract of the latter because of the certainty of produce.[25] *Audaka*, thus, could have been the same as the *adevamātṛka*. Similarly, *sthala* and *jāṅgala* were probably synonymous. Although not as desirable as the *audaka*, if cultivated properly and with suitable crop that did not require much rain for ripening (*sasyamalpavarṣapākam*), *sthala* could also yield a large harvest.[26] As *kedāra* is also mentioned in connection with the revenue to be collected from irrigation works, it appears that the term could also be used, apart from the land blessed with high rainfall, also for land receiving abundant water through irrigation.[27] The *Audaka, anupa, kedāra* lands seem to have been particularly suitable for the cultivation of paddy and the varieties like *sāli, vrīhi* were grown in such lands.[28] Muddy land, land with stone, saline and arid land were poorly thought of.[29]

But the real quality of the land, according to Kauṭilya, is not how it is naturally endowed with. It is the human enterprise and effort that goes towards its utilization that really determines the quality. He concludes an interesting discussion on the relative values of various kinds of lands by saying that 'land is what is made of it by man'.[30]

And a major means of making a piece of land valuable was to make arrangements for irrigating it. We have already seen that a land not dependent on rainwater, *adevamātṛka*, was regarded as more valuable than a land exclusively dependent on rain or vagaries of nature. In view of its obvious importance, irrigation was a major concern of the state. The state initiated and encouraged irrigation projects. In fact, some types of agricultural land for the purpose of revenue administration were grouped under the heading 'irrigation work'.[31] That practically all varieties of cultivable lands were put under this heading demonstrates how widely irrigation was used. It was natural, therefore, that the director of agriculture (*sītādhyakṣa*), a high state official, was expected to be an expert in the art of water divination (*śulba*).[32]

Although the terms *setu/setubandha* was used in a general way for irrigation works of all kinds, a few the different methods used for irrigation can be distinguished. In the context of *setu/setubandha nadī* (river), *sara* (lake), *taḍāka* (tank), *kūpa* (well), *utsa* (spring), *ādhāra* (reservoir), etc., have been mentioned.[33] All these sources were tapped for irrigation.

It has been said that waterless regions (*anudaka*) as well as regions with water supply (*sahodaka*) should be provided with appropriate irrigation facilities. The state officials were asked to set up wells waterworks, and spring.[34] The expression *kūpasetubandhotsān* may really signify spring wells rather than three different kinds of irrigation devices.

Two more terms, *sahodakasetu* and *āhāryadokasetu*, have been mentioned.[35] They may mean river canal and storage reservoir for collecting rainwater, respectively. *Sahodakasetu*, with perennial source of water supply (river), like *adevamatṛka* land, was regarded to be more dependable than *āhāryodakasetu* dependent on.[36]

In an interesting passage relating to water tax/rate (*udakabhāga*), four different methods of moving or lifting of water for irrigation are mentioned.[37] Since the rates differed for each, it can be logically surmized that the rates were linked to the efficiency of the methods. These four methods, arranged on a scale of ascending rates and efficiency, were: (i) *hastaprāvartim* or hand-moved; (ii) *skandaprāvartim* or shoulder-moved; (iii) *udghāṭam* or boosted; and (iv) *srotayantraprāvartim* or machine-moved. Hand-moved most probably meant manually drawing or throwing water from the source into the water channel running to the field. Various such manual devices are still practised. Shoulder-moved probably meant harnessing cattle for drawing water from the well. *Udghāṭam* must have meant some kind of boosting method similar to what later came to be stereotyped as the 'Persian Wheel'. *Srotayantra* may signify some kind of water lifting/moving mechanism driven by river currents.[38]

There were state-run irrigation works as well as non-government irrigation works. The non-government ones were either owned by private individuals or run by the community.[39] In case of newly settled villages, the government extended help in setting up irrigation works.[40] Water cess was collected from every user of irrigation facility—even one using one's own irrigation device was expected to pay.[41] However, private owners could offer the use of their facility to cultivators against payment of a part of produce.[42] Any case of non-maintenance or keeping a irrigation facility out of use could lead to the lapse of ownership right of private proprietors. How far these rules were actually enforced cannot be determined. But they do indicate the high stake that the state felt in the maintenance of irrigation facilities.

Like irrigation, in the land system too there seems to have been two primary categories: state land and private land. 'It is quite clear that all unoccupied land is supposed to belong to the king, i.e. to the state', writes Kangle.[43] However, it is not unlikely that a distinction was drawn between the state land and king's personal estate or crown land. The term *svabhūmi* used in the sutra 2.24.2 appears to refer to king's personal estate. Thus, the director of trade (*paṇyādhyakṣa*) was required to keep separate the produce of king's personal land (*svabhūmija*) from the produce received from other places (*parabhūmija*). He was required to make different arrangements for their sale.[44] Royal goods (*rājpaṇya*) from king's estate were to be sold at 'one centre' (*ekamukhaṃ*) and the goods from other places at many centres (*anekamukhaṃ*). In the light of this significant bit of information, it appears difficult to agree with the interpretation that *svabhūmija* means indigenous goods and *parabhūmija* foreign goods because it would imply that state trading involved more sale of imported foreign goods than goods produced indigenously. It must be remembered that the bulk of revenue was received by the state in kind and it is this scenario that necessitated their sale by state agency. It, thus, seems more likely that the expression *parabhūmija* signified goods received from lands other than those belonging to the king. In fact, we find two more terms in the context of goods *svadeśīyānāṃ* and *paradeśīyānāṃ.*[45] These two terms probably stood for indigenous and foreign goods, respectively. Moreover, foreign trade appears to have been mostly in the hands of private traders—the state, through the director of trade, exercised a supervisory and controlling role.[46] However, even if the state land was distinguished from the crown land, we do not have enough evidence to state whether any distinction was made in the manner of land utilization between the two categories.

The director of agriculture (*sītādhyakṣa*) was in charge of management and operation of state land; *dāsa*, *karmakara*, and *daṇḍapratikarti* were employed for tilling and cultivation. It is interesting that both *dāsa* and *karmakara*—and there is no doubt that *karmakara* were wage-labourers—were paid cash wages besides food.[47] The putting of *dāsa* and *karmakara* on the same footing even in regards to cash wage deserves examination. *Daṇḍapratikartri*, who seem to have been jailed convicts, received no wages other than their upkeep.[48]

It was perhaps not feasible to operate all the state land directly under the state's own management. Thus, state lands were also leased out. At least two categories of leasees are mentioned in this text: the *ardhasītika* and *svavīryopajīvin*. The former were clearly the well-known *ārdhika* or share croppers cultivating on a half-share basis.[49] The interpretation of the *svavīryopajīvin* is much more problematic.[50] Kangle is of the opinion that *svavīryopajīvins* too were a category of sharecroppers but they differed from the *ardhasītikas* in respect of certain terms and conditions. They paid three-fourth or four-fifth of the produce to the state because unlike the *ardhasītikas* who had to arrange for their own implements, equipment, seeds, etc., these were provided to the *svavīryopajīvins* by the state.[51] The exact nature of the rights and obligations of the *svavīropajīvin vis-à-vis* the state is not clear. However, from the fact that the term has a definite flavour of independence about it, it may be surmized that the *svavīropajīvins* enjoyed a greater degree of autonomy and better terms than the other categories of agricultural workers mentioned in the section. Moreover, the term reminds one of the expression *śūdraṁśca ātmopojīvin* mentioned in the *Manusmṛiti* which, according to the commentators, means independent śūdra labourers working for wages.[52] One gathers the impression from the section dealing with the functions of *śītādhyakṣha* that among the various categories of workers employed for the cultivation of state land, the *dāsa-karmakara* and *daṇḍapratikartṛi* were the least expensive. But this does not necessarily mean that they also accounted for the largest share of state-farming. In fact, it is not possible to form any clear idea about the relative quantity of state land cultivated directly under the state management and the land leased out on share-crop basis.

The extent of the direct state farming in the *Arthaśāstra* generally seems to have been often overestimated.[53] If the state had the requisite manpower and infrastructure for direct agricultural operation, then it would not presumably depend so heavily on private enterprise for bringing any virgin area under cultivation (*śūnyaniveśa*).[54] It may perhaps be surmized that the amount of state land leased out to cultivators could not have been very small. After all, the cultivation of the state land directly would require an extensive labour force of slaves (*dāsa*) and convicts (*daṇḍapratikarti*). Neither the spirit of our text (the *Arthaśāstra*) nor the contemporary accounts found in the writings of Megasthenes or the Mauryan inscriptions betray indications that slavery was extensive and crime was rampant. In fact, the evidence betray signs to the opposite direction.[55]

The *Arthaśāstra*, like some other contemporaneous evidence, reflects the extension of agriculture to virgin areas. The clearing of forest and the converting of the erstwhile forest land into agricultural land was an ongoing process right from the Mesolithic-Neolithic periods. The rise of organized states gave the process an added impetus. References in the *Manusmṛti* and *Milindapañha* reflect this process. The most extensive documentation of the process, however, can be found in the *Arthaśāstra*.[56] In a planned and systematic way at the initiative and under the supervision of the state, a policy of establishing new settlements was pursued. It was a better, organized and throught-out, scheme than random nibbling at the forest cover to create more agricultural land. The reclamation of virgin territory from under the forest cover, and presumably under the control of forest people (the *āṭavīkas*), was executed by rearing a carefully drawn support structure. It seems that a series of new settlements consisting of regular villages with an orderly administrative setup and arrangement for defence were raised in outlying regions abutting on forest areas.[57] These villages were far from sparsely populated huddles of a few huts in the midst of a sea of jungles. Kauṭilya's description leaves no room for doubt that these were planned as extensive settlements agog with life and bustle. The population ranged from hundred to five hundred families. It seems that distinct preference was given to the agriculturists and cultivators, to the hardy and industrious Śūdra and lower varṇa (*avaravarṇa*) cultivators (*karṣaka*).[58] Farmers with industry, commitment and initiative were protected and encouraged. Indolence and lethargy were discouraged.[59] Well-managed irrigation facilities with adequate scope for public participation in the management were provided.[60] However, despite the clear emphasis put on agriculture and farmers, care was taken not to make the settlements unidimensional in terms of occupational profile or cultural life. There were 'priests, preceptors, chaplains and Brāhmaṇas learned in the Vedas' along with Śūdra farmers as also physicians, horse trainers, elephant trainers, etc.[61] And, of course, there were the traders.[62]

The rules governing the settlement process shed light not only on the newly established villages but also on some general features of agrarian life and the agricultural policy of the state. In this connection, Kangle observed:

> It may be assumed that though villages in new settlements are to enjoy certain privileges and concessions in the initial stages, they would be expected to be treated, in course of time, in the same way as the other older villages.... Many of the rules of this Chapter (sūtra 25 onwards) appear to be intended for all villages in the state, and not those in the settlements only. For example, the rules about the guarding of the property of the minors and temples, the maintenance of orphans, etc., ... can hardly be understood for new settlements only. Moreover, the concluding stanzas of the Chapter (2.1.36–39) have undoubtedly the whole country in view and not only the new villages.[63]

In the *Arthaśāstra,* the expression *karṣaka* has been used as a general expression to signify a tiller of soil or cultivator without any specific status or property connotation. Unless the context specifies, merely from the expression *karṣaka,* it is not possible to determine whether a hired farm labourer or a slave or a tenant cultivator or an owner cultivator is meant. *Gṛhapatikavyañjana,* the seeming householder or a person, who acted and behaved as though he was a householder but was actually an employee in the Kauṭilyan elaborate espionage system, has been called a *karṣaka.*[64] The context would suggest that he was expected to look like an independent farmer with his own land. In connection with the maintenance of record of villagers by the revenue department, *karaṣkas* are distinguished from the *dāsakarmakaras.*[65] It is well known that *dāsakarmakaras* were usually landless and were often used as agricultural labourers. But from these facts alone we should not perhaps certainly conclude that the *karṣakas* were, in all cases independent peasant proprietors. In some places, we find the *karaṣaka* working for others against payment of wage (*vetana*).[66] We also find *karṣakas* as members of a collective body of labourers (*saṃghavṛta*) or a joint undertaking (*saṃbhūya samutthāna*) accepting employment against payment (*vetana*).[67] Clearly, the *karṣakas* referred to in these cases were wage earners. The possibility of their being landless seems quite strong. Even if they owned land, they had to seek employment as wage labourers to supplement the income from their own land.

Keeping in mind this neutral status connotation of the term *karṣaka* in Kauṭilya's *Arthaśāstra,* we may now sketch the overall agrarian profile of the 'new settlements' (*janapada niveśā* and *śūnya niveśa*), as depicted in the text.[68] Although these settlements obviously still bore to some extent the character of 'special areas', they may be taken as a fair index of the agricultural policy of the state and the general agricultural conditions in the country. The villages 'consisted mostly of Śūdra farmers'.[69] Arable fields were to be allotted to 'taxpayers alone'. And if a 'non-arable' field was being converted into an 'arable one', then it should not be taken away from the one who was making it cultivable. On the other hand, if a productive field was kept idle and uncultivated, then it should be taken away from the defaulting allottee and a fresh allotment of it was to be made. "Or village servants (*grāmbhṛtakas*) and traders (*vaidehakas*) should till it. Or those who do not till should make good the loss (to the treasury)".[70] However, *brahmadeyas*, exempt from taxes and fines (*daṇḍa*), were to be granted to the learned and religious Brāhmaṇas as "freeholds". Similarly, estates 'without the right of alienation' were granted to certain categories of state officials.[71]

The plan that seemed to have been envisaged by Kauṭilya for the newly established settlements was the blossoming of a multidimensional comprehensive society. The occupational and economic base was, of course, to be formed by agriculture and cattle-keeping and manned by people of lower (*avara*) *varṇas*, mainly farmers.[72] Suitability to agriculture was one of the principal considerations in the selection of areas for establishing new settlements.[73] The necessity of making arrangements for appropriate irrigation facilities for the growth of agriculture in the new settlements has been emphasized.[74] However, the merits of a settlement with diversified occupational profile in contrast to the one with a monolithic structure with an organized sector of occupation (*śreṇīmanuṣyā*) also have been highlighted.[75] The diverse occupational spectrum envisaged in the Kauṭilyan scheme of the new settlements is reflected in the various *sūtras* of the chapter on *janapadaniveśa*.[76] We find references to the learned Brāhmaṇas, who obviously ministered to the spiritual and religious needs of the villagers; to some categories of minor state officials residing in the settlements; to skilled professionals like physicians, elephant-trainers, horse-trainers; to traders; to workers in mines, factories; to cattle keepers; to ascetics and hermits; to a variety of entertainers; and so on.[77]

A number of scholars have noticed in Kauṭilya's advocacy of establishing settlements with lower *varṇas* constituting the majority, a conscious pursuance of a radical policy of land redistribution with the aim of giving landless Śūdra agriculturists the status of peasant proprietors. It has also been asserted that Kauṭilya discouraged landlordism.[78] The balance of evidence, however, does not support such a view.[79] There are indications—some explicit and some not so explicit—that landed estates, sometimes of substantial dimensions, were granted to learned Brāhmaṇas, certain state officials, members of royal family and other magnates and dignitaries.[80] It appears that apart from the state and crown lands, there were also private landlords with varying dimensions of holdings.

Landholders usually cultivated their land either by employing lablourers under their own direct management or leasing out land to tillers. The labourers could be either wage earners (*karmakaras*)[81] or slaves (*dāsas*). The normal wage given to the *karmakaras* was fixed at one-tenth of the produce. However, a different rate could be arrived at on the basis of mutual agreement between the employer and employee. We do not know at the bargaining counter the advantage lay with whom, the employer or employee. However, Kauṭilya refers to the collective bargaining effected by labourers' union (*saṃghabhṛta*); the union undertook to complete the work and distribute the wages to individual workers.[82]

There is no direct reference in the *Arthaśāstra* to the employment of *dāsas* in agriculture by private londholders. However, from the fact that *dāsas* were employed by the state in farming its land[83] and that the employment of *dāsa* in agricultural operation by private landholders was a common feature of the agrarian life in early India,[84] it may be assumed that private landowners also engaged *dāsas* as farm hands. It needs to be stressed, and stressed strongly, that the status of the *dāsas* were not the same as that of the slaves in ancient Greece or Rome or in the early Middle East. This fact has been noted by practically all scholars dealing with slavery in early India. In fact, the difference in the treatment of the *dāsa* in India from the slaves in the Hellenic, Hellenistic and Roman world must have been so great that it led Megasthenes to his famous statement that slavery was unknown in India.[85]

Landholders also obviously leased out lands. This is clear from some legal injunctions relating to lease agreements. Both the landowner and as well as the lesee had to abide by the provisions of the contract. The tenure of the contract, however, appears to have been generally of short duration; presumably these contracts covered a single crop season and the annulment was not very difficult. It was an offence for the *kṣetrika* (the owner of the field) to take away the field from the *upavāsa* (the leasee) and for the *upavāsa* to leave the field before sowing.[86] Does it indicate that the contract could be cancelled other than during the sowing season? The distinction made between the *kṣetrika* and *upavāsa* seems significant; it seems to imply that only a full title to ownership qualified a person to be called a *kṣetrika*.[87] *Svāmi* was possibly another technical term used for the owner of the land and his rights were more than the one who had mere possession (*bhuktika*).[88] The expression *bhuktika* might have signified the leasee as he enjoyed the right to produce. Although the nature and extent of the right of the leasee (*upavāsa*) cannot be precisely worked out, it appears that the landlord could not evict the leasee at will.[89] It is also not clear whether the term *bhuktika* also covered the sharecropper (*ardhasītika*) as well. The *Arthaśāstra* mentions sharecropping in the context of state farming; there is no clear and indubitable reference to sharecropping in the context of private farming. But in view of the wide prevalence of the practice in early India and from the fact that the practice is referred in the text in the context of state farming, it is safe to assume that in the agrarian framework reflected in the *Arthaśāstra* the practice was very much current in the domain of private farming as will. Moreover, a *sūtra* actually refers to the liability of the wives of sharecroppers for the debt they might have incurred.[90] It may be inferred that the responsibility to fulfil the obligations of the contract did not lie on the farmer alone but on his entire family. Since private irrigation facilities could be leased out (*prakraya*) or be given on hire (*avakraya*) or even on share basis (*bhāga*),[91] it may be assumed that the practice was prevalent in cultivation of land too.

Between the landless farm labours and big landholders lay the intermediate categories of owner cultivators and sharecroppers. It is not possible to form any idea regarding the extent of the land under the holding of these different categories (the big landholders, the owner cultivators and sharecroppers).[92] *Gṛhapatikas*, who were also cultivators (*karṣakas*),[93] and *Kuṭumbins*, who seems to have been 'medium farmers', are referred to in such a manner in the text that it leaves the impression that their number could not have been very insignificant. A normal *gṛhapati*, whose means of livelihood was not 'depleted' (*vṛttikṣīṇah*), appears to have been a prosperous landholding cultivator.[94]

Although specific reference to *Kuṭmbins* is found in the text in the context of city planning where the *Kuṭumbins* are found as city dwellers, the depiction leaves no doubt that generally, they were agriculturists of some means and stability.[95] If the term *kṣetrika* also signified farmers of this category, it may be further surmized that many of them might have got their land cultivated through *upavāsas*, sharecroppers and hired labourers.[96]

It is scarcely necessary to labour the point that the Brāhmaṇa recipients of gift land or the state officials who were granted land would normally employ others to cultivate their land. Perhaps they gave their land to tenants for cultivation on a temporary basis. Such temporary tenancy is hinted at in the following *sūtra*: 'if one does not till land that is inalienable, another may use it for five years and return it after receiving compensations for his exertions.'[97] In this connection, it is interesting to recall that the land granted to state officials was inalienable.[98] It is also noteworthy that some state officials during the reign of Aśoka used to go on long tours every five years.[99] It may perhaps be suggested that such touring officers used to give their land in temporary tenancy to cultivators. At the time of the termination of such a tenancy, some payment appears to have been made to the erstwhile tenant. The *sūtra* succeeding the one referring to temporary tenancy is also interesting. It reads: "Non-taxpayers living in a different place may live on the produce (of their fields)".[100] From all this information it appears that it cannot be claimed that Kauṭilya abrogated that absentee-landlordism.

Some restrictions seem to have been imposed on the free sale of land. A *brahmadeya* land could be sold only to a recipient of *brahmadeya* grant. It is clear that a taxpayer could not become the owner of non-taxpaying land. It is not possible to ascertain whether the restriction still remained even if the prospective buyer agreed to pay the normal land revenue. 'Kinsmen, neighbours and creditors, in this order, shall have the right to purchase landed property (on sale). After that, others who are outsiders (may bid for purchase).'[101] The purpose was to prevent outsiders from disturbing the cohesiveness of the holding.

NOTES AND REFERENCES

1. Text and translation of the *Arthaśāstra* we have used are R.P. Kangle, 1969, *The Kauṭīliya Arthaśāstra*, Pt. I, Second Edition, Bombay, Pt. II, Second Edition, Bombay, 1972. We have also extensively used Part III of Kangle's work consisting of his study of the *Arthaśāstra*, Bombay 1965.
2. *Artha.*, 15.1.1–2; Kangle, *op. cit.*, part III, p. 1, n. 1.
3. *Artha.*, 6.1.8.
4. *Artha.*, 2.1.21; 2.2.3–4. Some of these sanctuaries and parks were obviously reserved for royal sports and pleasures. An account of the royal hunting expedition has been recorded by Megasthenes, Strabo, XV.1.55, cited by R. Thapar, 1973, *Aśoka and the Decline of the Mauryas*, Delhi, p. 160.

5. *Śāntiparva*, 59. 100–106, 119–120; cf. *Viṣṇu Purāṇa*, I. 13; Romila Thapar, 1978, 'Origin Myths and Early Indian Historical Tradition' in *Ancient Indian Social History: Some Interpretations*, New Delhi, p. 294 ff.
6. D.R. Chanana, 1978, '*Uttari Bharat men Kṛshi ka Vistara*', in *Prachin Bharat* (Hindi), (ed.) D.N. Jha *et al.*, New Delhi, pp. 100–121.
7. The *Rākṣaśī pravṛitti* was usually identified with unmitigated surrender to the lower appetites. Ironically, however, gathering (*uñchavṛitti*) and not saving for the morrow was considered as the acme of virtue. In the portrayal of the unsullied golden age of virtue too can be discerned an image of the gathering economy.
8. *The Kauṭilya Arthaśāstra*, Part I, ed. R.P. Kangle, 2.1.6, 8.4.43, 9.2.18–20.
10. *Arthaśāstra* (henceforth Artha.), 2.4.24–25.
11. The sections dealing with the settlement pattern of cities and their administration in the *Arthaśāstra*, sections 2.4 and 2.36, do not contain any indication that agriculture was practised in cities. Amita Ray thinks that there was little economic difference between the city and the village—the city was larger in size than the village and was more prosperous, A. Ray, 1964, *Villages, Towns and Secular Buildings in Ancient India*, Calcutta, p. 24. But a perceptible difference in the economy between the city and the village seems to have emerged; the city was predominantly non-agricultural, A. Ghosh, 1973, *The City in Early Historical India*, Shimla, pp. 37, 41–42, 45–49.
12. *Artha.*, 2.21 and 2.24.
13. *Artha.*, 2.4.11 refers to *nagaradhānyavyavahārika.*
14. *Artha.*, 2.6.2–3 *rāṣṭra* referred to rural areas or countryside in contradistinction to fortified urban settlements. The corpus of revenue was first classified under six broad heads based on territorial distinctions or the nature of the source of income. These heads were forts, the country, mines, irrigation works, forests, cattle herds and trade routes.
15. *Artha.*, 9.1.17–19.
16. *Artha.*, 9.1.17–19; 2. 24.5.
17. R.P. Kangle, *Kauṭilīya Arthaśāstra*, pt. II, Bombay 1972, pp. 136, 149; Sibesh Bhattacharya, *Indian Economic and Social History Review*, Vol. XV, No. 2, p. 212.
18. *Artha.*, 2.24.6, cf. 2.24.10.
19. *Artha.*, 5.2.2, 6.1.1, 6.1.8.
20. *Artha.*, 2.24.5. Sixteen *droṇas* of rain, according to Kangle's calculation, would range between approximately 63 cm and 81 cm. The import of the concerned *Sūtra* is not clear. It may also mean that for sowing in the *jāṅgala* 16 *droṇas* of annual rainfall was necessary, where as for the *anupa* it was 24 *droṇas*, Kangle, *op. cit.*, pt. II, p. 49, n. 5.
21. *Maruprāya, Artha.*, 9.1.50.
22. *Artha.*, 2.6.5, 2.35.3; 3.9.26.
23. *Artha.*, 2.24.23; 7.11.16.
24. *Artha.*, 2.24.5.
25. *Artha.*, 7.11.3–5.
26. *Artha.*, 7.11.4; *cf.* Kulluka on Manu VII.69 regarding the high yield of *jāṅgala*.

27. *Artha.*, 2.6.5.
28. *Artha.*, 7. 11. 5; 2.24. 11–12.
29. *Artha.*, 6.1.8.
30. *Kṛtrimāhibhūmirguṇāh*, *Artha*, 7.11.19.
31. *Artha.*, 2.6.5.
32. *Artha.*, 2.24.1.
33. *Artha.*, 2.1.20–24, 2.6.5, 2.35.3, 3.9.28, 2.24.18, 2.34.8, 3.8.2, 3.9.3, 3.9.33, etc.
34. *anudake kūpasetubandhotsān sthāpayet*, *Artha.*, 2.34.8.
35. *Artha.*, 2.1.20, 7.12.4–5.
36. *Artha.*, 7.12.4–5.
37. *Artha.*, 2.24.18. For the interpretation of these terms and their implications, see Sibesh Bhattacharya, 'Land, Soil, Rainfall, Irrigation—Some Aspects of the Backdrop of Agrarian Life in the *Arthaśāstra* of Kautilya', *Indian Economic and Social History Review*, Vol. XV, No. 2, pp. 216–17.
38. The origin and introduction of the so-called Persian Wheel in India has been a controversial issue. For various views, see I. Habib, 'Technological Changes and Society: Thirteenth and fourteenth Centuries', Presidential Address, Medieval India Section, 31 Session, Indian History Congress, Varanasi, 1969; B.D. Chattopadhyaya, 'Irrigation in Early Medieval Rajasthan', *Journal of the Economic and Social History of the Orient*, Vol. XVI, pts II–III, pp. 303–304, n. 20–21.
39. *Artha.*, 2.1.20–23.
40. *Artha.*, 2.1.20–24.
41. *Artha.*, 2.24.18; *cf.* Kangle, *op. cit.*, pt II, p.151, note 18.
42. *Artha.*, 3.9.35.
43. Kangle, *Kauṭilya Arthaśāstra* part III, p. 168.
44. *Artha.*, 2.16.4, 2.16.8, 2.16.11, etc.
45. *Artha.*, 4.2.28.
46. *Artha.* 4.2.24–36.
47. *Artha.*, 2.24.2, 2.24.28–29.
48. For *daṇḍapratikartri*, see *Artha.*, 2.23.2, 3.13.18.
49. cf. *Manu*, IV.253.
50. *Artha.*, 2.24.16.
51. Kangle, *op. cit.*, II, p. 150, n. 16.
52. *Manu.* VII.138
53. For example, U.N. Ghoshal, 1966, *A History of Indian Public Life*, Vol. II, Oxford University Press, Delhi, Ch. V. Although Kangle has refuted the theory that all land belonged to the state, his treatment of the state economy leaves the impression that state agricultural operation under *sītādhyakṣa* accounted for the major share of revenue accruing from agriculture, Kangle, *op. cit.*, part III, ch. Seven.
54. *Artha.*, 2.1.
55. For a succinct discussion of the issues related to the theory of slave economy reflected in the *Arthaśāstra*, see Kangle, *op. cit.*, part III, pp. 186–87.
56. *Artha.*, ch. 2.1.
57. Kangle, *op. cit.*, part III, pp. 167–69.

58. *Artha.*, 2.1.2.
59. *Artha.*, 2.1.8–15.
60. *Artha.*, 2.1.20–24.
61. *Artha.*, 2.1.7.
62. *Artha.*, 2.1.11.
63. Kangle, *op. cit.,* part III, p. 169.
64. *Artha.*, 1.11.9–10, cf. 2.35.8.
65. *Artha.*, 2.35.4; they are also distinguished from *gorakṣaka, vaidehaka* and *kāru.*
66. *Artha.*, 3.13.27–29.
67. *Artha.*, 3.14.18–22.
68. *Artha.*, 2.1.
69. *Artha.*, 2.1.2; cf. 6.1.8, 7.11.21.
70. *Karadebhyah kṛtakṣhetrāṇyaikapuruṣikāṇi prayacchet, akṛtāni kartṛibhyo nādeyāni. Akṛiṣatāmācchidyāṅyebhyah prayacchet. Grāmabhṛtakavaidehakā vā kṛiṣeyuh. Akṛiṣanto vāvahīnaṃ dadyuh, Artha.*, 2.1.8–12.
71. *Artha.*, 2.1.7; from another *sūtra,* however, it appears that the provision about the right of alienation was not followed rigidly in all cases; it also appears that such land or landed property could be sold to a similar grantee of *brahmadeya* or other such grants, *Artha.* 3,10.9.
72. *Artha.*, 7.11.21.
73. *Artha.*, 7.11.3–5; cf. 6.1.8.
74. *Artha.*, 2.1.20–24.
75. *Artha.*, 7.11.18–20, see Kangle's remark regarding the interpretation of the term *śreṇīmanuṣyā*, Kangle, *op. cit.*, Part II, relating to *sutra* 7.11.18.
76. *Artha,* 2.1.
77. *Artha.*, 2.1.7,10–11,19,32–35, etc.
78. R.S. Sharma, *Sudras in Ancient India,* Varanasi 1958, pp. 147,173; R.S. Sharma, *Light on Early Indian Society and Economy,* Bombay 1966, p. 66; U.N. Ghoshal, 1966, *A History of Indian Public Life,* vol. II, Oxford University Press, pp. 88–89.
79. The issue has been discussed in detail by the author in 'Land-System as Reflected in Kautilya's *Arthaśāstra*', *Indian Economic and Social History Review,* Vol. XVI, No. 2, pp. 87–92.
80. *Ibid.*; cf. *Artha.* 2.1.7, 2.7.2, 2.35.3, 3.10.9, 7.16.16, 9.3.15–16.
81. *Artha.*, 3.13.26–29.
82. *Artha.*, 3.14.19–24.
83. *Artha.*, 2.24.2.
84. R. Fick, 1972, *Social Organisation in North-East India in Buddha's Time,* Delhi (reprint) p. 305; K.M. Saran, 1957, *Labour in Ancient India,* Bombay, p. 25; D.R. Chanana, *Slavery in Ancient India,* New Delhi 1960, p. 108.
85. The status and the conditions of *dāsa,* as reflected in Kauṭilya's *Arthaśāstra,* have been dealt with by Kangle and he too has underlined the difference in these regards between the *dāsa* and the *doulos* of ancient Greece and *helots* in Rome. The distinction between the unfree *daṣa* and the free wage earner *karmakara* in regard

to actual conditions was very thin. *Dāsas* were surely not well off but they were not treated as subhumans, Kangle, *op. cit.*, part III, pp. 186–87, 227–28.

86. *Artha.*, 3.10.8.
87. Kangle, *op. cit.*, pt III, p. 170.
88. Two sets of terms were used to mark this distinction in the nature of the holding; the terms like *svatva, svāmītva*, etc., indicated full unhindered possession, whereas terms derived from *bhuj* indicated the mere right to enjoy the produce, U.N. Ghoshal, 1973, *The Agrarian System in Ancient India*, Calcutta, Chapter V; cf. *Mitākṣarā* on *Yāj* II. 27. For the use of the expression *svāmī* in the *Arthaśāstra* in the context of landownership, see, *Artha*, 3.9.7,17. We find a clear use of the expression to indicate ownership of irrigation device (*setubandha*) in *Artha.* 3.9.32.
89. *Artha.*, 3.10.8.
90. *Artha.*, 3.11.23.
91. *Artha.*, 3.9.35–36.
92. A.N. Bose thinks that land was mostly held by the Brāhmaṇa landholders, *Social and Rural Economy in Northern India*, Vol. I, Calcutta 1970, p. 173ff, Vol. II, Calcutta 1967, p. 261. R.S. Sharma, however, thought that the peasant proprietors and well-to-do peasants had more land in their possession than the Brāhmaṇa and Kṣatriya landlords, *Light on Early Indian Society and Economy*, p. 62.
93. *Artha.*, 1.11.9–10, 1.12.22, 2.35.8.
94. *Artha.*, 1.11.9–10.
95. *Artha.*, 2.4.24–25.
96. In the Buddhist literature, *gahapatis* and *kuṭumbikas* appear to have been substantial landholders. *Gahapatis*, in fact, were both business magnates as well as big landholders, Sibesh Bhattacharya, 'The Meaning and Significance of the term Gahapati', *Archiv. Oriental. Praha*, Vol. 44, 1976, p. 149ff. However, it is not certain that the *gahapati*, of the Buddhist texts and early Brahmi inscriptions on the one hand and the *gṛhapatika* of the *Arthaśāstra* are identical. But the possibility appears quite strong.
97. *Artha.*, 3.10.13.
98. *Artha.*, 2.1.7.
99. Third Major rock edict.
100. *Artha.*, 3.10.15.
101. *Artha.*, 3.9.1–2.

CHAPTER 30

Agriculture as Reflected in the Dharmaśāstra Tradition

(With Special Reference to Manu and Yājñavalkya)

A.P. Ojha

There are two perceptions of the term Dharmaśāstra—restricted and wider. B.N.S. Yadava[1] has rightly pointed out that in a restricted sense, the term Dharmaśāstra was used by some[2] modern scholars for the versified *smṛti* texts of Manu, Yājñavalkya and others, which were designed as manuals of human conduct. But, in a wider sense, the term was also applied to the Dharmasūtras[3] which antedate the metrical *smṛtis* and postdate the early Gṛhyasūtras.

As the earliest sources of ancient Indian law, the early Dharmasūtras of Gautama,[4] Āpastamba, Baudhāyana and Vasiṣṭha have roughly been assigned[5] to the period from c. 500 to 200 BC.[6] The growth of the metrical *smṛtis* marks a distinct stage in the history of Dharmaśātra literature.

Among the mertical *smṛtis,* the development of *Manusmṛti* between c. 200 BC and 200 AD is most significant. The *Manusmṛti* or the *Mānava Dharmaśāstra*[7] is the oldest and the most well-known text of the Dharmaśāstra literature. It attempts to forge a synthetic common culture, rising above the various Vedic schools and ritual affiliations. It represents an advance on the Dharmasūtras in so far as it is detailed and more systematic. The *Yājñavalkya Smṛti*[8] (between AD 100 and 300) follows the *Manusmṛti* not only chronologically but also in point of prestigeousness. It produces its subject matter in a systematic manner under the heads of *ācāra, vyavahāra* and *prāyaścitta.* The *Viṣṇu Smṛti* (AD 100–300), *Nārda Smṛti* (AD 100–400), *Bṛhaspati Smṛti* (AD 300–500), and *Kātyāyana Smṛti* (AD 400–600) are the other well-known metrical *smṛtis.*

To begin with a chart of different terms used for agriculture and the agriculturists in the study of our period may be useful. It is prepared under the headings of the source of the term, its time and locale, the term itself and a column of remarks. We think that it will be more useful to start right from the Vedic age.

Source of the Term	Locale of the Source	Term/Terms Used	Remarks
Ṛgveda[9] (c. 1500–1000 BC)	Land of the Sindhu and its tributaries	Kṛṣi Kārṣman Kīnāśa Akṛṣīvala	Kārṣman denoting a 'furrow' found only in the *Ṛgveda* denotes 'not agricultural'
Atharvaveda[10] (c. 1000–700 BC)	Western UP	Kīnāsā Kārṣīvaṇa	 denotes a 'plougher'
Vājasaneyi Saṃhitā[11] (c. 1000–700 BC)	Western UP	Kṛṣi	denotes 'ploughing'
Taittirīya Brāhmaṇa[12] (c. 1000–700 BC)	Eastern UP & Bihar	Kṛṣi	plougher
Bhājā Cave Inscription[13]	Maharashtra	*Hālika*	a ploughman
Aṣṭādhyāyī of Pāṇini[14] (5th cent. BC)		*Hālika*	appears to have connoted an ox utilized for ploughing
Gāthāsaptaśatī of Hāla[15]	Andhra and Maharashtra	*Halio* *(Hālika)* *Pāmara*	
Manusmṛti[16] (2nd BC–2nd AD)		*Kṛṣīvala* *Kṛṣijīvī* *Kṣetrika* *Kṣetrina* *Kīnāśa* *Ārdhika*	denoting both peasant and ploughman share-cropper
Yājñavalkya Smṛti[17] (100–300 AD)	Mithila and Videha (Bihar)	*Kṣetrina* *Ardhasīrī*	
Arthaśāstra[18] of Kauṭilya		*Kuṭumbina* *Ardhasītika* *Karṣaka*	
Nārada Smṛti[19] (100–300 AD)	Nepal	*Karṣaka*	
Bṛahspati Smṛti[20] (300–500 AD)	North India	Sīravāhaka	
Spurious Gayā Copper-Plate Inscription of Samudragupta[21] (328–29 AD)	Bihar	*Kuṭumbī*	reference of *karada-kuṭumbī* (taxpaying cultivator) is significant

(Table contd.)

Source of the Term	Locale of the Source	Term/Terms Used	Remarks
Pañcatantra of Viṣṇuśarmā[22] (Gupta Period or a little earlier)		*Hālika*	ploughman or petty peasant performing manual work of tilling, etc.
Bṛhatkathā-śloka--saṇgraha of Budhasvāmina[23] (Gupta period)		*Hālika*	
Damodarpur Copper-Plate Inscription of the time of Budhagupta[24] (482 AD)	Bengal	*Kuṭumbina*	
Bhāruci's commentary on *Manusmṛti*[25] (500–650 AD)	Kashmir	*Kṛṣi* Karṣaka *Lāṅgala*	
Māliyā Copper-Plate Inscription of Mahārāja Dharasena II[26] (571–72 AD)	Gujarat	*Kuṭumbī*	
Daśakumaracarita of Daṇḍina[27] (550–650 AD)		*Hali* *Sīra*	
Amarakośa of Amarasingh[28] (c. sixth cent. AD)		*Kṣetrājīva* *Karṣaka* *Kṛṣaka* *Kṛṣīvala* *Kīnāśa* *Sairika* *Hālika*	except peasant these terms are also used for agricultural workers. appearas to have connoted an ox utilized for ploughing
Bṛhatsaṃhitā of Varāhamihira[29] (c. 7th cent. AD)	Madhya Pradesh	*Lāṅgala* *Karṣaka* *Kṛṣīvala* *Kṛṣījvī*	for plough
Harṣacarita of Bāṇabhaṭṭa[30] (c. 7th cent. AD)		*Kuṭumbī* *Kuṭumbika* *Kuṭumbina* *Kṛṣīvala*	*Kuṭumbī* is explained as *karṣaka* in the commentary of Śaṅkara[30A]

Source of the Term	Locale of the Source	Term/Terms Used	Remarks
Auśanasa Smṛti[31] (c. 600–900 AD)		*Kśetrina*	
Parāśara Smṛti[32] (c. 600–900 AD)		*Karṣaka* *Ārdhika*	also conceived as a separate mixed caste
Kāśyapīyakṛṣī-sūkti[33]		*Kṛṣīkāra*	
Medhātithi on *Manusmṛti*[35] (c. 9th cent. AD)	Kashmir	*Kuṭumbī* *Bhūmikarṣaka* *Vāha*	
Sāṇhyatattva Kaumudī of Vācaspati[36] (c. 9th cent. AD)	Mithila (Bihar)	*Hālika*	
Sārāvalī of Kalyāṇavarmā[38] (c. 9th cent. AD)	Gujarat	*Kṛṣīvala*	
Bhaṭṭotpala on *Bṛhatsaṁhitā*[39] (c.10th cent. AD)	Kashmir	*Lāṇgala* *Karṣaka* *Kṛṣīvala* *Kṛṣi* *Kṛṣijīvī*	
Kṛṣi-Parāśara[40] (c.950–1100 AD)	Bengal	*Karṣaka*	in the sense of peasant only
Kullūkabhaṭṭa on *Manusmṛti*[41] (c. 12th cent. AD)		*Karṣaka*	
Vijñāneśvara on *Yājñavalkysmṛti*[42] (c. 12th cent. AD)	South India	*Karṣaka*	
Dabok Inscription of the time of Dhavalappadeva[43] (c. 813 AD)	Mewāra (Rajasthan)	*Kṛṣaka*	
Haradatta on Āpastamba *Dharmasūtra*[44] (c. 12th AD)		*Kṣetrika* *Kṣetravāna*	in the verse of *Manusmṛti,* this term is used in place of *ārdhika.*
Parāśara-Madhavaḥ[45] (c. 14th cent. AD)		*Ārdhika*	

Agriculture is the source of perpetual creation on which a civilization depends. A fairly convincing picture of agriculture can be drawn on the basis of Dharmaśāstra literature. The Dharmaśāstra tradition brings into prominence the different aspects of agriculture corroborated by the other contemporary traditional (*śāstrīya*) and temporal (*laukika*) sources.[46] In attempt has been made to trace the history of agriculture in the pre-Dhamaśāstric period and to connect it with the post-Dharmaśāstric period so that a complete picture of agriculture may be presented.

CLASSIFICATION OF LAND AND SOIL[47]

Since the land and soil-fertility are the prime concern of agriculture, it should be dealt with at the very outset. We find a two-fold classification of soil, i.e. fertile and barren in the early Vedic period. In the later Vedic period, it is classified as *ūara* (alkaline) and *anūṣara* (non-alkaline).[48] A temporal (laukika) source of fifth century BC, the *Aṣṭādhyāyī* of Pāṇini, divides the land into cultivated (*karṣa*) and uncultivated. The later category is further divided into *ūara* (wasteland) and *gocara* (pasture).[49] In the *Arthaśāstra* also, we find several categories of land—*kṛṣṭa, akṛṣṭa sthala* (high and dry), *kedāra* (field sown with crops), and *mūlavāpa* (field of growing roots).[50] Kauṭilya also classifies the land/region on the basis of annual rainfall.[51]

In this connection, the Dharmaśāstra tradition also mentions twofold classification of land *kṛta* and *akṛta,* i.e., prepared or tilled and unprepared or untilled, respectively.[52] The gift of both these types of land was prevalent in the time of Manu, who declared unprepared fields to be less blamable than that of prepared one.[53] In other words, we may say that the gift of virgin land is preferred to that of furrowed land. The *jāṅgala* and *anūpa,* i.e. dry or arid and wet or marshy land, respectively, were the other classifications known to the Dharmaśāstra tradition.[54] Manu advises the king to reside in a country rich in forests and grain, where the land is charming and not marshy.[55]

The contemporary temporal sources belonging to grammatical and medicinal literature also throw light on the classification of land. Patañjali, the well-known grammarian of circa second century BC, refers to the twofold classification of arable and non-arable land as *kṣetra* and *gocara.*[56] He calls the cultivable land as *halya* or *sītya.*[57] The celebrated physicians of the first century AD, viz. Suśruta and Caraka, classified the land into *jāṅgala* (barren), *anūpa* (moist) and *sādhāraṇa* (ordinary). This threefold classification comes into view in the first century AD in the history of agriculture for the first time. The *jāṅgala* land has a flat surface with scanty growth of scattered thorny bushes. The *anūpa* land consists mostly of marshy areas thickly overgrown with forest trees. The *sādhāraṇa* land abounds with creepers, plants and trees. In addition, they make classifications based on the capacity of the land to yield plants of medicinal value.[59] It is stated in the *Suśrutāsaṃhitā* that the best soil for growing medicinal herbs and plants was 'glossy, firm, black, yellowish or red and not containing any sand, potash or any other alkaline substance.'[60]

FALLOW VIS-À-VIS PASTURE LAND

If the fallow land belonging to the village was being treated as the real wealth of the country,[61] the pasture land was also equally important in the eyes of Dharmaśāstrakaras.

A piece of land under cultivation was known as *kṣetra* or *kedāra.*[62] Apart from these two terms, ploughed land was called *Śūnāsīra,* after the gods *Śuna* and *Sīra,* respectively, 'wind' and the 'Sun'.[63] According to the Dharmaśāstra tradition every village should have a pasture. The Dharma-śastrakāras have given rules in connection with the maintenance of this type of land. Manu says that all around a village there should be a reserved area of one hundred bow-length *(dhanuḥaśataṃ).* or three shamyā-throws (shamyāpātastrayo), and three times that for a city.[64] Yājñavalkya also proposes that on the sweet will of the villagers or by the order of the king, pastures for the cows should be provided around every village.[65] To be more precise about the area, he further explains that one hundred, two hundred and three hundred bow-length land should be marked for pasture around each village, town and city, respectively.[66] Other Dharmaśāstrakāras like Viṣṇu,[67] Nārada,[68] Bṛhaspati[69] and Gautama[70] have also given such rules. Now it is obvious that forests and pastures were being maintained during the entire period of Dharmaśāstras. It can also be said that agriculture was the backbone of rural economy and villages were its nerve centres.

The Dharmaśāstra literature is acquainted with two kinds of properties, viz., *jaṅgama* (movable) and *sthāvara* (immovable).[71] The land falls under the latter category. Nārada and Bṛhaspati, conceding the right of sale, declare both movable and immovable properties as *paṇya* (vendible).[72] It is also interesting to note that two of Bṛhaspati's seven classes of legal documents are concerned with gift and sale of landed property.[73] Land has also been enumerated among the various articles of *ādhi* (pledge) in the Dharmaśāstra literature. To pledge or mortgage a chattel to the creditor himself is called *ādhi.* Nārada says, *adhikriyate iti ādhiḥ,* i.e. that to which a title is given, is *ādhi.*[75] Two kinds of pledges have been mentioned in the Dharmaśāstra tradition: one as *bhogya* (usable) and the other as *gopya* (unusable), of which the former comprises lands, houses, and so forth.[76] Nārada explains the former (*bhogya)* as 'a pledge which must be redeemed within a certain time' and the latter (*gopya*) as 'a pledge which must be retained till the debt has been discharged.'[77] Gautama[78] and Baudhāyana[79] include land in the list of objects that may be properly given away. But, at the same time Gautama, while mentioning the occupation of different castes in time of distress, quotes the opinion[80] of some who held land to be a forbidden article of traffic (*apaṇya*) for distressed brāhmaṇas. However, this restriction which evidently was not universally accepted even in Gautama's time; it was entirely done away with in the later Dharmaśāstras of Ápastamba[81] and Vaśiṣṭha.[82] On the basis of above pieces of evidence gleaned from the Dharmaśāstra literature, we may conclude that land as a property might be disposed of at will or given away as a gift.

LAWFUL MODE OF ACQUIRING WEALTH

Agriculture was one of the lawful modes of acquiring wealth (*vittāgama* and *dhanāgama*), according to the Dharmaśāstra tradition. Manu mentions seven modes of acquiring wealth in accordance with law (*dharmyā*).[83] One of these seven modes was *karmayoga,* i.e. the performance of work.[84] The very first commentator of the *Manusmṛti,* Bhāruci (500–650 AD)[85] who belongs to Kashmir, explains this *karmayoga* as 'agriculture or trade not carried on personally.'[86] The 9th and the 12th century AD commentators of the *Manusmṛti* Medhātithi[87] and Kullūkabhaṭṭa,[88] respectively, are also of the same view. In the very next metre, while describing the ten modes of eking out a living (*dasjīvanahetavaḥ*), agriculture was explicitly enumerated as one of them by Manu himself.[89]

We also find some contradictory views expressed by the Dharmaśāstra writers[90] regarding agriculture. It is found that at one place, agriculture has been extolled but at another, just opposite views have been forwarded. Manu has suggested to those Brāhmaṇās and Kṣatriyas who have been living by agriculture, to try hard to avoid farming.[91] Explaining why one should avoid farming, Manu states that because it tends to cause injury (*hiṃsāprāyāṃ*), is dependent upon others (*parādhīnāṃ*)[92] and damages the earth as well as creatures living within the earth by iron-tipped wood (kāṣṭhamayomukhama).[93]

In the fourth chapter, Manu describes different types of living for a Brahmin householder called *ṛta* (lawful), *amṛta* (immortal), *mṛta* (mortal) and *pramṛta* (deadly).[94] Concentrating on these four terms in the very next verse, it is said that subsisting by merely gleaning (corn) and gathering (single grains) would be known as lawful way of life;[95] living off unsolicited gifts would be immortal way of life;[96] begging for alms was the mortal way of life;[97] and farming the land was traditionally known as deadly mode of life.[98] Medhātithi (ninth century AD)[99] and Kullūkabhaṭṭa (12th century AD)[100] both are also agreeable in calling agriculture a deadly profession since it inadvertently causes the death of worms and insects in the course of ploughing. The next example of the contradictory view of Manu comes from the tenth chapter of his Dharmaśāstra when he says that some people think that agriculture is a virtuous profession (*sādhiviti*), but at the same time he also says that as a livelihood farming is blamed by good people (*sadbhirvigarhitā*).[101] A similar view has been forwarded by Parāśara in the second chapter of his text, viz., *Parāśara Smṛti.* On the one hand he proposes that agriculture causes a great sin (*mahādoṣa*)[102] to the Brāhmaṇas, but on the other hand he also proposes that it is the excellent (*praśasta*)[103] path of attaining heaven.

ATTITUDE TOWARDS AGRICULTURE

Agriculture was the backbone of society. It was on the surplus production that the kings and their subordinates, the priestly elite and a number of village artisans and servants[104] like washermen, barbers, sweepers, potters and carpenters, were living. But the attitude towards agriculture in the Dharmaśāstra tradition may be a point of discussion. In the time of Manu, it is said that agriculture was thought to be good by some, but as a profession it was blamed by all.[105] Bhāruci the 6th–7th century AD commentator of Manu, declares it as declamation (*arthavāda*)[106] intended to prohibit ploughing (*karṣaṇa*) without compelling reason on the maxim that 'one should not practise what is censured'.[107] There are also some other pieces of evidence. We find a term called *pramṛta* which, according to S.C. Banerji, is used by Manu for 'agriculture as a means of livelihood for a Brāhmaṇa.'[108] Medhātithi, the 9th century AD commentator of Manu, explains *pramṛta* occurred in connection with the duties of Brāhmaṇas, as a sin like death (*pāpīyah.*)[109] since plough driving (*lāṅgalākarṣaṇa*) is an act of carrying load (*bhāravāhatvaṃ*).[110] Kullūkabhaṭṭa, another commentator of Manu belonging to the 12th century AD, has also treated ploughing as a despised profession.[111] Thus, the Dharmaśāstric attitude towards agriculture seems to be critical. However, this aspect of the Dharmaśāstras needs further investigation.

NOTES AND REFERENCES

1. Pande, G.C. (edited), *Life, Thought and Culture in India* (from c. 600 BC to c. 300 AD), PHISPC, Vol. I, Part II, New Delhi, 2001, p. 254.
2. For example, Basham, A.L., *The Wonder That Was India,* Third revised edition, New Delhi, 2002, p. 112; Ghoshal, U.N., *Studies in Indian History and Culture,* Bombay, 1957, p. 473.
3. Winternitz, M., *A History of Indian Literature,* Vol. III, Reprint, 1985, pp. 575 ff; Kane, P.V., *History of Dharmaśāstra,* (hereafter *HOD*), Vol. I, Part I, Second edition, Poona, 1954, p. 14; Mehendale in *The Age of Imperial Unity,* editied by Majumdar, R.C. and Pusalkar, A.D., Bombay, 1953, p. 257.
4. The Dharmasūtra of Gautama is regarded as the earliest among these. As for as location is concerned Āpastamba and Baudhāyana have been placed in the Deccan. So has been the Grammarian Kātyāyana.
5. Pande, G.C., *Foundations of Indian Culture,* (hereafter FIC), Vol. I, Delhi, 1984, p. 229; However, Kane brings down the lower limit of the *Vasiṣṭha Dharmasūtra* to 100 BC *HOD,* Vol. I, Part. I, p. 105.
6. It is important to note that the Dharmasūtras look back to earlier traditions but, some later interpolations are also found in them. According to G.C. Pande, "...the Dharmasūtras present a more archaic social picture." *FIC,* Vol. II, p. 229.
7. Mandlik, V.N., ed., Mānava Dharmaśāstra, 2 Vols., Bombay, 1886; Jha, G.N. ed., *Manusmṛti,* Calcutta, 1932; Nene, G.S. ed., Kashi Sanskrit Series, Varanasi, 1970; Buhler, G., *The Laws of Manu* (translation of Manusmṛti), S.B.E., Vol. XXV, Oxford, 1886; Doniger, Wendy with Smith, Brian K., *The Laws of Manu,* Penguin Classics, 1991; Hopkins, E.W., *The Mutual Relations of the Four Castes according to the Mānavadharmaśāstram,* Reprint, Delhi, 1976.
8. *Yājñavalkya Smṛti* (hereafter *Yāj*), with the commentary of Vijñāneśvara, NSP, Bombay, 1926; Pandey, U.C. edited and translated, Varanasi, 1967; Yājñavalkya has generally been associated with Mithilā and Videha. But, according to K.P. Jayasawal, he probably wrote his work somewhere in Madhyadeśa (*Manu and Yājñavalkya,* Calcutta, 1930, p. 61).
9. Macdonell, A.A. and Keith, A.B., *Vedic Index of Names and Subjects,* (hereafter *V.I.*) Vol. I, Reprint, Delhi, 1995, pp. 151, 159, 181.
10. *Atharvaveda,* IV. 11, 10; VI. 30, 1; VI. 116, 1 vide *V.I.,* Vol. I, pp. 159, 181, fn. 7.
11. *Vājasaneyi Saṃhitā,* IV. 10; IX 22; XIV. 19, 21; *V.I.,* p. 181.
12. *Taittirīya Brāhmaṇa,* III. 1, 2, 5 etc.; *VI,* p. 181.
13. *Epigraphia Indica* (hereafter *EI*), vol. XV. p. 274 fn.

14. Agrawala, V.S., *Pāṇini-Kālīna Bhāratavarṣa*, Banaras, V.S. 2012, p. 156; Yadava, B.N.S., *Historical Investigation into Social Terminology in Literature: A Problem of the Study of Social Change (Mainly in the context of Early Medieval Northern India)*. Presidential Address (hereafter *PA*), (Fifty-Third Session Warangal), Allahabad, 1993, p. 7 fn. 3.
15. Basak, R.G., edited, *Gāthā Saptaśatī* of Hāla, Calcutta, 1971, 2.64, 69; 3.38; 4.24; 6.67 vide Yadava, B.N.S., *ibid.*, p. 7 fn. 5.
16. *Manusmṛti*, (hereafter *Manu*), III. 165; IV. 253; IX. 32, 38, 51, 52–54.
17. *Yājñavalkya Smṛti*, (hereafter *Yāj*) I. 166; II. 161.
18. Kangle, R.P., ed. & trans. *The Kauṭilīya Arthaśāstra*, 3 Parts, Second Edition University of Bombay, 1965, 1969, 1972, 2.4.24; 2.24.16; 3.11.23; 2.24.16; 2.35.4.
19. Jolly, Jullius, ed. *Nārada Smṛti* with the commentary of Asahāya, Calcutta, 1885; trans. *The Minor Law-Books*, Part I, S.B.E. vol. XXXIII, Reprint, Delhi, 1977.
20. *Bṛhaspatīsmṛti*, XVI, 1–2, cited in the Vyavahārā-Kāṇḍa of Lakṣmīdhara, ed. K.V.R. Aiyangar, Gaekwad's Oriental Series, Baroda, 1953, pp. 401 f.
21. Fleet, J.F. ed, *Corpus Inscriptionum Indicarum*, Vol. III. (hereafter CII) karadakuṭumbī-kārukādayan.... p. 257. l. 13.
22. Kale, M.R., ed., second edition, Delhi, 1969 part V, p. 27.
23. Agrawala, P.K., ed. *Bṛhatkathā-ślokā-saṅgraha—A Study* by V.S. Agrawal, Varanasi, 1974, 16.47.
24. Maity, S.K. and Mukerjee, R.R., *Corpus of Bengal Inscriptions Bearing on History and Civilization of Bengal*, Calcutta, 1967, p. 59, 1. 3; *EI*, XV p. 135; Sircar, D.C., *Select Inscriptions, Bearing on Indian History And Civilization* (hereafter *SI*), Vol. I, p. 333, 1.3; Yadava, B.N.S., *PA*, p. 14.
25. Derrett, J.D.M., ed. *Bhāruci's Commentary on the Manusmṛti*, 2 vols. (hereafter *Bhāruci*), Wiesbaden, 1975.
26. Fleet, J.F., ed. *CII*, Vol. III, p. 166, 1.25.
27. Kale, M.R. ed., *Daśakumāracarita*, fourth edition, Delhi, 1966, p. 120 (text), p. 76 (notes).
28. Sharma, H.D. and Saradesai, N.G., ed. *Amarkośa* with the commentary of Kṣīrasvāmina, Poona, 1941, 2.9.6; 2.9.64; 3.3.215.
29. Bhat, M. Ramakrishna, *Varāhamihira's Bṛhat Saṃhitā*, with Text, Translation & notes, Part I, Delhi, 1981, IV. 9; V. 29, 34; VIII. 52; XXXI. 4; XXXIII. 9, 21; XXXIV. 12.
30. Kane, P.V., ed. *Harṣacarita*, Reprint Delhi, 1973, 2.35, 1.20.
30a. Pathak, J. ed. *Harṣacarita*, Varanasi, 1972, pp. 131, 406, 408, 412; Yadava, B.N.S., *PA* p. 13 fn. 3.
31. *Auśanasa Smṛti* collected in Apte, V.G., ed. *Smṛtīnām Samuccayaḥ*, ASS, No. 48, Poona, 1929.
32. Tarkālaṅkāra, Chandrakānta, ed. *Parāśarā Smṛti*, Vol. I, Reprint Calcutta, 1974; ed. Sri Vāsudeva, Varanasi, 1968; ed. Sharma, Ram Chandra, Moradabad, 1925, 2.10–11.
33. *Kāśyapīyakriṣi-sūkti*, Wojtilla, G. ed. *Acta Orientalia*, XXXIII (2), Budapest, 1979 V. 308.
34. The core of the *Kāśyapīyakriṣi-sūkti* has been placed in the 8th–9th century, though the most of the material there in belongs to the early medieval times. Wojtilla,

Acta Orientalia Vol. XXXIX (1), 1985, p. 85 fn. 1, cited in Sharma, R.S., *Urban Decay in India,* New Delhi, 1987, p. 172 fn 31, Yadava, B.N.S. *PA*, p. 5, fn. 6.

35. Jha, G.N. ed., *Manusmṛti* with the commenatary of Medhātithi, Asiatic Society of Bengal, 1932; trans, Calcutta 1922–29, IV. 253 & IX. 150.
36. Jha, G.N. ed. & tr., Sāṇkhyā-Tattva-Kaumudī, Mumbai, 1896, p. 17 ll. 1–3; Yadava, B.N.S., *PA*, p. 8 fn. 4.
37. Tripathi, Shambhu Nath, ed., *Śākaṭāyana Vyākaraṇa,* Varanasi, 1944, p. 150, 2.1.96.
38. Chaturvedi, M.L., ed. & tr. Sāravalī, Varanasi, 1977, 21.34.
39. Tripathi, Avadha Vihari, ed., Commentary of Bhaṭṭopala on the *Bṛhatsaṃhitā* of Varāhamihira, 2 Parts, Varanasi, 1968, IV. 9; V. 29, 34; VIII. 52; XXXI. 4; XXX. 9, 21; XXXIV. 12.
40. Majumdar, G.P. and Banerjee, S.C., Kṛṣī-Parāśara, Calcutta, 1960, VV. 146, 163.
41. Nene, Pt. Gopal Śāstrī, ed, *Manusmṛti* with the commentary of Kullūkabhaṭṭa, Varanasi, 1970, VIII. 243; IX. 38, 150.
42. Śāstrī, Nārāyaṇa, ed., Commentary of Mitākṣarā on *Yājñavalkya Smṛti,* Varanasi, 1977, I. 166; II. 161.
43. Dabok Inscription of the Time of Dhavalappadeva: (Harsha) Saṁvat 207 (AD 813), *EI,* Vol. XX, p. 125, 1.13.
44. Commentary of Haradatta on *Āpastamba Dharma Sūtra* cited in *Dharmakośa-Vyavahārakāṇḍa,* Vol. I Part II, ed. Joshi, Laxman Sāstri, Wai (Distt. Satara), 1938, p. 842, 11.2; Yadava, B.N.S., *PA*, p. 15 fn 3.
45. *Parāśara-Mādhavaḥ,* Prāyaścittakāṇḍa, 11.24.
46. Temporal (*laukika*) sources include archaeological, epigarphical, grammatical, astrological, āyurvedic, numismatic and other such type of sources which fall under the traditional category of archaeological source. See for details, Ojha, A.P., *Prācīna Bhārata men Sāmājika Starīkaraṇa,* (hereafter *PBSS*), Allahabad, 1992, Chapter I.
47. Land and soil have been classified on the basis of geographical divisions or environs into dry and wet which form the twofold divisions of earth in the early historic period.
48. Ṛgveda, I. 127.6; *Āśvalāyana Gṛhya Sūtra,* IV. 7.9.
49. *Aṣṭādhyāyī,* III. 3.119; IV. 4.91, IV. 4.97 etc. Agrawala, V.S., *India as Known to Pāṇini,* Lucknow, 1953.
50. अक ष्यायां भूमौ...; स्थाल्याश्चानूप्याश्च...; केदारमूलवापा:..., *Arthaśāstra,* 2.2.1, 2.6.5, 2.24.23 & 7.11.7, respectively.
51. According to Kauṭilya, sixteen *droṇa* is the amount of rain in dry lands (जाङ्गलानां), one and half times in the wet lands (अनूपानां), thirteen droṇas and a half in the Aśmakas, twenty-three in the Avantis, unlimited in the Aparāntas and the snowy regions, and (unlimited) as to time, in lands where sowing are made with the help of canals— षोडशद्रोणं जाङ्गलानां वर्षप्रमाणम् अध्यर्धमनूपानां देशवापानाम्... कुल्यावापानां च कालत:, Arthaśāstra, 2.24.5.
52. अक तं च क तात्क्षेत्राद्गौरजाविकमेव च।
हिरण्यं धान्यमन्नं च पूर्वं पूर्वमदोषवत्। *Manu,* X.114
53. पूर्वं पूर्वमदोषवत्।। Ibid.
54. *Ibid.*, VII. 69. This classification had already been recognized by Kauṭilya. Supra fn. 51. The antiquity of this twofold dry and wet classification of land goes back at least

to the early historic period. See also Prana Nath, 'Date of compilation of Kautilya's Arthaśāstra', *Indian Antiquary,* Vol. 60, 1931, p. 111. Roy, Mira, 'Agricultural Practices in Ancient India' published in Chattopadhyaya, D.P. and Ravinder Kumar, *Science Philosphy and Culture* Multi-disciplinary Explorations, Part two, Delhi, 1977, pp. 164–201.

55. जाङ्गलं सस्यसम्पन्नमार्यप्रायमनाविलम्
रम्यमानतसामन्तं स्वाजीव्यं देशमावसेत्।। *Manu,* VII. 69

This classification of *jāṇgala* (dry) and *anūpa* (wet) were made on the quantitative scale of rainfall. The areas with sixteen *droṇās* of annual rainfall were called *jāṅgala* and the areas with twenty-four *dorṇās* of rainfall were called *anūpa.* Bhattacharya, Sibesh, 'Land Soil, Rainfall, Irrigation—Some Aspects of the Backdrop of Agrarian Life in the Arthaśāstra of Kauṭilya,' *The Indian Economic and Social History Review,* Vol. XV, No. 2, p. 213.

56. *Mahābhāṣya,* III. 3.19.
57. *Ibid.,* I. 1.72.
58. Raychaudhari, S.P., 'Agriculture in Ancient And Medieval India' published in Ray, Priyadaranjan and Sen, S.N. edited, *The Cultural Heritage of India,* Vol. VI, pp. 177–187.
59. *Ibid.*
60. *Suśruta-Saṃhitā,* Translated with commentary 'Āyurvedatattvasandīpikā' by Śāstri, Ambika Dutt, Varanasi, 1954, I. 37.2.
61. *Arthaśāstra,* 9.1.17–19.
62. धान्यानां भवने क्षेत्रे ख ा... केदाराद्या च, *Aṣṭādhyāyī,* V. 2.1 and IV. 2.40.
63. द्यावाप थिवीशुनासीर..., *Ibid.,* IV. 2.32.
64. धनु:शतं परीहारो ग्रामस्य स्यात्समन्तत:
शम्यापातस्त्रयो वा पि त्रिगुणो नगरस्य तु।। *Manu,* VIII. 237. शम्यापातस्त्रयो means three castes of a stock.
65. ग्राम्येच्छया गोप्रचारो भूमिराजवशेन वा। *Yāj.* II. 166.
66. धनु:शतं परीणाहो ग्रामे क्षेत्रान्तरं भवेत्।
द्वे शते खर्वटस्य स्यान्नगरस्य चतु:शतम्।। *Ibid.,* II. 167.
67. *Viṣṇu,* V. 147–148.
68. *Nārada,* XI. 40.
69. *Bṛhaspati,* XIV. 22.
70. *Gautama,* XII. 21.
71. *Nārada,* I 139; Bṛhaspati, qutoed in *Vivāda-ratnākara,* p. 189.
72. *Ibid.*
73. *Bṛhaspati,* VIII. 6–7.
74. Banerji, S.C., *A Glossary of Smṛti Literature,* Calcutta, 1963, p. 13.
75. *Nārada,* I. 124; Jolly, Julius, Translation, *The Minor Law-Books,* S.B.E. Vol. XXXIII, p. 72.
76. *Guatam Dharmasūtra,* XII. 39; *Vaśiṣṭha Dharmasūtra,* XVI, 18; *Āpastamba Dharmasūtra,* I. 20; *Manu,* VIII. 143–145; *Yāj,* II. 23, 25, 90; *Viṣṇu,* V. 181; *Nārada,,* I. 124, 139; आधिरपि द्विविधो-गोप्यो भोग्यश्च। Medhātithi on *Manu,* VIII. 143.
77. *Nārada,* I. 124; Jolly, *op. cit.,* p. 72.

78. भूमिस्तिलघ तमन्नमिति देयानि। *Gautama,* XIX. 17
79. *Baudhāyana,* III. 15; VII. 15.
80. *Ibid.,* VII. 15.
81. *Āpastamba,* I. 12.
82. *Vaśiṣṭha,* II. 24 ff.
83. सप्त वित्तागमा धर्म्या दायो लाभ: क्रयो जय:।
    प्रयोग: कर्मयोगश्च सत्प्रतिग्रह एव च।। *Manu,* X. 115.
84. कर्मयोगश्च..., *Ibid.*
85. Ojha, A.P., *PBSS,* p. 3-4.
86. कर्म-योग: क षि-वाणिज्ये स्वयं-क ते, Bhāruci on *Manu,* X. 115.
87. प्रयोगकर्मयोगौ कुसीदक षिवाणिज्यानि, Medhātithi on *Manu,* X. 115
88. कर्मयोगश्च क षिवाणिज्ये, Kullūkabhaṭṭa on *Manu,* X. 115.
89. विद्या शिल्पं भ ति:सेवा गोरक्ष्यं विपणि: क षि:।
    ध तिर्भैक्ष्यं कुसीदं च दश जीवनहेतव:।। *Manu,* X. 116.
90. *Manu,* X. 83–84; *Parāśara,* II. 2ff.
91. ...क षिं यत्नेन वर्जयेत्।। *Manu,* X. 83.
92. वैयव त्त्यापि जीवंस्तु ब्राह्मण: क्षत्रियो पि वा।
    हिंसाप्रायां पराधीनां क षिं यत्नेन वर्जयेत्।। *Ibid.*
93. भूमिं भूमिशयांश्चैव हन्ति काष्ठमयोमुखम्।। *Manu,* X. 84.
94. ऋताम ताभ्यां जीवेतु म तेन प्रम तेन वा।। *Manu,* IV. 4.
95. ऋतमु छशिलं ज्ञेयम...। *Ibid,* IV. 5.
96. अम तं स्यादयाचितम्। *Ibid.*
97. म तं तु याचितं भैक्षं।। *Ibid.*
98. प्रम तं कर्षणं स्म तम्।। *Ibid.*
99. कर्षणं तु मरणादपि पापीय:। Medhātithi on *Manu,* IV. 5
100. भूमिगतप्रचुरप्राणिमरणनिमित्तत्वाद् बहुदु:खफलकं प्रकर्षेण म तमिव प्रम तम्।। Kullūkabhaṭṭa on *Manu,* IV. 5
101. क षिं साध्विति मन्यन्ते सा व त्ति: सद्विगर्हिता। *Manu,* X. 84
102. ब्राह्मणश्चेत् क षिं कुर्यात्तन्महादोषमाप्नुयात्। *Parāśara Smṛti,* II. 8
103. प्रशस्तं स्वर्गसाधनम्। *Ibid,* II. 11.
104. There were a practice of paying shares of grain to rural artisans and servants. Among these artisans and servants mention may be made of *pañcakārukī's.* Thapar, Romila, *Ancient Indian Social History,* New Delhi, 1978, pp. 147, 151 fn 59. For the view that there were some elements of the *jajmānī* system in ancient India, see also, Rai, Jaimal, *The rural-Urban Economy and Social Changes in Ancient India* (c. 600 BC to 600 AD), Varanasi, 1974, pp. 4, 99 f. For *pañcakārukī* see Ojha, A.P., *PBSS,* p. 89–90.
105. क षिं साध्विति मन्यन्ते सा व त्ति: सद्विगर्हिता। *Manu,* X. 84
106. अर्थवाद:, Bhāruci on *Manu,* X. 84
107. "निन्दितं न समाचरेत" *Ibid.*
108. Banerji, S.C., *A Glossary of Smṛti Literature,* Calcutta, 1963, p. 78.
109. कर्षणं तु मरणादति पापीय: Medhātithi on *Manu,* IV. 5.
110. लाङगलाकर्षणं हि भारवाहत्वं..., *Ibid.*
111. प्रकर्षेण म तमिव प्रम तम्।। Kullūkabhaṭṭa on *Manu* IV. 5; सा पुर्नजीविका साधुभिर्निन्दिता..., Kullūkabhaṭṭa *on* Manu, X. 84. See also, Achhe Lal, *Prācīna Bhārata mein Kṛṣi,* Varanasi, 1980, p. 274.

CHAPTER 31

Problems and Perspectives of Agricultural Taxes in Early Northern India

O.P. Srivastava

INTRODUCTION

This chapter deals with agricultural taxes of early northern India. It is pertinent to point out that the topic is very comprehensive and a complex one. However, it is proposed here to focus on the gradual rise in incidence of agricultural taxes of early northern India.

REVIEW OF EARLIER WORKS

Prior to 1929, only a few scholars like N.C. Bandyopadhyaya, Magan Lal Buch and others referred to some kinds of agricultural taxes and their rate. But it was U.N. Ghoshal who, for the first time, made a serious study of the revenue system in ancient India in his book, *The Contribution to the Hindu Revenue System* (Calcutta, 1929). It was a piece of pioneer research on the subject, which dealt with the revenue system of India from the Vedic period to the twelfth century AD. The treatment of the revenue system of the Mauryan period is quite exhaustive but the coverage of the later period is partial and meagre. Another important work in this respect, though for a shorter period, is that of M.H. Gopal (*The Mauryan Public Finance,* 1935). In 1942, A.N. Bose presented *the Social and Rural Economy of Northern India* (c. 600 BC – 200 AD), Calcutta, where in he has devoted a chapter on the study of revenue system of the Mauryan and the Post-Mauryan periods. This aspect has also been discussed in *The Economic Life of Northern India in the Gupta Period,* 1957 by S.K. Maity. This subject has also been treated by R.S. Sharma in his work, *The Aspects of Political Ideas and Institutions,* 1959, Delhi. In the same year, Roma Niyogi's book, *The History of Gāhaḍavāla Dynasty,* was published from Kolkata. In it, she has discussed the revenue terms mentioned in the *Gāhaḍavāla* inscriptions. Puspha Niyogi's work the *Contribution to the Economic History of Northern India from the Tenth to the Twelfth century AD* was published in 1962 from Kolkata. She has also analyzed some revenue terms as well as the rate and mode of collection of taxes. Lallanji Gopal devoted the second chapter to revenue system in his work *Economic Life of Northern India* (Varanasi 1965) and the most interesting part from this point of view is his critical study of the revenue terms as found in inscription. In 1967, *The Revenue System in Post-Mauryan and Gupta Times* by D.N. Jha, Calcutta, appeared, which deals with land ownership and land revenue, its kinds and rates and also scrutinizes the view of previous authors on this point.

B.P. Majumdar, in his article 'Land Revenue in Early Medieval North India. AD 600–1200 AD' (in R.S. Sharma (ed.), *Land Revenue in India: A Historical Study* Delhi, 1971) discussed the meaning and significance of various fiscal terms and the mode of payment of agricultural taxes. In 1973, B.N.S. Yadava's book, *Society and Culture in Northern India in the Twelfth Century,* was published from Allahabad. In it, he has devoted a few pages to the study of taxation in the light of literary and epigraphical evidences. He has also thrown a welcome light on new dimensions of taxation such as oppressive features of taxation or over taxation, the machinery of collection, modes of payment, etc.

AGRICULTURAL TAXES

There are two main types of taxes: (1) Direct and (2) Indirect. Agricultural taxes come under direct taxes. In this chapter, I shall concentrate upon the rate of taxes and its implications from 600 BC to 1200 AD.

In the *Baudhāyana Dharmasūtra*[1], the *Gautam Dharmasūtra*[2], the *Arthaśāstra*[3], the *śāntiparva*[4] of the *Mahābhārata* and in the *Manusmṛti*[5], it is stated that 1/6th of the produce should be realized as land revenue. Kauṭilya[6] mentions at another place the expression '*ṣaḍabhāga*', i.e., one-sixth of grain produced as share taken by the king. Bhaṭṭasvāmin, commenting on a passage of the *Arthaśāstra,* explains *ṣaḍabhāga* in the general sense of royal share (*rajā-bhāga*). He further states that the term 'one-sixth' includes by implication other rates such as one-third, one-fourth or any other portion prevalent in different-different regions. However, this does not mean that Kauṭilya recognizes the uniform rate of one-sixth as land revenue in all the cases.[7] The same rate is mentioned by Nārada[8] and Viṣṇu[9] also. Megasthenes, who visited India during the Mauryan period, did not mention the portion of land revenue. It is not mentioned whether it was paid in addition to the $1/4^{th}$ of the produce of land.[10] But in the *Arthaśāstra,*[11] it is mentioned at another place that a cultivator had to pay water tax or irrigation taxes at varied rates from 1/3 to 1/5 of the produce, even though he had irrigated the field through natural resources.

In view of the above mentioned rates, M.H. Gopalan[12] has stated that under the Mauryan rule, the total burden of land tax was 1/6th or 16.66% of the produce. He has further stated that in addition to it, irrigation cess was also realized[13]. Thus, the revenue extracted by the Mauryans, according to him, 'was a little too heavy'[14]. But he has not explained the reasons behind the heavier levy on land. He has also not paid any attention to the total burden of revenue which includes the irrigation taxes also. Thus, the total burden of taxation ransges from 16.66% to 36.66% or 41.66% or about 50% (33+16.66 = 49.66%) and it appears that in the Mauryan times, the burden of agricultural taxes was heavier.

Another interpretation of the data mentioned above may be taken to calculate the land revenue as 1/6th of the produce from a non-irrigated field, plus 1/3, 1/4 or 1/5 of the remaining produce from the field irrigated by *srotayantra prāvartitam*[15] (a mechanism for letting water in channels leading to the fields), *udaghatum*[16] (to use the water wheel for raising the water) and *skandha prāvartitama*[17] (water carried on shoulders or on the back of bullocks), respectively, as is described by Bhaṭṭasvāmin, the commentator of the *Arthaśāstra,* who flourished in the twelfth century AD. Thus, the land revenue, according to the interpretation, ranged from 16.66% to 44.66% (16.66%, 32.57%, 37.5% and 44.66%). This interpretation is suited to the explanation of the commentator Bhaṭṭasvāmin when he states that the king was entitled to 1/3 to 1/5 of the produce as water tax over and above the revenue of 1/6 of the produce even when works were made by farmers themselves, because the king was the sole owner of all water as well as land in the Mauryan age.[18] Thus, the total burden was still too much on peasants and agriculturists. Some scholars think that the water cess was levied on 'crown land' as it has been dealt with under section, The Director of Agriculture.[18b] A close examination indicates that it applies to the 'state land'[18c], for the crown tenants were given 1/2 of the yield when they employed their own capital and 1/4 or 1/5 when the state supplied the implements and other necessities. Thus, the crown used to receive 1/2 or 3/4 or 4/5 of the grains produced and the tenant was left with 1/2 or 1/4 of 1/5 as the price of his labour. But in the passage in question, we find that irrigation levy was collected in addition to the normal rate of land revenue and tenants share of the produce was 67.43%, 62.5% and 55.34%. From this, it is explicit that the *udakabhāga* was referred to in the section as it was levied on state land.

The third interpretation of the evidence may be that the author of the *Arthaśāstra* appears to have meant to realize 1/6th of the produce from non-irrigated land and 1/5 , 1/4 and 1/3 from the produce of the irrigated field. If it is taken to be true, then it can be said that the land revenue in the Mauryan Period ranges from 16.66%, 20%, 25% and 33% of the produce. This rate does seem a little bit burdensome to cultivators if we take into consideration other different types of occasional demands (*Praṇaya*)[19], periodical taxes as it was termed *bhāga*[20], in the texts of the Mauryan and Gupta times, law digest and *Raghuvaṁśa*[21], *Abhijñānaśākuntalam*[22] and *Kādambārī*[23].

Throughout the Gupta period, the state claimed one-sixth of the produce of the land from the people in return for the protection given to the person and their property. This rate of taxation was so strongly entrenched in the society that Kālidāsa[24] states that for protecting the ascetics from obstacles and their wealth from robbers, the king was the claimant of one-sixth of their earnings and he was also made the enjoyerer of one-sixth of their religious merit (*tapasyā*)[25]. This statement by Kālidāsa is fully corroborated by Nārada[26], who states that one-sixth of the produce of the land forms the royal revenue, taken as reward by the king for protecting his subjects. Manu too supports this view with certain variations. According to him, a king should collect one-sixth, one-eighth or one-twelfth part of the crop according to the fertility of the soil[27a]. Kṣīrasvāmin, the eleventh-century commentator of the *Amarkośa* also refers to the *Arthaśāstra* to define *bhāga* as one-sixth share of the produce of king.[27b]

Bṛhaspati was the first law giver who lays down the tenet that the taxes varied from one-tenth to one-sixth (i.e. 10% to 16.66%). According to him, 1/10 on *Khil* land, 1/8 on the land exposed to rain and 1/6 of the crops harvested in spring season were levied[28]. In this context, D.N. Jha is of the opinion that these rates of land tax also varied for the reason that land exposed to dangers of rainwater make a rich harvest uncertain[27c], and the crops harvested in the spring season are least subject to risk[27d]. But this suggestion appears to be less required plausible as the danger of unseasonal excessive rains than the crops demanded hailstorm and attack of insects were also there in the spring season. The reason for the variation in rates of land revenue between the rainy season and spring season, most probably, seems to be the nature and property of the grains grown in these seasons. The most important grains of rainy season are paddy, *kodrava, syāmaka* and others which were unhusked and, therefore, could not be used as food in comparison with wheat, grama, *masūra,* etc., grown in the spring season. The rates of land revenue formulated by Bṛhaspati are also reiterated in other law digests of the early medieval period. Haradatta, who wrote the *Gautam Mitākṣarā,* a commentary on the *Gautam Dharmaśūtra* in the twelfth century AD takes 1/6, 1/8 and 1/10 rates to apply to different types of soil on account of their fertility.[29]

So far as the demand of 1/4 or 25% of the produced is concerned, it was realized in emergency only as is mentioned in almost all the texts on the *Dharmaṣāstra, Nītiśāstra* as well in other texts on the to political economy. For example, Manu has prescribed for a king to realize one-fourth of produce only in time of distress.[30] The commentators of Manu—*Govindrāja, Medhātithi* and *Kullūkabhaṭṭa* also agree with this opinion of Manu that a husbandman should give one-fourth of grain grown in emergency.[31] Lakṣmīdhara, who flourished in the twelfth century AD and served king Govindcandra of Gāhadavāla dynasty as his *sāndhivigrahika,* quoted with approval the views of Manu and recommended that one-fourth should be collected from the Kṣatriyas during the emergency.[32] Similarly, Kauṭilya had also recommended to realize one-third or one-fourth for replenishment of treasury.[33] In the *Mānasollāsa*[34] it is stated that the king should take 1/8, 1/12 and 1/6 of the produce according to the nature of the soil yield. But these rules could not have been followed in actual practice.

The *Dharmaśāstra* works lay down the law that the rates of taxes sanctioned therein were not to be altered and the imposts made of supplementary charges could be levied to meet the additional needs only, (*Rājadharmakāṇda* of *Kṛtyakalpataru*[35]).

In contrast, in the early medieval period, land tax became oppressive and arbitrary demands were put forth. The emergence of landed aristocracy involved a lord-vassal nexus, which was bound to, affect the taxation system of the early medieval North India. In spite of the general rate of tax, the taxes on the whole differed to such an extent from kingdom to kingdom and principality to principality,[36] that while explaining the term *deśadharma,* the *Mānasāra*[37] gives a list of increasing rates of revenue realized by the various categories of rulers and vassal chiefs in descending order, which reflects the actual condition of the age. They are:

1.	*Chakravarti* ruler takes	-	1/10
2.	*Adhirāja* or *Mahārāja*	-	1/6

3.	*Narendra*	-	1/5
4.	*Pārṣaṇika*	-	1/4
5.	*Paṭṭadhara*	-	1/3
6.	*Maṇḍaleśa*	-	?
7.	*Paṭṭabhāja*	-	?
8.	*Prahāraka*	-	?
9.	*Aṣṭgrāhiṇa*	-	?

What is important for our purpose is the different rates of royal taxes according to the position of the ruling class. The higher the authority, the lesser is the portion of revenue. In other words, the lower the privilege of the ruling aristocracy, the higher is the rate of land revenue. The exact rate of other kinds of the four rulers of *maṇḍaleśvara, Paṭṭabhāja, Prahārka* and *Aṣṭagrāhiṇa* are not mentioned. In this context, as R.S. Sharma[38] rightly says, half or even more of the produce was realized as land tax. Similar gradation or ruling class is mentioned in the *Aparājitapṛchhā* of Bhuvandeva, a text on architecture of the thirteenth century AD. They are *Mahīpati, Rājā, Narādhipati, Mahāmaṇḍaleśvara, Māṇḍalika, Mahāsāmanta, Sāmanta, Laghusāmanta, Caturānśika*, etc.[39] Although the text does not enumerate the rates of land taxes to be collected by those chiefs from the subjects, it may be taken to indicate the increasing burden of agricultural taxes on the peasants of the early medieval period. In the *Udaya-Sundarīkathā* of Śoḍḍhala, we find a chief of royal blood extracting money from rich people by imprisoning them.[40]

In the Mauryan period, the number of agricultural taxes[41] (*Sītā, Bali, Bhāga, Bhoga, Udakabhāga, Hiraṇaya, Kara, Viṣṭī, Udraṅga?, Ditya*) were nine or so.

These taxes continued upto the Gupta Period with a few additions like *cāṭa bhāṭa, uparikara, bhūtavāta-pratyāya, hālikakara, ditya, tulyameya, kḷpta, upkḷpta, viṣṭi*, etc.[42] In the post-Gupta period, the number increased to 18 types of agricultural taxes, as mentioned in the *Niśīthacūrṇi* (7th c. AD) and *Thāṇam*[43]. But, in actual practice, their number grew more than that. In the Gupta period, certain new land revenue terms appear in several inscriptions. References to these taxes in the contemporary record is usually followed by the word *ādi, etyādi*. For example, in a few Pallava charters, donees are granted land pieces with eighteen *parihāras* but their names are not mentioned. Similarly, the Basein grant of Vākāṭaka ruler Vindhyaśakti II refers to as many as fourteen *parihāras*[44].

In the inscriptions of the early medieval period we find the following terms for land revenue. These are as under: *bhāga, bhoga, hiraṇya, kara, bhūtavāta-pratyāya, uparikara, udraṅga, kḷpta, upakḷpta, nidhi, nikṣepa, upanidhi, jalakara, parṇakara, collika, paṭṭakila, tulyameya, tālārādāya, halasadi* (a tax on plough), *hasti-daṇḍa, cāṭa-bhāṭa, dāna, khalabhikṣā, khalayajña, halikākara, viṣṭi, ākaśotapatti, pātālotpatti, daśāparādha* (fine realized from villagers for ten offences), *Kalyāṇadhana, vāhoṭaka, dāni, khalakīya* (a threshing floor tax), *medhihāraka* (a special tax on the threshing floor), *māṅgalīyaka* (probably taxes collected on auspicious occasion the by royal family),[45] etc. Not only this even an item of very small value was taxed, as it seems from the term *akiñcitakaragrāhya* in the inscriptions.

Over and above, several other new taxes were imposed as per the will of the rulers. This is attested by the various terms such as *akiñcitapragrāhāyah.* In the Pāla and Sena inscriptions, *ādbhiḥ* in the inscriptions of Somavaṃsis of Orissa; *niyatāniyata— Samastādāyan* in the Gāhaḍavāla Charters, *Sarvādāya sametaḥ* in the Paramāra epigraphs, *ucitānucita* or *aparairapi* in the Candella grants and in the Kalacuri records *Samastarāja-pratyādāya,* etc.[46a]. We shall concentrate here on *khalabhikṣā, Khalayajña* and *Halikākara.* The terms *khalayajña* and *khalabhikṣā* are interchangeable[46b]. The term *Khalabhikṣā,* mentioned in Rajore records[47] of Madan Deva, of 960 AD, the Karitalai inscription[48] of Lakṣmaṇarāja II and the Baragaon stone inscription[49] of śabara has been described by some scholars such as Fleet[50], Ghoshal[51] and P. Niyogi[52], begging from threshing floor, consequently, may have denoted an additional tax in kind from the grains brought to the threshing floor by the state. But so was not the case. D. Sharma[53] takes it to mean two-third of the produce as the state's share from threshing floor and one-third as the share of the cultivator who received the straw also from the threshing floor. B.N.S. Yadava[54] has thrown a welcome light on the *Khalabhikṣā* in his presidential address of the fifty-third Indian History Congress (1993 session at Waranagal). He states that there is a reference in the *Lekhapaddati* that a tenant cultivator had to fulfil a number of obligations which include the payment of specified dues in kind to ruler (rāṇā)[55] and providing allotment of grain to five categories of village artisans (*pañcakārukī*), carpenter (*sūtradhāra*), iron smith (*lauhakāra)* and potter (*kumbhakāra*), barber (*nāpita*) and washerman (*rajaka)*[56]. The occurrence of these terms, the *grāmabhṛta* and *Pratibaddha–śilpś* indicate that payments were made to the village artisans.[57] It represents the antecedents of the *Jajamānī* system.[58]

Khalayajña does not mean the performance of ritual and sacrifices at the threshing floor[59], but it is mentioned by Mādhavācārya, (thirteenth c. AD) the commentator of the *Parāśara smṛti*[60], that one may, even Brāhmaṇas, can take up the vocation of cultivating the soil or being an agriculturist if he gives the following portion of revenue, i.e. 1/16 to the ruler, 1/21 of the produced as gift to the Brāhmaṇas and 1/3 of the same to the deity, respectively. In this context, B.N.S. Yadava[61] rightly says that the term *khalayajña* also signified the payment of royal dues in addition to gifts to deities and Brāhmaṇas out of threshed grain at the threshing floor.

The term *Halikākara,* B.N.S. Yadava says, occurs in the sense of a tax on sharecroppers in some Jain texts like *Deśīnāmamālā*[62] and the *pāi-sadda-mahāṇṇava*[63]. The epigraphic reference also points to the same conclusion. The Kāmana inscriptions[64] of Bharatpur of Gurjara Pratīhāra records refers to the donation of a piece a land ploughed (*Vāhayatva*) by a *hālika* in favour of a deity was made by a vassal chief, reducing him to the status of a *paradāsattān* (i.e. dependence of lesser degree). The term mentioned in Chamba records[65] for *āshṛta hālika* (ploughmen attached to the ground) appears to be the case. The *Hālikakara* shows a crystalization of a practice of squeezing unpaid services into tax. Y.B. Singh[66] thinks that it was a tax on plough or area under cultivation on one plough, as L. Gopal[67], Ghoshal[68] and S.K. Maity[69] also think. The *Hala* is obviously a plough, whereas *Hālika* refers to a ploughman. So, it appears to have been services rendered to lord in lieu of taxes. B.N.S. Yadava[70] takes it as tax on share cropping by the sharecropper or *karṣaka* attached to the land. There is an expression of the term in the context of sharecropping that *tasya grāmiṇaiḥ karṣaṇa halikā dīyante*[71], meaning thereby rendering a services by ploughing the field of chief or lord without getting any wage in lieu of his service.

Thus, the foregoing survey reveals rate and number of agriculture taxes from about 600 BC to 1200 BC and right from the Mauryan times to the twelfth century AD, the king's share of 1/6 of the produce of land had probably become the normal rate of the land revenue. But on account of various factors such as fertility of the soil, irrigational facility, manuring, flood and famine, emergency, etc., we find an enhancement in the rate of agricultural taxes; sometimes it had become too excessive. The study also shows that several additional taxes appeared in the early medieval period as compared to that of the Mauryan and Gupta period. These new imposts in the post-Gupta times appears to have been levied on account of the growing excessive demands of feudal lords and village chiefs. This, in turn, had certainly enhanced the miseries of the agriculturists of the early medieval period.

NOTES AND REFERENCES

1. Ed. G Buhler, 1882, Oxford, *Sacred Books of the East,* Vol. XIV, pt. 2, 10.81.1; see also *Baudhāyana Dharmasūtra* with the commentary of Govindsvāmin, ed. L. Srinivāsacharya, 1907, Mysore I, 10.81.1.
2. *Gautama Dharmasūtra,* Ed. A.S. Stenzler, London, 1876, With the Commentary of Maskarin, Ed. L. Srinivascharya, Mysore, 1917.X.24.
3. The Kautilya *Arthaśāstra,* Pts I–II, Tr. by R.P. Kangle, Bombay, 1969–1972, 2.15.3, *ṣaḍabhāga* 2.35; commentary of Bhaṭṭasvāmin, *JBORS,* Vol. XI, pt. III, p. 83.
4. *Mahābhārata* (*Śāntiparva* cr. Ed. S.K. Belvalkara, Pune, 1949–54, XII, 24.12 and 71.10.
5. *Manusmṛti,* VIII, 308.
6. *Artha.* 1.13.6 (*dhānyaṣaḍabhāgam*).

6b. *Artha.* 2.15 commentary of Bhaṭṭasvāmin, *Śaḍgrahṃam prayākatvadanyeṣāmapi yathā—deśa — prasiddhānāṃ — tritīyacaturthādibhāgānamupalakṣaṇāstham, JBORS,* XI, Pt. III, p. 83.

7. *Ibid.,* 2.35.
8. *Nāradā Smṛti* (*SBE*) Ed. J. Jolly Vol. XXXIII, pt. I, XVIII, 48.
9. *Viṣṇusmṛti,* Ed. by J. Jolly (*SBE*) Vol. VII, Oxford, 1880 III, 22–23.
10. Diodorus, ii, 40. In Strabo, XV, I, 40. But Strabo writes, 'The whole of the land belongs to the crown and the husbandmen till it on the condition of receiving one-fourth of the produced'.
11. *Artha.* 2.24.18. The line runs as *Sva-setubhyo hastaprāvartitamudakabhāgam—pañcaṃ—dadyuḥ, skandhaprāvartitam—caturthaṃ, Sroto-yantraprāvartitam ca tritīyaṃ, caturthaṃ nadīsarastāṭākakupodghāṇam.*
12. M.H. Gopalan, 1935, *Mauryan Public Finance,* London, Published by George Allen & Unwin Ltd., pp. 69–70.
13. *Loc.cot.,* p. 69.
14. *Ibid.,* p. 70
15. *Artha.* 2.24, 18.

16. *Loc.cit.*, p. 77.
17. *Ibid.*, p. 70, According to Bhaṭṭasvāmin *balivardaprāvartita yantrodakaseka nishpāditaṃ*, *Journal of Bihar and Orissa Research Society* (J.B.O.R.S.), Vol. XII part II p. 317.
18a. *J.B.O.R.S.*, Vol. XII, pt. II, p. 138.
18b. U.N. Ghoshal, 1929, *Contribution to the History of Hindu Revenue System (HRS)*, Calcutta, Calcutta University, pp. 30ff.
18c. M.H. Gopalan, *op. cit.*, p. 71.
19. *Arthaśāstra*, 5.2.16, 26.29; 8.4.23–24.
20. M.H. Gopalan, *op. cit.*, pp. 42, 59, 65, 127.
21. *Raghuvaṃśam* of Kālidāsa, 1929, Niraṇayasāgara Press, Bombay, II. 66.
22. *Abhijñaśākuntalam* of Kālidāsa, 1947, Niraṇayasāgar Press, Bombay, II.13.
23. *Kādambarī of Baṇabhaṭṭa*
24. *Raghu*, II, 66; V–8; XVII, 65; *Abh.Śāk*, II, p. 850.
25. *Rāghu*, XVII, 65.
26. *Nārad*, XVIII. 48.
27a. *Manu*, VII.130–131.
27b. *Rājagrāhyaṣaḍabhāgā dibhiḥ*, Kṣīrasvāmin's commentary on *Amarkośa*, II–8.28.
27c. *LRI*, p. 4
27d. *Loc.cit.*, p. 4.
28. *Bṛhaspatismṛti*, 1–43 vide *Land Revenue in India: Historical Studies* (*LRI*) Ed. R.S. Sharma, 1979, Delhi, Motilal Banarasidass, pp. 4–5.
29. *Gautama-Mitākṣarā* of Haradatta; X. 24.
30. *Manu*, X. 118.
31. On *Manu*, X.118, Medhātithi on *Manu*, X–118 and 120 and *Kullūka* on *Manu* X.118.
32. *Rājdharmakāṇḍa* of the *Kṛtyakalpataru* of Bhaṭṭa Lakshmīdhara, ed. K.V.R., Aiyangar Gaekawad Oriental Series Baroda NOC, p. 89.
33. *Artha.* 5.2.2.
34. *Mānasollāsa* of Someśvara, Ed. G.K. Shrigondekar, (G.O.S. Baroda, XXVIII) 1926, Vol. I p. 44.
35. *Rājadharmakāṇḍa* of *Kṛtyakalpataru*, pp. 89ff.
36. See for list of dynasty-wise taxes, L. Gopal, 1965, *Economic Life of Northern India* c. AD 700–1200 (ELNI), Varanasi, Motilal Banarasidass, pp. 32–70.
37. *Mānasāra* on Architecture and sculpture (Sanskriti Texts with critical 'Notes') Ed. P.K. Acharya, Oxford, 1933, VI 126, chs. 45 and 49 of the text.
38. R.S. Sharma, 1980, *Indian Feudalism* c. AD 300–1200 (*IF*), Delhi, The Macmillan Company of India, (Second edition), p. 167.
39. *Aparājitapṛcchā* of Bhuvandeva, Sed, P.K. Mankad, G.O.S., C.XV, Baroda, 1950, 81.2–10.
40. *Udaya-sundarī-kathā of* Soḍdhala: G.O.S. No. XI, 1920, p. 56.
41. See, D.N. Jha, 1967, *Revenue System in Post-Mauryan and Gupta Times*, Calcutta, Punthi Pustak, pp. 35–50; S.K. Maity, 1970 AD, *Economic Life In Northern India in the Gupta Period.* (cir. 300–350), Delhi, Motilal Banarasidass (second edn.) pp. 77–86; M.H. Gopalan, *op.cit.*, pp. 51–74; U.N. Ghoshal, 1929, *Contribution to the History of Hindu Revenue System* (*HRS*) Calcutta, Calcutta University, pp. 1–50.

42. *Epigraphia Indica* (*EI*), XXI p. 81f.; *EI*, XX, pp. 63f., 188, 122, 129; II, 166; XIII, p. 143; VIII, p. 287; J.F. Fleet, *Inscriptions of the Early Gupta Kings, Corpus Inscriptionum Indicarum* (*CII*), London, 1888, pp. 96, 98, 103, 112, 118, 122, 127, 131, 136, 173, 179, 257, 343, 346.
43. D.C. Sircar, 1965, *Select Inscriptions Bearing on Indian History and Civilization,* (*SI*), Vol. I, Calcutta, Motilal Banarasidass (second edn.); pp. 439–40; EI, XV, no. 11.
44. *S.I.,* pp. 408–409.
45. ELNI, pp. 32–70; B.N.S. Yadava, 1973, *Society and Culture in Northern India in the Twelfth Century* (*SCNI*), Allahabad, Central Book Depot., pp. 288–296; pp. 20–26; R. Niyogi, 1959, *History of the Gāhaḍavāla Dynasty,* Calcutta, Calcutta Oriental Book Agency, pp. 168–190; P. Niyogi, 1962, *Contribution to the Economic History of Northern India From the Ninth to the Twelfth Century,* Calcutta, Progressive Publishers, pp. 77–190.

    A.K. Majumdar, 1956, *Chaulukyas of Gujrat,* Bombay, Bhartiya Vidya Bhawan, D. Sharma, 1966, *Rajasthan Through the Ages,* Bikaner, Rajasthan State Archives, pp. 323–330; B.P. Majumdar, 1960, *Socio-Economic History of Northern India* (1030–1194 AD), Calcutta, Firma K.L. Mukhopadhyaya; D.C. Sarkar, 1966, *Indian Epigraphical Glossary,* Delhi, Motilal Banarasidass and idem, *Indian Epigraphy,* Varanasi, Motilal Banarasidass, 1965, *Lekhapaddhati* (LP) Ed. by C.D. Dalal and G.K. Shrigondekar, G.O.S., Baroda, 1925, pp. 3–54; etc.,

46a. *LRI,* p. 46.

46b. *Deśīnāmamālā* of Hema-Chandra; Ed. R. Pischel, Bombay, Sanskriti Series No. 40, 1923, VII, 89.

47. *EI,* Vol.III, p. 266, 1.11.
48. *LRI,* p. 24.
49. *Loc.cit.,* p. 24.
50. Fleet's view in *EI,* p. 264, fr.
51. *HRS,* p. 237.
52. P. Niyogi, *op.cit.,* p. 184.
53. D. Sharma, *op.cit.,* p. 184.
54. B.N.S. Yadava's General Presidential Address (GPA) entitled, 'Historical Investigation into Social Terminology In Literature: A Problem of the Study of Social Change (Mainly in the context of Early Medieval Northern India)', *Proceedings of the Indian History Congress (PIHC).* 53rd Session, Warangal, 1992–93, pp.11–15.
55. *Ibid.,* p. 13.
56. *LP,* p. 19.
57. *Kāśikā* of Vāmana and Jayāditya, A commentary on *Aṣṭādhyāyī* of Pāṇini, Ed. by A. Sharma, Hyderabad 1969, 5.4–95.
58. W.H. Wiser, *Hindu Jajamānī System,* Delhi, 1988 (reprint) Introduction.
59. Yadava, *GPA,* in PIHC, 1993, p. 11.
60. *Parāśarasmṛti,* 2.12–13.s
61. Yadava, *GPA* in *PHIC,* 1993, p. 11.

62. R. Pischel, *op.cit.*, pp. VI, 104.
63. *Pāia-Sadda-Mahāṇṇava*, V.S. Agrawal, Prakrit Text Society, 1963 (Second edn.), p. 943.
64. *EI*, XXIV, p. 366, II. 4–7.
65. Vogel, *Antiquities of the Chamba State*, Calcutta, 1911, p. 167.
66. Y.B. Singh's articles on, '*Hālikakara:* Crystalization of a Practice into a Tax', *Essays in Ancient Indian Economic History*, Indian History, Congress Golden Jubliee Year Publication, New Delhi, 1987, Vol. II, pp. 88–94.
67. *ELNI*, pp. 17, 22, 54, 61.
68. *HRS* pp. 213, 292.
69. S.K. Maity, *op.cit.*, p. 87.
70. B.N.S. Yadava *GPA* in *PHIC*, 1993, p. 4.
71. Devadatta Sastri (ed.) 1964, *Jayamaṅgalā* commentary on *the Kāmasūtra* of Vātsyāyana, Varanasi, Chaukhambha Vidya Bhawan, Varanasi, 5–5, p. 575.

SECTION FIVE

AGRICULTURE IN EARLY MEDIEVAL INDIA (FROM c. 600 AD TO c. 1200 AD)

CHAPTER 32

Some Aspects of Sharecropping in India in Ancient and Early Medieval Times

B.N.S. Yadava

Sharecropping has been found to be geographically widespread, and of great antiquity, varied manifestations and some persistence, especially in India, Persia and China.[1] In India also, it has existed from quite early times. The *Arthaśāstra* of Kauṭilya provides one of the earliest pieces of evidence bearing on sharecropping in the Mauryan period. The Director of Agriculture (*sītādhyakṣa)* is advised in this text to lease to cultivators, on crop-share basis, the crown or government farms which he has not managed to get cultivated through slaves *(dāsas),* labourers (*karmakaras)* and persons paying their fine by personal labour.[2]

Two main forms of share contract[3] are mentioned in the *Arthaśāstra* (2.24.16). The sharecroppers who impliedly provided their own seed, ploughs and bullocks are said to be entitled to half shares of produce (they are designated as *ardhasītikas*). Those who contributed only their labour (*svavīryopajīvino)* are said to be given a quarter or fifth share.[4]

We find an allusion to crop-sharing contract between the *kṣetrika* (owner of field) and the *bījī* (owner of seed) in the *Manusmṛti*[5] (between 200 BC and AD 200). As per the agreement, the *kṣetrika* is viewed as making over his field to the *bījī* for sowing it with the necessary seed provided by the latter in order to raise crop. In this context, the plough, other agricultural implements and the animal power necessary for cultivation also appear to have been provided by the crop-sharing peasant. Both the *kṣetrika* and the *bījī* are regarded as (equal)[6] sharers (*bhāginau*) of that (crop). Such sharecroppers are designated as *ārdhikas* (the peasants who work the land for half of the crop) in the *Manusmṛti* (IV.253). Such crop-sharing contracts appear to have been common in that age.[7] In the *Yājñavalkya-smṛti* (between AD 100 and 300) we find the term *ardhasīrī*[8] for *ārdhika.*

The evidence of the *Manusmṛti* as well as the *Yājñavalkya-smṛti,* noticed above, may be supplemented with that of a Pallava land grant[9] (third-fourth century) from Andhra, which mentions four sharecroppers (*addhikā = ārdhikāḥ*) who remained attached to the land even after it was donated to some Brāhmaṇas. These sharecroppers, who were tied to the land, appear to have been Śūdras.[10] Some inscriptions reveal that in the mountainous regions of Orissa and the Deccan sharecroppers and other peasants were specifically asked to stay with the land granted to the beneficiaries from the sixth century onwards.[11]

The emergence of landed intermediaries and a hierarchy of ruling landed aristocracy involved in the feudal relationship of domination and subordination, the weakening of the central authority, and a trend toward the insulation of self-sufficient local agrarian economy (especially for about three centuries after the downfall of the Gupta empire) led to the subjection and exploitation of a sizable section of peasants and field labourers in India in early medieval times (from the sixth-seventh to the twelfth century).[12] In this historical milieu sharecropping, as a mode of surplus appropriation, has been found to be associated with a class relationship to 'feudal intermediaries'.[13]

The evidence of the commentary (c. AD 900) of Medhātithi on the *Manusmṛti* (VIII.168) appears to be significant in this context. It indicates that sometimes *akṛta,* or *adatta* (voidable) gifts of unused (*anupayujyamāna)* fields and farms were forcibly *(balāt)* given to unwilling petty peasants for cultivation *(vāhanāya)*: the peasants were, thus, forced[14] to become sharecroppers or temporary tenants. The element of feudal coersion was involved, to some extent, in such a practice of leasing land to which the lessee—sharecropper or temporary tenant—could not legally claim any occupancy right even after working it for any number of years. However, such coersion could have been possible mainly on the estates of rulers, chiefs and big landlords. The evidence of the commentary of Medhātithi further suggests that in Kashmir (where the work was composed), as also in some other regions, land was more abundant than labour and the landlords resorted to the practice of forcing the unwilling peasants to cultivate strips of unfarmed land as sharecroppers or temporary tenants. This device was one of the factors leading to the expansion[15] of agriculture during the later phase of the early medieval period.

A model document for *bhūmisaṃsthā* in the *Lekhapaddhati*[16] reflects the subjection and exploitation of peasants involved in a form of sharecropping relationship as a more or less general phenomenon in some regions. A form of charter (*guṇapatra*) to be granted by a *rāṇā* (ruling chief of a region) to the inhabitants residing in the villages held by him, it asks the peasants (*kuṭumbikas*)[17] living in huts to cultivate the fields individually recorded in the register (*vahikāyām*) against their names (*nibandhabhūmi*), to suffer some penalty if they keep any part uncultivated and to render dues and services according to local customs,[18] including the payment of part of the produce to local officials, village artisans,[19] etc. It provides for the supply of the seeds of rice, wheat, barley, *ciṇā* and *lāṭa* from the barnyard or the threshing floor, the procurement of other seeds being the responsibility of the peasants. The peasants are required to carry two thirds of the grain to the chief's granary as his share and allowed to retain the remaining one-third and the grass and straw as their own share. A peasant found stealing grain is to be warned once, but if he persists, he is to be deprived of his share of the produce and, finally, to be turned out of the village. Complaints of peasants are to be entertained not individually, but only when four of them go together with the *guṇapatra.* The fields, grain, cattle and other property of a peasant who leaves the village and moves away elsewhere are to be taken over by the chief or ruler.[20]

All this may be taken to have been a regional practice in Western India and Rajasthan, for the text reflects the conditions of these regions. But the practice of sharecropping becoming widely prevalent during this period is clear from many pieces of evidence[21] and, above all, from the fact that the *ārdhika* (sharecropper) is mentioned for the first time as a separate mixed caste in the *Parāśarasmṛti*[22] (c. AD 600–900).

The form of document under consideration has been assigned a late date (Saṃvat 1407) in the text,[23] but it represents an earlier tradition. The evidence of the *Lekhapaddhati* may be put alongside that which we find in the *Ādi Purāṇa* of Jinasena (ninth century), which also reflects the conditions of the same region. The *Ādi Purāṇa*[24] asks the ruler to organize cultivation in his *bhaktagrāmas* (provision-villages) through the *karmāntikas* by providing them with seeds and by making other efforts.[25] The assignment of land is, of course, implied here. He is advised to take a just part of the produce from all of them.[26] The peasants in the *bhaktagrāmas* appear as sharecroppers or temporary tenants in a state of dependence on the ruler as well as the land.

The question of share for the sharecropping peasant is of considerable interest. The terms *ardhasītika*, *ārdhika* and *ardhasīrī*, occurring for the sharecroppers in the *Arthaśāstra*, *Manusmṛti* and *Yājñavalkya-smṛti*, respectively, indicate that a 50:50 share was the archetypal division between the sharecropper and the landlord. But, in spite of the fact that the term *ardhasīrī* occurs in a few minor *smṛtis* (c. AD 600–900)—*Vedavyāsa-smṛti*[27] (3.55) and *Aṅgiras-smṛti*[28] (verse 120), the share of the sharecropper generally ceased to remain as one half of the produce in the post-Gupta period. There were share contracts with unequal proportional divisions. The evidence of the *Lekhapaddhati*, as noticed above,[29] shows that in some regions of Western India and Rajasthan, the sharecropper was allowed to retain only one-third of the produce, along with the grass and straw, and the landlord, thus, claimed the greater share.[30]

In view of the non-uniform nature of share contracts in the post-Gupta period, Bhāruci (sixth-seventh century), while commenting on a verse of the *Manusmṛti* (VIII. 39), wrote: 'Entitled to half share means entitled to a share, for the word half is to be understood in the sense of a share, as for example, in the phrase sharecroppers (*ārdhikas*)....' .[31] What Bhāruci further says in this connection indicates that the calculation of the share was to be made according to the qualities of the sharecropper or at the pleasure of the ruler/landlord. This may suggest that the share contracts were considerably affected, in that age, by the distinctly unequal power relationship between the sharecropper and the landlord, leading to meagreness of the former's share of the crop and his exploitation. However, the quality of land provided by the landlord or tenure holder to the sharecropper and the extent to which the latter would supply the necessary instruments of production and inputs (plough, animals, seed)[32] may have been taken into account, to some extent, in the division of the harvest.

The evidence bearing on the recent survivals of leasing land on the basis of produce-sharing in eastern India shows that this practice, involving both economic and non-economic structures of subordination, was reinforced by indebtedness: it has been a common device for maintaining and reinforcing the 'web of dependency'[33] subsuming sharecropping and for transforming the free peasants into sharecroppers. In the light of this information, it may be surmised that the emphasis of Kātyāyana[34] (between AD 300 and 600) on the desirability of making the *karṣakas*[35] *et al.*, repay (a debt) by manual labour[36] has reference to the practice of making the poor indebted peasants toil as sharecroppers or agricultural labourers. The relevant line of Kātyāyana's verse has been cited in the *Smṛti-candrikā*[37] (AD 1200–1225) of Devaṇṇa-bhaṭṭa, which indicates that this practice remained prevalent, to some extent, throughout the early medieval period. The nature of economy in ancient and early medieval times suggests that loans in kind may have been more than those in cash.[38] However, the sharecropping system was not necessarily associated with the rural credit relations, even though they may have contributed, in varying degrees, to its growth at the lower levels in early medieval times.

NOTES AND REFERENCES

1. Cf. T.J. Byres, "Historical Perspectives on Sharecropping", *The Journal of Peasant Studies,* Vol. 10, Nos. 2 & 3, 1983, pp. 7 ff., 32.
2. *The Kauṭilīya Arthaśāstra,* Parts I (text) and II (translation), ed. & tr. R.P. Kangle, Second Edition, University of Bombay, 1969 & 1972, 2.24.2, 16.
3. Cf. Adrienne Cooper, "Sharecroppers and Landlords in Bengal, 1930-50: the Dependency Web and its Implications", *The Journal of Peasant Studies,* Vol. 10, Nos. 2 & 3, 1983, p. 229.
4. Those belonging to this category appear to have planned their process of cultivation on the lands leased to them, and, as such, they were sharecroppers and not simple agricultural labourers: the latter are looked upon as only assisting the landholders who plan and control the production process on their lands. However, the two categories cannot be said to be mutually exclusive. For the general distinction between sharecroppers and agricultural labourers, see B. Chaudhuri in *The Cambridge Economic History of India,* Vol. 2, ed. Dharma Kumar, Cambridge University Press, 1982, p. 163.
5. *Manusmṛti,* ed. G.N. Jha, Asiatic Society of Bengal, Calcutta, 1932, IX. 53.
6. Wendy Doniger with Brian K. Smith, tr., *The Laws of Manu,* Penguin Classics, 1991, IX. 53. G. Bühler, however, renders the term *bhāginau* as 'sharers' only, *The Laws of Manu,* tr., idem, Reprint, Delhi, 1964, IX. 53.
7. *Tasyeha bhāginau dṛṣṭau bījī kṣetrika eva ca—Manusmṛti,* IX. 53.
8. *Yājñavalkya-smṛti,* with the *Mitākṣarā* Commentary of Vijñāneśvara, ed. & tr. U.C. Pandeya, Kashi Sanskrit Series, Varanasi, 1967, I. 166.

   The terms *ārdhika* and *ardhasīrī* are viewed as synonyms by Medhātithi; commentary (c. AD 900) on *Manusmṛti,* IV. 253.
9. *Epigraphia Indica,* I, 1, l. 39.
10. R.S. Sharma, *Indian Feudalism,* Second Edition, Delhi, 1980, pp. 50–51.
11. Idem, *ibid.,* pp. 44 f.; *Early Medieval Indian Society,* Kolkata, 2001, pp. 26f., 189.
12. B.N.S. Yadava, Presidential Address (Section 1), *Proceedings of the Indian History Congress* (*P.I.H.C.),* Forty-first Session, 1980, pp. 24ff.; Presidential Address, *P.I.H.C.,*

Fifty-third Session 1992-93, pp. 8, 17ff.; *Society and Culture in Northern India in the Twelfth Century,* Allahabad, 1973, Chapter 3.

13. D.D. Kosambi, *The Culture and Civilization of Ancient India in Historical Outline,* New Delhi, 1970, p. 150. However, at the lower levels, sharecropping arrangements were also made by the individuals not belonging to the class of feudal intermediaries; infra.
14. As against this, the *Arthaśāstra,* (2.24.16), which throws light on the conditions of the Mauryan period, advises the ruler to lease unfarmed or unused crown or government farms to willing sharecroppers without resorting to any kind of coersion; supra.
15. Meenashri Yadav, *Agriculture and Peasants in Northern India from c. 600 to 1200* AD *(in Hindi),* thesis approved for the degree of Doctor of Philosophy by the University of Allahabad, 1992, pp. 215, 216, 287.
16. *Lekhapaddhati,* ed. C.D. Dalal, Gaekwad Oriental Series, Baroda, 1925, pp. 18f.
17. The terms *kuṭumbika* and *kuṭumbin,* as Sanskrit equivalents of *kuḍumbia* and *kuḍumbi,* respectively, are synonyms (*Pāia-sadda-mahaṇṇavo,* compiled by H.D.T. Sheth, Varanasi, 1963, p. 252, column 2). In earlier times, *kuṭumbikas* as well-off Vaiśya householders were associated not only with cultivation but also with trade and usury (cf. S.C. Bhattacharya, *Some Aspects of Indian Society from c. 2nd Century* BC *to c. 4th Century* AD. Culcutta, 1978, pp. 129 f., 248). But during the early medieval period, *kuṭumbin* as a generic term for peasants acquired a wide prevalence in Northern India (B.N.S. Yadava, *P.I.H.C.*, 53rd Session, 1992–93, pp. 5f.). The commentary of Medhātithi on the *Manusmṛti* (IV. 253) reveals that the usage of the word *kuṭumbin* in the specific sense of sharecropper was becoming common round about AD 900 (the time when the commentary was composed). Medhātithi has taken the word *kuṭumbī* as a synonym of *ārdhika* (sharecropper entitled to one half of the produce). Bhaṭṭasvāmin (twelfth century), a commentator on the *Arthaśāstra* of Kauṭilya, has designated the sharecroppers entitled to one half of the produce as *grāmya-kuṭumbinaḥ (Journal of the Bihar and Orissa Research Society,* Vol. 12, Part 2, p. 137; cited by Meenashri Yadav, *op. cit.*, p. 221 fn.5). In the *Lekhapaddhati* also, the term *kuṭumbika* has been used to denote a sharecropper.
18. *Lekhapaddhati,* p. 19.
19. We find the mention of five artisans *(pañcakāruka)* who may be the carpenter, the ironsmith, the potter, the barber and the washerman (*ibid.*, p. 108).
20. *anyatra praṇasyagatakuṭumbikasya kṣetrakhalakaḍhaura-prabhṛti sarvaṃ rājakule svādhīnaṃ kartavyam; ibid.,* p. 19. To threaten a fugitive with confiscation was one of the methods adopted by the lords in feudal Europe to prevent the migration of their subjects. They tried to retain their peasants, for under the condition of closed agrarian economy, it was useless to have an estate without labour to work it. But owing to the fragmentation of authority and the abundance of virgin soil, especially in certain regions of France, it was difficult to prevent desertions. Marc Bloch, *Feudal Society,* (London, 1965), p. 263.

21. A notable factor which could not but lead to the increase in the practice of sharecropping at the lower levels was the 'criterion of social evaluation', which tended to be 'deprecatory of physical labour', including of course labour involved in agricultural operations, in the post-Gupta period (G.C. Pande, *Foundations of Indian Culture,* Vol. II, Second Edition, Delhi, 1990, p. 240).
22. *Vaiśyakanyāsmudbhuto brāhmaṇena tu saṃskṛtaḥ, so hyārddhika iti jñeyo bhojyo viprairna saṃśayaḥ; Parāśara-smṛti,* ed. and tr. Shiva Dutta Mishra, Varanasi, 1969, XI. 25. *Ārdhikas,* or *ardhasīrīs* generally belonged to the Śūdra *varṇa* (R.S. Sharma, *Śudras in Ancient India,* Second Edition, Delhi, 1980, p. 257). However, the way in which their origin has been explained here suggests that a section of the Vaiśyas had also been reduced to the status of sharecroppers.

    The *Vedavyāsa-smṛti* (1.11) mentions the *Kuṭumbin* caste of the Śūdras, along with Vaṇik, Kirāta, Kāyastha, and Mālākāra castes (*Smṛtīnāṃ Samuccayaḥ,* ed. V.G. Apate, Anandāśrama Sanskrit Series, Pune, 1929, p. 357). As the term *kuṭumbin* had begun to be used for the sharecroppers also during the early medieval period (supra), it may be said that these may have represented a sizable component of this caste also at the formational stage.
23. *Lekhapaddhti,* p. 18.
24. Bhāratīya Jñānapīṭha, Kashi, 1951, 42. 139ff.
25. *Ibid.,* 42. 174-76.
26. *Ibid.,* 42. 177.
27. *Smṛtīnāṃ Samuccayaḥ,* ed. V.G. Apate, p. 365.
28. Ibid., p. 6.
29. Supra.
30. However, according to the Chinese traveller I-tsing, (AD 635–713), the Buddhist monasteries provided the monastic servants or some other (peasant) families with the land as well as the bulls for cultivation, and were 'responsible for nothing else'. This was also a form of sharecropping arrangement by which the monastery received only one-sixth of every product. But the division of the product was subject to modification according to seasons; J. Takakusu, *Record of Buddhist Religion as Practised in India and Malay Archipelago* by I-tsing, London, 1896/Delhi, 1966, p. 61.
31. *Bhāruci's Commentary on the Manusmṛti,* ed. and tr. J.D.M. Derrett, Vol. II, Wiesbaden, 1975, VIII. 39.
32. In the *Arthaśāstra* (2.24.16) of Kauṭilya also, these elements are taken into account while dividing the crop.
33. Adrienne Cooper, *The Journal of Peasant Studies,* Vol. 10, Nos. 2 & 3, 1983, pp. 227ff.; Bhaduri, *The Economic Journal,* Vol. 83, No. 329, cited in T.J. Byres, *op. cit.,* p. 8.
34. *Kātyāyana-smṛti-sāroddhāraḥ,* ed. P.V. Kane, verse 480.
35. During the post-Gupta period the term *karṣaka* signified not only dependent ploughmen, petty tenant-peasants and poor owner (in a restricted sense) cultivators but also sharecroppers in different contexts; B.N.S. Yadava, Presidential Address, *P.I.H.C.,* 53rd Session, 1992–93, pp. 6ff.

36. Bhāruci and Medhātithi (on *Manusmṛti*, VIII. 46) also prescribed that a poor debtor unable to repay his loan should be made to do manual labour. This is unlike what we find in ancient times when failure to repay debts led to slavery; B.N.S. Yadava, *The Indian Historical Review*, Vol. V, Nos 1–2, p. 41.
37. Mysore Govt. Oriental Library Series, III, p. 384.
38. Cf. R.S. Sharma, *Light on Early Indian Society and Economy*, Bombay, 1966, p. 127.

CHAPTER 33

Irrigation in South India (Upto 1300 AD) Techniques and Management

K.V. Raman

GEOGRAPHICAL BACKGROUND

The land south of the Vindhyas was known in ancient times as *Dakṣinapatha,* the 'Southern Region' in the later works as the Deccan. Geographically, it is the peninsular portion of India fitting into the Indian Ocean between the Arabian Sea and the Bay of Bengal and joining at Cape Comorin (Kanyakumari). But by contiguity, language, family, food habits and usage, the name southern India is applied for a more restricted area comprising the land south of the rivers Godavari and Krishna, i.e., the four states of Andhra Pradesh, Karnataka, Tamil Nadu and Kerala. We have the extensive Deccan plateau in the centre, a long mountain chain running from the north to the south along the west and east called the Western and the Eastern Ghats, the former in parts rising to the heights ranging from 4000 to 9000 ft and latter of considerably lower elevation. The centre of the peninsula consists of generally of undulating tablelands for 1000 to 5000 ft above the sea level. The land slopes gradually from the western hills towards the Bay of Bengal on the east, as indicated by its major rivers, the Godavari, Krishna, Tungabadra and Kavery, and also smaller rivers like the north Pennar (AP), south Pennar, Vaigai and Tambraparani. The deltas of the Godavari and Krishna (in Andhra Pradesh) and Kaveri (in Karnataka and Tamil Nadu) which alone can be called perennial, cover wide expanses of irrigated crops and, hence, form the most fertile region. The other rivers are seasonal and, so, limited in range as an irrigational source. A number of smaller rivers and as tributaries (e.g., *Kabbani, Emavati*) for Kaveri or as branches (e.g., *Kollidam*) have served as important sources of irrigation and attracted human settlements from very ancient times. The Godavari runs a course of 900 miles from hills in the Nashik district to the Bay of Bengal and drains an area of 1,12,000 sq. miles with its many tributaries and branches. The Krishna also has a long stretch of nearly 800 miles passing through Maharashtra, Karnataka and Andhra Pradesh. Kaveri, starting from the Coorg, waters the considerable portion of Karnataka and fertilizes the western districts of Tamil Nadu, Erode, Karur, Tiruchi and Thanjavur before its falls into the Bay of Bengal at the historic town of Kaviripumpattinam, covering a distance of above 500 miles. The Godavari, Krishna and Kaveri deltaic regions can indeed be called the granary of the south. Though much shorter in length and drainage, further south, the Tāmbraparni river formed the lifeline for agriculture in Tirunelveli district as the River Periar, the west-flowing river in Kerala. Apart from these perennial rivers and their tributaries and branches, there are a number of small rainfed local rivers, which often served as sources for filling the local lakes and ponds, e.g. the Pennar and the Palar.

South India receives the benefit of both the southwest monsoons (from June-July) and the northeast monsoons (from October–November), the former more copious than the latter. Kerala, Karnataka and western parts of Andhra Pradesh benefit from the southwest monsoons, whereas Tamil Nadu and the coastal districts of Andhra Pradesh receive rains under the influence of the northeast monsoons. But monsoon rains spread over only a short period in a year, besides being unpredictable, and are often found to be inadequate. Periodical failure of monsoons were recorded in more recent decades, and we know the consequent drought and famine conditions by way of literature, inscriptions and administrative reports (of the British period). Untamed rivers, during the times of excessive rains or floods, would damage the crops and cause scarcity of food grains. These factors necessitated the devising of ways and means to keep the supply of water to the fields through artificial means in order to ensure agricultural production. They fall under three main categories: tank, canal and well irrigation, which would include maintenance works like construction or strengthening of embankments, tank bunds, de-silting, etc. Though the history of agriculture in south India can be traced back to the Neolithic times, there is very little evidence of any irrigational activity for that period. By and large, it was dry and shifting cultivation. Organized irrigation or tank irrigation began probably during the Iron Age or the Megalithic period, around the middle of the first millennium AD. During that period, human settlements with burial monumental adjuncts proliferated all over the south. They are invariably found to occur in juxtaposition to irrigation tanks but very close to the arable lands, e.g., Rajakulipettai Megaliths close to the huge Kolava Lake in Chengalpattu and extensive urn burial site found at Adichanallur on the banks of Tambraparani River. The evidence of rice found in these and other sites attest to the practice of irrigated agriculture. Tank irrigation and lift irrigation seem to have been developed during this period.

Coming to the historical times, we have ample evidence of irrigation in south India from the literary, epigraphical and archaeological sources. An integrated account from all these sources is presented here.

IMPORTANCE OF IRRIGATION FOR AGRICULTURE

The importance of rains in their proper seasons was considered auspicious and absence of these was considered as the sign of the growing *adharma*, says the *Tirukkural.* The ancient Tamil work states: 'Neither alms giving nor ascetic practices will abide in the world, if the sky does not shower its gifts'. The constituents of a kingdom, according to it, are the two waters from above (rains) and below (springs). Construction of reservoirs and digging of canals were considered as acts of fame to a ruler by a Sangam poet. The *Amuktamālyada* written by Krishna Deva Rāya echoes the same opinion when it states virtue (*dharma*) and prosperity (*artha*) will increase only when tanks and irrigation channels are constructed (Canot IV, verse No. 236). Digging of a tank was considered as one of the seven meritorious acts of a person (*sapta-santanam*).

It was no wonder the ruling kings and chieftains vied with one another to undertake irrigational schemes to contribute to the agricultural wealth and increase water resources for the general welfare of the country. This ideal also inspired the landed gentry, merchant classes, agricultural guilds, village assemblies, temples and *maths* to undertake similar acts. One of the celebrated kings of the Sangam period, Karikala Chōḷa (c.150 AD) is accredited with leaving of the jungles and introduction of agricultural settlements, construction of embankments on both sides of the Kaveri River, construction of a dam (Kallanai now known as the Grand Anecut) across the Kaveri and also digging canals to harness it fully and prevent inundation, wastage or spillover. He was hailed as Peruvalattan, the great benefactor or fertilizer. The Sangam works provide clear evidence of agrarian settlements and irrigated rice fields in the valleys watered by Kaveri, Vaigai, and Tāmbraparani and besides evidence of man made canals and tanks. The classification of Tamil land into five ecological zones found in the *Tolkāppiam* and other works is also significant—the hilly regions (with waterfalls and springs), the forest area or the pastoral tract (*mullai* with wild streams), the riverine plains (*marudam* with rivers, wells and ponds), the coastal or littoral area (with sand wells and backwater) and the desert or arid tracts (*pālai* with barren or infertile soil). The riverine plains or the *marudam* area was considered the most suitable for human settlement, the home of the agriculturists (*Ulavar*). It is no surprise that the three ancient kingdoms of Tamil land rose up with their capitals located in this wet zone—Uraiyur, the capital of the Chōḷa on the banks of the Kaveri, Madurai of the Pāndyas on the banks of Vaigai and Karur-Vanji, the capital of the Cheras on the River Amravati, a branch of Kaveri. We can also add to this hot Kanchi city on River Vegavati, Tirukkovalur, and the capital of Malayamans on the River Peṇṇai.

References to tanks, minor dams and water lifting with basket-pails, the rivers overflowing the banks and filling up the ponds are found in the Sangam works. The method of construction of sluices and water-heads to control and regulate the Vaigai River are mentioned in the *Paripādal* (No. 20). The epic *Silappadikāram* (canto XI) describes how when a dam, tried to leap beyond, arrested the flow of the Kaveri and the noise it produced silenced the sounds of bucket or water-lifts, loud *picottah* or the palm-leaf used for irrigation.

Another ancient work, *Gathasaptasati* by the Satavāhanā king Hala (first century AD) contains interesting references to irrigation and agriculture. It makes a particular reference to a contrivance called *rahattagadia* for lifting water from the wells (Jogekar S.A. ed, *Saptasati*, Poona, 1912.). 'This contrivance was formed of a string of pots attached to a wheel in such a way that water lifted from the well and conveyed the fields through the channels' (Kotriah).

WATER RESERVOIR AND CANAL

Archaeological excavations have provided us with valuable material evidence of irrigational activities in the early centuries of the Christian era. We can draw attention to two such discoveries one from Tamil Nadu and another from Andhra Pradesh. At Vanagiri, a suburb of the ancient port town of Kaveripūmpaṭṭinam, a water reservoir made of earthen bund with a well-designed brick-built mouth or façade to receive the water was discovered. It has brick-built inlet channel to receive the water from a channel from the Kaveri. It is datable to first century AD. At Nagarjunakonda (Guntur district). In the ancient Vijaypuri the capital of the Ikshvakus on the banks of the River Krishna was discovered a canal about 50 ft wide and 6 ft deep with side-walls and floorings built of lime gravel mixed with it. A large brick-built rectangular tank was also exposed.

CANALS, TANKS AND SLUICES

Some early references to the Karnataka, artificial tanks and sluices are found, e.g. Kadamba King Mayurasarman (345–60 AD) and Kakustavarman (425–450 AD) are credited with construction of tanks, the latter at Talagunta in Shimoga district. A Ganga record of the ninth century lists the names of ten tanks around the place Marugara-Visaya. A record dated 714 lists the names of five tanks and a dam (*Sethu*) across the River Kilini. We also have a reference to the water from the Kaveri being channelized to a big lake appropriately named *Kaveri-Sāgara*. Tanks with crescent-shaped (*Balachandrakare-kola*) and halfmoon-shaped (*Ardhchandra-kola*) are mentioned. As has been rightly pointed out, their curved form must have been resorted to an account on the geographical factors at the site of construction. Tanks named after kings, princes, and private donors artisans are also plenty. Quite a number of records speak of the special honours bestowed on the donors of tanks and sluices like golden bracelets and titles like *Nolamba Doḍḍa* and *Sreemukha Doḍḍa*.

The tempo of irrigational activities steadily increased under the Chalukyas of Kalyāni and the Hoysalas. The famous encyclopaedic work *Manasollasa* by the Chalukyan King Somesvara (1126–1138 AD) reflects the trend of the times and lays down that it was the duty of the kings to construct and maintain wells, lakes, ponds and water-sheds as welfare measures. Somesvara II made a gift for construction of the Piriyakere Tank in 1071. Instances of taxes and levies collected for maintanence of a tank and numerous names of tanks can be culled out from the Hoysala records like *Dorasamudra* (Halebid), *Rudrasamudra, Gangasamudra, Achyutsamudra, Virasamudra* (the last four were made by Viraballala's minister in 1187). Contributions were made for the periodical maintenance of the tanks.

Equally important was the canal irrigation. In 1088, the excavation of a canal and subcanals from the river Tungabhadra after a preliminary survey has been recorded. Col. Mackenzie, the British surveyor, after observing the ancient channel for the *Yelachi* (or Yagachi) river, which supplied water to the Hoysala capital Dwārasamudra, covering a distance of 16 km, has recorded his admiration for the engineering skill and planning. From the Hoysala inscription we learn that there were hereditary professionals specialized in irrigation technology and they were called *tantrapalas* (C.T.M. Kotriah).

TAMIL NADU

As already noted, Tamil Nadu has preserved the evidence of irrigational strategies right from the early centuries of the Christian era. There are wet and dry zones and, hence, different methods had to be adopted. The valleys of the Kaveri, Tāmbraparani and, to a lesser degree, of smaller rivers canal irrigation supplemented by tank irrigation were resorted to. In drier zones like Tondaimandalam areas (northern districts), rivers like Palar and south Pennar are seasonal; canal irrigation was effective for shorter duration, but tank irrigation and well irrigation had to be extensively resorted to. The same was true of certain portions of Pāndya-nādu, especially the arid coastal districts of Ramanathapuram district.

But we have excellent epigraphical records to study the irrigation pattern. We get the names of innumerable distributive channels from the Kaveri that virtually honeycombed the Tiruchi–Thanjavur region during the Chōḷa times, making the area the most fertile rice-belt. In fact, the Kaveri deltaic region came to be called the land of floods *(Punalnāḍu)* and continuous efforts were made to regulate and tame it by digging canal networks for the distribution to irrigate the lands. Some important irrigational undertakings under the Pallavas may be mentioned here. They took a leading part in clearing jungles and arid lands into fertile tracts for which they earned the name *Kāduvetti.*

The Pallavas studded their country with large lakes called *ēri*, or *taṭākas*, many of which bear their names. For some of the natural lakes, they provided the bunds and the sluices. Some of these tanks are: Tiraiyan-eri at Tenneri in Kanchipuram district, Mahendra-taṭāka, Paramesvra-taṭāka, Vairamegha-taṭāka (near Uttiramērur) named after Dantivarman. Kaveripakkam tank, which is about four miles in length, was created by Nandivarman III (846–869); Marpidugu *ēri* (tank) was created by Dantivarman at Alampakkam near Lituchrapally. Village assemblies have specialized committees for the management of lakes called *ēri-vāriyam* (tank committee).

We also have evidence of canals (*vāykkāls*) dug extensively from the rivers like the Palar, Seyyar, Vegavati and Pennār. They were known as *ārrukāl* and the lands irrigated by them were known to receive *ārrukkal-pāyachal.* Inundation canals were known as *vellakkal* and the smaller branches leading to the fields were called the *Oḍai.* Canals were extensively used to feed the tanks or lakes.

The Pandyan country is blessed with two bigger rivers, Vaigai and the Tāmbraparani, both of which are seasonal but susceptible to floods during the heavy monsoons. Besides them, there are smaller streams like Gundaru, Veitaru, etc. It was on the banks of the rivers many ancient sacred centres (like the *Navativupatis,* nine Visnu shrines) on the Tāmbraparani and urban centres (like Madurai on the Vaigai) and numerous agrarian villages have risen. So, channels were drawn from the rivers for irrigation and also to feed the tanks. Examples of such tanks are Maranēri, Perumkuḷam, Maḍakkulam (near Madurai). They were also partly rainfed tanks. We have a record dated the seventh century AD, which has registered the construction of a sluice (*madagu*) and excavation of a channel (*kāl*) or (*Kālvai*) from the Vaigai. It was built by a Pandyan king Arikesari (alias Sendan), after whom, the sluice was named.

During the reign of Srimāra Srivallabha (811–857), a number of irrigation projects were undertaken, which included renewing existing tanks, canals and sluices and brick and granite blocks were used as building materials. The same process continued in later times too and we can cite a few references like the stone embankment named Kallanai, Parakrama-Pāndian Kallanai, Vira Pāndyan-Kal (channel) and water reservoirs like Sri Vallabha Pērēri (big lake).

MAINTENANCE AND REPAIRS

There was regular provision for the upkeep and maintenance and periodical repairs to bunds, sluices and channels by the collection of levies like *Vaykkal-pattam* (canal tax), *pasipattam* (fishing tax). A case of a fisherman being entrusted with the responsibility over the safety and supervision of an embankment (*aṇai*) and the connected channel is recorded at Suralur. He had to report to the village of any breach or leakage for urgent repairs. In return, the fisherman was granted tax-free land and also rights to collect the canal tax and fishing tax mentioned above.

Many of canals of the Pāndyan land continued to serve well even today. A British engineer named Horsley, in the last century, has recorded his admiration for the engineering skill of the Pandyan canal. 'The evidence of their skill and almost super-human perseverance as so marked that I have in carrying out the (repair) works felt contended and fully satisfied to follow in the footsteps of those whom I consider to have been masters in their art and irrigation engineering' (quoted by A. Appadorai).

CANALS AND TANKS CHŌḶA REGION

There was tremendous agrarian expansion during the rule of the imperial Chōḷa (c. 900–1270 AD) all over Tamil Nadu and more particularly in the Kaveri basin. Most of the distributory canals of the Kaveri, criss-crossing the deltaic region, belong to their period, e.g., Uyyakondan canal, Rajendran vāykkāl, Sembian Mahadegvi vāykkāl. There was a well-developed and highly efficient system of water management from their village level upwards. The increase in the royal patronage, and also the number of *devadana* and *bramadeya* lands which increased the role of the temples and village assemblies in the field. Committees like *ēri-vāriyam* (tank-committee) and *toṭṭa-variam* (garden committees) were active as also the temples with their vast resources in land, men and money.

The tanks that came up during the Chōḷa period are too many to be listed here. But a few most outstanding may be briefly mentioned: Rajendra Chōḷa I's huge tank named *Sōlagangam* in his capital town Gangaikonda Solapuram was described as the liquid pillar of victory (*Jalasthamba*). About 16 miles long, it was provided with sluices and canals for irrigating the lands in the neighbouring areas. Another very large lake of this period, which even today seems an important source of irrigation was the *Viranameri* near Kāttumannarkoil in South Arcot district; founded by Parantaka Chōḷa I. Other famous lakes of their period are Madurantakam, Sundra-chōlapērēri, Kundavai-Pērēri (after a Chōḷa queen).

ANDHRA PRADESH

In Andhra Pradesh, certain districts are extremely fertile and others extremely dry and arid. The districts watered by the Godavari and the Krishna and their innumerable channels and the portion of the Pennar basin are more fertile than the areas like the Rayalseema and Telengana, which are dry areas dependent on the annual rains, water tanks and wells. As already mentioned, irrigation works like digging canals, tanks and lakes are attested to the early centuries of the Christian era, e.g., Nagarju-Nakonda. A medieval inscription from Hanmakonda states that in *Andradesa*, 'there are hundreds of tanks and thousands of rivulets and they appear to be the ocean and its consorts, respectively'. We get the names of several tanks or Samudras. Speaking about the Kakatiyas, Professor Yazdani says, 'they inherited a system of irrigation which had been in use in the land from time immemorial. The main feature of the system was the existence of a network of tanks in which rainwater was caught and stored for utilization for the cultivation of soil'. We will mention a few examples of irrigation tanks that occur in the epigraphical records:

Kesari taṭākam (Motupalli inscription, eleventh century) excavated by one Prole and it is described as an ocean.

Sari-Samudra (Kazipet, by Prole II).

Hema Kunda taṭāka (an enormous tank in the centre of the city by Rudradeva in (1084).

Pakhal Tank—30 miles from Warrangal, capital of the Kakatiyas, by Kakatiya Ganapati (1199-1202).

Rāmappa and *Ghanpur* lake also by the same ruler.

Kumāra Samudram, Tirupura Samudram dug by one Sambadeva mentioned in a record, of 1295 from Tripurantakam in Kunool district.

SPRINGWATER TANKS, CANALS AND DAMS

Innumerable names of canals (*Kalva*), dams (*ana*) and sluices (*timu*) occur in the inscriptions, e.g., Rayashasramalla-kalva from river Cheyyaru. Springwater canals and tanks were called *uttu-kalva.*

An interesting instance of propitiating the raingod Varuna is available. *Varuna-pratishta* ceremony was performed after the construction of tanks to ensure copious and unfailing rainfall by a Kakatiya king.

As in Tamil Nadu and Karnataka, in Andhra Pradesh and Kerala too, the practice of entrusting the private individuals with the task of maintaining the public works like tanks and canals was popular. If it was called *Eripatti* in Tamil Nadu, here it was called *Dasavandha-manya.*

A piece of land or one-tenth share of the produce was gifted to as compensation. This practice was in usage in modern times also. In Karnataka, such grants were known as *Kodage.* Grants like these acted as a great incentive for public-spirited individuals to construct tanks, and channels and maintain them well.

From the foregoing account, it can be seen that south India provides a classic example of a land where systematic irrigational activities were widespread for well over 1500 years. The importance of evolving different methods of tapping water resources from the rivers, rainwaters, spring channels, tanks, and wells was widely recognized and practiced. Reclamation of fallow lands with an increase in population, and increased agriculture settlements or villages and towns, *devadānas, brahmadēyas* and the consequent agrarian expansion. The need for undertaking, irrigational networks was realised. In this vast undertaking, the state, royal officials, chieftains, village assemblies, merchants and merchant guilds, temples, private individuals have all played a part. Inscriptions provide thousands of such instances for clearing the jungles, reclamation of lands, river-basin irrigation including flood control, and cutting inundation channels canal-irrigation (*ātrukāl*), digging new channels (*kālvay*), bonding the vast natural rage in the catchment areas (*ēri* or *kulavāy*), digging smaller tanks to be fed by the rivers or canals (*kuḷam*), regulating the flow of the channels by dams (*anai*) and sluices (*madagu*), etc., go on to support the general conclusion that their was considerable skill, experience and knowhow in the management of water sources, not only creating new tanks and new channels but in the continuous maintenance and repair to such undertakings.

The villages on either banks of the River Kaveri and its numerous branches were to contribute for keeping the bund in good repair and constant care by paying cesses like *Kāverikarai viniyogam* and *āṭrankarai viniyogam.* For the care and maintenance of lake also, special cesses were collected like *ēri-ayam* and *nir-nilaikkasu.* Irrigation from wells was quite extensive by waterlifts or *picottas* (*ettam*) and a small cess was collected on them. We also get references to tank fund (*ēri-mudal*) to meet the expenses for maintaining tanks and lakes. The special watch and ward staff to inspect and report breaches was called as *Vettiyans* and *ēri-kaval,* etc. Lease for fishery (*eri-min-pattam or pasi-pattam*) brought the income from tanks to be used for their maintenance, desilting, repairs, etc. Often, those who took the fishing lease were also put in charge of the maintenance of the tanks. Experts in digging wells and other earth works were called *Upparas* and *Oddars.*

DE-SILTING OF TANKS

Periodical desilting of tanks as part of the regular maintenance of tanks is frequently mentioned in the inscriptions of all periods and the work was entrusted to the special squad on a lease basis. We get interesting details of the desilting operation done in a big tank. Six labourers, using a boat, removed baskets of silt in about 200 cubic feet of earth. A carpenter and a blacksmith were engaged to assist in the repairing the boat to keep it fit for the task (T.M. Srinivasan).

WATER-LIFTING DEVICES

We will now refer to some mechanical devices (*yantras*) used for lift irrigation. Already the use of bucket-wheel or water wheel was cited. Also of great antiquity were the *picottas* or water lever *ettam* (in Tamil) and *rātanam* (in Kannada). A giant metal bucket is attached to one end of a long pole, which is mounted across the top of a vertical pole close to the well. One man used to stand on the top to press the top cross pole towards the well so that the bucket sank in the water and then he lifted it to the top and emptied it to the mouth of the channel through a tube to supply to the fields.

The other method of water lifting was to attach the basket to a rope and a pulley wheel fixed to the top of the well. The rope was pulled forward and back by a pair of bullocks. By this way, water was drawn up for supply to the fields. This method is popularly known as *Kavalai* or *Kabila.* Both the *ēttam* and *kavalai* devices are popular even now, though electric pump sets are increasingly used for lifting water (T.M. Srinivasan, 1991).

INTER-LINKING OF TANKS

An interesting case of interlinking of tanks is available in the records of Karnataka. These tanks of different sizes were dug at three places at varying levels and at considerable distance from one another. The largest and the farthest one (*Pērēri*) occupied the highest level, the second was slightly smaller at a lower level and the third was the smallest at the lowest level (G.R. Kuppuswami). The principle behind the arrangement was that the bigger tank could feed the smaller ones and cover more villages. This system seems to have been followed in all parts of south India. This tank network had elicited the praise of the engineers and administrators of the British period.

An engineer survey report of 1899 in Pudukottai State in Tamil Nadu states 'that there was little or no scope for really fresh irrigation projects. The ancients have done their best and studded the country with the system of irrigation tanks, connected like the links in a chain, the surplus of one filling the next below so that not a particle of water could escape unutilized.'

British administrators such as Sir Thomas Munro, Crole, Major Handerson, Major Sankey and Horley have all recorded their admiration for the irrigation works of south India. According to them, there was very little scope for any new work and their main task was only to maintain them by regular repairs and renovations. Crole, in his Chingleput District Gazetteer 1879, wrote, 'almost every catchment basin, however small, still bears traces of having been bound across and, in many instances, this was done in order to secure a crop of paddy in a few acres of ungenerous soil to which all fostering care of British administration has failed to induce cultivation. They bear witness to the enlightenment of those Hindu kings, while the absence of scientific instruments in those remote times compels astonishment of the beholder.'

CONSTRUCTIONAL METHODS AND TECHNIQUES

Some scientific methods based on experiment and trial and error were evolved and laid down for construction of tanks. The field of hydraulics was recognized as a special field and was called *jalasastra* and *pathasastra.* The hydraulic engineer was called *jalasutrada* and water-diviners were called as *Kupadasakas.* There is evidence to presume that texts existed in the subject, though they are lost now. A late manuscript named *Kupasastra* in Tamil has been published by the Tamil Nadu Govt. Mss. library. Though it is a late mss. it seems to have been based on earlier works. But, fortunately, we have an excellent account of the method of tank construction preserved in the famous Borumamilla tank inscription dated 1367. Among the twelve requirements of a good tank, it includes experts learned in hydrology (*pathasāstra*), ground with hard clay, a river with sweet water with in three *yōjanas* distant, the hill and a dam of compact-stone wall, not too long but built between the hills, an extensive and deep bed, a quarry containing limestones, neighbouring fields, a water course or the sluicer and a gang of skilled workers. The record also gives a list of six defects or faults (*doshas*) in a tank like the presence of water-leakage, saline water, wrong situation, elevation in the middle of the tank bed, scanty supply of water and excess of water, (*Epigraphia Indica* XIV pp. 97 ff; also A. Appadorai). The inscription belongs to the early fourteenth century, and as such, can be taken to represent the cumulative expertise and knowhow born out of experience and experiments of the preceding times.

REFERENCES

1. Appadorai, A., 1956, *Economic Conditions in Southern India* (1000–1500), Madras.
2. Dikshit, G.S., Kuppusaami, G.R. and Mohan, S.K., 1993, *Gandhi Sahitya Sangha*, Bangalore.
3. Kotriah, C.T.M., 1995, *Irrigation Systems Under the Vijayanagar Empire*, Mysore.
4. Kuppusaami, G.R., 1975, *Economic Conditions in Karnataka*, Dharwar.
5. Sastri, K.A.N., 1955, *Chōḷa*, Madras.
6. Srinivasan, T.M., 1991, *Irrigation and Water Supply in South India*, Madras.
7. Tirumalai, R., 1995, *Collected Papers*, Madras.
8. Venkayya, V., 1903–04, Irrigation in Southern India in Ancient times, *A.S.I. Annual Report*.

CHAPTER 34

Some Aspects of Agriculture as Described in the *Lōkōpakāra*

A.V. Narsimha Murthy

INTRODUCTION

Lōkōpakāra[1] is a Kannada literary work written by Chāvuṇḍarāya II[2], the grandson of Pangauṇḍa Bhaṭṭa and the son of Bhaṭṭoparya and Malikabbe (mother). Chāvuṇḍarāya II was a Sarvādhikāri under Kalyāṇa Chālukya king Jayasimha II, also known as Jagadekamalla. He completed the above work in 1025 AD. Chāvuṇḍarāya II belonged to Mudgal in north Karnataka. He served the Chālukya king faithfully and rose to the position of Sarvādhikāri. Some inscriptions refer to him when he was a minister at Banavāsi. The author has called his work as *Lōkōpakāra* as he wanted to serve the people by writing this book. The book contains twelve chapters and each chapter is devoted to the study of a particular aspect. The sixth chapter throws light on many aspects of agriculture and horticulture. The seventh chapter, dealing with cookery, is also of some interest in this respect as it mentions various food articles that were used during the period. Though it is an encyclopaedia, its contents are taken to represent the practices of the early part of the eleventh century, coinciding with the period of the Chālukyas of Kalyāṇa.

Chāvuṇḍarāya calls the sixth chapter as *Vṛkṣāyurveda* and discusses the methods of growing various trees without any disease and to get maximum yield from them. Before sowing the seeds, an astrologer should be consulted regarding the suitability of the stars. *Rōhiṇi, Mṛigaśira, Makha, Hasta, Chitta, Viśākha, Mūla and Śrāvaṇa* are good and auspicious constellations for sowing the seeds. Then the soil should be tilled properly and the ground should be levelled. The farmer should select such a place where the subsoil water is available in plenty, or water should be made available through irrigation canals. Then pits should be dug, each pit measuring six feet in depth and three feet in breadth.[3] The distance from one pit to another should be 24 or 21 or 15 feet. Then the seeds, which drop out on their own, should be selected, mixed with cowdung and dried for five days under the sun. Then they should be soaked in milky water for five days. Afterwards, they should be treated in a solution of juice of *gullada* fruit and tender coconut water and later, administered the smoke of *vāyuviḷanga* and *ghee.* Such seeds are fit for sowing. The author further recommends the use of the flesh of pig with milk for immediate sprouting of the seeds.

Then Chāvuṇḍarāya describes the method of 'grafting the branches of certain, plants like pomegranate, jackfruit, *jambu, aśoka,* plantain, *mallige* (flower), *mādala* (fruit). A mixture of cowdung, milk, *viḷanga, ghee,* honey, *lāmancha* and sesame should be prepared and applied to the branches to be planted for better results. In addition to the usual stem grafting which is common, Chāvuṇḍarāya mentions cleft grafting, whereby the branch is vertically cut and a wedge put in between and the paste, as mentioned above, should be inserted within. Even modern horticulturists would recommend such a grafting.

The work *Lōkōpakāra* recognizes the importance of good manure for cultivation. But it is very curious that the author recommends the use of flesh of various animals as one of the components of the manure. Along with water, the elixir of goat flesh added to juice of *sabbasige* leaves should be given. Further, *māsha* and green gram should be boiled in goat's milk and that decoction, along with the elixir of the flesh of pig and deer with ghee and honey, should be placed around the roots of the plants.

Next, Chāvuṇḍarāya narrates the method of protecting the young plants from the insects and other diseases. He recommends the removal of weeds first and then an insecticide consisting of *ingu* (asafoetida), *baje,* pepper, *viḷanga* and cashew should be soaked in urine of the cattle and poured into the pit to destroy all insects. If a tree gets afflicted with any disease, which makes it lose its branches or leaves, such holes should be sealed by a mixture of hot ghee, *viḷanga* and black soil. Such a tree grows without any disease.

The next subject Chāvuṇḍarāya touches is the method by which one can get a very good yield of fruits like plantain, coconut, mango, jackfruit, grapes, pomegranate, orange, myrobalan, *mādala* (sweet lime), wood apple and tamarind. Secretion of the temples of an elephant should be put into the skull of a monkey or the tooth of a pig and it should be buried at the root of the plantain to get plantains as big as the ivory of the elephant. Sesame, linseed, *viḷanga, saindhava lavaṇa* should be poured and mixed with honey and liquor and this should be applied to coconut tree to get very good yield of coconuts. The yield of the mango could be substantially increased if cross-fertilization between mango tree and the creeper of *viḷanga* is done. Chāvuṇḍarāya has prescriptions for improving the quality of mango. If a decoction of the flesh of pig, deer, ox and fox along with the barks of fig, banana, jackfruit and pipal is fed to the mango tree, which was yielding sour fruits, thereafter it will yield sweet mangoes. In the same manner, if the decoction of *bhadramasthe,* cuscus grass, and *jambū* is fed into the roots of mango tree, the fruits will have sweet smell. Thus, by treating the mango trees with certain decoctions, its taste and smell can also be improved according to *Lōkōpakāra.* Another interesting point is to get the jackfruit at the lowest level possible by covering the tree with a special grass called *Nellu.* Feeding the trees with the manure of sheep and the flesh of goat can increase the size of the pomegranate.

Certain trees do not bear fruits at all. Chāvuṇḍarāya suggests medicines for such trees also. A decoction of jaggery, milk and flesh, if given to orange plant, it will immediately give good fruits. Similarly, if a branch of a myrobalan is cut vertically and honey, ghee and sesame are poured into it, the tree gives myrobalan in plenty.

That producing seedless fruit is not a modem concept is attested to by Chāvuṇḍarāya. A mixture of honey, sugar, *yaṣṭimadhuka, koṣṭha* and *ippe* flowers, fed to the root of any fruit tree, the result will be seedless fruits. Generally, trees and plants bear fruits during a particular season or seasons. But by feeding the juice of *ankole* seeds, *dhātakī* petals and flesh to certain plants, one would get the yield throughout the year. It is interesting to know that Chāvuṇḍarāya has a prescription to get brinjal from the cucumber plant.

Lōkōpakāra also prescribes certain treatment to get good flowers. If the dry grass is burnt and on that land *mallige* plant is grown, there will be very good yield of the flowers. If *devadāru* seeds, turmeric, molasses, *pippali,* areca, white sesame, neem seed, *baje* are made into a paste and given to the *champaka* tree, the result will be unprecedented growth of flowers. *Punnāga* flower can be obtained every day by applying a decoction of flesh, milk and *pancha pallava* water in the early mornings. It is interesting to note that the colour of the hibiscus can be changed. The plant of red hibiscus can be made to yield white hibiscus flower if a mixture of sugar, liquor and buffalo milk is given to it. Equally interesting is the method by which there could be exchange of smell from one flower to another. If a decoction *of agilu,* lotus root and its pollen are sprinkled over the *jāji* and *mallige* plants, there will be exchange of the smell of both the flowers.

Thus, many of these prescriptions of Chāvuṇḍarāya, as enunciated above, are beyond our understanding. But they cannot be dismissed as mere figments of imagination. Most of his prescriptions contain either vegetables or meat, both of which are considered to be good manure. Thus, by providing artificial manure in the form of flesh or vegetables and other articles, Chāvuṇḍarāya tries to obtain many results. In fact, many of these prescriptions are used even today in folk agricultural practices. Thus, the prescriptions of Chāvuṇḍarāya must have had a folk origin and hence their popularity.

Chāvuṇḍarāya's work *Lōkōpakāra* has another chapter (seventh) on the perfumery industry which incidentally refers to many crops that were grown during the period. He refers to *kesari,* clove, *baje, jāyikāyi,* jāyipatre tavakshīra, rēṇuka, cardamom, coriander and others. While making scented oils, he refers to *champaka, mallige, mṛgamada,* camphor, *pādari* and *jāji.* He also refers to almond, rose, *maruga, pacche, davana, khachora, kumkum, kastūri,* garlic, wood apple, etc. Thus, he possessed knowledge of all these crops.

The chapter (eighth) on *Sūpaśāstra* is also interesting from this point of view. In this chapter, while mentioning various food items, Chāvuṇḍarāya refers to different types of rice, millets, cereals and other items, of which he had a good knowledge. He refers to two types of rice, *tāvareakki* and *karaveakki,* the former being the sweet scented rice. He also refers to *(jave godhi)* a superior variety of wheat, which is highly nutritious. While referring to the popular south Indian dish *iḍḍali,* he mentions *mung dal,* cumin seeds, coriander and pepper. While preparing different types of sweets, he mentions various spices like camphor, saffron, dates, clove, cardamom, *patre* (nutmeg), etc. When preparing a dish from the flowers of *neem,* the other items he refers to are *methi,* mustard, ginger and chillies. He also refers to onion, garlic, *colotropis gigantea (ekka)* and a variety of green leaves, which are difficult to identify. He refers very often to wood apple *(kapittha),* which was specially grown for its medicinal qualities and also used as a drink.

Lōkōpakāra has a chapter (fifth) on water divination (*udakārgaḷa*). Perhaps the *Bṛhatsaṃhitā* of Varāhamihira should have been the source of Chāvuṇḍarāya for this aspect in addition to other sources. His main interest in this chapter is to help people by explaining the principles of identification of groundwater so that population gets enough water for their substenance. He particularly refers to the location of ponds and wells and the direction in which they should be built in a village. He also gives certain botanical and other clues to locate groundwater. He prescribes places where the thornless *Śami* tree grows as a sure sign of groundwater. Another clue is the location of the anthill. He also gives an estimate of the quantity of water as well as the depth at which water is found, based upon certain clues. All these show Chāvuṇḍarāya's anxiety of and interest in the water resources, which ultimately help mankind in many ways.

Chāvuṇḍarāya's work also has a chapter (ninth) on medicine *(narādi vaidya)*. This chapter is also useful to a student of agriculture because it mentions a large number of herbs and roots required for the purpose of preparing medicines like *cūrṇas* and *kaṣāyas*. Some of these are highly technical and cannot be identified. Such plants are *amṛtabaḷḷi, lingusticum ajwaen* (oma), *hippali, neladāḷa, aḷale* (ink nut), *āḍusoge, yaṣṭimadhuka, tārikāyi, doḍḍipatre, parpāṭaka, karibevu, ummatta* seed, *ankole* root, *bobbuli, atibaje,* etc. If some of them grew wild, at least others must have been grown for the sake of medicine. It is important to note that some of these plants are identified and their use and efficiency are confirmed by the people involved in medicines.

Considering all these merits, it has to be stated that Chāvuṇḍarāya's *Lōkōpakāra* is an interesting Kannaḍa work and throws some light on the practice of agriculture of the eleventh century AD, when the Chālukyas of Kalyāṇa were ruling this part of Karnataka.

NOTES AND REFERENCES

1. H. Sesha Iyengar (ed), 1950, *Chāvuṇḍarāya Lōkōpakāram,* Madras, Madras Oriental Library and S. Rangaswamy: 1996, *Chāvuṇḍarāyana Lōkōpakāra, Ondu Adhyayana,* (in Kannada language), Mysore.
2. There are many Chāvuṇḍarāyas in Karnataka. The first one is the famous Ganga minister who erected the Bāhubali image at Sravanabelgola. He is referred to as Chāvuṇḍarāya I. The author of the present work, who flourished at least a century later, is called Chāvuṇḍarāya II.
3. Actually, he gives the measurements in *hastas* which has been converted into feet.

   I *hasta* $= 1 - \frac{1}{2}$ feet.

CHAPTER 35

Agricultural Technology as Known from the *Kṛṣi-Parāśara*

D.K. Ganguly

INTRODUCTION

Notwithstanding the existence of a good number of Sanskrit works from the *Ṛgveda* to the sixteenth century jurist Raghunandana's *Jyotiṣa-Tattva*, embodying numerous, scattered references to agriculture and its allied issues, agricultural treatises, properly so called, are extremely rare in the vast mass of Sanskrit literature. Parāśara's *Kṛṣi-Parāśara*, variously called *Kṛṣi-Paddhati* and *Kṛṣi-Saṃgraha*, stands out prominently in the select group of three or four such works[1] that have been handed down to the posterity.

DATE OF COMPOSITION

Since the *Kṛṣi-Parāśara* is devoid of any direct reference to the date of its composition, any conclusion regarding the chronology of the work—as based on the meagre data at our disposal—is likely to be hypothetical. The text, no doubt, refers to Gārgya as an illustrious authority on agricultural science but since the date of Gārgya is not precisely known as yet, this reference does not offer us any help in fixing the chronology of the text. Since the author of a well-known *Smṛti*-text is called Parāśara, one may be induced to identify the two as one and the same person who may have flourished in the pre-Yājñavalkya epoch. There are, however, some difficulties that stand in the way of the acceptance of this view. The *Parāśara-Smṛti*, while dealing with the question of agriculture, does not contain any citation from the *Kṛṣi-Parāśara*, nor does it betray any acquaintance with the section on agriculture to be found in the *Kṛṣi-Parāśara*. The *Kṛṣi-Parāśara*, likewise, does not quote any verse from the latter text. This would certainly have been the case had both the works been composed by one and the same author. Secondly, whereas the *Parāśara-Smṛti* discusses at length the question of caste in relation to agriculture, the *Kṛṣi-Parāśara* passes over the issue of caste in complete silence. Similarly, there is hardly any agreement between the two works on the question of the categories of the bulls, considered unsuitable for use in agricultural operation. Moreover, while the *Kṛṣi-Parāśara* has attached a great deal of importance to the colour of the bulls, the *Parāśara-Smṛti* does not take any cognizance of the issue. Judged in this context, the theory ascribing the date of the *Kṛṣi-Parāśara* to the early centuries of the Christian era[2] does not sound probable.

Thus, while the internal evidence of the *Kṛṣi-Parāśara* does not help us at all in determining the chronology of the text, the sixteenth-century Bengali jurist Raghunandana's *Jyotiṣa-Tattva* fortunately provides us with a clue to the solution of the problem. Raghunandana ascribes some of the verses, found in the *Kṛṣi-Parāśara,* to the *Rājamārtaṇḍa* and Varāha. The *Rājamārtaṇḍa* has been identified with a text of the same name, composed by the King Bhoja of Dhārā, who is believed to have ruled in western India around the first half of the eleventh century AD. Varāha is believed to be the same as the illustrious astronomer Varāhamihira, who is generally assigned to the fifth-sixth centuries AD. This would lead us to place the *Kṛṣi-Parāśara* at a date from about the latter half of the eleventh century to the beginning of the sixteenth century AD.

IMPORTANCE OF AGRICULTURE

Parāśara speaks in glowing terms of the importance of agriculture which, as he suggests, has ensured the very existence of the people, and remained to be truly 'life itself' for the mankind[3] (*jantūnāṃ jīvanaṃ Kṛṣiḥ*). Rice, which is the staple food of the vast majority of the people all over the world, is produced from the agricultural crop paddy[4] (*annaṃ hi dhānya sañjātaṃ*). It is rice which, besides providing sustenance to the gods, demons and human beings (*dev-āsura-manuṣyāśca sarve c-ānn-opajīvinaḥ*) constitutes the very vitality (*annaṃ prāṇāṃ*), vigour (*balaṃ c-ānnaṃ*) and the means of the fulfilment of all the ends of life[5] (*annaṃ Sarvārtha-sādhanaṃ*).

Parāśara further asserts that rice is far superior to any invaluable metal, or learning or even any other prized possession that a man can boast of. A person howsoever richly embellished he may be with gold ornaments in his neck, ears and hands (*kaṇṭhe karne ca haste ca suvarṇam vidyate yadi*), when denied of rice, is inevitably threatened with starvation and death[6] (*upavāsas-that-āpi syād ann-ābhāvena dehīnām*). Likewise, Brāhmaṇa, though possessed of wisdom and the knowledge of the *Vedas* and the *Śāstras* (*catur-ved-āntago vipraḥ, śāstra-vādī vicakṣaṇaḥ*), when reduced to poverty (*alakṣmya gṛhyate s-o'pi*), is impelled to take recourse to solicitation in order to save his life (*Prārthan-aladharanvitaḥ).* Parāśara has accordingly enjoined the people to take up agriculture at the cost of everything else[7] (*tasmāt sarvān parityajya kṛṣiṁ yatnena kārayet).*

The importance of the agricultural produces in general, and of rice as a daily food in particular, so highlighted time and again by Parāśara, can hardly be undermined, for it is the agricultural crop, especially rice, that keeps the population alive by providing it with food and nourishment. Wealth and learning cease to be of importance for those who are faced with the threat of starvation. Parāśara's special emphasis on rice is quite in consonance with the great importance of the food crop which, at least in the Indian environs, continues to be practically the whole of the people's diet. Parāśara's statement about rice as providing sustenance to the gods compares favourably with an old Japanese notion[8], envisaging rice as the food of the gods and the *samurais.* It is worth remembering that Parāśara only speaks of the dietary usefulness of rice but is conspicuously silent on its importance as a source of employment for the millions of the Indian population over the centuries and its many other uses.

IMPORTANCE OF RAINFALL FOR CULTIVATION

Parāśara emphasizes the importance of rainfall for crops production in clear and explicit terms. 'All agriculture' says, he 'has rainfall at its root'[9] (*Vṛṣṭimūlā Kṛṣiḥ sarvā*). As is well known, water is the paramount factor for the growth of crops, and the author would have us believe that water, thus required for agriculture, is solely derived from rainfall. It is true that in a tropical country like India, the monsoonal rainfall from the end of May to mid-October is the principal source of water, needed for agriculture, but the onset, amount and distribution of rainfall being variable from region to region and year to year, the cultivation of crops is likely to be uncertain. Parāśara is fully aware of the unstable nature of rainfall and its impact upon crops, and has accordingly devoted a large section of his work to a fruitful discussion about the position of planets, methods of forecasting annual rainfall, courses of the wind, indications of immediate annual rainfall, measurement of rainfall, symptoms of drought and other related issues. Parāśara lays special stress on the preservation of water in the fields from mid-September to mid-November[10] (*Āśvine kārtike c-aiva dhānyasya jala-rakṣaṇaṃ Na kṛtam yena mūrkhena tasya ka śasya-vāsanā//*) and observes that water is as indispensable to crops as the women are to the families[11] (*Yathā kulārthī kurute kula-strī-parirak saṇaṃ/Tathā saṃrakṣayed vāri śaratkāle sam-āgate*). It is far from being precisely known whether the author here speaks of the water reservoirs, wells and tanks which were—to judge from the combined testimony of the literary and archaeological materials—widely excavated in ancient India for preserving and drawing water to distant fields. It does not escape notice that the rivers and canals do not find mention in Parāśara's scheme of water resources, which is extremely circumscribed by rainfed water.

FOREBODEMENT OF HEAVY RAINFALL

Parāśara, on a careful analysis of the courses of the wind and the position of the planets and stars, points out the following symptoms, anyone of which is said to be adequate enough to foreshow a heavy rainfall for the coming months of the year, resulting in floods and destruction of crops:

1. If the sky is whitened with dust by a violent wind and the western horizon is dazzled by flashes of lightning in the bright fortnight of the month of Pauṣa;[12]
2. If there be a shower or fog in the month of Pauṣa;[13]
3. If the sky is overcast with clouds in the bright fortnight of the month of Pauṣa;[14]
4. If it rains during the moon's stay in Pisces and Scorpio between the close of Pauṣa and the commencement of Māgha;[15]

5. If the first day of the bright half in Caitra be on Monday;[16]
6. If the moon be in the star Citrā in the middle part of Caitra;[17]
7. If the wind blows from the southwest during the full moon in the month of Āṣāḍha;[18]
8. If the Jupiter passes from the Zodiac to another;[19]
9. If the Jupiter passes through the stars Citrā and Svāti.[20]

FORECAST OF GOOD RAINFALL

Parāśara points out the following symptoms as foreshowing a good rainy season with a plentiful growth of crops:

1. If there is a shower and the sky is overcast with clouds on the seventh day of the bright half of the month of Māgha;
2. If there is a strong wind or a hailstorm on the following days:

   (i) the seventh day of the dark half of Māgha and Phālguṇa,
   (ii) the third day of the bright half in Caitra, and
   (iii) the first day of Vaiśākha.

3. If the first day of the bright half in Caitra be a Wednesday, Thursday and Friday;
4. If there be a shower in the stars Citrā, Svāti, and Viśākhā in the months of Jyaiṣṭha and Śrāvaṇa;
5. If the wind blows from the east during the full moon in Āṣāḍha;
6. If the wind blows from the north during the full moon in Āṣāḍha;[21]
7. If the wind blows from the northeast during the full moon in Āṣāḍha;[22]
8. If there be rain on the ninth day of the bright half in Āṣāḍha;[23] and
9. If there be a shower in the star Rohiṇī in the month of Śrāvaṇa.[24]

INDICATIONS OF DROUGHT

Parāśara points out the following indications of drought:

1. If the first day (*pratipadi*) of the bright half (*sitāyāṁ*) in the month of Caitra be a Saturday;[25]
2. If there be no rain in the star Rohiṇī (*Karkaṭe Rohiṇī kakṣe yadi Vṛṣtir-na Jāyate*) in the month of Śrāvaṇa during the sun's stay in the Zodiac Cancer;[26]
3. If Mars passes on to the stars Uttara Phālgunī, Uttara Āṣāḍhā, Uttara Bhādrapada, Śrāvaṇā, Hastā, Mūlā, Jyaiṣṭhā, Kṛttikā and Māgha;[27]
4. If the sun be at the back of Mars[28] (*kuja-pṛsṭha-gato bhānuḥ*);
5. If Mars be in conjunction with the sun[29] (*Sa eva ravinā Yuktaḥ*); and
6. If Mars be in Leo[30] (*Aṅgārako yadā siṁha*).

DISTINCTIVE SIGNS OF IMMEDIATE RAINFALL

Parāśara, in the section entitled *Atha Sadyovṛṣṭijñānaṃ* points out the following indications as foreboding immediate rainfall:

1. A waterspout (*Jala-hasta*) near (*nikaṭe*) or in the midst of water[31] (*Jalastho*);
2. The rising of ants (*pipīlikā*) from their holes with eggs[32] (*aṇḍam-ādāya*);
3. Croaking (*śabdāyate*) of frogs (*bhekah*) all of a sudden[33] (*akasmāt*);
4. Excited movement (*dhāvanti mattāḥ*) of cats, mongooses, serpents and of other animals living in holes[34] (*c-ānye* va *vileśayāh*);
5. Making of bridges with dust (*dhūlibhiḥ setu-bandhanaṃ*) by boys on the road[35] (*mārge*);
6. Dance of peacocks[36] (*mayūraś-c-aiva nṛtyanti*);
7. Sudden rheumatic pain in the body[37] (*nṛṇām-aṅge vyathā yadi*);
8. Ascent of a snake on the top of a tree[38] (*vṛkṣ-āgrarohaṇaṃ-c-aheḥ*);
9. Drying of the wings in the sun (*pakṣayoḥ soṣayaṃ raudre*) by aquatic birds[39] (*Khagānām-ambuvāriṇāṃ*);
10. Noise of crickets in the sky[40] (*Jhinjhīravas-tath-ākāśe*).

ROLE OF FARMERS

The success of the agricultural operation considerably depends on the attitude and qualities of a farmer whose negligence or lack of exertion is likely to jeopardize the prospect of a rich harvest[41] (*kṣaṇenaikena sīdanti muhūrtam-anavekṣaṇāt*). The farmer, with sufficient means and capacity, can conduct the operation successfully and the venture, undertaken with insufficient means and manned by incompetent workers, proves to be abortive, reducing everyone involved in the process to a state of extreme poverty.[42] Besides being ever-alert, resourceful and capable, a farmer should be attentive to cows (*gohitaḥ*), regular in duty (*kṣetra-gāmī*), diligent (*vitandraḥ*) and possessed of a sound knowledge of the seasons (*kālajña*) and seeds (*sarva-śasyādyaḥ*). Agricultural enterprises will be highly rewarding for those farmers who are endowed with all these qualities (*n-avāsīdati*).

ANIMAL POWER AND ITS PROPER UPKEEP

In those days when tractors were not invented, the animal was the principal tillage power source. Parāśara recommends the use of cattle in rice culture and is silent about the buffalo and other animals which were also employed sometimes. Cattle being of such paramount importance for cultivation, every effort should be made for their proper maintenance, which would ensure the smooth sailing of the agricultural operation and boost the prospect of a rich harvest.

Parāśara[43] advises the farmers not to subject the draught animals to pain or oppression (*bāhān-na pīḍayet*). Needless to emphasize, too much hardship adversely affects the health and longevity of the animals and proves to be detrimental in the long run to the interest of the farmers who are forced to look for new animals in order to keep the agricultural operation going. The animals should be provided with molasses (*guḍakaiḥ*), fodder (*yavasaiḥ*) and other kinds of nourishment (*anyār-api poṣaṇaih*), besides being allowed to graze freely in the morning and evening[44] (*sāyaṃ prātas-ca cāraṇāt*).

The cowshed (*go-śālā*) should be carefully constructed and properly maintained.[45] A good and well-maintained cowshed is conducive to the growth of the animals (*bāhā vivarddhante*). There should be provision for smoke in the cowshed to keep the place free from mosquitoes and other insects. A cowshed should be strongly built (*suhaṭā*) so that it can withstand the ravages of storm and heavy shower; it should be as spacious as measuring five steps (*pañca-padā*) to facilitate the free movement of the animals, besides being thoroughly clean (*śucih*). Cowdung (*go-śakṛt*) and cow's urine (*mūtra*), which are injurious to their health, should not be allowed to accumulate in the cowshed[46] (*varjitā*). The cowshed should be provided with a lamp in the evening[47] (*sandhyāyāṃ tu gavāṃ sthāne dīpo yatra na dīyate*).

Parāśara forbids in clear and explicit terms the use and observance of the following articles and practices in a cowshed:

(i) Bell-metal[48] (*kāṃsyaṃ*);
(ii) Water in a vessel made of bell metal[49] (*kāṃsy-odakaṃ*);
(iii) Hot scum of rice[50] (*tapta-manḍaṃ*);
(iv) Water with which fish has been washed[51] (*jhas-odakaṃ*);
(v) To card or comb kārpāsa-cotton[52] (*Kārpāsa-śodhanaṃ*);
(vi) Broomsticks[53] (*sammārjanī*);
(vii) To keep a pestle[54] (*musala*);
(viii) Remnants of food[55] (*ucchiṣṭaṃ*);
(ix) To cleanse cows' excrements by cow's urine[56] (*gomūtra-jālaken-aiva yatr -āvaskara mocanaṃ*);
(x) To keep goats (*aja-bandhanāt*);
(xi) To sell off cowdung on Tuesday, Saturday and Sunday[57] (*Ravi-Bhauma-Śaner-dine*); and
(xii) To construct a cowshed at the time of the sun's stay in Leo[58] (*Siṃha-gehe*).

Parāśara vehemently decries farmers who employ only two bulls (*dvi-gavāṁ*) in cultivation in utter disregard to their physical capacity. He condemns them as *gavāśin*, beef-eaters. Those who use four bulls are likewise blamed as being cruel to the animals (*catur-gavaṃ nṛśaṃsānāṃ*). Parāśara firmly believes that the ideal situation of cultivation would demand the services of eight bulls (*Halam-aṣṭa-gavāṃ proktaṃ*), although the employment of six ones may be conceded to be the normal practice of the day (*Ṣaḍgavāṃ 'vyavahārikaṃ*). What Parāśara intends to mean is that the bulls should not be continuously employed day after day but be provided with rest to rejuvenate them for being pressed into service once again.

USE OF CATTLE MANURE

Parāśara points out the usefulness of cattle manure for the steady and proper growth of paddy. 'The paddy plants', says Parāśara[59], 'without the application of manure, simply grow up but do not yield fruits' (*Vinā sāreṇa yadhā-nyaṃ varddhate phala-varjitaṃ*). Cattle manure contains considerable quantities of potassium, phosphorus and nitrogen which plants need for their nourishment. Parāśara prescribes the methods how dung is to be carefully preserved, stored, grinded and finally applied to the field. Too much shifting of a dung-heap is likely to reduce much of its nitrogen. It is obviously to minimize the loss of nitrogen that Parāśara enjoins that a dung-heap should remain undisturbed up to the month of Pausa but be raised (*Kudālais-tolayet*), dried up in the sun (*raudre saṃsoṣya*) and subsequently powdered (*tat sarvam kṛtvā gundaka-rūpiṇaṃ*) before it is finally stored in a pit in every field in the month of Phālguṇa[60] (*Phālguṇe pratikedāre sāraṃ garte nidhāpayet*) for being applied to the field at the time of sowing the seeds[61] (*tato vapana-kāle tu kuryāt sāra-vimocanaṃ*). The author is, however, silent about the easily available green manure which finds prominent mention in Śūrapāla's work *Vrksāyurveda* of about the tenth century AD as containing considerable quantities of plant food or about the organic manure like fish and bone of which the *Bṛhatsaṃhitā* speaks so much for its efficacy in the nutrition of crops.

THE PLOUGH

The plough, also called the indigenous plough or the country plough, which is mainly used for preparing and cultivating the rice fields, is described by Parāśara to be comprised of the following constituents:

(i) ***Īṣā:*** the beam of the plough
(ii) ***Yuga:*** the yoke
(iii) ***Niryola:*** the shoe and the body of the plough
(iv) ***Halasthāṇu:*** the handle of the plough or the piece of wood fixed to the Niryola, juxtaposed the ploughshare
(v) ***Phālaka:*** the ploughshare
(vi) ***Pāśikās:*** the iron angles that fix the ploughshare to the shoe of the plough
(vii) ***Addacalla:*** the pins of the yoke where the bull is tied
(viii) ***Śula:*** An extra piece of wood that firmly fixes the body to the beam
(ix) ***Paccanī:*** a goad for driving the bulls
(x) ***Yotra:*** the cord for fastening the bull to the yoke of the plough
(xi) ***Avaddha:*** the iron angle between the body of the plough and the beam

Parāśara's description would bring to light the following characteristics of the plough of his scheme:

First, Parāśara's description seems to imply that the shoe and body should be made of one piece of wood as is precisely the system in the states like West Bengal, Orissa, Bihar, etc., but in those parts of the country, where the soil is alluvial, the shoe and body are made as separate components to facilitate the replacement of the shoe when worn out.

Secondly, Parāśara's description would further suggest that the handle of the plough is to be separately made, although the ploughs with the shoe, body and handle being made of one piece of wood, are also used.

Table 1: Length and other Details of the Components of the Plough

Item		*Length*		*Materials*
Textual	*Modern Name*	*Textual Version*	*Conversion in Centimetre*	
1. *Īsā*	Beam	5 *hastas*	254	Wood
2. *Halasthānu*	Handle	5 *vitastis*	114.3	Wood
3. *Yuga*	Yoke	upto the ears of the bull	X	Wood
4. *Niryola*	Shoe and body	1.5 *hastas*	76.2	Wood
5. *Pāśikā*	Angle between Share and shoe	12 *aṅgulis*	22.86	Iron
6. *Aḍḍacallā*	Pin of the yoke	12 *aṅgulis*	22.86	Wood
7. *Saula*	Extra piece	1 *artani*, i.e., the distance between the elbow and the tip of the little finger 12.5/9 *muṣṭis*	X	Wood
8. *Paccnī*	Goad	12.5/9 *mustis*	635/457.2	Bamboo, iron tipped
9. *Āvaddha*	Angle between body and beam	54 *aṅgulis*	101.6	Iron
10. *Yotra*	Cord	4 *hastas*	203.2	X
11. *Phālaka*	Ploughshare	9 *aṅgulis*	17.78	Iron

It is worth remembering that Parāśara tells us about the length of the constituents of the plough only but does not spell out their respective dimensions which include, besides length, the breadth and thickness as well. This limitation apart, Parāśara's prescription about the measurement of the length of the various parts of the plough when compared with the dimensions of the components of a standard plough in modern India, as shown in Table 2, may be found to be fairly accurate.

Table 2: Dimensions and other Details of a Modern Plough

Item	*Dimensions*	*Material*
Share, cm × cm × cm	40 × 3.5 × 2.5	Mild steel
Shoe, cm × cm × cm	32 × 25 × 10	Wood

(Table contd.)

Item	*Dimensions*	*Material*
Body, cm × cm × cm	40 × 22 × 11	Wood
Handle, cm × cm × cm	59 × 9 × 6	Wood
Beam, cm × cm × cm	200 × 6 × 4.5	Wood
Angle between the shoe and body	120°	–
Inclination of the share to the ground	12°	–

OTHERS TOOLS AND IMPLEMENTS

Of the other agricultural tools and implements to be encountered *Kṛṣi-parasara,* mention may be made of the following:

1. ***Madikā:*** also called *Mayikā.* A ladder-shaped contrivance used for leveling rice-fields both before and after the sowing of seeds.
2. ***Viddhaka***: A 21-spiked hoeing implement for making furrows and loosening the soil.
3. ***Kodāla*:** A spade, mentioned in the text, as being used in lifting and throwing cattle manure. This hand tool is mainly used for pulverizing and breaking clods and removing weeds.
4. ***Khanitra*:** A harrow used for opening furrows and removing weeds, pebbles, etc.
5. ***Medhi*:** A threshing post, made of Nyagrodha, Saptaparna, Gambhari, Salmali, Udumbara or any other milk-exuding tree.
6. ***Sṛṇi*:** Sickles of various shapes and sizes primarily used in harvesting all the field crops.
7. ***Cālanī*:** A vessel for sifting.
8. ***Dhānyakṛt*:** A winnowing fan for separating the chaff from grain.
9. ***Sūrpa***: A winnowing basket.

COLLECTION AND PRESERVATION OF SEEDS

In the section entitled *Vīja-sthāpana-vidhiḥ,* Parāśara sheds interesting light upon the system of the collection and preservation of seeds. According to him, seeds are to be collected (*sarva-vijāni saṃharet*) either in the month of Māgha or in that of Phālguna (*Māgho va Phālgune māse*). Seeds are to be dried up thoroughly in the sun (*śoṣayed-ātape saṃyak*) and should not be kept on the floor so that they are not exposed to moisture (*n-aiv-ādho vi-nidhāpayet*). Proper care should be taken to cleanse the seeds by purging the chaff (*vidhānyaṁ tatra śodhayet*) and grass particles (*tṛṇam chinyāt vinir-gataṃ*). Parāśara points out that seeds mixed with husk are extremely harmful to crops (*vījaṃ vidhānya-sammiśraṃ phala-hānikaraṃ paraṃ*) and seeds mixed with grass particles are likely to yield grass instead of crops (*acchinna-tranake hy-asmin kṛṣih syāt tṛṇa-bhūṣita*). Seeds to be sown on the same field should all be of the same variety (*eka-rūpaṃ pcayatnena tasmād-vījam sam-ācaret*) for the seeds which are of the same class yield a rich harvest (*eka-rūpaṃ tu yad-vījam phalaṃ phlalati nirbharaṃ*). Seeds should be kept in a tight packet (*vījasya-putikām kṛtvā; suhadaṃ putakaṃ kṛtva*) and never be kept on an anthill (*na valmīke*), in the cowshed (*na go-sthāne*), in the place, where a woman has delivered a child (*na prasūta-niketane*) or in the house where a barren woman resides (*na ca vandhyāvati-gehe vīja-sthāpanaṃ-ācaret*). It is evident that some of Parāśara's suggestions are based on superstition rather than practical observations. Parāśara further enjoins that seeds should not be allowed to come in contact with the remnants of food (*n = occhiṣṭaṃ*), a woman in her monthly impurity (*sparśayed vījam na ca nāriṁ rajasvalāṁ*), a barren woman (*na vandhyā*), a woman in the family way (*garbhinī*), a woman who has, of late, delivered a child (*na ca sadyah prasūtikā*). Likewise, seeds should be kept aloof from ghee, oil, buttermilk, lamp and salt.[63]

Parāśara[64] quotes a few verses from Gārgya to show how essential the seeds are for a rich harvest, superseding the cultivators, manure, field, bull and the mass of clouds in importance. Parāśara further draws our attention to a statement, of Gārgya to the effect that it is 'the seeds which actually shape the quantum of harvest.'

TIME FOR SOWING SEEDS

Parāśara speaks of two different methods of rice cultivation—the broadcast method and the transplantation method.[65] The time for sowing seeds under the two schemes is obviously different. For the broadcast method,[66] the best period for sowing seeds is the month of Vaiśākha (*vaiśākhe vapanaṃ 'śreṣthaṃ*), while the month of Jyaiṣṭha is supposed to be tolerably good (*Jyaiṣṭhe tu madhyamam smṛtaṃ*). The month of Aṣāḍha is bad (*Āṣāḍhe c-ādhamaṃ proktaṃ*) for this purpose and Śrāvaṇa is considered to be the worst (*Śrāvaṇe c-ādhaṁ-ādhamaṃ*). This scheme of sowing rice crop, which conforms to the southwest monsoon, corresponds to the cropping season of the Aus paddy which starts with the onset of the pre-monsoonal rain. Similarly, the best period for sowing seeds for transplantation[67] is termed as *Śūci*, the exact meaning of which is far from being clear. It may signify the hot season in general, or, according to another interpretation, any of the two months of Jyaistha and Aṣāḍha (*Ropaṇārthaṃ tu vījānāṃ śūcau vapanaṃ-uttamaṃ*). The months of Śrāvaṇa and Bhādra are denounced by Parāśara as being unsuitable for sowing seeds (*Śrāvaṇe c-ādhamaṃ proktaṃ Bhādre c-aiv-ādhamaṃ*). This is precisely the time when the *Āman* crop is planted.

Parāśara, a staunch believer in the abiding influence of the planets and stars in the life and yields of the plants, points out that the stars like Uttara Phālgunī, Uttar Āṣāḍhā, Uttara Bhādrapada, Mūlā, Jyeṣṭhā, Anurādhā, Māghā, Mṛgaśiras, Rohiṇi, Hasta and Svāti, are propitious for the sowing of seeds. Parāśara further observes that the stars, including Śrāvanā, Pūrvā Phālguṇī, Purvā Āṣāḍhā, Pūrva Bhādrapada, Viśākhā, Bharaṇī, Ārdrā and Aśleṣa are so inauspicious that the seeds, sown on those occasions, would hardly allow the farmers to obtain more than what is barely needed for seeding. The farmers are advised to avoid the following days in sowing seeds:

1. Three days and a half[68] (*tridinaṃ sārdhaṃ*), both at the end of Jyaiṣṭha (*Jyaiṣṭhā-ānte*) and at the commencement of Aṣāḍha (*Āṣāḍh-ādau tath-aiva ca*).

2. Three days between the end of Jyaiṣṭha and the beginning of Āṣāḍha. It seems that this provision which calls for the avoidance of only three days for sowing seeds is meant for those farmers who cannot afford to lose six consecutive days in sowing seeds, as stipulated in the earlier provision.
3. Four days in Āṣāḍha from the tenth to the thirteenth of the dark half of the month when the earth is supposed to be impure.
4. Tuesday and Saturday, when there is every likelihood of the plants and crops being damaged by rats, locusts and insects[69] (*Vapane ropaṇe c = aiva vārayugmaṃ vivarjayet/Mūṣikāṇāṃ bhayaṃ bhaume mande śalabha-kīṭayoḥ*).

RITES AND METHODS OF SOWING AND TRANSPLANTATION

It is the owner who ceremonially inaugurates the process of sowing seeds by personally sowing, while meditating upon Indra (*Indraṁ citte sam-ādhāya*), three handfuls of seeds—(*svayaṁ muṣṭi-trayaṁ vapet*) moistened with cold water (*hima- vāri-niṣikta*). Next, the owner is enjoined to turn to the east (*prāṇ-mukhaḥ*) and invoke with a pitcher in his hands the Mother Earth to bless him and his associates with sound health, riches, paddy and prosperity (*Susthā bhavantu kṛṣakā dhana-dhānya-samṛddhibhiḥ*). This inaugural ceremony is called Dhānya-puṇyāha.[70] When the process of sowing is completed, all the labourers involved in the operation should be sumptuously fed with ghee and rice, boiled with milk and sugar[71] (*Kṛtvā tu vapanaṁ, kṣetre kṛṣakān ghṛta-pāyasaiḥ/Bhojayitvā su-bhojyena nirvighnā jāyate kṛṣiḥ//*). Afterwards, to facilitate the speedy and evenly growth of the paddy, the rice field needs to be levelled with a ladder, called *mayikā.*

Parāśara, likewise, prescribes the norms for selecting and transplanting seedlings. The seedlings to be transplanted should be tender for hard and full-grown plants do not yield high crops (*na phalanti dṛḍhāh sarve vījāḥ*). The plants, transplanted in the month of Śrāvaṇa (*karkaṭe*), should be one cubit apart from each other (*hast-āntaraṁ*); those planted in Bhādra (*siṁhe*) should be half a cubit distant and the ones, transplanted in Āśvina (*Kanyāyāṁ*) warrant a gap of four fingers between the two (*catur-aṅgulaṁ*).

THINNING OUT THE PADDY PLANTS

The diseased and worn out leaves and plants, which impede the growth of other leaves and plants in the field, are to be removed[72] from time to time. The process should begin with Āṣāḍha or Śrāvaṇa,[73] preferably before the onset of the rainy season (*avṛṣṭau*). If the process is commenced in Bhādra,[74] the yield would be half of the usual produce. Any further delay in the commencement of the operation would spell disaster for the crops (*phal-āśā n-aiva c-āśvine*).

Parāśara urges the farmers not to rest contented by thinning out the plants once for all but to repeat the operation thrice till the end of the month of Āśvina. The operation of thinning out paddy plants will go a long way in arresting the wastage of plant food and escalation of diseases, thus paving the way—as a consequence thereof—for the growth of a rich harvest.

WEEDS AND WEED CONTROL

Parāśara is aware of the importance of weeding out the rice field. It is admitted on all hands that the weeds, which occur in every paddy field throughout the world, are unwarranted, useless, prolific, competitive and often harmful to agricultural operations. They reduce the quantity and quality of crops by competing for nutrients, water and light and intensify the problem of diseases, insects and other pests by serving as their hosts, besides reducing the efficiency of harvesting. Parāśara, accordingly, advises the farmers to remove weeds and to ensure a conducive environment for the proper growth of crops. 'Therefore, with all care,' observes Parāśara,[75] 'render agricultural produce free from weeds; the agricultural produce from which weeds are removed, becomes wish-yielding to cultivators' (*Tasmāt sarva-prayatnena nis-tṛṇāṁ kārayet kṛṣiṃ/Nis-tṛṇā-hi kṛṣāṇānāṃ kṛṣiḥ kāmadudhā bhavet//*).

INSECTS, BIRDS AND ANIMALS AND THEIR CONTROL

Parāśara gives a fairly long list of insects, birds and anirnals[76] that cause severe damage to crops in various ways and spread viral diseases among rice plants. Of the insects damaging the crops, mention may be made of Vātā, Bhāmmā, Śāṅgī, Gāndhī, Pāṇḍaramuṇḍī, Dhūliśṛṅgī, Kumārī and Makā, some of which may be identified as stem borers, leafhoppers, planthoppers, gall midges, whorl maggots, leaf-folders, bugs, etc. The sparrow finds special mention as doing injury to corns. Among the animals, destroying crops, mention is made of boars (*śūkara*), deer (*mṛga*), rats (*mūṣika*) and buffaloes (*mahiṣa*).

Parāśara recommends the following measures to ward off the evils brought about by the harmful agents:

First, the roots of plants should always be kept submerged under water up to a point. The required quantum of water is to be strictly maintained; inadequate and unlimited water is likely to expose the plants to the threat of insects and diseases.

Secondly, poles are to be erected in rice fields with banners inscribed with the mantra *oṃ, āṃ, ghāṃ, ghīṃ, ghuṃ, ghaḥ*. This operation, it is claimed, would ensure the divine help of Rāma and Hanumant in driving away insects, birds and beasts from the rice fields. It is interesting to note that the practice of setting up poles of various figures and forms to ward off evils to the plants is still in vogue in some parts of the country.

OMENS AND PORTENTS AND THEIR CONSEQUENCES

Parāśara advises the farmer to be on guard when engaged in ploughing the field, against the occurrence of some omens and portents, and thereby ensure the safety of the operation and welfare of his family. The omens and portents and their consequences, as specified by Parāśara in verses 145–51 of the text, are enumerated below:

1. Raising of a tortoise (*kūrma*) by the plough, when driven, would subject the cultivator to the loss of his wife and the damage of his property by fire;
2. Falling of the ploughshare to pieces, while in operation, would lead to the desertion of his native land (*deśa-tyāga*) by the cultivator;

3. Breaking of the plough would herald the death of the owner of the land;
4. Splitting of the beam signifies the death of the cultivator;
5. Dissolution of the yoke signals the death of one's brother;
6. Disintegration of the *śaulā* spells the loss of one's son;
7. Tearing of the cord brings about diseases and damages to crops;
8. Falling down of the cultivator forebodes his harassment and detention at the hands of the king;
9. Dropping down of the bull in the course of ploughing is a warning for the loss of lives due to fever and dysentry;
10. Sudden excitement of the bull in the course of tilling the soil is a symptom of the failure of crops (*Kṛṣi-bhaṅgo*) and physical ailment (*pīḍā v-āpi śarīrajā*) of the cultivator;
11. Bellowing of the bull when engaged in ploughing the field or licking of its nose (*nāśa-līḍha*) means the four-fold increase of crops (*śasyam catur-guṇaṃ*);
12. Passing of dung by the bull, while driving the plough, signifies of increase of crops (*śasya-vṛddhiḥ*);
13. Passing of urine by the bull serves as a forecast for the advent of flood (*mūtre banyā prajāyate*).

AGRICULTURAL CROPS

Mention is made in the *Kṛṣi-Parāśara* of the three kinds of crops comprising sesamum, paddy and barley (*tila-dhānya-yavānāṃ*). This list of crops is, no doubt, far shorter than that is provided in the *Arthaśāstra* or the *Kāśyapīya-kṛṣisūkti*). The *Kṛṣi-Parāśara,* likewise, does not speak of the well-known categories of rice and their varieties but simply refers to the crop in a general way.

AGRICULTURAL RITES AND CEREMONIES

Parāśara believes in the efficacy of certain ceremonies and festivities to ward off the maladies that are likely to plague the growth of crops. One of these ceremonies is called *Go-parva* which is believed to invest the cows with sound health and 'freedom from various diseases for a year'[77] (*sarvā gojātayaḥ susthā bhavanty-etena tad-gṛhe/nanā-vyādhi-vinir-muktā varsam-ekaṃ na saṃśayaḥ*). The ceremony is to be observed on the first day of the month of Kārtika when the *śyāmā*-creepers are to be tied on the horns of the cows, their bodies are to be besmeared with turmeric oil, saffron and sandal paste and the animals are to be decorated with ornaments, cloths, etc., while the chief among them is to be taken round the village with the accompaniment of vocal and instrumental music and the farmers holding clubs in their upraised hands. The cows are to be marked with heated iron and a portion of their tails, hair and ears is to be cut off.

Parāśara, likewise, insists on the performance of *Hala-prasāraṇa,* the ceremony of driving the plough to ensure the increase of crops[78] (*hala-prasāraṇaṃ kāryaṃ kṛṣakaiḥ śasya-vṛddhaye*). This is obviously the ceremony to be performed by the farmers at the commencement of the harvesting season. Monday, Wednesday, Thursday and Friday are considered to be highly suitable for the performance of this festival. The farmer, after having bathed and put on a pair of white cloths, has to besmear the tip of the ploughshare with honey, apply butter or ghee to the faces of the cows which should be black, red or red-and-black, duly propititate the earth, planets, Pṛthu and Prajāpati and finally offer milk, white flowers, curd, and condensed milk to Indra, the god of rains, who on being appeased, is supposed to release a good rainfall, so essential for the growth of crops.

Parāśara also refers to a festival called *Muṣṭi-grahaṇaṃ*, 'taking handfuls of paddy'[79]. in the month of Agrahāyaṇa, which is believed to give a boost to the production of crops. It is on an auspicious day that the farmer should duly worship the paddy plants with sandal paste, flowers, incense and offerings and shear off two and a half handfuls of paddy from the northeastern corner of the field. Then the farmer, carrying the bundle on his head with stalks turned should start to his home silently without touching anybody, and on entering the main room and walking seven steps, should keep the bundle to the east and duly worship it. Parāśara enjoins the farmer to observe the ceremony and the non-compliance of the injunction as we are informed would subject the latter to constant failures in his enterprise[80] (*pade pade viphalatā tasya dhānyam kuto gṛhe*).

The *Kṛṣi-Parāśara* refers to a ceremony called Puṣya-yātrā[81], which is to be performed in the month of Pausa before the onset of the harvesting season. On this occasion, the people should put on new garments, assemble themselves near the field, besmear each other with sandal paste, *catubhasma* and perfumed oil, decorate themselves with flowers and perform great merriment with vocal instrumental music (*gītair-nrtyais-ca-vādyaiś-ca*), besides being sumptuously fed by the owner of the field.

At the end, the people thus assembled, are to offer a prayer to the solar deity for the increase of paddy and fame (*dhānya-vṛddhir-yaśo-vṛddhiḥ*), welfare of the king and their own families (*Pravṛddhir-dāraputrayoḥ rāja-sammāna-vṛddhiś-ca*) and increase of cattle (*gavāṃ vṛddhiḥ*). The people are advised not to take any more meal on that day.

Parāśara's clear and categorical insistence on the farmer's observance of so many rites and ceremonies is quite in keeping with the fond belief among the Indian peasants that divine help goes a long way in combating drought, flood, damages, caused by birds, insects, etc. and other maladies that are likely to impede the growth of crops and jeopardize the prospect of a good harvest.

NOTES AND REFERENCES

1. To this category of works may be assigned the *Kṛṣiśāsana* and the *Kṛṣiśāstra*, which are of little historical importance.
2. That was precisely the epoch of Parāśara. This is proved first by the fact that *Yājñavalkya*, who is assigned to the fourth century AD by Jolly and to 100–300 AD by *Mahāmahopādhyaya* P.V. Kane, mentions Parāśara as an author of *Dharmaśāstra*-text, and secondly, by the fourth century AD, Bower manuscript which refers to a number of earlier authorities, including *Parāśara Hārita* and *Śuśruta*.
3. Verse 8.

4. Verse 7.
5. Verse 6.
6. Verse 5.
7. Verse 7.
8. S.K. De Datta, *Principles and Practices of Rice Production* (Toronto 1981) P.I.
9. Verse 10.
10. Verse 196.
11. Verse 197.
12. Verse 34. *Dhūlibhir-eva dhavalīkrtam-antarīkṣam*
Vidyucchatācchurita-vāruṇa-dig-vibhāgam
Pauṣe yadā bhavati māsi site cāpakse
Toyena tatra sakalā plavate dharitrī
13. Verse 35. *Pauṣe māsi yadā vṛṣtiḥ kujjhatir-vā yadā bhavet*
Tad-ādau saptame māsi vāripūrnā bhave-namahī
14. Verse 36. *Yadā Pauṣe site pakṣe nabho mech-āvṛtam bhavet*
15. Verse 37. *Mīna vṛṣcikayor-madhye yadi varṣati vāsavaḥ*
16. Verse 44.
17. Verse 46.
18. Verse 59.
19. Verse 71.
20. Verse 73.
21. Verse 59.
22. *Ibid.*
23. Verse 60.
24. Verse 62.
25. Verse 45.
26. Verse 63.
27. Verse 75.
28. Verse 76.
29. Verse 76.
30. Verse 77.
31. Verse 65.
32. Verse 66.
33. *Ibid.*
34. Verse 67.
35. Verse 68.
36. Verse 68.
37. Verse 69.
38. *Ibid.*
39. Verse 70.
40. *Ibid.*
41. Verse 91.

42. *Ibid.* The term *samm-arthaḥ*, as used by Parāśara in verse 92, means one, resourceful in physique and money.
43. Verse 84. Parāśara discusses the issue at length in the section entitled *Bāhana-Vidhānam.*
44. Verse 86.
45. Verse 87.
46. Verse 88.
47. Verse 95.
48. Verse 90.
49. *Ibid.*
50. *Ibid.*
51. *Ibid.*
52. *Ibid.*
53. Verse 91.
54. *Ibid.*
55. *Ibid.*
56. Verse 92.
57. Ibid.
58. Verse 89.
59. Verse 111.
60. Verse 110.
61. Verse 111.
62. The table is based on the verses which read as: *Pañca-haśtā bhaved-Īsā sthāṇuḥ pañca-vitastikāḥ/Sārdha-hastas-tu niryoloyugaṃ Karṇa-samānakaṃ* (Verse 113), etc.
63. Verse 163. *Ghṛtaṁ tailaṃ ca-takraṃ ca pradīpaṁ lavaṇaṃ tathā, Vij-opari bhrameṇ-āpi kṛṣako n-aiva kārayet* !!
64. Verse 166. *Kṛṣān-sāra-kedāra-vṛṣa nīrada-sañcayāh/sarve eta bandhyatāṁ yānti vije bandhyatām-āgate.... Vīja yatnamataḥ kuryāt vījamūtāḥ phalasṛiyaḥ* !!
65. This is evident from the use of the expressions *vapaṇaṃ* and *ropaṇ-arthaṃ* in verses 168–69 of the text.
66. Verse 168.
67. Verse 169.
68. Verse 174.
69. Verse 172.
70. Verse 178.
71. Verse 181.
72. The process is described in the *Kṛṣi-Parāśara* as *Dhānya Kaṭṭana-vidhi.*
73. Verse 186.
74. Verse 187.
75. Verse 192.
76. G.P. Majumdar and S.C. Banerji edited *Kṛṣi-Prāśara* (Calcutta, 1960), p. 48.
77. Verse 104.
78. Verses 121–40 give a detailed description of the ceremony.

79. Verses 206–13.
80. Verse 211.
81. Verse 221–36.

CHAPTER 36

Agriculture as Known from *Khanā's Vacanas*

Rita Chaudhuri

INTRODUCTION

"Blessed is agriculture, holy is agriculture and agriculture is the life of creatures," this statement in the *Kṛṣi-Parāśara*,[1] echoes the sentiments of generations of Indians who resorted to cultivation, with the transition of life from a nomadic stage to a more settled existence. Nature's bounties, coupled with the ingenuity of our ancient ancestors, have enabled India to build up a sound agrarian system, features of which exist even today as some of our important legacies of the past.

Perhaps, the greatest contribution made by the River Ganges in moulding the destiny of the Indian civilisation was the creation of the broad belt of alluvial plains which skirts the Himalayan foothills and is bounded on the west by the semi-arid Punjab and Rajasthan plains, on the southeast, by the Bay of Bengal and on the south by the Central highlands. Endowed by nature with all the prerequisites conducive to human settlement and ranked amongst the world's most productive regions, the great northern plains as it is geographically designated, nurtured the growth and subsequent flourishing of Indian culture. Especially, in the lower reaches of the Ganges delta, which cover a major portion of southern Bengal, 'the traveller passes through a wide area of crop producing land broken by clustering groves of mango, tamarind and other trees, giving place gradually to long lines and avenues of palms, bordering the fresh verdure of irrigated rice fields'.[2]

It is indeed surprising that in a country like India, there should be a dearth of ancient works bearing on agriculture and agricultural operations. Countless such references are to be found scattered in various works starting from the *Ṛgveda*. Unfortunately, specialized works on the subject are rare. The *Kṛṣi-Parāśara*, as mentioned above, is the only work in Sanskrit, hitherto known, devoted exclusively to various aspects of agricultural pursuits. The work entitled *Kṛṣi-śāsana* is merely a compilation of passages culled from different sources, referring to agriculture. The practical value of the *Kṛṣi-śāstra*[3] is negated by the fact that it deals only with the time suitable for some of the items of agricultural operations.

By itself not an exhaustive work, a major part of he *Kṛṣi-Parāśara*, bears a striking affinity with the agricultural proverbs popular all over India. In Bengal, such a collection goes by the name of *Khanār Vacan.*[4] Incidentally, identical ideas expressed in the two works, regarding certain customs, superstitions and other aspects of agriculture, along with certain philological affinities, has led S.C. Banerji[5] to tentatively postulate a Bengali provenance for the *Kṛṣi-Parāśara.* He is, however, aware of the fact that there is no conclusive proof and that it would be feasible to conclude that both originated in northern India, wherein the fields were extensively cultivated, particularly in Bengal, from very early times.

In seeking to trace the rudimentary beginnings of village life based on agriculture in Bengal, it is reasonable to accept S.K. Chatterjee's contention, that the pioneers were the Austric tribes of India, who generated the process in the pre-historic period. They brought with them a primitive system of agriculture in which a digging stick was employed to till the hillside. Terrace cultivation of rice on hills and plains cultivation of the same were, in all likelihood, introduced by them. They brought, as the names of their language would suggest, the cultivation of the coconut (*nārikela*), the plantain (*kadalā*), the betel vine (*tāmbula*), the betel nut (*guvāka*), probably also turmeric (*haridra*) and ginger (*śriṅgavera*) and some vegetables like the brinjal (*vatingana*) and the pumpkin (*alābu*).[6] Interestingly, all these crops are widely cultivated in various parts of Bengal even today. It must be remembered that the rural folk in India had always been more or less ignorant of the science but proficient in the art of agriculture. They reposed a great deal of faith on traditions—oral or otherwise—which provided them with a set of systematic rules to follow. Herein lies the importance of such literary compositions as the *Khanār Vacan.* Written in a language understandable to the masses, and consisting of short rhyming sentences, easy to memorize, it embodies the experience of ages. Manifesting a keen observation of nature and a strong bond with the soil of Bengal, it is still regarded as an indispensable manual for domestic and agricultural purposes.

The identity of the lady named *Khanā,* who has so endeared herself to the masses, is shrouded in mystery, being submerged in innumerable dubious traditions. Scholars have suggested that the name is derived from the fact that the sayings deal with *kshana* or right time for performing a particular kind of activity, including agricultural operations.[7] Popularly, she is often considered to be the daughter-in-law of Varāhamihira, who was a prodigy in astronomy in the days of King Vikramāditya (c. 380–413 AD) of Ujjayini. In fact, some Sanskrit formulae in the *Bṛhat Saṃhitā* of Varahamihira, though applicable in general to conditions prevalent anywhere in India, are very similar to some of the Bengali sayings of *Khanā.* However, the very fact that the latter deals with subjects exclusively concerned with Bengal and is composed in the Bengali language, disputes the contention that *Khanā* could have been a native of anywhere else except for Bengal.

Dr. D.C. Sen[8] draws our attention to the village of Deuli, some miles to the east of Barasat in Bengal, where many legends about Khanā and her proficiency in agriculture and astronomy are prevalent. Herein can be witnessed the ruins of what was once the kingdom of Rājā Chandraketu, known as Chandrapur. A little further to the west of the now dilapidated palace can be witnessed a mound of earth, full of antiquarian remains, which is reputed among locals to have been the place where *Khanā* and her husband Mihira once resided. The coincidence of similar names and their association with astronomy may have culminated at a later age in this Khanā being identified with an actual or fabled heroine associated with the court of Vikramāditya. There are scholars who consider her to be a mythical person; rather a personification of the sayings.[9]

Superstitions and obscure traditions have made it difficult to unveil the truth. It is reasonable to conclude that the people of Bengal have, through the ages, contributed to these aphorisms, which now pass under the name of Khanā. However, though simplified and altered, the presence of old and antiquated forms of expressions in it, remind us that the sayings must be traced to an early age. Sen assigns it to the pre-Muhammadan period, preferably between 800–1200 AD.[10] The subjects treated in the work cover a varied field but by far the greater portion of them is devoted to agricultural matters, a discussion of which will now be made under specific categories. It is necessary to remember in this context, that no single compilation contain all the *Vacanas* of *Khanā*. Various editions ascribe to her the authorship of different verses, which, moreover, often render the same verse with slight variations. Consequently, for an overall picture, it is necessary to consult not only various editions of the text, but other relevant literary sources.

SOIL

All cultivable land are not fit for producing every kind of crop. It should be remembered that though the soil of Bengal is essentially alluvial, its natural composition differs a great deal in various regions, depending on the working of natural forces (rainwater, rivers, wind), the action of soil organisms and the interaction of various chemical substances present in the soil. The oldest classification of arable lands were simply into those in which crops had to depend entirely on rainfall (*devamātrikā*) and those which were inundated by or could be irrigated from rivers (*nadimātrikā*). Later texts, however, reveal a profound knowledge of the varied nature of the soil (as to whether it was sandy, clayey or loamy) and the kinds of crops that could be grown on each specific type. Thus, addressing herself to the son of ploughman, Khanā advises them to plant patol (*Trichosanthes dioica*) in a sandy soil for the fulfilment of their expectations:

Śunre bāpu cāṣār beṭā
māṭir madhye bele (sandy) yeṭā
tāte yadi bunisa patol
tātei tor āśār saphal[11]

The sandy banks of rivers with silt deposits are ideal for the cultivation of Āuś paddy (*oryza sativa*), while the soil for jute (*corchorus capsularis* and *C. olitorius*) should be clayey[12] (*āṭāle*). The cultivation of raddish (*mūlā-raphanus sativus*) requires a soft and fine-textured soil (*tūlā*), while that of sugarcane (*saccharum officinarum*) should contain lime (*calcium*).[13] Tobacco (*Nicotiana tabacum*) flourishes on soil which is loosened and grounded before the seedlings are planted,[14] presumably an indication of the coarse, gravelly nature of the soil. Coconut trees (*cocos nucifera*) planted on saline soil will bear fruit quickly.[15] *Khanā* further reveals a remarkable knowledge of a wider variety of crops, grown under the influence of other congenial conditions, on her native soil. This will be dealt with subsequently in this chapter. It will suffice here to observe that the text provides the reader with a panoramic view of a land, fertile and well-cultivated, corroboration of which is to be had from the works of writers from Bengal and outside.

'A low and moist country', wherein 'crops are abundant' and farming operations were regular' is how Hiuen Tsang describes Bengal, as he saw it in the seventh century AD.[16] Varendri (north Bengal), according to Sandhyakara Nandi[17] (second half of eleventh to first half of twelfth, AD), had "all its important regions filled up with crops and water." She was "the sparking crest jevel of the earth;" on account of her being adorned with, "...paddy plants of various kinds, which was further spread over by the fine bamboo clumps and which had (as additional charm) the sugarcane plants that were flourishing excellently there." This reminds us of the claims of classical writers, that Indians near the Ganges used to quaff sweet juices from tender reeds.[18] Furthermore, Suśruta refers to a kind of sugarcane called *pauṇḍraka.*[19] Commentators agree that it was named because it was grown in the *Pauṇḍra* country, originally parts of north Bengal; During the early medieval times it came to include extensive areas stretching from the eastern side of the Bhāgirathi to Chittagong and Sylhet.

Assuming that the author of the *Kṛṣi-Parāśara*[20] was closely acquainted with conditions in Bengal, we may note here an interesting passage in the text purporting to describe the changes in the nature and productive capacity of the soil at various times of the year. Thus, it is like gold in *Māgha* (January-February), silver in *Phālguṇa* (February–March) copper in *Caitra* (March–April) and like paddy in *Vaiśākha* (April–May). The soil should be known like the soil itself in *Jyaiṣṭha* (May–June), full of mud in *Āṣāḍha* (June–July) and that (soil) which is raised by ploughs in *Srāvana* (July–August) is barren. Such statements, embodying the theory of tillage, are applied to practice by our cultivators, who traditionally follow the guidelines provided by these time-tested principles.

It is significant that the *Kṛṣi-Parāśara* speaks of barren soil, for all the soil of Bengal could not have been productive. Terms such as *jaṅgala patha,*[21] where there was no water, *avaskara-sthana,*[22] denoting refuse land and *ushara*[23] meaning sterile or barren soil, have often been used in inscriptions, found in various parts of Bengal, to describe land fit neither for cultivation nor habitation.

WEATHER LORE AND METEOROLOGY

Weather lore has always been popular with the masses; the more so, because the farmer's fortunes invariably depend in a large measure on the vagaries of nature. Statements like 'all agriculture has rainfall at its root' (*Vriṣṭimulā Kṛṣi Sarva*)[24] amply illustrates the fact. The author of the *Kṛṣi-Parāśara* further claims that "life too has rainfall as its source. Therefore, at the outset, one must acquire knowledge of rainfall very carefully."

Interestingly, in Bengal, the twelve subdivisions of the year manifest varied weather conditions which exert a profound influence on the life style of the people. The *bāramāsi* or a lyrical description of the peculiar characteristics of the twelve months of the year is a favourite subject with our ancient writers. In the *Khanār Vacan,* there are frequent references to conditions of weather, especially to the effect of rainfall, occurring at different times of the year, on the prospects of paddy. The observances are given as much importance today as in the days of yore. To quote some of them: if it is dry in the month of *Jyaistha* (May-June) and there is heavy downpour in the month of *Āṣāḍha* (June–July), the earth will be burdened with crops:

Jyaiṣṭhe śukna, Āsāḍhe dhārā
dhara habe śaṣyer pasrā[25]

If there is heavy rainfall in cancer *Śrāvaṇa* (July–August), dry in Leo *Bhādra* (August-September), abundant rainfall in Virgo *Āśvīna* (September–October), and windless accompanied by a light drizzle in Libra *Kārtik* (October–November), then there will be a good harvest of paddy:

karkaṭ (cancer) charkat, siṅgha (Leo) śukā
Kanyā (Virgo) kāne kān
binā bāya varṣe Tūlā (Libra)
Kothā rākhbi dhān[26]

The following four verses appear to be more popular with the young and the old alike in Bengal who are often heard rendering them under similar weather conditions. It is, indeed, amazing as to how observances of such antiquity could have been so accurate:[27]

If it rains in the month of *Agrahāyaṇ* (November–December), the king goes begging:

Yadi vare Āgane
rājā nāmen māgane

In other words, the country will be in dire straits, as untimely rains might harm the already ripened crops.

If it rains in the month of *Pauṣa* (December–January), money may be had even by selling the chaff:

Yadi vare Pauṣe
kaḍi (cowry) hay tuṣe

By the month of *Pauṣa*, the winter crop has been harvested, and one can be free from anxiety.

If it rains at the end of the month of *Māgha* (January–February), the king and his country become blessed:

Yadi vare Māgher śeṣa
dhanya rājār pūṇya deśa

Rainfall, in accordance with this saying, will ensure an abundant growth of various kinds of crops as the following verse indicates:

If it rains in *Phālguṇa* (February–March), the millet (*cīnā- kāon panicum miliaceum*) grows abundantly:

Yadi vare Phālgune
Cīnākāon hay dviguṇe

These verses reveal an inherent truth about Indian economic life—the anxious waiting and precarious dependence of the agriculturists for and on seasonal rainfall for the success of their endeavours. In fact, the most important feature of the Indian climate is the monsoon which arrives in June (*Āṣāḍha*), heralded by dark moisture-laden clouds and much thunder and lightening. Preceding this, northern India in general, experiences a hot spell with a rapid rise in temperature to as high as 40°C in the plains. Some relief is provided by the local storms called the *Kālvaiśākhi* in Bengal (meaning calamity of the month of Vaiśākha) which is accompanied by violent winds, heavy rain and hail. It wreaks great havoc, destroying lives and damaging crops.

The monsoon showers continue intermittently till about the month of September when the temperature gradually begins falling. The cold weather season commences early in December and extends to the end of February. Fine weather conditions are sometimes broken by cyclonic disturbances; our agriculturists usually regulate their work according to expected weather conditions and even raise a winter crop by irrigation.

However, nature is not always predictable. Unforeseen conditions occur, rendering the farmer helpless in the face of what seems to be nature's wrath. *Khanā* strongly contends that nature herself provides indications about the future that lies ahead with regard to climatic conditions and the consequent agricultural prospects. Her prophesies on inclement weather conditions such as the occurrence of storms, excessive rainfall and floods, mists, droughts, earthquakes, and plagues are obviously a result of a keen observance of the workings of the various natural phenomena and their consequences. Some of the verses may be taken into account to illustrate the point:

If a southern wind blows in the month of *Āṣāḍha* (June–July), there will be flood in that year:

Pūrna āsādhe dakhinā bāy
Sei bachare vanyā hay

It may be noted herein that people in Bengal even today regard the blowing of a southerly wind in *Āṣāḍha* as an indication of impending floods.

If in *Pauṣa* of (December–January) there be heat in the atmosphere and cold in *Vaiśākha* (April–May) in that year, heavy rainfall will commence from the first part of *Āṣāḍha* (June–July).

Pauṣe garmi vaiśākhe jādā
pratham āṣāḍhe bharbe gāḍhā

If the sky is covered by mist in *Caitra* (March–April) and there be plenty of paddy in *Bhādra* (August–September), the earth is afflicted with plague and other disasters of that sort

Caite kuyā, Bhādre dhān.
narer muṇḍu gaḍāgadi yān[28]

If there is no rainfall in the months of *Śrāvaṇa* (July-August) and *Bhādra* (August–September) and if there is rainfall in the morning and the night skies are clear, then the earth will be inflicted with sorrows

Khanā bale śuno vānī
Śrāvaṇe Bhādare nāiko pāni
dine jal, rāte tārā
tabei to dukhinī dharā

A rainbow visible in the west indicates drought, while one in the east is the harbinger of rainfall;

Paścime dhanu nitya kharā
pūrve dhanu varṣe dhārā[29]

The following verse referring to dark moisture-laden clouds frequenting the skies along with other indications, heralds the advent of the monsoons.

If the clouds take the form as if cut by spade and axe and the wind blows off and on, it should be understood that rainfall will commence in a day or two. O, my peasant friend, do not waste time in such weather, but busy yourself in constructing a ridge:

Kodāle kuḍule megher gā
mājhe mājhe dicche bā (wind)
bal cāṣāre bandhte āl I
vṛṣṭi habe ājkāl[30]

Interesting too, are the verses which give indications of imminent rainfall, which the people of Bengal believe in, even today. The most common of them is the croaking of frogs.

Byāṅ ḍāke ghana ghana
śīghra vṛṣṭi habe jeno[31]

Moreover, if the sky is overcast in the month of *Bhādra* (August–September) and the wind blows in the opposite direction, there is a possibility of rainfall on that day.[32] *Kṛṣi-Parāśara's* claim that amongst other indications there will certainly be rainfall when 'ants rise (from their holes) with eggs as also when peacocks dance' and there is noise of crickets in the sky[33] are other methods popular with the masses in Bengal of ascertaining the possibility of rainfall either, instantaneously or in the near future. Sandhyākara Nandi, aptly described Varendrī, (north Bengal) as a country 'where large and swiftly moving clouds gave abundant rain' and where the flow of water ... 'came from large clouds'.[34] However, Varāhamihira refers to a great advancement made in the field of meteorology. He mentions the use of the rain gauge for measuring rainfall. The text further gives statistics of the quantities of rainfall in the *droṇa* measures together with carefully forecasting excessive, sufficient or scanty rainfall in the light of astronomical and meteorological data.[35]

CUSTOMS AND BELIEFS REGARDING THE COMMENCEMENT OF PLOUGHING AND OTHER AGRICULTURAL OPERATIONS

Around the world, the agricultural cycle is hedged around with ceremonious acts directed towards furthering the powers of fertility manifest in various crops and to ensure a good harvest. *Khanā* emphasizes on the necessity of setting out to the fields for ploughing on an auspicious moment. Even to receive any disheartening news, while on the way, is considered to be an ill omen. On entering the field, the farmer should ascertain the directions and start ploughing from the east, as this would ensure a good field. Agricultural operations should be carried out by the owner of the field himself with the assistance of his son and, in the absence of the latter, by a brother of the former.[36] Probably, *Khanā* herein refers to small holdings of peasants, which were cultivated by the owner and his family, as is the case even today. In fact, in some inscriptions, lands owned by private individuals are mentioned in connection with the demarcation of the boundaries of donated lands.[37] Fields which were owned by cultivators themselves are generally described as *kauṭumba-kṣetras.*[38] Often the term *satka* is mentioned to convey the idea of ownership.[39]

However, there were also larger holdings, which were not farmed by the owners personally. Thus, the Āśrafpur plate of *Devakhaḍga*[40] mentions a *pāṭaka* of land enjoyed by one Sārvāntara, but cultivated by Śikhara and others. Eleven *pāṭakas* of land are mentioned in the Gunāigarh plate,[41] where the owners engaged hired labour for their cultivation or let out the land to sharecroppers. Not surprisingly, we find that our law givers have prescribed certain rules to safeguard the interests of both the landowners and hired labourers. Some of our *Dharmasāstras* contain deliberations on the wages or the grain-share to be paid to the hired labourer, the care to be taken by the latter for the implements of work provided to them and the penalties to be paid for any breach of contact by both the parties.[42]

Khanā,[43] further recommends or prohibits certain days of the month, lunar days or lunar mansions (*nakṣatras*) for the commencement of ploughing, as they are considered to exert a strong influence on agricultural prospects. Thus, new moon (*amāvasyā*) and full moon days (*pūrṇimā*) are considered to be inauspicious and will only bring sorrow to the farmer. His bulls will be rendered useless, being struck with gout (bāt). Herein, perhaps, we get an indication of the concern felt by the people for the animals on whom they were dependent for their subsistence. The *Bṛhat Saṃhitā*,[44] explicitly prohibits the use of cattle for cultivation, which were lean, very old, small, diseased, apt to run away, blind of one eye or lame. Recognizing the indispensability of bulls in agricultural operations, the *Kṛṣi-Parāśara*[45] ordains great care and humane treatment towards them. Certain rites in the month of *Kārtika* (October–November) have been enjoined as they are supposed to be conducive to the health of cattle.

Khanā[46] is, however, emphatic on the fact that a person who is in possession of bulls should make use of them or else he is destined to suffer from despair throughout the year. In the *Amarakosa*[47], there is a reference to the idle cultivator who ploughed his field very roughly and then sowed his seed and consequently could not expect a good yield.

Khanā pronounces prognostic statements regarding the efficacy of planting seeds under the influence of certain stars and planets, a practise followed by peasants even today. Thus, she contends, that the year in which Saturn (*Śani*) is the king and Mars (*Mangal*) the minister, is not a good year for paddy[48] '...fight, storms but scanty rainfall, constant wind, disease and calamity,' are the consequences which *Kṛṣi-Parāśara*[49] apprehends, would be the outcome of such a year. Khanā recommends the year wherein Mercury (*Budha*) is the king and Venus (*śukra*) the minister as being the time for reaping a good harvest.[50] Such a year, in the opinion of the *Kṛṣi-Parāśara*[51], will be a beneficial one, there being '... freedom from disease, easy movement and easy availability of aims on the earth,' thus indicating a profitable year, free from calamities. We are reminded herein of Kauṭilya's[52] reference to the practice of forecasting the monsoons by observing the position of Venus (*śukra*) and Jupiter (*Bṛhaspati*) and the usual and unusual appearances of the sun.

According to Khanā, if on the seventh or eight days of the month of *Phālguna* (February–March) in either the *Śuklapakṣa* or the *Kṛṣṇapakṣa*, the moon resides in the lunar mansion (*nakṣatra*) of Rohini, then there is sure to be a good harvest of crops. If it is the ninth day, there will be inundation and if it is the tenth day, there will be total destruction. A failure of crops is also to be apprehended if, on the thirteenth day of the month of *Caitra* (March–April), the planet Saturn (*śani*) reigns supreme.[54]

Khanā is firm in her belief that from the arrangements of the days in a particular month, one can foretell inclement or favourable weather conditions, a theory which receives acceptance even today. Thus, it is said that a month having five Sundays will witness storms and droughts.[55] If the first day of the month of *Caitra* (March–April) is a Sunday, there will be scanty rainfall that year; if it is a Tuesday, then there will be a good amount of rainfall. Famine is forecasted if the first day is a Wednesday. It is good news for the growth of crops if it is either a Monday, Thursday or a Friday and a year wherein the month of *Caitra* has five Sundays, will be visited by plague, which will be fatal for man and animals:

Madhumāse (caitra) pratham divase hay yei var
Rabi śese, Mangal varse, durbhiksa Budhvar
Som, Śukra ār guruvār

pṛthvī nā say sasyabhar
pañca, Śani pāy Mīne
Śakuni maṃsa nā khāy ghṛṇe[56]

Khanā has great foreboding regarding a year in which in the month of *Bhādra* (August–September), along with rainfall there is earthquake; such a year is bound to be disastrous for the kingdom, crops and cattle as well as for all human kind:

Khanā bale śuno śuno, ogo patir pitā
Bhādra mase jaler madhye naden Vasumata
Rājyanāśa go nāśa, ār atiśay bān
hāte kāṭhā grihī phere, kinte nā pāy dhān[57]

As to the reason for choosing particular dates for specific agricultural operations, which are often accompanied by elaborate ceremonies, we do not possess adequate knowledge. However, we may be sure that they formed an indispensable part of an old agricultural calender which, given the inherent Indian love for tradition, find many faithful followers even today. Albeit, most of the observations made by Khanā, in the previous passages, are speculative and one may legitimately question their accuracy. However, one should not reject them outright, so long as they have not been tested by competent persons in the proper scientific manner. It should be remembered that the modern agriculturist, though often acquainted with the advanced techniques made in the field, prefers to depend largely on the traditional methods of their forefathers. Obviously, the time-tested manuals are used as almanacs by the farmers and are often referred to.

SUPERVISION

For the success of the whole process of agriculture, it is imperative that the farmer should personally involve himself and supervise the work. Khanā thus opines:

Khāṭe khāṭāy lāber gānti
tar ardhek kāndhe chati
ghare vase puche bāt
tār ghare hā bhāt[58]

One who personally toils on his fields, along with his assistants, is amply rewarded with a good profit. One who makes a leisurely visit to the fields, just to see the work being done, only gets half the profit, while a person who merely sits at home and enquires about the progress of the work will have no harvest to reap and, consequently, will have to go hungry. The *Kṛṣi-Parāśara* reiterates the view when it states that agriculture, when supervised, yields gold; when not supervised, it indeed causes poverty. The text further states that, 'A farmer doing good to cows, regularly going to fields, knowing the proper season, mindful of seeds, free from lethargy (becomes rich) with all crops and does not suffer.'[59] An excellent example of personal involvement in the work of agriculture is to be had from a passage in the *Raghuvaniśa*[60], which refers to the wives of cultivators sitting in the shade of sugarcane plants, guarding the *śāli* paddy.

MANURE

Fertiliser technology appears to have had a long history in India. Ancient Indian texts referring to agriculture allude to the use of farmyard manure to enrich the soil by providing it with additional nutrients and to rejuvenate sick plants. *Khanār Vacan* provides illuminating information in this respect. However, with great foresight, she cautions against the harmful effects of fertilisers on human health:

> *Mānusa mare yāte, gāchla bānce tāte*
> *pachlā sarāy gāchlā vāḍe, gondhla diye mānuṣa māre*[61]

Thus, it is said, that those things (e.g., cowdung and decayed remains of plants and animals) which injure man (as the gas that it emits is poisonous) cure the plants. This verse, however, makes it obvious that Khanā recommends the use of indigenous methods of providing the soil with manure, which will not be difficult for the Indian farmer to procure. Thus, in accordance with her advice, 'if the water in which fish is washed is poured at the root of a bottle gourd (*lāu-lagenaria sicearia*) plant, it will be strengthened'; 'the land which contains rotten paddy as manure is fit for the rearing of chillis (*Jhāl-laṅkā-capsicum annuum*)', 'the smut of corn (paddy) should be thrown into the bamboo grove (*bānśa-bambusa tulda*)'. 'If one does so near the root of the shrubs, they will soon cover two *Kuḍas* of land' (174 cubits square); 'Earth should also be thrown beneath a bamboo grove to serve the purpose of land manure'; 'betelnut plains (*śupari–Areca catechu*) require cowdung manure for their growth'; pieces of rotten straw or chips of wood should be used as manure at the roots of arums (*ola-amorphophallus campanulatus*); and 'edible arums *(mankacu-alocasia indica*) will thrive if ashes are used at their roots'; 'toro (*Kacu-colocasia esculenta*) too flourishes; when ash is used as fertiliser and when it is planted by the riverside'; 'Along with areca-nut trees (*śupari*), the coral tree (*māndār-erythrina variegata*) should be grown in the same plantation, as the leaves of the latter, falling at the root of the former, enable it to grow faster'; 'Coconut palms (*nārikel-cocos nucifera*) flourishes well when the smut of corn is used as fertiliser'; 'potatoes (āloo-solanum tuberosum) should be planted in a bamboo grove'—this will enable it to thrive.[62] Obviously, all these were a result of experimentation over a period of time with things easily available at hand, viz. cowdung and organic refuses—to make up for soil deficiency. It is indeed heartening to realize that our ancient ancestors were successful in ensuring for the peasant farmer for times to come, inexpensive means to achieve better crop production.

The *Kṛṣi-Parāśara* too recognizes the usefulness of the use of manure for crops. It clearly states that without manure, 'the paddy plants grow up bereft of fruits.'[63] Cowdung, as a manure has been highly extolled in the text. The *Arthasastra,* too mentions the manuring of cotton trees with bones and cowdung.[64] The use of similar ingredients as fertilisers has been mentioned in other ancient texts referring to agriculture.[65]

CONDITIONS INFLUENCING AGRICULTURE AND CERTAIN RULES TO BE FOLLOWED WHILE PLANTING VARIOUS CROPS

Certain specific natural conditions, influence the growth of crops and enable them to flourish. To illustrate the point, we quote some of Khanā's sayings:

> ... paddy thrives in the sun and the betel (pan-piper betel) in the shade

> *Khanā deke bole yān*
> *Rode dhān, chāyāy pān*

If the paddy gets profuse sunshine by day and showers by night, it rapidly develops. Khanā says the drizzling rain in the month of *Kārtik* does immense good to paddy.[66]

> *Dine roda, rāte jal*
> *Tāte vāḍe dhāner bal*
> *kārtiker uṇa jale*
> *Khanā bale duno phale.*

Khanā further advocates certain rules of agriculture which, if followed by the farmer, leads to the healthy growth of trees and plants. Thus she says:

> *Dātār nārikela, vakhiler baṃśa*
> *kame nā vāḍe nā vāro māsa*

The fruits of a coconut palm belonging to a generous person do not decrease, just as the bamboos in a grove of a miserly person do not grow.[67] The idea inherent here is that a wise man will never let his trees to be overburdened with fruits, for fear that they will lack nourishment. So he will pluck some from time to time and give them away. Likewise, if a person does not allow his bamboos to be cut, he will not profit, as bamboos grow faster when cut frequently. The same principle is applicable to the following:

Hemp (*śoṇ-crotalaria juncea*) should be cut when it flowers. Jute grows faster when the tree is cut:

> *Hale phul kāto śon*
> *gāch kāṭle pāṭ dviguṇ*[68]

Khanā offers interesting advice on the spacing of various kinds of trees which are indispensable for their healthy growth. Thus: 'sow the seeds of mustard (*sariṣā-Brassica campestris*) and those of rye (*sinapis ramose*) at some distance from one another. Cotton plants (*Kāpās-gossypium* sp.) should be put at the distance of a leap from one another and jute (*pāt-corchorus capsularis* and *C. Olitorius*) should by no means be planted near them, for cotton plants will perish if they come in contact with water from the jute field'.[69]

Regarding the cultivation of 'Āuś' paddy, Khanā states that 'the grains are to be planted at some distance from one another to ensure a thick growth and, consequently, Goddess Lakshmī will reside therein'. In other words, the farmer can expect a good profit from a rich harvest:

Kol pātlā dāgar guchi
Laksmī bale hethāy āchi[70]

In fact, the seedlings of paddy are usually planted 15–23 cm apart, in small bunches of three to five, or even singly. Sometimes, however, they are planted in large bunches of fifteen or so in flooded areas.

A similar set of rules has been ordained for the cultivation of fruit trees. Thus, 'if you want better fruits you have to plant coconut plants twelve cubits (approx. 18 ft) and betel-nut palms eight cubits (approx. 12 ft) apart and if they are not planted according to principle, you have to reduce them.'

Nārikela vāro, śupāri āta
era ghana takhani kāṭo[71]

It may be mentioned herein that this spacing is popular in Bengal even today.

Mango (*ām-mangifera indica*) as well as jackfruit trees (*kānthal-artocarpus heterophyllus*) should be planted at a distance of 20 cubits from each other. A dense growth is unfavourable for the growth of fruits. If there is no rain in the month of *Agrahāyaṇa* (November and December), jackfruit trees do not bear fruit.[72]

Plantain trees (*kalā-Musa paradisiaca*) should be planted at a distance of seven cubits from each other. As even the leaves of the plantain are useful, it is explicitly forbidden to cut them, for it will provide food and clothing for the family.[73]

Regarding Palmyra trees (*tāl-Borassus flabellifer*), Khanā says that planted by one generation and tended with care by the next, this tree will bear in the third generation. If the tree and its leaves have not been chewed by a cow, it will bear fruit in twelve years.[74] The usefulness of the palmyra tree has been recognized through the ages. In ancient times, its leaves were used as writing materials. Among the Bengal peasants, its fermented juice is popularly known as *tādi*, which is an intoxicating drink.

The wide cultivation of some of the fruits mentioned above is attested to by a large number of Pala and Sena inscriptions, besides ancient texts. The Āshrafpur Grant of Devakhaḍga states that the donee should enjoy the land donated to him after cultivating betel-nut palms and coconut trees.[75] A mango track, forming the boundary of a village is referred to in an inscription from Bengal.[76] Amongst the constituent elements of a donated plot mentioned in the Candravati plate of Candradeva[77] (Samvat 1148) are '.... groves of madhuka and mango trees ...' The plantain tree is frequently depicted in he Paharpur terracotta plaques (Paharpur, 70). It also occurs in such sculptures as the Caṇḍi images of the Rājshāhi museum.[78]

AGRICULTURAL PRACTICES AND CROPS GROWN

We may now attempt to form a comprehensive picture of the actual practice of agriculture as prevalent in Bengal and conveyed by the sayings of Khanā. For agriculture to be successful, the fields have to be made ready by ploughing and cross ploughings. The time taken for this varies with the type of crop to be cultivated. Thus, it will take sixteen days for raddish, eight days for cotton and four days for paddy. However, the cultivation of betel does not require any elaborate preparation:

ṣola cāse mūlā
tār ādhā tūlā
tār ādhā dhān
binā cāṣe pān[79]

Rice

Rice (*oryza sativa*) cultivation seems to have been popular all along the Gangetic plains, with the cultivators being conversant with a varied type of the food crop. The inscriptions of the Sena kings refer to 'smooth fields growing excellent paddy.'[80] The Anulia copper plate of Lakṣmaṇa Sena refers to the gift of myriads of villages, consisting of land growing paddy in excessive quantities.[81] The Rājavāḍi (Bhāwāl) plate of Lakṣmaṇa Sena[82] describes the capital (dharyya-grama) as a place where '... the thrill of victory (prosperity) of the earth is expressed in the guise of (shivering) corn (paddy) plants...'

Khanā is of the opinion that for the sowing of seeds of paddy, *Vaiśākha* (April-May) is the best month.

Vaisākher pratham jale
āus dhān dugun phale

Āuś paddy thrives well when seeds are planted with the first showers of *Vaiśākha*[83]. Agreeing with this statement, the *Kṛṣi-Parāśara* further claims that for sowing, *Jyaistha* (May-June) is tolerable, *Āṣāḍha* (June–July) is bad and *Śrāvaṇa* (July–August) is the worst.[84]

However, it must be remembered that the Greek writers had earlier claimed that India has a double rainfall and that Indians generally gather two harvests.[85] The *Bṛhat Saṃhitā* speaks of summer crops, autumnal crops and vernal crops.[86]

Khanā too refers to two kinds of paddy—Āuś, which ripens in the rainy season, and *śāli,* which can be harvested in winter. She, thus, professes knowledge of both a khārif and a rabi crop. This is reinforced by the fact that besides speaking of the beneficiary effects of rain in summer, she claims;

Māgha māse varṣe debā
rājā chāḍe prajār seva[87]

If it rains in the month of *Māgha,* the rabi crops thrive and the king need not have any apprehension regarding the future of his subjects.

Another class of rice is known to have been grown in Bengal. In the Maināmati plates of the Candra Kings[88] occur such names of villages as Bālesvara-varddaki voraka and Brāhmana-deva-voraka grama. The work *Voraka* seems to be akin to the Bengali word *boro* which essentially means a sort of rice sown on low swampy ground or near the bank of rivers. In all probability, the term denotes land fit for boro cultivation. Two croppings of this is even now practised in Bengal—*kharif* or rain crop sown in *Āṣāḍha* and harvested in *Āśvina* and *rabi* or winter crop sown in *Kārtika* and harvested in *Caitra.* Apparently, even in the ancient period, rice was grown in some form or another throughout the year.

In Bengal, both the systems of sowing seeds directly into the soil and of growing rice in nursery beds and replanting seedlings in flooded fields were known. The *Raghuvaṃśa*[89] of Kālidasa affords us a glimpse into this method of rice cultivation. Describing Raghu's conquest of the Vaṅgas, the poet remarks that the conqueror uprooted and replanted them (*utkhāta pratiropita*) like rice plants. Peasants are further referred to as replanting the seedlings of *śāli* paddy. It is for this specific purpose that Khanā advises agriculturists to prepare the fields by buildings ridges around them in order to preserve rainwater:

Āge bāndhi āli
Ropa tāte śāli[90]

The *Kṛṣi-Parāśara*[91] emphasizes on the need of releasing excess water in the month of *Bhādra,* through an outlet made in the field. Only as much water as is necessary for dipping the roots of plants should be allowed to remain; otherwise the plant will be inflicted with diseases.

The seasons for transplantation have also been identified by Khanā, thus:

Vaiśākha vona, Āṣādhi royā
Jaygā nā hay dhān thoyā[92]

Āuś paddy sowed in the month of *Vaiśākha* (April–May) should be transplanted in *Āṣāḍha* (June–July). This will yield such an excessive harvest that there will not be enough place to store the grain.

For other seasonal varieties, the time for transplantation was between the month of *Śrāvana* (July–August) and the twelfth day of *Bhādra* (August–September):

Śrāvaner puro, Bhādrer vāro
roo er madhye yata pāro[93]

According to *Kṛṣi-Parāśara,*[94] seedlings transplanted in *Śrāvana* should be one cubit apart from one another; in *Bhādra,* half a cubit, and in *Āśvīna,* they should be four fingers apart *Ropoṇa* is forbidden in lowlands.

Kaṭṭana of Paddy

For *kaṭṭana* or 'thinning out' and levelling the field after the seeds have sprouted, the months of *Āṣāḍha* (June–July) and *Śrāvana* (July–August) are the most suitable:

Āṣāḍhe kāḍan nāmke
Sravane kadan dhanke
Bhādre kādān śīṣke
Āśvīne kādān kiske[95]

The *Kṛṣi-Parāśara* warns that the paddy of which *kattana* has not been done remains like seeds[96]. Both the texts agree that there is indifferent and poor growth of paddy in the absence of *kattana.*

Weeding

Khanā ordains that the period extending from the month of *Āṣāḍha* (June-July) to *Bhadra* (August–September) is the best time for the removal of weeds from the soil.[97] Confirming this, the *Kṛṣi-Parāśara*[98], cautions that 'agriculture suffers loss due to weeds.' Conversely, 'the agricultural produce from which weeds are removed becomes wish yielding to cultivators'.

Harvesting

Some of the sayings of Khanā enable the lay reader to trace the gradual growth of the paddy plant till it is ready to be harvested:

Thor tirise, phulo bise
ghoda mukhe tera din
sisa dekhe bisadin
kātte Māḍte daśdin[99]

Thirty days after being planted the seeds begin to sprout. The stalks that appear begin to flower, twenty days thereafter. Gradually, the pannicle of spikelets, constituting a paddy plant, comes to resemble the head of a horse. After a lapse of twenty days thereafter, it is time to reap and thresh the corn. This will take at least ten days time. The *Rāmacarita*[100] refers to the threshing floor, where the 'reaped crops were spread out and threshed by means of bullocks that went round and round over them'. The *Kṛṣi-Parāśara* recommends an elaborate ceremony called "*pusyayātra*" in *Pauṣa* (December–January) before the harvesting of obviously, the winter crop. Thereafter, having harvested the crop, one should store it in granaries and worship the Goddess of Wealth.[101]

The aphorisms of Khanā, make it apparent that the farmers were well acquainted with the fact, that weather and soil conditions in Bengal, permitted the growth of many other kinds of crops, besides paddy. Thus the seeds of Brinjal (*Begoon-Solanum melongena*) could be sown throughout the year, except for the months of *Caitra* (March–April) and *Vaiśākha* whilst *Vaiśākha* and *Jyaistha* (May-June) were suitable for the cultivation of turmeric (*Halood-Curcuma longa*).[102]

It was believed that Betel (*Pan-piper betel*) cultivated in the month of Śrāvana (July–August) yielded a good harvest. However, the crop needed three years to mature.[103] Interestingly, *barajas* or betel-leaf plantations, yielding a high revenue are included within the confines of a donated plot as recorded in the Calcutta Sahitya Parishat copper plate of Visvarupasena.[104] An image inscription[105] attest to the fact, that in the villages of Pāikpārā and Betkā, lived a class of people known as 'Barai' who derived great wealth from the sale of betel-leaf cultivated by them.

Towards, the end of the season of Śarat (August-September-November), mustard (*Sarisa-Brassica campestris*) and raddish (*mūlā-Raphanus sativus*) could be cultivated.[106] Incidentally, the Vappaghoshavāṭa grant of Jayanāga (7th AD) states that a '*sarṣapa yānaka*' (mustard channel) formed one of the boundaries of a village in the Audambarika visaya of Karṇasuvarṇa.[107]

Between the last four days of *Bhādra* (August–September) and the first four days of *Āśvīna* (September–October), black gram (*kalāi-phaseolus mungo*) should be cultivated. The seed of the chilli (*Lankā-capsicum annum*) should be sown at the same time. Any delay in maintaining the time schedule would lead to the crop being inflicted with disease; avoiding the last nine days of *Āśvīna* and the first nineteen days of *Kārtika* (October–November), the pea (*maṭar-pisum sativum*) could be cultivated; white tila (*gingelly-sesamum indicum*) could be sown in the month of *Kārtika* (October–November) and black tila in *Māgha* (January–February); the month of *Phālguṇa* (February–March) was conducive for the cultivation of *patol* (*trichosanthes dioica*) and the esculent plant ola (*amorphophallus campanulatus*) and in the month *Caitra*, maize (*bhuṭṭā-zea mays*) could be grown.[108]

CROP ROTATION AND DOUBLE CROPPING

Khanā's Vachan, further reveals knowledge of the practice of growing different crops in succession in different seasons on the same piece of land as well as of growing two crops thereon, at the same time. Thus, it has been claimed the chillies will grow well on the land where paddy has been cultivated:

Dheno māṭite jhāl prabal[109]

The prevalence of the system of double cropping is amply revealed in the following verses:

Sariṣā bune kalāi, mugh
bune beḍāo cāpḍe buk[110]

Sow the seeds of kalai (black gram—*phaseolus mungo*) and *mugh* (green gram—*phaseolus aureus*) on the same piece of land. The yield will make the farmer proud. The cultivation of the coral tree (māndār) along with betal nut palms in the same grove has already been referred to. Herein, the reason is that the leaves of the former will provide manure to the latter.

Interestingly, Khanā refers to a practice prevalent in rural Bengal even today—that of growing vegetables in the plot of land at the disposal of the farmer, however small that may be. Thus she maintains:

Uṭhān, harā lāu śaśā
ghare tār Lakṣmir dasā[111]

A person who cultivates the bottle gourd plant (*lāu-legenaria siceraria*) and cucumber (*cucumis sativus*) in his courtyard will be blessed by the Goddess Lakṣmī.

Cāl bharā kumḍa pātā
Lakṣmī balen thāki tathā[112]

'The Goddess Lakṣmi resides in that home where on the rooftop there is an abundant growth of pumpkins (*kumḍa–cucurbita pepo*).'

Significantly, the cultivation of all these food crops do not involve much expense. It is a common sight in almost every house in rural Bengal to spot these tendril climbers being provided with supports, entwining which they climb to the rooftops, where the vegetables grow. The *Subhāṣitaratnakosa* of Vidyākara, which contains a number of verses reflecting the poverty of the villagers, refers to a famished householder, consoling his wife with the wishful thinking that with their pumpkins, which they will grow without labour by their hovel at the end of the summer season, they will feed themselves like a king. In another verse, a housewife relies only on pumpkins for her family, but does not get a bounteous crop, because the water which she borrows from well-wishers leaks through the cracked pot and she cannot water them.[113]

In fact, the vegetables grown in the courtyard of the householder not only provide his family with food but often provide the only means whereby the peasant farmer can earn a living.

AGRICULTURAL IMPLEMENTS

Unfortunately, regarding this aspect of agriculture, Khanā is not very informative. She merely refers to the plough (*hāl*) drawn by bullocks and the spade (*kodāl*) used for digging up the soil.[114] However, the very fact that she refers to various kinds of agricultural operations indicates the use of some other implements other than those mentioned above. We may utilize data from some other sources, to form a more concrete picture. The *Kṛṣi-Parāśara* mentions eight constituent parts of a plough, 'conducive to welfare in all operations' and 'should be made very strong by cultivators.' According to the author, the *phālaka* (i.e., the ploughshare) resembles an *arka leaf* (*madar, calotropis gigantea*, Br-*Akanda* in Bengali).[115] In fact, in India, there are ploughshares consisting of a pointed piece of flat iron side by side with broad-lance shaped ones. There are sufficient and scientific reasons, such as differences in the nature of the soil, subsoil, crop, season and manure, for the choice of the former for soils where the latter would be out of place.

The *Kṛṣi-Parāśara* further refers to the *madikā* or *mayikā,* a word not found in standard Sanskrit lexicons, indicating its provincial origin. Its equivalent used in Bengali is the *mai,* a ladder-shaped contrivance used for levelling the rice-fields. In fact, the *Kṛṣi-Parāśara* contends that 'in the absence of that (application of the *mayika*), there is no even growth of the seeds'. The *viddhaka* is designated as the *bida* in Bengali and refers to the Indian harrow, used for breaking up clods of earth.[116]

The threshing floor, where on the reaped crops are stocked, has already been referred to. The *Kṛṣi-Parāśara* refers to the planting of the *medhi* or a pillar in the middle of a threshing floor to which the oxen are bound, in the month of *Agrahayaṇa* (November–December) which is attended by elaborate ceremonies.[117]

It is obvious from the above account that compilations like the *Khanār Vacan,* help greatly in filling the void created by the paucity of ancient works on agriculture. This accounts for their indispensability, in spite of certain defects, in the reconstruction of the economic life in ancient India, wherein agriculture constituted the mainstay of the people from times immemorial. It has often been claimed and, to some extent rightly, that a bounteous nature has encouraged a life of ease and comfort and a tendency towards fatalism amongst the Indians. However, we have enough testimony to prove that our ancestors were never negligent of the actual needs of the society, their no mean achievements being reflected in Indian life, even today.

NOTES AND REFERENCES

1. Majumdar, G.P. and Banerjee, S.C., ed. and trans, 1960, *Kṛṣi-Parāśara,* (Bibliotheca Indica Series No. 285) Kolkata, The Asiatic Society, verse 8, pp. 2,63 (henceforth ref. to as *KP.*). The authorship and date of the text can only be conjectured from certain data available in the work. In spite of the fact that both the title of the work and its colophon associate the name of Parāśara with it, there is no conclusive evidence to prove the identity of the authors of the *Parāśara-Smṛti* and the *KP.* Some verses of the *KP* have been ascribed by scholars to the *Rajamartanḍa,* a book by King Bhoja of Dhara (1,000–1055 AD) and others to Varāhamihira (sometime between the fifth and the sixth centuries AD. It is inconclusive as to whether the author of the *KP* was the borrower or *vice versa.* It is perhaps feasible to conjecture, that the author was earlier than the sixth century AD and by no means later than the eleventh century AD (*KP* v–ix). Ray J.C. (*Ancient Indian Life,* Cal. 1948), however, places Parāśara's work on agriculture between the sixth and the eight centuries AD. Chap. I, p. 30.
2. *Imperial Gazetteer of India,* new edition (Oxford 1907), vol. I, p. 22.
3. *Triennial Catalogue of Mss. 1925–26 to 1927–28* (Government Oriental Mss. Library,

Madras), Vol. I pt. I (Sanskrit), pp. 7277–78, Ms. no. 5276. fol. 16.

4. *Khanār Vacan* published by Visvadeva Mukhopadhyaya, Calcutta. The numbering of the chapters, verses, and pages in the present article has been made in accordance with the Pandit Rampada Bhattacharya's compilation of the *Khanār Vacan*, Kolkata Rajendra Library (not dated) (henceforth referred to as *Khanā*).
5. *KP.* Intro., pp. ix–xi.
6. Cf. Majumdar, R.C. (ed), 1943, *The History of Bengal, vol. I (Hindu period)*, Dacca, p. 562.
7. Ray, J.C., 1948, *Ancient Indian Life*, Kolkata, p. 29.
8. Sen, Dinesh Chandra, 1914, *Vaṅga Sāhitya Parichay (in Bengali) or Selections from Bengali Literature (From Earliest Times to the Middle Nineteenth Century)*, Kolkata, (University of Calcutta), part I. pp. 10–11. (Henceforth referred to as *Vaṅga Sahitya*); Sen, D.C., 1954 (Second Edition) *History of Bengali Language and Literature* (Kolkata, University of Calcutta), pp. 22–23, (henceforth, *Hist. Beng, Lang.* .....)
9. Ray J.C., *op. cit*, p. 29.
10. *Vaṅga Sahitya* ... p. 11.
11. *Khanā*, II, verse 87, p. 13.
12. *Ibid.*, II, verse 39, p. 6.
13. *Ibid.*, II, verse 65, p. 10, The word used for sugarcane soil in the verse is *dhūlā*, which generally means dust. A similar sense is conveyed by the Bengali word *curna*, which in turn, is synonymous with *chūṇa* meaning lime (calcium). The latter meaning is applicable in this context as sugarcane soil, in fact, contains as high as 25–30% of lime.
14. *Khanā*, II, verse 97, p. 14—*tāmāk buno guḍiye māti, vīja ponto guṭi guṭi*—the last line refers to an important means of propagation of plants. *Guti* or *gula* refers to the ball of clay applied round the stem, for striking the root.
15. *Ibid.*, II, verse 103, p.15—*nārikela gāche nune* (saline) *māṭi/śīghra śīghra bandhe guti.*
16. Watters Thomas, 1905, *On Yuan Chuang's Travels in India*, London. R.A.S., Vol. II, pp. 184–185, 187, 189–191.
17. Sandhyākaranandin, *Rāmacaritam*, ed. Hm. Haraprasad Sastri, revised with Eng. trans and notes by Dr. Basak R.G., 1969, (*Memoirs of the Asiatic Society of Bengal*, Vol. Il, No. 1, Kolkata), Chap, III, verses 17–20, 67 ff–67; ed. Majumdar, R.C., R.G. Basak, Banerji, N.C., (Rajshahi, Bangladesh, 1939), verses 17–20, 31 ff.
18. McCrindle J.W., 1877, *Ancient India as Described by Megasthenes and Arrian.*
19. Bhisagaratna, Kunuja Lal, (Tr.), 1963, *The Sushruta Saṃhitā*, Varanasi, Chowkamba Sanskrit Series Office, Vol. I, pp. 45, 138–40.
20. *KP.* verses 154–155. pp. 39–78.
21. *Epigraphia Indica* (Archaeological Survey of India, New Delhi), (henceforth *EI*) Vol. VI, 203–07. cf. Sen, B.C., 1942, *Some Historical Aspects of the Inscriptions of Bengal*, (Pre-*Muhammadan epoch*) Kolkata, p. 59 (henceforth *SHAIB*).
22. EI, XXII, 150 ff.
23. Majumdar, N.G., 1929, *Inscriptions of Bengal* (Rajshahi) Vol. III, 99 ff. (henceforth *IB*).
24. *KP.* verse 10, pp. 2, 63.

25. *Khanā,* III, verse 140, p. 19.
26. *Ibid.*, II, verse 24, p. 5.
27. *Ibid.*, III, verses 135–138, pp. 18–19, Trans. of these verses in *Hist. Beng. Lang...*, pp. 20–21.
28. *Khanā,* III, verses 118. 120, 121. p. 17. Trans of these verses in *Hist. Beng. Lang...*, p. 26.
29. *Khanā,* III, verses 122, 123, 129. pp. 17, 18.
30. *Ibid.*, verse 128. p. 18,
31. *Ibid.*, verse 131, p. 18.
32. *Ibid.*, verse 130. p. 18.
33. *KP.* verses 65–70, pp. 17–19, 70.
34. *Rāmacarita,* III, verse 26 (B) p. 70; V, v. 4(B), pp., 79, 80.
35. Varāhamihira, *Bṛhat Saṃhitā,* trans. Kern H. (Cal.1865) XXI, pp. 32, 34 ff; XXIII, 6–9 (statistics of rainfall); XXIII, 2 (rain-guage). (henceforth *Bṛhat.*).
36. *Khanā,* II, verses 30–34, p. 5.
37. *EI,* Vol. XX, 123–125. Vol. XXI, 172; Vol. XXXIII, 236–7.
38. *Corpus Inscriptionum Indicarum* Vol. IV. No. 22, lines 20. 24; *EI.*, Vol. XXII, 115–20, Vol. XXI, 183 f; Vol. IV, 76–81; cf. Gopal, L. 1965, *Economic Life of Northern India,* Varanasi, Motilal Banarsidass, pp. 5–7, for more details.
39. *EI,* XXIII, 236 f, XXI, 172, 183 f; 210 f; IV, 76–81; XXII, 115–20.
40. *Memoirs of the Asiatic Society of Bengal (MASB),* I, pp. 88, 90.
41. *Indian Historical Quarterly (Calcutta),* VI (1930) 53–56.
42. *Nārada Smṛti,* trans, by, Joly J., (Sacred Books of the East, xxxiii, Oxford, 1889), Chap. VI, 3, 4, 5; Chap. V, p. 23; *Bṛhaspatismṛti,* ed. A.K.V. *Rangasvami* (Baroda, 1941), Chap. XV, pp. 13, 9, 14; XVI, pp 1–3, 5.
43. *Khanā,* II, verse 35, p. 6.
44. *Bṛhat,* XIII, p. 31.
45. *KP.* verses 84–104, pp. 23–27. 72, 73.
46. Khanā, II verse, 88, p. 6.
47. *Amarkosa* (with Kṣirasvāmi's commentary) ed. Sharma, H.D. and Sardesai, N.G. (Pune, 1941), 9, 8, p. 202.
48. *Khanā,* II, verse 52, p. 7.
49. *KP.*, verse 21, pp. 5, 65.
50. *Khanā,* II, verse 54, p. 8.
51. *KP.* verse 18, pp. 4, 64.
52. Kauṭilya, *Arthaśāstra,* trans. by Dr. R. Shamasastry (Mysore, 1967). Book II, Chap. XXIV, p. 130 (henceforth, *Artha*)
53. *Khanā,* III, verse 147, p. 20.
54. *Ibid.*, verse 14B, p. 20.
55. *Ibid.*, verse 145, p. 19.
56. *Ibid.*, verses 149–151, p. 20.
57. *Ibid.*, verse 146, p. 20.
58. *Ibid.*, II, verses 56, 57, p. B.

59. KP. verses 79, 83, pp. 22, 23, 71, 72.
60. Kalidāsa, *Raghuvaṃsā* (Mallinatha's Comm.) ed. Velankar, H.D., Bombay, 1948, IV. p. 20. (henceforth *Raghu*).
61. *Khanā,* II, verse 40, p. 6.
62. *Ibid.*, II verses 72, 92, 93, 99, 100, 101, 108, 109; pp. 12, 14, 15, 16. cf, Das Gupta, Tamonash Chandra, 1935, *Aspects of Bengali Society from Old Bengali Literature,* Kolkata (University of Calcutta), pp. 236–239.
63. *KP.*, V. III, pp. 29, 74.
64. *Artha.*, II, p. 24.
65. *Brhat.*, LV. pp. 4–7; *The Sūkra–Nīti,* ed. Major Basu B.D., trans, Sarkar B.K., (*The Sacred Books of the Hindu Series. Vol. III*) (Allahabad, 1914), Chap, IV, Sec. IV, lines 91–93, p. 165.
66. *Khanā,* II verses 26, 27, p. 5. cf. *Hist. Beng. Lang.*, pp. 20, 21.
67. *Ibid.,* II, verse 106, p. 15.
68. *Ibid.*, II verse 75, p. 11.
69. *Ibid.*, II, verses 69, 70, p. 10; *Hist. Beng, Lang.*, p. 21.
70. *Ibid.*, II, verse 42, p. 6,
71. *Ibid.*, II, verse 102, p. 15.
72. *Ibid.*, II, verses 114–116, p. 16.
73. *Ibid.*, II. verses 78–79, p. 12.
74. *Ibid.*, II. verses 111–113, p. 16.
75. *Memoirs of the Asiatic Society of Bengal,* Vol. I, p. 90.
76. *SHAIB.* p. 99.
77. *EI,* IX, 302 ff.
78. Majumdar, R.C. (ed), 1943, *The History of Bengal,* Vol. I *(Hindu Period)* Dacca, p. 451.
79. *Khanā,* II, verses. 21, 22, p. 4.
80. *IB.*, p. 129.
81. *Ibid.*, 86, 89–90.
82. *Journal of the Royal Asiatic Society of Bengal (Letters),* Vol. I (1942), p. 1 ff; *EI,* XXVI, 1 ff.
83. *Khanā,* II, verse 28, p. 5.
84. *KP.*, verses 159–177.
85. McCrindle, J.W., *op. cit.*, p. 54.
86. *Bṛhat.*, verses 21, 27, 78 and 90; VIII., 47; IX. 42, 43; X. 18; XXV. 2; XXVII, I; XL, 2–3.
87. *Khanā.* III. verses 139, p. 19.
88. *EI.* XXX, 197 ff.
89. *Raghu.*, IV, 37.
90. *Khanā.* II, verse 53, p. 8.
91. *KP,* verses 193–194, pp. 48, 82.
92. *Khanar Vacan,* referred to in KP, p. XV.
93. *Khanā,* II, verse 20. p. 4.
94. *KP,* 185. 190, pp. 46, 47, 81, 82.

95. *Khanā,* II, verse 43, p. 7.
96. *KP,* verses 186–188, pp. 46, 82.
97. *Khanā,* II, verse 95, p. 14.
98. *KP.* verses 189, 192; pp. 47, 82.
99. *Khanā,* II, verses 58, 59; p. 9.
100. *Rāmacarita,* verse 176; p. 91.
101. *KP.* verses 221–237; 241–243, pp. 55–60, 86–88.
102. *Khanā,* II. verses 88, 94, pp. 13, 14.
103. *Ibid.,* II. verse 49, 62. pp. 7, 9.
104. *IB,* p. 140 ff.
105. *El,* XXVII, 26 ff. cf. Niyogi, P., 1962, *Contributions to the Economic History of Northern India,* Kolkata, Progressive Publishers, p. 33.
106. *Khanā,* II, verses 63, 65, pp. 9, 10.
107. *EI,* XVIII, p. 61.
108. *Khanā,* II, verses 64, 67, 68, 73. 74, 86, 90, pp. 9, 10, 11, 13.
109. *Khanā,* 1, II, verse 72, p. 11.
110. *Ibid.,* II, verse 66, p, 10.
111. *Ibid.,* II, verse 85, p. 13.
112. *Ibid.,* II, verse 84, p. 14.
113. Vidyākara, *Subhāṣitaratnakosa,* ed. by Kosambi, D.D., and Gokhale, V.V. (Harvard, 1957), verses 1306, 1315; cf. *The Economic Life of Northern India,* (Varanasi, 1965), p. 244.
114. *Khanā,* II, verses 35–37, 67, pp. 6, 10.
115. *KP,* verses 112, 113, 117, pp. 49, 75.
116. *Ibid.,* verses 118, 182, pp. 45, 30, 75, 81.
117. *Ibid.,* verses 214, 220, pp. 53–55, 85–86.

CHAPTER 37

Vṛkṣāyurveda in Ancient India

Brajdeo Prasad Roy

DEFINITION AND SCOPE

In modern times, the term Botany is used to mean plant science but, within its scope, Biology is also included. Biology is a term of modern origin. In ancient India, there were concepts of these sciences. There are evidences which prove that the Indus Valley people had the knowledge of several plants and creepers. The *Ṛgvedic* seers were acquainted with plants, plant life, their medicinal use and the gradual process of their germination, growth and decay.[1] The *Atharvaveda* describes in detail the characteristics and medicinal properties of several plants.[2] But these texts do not refer to the term *Vṛkṣāyurveda*, meaning, plant science. The physicians were the persons who had knowledge of this branch of science as they had to deal with the plants and their medicinal properties.

Both Caraka and Suśruta define *Āyurveda* as the science which protects and increases the span of life of a person, keeping him physically and mentally healthy.[3] It cures their ailments by using several remedial methods. In the same sense, *Vṛkṣāyurveda* is the branch of knowledge which enables the plants to grow properly and yield desired roots, flowers, fruits and grains. It deals with their medicinal care and method of their application in order to cure them of their diseases. For the first time, Kauṭilya refers to the term *Vṛkṣāyurveda*. According to him, a *Vṛkṣāyurvedajña* was an expert in plant science.[4] After him, Varāhamihira[5] and the *Agnipurāṇa*[6] refer to this term. Both these texts contain sections dealing with the topics of *Vṛkṣāyurveda*. This section in the *Agnipurāṇa* is nothing but a popular encyclopaedia of all kinds of knowledge of plants. The *Bṛhat Saṃhitā* is a manual containing directions for the applications of the knowledge of astronomy and astrology in practice. All these three texts are non-Botanical treatises, but in each of them we come across a section dealing with the subject of *Vṛkṣāyurveda*. Kauṭilya refers to *Vṛkṣāyurveda* without discussing its central point, normally the treatment of plant diseases and prescriptions for their remedies. Mainly, Varāhamihira and the author of the *Agnipurāṇa* have discussed this point. It appears that in Kauṭilya's view, *Vṛkṣāyurveda* was a section of the *Kṛsitaṃtra*, obviously signifying a treatise on agriculture.

In the other two texts, namely, the *Bṛhat Saṃhitā* and the *Agnipurāṇa*, this term looms large while *Kṛṣitaṃtra* does not find any mention. But in them the purpose is just to indicate the application of the knowledge relating to the ancient science of plant life for agriculture, horticulture and for irrigational and economic purposes.

All these texts furnish an impression that the subjects within the scope of the ancient science of the plant life consisted of the collection and selection of seeds, their sowing, germination, grafting, cutting, planting, nourishing, selection of soil, manuring, irrigation, protection and harvesting.

It appears that there were independent texts pertaining to plant science, which contained sections on the art of agriculture before it gained an independent status of its own. At present, those independent treatises on agriculture are not available. Bhattopala, a commentator on the *Bṛhat Saṃhitā,* while commenting on the section of *Vṛkṣāyurveda,* cites quotations from the ancient authors, such as, *Kaśyapa, Parāśara* and *Sāraswavata,* whose treatises were primarily concerned with *Kṛṣi,* the art of agriculture. At present, we have only one text on agriculture entitled *Kṛṣi-Parāśara.* However, the term *Vṛkṣāyurveda* also indicates the close association of the knowledge of plants and plant life with the art of healing, references to which are numerous in ancient Sanskrit texts. The close relationship between the medical science and plants may be guessed from the fact that medicinal plants have been termed as *auṣadhi* (medicine). The term *bheṣaja* means medicine or medicinal plants. This fact may be corroborated by Caraka's opinion that an expert physician, who is well acquainted with the name of external features of plants and is able to use them properly and according to their properties.[7] *Bheṣaja Vidyā,* the knowledge of healing using medicinal plants signifies plants, and plant life with special references to medicinal plants and their uses.

In some of our Sanskrit texts, we find theories about the evolution of plants, their position in the entire scheme of nature and development, along with various philosophical speculations. As in case of the human beings, *Ayurveda* is a very comprehensive science which deals with the human life from conception (*garbhasthāpana*) to death. Similarly, *Vṛkṣāyurveda* also covers the complete process of the life of a plant, beginning from sowing of seed to the harvesting of crops. In its comprehensive scope, it comprises all aspects of plant life—germination of seeds, morphology, external and internal physiology, ecology, and taxonomy of plants.

SOURCES

For the study of plant science in ancient India, one has to utilize two kinds of sources, namely, the literary and archaeological. The earliest literary source is the *Ṛgveda* (1500–1200 BC) but the *Atharveda* (1200–1000 BC) is of utmost importance, which directly deals with the characteristics and medicinal properties of near about a hundred plants. *Āyurveda* is considered as a part of this Veda, which also describes the proper use of plants for curing ailments. The *Yajuṣ Saṃhitās* (1000–600 BC) also throw light on this science. The Vālmīki *Rāmāyaṇa,* and the *Mahābhārata* along with some of the *Jātaka* stories (500 BC–200 AD), are very helpful in this regard. For the first time, the Kauṭilya *Arthaśāstra* (fourth century BC) discusses several aspects of plant science though the *Aṣṭādhyāyī* of Pāṇini (pre-Kauṭilyan), the *Mahābhāṣya* by Pātanjali and the *Millindapañho* by Nāgsena (second century BC) also throw considerable light on plant science.

The Caraka and Suśruta *Saṃhitās*, which were finally compiled in the second century AD, are also important sources. Several texts of Kālidasa and *Amarakoṣa* of Amarsiṃha (Gupta period) are important sources for the study of *Vṛkṣāyurveda*. In a direct way, Varāhamihira in a section on *Vṛkṣāyurveda* in the *Bṛhat Saṃhitā* (Chapters 24, 40, and 55) discusses matters of plant life in a comprehensive way. He flourished sometime in the late fifth or the early sixth century AD.[8] The *Kṛṣi Parāśara*, an informative text of uncertain date but composed not later than the fifth century AD, is devoted mainly to the plantation of paddy and throws welcome light on other aspects of agriculture.[9] It is an all-comprehensive book full of a large number of pregnant aphorisms relating even to the minutest particulars of agricultural processes. Similarly, the *Kaśayap Saṃhitā* is a text on agriculture, which draws much from the *Bṛhat Saṃhitā*. Though several *Purāṇas* contain information about plant science, but the *Agnipurāṇa* surpasses all of them concerning the *Vṛkṣāyurveda*. It appears to have been finally composed somewhere in the northeastern India in early medieval times, but not earlier than the tenth century AD.[10]

Among the foreign literary sources, the accounts of Greek writers[11], namely, Megasthenes, Arrian, Strabo, Plutarch, Nearchus, etc., and the Chinese travellers[12] such as, Fa-hien and Hiuen-Tsang also throw considerable light on the fertility of the Indian soil and several crops of different regions of India.

Some of the archaeological sources help us understand *Vṛkṣāyurveda*, such as the seals and potteries of the Harappan civilization which contain the figures of plants and creepers. The punch-marked, Indo-Bactrian, Tribal and the Gupta coins also have depictions of plants, creepers, flowers and fruits. The early Indian classical art, and the Kuṣāṇa and the Gupta schools of art also contain depictions of plants, creepers, flowers and fruits. Several specimens of remains of plants and food grains have been unearthed from various archaeological sites, which are also useful sources for the study of plant science in ancient India.

IMPORTANCE OF PLANTS

In pre-agricultural times, the people were pastoral and migratory ones, who grazed their cattle in pastures and gathered their food from roots and plants. Hence, naturally, they realized their importance. The Vedic people, with their spread in the Saraswatī and the Gangetic valleys during the compositional period of the *Saṃhitās*, had to clear the forests for settlements. They availed of roots, fruits and grains both from cultivated and wild plants. The terms *vṛkṣya* and *vānaspatya* were used to denote the fruits of trees. On account of getting fruits from plants, the fruit-bearing trees were rightly considered as of greater importance than those which did not yield fruits. For possessing fruit-bearing trees, a certain rite was performed in which the sacrificer had to eat the fruits which he desired and had to offer the wood pieces of such trees, which he wanted to obtain. Different kinds of fruits were obtained from the trees, which they used as the items of their diet besides they obtained honey as well, which was used as a medicine as well as a drink. They cultivated different kinds of plants, which yielded cereals; hence, naturally, they realized the importance of plants in their economic life.

The plants were also useful in providing the people of ancient times with shelter. In villages, most of the houses were built of grasses and bamboo along with the wood pieces. Archaeological excavations have revealed the use of wood and grasses for building the houses. At Atranjikhera, third period, representing the P.G. Ware culture, the hearths are circular in shape. The presence of postholes over thick and mud floor suggests thatched roofing over bamboo or wooden posts. Some houses were built of wood and were comparatively durable. It was also easy to build houses of wood, as this material could be easily procured.

In the life of the people, chariots played a significant role and were required on battlefields. Chariots were the most comfortable and swift means of communication and transport. Small agricultural tools, such as, hoe, handles of spades, sickles and others were made of wood or bamboo. All these necessitated realizing the importance of plants.

In the domestic life of the people, utensils made of wood and clay were used. The poorer sections of society could not afford metallic utensils, hence, wooden containers, dishes, lids, jars, and pestles were used. The *Brāhamaṇa* texts prescribe specific timbers for making different kinds of royal thrones to be used on specific occasions.

To cook food, fuel was an essential item, which was easily obtained from plants. Dry and hard stems of cereal plants were used as fuel but bigger plants were also used for this purpose.

The life of people was so closely associated with the vegetation world that they attached religious sentiments with them to a great extent. They thought that the plants were the givers of their life and truly, they were their sustainers. On account of this, they rightly looked for the abode of the immortal gods in plants. The people considered some plants as gods and others to be their abodes.[13] The *soma* plant was considered as the very embodiment of god and several hymns have been dedicated to it in the Vedic texts, as its juice was intoxicating, life giving and thought provoking. *Soma* was also offered to the Gods. The *aśvattha, nyagorodha, palāśa, khadira* and other trees were considered as the abodes of gods. In sacrifices, sacrificial offering of the pieces of wood were offered to different gods. Sacrificial posts of several shapes and sizes were made and erected on the sacrificial grounds in different kinds of sacrifices. The *Yajus Saṃhitā* and the *Brāhamaṇa* texts prescribe specific trees to be cut down for making specific kinds of *yūpas* to be erected in particular sacrifices. *Vānaspatya,* Araṇya, Aranyānī, Soma, Marut and Rudra were presiding gods over the vegetation world, in whose honour prayers have been dedicated in the Vedic texts.

Man is highly indebted to the plants for their utilitarian and aesthetic values. In every age and stage in civilization, men and women have been fond of flowers and garlands for their personal embellishment. Their passionate love for flora is apparent through fascinating and gorgeous descriptions in literature. The lotus flower has been usually mentioned in the texts in the contexts of sacrifices and personal decoration. Various kinds of cosmetics were prepared with flowers and roots, which were used for enhancing physical beauty.

The utility of plants made the people to associate closely with them. The importance of wood was so high that in society, special terms were coined to denote the persons engaged in wood-works. The woodcutters who adopted plant cutting as the means of their livelihood. *Rathakāra* was the person who engaged himself in the specialized art of chariot making. The *bardhakī* was the person who worked in wood and made various kinds of objects to be used by people in different ways.

FORESTS

In the economic life of the people, the forests proved to be very useful. In literature, different terms for forests have been mentioned such as *Araṇya,* which means far away from the village or that which is not worth in which one may move (*a-ramaṇa*). *Araṇya* is contrasted with settlement and agricultural land. On account of its dense trees, the forest was a natural and the most suited shelter for the thieves. These facts reveal that forests were far from settlements and it was fearful to settle in dense forests and practise cultivation.

The people did not like to establish their settlements amidst deep and dense forests for fear of beasts and dangerous reptiles, which destroyed their life and cattle, as also their standing crops. It was time consuming and laborious work to clear the forest for settlement and agricultural operations. In spite of this, people had contact with the forests. The retired persons went to the forests wherein they meditated and thought over the vital problems of life and metaphysics. The contact was so intimate that the retiring persons preferred to go to the forests to pass the rest of their life (*vānasprastha*). They also composed the *Āraṇyaka* texts in the forests, which played an important role in the social and religious life of the people.

There are ample evidences to show that originally, the Yamuna-Gangetic Valley was a thick monsoon-fed forest. The present alluvial plains are a result of heavy de-forestation at the hands of man. There are records of its being covered at one time by vast forests. This region, with its forests and the hard calcareous swirl, was not a congenial place for settled agriculture in the Copper-Bronze Age. Archaeologically, it is obvious that before the P.G. Ware presents the picture of village life with a limited agriculture. Only in the N.B.P. times do we have the cities coming up as the result of a large-scale agriculture, which was possible by the abundant iron implements. The Gangetic Valley needed heavy ploughshares to produce the agricultural surplus. Undoubtedly, the P.G. Ware people started colonization of the Gangetic Valley with iron tools, but large agricultural surplus was possible only in the NBP times with the availability of iron from Bihar.

The people conceived the idea of Araṇyānī, the goddess presiding over the forest for their protection in dense forests from beasts. They sought her help in finding out troubleless routes in their travels. The state also came to realize the importance of forests and adopted adequate steps for their protection from devastation. The *Vājasaneyī Saṃhitā* refers to the *Vanapa,* an official of the state who was in charge of guarding the forests from being devastated or destroyed or misutilized by human agencies.[14] With the growth of population, the forests were being encroached upon which were the property of the state; so the appointment of *Vanapa* was a step in the right direction. In spite of this, some natural agents caused devastation to the forests, such as, *dāva* or the forest fire, which caused conflagration in them especially during the summer. Kauṭilya advises the king to appoint officers for protecting the forests and utilizing their products properly to enrich the treasury.[15]

The forests were classified into two classes on the basis of their extent. *Araṇya* was the common term to denote the forest in general but *dīrghāraṇya* referred to extensive and vast forests. Sometimes, these terms were substituted with *vana* and *mahāvana*, respectively. During the compositional period of the *Aitareya Brāhamaṇa*, there were numerous villages which were settled close to each other in the eastern region but in the western region, the signs of villages were a few as it was covered with dense forests. In the Yamuna-Gangetic basin, *Naimiṣāraṇya* was a famous forest, which was the most famous seat of learning. Several sages had established their *āśramas* in this forest. *Naimiṣīya* denoted the inhabitant of that forest. This forest is identified with Nīmasāra situated on the banks of the River Gomatī at a distance of 45 miles northwest from Lucknow. An important forest was the *Khāṇḍava*, which marked the boundary of the Kuru country. The Purāṇic traḍitions locate this forest on the River Aśvarathā or the Yamunā and sometimes regard *Khāṇḍava Prastha* as identified with Indraprastha though the Khāṇḍava is supposed to have extended from the present Bulandsahar to Saharanpur in U.P. But really, Khāṇḍava was to the south of Kurukṣetra. The Jātaka stories and the Kauṭilya *Arthaśāstra* throw considerable light on the economic utility of the forests.

SEEDS AND THEIR GERMINATION (*AṄKUROBHEDA*)

The first stage in the growth of plant is germination, which means awakening of life, latent in the seed under certain given conditions, such as, supply of air, water and heat. Suśruta, elucidating this point (sprouting) in detail, remarks that for germination of first century AD strong undiseased sprouts, some factors are essential, such as, proper season (*ṛtu*), good soil (*kṣetra*), water, vigorous seed (*vīja*) together with proper care.[16] These points have been emphasized by Bālāditya, the author of the Gwalior inscription of Mihira Bhoja that from the fresh seed taken from a fruit in a garden and shown in a natural fertile soil, grows having strong roots of superior kind from which again spring fresh plants.[17]

One of the ways of cooperating with nature for increasing yield was to use an improved variety of seeds. The seed production was not merely a commercial activity but a highly technical job involving quality control, seed testing and safe storage. The ancient Indian agriculturists were well acquainted with the fact that good yield of grains depended upon good seeds. So, emphasis was given on good quality of seeds for sowing.

The seed, prior to sowing, required modification for easy germination. Modification included selection of good quality of seeds, scrutiny of healthy ingredients, removal of worn out grains, and keeping them in water to cause early germination in the field when sown. It was rightly observed that the seed, which is poured into the womb (furrows), becomes generative.

Collection of better quality of seeds was an important work for their proper germination and growth of healthy plants. So, the cultivators as well as *Sītādhyakṣa* collected in the proper season seeds of all kinds of grains, flowers, fruits, vegetables, bulbous roots, creepers, flax and cotton. Different kinds of seeds required different kinds of soil, such as, the seeds of *vaṭa, pippala, nimba* and the rest, sprout during the rainy season under the influence of dew and air.[18]

Prior to sowing the seeds, it was essential to treat them so that they could germinate properly. Different varieties of seeds of grains needed separate treatment. Kauṭilya suggests that the seeds of *dhānya* (grains) should be left in the open for seven days and nights, soaking in dew by night and drying in heat by the day, and for three days and nights or even five, in the case of seeds of pulses (*koṣa dhānya*). In the case of *kāṅdavīja,* such as, sugarcane, mixture of honey and ghee should be smeared on both the ends. Smearing should be made with cowdung in the case of stone-like seeds (*asthivījja*) and in the case of trees, burning in the pit and fulfilment of the longing with cow bones and cowdung at the proper time. Kauṭilya thinks that at the first sowing of all kinds of seeds, the cultivator should sow the first handful after it is immersed in water containing gold and then offer prayers to Kaśyapa, the lord of creation and the God of rain and should invoke the divine Sītā for successful harvest.[19]

The seedling is called *aṅkura* which etymologically means that the plant becomes visible for the first time by it. Scientifically, this term is more accurate than the English synonyms 'seedling'. The ancient Indians knew of the process of germination accurately. The term uttānapāda (*urdhvamula*) indicates that the primary root comes out first (*mūla*) regardless of whatever position the seed is placed.

Kauṭilya suggests that seeds of various grains should be sown according to the availability of water in different regions of the country. The cultivators should decide on wet crops (*kaidaraṃ*), winter crops (*haimanaṃ*), or summer crops (*graiṣmikaṃ*). *Śāli* and others are the best crops, vegetables middling but sugarcane is the worst as these are fraught with many dangers and require much expenditure.[20]

On this point, Kauṭilya says that a region where the foam strikes the banks is suited for creeper fruits (*vallīphal*); regions on the outskirts of overflows are suitable for long pepper, grapes and sugarcanes; those on the borders of moist beds of lakes for green grasses; ridges are suitable for plants reaped by cutting, such as, perfume-plants, medicinal herbs, *usira* grass, *piṇḍaluka* and others. On lands suitable for each, he should raise plants that grow both on dry lands and in wetlands.[21]

Kauṭilya suggests that in conformity with the rainfall, crops should be shown requiring both plenty or little water. *Śāli* and *vrīhi* varieties of paddy, *kodrava, sesame, priyangu, udaraka* and *varaka* should be shown first (*pūrva vapaḥ*). Then *mudga māṣa* and *saimlya* should be sown (*madhyavapa*). Lentils, tulottha barley, wheat, *kalaya,* linseed and mustard are the last showings (*paścādvapaḥ*) or the sowing of such seeds should be in conformity with the season.[22]

Ploughing of plots was considered as an essential condition for sowing seeds. It has been observed that furrow is like the womb of the seeds, and if one casts seeds into unploughed field, it is just as if one were to sow seeds elsewhere than into the womb.[23] A somewhat similar view has been mentioned in the *Avesta* as well that he who would till the earth, unto him will she bring forth plenty. Like a loving bride on her bed unto her beloved, the bride will bring forth children, the earth will bring forth plenty of grains and fruits.[24] The earth is like a maiden who wants a good husband. Hence, attention was given on good ploughing. Mistakes were but natural in ploughing the fields, which were rightly corrected and this act has been compared with the correction made by the *hotṛ* reciting the Vedic hymns.[25] On account of the special importance of plough for ploughing the land, it was also asked as a boon. The field for sowing seed was prepared by the cultivator with the help of oxen drawing the plough. Hence, prayers were offered so that the ploughshare (*phāla*) may plough the field properly.[26] Kauṭilya says that the *Sītādhyakṣa* should cause the seeds to be sown in land suitable for each, which has been ploughed many times. He should cause no delay in this respect on account of ploughing implements and bullocks.[27]

After the preparation of soil, seeds were sown into the field. This process was very important; hence prayers were offered to different gods for healthy germination of seeds and growth of plants. *Kṣetrapati* was the presiding deity over agriculture and was worshipped for increased yields of grains. Soma, the god of plants, Pūṣan, the god of fertility, and Indra, the god of rain, were worshipped for letting out the tender shoots and further growth of plants. Indra was the lord of the fallow land. He was the lord of tilth and, as such, he has been prayed to sink the furrow as Pūṣan guides it. He was the god of plants and grains; sowing operation was viewed to be vitally dependent upon magical observance. Magical beliefs acquired importance at an early agricultural stage. The Vedic *Saṃhitās* contain chants which were recited while sowing the seeds to insure good germination and harvest. The Brāhamaṇa texts and the *Śrautasūtras* also prescribed rites, which were to be performed on the occasion of sowing seeds. The works of tilling, sowing and reaping were slow and arduous, requiring patience and faith. Accordingly, they depended upon fate, so for acquiring divine grace, they took resort to magic.

Udvija was one of the three kinds of creation, which meant something produced from sprout or germination. For securing different kinds of grains, several kinds of seeds were sown in the field. It was observed that from the right side the seed is infused into the womb (furrows); water heals what is injured or torn in the earth and from the seed, lands spring forth full buried as these spring forth from the seal of the waters. They scattered the seeds along the furrows for germination. After a few days, the sown seeds grew. The shoots of rice, panic and barley have been referred to in the texts.

After sowing the seeds in furrows, these were covered by ploughing the land once again. But in some cases, it was not essential to plough the land for sowing seeds. A perusal of the process of the performance of the sacrifice shows that grains were produced both on the ploughed and unploughed land, and for producing them on both kinds of land, special rites were prescribed. Even now, most of the seeds are sown in ploughed fields but some are sown in unploughed fields, such as, gram and others. The *Śatapatha Brāhamaṇa* describes the persons carrying seeds on their heads for sowing their fields, but bringing the harvested crop on the cart.[28] Digging stick was one of the earliest instruments of food gatherers with which they made holes for sowing seeds, and sometimes for breaking clods, and occasionally a wooden blade to chop down the top of weeds. The ancient Indians had minutely observed the process of germination of seeds as they rightly thought that these were impregnated in front behind. In the operation of sowing, the sower played an important role because the success of agriculture depended upon him also, so in the *rājasūya* he has been dedicated to welfare.[29]

The cultivators were acquainted with the importance of the rotation of crops,[30] which was an important advancement in agricultural knowledge. Continuous cropping was the usual practice but the crops were also grown in the same field by rotation and the system of following was known.[31] In the course of a year, two crops were harvested from the same field.[32] Barley was sown in winter and ripened in summer, whereas rice was sown during the rainy season and it became ripe by autumn. Beans and sesame were sown during the autumn and they became ripe during the winter season.[33] The winter crop ripened by the month of *Caitra.* The ripening of crop in winter was so important that the name *Śarada* was ascribed to this. This season was looked upon as an important one because not only the rice ripens in this season, but it was also harvested during the same period.

In the post-Vedic times also, two crops were produced in a year, which was a source of great wonder to the Greek travellers who came to India. In winter in parts of northern India, two crops grow without irrigation while in the plains, a summer crop of rice grew during the rainy season and a second crop in the dry season.[34]

MORPHOLOGY (FEATURES OF PLANTS)

It is essential to have knowledge of the external and internal features of plants, which help us in identifying them for their specific usage. The ancient texts describe these features of different plants in detail right from the Vedic times. These contain terms and expressions covering all the features of plants. In external morphology, the roots, stems, leaves, flowers, fruits and seeds have been included.[35] Every plant is broadly divided into two parts, namely, the root (*mūla, pāda*) and shoot (*vistāra*). *Mūla* is expressive of its functions and locations; by its means the plant is fixed in the soil. The plant is called *pādapa* because of its *pādas* (*mūla*). It observes water (*rasa*) from the soil. Adventitious roots were called *Śākhāsipha* (fibrous roots springing from the branches). Fibrous roots were called *sipha* or *jaṭā.* Adventitious roots hanging from spreading branches were called *avaroha,* meaning that which goes down, for example, *nyagrodha* tree (*Ficuous Indica*). Roots of some of the plants were used as food items.

The shoot (*vistāra*) is divided into stem and leaves. The main stem (trunk) was called *prakāḍa,* meaning the part which is between the main root and the place from which branches originate. It is called *skandha* as it bears the head or crown. Strong stemmed plants were called *vanaspati* because they stood erect but the weak-stemmed plants cannot support themselves and according to their habit, they are distinguished into a creeper and a trailer. The creeper (*vallī, vratati* and *latā*) includes both a twiner and a climber. *Latā* means that which embraces or twines. It goes from the root to the top of a tree. The other kind was called trailer (*pratanin*) meaning spreading on the ground.

The stem may be plain or jointed (*parva*). Each joint.is called a *parva* or *granthi,* from which the leaves spring. Plants may be with stem or as a stemless plant. Stemmed plants are called *sakāḍa* and stemless as *aparkāda* or *stamba.* The primary branches are called *skandaśākhā* and secondary ones are known as *praśākhā, pratisākhā* and *anuśākhā.* Branchless stems are called *sthāṇu* or *śaṅku.* Apex of the tree or the treetop is called *śiras, agra* or *śikhara.*

Patra (leaf) is so called because it falls very soon. It is also called *parṇa* because of its green colour. The stalk of the leaf, when present, is called *bṛnta.* New leaves are called *pallava* or *kisalaya.* The number of leaflets, namely *dvipatra, tripatra, saptparṇa* and so on describes compound leaves. Leaves are also described by their shapes, such as, *aśvaparṇaka,* meaning resembling the ear of the horse and *mūṣikaparnī,* resembling the ear of a mice.

Puṣpa (flower) was called *sumana,* meaning that which pleases the mind and *puṣpa* also because it opens. It is also called *prasūna* because it is born from plants. *Kalikā* is the bud (unopened flower). An opening bud is called *mukula* and full-blown flowers are called *vikaca* and *sphuta,* and the bunch of flowers is called *stabaka.* Petals are called *puṣpadala* and stamens are known as *keśara.* Pollen grains are called *keśavareṇu* or *parāga.* The *Dhammapada* describes different characteristics of flowers in detail while narrating the virtues of a good person.[36]

Phala (fruits) means the result of a previous process. Green fruits are called *salatu,* dry fruits are termed *vana* and fleshy fruits are known as *jalaka.* There were several kinds of fruits which were mainly known after the names of particular trees. A legume or pod is called *sanai* and the seeds in it are known as *śamī-dhānya.*[37] The juice of a fruit is called *rasa.*

Vīja (seed) means that out of which something grows. A seed was enclosed in a vessel called *vīja koṣa* (seed vessel). Sometimes, seed vessels are identified with the ovarian wall. The kernel of the *vīja* has been called *sasya* (endosperm) and the cotyledon is known as *vījapatraṃ.*

Besides knowing the external features of the plants, the ancient Indians had the knowledge of their histology, which dealt with their internal morphology. For example, the stem was divided into two parts, namely, *twaca* or *valkala,* meaning skin and inner part of wood or essence was called *sāra,* meaning, that which lasts till the end of time. It included the *majja,* the pith. Thus, the stem is internally divided into three parts, such as, an outer skin (*twaca*), and the inner, the wood between which stand the softer part with strong fibres corresponding to the human flesh along with nerves. The wood encloses the soft pith. A more elaborate description of this has been given in the *Bṛhadāyraṇyaka Upaniṣad,* which mentions that the plant is similar to a human body. Like the hair and skin in the human body the plants have the leaves and bark. As blood rushes out from the human body, so does sap flow from the tree. When struck, the human body lets out a stream of blood. This happens also to the plants. In the human body, there are several layers of flesh; likewise in the plant there are layers of wood. The muscle in the human body is similar to the fibre in the plants. They are like the wood within, and the marrow resembles pith.[38] A curious advance in the knowledge of plant life is displayed in the Śaṅkara Miśra's *Upasakara,* wherein he notes the growth of organs or tissues by natural recuperation after laceration. The closing up of fracture in plants is manifest means of fruitation. In *Gunaratnas* commentary, there is a reference to healing of wounds in plants.[39]

PLANT PHYSIOLOGY

In this aspect of plant science, the influence of soil and food upon the vegetation system has been studied. These things ensure the health of plants and remove their diseases. In the nourishment of plants, the most important factor is soil, the principal source of substance, which is roughly classified into two classes, namely, fertile (*urvara*) and barren (*usara*). The fertile soil is capable of supplying the plants with their necessary food materials while the barren does not have this capacity. The plants draw food materials from the soil through the help of the root. The *Mahābhārata* describes in detail that plants drink water through root with the help of air.[40] The root of a plant is the mouth for sucking the watery substance. Energy (*agni*) and air (*vāyu*) help in the digestion of water, which is absorbed through the roots of the tree, which is then conveyed to the leaves. As a result of this process, the plants undergo development and become graceful. The leaves are very important parts in a plant and, obviously, act as a kitchen where the food of the plant is prepared. Without leaves, a plant cannot grow. During the rainy season, more solar energy and water are available; consequently, there is more storage of food in plants. But betel is a shade-loving plant and has larger leaves. The cause of this is absence of sunlight. Water, going upto the leaves, circulates all over the plant.[41]

PLANTING

The science of planting is very ancient in India. Its full-fledged development has been mentioned in the *Arthaśāstra*, which states that the *Sītādhyakṣa* must be possessed of knowledge of the science of agriculture dealing with the plantation of plants being assisted by those who are trained in this science.[42] Varāhamihira also refers to different kinds of plants to be planted in a garden or in the house compound for welfare of the residents.[43] Kaśyapa innumerates plants which were to be planted in temples, gardens, houses and kitchen gardens. The *Agnipurāṇa* mentions the trees to be planted all over the four directions of the houses.[44]

In addition to ordinary methods of propagation by seeds, cuttings and graftings were also in practice. The *Bṛhat Saṃhitā* mentions several trees which were to be propagated by means of cuttings (*kāṅḍaropaṇa*), being smeared with cowdung.[45] Grafting was a better of propagation of plants which was done in two ways. The first was *mūlakṣeda*—uprooting a plant and inserting on the root of another and the next was *skandharopaṇa* —inserting the stem of another plant.[46] While planting the plants, the planter had to keep in mind the ideal distance of plants from each other as it was best plant trees at the intervals of 20 cubits, next at 16 while 12 cubits interval was the minimum that could be prescribed.[47]

MANURING

Only good ploughing was not an essential condition for a good harvest. The soil was one of the most important factors, which played a vital role in the crop production. It provided strength to the standing crops and acted as the storehouse for water and mineral nutrients, which in turn, were vital factors for controlling the rate of growth of plants and the final yield. In order to increase the rate of fertility of the soil, manures were mixed with soil in fields. Every crop-plant required nutrients. Some of these were available to the plants from water and atmosphere and some were taken up by plants root from the soil. In intensive cropping, crops removed large quantities of these nutrients from the soil, and thus, the nutrient reserve of the soil become exhausted. In order to maintain the balance of nutrients, it was essential to preserve them through the use of manures. The purpose of using manures is to add to the soil the mineral nutrients for maximum growth. The term for fertilizer was *purīṣam*, which has been derived from the root 'pṛ', meaning, to fill. It fills the exhausted fertility of the soil. Fertility was considered to be an aspect of Rudra who was associated with the commoners. The Vedic Āryans were well acquainted with the fertile nature of cowdung, and they used it as manure in their fields to increase the fertility of the soil. Rotten cowdung was spread over the fields.

The *Atharvaveda* mentions the making of manure with the straw of the brown-whitish jointed barley with the stalk of sesame etc.[48] The *Bṛhat Saṃhitā*[49] and the *Agnipurāṇa*[50] mention that to increase the production of flowers and fruits, one should sprinkle ghee mixed with cold milk. A mixture of sesame, excreta of goats and sheep, barley powder and leaf thrown into water and left standing for seven nights, should then be poured round the roots of plants. Different kinds of plants required various kinds of manures for a specific period, which have been described by both the texts.[51] The idea of pouring fish washing as a means of helping development of fruits has been mentioned by Varāhamihira.[52] Cakradatta mentions a recipe for the preparation of an oil, which when applied or sprinkled at the root of a barren tree, it becomes full of flowers and fruits, graceful and strong.[53]

TREATMENT OF PLANTS

The science of the treatment of diseased plants has been described both by Varāhamihira and the *Agnipurāṇa*. These texts mention that similar to the human beings and animals, the plants also suffer from different diseases, such as, inception of diseases, displacement of flowers, fruits, leaves and barks.[54] By applying appropriate medicines, unnatural growth, deteriorations, wounds and fractures may be cured. Śaṅkara Mishra, in his *Upasakara*, notices the application of drugs in plants. Varāhamihira mentions some of the causes of the diseased condition of plants (*taruroga*); namely, cold climate, wind and the sun. The leaves become yellow, buds do not develop or their growth is arrested, branches become dry and the sap exudes.[55] To cure the ailments of plants, some prescriptions have been made which are both preventive and curative. Varāhamihira suggests that mud kneaded with ghee and *vidaṅga* should be applied to the roots, after which milk diluted with water should be poured.[56] The *Agnipurāṇa* also suggests that *vidṅga* mixed with rice, fish flesh, mixed together, constitutes a remedy invigorating to the plants and curative of their diseases.[57] Varāhamihira further suggests curative prescriptions for the most incurable diseases, such as, barenness. A hot decoction should be made of *kulattha, māṣa, tila* and *yava* which, when cooled, should be poured round the roots.[58]

SENSITIVITY

The phenomenon of the movement of plants, their capacity to sleep, their sensitiveness to touch, heat, wind, and noise were noticed by the ancient Indians. The *Mahābhārata* refers to this sensitiveness in plants.[59] A Buddhist text, *Nyāyavindu Tīkā* by Dharmottara refers to sleep contraction of leaves in the night in certain plants.[60] Udayana, in his *Kiraṇāvalī*, notices in plants the phenomenon of life, death, sleep, waking, disease, drugging and transmission of specific characters by means of ova, sensitiveness towards what is favourable and avoiding that which is unfavourable.[61] *Guṇaratna* enumerates the characteristics of plant life, such as, various kinds of sensitiveness or actions connected with sleep, waking, expansion and contraction in response to touch and also movements towards a support. He notices sensitiveness to touch in plants like *lajjāvatī*, which manifests reaction in the form of contraction (*patrasaṃkocan*). He states that the lotus opens with the sunrise, *ghoṣatakī* and others in the evening and lily with the rise of the moon.[62] Śaṅkara Misra, in his *Upasakara*, describes plants as characterized by approaching the agreeable and avoiding the disagreeable. In this connection, the name of the *sūryamukhī* (sunflower) may be mentioned which always faces the rising sun.[63]

Sensitiveness of plants may also be proved by the stages of infancy, youth and old age. A tree undergoes through stages of sprouting, seedling with new leaves, branching and so forth. The conditions of growth were of different kinds, such as, food, water and light. It has been said that water is a condition of the growth of plants, light by day and water by night lead to the growth of strength in plants. Light is an essential condition for the growth of plants, but some plants, such as, betel leaves require shade. The sensitiveness of the plants may be ascertained by their age and death. The main cause of the death of plants depends on suitable and unsuitable food and conditions of growth.

Consciousness in plants (*caitanya*) was well known to the ancient Indians. *Vṛkṣādevatās* (tree spirits) were worshipped in ancient India. The *Jātaka* stories frequently refer to these plant spirits that indicate that the ancient Indians believed the tree as being enumerated as the spirits. The *Ṛgveda*[64] and the *Atharvaveda*[65] address the plants, evidently referring to their possession of the gift of hearing. Mahidasa Aitareya includes the herbs and trees along with animals in the organic world. The plants belong to the last of the four classes of beings; namely, propagated from germs. In his theory of gradual development of the soul, he says that in herbs and trees, for example, life only is seen but thought (*citta*) in the widest sense is in the higher forms of life.[66] *Uddālaka* is of the same opinion as he thinks that the living principle is the potentiality of living bodies of the real seeds of things. Furnishing its complete illustration, he thinks that there is potentiality or vitality in a small seed from which a large *Nyagrodha* tree springs into existence. It is the spirit which animates all the parts of living beings. When this spirit departs from any branch of a tree, that withers. When it leaves another branch, that too withers and when, in this way, it leaves finally, the whole tree withers and perishes but the living principle never dies.[67] Manu also is an advocate of life spirit or consciousness in the plants as he states that the trees are trees on account of their being under the influences of *tamoguṇa* but they possess a kind of dormant and latent consciousness and are capable of realizing pleasure and pain.[68] The *Mahābhārata* presents an analytical exposition of the theme of the consciousness in plants. In a conversation between Brahman and Bhāradvāja, it has been discussed as how one may doubt the existence of the gift of touch in the plants seeing that their leaves, bark, fruit and flowers wither under the influence of heat. The plants have unconsciousness and on account of this they grow, bear flowers and fruits and decay.[69] The *Bhāgavata Purāṇa* (3.10.19–20) very distinctly points out the peculiarities of plants, namely, their process of taking food from below upwards, dormant consciousness, and the possession of the sensation of touch.[70] Chakrapani, in his work *Bhānumatī*, states that the consciousness of plants is a sort of stupefied consciousness.[71]

The ancient Indians firmly believed that there is *prāṇa* (vital life force) in the plants. They have feelings of happiness and sorrow. But the modern botanists did not accept this theory. Jagadish Chandra Bose rediscovered this important aspect of plant life in 1901. On the basis of his scientific experiments on plants, he established his theory that they also have life force. He demonstrated that response given by the plant was physiological and that it afforded an accurate index of the vital activity of the plant. The test applied by him in order to discriminate as to the physiological nature of the response consisted in observing the effects of anesthetics, poisons and exceedingly high temperatures, all of which are known to depress or destroy the activity of life. After his experiment, he observed that the electric response is a faithful index of physiological and that such a response is given by all plants and by their different organs. His theory was much appreciated and recognized by the botanists who played an important role in the development of botany.

SEXUALITYAND REPRODUCTION

In ancient India, the idea of sexuality in plants was vague and inaccurate. The general idea of division of the plants into male and female based on human analogy was formed. The *Hārita Saṃhitā* records possibly the earliest germs of sexuality in plants in which it has been said that the seed is produced by the co-operation of different sexes among the creeping and the fixed plants, its quality varying according to the sperm (*dhātu*). All the plants are endowed with Śiva and Śakti, i.e. the male and female procreative energies. The functions of the male and female arise from their combination. The mango flowers, fruits, and stones (embryo within), likewise, are endowed with *śukra* (the generative power).[72] Caraka is more precise than Hārita who remarks that *vetasaka* that bears white flowers, large fruits and tender leaves belongs to the category of males, and that which bear red and yellow flowers and small fruits, small stalk and whose colour is green belong to the category of females.[73] Other plant scientists now advanced this aspect of sexuality of plants. They divided the plants into male and female classes based on the differences of their attributes and characteristics. *Amara* represents this observation as going a bit further. He confounds the pollen of flowers as corresponding to the female menstrum.[74] Some of the foreign travellers also knew of the concept of Indians about plants. Pliny mentions that the Indians are of the opinion that the plants have the distinction to male and female, the body of the male being more compact, and that of the female of greater amplitude.[75] To get the differentiation of sexes in plants based on actual observation, the example of *Ketakī* may be cited which was considered as a couple, the male one being designated as *siaketaki* and female one as *svarṇaketakī*. The male *ketakī* was known as *ketaka* as also *viphala*, not producing fruits, having flowers with dust, meaning, with only pollen grains (*dhūlapuṣpika*).

Various methods of propagation of plants were in practice, such as, by fruits and seeds (*vijaruha*), by roots (*mūlaja*), by cutting (*skandhaja*), by graftings (*skandharopaṇīyana*), by apical portions (*agravija*) and leaves (*parṇoyoni*). A method of self-layering was in practice, namely, the Nyagrodha tree from which branches grew out to the length of even 12 cubit. These grow downwards and touch the ground. They penetrate into the soil and take root like shoots that have been planted. Then, they spring upwards and form a trunk. Strabo mentions this kind of propagation with regard to the Indian fig free.[76] In the *Brahmajālasutta* also, these methods of propagation of plants have been mentioned.[77] In his commentary on the *Dīgha Nikāya,* Buddhaghoṣa mentions these elaborate methods in detail.[78]

GENETICS (HEREDITY AND PLANT BREEDING)

The principle of heredity plays an important role in modern biology. Some ancient Indian biologists knew of this important problem, which deals with transmission of specific characters that parental characters are transmitted to their offsprings. The question is raised by Caraka[79] but, earlier to him, some of the *Brāhamaṇa* texts also discuss it. These describe as how specific characters are transmitted, why the offspring is of the same species as the parental organism. Caraka, Suśruta and Dhanavantarī hold that the foetus or rather, the fertilized ovum, develops by polingenesis. All the organs are potentially present there in at the same time and unfold in a certain order. As the sprouting bamboo seed contains in miniature its entire structure; similarly, the mango blossom contains the stone, pulp, and the fibres which appear to be separated and distinct in the ripe fruit, though from their excessive minuteness they are undistinguishable in the blossom, such is the case with the fertilized ovum.[80]

The inheritance of specific characters explained in accordance with this view of Caraka that the sperm cell of the male parent contains minute elements derived from each of its organs and tissues.[81] The great Indian philosopher, Ādiśaṅkara states that the sperm cell, the seed in the case of plant, represents in miniature every organ of the parent organism and contains in potential the whole organism that is developed out of it.[82] But the congenital deformities of the parent do not necessarily produce corresponding deformities or infirmities in the offspring in all cases. It cannot, therefore, be said that the fertilized ovum represents in miniature every organ or tissue of the parental organism. Ātreya, a predecessor of Caraka, had found its solution. He states that the fertilized ovum is composed of the elements, which arise from the whole parental organism (*samudayaprabhāva*). But these are not the developed organs of the parents with the acquired characters that determine or contribute the elements of the sperm cell (seed). In fact, the parental *vīja* (seed) is an organic whole, independent of the developed parental body of its organs. Ātreya and Caraka emphasize the influence of the defective nutrition and the constituents of the food on the characters of the *vīja* in the reproductive tissue, especially with regard to the sexual character, the stature and the colour pigment (*varṇa*) of the offspring. They think that the true congenital variations are accidental.[83]

ECOLOGY (STUDY OF PLANTS IN THEIR NATURAL SURROUNDINGS)

For the proper growth of plants, it is essential to have suitable ecology around them. It mostly depends on the types of land on which the plants are planted or grow naturally. Caraka divides the land into different regions according to the nature of the soil and climate that determine the characteristics vegetation. He mentions some of the plants, which grow in particular regions, such as, *jaṅgal* (dry deserts), *anūpa* (abounding in water), and *sādhāraṇa* (ordinary land).[84] According to Suśruta and Caraka, in the *jangala* region *khadira, aśvakarṇa, dhava, śāla, vadara, āmalakī, vaṭa, śamī, arjuna* and *siṃsapa* grow. In anūpa region, *vanira, hintala, lamala, tāla, nārikela* and *kadalī* grow. Varāhamihira furnishes the names of some of the plants, which grow in *anūpa* region, such as, *jambū, vetasa, kadamba, vījapuraka, drākṣā, timira,* etc.[85] Amarsimha also refers to some aquatic plants, namely, *saugandhika, utpala, indibara, kumuda, padma, vārīparṇī* and *śaivāla.* In *sādhāraṇa* region, according to *Amarakoṣa, mandāra, pārijātaka* and *haricandana* grow.[86]

In some places, plants of a particular species predominate over the others and the place is given the name after the dominating plant, such as *kumudvatī,* where white water lilies predominate. A place abounding in lotus is called *Padminī* and *Śādvala* is the place were green grasses abound. Even some rivers are given names after the plants or creepers which grew predominantly on their banks, namely, Sarasvatī—abounding in *sara* (reeds) and Vetravatī abounding in *vetasa* or *vetra.* The names Kuśadvīpa, and Śālmalidvīpa and a few others, as mentioned in the *Purānas,* may be identified with particular places abounding in corresponding plants.

TAXONOMY (CLASSIFICATION AND NOMENCLATURE OF PLANTS)

The Vedic people had classified the entire creation into three classes, namely, the *aṇḍaja,* born from the egg, *jīvaja,* born from the living beings and *udbhija,* born from a sprout.[87] The plants were included in *udbhija* class, which was divided into four subclasses, such as, *auṣadhi, vanaspati, latā* and *tṛṇa.*[88] The *ausadhi* included all the plants possessing healing powers or medicinal ingredients. Those plants had bulb and flowers. The *auṣadhi* (herb) is so called because it sucks the burning element or because people suck it when something is burning on the body.[89] The *Ṛgveda* contains numerous references to at least 107 species of herbs.[90] *Vanaspati* was the generic name of the trees. *Latā* included all the creepers and *tṛṇa* denoted all grasses.

Again, the vegetation world was divided into two classes, namely, the *grāmya,* cultivated and *āraṇya,* the wild plants.[91] These terms signified the plants, which were cultivated and planted, such as, the cereal plants and the plants of the roots and fruits, and the plants, which grew wild, respectively, but were all useful.

Further, on the basis of the characteristics of plants, they have been classified into several classes, such as, those which expand, those that are bushy, those having only one seath, those that creep, those having numerous stalks, those that have numerous joints or are knotty, and those having spreading branches.[92]

Kauṭilya furnishes an exhaustive list of plants divided into several groups, which include agricultural and forest products.[93] Manu classifies vegetation into eight subgroups on the basis of their characteristics.[94] Caraka classifies plants into four groups, namely, *vanaspati, vasaspatya, virudha* and *auṣadh*.[95] Again, these have been classified into 10 *vargas* (groups) in which more than a thousand have been enumerated.[96] The classification of Suśruta is very minute, with 37 sections (*vargas*).[97]

AUṢADHI

Apāmarga was used for curing different kinds of ailments. The *Atharvaveda* contains a *sūkta* devoted to it in which its characteristics have been discussed along with its medical value.[98] It is identified with *chirachira* in dialect in botany and known by the name of *achyranthes aspera*. Its branches tend to spread in one direction only and the fruit in the another. It is regarded as a symbol of vigilant circumspection.[99] It was used as medicine in different diseases. Its use in witchcraft also has been prescribed. In rājasūya, it was used in the *apāmarga*[100] offering. Till now, it is used for curing wounds and purifying blood. In taste it is bitter.

Arka has been identified with *madara* in vernaculars. In botany, it is known as *Calotrapis gigantea*. It grew wild and its fruits and flowers were used in sacrifices. It was considered as growing and spreading from the resting place of Rudra, on account of which, till now, its flowers and fruits are offered to Him.[101] Its juice is white and is used in curing diseases of the eyes.

Jaṅgiḍa[102] was used as a medicinal herb, which generally is identified with *vacā* in vernaculars. It was used to cure diseases like fever, rheumatic pain, consumptive cough and pleurisy.[103] It was so useful and effective that it was considered as the most excellent herb. The *Atharvaveda* euogolizes it in these words 'the ancient herbs surpass thee not, nor any herb of recent days'.[104] It was considered as a very useful herb, protecting a person from disease first as a treasury secures the wealth.

Kuṣṭḥa is identified with *gostus-speciosus* or the *arabicus*. It grew on the high peaks of the Himalayas and was used for curing numerous diseases. Hence, it was considered as a medicinal herb possessing divine powers.[105] On account of its utility, it was considered as one of the most popular and useful medicinal herbs as it possessed properties which were useful in several ailments of general type.[106] This important herb was discovered by Ikṣvāku.[107]

Pṛṣniparṇī generally is identified with *pithavana* in vernaculars.[108] It contained ingredients, which were used in removing even dangerous diseases.

Rohiṇī was very useful in the diseases of bones. It was used as a binding material over broken bones in the body and was supposed to join them together. In other ailments also, it proved beneficial.[109]

Sahasraparṇī was supposed to bestow vigour and used as a medicinal herb in numerous diseases. It was a widespreading herb, which had thousands of leaves. It contained at least one hundred tendrils and 33 descending shoots. In colour, it was brown.[110]

Śatavāra as a medicinal herb contained properties, which were used to remove several ailments. It was a special medicine to be used to remove the diseases caused by dogs. Its tops were yellowish in colour. Griffith thinks that *śatavāra* was not a medicinal herb but a powerful amulet made apparently of a great number of medicinal herbs and particularly of *darbha* grass.[111]

Śītikā, as the name suggests, was a medicinal herb, which contained cooling properties. It was full of fresh juice; hence was considered as a refreshing herb. It was applied on burns caused by fire and the victim soon realized relief.[112]

Soma is the most discussed plant in the entire range of the plants and herbs mentioned in the Vedic texts. It remains as a topic of discussion among the scholars. It was the most wonderful among all the plants. It contained several ingredients, which were used in different ways for gaining strength and removing several ailments. On account of this, it was considered as the *samrāṭ* among herbs.[113]

Both the Vedic Soma and the Avestic Hoam are derived from the root *su (zend Hu)* meaning to press or produce. It denotes a spirituous liquor-extracted plant, which grew on the mountains. The *Ṛgveda* and the *Avesta* distinguish between an earthly *soma* and the world of men forms the central element of the *soma* myth. The potent juice endowed the feeble mortal with godly powers and, for a time, freed him from the earthy cares. The divine personality of the *soma* has been much praised. The *soma* plant was a source of light and life and in the Brāhamaṇa texts, it has been identified with the moon. Such identification is observed in the *Ṛgveda,* which later on, has been developed in the later Vedic texts.

The exact identity of the *soma* plant is somewhat difficult and doubtful. Several historians and botanists have attempted to identify it, among whom, Roth, Maxmuller, Baker, Dyer and Schindler were prominent. Schindler attempted to identify it with the help of the Parsis in Iran who prepared *soma* juice of the *Avesta.* According to this description, this plant grows to the height of four feet and consists of circular fleshy stalks of whitish colour and has a sweetish taste. After being stored for a few days, it turns sour, and its stalks become yellowish brown. The stalks break easily at the joints and then form small cylindrical pieces. In this condition, they lose their leaves, which are said to be small and formed like those of the *jessamine.* This description indicates that *soma* is a plant whose identification may be established with the *sarcostmma* (Milkweed) or some other group of *asclepis,* such as, the *pariploca-aphylla,* which has been traced by Haussknecht to height of 3000 in the mountains of Iran and is common in Afghanistan.[114]

In fact, several varieties of *sarcostamma* or *asclepidas*—somewhat different to that of Iran and Afghanistan, which are not to be found so far south—seem to have been found, and, indeed seem still to be made use for the *soma* sacrifice.

The Vedic texts inform that the *soma* plant grew on the mountains and the *Munjavanta* was renowned for it. The *Taittirīya Saṃhitā* mentions that the *soma* plant grows in the mountainous region; hence rightly stone, is its pressing object.[115] Generally, it has been held to be the *sarcostemma viminale* or the *aselepias acida.* The *sarcostemma acidum* meets the requirements of the case. Watt thought that Afghan grape was the real *soma* but Rice opines that a sugarcane might be meant, while Maxmuller and Rajendralal Mitra suggest that the juice was as an ingredient in a kind of beer. In their view, the *soma* plant was a *hop.* Hilerbrandt considered that neither hops nor the grapes can explain the references to *soma.* In the *Yajuṣ Saṃhitā,* it has been maintained that it grew on the mountain and could not be obtained by ordinary people.

It appears that *soma* was the product of the west and not of the east; hence it was generally brought in the east from the persons of the west. Mostly, its plant was used in sacrifices for extracting juice to be offered to the gods. Religious and spiritual sanctity was accorded to it and it was above commercial transaction.[116] The priest doing so was considered to be a degraded person. However, in certain cases, he was allowed to accept *dakṣinā* in the form of gold, cow and goat, which were considered as sacred objects.[117] *Soma* plants were brought to the sacrificial grounds on bullock carts. In some cases, the *soma* plant could be exchanged with a ruddy and yellow-eyed cow, as that was the form of the *soma* plant, and in this way, it was exchanged with its own form and deity. Besides, it could be exchanged with the she-goat, milch cow, and bull, a draught ox able to carry load and even with a garment.[118]

The twigs of the *soma* plant were ruddy[119] (*aruṇa*) or tawny[120] (*hari*), and its other parts have been described in the Vedic[121] texts. The *soma* plant grew wild and was not cultivated in fields by adopting artificial methods. It grew in stony unfertile soil on the hills in the bright half of the month, and ceased to grow in the dark half. Its bundles were brought to the sacrificial ground on bullock carts and its juice was prepared, which was of different tastes.

In the eastern region where the *soma* plant was unavailable, its various substitutes were prescribed which could be used for extracting juice to be offered to the gods in different sacrifices. Such substitutes were the *praprotha, adara, usana* and the *pṛṣniparni.*[122] The last one had speckled leaves and its wood was used for protection from the negative effects caused by evil spirits.[123] It has been identified with *lakṣamaṇā*, a herb curing infertility but according to some others, it should be identified with *glycine-debi. Putika* was also a substitute of *soma* plant whose wood was used for making churners to churn milk.

Taraka[124] was used to cure different kinds of old and chronic hereditary diseases. *Vaiṣṇavī* was also such a medicinal herb but it is difficult to identify both of them.

Auṣadhis numbered in thousands and it is not useful to enumerate and identify all of them. It goes to show that the people were acquainted with numerous plants, which were used as the *auṣadhis* as they also contained medicinal ingredients.

VṚKṢA

Āmalaka has been mentioned as the tree, *emblica officinals.* In vernaculars, it is known as *amala.* Its fruits were also known by the name of the tree which has medicinal properties—which are digestive and cause vigour in the body.[125] Till now, its fruits are used for preparing *chutney*, pickles, preserves and oil.

Aśmagandhā[126] literally means rock smell. It is identified with *aśvagandhā*, which in botanical term, is known as *Physalis flexuosa.* It contains medicinal properties. Until now, it has been used for preparing different medicines.

Aśvattha was considered as one of the most sacred trees. Even before the Vedic period, this tree had attained religious sanctity. The Indus seals[127] contain the figures of the leaves of this tree. On the Indus potteries as well, its trees and leaves have been depicted. It appears that the *Ṛgvedic*-Āryans borrowed the *asvattha* tree and attached religious sanctity to it, like their Indus counterparts. Its literal meaning is the place for stabling the horse.

The Vedic texts refer to this tree in different contexts. Sacrificial vessels were made of its wood. Its hard wood formed the upper piece used for kindling the sacrificial fire. Besides, in certain sacrifices, its wood was used for offering to different gods. It was considered as a sacred tree under the shadow of which gods were supposed to rest. It contained medicinal properties as its berries were used for curing wounds.[128] The *aśvattha* grown on the *śamī* tree was considered to have medicinal properties, which were supposed to ensure the birth of a male child.[129] Amulets also were made of its wood, which were donned to remove the effects of evil[130] spirits. Until now, in villages, the leaves and the wood of this tree are put round the neck or on the arms for the same purpose. Its fruits were known as the *supippala* and were favourite edibles of birds. When ripe, they became reddish and blackish and were consumed even by human beings.[131]

On account of its religious importance and other considerations, *aśvattha* was given the status of the Kṣatriya among the vegetation world. The *Aitareya Brāhmaṇa* states that as to the fruit of the *aśvattha* tree which sprang out of lustre and which has the sovereignty over the trees, the Kṣatriya, when drinking the juice prepared from its fruits, places in the Kṣatra lustre and the sovereignty over the trees. Its *kṣatriya* status or hardness may be well assumed from the fact that it was called burster because it split asunder the wood or stone in which crevices its seed has germinated and grown.[132] Till now, the Hindus consider the *aśvattha* tree as the most sacred.

Bilva has been mentioned in the Vedic texts since the later *Saṃhitās* onwards.[133] In botanical terms, its name is *Aegle marmelos.* Its leaves and fruits were supposed to contain medicinal properties. These were used in several ailments and had a cooling effect. Sacrificial posts were also made of its wood, and its leaves and wood were offered to different gods.[134] Its amulets also were made and worn for removing the evil effects caused by the evil spirits and diseases. The *bilva* tree was regarded as the symbol of fatness and fertility because it increased in size from the roots upto the branches and it yielded fruits every year. So, the sacrificer wishing property was advised to erect a post of *bilva* wood on the sacrificial[135] ground. Its fruits, when ripe, are yellowish and very tasteful. Generally, its fruits are ready during the summer and are used for preparing delicious juice for a cooling and soothing effect.

Harītaki[136] was noted for its fruits of the same name. Its berries were used as medicine, which was conducive to digestion. Till now, these are used for the same purpose.

Karkandhu is the *jujube.*[137] It originated from the mouth of Agni. Its tree and fruits are known by the same name. In vernaculars, it is known as *jamun.* It grows wild. Its fruits are red when ripe. *Badara* was a variety of *jujube.* In ancient times, it grew wild and was domesticated as well. Its fruits were used in sacrifices for offering to the Gods.[138]

Karīra was a leafless tree and its fruits also were known by the same name.[139] In vernaculars, it is known as *karila.* It grows wild in forests.

Kasya was the tree, which sometimes was the synonym of the *śāla* tree. In northern and eastern India, there were extensive forests of this tree. It was a tall tree and its wood was famous for its durability and was mostly used for building houses and furniture.

Kurcī was a very useful plant. It possessed medicinal properties, which were used for removing troubles caused by dysentry. Its beads and amulets were put on the body as medicinal charms. This tree has affinity with the family of *pocynaceae.* Its remains have been found from Hastināpur.[140]

Kārṣmārya has been identified with *gmelina orborea.* It was considered of divine origin. Its wood was prescribed to be placed to the south of the sacrificial altar to ward off the devils.[141]

Khadira is one of the most sacred and hard woods. In botany, it is known as the *mimosa* or a *cacia catchu.* In vernaculars, it is known as *khaira* from which *kattha* is made to be taken with betel leaves. The *khadira* tree had religious sanctity even in the Harappan culture as on some of the Harappan seals depict a railing which encircles the figure of this tree and on its base a platform has been raised.[142] The *khadira* wood was very strong (*bahusāra*) so the bolts of the axles of cart were made of it. Different sacrificial utensils, such as, jars, lids, and ladles were made of its wood. Amulets made of its wood also were put on the body for warding off the effects of evil spirits.[143] On the occasion of the *rājasuya,* the king sat on the throne made of its wood as it was supposed to ensure the durability of the king's reign.[144]

Kharjūra was a plant and its fruits also were known by the same name. The faience sealing in the shape of date palm seal has been found from Harappa and some badly carbonized date seeds from Mohenjo-daro also have been discovered. These findings obviously indicate that *kharjūra* grew in the Indus valley.[145] In eastern India, it grows wild.

Kṛmuka was considered as a very sacred plant. Its wood was used as sacrificial oblation to the fire. It was sweet as it had vital essence in it. In colour, it was red; hence it was considered the very embodiment of[146] Agni.

Nalada is identified with *narad stachys,* which in vernaculars, is known as *jaṭāmāsī.* Its flowers are red and knotty and used for making garlands.[147]

Nyagrodha, like the *aśvattha,* was a very sacred tree. In botany, it is termed as *Ficus indica* and in vernaculars it is known as *baṭa* or *vada* or *bara.* In the Vedic texts, it has been mentioned from the *Atharvaveda* onwards. According to the authority of the *Aitareya Brāhamaṇa,* it grew abundantly in the region of the Kurukṣetra, which was the place of its origin.[148] The Chāndogya Upaniṣad presents the simile of *nyagrodha* seed developing into a vast tree to exhibit the growth of the universe from the Ultimate Spirit who is smaller even than the *nyagrodha* seed.[149]

This tree was termed *nyagrodha* on account of its downward growth (*nyagrodha*). This tree has the characteristic of bending its branches down to the ground and developing new secondary trunks so that a single tree may, in course of time, form a large grove. In order to justify its downwards and lateral growth, the *Śatapatha Brāhmaṇa* innovates a story that when the gods were performing a sacrifice, they tilted over *soma* cups, and turned them downwards, those took roots.[150] Besides, it also was noted that the *nyagrodha* tree never grows straight and slightly inclines to one side.

Perhaps this story was innovated to attribute divine origin to the *nyagrodha* in order to make it a substitute of the *soma* plant, which was not available in eastern India. It has been asserted that the kṣatriya has not been allowed to drink the *soma* juice in the sacrifices. However, he may take the juice extracted from the descending roots of the *nyagrodha* together with the fruits of the *udumbera, aśvattha* and *plakṣa* trees. It is remarkable that the airy descending roots of the *nyagrodha* are also reddish, which is the colour of the top of the *soma* plant. So it was an appropriate substitute for the real *soma* plant. The drinking of *nyagrodha* juice by the king has been justified on the grounds that the kṣatriya in the human society occupies the same place as the *nyagrodha* among the trees because the *kṣatra's* power alone is spread here as being alone invested with sovereignty. The *nyagrodha* is, as it were, firmly established in the earth, and by means of its descending roots, expands on all directions and, therefore, is a sign of the great extent of the *kṣatra* power over the entire earth. When the *kṣatriya* performing a sacrifice enjoys the juice squeezed out of the descending roots of the *nyagrodha* tree and its fruits, he places in him the royal power exercised by the *nyagrodha* over the trees. Just as the *nyagrodha* tree has, by means of its descending roots, a firm footing on the earth, the royal power of a kṣatriya who enjoys while sacrificing this portion of food has a firm footing and his rule cannot be overthrown.[151]

In sacrifices, the sacrificial posts and utensils made of the *nyagrodha* wood were used. Its milky juice is applied in the eyes to cure certain ocular diseases. Besides, its airy descending roots are also used as medicines in certain ailments. Aśoka informs that he had planted *nyagrodha* trees on roadsides in his empire for providing shades to human beings and animals.[152]

Strabo (first century BC), on the testimony of Megasthenese, informs that in India, numerous strange trees grow among which is the one which branches bend downwards and whose leaves are no smaller than a shield in the country musicians, which is the most southerly part of India. These are great trees whose branches have first grown to the height of 12 cubits, and then have grown downwards as though bent down till they have touched the earth and that they then just distributed, have taken root underground like layers and, then growing forth, have formed trunks. The branches of these trunks again, likewise, bent down in their growth. They form another layer and then another and so on successively. Thus, that forms only one tree there is formed a vast sunshade like a tent with many supporting columns. He writes also of the size of the trees that five men could hardly embrace their trunks. On the testimony of Aristobulus, Strabo states that in Punjab also, there are such trees that have their branches but downwards and such size that even four hundred persons can pass the noon in shade under one tree.[153]

Palāśa (*butea frondosa*) was considered to be one of the most sacred trees and its wood was used in sacrifices for different purposes. Among the vegetation world, it was accorded the status of the brāhmaṇa.[154] The sacrificer wishing to obtain spiritual knowledge and lustre was advised to erect the sacrificial post made of the *palāśa* wood on the sacrificial ground as this was a symbol of beauty among the trees and considered as the womb of all the plants.[155] This tree grew wild in forests. Its amulets were worn for gaining health and material prosperity.[156] Different kinds of sacrificial ladles, lids, cups and utensils were made of its wood. It was considered to have originated from the flesh of Prajāpati, so its juice was red.

Plakṣa is a large and beautiful fig tree. Its fruits are white. Its mythological origin has been narrated in the Vedic texts. It has been maintained that gods seized an animal for sacrifice. Its sacrificial essence flowed down and on that place a tree sprang up. The gods beheld it, wherefore, it was called *prakhya* (visible). *Plakṣa*, doubtless, is the same as the *plakhya.* With that same sacrificial essence, he who completes it makes it whole; hence, there are *plakṣa* branches as an upper covering.[157] Like other trees, it also had its sacrificial importance; thus, sacrificial implements made of its wood were used in different sacrifices. The simile of the fig seed has been sited in the *Brāhamaṇa* texts. It has been stated that as to the fruits of the *plakṣa* tree, which sprang up from glory and in which there are indepedence and brilliance of trees, the Kṣatriya places in this way the independence and brilliancy in kṣatraṃ.[158]

Pilu was an evergreen tree of the same nature as of *śamī.* It grew abundantly in the *Vāhika* country. Its berries were small and called as *pilukuna.*[159] It abounded in the Punjab.

Putudru has been identified with *cedrus deodar.* It is a very ancient specimen and even the people of the Harappa culture used its wood. Some of the specimens have been discovered from Harappa.[160] In the Vedic texts, it has been mentioned as a sacred tree. Its wood is very light. Sacrificial posts were made of its wood. It also had religious sanctity and divine origin was attributed to it. It was characterized by a sweet smell and was easily inflammable.[161]

Śālmali (*salmali malabarica or hippamone mancolla*) was one of the useful trees. In vernaculars, it is termed as *semara.* The *Ṛgvedic* people were well acquainted with it and they thought it to possess poisonous ingredients.[162] It is interesting to note that poison extracted from this tree is used for tipping arrows even now. Its flowers are red, which look very beautiful and, when ripe, these yield soft cotton. Its seed are black and contain poison. Its wood is very light so these were easily portable on bullock carts. This tree was famous for its tall growth and rightly has it been mentioned as the tallest among the trees.[163] The tree *śālmali* was so called because it is easy to pierce or because it bounds in pricking thorns.[164]

Śamī (*prosopis spicigera or cacia suma*) is one of the most sacred trees of the Vedic Times.[165] One of the pairs of the fire sticks (*araṇi*) was made of its wood to produce a sacrificial fire. Its wood was very hard. Its stick, fitted into a hole in a lower block, was twirled like a carpenter's brace in order to produce fire by friction. On account of this, it was supposed to contain fire in its womb. This tree grew abundantly in the dry arid zones of the Punjab, Rajasthan and Sindh. Its tree had hundreds of branches and its leaves were large.[166] The juice of the *śamī* tree was used as a medicine for causing the lustrous growth of hair.[167] Its fruits were termed as *śamīdhānya.*

Sidharaka was a tree that yielded very hard wood; hence, on that account, it was known as *sāravṛkṣa.*[168] Its wood was used for making furniture.

Śiṃśapa (*darbergia sisso*) is a very useful tree. In ancient times also, its wood was used for making houses, furniture and domestic utensils. Carts also were made of its wood.[169] It grew wild and its branches were used for fuel. The Harappans were acquainted with this tree, as an interesting representation of its leaf has been found on a pottery sherd from Harappa.[170]

A perusal of the *śiṃśapa* tree shows that it grew in western India, especially in the regions of Gandhāra, Sindh, the Punjab and Baluchistan. It grew wild in the eastern regions as well and its remains have been discovered from Hastināpur.[171] It appears that India is the place of its origin, wherefrom it spread to the west. The inscriptions of Darius I prove that the wood of *yaka* tree, was used in building his palace at *susa* which were brought from Gandhāra and Karman.[172]

Tilavaka tree was not so important for the purpose of making furniture or houses but the sacrificial pillars made of its wood were erected on the sacrificial grounds in certain sacrifices. It was considered as sacred and dead bodies were not disposed of near it.[173]

Udumbara (*picus glomerata*) is identified with *gullara* in vernaculars. It has not been mentioned in the *Ṛgveda* but has been referred to in the later Vedic texts. It grew wild and sometimes its forests also have been mentioned. Its wood was very durable. Its flowers were not seen and were a simile for non-existent things but its fruits, when ripe, were sweet like *madhu.*[174] It had the characteristics to yield fruits thrice a year.[175] The *udumbara* tree was considered as the most excellent among the plants on the grounds that 'it contains whatever pith and vital sap there is in the trees, hence it is always moist and is full of milky juice so it is also food. The sacrificer gratifies Agni by every kind of food and kindles him by all kinds of wood.'[176] Its use in sacrifices was popular as it was the symbol of strength and a sacrificer, having erected a sacrificial pillar of its wood on the sacrificial ground, ensured for himself strength and vigour, just as *udumbara* signifies strength.[177] Considering it as the embodiment of strength, the royal throne in the *rājasūya* was made of its wood and it was thought that this wood bestows strength on him. Sacrificial utensils like jars, lids, ladles and others were made of timber. Amulets made of the *udumbara* wood were donned to beget material prosperity and ward off evils.[178]

On the occasion of the performance of the *rājasūya*, the king designated, going through the ceremony, enjoyed the juice prepared with the fruit of *udumbara* as it was thought that the *udumbara* tree has the vigour of all the plants furnishing nourishment. This tree means sustenance. An *udumbara* container was used for mixing different juices in sacrifices.[179]

Vibhīdaka (terminalia bellerica) was known to the *Ṛgvedic* people. Generally, it grew wild on the slopes of the mountains and *akṣas* were made with its seed.[180] Its fruits were known by the same name and used as medicine conductive to digestion. In vernaculars, it is known as *bahera.*

Vikaṇṭaka (flacourita sapida) was a sacred wood and in some of the sacrifices, it was offered to Agni. Divine origin was attributed to it. As the name suggests, it had no thorns in it and was very soft.[181]

Besides the above-mentioned trees, the texts under review contain references to some other plants but they were not so important, such as the *rajjudala* (*cordia myxa*), *varaṇa* (*crataera roxburghii*) and *sphūrajaka* (*diospyros embryopteris*).

TṚṆA

The literary and archaeological sources prove the existence of several domesticated and wild grass, which were used for different purposes. A favourite term for grass was *barhī,* which shows the general characteristics of the grass, i.e. one which grows rapidly. Grasses like bamboo, *muñja*, *sara* and others were used for building huts as there were easily available. The usefulness of grass as fodder for the animals can be assumed from the term *atṛnavatsa*, which denoted the calf that yet had not started to live on fodder. *Ghasa* denoted the green grass used as fodder for the animals. The cattle yielded much milk when served with green grass. Some grass had got religious sanctity as kuśa was used in sacrifices and gridles for the *brahmachārins* were made of it. Strings were also made of it and used for different purposes. Besides household materials, namely, containers, mats, baskets, etc., were made of them. A few types of grasses had medicinal properties and were used as medicines for curing certain ailments. Though the texts contain references to hundreds of grass names, some of them are very useful.

Aśvavalā was a wild cane and a species of reed of the class of *saccharum spontaneum*. Its remains have been discovered from Hastināpur.[182]

Bhāng was hemp. In the *Ṛgveda*, it is an epithet of the *soma* plant, perhaps, in the sense of intoxication.[183] In vernaculars, also it is known by the same name. It grows wild but in some regions, it is cultivated. Its plants are thin and long and the leaves are green. Its dry leaves are crushed and burnt and the smoke is inhaled. Its powder is taken with sweet juice.

Darbha was considered as one of the most sacred grasses and it was used in sacrifices in different ways. Its characteristics have been mentioned in the texts, such as, it spreads rapidly and re-roots itself, hence, it is known as having a thousand joints, several roots (*bhūrimūla*), thousands of leaves (*sahasraparṇa*) and hundreds of stalks (*śatakāṇḍa*). *Darva* was used as a medicine for lessening anger and had cooling effects on the mind and body. It was used as an amulet, ensuring the proper growth of hair and protection against their scattering.[184] Among the herbs, it was considered as the excellent.

Darbha is identified with the *kuśa* grass. It was used in sacrifices for covering the ground for placing oblations and sitting purposes. This grass is said to have sprung from the heaven. Various sanctifying and other far-reaching supernatural powers have been attributed to it. Sometimes it has been equated with *soma*. On account of its importance, divine origin has been attributed to it.[185] It has been asserted that *darbha* contains both kinds of foods as it is both water and plant. It has sprung up from the water released from the body of Vṛtra. This grass was the very embodiment of vigour (*ojas*) and its roots were so deep rooted that it was difficult to eradicate them.[186]

Dūrva (*panicum dectylon*) was considered as a sacred grass. It grew on damp soil. Like the *darbha*, it also spreads rapidly on the ground and even during the summer it does not dry. Its different varieties were known to the Vedic people, such as, the *sāṇḍa dūrvā*, whose roots were like eggs. It may be identified with *motha* in vernaculars, whose egg-like round roots are used as medicine in different ailments, specially fevers. *Pāka dūrvā* grown on the spot where the dead bodies were cremated. Sada was also a variety of *dūrvā*, which gave birth to the term *sādvala*, meaning a green piece of land. *Sahasrakāṇḍa* was a kind of *dūrvā* and in its glorification, a complete Sūkta has been dedicated. It was taken as a medicine for its cooling effects and removing numerous physical disorders. On account of its medicinal properties and importance, it was considered to be of divine origin. It was used in sacrifices and at present also, this practice is in prevalence. On account of its vigour, it was accorded the status of the kṣatriya among the grass. The kṣatra is represented by it as it spreads everywhere.

Ikṣu (*saccharum spontaneum*), as mentioned in the Vedic texts, is the most useful variety of grass, and of importance in the economic prosperity of the people. It is identified with *ikha* in vernaculars. For the first time, it has been mentioned in the *Artharvaveda*,[87] wherein an independent *Sūkta* has been dedicated to it in which its characteristics have been described in a systematic manner. Sugarcane was considered as the very embodiment of sweetness and was cultivated with a desire to get sweet juice. Its shoots have been compared with the eyelashes of Prajāpati and its pieces with his eyelids.[188] Till now, sugarcane is cultivated in a large region of northern India. The credit of its discovery and first cultivation has been given to Ikṣāvaku, the king of Ayodhyā.

Iṣikā was a reed-like grass, which may be identified with *sikki* in vernaculars. It has its characteristics of growing rapidly during the rainy season and is ready in the autumn. It was a useful grass and used for making *sūrpa* for winnowing grains.[189] The roof of a house was also covered with it. It was regarded as the symbol of fragility. Toys and other household objects were also made of it. Till now, the region of Mithila is famous for *sikki* work, wherein it is practiced as a cottage industry. *Kattṛṇa* was a fragrant grass, which may be identified with *sugandhitejan.*[190]

Kasa was a kind of wild cane and a species of *saccharum spontaneum,* which was used for making mats.[191] It had sanctifying qualities; hence in sacrifices, it was used for different purposes. Archaeological excavations at Hastināpur prove that it was used to give strength to the mud wall, which was further enforced by plaster of mud mixed with rice husks.[192]

Kūśa was considered as a very sacred grass, used in sacrifices in several ways. It had sanctifying qualities; so it was spread on the sacrificial ground on which oblations were kept and the priests also sat on them.[193] Mats were also made of it, sitting on which persons performed sacrifices or other kinds of religious acts. Till now, this grass occupies the same place in the sacrificial rites.

Mūñja was a variety of grass, which grew wild so abundantly in the western mountainous regions that on account of this, a particular mountain was named as the Mūnjavanta, meaning, abounding in the *mūnja* grass.[194] *Mūñja* is so called because it is thrown out by a kind of rush. It was used in sacrifices for securing protection because it was considered as the womb and it was further thought that the womb does not injure the child.[195] Strings made of *mūñja* grass were used for making girdles for the Vedic students and the ascetics. Cots and some furnitures were woven with the strings made of this grass. It was used to filter the *soma* juice. Its stalks were used for making baskets. This grass was light in weight, so it is known as *suṣira,* meaning, hollow. It was used for making the plaited part of the throne. It grows wild in rural areas and is used for several purposes.

Naḍa was a variety of reed, which grows on the banks of lakes and tanks or on moisturous places.[196] It grows abundantly during the rainy season; so it was termed as *vārṣika.* It grew profusely and increased the hair on the head. Its plants were used for making mats. Mainly women were engaged for making mats and other objects of this grass.

Sometimes, its plants were used for making big mats, which were spread on cot or *chauki*, and such a bed was known as *naḍvala*. This grass is used for making mats and popularly used by the poorer sections of society. It is known as *narkaṭa* in vernaculars.

Sara was a variety of reed which grew profusely in watery regions; hence, it was rightly considered as the flower of the waters.[197] Its plants were very weak and fragile and were easily to be broken. It was very light and on account of this, it was used for making arrow-shafts. Sometimes, girdle was also made of this grass and used in sacrifices in several ways.

Vaṃśa was the strongest and the longest variety of grass, which grew wild in forests and villages. It was so called because it grew in a forest (*vanasaya*) or it could be divided into different parts.[198] Its charred remains have been discovered from Harappa.[199] In rural areas, it is used mostly as a material for house building. In ancient times, it was used for the same purpose. Carts, baskets, chairs and fences around the fields were made of it. Personal names also were kept after it, such as, Prācīnvaṃśa. It was very hollow and also known as *veṇu*. It was used for making shafts of bows as it was flexible.

LATĀ (CREEPERS)

Creepers mostly grew during the rainy season and were mainly used for fodder, medicinal purposes, vegetables and decorative purposes. Generally, creepers were of three varieties, namely, those which spread on the ground, those which grew and spread on water; and lastly those which grew and climbed trees. *Latās* yielded flowers and fruits and were utilized by the people for different purposes.

Avaka was a water creeper, which has been usually mentioned in the *Brāhmaṇa* texts, and in later texts, it has been mentioned[200] as *śaivāla*, which in vernaculars is known as *sewāra*. It spread on the surface of water in tanks and lakes and protects the fish. The old term for *śaivāla* is *śipāla*, the botanical name for which is *blyxa octandra*. Hence, a tank or lake overgrown with it has been termed as *śipalya*. *Avaka* (*śaivāla*) was the symbol of durability and growth, so it was prescribed to be used in certain sacrifices for ensuring longevity and prosperity of the sacrificer.

Bhūmipāśa was an insignificant creeper, which spread on the earth like a net and bound it firmly.

Kiyamu grew on the spot where the dead bodies were buried. Its literal meaning is having some water. It was considered as an inauspicious creeper.

Kumuda[201] was a water creeper. In its appropriate literal sense, it represents white water lily. Its small plants and flowers were termed as *kumudinī*. It grows in tanks during the rainy season and dries up during the summer. Its roots and fruits are consumed by the poor people and its flowers are used for making garlands.

Puṣkara is a very famous and useful creeper, which has been mentioned in the texts in different contexts. In its true literal sense it is a blue lotus. Its sweet fragrance has been beautifully described. Generally, *puṣkara* grew wild in tanks and lakes and so had the name of *puṣkaraṇī*. The leaves of this creeper are green and so are its stalks. Since the earliest times, *puṣkara* has been used as an item of physical decoration (*vapuskarmaṇ*).[202] The Aśvinikumaras have been mentioned as Puṣkarasarajas, wearing garlands of lotus flowers. The bowl of the ladle was termed as *puṣkara,* as its shape resembled the lotus bud. It contained sweet fragrance; so usually bees sat on it were known as *puṣkarasād.*

Puṇḍarika was the white species of lotus and garlands were made of its flowers. In various texts, the shape of the human heart has been compared with the shape of a *puṇḍarika.* The people liked to have tanks with its plants in the vicinity of their houses to beautify the surrounding. The Vedic seers practically overlooked the aesthetic aspect of the lotus and the lotus pond. Only once in the *Ṛgveda,* the lotus pond has been mentioned as the most beautiful place and has been kept at par with the place of the gods regarding its picturesqueness. In the *Brāhamaṇa* texts, lotus has been symbolized as an immortal element.[203] In the Vedic philosophy, the lotus has a deep mystical significance as it symbolizes the cosmic waters.[204] On account of this, its flowers and leaves were prescribed to be used in sacrifices. It was the symbol of prosperity and immortality so it was put on different auspicious occasions for ensuring prosperity for the sacrificer. It appears that even the Indus Valley people were also acquainted with this flower as a representation of lotus fruit in faience has been discovered from Harappa.[205] In the post-Vedic texts, *puṣkara* became the usual simile of love, beauty, tenderness and immortality. During the times of the Buddha, several *Puṣkarṇis* existed in different localities. The Abhiṣeka Puṣkarṇī at Vaiśalī was very famous and only a few privileged ones were allowed to touch its waters. Till now, the lotus flower has retained it importance and remains a coveted object for the connoisseurs of refinement and beauty.

Urvāru is the cucumber and its fruit was termed as the *urvāruka,*[206] which is *kakadī* in vernaculars. In the Vedic texts it has been a frequent simile to denote the release of man from the bondage of death.

Vetas was a water plant or creeper of the class of calamus rotang or a similar reed. On account of its growth in water, it was termed as *apsuja.* Its colour was yellowish; hence, sometimes it has been termed as *hiraṇya.*[207] *Vetasavanta* denoted the person who possessed abundant of *vetas.* It was elastic and so chairs, baskets, sacrificial thrones and shafts of bow were made of it. It was used in sacrifices and was the symbol of elasticity and durability.

CULTIVATED CEREAL PLANTS

After agriculture was once established, attention was directed towards the cultivation of more suitable food grains and this gradually resulted in the production of grains differing markedly from the original wild forms. The reward of cereal cultivation is greater than that of any other form of agriculture, the gathering and threshing of grains are less laborious than the digging of roots. Further, the products can be stored easily and will remain for long periods. Surplus also can be accumulated and this may help in protecting the dense population against crop failure.

In the *Ṛgeveda,* the only name of cereal is *yava,* but it appears to have been a generic term to denote the grains in general. The archaeological evidence shows that other cereals were also cultivated but they have been not mentioned by their particular names. The later *Saṃhitās* show a sign of change and for the first time, a systematic enumeration of 12 different varieties of cultivated cereals have been furnished by the *Yajus Saṃhitās* and the Upaniṣads. This proves the growing importance of agriculture.

VRĪHI (PADDY)

Paddy is the most favourite food grain of the tropical regions. It grows during the rainy season under swampy conditions. On account of this, it has been rightly termed as *varṣavṛddha*, meaning that which grows during the rainy season.[208] The botanists think that for the first time its cultivation began somewhere in the region of southeast Asia, including India. The distribution of cultivated plants and their wild ancestors prove that rice was cultivated in a large part of India, leaving a strip in the extreme northwest. The Neolithic rice discovered from Orissa has been referred to as *oryza sativa*, which also grows wild in the marshes of Rajasthan, Bengal, Assam and Central India. It is, of course, easy to presume that it was this wild species which had been brought under cultivation. During the Chalcolitihc period, the pattern of plant economy undergoes a considerable change.

Some historians think that in the *Ṛgveda*, there is no mention of paddy; hence, the early Vedic Āryans were not acquainted with it. But it is very difficult to accept this view as the two terms, namely, *dhānyavīja*[209] and *kṣirapākaṃ*[210], *odanaṃ* refer to the existence of rice. The former, in its corrupt form *dhāna*, still survives in vernaculars for paddy and, generally, *odana* is taken in the sense of boiled rice.

Recent archaeological excavations prove that the cultivation of rice in western India was in practice even before the date of the *Ṛgveda* because rice husks have been found at Rangapur[211] (pd. 2nd A, 2000–1500 BC). Of particular interest is the discovery of rice and spikelets embedded in clay and pottery at Lothal.[212] The excavations at Ahar (Rajasthan) have brought to light some sherds, which bear the impression of rice husks. Rice husks have been found from Baidipur in Orissa whose scientific analysis proves that it was not wild rice but a cultivated one. Even as early as 1200 BC, rice was cultivated in the Narmada valley. So these evidences go a long way to show that the cultivation of rice was in practice in India ever since 2000 BC.

The later *Saṃhitās*, the *Brāhamaṇas* and the *Upaniṣads*, along with the *Āraṇyakas*, preferred *vrīhi* and described rice preparation both for domestic use as well as for sacrifices. These texts were composed in the Yamuna-Gangetic valley, which was mostly the rice-growing zone. Hence, references to it were but natural. For the first time, the term *vrīhi* occurs in the *Yajus Saṃhitās*[213] and since then, has become its popular name. Excavations at Chirand and Hastināpur[214] indicate that paddy was cultivated in the Yamuna-Gangetic valley (1800-800 BC), and rice husks were used as a binding material in mud plaster of houses at the later place.

The *Yajus Saṃhitās* inform us that different varieties of paddy were cultivated and each of them had its own specific characterstics, such as, the *kṛṣṇavrīhi*[215] (black paddy) and *śuklavrīhi*[216] (white paddy). *Āśudhānya* was ready for harvest within a short period of two months[217]. Such a variety is cultivated even now and is known as *sāṭhī*, as it is ready within 60 days after sowing. *Hāyana*[218] was red-husked paddy, which took one year to ripen. It is the same, which during the post-Vedic times, was known as the *saṃvatsara-pakva-vrīhi.*[219] The *masūsya* and *priyangu* were two other varieties, the former being cultivated in the north and the latter in the south, where it was famous for its gold-like shining husk.

Besides these general varieties, *mahāvrīhi* was the most important variety of paddy. The term indicates that in size, its grains were comparatively larger than others and its taste was also superior to other varieties. So, on this account, it had been considered as the *samrāt* among the cereal plants.[220] It was known by the name of *mahāśāli* as well because it was a special food item used specially by the rich peasants or the nobles whose household establishments were extensive. It has been maintained that *vrīhi* represents the *kṣatra* and by bringing sprout of such grains, the priest places universal sovereignty in the king.[221]

Patañjali[222] informs that *mahāśali* was a speciality of Magadha and was cultivated in that region. This fact has been corroborated even by the foreign accounts. In the seventh century AD, *mahāśali* had attracted the attention of Hiuen-Tsang, who has clarified its meaning. He mentions that it was for the use of persons of upper classes and was a special variety of rice grown in Magadha. Its grains were larger, scented and an exquisite taste and were famous for their shining colour.[223] Hwui Li, the biographer of Hiuen-Tsang informs that *mahāvrihi* was grown in Magadha.[224] In Chinese, it was known as *kung-ta-jin-mai,* meaning the rice offered to the great householder.[225] On all these consideration the mahāvrīhi should be identified with *basmati* or patna rice, which is famous for its shining quality, larger grains and excellent taste.

Besides these cultivated varieties, some wild varieties were known, such as, *nīvarā.*[226] It was grown in shallow tanks. Its cultivation was very easy and it grew wild hence, it was used by the landless people. Till now, *nīvarā* is considered as a sacred grain and is used as a non-cereal grain to be consumed on fasting days. Cake of wild rice[227] was prepared for offering to Bṛhaspati in *vājapeya.*

YAVA (BARLEY)

Barley was also a favourite food of the people and the evidences prove that even the Indo-Europeans were acquainted with this cereal. In different Āryan languages similar terms have been used. Its original form was *yevo,* which became *yava* in the *Vedas* and the *Avesta* and Jea in Greek.[228] In the epics of Homer, barley has been mentioned as a favourite foodgrain.[229] In the *Ṛgveda,* it is the only foodgrain which finds mention worth the name,[230] but doubt has been expressed whether it denoted the true cultivated barley or was a generic term to denote all vegetation. Like *dhānya,* it was a generic term for all the foodgrains.[231] In the *Ṛgveda* and the later *Saṃhitās, yavas* denoted the grass on which animals subsisted. It is interesting to note that Homer has also mentioned *yava* as fodder for horses. In India till now, jayee is a favourite fodder for horses, which is of a wild variety.

In the Vedic texts, different terms derived from *yava* have been mentioned, such as Yavyāvatī, which was the river, and its valley was famous for the cultivation of bumper crop of barley.[232] Yavaśira was an epithet of *soma* because the powder of *yava* grains was mixed with it when its juice was prepared. *Yevasa* or *yavasa* was a destructive insect, which destroyed the barley crop.

Since the times of the later *Saṃhitās*, *yava* has been used in the strict sense of barley grains. *Yava* was the stalk of the barley plant. *Yava* was considered as a sign of prosperity and the person possessing much barley was distinctly known as the *yavamanta*[233]. The eyebrows of Agni have been described as made of barley grains.[234] Its grains were considered as very strong and nutritious. Its stalks were subjects of special attraction on account of being tawny-brown in colour and with silvery ears.[235] The Sarasvatī valley was famous for the rich harvest of *yava.*[236] Its plants have been considered to be more moist than other plants hence these thrive lustily where other plants wither.[237] Its grains also were hard and in skill and strength were compared with a military commander. The priest, on the occasion of the king's *mahābhiṣeka,* brought their sprouts to place on his head and thus bestowed on him the same skill and strength.[238] This tradition is maintained even now on certain religious occasions and specific festivals. People place its sprouts on their heads. *Yava* and paddy were regarded to be so important that they have been considered as the two sons of the god.[239] Different preparations were made of barley. *Upa-vāk* was a variety of barley, which later on was known as *Indra-yava.* It formed the essential element of gruel. *Upa-vāk* groats are mentioned in the *Śatapatha Brāhamaṇa.*[240]

Records of barley are not only scarce but are represented by a few grains. *Hordeum vulgare var. nudum,* a variety of barley, is known from Mohenjo-daro and *H. Var. hexastictum* from Harappa[241] (2300–1650 BC). Kalibangan is the only site where abundance of barley is noted and at Lothal, only rice has been found. Again, barley appears at Atranjikhera[242] (1000 BC). However, it appears that barley was cultivated in an extensive area.

MĀṢA

Māṣa appears to have been a grain of the Indo-Iranian period.[243] In vernaculars, it is known as *urada.* Like barley, *māṣa* was also considered as a very sacred grain and its preparations were offered to the gods and departed souls. It was sown in the *Hemanta* and harvested in the *Śiśira.*[244] Its inferior variety was known as *kulmāṣa* and it was disliked for consumption; hence, it was wasted away in families and on account of this, it received its nomenclature.[245] Mostly, *māṣa* was used as a pulse and its powder was consumed by the people of the lower sections of society. The grain of *māṣa* was so beautiful and uniform in size that it was adopted as the unit of weight. Archaeological excavations prove its cultivation in the Narmada valley as early as 1200 BC.[246]

TILA (SESAME)

Sesame was a favourite grain and used in sacrifices as well. Its grains were offered as oblations to the gods and departed souls. *Dhānya* and *tila* have been compared with the cow and the calf.[247] *Tila* was mostly used for extracting oil and it gave birth to the term *taila,* meaning, oil in general. *Tila* has been mentioned in the *Ṛgveda* but from the *Yajus Saṃhitā* onwards, it occurs frequently. Like *māṣa,* it was also cultivated during the *Hemanta* and harvested in the *Śiśira.*[248] It had its two varieties, namely, the cultivated and the wild. Its buds were considered as very beautiful and its dried plants were useful for fuel. *Tilaudana* was also taken on special occasions.[249] *Jaratila* was a wild variety of sesamum. Its use in sacrifices were prohibited.[250]

Archaeological excavations in the Indus Valley prove that the people cultivated sesamum and used it in their diet.[251] It has been found at Harappa.[252] In Sumer also, it was cultivated at least in the beginning of the third dynasty at Ur (2350 BC).[253] Perhaps, it spread there through India or had come to the Indus valley from there, as both these places had commercial and cultural relations. The problem of the first cultivation of sesamum has been studied by historians and they come to the conclusion that Africa was the original home of sesame. For the first time, it was domesticated near the headwaters of the Niger river.[254] The African people were responsible for its spread in Sumer, wherefrom it spread in the Indus Valley. The earliest waves of people to India were Negroids from Africa.[255] Human skeletons showing Hamitic Negroid features have been found associated with Langhnaj (Gujarat) microlithic cultures.[256] This may be the story of sesamum, which spread from Africa to Sumer and India along with the migration of the people. Certainly, it was cultivated in the Indus Valley on the eve of the arrival of the Āryans who also adopted its cultivation and included it in their diet. It gained religious importance and now, many of us hardly feel that it came from Africa.

MUDGA

Mudga was a kind of bean (*phascolus mungo*), which occurs in the list of vegetables in the *Vājasaneyi Saṃhitā*.[257] In Hindi dialects, it is known as *munga*. Its grains were consumed in a boiled form (*mugdodana*). It is used for the purpose of preparing pulse, which is considered to be delicious and light. It appears to have been a corn of some importance as personal names also were current after it, such as, *Mudgala*, meaning the person who possesses *mudga* or swallowed[258] it. The descendants of *Mudgala* were known as *Maudagalyāyana*.

KHALVA (CICERRICTINNUM)

Khalva was a minor cultivated grain but still very useful. It has been identified with gram. It was reduced to powder with the help of pounding stones.[259] Its powder (*saktu*) was considered to be very cool with medicinal ingredients. The *Yajus Saṃhitās* mention it as a sacred grain. Archaeological excavations prove that gram was cultivated in the Narmada valley[260] in 1200 BC. The discovery of this grain from Amrawati (Maharashtra) is an interesting addition to the ancient plant economy of India but those are late in history. Gram is also associated with horses, which characterize the Āryans. We should find out the records of gram from the earlier strata and use them as a time maker for the advent of the Āryans in India.[261]

Similar to *khalva, khala-kula* was also a cultivated grain which is identified with *Kulthī*. It was considered as a sacred grain[262] to be used in sacrifices.

PRIYAṄGU

Priyaṅgu was an inferior variety of corn, which has been identified with *Panic* or *Setaria Italica*. In vernaculars, it is known as *keoni*. In the Vedic texts, *priyaṅgu* has been mentioned as having represented the enjoyment of pleasure among the herbs. On account of this, it was used in sacrifices and on the occasion of the king's coronation, its sprouts were placed on his head.[263] *Priyaṅgu* is of common cultivation in southeast Asia. The Chinese records show that the history of its cultivation in that country goes back to the third millennium BC and it appears to have spread in the west through China and India.[264]

AṆU

Aṇu is a comprehensive term, which includes several grains of considerably small size. It consists of plants of different genera. These cultivated, tall cereal plants have several characteristics in common. They yield grains, which are both smaller and less nourishing, but they can be cultivated in climate considered as too tropical for wheat or barley cultivation, and also too dry for rice. They can resist humidity and severe drought during some part of the growing season. They can also be cultivated under widely diverse conditions. This makes them of a prime importance in many regions. The botanical evidences suggest that the areas of first cultivation of true millet lay in the same region as that of wheat and barley. Archaeological excavations in west India have brought to light some grains termed as *aṇu* and it appears that in Rajasthan, these were cultivated between 1800–300 BC. Some potsherds found from Ahar[265] (1800–300 BC) bear fix form impressions in the form of cavities showing the surface and the lateral views of these grains. An analysis of these impressions shows them to belong to millets, such as, *panicum millaceum*, in vernacular known as *sāwān* (or Chīnā), *sorghum vulgare* (*jowar*), *pennisetum* (*bājarī*), *setaria* (*kaun*) and *leusine coralana* (*torī*), etc.

A wild variety of jowar originally belongs to the island of San Antonio in Africa, from where it was introduced and domesticated in Egypt by the prehistoric people and its earliest domesticated remains have been found from the tombs of ancient Egypt (2200 BC). Possibly, it was introduced in Mesopotamia through Egypt and from there to India via the Indus Valley and Rajasthan.[266]

ŚYĀMAKA

Śyāmaka belongs to the *aṇu* group of grains but on account of its importance, it has been mentioned as an independent variety of grain. It is very small in size and light in weight. It was so light and so small in size that it was used as a simile for showing lightness of things.[267] One of its characteristics is its speedy growth hence, in the *rājasūya*, its cake Vanaspati was offered to *Soma*.[268] It was the favourite food of pigeons.

GODHŪMA

A study of the Indo-European group of languages proves that the early Āryans before their dispersal to different regions were acquainted with barley, but not with wheat. It appears that only the Indo-Iranians came in touch with this cereal so they had a common term for this, i.e., *godhūma* in the *Vedas* and *ganduma* or *gantuma* in the *Avesta*. On the contrary, Schrader thinks that wheat was known in prehistoric European culture. For it, he finds the word *hordeum*[269] which may be equated with Sanskrit *godhūma* and *zend godhūma.* But in the *Ṛgveda,* there is no mention of wheat, though the early Āryans were in contact with the wheat-producing people in the Punjab. *Godhūma* has been mentioned in the later *Saṃhitās* and the *Brāhamaṇa* texts. It is interesting to note that *yava* has been prescribed as an offering to the gods and departed souls but the use of wheat in sacrifices has not been mentioned. It appears that the Vedic priests considered it as impure on account of its association with the Asuras. It has been suggested that the Indo-Iranians borrowed wheat from the Assyrians who called it as *sheum.* It also appears that the Āryans spread the cultivation of wheat in eastern India during the times of the later *Saṃhitās.* Different preparations of wheat were prepared and relished.

A wheat superior to *emmer* was developed in early times. This is known as bread wheat (*triticum vulgare*), whose improved variety provides the world's supply of wheat today. On the evidence of carbonized remain of grains, it has been proved to be cultivated by the later prehistoric people of Mesopotamia. It was certainly grown by the Indus Valley people.[270] Remains of *triticum compactum* (club wheat) have been discovered from Mohenjo-daro and Harappa. The *Triticum sphaerolocum* variety has been unearthed, which is thick and short.[271] It is known as dwarf wheat. These two varieties are still cultivated in the Punjab and adjoining regions. It is difficult to ascertain whether wheat cultivation began in the Indus valley through Mesopotamia or was begun independently. The wild variety of wheat grew in the northwestern India and in Kashmir as its remains have been discovered from the Neolithic site Burzahom.[272] During the Chalcolithic period, the cultivation of wheat spread to the entire north and central[273] India. It appears that the cultivation of wheat was begun by the Āryans in the Ganga-Yamuna Valley. At Atranjikhera, for the first time, remains of wheat were discovered from the PGW levels along with paddy and barley.[274] In this context, it is interesting to note that remains of wild wheat have been discovered from the Neolithic site at Mahagara in the middle Gangetic Valley.

IRRIGATION

For the proper growth of plants, it was essential to provide them with moisture. This process is called irrigation. The *Ṛgveda* refers to some of the means of irrigation, but the people depended more on the grace of nature. They had developed very primitive and crude artificial means of irrigation. The later Vedic sources throw considerable light on the means and methods of irrigation adopted by the people. They were acquainted with different sources of water unknown to the early Vedic people.[275] Besides, they were familiar with waters of different kinds.[276] These things indicate that much progress was made in the field of irrigation, both artificial and natural. It also shows the growth of agriculture in place of pastoralisim.

Water was considered as the vital element or the existence of all, because everything that exists is produced by water. It was held that the essence of plants is water. So, it was rightly thought to be an essential element of their growth.[277] The necessity of irrigation may be assumed from the fact that in a process of the preparation of the site of the *āhavanīya* sacrifice, five jars filled with water were poured over the ploughed as well as unploughed ground which promoted the growth of vegetation.[278]

The success of agriculturists depended upon nature or irrigating the fields; hence, the people prayed to *Parjanya* and Indra to pour adequate rainwater for obtaining a rich harvest. The three hymns of the *Ṛgveda* dedicated to *Parjanya* and the frogs were clearly intended in rain-charms.[279] Prayers were offered to Varuṇa, Indra and *Parjanya* to pour adequate waters in time for a good harvest.[280] It is interesting to note that the Vedic people were well acquainted with the scientific process of rain. They knew all about the heating of the oceanic water through the sun rays—how it changed into vapour, how it went high up in the sky, and how poured rains.[281]

The later *Saṃhitās* prescribe rituals and prayers for adequate rainwater. The *Chāndogya Upaniṣad* describes the necessity of rain for the success of agricultural operation. It mentions heat as the original matter which emits water. Water wished to be many—it procreated itself and emitted food. Therefore, whenever it rains, there is abundant food. It states that water is more important than food; therefore, when there is no adequate rain, living creatures become sick with the thought that food will now become scarce, but when there is rain they become happy believing that food will become abundant.[282] Prayers were offered to rainwater for causing material prosperity to the people in the form of good harvest and abundant fodder.[283]

Sometimes it did not rain or it was inadequate. So the plants did not grow, resulting in famines. Some of the *Jātaka* stories mention such famines. Kauṭilya describes the measures of rainwaters essential for successful harvest of different crops in several regions of India.[284] Megasthenes mentions adequate rains twice in a year in India, causing a rich yield of crops.[285] The major areas of northern India were covered with a network of rivers originating from the Himalayas and flowing towards the east and the west.[286] These rivers, during rainy season, flooded their valleys, deposited fertile siltation in the fields and watered the plants. Floods were useful in two ways—first, they watered the standing crops and spread a fertile layer of mud on the fields, which provided the plants with fertile manure. Megasthenes also notes the spread of alluvial soil by the rivers in northern India, making the region very fertile.[287] On account of this, the rivers thus helped in obtaining a rich harvest and that is why the people revered them as their mothers. But, similar to the rain irrigation, flood irrigation was also uncertain and sometimes it caused devastation as it washed away standing crops, domestic animals and even settlements. It caused heavy damage to the fields situated on the banks of the rivers. Evidently, sometimes, the people constructed embankments on the rivers which were strong and beautiful.[288] In the post-Vedic times, Kauṭilya furnishes more details about the construction of *setu* (embankments) for utilizing the river water for irrigating the fields.[289] Rain and floods were the natural means of irrigation but these were not dependable; hence, artificial means were also adopted. *Khanitrim*[290] denoted artificial means for obtaining water, such as, the wells, tanks and canals, which are of remote antiquity in India. The failure of monsoon and uneven rainfall were some causes which compelled the people to adopt these means and methods for irrigating their fields.

Wells appear to have been the most convenient and useful means of artificial irrigation as these could be easily dug out and then contain sufficient water throughout the year. *Avata*[291] was an artificial hollow in the earth, containing water. It may be compared with the present days' temporary *kacca* wells. It was in contrast with the *utsa,* meaning a natural spring.[292] The well has been described as an unfailing source of water. *Kakaṭa* was the well having less water and *reṇukakaṭa* was the one having sandy water, hence considered useless.

The cultivators had invented the means and methods for lifting water from the wells. It was lifted by means of wheels, a strap and water pails.[293] It also might have been raised by a bucket tied to a rope, the other end of which is tied to a wooden pole with the fulcrum near the other end of the pole that carried a heavy weight. The same method is still prevalent in some parts of India. A device known as *macakra* was used for lifting water.[294] Its exact functioning is not known but it may be conjectured on the basis of its description in the Vedic texts. It appears to be a wheel made of stone and water was raised in a pail (*koṣa*) by means of a leather strap (*varatra*). Kauṭilya advises the king to cooperate with the farmers by digging wells[295] and Varāhamihira discusses in detail the signs, means and methods for digging wells so as to obtain sufficient water.[296]

Tanks were useful means of irrigation, which were of different sizes. *Hrada* was a big tank or lake. *Vesanta* was a tank of bigger size, but Vesantī was small in size.[297] *Sara* also denoted a tank. The *Ṛgveda* classifies tanks into two categories, namely, those of bigger size with much water in which one could take bath and the other was with less water.[298] During rainy season, much water would be stored in tanks. The same water was utilized for irrigating the vast tracts of agricultural fields, which the wells could not irrigate with their limited water. In rural areas, tanks still play a vital role in irrigating the fields. The method of lifting water from the tank was the same as is prevalent in northern India at present. The ancient Indian kings were very careful towards their farmers and provided water for irrigation by constructing huge lakes like *Sudarśana* and repairing them for centuries, as was done by Chandragupta Maurya, Aśoka, Rudrādaman and Skandagupta.[299]

The question of long canals in ancient India does not arise. However, there are references to *kulyās*,[300] which are taken in the sense of canals. *Kulya* was an artificial water course flowing into a reservoir. It was cut from the river. Allegories referring to the former as a calf and the latter as a cow indicate that rivers were the source of canals.[301] Rituals were prescribed which were to be performed on the occasion of the opening ceremony of letting out the river waters flow through the canals.[302] Possibly, small canals were dug out to irrigate the fields during off seasons.

To some extent, rivers were also the means of artificial irrigation of fields situated on their banks. It was very easy to lift water from them, as it did not involve much labour to cost. Sometimes, dams were built to utilize the water of rivers for this purpose. The Śākyas and the Koliyas utilized the water of the River Rohiṇī for irrigating their fields.[303] *Rodhas* was the term to denote a dam which restrained the flow of the stream or the river.[304]

PLANT PROTECTION

The cultivators hoped that they would get corns in such quantity that their earthen storage jars might burst and the heaps of grains might become exhaustless like the ocean.[305] They had to pay attention on plant protection for a good harvest. Natural calamities like drought and lightning caused damage to the standing plants, so it was essential to protect them. For this purpose, prayers were offered to Indra.[306] Most of the fields around the villages were surrounded by forests and wild animals like elephants, deer, hare, monkeys, and domestic animals who also damaged the crops. In order to ward them off, people raised fences of bamboo and other materials around their fields. Kauṭilya advises the king to engage hunters for killing wild animals and punishing the animal owners causing damage to crops.[307] Birds were also the cause of damage to standing crops, so different methods were adopted to ward them off from the fields, such as, clapping or verbal sounds.[308] Reptiles were harmful to crops. They cut the plants, stole away grains and stored them in the holes made into the earth. On account of this activity, they were rightly called *mūṣa*, meaning, one who steals.[309] It is an Indo-European word which survives in English (mouse) in the same sense. Kauṭilya considers the menace created by the rats as one of nature's calamities and so indicates methods for their eradication.[310]

Wild grass also hampered the growth of crops; so these were weeded out with the help of weeding tools made of wood, copper or iron. The people invoked the divine grace for protecting their plants. In the later *Saṃhitās* and the *Brāhamaṇa* texts several spells and rites have been prescribed to ward off the elements, causing harm to the plants.[311] In spite of these precautions, sometimes, crops were badly damaged, which resulted in famine. The people of the Kuru country had to suffer much as their crops were damaged by the locusts.[312] The early Pāli texts also throw considerable light on the problem of plant protection. The people adopted several methods for this, viz., digging pitfalls for capturing animals, fixing stakes and setting stone traps.[313] There were watchmen in fields who protected crops from animals and birds.[314] This practice is still surviving in rural areas.

HARVESTING

In due course, different crops became ripe and were ready for harvest. Seasons played an important role in ripening crops. The later *Saṃhitās* and the *Brāhamaṇa* texts furnish interesting information regarding the fact that barley ripens in the summer, rice in the autumn and beans and seasme in the winter.[315] The seeds sown in the winter were ripe by *Caitra* (March–April).[316] These indicate that the people were well acquainted with different crop cycles.[317] Generally, two crops were harvested in a year, one in the summer and the other in the winter.[318] Pāṇini, Kauṭilya and Megasthenes refer to harvesting of two crops, namely, the winter and the summer crops in a year.[319]

With the penetration of the people into the Deccan, their agro-geographical knowledge also increased, as the *Aitareya Brāhamaṇa* informs that the crops are first ripened in the southern countries so one who wishes for food should turn towards the south and one who goes there enjoys much food and becomes the master of food.[320] In the north, mainly barley and wheat were cultivated, which were ready for harvest in *Caitra*, but in the Deccan, mainly rice was cultivated, which was ready for harvest in *Kārttika-Mārgaśīra* (November–December).

The standing crops were cut with sickle[321] (*dantyā*) made of copper or iron. It was known as *dātra* in the north but in the east, it was termed as *dātī*. It appears that sickle was evolved even during the Indo-Iranian Period. *Dātra* was the common term to denote it, which still survives as *basa* (modern Persian). When the crop became ripe, people went to the fields with sickles in their hands and cut the standing plants upto the roots above the ground. The sickles were the most favourite tools of the farmers. The use and existence of iron sickle is attested by the find of an iron sickle blade from Hastinapur[322] Pd. III.

The harvested crop was collected and bundles were made out of it. Those were carried to the threshing ground. If the bundles were few, they were beaten out on the floor.[323] Oxen were also used for separating grains from the stems by treading them with their feet.[324] After this, the grains were separated from the chaffs and refuse with the help of a winnowing fan (*sūrpa*)[325] or a winnowing basket (*śyama*).[326] Winnowing was performed in the wind to clear the grains from chaffs and light grains.[327] The winnower was termed as *dhānyakṛta* because he separated the grains from the chaffs. Kauṭilya also describes the harvesting of crops, their collection on threshing ground (*khalva*) for separating corns from plants.[328]

STORAGE

The remains of a granary at Mohenjo-daro prove the storing skill of the Indus Valley people. The cleared grains were measured and stored in granary (*sthīvī*). *Sthīvīmant* was the person who possessed storage jars in considerable numbers and it was the mark of material prosperity in the form of corns.[329] In villages, the term *sthīvī* survives in the form of *theli* in dialects. Barley has been mentioned as coming out of it.[330] *Urdaraṃ* and *Kṛdaraṃ* were also storage devices.[331] The former was cylindrical in shape with perforation on the top, and had an opening on the upper part. The latter was a storage device into which a hole was bored on its lower portion to take out the grains. Both the openings were covered and removed when desired.[332] Their description shows that those were nothing but the *deharis*, which are still used for storing grains in villages. Kauṭilya furnishes more details about storing chambers of foodgrains[333], and the Sahagaura copper plate inscriptions confirm its structure.[334]

NOTES AND REFERENCES

1. Roy, B.P., 1984, *The Later Vedic Economy*, p. 168–200, Janaki Prakashan, Delhi/ Patna.
2. Roy, B.P., *loc. cit.*, pp. 168–202.
3. *Caraka Saṃhitā*, *Sūtrasthāna*, 1,40–44 Suśruta, *Sūtrasthana*, 1.4–15.
4. *KA*, 2.24.
5. *Bṛhat Saṃhitā*, Eng. tr. Pt. I, Bhat, M. Ramkrisna, 1997, Delhi, Motilal Banarasidass, p. XI, also ch. 55.
6. *Agni Purāna*, Ch. 281.

7. *Caraka Saṃhitā*, i, ch. 30. 8–9.
8. *Bṛhat Saṃhitā*, as ref. no. 5.
9. Majumdar, G.P., 1927, *Vanaspati*, Calcutta, Calcutta University, p. 207.
10. Mishra, B.B., 1965, *Polity in the Agni Purāna*, Calcutta, Punthi Pustak, pp. 22–26.
11. Majumdar, R.C., (ed), 1960, *The Classical Accounts of India*, Calcutta, Firma K.L.M. Pvt. Ltd.
12. Samuel Beal, 1994, *Buddhist Records of the Western World*. Delhi, Motilal Banarasidass.
13. Sinha, B.C., 1960, Tree Worship in Ancient India.
14. *VS.* 30.19.
15. *KA.*, 2.17
16. Suśruta, *Sarīrasthāna*, 2.33., *Mbh.*, Karṇa, 48.6.
17. Sircar, D.C., 1983, *Select Inscriptions*, pt. II, p.243, Delhi, Motilal Banarasidass.
18. *Guṇaratna's Commentary on Ṣaḍadarśana Samuccaya*, śloka. 49.
19. *KA*, 224.
20. *Ibid.*
21. *Ibid.*
22. *Ibid.*
23. *ŚB*, 7.2.2.5.
24. Vide Dange, S.A, 1971, *Vedic Concept of Field and the Divine Frutification*, Bombay, pp. 68–82.
25. *AB.* 3.3.38.
26. *ŚB*, 7.2.2.9.
27. *KA*, 2.24.
28. *ŚB*, 3.3.3.17.
29. *VS.* 30.7.
30. *ṚV*, 10.3.2.
31. *Ibid.*, 10.31.2.
32. *TS*, 5.1.7.1–3.
33. *Ibid.* 4.2.7.2–10.
34. Basham, A.L., 1967, *The Wonders That was India*, p. 195, London, Sidgwick and Jackson Ltd. *Ancient India as Described by Megasthenes and Arrian*, Calcutta, Mc Crindle. J.W., (reprint), pp. 52–53, Majumdar. R.C. *op. cit.*, p. 233.
35. *TS*, 7.320.l; vs 22.28; AV,1.7.33r *Viṣṇu Purāṇa*, 7.37–39.
36. *Dhammapada* 4.1–59
37. Amarakośa, Vāmanajodhivarga.
38. *BU*, 3.2.28.
39. Upaskar on Vaiśeṣika Philosophy, 4.2.5.
40. *Śāntiparvan*, 184.
41. *Bhāgavata Purāṇa*, 3.10.20. Cf Kaṇād, 5.2.7.
42. *KA*, 2.24.
43. *Bṛhat Saṃhitā*, 55.3.
44. *Agni Purāṇa*, Ch. 281.
45. *Bṛhat Saṃhitā*, 55.4–5.

46. *Ibid.*, 55.6.
47. *Agni Purāṇa*, 281.8–9.
48. *AV*, 2.8.3.
49. *Bṛhat Saṃhitā*, 55.17–18.
50. *Agni Purāṇa*, 2.81.11–12.
51. *Bṛhat Saṃhitā*, 55.2–26, *Agni Purāna*, Ch. 194.
52. *Bṛhat Saṃhitā*, 55.21–26.
53. *Cakradatta, Cikitsā Saṃgraha*, 86–87.
54. *Bṛhat Saṃhitā*, 55.15., *Agni Purāṇa*. 281.10–13.
55. *Bṛhat Saṃhitā*, 55.14.
56. *Ibid.*, 55.15.
57. *Agni Purāṇa* 281.13.
58. *Bṛhat Saṃhitā*, 55,18 and 20; cf. *Agni Purāṇa*, 281.10.
59. *Śāntiparvan*, ch.184.
60. *Bibliotheca Indica*, ch.II, p. 23.
61. *Ibid.*, new series (1912), p. 134.2.
62. *Ibid.*, (1907), p. 1151.
63. *ŚB.* II, vol. 4.
64. *ṚV.* 10.97.21.
65. *AV.* 11.6.10.
66. *Aitareya Āraṇyaka*, 1.51.9.,2.6.1.5.
67. *CU*, 6.12.1.2.
68. Manu., 1.40.
69. *Śāntiparvan*, Ch. 184.
70. *Bhāgavata*, 3.10.19–20.
71. Vide, B.N., Seal, 1959, *Positive Science of the Ancient Hindus*, Motilal Banarasidass, Delhi, p. 175.
72. *Hārita Saṃhitā*, Śarīrasthāna, ch. 1.
73. *Kalpasthāna*, 3.5.
74. Vanauṣadhi Varga, *Amarakośa*, 50.
75. Vide, G.P., Majumdar, op. cit., p. 65.
76. Majumdar, R.C., *op. cit*, p. 253.
77. *Brahma Jālasutta*, II.
78. *Sumangala Vilāsinī*, pt, I, p. 81.
79. *Sarītasthāna* 3.22.
80. *Suśruta Saṃhitā, Śarīrasthāna*, 3.18
81. *Caraka Saṃhitā, Sarīrasthāna*, ch. 3.
82. *Ādi Śaṅkar on Bṛhadāraṇyakopaniṣad.*
83. *Caraka Saṃhitā, Śarīrasthāna*, 3.22–267; *Mbh, Śānti*, 204.2–16.
84. *Ibid, Kalpasthāna* I.
85. *Bṛhat Saṃhitā*, 55.10–11.
86. *Amarkośa, Svargavarga.*
87. *AV.* 11.7.21., cu.6.3.1.

88. *TS.* 2.5.3.2.
89. *Nirukta,* 9.27.
90. *ṚV,* 4.97.1, *Nirukta,* 2.28.
91. *ŚB,* 11.1.7.2.
92. *AV,* 3.7.1.
93. *KA.,* 2.15; 2. 17.
94. Manu., 1.46–48.
95. *Sūtrasthāna,* 1,36–37.
96. *Ibid.,* 4.1.
97. *Suśruta, Sūtrasthāna,* 38.166.
98. *AV,* 4.17.166.
99. *Ibid.* ,7.65.1–3.
100. *ŚB,* 5.2.4.15.
101. *Ibid.,* 9.1.1.4.
102. *AV,* 2.4.1,19.34–25.
103. *Ibid.,* 19.34.10.
104. *Ibid.,* 19.34.7.
105. *Ibid.,* 19.50.1–8.
106. *Ibid.,* 19.38. 8–9.
107. *Ibid.,* 19.3 9.9.
108. *Ibid.,* 2.25.1–5.
109. *Ibid.,* 4.12.1–7.
110. *Ibid.,* 6.134.1–3.
111. *Ibid.,* 19.3.4–5
112. *Ibid.,* 18.3.60.
113. *Ibid.,* 6 96.16 6.
114. *SBE,* XXVI, pp. XXIV–XXV.
115. *TS,* 6.1.11.4.
116. *AB.* 1.3.12.
117. *ŚB.,* 3.3.3.7.
118. *TS,* 6.1.6.
119. *ṚV,* Maṇḍala, IX.
120. Ibid., 9.21.1.
121. VI *I.*474.
122. *ŚB,* 5.410.4.3.4.13.; 7 13.8.1–16.
123. *AV* 2.25.1.
124. *Ibid.,* 2.8.3–5.
125. *CU,* 7.31, *JUB.,* 1.386.
126. *ŚB,* 8.8.1.16.
127. Vats, M.S., *Excavations at Harappa,* p. 333.
128. *RV,* 10.97.5; AV. 6.110.1.
129. *BU,* 6.11.1.
130. *Ibid.,* 2.6.8.

131. *ṚV.*, 1.164.20–22; VS, 28.20.
132. *AV*, 3.5.6, 8.8.3.; AB. 7.5.32.
133. *AV*, 20138.13; *ŚB*, 13.4.4.8; AB., 2.5.
134. *TS*, 2.1.8.1–2; *ŚB*, 1.3.3–20.
135. *AB*, 2.1.1.
136. *PGS*, 4.1.41.
137. *VS*, 19.23; *ŚB*, 5.5.4.10.
138. *VS*, 21.30–31.
139. *TS*, 2.4.9.2; *ŚB*, 2.5.2.11
140. *AI*, 9–10, pp. 130–133.
141. *TS*, 5.2.7; *ŚB*, 3.4.1.6; 7.4.1. 57–61.
142. *Vats*, *op. cit.*, pp. 330–331.
143. *AV*, 3.6.1; TS, 3.5.7.1.
144. *ŚB*, 13.4.4.5–9.
145. Vats *op. cit.*, p. 467; Marshall, *MIC*, II p. 587.
146. *ŚB*, 6.6.2.11.
147. *AV*, 6.102,3; *AB*, 2.2.4.
148. *AV*, 6.102.3; *AB*, 7.3.30.
149. *CU.*, 6.12.1–3.
150. *ŚB*, 5.3.5.15; 13.2.7.3.
151. *AB*, 7.5.30–31; *ŚB*, 13.2.7.5.
152. PE.VII
153. R.C., Majumdar, R.C. *op. cit.*, p. 253.
154. *ŚB*, 6.6.3.7.
155. *AB*, 2.1.1.
156. *AV*, 3.5.4; 18.5.53.
157. *TS*, 6.3–10.2, *ŚB*, 3.8.52.
158. *AB*, 7.5.32; 8.3.18.
159. *Aṣṭa*, 5.2.24.
160. Chaudhary and Ghosh, 'Plant Remains from Harappa', A 9 No.7. p. 17.
161. *KS*, 2.5.6: *AV.*, 8.2.28; *ŚB*, 13.4.4.5–7.
162. *ṚV.*, 7.17.3
163. *VS*, 2.3.13: *ŚB*.13.2.7.4.
164. *Nirukta*, 12.8
165. *ṚV*,10.97.5: *AV*,6.11.1; *ŚB*,11.5.1.15.
166. *AV*,6.30.2–3 .
167. *Ibid.*
168. *TB*, 3.4.10.
169. *ṚV*, 3.15.19.
170. Vats, *op. cit.*, p. 468.
171. *AI*, Nos. 10–11, p. 124.
172. Kent, R.I., *Old Persian*, DSF. 3451.
173. *MS.*, 3.1.9; *ŚB*, 13.8.1.16.

174. *AB,* 7.15.
175. *Ibid.*, 5.24.
176. *ŚB,* 6.6.5.2–3.
177. *TS,* 5.2.8.7; *ŚB*.6.1.2.
178. *AB,* 8.2.8; AV, 19. 13,1–14.
179. *AB,* 7.5.32; *ŚB,* 5.3.4.2.
180. *ṚV,* 10.34.1.
181. *ŚB.*, 6.6.3.1; 1.4.2.5.
182. *Ibid.*, 3.4.1.7; *AI.*, Nos 9–10, pp. 126–130.
183. *ṚV,* 6.6.1–13; *AV,* 11.6–15.
184. *ṚV,* 8.7.20; 6.43.2; 6.4.3; 19.32.2.
185. *AV,* 1.34.166; *TS* 6.2.1.5.
186. *AV,* 7.58.4. *ŚB,* 1.1.4.19.
187. *Ibid.*, 2.7.1 ff.
188. *AV,* 1.34.166: *TS.* 6.2.1–5.
189. *AV,* 7.58.4. *ŚB*.1.1.4.19.
190. *TS,* 6.2.8.4.
191. *ṚV,* 10.100.10.
192. *AI,* nos. 9–10, pp. 152–153.
193. *ṚV,* 10.134.5: *TB.* 1.5.10.–1–7: *ŚB,* 2.5.2.15.
194. *ṚV,* 1.161.8: *AV,* 1.2.4.
195. *ŚB,* 6.6.1. 23.
196. *AV,* 4.19.9.
197. *TS,* 4.42: *AV,* 1.2.1.
198. *Nirukta,* 5.5.
199. Vats, *op. cit.,* p. 467.
200. *ŚB.* 7.5.1.11.
201. *AV,* 4. 34. 5.
202. *Nirukta,* 5. 14.
203. *ŚB,* 10.5.1.5.
204. *TS,* 5.6.4.2.7.
205. Vats, *op. cit.,* p. 467.
206. *AV,* 14.1.17. VS, 3. 60.
207. *TS,* 5.5.12.2
208. *VS,* 1. 16.
209. *ṚV,* 5.53.13.
210. *Ibid.,* 8.77.10.
211. *AI,* Nos. 18–19, p. 172.
212. Allchin, *BIC,* p. 259.
213. *TS,* 2,10. *KS,* 10.6. VI. I. 345; *Mbh, Āraṇayaka,* 246.1–13.
214. *Al,* No. 10–11, pp. 14, 120–24: S.S. Ghosh, 'Further Records of Rice from Ancient India, Indian Forester', Vol. 87, No. 5, p. 296.
215. *TS,* 1.8.10.1.

216. *CU.* 5.10.6.
217. *ŚB,* 5.3.5.2.
218. *Ibid.,* 5.3.3.6.
219. *Aṣṭa,* 3.1.48.
220. *AV,* 11.4.13; *AB,* 8–18 Cf. *Aṣṭa,* 62–38.
221. *AB,* 8.3.16.
222. Puri, B.N., Vide, *India in the Times of Patañjali.,* p. 124.
223. Beal, S., *Si–Yu–Ki,* pt II, pp. 82–109.
224. Sankalia, H.D., *Nalanda,* pp. 112–113.
225. Gode, P.K., 'Studies in the History of Indian Plants', *New Indian Antiquary,* Vol. 6 to 12, p. 267.
226. *VS,* 18.12; *ŚB,*5.1.4.14.
227. *ŚB,* 5.1.4.14.
228. O, Schrader, 1962, *Prehistoric Antiquities of the Āryan,* Peoples, Delhi, p. 282.
229. Homer, Ibiad, Book, 18, p. 353; Odyssey, Book 9, p. 129.
230. *ṚV,* 1.23.15; 8.2.3.
231. *VI,* II, p. 187.
232. *ṚV,* I.94.
233. *VS,* 19.5.
234. *Ibid.,* 19.91.
235. *ṚV,* 2.8.3
236. *Ibid.,* 7.30.1.
237. *ŚB,* 3.6.1.10; *Mbh, Śānti,* 44.8–9.
238. *AB,* 8.3.16.
239. *TS,* 7.2.10.2.
240. *ŚB,* 12.9.1.5.
241. Vats, M.S, *op, cit.,* p. 467; Allchin, *BIC,* pp. 258–59.
242. Mittra, V., Status of Barley in Indian Archaeology, *Purātattva,* No. 3, pp. 1–2.
243. O., Schrader, *op. cit.,* p. 284.
244. *TS,* 7.2.10.2.
245. *CU,* 1.10.2.7; Nirukta, 1.4; Cf. Aṣṭa, 4.4.4.
246. Sankalia, H.D., 1958, *Excavations at Maheshwar and Naradatoli,* Pune, p.332.
247. *AV,* 18.426.
248. *TS,* 7.2,10,2.
249. *BU,* 6.4.16.
250. TS, 5.4.3.2; Mbh, Sabhā., 68.13–14.
251. Piggot, S., *Prehistoric India,* p. 153.
252. Vats, M.S., *op. cit.,* p. 467.
253. Burkill, I.H., Habits of Man and the Origin of the Cultivated Plants of the World, *Proc. Soc. Land.,* Vol. 164,12.
254. Murdocks, G.P., Africa: Its people and Their Cultural History, New York (1969).
255. Chatterjee, S.K., Race, 'Movements and pre-historic culture in the Vedic Age' (*In the History and Culture of Indian People*), pp. 143–66.

256. Sankalia, H.D., 'From Food Collection to Urbanisation in India', *Indian Anthropology*, (1962).
257. *VS*, 18.12; Cf. *Mbh, Anuśāsana*, 112.62–63.
258. *ṚV*, 10.102.; *Nirukta*, 9.24.
259. *ṚV*, 5.23.8.
260. Sankalia, H.D., EMN, p. 332.
261. Mittre, V., Inter-relations between archaeology and plant sciences, Purātattva. No. 1, p. 8.
262. *BU*.1, 6.3.22.
263. *AB*, 8.3.16.
264. Mittre, V., Remains of rice and millet in excavations at Ahar, pp. 2.31.
265. *Ibid.*
266. *Ibid.*
267. *AV*, 10.50.4: CU, 3.14.3.
268. *ŚB*, 5.3.3.4.
269. Schrader, O., *op. cit.*, p. 284.
270. Marshall, *MIC*, II, p. 585.
271. Vats, M.S., *op, cit.*, p. 466.
272. Mittre.V., *Purātattva*, No. 1, p. 11.
273. *Ibid.* p.8.
274. Chaudhary, A.K., *Ancient Agriculture and Forestry In North India*, pp. 75–77.
275. *VS*, 16.36–38.
276. *Ibid.*, 22. 25.
277. *ŚB* 7.4.1.6: 3.6.1.7; *TS*, 5.1.3
278. *ŚB*, 7.2.4.1–12.
279. *ṚV*, 7. Sūktas, 101–3.
280. *Ibid.*, 10.104.8; AV, 3.17.7; TS, 2.4.10.
281. *ṚV*, 4.77.10; AV, 4.15.5; CU, 7.11.1.
282. *CU.*, 6.2.1–4; 7–10.1
283. *AV*, 3.17.1; 3.31.11; *Mbh, Ādi*, 163. 15–20.
284. *KA*, 2.24.
285. McCrindle, *J.W.*, *op. cit.*, p. 31.
286. *AV*, 6.2.4.1.
287. McCrindle, *J.W*, *op. cit.*, pp. 30–31.
288. *ṚV*, 1.3.11.
289. *KA*, 2,24.
290. *ṚV*, 7.4.92; *AV*, 1.6.4
291. *ṚV*, 1.105.19.
292. *Nirukta*, 5.26; *Mbh, Ādi.* 23.15–18.
293. *ṚV*, 8.69.12.
294. Raghavan, D., 1964, *Agriculture In Ancient India*, Delhi, p. 13, *ṚV*, 10.101.6.
295. *KA*, 2.1.
296. *Bṛhat Saṃhitā*, 54.1–125.

297. *AV*, 1.3.7.
298. *ṚV*, 10.71.7.
299. Junagarh Rock Inscriptions of Rudradāman I and Skandagupta, respectively.
300. *ṚV*, 10.437, AV, 20.17.7.
301. *AV,1.* 6.4; 3.13.7; Sharma, 'Ancient Canals', *JIH*, Vol. 41; pt. 2, pp. 202–3.
302. *Kauśika Sūtra*, 40.1–10.
303. *Kuṇāla Jātaka.*
304. *Nirukta*, 6.l. *Mbh. Droṇa*, 62.2; *Śānti.* 120.9
305. *AV*, 6.142.l–3.
306. *Anguttara Nikāya*, III,104, AV, 7.11.1 ff:
307. *KA*, 2.24.
308. *ṚV*, 10.68.1
309. *Nirukta*, 4.5
310. *KA*, 4.3., *Mbh. Śānti*, 221.58.
311. *AV*, 6.50.1,6.11.1.
312. *CU*, 1.10.1–3.
313. *Jātaka*, i,145.
314. *Ibid.*, II, 110; III.52.
315. *TS*, 7.2,10.2
316. *KB*, 19.3.
317. *MU*, 1.1.7; KA,1.6.
318. *TS*, 5.1.7.3.
319. *Aṣṭa*, 3.l.ll4; 4.3.454; KA, 2.24; J.W. McCrindle, *op. cit.*, p. 31.
320. *AB*, 1. 2.7–8.
321. *VS*, 10.32; Nirukta, 2.2.
322. *AI*, No.10–11, p. 98.
323. *ṚV*, 10.131.2.
324. *Ibid.*, 10.48–7; Aṣṭa, 4.2.50–51.
325. *Ibid.*, 10.71.2; AV, 12.3,18.
326. *Nirukta*, 6.9. *ŚB* 1.1.4.19
327. *ṚV*, 10.3.2; *Mbh, Śānti*, 174.7.
328. *KA*, 2.24
329. *AV*, 10.27.15.
330. *Ibid.*
331. *ṚV*, 2.14.11; *Nirukta.* 3.20,
332. *Nirukta* 3.20.
333. *KA*, 2.24.
334. Sircar, D.C., *Select Inscriptions*, I, p. 82.

ABBREVIATIONS

AĀ	*Aitareya Āraṇyaka*
AB	*Aitareya Brāhamaṇa*

AI	Ancient India
Aṣṭa	*Aṣṭādhyāyī*
AV	*Atharvaveda*
BIC	Birth of Indian Civilization
BU	*Bṛhadāraṇyaka Upaniṣad*
CU	*Chāndogya Upaniṣad*
EMN	Excavation at Maheshwar and Navdatoli
JIH	Journal of Indian History
KA	Kauṭilya Arthasastra
KB	Kauṣītaki *Brāhmaṇa*
KS	Kāṭhaka Saṃhitā
Mbh	Mahābhārata
MIC	Mohenjo-daro and the Indus Civilisation
PGS	Pāraskara Gṛhyasūtra
ṚV	Ṛgveda
ŚB	*Śatapatha Brāhamaṇa*
SBE	Sacred Books of the East
TB	*Taittirīya Brāhmaṇa*
TS	*Taittirīya Saṃhitā*
VI	Vedic index
VS	*Vājasaneyī Saṃhitā*

CHAPTER 38

Artificial Means of Irrigation in Early Medieval Northern India with Special Reference to *Arahaṭṭa*, or *Araghaṭṭa*

Meenashri Yadav

The early medieval period of Indian history is generally taken to extend from about 600 to 1200 AD.[1] This phase of Indian history witnessed considerable improvement in the techniques of agriculture, leading to agricultural expansion and development.[2] In the present context, it is proposed to deal with the artificial means of irrigation with special reference to *arahaṭṭa*, or *araghaṭṭa*, which have been viewed as essential for the protection of crops against the vagaries of monsoon.[3]

The early medieval period (roughly speaking, the post-Gupta period) is found to have witnessed the widening prevalence and proliferation of the artificial sources and means of supplying water to arable lands and crops.[4] However, in addition to these, there were also the natural[5] sources of water, e.g., rivers, natural ponds., etc.

The *Aparājitapṛcchā*[6] underscores the need of expanding the sources of irrigation. The digest-writer Lakṣmīdhara[7] (twelfth century) went beyond the Dharmaśāstric convention in so far as he brought under *dāna* (gifts of spiritual merit) the excavation of wells and tanks, and the construction of reservoirs (*dvārībandhas*) for irrigation by damming mountain springs. This text as well as the *Kāśyapīya-kṛṣi-sūkti*[8] indicates that wells were the most important source of irrigation in Northern India. As many as ten types of *kūpas* (wells), varying in the diameter of mouth from four to thirteen *hastas* (hands) and two types of *kūpikās* (small wells), measuring two or three *hastas* in the diameter of mouth, are enumerated in the *Aparājitapṛcchā*[9] (p. 183). However, this text is of prescriptive nature and the elaborate classification given in it may not have always been followed in actual practice. The *Kāśyapīya-kṛṣi-sūkti* (vs. 145 f.) divides wells into three broad categories—small well, well and large well (*mahākūpa*)—and advises the ruler to get the wells lined with bricks dug in his kingdom (verse 161). But brickless wells may also have existed in a large number.

While prescribing for the construction of water reservoirs, the *Aparājitapṛcchā* (p. 183 ff.) classifies them into three categories—*vāpīs, taḍāgas* (tanks) and *kuṇḍas.*

These are subdivided into four, six and four types, respectively, according to their dimensions and some other features. The water reservoirs, as also the wells, were used for irrigation and other purposes.

We also find references to canals as a source of irrigation. The term *kulyā*, defined as a small artificial river or stream (*alpā kṛtrimā sarit*) in the *Amarakośa* (1.9.34), appears to denote an irrigation canal also. The *Kāśyapīya-kṛṣi-sūkti* (verse 119) mentions *kulyās* ranging in width from four to ten *hastas*. In some other sources,[10] the term *sāraṇī* occurs in the sense of canal.[11]

The literary and epigraphic evidence[12] of the post-Gupta period reveals the widespread practice of excavating water reservoirs and wells by kings and members of their families, ministers, feudatories, and private individuals in Kashmir and in the kingdoms of the Cāhamānas of Naḍḍūla, the Caulukyas of Gujarat, the Paramāras of Malwa, the Kalacuris of Central India, the Candellas of Bundelkhand, the Pālas and Senas of Bengal, etc. The *Rājataraṅgiṇī*[13] of Kalhaṇa informs us that in the ninth century the engineer Suyya constructed a number of dams and irrigation canals, which gave a boost to agricultural production in Kashmir. We also find evidence[14] of some special attention paid to the protection and maintenance of water reservoirs and to stepping up their efficiency by clearing them of mud (*paṅka*).[15]

The literary and epigraphic sources throw some light on the methods of raising water. *Dheṅkā*,[16] or *kūpatulā*,[17] was a contrivance for raising water from a well, in which a horizontal beam, with a bucket hanging from one end thereof, sea-sawed on a vertical post. It appears to have been commonly used in Rajasthan, Malwa, Gujarat, etc.

An inscription[18] (946 AD) from South Rajasthan refers to a field irrigated by a *kosa*. This term appears to have been used in the sense of *camma-kosa*, which occurs in some Jain texts[19] for a leather bucket. The method of raising water from a well in a leather bucket drawn by bullocks walking down an inclined plane had been a common sight in many regions of Northern India till recent times.

A significant water-lifting device during this period was the *arahaṭṭa*, or *araghaṭṭa* (a wheel, or a machine for raising water generally from a well). Though not unambiguously lexiconized in the *Amarakośa*,[20] it finds mention in the lexicons,[21] other literary works, and inscriptions of the post-Gupta period. The framework of *araghaṭṭa*, as indicated by the *Gauḍavaho*[22] of Vākpatirāja (8th century), consisted of wooden spokes and a component part made of iron (*loha*). But, according to the commentary[23] of Prabhānanda Sūri (before the twelfth century) on the *Ṛṣabha-pañcāśikā* of Dhanapāla (tenth century), *araghaṭṭa* was a wooden[24] (*kāṣṭha-ghaṭita*) contrivance, with a string of earthen pots (*ghaṭīmālā mṛṇmayā*), which was fitted to a well (*kūpa*). This appears to reflect the regional variation in respect of the materials used for the construction of *arahaṭṭa*.

There is much controversy about the nature of *araghaṭṭa*, or *arahaṭṭa*. Irfan Habib[25] has equated it with noria (gearless wheel carrying pots and buckets fixed on its rim), which could be operated on an open surface (stream or reservoir) for raising water of which the level was quite high. As pointed out by him, the geared Persian wheel, with a bucket chain, which was moved by bullocks and could raise water from a deep well, was introduced in India in the period of the Turkish conquests and the two subsequent centuries (thirteenth and fourteenth centuries).

L. Gopal[26] has, however, questioned the tenability of the above view. On the basis of a wide range of evidence adduced by him, he has suggested a probable date 'in the fourth century or a little earlier' for the introduction of the *araghaṭṭa* device, which, according to him, was a geared wheel device for raising water. In his opinion, the designation 'Persian wheel' is a misnomer for the device which was used in India even before its earliest known use in Persia.

A piece of fresh evidence from the Jain *Mahāpurāṇa*[27] of Puṣpadanta (tenth century), has a significant bearing on the nature of *arahaṭṭa.* It gives a metaphorical picture of *rahaṭṭa (arahaṭṭa)*, having a long chain of pots and being moved by two bullocks. The chain of pots *(ghaḍimāla)* is conceived here as going round and round, the filled up pots coming up out of the water and emptying themselves. In this context, we do not find any reference to the gearing mechanism. But the conception of the sun and the moon as two bullocks moving the *arahaṭṭa* round and round clearly implies the existence of the same (gearing mechanism) as a component part of the device. Thus, the occurrence of this evidence in a Purāṇa written in Apabhraṃśa points to the fact that *arahaṭṭa*, with a gearing mechanism, was a well-known water-raising device in India in the tenth century,[28] which was before the Turkish conquest of Northern India. This piece of evidence strengthens the probability that the so-called Persian wheel was, in fact, an indigenous device. But there is no clear evidence of the prevalence of the *arahaṭṭa* with a gearing mechanism in India prior to the tenth century AD—the time when the *Mahāpurāṇa* of Puṣpadanta was composed.

However, some sources indicate that *arahaṭṭa* was operated by human energy also,[29] and this practice appears to have been in vogue even before the early medieval period. According to the *Kāśyapīya-kṛṣi-sūkti*,[30] which contains the traditions of the early medieval period, *ghaṭīyantra*[31] (*araghaṭṭa*, or *arahaṭṭa* in this context) of only the inferior type was operated by human energy (*naravāhyakam*): that of the superior type was moved by two bullocks.

The use of *arahaṭṭa* with a bearing mechanism must have tended to increase the efficiency of irrigation, which in turn, was bound to step up intensive cultivation and also to help in the extension of arable land. But the literary and epigraphic evidence suggests that the use of this water-raising device was confined mainly to kings, feudatories and big landholders.[32] The setting up of this device appears to have been a costly affair, not within the reach of the medium cultivators and small holders. However, a clear indication of its growing prevalence among the peasants on the higher echelons, in the second phase of the early medieval period, is found in the *Jyotirvidābharaṇa*,[33] which reflects the tradition of that phase. It prescribes, for the first time in the astrological tradition, auspicious *nakṣatras* (constellations) for the installation of *jalayantra*[34] (explained as *arahaṭṭa* by the commentator Bhāvamuni). However, the choice of auspicious constellations was obviously in respect of the superior[35] as well as the inferior[36] (operated by human energy) type of *arahaṭṭa.*

The vogue of *arahaṭṭa*, operated by human energy, also increased in the early medieval period. This fact is borne out by the occurrence of specific terms—*arahaṭṭiya-nara*,[37] *arahaṭṭīya*,[38] etc., for the skilled workmen who operated the *arahaṭṭa*, in the contemporary literature. The *araghaṭṭas*, or *arahaṭṭas* set up on the banks of rivers[39] to raise water for the purpose of irrigation appear to have been operated by such workmen.

The growing popularity and use of *arahaṭṭa* in the second phase of the early medieval period is further indicated by its depiction in art. A panel[40] (tenth century AD) from Maṇḍor near Jodhpur (Rajasthan) depicts an *arahaṭṭa*—wheel with a chain of terracotta buckets (Fig. 1). According to Irfan Habib,[41] the device shown is only a noria-wheel carrying pots, without any gearing mechanism, and worked manually. But, in the opinion of M.S. Randhawa,[42] this sculpture shows an *arahaṭṭa* in profile and, as such, the gearing mechanism would be on the other side. Another panel[43] (eleventh century) depicting the *arahaṭṭa*—wheel with a chain of terracotta buckets (Fig. 2)—has been discovered in the Jogeśvara temple situated near Sādarī town of District Pali (Rajasthan).

Fig. 1: Panel depicting *arahaṭṭa* from *Maṇḍor* (Rajasthan)

Fig. 2: Panel depicting *Arahaṭṭa* in Jogeśvara temple (Sādarī, District Pali, Rajasthan)

Agriculture continued to be viewed as dependent on rains to a considerable extent.[44] A verse in the *Subhāṣita-ratna-koṣa*[45] of Vidyākara (twelfth century) shows how the people, anxious for crops, used to wait for the first rainfall of the rainy season for the purpose of starting their agricultural operations. However, for protection of crops against *anāvṛṣṭi* (drought) and *svalpavṛṣṭi* (scanty rainfall), the importance of the protective role of artificial means of irrigation was increasingly realized[46] by the thinking people and the rulers of the age.

REFERENCES

1. Meenashri Yadav, 1992, *Agriculture and Peasants in Northern India from* c. *600 to 1200* AD (in Hindi), thesis approved for the D. Phil. Degree of the University of Allahabad, Introduction, p. 2; B.N.S. Yadava in *Proceedings of the Indian History Congress,* 53rd Session, 1992–93, p. 2.
2. Meenashri Yadav, *op. cit.,* pp. 272 ff.
3. *Aparājitapṛcchā* of Bhuvanadeva (twelfth century), Gaekwad Oriental Series, No. CXIV, Baroda, 1950, p. 188, verses 32 ff. The text was composed in Gujarat.
4. Cf. Meenashri Yadav, *op. cit.,* pp. 59 ff.
5. Kṣīrasvāmin (eleventh century) on *Amarakośa,* ed. H.D. Sharma and N.G. Sardesai, Pune, 1941, 1.9.27; *Maṅkhakośa* (composed in Kashmir in the twelfth century), ed. T. Zacharia, Varanasi, 1972, verse 367 (*devakhātaṃ hrado....*).
6. *Aparājitapṛcchā,* p. 188.
7. *Kṛtyakalpataru–Dānakāṇḍa,* ed. K.V.R. Aiyangar, Baroda, 1941, Int., p. 114.
8. *Kāśyapīya-kṛṣi-sūkti,* ed. G. Wojtilla, *Acta Orientalia,* XXXIII (2), Budapest, 1979. This text acquired its present form in the medieval period, which followed the early medieval period. However, it contains the essential agricultural tradition of early medieval period; Meenashri Yadav, *op. cit.,* Introduction. Cf. G. Wojtilla, *Acta Orientalia,* Vol. XXXIX (1), 1985, p. 85 fn. 1; R.S. Sharma, *Urban Decay in India,* New Delhi, 1987, p. 172 fn. 31. The core of the *Kāśyapīya-kṛṣi-sūkti* has been placed in the eighth-ninth century by G. Wojtilla (*loc. cit.*).
9. B.N.S. Yadava, 1973, *Society and Culture in Northern India in the Twelfth Century,* Allahabad, p. 258.
10. *Viśvaprakāśa-kośa,* ed. Śrī Śīlaskandha Sthavira, Varanasi, 1983, p. 53, verse. 78; *Aparājitapṛcchā,* p. 188, v. 36.
11. However, the term *sāraṇī* was also used in the sense of a channel for irrigation; Meenashri Yadav, *op. cit.,* p. 71.
12. L. Gopal, 'Technique of Agriculture in Early Medieval India', *University of Allahabad Studies* (Ancient History Section), 1963–64, pp. 2 ff.; Meenashri Yadav, *op. cit.*, pp. 100 ff.
13. *Rājataraṅgiṇī,* ed. & tr. M.A. Stein, V. 109.
14. Meenashri Yadav, *op. cit.,* pp. 100f.
15. E.g. Gwalior tank inscription (1194 AD), *Epigraphia Indica (E.I.),* XXXVIII, p. 134.

16 & 17. *Deśīnāmamālā* of Hemacandra (twelfth century), ed. R. Pischel, 1938, second edn., Vizianagaram, 4.17, Glossary (p. 42). As many as six more terms are given in the lexicon for this contrivance; L. Gopal, *op. cit.*, p. 9. The words *ḍheku, ḍhikuyau* and *ḍhiku* occur in some inscriptions for the same water-lifting device; Meenashri Yadav, *op. cit.*, pp. 77 f.; B.D. Chattopadhyaya, 'Villages, Wells and Rulers in South-

eastern Marwar', paper presented at the Rajasthan Studies Conference (December, 1987), Jaipur, pp. 7f.

18. *E.I.*, XIV, p. 102.
19. *Pāia-sadda-mahaṇṇao* by H.D.T. Sheth, Second Ed., Varanasi, 1963, p. 319; cited in Meenashri Yadav, *op. cit.*, p. 79 fn. 2.
20. Only the term *ghaṭīyantra* occurs in the *Amarakośa* (2.10.28).
21. *Halāyudhakośa*, ed. J.S. Joshi, Lucknow, 1967, vs. 684f.
22. Ed. & tr. N.G. Suru, Ahmedabad, 1975, verse 685.
23. On *Ṛṣabha-pañcāśikā*, ed. H.R. Kapadia, Surat, 1933, p. 93, verse 30.
24. The use of *arahaṭṭas* (commonly known as Persian wheels) turned by 'rough wooden' gearings worked by cattle has been noticed in some regions of Uttar Pradesh even in the early decades of the 20th century; W.H. Moreland, *The Agriculture of the United Provinces*, Allahabad, 1912, p. 86.
25. Presidential Address, Medieval India Section, *Proceedings of the Indian History Congress*, 31st Session, Varanasi, 1969, pp. 149–53. According to him, the earliest description of the Persian wheel in India is the classic one by Babur in *Babur-nama* (tr. A.S. Beveridge, Vol. II, London, 1921, pp. 486 f.).
26. *Aspects of History of Agriculture in Ancient India*, Varanasi, 1980, p. 165.
27. *ai somasahāba mahāraudda*
   sañcalliya cala sasi ravi baladda.
   kiha vañcai kālarahaṭṭa cāru
   ghaḍimālai laṅghiṃ āuṇīru.

   Mahāpurāṇa of Puṣpadanta, ed. P.L. Vaidya, tr. D.K. Jain, Bhāratīya Jñānapīṭha Prakāśana, Parts 1 and 2, 1979, 27.1.1-2; cited in Meenashri Yadav, *op. cit.*, p. 85. Puṣpadanta flourished in the kingdom of the Rāṣṭrakūṭas, and he completed this work in 965 AD.
28. The evidence of the *Upamiti-bhavaprapañcā-kathā* (AD 906), which alludes to sixteen bulls moving an *arahaṭṭa*, also suggests the same; *ibid.*, p. 985, cited in L. Gopal, *op. cit.*, p. 167.
29. Meenashri Yadav, *op. cit.*, pp. 87 ff.
30. Verses 167 f.
31. For the connotations of *ghaṭīyantra* see L. Gopal, *op. cit.*, pp. 134, 142 ff.
32. cf. L. Gopal 'Technique of Agriculture...', *University of Allahabad Studies (Ancient History Section)*, 1963–64, p. 10.
33. Ed. & tr. Ramachandra Pandeya, Delhi, 1988, p., 576, vs. 31, 32; p. 577, verse 33.

   The text is ascribed to Kālidāsa, but the internal evidence clearly reveals that it was composed in Śaka Saṃvat 1164/AD 1242. The necessity of prescribing auspicious *nakṣatras* for the installation of *arahaṭṭa* must have arisen before this date in the wake of the growing prevalence of the device.
34. The term *jalayantra* occurs in the *Bṛhatsaṃhitā* (ed. A.V. Tripathi, Varanasi, 1968, Chapter 2, page 39) of Varāhamihira (fifth-sixth century AD) as well, which is a well-known earlier text on astrology. But there it means a water-clock used for measuring time, and not *arahaṭṭa* (A.M. Shastri, *India as Seen in the Bṛhatsaṃhitā of Varāhamihira*, Delhi, 1969, p. 256).

35. Supra.
36. Supra
37. *Ākhyānaka-maṇi-kośa* (twelfth century), Prakrit Text Society, Varanasi, 1962, p. 146; cited in B.N.S. Yadava, *Society and Culture...*, pp. 259, 305.
38. *Kumārapāla-pratibodha* of Somaprabha, cited in *Pāia-sadda-mahaṇṇavo* of Hargovind Das T. Sheth, Second Edition, Prakrit Text Society, Varanasi 1963, p. 71.
39. King Lalitāditya (eighth century) of Kashmir is said to have set up a number of *arahaṭṭas* at Cakradhara (modern Tasakdar) for raising the water of the river Vitastā in order to supply it to various villages; *Rājataraṅgiṇī* of Kalhaṇa, ed. & tr. M.A. Stein, 4.191. However, wells were also fitted with *arahaṭṭa* device. A clear evidence for this is found in the maxim *kūpa-yantra-ghaṭikā-nyāya* occurring in the Ṣanskrit drama *Mṛcchakaṭika* (X. 60) of Śūdraka (Gupta period). Skilled (*dakṣa*) workmen, moving fast (*vegena*) the mechanical device (*arahaṭṭa*) fitted to a well (*kupāntarvarti-yantra*) are mentioned in the commentary of Pūrṇakalaśa Gaṇi on the *Kumārapāla-carita* of Hemacandra (twelfth century); *Kumārapāla-carita* (with the commentary of Pūrṇakalaśa Gaṇi), ed. S.P. Pandit, Bhandarkar Prācya Vidyā Saṃśodhana Mandira, p. 124, commentary on verse 29. This shows that *araghaṭṭas* fitted to wells were not only moved by two bullocks but also operated by skilled workmen.
40. My thanks are due to Sri Jai Ram Pandey, Regional Forest Officer, Jhalawar, Rajasthan, for making available to me the photograph of the panel from Sardar Museum, Jodhpur.
41. Idem, 'Jatts of Punjab and Sind,' *Punjab Past and Present,* Essays in Honour of Dr. Ganda Singh, pp. 97, 98.
42. Idem, *A History of Agriculture in India,* Vol. 1, Indian Council of Agricultural Research, New Delhi, 1980, p. 478.
43. I am thankful to Sri Jai Ram Pandey (vide footnote 40 above) for the photograph of this panel in the temple.
44. *Bṛhatsaṃhitā* of Varāhamihira (with the commentary of Bhaṭṭotpala), Part 1, ed. A.V. Tripathi, Varanasi, 1968, 21.1; commentary of Bhaṭṭotpala on *ibid.* Varāhamihira is generally assigned to the fifth-sixth century AD; A.M. Shastri, *India as Seen in the Bṛhatsaṃhitā* of *Varāhamihira,* p. 16, However, there are some interpolations in his *Bṛhatsaṃhitā*; cf. B.N.S. Yadava, in the *Proceedings of the Indian History Congress,* 1980, p. 11. Bhaṭṭotpala, the commentator of Varāhamihira's works, flourished in Kashmir in the 10th century.
45. Ed. D.D. Kosambi and V.V. Gokhale, 1957, Harvard Oriental Series, Vol. 42, p. 46, verse 29.
46. E.g., *Aparājitapṛcchā,* (twelfth century), p. 188, verses 32 ff.

CHAPTER 39

Role of Forced Labour in Ancient Indian Agriculture

G.K. Rai

INTRODUCTION

In the following pages, we wish to examine the role of forced labour in ancient Indian agriculture. In reviewing the applicability of Marxist models to early India, D.N. Jha observes: '...as compared to Europe, forced labour was less common. In other words, the incidence of labour rent, a prominent feature of European feudal economy, *was much less in India'* a fact which can be appreciated against the background of the varying agrarian organizations prevailing in the two regions[1]. Similar is the observation of R.S. Sharma: '...the practice was hardly different from what we find in Europe....But in other respects, the position was different. Probably those who were subjected to forced labour were not compelled to work on the fields of the grantees to the same extent as peasants in medieval European manors. Furthermore, the area of land under the direct cultivation of the grantee was limited... consequently there was little occasion for such work, and its scope was probably restricted'.[2]

Such an answer, in reality, is no answer at all. It is not enough to identify some variable of the feudal syndrome. What is more important is to determine the limits of permissible variability within which the feudal essential resides. It is methodologically improper to argue that the practice of forced labour in ancient India was similar to European practice but *much less in degree.* 'How much less is the relevant question.' And the answer to this will determine the nature of the practice. We cannot go on diluting the nature of the practice *ad infinitum;* there comes a stage when further dilution results in the change in the very nature of the practice.

The question of the role of forced labour in ancient India has recently come under scrutiny. Mukhia feels that "the nature of forced labour in India—of which there is considerable evidence throughout her history—is in its very essence different from the one in Europe, for in India, it was very rarely used for purposes of production... *Forced labour in India remained, by and large, an incidental manifestation of the ruling class' political and administrative power than a part of the process of production."*[3] On the other hand, B.N.S. Yadava believes that the practice was *not only essential but more or less a specific element of the feudal complex.*

It has been supposed by R.S. Sharma that the passage of *Kāmasūtra* gives an idea of the nature of the work included under *viṣṭi*[4] and that 'the inclusion of working in the fields of the headman in the list of labour services catalogued here marks the beginning of an important feudal practice.'[5] The passage reads: 'Sexual connection with these women may be had on those occasions, when they are employed for *viṣṭi* (unpaid work), such as filling up of the granaries, taking things in and out of the house, cleaning the houses, working in the fields, spinning yarn of cotton, wool, flax or hemp, and the purchase, sale and exchange of various articles.'[6] In fact, the passage should be correctly translated as: 'Sexual connection with these women may be had on those occasions when they are employed in *visti* and other works such as filling up of the granaries...etc.'[7] The expression *teṣu teṣu ca* leaves no doubt that *viṣṭikarma* was viewed as different from other works enumerated in the same passage. This is further confirmed by the *Jayamaṅgala-tīkā* of the *kāmasūtra,* a work of thirteenth century AD.[8] The *tika* comments on *viṣṭi* separately. It shows that the commentator does not take for granted the fact that all the works enumerated above constitute *viṣṭi.* He explains *viṣṭikarma* as '*bhaktamatrena*[9] *yani pesanakuttanarandha-nadikaryani tani viṣṭikarmāni.*' The *Kāmasūtra,* therefore, does not refer to working in the fields of the headman, i.e. unpaid labour was not utilized in agriculture. Still, the passage is significant in at least one respect; it shows that by sixth century AD,[10] the village headmen and other petty officials could exact forced labour for their personal use.

B.N.S. Yadava has made an attempt to bring into relief the role of forced labour in agriculture.[11] He suggests that even though there is no direct reference to the use of forced labour in agriculture except in the *Bhāgavata Purāṇa*[12] (which cannot be later than AD 800), there is some evidence to believe that forced labour was used in agriculture as early as fourth-fifth century AD.[13]

First, we will examine the earliest direct reference to the use of forced labour in agriculture. This conclusion is based on 5.9.9. and 5–9.11 of the *Bhāgavata Purāṇa.* The passages under review read:

> *karmāni ca sa kāryamānah parecchayā karoti vistito vetanato vā yācnyayā yādṛcchayā vopasāditamalpaṃ bahu mṛṣṭam kadannam vābhyavaharati paraṃ nendriyaprītinimittaṃ* (5.9.9)[14] *and yadā tu parata āhāraṁ karamavetanata īhamānaḥ svabhrātṛbhirapi kedārakaramaṇi nirupitastadapi karoti kintu na samaṃ viṣamaṃ nyūnamadhikamiti veda kaṇapinyakaphalīkaranakulmāsasthāli puriṣādinyapyamṛta-vadabhyavaharati.* (5.9.11).[15]

The passages read:[16] 'Bharata, being made to engage in work by the instance of others, by *viṣṭi* (forced labour) or by wages, he ate whatever was obtained by way of alms or by chance, much or little, sweet or foul, but never for satisfaction of senses (5.9.9.)... when he wished to have food (for his livelihood) in return (as wages) for work for others, he was asked even by his brothers to work in the fields and he did it, but he knew not whether it was level or uneven, whether he did more or less (in proportion to the wages). He swallowed, as if it were nectar, whether it was broken rice or other grain: oil-cake, husk, worm-eaten grains and charred rice sticking to the bottom of pot' (5.9.11).

Both of the above passages have been incorrectly linked up to reach the conclusion that *viṣṭi* was used in agriculture,[17] whereas a perusal of the above passages shows that use of forced labour in agriculture has not been mentioned at all in the *Bhāgavata Purāṇa.*

Yadava further argues that 'substitution of *vṣṭikarts* for *bhṛtijīvins* leaves little room for doubt that forced labour in agriculture is also implied in *viṣṭi,* especially when we view it in the light of the evidence of other texts noticed before.' In support, he cites passages from the *Vṛddhayavnajātaka (VYJ)* and the *Jataka.* These passages are crucial for our discussion; therefore, a detailed examination of these passages is in order.

Yadava bases his conclusion on *VYJ,* XXXIII.3 but for discussion, we cite *VYJ* from XXXIII.1 to XXXIII.3:

karmāśṛtastīkṣṇakaraḥ svatuṅge
karoti karma vratakrtyameva
Tadaṃśakastho vadhabandhajaṃ ca
ṣaḍvargaśuddhaḥ sakalaṁ kalājaṃ XXXIII.1
nicaśṛtastikṣṇakaro'ambarastho
nṛṇām vidhatte bahudāsakarma
tasyaiva bhāge vaṇijodbhavaṃ ca
vargottamastho vaṇijaprasūtum XXXIII.2
mitrāśrayastho dinakṛt khasaṃsthaḥ
karoti pumsāṃ kṛṣijaṃ ca karma
mitrasya bhāge kunṛpasya sevajaṃ
pāpasya varge vadhabandhajaṃ sadā, XXXIII.3

Translated,[18] they read:

If the Sun is in the tenth house and in exaltation then the subject does ritualistic work.
If it is in that *amsa* then it leads to chastisement and imprisonment
If the Sun is pure in *śaḍvarga* then it leads to all kinds of crafts, XXXIII.
If the Sun is in tenth and in debilitation, then it causes multifarious servile work.
If it is in its own portion, then it leads to work arising out of trade.
If the Sun is *Vṛgottam,* in the same *rāśi* and in the same *aṃśa,* then it leads to commercial work, XXXIII.2.
If the Sun is in the tenth house and in a friendly sign, it leads to agricultural work.
If it is in the portion of friendly planet, then it leads to the service of an evil ruler.
And if it is in the portion of malefic planet, it leads to constant chastisement and imprisonment, XXXIII.3.

From this verse (XXXIII.3), can it be concluded that forced labour was used in agriculture? Such an inference is possible if it is assumed that the result of a planetary position in the natal chart in a friendly sign governs and determines the result, if in this sign, then planet is in the portion of a friendly planet or in the portion of the malefic planet, i.e., to say that results in the portion of a friendly or malefic planet will be different aspects of the same result. On the basis of this logic, Yadava argues that *kuṇṛpasya sevajaṃ (sevanaṃ)* does not simply mean service of an evil ruler but 'agriculture service to an evil ruler'. Similarly, *vadhabandhajam (vadhabandhanam)* means 'agriculture work caused by direct coercion-corporal punishment and bondage, or dependence'.[19]

But such a logic has neither been stated nor followed in the text (see, e.g., xxxv.3).[20] Moreover, even if we accept the logic, *vadhabandhajam* cannot imply forced labour as is clear from the following lines of the *Mahābhārata*:

> *Mānuṣā mānūṣāneva dāsabhavena bhuñjate*
> *Vadhabandhanirodhena kārayanti divaniśam*
> *Ātmanaścapi jānāti yadduhkhaṁ vadhabandhane.*[21]

The other passage on which Yadava bases his conclusion regarding the use of forced labour in agriculture in the sixth century AD is from a *jātaka* tale. The account runs as under:

> For in days to come, the world shall decay; the kingdom shall grow weak, its kings shall grow poor and niggardly; the foremost among them shall have no more than one thousand pieces of money in his treasury. Then shall these kings in their need set the whole of the country-folk to work for them; for the kings' sake shall the toiling folk, leaving their own work, sow grain and pulse, and keep watch and reap and thresh and garner; for the kings' sake shall they plant sugarcanes, make and drive sugar-mills, and boil down the molasses; for the kings' sake shall they lay out flower gardens and orchards, and gather in the fruits. And as they gather in all the diverse kinds of produce, they shall fill the royal garners to overflowing, not giving so much as a glance at their own empty barns at home.
>
> Kings shall...be fierce and cruel, amassing wealth by crushing their subjects like sugarcanes in a mill and by taxing them even to the uttermost farthing. Unable to pay the oppressive tax, the people shall fly from village and town and the like.[22]

The period for which the society the *jātakas* serve as documents, is highly problematic.[23] Even if we agree for a moment that the above account pertains to sixth century AD, the question still remains as to whether the conditions described pertain to Ceylon or India. The nature of the evidence is so general, the dates so uncertain, that unless more evidence is forthcoming, one feels great hesitation in concluding on the basis of this evidence alone that forced labour was used in agriculture. Yadava himself admits that the account is, of course, exaggerated.[24]

Some information regarding the practice of forced labour can be gleaned from the works of *Varāhamihira*, who is generally assigned to the last quarter of the fifth century AD and the first half of sixth century AD.[25]

In the *Bṛhat Saṃhitā, viṣṭikaraṇa* has been deemed so inauspicious a time as to be avoided for the start of such operations as ploughing, sowing, trading, etc., but has been regarded as suitable for attaining success in activities characterized by violence and outrage, such as poisoning, burning, killing of enemies, etc.[26]

After reviewing the evidence, we return to the debate again. Among the ideological, technical, organizational, and social features that appear in any given society, some are essential for the society's proper functioning, while some are not. Among the essential features, some are specific, some are not. A third group is neither essential nor specific.

According to Karl A. Wittfogel, 'corvee labour is an essential element of hydraulic and feudal societies. However, it is essential to more than one type of society and specific to none. The corvee is not confined to hydraulic societies; forced labour of non-slaves also appears in other social types. It is specific in that in agro-hydraulic civilizations, different from feudalism, corvee labour is imposed on the mass of the population by the state.'[27]

In case of Europe, all the estate descriptions emphasize the importance of labour obligations from both free and servile peasant holdings, so that although there were rents both in kind and in money, labour rent was apparently predominant.[28]

Witold Kula's hypothesis suggests the following limiting case for the feudal system: 'a small peasant plot, operating according to a completely natural economy, capable at most of simple reproduction, and obliged to provide the lord only with the corvee.'[29]

We do not find any parallel in case of India. In our survey, we have not come across any evidence which may indicate the use of forced labour in productive activities. In Nepal, where the practice continued to exist till modern days and where it flourished in its most evolved and widespread form, forced labour was not utilized in agriculture. In this connection, Fukazawa's survey—even though it does not fall in the period of our study—is of great interest. As an indicater, he carried out an indepth survey of the position of forced labour as it was practised in the directly administered regions *(svarājya)* of the eighteenth-century Maratha confederacy on the basis of about fifty contemporary records.[30] Out of thirty records examined by him, he found: (a) Seven records pertaining to corvee exacted for construction or repair of forts *(killā),* police stations, *(thānā),* residences of local bureaucrats, as well as the dams *(dhāam)* for irrigation; (b) Six records regarding porterage *(hājir begār)* of grains, timbers and other goods of government; (c) Five records pertaining to corvee in cutting fodder *(gavat)* at government meadows *(sarkārcha kuraṇ)* scattered in various places; (d) Five records concerned with miscellaneous labour *(rābṇūk* or *rābate)* at the local as well as the central government offices; (e) Two records related to miscellaneous labour and saddlery at government stables *(hujūr pāgā);* (f) One record on free service of watchmanship *(baiṭhak veṭhbegār or choki veṭhbegār)* at market places *(bājār);* (g) Four records on forced labour in *inām* villages; three of them on construction of houses and one on porterage for the sake of *ināmdar.* 'Above itemization clearly shows that the government imposed various kinds of forced labour upon the people, both in the centre and in the countryside. It also may suggest that the corvee was hardly used for the cultivation of state lands as well as the directly managed lands or demesne of *ināmdar* in his *inām,* village. Such lands, were being usually cultivated either by sharecroppers *(vātgkari, ardhelī,* etc.) or on a fixed rent.'[31]

Even though the relation between the emergence of forced labour and decline of slavery has not been stated in so many words, it has been asserted that the period which saw the emergence of forced labour was also marked by the beginning of the loosening of the bonds of servitude. Proceeding further, we are told that in the post-Gupta period, in which the practice of forced labour was increasing, slavery declined.

'Decline', as Finley suggests, is a dangerous word. Slavery was an institution performing various functions. So long as that labour is needed, slavery cannot decline, it has to be replaced. If a change-over was presumably taking place in the status and organization of labour, the question arises: where and and in which section or sectors of the labour force?

What to say of the relationship between the emergence of forced labour and the decline of slavery, when even the fact of decline of slavery in post-Gupta period is disputed.

The Marxist trend in historiography envisages the view of slavery as giving way to feudalism in the Indian context also. The picture of social and economic change in this country does not really appear to have been so simple. The mention of *kṛṣikarma* or *kṣetrakarma* as a duty of female slaves in the *Lekhapaddhati,* and the evidence in the *Likhanāvali,* of the extent of *sūdra* slavery as well as the employment of slave-labour in agricultural production, of these suggest that the situation here was extremely complex, and even complicated.[32]

Another European parallel of the rise of the money economy leading to the disappearance of forced labour is also belied by the facts. In sum, **we find the practice of forced labour neither essential nor specific to the socio-economic formations of the early medieval India.**

NOTES AND REFERENCES

1. Jha, D.N., *Presidential Address,* History Section, I.H.C. XI Session, Waltair, 1979, p. 26.
2. Sharma, R.S., *Indian Feudalism,* I.H.C. p. 53.
3. Mukhia, Harbans, *Presidential Address,* I.H.C., 249.
4. Gopal, L., *The Economic Life of Northern India,* pp. 27-28.
5. Sharma, R.S., *op.cit.,* p. 52.
6. *Kāmasūtra,* V. 5-6.
7. *Ibid., tābhissaha viṣṭi karmasu koṣṭhāgārapraveśe...tesu teśu ca karmasu samprayoga.*
8. Winternitz, *History of Indian Literature,* Vol. III, Part II, p. 624.
9. Cf. *Agni Purāṇa,* 223–33: *bhuktamātreṇa ye cānye svaśarīropajīvinaḥ.*
10. *Ibid.,* p. 623: Vātsyāyana wrote the *Kāmasūtra* in any case before 7th century AD.
11. Yadava, B.N.S., *Presidential Address,* I.H.C., 1980, pp. 16–20.
12. *Ibid.,* p. 20 fn. 3.
13. *Ibid.,* p. 16.
14. *Bhāgavata Purāṇa* (Geeta Press edition, Gorakhpur) 5.9.9.

15. *Ibid.*, 5.9.11.
16. English rendering of 5.9.9. and 5.9.11 is based on the translation of S. Subbarau and J.L. Shashtri.
17. The present author has also contributed greatly to this confusion, Rai, G.K., *Forced Labour in Ancient and Early Medieval India,* IHR, Vol. III No. I, p. 33.
18. I owe this translation to Professor G.C. Pande.
19. Yadava, B.N.S., Presidential Address Indian History Congress, 1980, p. 16.
20. In Vṛdha-Yavana Jātaka, XXXV.3, three results for corresponding configuration are expenditure incurred due to friend's work, on bad women and due to untouchables. All these results are unrelated.
21. *Mbh,* XII.261.38 f.
22. *Jātaka,* Vol. I, tr. Robert, Chalmers, p. 190.
23. Though they cannot serve as documents for the social conditions at the time of Buddha, but at the most, for the period of third century, BC and for the greater part, especially in their prose, only for the fifth or sixth century AD, yet so much has remained unchanged in India throughout countries, that the picture of civilization in the *Jatakas* may nevertheless be regarded as very 'ancient', Winternitz, Maurice, *History of Indian Literature,* p. 156.
24. Yadava, B.N.S., *Kali Age, IHR,* Vol. V, Nos. 1–2, p. 55.
25. Shastri, A.M., *India as Seen in the Bṛhat Saṃhitā of Varāhamihira,* p. 16.
26. Varāhamihira, *Bṛhat Saṃhitā* (with the commentary of Bhottotpāla), 99.4.
27. Wittfogel, Karl A., *Oriental Despotism,* p. 415
28. Hilton, Rodney, Introduction to *The Transition from Feudalism to Capitalism,* p. 16.
29. Kula, Witold, *An Economic Theory of the Feudal System,* p. 17.
30. Fukazawa, H., 'A Note on the Corvee System (Veṭhbegār) in the Eighteenth Century Maratha Kingdom', in *Science and Human Progress;* Essays in honour of Late Prof. D.D. Kosambi, pp. 117–130.
31. Fukazawa, H. *op.cit.,* pp. 120–121.
32. Negi, J.S., 'Some Light on the Institution of Slavery from the Likhanāvali of Vidyapati' in *K.C. Chattopadhyaya Memorial Volume,* Allahabad, pp. 93–96.

CHAPTER 40

Peasant Differentiation and Categorization in Northern India in Early Medieval Times (Sixth–Seventh to Twelfth Century AD)

Meenashri Yadav

During the earlier phase of the early medieval period extending for about three hundred years, the decline of trade, towns and money economy led to the insulation of the local agrarian economy tending towards self-sufficiency.[1] The entrenchment of local villageholders on account of religious and secular land grants, the weakening of central authority paving the way for the rise of local chiefs, the transformation of hereditary village headmen into village lords—all these led to peasant subjection and extensive emergence of dependent peasantry.[2] Thus, we come across a socio-economic milieu characterized by the *sāmanta* system, or Indian feudalism.[3]

The expansion of agriculture and the ranks of peasants was furthered[4] in the early medieval period by the increasing spread of the use of iron implements; the improvements in the artificial means of irrigation and other techniques of agriculture; the transformation of Śūdra servants and slaves, to a marked extent, into peasants (mostly dependent ones); and the absorption of some aboriginal tribal groups[5] in society, that began to live by agriculture.

A considerable degree of differentiation existed among peasants in Northern India in early medieval times. We find evidence of more than one criterion of such differentiation. The mention of two methods of organizing productive activity in agriculture in the commentary (c. 900 AD) of Medhātithi on the *Manusmṛti*[6] (III. 155) reflects the stratification of peasantry into two broad segments. The term *kṛṣijīvī* has been defined by Medhātithi as one who lives by agriculture through self-employment *(svayaṃ kṛtayā kṛṣyā)*[7] and also one who obtains his livelihood from agriculture without getting engaged in actual agricultural work (*asvayaṃkṛtayā*...). It is clear that the former was small-holding peasant who cultivated his fields through self-employment or family labour and the latter was a big landholder who engaged labourers for agricultural work on his fields and/ or leased his unfarmed fields to sharecroppers. However, these two methods of organizing productive activity—suggesting two broad categories of peasants, cannot be said to have been mutually exclusive. Many peasants belonging to the middle level of peasantry may have resorted, in different degrees, to both these methods. Then again, many Brāhmaṇa and Kṣatriya peasants of even the lower level, for whom agriculture as a regular occupation was interdicted by the earlier Dharmaśāstra tradition,[8] may have desisted from actual agricultural work.

The evidence of the *Kathākośa-prakaraṇa*[9] of Jineśvara Sūri (eleventh century) indicates that the status of the peasants who did manual labour in agriculture, and, thus, lived by agriculture through self-employment, was viewed as lower than that of the peasants who were big landholders and engaged labourers for their agricultural works. But generally, only the village headmen/lords and a few men who were members of village councils may have been big or rich landholders in villages.[10] In an *ekabhoga grāma*[11] (village enjoyed by an individual), only the village lord who may have carried on direct cultivation by engaging workmen and also leased his unfarmed land to sharecroppers was a big landholder: the other residents of the village were dependent peasants, agricultural labourers and servants.

There is a noticeable evidence of two broad divisions of peasants on the basis of land rights and land tenure. The commentary of Haradatta (twelfth century) on the *Āpastamba Dharmasūtra*[12] (2.11.2) refers to the *svatantra-kṣetravān*, who was regarded as free by virtue of his land rights and could punish his ploughmen if they left the agricultural work. The existence of *asvatantra kṣetravān* is also implied here, who had no right to punish his ploughmen for such a default: only the ruler could punish them. Obviously, the *svatantra-kṣetravāns* were *kṣetra-svāmins* (owners of plots), and the *asvatantra* (unfree) ones were dependent peasants (tenant-peasants, sharecroppers, etc.). The term *karada-kuṭumbī* (tax-payer peasant), which occurs in the spurious Gaya Plate[13] (probably the beginning of the eighth century), appears to refer to the free owners of plots (*svatantra-kṣetravāns)* mentioned above.

The number of ploughs owned by a peasant was a rank symbol mainly evidencing his economic status based on the area of land under his cultivation. According to the *Kṛṣi-Parāśara*[14], there is Lakṣmī (goddess of wealth) in ten ploughs, wealth (*dhana*) in five, and food (*bhaktam*) in three. It is further stated that with two ploughs there is merely the maintenance (*poṣaṇa*) of one's own self and that a peasant possessing only one plough remains so poor as to get into debt (*ṛṇam)*.

Some other sources indicate that big landlords resorting to direct cultivation, on a large scale, possessed a large number of ploughs. In a story of the *Daśakumāracarita*[15] of Daṇḍin (AD 550–650), we find reference to a *gṛhapati janapada-mahattara* in the region of Mithilā and Puṇḍra-deśa, who has been designated as '*śatahali*'[16] (one possessing 100 ploughs). At another place in the same text, a *grāmaṇī* (*grāmapati*—village headman/ lord) has been called *anantasīra*[17] (one possessing innumerable ploughs). These *grāmaṇīs* may be compared with the village headmen or *pañcas* of the Mughal period, who were called *muqaddams.*[18] They wielded considerable power over their villages and were big landholders who engaged labourers for their agricultural work.

A rough and ready indicator of the upper, middle and lower ranks of peasants was the type of storage in which the grain produced in agriculture was kept by the peasants. In the *Brahmapurāṇa*,[19] the peasants of the upper, middle and lower ranks are called *gartadhānyadhanāḥ* (those who stored their grain in large underground chambers), *kusūla-dhaninaḥ* (those who stored their grain in granaries), and *ghaṭakṣipta-dhanāḥ* (the peasants who kept their grain in earthen pots), respectively. The peasants belonging to the last category were obviously poor peasants who constituted a noticeable segment of peasantry.

In the *varṇa-jāti*—divided hierarchical society the ritual status was obviously an important basis of the gradation of peasants. According to the ancient Dharmaśāstra ideology, agriculture was the proper occupation of only the Vaiśyas (*Manusmṛti*, 1.90; 10.79), and that it could be only *āpaddharma*[20] (occupation to be temporarily resorted to in distress) for the first two *varṇas*. But it has been sanctioned for all the *varṇas* in the *Parāśara-smṛti*[21] (AD 600–900).[22] This may be taken to reflect, *inter alia*, the comparatively increasing preponderance[23] of agrarian concern and expansion of the ranks of peasants during the early medieval period.

The literary and epigraphic evidence reveals that the Brāhmaṇa peasants, having the same ritual status, belonged to different socio-economic strata from the material point of view.[24] The same was the case with the peasants of Kṣatriya[25] and Vaiśya *varṇas*. The Śūdra peasants mostly belonged to the strata of dependent peasantry[26]: a large number of Śūdras were agricultural workers[27] also.

As noticed above, peasantry was differentiated internally in terms of certain significant variables. Thus, all peasants cannot be aggregated together under a homogenous, monolithic category. However, in some contexts, the commonalty of peasants—mostly including those of lower strata, who lived under similar agrarian economic conditions of existence, were seen as being a category within the wider hierarchical social system.

In his *Bṛhatsaṃhitā*[28], Varāhamihira has mentioned the following *varṇas*, occupational classes and social groups in course of associating different constellations (*nakṣatras*) with each of them: Brāhmaṇas, Kṣatriyas, *kṛṣīvalas* (peasants), *vaṇiks* (merchants), people belonging to Ugra *jāti*, *sevājana* (menials), and Caṇḍālas. The trend of the toiling peasantry emerging as a socio-economic category cutting across the *varṇa* structure is, to some extent, reflected in the muddled account of the distinctive sections of society in the *Nītivākyāmṛta*[29] of Somadeva (tenth century). We find here the separate mention of *kṛṣīvalas* along with Brāhmaṇas, Kṣatriyas, Kirāṭas, *vaṇiks*, and Vaiśyas.

In the twelfth century, Bhuvanadeva prescribed in his *Aparājita-pṛcchā*[30] the *vedis*[31] of different lengths (in *hastas*–cubits) for the main sections of the hierarchical society as follows:

(1)	Brāhmaṇas	7 *hastas*
(2)	Kṣatriyas	6 *hastas*
(3)	Vaiśyas	5 *hastas*
(4)	Śūdras	4 *hastas*
(5)	*Karṣakas*[32]	3 *hastas*
(6)	*Prakṛtis* (artisans and craftsmen)	2 *hastas* or 1 *hasta*

In this hierarchy, the *karṣakas* (lower peasantry including the ploughmen) have been assigned the fifth place in descending order. Here we find a clearer evidence of the emergence of lower peasantry as a socio-economic class cutting across the *varṇa* structure of society. There is also some indication of its position relative to other sections of the social system.

The mass of toiling peasantry is viewed as a distinct class in some other sources[33] of the early medieval period also, and there are some pointed references to the antithesis[34] between the ruling landed aristocracy involved in the *sāmanta* system, reflecting feudal relations, and the peasantry. But the existing state of peasant differentiation, as noticed above, tended to preclude the emergence of a high degree of class consciousness. As such, the peasantry cannot be said to have been a class for itself. Nevertheless, some pieces of evidence[35] clearly reveal that, especially during the later phase of the early medieval period, the exaction of exorbitant taxes and forced labour, and other kinds of oppression at times led peasants to unite locally for protesting against oppressive measures or offering resistance to their oppressors.

NOTES AND REFERENCES

1. Cf. B.N.S. Yadava, Presidential Address, *Proceedings of the Indian History Congress*, 53rd Session (1992–93), p. 8.
2. Idem, *op. cit.*
3. Cf. R.S. Sharma, 1980, *Indian Feudalism,* Second Edition, Delhi, pp. 213 ff.; B.N.S. Yadava, 1973, *Society and Culture in Northern India in the Twelfth Century,* Allahabad, Chapter 3.
4. Meenashri Yadav, 1992, *Agriculture and Peasants in Northern India from* C. *600 to 1200* AD (in Hindi), thesis (unpublished) approved for the D.Phil. degree of the University of Allahabad, pp. 238f.; cf. B.N.S. Yadava, *Proceedings of the Indian History Congress,* 53rd Session (1992–93), p. 8.
5. They are also referred to as *kṛṣijīvins*; *Kāśyapīya-kṛṣisūkti*, ed. G. Wojtilla, 1979, *Acta Orientalia,* XXXIII (2), Budapest, verses 187–88. This text acquired its present form in the medieval period. However, the core of it has been placed in the eighth-ninth century by G. Wojtilla; *Acta Orientalia,* XXXIX (1), 1985, p. 85 fn.1.
6. *Manusmṛti* with the *Manubhāṣya* of Medhātithi, Vol.I, (ed.) G.N. Jha, Calcutta, 1932.
7. Medhātithi on Manu, III. 155.
8. Infra.
9. Bhāratīya Vidyāpīṭha Bombay, 1949, p. 115; B.N.S. Yadava, Presidential Address, *Proceedings of the Indian History Congress,* 53rd Session (1992–93), p. 7.
10. Cf. B.N.S. Yadava, *loc. cit.* Some inscriptions of the early medieval period mention the *pradhāna* (principal) residents of villages; e.g. Copper Plate Grant (C. AD 1142) of a ruler of Kāmarūpa; *Epigraphia Indica,* Vol. II, p. 353, lines 49f.
11. K. Rangachari, "Town Planning and House Building in Ancient India according to *Śilpaśāstras*", *Indian Historical Quarterly,* Vol. 3, pp. 826f.; Meenashri Yadav, *op. cit.,* pp. 261f.

12. *Āpastamba Dharmasūtra,* Varanasi, 1969; *Dharmakośa–Vyavahāra-kāṇḍa;* ed. Laxman Joshi, Wai (Dist. Satara), 1938, p. 842.
13. *Corpus Inscriptionum Indicarum,* III, (ed.) J.F. Fleet, 1888, London, p. 257, line 13.
14. *Kṛṣi-Parāśara,* ed. & tr. G.P. Majumdar and S.C. Banerji, The Asiatic Society, Calcutta, 1960, verses 97, 98. The work has been assigned to the middle of the eleventh century by L. Gopal; idem, *Aspects of History of Agriculture in Ancient India,* Varanasi, 1980, p. 30.
15. Ed. M.R. Kale, 1966, Fourth Edition, Delhi.
16. *Ibid.,* p. 120 (main text), 76 (notes).
17. *Ibid.,* p. 120.
18. Irfan Habib in *The Cambridge Economic History of India,* Vol. I, ed. Tapan Raychaudhuri and Irfan Habib, 1982, Cambridge University Press, p. 221; idem, Presidential Address, *Proceedings of the Indian History Congress,* Kurukshetra Session (1982), pp. 31f.
19. Ed. Taranish Jha, 1976, Hindi Sahitya Sammelana, Prayag, 80.58.
20. *Manusmṛti,* 10.82 ff.
21. *Parāśara-smṛti,* ed & tr. Shiva Dutta Mishra, Varanasi, 1981, 2.2, 6, 18, 19; cf. L. Gopal, *Aspects of History of Agriculture in Ancient India,* p. 27.
22. P.V. Kane, 1977, *History of Dharmaśāstra,* Vol. V, Part II, Second Edition, Poona, Chronological Table, p. XIII.
23. Cf. B.N.S. Yadava, "Dissent within the Fold of the Dharmaśāstra. Tradition as Reflected in the *Parāśara-smṛti", Journal of the Ganganatha Jha Kendriya Sanskrit Vidyapeetha,* Lii-Liii, January–December (1996–1997), p. 40.
24. Meenashri Yadav, *op. cit.,* p. 234.
25. *Ibid.,* p. 235.
26. *Ibid.,* p. 235.
27. *Kāśyapīya-kṛṣisūkti,* verses 211-12.
28. *Bṛhatsaṃhitā* (with the commentary of Bhaṭṭotpala), Parts I & II, (ed.) A.V. Tripathi, Varanasi, 1968; 15.28–30. Varāhamihira may be assigned to the last quarter of the fifth and first half of the sixth century AD; A.M. Shastri, 1969, *India as Seen in the Bṛhatsaṃhitā of Varāhamihira,* Delhi, p. 16.
29. Ed. Sundarlal Shastri, 1976, Varanasi, 7.32-34, 38ff.
30. *Aparājita-pṛcchā,* ed. P.A. Mankad, 1950, Baroda, 77.15–16; L.M. Dubey, 1987, *Aparājita-pṛcchā—A Critical Study,* Allahabad, p. 452.
31. A kind of balcony (M. Monier-Williams, Sanskrit-English Dictionary, p. 1017, col. 3) or balustrade (L.M. Dubey, *op. cit.,* p. 179).
32. By the twelfth century, the term *karṣaka* began to denote in its wider sense a broad, though not clearly defined, socio-economic category including sharecroppers, petty tenant-peasants, poor owner (in a restricted sense)-cultivators, and dependent ploughmen; cf. B.N.S. Yadava, Presidential Address, *Proceedings of the Indian History Congress,* 53rd Session (1992–93), p. 7.

33. *Padma Purāṇa,* of Raviṣeṇa (seventh century), (ed.) Panna Lal Jain, 1958–59, Bhāratīya Jñānapīṭha, Kāśī, 11.350.
34. *Ibid.*, 31.70 ff.; B.N.S. Yadava, 'Problem of the Interaction between Socio-Economic Classes in the Early Medieval Complex', *The Indian Historical Review,* Vol. III, No. 1, p. 46.
35. Meenashri Yadav, *op. cit.,* pp. 268 ff.; R.S. Sharma, *Indian Feudalism,* p. 220.

CHAPTER 4i

The History of Agriculture in South India

Vijaya Ramaswamy

INTRODUCTION

Agriculture constitutes the lifeline of the Indian economy. Oral traditions in south India, especially the folk songs, drew heavily upon the images and metaphors of the plough, the wet clay of *nanjai* (the term used for fertile crop area in Tamil) and the young grain shoot swaying gracefully in the breeze. In fact, both men and women sang while engaged in the task of irrigating the fields and this created the genre of folk music known as *temmangu*. In the pre-industrial era, not only the villages but even the towns and cities were sustained by the agricultural surplus. It was pointed out by the renowned historian Frank Perlin that the townships in early India should not be characterized as urban but rather as *rurban* (a portmanteau word which is a combination of 'rural' and 'urban'). In the context of south India, R. Tirumalai in his book: *Studies in the Ancient Townships of the Pudukottai State* (Govt. of Tamil Nadu, Madras, 1981) makes the following observation:

> The term 'township' needs an explanation. In modern usage, it has come to be appropriated by a part of human settlements in an urban concentration. Etymologically, it is derived from the German word *Tun* meaning a hedge. It not only refers to a marked territorial area but also the community—the *communitas* inhabiting it. *Its connotation is agrarian and it denotes land...* (emphasis mine). It aptly denotes the area of the community's interests and activities, their agricultural pursuits and identity.

Therefore, a study of the history of agriculture in south India involves not merely the study of the eco-regions, cropping patterns and irrigation but also the over arching agrarian structure, questions of rights over land and the development of relations in land over historical space and time.

The broad framework of this study will comprise more or less the geographical territory known today as the Tamil country, although much of the old Tamilaham which constituted the boundaries of the Sangam age, included most of southern India with the 'Dravidian' culture as its core. The present study will, of course, also cover the history of agrarian conditions under dynasties such as Vijayanagara, which stretched far beyond the Tamil country.

AGRICULTURE IN THE SANGAM AGE

The ancient Tamils divided the landscape of Tamilaham into five eco-zones which they called *tinai.* These were:

Kurinji–Hilly regions inhabited primarily by hunter-gatherers. The archaeological artifacts belonging to this region such as axe-heads, spears, choppers and scrapers have been excavated from Cuddappah, Nellore, north Arcot, Chingleput and other regions having this type of eco-landscape. The *kudi* (generic term referring to 'inhabitants') of Kurinji were known as Kanavar, Kunruvar, Vettuvar/Vedan and Kuravar.

Although the Kurinji *kudi* were primarily hunters, there are evidences that they cultivated the hill slopes and practiced agriculture. The *Madurai Kanchi (stanza:* 286–293) states that a few varieties of rice, mustard, ginger, pepper, turmeric and a type of beans (*avarai*) was cultivated in the Kurinji region. The text makes the interesting comment that sandalwood trees were cut down to clear the land and make it arable! Another Sangam text *Malaipadukada* (lines: 102–117) also refers to the cultivation of millet (*varagu*), different varieties of yam plantain as well as sugarcane. There are innumerable references in these texts to the guarding of the crops by young girls with the help of rattles and circular devices called *kavan, tattai* or *tazhal.* Such times also became occasions of romantic dalliance and text like *Tirumurugarrupadai* would talk about Valli, the Kuratti maid from the hills meeting the divine Murugan while she was engaged in guarding the crops with the *tazhal* and eventually becoming Murugan's consort.

Mullai *tinai* referred to the pastoral tracts and the main occupation here was cattle rearing, husbandry and dairy farming. The *kudi* of Mullai were known as Kovalar, Ayar, etc. P.T. Srinivasa Iyengar, in his book *History of the Tamils* (Reprint, Asian Educational Services, New Delhi, 1992), suggests that since cattle wealth tended to multiply very fast, the first, evidence of private property and the "fission of tribes into families" happened in this region. Rajan Gurukkal stretches this argument to say that the beginnings of state formation took place in the Mullai rather than in the Marudam zone.[1] To quote Gurukkal, 'It was dryland agriculture accompanied by herding that dominated the economy as evidences by the predatory politics, re-distributive society and heroic culture' (Dev Nathan, 1997: 213). The evidence for the existence of agriculture in the pastoral tracts come from texts like the *Perumpanattrupadai* (verses 147–205). *Varagu* or the finger millet seems to have been the main crop of the region. The above verses from *Perumpanattrupadai* state that sturdy bullocks were yoked to the plough (*kalappai*) which had an iron tip. The term used for the iron tip is *perumkozhu.* This evidence clearly contradicts the assertion of some historians that plough agriculture did not have much application outside of the riverine tracts, i.e., the Marudam region (for instance, the arguments of Gurukkal *vide* Dev Nathan, 1997: 213).

Neydal region constituted the seaboard of the Tamil country. The inhabitants of the littoral tract were called Paradavar and the main occupation of the region was fishing and salt making, products which were then exchanged for rice. There is no evidence of reclamation of land and of agricultural production in the Neydal region.

The Palai *tinai* was arid, where not much kind of cultivation was possible Nevertheless a kind of coarse grain was produced here for local consumption. The original *kudi* of this region, the Maravar and the Kallar, were militaristic and nomadic. At some point, they seem to have acquired reputation as thieving communities and that is an image that has stuck to them till today. In fact, the term *kallar* is now used as a generic term for 'thief' in Tamil.

SETTLED AGRICULTURE IN THE MARUDAM TINAI

The most important zone for the development and spread of agriculture was the Marudam zone. P.T. Srinivasa Iyengar (1989: 12) argues that this was the last *tinai* which was settled by the ancient Tamils towards the close of the Paleolithic period. The domestication of crops in the region occurred in the beginning of the Neolithic period. There are extensive literary references to agricultural techniques, cropping patterns and the domestication and spread of crops in the Sangam literature, especially the *Madurarikanchi, Perumpanattrupadai* and the *Malaipadukadam.* In these works, the Marudam land is also referred to as *panai,* which means fertile tract. The *Perumpanattrupadai* provides a detailed description of the various processes involved in agriculture such as getting the land levelled by the use of bullocks and then smoothening out the soil by stamping upon it in a systematic manner (verses: 209–211). It refers to the removal of wild weeds with the help of a tool called *toduppu* (verses: 201–202). It further describes processes like sowing of seeds (this is the earliest evidence of sowing in any Sangam text) and harvesting, and says that the gods were first placated with offering of paddy (verses: 228–240). The area where the paddy was husked is referred to as *kalam.* What is significant about this evidence is the absence of reference to the plough. The mention of the plough in certain contexts and its absence in others is intriguing.

It is clear that irrigation was practiced. There is evidence of both tank and well irrigation. The *Pattinappalai* anthology contains a poem by Ruthiran Kannanar, who lauds Karikala Chola II for his contributions to the construction of irrigation tanks (verse 284). The best known is the grand anicut on the Kaveri River constructed in the second century AD. Irrigation technology was simple. The farming people would stand in a long row and pass the water from hand to hand till it reached the fields. The *Maduraikanchi* states that they sang as they worked rhythmically. The *temmangu* genre of folk songs is directly linked to irrigation and involves the singing of romantic duets (sometimes in the form of a dialogue) between young courting couples.

The main types of grains grown in the Marudam region were red grain (actually grain with an yellowish tinge) and white grain which was initially undomesticated. In the *Perumpanattrupadai,* a *Panan* (who was a wandering minstrel) says that in the Marudam region, he was fed with white rice and chicken (verse: 254–56). Apart from rice, the other staple crops of the region were finger millets, pearl millets (*ragi* or *varagu,* especially kezh *varagu*) which were largely consumed by the lower classes and the cultivation of maize (*cholam*). Sugarcane was extensively cultivated. We have reference to the cultivation of plantains, jackfruits, ginger, turmeric, castor seeds (possibly for the extraction of oil) and varieties of gram like black gram and horse gram. These were consumed by the classes.

The social transition from *kudi* to *jati*, i.e., from occupational differences to a caste-based society, occurred in the Marudam *tinai*. The importance of the Marudam *tinai* to early agrarian societies stemmed from its location in the river valleys and deltaic regions, the best known being the settled agrarian culture of the Kaveri delta, which led to the state formation in the region and imperial polity of the Cholas. Sangam literature, in dealing with the Marudam region (and to some extent the Mullai region), deals with a proliferation of specialized groups. The development of economic and social stratification is most clearly to be perceived in the Marudam. It is noteworthy that the king in ancient times was also called *Vedan*, the name of the patron deity of the agricultural tract (who came to be known as 'Indra' in the Sanskritic pantheon).

The dominant *kudi* in the Marudam region were obviously the agriculturists. The term used for a farmer is *ulavan*, literally 'one who ploughs' the *ulatti* is its feminine from (*Tolkappiyam*: Porul Adikaram: 20). The term *Vellalar* for the farmers itself occurs in the Sangam anthology as *Paripadal*, which falls under *Ettutogai* anthology which could not have been composed before the third century AD. According to P.T. Srinivasa Iyengar, the Vellalar derived their name from the Tamil *vellam*, meaning flood, since they specialized in the channelizing of floodwaters for land irrigation (1989: 13). The term used in *Paripadal* for plough is *er*. In the period, around the seventh century, with the growing system of land grants under the Pallavas and the Cholas, the Vellalas emerged as the dominant caste in large parts of Tamilaham. In conjunction with the Brahmins, they became the main upholders of patriarchy and a stratified caste society. The term *ulavan* in, course of time, no longer denoted any specific community but became a generic term for a farmer.

Notions of patriarchy also seem to have emerged in the Marudam *tinai*. The concept of the family with the man as the patriarch developed alongside notions of land ownership. However, there can be no one-to-one equation between the fall in the position of women in the Marudam region with the domination of plough agriculture, a logical postulate which has been made in the context of northern India. Women were crucial to all agricultural activities even after the shift from the hoe cultivation (where women have been known to be dominant) to plough cultivation (which is said to be male dominated due to the sheer weight of the plough, which women were believed to be incapable of managing). Sangam literature clearly shows that even if men did the ploughing, women were crucial to agricultural activities such as planting the seed, guarding crops, husking paddy and pounding grain besides sharing in the work of irrigation. Ritually also, women were crucial to all agricultural activities. Goddess Korravai, who was the diety of the Palai region, was worshipped in all the five eco-zones. The origin of Korravai worship lay in the adoration of the mother goddess as a symbol of fertility. In fact, the ancient custom of plucking the heads of corn and offering them to the mother goddess as a ritual sacrifice marks the beginning of this worship. I have argued elsewhere[2] that with the growing influence of Sanskritization and the transformation of Korravai into the Brahmanical Durga Paramesvari, the status of the women in the Marudam region declined. Therefore, there is no logical nexus between settled peasant agriculture and patriarchy. The fall in the social and ritual status of women in stratified peasant societies is to be looked at in terms of the emergence of the Brahmadeyas (land grants to Brahmins) and the process of Sanskritization in Tamilaham.

AGRICULTURE IN THE PALLAVA-CHOLA PERIOD

The Sangam age was followed by the historical phase popularly known as the Kalabhara interregnum. The Kalabhra chieftains came into Tamil Nadu from the Andhra region and were followers of Buddhism. Settled agriculture was adversely affected by the invasion of the Kalabhra tribes. Agriculture revived only with the expulsion of the Kalabhras from the Tamil country around the sixth century AD.

Epigraphical records provide the most detailed picture of cropping patterns, land surveys and revenue administration during the period characterized by the presence of the Pallavas in Thondaimandalam, the Cholas in Cholamandalam and the Pallavas in Pandyamandalam.

Most of the agricultural lands were used for cultivation of rice which was and continues to be the staple food of south Indians. Rice also constituted the principal item of barter. Pallava inscriptions refer to some specific varieties of paddy. For example, an inscription of Nandivarman III [Annual Report of Epigraphy (henceforth A.R.E.) 73 of 1900] refers to tellunellu. According to C. Minakshi, *tellu* meant a piece of land prepared for immediate cultivation and *nellu* refers to paddy. This was similar to *irainel* and indicated a kind of coarse grain for the consumption of the lower classes. An inscription of Nripatunga (A.R.E. 122 of 1929) has a reference to *iennel*, as a variety of reddish paddy. Paddy used in raising tender transplantation was called *narnel*. *Pala arisi* or old rice was regarded more highly than new rice, because it was refined by being pounded a number of times.[3]

There is no clear evidence of the land tenure system under the Pallavas but a ninth-century inscription from the reign of Kampavarman indicates that tenure patterns were also derived from the centrality of *nel* or paddy. The Kilputtur inscription, dated in the eleventh year of the king, records a unanimous agreement among the villagers to supply annually one *Kadi* of paddy on every *patti* of cultivated land (*vilainilam*) as *erikadi*, i.e., 'tank duty' to Madevanar in return for the gold received from him to construct the tanks.[4] The cultivable lands of the village seems to have been held under different tenures. The records refers to *payalnilam*, which Dr. Minakshi interprets as referring to the landlord-cultivator share, as being fifty-fifty. The term *adainilam* is interpreted by her to refer to the king's share of the land produce, while the term *karainilam* is defined by her as those lands which were subject to periodical redistribution among the tenant peasantry. Her interpretation has been questioned by Tamil scholars like Natana Kasinathan, who argues that *adainilam* refers to land that has been given out on lease (from the Tamil word *kuttagai*) and *karainilam* indicates the growing of *kar* crops rather than any periodic re-distributive system.[5]

A major aspect of agriculture under the Pallava kings was the turning of forest lands and cultivable wastes into arable lands. It is significant that the Pallava kings enjoyed the title *kadu vetti*—literally, 'one who felled forests'—obviously for purpose of cultivation.[6] The reclamation of land to cultivate crops is a recurrent theme in many inscriptions from the Chola and Pandya areas as well. A record from the reign of Kulottunga Chola III praises the king for making wastelands arable by felling trees (*kadu vetti*) and smoothing the soil.[7] The clearing of lands and the gifting of ploughs was also apparently considered as an act of great merit on the part of kings along with the performances of the well-known sacrifices like *Ashwamedha* and *Rājasuya.* A record from Hirahatahalli refers to some king as one who gave hundred thousand cows and hundred thousand ploughs in charity.[8]

Agricultural crops were divided into two categories—*van payir* and *pun payir* plantains, sugarcane, ginger, turmeric; fruits like mango and jackfruit; cotton seeds and castor seeds (from which castor oil was extracted) are referred to as *van payir and* paddy (*manavari puzhudi nel,* probably indicating a coarse variety of paddy), millet and sesame as *pun payir.* Coconut was in great demand for religious as well as domestic use and the donees of Brahmadeya and Devadana villages secured a special permission to cultivate coconuts. The making of toddy from both coconuts and palm fruits seems to have been a popular local practice and Minakshi refers to the tax *ilam putchi* in the Pallava inscriptions as a tax on toddy makers. It is interesting that the coconut palms in the Brahmadeya and Devadana villages were out of bounds for toddy drawers who were prohibited from climbing those trees. This was due to the moral censure attached to toddy drinking amongst the Brahmin community.[9]

The importance of irrigation to the south Indian agricultural system can be seen in the following verse from the Sangam anthology *Purananuru:*

> Verily he who has turned the bent (low) land into a reservoir to arrest the flow of the running water is one who has established a name in the world.[10]

The twelfth-century canonical text *Mitakshara* of Vijnanesvara has this to say:

> The construction of a dam (*sethu*) to a water flow should not be stopped by the owner of the field, even though it destroys his land: provided it causes little injury and is productive of much benefit to many. A well, moreover, as it occupies a small portion of land, and is beneficial on account of the abundance of water in it and so its construction should never be stopped.[11]

The *Er Elupati,* meaning 'Seventy Verses in Praise of the Plough' was composed by Kambar, the court poet of the Chola kings in the thirteenth century. This is an invaluable source in understanding the agrarian structure and agricultural technology in the Chola-Pallava times. Stressing the importance of irrigation for the economy, Kambar writes:

> Let not rain fall, let not kings rule the kingdom with justice. Let not each do his respective duty in this world. Let there be famine everywhere. Even if all

> these adverse circumstances happen at the same time, if only the peasants will do their duty and water their corn from wells by means of *ettam* (water life), the corn will grow and there will be no hunger in the land.[12]

The various theoretical models which have been used to explain the nature of society and economy in south India have given primacy to the role of irrigation—the technology, installation, control and distribution of water-raising devices in a societal fabric based on agriculture to a large extent. The best known of such models is Witfogel's theory of the hydraulic state. Without subscribing to any of irrigation centred theoretical models, it is possible to perceive the seminality of the construction and maintenance of water works in the Chola-Pallava times.

The early medieval references to irrigation begin with the Pallava records. Thondaimandalam was a draught-prone area and the hymns of the Nayanar saints refer to the cruel effects of famine on the land. The Pallava kings and their feudatories took a lead role in channeling rainwater and providing a network of canals, tanks, sluices and wells. Royal tanks were called 'Rajatataka' and were prefixed by the name of the king/ queen who had commissioned them. The Gunapadeyam copper plate inscription of Charudevi (included in the 'Thirty Pallava Copper Plates') refers to one such royal tank. The famous Kasakkudi copper plates[13] of Nandivarman Pallavamalla refer to the Tirayaneri (the term *eri* in Tamil stands for a water tank) located ten miles to the east of Kanchipuram. The Sanskrit portion of the copper plate refers to it as *Tiralaya tataka.* This is probably the same tank which is known today as Tenneri. The Kuram copper plate inscriptions record the construction of the Paramesvara tataka by the Pallava King Paramesvaran I, the subject of these set of inscriptions. The record states that a new Brahmadeya village called Paramesvaramangalam was established by him and the tank was constructed in this connection.

It is noteworthy that the setting up of these tanks was considered an act of great merit. The association of merit with the setting up of waterworks (a modern parallel being the free providing of drinking water in memory of loved ones) accounts for the abundance of tanks at ancient pilgrimage sites. The holy town of Tirvuvannamalai is a major example of different dynasties leaving behind memorials in the form of irrigation tanks. The best known of these, some of them dating back to the Pallava time, are: Pali Theertham Simha Tirtham and Nirutti Theertham.

Historical evidence for irrigation tanks can be found in the context of the Kaveri, Tamraparani, Vaigai and Pennar rivers. The Uttiramerur inscription refers to a special 'tank committee' called the *eri variyam* and it is noteworthy that the Vairamegha tataka built by the Pallava kings was located in Uttiramerur. Inscriptions refer to the Chola Varidhi of Sholingur, the Kaliyaneri near Anaimalai in Madurai, the Kallimangaikulam at Cholapuram and the Rajendra Chola Periya Eri at Punganur. It was the task of the tank committee to see to the repair and maintenance of these tanks. The records from Bahur and Tribuvani mention a tax called *eri ayam* at the rate of one *padakku* of grain per *ma* of cultivated land.[14] Water lifts called *ettam* which, in later times, came to be known as *picotah,* were used to take the water. In AD 101, the Brahmadeya Sabha of Nemali (in south Arcot district) set apart certain incomes as *eri ayam* for the maintenance of the local tank including one quarter *Pon* (gold coin which was a unit of currency under the Cholas) on the Brahmin community in the form of death duty. Hundred years later, there is record of an individual called *Araiyan* (usually refers to a chief) repairing the breach in the tank at Tirukkanji and constructing a stone revetment to the bank (the term used is karpadai). In 1188, a merchant named Dasi Chetti, for the merit of his son, enlarged the tank at Banavur and made a sluice and built two new tanks.[15] Similarly, a tank was built by Lalamadeva and his brother in 1285 in the memory of their mother. At Tirukachchur, it was the local temple which constructed fresh sluices to ensure that the temple lands were well irrigated, although the tank itself was constructed by and belonged to the people of Sengunram. The water in the tank was accordingly to be shared between the villagers and the temple in the ratio of their holding.[16]

The principal of the *anicut* can be said to be the original contribution of the Cholas to irrigation technology. Their construction of tanks essentially involved the idea that a river is best controlled at the head of its delta. The grand anicut over the River Kaveri is said to have been the work of Karikala Cholan, while the sixteen-mile-long reservoir with substantial sluices was built under the patronage of Rajendra Chola at Gangaikonda Cholapuram. Any account of the promotion of agriculture and the setting up of tank irrigation under the Cholas would be incomplete without a special reference to the Uttiramerur inscriptions in the twefth and the fourteenth regnal years and the Karandai plates of Vira Narayana Parantaka Chola I (AD 907–9055). The Uttiramerur Sabha seems to have taken a seminal interest in irrigation, as can be seen by the creation of the *eri* (tank) *variyam* (committee). The record further stipulates that the member of the tank committee will enjoy a special status within the decision making body of the Sabha.[17] The Karnadai plates also deal with the setting up of irrigation tanks all over Cholamandalam by Parantaka I.[18]

Evidence of ancient irrigation works has also come down to us from the Andhra and Karnataka regions. Water harvesting in the Karnataka region was through riverfed tanks (called *kere* in Kannada) over the Kaveri, the Hemavati, Swarnavati, Bhadravati (Shimoga in particular and generally the Malnad area) etc. It is noteworthy that the tanks were built in a series usually situated merely a few kilometres apart. Irrigation was also by wellsprings which are referred to as *talpariges.*[19] The Munirabad stone inscription of the thirteenth year of Vikramaditya VI's reign is of great significance in this context. The record provides proof of the early water harvesting techniques over the Tungabhadra. One Aditya Bhatta, the father of Somanatha, is said to have obtained the village of Pulige from Trailokyamalla (Somesvara I) for his own maintenance. The inscription states that he surveyed the banks of the Tungabhadra river and founded a canal with its subchannels like 'threads drawn out cutting asunder the stalk of a lotus...the mass of plantain trees growing there increased and satiated (the people) with water'. Even more significant is the reference in this inscription to lower level and higher level canals—*palla vaykalu* (*vaykalu* means canal) and *uddiya vaykalu,* among the boundaries of the donated land.[20] A Hoysala record of circa 1300 AD, found near Halabid, speaks of a channel drawn off from the Elachi River. It is said that considerable engineering skill was employed in drawing the river water which flowed at Belur into Halabid. The epigraphical record is supported by the remains of a deep cutting near the 16th milestone of the Hassan-

Belur road.[21] The evidences for tank construction (both riverfed tanks and storage tanks) along with their maintenance and repair by individuals, the village community or by the state machinery/royalty are so prolific (*The South Indian Inscription* series are replete with the requisite details) that it is sufficient to provide a few examples here to indicate their central role in the agrarian economy.[22]

AGRARIAN RELATIONS

As Nilakantha Shastri puts it in his classic work: *The Cholas*, '...the village was primarily a settlement of peasants and the village assembly an association of landlords.'[23] There was individual ownership of land (*Ekabhogam*) as well as collective ownership (*Ghanabhogam*). The wastelands, forest lands and pasture lands seem to have been held in common by the village community. An inscription of the reign of Sundara Chola from Madurantakam records the sale by the Sabha (Sabhai-vilai) of some land described as being a part of the common land of the village.[24] Evidence of communal ownership is provided by the recurrent use of terms like *Ur-manjikam* and *Ur-podu*, as Nilakantha Shastri points out in his chapter on land tenures (*Colas:* 1975: 567). Chola inscriptions also use the term *karayidu*, which seems to indicate periodic re-distribution of land.

Individual ownership of land was clearly recognized in the innumerable instances of alienation by sale or gifts of the absolute proprietorship of the soil by individuals. The inheritance of such property from a father to his son can be gathered from numerous Chola inscriptions. One of the most recent books which discusses the whole issue of property relations in the Chola period with the focus on the development of *kani* rights is *Gifts of Power: Lordship in an Early Indian State* by Heitzman.[25] The term *kani* was originally interpreted by scholars like Karashima to mean *asthabhoga* rights, i.e., the right to absolute ownership over land, inclusive of the right to purchase, sell or mortgage, as also mining rights. Heitzman, however, shows *kani* rights to indicate a trope ranging from *nilakani* to *uzhavukani* to even unusual rights like *patavya kani*—literally, *kani* to be sung meaning that here *kani* indicated a right as well as a responsibility.[26] Noboru Karashima was among the first to evolve a sustained pattern of agrarian relations on the basis of thirty inscriptions from Tiruchirapalli. He contends that a new agrarian order comprising powerful individual landholders emerged in the lower Kaveri valley between the third and the fourth decade of the thirteenth century. The historical stages of this socio-economic transformation is set forth in four phases of Chola rule—phase I from Vijayalaya to Uttama (AD 846 to 985); phase II from Rajaraja I to Adirajendra (AD 985 to 1070); phase III from Kulottunga I to Rajadiraja II (AD 1070–1179) and phase IV from Kulottunga III to Rajaraja III (AD 1179 to 1279). The development of individual landholding is ascribed by Karashima to the fourth phase. The following historical processes are identified as the basis of this development:

a. Accumulation of wealth in the Kaveri valley, the heartland of the Chola country as a result of the imperialist expansion by Rajaraja I and his son Rajendra I.
b. The emergence of military holding or padaipaṟṟu in the region. The author argues that these chieftains began to acquire more and more land for themselves during the time of political disorder towards the middle of the thirteenth century.

c. An increase in agricultural productivity by way of improved irrigation by means of canals and tanks.

d. The role of the brahmadeyas in social integration and the spread of superior agricultural techniques in non-brahmadeya villages.

Karashima focuses on the twin villages of Allur and Isanamangalam, the first aũr (non-Brahmin) settlement and the latter a brahmadeya, to prove the emergence of stratification and landlordism in the south.

Several interesting concepts emerge in the context of the emergence of big landholders in the Kaveri delta. Some of the records which register a land sale deed use the term *Kudineekki* (evicting the resident cultivators), while others stipulate that the transaction should be *Kudineekka*, i.e., without any dislocation being caused for the cultivators.[27] Y. Subbarayalu, who has closely collaborated with Karashima in many of his research projects, takes up the same theme for a closer appraisal in his recent article, 'Property Rights in Medieval Tamil Nadu as seen from the sale deeds in Chola Inscriptions.'[28] Subbarayalu identified 276 inscriptions that deal directly with sale transactions involving agricultural land as well as house-sites. Using the same quantitative analysis method and following the same periodization as Karashima, Subbarayalu conclusively proves the predominance of individual landholders in the brahmadeyas in the early phase followed by the decline of sale and purchase of land by Brahmins in the latter phase. The Brahmin Sabha accounted for about 52.5% of the transactions (roughly 139 sales) in the first period, while sale and purchase by individual Brahmins accounted for 29%. The lands sales by the Brahmana assemblies are said to be from their village common, while Brahmin individuals sold lands in their possession showing the prevalence of individual ownership of land in the brahmadeyas. In a large number of land transactions in the Chola period, the temples seem to have been the beneficiaries/donees though they themselves hardly ever felt the need to sell land.[29] Both Karashima and Subbarayalu postulate the thesis that landownership by individuals in the Ur or Vellanvagai villages increased during the twelfth-thirteenth century. The emergence of landlordism in a big way in these non-Brahmin villages was simultaneously marked by the decline of the brahmadeyas.[30] The issue of 'Landownership and Succession in Medieval Karnataka' has been discussed by G.R. Kuppuswamy in a recent article where, along with individual ownership, he has also discussed 'common ownership and part-rotation or periodic re-distribution called *Tattu*.[31]

TAXATION AND PROTEST

Land was the mainstay of the economy under the Chola-Pallavas, first as it was in early and medieval India everywhere. Land tax collected, whether in cash or kind or a combination of both, was the chief source of revenue for the state. *Nilavari* or *nilaigai* are the most common terms used for land tax in the inscriptions. The land revenue department of the Chola state was called *Puravu vari tinai kalam* and was a highly hierarchical department with specialized officers in charge of land measurement, assessment and collection. While land dues collected in kind seem to have borne the suffix *kadamai,* all cash payments were called *antarayam.* The rate of taxation varied according to the soil fertility, irrigation facilities and even the number of bullocks used in ploughing the land. Inscriptions show that lands were surveyed and re-surveyed in the Chola period and perhaps the Valanadu was created as a geographical-cum-administrative unit in order to facilitate the task of land revenue collection. N. Karashima and B. Sitaraman have documented the innumerable terms that appear in connection with land tax in the Chola inscriptions.[32] Heitzman has estimated that as many as 95 individual tax terms appear in the Chola heartland, which he has studied in detail.[33] Echoru seems to have been a special levy on the peasants to meet the hospitality expenses of the revenue officials. Some inscriptions suggest a nexus between the temple and the Chola state in the process of tax gathering. For example, in Pandanallur (located in Tanjavur), certain professional communities including the *Kaikkolar* (weavers) could not pay land tax on their *kani* lands. The land revenue department confiscated their lands and sold the lands to cultivators who could till them, with the sale price exceeding back taxes owed to the temple.[34] Another inscription dated 1001 AD from Tiruvidaimarudur (also in Tanjavur) says that tax defaulters were made to lie in water or stand in the blazing sun for hours! Unable to bear the torture, the Sabha of brahmadeyam Mahendramangalam protested and took the matter to the king. Heavy taxation led to protests and the desertion of lands by the peasantry. In the years 1238–39, two records from Mannargudi, again the heartland of the Chola country, record the lament of the peasantry against arbitrary taxation and the unbearable punishment meted out to them by the tax officials. The peasants resolve to abandon cultivation unless and until their grievances were redressed.[35] Desertion of lands by the peasantry had a fallout of two kinds: the peasantry protested actively, resulting usually in the remission of taxes or they deserted their lands which sometimes led to the auctioning of their land by the state. Nilakantha Sastri, in his *Colas,* provides a number of instances of desertion of lands due to the inability to pay the taxes and the subsequent auctioning of these lands.[36]

AGRARIAN CONDITIONS UNDER THE VIJAYANAGARA EMPIRE

Historians of south India concur that the post-Chola period witnessed the decline of the brahmadeyas and the replacement of the *nadu* by the *periyanadu.* The Brahmins had played a crucial role in the process of state formation in early south India. Brahmins provided ritual sanction to the imperial authority of the Pallavas and the Cholas and officiated at Vedic sacrifices such as the *Ashwamedha,* the *Rajāsuya* and the *Vajapeya yajña.* The kings acknowledged their gratitude through generous endowments to temples and Brahmins and the donating of brahmadeya lands in the most fertile areas.[37] These had emerged as urban centres in course of time. For example, Madurantaka Chaturvedi Mangalam, a Taniyur in Kalattur Kottam in Jayangonda Cholamandalam, became important because Parantaka Chola I had attached to it the Nagaram Viracholapperum Angadi (the term *perum angadi* refers to a major marketing centre). Uttiramerur in Chingleput district, Tribhuvani in South Arcot district and Mannarkoyil (originally called Rajaraja Chaturvedi Mangalam) in Tirunelveli district are a few examples of prominent brahmadeyas in the Chola-Pallava period. The decline of the brahmadeyas and the fall in the percentage of individual Brahmin landholders can be seen in the post-twelfth century inscriptions. A major factor which shaped the history of the Tamil country in the post-Chola period was the emergence of new elements in this macro region and the expansion of the Telugu chieftains into the Tamil region as well as the establishment of the Deccani Sultanates by dissident Muslim Governors. Stein, in fact, argues that with the Vijayanagara rulers who were of Telugu origin came a number of influential and not so influential migrant groups in search of land and opportunities. These men included mercantile guilds, artisans, cultivators, Telugu Brahmins and motley groups of landless labourers.[38] All these factors led to changes in the agrarian landscape in the era of the Vijayanagara empire.

The post-Chola period witnessed the acquisition of land by non-Brahmins. Karashima, in his specific analysis of inscriptions from the Tiruchirapalli and Tanjavur districts, observes that communal holding in the Chola period gave way to individual landholding in the lower Kaveri valley. He examines seven inscriptions relating to Ukkal, a famous brahmadeya of the Chola period located in the North Arcot district and also an inscription from Kaveripakkam, another brahmadeya from the same region. The earliest record from the region, dated 1241 AD, records a sale of land by the Mahasabhiyar of Ukkal to a non-Brahmin called Pattiyar Poyan Perungan Devapperumal. The sale deeds indicate that the emerging non-Brahmin landlords in the area held titles like Udaiyan, Kilavan, Araiyan and Muvendavelan. Like the hamlet Madaganmedu which changed hands, the village of Viranakkurichchi and the pidagai of Arasanipalai also changed hands from the Brahmins to the non-Brahmins, predominantly Vellalas.[39] These sale deeds broadly suggest that loss of patronage, leading to poverty and indebtedness among the Brahmins, and causing the distress sale of brahmadeyas.

The other major development in agrarian relations in the Vijayanagara period was the shift of landed power from the hands of the Nattar to those of the Nayakas. The Nayakas were leaseholders of lands referred to in the Vijayanagara inscriptions as *amaram* lands. While the other two major land categories consisted of the *bhandaravada*, i.e. crown lands and *manyam*, i.e. religious or charity grants, the *amaram* was the largest category, amounting to three-quarters of all the villages. *Amarakani* was not as absolute proprietary right to land but implied one-sixteenth share of the entire agricultural income/usufruct. The Nayanakara system implied not only control of an external authority over the locality land system but the combination of bureaucratic powerbacked immense military capability. The militarization of the state and its representative authorities meant high potential for conflict. It is in the context of the Amaranayaka system that Karashima perceives the features of feudalism, while Burton Stein postulates further segmentation and the emergence of 'colonial' states, which replicated the segmentary model of the Vijayanagara state.[40] The Nayakas, whom Stein calls 'supralocal chieftains', served initially as extended arms of the Vijayanagara empire in its peripheral areas of control.[41]

AGRICULTURAL PATTERNS AND PRACTICES

Domingo Paes, a Portuguese merchant, visited the kingdom of Vijyanagara in circa 1520 and left extensive records of his travels. He writes:

> This is the best provided city in the world and is stocked with provisions such as rice, wheat, grains, jowar and a certain amount of barley and beans, pulses especially *moong*, horse gram and many other seeds grown in this country which are the food of the people, and there is a large store of these and very cheap; but wheat is not so common as the other grains since no one eats it except the Moors.[41] Fernao Nuniz, who visited the country just a decade later, comments on the abundance of rice, gram and millet and says that 'millet was the grain which was most consumed' by the common people.[42]

The importance of rice as a principal monsoon crop of the wetlands (*Nanjai*) of south India continued into the Vijayanagara period. While white rice was consumed by the richer classes, coarse red rice was consumed by the poorer classes. Barbosa says that 'black rice' (the term he used for 'red rice') was healthier and more wholesome than the refined white rice. He mentions the following varieties of rice: girasal, asal, auavagas and pachcha.[43] Another variety referred to in inscriptions is 'kuruvai'.

The most productive method of cultivating paddy was through the transplantation of seedlings. 'Seeds were sown densely in seed beds. In a period of eight weeks, the field was ploughed four times after which manure was put in and the seeds transplanted. The rice fields were kept under water. Harvesting took place after four months. Most regions were two crop areas: *kuddapah-kar* and *samba-pashanam.* The deltaic areas produced a third harvest called *kadaipu.* Among the other crops were varieties of cereals such as barley, pearl millets and *tinai* besides a variety of pulses such as black gram, horse gram, red gram and beans. Like Nuniz, Paes also emphasizes the importance of millet as a staple food, which he calls 'Indian corn'. Considering the glorious history of cotton textiles in south Indian craft production and export, the remark of Paes that there is an 'infinity of cotton' in south India, assumes great significance.[44]

The medieval peasantry were familiar with the usefulness of crop rotation, manuring and of fallows. Cowdung was the most common natural manure and it was estimated at a *kani* of land (one *kani* is roughly equivalent to 1.32 acres) might be regarded as well manured if a flock of thousand head of cattle spent five or six nights in it. The other natural manure which were used were ashes, household refuse, etc.

One very interesting aspect of agricultural technology in the Vijayanagara period pertains to the use of the *kozhu* or iron-tipped plough. It was conclusively proved in the course of the nineteenth-twentieth century when the English East India Company Agents tried to introduce the coulter into Indian agriculture. Its use was found to be by and large unsuitable to the south Indian soil in most places. In Malabar, for instance, the use of the iron-tipped plough disturbed the rich surface soil, bringing inferior soil on top and reducing the productivity. The same point about the inappropriateness of applying the *kozhu* to all arable regions seems to have been realized by the Vijayanagara state. The state, however, chose to couch its injunction in moral terms rather than in the language of economics and appropriate technology which have made no appeal to the mind of the peasant. Thus, Madhavacharya, said to have been the minister of the Vijayanagara King Krishnadevaraya, wrote in the *Parashara Madhaviyam:*

Loha sahitena Langala mukhena praninam chinnavadho bhavanti iti matsyavadhat papadhikyam uktam.[45]

Using an iron-tipped plough causing many to die (by farming) one commits a greater sin than killing of fish (by a fisherman).

The economic rationale behind a seemingly religious/injunction was to prevent the erosion of the rich surface soil and the destruction of alkaline substances and nitrogen, which was beneficial to crop cultivation.

A major development in the agricultural period under the Vijayanagara kingdom and the Deccani Sultanates was the growth in the production and export of spices. Spices were produced largely in the Malabar region and consisted of pepper (both ordinary and long pepper), cloves, ginger (both green and dried), nutmeg, cardamom and cinnamon. Barbosa writes that these products were exported to foreign countries where there was great demand for them.[46] Coconuts and betel were two other major products of south India. While coconut groves were to be found along the coast, the betel leaf and the betel nut were produced in all parts of the south. Abdur Razzak, the fifteenth-century traveller to south India writes:

> This mastication lightens up the countenance and excites an intoxication like that caused by (drinking) wine. It relieves hunger, stimulates the organs of digestion, disinfects the breath and strengthens the teeth. It is impossible to describe its aphrodisiac virtues.[47]

IRRIGATION AND THE ROLE OF THE STATE

Tank irrigation continued to dominate the economic landscape of south India during the Vijayanagara King Krishnadevaraya, states:

> When a state is small in extent then both dharma (virtue) and artha (material prosperity) will increase only when tanks and irrigation channels are constructed.[48]

An early description of tank construction in the Vijayanagar a period by Bhaskara (*alias* Bhavadura), a prince of the First Vijayanagara dynasty, occurs in the celebrated Porumamilla inscription from Badvel taluq, Cuddapah district dated 1369 AD.[49] After quoting the *Vedas* and the *Shastras* to proclaim that tank construction is an act of great merit, the inscription states:

> A shed for distributing water (*prapa*), a well (*koopa*) and a reservoir, canal and a lotus tank; the merit of constructing them is millions and millions (of times) higher in succession. As the water from the tank serves to nurture both animate and inanimate creation on this earth, even the lotus-faced Brahma is unable to recount the fruits of merit arising from it.(verse: 26)

The record goes on to provide one of the most detailed descriptions of the process of construction. The tank is seven miles long and two and a half miles wide. The bund was surrounded by four natural hills, connected by three dams, which were made of mud and riveted with Cuddapah slabs. The artificial bund is said to have been five thousand *rekhadanda* long, eight *rekhadanda* wide and seven *rekhadanda* high (one *rekhadanda* being roughly equivalent to one and a quarter yards. Interestingly, the record states that the science of *jalasutra* (hydraulics) was well applied by the builders of this great tank. A labour force of thousand is said to have been employed in its construction. It is noteworthy that the Porumamilla tank continues to provide irrigational facility to farmers in the Cuddapah region.

Lengthy descriptions of the construction of irrigation works comes from the accounts of both Fernao Nuniz and Domingo Paes. Peaes described the manner in which a tank was constructed at the new city of Nagulapura at the mouth of two hills. Water coming from either hill would collect there besides which pipes were laid out which ran along the range for three leagues and brought water from an artificial lake. Nuniz, in his chronicle, says that since no mason in his country felt competent to undertake the task of creating this lake, the king commissioned one Joao della Pone,[50] 'a great worker in stone' who undertook the task. Nuniz narrates a gory tale of blood sacrifice of the state prisoners which was considered imperative to please the gods and complete the construction successfully! Nuniz also observes that when della Ponte asked for huge supplies of lime the king belittled his idea since Indians only knew the use of stones.[51] Perhaps it was this factor rather than any supernatural force which caused the construction to collapse initially.

The story of human sacrifices in the context of the construction of irrigation works extends to other sites as well. The Madag reservoir built in the Vijayanagara period irrigated the agricultural lands of Dharwar. The gap between the hills through which the Kumadvati, a feeder river of the Tungabhadra flowed, was closed by a bund 30.5 m high. Two other embankments were also built to fill the gaps in the hills on either side of the Kumadvati valley as water catchment areas. The embankments were provided with sluices. It is said that when the king tried to raise a victory pillar at the site of the reservoir, it collapsed repeatedly. The soothsayers believed that only a maiden's sacrifice would solve the problem and accordingly, Lakshmi, the daughter of the chief *Bedar* or pond digger offered herself at the site. The pillar came up and the sluice became a temple in honour of Lakshmi. It is curious that the Indian tribes of Canada narrate a similar story about a young maid's sacrifice at the site of the Niagra falls! Another maiden of Madag, Kenchava, is said to have desisted the advances of the king of Vijayanagara and drowned herself in the Madag reservoir and the local peasants worship her by offering nose-ring and gold ornaments into the lake.[52] The popular belief is that she blesses them with good rains.

As in the Chola-Pallava times, the state was not the only institution involved in the active patronage of irrigation works. Private individuals often constructed wells or tanks to commemorate births or deaths or to expiate for a crime. Rayasam Kondamarasayya, the minister of Krishnadevaraya, constructed two tanks in the Kondavidu province. Penugonda Viranna, the brother of Virupanna, the *talar* of Vijayanagara dug up a spring and irrigation canal called 'Nutana Tungabhadra' at the village of Modaya. In 1441, an ordinary individual called Mallanary of Udayagiri built a tank at the village of Mandanapati.[53] Among private agencies in addition to individuals, sometimes the village community jointly also built tanks or embankments. In 1486–87, the residents of Tiruvamattur sold portions of their lands to the local temple treasury to dig a channel from the river, leading to the irrigation tank of the village. Individual/community involvement in irrigation led to the emergence of a unique system of land rights. The same rights were called *kattu-kodage* in the context of Karnataka. The creation of these special rights in land was the state recognition of the contribution of non-governmental agencies in the construction, maintenance and repair of irrigation works. The person/persons who undertook such work were given a piece of tax free land watered by that particular well, tank or canal.[54] A lengthy inscription from Mulbagal taluqu, Kolar district (in Karnataka) spells out the implications of the grant of *dasavanda* rights by the state. In AD 1496, a person constructed a tank in a *devadāna* village (attached to the temple) under an arrangement with the head priest of the Narasimha temple. It was stipulated that the builder of the tank would be entitled to three-tenths of the produce from the rice land watered by the tank. The builder was additionally granted income shares from the cultivation of *ragi*. From his share, it was provided that he would be responsible for repairing and maintaining the tank under the penalty of having to make a special payment to the temple in case he failed to do so.[55] The emergence of *dasavanda* rights under the Vijayanagara empire is unique to this period and has no parallel in the agrarian system of the Chola-Pallava period.

CONCLUSION: AGRARIAN CONDITIONS IN THE PRE-COLONIAL ERA

The battle of 1565 involved most of the southern states in an expensive and destructive war. While technically the alliance of the Sultanates of Golconda, Bijapur, Ahmednagar, Berar and Bidar won a 'great' victory against Vijayanagara, in effect, no Deccani state recovered from it completely. Added to political debility was the onslaught of the Marathas on the one hand and the Mughals on the other. The Dutch and the English East India Companies aided any side which was likely to grant them commercial favours in the region. In a situation of political uncertainty and economic insecurity, agriculture suffered and irrigation facilities fell into disarray for lack of maintenance. David Ludden, in his study on the cultural economy of irrigation in Tirunelveli, makes a very interesting observation that in the pre-colonial south Indian agriculture, tank irrigation had played a key role in the ritual-based polity and economy of the Tamils. To quote Ludden: 'It was this system—within which irrigation facilities were constructed, maintained, and regulated by the same organizational units that controlled cultivation processes as a whole—that confronted British administration...'[56] The establishment of the English East India Company brought with it the British anxiety about private property. The notion of periodic redistribution of land, which seemed to defy the laws of private property, were beyond the comprehension of the British. In land ownership, the Company servants replaced the loosely structured *kani* rights with *mirasi* rights. The Board of Revenue records describe the slow and painful process by which the Company arrived at a limited understanding of the agrarian system of south India and devised the raiyatvari system in which 'each field was to be surveyed, its output estimated and then converted into each. Each field was registered in the name of a *raiyat* sometimes referred to as the pattadar, who was directly responsible for payment of revenue to the government.'[57] Once a servile landlord class had been created, it was fairly easy for the Company to get the peasants to shift from rice to the production of cotton which could be exported to England to supply the needs of the textile manufacturing centres. The mangling of the South Indian agrarian structure in order to suit imperial needs resulted in a series of famines which devastated the southern countryside in the eighteenth century. Partial recovery came with the emergence of powerful regional potentates like the Odaiyars of Mysore who restored patronage and state care to the agricultural sector.

REFERENCES

1. Rajan Gurukkal, 1997, 'From Clan and Lineage to Hereditary Occupations and Caste in Early South India', in Dev Nathen (ed.) *From Tribe to Caste,* The Indian Institute of Advanced Study, Shimla, pp. 205–222.
2. Vijaya Ramaswamy, 'The Kudi in Early Tamilaham and the Tamil Women from Tribe to Caste' in Dev Nathan (ed.) *From Tribe to Caste op. cit.* pp. 223 to 246. See especially pp. 232–33 for the details of this argument in the context of social transformation in early Tamilaham.
3. Inscriptional references to paddy and rice and an analysis of the terms are to be found in C. Minakshi, 1977, *Administration and Social life Under the Pallavas,* University of Madras, Madras, pp. 380–81, Note II.
4. *Annual Report of Epigraphy* henceforth A.R.E. 116 of 1923 cited in *Ibid.* p. 169.
5. Natana Kasinathar, 1973, *Vellan iyal* in K.P. Aravanan (ed.) *Aivu koththu,* 'Araichi Pervai', Chennai, p. 15.
6. R. Nagasami, 1973, *Yavarum Kelir,* Bookventure, Chennai, p. 72.
7. *Ibid.,* p. 147.
8. *Ibid.,* p. 72.

8a. *Brahmadeya* refers to land grants made to Brahmins while the term *Devadana* refers to land grants made to temples.

9. C. Minakshi, *op. cit.,* p. 171.
10. Irrigation in South India (1000–1800 AD) a working paper by T.M. Srinivasan, Indian Institute of Chemical Technology, Hyderabad.
11. *Mitakshara* of Vijnanesvara, Commentary on the Smṛti of Yajñavalkya, tr. By J.R. Gharpure, Bombay, 1936–42, II, 126 (tr.) p. 1165 cited in S. Gururajachar, 1974, *Aspects of Economic and Social Life in Karnataka,* Prasaranga, Mysore University, p. 53.
12. T.M. Srinivasan, *op. cit.,* pp. 6–7.
13. *South Indian Inscriptions* (S.I.I.), Vol. II pt. III, pp. 351 ff.
14. Annual Report of Epigraphy, 178 of 1902 and 192 of 1909 cited in Nilakantha Sastri, 1975, *Cholas* University of Madras, Chennai, p. 583. The exact measures represented

by 'ma' and 'padakku' is not clear. The *Pre-Pallava Tamil Index* compiled by N. Subrahmanian (1990, 2nd ed), University of Madras, states that a *ma* equals 100 *kuli* of land.
15. *Epigraphica Carnatica,* Vol. V, Arasikere 22 cited in A. Appadorai, *op. cit.*, p. 206.
16. A.R.E. 295 of 1909 sited in Nilakanta Sastri, *op. cit.* p. 584. The reference to the Tirukkanji eri (A.R.E. 215 of 1919) is also to be found on the same page.
17. A.R.E. 1 and 2 of 1898 but the whole text can be found in the Archaeological Survey of India, Annual Report, 1904–05, pp. 136–38 (Part I of the record) and 139–42 (part II of the record).
18. *South Indian Inscription.* II pp. 383ff cited in Nilakantha Sastri, *Cholas, op. cit.*, p. 135.
19. A bird's eye view on the theme of 'Water harvesting in South India' is to be found in a newspaper article by N. Ramachandran, titled 'Dying Traditional System' (*National Herald:* New Delhi: Tuesday, Oct. 27th 1998).
20. Hyderabad Archaeological Series Nos. 5 and 6 on pp. 10 and 11 cited in S. Gururajachar, *op. cit.*, pp. 53–54.
21. Mysore Archaeological Report, year 1911, p. 51 para 110 cited in *ibid.*, p. 54.
22. A most useful work on the theme of traditional water harvesting systems is the report: *Dying Wisdom: Rise, fall and potential of India's traditional water harvesting systems,* State of India's Environment: A Citizens Report, edited by Anil Agrawal and Sunita Narain, 1997, published by the Centre for Science and Environment, New Delhi.
23. Nilkantha Shastri, 1975, *Chola,* Madras University, Chennai, p. 567.
24. A.R.E. 396 and 157 of 1922 cited in *ibid.*, p. 567.
25. James Heitzman, 1997, *Gifts of Power: Lordship in an Early Indian State,* OUP, Delhi.
26. *Ibid.*, see pp. 54–72 for a discussion of the term *kani.* The term *patavya kani* occurs on p. 59.
27. Many of the issues relating to agrarian structure and the changing nature of agrarian relations were raised by Noboru Karashima in his book, *South Indian History and Society: From Inscriptions,* AD 850–1800. I had commented on the analysis and theoretical formulations presented in this book in my review titled 'Confined Conclusions', *The Book Review,* Vol. IX No. I, 1984.
28. Bhairabi Prasad Sahu, 1997, *Land System and Rural Society in Early India,* Manohar Publications, New Delhi, pp. 151–162.
29. Y. Subbarayalu, *Ibid.*, pp. 154–155.
30. *Ibid.*, p. 155. While the first period identified by Subbarayalu relates to the ninth-tenth centuries and here he logically locates the growing importance of brahmadeyas, his subsequent periodization appears to be quite ambiguous which becomes a major drawback in accepting some of his conclusions.
31. G.R. Kuppuswamy in Bhairabi Prasad Sahu edited *Land System and Rural Society in Early India, op. cit.*, pp. 163–177.
32. Revenue Terms in Chola Inscriptions', *Journal of Asian African Studies,* 5, 1972, pp. 87–117.
33. Heitzman, *op. cit.*, p. 161.

34. A.R.E. 115 of 1931–32 cited in *Ibid.*, p. 156.
35. 159 of 1895 from Pandanallur and 97, 98 and 104 of 1897 from Mannargudi vide Nilakantha Sastri, Cholas *op. cit.*, pp. 537–38.
36. *Ibid.* p. 540.
37. An interesting paper on this subject is Burton Stein's 'All the Kings' Mana—Perspective on Kingship in Medieval South India which first came out in 1977 in J.F. Richards (ed.) *Kingship and Authority in South Asia,* University of Wisconsin Press, Madison. The most recent book which deals with this and related issues is Heitzman's book, cited earlier.
38. Burton Stein, 1980, *Peasant State and Society in Medieval South India,* OUP, Delhi. This point is also specifically made by Stein in his article Vijayanagara c. 1350–1564 in Tapan Raychaudhuri and Irfan Habib edited, 1982, *The Cambridge Economic History of India,* Orient Longman with CUP, pp. 110–111.
39. The changing nature of agrarian relationships and proprietary rights in the Vijayanagara period has been discussed at some length by Noboru Karashima in his recent book, 1992, *Towards a New Formation: South Indian Society Under Vijayanagara Rule,* OUP, Delhi, See Chap, 'Emergence of New Groups of Landholders', pp. 117–130.
40. An in depth analysis of the various arguments presented by Stein and his followers on the one hand and Karashima and his school on the other hand albeit being crucial to South Indian history and historiography would be out of place here. However, some of their most important conclusions are presented in: Burton Stien: *Peasant State and Society in Medieval South India, op. cit.,* pp. 374–80ff and his *Vijayanagara,* 1989, The New Cambridge History of India, Orient Longman with CUP, and Noboru Karashima, 1992, South Indian Society under Vijayanagara rule, OUP, Introduction, chapter 7 and passim. For the earlier period, a sound discussion on this theme is to be found in R.N. Nandi, 1984, 'Feudalization of the State in Medieval South India', *Social Science Probings,* Vol. I, No. I, March 1984 and, more recently, in Kesavan Veluthat's book, 1993, *The Political Structure of Early Medieval South India,* Orient Longman, New Delhi. This debate was joined by Brjadulal Chattopadhyaya, 1994, with *The Making of Early Medieval India,* OUP.
41. Robert Sewell, 1962, *A Forgotten Empire: Vijayanagara* (Indian Reprint), National Book Trust, New Delhi, pp. 345–347.
42. *Ibid.*, pp. 279–376 records the chronicle of Nuniz.
43. *The Book of Duarte Barbosa,* Two volumes, edited and translated by M.L. Dames, Hakluyt Society, London, 1918, Book I, p. 192 and pp. 195–96.
44. The Narrative of Paes in Sewell, *op. cit.,* p. 230.
45. *Parashara Madhaviyam* of Madhavacharya Book I, p. 432 cited in A. Appadorai, 1990, *Economic Conditions in Southern India* (1000–1500) Vol. I, University of Madras, Chennai, reprint, p. 235 footnote 616.
46. Duarte Barbosa, *op. cit.*, Book 1, p. 203; II, p. 215, etc. vide T.V. Mahalingam, 1975, *Administration and Social Life Under Vijayanagara,* Madras University, Chennai, p. 81.
47. Cited in M.S. Randhawa, 1982, *A History of Agriculture in India,* Vol. II, Indian Council of Agricultural Research, New Delhi, p. 104.

48. A Rangasvami Saraswati, 1925, 'Political Maxims of the Emperor-Poet, Krishnadeva Raya.' *The Journal of Indian History,* No. 4, pp. 61–88. The above maxim is to be found on pp. 68–69.
49. *Epigraphica Indica* vol. XIV, pp. 94–97 ff. The record is discussed at some length in A. Appadorai, *op. cit.*, pp. 200–201.
50. Robert Sewell in his book *The Forgotten Empire, op. cit.*, speculates that Jodo della Ponte belonged to the celebrated Ponto family of engineers from Venice, Italy (see footnote 2 on p. 345).
51. *Ibid. The account of Nuniz,* p. 346.
52. Some of these fascinating legends regarding the constructing reservoirs and dams by the Vijayanagara kings are to be found in Anil Agarwal and Sunita Narain edited *Dying Wisdom, op. cit.* pp. 205, 210 ff.
53. A.R.E. 336 of 1915, 68 of 1912 and 269 of 1905 *vide* T.V. Mahalingam, *op. cit.,* p. 85.
54. A.R.E. 7 of 1922 from North Arcot district, Several instances of dasavanda rights are given in T.V. Mahalingam, *op. cit.*, p. 86ff.
55. Epigraphica Carnatic, Vol. 10, No. 172 from Mulbagal vide Burton Stein, *Peasant State and Society in Medieval South India, op. cit.*, p. 427.
56. David Ludden, 1979, 'Patronage and Irrigation in Tamil Nadu, A Long-term View, *The Indian Economic and Social History Review,* Vol. XVI, No. 3, 1979. See also his more recent publication on the broader theme of *Peasant History in South India* brought out by Princeton University Press in 1985.
57. Dharma Kumar, 1965, *Land and Caste in South India,* Cambridge University Press, p. 21.

CHAPTER 42

Expansion of Agriculture in Ancient Bengal

Puspa Niyogi

EXPANSION OF AGRICULTURE IN ANCIENT BENGAL

Agriculture was the principal occupation of the people of ancient Bengal. A large section of her population was dependent upon it for their very existence and livelihood. In fact, agriculture plays an important role in the economic set up of a country, which is mainly agricultural. It is a source of perpetual creation on which our civilization depends.[1]

A close study of agriculture is expected to furnish important data relating to soil, rain; manure; seeds, sowing, transplantation, removal of weeds, rotation of crops and double crops; harvesting and crop processing, storage, agricultural implements, irrigation, and agricultural products. These subjects have not yet been treated in a systematic or comprehensive manner based on an intensive study of the available materials but as part of a general narrative of the past.

Also, archaeological information from this part of India is still incomplete and comparatively scanty. We have no positive idea of how the transition from pastorals to crop farming took place. Neither do we know how the first farmer settled down to cultivate food crops or how the first piece of land was brought under cultivation in Bengal and by whom. Very little is known about the agricultural communities that produced a surplus of agricultural products regarded as sufficient for the population and also how it filled the granaries for emergencies.[2] In spite of the drawbacks, it will be absurd to minimize the importance of the matters embodied in the numerous land grants and other documents which either exist or are studied in detail.

ORIGIN

Scholars are divided about the time and place of the origin of agriculture which has no single, simple origin, but its adaptation took place, no doubt at an early age in this part of the subcontinent.[3] About the beginning of agriculture in Bengal, S.K. Chatterji states: 'The Austric tribes of India appear to have belonged to more than one group of the Austro-Asiatic section—to the Kol, to the Khasi, and to the Mon-Khmer groups. They were in the neolithic stage of culture and perhaps in India they learned the use of copper and iron. They brought with them a primitive system of agriculture in which a digging stick (*lag, lang, ling, lak*) was employed to till the hillside. Terrace cultivation of rice on hills and plains cultivation of the same grain were in all likelihood introduced by them. They brought the cultivation of cocoanut (*narikela*), the plantain (*kadala*), the betel vine (*tambula*), the betelnut (*guvaka*), probably also turmeric (*haridra*) and ginger (*sringavera*), and some vegetables like the brinjal (*vatingana*) and the pumpkin (*alabu*). They appear not to have been cattle-breeders but they were probably the first people to tame the elephant, and to domesticate the fowl. The habit of counting by twenties (Hindi *kodi*, Bengali *kudi*) appears to be a relic of an Austro-Asiatic habit. The later Hindu practice of computing time by days of the moon (*tithis*) seems also to be Austric in origin.'[4]

S.K. Chatterji[5] further points out that the Alpine race, which forms the main element in the composition of the present Bengalis, contributed much to our material culture such as the cultivation of some of our most important plants like rice and some vegetables and fruits like the *tamarind* and the *cocoanut*, etc., the use of the betel leaf in Hindu life and Hindu ritual, and many other things would appear to be a legacy from our pre-Aryan ancestors.

Like S.K. Chatterji, some scholars like Przyluski, Bloch, Levi, Bagchi revealed that the Austro-Asiatic language-speaking peoples of South-east Asia were the torch-bearers of the Neolithic revolution, and as such, they were primarily responsible for the introduction of cultivation and agriculture in North-Eastern India, Bengal and the adjoining region.[6]

India's most important contribution to world agriculture is rice, the staple food of Bengal and the East.[7]

Agriculture and civilization appear to be corelated. It may be presumed that before man became settled, civilized and practised agriculture, he passed through some stages of development; starting from the savage stage, when all men were hunter-gatherers; the second step was nomad stage (i.e., herdsmen) when man domesticated some animals and finally, man reached the farming stage when he started agriculture and became civilized. In fact, the Nishadas or Austric-speaking people led life based on agriculture the foundation of civilization of India.

However, it seems probable that in the beginning, both agriculture and settlements followed the courses of the great rivers, which acted as powerful agents of the soil in their neighbourhood. During the first millennium BC, the Gangetic valley became the principal seat of Indian civilization, and the rich alluvial plains of Bengal, watered by the Ganges and issuing from the Himalayas, is the greatest of all Indian rivers and worthy of comparison with the River Nile. The farmers took advantage of the alluvial soil and began to carry on agriculture in the Gangetic Delta region. In course of time, the Gangetic valley became the birthplace of agriculture-based civilization. We may safely conclude that it was the place of origin of agriculture in ancient Bengal. From here, it spread throughout the country, specially Bengal and Bihar and the East, but it failed to penetrate to some interior regions, of which too, there exist exceptions.[8]

Agriculture is venerated in ancient Bengal for many reasons. Firstly, it is regarded as a source of wealth; secondly, it provides employment; and lastly, it supplies food and raw materials for the growth of our industries. Apart from these material considerations, it is the way of life, special and indispensable in human values. These factors find ample representation in our sources. *Baudhayana* suggests a link between the study of the Vedas and the pursuit of agriculture.

Agriculture, which was known to the people from a very early age, expanded with the progress of time. By the time of the Guptas, demand for land increased to such an extent that both cultivable and uncultivable land was gradually brought under the plough. This extension of cultivation is amply borne out by the copper plate inscriptions of Bengal from the fifth, sixth and seventh centuries AD. Most of these records deal with the purchase of land, which are described as *aprada, aprahata* and *khila* (unsettled, uncultivated and fallow). It may be supposed that the buyers wanted to buy the waste land perhaps because of its low price. Moreover, religious donations to the highest caste (i.e., the Brahmins) of untilled land were free from tax. Thus, in one of the Damodarpur Plate (I) of the time of Kumaragupta I of the Gupta year 124—(443–44 AD) we come to know that a Brahmana named Karppatika made an application before the local government for a permanent grant to him, of one *kulyavapa* of uncultivated (*aprada*) land, unploughed and not given to anyone, for the performance of his agnihotra rites.[9] Similar information are found in some more Gupta copper plates.[10]

In all these copper plates, small areas were transferred, consisting of uncultivated and fallow lands.

Later, the *Gunaighar* inscription of *Vainyagupta* dated in the Gupta year 188 (508 AD) records the gift of some land to a congregation of Buddhist monks residing in a monastery built by Mahārāja Rudradatta, at whose request, the grant was made. It consisted of eleven *patakas* of *khila* land[11] in five plots, in the village Kantedadaka in Uttara Mandala. The gift land is described as *Sunya - pratikara hajjika - khila bhumi* (waterlogged, wasteland, exempt from any tax). From a somewhat later inscription of *Lokanatha*[12] it is clear that lands in the forest region in *Suvanga - visaya* were given to the Brahmanas for their maintenance. 'It was outside the pale of human habitation, where there was no distinction between natural and artificial, infested by wild animals and poisonous reptiles and covered with forest out-growths.'[13] From the above it follows that both cultivable and uncultivable lands were brought under the plough, and nobody wanted that any land to remain fallow and uncultivated.

The wastelands, covered with dense forests, were gifted to the donees, so that they may bring them under tillage and thus make these regions productive and fit for cultivation. They were exempted from paying any tax. *Narada* recommends that 'when the possessor of a field is unable to cultivate his land for some reason or other, any stranger who undertakes its cultivation, should be allowed to keep the produce.[14] But when the owner returns, and the stranger is still engaged in cultivating the field; in that case, 'the owner is expected to pay the stranger the whole expense spent in cultivating the land, after which he recovers his land.'[15] Thus it appears that land was not allowed to remain untilled for an indefinite period.

Sukra[16] lays down certain good laws for restoration of the barren lands; it states that 'if somebody starts new industries or cultivates untilled lands, and dig tanks, canals, wells, etc., for their betterment: the king should not claim anything from them, until they realise profits amounting to twice their expenditure.' Moreover, the construction of irrigation works in many parts of North India, gave a further impetus to the development of agriculture, even in the dry and sterile lands. The lands were, step by step, brought into cultivation by a number of people, besides the State, and the individual cultivator. By their joint exertion, new abodes had increased and the forests were cleared for cultivation and the state most gladly lent their kind favour by pardoning royal dues.

From the above it follows that three centuries from the fifth to the seventh century AD postulates a systematic continuation of agriculture, progress and settlement. As pointed out earlier, agriculture in Bengal began much earlier and continued from century to century and from region to region. P.C. Chakravarti clearly traces the causes of the progress of agriculture as follows: 'The pressure of a growing population, the growing desire of priests for material prosperity and religious zeal of the kings, all served in various ways to organise a widespread attack on some of the "negative lands" of the province, which settlement and agriculture had at first avoided.'[17]

There may be other causes, which helped in the expansion of agriculture in ancient Bengal. After the Guptas, there was a marked decline of trade and commerce resulting in complete dependence on agriculture.

By the seventh century AD, there was a distinct sign of an expansion of agriculture in ancient Bengal. Most of the people took up cultivation for their livelihood. This is clearly reflected in some contemporary works and inscriptions, which describes the state of agriculture in the different parts of ancient Bengal.

The Chinese Pilgrim, Hiuen Tsang, who visited the different parts of ancient Bengal and noticed the cultivated fields and characteristic products of the regions, gives an eyewitness account. He specially mentioned a dozen states, which were remarkable for their fertile soil, good farming and rich crops.[18] His account is corrobarated by Sandhyakara Nandi in his *Ramacharita*.[19] He described Varendri as follows:

> 'The land which had all its important regions filled up with crops and water and had as their ornaments the groves ... it had elevated lands bearing excellent flowers.' Varendri as described by him, had large marshy lands besides, lands in which paddy plants of various kinds grew. It abounded in 'sugarcane and bamboo and there were also vast fields for growing fine plants.'

The *Saduktikarnamrita* depicts the picture of plants and prosperity in villages after the harvest is over.[20] It appears from the *Krsiparasara* that paddy fields were cultivated in Bengal.[21]

Ibn Battuta states that 'Bengal is a vast country and abounds in rice.'[22] Some terracotta plaques found in Paharpur also depict paddy fields.[23]

Inscriptions, particularly those of the Senas, give beautiful description of paddy fields. Thus, the Anulia copper-plate of Laksmanasena speaks of the harvest of *sali* rice in autumn. And in another verse, we are told that the gift land consist of 'myriads of excellent villages', consisting of lands excessively growing paddy and beautifully interspersed with gardens.[24]

Another reference is found in the Edilpur copper plate of Kesavasena which mentions 'villages with smooth fields growing excellent paddy.'[25]

The Mainamati plate of *Ranavankamalla Harikaladeva* also refers to "agriculture rich with harvest."[26]

An important fact is gleamed from the study of some of the Sena inscriptions, ranging from the eighth to the thirteenth century AD. It shows how slowly and steadily the agricultural community became prominent and influential in the society. Some of these inscriptions mention the cultivators (kshetra–karah or karshakah) as a distinct class apart from 'the officials, Brahmanas and others'. No longer were the cultivators neglected as before but they were informed of all land transactions in the State. In short, they enjoyed equal status with those of the higher classes without prejudice to their occupations. They, thus, played an important role in the socio-economic set up of the community and enjoyed considerable position in matters of local administration.

The great importance of agriculture in our period is reflected in our inscriptions in yet another way. The land is always measured according to one or more standards, which are associated with agriculture. Agricultural terms such as the *hala, kulya,* etc.,[27] are frequently used signifying a close associate of agriculture with land. All these facts clearly prove that slowly and steadily, agriculture expanded in ancient Bengal.[28]

It is interesting to note that as early as the time of Kalidasa (fifth century AD), agriculture was recognized in both religion and secular society. According to Kalidasa, even the *risis* (hermits) utilized lands for the purpose of cultivation and produced different kinds of crops in the fields adjoining their *asramas* for their own use.

2. SOIL, NATURE IN AGRICULTURE, MANURE, SEEDS AND SOWING, TRANSPLANTATION, CROP ROTATION AND PROTECTION, HARVESTING, IRRIGATION, IMPLEMENTS

The agriculture of a country is dependent to a large extent on the nature of its soil, which is influenced by climatic factors. The soil is of utmost importance in agriculture and the role of nature in helping the soil to become fertile. The alluvial soil is most important for agriculture. This is formed by the constant flowing of the rivers, which swell during the rains, carry down part of rocks with them. Thus, the river beds in higher altitudes are found covered with boulders and gravel in winter season. By a continual grinding action, these boulders and gravel become smaller and smaller as they come down to the plains where they are found in the form of sand. Thus, the soils we see on both sides of a river, are alluvial soils, which consist of sand and silts containing organic matter, and are the most fertile tracts for the growth of various cultivated crops. This kind of soil is found is Eastern India in the regions washed by the rivers Ganges and Brahmaputra and their tributaries. This fact is attested by earlier Classical Writers[1] and even Chinese pilgrims who visited India in the seventh century AD.[2] Contemporary writers and many others also referred to the fertile nature of the soil of Bengal. Thus, Sandhyakara Nandi describes Varendra (North Bengal) as 'the land which had all its important regions filled up with crops and water and had as their ornaments the groves .... it had elevated lands bearing excellent flowers', he further adds, 'Varendra ... had large marshy lands, besides land in which paddy plants of various kinds grow'. It abounded in sugarcane and bamboo and there were also vast fields for growing fine plants.[3] There are other inscriptional and literary evidences in support of this. The earliest reference to a rice granary in Bengal is found in the Mahasthan Brahmi inscription of the Maurya period. The inscription records relief measures adopted by the State to relieve the people from the distress. Paddy was distributed from the royal granary (storehouse). It was located in Pudanagala (Pundranagara).[4] In support of this, Hiuen-Tsang states : 'the soil is flat and loamy, and rich in all kinds of grain produce' ... 'it is regularly cultivated, and is rich in crops' ....[5] Kalidasa hints at transplantation of paddy plants.[6] Some of the poems in the *Saduktikarnamrita* depict the picture of plenty and prosperity in the villages after the harvest is over.[7] Many of the Sena inscriptions also indicate the prosperous condition of farming. The Anulia copper plate of Laksmanasena refers to 'myriads of excellent villages consisting of land excessively growing paddy and beautifully interspersed with gardens'.[8] In the Edilpur copper plate of Kesavasena there are references to villages with smooth fields, growing excellent paddy.[9] The *Mainamati* copper plate of *Ranavankamalla Harikaladeva* refers to agriculture rich with harvest.[10]

Although there were extensive stretches of cultivated fields, it may be noted that all land in ancient Bengal was not fertile. Side by side of cultivable land there were *usara* and sterile land. The term *usara*, used along with *gartta* in the land grants means barren and infertile or infecund land. Medhatithi explains *usara* as part of land in which seed do not sprout on account of some shortcomings in the soil (*usaro bhumibhaga uchyata*).[11] *Kulluka* also bear with Medhatithi by stating *usara* is that where seeds, when cultivate do not shoot up and grow.[12] According to Halayudha, *usaram is trinam.*[13] Monier-Williams states that *usaram* means saline or sterile soil.[14] Thus it is clear that *usara* is a barren piece of land not at all fit for cultivation or vegetation. It may also be presumed that the phrase *sa-gartta-asara* is a part of the conventional phraseology and need not necessarily mean that every village had *usara* land.

Side by side with fertile land, there were sterile regions in Bengal. In this connection, reference may be made to a sterile region in Bengal. The Bhubaneswar Prasasti described a place in Uttara-Radha (West Bengal) as *jangala-patha* where there was no water (*Radhayam-ajallasu jangala-patha*).[15]

Inferior land and land used for purposes other than those of cultivation are mentioned in some inscriptions although rarely. *Apakrsta-bhumi* means inferior land which was comprised in the fertile area (*ksetra*); gourds grew in these land.[16] Refuse land is indicated by the term *avaskara. Avaskara-sthana* of the Irda plate of Nayapaladeva.[17] means a place where sweepings are thrown. Such land was not used for cultivation.

River-side land is probably indicated by the expression *kachcha-bhumi.*[18] Barnett suggests that the word *kachacha* means river side.[19] It may be noted that bosporum (i.e., small grains) used to grow on lands situated between the rivers.[20] Kautilya states that river banks and sea beaches (i.e., land beaten by the foam) were suitable for growing creeper yields (*valliphalam*) such as goard, pumpkin, and the like.[21] We are told by Patanjali that the banks of the river Devika provided grounds for sowing paddy crops.[22] From this, it follows that river-side lands were also useful for cultivation.

Mention may also be made of another type of land known as *vagara-bhumi.* It is mentioned in the above mentioned inscription. It bears the same meaning as Hindi *bagar* which stands for a hedge.[23]

Besides the above there are two kinds of tropical soils (not directly referred to in the land grants) in relation to water supply, both of which are important in our crop production. They may be classed as (1) dry soil and (2) water logged soil. The first kind of soil is situated in the slopes of a hill or on a higher level. It is always well drained. Such a soil is very suitable for fruits and other crops, such as wheat, barley, mustard, tea, sugarcane, cotton, etc.

The soil in lower levels, where there is no outlet for water, always remains waterlogged. Apart from paddy and jute, most of our cultivated crops do not grow in such a soil. Dry soil is not found in Bengal. Typically dry soil is found in Punjab, U.P. and Sind; where the rainfall is very low.

In the Amarkosa, twelve types of land are mentioned—*urvara* (fertile), *usara* (barren), *mara* (desert), *aprahata* (fallow), *sadvala* (grassy), *pankila* (muddy) *jalaprayamanupam* (water or wet land), *sarkara* (land full of pebbles and pieces of limestone), *sakravarti* (sandy), *nadimatrika* (land watered by rivers), *devamatrika* (land watered by rain).[24]

From the *Abhidhanaratnamala*, we come to know that village lands were classified 'as *urvara* (fertile), *irina* (barren), *sadvala* (grassy), *nadvala* (filled with reeds), *khila* (fallow), *maru* (desert), and those, which were black or yellow, and those, which owed their fertility to rivers or rains.'[25] It is further stated in the same work that there were different kinds of fields for different classes of crops, such as fields for producing varieties of rice, beans, oilseeds, hemp, barley, vegetables, etc. The same idea is also preserved in the *Vaijayanti* of Yadavaprakasa where we find terms for fields named after the crops grown viz., *mudga, kodrava, yava, vrihi, sali, sasti, tila,* etc.[26] Panini divides agricultural fields into two classes on the basis of crops grown in them and in accordance of the quantity of the seed required for sowing in them, e.g., *vraihaya* in which *vrihi* is sown in them. *Saliya* in which *sali* is grown.[27] T.C. Dasgupta,[28] on the observation of Khana, gives us proper account about the various crops and the types of soil required for each of them. He states, the *Aus* paddy requires sandy soil and the jute crop requires clayey soil. He further continues, 'One should sow *patal* (*Trichosanthes dioeca*)... in a soil which is sandy'. *Arum* should not be shown in a shady place...when sown by the river side, will have the proper growth. It is interesting to note that soils of some places are suitable for certain crops such as Badhsabogh rice grow specially in Burdwan, Bankura, Hooghly and 24 Parganas. This is true of other varieties of rice also.[29]

There is another class of land in Bengal known as cultivable waste. These land may be brought under cultivation with the help of irrigation. From the agricultural point of view, lands have been classed as *devamatrika* and *nadimatrika.* A third class of land was near the sea, and in them agricultural operations were carried out without hardship.[30]

Soils of different colours are also noticed (such as black and yellowish). The *Abhidhanaratnamala* refers to countries with red, yellow, sandy soil etc.[31] The author of the *Krsiparasara* put forward his suggestions regarding the suitability of the soil for cultivation in different months. Thus, 'the soil is said to be like gold in Magha, silver in Phalguna, copper in Caitra, and so on'; it is further stated that 'cultivation in the dew season (*hemanta*) is held to produce the richest yield, while at the advent of the rains (*ghanagama*) it results in dire poverty.[32]

ROLE OF NATURE IN AGRICULTURE

Role of nature and weather information is of utmost importance in Indian agriculture. It remains an important limiting factor because agriculture is dependent on rainfall.[33] The monsoon rains are of great advantage to the agriculturists. The *Krsiparasara* consider agriculture as depending wholly on rainfall (*vrati - mula - kraih - sarva*). It gives details about rainfall in the different months of the year; about immediate downpour; at the passing of planets (from one Zodiac to another). The influence of planets on agriculture and rainfall is also taken up in some details. But the *Krsiparasara* does not give us an idea about total rainfall in different parts of India, including Bengal.[34]

The clouds have been divided into four types, viz., Avarta, *Samvarta, Puskara* and *Drona,* and effect of each are described.[35] Some interesting methods have been cited by Parasara about the annual rainfall. We get some idea about the technology of weather and knowledge about rain in his book. Parasara hints at the behaviour (conduct) of some of the animals which may help to forecast immediate rainfall: 'the rising of ants from their holes with eggs, sudden croaking of frogs, ...boys making bridges with sand, dancing by peacock, ...snakes climbing on the tops of trees,' etc.[36]

The *Krsiparasara* gives a detailed account about rainfall on the basis of observing the courses of the wind in the different parts of the year and the position of the *naksatra* Citra in *Caitra* which hint at the amount of rainfall; another device is to put a marked stick into a river in the month of *Vaisakha* to confirm the quantity of rain and flood.

The nature of rainy season is indicated by the downpour under certain *naksatra* in the months of *Jyaistha* and *Sravana* and the course of the wind during the full moon in *Asadha* can also foretell the weather in the rainy season. Rain on the ninth day of the month of *Asadha* foretells the rainy season. On the basis of the *naksatra* the nature of the rainy season is decided in the months of *Sravana* and *Bhadra.*[37]

The people of Bengal were fond of weather forecasts, which were based on practical observations and cultivator's manuals, as a result, the people of Bengal in general became trained in the study of weather condition. The peasant-folk got by heart the saying of the cultivator's manual—the *Khanar Vacan*—and became accustomed to quote the lines about the observation of nature in their every day life or field work.[38] In this way, it passed on from generation to generation. Khana emphasized how the clouds, winds, rainbows and rain forecast the weather.

Some of the sayings of Khana are given below:

(1) 'If rainbow is seen in the eastern sky during the rainy season, surely there will be too much rain and consequent over-flooding of the land.'

(2) '....the patches of clouds that are floating, with intermittent breeze at the same time, will tell you about the imminent rains.'

(3) 'The rainbow in the western sky will bring drought but the rainbow in the eastern sky will fetch rains.'

(4) 'The frequent croaking of frogs is a sure sign of immediate rains.' Parasara makes the same statement.

(5) 'When in the beginning of the Bengali year *Baisakh* (April-May), the wind blows from the northeast, you may be sure of heavy rains that year...'

(6) '...when there is rain on Saturday, it will continue for a week, when it will rain on Tuesday it will go on for three days. In other days, it will end on the very day when it will begin to rain.'

(7) If in any day of the month of *Bhadra* (August–September), there are clouds moving in the sky and the wind blowing at random from different directions, then, on that day heavy rains are expected.

There are another group of observations of Khana as advice to the cultivators about the effect of rain on agriculture. They are:

(1) 'If there be rains in the month of *Agrahayana* (November–December), then famine is apprehended due to the destruction of crops...'

(2) 'If there be rains in the winter months of *Pausa* (December–January), the paddy falls off early from the stalks, as a result of which so much scarcity prevailed.'

(3) 'If there may be rains in the latter part of *Magha* (January–February), praise be to the king and his family and his happy land as the rabi crop will grow abundantly.'

(4) 'If there be shower in *Phalgun* (February–March), then the crops '*Chira* and *Kaon* are abundantly expected.'

(5) 'If there be drought in the month of *Jyaistha* (May–June) and heavy rains in *Asadha* (June-July), surely there will be crops to such a quantity that the earth will hardly be able to sustain the burden.'

(6) 'If it rains in *Magha* (January–February), the people become very rich as the rains are conducive to the growth of crops.'

(7) 'The drought in *Jyaistha* (May–June), and the flood in *Asadha* (June-July) make the land bear abundant crops.'

(8) 'If it rains in *Magha* (January–February), even the unfertile uplands will bear ample paddy crops.'

(9) 'If there be showers in *Chaitra* (March–April), it is quite favourable for the growth of paddy.'

(10) 'If the sky of the full-moon night of *Kartik* (October–November) is clouded and there is strong wind, the *Rabi* crop is expected in abundance. Cloud and rains in that night indicate injury to the crop altogether.'

(11) 'If in the rainy season, the weather is cloudy at night, and water increases at the time, the paddy crop will be destroyed through excessive water.'

(12) 'If in the rainy months of *Asadha* (June–July) of any year heavy shower occurs in the ninth quarter of the full moon, there will be drought in that year. If there be very little rain during the month, fishes may be had in abundance. If there be moderate rains in that month, surely the earth will bear good crops. If in any

evening of *Asadha* (June–July), the sun goes down shedding brilliant lustre, no crop is expected in that year.'

(13) 'If the southern wind blows throughout the month of *Asadha* (June–July) from the beginning, no doubt the year will see much flood.'

(14) 'When in one year there will be mist in the month of *Chaitra* (March–April) and flood in the month of *Bhadra* (August–September), the death rate of the people will appreciably increase.'

(14) The Brahmanas, the rain and the flood—all go away on getting *daksina* (fees). In plain language, the Brahmans leave a house after getting their priestly fees while the rain and the flood leave the land of Bengal on the approach of the southern wind.

These are some of the maxims put forward by Khana, and the lesson derived from them is of great advantage to the cultivators. In ancient Bengal, the farmers made the best use of the sayings of Khana in regard to sowing, ploughing, etc. of their fields, and in their everyday life.

Before concluding this section, it may be pointed out that the ancient people had some knowledge about rain-gauge. As early as the fourth century BC, Kautilya[39] refers to a bowl (*kunda*) with its mouth as wide as an *aratni* (24 angulas) as a rain guage (*varshamana*) used for the measurement of rainfall. The *Krsiparasara*[40] refers to an inelegant way to ascertain rainfall. It explains *adhaka* as a vast expanse of water as a hundred *yojanas* wide and thirty *yojanas* deep. The *Meghamala*[41] states that continuous rain for seven nights together was called *drona*. Varaharmihira, in his *Brihat Samhita* deals with the metreological observations, about the different quantity of rainfall and its results.[42]

MANURE

Soil fertility is the capability of the soil that helps the development of plants; where the soil is not fertile or productive, materials natural or manufactured material or creative substance may be applied to the land to supply the plants nutrients. These are called fertilizers or manures.

The ancient Indians apparently did not know the use of chemical fertilisers but they used some indigenous materials such as bones, flesh of animals, water from washing fish, vegetables and animal product, etc. The manure that was mostly used consisted of cow-dung and excreta of various other animals. The ancient people knew the fertilising property of the cow dung although they may be ignorant about their chemical composition. They also knew about the physical effects of fertiliser on the soil. They also knew that the addition of fertiliser to the soil will increase yields. But the overdose of the same may be harmful.

We have seen before that there were fertile regions and the fertile lands do not remain the same all the time. Due to continuous cropping the natural fertility of the soil is reduced, the cultivators has to resort to the need of the land; the degree of deficiency must be determined and the amount and kind of fertiliser needed for a given yield must be found. The Indian farmers knew about their land so well that no scientific examinations were required for detailed examination of plants and soil conditions in the field. They had their own formulae about the manure. Thus, *Vārahamihira* says,[43] to promote inflorescence and fructification, 'a mixture of one *adhaka* (64 *palas*) of sesame, two *adhakas* of excreta of goats or sheep, one *prastha* (16 *palas*) of barley powder, one tula of beef thrown into one *drona* (256 *palas*) of water and standing over for seven nights should be poured round the roots of the plant.' He further prescribes that the seeds before sowing should be treated as follows: 'They should be taken up in the palm greased with ghee thrown into milk; on the following day, the seeds should be taken out of the milk with greased fingers and the mass separated into single seeds.' This process is to be repeated for ten successive days. Then, the seeds are to be carefully rubbed with cowdung and steamed in a vessel containing pork or venison. Then the seeds are to be sown with the above-mentioned flesh and lard in a soil where previously sesame was sown and dug up, or trodden down and then water mixed with *ksira* must be sprinkled daily.[44]

'To ensure the growth of *ballaris* (i.e., sprouting and the growth of luxurious stem and foliage), the seeds should be properly soaked in an infusion of powdered paddy, *masa* (bean), sesame and barley mixed with decomposing flesh and then steamed with *Haridra* (turmeric). This process will succeed even with *tintidi* (*tamarindus indica*). For the *kapittha* (*Feronia elephantum*) the seeds should be soaked for about two minutes...in a decoction of 8 roots—*Asphota* (Jasmine), *Amalaki* (*Phyllanthus embellicus*), *Dhava* (*grislea tomentosa*), *vasika* (*tustica guarderussa*), *vetula* (*calamus rotung*), *suryavalli* (*gynandropsis pentaphyla*), *syama* (*echites fructescens*), and *atimuktaka* (*aganosma caryophyllata*) boiled in milk. The seeds then should be dried in the sun. This process should be repeated for 30 days. A circular hole should be dug in the ground, a cubit in diameter and 2 cubits deep, and this should be filled with the milky decoction. When the hole dries up, it should be burnt with fire and then pasted over with ashes mixed with ghee and honey. Three inches of soil should now be thrown in, then the powder of bean, sesame and barley, and then again three inches of soil. Finally, washings of fish should be sprinkled and the mud should be beaten and reduced to a thick consistency; then the seeds previously prepared should be placed in the hole under three inches of the soil and fish washings (with fish). This will lead to luxuriant ramification and foliage, which will excite wonder.'[45]

The *Agni Purāṇa*[46] gives us another formula about manure. It is as follows:

> A tree becomes laden with flowers and fruits by manuring the soil with powdered barley, sesamum and the offal matter of a goat mixed together and soaked in washings of beef for seven consecutive nights. A good growth of trees is secured by sprinkling them with the washings of fish.

Kautilya[47] states 'the seeds of grains are to be exposed to mist and heat (*tusarapayanamusnam ca*) for seven nights; the seeds of *kosi* (such as mudga and masa, etc.) are treated similarly for three nights; the seeds of sugarcane and the like (*kandabijanam*) are plastered at the cut end with the mixture of honey, clarified butter, the fat of hogs, and cowdung; the seeds of bulbous roots (*kanda*) with honey and clarified butter; cotton seeds (*asthibija*) with cowdung and water pits at the root of trees are to be burnt and manured with bones and dung of cows on proper occasions.'

'The sprouts of seeds, when grown, are to be manured with a fresh haul of minute fishes and irrigated with the milk of *snuhi* (*euphorbia antiquorum*).'

The *Krsiparasara* recognizes the usefulness of manure for crops. It clearly states that without manure, 'the paddy plants grow up bereft of fruits'.[48] It also describes how the chief manure (cowdung) was made and used in India. It states, 'having worshipped the heap of cowdung in *Magha* one, with reverence, should lift it with spades on an auspicious day and *naksatra*.' The *Krsiparasara* then continues, 'having powdered all that and dried it up in the sun, throw the manure into a pit in every field in *Phalguna*. Then, at the time of sowing, take out the manures (and spread over the fields)'. Lallanji Gopal expresses his opinion on the sayings of Parasara about manure. He states: 'that the passage implies a knowledge of the component elements in which modern science analyses cow-manure, we can hold that it indicates an awareness of its fertilizing property and an appreciation of the way in which this property could be preserved and augmented.'[49]

Manure must be carefully stored to minimise loss of nutrients, particularly nitrogen. R. Gangopadhyay rightly points out that the loss can be minimised only if the dung heap is not disturbed; for any disturbance causes rapid fermentation of the liquid portion of the manure, viz., urine with a consequent increase in the evaporation of ammonia. The chief fertilizing element, that of drying dung into balls also 'reduces ammonia, which would otherwise be injurious to the seeds and the tender roots of plants'. The placing of the dung balls in pits increases the humus that contributes greatly to the fertility of the soil. 'Knowledge of manuring at the time was probably a result of extensive practical and not scientific observations.'[50]

From the above information, it follows that farm manure is of major importance during this period. Manure is understood to mean the refuse from farmyard and stables including both excreta and straw. Livestock produces a large amount of manure, specially cowdung; is valuable in maintaining and improving soil because of the plant nutrients, humus and organic substances contained in it. Manure must be applied to the right kind of crop at the proper time. The main benefit of manure are many. It provides humus which improves the soils material character. Manure can be applied as liquid or solid. When accumulated as a liquid from farmyard it may be stored, liquid manure reduces labour but the noxious odours are highly objectionable. The solid manure is to be spread uniformly over the land. The process can be carried out during convenient time but not when the crop is growing.

T.C. Das Gupta[51] gives an account of some of the manures for crops on the basis of Khana. Khana advices the cultivators thus: 'O worthy cultivator, for a vigorous growth of bamboo the smut of paddy should be thrown into the bamboo grove; the edible arums will thrive if ashes are used at their root, and so on.'

SEEDS AND SOWING

The quality of seed is of vital importance in agriculture. Seeds are of two kinds—one for sowing and the other for planting. Only healthy seeds are used for sowing. In agriculture, much depends upon the property of the seed sown and its preservation. The *Krsiparasara*[52] lays down rules about the preservation of seeds, procedure of sowing and its collection. All seeds are to be collected and dried in the sun in the month of *Magha* or *Phalguna*. They are to be kept in small bundles (*putikas*). Seeds, uniform in shape, and free from chaff or other kinds of grain, yield a rich harvest; so care should be taken to keep them together; the bundles should be free from grass, tightly tied up. Next, they should be kept in sacred and clean place. Parasara further warns the cultivators about the sanctity of the seeds. They should not be kept 'on an anthill, in the cowshed, the place where a woman has delivered a child, nor also in a house having a barren woman in it'. He further continues, 'do not allow the seed to come in contact with remnant of food, a woman in her monthly impurity, a barren woman, a woman in the family way and a woman just delivered of a child.' 'The cultivator should not...keep on seeds ghee, oil, butter-milk, lamp and salt.' The same authority on seeds also furnishes more details. He continues, 'the seed that has come in contact with a lamp, fire, smoke, that has been damaged by rain, and has been covered up in holes, is always to be rejected.' He further adds, 'never,...sow the seed that is kept underground, and is mixed up', 'the seed that is a chaff or mixed with particles (of grains) becomes barren.' 'The wealth of harvest depends on seeds.' There are some others who hold similar views about seeds. Thus, from Medhatithi we learn that the sowing of unripe seed may harm the crops,[53] and from Merutunga we come to know that seeds which have been burnt, do not grow or breed.[54] Mention may be made of seeds , which under special circumstances do not sprout, viz., *pulaka* grains cannot grow owing to excessive heat of the soil and *kakayava* is the term for seeds which are disjoined from the covering skin and thus become barren or useless.[55]

The time of sowing of seeds (for paddy cultivation) is 'best in *Vaisakha*, of middle quality is *Jyaistha*, bad in *Asadha* and worse in *Sravana*'. It also mentions the constellations (*naksatras*) which are good and bad. In the Krsiparasara, it is further stated to avoid two days (Tuesday and Saturday) and certain *tithis* and he also gives some prohibited period when sowing should not take place. Thus, he states, '...should avoid sowing of all crops in the three days and a half at the close of Jyaistha as well as in the beginning of *Asadha* he further continues, 'During the three days between the end of Taurus (i.e., *Jyaistha*) and the beginning of Gemini (i.e., *Aṣadha*)...' So also in the period between the cessation of *Mrgasiras* and the first quarter of *Ardra Ambuvaci* (when the earth is supposed to be unclean) takes place.

After the prohibited period is over, the farmer or owner of the farm should 'sow three handfuls of seed moistened with cold water after meditating upon Indra. Next he should pray to the goddess and lastly feed the cultivators.'[56]

From Khana, we get the following about sowing: '...it is believed that sowing paddy seeds within the first five days of the month of *Asadha* (May–June) will yield much crop' '...sowing is recommended throughout the month of *Sravana* (July–August) and the first 12 days of the next month. Simliarly, the pulse *kalai* is to be sown either in the last four days of *Bhadra* (August–September) or within the first four days of *Aswin* (September–October) which are taken to be auspicious.' Another variety of *matar-kalai* is to be sown throughout the month of *Aswin* (September–October) and *Kartik* (October–November) except the last 19 days of the former and the first 19 days of the latter month for getting a full harvest. *Tila* (sesame) is to be sown either within the last eight days of *Phalguna* (February–March) or within the first eight days of *Caitra* (March–April). He further adds 'he who cultivates the soil in the days of the full moon and the new moon is sure to suffer misery'.[57] Khanā have more to say about this[58] and it will be harmful if we ignore the sayings of both Parasara and Khanā. We must admit that the people mostly connected with land and cultivation arrived at these conclusions after a close observation of the conditions of the soil, in different weather and in different seasons. Khanā do not hesitate to advice the peasants to 'start for the fields on an auspicious day'. 'Just reaching the field they have to enter it by the eastern side and begin ploughing from thence.' Similarly 'in collecting the harvest, the southern side should be taken as the side for beginning work.' The ceremony of harvest-gathering is called *navanna*; it is celebrated in the month of *Agrahayana* (November–December) or *Pausa* (December–January). Besides this there is another ceremony in connection with agriculture. It is known as *hala-pravaha*.[59]

It is understood that sowing has to be done on the fixed date. Medhatithi[60] states that untimely sowing may be harmful, it is required to know how a particular seed is to be sown—closely or scattered. Generally, the Indian farmers scatter the seeds on the fields by hand.[61] From *Medhatithi*, we come to know that after the field had been ploughed the seeds were sown; and seeds were sown with the help of the plough, etc.[62]

Varāhamihira advises that before a seed is sown it should be kept inside milk and then handle it with hand, greased with ghee; next it should be rubbed with cowdung and mixed with flesh of hog and deer...then it should be planted in prepared soil, after ten days. Varahamihira gives us further details on the subject, and further advices to plant trees under the influence of auspicious stars and planets.[63]

Besides, a cultivator should have knowledge about the need of his field as regards the seeds. This is best illustrated in some of the land grants of the period. The extent of land concerned is indicated by a term denoting the measures of seed which could be sown on it.[64] In some of the Gupta records, the expression *kulyavapa* is used as a denomination of land measure denoting the area of land, possessed of the capacity of bearing one *kulya* of seed, The element *vapa* in the compound *kulyavapa* derived from the root *vapa* to sow definitely establishes this interpretation. In later times land measures allied to the *kulyavapa*—measure, e.g. *dronavapa, adhavapa,* etc. become widespread. It is to be noted that the use of the word *vapa* in these expressions clearly points to the extent of the seed bearing capacity of the land covered. Other seed measures such as *adhaka, pataka. unamana, kaka, kakanika.*, etc., is found in Bengal inscriptions. These along with *kulyavapa, dronavapa, adhavapa,* etc. constitutes different grades of measure, based on a common unit, which has to be ascertained.

The *farmers* of ancient Bengal also had experience of seed-bed preparation, which mostly depended on the condition of the soil. They were also familiar with cropping system i.e. the kind and sequence of crops grown during a fixed period of time on a fixed area of land is described as the cropping system. It may be either growing a single crop, year after year on the fixed land or it may be a pattern of regular rotation of different crops.

TRANSPLANTATION

The *Krsiparasara*[65] gives us an account of transplantation of seedlings. They should be uprooted when young, tender and soft. The same idea is found in the *Brihati Saṃhita*,[66] which states that young trees without branches must be transplanted in the (dewy) autumn season. Strict rules are laid down, which are to be followed during transplantation. From the month of *Sravana* to *Asvina* transplantation of seedlings takes place; in *Sravana* the seedlings are planted at a distance of one cubit each. In *Bhadra* at half a cubit and in *Asvina* four fingers apart.[67] It may be noted that transplantation for *Jyaistha* or *Asadha* is the best for the seed sowing. *Sravana* and *Bhadra* are termed as bad and worse respectively. From the same authority we learn about the days and stars under which transplantation is to take place.

KATTANA OF PADDY

Kattana means 'thinning out'. The *Krsiparasara* advices the farmers about levelling the field after the seeds have grown, the months of *Asadha* and *Sravana* is the most appropriate time for it. The following two months are not good; if done in *Bhadra*, only half the harvest is acquired and, in *Asvina*, nothing; low land should be avoided, and grasses to be weeded out.[68] Preservation of water in the field is necessary.

REMOVAL OF WEEDS FROM PADDY

Weeds can create havoc in agriculture. The ancient people realised the many difficulties of the weed. It does not help the growth but lessens yields; increases the production cost, it intervenes with harvest and at the same time the quality is lowered. Moreover, its growth obstructs irrigation water flow, so their removal from paddy fields was regarded as essential. Parasara is very serious about it and speaks of removal of weeds from paddy fields. The technical term *nistrnikaranam* is used by the *Krsiparasara* to mean removal of weeds from paddy fields.[69] Paddy trees are surrounded by grass and weeds. If these trees are not made free from weeds it will not bear sufficient fruits. Weeding is generally done in the months of *Asvina* and *Kartik*.

In ancient times, some methods were practised for weed control. It included, sowing, cultivating, smoothering, burning and crop rotation. The ancient people applied these methods and removed the weeds and thus safeguarded the paddy and other plants.

Care was taken for the growth of paddy plants by allowing preservation of water up to the roots in the month of Bhadra and excess water was let out of the fields, but in *Asvina* and *Kartik*, water was stored in the fields.[70]

CROP ROTATION AND DOUBLE CROP

It is interesting to note that the ancient people knew the method of improving the fertility of the soil by what may be called rotation of crops. Such a system has some advantages.

First, it maintains productivity; secondly, it helps in keeping the soil structure favourable, and lastly, it tends to reduce erosions.

In fact, the decision of rotation of crops depends upon whether the crops compete with each other or complement to each other (i.e., by growing one crop lowers the yield or increases its production). In short, it is important in practical agriculture that the same crop should not be grown in a field successively in every season but it should grow only once in the course of two or more crop seasons in rotation with other crops. The reason may be that different crops have different root systems and secondly, the plant foods are better distributed in a rotation and lastly, pests and diseases peculiar to one crop tend to damage the same crop if grown without rotation.

Crop rotation was known from early times and sowing of different items, at stated intervals, was also practised.[71] Kautilya refers to the growing of different crops in succession in the different seasons.[72] In the *Mahābhāsya* of Patanjali, it is stated that the land was ploughed in conformity with the need of the main crop, for example, sesame was sown with beans, the later is considered as the main crop and the other its subsidy. (This system was also known as mixed cropping.) Patanjali further points out that the land was ploughed in conformity with the need of the main crop and the seeds of the minor crop were sprinkled at sowing time.[73] Such union of both fully use the power of the soil and also increase the total yield by raising different crops. The *Yuktikalpataru*[74] is of the opinion that the system was not much in practice. It states that cultivation from year after year results in loss of fertility. In such cases cultivation should be carried on in a different place or field.

Besides rotation of crops, India had a double rainfall and Indians generally gathered two harvests. The Classical writers also held the same view.[75] Eratosthenes states: 'India is watered by the summer rains, and the plains are overflowed. During these rains, accordingly, flax is sown and millet, also sesame, rice and *bosmonum* and in the winter time, wheat, barley, pulse and other esculent fruits unknown to us.'[76]

In the *Arthasastra,*[77] there is clear reference to three harvests: the raining crop, autumanal crop and the winter crop. The *Milindapanho* speaks of a third monsoon, besides the regular rains.[78] The *Brihat Samhita* refers to summer crops, autumnal crops and vernal crops.[79] Thus, the crops, which were sown in summer, would be duly ripened in *Sravana* or in the early part of autumn; the grain sown in autumn was known as spring crops. Lastly, the crops sown in early spring, would ripen in *Chaitra* or *Vaisakha* (April–May).

MONOCULTURE

The practice of growing the same crop year after year, on a given acerage is known as monoculture. It prevailed in ancient Bengal. It had got its advantages and disadvantages.

CROP PROTECTION

Crops are exposed to attack by various agencies, as follows, and therefore, need to be protected:

(a) Nature and Natural Calamities

Ancient Bengal suffered from numerous natural calamities, such as excess of rain (*ativrsti*), flood, cyclone, drought (*anavrsti*); occasional and inundation which swept away the seeds; lightning, etc.[79A] It is difficult to take protective measures against these natural calamities; but inspite of that, some attempts are made to combat these enemies.

(b) Protection from Animals and Birds

Besides natural calamities, animals and birds damaged standing crops; insects, for example, can destroy a crop in a relatively short time. It may be noted that protection of crops was more or less the same everywhere in India. East, west, north and south, all followed the same method. Standing crops were damaged and destroyed by a large number of animals and birds. In some parts of North India deers and hares perished rising buds of the sugarcane plants.[79B] A verse of Yogesvara refers to pigeons eating up *kodrava* corns.[80] There is also reference to deer and parrots consuming corn.[81] Besides, the above, stray cattle and wild animals, both large and small such as jackals, wild pigs, monkeys and even elephants caused great damage to crops. The *Krsiparasara* gives a list of animals as destroyer of crops. They are goats, swine, deer, buffalo and birds like sparrows, parrots, etc.[82] Not only animals but even crops are attacked and damaged by insects, pests and other diseases.

Crop protection and control measures have engaged the attention of farmers from ancient time, yet success has not been achieved, and the battle goes on. Several hundred species of insects are harmful and requires some degree of control. The insects destroy food and grain and in addition they carry and transmit many diseases.

To put a check on crops and protect them, Brihat Parasara advocates the construction of fences[83] of various kinds. Fences in the form of quick hedges of thorny plants like milk hedge, pricky-pear-stripper-thorn, milk-bush (Euphorbia tirucalli, Pithecolobium, Cuphrbia antiquorrum. Opuntia clatier, etc.) were cultivated by the sides of the boundary walls, so that when grown up, animals will not be able to penetrate inside. This is confirmed from the Prabandhacintamani where we come across the watchmen collecting the branches of thorny shrubs[84] for the purpose of fencing the fields.

The second method adopted was the use of scarecrows. Govardhanacharya refers to such scare crows armed with bows and arrows made of straw or grass (*trṇabānastṛṇadhanvā*).[85] Use of scarecrows in the form of forged (fictitious) figures made of grass or straw is mentioned by Yogesvara.[86] Vardhamana who calls them Cancā.[87]

Bana refers to buffalo-skeletons.[88] Hemachandra gives names of grass man as *avadao, jharamko,* and *jharamto*[89] there are other references to scarecrows.[90]

Another device of protecting crops is by appointing watchman. This is referred to in the *Subhasitaratnakosa.* Hemachandra also refers to it.[91] We also hear of girls belonging to the cultivator's family keeping watch over the standing paddy crops. The *Aryasaptasati* also refers to them.[92] According to Hemachandra the expression *hindolam* is used in connection with a craft to frighten and drive away the birds from the fields.[93]

Certain other practices can prevent or reduce insects from crop damage. These include destruction of crop residues, crop rotation, etc. Such practices are useful but cannot be relied upon.

Insects, of course, are not the only agents hazardous to crops. Plants diseases and microscopic worms have the power of creating total destruction of crops. There is hardly a single crop which is not subject to these attacks. Crops are attacked in all stages and nearly every part of if is subject to attacks.

Plants are to be saved from infection, by cleanliness or plant sanitation. The farmers can prevent infection by careful scrutiny of the seeds before planting. The farm should be kept clean by cutting down weeds and wild grasses from which diseases are likely to be spread.

Field pests and storehouse pests of crops were common and alarming enemies of the farmers even in the Vedic times, and later on, large and visible pests like beetles, bugs, caterpillars and invisible foes like fungi and bacteria attacked crops on the field and in the granaries. Rats and sparrows, moths and other enemies caused havoc to crops.

The farmers tried their best to protect the crops in and outside, but even with the best effort, it is not always possible to keep pests and diseases out. It seems that the Vedic people had much faith in the power and potency of *mantras* and incarnation through which they invoked the help of gods like Asvins and others to drive out these agencies of crop destruction.[94] Even in a later period, we find such practices in vogue. Thus, in the *Sarngadharapaddhati*, it is said that if a pot full of sand is empowered with *mantras*, and is thrown in the field, all crop destroyers, viz. insects, cranes, parrots, boars, deer, rats and hares will keep away.[95]

Parasara has a strong faith in *mantras*. The force of *mantra* written on the leaf of *ketaki* and tied in the North-east corner of the field will help and protect the crops from diseases, insects, and animals. It is suggested that it should be written with lac dye, tied in the midst of the crops.[96] There are some herbal remedies for protection of crops also.[97] Lastly sun's rays destroy all types of worms when they are directly exposed to their powerful radiation. The farmers should be cautious and take care in planting trees in the northern and southern directions for the sun's rays.

NALA-ROPANA

The *Krsiparasara*[98] further advice the farmers that on the occasion of the *Kartika-Samkranti* cultivator should plant a *nala*[99], along with leaves, on the North-east corner of the field, and carefully worship it and pray for the uniform growth of all paddy plants and increase of the same.[100] It is expected to avert all evils to paddy.

PREHARVESTING RITUAL IN AGRAHAYANA

The *Krsiparasara* is very particular of the ritual called *musti grahaka* in *Agrahayana* (November–December). He advices that on an auspicious day in the month of *Agrahayana* the farmer should shear off two and a half handful (*musti*) of paddy to the North-eastern corner of the field after duly worshipping the paddy plants. Then carrying that (bundle) on the head with the stalks turned forward, one should go home peacefully. 'Then entering the main room and walking seven steps, one should place it, well worshipped, to the east.' Parasara advises not to touch the paddy before *Agrahayana.*

This ritual is beneficial to the farmer and will be gifted with paddy, wealth and happiness. Non-observance will not be rewarded.[101]

HARVESTING: PLANTING OF MEHDI-ROPANA

The *Krsiparasara* states that the harvesting of paddy should take place under a good star (*naksatra*)[102] in the month of *Pausa*[103] (December–January). When the crop is ripe, one should thresh the corns properly and weigh with an *adhaka.*[104] The *Agni Purāṇa* also mentions auspicious occasions for harvesting.[105]

The *Krsiparasara* next mentions the threshing pillar on post called *Medhi,*[106] which was fixed in the centre of the threshing floor, known generally as *khala.*[107] The medhi was planted in the pit in *Agrahayana* (November–December) on an auspicious day, and the *Krsiparasara* is very serious about the rites to be performed on that day. From the *Abhidhaharatnamala,* we come to know that bullocks were tied to the medhi to thrash out the grains.[108] In the *Ramacharita,*[109] there is reference to the threshing floor on which the reaper corns were scattered in all directions and thrashed by bullocks.[110] Next the corns were measured by the measuring vessel called *adhaka.*

The *medhi* should not be planted in an inauspicious day and also in the *naksatra* (star) called *Sravani.*[111]

PUSYAYATRA

The *Krsiparasara* gives a description of an interesting ceremony called *Pusyayatra.* It takes place in the month of *Pausa* (December). The people together should perform it near the field on an auspicious day. The *Krsiparasara* attaches great importance to the ritual, which is to be carefully performed according to rules, "for the removal of all obstacles and the increase of crops", and the ritual is concluded by a feast and a prayer.[112]

The *Amarkosa* gives very interesting details about the husking of corn. For dividing the grains from the husk the cultivator should carefully shift the superior grain from the inferior staff by means of the winnowing fan. A pestle was used for pounding corn and then husk was separated from the grain by a winnowing basket.[113]

References to distinct terms for unhusked rice, husk, eye of corn, water and rice-powder are noticed in some other work.[114] In the *Desinamamala,* we also find terms for pestal, mortar, etc.[115]

STORAGE

The *Krsiparasara* requires that after all the formalities have been done the grains are to be measured and kept in the granary (*dhanyagara*) on an auspicious day[116] and star, together with the crops, two *mantras* are to be written and kept in the granary for prosperity. Then one ought to worship this goddess of wealth very well.[117] The advice of the *Krsiparasara* is also echoed in the *Agni Purāṇa.*[118]

IRRIGATION

India being mainly an agricultural country, the importance of irrigation and drainage cannot be overlooked; irrigation is the artificial application of water to soil for the production of crops. Drainage is the artificial way to draw out water from land. Again irrigation may be of two categories: natural and artificial. Natural irrigations are those, which are nature's gifts, and artificial irrigation is the efforts of man's labour.

Irrigation means the application of water to cultivate land for the production of crops. The sources of irrigation water, that is surface water and ground water were provided by rivers, canals, lakes, wells, ponds, pools, artificial reservoirs of many varieties, great and small, public and private, to which there are numerous references in the sources, both literary and archaeological. Thus the whims and failure of rainfall were met by one or other means of irrigation.

Fortunately Eastern India is blessed with great rivers, such as the Ganges, the Brahmaputra and their branches and tributaries. Besides, North-eastern India is plentifully watered by two monsoons. Above these, there are other means of irrigation in this part of the subcontinent.[119]

The origin of artificial irrigational schemes is very old. Although no one really knows when or where irrigation was made use for the first time.[120] The farmers realised that they cannot wholly depend on rain, so irrigation was absolutely necessary. This is emphasised in the Epics.[121] It was considered to be meritorious work to dig wells and tanks, which caused kings and aristocrats to engage themselves in such activities.[122] Not only Epics but others also spoke about irrigation. *Kamandaka*[123] speaks in favour of irrigation and Manu directs the kings to undertake the work of excavation of wells, tanks and canals.[124] Kautilya refers to two types of *setu*.[125] Narada[126] classifies the dikes into two classes: *kheya* (which is dug into the ground) serves the purpose of irrigation and *bandhya* (which prevents the access of water) protects the field from excessive water.

DIFFERENT KINDS OF IRRIGATION

Artificial irrigation was carried in Eastern India and outside from wells, tanks, ponds, pools, lakes, reservoirs, canals, etc. When there was scarcity of water, people had to depend on water of wells. The same is true of Bengal. Some of the villages depended on water from wells. The *Ṛgveda* contains references to the water of the wells, being used for watering the fields and we have repeated mention of the word *avata* meaning a well.[127] Kautilya states that tanks and wells should be built in barren and less fertile parts of the country.[128]

Other authorities also refer to tanks, wells, canals and reservoirs. The *Amarkosa* refers to wells and it is regarded as the most popular form. A verse in the *Baudhayana Dharmasutra* shows that some of the villages entirely depended on wells for their water requirements.[129] Besides, the water of the well remains the same as the time of the drought, when tanks are dried up. The *Aparajitaprecha*[130] refers to ten categories of wells, viz., *srirmukha, vijaya, pranta dundubhi, manohara, cudamani, digbhadra, jaya, nanda* and *sankara,* which differed by four to thirteen hands in length, each being one hand long in due order, i.e., the second is one hand longer than the first. The same authority also mentions four categories of *vapis* and *kundus*. Next, it refers to six types of tanks, namely *sara, mahasara, bhadrakas, subhadra, parigha* and *yuguaparigha*. They are further classified according to their length and width of their fences (*pati*).

Tanks are denoted by the term '*vapi*' which is a tank of the smallest size, *tadaga* implies a large tank and *dirghika*, a tank of oblong shape and a tank of a very large size.[130A]

The *Abhidhanaratnamala*[131] classifies reservoirs, ponds, tanks and springs accordingly. Thus a reservoir is distinguished from a deep-water reservoir, similarly, a pond is classified into a small pond, a natural pond and an artificial pond, and tanks are also classified as a small tank, a large oblong tank and a dug spring, and natural spring.

The *Upamitibhavaprapancakatha*[132] mentions some reservoirs, viz., *dirghika, gunjalika* and *yantravapika*. A steep well and a well in a high place is also distinguished by Hemachandra.[133] Mention may also be made of a ring well. In the ring wells instead of bricks, terracotta rings were used. Ring wells appeared in India in the sixth century BC, and continued till about the second century AD. According to Y.D. Sharma, ring wells can be seen even at present in Orissa and Bengal. Their present use is for drinking water.

Epigraphic records also refers to the excavations and gifts of tanks, wells, etc. by different rulers and individuals. It was considered the most meritorious work to make provision for supplying water to land and its people. Accordingly, we find a minister undertaking this type of work with his resources. Bhatta Bhavadeva, the minister of king Harivarman of the Varman dynasty of Bengal, performed the excavation of a tank before the temple of Visnu in Radha (West Bengal). The Bhuvanesvar inscription further claims that in Radha, in the waterless (*jangalapatha*) boundary lands abutting on a village situated in an arid region, has been made (i.e. excavated) by him a reservoir of water (or tank) which gladdens the soul and mind of the company of tourists, sunk in fatigue, and whose beds of lotuses have become devoid of bees.[134]

The Deopara inscription of Vijayasena records the construction of a temple by Vijayasena and excavated a tank near it.[135] Yaksapala executed one tank and named it "Uttara-manasa"; many similar examples exist. Thus, in the *Tarpandighi* copper plate of *Laksmanasena*, we have names of tanks, which formed boundaries of gift villages. The boundaries of the gift villages is given as follows, the south boundary is formed by the *Nichadahara's* tank, the west, by the *Nandiharipa kund* (or the tank belonging to Nandiharipa) and the north, by *Mollana-Khadi* (or the ditch belonging to Mollana)[136]. Many tanks and wells built by private individuals and named by them are found but records of which are not preserved. It appears that wells and tanks are popular means of irrigation. Kautilya[137] believed that irrigation work having a perennial supply of water is better than that which is fed by water drawn from other sources. In the *Mahabharata* we are told that a tank is hundred times more important than a well.[138] In spite of this, wells have been an important means of irrigation.

CANALS

In addition to wells, tanks, etc., mention may also be made of canals. It is interesting to note that the *Veda* contains references to canal digging and in the *Atharvaveda* we find description of canal digging.[139] In Buddhist literature in the time of the Buddha, the Khatlas of Magadha were intersealed by a network of canals and ridge which formed the boundaries of the fields.[139A] The process meant to regulate the inflow and outflow of water from the fields.

It may be noted that volume of rich water in the rivers of Bengal (the Ganges, the Brahmaputra and their tributaries) increased greatly during the monsoon. So canals were dug to divert waters from the rivers for irrigating the fields. In fact, there were a number of channels, large and small, all fed from the Ganges, which watered the vast plains of Eastern India and around. They also knew the process to regulate the flow of these fertilising waters of the rivers with the poor water of the monsoon rainfall or to make these canals overflow and to wash the rice fields. This system of irrigation is defined by William Willcocks as a 'overflow irrigation'. The main features of these types of irrigation are the following:[140]

(1) The canals are broad and shallow, carrying the crest waters of the river floods, rich in fine clay and free from coarse sand;
(2) The canals are long and continuous and fairly parallel to each other, for purposes of irrigations;
(3) The irrigation was performed by cuts in the banks of canals, which were closed when the flood was over.

A.L. Basham holds the same idea when he states: 'In the flat plains, the land was cut by canals running from the great rivers and dotted with artificial reservoirs, which were made by and fed its smaller channels, which watered the fields.[140A] This water contained a great deal of silt, which helped the soil and the crops. This soil is known as *nadi-matrika.* This is emphasised in the *Tilakamanjari* [141] where there is reference to the fertile banks of the river Saraswati, whose water was sent to the neighbouring regions. This view is upheld by Dio Chrysoslom as early as the first century AD.[142] Cultivated fields irrigated by canals is referred to by Kautilya and *Amarkosa* (*jalamirganah*); Patanjali also refers to canals.[143]

LAKES

A lake is a large body of water within the land. A lake is probably bigger than a tank, pool, pond, well, etc. It may be compared to a big reservoir. In ancient Bengal, there were some lakes but specific references about them are not available. The *Ramapalacarita* refers to the reservoirs constructed by King Ramapala.[144] It may be a lake. Outside Bengal, there are many lakes like the *Sudarsana* lake lying on the foot of Mount Girnar in Surastra. From the Junagarh Rock Inscription[145] (c. 150 AD) we learn that the lake in question was originally built during the reign of Chandragupta Maurya. The dimension of the dam (420 cubits and 75 cubits) gives an idea about the vastness of the reservoir. This lake had strong embankments and was provided with drains for proper distribution of water to the cultivated fields. In the time of Rudradaman, there was a flood, resulting in a damage of the lake. The repair work was carried out by the State after two months efforts and an expenditure of huge money. It is doubtful whether the lake of this kind is known to ancient Bengal. It may be presumed that there were many lakes in ancient Bengal but they may not be owned by the State, so references about them are still wanting.

More important was the achievement of Jayasimha, who constructed the Sahasalinga lake of Anahilapura. An old inscription states that the water of the river Saraswati was canalised to fill up the lake with its water.[146] The lake has been referred to by Hemachandra in his *Davyasraya Kavya.*[147] Merutunga refers to the excavation of the famous Karna Sagara lake by king Karna, [148] and *Kalhana* refers to the Pamba lake executed by king Harsa (1089-1111 AD) of Kashmir, which has been identified by Stein with the lagoon called new Pamba Sar.[149]

It is important to know something about the many irrigational works and projects that existed in India, outside Bengal. References to some of them are already mentioned. It is not possible to notice all such works and projects; only references to some important projects may be stated here.

The most important irrigational project is in Kashmir. Kalhana's *Rajatarangini* gives a picture of Kashmir suffering from famine under devastating floods of the Mahapadma lake. Suyya constructed a stone dam and built stone embankments along the river banks. He thus saved Kashmir. King Lalitaditya Muktapida effected improvements by distributing the water of the rivers to various villages, by constructing a series of water wheels, as a result of this the produce of the soil increased. Shortly afterwards the country was again overtaken by flood and famine. These natural calamities were completely averted during the reign of *Avantivarman* (c. AD 855–883) and others.[150] Besides Kashmir, other dynasties also constructed irrigational projects in their respective kingdoms. Thus, the Chalukyas of Gujarat, the Chahamans, the Kalachuris, the Candellas, and the Paramaras excavated and constructed tanks, wells and other irrigational works.[151]

During the reign of Jayapala of Kamarupa, a Brahmana named Prahasa executed a tank.[152] In Bihar, Gangadhar executed a tank in the Gaya district sometime before AD 1137.[153]

Raulavallaladevaka, a feudatory of *Narasimhadeva,* constructed a *vaha,* i.e. a water channel some time before 1158 AD.[154] Another feudatory, Malayasimha excavated a tank in Rewah in 1192 AD at the cost of 1500 *tankas.*[155] There must have been many other tanks and wells excavated by private persons in North India.

Not only the ancient people excavated irrigational works and projects, but also they knew how to operate them. The *Abhidhanaratnamala* describes the different parts of the wells and the methods used for its operation.[156] It also refers to *araghatta,* i.e. the wheel or machine for raising water from a well.[157] It is interesting to note that as early as the Vedic period water seems to have been raised from wells by means of a wheel (*cakra*) to which baskets of wood were tied. The evidence of another passage in the *Ṛgveda* shows that sometimes this water was poured into channels and set to different parts of that field.[158] *Muir* took the word *kulya* to mean artificial waterways which carried the water into reservoirs.[159] In this connection, mention may be made of *utsecane stambha.* This phrase probably refers to the water lifting apparatus (Tamil *erram*) familiar in South

India. It is like a full grown stump of a tree at the end of which was attached the pitcher to lift water from well.[160]

The *Desinamamala* explains some technical terms such as *aqaati, ukka, ukkamdi, dhemka* etc. in connection with raising water from a well,[161] strong machines were also used.[162] The machine known as *araghattaka* was not used by everybody but by people of the higher status of the society. It was expensive.[163]

Medhatithi refers to a term *yantra* to mean building embankments for regulating the flow of water.[164] He also tells us about water being drawn from the well or the tank and kept the water safely in a reservoir.[165] Four water-machines (*vari-yantra*) are mentioned[166] to bring water down, to raise it first and then to bring down again and then to raise it. How far they were used for ordinary purpose is not known.[167]

CONSTRUCTION AND PROTECTION OF IRRIGATION WORKS

Different kinds of irrigation works were undertaken by the State, such as big irrigational projects, wells, tanks and canals; they depended to a great extent on the joint labour and cooperation of the villagers.[168] Lastly, individuals undertook (sometimes) the excavation of wells, tanks, etc.

Digging of wells, tanks were considered to be meritorious works, which induced king and rich peoples to engage themselves in said activities. Very hard labour is required to engage in irrigation work. From the *Mahabhasya* of Patanjali, we come to know about a well digger who had to labour very hard to fulfil his objectives.[169]

It may be presumed that during the historical period in India, both North and South followed the same principles regarding major issues, and projects. The State took up full responsibility for the development of irrigation, as required. The State also granted many facilities for undertaking new irrigation works or repairing the old ones. On the other hand, the State was serious against those who violated the minimum norms set forth. Moreover, rules were framed by the ancient jurists for the protection of irrigation works. They are the following:

1. (a) According to Kautilya, a tank must not be made below one already existing in such a way as to submerge an irrigated field.
   (b) The flow of water from higher tank to a lower one shall not be stopped unless the later has been out of use for three consecutive years.
   (c) The emptying of a tank is punishable with fines.[170]

2. The ownership of any private irrigation work (*setu bandha*) which has been neglected for five years shall lapse, except in the case of public calamity.[171]
3. Persons making use of irrigation work belonging to others on payment of a lump sum, as annual rent, or share of the produce shall keep the works in repair or pay twice the cost of putting them in repair.[172]

4. (a) The penalty for letting water out of a tank otherwise than by the proper sluice (*apara*) or obstructing a sluice (*para*) through negligence is six *panas.*

(b) For wilfully obstructing irrigation or the flow of water from the sluice gate of tanks shall pay the same fine (*purvasahasadanda*).[173]

5. Persons who cultivate the lands below tanks, etc., of others at a stipulated price (*prakraya*) or for annual rent (*avakreya*) or for a certain number of shares of the crops grown (*bhaga*) or persons who are permitted to enjoy such lands free of rent of any kind, shall keep the tanks in good repair, otherwise they shall be punished with a fine of double the loss.[174]
6. Manu lays down that a person breaking (the dam) of a tank should be drowned in the water or he may be asked to repair the damage and to pay a heavy fine.[175]
7. Stealing of water from tanks and wells as well as the sale of tanks or wells was considered heinous offence according to Manu.[176]
8. It is further ordained that the people who fail to help the king when a dyke, or embankment is demolished by burglar (thieves) or gangsters (*hitabhanga* according to Kulluka) should be banished with their goods and chattels.[177]
9. Visnu prescribe death punishment for the breakers of dykes.[178] Brhaspati is more lenient and lays down that he who destroys an embankment shall be fined 100 panas or more according to the nature of offence.[179] Protection of irrigation works is also recommended in the Great Epic.[180]

From the above data, it follows that in ancient India great importance was attached to irrigation works and projects. Any harm done to it was a serious concern of the State which is clearly reflected in the Great Epic, in the *Arthasastra*, and in writings of Manu, Visnu, Brhaspati and others.

Not only irrigation works were protected, but several ceremonies in connection with it were held. One of the Sutra[181] describes a ritual performed at the time of making a river flow in a particular direction, the proposal seems to be irrigation of agricultural lands.

STATE FUNCTIONARIES

We know from the *Arthasastra* of Kautilya and others that administration of irrigation was carried by a number of departments of the government, each organised under an *Adhyaksa* or Superintendent. The list of officials mentioned in the Pala records points to a similar plan. This is more definitely established by the facts mentioned in the Irda copper plate of the Kamboja King Nayapala, which throws light on the organisation of administration. It includes in the list of officials, 'the heads of Departments (*Adhyaksha-vargge*) along with the clerks (*Karana*).[182]

Strabo speaks of a class of royal officers-in-charge of irrigation and its improvements. They may be compared with the officers called Nadipala[183] in the *Arthasastra*. Kautilya further tells us that the Superintendent of Agriculture (*Sitadhyaksa*) had the general control of State owned irrigation works. His duties consisted of "letting in water from river (*nadi*), lake (*sarah*), reservoir (*tataka*) and wells (*kupa*) by regulation of sluice gates (*udgatam, udghatyate, nissi, ryate, jatam, anenetl, udghato, araghattakadi, yantram*).[184] From the *Arthasastra*, we further learn that 'letting out water and receiving it out of turn (*vara*) is punishable offence.'[185] Cultivated fields irrigated by wind power is referred to in the same work.[186]

Secondly, there is a reference to a process of irrigation called *srotoyantra pravartimam* which means the arrangement of appliance by which the irrigation officer distributes the water among the fields by bringing it from flowing streams.[187] As regards the statement of Megasthenes about distribution of water, should be such that 'every-one may have an equal supply of it'. Kautilya also ensures this. All are taxed equally for the water they use for their respective irrigation needs. The irrigation officer is empowered to settle dispute as to distribution of water among different fields (*kedara*) situated at different levels.[188]

Nor should the flow of water from a new and higher to an older and lower tank be stopped.

Irrigation works like the construction of wells, tanks, reservoirs, etc., depended on the joint labour and cooperation of the villagers and the State. Moreover, rules were made for the protection of irrigation works. The State tried its best to develop irrigation. In course of time irrigation became a major industry in Bengal as also outside. It greatly helped agriculture.

OWNERSHIP

There is a long standing controversy regarding the question of the ownership over irrigational projects and works in North India.

There is a view that the king was the owner of the soil and the irrigation works, such as wells, tanks, etc.[189] But this view is not widely accepted. Moreover, there was a distinct attempt on the part of the ruling power to establish their ownership right over irrigation projects. In theory, at least the king may have been regarded as the ultimate owner of all lands and irrigation projects in his kingdom, though in practice private property in land was duly recognized.

In support of this, a passage from Manu[190] is quoted to prove private ownership. The passage says that the first clearer of forest, is the owner of the land thus reclaimed.

Kane[191] has collected valuable materials, from *Smrtis* and their commentaries and other allied texts, which in his opinion go to establish the theory of private ownership on an indisputable basis. It may be noted that what is true of land is also true of irrigation works. It is clear that the man who had a well dug on his own land was its natural owner.[192]

Medhatithi[193] suggests that for all practical purposes, the claim of the State to the ownership rights over natural irrigation works (such as ponds, lakes, rivers, streams, etc.) situated in the village, should be vested in the village as a whole which engaged it on behalf of the king.

Land grants issued by kings are found conveying right to receive the royal share of the produce. Thus, for example, when the grant of a village is made, it means a transfer to the grantee of the income, derived by the king from that village, there is no evidence to show that such a grant affected the right of private ownership, already possessed by its inhabitants in respect of the landed property situated in the village concerned. A grant in such a case did not affect any change in the existing rights enjoyed by the residents of the village, of course, there is no mention of such rights being allowed to continue unaffected.

IMPLEMENTS

Archaeological information about agricultural implements is comparatively scanty. It does not reveal when and how the first farmer settled down to use plough and other agricultural implements to cultivate food crops. No archaeological or pictorial evidence of farm implement has survived of the Harappan people. In fact, written records available from c. 1200 BC are more reliable.

In the Vedic literature, we have repeated mention of the plough (*langala, sira*). It was used for tillage. But very little is known about its construction and shape. We only know that it was large and heavy and required four or more oxen to draw it. In the *Atharvaveda,* the number of oxen used is increased to eight or twelve.[194] The ploughshare was called *phala.* The plough was sharp and pointed with smooth handle (*tasru*).[195] In the *Atharvaveda* and other Samhitas a heavier plough was used, and also other implements, such as *khanitra* (shovel), *datra* and *sine* (sickle), *titan* (sieve) and *surpa* (winnowing fan).

According to another scholar, the mattock and the hoe were also used. The grain measuring vessel known as *urdara* is also mentioned.[195A] The list supplied by the Vedic literature is supplemented by others. Panini gives a list of agricultural implements. He mentions *sira* (VI.2.187), *hala* (III.2.183), *hali* or *titya* (large plough) (III.2.184), *khanitra* or spade (III.2.184) *stambaghna* or the hoe for uprooting weeds (III.3.83) *datra* or *lavitra* meaning sickle (III.2.184).

Panini also gives us the list of the different parts of the plough. They are: a (pole), *potra* (bend in the middle), *kusi* (plough-share), *voga* (yoke), *varatha* (long rope). The plough appears to have been drawn by bulls called *halika* or *sairika* (III.1.81), *sita* (furrow).

In the Great Epic also, a list of agricultural implements are found. Mention may be made of the plough (*langala* 254.17)[196], and *ayomukha, kastha* or a piece of iron tipped with wood.[197]

The plough was used for tilling the soil. First, to break up the soil until it became an aggregate of particles out of which a well tilled soil is composed. Secondly, the plough is used to remove roots, weeds, seeds, and other sprouting material and clear the soil and lastly, to destroy insects, pests, etc., which may be in the soil. The farmer ploughed the field from one end of it to the other when the field is filled by repeated ploughing, the seed-bed is properly prepared.[198]

The *Krsiparasara*[199] held the plough in high esteem and recommended the construction of strong ploughs; otherwise cultivators would face problems at each and every stage. He described the different parts of the plough and, subsequently, has given the description and measurements of the different parts as follows:

Sr. No.	*Names*	*Description*	*Measurements*
1	*Isa*	the pole of the plough	8 cubits long
2	*Sthanu* or *hala sthanu*[200]	the handle of the plough	5 *vitastis* (1 *vitasti* = 12 *angulas*) Monier Williams.
3	*Niryola*	rod of the plough exclusive of the pole or share	one and a half cubit
4	*Yuga*	Yoke	upto the ears of the animals
5	*Niryola-Pasika*	the plates that fix the plough share to the *niryola*[201]	twelve fingers
6	*Addacalla*[202]	pins of the yoke where the bullocks are tied	twelve fingers
7	*Soula*	an extra piece of wood to fix *niryola* to pole	one *aratni*[203]
8	*Paccanika*	made of bamboo, hard, iron tipped	12½ *mustis* or 9 *mustis* long[204]
9	*Abaddha*[205]	circular in shape. A rod of iron which prevents the *niryola* from getting out of the pole	fifty-four fingers
10	*Yotra*	yoke tie	4 cubits long
11	*Rajju*	Rope	5 cubits long
12	*Phalaka*	plough-share resembling an arka-leaf	9 fingers long or one cubit and five fingers
13	*Viddhaka*[206]	a kind of harrow	21 spikes
14	*Madika*[207]	*mai* Bengali	9 cubits in length

The *Abhidhanaratnamala*[208] gives the names of the following implements:

Sr. No.	*Names*	*Description*
1	Ploughshare	
2	Yoke tie	
3	Pin of Yoke	
4	Pole	
5	Goad	
6	*Matya*	Roller to level the sown field
7	*Kottisa*	Harrow to break clods of earth
8	*Khanitra*	
9	*Avdarana*	Spade on hoe
10	*Godarana or kundala*	Scraper or shorel
11	*Abhrih or Ksuh*	
12	*Datra or Srni*	Sickle
13	*Laritra or Asida*	Sickle
14	*Vantaka*	
15	*Kuddala*[209]	The digging hoe
16	*Samya*[210]	Same as *Addaccala* of *Krsiparasara*

In this work, the measurements are not given.

Names of some implements are also noticed in early Bengali literature. *Ramai Pandit* and *Rameswar* give a list of some implements. They are:

(a) Yoke, (b) a pin or nail, (c) a ploughshare, (d) cleaver, (d) sickle, (f) frame, (g) ladder, (h) stick, (i) a rice-husking pedal and many other implements. In the *Chasapala* of Rameswar, a vivid description of making of these implements is found.[210] The plough was held in high esteem and was regarded as an auspicious article.

3. PRODUCT

Our main drawback is that we have no comprehensive account from which we can compile a fairly complete list of agricultural products of ancient Bengal. Some materials, however, lie scattered in the different inscriptions and also in contemporary literary and historical accounts, including those left by Classical, Chinese, Arab and Muslim writers.

A detailed study of the different sources clearly reveals the fact that the crops of India as a whole, during the period, were not very much different from those grown in earlier times. North India (and Bengal) in our period produced most of the crops which continue to grow even now.[1]

A list of Indian grains is found in the *Abhidhanachintamani* of Hemachandra.[2] In this work, the author mentions seventeen kinds of *dhanyas* (rice). The word *dhanya* in this work is used in a wider sense and signifies grains in general and not merely rice. The list includes *vrihi, yava, masura, godhuma* (wheat), *mudga* (green gram), *masa* (black gram), *tila* (sesame), *chanaka* (Bengal gram), *anava* (the meaning of the word is not clear). Hemachandra defines it as *salibhada.* Probably, *Sali* here indicates grain in general. Madhavakara[3] has given a detailed list of the varieties of agricultural products in his book entitled *Paryayaratnamala.* Sarvananda Vandyaghatiya has referred to this work many times; in this work, the names of various crops are found. Thus, *vrihi, asuvrihi, jombula* or *devadhanya, priyangu* or *kangami, godhuma* or *wheat.* Names of some pulses are also found, e.g., *mudga, kalaya, kulattha* (horse gram).[4] Some vegetables are also mentioned in this work. Two lists of grains are also found in Nemichandra's *Prava-chanasaroddhara,*[5] a work antedating Hemachandra, with a commentary belonging to the twelfth century AD one of these mentions twenty-four kinds of grains and the other twenty-five.

The *Pravachana's* list includes *anava, chana and mayushthaka.* The additional items furnished in one list are *yavavava* (a kind of barley), *cavalaka* (cow-pea), *Krsnacanska* (black chick pea), *valla* (climbing bean), *atasi* (common flax), *latva* (sun flower), *Koradushaka, barathi, siddhartha, mulaka* (radish), *ralaka.* In the other list are to be found *sastika triputaka* (chickling vetch), *iksu* (sugarcane), and *dhanyaka.*

These lists, together, give us some idea about the common agricultural products of India.[6] It may also be mentioned that earlier literary sources throw important sidelight on the subject, although all the names of grains are available are not furnished in the inscriptions of the period.

Besides the above data, we have got other sources, which throw important light on the subject. The *Abhidhanaratnamala*[7] gives the names of a variety of cereals and other foodgrains. The list includes *sali* rice of three varieties, course grain (*kodrava*), mustard (*sarshapa*) of two varieties, long pepper or saffron or Italian millet (*priyangu*), wild sesame (*jartila*), wild rice (*nivara*), as well as pulses of four different kinds (*masura, kalaya, ralla* and *adhaka*). A list of seventeen articles (including rice and barley) included by the *Smrtis* in the category of grain (*dhanya*), as noted by Medhatithi.[8] Similarly, Vishnudhanmottara, quoted in Mitramisra's *Paribhashaprakasa*, gives a list of five kinds of grains (*dhanya*), viz. barley, wheat, paddy, sesame and edible grains of two varieties, namely, *kangu* or *syamaka* and *chinaka*.[9] Among other agricultural products camphor and aguru are given by Medhatithi[10] as examples of costly articles. The *gosirsha* variety of sandalwood is included in a list of specially precious products in *Upamitibhavaprapanchakatha*.[11]

On the evidence referred to above, we may classify the agricultural crops of Bengal as follows:

RICE (*DHANYA*)

Rice is the staple food of Bengal and India's most important contribution to world agriculture (*oryza sativa*). It was widely cultivated not only in Bengal but in many regions of India from very early times.[12] Bengal was another important rice-producing area.[13] The earliest reference to a rice granary in Bengal is found in the *Mahasthan Brahmi* inscription of the Maurya period. It was located in *Pudanagala* (Pundranagara).[14] From one of the *Jatakas* we learn that a brahmana landowner grows only rice and no other crop on this large holding located in Magadha.[15] Kalidasa's *Raghuvamsa* contains an indirect reference to the method of rice cultivation in Bengal. Describing Raghu's conquest of the Vangas, the poet remarks that Raghu uprooted and replanted them (*utkhala-pratiropita*) like rice plants.[16] Patanjali praises the *sali* rice grown in Magadha.[17] According to Apararka, Magadha produced abundant paddy.[18] He also refers to another kind of rice called *sastika*, which used to ripen in sixty days;[19] he also refers to bumper harvest of *vrihi* rice.[20] The *Ramacarita* mentions paddy plants of various kinds grown in Varendra.[21] The inscriptions of the Sena kings mention smooth fields growing excellent paddy and 'myriads of villages consisting of land growing paddy in excessive quantities'.[22]

The *Paryayaratnamala*,[23] the *Danasagara*,[24] the *Aryasaptasati*,[25] the *Abhidhanaratnamala*[26] and the *Krsiparasara* tells us about paddy cultivation in Bengal. According to one scholar, 119 varieties of rice grew in the 24 Parganas.[27] Ibn Batuta states that Bengal is a vast country and abounds in rice.[28] Some terracotta plaques found at Paharpur depict paddy fields.[29]

It is interesting to note that varieties of rice are grown in Bengal. The *Amarkosa*[30] refers to three varieties of paddy—*kalama, sastika* and *asuvrihi* (also called patala). The *Abhidhanaratnamala* of Halayudha[31] mentions three varieties of rice, viz., *raktasali, mahasali and kalamasali.* They are also mentioned in other works.[32] All the varieties mentioned above were not grown in Eastern India. Varahamihira[33] adds four kinds of paddy, viz., *kalama*, red rice, yellow rice and hog's rice. Varieties of wild paddy were found in Bengal. *Asu vrihi*, mentioned in the *Paryayaratnamala*,[34] may be identified with modern *aus* paddy of Bengal.

Hiuen-Tsang mentions the characteristic products of the territories and regions visited by him. He specially mentions a dozen states, which were remarkable for their fertile soil and rich crops. Paryatra (Bairat) produced a variety of rice, which was ready for harvesting in sixty days, while Magadha grew another variety with large grains, of extraordinary fragrance, which was called rice for granders. Shaman Hwui Li states in the biography of Hiuen-Tsang about the reception he (i.e., Hiuen-Tsang) received from the authorities of the Nalanda monastery. During his stay there, Hiuen-Tsang was provided with daily offerings consisting of many things, including *mahasali* rice, an exceptional variety grown in Magadha only, and exclusively reserved for kings and religious leaders of outstanding distinction.[35]

Different species of paddy grew in Bengal. A list is furnished in the *Sunya Purana* of Ramai Pandit and in the *Sivayana* of Rameswar. Some names are common in the two lists. It is believed that some of these paddies are cultivated in various parts of Bengal.[36]

SUGARCANE (*IKSU*)

Another important agricultural product was sugarcane which was produced in some parts of Bengal and referred to by *Sasruta* as *Paundraka.*[37] Commentators agree that it was so named because it was grown in the Paundra country (North Bengal). The *Ramacarita* also refers to the cultivation of sugarcane plant of Varendra (North Bengal),[38] the *Amarkosa* mentioned *kantara* as *Kantaraka* and also refers to raw and refined sugar.[39] It is mentioned by Lucan that the Indians near the Ganges used to 'quaff sweet juice from tender reeds' and the Classical writer Aelian mentions a kind of honey drawn out from the reeds which grew among the *Prasioi.*[40] Sugarcane also appears in the list of export to African ports from Barygaza in the Periplus.[41] In some later works there are references to the making of molasses (*guda*) from the juice of sugarcane and the machine (*iksu-yantra*) used for pressing sugarcanes. The *Aryasaptasati*, the *Saduktikarnamrta* and the *Subhasitaratnakosa* also refer to sugarcane.[42] All these facts point to the conclusion that certain kinds of sugarcane were cultivated in ancient Bengal.

It is interesting to note that there were many sugarcane producing areas in India besides ancient Bengal. *Bana* refers to the unbroken rows of fields of the Paundra variety of sugarcane in the Srikantha *Janapada* and in the neighbourhood.[43] The *Gaudavaho* refers to sugarcane produced in the Vindhyan forests.[44] In the inscriptions of the Chandellas in particular, there are references to *iksu* (sugarcane), which point to the fact that it grew in the land of the Chandellas between the Jamuna and the River Narmada. It is still a sugarcane producing area in India.[45] It is believed that some amount of sugarcane grew in Gandhara.[46] In the *Kalika Purana*, it is stated that sugarcane was produced in Assam in the tenth-eleventh centuries.[47] Sugarcane matures in the winter months along with winter paddy. Mention may also be made of *Dharmasvamin*, the Tibetan pilgrim to Bihar. He noticed standing sugarcane plants of *Tirhut* from which raw sugar (*bu-ram*) was produced.[48] References to sugarcane fields are found in some Muslim accounts,[49] and in some literary texts already referred above. The *Milinda Panho*[50] refers to machines for pressing sugarcane. Sugarcane stalks were used for fuel.

It is interesting to note that there were varieties of sugarcane such as *paundra, kantara, vamsaka, bhiru, kosakara* and a black type of sugarcane. This crop was an important industry of Bengal.

We know from Kautilya that sugarcane crop (*iksu*) were very difficult to grow because, besides subject to various evil, they required much care and expenditure to reap. Moreover, the term *ksara* used in the *Arthasastra* may be taken to mean 'jaggery', sugar candy, etc.

PULSES

Scholars are divided in their opinion about pulse cultivation in Bengal. The *Paryayaratnamala* of Madahavakara testifies to the production of *mudga* in Bengal. The *mudga* pulse produced in Bengal was of two varieties black and yellow. The former was known as *krsna-muga* and the latter as *sona-muga.*[51] The *mudga* pulse may have also been cultivated in some places outside Bengal. The *Abhidhanaratnamala*[52] refers to *mudga* fields, and also to four different kinds of pulses (*masura, kalaya, ralla* and *adhaka*). Probably, these four kinds of pulses grew in Bengal. *Masa* is the name of a grain used for pulse. It is referred to in the *Krsiparasara, Abhidhanaratnamala* and some other works.[53] Some other pulses also grew in Bengal.

The *Paryayaratnamala* mentions *matara* (peas).[54] Mention may also be made of other pulses which may have grown in Bengal, such as *canaka, kulattha,* (horse gram, modern *kurthi*), *khasada* and some others. Khana observed that peas (*kalai*) or kidney bean (i.e. *mug*) may be sown in the same field where mustard was first sown. Both will then grow in abundance to the joy of the cultivators.[55]

Some are of the opinion that pulses did not grow in Bengal and the people of Bengal refrained from taking preparation made out of pulse and 'the practise of taking pulse in Bengal are gifts of Aryan India'. How far these observations are acceptable is questionable. The same author speaks about the use of pulse by the people of Bengal, those who are above the poverty line. From this contradictory statement, it appears that the learned author[56] is ignorant of the subject.

OILSEEDS

Oilseeds of various kinds were known and cultivated in different parts of ancient Bengal. In the *Jātakas,* there is mentions of oilseeds known as sesame.[57] It is also referred to in a verse of Yogesvera,[58] which refers to the maturing sesame in winter. It is also noticed and mentioned in *Kalaviveka,* the *Ramacarita* of *Sandhyakaranandi,* the *Krsiparasara,* the *Brahmana Sarvasva* and in many other works.[59] Sesame fields are mentioned in *Abhidhanaratnamala.*[60] From all these references, it is clear that sesame grew considerably in Bengal.

Castor was another important oilseed but it was not popular in Bengal. Mention may also be made of *tila* oil and linseed (*atasi*).[61] The *Amarkosa*[62] mentions sesame, linseed, and black mustard. Beside all these oilseeds, mustard is frequently referred to in many works. The *Saduktikarnamrta* gives a poetic account about it during winter (*hemanta*);[63] whereas from another work we learn that 'the proper time for sowing mustard are at the end of autumn'. The mustard should be sown very close to one another, the rape seed (*sinapisracemosa*) should be sown rather apart.[64] It was known from very early period, the *Arthavaveda* referred to it.[65] It is also mentioned in the *Sadhanamala, Arya-Manusrimulakalpa, Krsiparasara,* and many other works.[66] The *Abhidhanaratnamala* states that there were only two varieties of mustard. One was *sarsapa* of white colour and the other was *rajika* or *krstika* of black colour.[67] Not only do we have literary evidences about mustard. We also have epigraphic evidence in support of the theory regarding mustard cultivation in Bengal. Thus, the Vappaghosavata grant of Jayanaga (seventh century AD[68] refers to a '*sarsapa yanaka*' (mustard channel) in the Andumvarika-visaya of Karnasuvarna (West Bengal).

Not only do we have references to oilseeds but also to oil mills (*tilapidana-yantra*). These oilmills were sometimes subjected to the payment in kind, of a rent in the nature of a religious cess, as mentioned in some inscriptions.[69] It is interesting to note that oil mills working in villages are mentioned in some cases to have contributed portions of their output, to religious or other endowments. But this is not very common in Eastern India.

WHEAT, BARLEY, MAIZE AND OTHER GRAINS

In some literary works, there are references to wheat, barley, maize and other crops grown in ancient Bengal. In the *Yajurveda Samhita* and *Brahmanas,* mention is made of wheat (*godhumah*), barley (*yavas*) and many other products. References to these are also found in the *Brhadaranyaka Upanisad.*[70] The *Paryayaratnamala* mentions wheat (*godhuma*). The *Saduktikarnamrta* suggests that barley, a winter crop, grew in the region stretching from Uttar Pradesh to Bengal. A poem quoted in the same book states that the hairs of the barley corn growing in the fields were as delightfully green as the blue lotus growing in the pond at the boundary of the village.[70A] Barley fields are referred to in the *Abhidhanaratnamala* and some other works.[71] Yava is referred to as the earliest cultivable grain in India and it is mentioned in the *Ṛgveda,* the *Jatakas* etc. It is still doubtful whether it was cultivated in ancient Bengal.

References to maize fields are found in the inscriptions of Assam (*kostha makkhi-yana*). It is interesting to note that maize was once sold at a very high price in Bengal. Khana knew about it and said if you want to earn money and become rich, then sow maize in the month of *Chaitra* (March–April).[72]

FIBROUS CROPS

Next in importance to the foodgrains, we have important fibrous plants. These were of great service to humanity, since they supplied man with materials for clothing. Chief among the fibrous plants of India are the following:

COTTON (*KARPASA*)

Cotton was an important product of ancient Bengal and also of India. The use of cotton in manufacturing clothes was known in the period of the Indus Valley culture. Weaving was a common industry among the Ṛgvedic Aryans. Mention of cotton is found in the *Asvalayana Srauta Sutra.* Manu states that the sacrificial thread of a Brahmana must be made of cotton. Manu further states that for stealing thread, raw cotton, and such items, the fine should be thrice the value of the article stolen.[73] *Herodotus* is perhaps the first foreign writer who seems to refer to cotton fibre. In his account of India, he writes, 'The wild trees of that country bear fleeces as their fruit, surpassing those of sheep in beauty and excellence and the Indian use cloth made from these trees'. The Greeks of the days of Alexander learnt of its use and the name of *karpasa* is actually mentioned in the accounts of the Greek writers. In the beginning of the Christian era, cotton—both raw and manufactured—formed one of the staples of trade between India on the one hand and Egypt and the Greek world on the other. According to Arrain, it appears that in his time Indian cotton used to be exported to Adulie, a port on the Red Sea and that a trade was established with Ariake, Barygaza and some other places. It is mentioned in the *Jatakas* and also in the *Arthasastra* of Kautilya.[74] Fa-hien calls the cotton fabrics of the country (*po-tie*). Besides we have more literary evidences about cotton, thus in the *Amarkosa* mention is made of two varieties of cotton.[75]

The *Paryayaratnamala* refers to *karpasa* and the *Krsiparasara* to *karpasasodhana* (carding of cotton) appears to be important and may suggest cotton cultivation in ancient Bengal.[76] A verse quoted in the *Saduktikarnamrta*[77] mentions the courtyard of the house of a poor brahmin, which is full of the seeds of cotton (*karpasasthi*). Cotton, along with other products, was included in gifts of villages as learnt from inscriptional documents mostly belonging to the Chandella dynasty. In an inscription of Vijayasena of Bengal it is stated that the ordinary rural folk were familiar with seeds of cotton (*karpasabija*).[78] The early *Charya-padas* also refers to cotton cultivation.[79] Marco Polo states, 'They (i.e. the Bengali) grow cotton in which they derive a great trade'.[80] From all these evidences it is clear that cotton grew in ancient Bengal. Cotton grew not only in Bengal but also in many parts of India including Assam, Gujarat and many other places in Northern India. It may be presumed that not only a simple or ordinary class of cotton were produced in Bengal but also different varieties of cotton grew in Bengal including silk cotton trees (*salmali*). There are two types of cotton, one with plaited seeds and the other having single seeds. Both are known by the name *baiga*, which suggests that this type originally belonged to Vanga (Bengal) from where it went to Bihar. *Baiga* is the most popular type of cotton grown in Bengal.[81]

From the observations of Khana, we learn that the cotton plant should remain wide apart in such a way that cotton may be plucked easily standing in the midst of them or the plant may be crossed at pleasure. Khana further states: 'The jute plants and the cotton plants should not be planted in the same field as the former require sufficient water for its growth while the latter will die out if water reaches them.'[82]

Cotton was an important product of India and it grew in many places of North India, such as Bihar, United Province, Gujarat and Assam and also in some other places.[83]

HEMP (*SANA*) AND JUTE (*PATA*)

We have references to the production of hemp and jute in Bengal. The *Aryasaptasati* describes the rows of hemp plants (*sana-sreni*) standing in a field with yellow flowers.[84]

Jute was also cultivated in Bengal and Assam. Khana observes: 'the jute plants and the cotton plants should not be planted in the same field as the former require sufficient water for its growth while the latter will die out if water reaches them.' The jute crop requires clay soil.

In the *Chandikavya* of Mukundaram, there is a reference to *sadis* and they were called *kala pat-sadi, Agun-pat-sadi.* In short, these were known as *patsadis.* The term *pat,* in modern Bengali, means jute. Thus, though we have no definite proof of its cultivation in our period, it is perhaps not quite unreasonable to infer that jute was known and cultivated in ancient Bengal.[85]

BETEL NUTS

Inscriptions of Bengal, in particular, refer to betel nuts and it seems that betel trees grew abundantly in Bengal. A copper plate grant of Visvarupasena states that the donee named Halayudha had, among other things, gardens of betel nut trees, which were regarded as a source of wealth. It may be presumed that betel nuts were sent to other parts of India from Bengal and thus a lucrative business was carried on in the article. Khana observes: 'that betel nut plants require cowdung (liquid manure) as manure for their growth. In another place, it is said the plants of *mandar* (Erythrine Fulgens) should be reared in orchard of the betel nut plants. The leaves of the former, falling beneath the latter, help the rapid growth of the nuts. It is further stated by the same authority that if the young plants of betel nut are removed from the place where they were first planted, these plants get more strength and grow rapidly.'[86]

BETEL LEAF (*PAN*)

The cultivation of betel leaves was in the hands of a class of people known as *barai.* They lived in the village of Paikpara and neighbouring village of Belka (Bangladesh) as evidenced by an image inscription.[87] The betel leaf cultivators lived there from very early times, and derived their wealth from the sale of betel-leaf grown by them. The copper plate grant of Visvarupasena, to which we have already referred, shows that large plantations of betel leaves existed in Bengal, during the Sena period, augmenting the wealth of her people.[88] The village of Baraipada situated to the south of Pinjakasti in the Vikramapura section of Vanga within the Pundravardhana *bhukti* was also a settlement of the *varajivis* (betel-vine growers).[89] It is said in the *Rajatarangini,*[90] that foreign traders sold betel leaves in Kashmir. A king of Kashmir, liked betel so much that he is said to have gone to the extent of spending his whole revenue in procuring the commodity. This sounds fantastic, but it must have been costly and regarded as a rare luxury.

Bengal and Malabar being the most important betel leaf growing regions in India, it may be surmised that the commodity went to Kashmir from these lands and sold at an exorbitant rate. Betel leaves were to be carried over a long distance and particular care was to be taken by means of quick transport and other devices to preserve their freshness when they arrived for sale. All this must have accounted for the high price, which was to be paid by the Kashmirians for this article. It is interesting to note that no ploughing is required for betel leaves and it grows well in a shady place. The betel vines should be planted in the latter part of July.[91] (After two years one may expect plentiful leaves of the plant.)

BHĀNG

Cakrapanidatta (eleventh century AD), author of *Sabdacandrika,* mentions the *bhāng* plant.[92] Grierson calls it hemp. The leaves of this plant are generally used for preparing a particular type of drink. The *Abhidhanaratnamala* speaks of fields in which *bhāng* was grown.[93]

FRUITS AND FLOWERS

In the *Ramacharita,*[94] names of many trees are found, such as *asoka, amalaki, karuna, pariyala,* etc. References to gardens abounding with fruit-trees and flower-trees occurs in many foreign accounts.[95] Certain kinds of fruits grow only in particular localities, as will be mentioned later. The fruit trees include cocoanut (*narikela*), mango (*amra*), palm, date, figs, madhuka,[96] jackfruit (*panasa*), walnut,[97] gourd (*labakuti*[98]), grape,[99] lemons,[100] monj,[101] etc. Cocoanut (*narikela*) is frequently referred to in the inscriptions of Bengal. The *Ramacharita* refers to Varendra as 'the congenial soil for coconut trees in the world'.[102]

Mango (*amra*) and jackfruit (*panasa*) are mentioned in large number of Pala and Sena inscriptions.[103] Hiuen-Tsang refers to the growth of *panasa* in Pundravardhana and gives a detailed account of the fruit, which was well known.[104]

GRAPES (POMEGRANATES) AND BANANA (PLANTAIN)

Pomegranates grow in different parts of Bengal. We learn from the Govindapur Copper plate of Laksmansena that the western boundary of the gift village of Viddara-Sasana in the subdivision of *Vetadda-Chaturaka* (modern Bator in the Howrah district) was formed by an orchard of pomegranates (*dalimvakshetra*). Pomegranate, grapes, plantain all these fruits are mentioned in other literary works.[105] Ibn Batuta says that the grape trees bear fruits twice a year.[106]

A well-known fruit is the plantain. The plantain tree is frequently depicted in the *Paharpur* terracotta plaques.[107] It is also noticed among the sculptures in the Candi images of the Rajshahi museum.[108] It figures very prominently in the sayings of Khana: 'the banana should be planted by the mother and her sons. One should not cut the leaves from the young plants. It will then yield abundant bananas...' Khana advises, 'cut the thick banana root and plant the same in rows in the month of *Phalguna* (February–March). Then plants will grow from the offshoots in large numbers yielding a large quantity of fruits. The time for planting is the rainy months of *Asadha* (May–June) and *Sravana* (June–July).[109]

According to others, the best month for planting bananas is the month of *Phalguna* (February–March). The banana plants should be given first preference in making orchard.[110] Bananas are valued for many reasons. Even the skin is used for cleaning purposes. Moreover, the Hindus hold the plant auspicious on religious and festive occasions. Plants are mentioned in many other works.[111]

ARECA (GUVAKA)

Another valued product of ancient Bengal was the areca (*guvaka*). The earliest reference to its cultivation is found in the Ashrafpur copper plate of Devekhadga, which states that the donee should enjoy the denoted lands by the cultivation of areca nuts.[112]

The inscription of Bengal frequently refers to it. In one copper plate of *Visvarupasena* it is stated that the donee was to enjoy the land together with the price of the areca nuts. The *Ramacharita* and the *Abhidhanaratnamala*, refers to the cultivation of areca nuts in Bengal.[113]

MANGO

Mango was most extensively cultivated in Northern India, also including Bengal. It is mentioned as a village product in inscriptions from different parts of the country. The *Gaudavaho* and some other works refer to mango trees of Bengal; mango is also mentioned in large number of inscriptions of Bengal. There is a reference to a mango track forming the boundary of a village of Bengal.[114]

JACKFRUIT (*PANASA*)

Jackfruit appears frequently in the inscriptions of Bengal. It is also noticed in some of the literary works and inscriptions.[115] Hiuen-Tsang[116] refers to the abundant growth of *panasa* in Pundravardhana and gives a detailed account of this fruit which was 'highly esteemed'. The Paryayaratnamala mentions other fruits such as *amala, beheda.* palmyra, jujube and *amada,*[117] and the *Ramacharita* refers to the growth of many fruits trees including *Lakuca, bela,* orange, and *mahua.*[118] Certain verses of the *Subhasitaratnakosa* refer to *bela, mahua* and jujube as growing in the villages of Bengal.[119] The *Krsiparasara* and *Gitagovinda*[120] states that palm fruits grew in Bengal. Kalidasa writes that the Eastern parts of India are darkened by the excessive growth of palm trees.[121]

In addition to the above references, we have got an exhaustive list of trees, described by Sukra as[122] fruit bearing, which comprise an *udumyara* (ficus glomerata), etc. Some of the trees are also found in the inscriptions of Bengal. Thus, the Khalimpur plate mentions *vija* (i.e., citra) and *khajura* (date); the Kotalipada plate mentions *parkati* (figs) and the *Ramganj plate* and the *Ramacharita* refers to madhuka (bassia latifolic).[122A]

The *Ramacharita* describes Varendra as a land of excellent flowers, where grew *malati, nagakesara, kesara* (i.e., *vakula*), *madhu* (asoka), *parijata kanaka* (i.e., *champaka*), lotus and *ketaka.* The author describes Varendra as a land suitable for the growth of flowers of endless varieties, where blew a breeze, which was cool and fragrant, because of the presence of water with plenty of lotuses, red and blue.[123]

Honey was also manufactured in some parts of India.[124] The *Ramacharita* also furnishes names of trees which are used for medicine, viz., *amlaki, triphala, haritaki* etc. They were cultivated in ancient Bengal.[125]

BAMBOO

Bamboo is an important product of ancient Bengal. It is used mainly for construction of houses in the villages. Many other items are also produced from bamboo, viz., baskets, mats, fans, even furniture, winnowing baskets, etc. Khana is very conscious about the bamboo and its usefulness. She suggests that 'the smut of corn (paddy)' should be thrown into the bamboo grove. If this is done the result will be according to Khana, very rapid increase of the grove, 'earth should also be thrown beneath a bamboo' grove to serve the purpose of manure.[126]

VEGETABLES

A great variety of vegetables also grew in ancient Bengal. The *Paryayaratnamala* refers to many vegetables, viz., two varieties of garlic; onion of the two kinds—red and white; *patola (paravala)*, (*Trichosanthes dioece*), *olla* (Arum indicum); sobhanjana Moringa *pterygosperma* (modern sajina), *kecuka* (*kacu*, a variety of *kanda*); *sthalakanda* (Vena olakacu), *ervaruka or karkatika* (*kahkuda* modern *kaumkari* or *khekhasa*) beans—white and black (*simba*), pumpkin, *karavellak* (bitter ground, modern *kerala*, several types of *sak, kusmandake* (a kind of gourd) and *vartaku* (brinjal), etc.[127]

The *Kalaviveka* also mentions pumpkin, bean (*simbika*) and brinjal.[128] The *Abhidhanaratnamala* refers to the fields of sake[129] and the *Ramacharita*[130] mentions closely planted rows of *kanda* or *kanda.* Reference to *mulaka* (radish) is also found. Thus it appears that among the vegetables pumpkin, gourds, sakas, brinjal were much popular and they were widely grown in North-eastern India. Hiuen-Tsang included pumkin in his list of the agricultural products of India.[131] From the *Amarkosa,*[132] we learn that the following vegetables were cultivated by the villagers, viz., cucumber, onion, garlic, pumpkin and gourd. The daughters of rishis used to water the trees, plants and vegetables with the help of small pitchers.[133]

Khana made some observations:

According to Khana, the radish should be ploughed at least sixteen times to grow well. The soil should be powdered to dust by repeated ploughing to grow the radish. But in the case of *arum indicum* (*man*) and sesame, the ground should be prepared with a spade. But for growing the sesame, regular ploughing will be required. The time for sowing the sesame is also recorded. The white sesame should be sown in *Aswin* (September–October) or in Kartik (October–November), but the black one is to be sown in *Phalguna* (February–March).[134]

Khana[135] further adds, the *Trichosanthes dioeca* (i.e., *patal*) if sown in a sandy soil it will yield rich crops. The *arum* should be sown in *Phalguna* (February–March). It is advisable not to sow it in a shady place near the river side. The gourd should be reared in a place where ashes are thrown.

The brinjal may be cultivated throughout the year, except for *Baisakh* and *Jyaistha* (April–May and May and June).

SPICES

Besides the above, we have reference to spices of various kinds, grown in Bengal. Thus, the *Paryayaratnamala* mentions some of the spices grown in ancient Bengal. They are the following: cumin seeds of three varieties, black, white and small; cardamom of two varieties; turmeric (its many synonyms indicate its wide popularity and diverse use); javari and corriander.[136] The *Saduktikarnamrta*[137] refers to *javani, satapuspika* (aniseed) and *kustumbari* (corriander crops), cardamom (*ela*) was referred to in the *Gaudavaho*[138] and *Ramacharita.*[139] Also, black pepper (*merica*) and ginger (*ardraka*) were recorded by Marco Polo.[140]

Khana tells us something about the time and manure to be used. Thus, chillies should be planted in October when they are very young. But if planted later, 'the plants become too big and liable to be eaten up by worms.' She further adds, the land which contains rotten paddy as manure is fit for the rearing of chillies. Similarly, turmeric should be planted in *Baisakh* (April–May), and *Jyaistha* (May–June). 'Clear the soil of weeds in *Asadha* (June–July) and *Sravana* (July–August) and then sow in *Bhadra* (August–September).'[141]

Thus, we have many references to spices, oil crops, medicinal herbs, etc. The farmers produced these crops mostly in the villages. Besides, there was a heavy demand by the foreigners (specially the Romans) for our spices, aromatics and other plants, which compelled the people to produce them for export. Thus, it also seems that there are numerous references to successful agricultural growth. But India was not free from famine, drought, inundations, bad harvest, failure of crops, etc. Inspite of adverse conditions, Bengal in those days was not only self sufficient with regard to production of foodgrains but could also afford to export the surplus to other countries. Only pepper and ginger, according to Pliny, fetched a price of a hundred times their original value, let alone incense, spices, arometics and other food grains.

NOTES AND REFERENCES

1. Historical evidence prove that among pastrol people or even semi-savages, agriculture in some form or other has been practised. From the evidence of the Vedic hymns we may safely draw the conclusion that by the time of the earliest hymns, the Aryan masses had settled down to a peaceful agricultural life though the *Vratyas* (one section) retained their wandering nomadic habits (Panca. V; Brahmana, XVIII). Agriculture figures prominently in the *Ṛgveda* (3. II. 6, 6; 1, 179, 6; II, 14, II, 15, 12; IV, 57, 5, 6, 8; VII, 79, II; X, 71; 2, etc.). The word *pancakrsti* (RV. IV.38:10) though explained differently by Sayana, is taken by some to mean either rotation of five crops in the same field or the growing of five kinds of crops in different fields, cf, S.C. Banerji, 1993, *Society in Ancient India,* D.K. Print World Pvt. Ltd., New Delhi, pp. 230; N.C. Banerjee, 1945, *Economic Life and Progress in Ancient India,* Calcutta, University of Calcutta, p. 126 *Pancakrstayah* denotes great tribes. There is much evidence about agriculture as the principal occupation in the age of the Brahmanas

(Sata. 16.13; X. 11.1. 72; II. 2.7.32;). Agriculture finds a prominent place in the society reflected in the *Astadhyayi* of Panini where mention is made of agricultural operations and implements. Agriculture is the principal occupation in the age of Patanjali also. Agricultural land was called *halyas* or *sitya* (1.1.72.) and cultivator's *langalagrah.* Agriculture was the main source of livelihood of the people of the time of the *Rāmāyana.* In the *Mahābhārata,* agriculture figures very prominently. This book requires the king to do many things for the development of agriculture.

In the *Arthasāstra* of Kautilya, the main occupation of the people appear to have been agriculture, cattle rearing and trade. The three being jointly called *varta* (1.41). Agriculture was under the charge of the *Sitadhyaksa.* Kautilya refers to sudra agriculturists. *Kamandaka* states that the subsistence of a vaisya is cattle rearing, agriculture and trade. Those who practised *varta,* according to *Kamandaka,* 'can be assured of freedom from want'. (II. 14).

Agriculture is a prominent occupation in the *Kalpasutras* too. The *Srauta* and *Grhya-sutras* refer to several ceremonies in connection with agricultural operations and cattle-welfare (*Kans.,* S. 24., 1–2; 51. 16. 22. Kath *Gr. Sutra,* 71. 8, etc.). *Raghuvamsa* contains references to rice cultivation in Bengal (Raghu. IV.37).

2. *El,* XXI, pp. 83 ff.
3. The Chalcolithic stage of culture began with the use of copper and bronze and also record of production of foodgrains. Reference may be made to the results of excavations at *Pandurajardhibi* (Burdwan district. West Bengal) which, in 1962 yielded records relating to the oldest inhabitants at the site who appear to be the simple agricultural people, living in mud houses and making hand-made pottery. Near the site where pottery was made, paddy-husks impression were also noticed (P. Dasgupta, Excavations at Pandu Raja Dhibi, pp. 14, 16–20). Scientific analysis of the paddy-husk impression has shown that it is a variety of cultivated paddy, known as *oryza sativa* (*ibid*). Another site in the district of *Birbhum,* namely, *Mahishdal* has yielded carbonised rice assigned to 1380–855 BC. These are the two oldest records of produced foodgrains in Bengal (Cf, IHC, 1981, pp. 100 ff.). Vivilov describes India and Central Asia (including Pakistan) as two separate zones amongst 'seven world centres of origin of cultivated plants, and further holds that these centres have given rise to the whole world agriculture' (IHC, 1981, pp. 42 ff.(N.I.) Vivilov, *The Origin, Variation, Immunity and Breeding of Cultivated Plants,* Trans. from Russian by K. Starr Chester, 1951, *Chronica Botanics,* XII, pp. 1–366.
4. S.K. Chatterjee, *Indo-Aryan and Hindi,* p. 35 ff.
5. *Ibid.,* pp. 31 ff.
6. Digging by stick or stone hoe axe was gift of the Austric people. Reference may also be made to Jhum or shifting cultivation widely practised in North-East India by the primitive peoples who are mostly of Austro-Asiatic origin. P.C. Bagchi points out that cultivation by plough (*langala/langula*) can be traced among the Austric peoples (P.C. Bagchi, *Pre-Aryan and Pre-Dravidian in India.* Levi, Sylvain, Jean Przyluski and Jules Bloch, Trans. from French by P.C. Bagchi, Calcutta, 1975; pp. 8–15); besides plough, even the various words denoting agricultural products

like *dhanya* (paddy/rice), *narikela* (coconut), *quvaka/supari* (betelnut), *pan* (betel leaf) are all of Austric derivations (Cf. P.C. Bagchi, pp. 19; Hutton *Census Report of India,* Vol. I, Part 1, 1931, p. 360). A critical study of folk religious rites and practices of Bengal related to agriculture are of Austro-Asiatic origin (S. Das. *Folk Religion of Bengal,* pp. 12–15; 24–33).

7. Vivilov (1951) described India and Central Asia (including Pakistan) as two separate zones amongst seven world centres of origin of cultivated plants and further holds that these centres have give rise to the whole world agriculture, Vivilov, N.I, *op. cit.,* pp. 1–366).
8. Cf, The Tippera Ins. of Lokanatha furnishes some important information regarding the progress of Aryanisation and cultivation in remote and inaccessible parts of Bengal, which was covered with dense forests.
9. *El,* XV, pp. 113 ff.
10. Cf. The Baigram CP. inscription (*El,* XVII, pp. 45 ff.). The Jagadishpur inscription (*JVRM,* 1, 1972, No. I, pp. 23 ff.).
11. *IHQ,* VI, pp. 45–60.
12. *El,* XV, pp. 307 ff.
13. *Ibid.* The boundaries of the gift land are given. This inscription has some special features, which may be noted here. In the first place the king does not make the grant directly to the brahmin settlers whose names are recorded in the text. Originally, he gives some land to 'a high administrative official who is a brahmin by caste, named Mahasamanta Pradosa-Sarman, son of Tosa-Sarman. On the land received from the king, Pradosa-Sarman constructed a temple of Ananta-Narayana and settled there more than one hundred brahmins, by allo ting to them specific shares of land.' (*El.* XV, pp. 302 ff.).
14. Nar XI, 23.
15. *Ibid.,* XI, 24.
16. Sukra, IV, II. pp. 242–44.
17. R.C. Majumdar, 1963, *The History of Bengal,* Vol. 1, The University of Dacca, Dacca, 2nd Ed. pp. 649.
18. S. Beal, *Si-yu-ki. Buddhist Records of the Western World.* Tr. from the Chinese of Hiuen-Tsang by S. Beal, Delhi, Vols. I and II, pp. 191, 194, 199, 200–201.
19. *Ramacharita* of Sandhyakara Nandi, 1939, ed. by R.C. Majumdar, R.G. Basak and N.G. Banerjee. Published by VRS, pp. 2 ff. Motilal Banarsidass, Rajshahi.
20. *Saduktikarnamrta* by Sridharadasa ed. by R. Sharma, 1933, Lahore, 2.136.5.
21. *Krsiparasara* ed, and tr. by Girija Prasanna Majumdar and Suresh Chandra Banerji, Calcutta, 1960.
22. *The Relha of Ibn Battuta,* trans. by H. Hussain, GOS, Baroda., 1953, p. 234.
23. *MASI,* Vol. 55, p. 57 and plates Xld–4, XI, II, a 5.
24. *IB,* III, pp. 85 ff.
25. *Ibid.,* pp. 124 ff.
26. *IHQ,* IX, pp. 282 ff.
27. Cf, Puspa Niyogi, *Economic History of Northern India,* Chapter, Land Measurement.
28. The *relha* of Ibn-Batuta, trans. by H. Hussain, VOS, Baroda, 1953 in p. 234.

2. NOTES AND REFERENCES

1. Megasthenes—India has many vast plains of great fertility—more or less beautiful ... (pp. 29–30); 'The magic soil of India produced two or three bumper crops' (Diod.: 11.16.5; II, 35.6; Strabo, XV, 1.20; the fertility of the Gangetic plain was an object of amazement for the Greek and Latin writers (Diod. 11.35.36); Pliny, VI. cl7 (2L); *Aelian History*, XII, XXXIII).
2. Hiuen-Tsang bears testimony to the fact that in all parts of the province he noticed regular and intensive cultivation of the land and the production of grains, fruits and flowers in abundance—Beal, *Si-yu-ki, Buddhist Records of the Western World*, Tr. from the Chinese of Hiuen-Tsang by S. Beal, Calcutta, 1958, two Vols., T. Watters, 1961. On *Yuan Chwang's Travels in India*, Munshiram Manoharlal Publishers Pvt. Ltd., Delhi, two vols.
3. Sandhyakara Nandi, *Ramacharita*, by R.C. Majumdar (Ed) R.G. Basak and N.G. Banerji, 1939, Rajshahi (Bangladesh), v., 17–20, pp. 91 ff.
4. *EI*, XXI, pp. 83 ff.
5. Beal. *Record*. II. pp. 191, 194, 199, 200–201.
6. *Raghuvamsa*, IV, 37.
7. *Sridharadasa, Saduktikarnamrta*, by R. Sharma, (Ed.) 2.136, 5.
8. *IB*, III, pp. 85 ff.
9. *Ibid.*, III, pp. 124 ff.
10. *IHQ*, IX, pp. 282 ff.
11. *Medhatithi on Manu*, 2. 112.
12. *Kullaka on Manu*, 2.112.
13. *Abhidhanaratnamala* of *Halayudha*, T. Aufrecht; (Ed.) 1891, (London); reprinted Lahore, 1928, 2,4.
14. Monier-Williams, *A Sanskrit-English Dictionary*, Reprinted, 1956, p. 223.
15. *EI*, VI, pp. 203 ff.
16. *JASB*, LXVII, pp. 99 ff.
17. *EI*, XXII, pp. 150 ff.
18. *Ibid.*, XIII, pp. 216 ff.
19. *Ibid.*, p. 217 n.
20. *Arthasastra*, II, p. 129.
21. *Ibid.*
22. Patanjali, *Mahabhashya*, III, 360.
23. *JASB*, LXIII, pp. 99 ff.
24. *Amarkosa (with Kshirasvami's commentary)*, (Ed.) H.D. Sharma and N.G. Sardesai, Poona, 1941, Chapter I, pp. 5 ff; 70 ff.
25. *Abhiddhanaratnamala of Halayudha*, Ed. T. Aufrecht, London, 1861, Reprinted, Lahore, 1928, Chapter II, pp. 3–9; The Namalinganusasana classifies land as follows: (i) *urvara* (fertile), (ii) *usara* (barren), (iii) *maru* (desert), (iv) *aprahata* (uncultivated), (v) *sadhala* (grassy), (vi) *pankhila* (loamy), (vii) *anupa* (marshy), (viii) *kaccina* (in proximately of water) (ix) *sarkara* (full of pebbles and limestone),

(x) *sakravarti* (sandy), (xi) *nadimatrika* (irrigated by river water), (xii) *devamatrika* (depending on rainfall).

26. *Yadavaprakasa's Vaijayanti,* Ed. G. Oppert, Madras, 1983, pp. 36 ff. (38.1.89).
27. Panini, 2, 1–4.
28. T.C. Das Gupta, 1935, *Aspects of Bengali Society From Old Bengali Literature,* Calcutta University, Kolkata, pp. 222 ff.
29. *Ibid.*, pp. 254 ff.
30. Cf, *Society in Ancient India,* pp. 247 ff.
31. *ARM,* 160.
32. *Krsiparasara* pp. 78, vv. 154, 156.
33. *Ibid.*, vv 30–78
34. *Ibid.*, pp. 70ff, vv. 71–74; cf, Ramayana, Ii, 112, 12, 32.
35. *Ibid.*, pp. 65, vv. 23–25.
36. *Ibid.*, vv. 65–70.
37. *Ibid.*, vv. 32–64.
38. Das Gupta, *op. cit.*, pp. 222 ff.
39. *Arthasastra,* II, 5.
40. *Krsiparasara,* vv. 26–28.
41. *Journal of the Oriental Institute,* IX, p. 418.
42. Cf, S.K. Maity, 1957, *Economic Life of Northern India,* The World Press Pvt. Ltd., Calcutta, see Appendix 1.
43. *Brihat Samhita,* LV, 17–18.
44. *Ibid.*, 19–20.
45. *Ibid.*, 21–26.
46. *Agni Purāna* Vol. II, 1038.
47. *Arthasastra,* II, 24, p. 132.
48. *KP,* VV. III, p. 74.
49. Gopal, Lallanji, 1989, *The Economic Life of Northern India, Delhi,* Motilal Banarsidass Publishers Pvt. Ltd., 2nd Ed. Appendix II, pp. 298 ff.
50. *IHQ,* VIII, p. 200.
51. Dasgupta, T.C., 1935, *Aspects of Bengali Society,* University of Calcutta, Kolkata, Chapter XIV, pp. 222 ff.
52. *KP.* VV. 157–167.
53. On Mann, VIII, 243.
54. *Prabandhacintamani of Merutunga,* 1901, Tran. by C.H. Tawney, Kolkata, p. 82.
55. *Commentary of Nilakantha,* on XII 187, 7.
56. *KP.* VV. 168–181.
57. T.C. Dasgupta, *op. cit.,* Chapter XIV.
58. See Chapter on Products.
59. T.C. Dasgupta, *op. cit.,* Chapter XIV.
60. On Manu, IX. 243.
61. *Medhatithi on Manu,* IX. 350.
62. On Manu II, 112.

63. *Brihat Samhita.* LV, 19–23.
64. Puspa Niyogi, 1962, *Contributions to the Economic History of Northern India from the 10th to the 12th century AD Calcutta,* Progressive Publishers, pp 89–95; cf; *IB,* pp. 68 ff 169 ff: *El*,XIV PP 156 ff; other seed measures, *dronika, khadika El,* XXVI, pp. I ff; *IB,* III, pp. 106 ff, 169 ff.
65. *KP. VV.* 183–185.
66. *Brihat Samhita,* I, VV. 8–9. It states that trees, when being transplanted, should be smeared all over the stem down to the root with ghee, oil, wax, milk and cowdung, (LV–7). The planter should pay honour to the tree by washing it before planting. When the trees have been planted, they need to be watered, morning and evening in summer, during the day in the cold season, and whenever soil is dry during the rains (*Ibid.,* 1. V. 7).
67. *KP, VV.* 185.
68. *Ibid.,* VV. 186–188.
69. *Ibid.,* VV. 189–192; the *Bhattikavya* (Pt. l. 2nd Edition. Banaras, 1951)2, 13; refers to the rows of paddy plants, clearly weeded out in autumn in the Eastern part of Uttar Pradesh.
70. KP. VV. 193–194; 196–7.
71. The *Yajurveda* (vv. 18.12) mentions a number of crops, such as rice, barley, beans, sesame, millet, wheat, lentil, etc., which were produced and harvested in different seasons, and we may infer that they were produced in rotation on the same cultivated plot or cultivated field; cf, Gupta, K.M. *The Land System and Agriculture of Vedic Age.* Sir Asutosh Silver Jubilee Volume, Vol. III, Pt–II. *Taittiriya Samhita* (V. 1.73) states that there were two harvests: barley ripens in summer, medicinal herbs, in rainy season, vrihi (rice) in Sarat, beans and sesame in *Hemanta* and *sisira* (winter). Apparently, there was a rotation of crops. *Yava* (barley) was followed by *vrihi* (rice) beans and *tila.*
72. *Arthasastra,* II, 24.
73. *Mahabhasya,* VIII, 4, 13; cf, V.S. Agrawala, India as known to Panini, Lucknow, 1953, p. 200.
74. *Yuktikalpataru of Bhoja, Isvara Chandra Sastri,* Calcutta, 1917, p. 6.
75. *McCrindle, Megasthenes and Arrian,* p. 53; Megasthenes, witnesses the sowing of wheat in early winter rains and of rice *bosporum,* sesamum and millet in the summer solstice (*Diodorus,* II, 36.).
76. Cf, *Strabo,* XI, 1, 13.
77. *Arthasastra,* II, 24.
78. *Milinda,* p. 114.
79. *A Brihat Samhita,* XL. 2–3; XXV, 2; IX, 43; VIII, 47 – summer crops, XXVII, 1; XXV, 2; X, 18; IX, 42 - autumn crops; XXVII, I – vernal crops.

79A. Cf. The *Ramayana* refers to the traditional six impediments, in agriculture, viz., excessive rainfall, drought, locusts, rats, parrots, foreign invasions (Ram, III, 117, 9; VII, 84, 4–6) The *Aryamanjusrimulakalpa* speaks of *ativrsti, mahavrsti. anavrsti, durbhiksa* and *mahadurbiksa,* (pp. 657, 555, 557, 207, 209, 220, 221, 557, 657, 637) There are other references to drought, famine, etc.

79B. *Harsacarita of Bana.* Trans. E.B. Cowell and F.W. Thomas, Delhi, 1996, p. 329.

80. Cf; *Subhasitaratnakosa of Vidyakara,* Ed. by D.D. Kosambi and V.V Gokhala, 1957, Harvard University Press, V. 264; *Saduktikarnamrta of Sridharadasa,* Suresh Chandra Banerjee, (Ed.) Calcutta, 1965 p. 2, 162, 1.

81. *Aryasaptasati of Govardhanacharya* Ed. by Pandit Durga Prasad and Kasinath Pandurang Parah, 2nd Bombay, 1895, VV. 101, 193, 346.

82. *Krsiparasara.* Ed. and trans. by G.P. Majumdar, and Suresh Chandra Banerjee, Calcutta 1960, pp. 48. The *Krsiparasara* mentions their names as, *vata, bhambha, bhanti, sankhi, gandhi, pandara, mundi dhuti. Sringari, kumari, madaka,* etc. (KP, p. 48) some of them may be identified.

83. *Brihat Parasara in Dharmasastra-Sangraha,* Jivananda Vidyasagara Bhattacharya, (Ed.) Calcutta. 1876. Chap. III.

84. *Prabandhacintamani* of *Merutunga,* Jivananda Muni. (Ed.) Eng. Trans. by C.H. Tawney, B.I. Calcutta, 1894–1906.

85. *Aryasaptasati* of *Govardhanacarya.* Ed. by Pandit Durga Prasad and Kasinath Pandurang Parah, 1895, 2nd revised Ed. Bombay, V. 192.

86. cf. *Subhasitratnakosa, op, cit.,* p. V–264; *Saduktikarnamrta, op. cit.,* p. 2, 162, 1.

87. *Ganaratna mahodadhi of Vardhamana,* 1963, (Ed.) Julius Eggeling, 2nd Edn. Delhi, p. 227.

88. HC *op. cit.,* p. 229.

89. *Desinamamala, of Hemachandra,* R. Pischel, Ed. 2nd Ed. by P.V. Ramanujaswami, 1938, Vizianagaram, III. 55.

90. *SRK, op. cit.,* V. 264.

91. *Ibid;* cf, *Desinamamala, op. cit.,* VI. 26.

92. *Arya.* V. 101 and 346.

93. *Desinamamala* VIII, 69.

94. *AV,* 5–23; 4–10 2–31; 1–2,5; 2–32 2–3 cf. *KP,* p. 48. It is interesting to note that there is a reference to Rama directing Hanuman to drive away from the field with his tail, the insects, birds, and animals which destroy the crops.

95. *Sarngadharapaddhati,* Pp. Peterson. (Ed.) Vol. 1. Bombay, 1888, vv. 3017–18.

96. *KP,* VV. 194–195.

97. In the *AV,* there are hymns devoted to prevention and cures of plant diseases. Thus the seer Badarayani (*AV,* 4. 37) refers to some of the preventive and remedial herbs and trees, viz. *ajasrngi, guggula, pila nalata, anksagandhi, pramandini, asvattha, nyagrodha, mahavrksa, sikhandi, haritaka, arjuna, apamrga, karkari, ayammayi, hiranyaparna and muska.*

98. *KP,* VV. 198–200.

99. A kind of reed, cf, *KP,* p. 83, n. 4.

100. *KP,* VV. 201–05. Those who do not implant nala, do not get the proper yield.

101. *Ibid.,* VV. 206–213.

102. *Ibid.,* VV. 212–213, harvesting of paddy should take place under the following stars (*naksatras*) viz, Ardra, Magha, Krttika, Citra, (or any one of the following: *Mrga-siras, Anuradha and Revati*), *Pusya, Hasta, Svati, Uttara-phalguni, Uttara-asadha,*

Uttara-bhadrpada, Mula and Sravana. Parasara advices to avoid the following viz. *Vyatipata, Visti, Rikta* and *Vaidhrti* and also Tuesdays, Wednesdays and Saturdays.

103. *KP,* V. 237.
104. *Ibid.,* V. 237–240. A measuring utensil for grains: made of wood (*adhaka* is also used as a measure of water, cf, *Ibid.,* V.II). It is twelve fingers in size; the weighing should be from left to right.
105. *Agnipurana,* CXXI, 50 cf, *Medhatithi on Manu,* VIII, 243.
106. KP, VV. 214–220 the *medhi* is a piller, (post) in the middle of a threshing floor to which oxen are bound: (M. Williams). It is to be made by farmers with feminine name, when the sun is in Scorpio, it is to be made of *Nyagrodha* (Ficus Indica), *Saptaparna* (*Bengali-Chatim*), *Gambhari* (*Gmelina arborea*), *Salmali* (*Bengal Simul*), particularly *Udambara* (*Bengal dumur*) or any other milk-exuding tree should be used. It is protected with leaves of neem and mustard (neem and mustard drives away all evil spirits and infection from any source) filled with stalks of paddy, grass and *karkataka* (a kind of plant). *Medhi* should never be made of *Kapittha* (Bengali *kayad bel*) *Bilva* (*Aegle marmelos*), Bamboo and *Trnaraja* (may mean cocoanut tree, a kind of bamboo or sugarcane). Medhi is also known as *Methi* (*ARM.*2.423); or *Mathi* (Vaij, p.125).
107. *ARM.* 2.423; *Arya,* VV. 58 and 468; *HC,* p. 49 *khala* or *khaladhana* was prepared in the month of *Agrahayana.*
108. *Abhidhanaratnamala,* V–578.
109. *Ramacharita,* p. 13.
110. Cf, *Desinamamala* of Hemachandra, III, 37; VI, 34.
111. *KP,* V. 219.
112. *KP.* VV. 221–237.
113. *Amarkosa,* IX, pp. 8–26 pp. 202–205.
114. *Vaijayanti,* p. 164, II, 130–31;
115. *Desinamamala,* 1, 26, 32, 74; II, 56, III, II; V, 23; VI, 15; VII, 94.
116. *KP.* VV, 241–42, the auspicious *naksatras* are *Hasta, Sravana, Dhanistha, Satabhisha, Pusya, Revati, Rohini, Bharani, Mula, Uttara-Phalguni, Uttara-Sadha, Uttara-bhadrapada, Magha* and *Punarvasu.* Thursday, Monday, Friday are good. Saturday is to be avoided.
117. *KP,* VV. 241–242.
118. *Agnipurana,* CXXI, 52–54; Cf, the *Vaijayanti* (p. 104, 11.127–28) mentions sacks, boxes and baskets together with the granary.
119. The *Manasollasa of king Somesvara* G.K. Shrigondekar (Ed.) 2 Vols, Baroda, 1925, 1930. Vol. 1, p. 8, VV. 71–72, The list of reservoirs, includes *kupa* (well), *vapika, puskarini, dhirghika and tadaga.*
120. Prehistoric tablets and carving point out that ancient civilization were born near rivers that supplied irrigation water to the fields. Reservoirs also developed in the ancient past. The *Ṛgveda* contains references to the water of wells being used for watering the fields and we have repeated mention of the word *avata* meaning a well (see *R.V.I,* 85, 10; II, 116, 9; IV, 17, 16; VII, 49, 6; X, 25, 4): It has been pointed

out by one scholar that the most ancient centres of irrigation were the seats of the oldest civilization: the Egyptian introduced system of basin irrigation in the Nile valley; at the same time the Babylonians began their system of perennial irrigation in the joint delta of Euphrates and Tigris: the 'overflow irrigation' evolved by the rulers of ancient Bengal some 3000 years ago-William Willcock, *Ancient System of Irrigation in Bengal*, Calcutta 1930. pp. 166.

121. In a well-known verse in the *Mahabharata* (2.5.67), the sage Narada tells Yudhisthira: *Kaccid rastra tadagini, purnahi ca brhmi ca bhagaso vinivistani na krsi devamatrika* (I hope, that in your domain the tanks are full and increasing in number and the water is divided proportionately). Agriculture should not depend on rainfall. Irrigation was also known in the *Ramayana* (2.74, 10–11). It is stated that while Bharat was marching to meet Rama in the forest his men erected dams wherever it was possible and excavated water course to carry of the excess water (*parivahana bohudakan*) and created many large lakes. They also excavated various kinds of wells; Rama praises the lord of Kosala as *adavanatakah,* i.e. not depending on rainfall but on irrigation (*Ramayana,* II.100, 46) and the *Arthasastra* uses the same ephithet to describe the qualities of a good country (*Arthasastra* VI, I). According to Diodorus (II, 92) a large part of the soil was under irrigation.

122. Cf. A.K. Majumdar, 1977, *Economic Background of the Epic Society*, Progressive Publishers, Kolkata, pp. 135–136.

123. *Kamandaka,* IV, 51–52.

124. *Manu,* VIII, 248; cf, *Medhatithi on Manu* (1, 30).

125. Kautilya refers to two types of *Setu* (or embankment or dam), which is built for holding waters. The *sahatlekasetu* and the *aharyadakasetu.* The former appears to refer to tanks, wells, etc., which are fed by natural springs of water, while the later seems to imply the storing of water in reservoirs by means of embankments (2.1,20) of the two, the former is preferable to the latter—(7–12.4–5): of irrigational work that which is perennial water is better than that which is fed with water drawn from other sources and of works containing perennial water that which can irrigate an extensive area is better (VII, 12, 4–5. p. 331).

126. *Narada.* XI. 18.

127. *Ṛgveda,* 1, 85. 10; 1, 116.9; IV, 17.16; VIII, 49,6; X, 25,4.

128. *Arthasastra,* II, 34. It may be noted that it was followed in ancient Bengal also.

129. *Amarkosa,* 9–10, p 63; *Baudhayana,* II, 3, 6, 32; Gautama, IX, 10.

130. *Aparajitaprccha,* of *Bhuvanadeva,* P.A. Mankad (Ed.) Baroda, 1950, pp. 183–185; VV. 2.4–36; Cf, L. Gopal, *op. cit.,* pp. 288–289.

130A. Cf. J.F. Fleet, *Corpus Inscription Indicarum,* Vol. III, London, 1888. Reprinted, pp. 75, 166, 199; *EI* XI, pp. 107. ff; *Amarkosa,* 9.28.66; The Sasanka dighi of West Midnapore district, which is popularly known as excavated by *Sasanka* king of *Gauda* is about 1/3rd mile in length and a little less in breadth; and there are also many other *dighis* of very big sizes in many other parts of ancient Bengal.

131. *Abhidhanaratnamala,* VV. 675–677.

132. *Upamitibhavaprapancakatha,* p. 56.

133. *Desinamamala of Hemacandra.* R. Pischel (Ed.), 2nd Ed. by P.V. Ramanujaswami Vizianagaram, 1938; 1, 94,89.
134. *IB*, III, pp. 25–41; Gamiyya, a minister of King Hemadrideva of the Berar region, also excavated a deep tank and a well. EI, XXI, pp. 132ff.
135. *IB*, III pp. 44 ff.
136. *Ibid.*, pp. 99 ff.
137. *Arthasastra,* VII, 12, p. 330.
138. *Mahabharata,* Adi Parva, 69.21. (Ed. S.Sukthankar).
139. *Ṛgveda,* 1, 85, 10; 1, 116, 9; IV, 17, 16; VIII, 49, 6; X, 25, 4.
139A. Mahavagga, A., VIII,12,1–2.
140. William Willcocks, 1930, *Ancient System of Irrigation in Bengal,* University of Calcutta, Kolkata, , pp. 5 ff.
140A. A.L Basham, 1967, *The Wonder That Was India,* Sidgwick and Jackson, London, p. 102.
141. *Tilakmanjarikatha* of *Dhanapala,* N.S.P., Bombay, 1903, p. 181; cf, *Aparajitprcchha,* p. 188, v. 37.
142. McCrindle, J.W., 1901, *Ancient India as described in Classical Literature Westminster,* p. 175 Strabo (XV.1.50) refers to the embanked canals from which water was distributed into channels so that all could be benefitted. Patanjali refers to canals for irrigation (*Mahabhasya* 1.1.24; cf 1.82.27); The Hathigumpha inscription (second half of the first century BC) (EI, XX, pp. 72ff) records that King *Kharavela* of the Ceta dynasty caused a canal opened to be brought to his capital from Tanasuli (Tosoli). From the account of Apollonius of Tyana (first century AD) (R.C. Majumdar, *The Classical Accounts of India,* Calcutta 1960, p. 395) we learn that dykes roving from place to place, served the purpose of canals for irrigation. Excavations have brought to light some important canals in different parts of India. e.g *Besnagar* yield a canal 185 feet long (300 BC). Another canal was dug out from Kumrahar (Patna) (*Indian Archaelogoy: A Review* 1954–55 pp. 19ff): the Chola King *Karikala* (first century BC) built embankments along the Kaveri river to control its water and distribute them to irrigate the barren regions (IA, 1908 pp. 233 ff)—The Age of Imperial Unity, Ed. by R.C. Majumdar, Bombay, 1953, p. 231). The Excavation at Nagarjunikonda revealed a huge construction of an ancient canal for irrigation belonging to the time of the Ikshvaku dynasty (*Indian Archaeology—A Review,* 1958–59, pp. 22–3; *JIH,* XI, p. 309).
143. *Arthasastra,* 10, 24. *Amarkosa,* 9.1, pp. 62 ff; 9.36. pp. 6–7; *Mahabhasya,* 1, 1024.
144. *Ramapalacarita,* III, 42, p. 108.
145. *EI,* VIII, pp. 42ff; *IA,* VIII, pp. 257 ff.
146. Annual Report, Baroda State, 134–35, pp. 18 ff.
147. *Dvyasrayakavya,* Cantos, XV, 120–121; PIHC, (Calcutta) 1939, p. 479.
148. *Prabandhacintamani,* p. 79; *Ras Mala,* 1, 104.
149. *Rajatarangini,* VII, 1940.
150. *Ibid.,* IV, 191; V, 84–121.

151. *EI*, I, pp. 122 ff; II, pp. 438 ff; XVIII, pp. 213 ff; XIX pp. 198 ff; pp. 298 ff; XXI, pp. 132 ff; pp. 150 ff; pp. 164 ff; pp. 198 ff; pp. 298 ff; XXVI, pp. 261 ff; XXVII pp. 283 ff, etc.
152. *Ibid.*, XIII, pp. 283–295.
153. *Ibid.*, II, p. 358.
154. *IA*, XVIII, p. 213.
155. *EI*, XIX, pp. 298–299.
156. *Abhidhanaratnamala*, vv. 684–85; *Vaijayanti*, pp.154–55; II, 16–18; 40–42.
157. *Ibid.*, V.685, *Vaijayanti*, pp. 155. 1.42; (water was also raised by baskets tied by ropes to one end of a long wooden pole. Another method was by large leather bucket, tied to a couple of ropes mounted on a wooden pully and drawn by the bullocks).
158. *Ṛgveda*, VIII, 6.9.12; Vedic Index, 1, 39.
159. Muir, Sanskrit Text, V, 465–66.
160. Cf. A.K. Majumdar. *Economic Background of Epic Society*, p. 136.
161. *Desinamamamala*, 1.63, 87; IV, 17, 44; Vll, 36. *Kupatula*.
162. *Abhidhanaratnamala*, V, 685; *Vaijayanti*, p. 155.1.42.
163. Cf, L. Gopal *op. cit.*, 290. *araghattaka* is the name of wheel for raising water from a well (*Abhidhanaratnamala*, V. 685).
164. *Medhatithi on Manu*, XI, 62.
165. *Ibid.*, XI. 162.
166. *Samaranganasutradhara*, I, pp. 178–79; vv. 109–114.
167. *Ibid.*, p. 175, vv. 75–76; for a device to convey water into a field and also to pour it out from it.
168. In Buddhist literature (*Jataka*, 1, 199 ff; 337), there are numerous references to the digging of canals, tanks etc., by joint efforts as well as by the co-operation of the villagers. The Great Epic (*Santi Parva*, 65. 19) informs us that tribal races were encouraged to dig wells for public use. The author of the *Arthasastra* lays down certain rules about the subject. He states that persons refusing to take part in any joint irrigation work (*Sambhuya-setubandha*) should be punished by being charged with the cost of labourers and bullocks employed on that account, and deprived of their share of benefit of the work.
169. *Mahabhasya*, 1.11.
170. *Arthasastra*, III, IX, p. 195.
171. *Ibid.*
172. *Ibid.*
173. *Arthasastra*, III, IX, p. 196.
174. *Ibid.*
175. Manu, IX, 279; *Yajnavalkya*, 11.278; Kautilya holds a similar view. He states 'when a person, breaks the dam of a tank full of water, he shall be drowned in the very tank; of a tank without water, he shall be punished with the highest amercement; and of a tank which is in ruins owing to neglect, he shall be punished with the middle most amercement', *Arthasastra* IV, 11, p. 260.
176. *Manu*, IX, 281, *Visnu* 1, 11.8.
177. *Manu*, IX, 274.

178. *Visnu* V. 15.
179. *Brihaspati,* XXIII.5.
180. *Mahabharata* (Santi Parva) 120.8.
181. *Kaus,* 5.40.1.10.
182. *HB,* Vol. I, p. pp 276 ff; *El,* XXII, pp. 150 ff.
183. *Arthasastra,* II. 6.
184. *Ibid.,* II, 24; a Superintendent of water house *paniyagharika* and Superintendent of works (*karman-tika*) Luder's list, Nos. 1270, 1186: cf. IA, XIV, p. 331; These designations appear to have some connection with the construction of irrigation works.
185. *Ibid.,* III, 9, p. 195.
186. *Ibid.*
187. *Ibid.,* II, 24, p. 131.
188. *Arthasastra.* III. 9, pp. 195–196.
189. *Ibid.,* II, 24; *JBORS,* Xll, p. 139; Bhattasvamin believe that the king was the owner of all land and water.
190. Manu, IX, 44.
191. P.V. Kane, *History of Dharmasāstras,* Vol. II, Part II Chapter, XXV, pp. 865 ff.
192. Cf. *Agni Purana,* CCL VII, 8.
193. *Medhatithi on Manu.* 1, 21.
194. It is believed that the Sumarians was the first to develop the plough about 2700 BC and the Harappans learnt the use of the plough from the Sumarians. Primitive ploughs were made of wood, a perishable material. Hence no wooden plough was found in Harappa. However, a terracotta model of a plough has been discovered from Mohenjo-Daro. Its shape is rather peculiar. Apart from the toy plough referred above; there is an indirect evidence of the existence of the plough during the Harappans time. To the Southeast of the pre-Harappan settlements, a ploughed field was discovered by B.B. Lal and B.K. Thapar. This is the earliest ploughed field so far excavated. It showed a grid of furrows, with one set more closely spaced, running east-west and the other widely spaced running north-south.
195. *AV.* 3.17.3.
195A. *Ibid.*
196. *Mbh, Vana,* 254.17.
197. *Ibid., Santi,* 261–46.
198. *RV, IO.* 117–7; I.23, 15; I.117. 21.
199. *Krsiparasara,* vv. 112–120.
200. Cf. *Desinamamala,* III, 41. Uses the term *javnam* for *halsathanu.*
201. Monier–Williams explains it as a strap of leather on a plough.
202. *The Abhidhanaratnamala,* use the term *samja* in place of *addacalla* (v. 575; *Vaijayanti,* p. 135. 1.55).
203. *KP,* p. 75.
204. *Musti* is equivalent to one cubit measured with the fisted hand—cf, *KP,* p. 75.
205. The word is read differently as *abandha,* and *abandha* in different Mss. cf, *KP.* p. 75, n. 4.

206. Cf, Monier Williams; R.C. Ganguly, 1932, *Some Materials for the Study of Agriculture and Agriculturists in Ancient India*, Serampore, West Bengal—a *viddhaka* was a big hoeing instrument.
207. Probably same as *mai*, a Bengali word meaning a ladder-shaped contrivance for levelling rice fields—KP glossary.
208. *Abhidhanaratnamala*, vv. 575–77; *Vaijayanti* p. 125, II, 56–60.
209. *Prabandhacintamani*, p. 108; II, 1–2; p. 109. cf. *Medhatithi on Manu*, IX, 293.
210. T.C. Dasgupta, 1935, University of Calcutta, Kolkata, *Aspects of Bengali Society*, pp. 229 ff.

3. NOTES AND REFERENCES

1. Cf, Sir G. Watt, *The Commercial Products of India*, 1908; *Dictionary of Economic Products of India*, 6 Vols. Govt. of India, Department of Revenue and Agriculture, 1893; Cosmo Publications, Delhi, 1972; A.H. Church, *Food Grains of India.*
2. Hemachandra, *Abhidhanachintamani*, Ed. By Boht-Iingk and Rieu, St. Petersburg, 1847, IV, 233; cf, *JAOS*, Vol. 61 (1941), pp. 167–71.
3. Madhavakara is the famous author of *Nidana*; he was born in *Silahrada*, probably in Eastern India in the ninth century AD, cf, Patna *University Journal*, Vol. II, 1945–46.
4. Cf, B.P. Majumdar, 1960, *Socio-Economic History of Northern India*, Firma K.L. Mukhopadhyaya, Calcutta, pp. 177–178.
5. Cf, *JAOS*, Vol. 61 (1941), pp. 167–71.
6. The *Amarkosa* (ii. 4) has a whole section dealing with agricultural resources, such as trees, plants, forests, gardens, etc. (As regards the cultivated grains of the earliest period, the Ṛgveda mentions only the *yava* and the *dhanya*. In the *Atharavaveda, vrihi* is repeatedly mentioned. Speaking generally, in the *Atharavaveda* or other later *Samhitas* we find a gradual development of agriculture and copious references to various crops.) The *Vajasaneyi Samhita* (18. 12) contains a list of twelve different kinds of crops such as rice, wheat, *masa, yava, mudga, masure*, etc. The *Vrhadaranyaka Upanisad* (VI. 3. 12) enumerates ten kinds of crops (*gramyani dhanyani*). The *Arthasāstra* of Kautilya gives a list of crops, vegetables and fruits, etc., viz., rice of different varieties, *kodrava*, course grain, *tila* (sesamum), *priyamgu*, pepper and saffron; pulses like, *mudga, masa, masure; kulattha, yava, godhuma* (wheat), *kalaya, atasi* (linseed), *sarshapa* (mustard), etc. some vegetables and *iksu* (sugarcane), etc. (*Artha*, II, 24). Megasthenes also furnished some information about the products of India. See, *Megasthenes Fragments*, I.; Hiuen-Tsang gives an account of the varieties of agricultural products according to the different regions where they were grown. This list includes rice, wheat, ginger, mustard, pumpkin, tamarind, etc. While giving a detailed account he also mentions the characteristic products of the regions visited by him. From this, it appears that almost the whole of cultivable land in India was cultivated. I-Tsing mentions non-glutinous rice, sweet melons, sugarcanes and fruits of various kinds were all-abundant in the country. But millet was scarce.

7. Halayudha, *Abhidhanaratnamala*, Th. Aufrecht, (Ed.)London, 1961, II, 425–29.
8. VIII, 321.
9. *Paribasha*, 115.
10. *Medhatithi*, VIII, 321.
11. *Upa*, 420.
12. In the *Ṛgveda dhanya* is taken to mean grain in general (VI, 3–4); though in later literature it means rice. In the *Atharvaveda* rice is meant by the word *Vrihi* (XII. 2. 54; XVII, 4; XVIII, 3, 6–9); *Taitt. Sam.* distinguishes between black and white rice and speaks of *asu-dhanya* and *maha-vrihi* (II, 3. 1. 3; also *Taitt. Br.* 1. 7, 3, 4). The *Jatakas* contain numerous references to rice (I, pp. 429, 484; II, pp. 135, 378, III, p. 383; p. 276, etc.). References to paddy are found in many works, specially, in Patanjali (*Mahabhasya*, I, 230); *Aryamanjusrimulakalpa* (pp. 34, 170, 534); *Gaudavaho* (VV. 273, 277, 279); *Kathasaritsagara*, (82, 22); Kautilya (*Arthasastra*, 24); Kālidāsa (*Raghuvamsa*) (IV, 37); *Varahamihira*, (*Brihat Sm.* XIX, 4–6; XXIX, 2).
13. The district of the whole of Bengal proper or the great alluvial and Deltaic plain between the Himalaya and the Bay of Bengal and the province of Orissa are the alluvial territory between the hills and the sea, connecting Bengal with Madras—a level area of nearly one hundred thousand square miles constitute the great rice-producing area of India, which is ordinarily much more than self-supporting. *The Administrative Report for 1882–83*, p. 12; cf, Watt, Vol. V, pp. 518 ff.

    According to Sankalia, the whole of Eastern India, comprising Assam, Bengal and Bihar, is pure Neolithic culture region. A fairly extensive, early rice-growing, riverine culture once existed in West Bengal. Pandu Rajar Dhibi, situated on the bank of the river Ajay, is an agricultural village of the Chalcolithic period. The impression of paddy husks from Pandu Rajar Dhibi were identified by the Economic Botanists as those of cultivated paddy *oryza sativa* (H.D. Sankalia, Bombay University, Bombay, 1962. The Prehistory and Protohistory of India and Pakistan, Bombay University, Bombay, 1962, pp. 309–312).
14. *EI*, XXI, pp. 83 ff.
15. *Jat.*, IV, pp. 272 ff.
16. *Raghu.*, IV, 37.
17. *Mahabhasya*, l. 19. H.
18. *Aparka on Yaj.*, 1. 212.
19. *Ibid.*, II, 360.
20. *Ibid.*, I, 230.
21. *Ramacharita*, V, 176, p. 91.
22. *IB*, III, pp. 89 ff; pp. 129 ff.
23. *PRM*, II, 696 and 702.
24. *DS*, pp. 65, 503.
25. *Arya*, VV. 101, 130, 192, 346.
26. *ARM*, 2, 425.
27. Lal Bahadur Dey, 1926, *The Bengal Peasant Life*, Macmillan and Co. Ltd., London, New Edition, Reprinted 1926, p. 201.

28. The *Relha* of Ibn Batuta, Trans. by H. Hussain, GOS, Baroda, 1953, p. 234.
29. *MASI*, Vol. 55, pp. 57 ff and plates.
30. *Amarkosa*, 2.9.15; 2.9.24.
31. *ARM*, 2. 427; *raktasali* was red in colour; *mahasali* was large or long in shape and *kalamsali* was sweet smelling paddy. The Jaina canon testify to the cultivation of the last mentioned variety of rice in Eastern India (J.C. Jain, *Life in Ancient India* as *Depicted in the Jaina Canon*, Bombay, 1947, p. 90). It is also mentioned in other books. This book also mentions, *kangu, nivasa* and *syamaka* varieties of rice found in Bengal. (ARM, 2, 427, 429).
32. The *Manasollasa* of King Somesvara, a twelfth-century work from Western India refers to eight varieties of paddy distinguished by their colour, odour, size etc. The *Susruta Samhita* enumerates various types of paddy (I, p. 207 ff).
33. Varahamihira, *Brihat-Samhita*, XIX, 4–6; XXIV, 2.
34. *PRM*, II, 6–6,702.
35. Shaman Hwui Li, *The Life of Hiuen Tsang*, Trans. by Beal, New Delhi, 2nd, Ed., 1973, pp. 110; Thomas Watters, *On Yuan Chwang's Travels in India*, London, 1905, two Vols—Vol. I, pp. 184 ff. *Si-yu-ki*. Vol. II, pp. 83 ff.
36. Dasgupta, pp. 248 ff; see Appendix for the List of rice. Some names are common in the lists. It is believed that some of these paddies are still cultivated in West Bengal.
37. *Susruta*, 45, 138–140.
38. *Ramacharita*, V, 17b, p. 91.
39. *Amarkosa*, 2, 4, pp. 163: 9, 43, p. 200; 3, 42, p. 279.
40. J.K. McCrindle, Indian Antiquary, 1876–77: Reprinted, Chuckerverthy, Chatterjee and Co. Ltd. Calcutta, 1926, *Ancient India*, Calcutta, 1877, p. 122 n 3.
41. *The Periplus of the Erythraean Sea* Trans. By W.H. Schoff, Oriented Books Reprint Corporation, 1974, New Delhi, 1974 (2nd Edn.).
42. *Saduktikarnamrta* of Sridharadasa ed, by R. Sharma, 2, 176, 3; 2, 177; *Subhasitaratnakosa* of Vidyakara ed, by D.D. Kosambi, and V.V. Gokhala, Harvard, 1957, VV. 282, 316; *Aryasaptasati* of Govardhanacharya Ed. by Pandit Surja Prasad and Kasinath Pandurang Parab, 3rd, Edn. Bombay, 1785, vv. 110, 124, 125, 244; cf, *Raghuvamsa* refers to the wives of the cultivators sitting in the shade of the sugar plants, are mentioned as guarding the *sali* plants. This shows that sugarcane was harvested in those areas where rice was cultivated (*Raghu*, IV, 30). The sugarcanes were generally matured during the winter along with winter paddy.
43. *Harsacarita* of Banabhatta, Trans. by E.B. Cowell and F.W. Thomas, Delhi, 1961. pp. 94, 229.
44. *Gaudavaho* of *Vakpatiraja*, S. P. Pandit, (Ed.) v. 392.
45. *EI*, XVI, pp. 272 ff. *IA*, XXV, pp. 205 ff; cf, Watt, *op.cit.*, vol. VI, Pt. II, pp. 58 ff.
46. *JBORS*, IV, pp. 437 ff.
47. *Kalika Purana* (Ed.) P. Tarkaratna, Calcutta, pp. 62, 67, 74, 87; *Harsacarita*, pp. 217, 218.
48. *Biography of Dharmasvamin*, Patna, 1959, p. 58.

49. Sir H.M. Elliot and John Dowson, *The History of India as Told by its Historians*, 8 volumes, 1866–77, Vol. III, pp. 428 ff.
50. *Milandapanho,* p.166.
51. *PRM,* II, 693–94; a third variety was known as *vanmuga.*
52. *ARM.* 2, 7.
53. *Krsiparasara,* V, 191; *ARM,* 2, 8; *PRM,* 1, 427, 478, 687, 703.
54. *PRM,* 1, 478, 685, 687.
55. T.C. Dasgupta, *op. cit.,* p. 242.
56. N.R. Ray, 1949, *Bangalira Itihasa,* (in Bengali), Calcutta Book Emporium. Calcutta, pp. 542 ff; cf, Laksmidhara does not favour the taking of *mudga* and *masure.* Kalhana considered *mudga* as a food for poor people (*Rajatarangini,* VII, 758).
57. *Jat.,* VI, pp. 335.
58. *Sad.* 2. 175. 5.
59. *KV,* 378–80; RC, 1, 148; KP, v, 167; HBS, pp. 32 ff.
60. *ARM,* 2. 428.
61. *Artha.,* 2. 24, 12–15.
62. *Amra.* pp. 4 ff,
63. *Sad.* 2. 177. 4.
64. Observation of Khana, cf, T.C. Dasgupta, *op. cit.,* pp. 241.
65. *Arthravaveda,* xii. 2. 54; xvii, 3, 8–9; xvii, 4.
66. *Sadhanamala,* II, pp. 463, 568, 589; *AMMK,* III, pp. 673, 677, 684, 704; KP, v., 217.
67. *ARM,* 32, 426.
68. *EI,* XVIII, pp. 60 ff.
69. *EI,* I, pp. 97 ff; 287 ff; XIV, pp. 176 ff; XLI, pp. 85ff.
70. *Vr. Upa.* VI, 3. 12.
70A. *Sad.* 2. 163; 2. 173; 1.2.176; 5.2.177.
71. *ARM,* 2, 8; *KP,* v. 167; *HBS,* pp. 33, 34, 52.
72. *JASB,* LXVI, pp. 113 ff; cf, T.C. Das Gupta, op. cit., p. 243.
73. Manu, 11, 44; VIII, 326–329.
74. *Jat,* III, 286; V, 343; VI, pp. 46 ft. The *Arthasastra* states that Kasi, Vatsa and Vanga were noted for their high-class cotton fabrics (*Kasikam vangakam vatsakam ....ca karpasikam sresthamiti*)—Bk II, XI. p. 84.
75. *Amarkosa,* 4, 47–48, pp. 92.
76. *PRM,* 1, 590; KP, V. 90.
77. *Sad.* 3, 17, 2.
78. *IB,* III, pp. 42 ff (verse 23).
79. R.C. Majumdar, *History of Bengal,* Vol. I, pp. 651 n. 4.
80. Sir Henry Yule, (Ed.) *The Book of Ser Marco Polo,* two vols. 3rd Ed. London, 1903, pp, 115 ff.
81. *PRM,* 1, 330; *Arya,* v, 541; *KP,* v, 216; A.K. Choudhury, *Early Medieval Villages in North Eastern India,* p. 169, Punthi Pustak, Calcutta, 1971.
82. Cf, T.C. Dasgupta, *op. cit.,* pp. 243.
83. Watters, T, *On Yuan Chwang's Travels In India,* two Vols. London. 1904–05, Vol. I, pp. 301; HC, pp. 238 ff.; *JASB.* LXVI. Pp. 285 ff.; *El.* XIV, pp. 295 ff.

84. *Arya*, v. 476.
85. T.C. Dasgupta, *Aspects of Bengali Society*, p. 243 pp. 270–277 ff.; Hemachandra mentions seventeen varieties of grains including *sana*. It is strange to find that the great *Jaina* scholar has included *sana* as a generic expression for grain (*JAOS*, Vol. 61 (1941), pp. 167–71.
86. *IHQ*, II. pp.77; *IB*, III pp. 140 ff; cf, T.C. Das Gupta, *op. cit.*, pp. 237–239.
87. *El*, XXVII, pp. 26 ff.
88. *IB*, III, pp. 140 ff.
89. *Ibid.*, pp. 132 ff.
90. *Rajat.*, Vll, 194.
91. T.C. Dasgupta, *op. cit.*, pp. 242–43.
92. *IA*, XXIII (1894) pp. 260 ff.
93. Grierson 'The Hemp Plant in Sanskrit and Hindi Literature', *IA*, XXIII, (1894) pp. 260 f; *ARM.*, 1, 8.
94. *Ramacharita*, p. 90.
95. Megasthenes states that 'the soil bears on its surface all kinds of fruits which are known to cultivation. India has many huge mountains which abounds in fruit trees of every kind they are of spontaneous growth' (*Ancient India as Described by Megasthenes and Arrian*. Trans. by J.W. McCrindle, London 1877; Calcutta, 1926 pp. 28–29). Hiuen-Tsang, 1-Tsing, Kautilya, etc., refer to different kinds of fruit trees.
96. It occurs frequently in inscriptions of North India; it was used for the manufacture of liquor.
97. *JASB*, LXVI, pp. 113 ff; LXVII, pp. 120 ff.
98. *Ibid.*, LXVII, pp. 90 ff.
99. *HIED*, I, pp. 39, 67.
100. *Ibid.*, I, p. 38; Index, p. 36.
101. Index, p. 59.
102. *Ramacharita*, V, 196, p. 93.
103. *GLM*, pp. 9, 33, 55, lB, III, 1, 14, 57.
104. Beal, *Records*, II, p. 194.
105. *IB*, III, 92 ff.; *PRM*, 1, 656, 657, 661; *El*, IV, No. 34, (Khalimpur plate).
106. *RIB*, p. 18.
107. *MASI*. No. 55 plate IVa.
108. Cf. *HB*, p. 451 and plate LXXVII, No. 181.
109. T.C. Dasgupta, *op. cit.* pp. 246 ff.
110. *Ibid.*, pp. 246–48.
111. *PRM*, 661; *ARM*, 2, 45.
112. *MASB*, Vol. I, pp. 85 ff; *IB*, III, p. 178 ff.
113. *Ramacharita*, III, V, 19 p. 931 *ARM*, 2, 45.
114. B.C Sen, *Some Historical Aspect of the Inscriptions of Bengal*, Calcutta, 1942, p. 99.
115. *IB*, III, pp. 14 ff; 42 ff.
116. Beal, Records, II, p. 194.
117. *PRM*, 1, 219; 656; 662; II, 666 and 667.

118. *Ramacharita,* 1, 3, 12, 13, 19, 21 and 42.
119. *SRK,* VV, 321, 316, 1191.
120. *KS,* V. 200. *Gitagovinda,* chapter 9, pp. 47 ff.
121. *Raghu,* IV, 34.
122. *Sukra,* IV, 4, 95–103, pp. 165.
122A. *IB,* III, 154, 156; *Ramacharita,* III, 21.
123. *Ramacharita,* V, 22B, p. 95.
124. *HIED,* I, p. 38; Index, pp. 24, 56.
125. *Ramacharita,* III, pp. 12, 16.
126. Cf. Dasgupta, pp. 237.
127. *PRM,* 1, 361–375; II, 89–90; 635–38.
128. *KV,* p. 338.
129. *ARM,* 2, 9.
130. *Ramacharita,* III, 12.
131. Watters, I, p. 178.
132. *Amar,* 4, 118, p. 110; 4, 120, III; 4, 148–149. p. 118; 4, 156–157, P. 120; *Raghu,* IV. 42.
133. *Raghu,* 1. 50–53.
134. Dasgupta, pp. 242 ff.
135. *Ibid.*
136. *DRM,* II, 229–231, II, 246–47; *PRM,* II, 127–28; *PRM,* 1, 221; II, 107–8.
137. *Sad.,* 2, 177, 1.
138. *GV,* V. p. 417.
139. *Ramacharita,* 3, 18.
140. Yule, *Marco Polo,* II, p. 115.
141. Cf. Dasgupta, pp. 237; p. 244–45.

APPENDIX I

Ramai Pandit gives the names of different kinds of paddy in the *Sunya Purāna*:

1. Joudhan
2. Jhijhira
3. Amla – Myrobalan
4. Alachit
5. Phaphari – the black rice
6. Sanakharki – straw like
7. Durga bhog – offering for the Goddess Durga
8. Angikal
9. Muktahar – the pearl necklace
10. Kalmugra – the black club
11. Nagaryuyan – youthful lover
12. Tulasali – cotton like *sali*
13. Asati – unchaste
14. Bak-cowrie – the white cowry
15. Gotampalal
16. Pangu-sia – the faded
17. Bhad-mukhi – the bhadoi
18. Tulsa-dhan – the cotton white paddy
19. Dudhuraa – milk white
20. Gujura – Gujarati
21. Yoja-ali – yoke-shaped
22. Dar – the oar-shaped
23. Hathi-Panjar – the elephant's rib (in hardness)
24. Bura Matha – the ripe Datura
25. Hatia
26. Hutia
27. Kaa
28. Tilsagari
29. Lata-mou – honey of a creeper
30. Mau-Kalas – the jar of honey
31. Khajur-churi – the bunch of dates
32. Parvat Jira – the cumin seed of the mountain
33. Gandha-Tulsi – the holy basil (*ocymum sanctum*)
34. Dala-guri
35. Bandhi – the prisoner
36. Basgaja
37. Sitasali – the *sali* rice bearing the name of *Sita*
38. Huccoli
39. Harikali

40. Kusum *sali* – the *sali* rice resembling the beauty and scent of a flower
41. Rakta-*sali* – the red *sali*
42. Chandansali – the sandal-scented *sali*
43. Rajdal – the vanquisher of kings
44. *Ura-sali* – the flying *sali*
45. *Vindhya-sali* – the sali rice of the Vindhya region
46. *Lau-sali* – the *sali* rice resembling a long gourd
47. *Ajanasali*
48. *Kalakartik* – the black beauty
49. *Maghi*
50. *Khirkamba*
51. *Pachal*
52. *Rasaja* – best taste
53. *Kamad* – bestower at will
54. *Khud-dudhu-raj* – tiny milk-white grains
55. *Javana* – coming from Java
56. *Baki* – the bent
57. *Mula* – the radish
58. *Pipirah* – the ant
59. *Kakachi*
60. *Madhavalata* – the Modhavi creeper
61. *Bagunbichi* – the brinjal seed
62. *Kotamata*
63. *Raigarh* – coming from Raygarh
64. *Tojana*
65. *Arbor*
66. *Kumar-bhog* – a prince's meal
67. *Jalarang* – the rice which makes the water red (perhaps after being boiled)
68. *Kanakchar* – yellow coloured (lit. thief of gold)
69. *Lalkamini* – the red-complexioned girl
70. *Sholpona*
71. *Pachaha-bhog*
72. *Bukhi*
73. *Ajana Lakshmi* – the unknown prosperity
74. *Basmati*
75. *Pashi*
76. *Kad* – bestower at will
77. *Gandha-malati* – name of flower Gandhamalati
78. *Ampaban*
79. *Gayabali* – the sand of Gaya
80. *Pathra* – stony
81. *Masilot*
82. *Jhingasal*
83. *Samadhuna*

84. Hari – bearing the name of the god Hari (Vishnu)
85. *Tangan*
86. *Su-asa* – the comfortable seat
87. *Mahipala*
88. *Baksal* – the bent *sali* riches
89. *Mangalan* – the bestower of bliss
90. *Bakchor*
91. *Puan* – the eastern
92. *Bidi*
93. *Gari* – the snail
94. *Gopala*
95. *Hura*
96. *Baskata*
97. *Marich*
98. *Ajay* – the unconquerable
99. *Amol*
100. *Palia*
101. *Dawa* – obligation
102. *Bira*
103. *Boy*
104. *Latra*

APPENDIX II

In Rameswar's *Sivayana*, we find the following names of some paddy:

1. Harisankar
2. Hatipanjar
3. Hura
4. Harkuli
5. Hatinad
6. Hingtsha
7. Haludguda
8. Kala Kanu
9. Kalajira
10. Kalia-kartika
11. Kayakaccha
12. Kasiful
13. Kapat-kuhintika
14. Kalindi
15. Kataki
16. Kusum-sali
17. Kanak-chura

18. Dudhraj
19. Durgabhog
20. Pardeshi
21. Dhustur
22. Krishna-sali
23. Kongur-bhog
24. Kongor
25. Purnima
26. Kalmilata
27. Kanaklata
28. Kamod
29. Garima
30. Khajur-dhupi
31. Khayer-sali
32. Khem
33. Gangajal
34. Gaya-bali
35. Gopal-bogh
36. Gouri-kajal
37. Gandha-malati
38. Guya-dhupi
39. Gunakar
40. Chamardhali
41. Chandan-sali
42. Chattra-sali
43. Jata-sali
44. Jagannath-bogh
45. Jamailar
46. Jalarangi
47. Jhinga-sali
48. Balai-bogh
49. Dhuna
50. Nimui
51. Nandan-sali
52. Rupnarayan
53. Patasa-bogh
54. Pairaras
55. Pipirabak
56. Tilsagari
57. Bak-sali
58. Bakai-buyali
59. Darbangi
60. Bakchur
61. Bura-matra

APPENDIX III

A classified list of paddies cultivated in different parts of ancient Bengal.

Sr. No.	*Names*	*Districts*	*Aus*	*Aman*	*Type*
1.	Parbatgira	Maldah and neighbouring districts	—	Aman	—
2	Badhsabhog	Burdwan, Bankura, Hooghly, 24 Pargamas	—	Aman	Fine
3	Nagra (Nagar-yuan)	Bankura, Hooghly, Burdwan	—	Aman	Medium
4	Elai (Elachita)	Hooghly, Maldah, Rajshahi Dist. Mymensingh Dist.	—	Aman	—
5	Kaya (Kaikachha)	Some parts of West Bengal, Burdwan, Bankura	—	—	Inferior
6	Hookooli (Hoorkosali)	Bankura and surrounding districts	—	Aman	Ordinary
7	Madhablata (Madhabi)	Mymensingh Dist	Aus	—	Ordinary
8	Gopalbhog (Gopal)	Murshidabad, Burdwan Noakhali, Bankura, Maldaha Mymensingh, Dinajpur, Backerganj, Dacca (Dhaka)	Aus	Aman	Fine
9	Jhingasal	Backerganj, Bankura, Murshidabad, Nuddea, Jessore, Pabna, Midnapur Faridpur, Hooghly, Birbhum Burdwan, Maldah, Rajashahi, Dist. Hajaribagh Lohardaga and Santhal Parganas outside Bengal	—	Aman	Fine and Coarse
10	Marichbuti (Marich)	Rajshahi Dists	Aus	—	Medium
11	Kataktara (Kateki)	Murshidabad Hooghly, Dacca, Barisal, Mymensingh, Rangpur, Bogra, Comilla, Jessore	Aus	Aman	Medium Fine
12	Ramsail	Bankura	—	Aman	—
13	Kalmilata Kalamtal or Jatkalma	Hooghly, Burdwan	—	Aman	Fine
14	Kanakchud	Bankura, Mahdah	—	Aman	Fine
15	Dudhkalma (Dudhraj)	Bankura, Hooghly, Burdwan Dinajpur	Aus	Aman	Fine
16	Kelejira	Bankura	—	Aman	Fine

APPENDIX IV

Another list of paddies grown in ancient Bengal (some of these have evidently changed their names or are known differently in different parts by local names).

Sr. No.	*Names*	*Districts*	*Aus*	*Aman*	*Type*
1	Dudhsar	Maldah, Rajashahi, Hooghly, Dacca, Mymensingh	—	Aman	A type of transplanted paddy heavy yielder
2	Dudhsali	Bankura	—	Aman	Medium
3	Sindurmukhi	Bankura	—	Aman	—
4	Hatisal	Hooghly other parts of West Bengal	—	Aman	—
5	Hatipanjar	—	Aus	—	Coarse
6	Tilakphul	Maldah	—	Aman	—
7	Gaurangsali	Bankura	—	Aman	Medium
8	Dadkhani	Bankura, Maldah, Rajshahi, Dacca	—	Aman	Fine
9	Bhramarkandi	Burdwan, Bankura		Aman	—
10	Donarguda	Bankura	—	Aman	Very Fine
11	Chatninakhi	Bankura	—	Aman	more fine than Donarguda
12	Nona	Many dists. of West Bengal	—	Aman	—
13	Kathianona	Bankura	—	—	—
14	Lakshmansali	Bankura	—	Aman	Medium
15	Newali	West Bengal	Aus	—	Ordinary
16	Badkalamkati	West Bengal	—	Aman	Ordinary
17	Barankalamkati	Burdwan	—	Aman	Ordinary
18	Mahipal	—	—	Aman	—
19	Indrasali	Dacca, Rajashahi Bankura, Jessore	—	Green Aman	Coarse
20	Manik Kalma	Birbhum, other parts of West Bengal	—	Aman	—
21	Bhasmanik	West Bengal	—	Aman	Medium
22	Kajalsali	Hooghly, Chinsura side	—	Aman	Medium
23	Kataribhot	Maldah, Hooghly	—	Aman	Fine
24	Radhunipaged famous for excellence and fragrance	Burdwan, Nadia, Murshidabad	—	Aman	Very good and fine
25	Joshabalam	Rajashahi, Hooghly	—	Aman	—
26	Gazia	Rajashahi, Hooghly Mymensingh	—	Aman	—
27	Suryamukhi	Burdwan, Nadia, Backerganj (grown in Highlands,	Aus	—	Medium fine

(Contd.)

Sr. No.	*Names*	*Districts*	*Aus*	*Aman*	*Type*
28	Charnock	Jessore, Murshidabad	Aus	—	Very fine
29	Malati	Dacca	—	Aman	Very popular
30	Bansphul	Many localities of Bengal	—	Aman	Fine
31	Jethi	Maldah	Aus	—	—*

* Cf. Dasgupta, Aspects of Bengal Society.

ABBREVIATIONS

Agni Purana	–	*Agni Purāna.* Tr. M.N. Dutt Calcutta 1903
Ait. Br.	–	*Aitareya Brahmana.*
AIU	–	*The Age of Imperial Unity,* Eds R.C. Majumdar and A.D. Pusalker, Bombay, 1953.
Amarkosa	–	*Amarkosa with Ksirasvamin's Commentary,* Ed. K.G. Oak, Pune, 1915.
Aparka	–	Aparka's commentary on Yajnavalkaya; two vols., Poona, 1903, 1904.
Aparajitaprecha	–	*Aparajitaprecha of Bhavanadeva,* Ed. P.A. Mankad, Baroda, 1950.
ARM	–	*Abhidhanaratnamala of Halayudha.* Ed. T. Aufrecht, London, 1861. Reprinted Lahore, 1928.
Arthasastra	–	*Arthasastra of Kautilya,* English trans. by Shamasastry, Mysore, 8th Edition 1967.
Arya	–	*Aryasaptasati of Govardhanacarya,* Ed. Pandit Durga Prasad and Kasinath Pandurang Parah. 2nd revised edition, Bombay, 1895.
ASI	–	Archaeological Survey of India.
ASR	–	*Archaeological Survey of India Report by Sir A. Cunningham.*
AV	–	*Atharva Veda.*
Beal, Records	–	*Buddhist Records of the Western World.* Trans. from the Chinese of Hiuen-Tseug by S. Beal (Si-yu-ki). Two Vols, 2nd. Ed. New Delhi, 1983.
Brihat Samhita	–	*Brihat Samhita* of Varahamihira, Varanasi, 1959.
Dasgupta	–	*Aspects of Bengali Society,* T.C. Das Gupta, University of Calcutta, 1935.
Desinamamala	–	*Desinamamala of Hemachandra.* Ed. R. Pischel 2nd Ed. by A.V. Ramanujaswami, Vizianagaram, 1938.

EI	–	*Epigraphia Indica.*
Gaudavaho	–	*Gaudavaho of Vakpatiraja,* Ed. S.P. Pandit, Bombay, 1887.
GLM	–	*Gauda-lekha-mala (in Bengali),* Akshaya Kumar Maitra, Rajshahi 1319, (B.S.)
Harsacarita	–	*Harsacarita of Banabhatta,* English trans. by H.C. Cowell and F.W. Thomas, Delhi, 1961.
HBS	–	*Brahmanasarvasva of Halayudha,* Ed. Tejaschandra Vidyananda, Calcutta, 1924.
HIED	–	*History of India as told by its own Historians,* Ed. H.M. Elliot and J. Dowson. 8 Vols., London 1866–77: Vol. II. Ed. by Habib, Aligarh, 1952.
IA	–	*Indian Antiquary,* Bombay.
IB	–	*Inscriptions of Bengal,* Vol. III, N.G. Majumdar, Rajshahi, 1929.
IHQ	–	*Indian Historical Quarterly,* Calcutta.
Index	–	*Biblographical Index to the Historians of Muhammadan India,* Vol. I, Ed. By H.M. Elliot.
JAOS	–	*Journal of the American Oriental Society.*
JASB	–	*Journal of Asiatic Society of Bengal, Calcutta.*
Jat	–	*Jatakas.* English Trans. E.B. Cowell.
JBORS	–	*Journal of the Bihar and Orissa Research Society,* Patna.
Kamandaka	–	Kamandaka–nitisara, Trans. by M.N. Dutta, Calcutta, 1896.
KP	–	*Krsiparasara,* Eds. G.P. Majumdar and S.C. Banerji, Calcutta 1960.
KS	–	*Kamarupa-Sasanavali (in Bengali),* P.N. Bhattacharya, Rangpur Sahitya Parisad, 1338 Sāla (AD 1931).
Kullka	–	*Kulluka's commentary of Manusmrti,* Ed. N.R. Acharya Bombay, 1946.
Manasollasa	–	*Manasollasa of King Somesvara,* Ed. Shrigondekar two Vols. Baroda, 1925, 1930.
Manu	–	*Manusmriti.*
MASB	–	Memories of the Asiatic Society of Bengal.
McCrindle	–	*Arrian's Account of India by McCrindle.*
Medhatithi	–	*Medhatithi's commentary on Manu.* Ed. G. Jha, Calcutta, 1932-39. Trans. G. Jha, Calcutta 1922–29.
Nar	–	*Naradasmriti.* Ed. J. Jolly, Calcutta, 1885.
Panini	–	*India as known to Panini* by V.S. Agrawala, Lucknow, 1953.

PHB	–	*Early History of Bengal*, P.L. Paul, Calcutta, 1939.
PRM	–	*Paryayatnamala* Madhavakara, Ed. T. Chowdhury, *Patna University Journal*, Vol. II, 1945–46, Nos. 1, 2 and 3.
Ramacharita	–	*Ramacharita* of Sandhyakara Nandi Ed. R.C. Majumdar, R.G. Basak and N.G. Banerji, Rajshahi, 1939.
Raghu-Raghuvamsa	–	*Raghuvamsa of Kalidasa*, English trans. by G.R. Nandargikar, 3rd Ed. Bombay, 1897.
Rajat	–	*Rajatarangini of Kalhana*. Ed. and trans. M.A. Stein.
RIB	–	*The Relha of Ibn Battuta*. Trans. by M. Hussain Gus Baroda, 1953.
RV	–	Ṛgveda.
Sad	–	*Saduktikarnamrta* of Sridharadasa, Ed. R. Sharma, Lahore, 1933.
SRK	–	*Subhasitaratnakosa of Vidyakara*, Ed. D.D. Kosambi and V.V. Gokhale, Harvard, 1957.
Sukra	–	*Sukraniti*, Ed. G. Oppert, Madras, 1982. English Trans. B.K. Sarkar.
Vaijayanti	–	*Vaijayanti of Yadavaprakasa*. Ed. G. Oppert Madras, 1893.
Watters	–	*On Ywan Chwang's Travels in India*, T. Watters, Two Vols. Eds. T.W. Rhys Davids and S. Bushell. Munshiram Manoharlal Publishers Pvt. Ltd., Delhi, 1961.

CHAPTER 43

Agricultural Economy of Gujarat: Aspects and Appraisal Upto 1300 AD

Rasesh Jamindar

INTRODUCTION

Gujarat has been a familiar name among tourists in India for quite a long time. Situated between 20°1′ and 24° 7′ latitude and 68.4 and 74° 4′ longitude, the present area of Gujarat consists of 1,96,024 sq km. Bordered by Marwad in the north, Mewad in the northeast, Malwa (of M.P.) and Khandesh (of Maharashtra) in the east, Nasik district of Maharashtra in the southeast, Konkan in the south. Arabian Sea in the west and Sindh (of the west Pakistan) in the northwest. Geographically, Gujarat consists of the three divisions; viz., the peninsula of Kachchha, the peninsula of Saurashtra and the mainland Gujarat, (i.e. the territories between the rivers Banās in the north and Damaṇ-gaṅgā in the south). Watered by seas in the south and west as well as bordered by the great Rann in the north and a small one in the east and southeast, Kachcha is geologically, geographically and physically an area of vital interest. Earlier known as Surāshtra and Kāthiāwāḍ the present-day Saurashtra region of Gujarat, both historically and culturally, is very interesting. As the tropic of cancer passes through Gujarat's northern border, it has an intensely hot or cold climate. But, as noted, both the Arabian Sea and the Gulf of Khambhāt washing its western coast reduce the temperature and render the climate pleasant and healthy. Rivers like Banās, Sabarmati, Mahisagar, Tapti and Narmada on the coastal plains and Bhādar, Shetrunjee and Bhogāvo on the peninsular plains keep her land green and fertile. The presence of the forestcovered rugged hills and mountains on the eastern boundary also help to reduce the intensity of climate. Gujarat's mountains are rich in scenic beauty and have been closely associated with the religious and historio-cultural currents and contours of Gujarat's life. Most important among these are the Girnar, the Shetrunjo, the Choṭilo, the Barḍo, the Ārāsura, the Pāvāgadh and the Tārngā hills. Of these the Girnar, near Junagadh, has a flame-like appearance and many a romantic tales and religious events are associated with it; Pavagadh near Vadodara is famous for one of the *Shakti-peeths* known as the residence of Mata Kali; the Shetrunjo is ornamented with the famous Jaina temples of marble and is known as a temple city on a mountain.[1]

THE NAME OF THE REGION

Generally Gujarat represents the territories that actually comprised the 'Gujarat' from the Chaulukya period and the territories known in the earlier period under different names such as Ānārt, Lāṭa, Aparānt, Surāshtra and Kachchha. Ānart forms the present-day north Gujarat, while Lāṭa that of central and southern parts of Gujarat. These boundaries of Gujarat have been found varying throughout the course of history.

Gujarat derives her name from the Prakrut *Gujjaratt,* i.e. Gurjara Rāshtra in Sanskrit, meaning thereby the 'land of Gurjaras.' The Gurjaras are believed to be an immigrant tribe who entered India along with the Hunas and settled in Rajasthan. The Gurjaras, since called Arbudes, passed through the Punjab and settled in some parts of Mt. Arbudā. These areas came to be known in due course as Gujarat, a name which became popular around the tenth century AD. It is noteworthy, however, that no Gujarat chronicler with the exception of Hemachandra and his commentator Abhayatilaka Gani, used the word Gurjaratra. The familiar terms in the chronicles are *Gurjarabhūmi, Gurjaramaṅḍala, Gurjaradharitri* and *Gujaradharā,* the last one being used even during the Muslim period by the Sanskrit writers. It may, therefore, be concluded that during the twelfth century the word *Gurjaratra* and some of its variants were in use to denote a part of Gujarat, but the only form sanctioned by Hemachandra was *Gujaratra.*[2]

THREE MAIN KINGDOMS

The period chosen here for this chapter, i.e. from the early times through the thirteenth century AD, covers a vast tract of years. Without making any reference to the pre-history and proto-history of Gujarat, within this long period, lie three main independent kingdoms in the political history of Gujarat:

(1) The Western Kshatrapas of western India from 23 to 415 AD,
(2) The Maitrakas of Valabhi from 470 to 788 AD, and
(3) The Solanki dynasty (otherwise known as Chaulukyas of Gujarat) from 942 to 1244 AD.[3]

BRIEF HISTORY OF GUJARAT

The story of early man in Gujarat begins with the Palaeolithic Age, which has left various stone tools in the valleys of the Sabarmati and other rivers. Their antiquity goes back to 150,000 to 200,000 years. The pre-historic stone ages were followed by the proto-historic copper age representing the southern extension of the Harappan civilization. Archaeological excavations at *Lothal* (Ahmadabad District) and *Dhoḷāvīra* Kachchha District) have revealed late and pre-Harappan townships, which have not only yielded a number of seals, sealings and other important artifacts, but also cradled one of the oldest and the most sophisticated civilizations with oldest and largest reservoirs. These have upturned the most popular truth of totally riverine civilization as they are located on the barren area and in the middle of Raṇṇ, respectively. Hence, both these sites are the most important of all Harappan sites.

EPIC AND PURANIC TRADITION

The proto-history of Gujarat is supplemented by the Epical and Puranic traditions, giving an account of the Sharyātas in Saurashtra and the Bhrugus in south Gujarat. The Yādava immigrants from Mathura renovated the old deserted capital of the Sharyātas and settled there, giving the new name Dwārakā to their capital. Among the Yādava heroes, Kriṣṇa Vasudeva played a prominent role in the political history of the country, while Neminātha emerged as a Tīrthaṅkar of the Jainas.

MAURYAN PROVINCE

Though the documented history of Gujarat commences with the findings of inscribed seals from Lothal, but as they are not yet fully deciphered, the real documentary history begins with the reign of Chandragupta Maurya (322–299 BC) whose governor constructed a dam and built a beautiful reservoir at Girinagar (modern Junāgaḍh). The local version of Asoka's 14 rock edicts is inscribed on a rock in the vicinity of the provincial headquarter. That the early history, especially for our purpose here, of Gujarat is full of the imperial grandeur of Chandragupta Maurya whose powerful forces conquered the earlier states of Gujarat. It is said that a Vaiśya called Pushyagupta ruled at Junagadh as the governor of the Mauryan emperor in the last years of the third century BC. It was Aśoka, the grandson of Chandragupta, who not only left an everlasting memorial of his spiritual empire as mentioned in his famous edicts engraved on the rock on the bank of the lake near Junāgadḥ; but also improved the lake with canals.

INDO-GREEKS

Gujarat then seems to have passed under the sway of the Indo-Greek kings. The coins of Kings Menander and Apollodotus (second century BC) were in circulation here for several centuries, thereby proving her contacts with the Hellenistic world. The finds of numerous Greek and Roman coins in this region evidently prove the commercial relation of Gujarat with the western world.

FIRST INDEPENDENT STATE OF GUJARAT

In the first quarter of the first century AD, the Western Kshatrapas of the Śaka origin and of the Kshaharāta family established their rule over Gujarat, which was continued mainly by the kings of the Kārdamaka family. The post-Independence discovery of the early dates of King Chāṣṭana, i.e. (inscription dated S.E. 6 from Daulatpur in Kachchha district) has enhanced confirmation of the foundation of ascribing the origin of the Śaka era to this founder of the Kārdamaka family of the Western Kshatrapa dynasty. His grandson, Rudradaman, also engraved an inscription on the same rock. It has a unique place in the history of the world epigraphy. Their silver coins held a unique place in Indian coinage.

The Western Kshatrapa regime was replaced by the Guptas in about 415 AD. Kumaragupta I annexed Gujarat into the Gupta Empire. Skandagupta has left an inscription (452 AD) on the same famous rock of Aśoka.

MAITRAKA KINGDOM

As the Gupta power declined around the middle of the fifth century AD, Senapati Bhaṭṭarka, the Maitraka general of the Guptas, established himself in Saurashtra with his new capital at Valabhī (now vālā or Valabhīpur). Maitraka kings of Valabhī were very powerful and dominated large parts of Gujarat and even Malwa and more so they rule over Gujarat for a pretty long period as noted earlier. Valabhī not only became famous as the seat of a powerful kingdom, but also could boast of a well-known university, which could be compared with Nalanda. The Maitraka dynasty was uprooted by the Arab attack from Sindh in 788 AD.

CHAVADA RULE

Chāvaḍā (880–942 AD) who were the vassals of the Maitrakas, held sway over some part of north Gujarat and became independent with the fall of Valabhī. Of the eight rulers of the Chāvaḍā dynasty, the name of Vanaraja, the founder of the dynasty and Aṇahilwāḍ Pāṭaṇ, stands out most prominently. He thus established his rule in the northwest part of Gujarat, which was destined to play a great role in the history of Gujarat. The power of the Chāvaḍā dynasty was lost to Prince Mularaja of the Solanki family in 942 AD.

CHAULUKYAS OF GUJARAT

The Solanki dynasty gradually consolidated the whole of Gujarat under its sway and gave her a glorious place in the politico-cultural map of India. Mularaja gave generous grants to learned *Brāhmaṇas* who were invited from different parts of the country, especially from the north India and were got settled in Gujarat. He established his complete hold over Surashtra and Kachchha by defeating Graharipu of Junāgaḍh and Lākho Fulāṇi of Kachchha. Of the other ten rulers of the Solanki dynasty, the names of Siddharaja Jayasimha (1094–1143 AD) and Kumarpala (1143–1174 AD) stand out prominently. The story of Siddharaj Jayasimha besieging the Junāgaḍh fort and ultimately capturing it along with Ranakdevi, the wife of the ruler, Rā Kheṅgār, who became Sati at Vaḍhvāṇ—though not supported by history—has become popular legend of the bards.

Lord Somanātha at Prabhāsa-Pāṭaṇa was the chief god of the Solankis, the temple of which was first built during the Western Kshatrapa dynasty. The temple of Somanātha was sacked twice during the Solanki regime and it was Kumarpala who reconstructed it. It is now rebuilt on the old site during the years 1950 onwards.

VĀGHELĀ REGIME

Vāghelās (1222–1298 AD) who were formerly in the service of the Solanki rulers founded a powerful dynasty after the decline of the Solanki dynasty. Of the eight Vāghelā rulers, Vīradhaval (1233–1238 AD) and Vishaldeva (1243–1261 AD) were mainly responsible for stabilizing the prosperity of Gujarat after the fall of the Solanki rule. The regime of Vīradhaval is noted for the successful administration of his two most distinguished *amatyas*, namely Vastupala and Tejapala, who built the magnificent temples at Abu, Kumbhariya, Girnar and Shetrunjaya. Karana Vāghelā was the last ruler, who forever lost the Rajput hold over Gujarat against the superior forces of Sultan Allauddin Khalji.[4]

CULTURAL PERSPECTIVE OF GUJARAT[5]

The tool technology of Gujarat of the prehistoric time links it with South India on one side and the Himalayan traditions on the other. If similar techniques and shapes reflect relationship of thoughts and thinkers, the Pan-Indic nature of the cultures of Gujarat was established in the remote prehistoric days. But these are worked on the local materials.

A perusal of these cultural traits in the later period in other technologies of copper, clay, stone, etc., is seen. It is interesting to note that anthropomorphic, copper celts, terracotta figurines, seals and sealings of steatite show strong affinities with those associated with the cultures of Sindh and Punjab. The ceramics also indicate the relationship with these parts as well as with Rajasthan and Malwa.

The Vedic tradition continued unabated in Gujarat throughout the centuries. The inscriptions have many references to different Vedic *Sākhās* and the *Brāhmaṇas* who followed them. The Vedic tradition was well established and Vedic scholars were found to be active in Jambusar, Vaḍanagar, Valabhī and other centres. The Vedic scholars coming from different parts of India, e.g. Shrimāl, Gauḍa Sāraswata and Dravida regions maintained this tradition in Gujarat.

This tradition had developed different schools of philosophy. Of them, the *Vaisesika* system of Kaṇāda has its origin, according to the tradition, from Prabhāsa Pāṭaṇ as *Āchārya* Kaṇāda hailed from this place. Then, in the world of philosophies, Gujarat had contributed its might in the earlier times and different philosophies were taught in the centres of Gujarat, viz., Valabhī provided if one relies on Yuan Chwang (or Hiuen Tsang).

Many Vaiśeṣikas were the devotees of Śiva. In Gujarat, Śaivism was fairly strong as it is evident from the temple of Somanātha. The discovery of Shivalingas too, of the early period from Valabhī and the names of the Kshatrapa rulers beginning with *Rudra* suggests its popularity.

It is, therefore, very interesting to note that the Śaiva tradition mentions that the 24th incarnation of Śiva, i.e. Lakuliśa, was from Gujarat. Ulkāgrāma, the place of his birth, Kāyāvarohaṇa, the place where he merged himself with Śiva are existing in the Vadodara district near Narmada. The spread of Lakuliśa Paśupata sect in and outside India and its tradition coming upto at least the thirteenth century AD in Gujarat, the work of Sarvajna about this sect are significant contributions of Gujarat to the Pan-Indic culture.

It is also notable that one of the greatest Vaisnava centres, i.e. Dwārakā, is in Gujarat. This centre of Krushṇa attracts pilgrims from all over India. Early Vaisnava images were obtained from different parts of Gujarat. Moreover, the founder of the *Mahānubhāva* sect, i.e. Chakradhar, was from Bharuch in south Gujarat. He preached in Maharashtra and his followers went upto Punjab and their contribution to the understanding of Mahānubhāva school is significant.

The Puṣṭimārga of Vallabhacharya has a great impact in Gujarat. The Vaisnava *Bhakti* is deeply rooted in the culture of Gujarat. Its great exponent from Gujarat was the 15th century poet Narasimh Maheta from Junagadh. *Padas* and *Bhajans* attributed to him form a rich heritage of Gujarat.

Such features of Pan-Indic phenomenon could be traced in the Śākta worship also. Gujarat had established Śākta centres at Kāravaṇ and Pavāgaḍh in Vadodara district, Bahucharaji in Mehasana district and Ambaji in Sabarkantha district, which were recognized by the Śāktas. The centre of Ambaji also became famous.

This perusal, of course in brief, of a long span of cultural history in different facets like religion, technology, language, literature, and migration of peoples indicates that Gujarat has received not only Pan-Indic influences but also those from other parts of the world. However, it has used the local materials and talent that has contributed to its might in the cultural movements of India and played its own part in the world.

LAND AND PEOPLE

The trans-Mahi area, now known as Charotar (and earlier Chāruttara) and one of the historic divisions of Gujarat, is noted for the high fertility of its soil. Charotar is the home to the Leuva and Kaṇabi Pāṭīdārs, the agricultural aristocracy of Gujarat, then and now. It is here that agriculture is most intensive and population thickest. Adjoining to Charotar, there lays the soil known as Kānam and Vākaḷ (which has half-black and half-red soil). The bulk of the population in both these areas is typically agricultural. The soil of Kānam is black, easily tilted, of great depth and containing a fair proportion of organic matter. Hence, it hardly requires manure. On the other hand, the *Gorāḍu* (sandy and reddish) soil of Charotar requires careful cultivation and abundant manure to produce good crops, but with efforts and proper irrigation, is capable of producing finer crops and a greater variety of them than the best black soil. Tracts like Charotar and Vākaḷ offer greater attractions to settlers, particularly to the superior type of agriculturists, than even the best of Kānam. In north Gujarat, the land everywhere interspersed with large patches of incurably barren salt-land and the villages mere clusters of miserable hovels built of mud and straw, more like pigsties and beehives than human abodes. The half of the land of south Gujarat is fertile, while the rest is notoriously unhealthy and their water is full of organic matter and not fit to drink especially in the Rāṇi Mahal areas. The structure of Saurashtra is the result of its geological formation. The central tract belongs to the 'Upper Cretaceous' and 'Deccan Trap' period, while the narrow strip of coastline all around belongs to the 'Post-Tertiary' period. Consequently, the surface of the land in the interior is undulating. The northern portion is flat while in the south is the Girnar, apart from a few solitary hills. The coast region is an alluvial plain.[6]

A special note on *Bhāl* region is worthy of mention here. Literally, *Bhāl* means forehead but geologically it means alluvial soil. This region, situated around Dhoḷkā in the Ahmadabad district, has a scarcity of good drinking water. Generally in this area the underground water is highly saline. Therefore, parts of Guṅḍi, Dhandhukā, Dholerā Talukas are known as waterless *Bhāl* (i.e. *Na-Pāṇiyo-Bhāl*).

All these features of the land surface and even underground water are highly suggestive of the existence of marine transgression in this area. If this transgression be assumed to be a reality, two points then require further analysis:

(1) Its shoreline; and
(2) Its extent.

An examination of the area beginning from Sarkhej on the west and Vaṭvā on the east (both on the outskirts of Ahmadabad) indicates the differences in levels of about 12 mts. The northern side is on an elevation of average 50 mts. contours and the southern line is of about 38 mts. contours. Between these two contours at 47 mts. height and 42 mts. height, two distinct terrace formations exist. They are suggestive of the tidal zones with the maximum limit of about 47 mts. and lower limit of about 42 mts. with MSL at about 38 mts.

The absence of the similar line almost upto the sea indicates a recession of the marine transgression on the south. In the west, the probability of its extent upto Saurashtra seems to be a possibility. Thus, the vast terrain of the parts of Ahmadabad district and the Bhāl area was submerged by marine transgression prior to the emergence of this land. One has to consider the chronological parameter of this phenomenon. For this, some indicators are available but further study is necessary for closer chronology.

The sand dunes at Ahmadabad have given the proofs of the existence of the late Stone Age industries at Vaṭvā, Shreyas, Solā, Thaltej and other sites. These industries are ascribed to a period of about 2500 BC to about 1300 BC from various parts of India.

Some studies on the Sabarmati, Narmada and other sites suggest that the present soil in this region had already formed by about 10,000 years from now.

Taking these two aspects into consideration along with the existence of Chalcolithic sites on this landform indicates that the land had already formed before the arrival of these groups in this region.

Significantly, some of the village settlements of these groups have been traced almost upto the present-day shoreline in Kheḍā, Ahmadabad and Bhāvnagar districts.

Thus, there is a strong evidence to suggest that the present-day landscape had become fit for human occupation at least by abut 3000 BC and therefore, the sea had already receded from this area before this time.

After the recession of the sea the rivers like Khārī and Sābarmatī; possibly Meshvo and others like Bhogāvo extended their beds in their sluggish flow in their new land.[7]

Hence, such phenomena have bearings on the agricultural growth in this region, then and even now.

While concluding, we may quote Sankalia here: The position and structure of Gujarat are mainly responsible for its complex culture. Its seafront was the gateway from the earliest times through which ideas, influences and cultures passed to and fro as a result of commercial and other intercourses. A succession of influences entered Gujarat in the shape of invaders, refugees and immigrants from its other borders, never truly insurmountable. And the nuclei of all these varied influence, political and cultural, were the fertile Doabs and river tracts in the coastal plain, for instance, the Mahīkānthā and the Sāraswat Maṅḍala boasting of good and prosperous ports; in the peninsula, the foot of the hills and the alluvial strip possessed of a few good harbours. On this alignment grew the kingdoms of Mauryas, Kshatrapas and Guptas in Gujarat with Girinagar as the capital, Maitrakas in Gujarat with Valabhī as capital; of the later dynasties, Gujjaras, Chaulukyas and the Rashtrakutas in the southern coastal plain around Khéḍā, Bharucha and Navasārī; of the later Chalukyas on the Doabs of the Saraswati and its tributaries, the Rupen and the Pushpavati, round Aṇahilavāḍa and Siddhapura; while the famous harbours of Bharucha, Khambhāt and later Surat and Valabhī; Somanātha and Dwārkā owed no less their political or commercial and religious importance to their position.[8]

JOTTINGS OF FOREIGN TRAVELLERS

Yuan Chwang noted that the land of *Malava-desha* was very good for producing winter crop such as wheat.[9] But we hardly get any in Saurashtra and north Gujarat at this time. Arab pilgrims of ninth and tenth century AD have described the land of Gujarat as fertile and agriculture as prosperous. Both crops and fruits were grown in plenty. Mangoes and coconuts were of high quality and the production of lime and honey were good. Sugarcane was also grown at some places.[10]

Regarding the fertility of the soil of Gujarat, we have an excellent testimony left by the Muslim writer Abdullah Wassaf (1238 AD), who finds no adequate words for describing the nature and beauty of Gujarat. He writes: 'In the course of the four seasons[11] of the year, seventy different species of beautiful flowers grow within that province. The purity of the air is so great that if the picture of an animal is drawn with the pen, it is life like. And it is another matter of wonder that many plants and their herbs are found well grown and cultivated there. One may always see the ground full of tulips even in the winter season. The air is healthy and the earth picturesque, neither too warm nor too cool, but in perpetual spring the strength of the soil is so great that the cotton plant spread their branches like willows and plane trees and yield produce for several years. Had the author full leisure to express fully the circumstances of that country (i.e. Gujarat) and to explain them, still he would not be able to record even a portion of the marvels and excellencies of that country.'[12]

SOURCE

From the extremely meagre data that we possess for the period under review, it is very difficult to give any adequate idea and complete picture of the agricultural activity of the people of Gujarat. We shall, therefore, merely present the available information.[13] Some archaeological findings, stone-inscriptions, few travelogues[14] and some literary works[15] help us to draw a gloomy picture of agricultural economy. The copperplates of Maitrakas, available in good numbers, come to our rescue to help design the status of agriculture in Gujarat.[16] The Prakrut work *Aṅgavijjā* also helps understand the different aspects of agriculture.[17] The Girinagar inscription of the Western Kshatrapa King Rudradaman, who engraved his information on the Aśokan Rock, sheds some light on the nature of monsoon and necessity of a reservoir for irrigation purpose.[18]

AGRICULTURE: A MAIN OCCUPATION

It is a known fact that the bulk of the people in India has always been cultivators and undoubtedly, Gujarat is no exception to this during the period under review. The bulk of the people were employed in the cultivation of land. In other words, agriculture appears to have been the chief occupation of the people of Gujarat during the periods under review. Nevertheless, there were in the rich classes of traders and merchants.

Gujarat was ranked among the most important province then.[19] This is specifically evident from the Rock edicts of Girinagar, whose main aim was to note the importance of reservoir and irrigation for agricultural activity. From the study of the copper plates of the Maitraka kings of Valabhī, it becomes clear that the land was the prime thing both for the society and economy. Inscriptions of the Solankis tell the same story.

EPISODES PERTAINING TO LAND

It will not be out of place to narrate here couple of episodes regarding land cultivation in India and Gujarat for showing the importance of land in the life of the people of our country.

The episode of Pruthu, a king of *Bhāratvarsha*, if explained in terms of ordinary life, marks the advent of cultivation in India, when the earth was first brought under cultivation. He removed the unevenness of the earth with the end of his bow (i.e. with ploughshare) and then milked the earth in the form of a cow, and, it is this probably that gave him the title to a king's one-sixth share in the produced of the land in order to protect the fruitful cow. Another such example is that of Viswamitra, who is said to have created a new world of his own, perhaps a new settlement, which he cultivated and where new crops were first introduced by him.[20]

Here is an example from the history of Gujarat. Lākho Fulāṇī, defeated and slain by Mulraj, a Solanki King, in the tenth century AD, is said to have first imported the grain, called Bājarī, to Saurashtra from distant eastern country.[21] In that country, it is said the grain was called *Khaḍ-Dhānya* (coarse grain obtained from grass).[22] This may suggest the date when some portions of the interior of Saurashtra were first brought under cultivation.[23]

VILLAGE PROFILE AND AGRICULTURE

The typical Gujarati village is essentially agricultural. Hence, it seems, the main occupations in the then Gujarat were agriculture and allied activities. While referring to the village of his times, Kautilya uses the terms consisting mostly of Sudras and farmers (*Sudrakarshaka Prāyams*). The usual type of village in Gujarat has been of the same type, as viewed by Chanakya, and that description more or less still holds true; for even now the immense majority in the Gujarati village consists of non-Brahmans and their profession is either agriculture or cattle domestication or some other pursuit subsidiary to it.

It seems from the Asokan edicts at Girinagar that Gujarat enjoys less rain and that too irregular. Hence, water storage is the prime factor for Gujarat, and therefore, the activity of preparing reservoirs in Gujarat is the prime concern, which is evident from the epigraphical evidences. Therefore, near the sea and rivers, we see the settlements, which were dependent on agriculture.

A farmer can hardly proceed with his work without the assistance of a smith or a carpenter, as he needs an axe blade and the ploughshare, the plough and the cart. The smith and carpenter alone can supply these articles to him. Then apart from his agricultural requirements he has also some other in common with the rest of the population. There can hardly be any civilized community, which can dispense with the services of a barber or a shoemaker, a potter or a washerman. All these and similar other artisans have been existing in villages from times immemorial but they exist merely to serve the needs of the agricultural-community.

As the tradition goes, each farmer paid a certain grain share every year to all the village artisans at the time of the annual harvest. Payment was not made in cash; nor was his payment in kind made on each occasion the service was rendered, but annually at the harvest time whether each farmer required the services of the village servant or not.[24]

There is a Gujarati proverb: *Uttam Khétī, Madhyam Vépār Ané Kaniṣṭa, Nokarī.* This proverb indicates that agriculture is the best occupation. Business is mediocre and that the service is mean. This proves that in Gujarat, agriculture was and is the main occupation. Kautilya has to say the same thing. He said that the people of Surāshtra region were known as a race of agriculturists and of warriors.[25] What he says for Surāshtra is equally good for entire Gujarat.

OWNERSHIP OF LAND

It is not unnecessary for our present purpose to discuss whether the entire land in earlier India belonged to the king. But as pointed out by our celebrated seer Yajnavalkya: the king, being the sovereign lord of the state, and all land being vested in him as state property, could make gifts of it to whosoever he desired (III, 401). This suggests that the king was the owner of the land. Does this apply to Gujarat also?

We do not have any references to show what was the position regarding the land-ownership in Gujarat during the pre-Mauryan, Mauryan and post-Mauryan periods. Viewing through the Yajnavalkya's statement we may presume that the ownership of land in the then (i.e. during above-mentioned periods) Gujarat was vested in the king as he being the sovereign head of the state.

But coming down to the beginning of the Christian era, there seems to have been a somewhat different picture. The Western Kashatrapas were ruling over Gujarat from 23 AD to 415 AD. Here we get one epigraphic evidence which shows that Ushavadatta, the son-in-law of Nahapana, who was the most powerful and warlike king of the Kshaharāta family and husband of Dakshamitra, purchased a piece of land worth 4000 Kārshāpaṇas just for donating the same to sadhus.[26] Being the member of the royal family he must had enjoyed the influential status and by virtue of that he and his wife seem to have engraved many stone inscriptions as records of donation of rock caves presented by them,[27] and might have had some say in the state-craft. In spite of these things that he had purchased a piece of land for donation. Hence, we may certainly infer from this evidence that during the reign of the Western Kshatrapas, the ownership of land in Gujarat might have been vested in individual who was cultivating the land and that the proprietorship of the state would have been limited to only fallow land and waste land.[28]

The picture regarding the ownership of land during the Maitrakas seems altogether different. As already observed, Maitrakas of Valabhī ruled over Gujarat from fifth through eighth centuries after Christ. Maitrakas have issued Dānaśāsanas and we come across many such copperplates throughout their rule.

In the sphere of *Dharmadāna,* the *Bhūmidāna* holds great importance. Not only this but for the receiver of *Bhūmidāna,* it becomes a safe source of permanent livelihood. For these two reasons, the prevalence of *Bhūmidāna* during the *Maitraka* administration was most eyecatching. The technical terms used for the copperplate grants were *Tāmraśāsana* or *Dāna-patra.* These two terms are actually used in the records of the Maitrakas of *Valabhī.* Thus of the two terms, the former, i.e. *Tāmraśāsana* indicates the material (*Tāmra* = copper) used for inscribing the grants, while the latter term (*Dāna* = grant and *Patra* = plate) suggest it to be a plate of grant. With the single exception of the earliest known grant of the Valabhī kings, which consist of only one plate[29], all the other Maitraka grants were written on two plates linked by two copperrings upon which was engraved the royal seal where two ends meet proving, thereby, their authenticity. These plates are incised on one side only. Their edges are slightly raised in the form of a rim so that the incised side may not damage by frequent use or contact with the other place.

Bhūmidāna being the main aim of the Maitraka *Dānaśāsanas,* it contains many references pertaining to the title of the land. The ownership of all the lands of the state was vested in the king, as it is evident from the grants themselves. The kings gave all these grants of the land themselves in donation. This evidently proves that the right of donating the land was with the king proving thereby that he was the sole proprietor of the land.[30] On the strength of its possession, gift or purchase, the title of it rightly was in the hands of the landhholder but in actual practice the owner of the land was that person who rightfully enjoys the possession of land. Hence, in these Maitraka grants, we sometimes come across the usage of such words: *Saṅghakshetra, Brāhmaṇskandsatkshetra* and *Kuṭumbisūryyakpratyayakshetra* for showing the ownership relation. Sometimes the *Kshetra* was found mentioned in the name of the filler (i.e. one who is filling the land), e.g. *Kuṭumbisamudraprakrushṭakshetra, Dāsakaprakrushṭakshetra* etc. We know that the holder of the Kshetra was free to legally sell his *Kshetra* (i.e. field) through the state during the Gupta period.[31] In a nutshell, we say that the ownership of land was the prerogative of the state, and hence of the king.

Such *Bhūmidāna*, from the viewpoint of *Dharmadeya*, was normally given either to religious men or religious institutions. Generally the *Dāna* of a *Kshetra* or a village was given as *Bhūmidāna* and sometimes that of *Vāpī* or *Kūpa* (i.e. well and stepwell). During Maitrakas such type of *Bhūmidāna* were given on the basis of *Bhūmichchhidranyāya*; and hence the enjoyment or possession of the land was vested not only in the person who gifted the land but its enjoyment was hereditary (i.e. the possession was passed onto from one generation to the next one and so on). Its ownership remained as long as the sun and moon endure, i.e. forever or eternal possession. Thus the main aim of the *Dharmadeya* was to encourage the receiver economically and hence such land was tax-free or revenue free.[32]

As we do not get the direct reference pertaining to the ownership of a land during Solanki period; at least we may definitely infer from the position regarding the ownership of land during pre-Solanki periods that even during Solankis and Vaghelas the ownership of land might have been vested in the state, i.e. the king might have enjoyed the possession of land that comes under his sway.

We shall mention here in this regard two documents discussed in the *Lekhapaddhati*, both dated Vikrama Samvat 288 (LP. 23), which shows that in case of doubtful proprietorship the land was escheated. The owners in such cases had to go to the court (*Shreedharmādhikaraṇa* = literary meaning—ministry of justice but in the present context it is used in the sense of court); where he had to establish his right and titles with the half of witnesses. If the decree was in their favour, the *Mahāmātya* in charge of the *Dharmādhikaraṇa* reinstituting to them their property issued an executive order.

That procedure of escheat confirms to a provision of Manu (VIII, 200) which lays down that 'Where possession is evident but no title is perceived, there the title shall be a proof of ownership and not possession', which means that mere possession without titles shall not be any proof of ownership at least where the state is adversely affected[33] villages were also confiscated for its owner having taken part in a rebellion. This we learn from a document in the LP dated vs 1288 (1288 (LP 25). Hence, it is possible that Raja Nagapala, whose village is said to have been confiscated, was one of the *Māṇḍalikas* who rebelled against Bhima.[34]

AGRICULTURE: A STATE AFFAIR

It has already been noted that agriculture was the main occupation of the people of Gujarat during the period under review. Not only this but for the progress of agriculture the state was giving unstinted attention as it is evident from the construction of the lake, viz., Sudarshana, in the foothills of the mountain Girnār, by the Mauryan King Chandragupta mainly for irrigation objective; the building of canals from it by Asoka Maurya; the repairs of the dam and canals undertaken vigorously and vigilantly and more so spending money from the state treasury by Western Kshatrapa King Mahākshatrapa Rudradaman after they were damaged by the heavy floods in the adjoining rivers. It was pleasurably looked after by the Gupta King Skandagupta who also built a Viṣṇu temple on the bank of the said reservoir, etc. It appears from these evidences that the development of the agriculture pursuits was the responsibility of the state in Gujarat.[35]

PROCEEDS OF THE AGRICULTURE

It is very difficult to say anything about any sort of agricultural activities that had taken place before the fourth century BC, as we do not have any source or record to name one in this regard in Gujarat. So, we have a very scanty picture upto the fourth century after Christ.

The Asokan Rock Edict at Girinagar, which contains three inscriptions engraved on that rock by three different kings of three different dynasties at three different periods of the history of Gujarat but with only one aim, helps us to formulate the design of agricultural activities in Gujarat from the beginning of the Mauryan dynasty through the end of the Gupta dynasty. This rock edict is a specific specimen to know something about the agriculture, the rainy season as well as state of the resources for the requirement of water in relation to agriculture. Taking into consideration the aim of this rock-dict three points require some elucidation:

1. That Gujarat had experienced and is experiencing the uncertainty of rain. Hence, water storage is a deadly necessity.
2. That she is prone to less rain, i.e. she is situated in an area, which receives less rain, at least from the fourth century BC.
3. That the agriculture of Gujarat is not an *Ākāshiyā Khetī* (i.e. whose crops do not depend for water only on the rains) but it is *Siṅchāi-Pradhān Khetī* (i.e. whose crops mostly depend for water on irrigation).

Hence, all the three dynasties (Maurya, Western Kshatrapa and Gupta) had taken too much care for the growth of agriculture in Gujarat by providing her a reservoir in the foothills of the mountain Girnār and on the banks of the rivers *Suvarṇshiktā* (Sonarekha now) and *Palāshinī* (now Palāsio) whose very name (Sudarshan) suggests its utility, vastness and appearance.

During this period, it may certainly be inferred that the main occupation of the people of Gujarat was agriculture. During this period in the Surāshtra and Bharukachchha regions of Gujarat, the land was fertile and vital. Says the author of the Periplus: The interior parts of the Barigaza and Saurasterene produced abundantly corn and rice, the oil of sesame, purified butter, muslins and coarser fabrics. It has also numerous herds of cattle.[36] From the quotation of Periplus, two points come to our mind:

(1) The regions of Saurashtra and Bharuch, being mainly agricultural, do not seem to have produced cloth for foreign commerce, hence its manufacture is mentioned

as being of a coarse texture.

(2) The domestication of cattle was the main supporting occupation.

It is observed that the business of agriculture keeps the people busy around six months throughout the year and hence they need a supplementary occupation to meet the household expenses and use the spare time for the rest of the period. Two things had come to their rescue as revealed from the Periplus:

1. Domestication of cattle; and
2. Spinning of yarn.

In rearing the cattle—both cows and buffaloes—gave the agriculturists supporting income not only in the spare time but almost throughout the year by selling milk and its products and more so by collecting and depositing the farmyard manure in the form of debris for using it into the fields as and when needed, and so also by rearing oxen for agriculture purpose. Though it brought in little money, the spinning wheel craft also supplied them some income. Especially, the disabled and old ones helped the family monetarily by working on spinning wheel at home almost round the year. Fishing is also one of the occupations of the people of Gujarat living near the seashore: but we do not get any direct evidence for it. It is simply presumed from the use of fish as food.

FOOD DURING THE WESTERN KSHATRAPA PERIOD

Keeping in view the agricultural proceeds, we may say that the people of this period were using wheat, Bājari, rice, and java for food. Milk and ghee were also used as food. Those who were living near the sea shore probably used fish for food. Paddy and *Kodara* (a kind of cheap coarse corn) were also used for food as we have had evidences from Nagarā (now a small village near Khaṁbhāt but in earlier times a very flourishing centre of civilization) excavation.[37] Sakas were fond of onions which is evident from its mention by Vagabhaṭṭa in his work *Astāṅgahridaya.*[38] As per *Charaka-Samhita,*[39] they were eating meat and wheat and drinking wine (made of Madhvīka, i.e. Mahuḍā = Mowra tree = *Bassia latifolia*). It seems that meat was eaten in Gujarat, most probably in royal household as they came from a foreign land. Different types of food items were prepared from rice as is evident from the Prakrut work the *Aṅgavijjā* such as *dūdha-bhāt* (milk and rice) *ghee-bhāt* (rice with ghee), *bhātnī-khīr* (rice cooked with milk), *bāfelā-bhāt* (boiled-rice), etc. Four types of food-preparations were in vogue as narrated by the *Aṅgavijjā*: boiled food, spicy food, stone roller food and steam prepared food. Besides, soup and juice were also taken as drink.[40]

FOOD DURING MAITRAKAS

स धनी यस्य भूभागः says Neeti, meaning thereby that the rich or wealthy is one who has the land. Hence, land donation was the important gift. The important property during this period was the agricultural land, and that the agriculture was the main occupation. The information regarding land grant from the copper plate grants of the Maitrakas indicate the same thing. Generally, we found the system of measurement, limit, place, location, names of the owners and farmers in these *Dānapatras* but we hardly get any information regarding the pattern of crops during the Maitrakas of Gujarat. Whatever little information we gather from them is: The land of *Kheṭaka Āhār* (now Kheda district) and Bharukachchha region was suitable for paddy growing. It is not possible to get clear details of crops of both north Gujarat and Saurashtra regions.[41]

The Arabian travellers of the ninth and tenth centuries AD say, that the land of Gujarat was very fertile and that the agriculture was very prosperous. There is also a reference that the harvest of both grain and fruits was good. Gujarat also produced mango, coconut, lemon and honey abundantly. Sugarcane farming was also in existence in some parts of Gujarat.[42]

A study of place names helps to identify the source from which the names were derived. Hence it is one of the source materials for writing history. Here, we use that source. *Dānapatras* also refer to many place names from which we are in a position to draw some information regarding trees that were growing in Gujarat such as mango, *udumber* or *umarḍo* (fig tree—*ficus glomerata*), *kadamb, shireesh,* (a kind of flowering plant—*albizzia lebbela*), *kavitthika-kothī* (wood-apple tree), *khajjurī* (date palm), *jambu-jambuḍo* (roseapple—*eugenia jambulona*), *nimba-limaḍo* (melia aza—*dirachta*), *palāsha-khākharo* (*buteafrontusa*), *pippal* (*ficus religios*), *Bilva* (*eagle marmaloss*), *vāmsha-vānsa* (bamboo) *shami-samadi* (*Prosopis spicigera*), etc. These are just some of the specimens. We have more suggestions for some more trees.

In Maitraka's *Dānaśāsanas,* there is one word we often come across, is *vrīhipiṭaka.* This term is of much importance for knowing a particular crop harvested in particular regions. Here, *Vrīhi* means paddy and *Piṭaka* means basket. This is a kind of measurement, about which we will have a separate discussion later on. That the *Vrīhipiṭaka* measurement was in use in both Kheṭaka and Bharukachchha. The usage of this term helps to know that paddy was cultivated in Kheṭaka and Bharukachchha areas of Gujarat during the Maitrakas of Valabhī.

SOLANKIS AND VAGHELAS

We get a little clearer picture of crops during these periods, though the sources for that are very scanty.

The principal crop of Gujarat at this time was probably rice, which seems to have formed the staple food for the people. In the *Nābhinandan-Jirṇoddhār Prabandh*[43] there is a specific mention of some crops such as *Tuver* (pigeon pea), *Aḍad* (black-beans) wheat, paddy and *juwār* (kind of corn). This work refers to following fruits: orange, lemon, *jambul* (rose apple—*egenia jambulona*), *kelā* (plantain tree), *kothuñ* (wood apple tree fruit), *Karmadāñ* (Corinda fruit *arissa corundo*), *chārolī* (kind of dry fruit, *chirongia sapida*), *pīlu* (kind of frui—*solanum nigrum*), mango, *sitāfaḷ* (custard apple—*annona squamosa), Bijoroñ* (*Citron-citrus medica*), *khajūr* (date palm: kind of sweetmeat); grapes, sugarcane, *Faṇasa* (Jackfruit—*artocarpus integrifolius*). The Prakrut work *Dvyāshraya* notes these fruits: *Sopārī* (betel nut), *Shreefaḷ* (coconut), *Dāḍam* (pomergranate: *punica granatum*), Āmbalā (hog plum—*hyllanthus embellica*), etc.[44] From other sources, we also come to know about the following crops: *masur* (lentil, sort of pulse), *chaṇā* (gram), *vaṭāṇā* (field pea/pisum), *tuver* (kind of pulse—*cajanus indicus*), *java* (barley), (*juwar* kind of corn), *til* (sesame seed indicum), *bājarī* (a kind of cereal) millet, *kodarā* (kind of cheap coarse corn), etc. Besides the cultivation of *Sheraḍī* (sugarcane), *gaḷī* (*Indigo-feratinctoria*), *kapās* (Cottonseed, cotton plant), etc., were also there. *Nāgarvel—pān* (leaves of betel plant) in the *Māngrol-Chorwād* area and coconut on the sea coast were also growing. The sugarcane crop was abundantly growing and hence the business of making jaggery, sugar and lump sugar from it was very brisk. *Varṇaksamuchchaya* notes the nine types of jaggery, fourteen varieties of sugar and seven kinds of *sākar* (sugar candy or lump sugar). *Kothuñ* was the type of the sugarcane press machine.[45]

But the principal crop, as has been mentioned earlier, was probably rice, which seems to have formed the staple food of the people, not only during Solankis and Vaghelas but also even during Mauryas, Kshatrapas, and Maitrakas. In the *Deshīnāmamālā,* Hemachandra included several local words all meaning nothing but paddy. These are *Aṇuo, Joṇṇaliyā*[46] that, according to Hemchandra, means Jowārī or *Dhānya.* Though, in the same verse he mentions Jowari as another local word. Hemchandra has also mentioned *Shālidhānya,* which was harvested during the autumn and was probably of two kinds: one is called *Talapphalo* and the other one *Talahalo.*[47] It is difficult to say whether all these local words mentioned here meant different varieties of rice or were merely synonyms. For the present, one may accept that the terms denoted different kinds of rice.

Meagre though it is, the only information of the rotation of crops is supplied by Wassaf. According to whom the winter cultivation was brought about only through the moistness of the dew, called *barasi.* When the harvest is over they (i.e. farmers of Gujarat) begin summer cultivation, which is solely dependent upon the influence of rain. The vineyards in this country bring forth blue grape twice a year.[48] But the main harvest, particularly of paddy, was collected during autumn.[49] According to Al-Idrisi (1154 AD), wheat was grown in the region of Khambhāt[50] and the document in the *Lekhapaddhati* (p. 21) mentions the harvesting of wheat in the month of Jyeshtha. Thus it appears that the second principal crop was wheat which was cultivated in Gujarat and harvested during the summer.[51]

In the *Deshīnāmālā,* Hemachandra mentions *Udido* for which he gives as Sanskrit equivalent the word *Māsha-dhānya* (1, 98). Here, *Māshā* means a bean; hence *Māshā-dhānya* probably means some kind of leguminous plant such as black gram as suggested by Pischell. He also suggested that the word *Udido* was derived from Tamil *ulud* and Canarese *uddu,* so that the plant was probably borrowed by Gujarat from her neighbouring provinces, though the process might have been exactly the reverse.[52] *Chaṇā* (gram) also seems to have been grown for which we have the local word *Anuio.*[53] We also come across another local word which Hemachandra explains as *Masur-Adinama-Pishtam* (DN, 11, 110). Evidently, some pulses other than *Masur* was intended by the word Adi, but Hemachandra does not mention them in the DN as to which were they.

He, however, mentions 17 kinds of grains in a quotation in his commentary to *Abhidhāna Cintāmaṇi* (IV, 233). As usual, he used the word *Dhānya* in its widest sense and included not only pulses and grains but hemp and sesame also. The 17 enumerated *Dhānya* are thus: *Vrīhi* (rice, paddy which ripens during monsoon), *yava* or *Java* (barley), *Masura* (lentil), *Godhūma* (wheat), *Mudga-munga-mag* (kidney bean), *Māsha* (black gram), *Til* (sesame), *Chaṇak* (chick pea), *Aṇava* (great millet), *Priyangu* (Italian millet), *Kodrava-Kodarā* (kodon millet), *Mayushṭhaka* (*phaseolus aconitifolius*), *Sāli* (rice harvested in autumn), *Āḍhaki* (pigeon pea), *Kulattha* (horse gram), *Kalay* (pea) and *Sliana* (hemp). The list by Hemachandra probably represents a fairly accurate roll of the principal crops grown in Gujarat during his time.[54]

Sheraḍī (sugarcane) was the very important agricultural product during this period. Hemachandra, in his work DN, mentions three different words for *Sheraḍī*: *Ingali*, *Angaliam* and *Gandiri* (1, 28, 79; 11, 2, 82). It is not possible to say whether these three words denote three types of sugarcane or are merely dialectical variations of the same name. He also gives one more word for sugarcane field: *Uchchhusranam* (or Ikshuvathah) DN, 1, 117). In his other work, namely, *Trishashṭishalākāpurushacharit* (Hence forth T.S.P.), he states the growing of purple sugarcane plantation guarded by milkmaids while singing (IV, 14). The *Nābhinandanoddhārprabandha* also mentions the cultivation of sugarcane. Hence, it seems, after rice and wheat, the sugarcane was the third main agricultural product. Thus, the sugar industry played an important role in the economic life of Gujarat.

Cotton is another important product of Gujarat during Solankis and Vaghelas. In DN, there is a word *Karaini* for the cotton tree (11, 18) that is also called *Shālmalī-taru* or *Shīmaḷo* (silk cotton tree, *bombax mulbericum*). It may, therefore, be infered that during this period *Shālmli* cotton was used for making mattresses, quilts and pillows. Besides, cotton was also used for weaving cloth. The author of the Periplus, Marco-Polo (1294 AD) and many Muslim travellers and geographers testify to the growth of cotton in Gujarat. Some of them have praised the excellence of the textile products of the country. According to Marco Polo, Gujarat produced plenty of cotton in very big trees which reached six paces in height when 20 years old were used for quilting and stuffing and not for spinning: only cotton gathered from trees less than 20 years old was used for spinning. Macro Polo does not state how the cotton grown on trees more than 20 years and less then twenty years was used.[55]

Among arts and craft, which gives us an insight in the socio-economic life of the people of Gujarat; that of cotton manufacture is the most significant with its deep-rooted effect upon their national character. With the evaluation and progress of cotton manufacture in Gujarat, the Greek historian Herodotus (445 BC), being struck with one peculiarity that distinguished India from other countries, records that cotton was the customary and common wear of the Indians in general and Gujaratis in particular. They possess likewise a plant. Which instead of fruit, produces wools of a finer and better quality than that of a sheep; of this, the Indians make their cloths.[56]

The author of the Periplus mentions that the Parthian kings held under their sway the whole of the northwest area, including Sindh, Saurashtra and northern Gujarat as far as Bharuch. In spite of this foreign domination, the province of Sindh was a flourishing one in the matter of manufacture and commerce. Amongst various commodities that this priovince supplied for export to foreign lands, the author notices cotton cloth, silk yarn and indigo. Proceeding further he mentions the coast of Syrastrene as being a very fertile province, which produced, among other things, 'cotton and Indian cloths made therefrom of the coarse sorts'. The region of Saurashtra, being mainly agricultural, does not seem to have produced cloth of foreign commerce; hence its manufacture is mentioned as being of a coarse texture.[57]

The *Paṭoḷās* of Pāṭaṇ have been famous since the days of Solankis and even earlier. In preparing the *Paṭoḷās,* bundle of multi-colour warp and weft are so arranged that when woven, each colour appears at the exact place where it is required by the exigencies of the pattern and what is most curious is that there is no 'wrong' side to the completed fabric. The lovely wedding sarees of Gujarat are made of *Paṭoḷā* silk. The silk sarees of Saurashtra, especially the embroided ones, can hardly be equalled throughout the length and breath of India for finish and beauty of colouring. We can learn from contemporary poets that silk was made in Pāṭaṇ and rough quilts and blankets of wool in Sorath. The popular saying, *Paḍī Paṭoḷé Bhāt Fāté Paṇ Feeté Nahiṇ* shows that colour printing on *Paṭoḷā* silk never fades away even though the silk (i.e. cloth) itself be decayed.[58]

Cultivation of Indigo was known in this part of Western India from the very early days. The author of the Periplus states that the indigo was cultivated in Barbaricum (p. 38) and Marco Polo mentions that there was an 'abundance of Indigo' in Gujarat.[59] It may be concluded from these statements that indigo was grown in large quantities in Gujarat. *TSP* of Hemachandra mentions the colour of Indigo. Both Mandelslo (1638 AD) and Thevenot (1666 AD) also say that the indigo of Sarkhej (now a subrub of Ahmadabad) was excellent in all kinds of indigo[60] and that indigo was exported from Ahmadabad in a great bulk (or quantity).[61]

The crop of sesame was cultivated and oil was produced from it in Gujarat, according to the Periplus (39). There is a word *Til* for sesame in *Dvyāshraya* (XVIII, 18–19). It also mentions two more words for this crop: *Urnayu* and *Vati* (*ibid.*, 73). There is a story that king Arnoraj (grandson of the King Vigrahraj, the 3rd of Shakambhari and contemporary of Solanki King Siddharaj) always considered his enemies like *Urnayu* and *Vati* and destroyed them. From this, it seems that the sesame product was harvested abundantly in Gujarat.

According to Hemachandra, the local word *Umato* (*Ummatto, Dhatturaka*) meant Éraṅḍ, (*resinus communis*) which is caster-oil plant (*palma christi*).[62] The same word also meant thorn apple. Probably it was cultivated in Gujarat. The word *Éraṅḍā* is mentioned in *DN* (1, 89). It is a seed from which the caster oil is produced. In *Dvyāshraya* (XIX, 79), we also find mention of *juwār* (kind of corn), *sarsava* (rapeseed or mustard seed), *uma* (i.e. turmeric) and *gudadhāṇā* (coriander seeds). We may thus infer that all these crops were produced in Gujarat.

An inscription from a stepwell, known as Soḍhalīvāv in Mongrol, mentions the best production of betel leaves and hence a tax of one *kārshāpaṇa* was levied to each field.[63] Hence, we may certainly say that the crop of *nāgarvelpān* was cultivated in great quantity in Gujarat during Solankis. *Devapattan–Prasasti* from Sintra states that the leaves of *nāgarvel* (betel plant) were use in Somanath Mahādeva temple for worship.[64] *Sopārī* (areca of or betel nut) is also mentioned in the same inscription from Mongrol wherein betel-plant leaves are mentioned and *Sintra-Prashasti* also states that betel nuts were also used for worship at the *Somnath* temple. *Nābhinaṅdanoddhārprabandha* does refer to *sopārī*. This was written immediately after the fall of Vaghela's reign. All these evidences prove that the crop of *sopārī* was abundantly cultivated in Gujarat during this period.

Marco Polo mentions that quantities of pepper and ginger were exported from Gujarat (p. 332), but it is not clear whether these were local products or imported from neighbouring areas and then exported.

Various fruits and vegetables must have grown in Gujarat during this period. In *Nābhinandanoddhārprabandha*, we come across the names of the following fruits: *nārangī* (orange tree), *limbu* (lemon), *jāmbu* (rose apple), *kelā* (plantain), *kothuñ* (wood apple), *karmaḍoñ* (corinda plant), *chārolī* (chirongia sapida), *pilu* (*solanum nigrum*), *keri* (mango), *sitāfaḷ* (custard apple), *bijoron* (citrus medica), *khajūr* (date), *drāksha* (grapes), *faṇas* (jackfruit = *artocarpus interifolius*), etc. Hence, all these fruits were produced in Gujarat during the Solankis and Vaghelas. Hemachandra also mentions in the *Dvyāshraya*, the names of *sopārī* (betel nut), *shreefaḷ* (coconut), *amajon* (*embellic myrobalan*), etc. In *DN* (1, 12, 15; 111, 7, VI, 48), there is a reference to *dāḍam* (pomegranate). It seems, the Gujarat was very rich in growing these fruits. Regarding fruits, Abul Fazl mentions that in the country between Surat and Bāglāṇ (i.e. the modern Navasari district), pineapples, pomegranates and oranges used to grow.[65]

It is difficult to say that all these fruits were cultivated in Gujarat during pre-Solanki times. It is also not known whether some of them were introduced in the country by her Muslim conquerors.

AGRICULTURE METHOD

Nothing comes to our knowledge so far as the methodology of agriculture is concerned. Evidences do not say anything directly in this regard. Nevertheless, as the main occupation of the people of Gujarat from the very early time through the thirteenth century was agriculture, it cannot work without any methods or systems of any sorts.

We certainly get some scanty information in this matter during the Solanki period. Though from the fact that indigo was cultivated it may be presumed that expert knowledge of agriculture was not wanting. We learn from a verse in *Dvyāshraya* that iron shod ploughs were in use (XIX, V, 37) and from *DN*, it is learnt that grains were crushed underfoot, a process known as *pamadda* (VI, 40) पामडा: पादाभ्याम धान्य मर्दनम्. In this system, as at now, it seems that both oxen and labourers were used. Hemachandra gives the meaning of the word *padiyantau* as *karmakar*, i.e. a hired labourer, presently called *khetmajūr*. Though it is not stated that *padiyantau* was an agriculture labourer,[66] we may tentatively accept Hemachandra's reading as such.[67]

MEASUREMENT

Of the many systems we have had at present employed in agriculture in Gujarat, little is known as to which of the methods was used in earlier times. We have noted that the son-in-law of the WK Kshaharat King Nahapana, Ushavadatta had purchased a piece of land by paying 4000 *kārshāpaṇas*. But nothing has been said about the area of the field or neither place nor we get any information regarding the size of the land, i.e. measurement of the land. Though some of the systems would have been put into practice, without which one cannot buy or sell a piece of land, little is known to us. But during the Maitrakas of Valabhī, as we know, a system of *bhūmidāna* was very much in vogue. These *bhūmidānas* or *tāmrashāsanas* had noted the *parimāṇ* (measurement) either of *kshetra, vāpi* or *kūpa* (i.e. field, stepwell and well, respectively). From the details of the measurement employed then, we may infer that the measurement might have been worked out in a very accurate manner.

In Gujarat, in the times of Maitrakas, different measurements were prevalent in different regions as it is revealed from the system of collecting the land revenue. The land revenue must have implied an elaborate system of land tenure with precise measurements of plot over which the state levied the tax. We know from the land grants that they give exact extent of the plots of the land with their four boundaries (*āghāṭanani*) specifically mentioned. For instance, in the Vala grant of Dhruvsena II a plot of land is described as the (third) plot lies on the north boundary and measures ten *pādāvartas* of land, to its east (lies) the field tilled by Adityadatta; to the south the field tilled by Sangamdinna, of the west the field tilled by Dasak and to the north field tilled by Dasak.[68]

Thus, the precise description of the plot or well was given as in our own times. In every case a plot of land must have been described in the records by their limits bounded by another plot of land, a well, a road, a border of a particular village, a river, a pond, a tree and so forth thus showing that Valabhī kingdom possessed an elaborate and efficient system of land survey and measurement.[69]

In the measurement of land, some kind of unit must have been used. The standard unit of the Maitrakas as seen from their records was *Pādāvarta.* As to the extent of this unit, the copper plate grants tell us that a plot of land given in gift measured from 10 to 7 hundred *pādāvartas,* whereas a Vapi or a stepwell measured from 12 to 55 padavartas. This shows that *pādāvarta*[70] cannot be a small unit as a square foot as explained in Monier Williams's dictionary; for according to this explanation, a plot of 10 *pādāvartas* measures only about 10 sq ft in all, which would be too small an area to be recorded as a gift on a copper plate grant. And more so in such a small plot of land the ploughing or cultivation casually would not be economically viable and in terms of expenses would be more expensive and enjoyment would be minimum.

On the other hand, Fleet's suggestion seems to be nearer the mark. According to him, a plot of one hundred *pādāvartas* may be taken to measure hundred feet each way, i.e. ten thousand square feet.[71] Thus, *pādāvarta* means the area covered by one foot each way.[72] It seems to have been the standard unit of measurement. The extent of the *kshetra* given as *dāna* was probably either 100 *pādāvartas* or 50 or even 120. Sometimes, the *dāna* was given upto an extent more than 120 *pādāvartas,* reaching upto 200, 300 or 700 *pādāvartas.*[73]

However, there were other measurements in use in other parts of Gujarat as in Saurashtra. One such unit was *bhukti* and the other one was *vrīhipiṭaka.* In Malavaka, the land was measured in *bhukti* while in Kheṭaka, it was measured in *vrīhipiṭaka.* It is difficult to ascertain the extent of these measurements. But it must have been the standard in Malava, for we find among the Maitraka records two known land grants in Malava measured in *bhukti.*[74]

This measure seems, however, not much or smaller than a square foot. We next have a unit known as *Vrīhipiṭaka* (i.e. basket of paddy). This unit was probably used for measuring land in the *kheṭaka Āhār.* From the term *vrīhipiṭaka,* it is evident that one unit represented a plot of land for which one standard sized basketful of corn (i.e. vrīhi-paddy) could be sown. The plots recorded to have been gifted measured one, two or even four *vrīhipiṭaks.*[75] That the *vrīhipiṭaka* was in use in Kheṭaka is farther corroborated by the expression *kheṭaka manenam.* In Bharukachchha district also, the unit of land measurement was *vrīhipiṭaka* which, however, was probably not the same as the *kheṭaka* standard unit, but as we find no specific reference given of the Bharukachchha *vrīhipiṭaka,* it must have been of the general standard in use elsewhere among the neighbouring district. Thus the exact extent of one *Vrīhipiṭaka* plot of land cannot be made out from the extant evidence.[76]

In south Gujarat, to measure the land the measurement was known as *Nivarttana.* Like *pādāvarta,* the *Nivarttana* was also a two-sided measurement. One *Nivarttana* is equal to 30 *daṅḍas* long and 30 *daṅḍās* wide.[77] Literary, *Nivarttana* means a piece of land, which could be tilled during a day and then retired.[78]

Both in Kheṭaka and Bharukachchha, there was another system of measuring a land, especially a *Kedār* land. *Kedār* is a field of rice or it is a lowland where water may be accumulated. It is also called a *kyāraḍī* or *kyāraḍu* or *kyārī.* In this system, calculation was being worked out on the basis of how many *piṭakas* full of rice or paddy were sown in that land. That is a piece of land which required two basketful (*piṭakas*) of paddy or four basketful of rice. But it is difficult to say for sowing one basketful of paddy how long the *kyārī* land was required.[79] The *Dānashāsanas* help us to know that the special measurement was in vogue and it was known as *vrīhipiṭaka.* The same was used to in Saṅgama Kheṭaka, an area in Vadodara district.

The smaller than *piṭaka* measurement was *prastha.* One *prastha* considered equal to 64 tolas, i.e. equal to 652.96 gms. But, surprisingly, it is not known that one *piṭaka* is equivalent to how many *prasthas.* Probably, it is as good as *prasthas* a like one *adhaka.* It is a wooden measure of four sher, i.e. 3.72 kg of corn or 16 *prasthas* is equal to one *droṇa.* It is a kind of old measurement.

In eastern India, a system prevalent for measuring a land was known as *kulayavāpa,* wherein, one *kulaya* (scuttle from winnowing basket) paddy can be sown in one acre of land. Accordingly, for sowing, one *piṭaka* (basket) corn required a piece of land little bigger than this.

Thus, during Maitrakas of Valabhī *pādāvarta, bhuktī, piṭakvāp* and *nivarttana* were the different systems prevalent in Gujarat for measuring a land.[80]

During Solankis and Vaghelas, we do not come across significant evidences to show what system of measurement for land was prevalent. The main reason is that unlike Maitrakas of Valabhī, Solankis and Vaghelas do not seem to have extended much activities towards donation of land. Though we certainly get some copper plates of these periods they do not help us in supplying any sort of material in this matter. But one thing is certain that when these copper plates meant land grants, there must have been some sort of system for measuring the land; without which the donation of land will be impossible. But we hardly get such information excepting the grants given during

Bhimadeva, the second (twelfth and thirteenth centuries) and Visaldeva (thirteenth century). The *Dānaśāsanas* of these kings mention that the system of measuring the land was known as *halavāha*. The meaning of this term is the measure of a land.[81] During Harshavardhana's times, the land was measured as to the length of the plough.[82] But the length of the *hala* has not been mentioned. It seems that this system did continue even after Harsa's times during (Chalukyas and Vaghelas) this system was used in Gujarat. Other than *halavāha* another system was also in vogue. The copper plate of Mehar Jagmalla found at Timana village mentions the word *patha*.[83] This may also be a kind of measure for land. We also get *vishopaka* from LP (106), which means a *vighuṅ*, i.e. equivalent to 5/8 acre in Gujarat. (one acre = 4046 sq metres and one *vighun* = 2530 sq metres).

We do have two more literary evidences in this matter. They are *Gaṇitsāra* of Shreedharacharya (dated V.S. 1449, AD 1393) and *Rājvallabhha* of Mandan (dated 1911). Rajkirtimishra has written a commentary on it in old Gujarati. The first manuscript was specially writen for the children of the families of *Moḍha Vaṇik* caste of *Pāṭaṇ* for their learnng and hence its practical utility is very clear. In this work, we get the following information in matter of the measure of land:

8 sarsava = 1 java (mustard seed and barley)
6 java = 1 aṅgula (barley and finger)
24 aṅgula = 1 hasta (finger and hand)
4 hasta = 1 daṅḍa (hand and stick)
2000 daṅda = 1 kosha *(gāu)* (stick or staff and gau 3.2 kms)

4 Kosha = 1 yojan (3.2 kms 24 miles

3 Daṅḍa = 1 Vāṁsa (i.e. stick 2 bamboo = 10 to 12 feet (2)
14 vāṁsa = 1 netan (bamboo)
20 netan = 20 halvāha
1 halvāha = 3360 hasta
= 80640 aṅgula
= 483840 java

In *Rājvallabha*, we find the following particular system of weights and measures for crops:

2 paṇa = 1 prasūti
2 prasūti = 1 kuḍava
4 kuḍava = 1 prastha = 64 tolas
4 prastha = 1 āḍhaka (a wooden measure of 4 seer or 3.72 kg)
4 āḍhaka = 1 droṇa (3.72 kg and kind of measure)
16 droṇa = 1 khārī
4 pawālā = 1 pālī (a measure of corn and a measure of corn)
4 pālī = 1 māṇaka
4 māṇaka = 1 seti (= mana)

6 seti = 1 hārī
4 hāri = 1 mānī
16 seti = 1 kaḷashī (= 16 maunds)
10 kaḷashī = 1 mūḍo

Many of these measures are still used in villages of Gujarat, especially by farmers.

LAND REVENUE

The major source of income for the state, now and then, is nothing but the land revenue. Direct or indirect evidences to know something in the subject in pre-Mauryan times of Gujarat are very difficult. None of the sources helps us in any way.

Even during Mauryas and pre-Maitraka times, we do not come across any epigraphic or archaeological sources regarding land revenue. Among literary sources the most authentic is the evidence of *Arthaśāstra* of Kautilya. Though the information about land revenue in it pertains to Pataliputra in particular, general references may be followed on the basis in respect of Gujarat in the Mauryan times.

There is a reference in the Kautilya's *Arthaśāstra* regarding the land revenue.[84] Accordingly, the 1/6th of the produce (i.e. *bhāg*) of the grain and taxes (*kar*) in relation to fruits and trees were to be given to the king. Again, the same work[85] deals with agricultural activities in details, for which the chapter has been named as *Seetādhyaksha.* Accordingly, the state was collecting 1/5th from the farmers of the produce with the help of irrigation water from the well or tank made of by their own expenses; while the farmers had to pay 1/4th of the produce if they had utilized the water for irrigation from the government tanks or wells. When the fields were irrigated by making small canals or drainages the farmers were supposed to pay 1/3rd of the produce to the state, if the water is fetched for irrigation from the river, reservoir, tank and well by using waterwheel then the farmer had to pay 1/4th of the production.

Taking this reference into account, we may very probably say that the same type of revenue systems were in vogue in Gujarat from the 4th century BC till the end of the fourteenth century AD.

We get more perfect knowledge in this regard from the beginning of Maitraka's rule till the end of Vāghelā's rule; for the sources of this period include mostly *dānaśāsanas* or *bhūmidāna* or land grants or copper plate grants. In many of the grants of earlier times, generally, we come across two terms for revenue officer in the early part of the administration. They were *Āyuktaka* and *Viniyuktaka.* The former one, i.e. *Āyuktaka* was the main officer of the revenue, accounts department,[86] while the latter one was the assistant officer in the same department. *Dhruva* was another revenue officer who was responsible for collecting state-share of the produce of the land.[87] The post of *Dhruva* as an accountant of the revenue is still prevalent in the Kachchha district of Gujarat and the same term is still preserved as a family name in certain families of Gujarat.[88] There was another revenue collecting officer, named, *Anutpannādānasamudgrāhaka,* who was responsible to collect those revenues which were not remitted at proper time. (We have already noted that the main officer in the revenue department was called *Samāhartta,* which during British period was known as collector and in the post-independence Gujarat *Samāhartta* is again in use.) Unfortunately, the names of these officers do not figure in the Maitraka grants. But we do get other terms such as *uparika* and *bhogika* in the land grants of the Maitrakas as the superior officers in the revenue department.[89]

Various taxes that were waived and mentioned in the *Dharmadeya* help us to understand the different types of land revenues. For revenue, the term that was used is *ādeya* or *ādāna*. Many types of *ādeyas* are enumerated in *Dānaśāsanas: udraṅga, uparikara, dhānyādeya, hirṇyādeya, bhāga, bhoga, vāt-pratyāya, bhūtapratyāya, vishṭi* and *dashāparādha.* The grantee of the land given as *Dharmadeya* was also endowed with the right of receiving all sorts of revenue from it as mentioned in the expression *Sarwa Adana Samgrahya* in the Katachchuri grants.

Udrṅnga and *Uparikar* were the revenue taxes mentioned in all most all the *Tāmraśāsanas* of the Maitraka kings and kings of the other contemporary states of Gujarat.

Uparikar tax seems to have been levied on cultivators who had no proprietary rights in the soil as may be gathered from Fleet's interpretation of the term.[90] It is also interpreted to mean a tax imposed on those who were literally above the right of proprietorship, i.e. who had no proprietary right in soil.[91] Evidently, it was a tax contrary to the *Udraṅga.* Thus the land tax paid by the tenants of the soil was called *Uparikar,* while that paid by the proprietors was known as *Udraṅga.*[92] This was the tax lavied on the produce of the land from those who had a proprietary rights in the soil. Tenants had to pay this tax to the Zamindars who was entitled to collect it on behalf of the state and in turn supposed to pay the total revenue of the village in gross. It must probably have been levied on those who were granted land with a permanent proprietary right to it.[93]

Hence, it seems that during the Maitraka's times, the right of the possession of land was more current than that of the ownership of land and, therefore, in comparison to *Udraṅga* tax the income from *Uparikar* tax was evidently much more. If the authority of *Uparik* officer was to collect the *Uparikar* tax, then it implies that the special superior officer was appointed for that.

The revenue paid as certain part of the corn (i.e. in kind) was called *Dhānyādeya,* literary meaning corn, while the one paid in the form of gold (i.e. cash) was called *Hiraṇyādeya.* The one paid by the proprietors or by the permanent tenants in a settled or fixed royal form of the produce of the land was commonly called *bhāga.* The term denotes the fixed royal share of the land produced and generally this share was 1/6th of the total produce.[94] Manu informs us that the said *bhāga* was 1/6th of the total (VIII, 130). But there is no reference we ever come across about how much *bhāga* was levied either in the Maitraka rule or in the other contemporary states. Nevertheless, the Maitraka rulers were described as having been true followers of the *Smritis* must have been very careful regarding the collection of this tax for the *Manusmriti* lays down, as already noted, 1/6th of the total as the maximum that may be collected while 1/8th and even 1/12th are recommended to the king (VII, 130).

From records of the Dharasena IV we are in a position to know of a reference of *Mridukargrahaṇa,* which indicates that the levy might have been low meaning, thereby that he claims to have reduced the usual rates of tenure but unfortunately, the records do not give us any figures.[95]

There is another term, of course, other than *bhāga,* called *bhoga.* It means an income (or a petty tax payable in form of) or through the daily gifts in the form of vegetables, fruits, flowers, etc.[96] the authority of enjoying the possession of such things was legal.[97] In other words, the *bogha* was a due (*ādeya*) and it was taken either in cash or kind. An officer called *Bhogdharṇika* or *Bhogika* was in charge of the collection of this tax.

We may, therefore, conclude that the terms *bhāga* and *bhoga* covered all sorts of land revenue.

Besides, these varied taxes levied on a land as revenue, sorts of other *pratyāya* (i.e. *Śulka* or custom duty) were also current in Gujarat. They were *Vātapratyāya* and *Bhūtpratyāya.* Of these two the one, that is *Vāta,* as a term, denotes the things produced in a village (i.e. local things).

The term *Daśāparādha* denotes the income taken in the form of a penalty on 10 types of offences or crimes. There is another term worth interesting. It is *Utpadyamānviṣṭi.* This term indicated the right of having work without payment and against one's will (or unpaid forced labour is *Viṣṭi*) as and when the requirements arise.[98] This is also a sort of land revenue.

Along with the *vṣṭti* there is another reference to two terms namely, *ditya* and *pratibhedana.* The meaning of the roots of these two words is almost the same. Hence, it is difficult to understand the difference between the two. In spite of that the meaning intended thereby (or referred to) is the customary distribution after having the divisions of the yields thus earned from in the form of cutting of grass (i.e. *ditya*) and harvesting of corns (i.e. *pratibhedana*). *Shaibar* was another kind of tax levied, whose meaning is not clear.[99]

The main source of the state income must have been then, as now, the land revenue during the reigns of both Solankis and Vaghelas.

As explained by Abhaytilakgani in his commentary on the *Dvyāsrayamahākāvya,* king's officer mainly collected the land revenue from the villagers (i.e. farmers). In our own times, the revenue was collected from the farmers in this manner. The same work also helps us to know that after the harvest was over or at the time of harvest in autumn a share of the produce was collected by the village landlord (i.e. *grāmapati*) and another share by the king himself through his officers (111, 18, 2). It is possible that both Hemachandra and Abhaytilakgani were giving illustrations of two different systems current then as discussed just now. Accordingly, in the first system of the two, there was a landlord in a village who collected the shares both for himself and the king, while in the other cases, the king directly collected his share of the produce through his collectors, presumably without the intervention of any landlord.[100]

But from the documents given in the *LP,* it appears that there were other systems for collecting revenue than those mentioned here earlier. The *LP* (p. 8) has texts of two land settlement-deeds (*grāmapattaka*), both of which are agreements between a private person and the government for payment of yearly revenue, what would be today called *Ryotwārī* settlement. Both these deeds make it clear that the agreements are for one year only. The first document is dated V.S. 802 and the second document V.S. 1288;

but the one dated in 802 must have served as a model in the Chalukya period as well, otherwise its retention in a compilation of this nature cannot be explained.[101]

Since both these documents are similar for all practical purposes, we shall consider here the first one only. This was an agreement between the *Pañchakula* and a man named Goda, who agrees to pay 216 *drammas* as the perquisite of the *Pañchakula*.[102]

The first document has *Pañchakulasya Karpatbhavye Deya Dramma* 216, while the second document is very clear and states *Pañchakulasya Bhavye Deya Dramma* 216. Goda, one of the agreementors, further agreed to pay 40 *drammas* for miscellaneous expenses (i.e. *vikarpad*). In addition, he was found to pay by agreement the following:

(1) any increase in rent called *chatapak* or *chadvo*;
(2) another tax which was to be utilized when the whole village wanted to show respect to a particular man. This tax was called *malmargana* = *mal-viro*;
(3) another revenue which seems to have been collected on some auspicious occasions called *Māṅgliyaka*;
(4) one more tax, Goda agreed to pay, was collected probably for maintenance of the police station called *chaturak* (*chattvar, choro*), i.e. a public place in a village, a police station.[103]

Further clauses provided that Goda was to pay the land revenue of 3000 *drammas* to the treasury in three equal installments called *skanda*; the first in *Bhādrapada*, the second in *Mrugśīrṣa* and the last one on the *Akshayatrutīyā* day. It seems that this practice of paying the land revenue in three installments was current in Gujarat since time immemorial: for in a grant of Krushna II,[104] the Rashtrakuta King of Gujarat, we find that the money, apparently the revenue, was to be paid in three installments—द्रम्मतित्रिलि the first installment in *Bhādrapada*, स्कंदकरै देय the second one in *Kārthika* and the third in *Māgha*.

Thus, we see that the payment of government revenue in three installments in cash was a very old custom in Gujarat.[105]

The *LP* further states that after each payment was made, the *Sreekarana* (i.e. the finance department of the state) was to be informed. It is not mentioned which *Sreekarana* is meant, but it must have been that of the *maḍal* or *pathak*. The system undoubtedly was devised to act as a check against dishonest collectors and revenue officials, and more so the state gets the exact idea of the income.

In the second deed, there is an additional clause under which the tenant was obliged to maintain the road in his area in good repairs. As this clause is absent in the first document, we presume that this innovation was imposed by the Chalukya kings who were always under the necessity of moving troops from one part of the country to another and probably wanted their roads to be kept in good condition.[106]

According to the *LP* (p. 10), there was another system of revenue in force under which the rent of a village was fixed and the villagers were free from having to pay any additional imposts other than the annual revenue mentioned in the deed called *Samkar-Yuddha-Gramma-Pattak.* As per this the villagers had to pay revenue of 4000 drammas and nothing else. This agreement, which was in the nature of a permanent settlement, was between a person and the *Pañchakula* working under a *Mahāmanḍalesvar.*

Yet another system of land revenue has been recorded in *LP* in which the government fixed different rates of rent for different types of land, depending on their productivity. This seems to be according to the *Śukranīti* (IV, 2,124–5), which prescribes that the king should assess land revenue after causing the land to be classified according to their fertility.

It is apparent from the *LP* that under this system the rent was collected through the king's officers (p. 16), as mentioned by Abhaytilakgani, noted earlier.[107]

There are two specimens of this type of document called *Grāmasaṁsthā* in the *LP* one of which is dated 802 V.S. and the other is undated. But for the reasons already stated, it is presumed that this system continued to be in force during the Chalukyas.

The document of the *LP* dated V.S.802, interestingly, states that the rates of the revenue referred to in it will be applicable in all the villages of the *Vishayas* under *Lāṭpalli,* whereas the second document simply states that the rates will be current in all the villages of the *Lāṭpalli.*

Apparently, the system of subdividing into *vishayas* which was customary in V.S. 802 was given up by the time the second document was prepared; so that while the latter was in many ways a copy of the first one; it has to be modified and brought into conformity with the later usage when the country was no longer subdivided into *vishayas.*

The *Grāmasaṁsthā* was more in the nature of a notice than an agreement. It was issued by the *Adhikārī* and his *pañchakula* and fixed the land revenue of an entire area. The rate of the revenue was thus as per the *LP* (p. 16):

1. In every village, there was some land, called *saṁkarbhūmi,* of which the revenue was fixed permanently at 24 *drammas* per *viṁshopaka.*
2. There was some weak and soft land called *pochalbhūmī* the revenue was fixed at 20 *drammas* per *viṁshopaka.*
3 For the uncultivated, highland, which is called *uddhakhilbhūmi,* one had to pay 16 *drammas* per *vimshopaka. Khil* or fallow land is also mentioned in the Damodarpur copper plate of Buddhagupta[108] and a comparison of these plates shows that the *Khil* was assessed at a lower value than ordinary land.
4. According to the document of the *LP* dated V.S. 802 if such fallow land was put under cultivation by a farmer who came from outside he had to pay only 10 *drammas* per *viṁshopaka* (नव्य-समायात कुटुम्बिकैः).
5. The undated document, however, makes no such distinction, but states that for *uddhakhilbhūmi,* one had to pay 16 *drammas* per *viṁshopaka* and for *khilbhūmi,* 10 *Drammas* per *viṁshopaka.*

This difference in these two documents, otherwise similar, may be due to the fact that in the ninth century of the *Vikram Samvat,* some inducement was necessary to bring farmers from outside and put the fallow land under cultivation. In due course of time, this necessity was no longer required and people began to cultivate even the low fallow land which is probably the meaning of the *khilbhūmi* as opposed to *uddhakhilbhūmi,* though the *khilbhūmi* is not even mentioned in the document dated V.S. 802.[109]

The *Nadiyaka* or *Anadiyaka* was another class of land for which the revenue was just three *drammas.* The place where the carts stood for disposing of goods is *Nadiyaka* or *Anadiyaka* as these terms indicate.

Apparently, the revenue for this and the following types of land was not fixed according to the area but to the use to which the land was put. The settlement officers could probably form a rough guess as to the area, which such lands would occupy in a village[110]:

(1) For the land determined to graze cows and buffaloes, two *drammas* were charged.
(2) In case of the land reserved for the grazing for oxen, only one *dramma* had to be paid.
(3) The land earmarked for grazing for ramps and sheeps charged 1/2 *dramma.*
(4) In the grazing land determined for bullocks used in ploughing lands, no tax was levied (वहमान-हल-बलीवर्दानाम् गोचरों मही).

Gochar as a tax is mentioned in the *Kadi* grant of the time of Mularaja. We also learn from Abhayatilakgani, that the villagers had to pay tax on grazing land for animals and those taxes were collected by the tax collector.[111]

The same deeds then proceed further to provide for fines to be realized for the here-mentioned offences:

1. To cultivate the land belonging to someone else the fine fixed was 6 *drammas.*
2. To allow the cattle to graze on cornfields the fine determined was 1 *dramma.*
3. For disobeying the state orders the fine settled was 5 *drammas.*
4. For the theft of hide (चर्म चौरिकायाम्) 25 *drammas* was fixed as fine.

It is to be noted here that Gujarat during this period used to export large quantities of leather goods for which she was famous. Hence precaution probably had to be taken against the theft of hide which must be kept during certain process of tanning in an open space.

Mention made of these fines in a deed of land settlement is peculiar and this aspect of the deed shall be discussed along with that *Daśāparādha.*

It has been already shown that in such expressions as सहिरण्य-भाग-भोग and भाग-भोग-कर when occur in land grants. भाग-भोग-कर has been taken to mean the king's share in produced grains or contribution in kind to be paid by the villagers and *hiraṇya* to be tax levied upon certain special kinds of crops which was of a nature difficult to be divided into share. It has also been suggested that *kara* (कर) was a periodical tax levied primarily upon agricultural land over and above the king's normal share of grain.[112]

Though according to the *Smritis, kara* meant the tax on merchant.[113] According to Altekar, *bhāgakar* was the land revenue whereas *bhogakar* represented the petty taxes in kind to be paid to the king everyday and usually assigned in practice to local officers as part of their income.[114]

The term *kara* is, however, explained by Abhaytilakgani, as used by Hemachandra, in the *Dvyāshraya* as thus कर: क षि पशुचा रणादि कृत राजकीय-भुमयु भाग हेतुको-राज गाह्य भाग:! (111,18).

The meaning of this that it is the share due to the king by one who has enjoyed the fruit of the royal agricultural land and pasture. That the farmers paid *kara* during the harvesting season in corn (i.e. kind), states Abhaytilakgani definitely. But this might have been the part due for using the agricultural land; for pasture, the tax was probably paid in cash.[115]

When, due to a failure of crops, the farmers were unable to pay their revenue, Merutungacharya while narrating an incident that happened in the regime of Bhima, the first, uses the expressions राजदेयविभाग and सस्य: निदानी भूतदानी-समबन्ध![116]

The second expression has been correctly translated by Tawney as the king's share that was to be deducted from the grain, and it seems that *bhāga* and *dānī* were synonymous terms and that both dented the tax payable in kind. Thus, we see that part of the land revenue such as *hiranya* and *gochar* (the grazing tax) were paid in cash and the rest of the land revenue in kind. But probably, with the exception of land revenue, all the other taxes were paid in cash.[117]

The fact, that with the exception of one document, i.e. supposed to be a copy of a land grant issued by Lavanyaprasad during the reign of Bhima II in V.S. 1288—all the other documents in the *LP* mention taxes in cash only requires to be pointed out here. This was probably due to the fact that the taxes in cash and kind were sanctioned by long usage (i.e. by tradition) and separate documents were not necessary to collect them.

If cultivators failed to pay these dues, they were probably put under arrest, as related by Merutunga or distained in some other way till they paid. These royal dues, sanctioned by age-old usages, were not, therefore, subject to any written agreement between the king and the farmer. But other arrangements, which completely differed from the customary one, required written documents for their validity and enforcement, and it is these documents which were collected by the compiler of the *LP*.

They show a distinct tendency to collect land revenue from one person. This had the advantage of lowering the collection charges while it ensured the state a fixed revenue every year, irrespective of the failure of crop or any other misfortune.

In short, it gave the state many advantages, which later on induced Lord Cornwallis to introduce permanent settlement in Bengal during British rule. But the arrangements in the Gujarat during Chalukyas was better than permanent settlement, because the agreement generally was for one year only with distinct understanding that the revenue was liable to be increased, thereby fully protecting the interest of the state.[119]

WATER RESERVOIRS

The detailed description about the establishment of reservoirs and different types of *Ghāṭs* and often tank with steps leading to the water and their importance in the planning of cities is found in our books on *Vāstuśastra,* i.e. writings on science of building houses.

Generally, water reservoirs are those constructions having the entire four sides bounded with high raised earthen work. Likewise, *kunds* are also water reservoirs having all the four sides bound with the construction of stone steps leading to the water. Similarly, the specially constructed *vavs* have also had its place of importance.

All these types of water reservoirs have had their important place in the growth of cultures in Gujarat since time immemorial as she is located in the less rainfall area and temperate zone.

Fortunately, we have authentic information regarding historical water reservoirs in Gujarat. It is the established fact that the civilization of Gujarat is known as reservoir-borne civilization, unlike the river civilization of India, as has been proved by the evidences brought to our knowledge by the excavations carried out in Gujarat since the Lothal and Dhoḷāvīrā cultures down to the present-day *Shyama Sarovar* built by the Government of Gujarat in the latter part of sixties on the famous and historical site located on the Meshvo River and known as Devni-mori Bauddha site near the famous religious *Toraṇa* of Shāmaḷājī in the Sabarkantha district of Gujarat.

The moot problem of the Lothal reservoir is still to be settled as it has brought about heated discussions, whether it is a sea wall or an embankment of a reservoir. Hence, we leave these things for future and further researches. Though the excavations at Dhoḷāvīrā in the Kachchha district is still continuing; it has no doubt brought to our knowledge by the 5000 year-old remains of one of the biggest manmade reservoir. As this area had no known river around it, the dam seemed to have served the requirements of the farmers for irrigation purpose. The dam also indicated the technology developed for preserving rainwaters. Therefore, we may certainly say that the growth of the cultures of Gujarat in prehistoric period started taking place on the bank of Dhoḷāvīrā *Sarovar* in Kachchha, reached her prosperity on the bank of *Sudarshan Sarovar* at Junagadh. Culminating later on the bank of *Sahasralinga Sarovar* at Pāṭaṇ and found her extremity at *Shyāmsundar Sarovar* near Shāmalāji. Other known reservoirs of the period under review are: *Meenal Talāv* at Dholaka, Munsar *Talāv at* Viramgam, *Suryakund* at Modhera, *Rankivav* at Pāṭaṇ and Aḍālaj-vav near Ahmadabad, to name a few.

All these historical reservoirs were manmade and digging the earth to an extent that made them give it a form of a reservoir. Naturally, while digging the site or a place, the dugout earth was used to construct a high-raised boundary on all sides. Thus the embankment was prepared with the use of dugout earthen material. Excavations have brought to our knowledge numerous such historical earthen embankments. The most known and famous one, but not in existence, is found at Junagadh, as has been noted earlier, which was built by the Mauryan king Chandragupta just for agricultural purpose.

We may now take stock of some of the historical water reservoirs of Gujarat.

The most known and historically famous one is *Sudarshan* Lake at Junagadh on the way to Girnar mountains. Built by Chandragupta, the Mauryan king with the help of his *Rāshtrīya* (i.e. governor) of Gujarat province, *Vaishya* Pushyagupta—as Gujarat was one of the very few richest provinces of the Mauryan empire—the earthern-built embankment of this lake was very strong and solid. The entire construction was sound and that too without any joint or patch. The natural dam was also in the middle of the reservoir. It contains at proper places many *praṇālis* (culverts), *parivāha* (overflowing outlet or channels) and *miḍhavidhān* (Strainers) for excess water to flow out.

On behalf of Asoka, his *Rashtriya, Yavanraj* Tushaspha, in charge of Surashtra, had decorated the very lake with culverts. He also added on the embankment *aṭṭālak* (balconies), *upatalpa* (an umbrella), *dwara* (gates), *sharana* (rest house) and *uchchhaya* (arched gate, i.e. *Toraṇa*). The entire construction was so strong and sound that no damage took place from c. 320 BC through 150 AD, i.e. nearly for five centuries.[120]

In 150 AD, when the then two rivers, namely *Suvarṇaśiktā* and *Palāshini* (now *Sonarekha* and *Palāsiyo* respectively and presently both are just dry riverbeds flooded due to excessive and unusual rainfall in the month of *Mrugashīrsha* has caused the *Setu* (*Bandha* = embankment) much damage in the reign of the *Mahakshatrapa* Rudradaman and again in 455 AD, during the reign of Gupta King, Skandagupta, the *setu* was damaged. At both these catastrophic occasions the then kings, under the direct supervision of *amātya* (governor) Suvishakha Pahlava and *gopta* (governor) *Parnadatta* (along with his son Chakrapalit, a Mayor of the city of Junagadh) without any delay and more so without levied any taxes on and taking any *visti* (labour) from their subjects, had gotten repaired the *Setu*. After the fifth century what had happened to its existence is anybody's guess, except that we have only at our disposal the remnants of it.[121]

The objective, obviously quite known in the construction of this lake, viz., *Sudarshan* and extended channels and drainage made from it, was none other than the growth of agricultural activities. These facts are known especially from the inscription of *Mahakshatrapa* Rudradaman, who engraved his noble writings on the western side of the famous Asokan rock edict. This is the earliest known available epigraphic historical record of India. On the eastern side of this rock Asoka had engraved his royal charters, of course, not political but religious-spiritual. He does not say anything regarding this reservoir. Skandagupta had inscribed his feelings on the northern wall of the rock. It seems from these edicts that Gujarat was prone to less rainfall area and that these kings were very noble and benevolent and more so they were very concerned for the growth of agricultural production. Hence, this *Sudarshan* reservoir and its importance for the agricultural development.[122]

It is noted in the rock edicts that this *setu Mruttikopalamay* was constructed with stone and earth. It was one km long, its base was hundred metre wide and the upper part was eleven metre broad. Its upper surface was evenly done. Though this *setu* was 17 metre high keeping in view the height of the land, its height was varied at different places. The entire construction was done using nearly 9,43,500 cubic metre of earth. When the *setu* was damaged in the time of Rudradaman, about 26,225 cubic metres of the earth's portion was washed away. During Skandagupta's reign, 10,400 cubic metre earth portion washed away. Precaution was properly taken by putting necessary strainers so that the reservoir does not get filled with earth brought in by the incoming waters. Such *setus* are found in Gujarat at various places such as Kheda, Sabarkantha, Ghumali, Godhara, etc. The tradition of constructing reservoirs was continued for long.[123]

We most probably do not find such lakes during the Maitrakas of Valabhī. But it seems that the use of *Sudarshan* Lake might have continued during this period as well.

The capital city, *Aṇhilwāḍ Pāṭaṇ*, of the Chaulukyas is situated on the border of the small desert of Kachchha, and the rivers, Banas and Saraswati, were flowing by it. Therefore, for utilizing the waters of these two rivers, the Chalukya King Durlabharaj, the predecessor of famous king Siddharaj, had constructed a lake named after him, i.e.

Durlabh Sarovar. Siddharaj, at a later stage, renovated it and named it as *Sahasralinga Sarovar.* One of the currents of the *Saraswati* was directed into this lake for making it prosperous and beautiful. But it is not known whether it was used for agriculture purposes. Historical data shows that Karnadeva, the illustrious father of the illustrious son Siddharaj, is also known to have built some extensively large and beautiful lakes. Available evidences prove this. He made large reservoir near Chanasma, in Mehsana district, namely, *Karṇasāgar,* the evidences of which have come to the knowledge. Another such lake he constructed near his military camp, viz., Karṇāvatī, which is now known probably as *Chandola Talāv,* presently located in the southeastern part of the modern city of Ahmadabad as has been believed by archaeologists.[124]

Two reservoirs are associated with Minaldevi, the mother of Siddharaj. They are *Munsar Talāv* at Viramgam, a miniature copy of the *Sahasralinga* and *Malāv Talāv* at Dholaka: This second one is bounded by on all sides with the stone steps and ghats. There is a popular Gujarati saying attached to it. It runs thus: *Nyāya Jovo Hoya to Juvo Malāv Talāv,* i.e. 'For justice see this lake', which is wrongly associated with Minaldevi's acts and construction of this tank.' Exploring and studying this lake R.N. Mehta has disproved this saying, and considered its construction on the basis of architectural principles.

BHŪMICHCHHIDRA

Kautilya refers to the system of *bhūmichchhidra Dharma* (11,2) with elaborate details. Land could not grow any crop is called *bhūmichchhidra.* Such land normally is saline. He further states that the land, which is not cultivable, is to be reserved for making pasture grounds where cattles may graze with ease.[125]

The land given in donation as *dharmadeya* was called *bhūmichchhidra* as mentioned in the *Dānaśāsanas* of Maitrakas.[126] Examination of this term referred to in *Dānaśāsanas* indicate that the original meaning of *Bhūmichchhidra* is pasture land or fallow land and such land, if given for cultivation, is not to be taxed.[127] Thus *bhūmichchhidranyāya* means fallow land in accordance with justice. Hence such land was made free from all sorts of revenue.[128] Not only this but such fallow land was given forever or permanently. For this *Dānaśāsanas* refer this custom as आ चन्द्राक्कर्णिर्णवक्षिति सरित्यर्व्वत समकालीनम. The Aliṇā copperplate grant of Siladitya VII dated 766–767 AD reveals how the village of Mahibali in the *Upahilpathak* in the famous *Khetaka Āhār* was granted in accordance with the custom of *bhūmichchhidra.*[129]

There is a reference of *Akshayanividharma* in the Bhamodarawala copperplate grant of the time of Maitraka King Dhruvasena Baladitya. According to this reference in this type of possession, right the real enjoyment is of the products of the land only, and could be enjoyed only by the receiver and its right of inheritance is not given to the receiver's descendant.[130] But the land donated on the basis of *bhūmichchhidra* system could be possessed by the descendants generations of the receiver as conformed by the *Dānaśāsanas* of the Maitrakas (पुत्र पौत्रान्वय भोग्यम). Such land receiver had the right to possess it, cultivate it and entrust it to his progeny.[131] Hence, such authority was eternal. It seems that the objective for referring *bhūmichchhidranyāya* is this. Besides this, we come across to the reference such as in some of the *Dānaśāsanas* of south Gujarat meaning thereby that the negotiations and removal of penetration was the right of the receiver.

Parihar means exemption from state revenue. As the land of *Dharmadeya* was tax-free, any royal person or personnel enacted prohibition for any sort of interference. Even *bhaṭs* (warriors) and *chāṭs* (greedy) were not entitled to enter in such land. Therefore, the receiver of the land had all the rights, at hand, of possession, cultivation, to be tilled and transfer. In this sort of enjoyment, the receiver had not only the right of enjoying products from such land but also included all kinds of *ādāna* or *ādeya*. Such *ādeyas* included *udraṅga* and *uparikar* as has been mentioned in all the *Dānaśāsanas* of the states in Gujarat.

Over and above these dictums, the king was entrusting also the revenue product to the donee such as *bhūtpratyāya,*[132] *vāt-Pratāyāya,*[133] *dhānyadeya,*[134] *hirṇyadeya,*[135] *bhāga,*[136] *bhoga,*[137] *utpadyamānviṣṭi,*[138] *dashāprādha,*[139] etc.

If these revenues are to be collected from the land receiver and, in turn, they are to be given back to the donee, in practice, this sort of procedure is nothing but freedom from paying taxes.

In the *Dānaśāsanas*, we commonly find the references to the statement regarding the land grant and about the owners of the land, which is being given away to the recipients. This helps us to say specifically that the land was indeed donated.

Of course, the reference pertaining to *ādeyas* (dues) actually refers to the village itself. Though the land being given in donation, the possession of the land of the village remains permanently with the original landholder. But all the revenues levied on it were to be entrusted to the donee instead of the king who was the real rightful taker of the revenues.

But in the case of *kshetradāna*, the real possessor of the field is the donee himself, he is not entitled to have any taxes from it. In the *kshetradāna*, especially, the question arises is that of the exemption of the taxes and not of the collection of it. It seems that the advantages the king was entitled to get from the land of the village in terms of revenues and share from the crop product was automatically entrusted to the donee as an economic advantage, which could be viewed from the references about *parihāra* and *ādeya* we come across in the village grants.[140]

FLOOD

The evidences of floods are evidently brought to our knowledge by archaeological studies at many places in Gujarat. The construction activities of building water reservoirs were also carried out with much vigour for warding off the water scarcity in Gujarat. The *Sudarshan* lake at Junagadh, *Sahasralinga* at Pāṭaṇ, big one at Kheda, and many such around Shāmaḷājī etc. prove the utility and usefulness of hoarded water. This activity of constructing lakes helps us to understand the arrangement and protection of them and in turn makes us aware of the water scarcity and irregularity of rains in Gujarat.

The excavation carried out at places such as Akoṭā in Vadodara and the studies brought out by archaeologists prove that the rivers of Gujarat often get flooded and damages caused by it and calamities experienced by the people as a result of the floods. Viewing and observing such evidences and more so from the evidence of the great wall built up on the Meshvo river near Devni Mori Stupa in Sabarkantha district indicate that the river floods in Gujarat had done major damage. Hence this has a bearing on the development of agricultural activities.

The earliest epigraphic record of a flood in Gujarat is preserved in the inscription of *Mahākshatrapa* Rudradaman. The event was the bursting of the great dam of the *Sudarshan* lake in the foothills of the Girnar mountain in 150 AD. The floods in the time of Rudradaman seems to have been a truly terrible one. In spite of all possible precautions to avert imminent danger, the dam did give way and the unprecedented devastation that followed can better be imagined than described. Heavy rain and that too unusually in the month of *Mārgaśīrṣa* (i.e. in autumn season) had turned the whole earth into a lake and had converted the rivers *Suvarṇasiktā* and *Palāshinī* as well as other streams which flowed from mount Girnar into the lake, into powerful torrents, beyond the strength of the dam to resist. Furious onslaughts of wind and rain completed the havoc. The wind was so strong to uproot and hurl into the lake everything that stood in its way, from the trees that crowned the summit of Girnar to the houses and villages that nestled at its base. The dam, then, burst and the newly liberated waters rushed headlong to the sea carrying all before them. It was an overwhelming calamity, with which (so say the inscription) all the courage and administrative skill of the official on the spot were powerless to deal. The record registered that the double breaking and the double mending of the dam of the lake *Sudarśan* by the state without charging any extra taxes and taking labour from the people is an important landmark not only in the early political history of Gujarat but in the early history of India too.[141]

There might have been floods in the post-Kshatrapa periods in Gujarat but we do not come across any evidences as such. Hence description of floods till the 1300 AD is not possible.

FAMINE

Next in importance and influence to floods and years of scarcity are the devastating forces of epidemics due to the calamity of famines.

The earliest extant historical references to such agricultural calamities in Gujarat is of a famine in the reign of Bhimadeva I (1002–1064), when the rains had failed round about Palanpur, the cultivators were unable to pay the king's share of produce. They were brought before king Bhimadeva; but the Prince Mularaj, in great pity, begged the king to remit the revenue in lieu of a boon that had been promised him. The king consented. In the third year after this, Prince Mularaj died. But the next year being of plenty, of course, in agricultural produce, those poor but honest cultivators came to pay to the king his due for the past and the present years. Bhimdeva refused to take the arrears but, to please the cultivators, spent them later on the charity. A truly noble episode, reflecting credits alike on the ruler and ruled.[142]

The second rigorous famine we came to know took place in the reign of Visaladeva Vāghela, which was successfully combated by reason of the magnificent munificence of the philanthropic Jaina trader Jagḍushā of Mahuva in Sundernagar district. He was forewarned by a Jaina Muni in V.S. 1312 about a great famine that would last for three years and to be known as *Pandrotaro* (V.S. 1315). After having such enlightenment from a Muni, he accordingly forearmed himself by hoarding a huge amount of corn brought from distant countries by his aiders and abettors. The famine came as severe as had been predicted. As cautioned, it lasted for long. In two years, even the royal (or the state) granary was empty. Grains became so scarce that only 13 gms could be had for a *dramma.* It was then that Jagḍushā started to distribute freely the grains he had collected earlier. Later on, he was requested by the minister of Visaladeva to give a loan of corn to the king himself; but the brave, philanthropic and kindhearted as he was, this merchant refused to obey royal request by pointing out that a copper plate inscription in his granaries declared that 'these grains or corns are collected only for the poor.'[143] After many entreaties, however, Jagḍushā gave away 8000 *muḍās* (equal to 320000 *maunds*) of corn to the king.[144] This episode suggests three things:

(1) Precise forewarning of muni. That muni was kind enough to let it be known to his disciple.
(2) The philanthropic approach of a Jaina merchant.
(3) The king had also cared for his subject.

TRADE ROUTES

Trade routes, though not directly related to the growth of agriculture, play an important role in transport of grains, manures and seeds. Hence, a brief description of it is desirable.

The chief trade route in earlier times in the region followed the coast from Goghā southwest to Somnāth (the famous place commonly known as Somanāth Pāṭaṇ was one of the twelve Shaivite temples is situated) and hence northwest to Dwārkā (another the famous place commonly known as Dwārkā of Lord Krishna). The chief land routes were those which connected the peninsula (i.e. Surashtra region of the present-day Gujarat) with the main land and of these the most frequented passed by Zinzuwāḍ Pāṭaḍi to Vadhavāṇa and by Viramgam to Valabhi were also in common use.

Again, there seems to have been a road joining Valabhī and Junagadh and Vanthali; but, as the greater part of it lay in forests and thinly populated area, the trade route followed the coastline. Later on, in the twelfth century, the Solankis of Aṇahilwāḍ-Pāṭaṇ constructed a military road, from Vadhavāṇa to Junagadh via Shāyalā, Sardhār, Gondal, Virpur and Jetpur, which was used as a trade route for many years. It also became the regular road for travellers and the coast route was gradually neglected save for religious pilgrimage.[145]

As to the state of communication in the mainland under strong rule of the Aṇahilwāḍ dynasties (943–1300 AD), the primeval forest that stretched from the hills of Modasa to the south of Sabarmati and the fringes of the Rann, was gradually converted into cultivated land studded with wealthy and populous towns, whose very existence testified to the development of flourishing trade. Its chief inland mart was Dholka. Al Idrisi (1150 AD) speaks of Dholka-Ashawal (on the present site of Ahmedabad) and

Shihor, the chief town of Chuval, as places of good trade.[146] Hence, the growth of agricultural activities had its role in the building up of trade routes. Thus both were to some extent interdependent.

Gujarat was connected with her neighbours by five earlier overland routes. Of these, the most important and more so from early times very active was the northern route connecting Gujarat with Sindh and Punjab via Rajputana. Though the greater part of Rajputana is today a desert, this region supported what seems to have been, flourishing settlements in prehistoric times. The explorations and excavations carried out in the decade that followed our independence in Bikaner in Rajasthan and at Rangpur and Lothal in Gujarat have revealed not only an extension of the Indus valley cultures but also its contacts with Gangetic valley and north Gujarat, which seems to have been the meeting place in the prehistoric times, of the stone age cultures of Punjab and south India.[147] Probably, the earliest large scale movement alongside this route was that of the Sakas and the Huna (BC 200) followed by the Gurjaras.[148]

The eastern routes connected Gujarat with Malva and the Gangetic basin. A northerly route following the river Meshvo and the upper reaches of the Hāthamatī, through Idar, Khedabrahma, Harsol and Shāmaḷājī was extensively used by the Mauryas, the Guptas and the Solankis. The other was the waterway provided by the rivers Narmada and Tapti in the south Gujarat since the greater part of this lay between the forested hills of Satpuda ranges, it is probably that it was extensively used although the Marathas entered Gujarat in early eighteenth century through this route.[149]

Then a main northeasternly route from Surparak (modern Sopara in the Nasik District of Maharashtra) passed north via Dahod-Ratlam pass. With this was linked up the old onyx route from Ujjain to Bharucha near Vadodara. Finally, the main north-south route along the north Konkana coast was linked up at Sopara with the two routes from the Gangetic basin and from the Deccan. The Chalukya and Rashtrakuta invasions seem to have occurred along with route.

The Gujarat was open to culture-contacts and migrations from Sindh and Punjab via Rajputana in the north, from Malva and Gangetic basin in the east and from Deccan in the south, not to speak of the contacts from the sea-routes,[150] and more so helping the advancement of agricultural activities.

MISCELLANEOUS

In one of the Maitraka *Dānaśāsanas*, there is a reference about a donation of *Puṣpavāṭikā* (garden) along with *kshetra* (field), *vāpi* (stepwell) and *kūpa* (well). Regarding the tradition of *Purtakārya*, there is a statement in *Dharmaśāstras* that of constructing and maintaining of *ārāms* (i.e. garden), in the light of this that we come across such references. From another such grant, we know that a donation of a *Puṣpavāṭikā* was given along with three fields, two *vāvs* and four *kūpas.*[151] In varied sorts of donative materials mainly include cow, land, learning, etc. These were believed as *atidāna* (i.e. extreme kind of donation) and among them sometime the land grant was considered best of all.[152] All these sorts of donations have a direct bearings on the growth of agriculture in Gujarat.

All the agricultural land of the state was taxable (for which the term *karad* is used in the grants). But the land given away in donation as *dharmadeya* was considered non-taxable (i.e. *akarad*). The possession of this type of land was called *dharmadeya sthiti.* If such *dharmadeya* has been given to any temple then it was called *devadeyasthiti,* and brahmanas it is *brahmadeyasthiti.* Maitraka *dānaśāsanas* mention other types of enjoyment right of *dharmadeyabhūmi.* There is also a mention of many *brahmadeyabhūmi* in the boundaries of the *devadeyabhūmi in Tāmraśāsanas* (e.g. ब्राह्मण शंकर सत्यकब्रह्मदेयक्षेत्र) such *brahmadeyabhūmi* was also called *agrahāra* (i.e. grant of a land given by a king either to a *brāhmaṇ* or a temple). This term often used, especially for the *brahmadeyas* of the villages, in relation to this the application of the term *agraharsthiti* was also in vogue for *brahmadeyasthiti.* The terms such as *devikshetra, saṅghkshetra, ādityadevasatkvāpi,* etc., indicate the probable meaning of *devadeya* along with the possession and enjoyment of the temple.[153]

The donation of immovable land was considered eternal and, hence, the written document was always must for such land grant. Therefore, such a document was required to be engraved on copper plates. The grants being incised on *tāmrapatras* suggest the existence of information about land donation as early as *Maitraka* rule.

Often the land was recognized either in the name of its owner or in the name of the main tree on the field. For instance, *Saṇghkshetra, Siddhakshetra, Devīkshetra, Shamikedarkshetra, Kāvitthikakshetra,* etc., likewise the stepwells were known by names, e.g. *Shrāvakavāpī* (stepwell of Jaina lay man), *Mallavāpī* (wrestler's stepwell), *Pipplavāpī* (stepwell having a pipal tree nearby), *Shireeshvāpī* (stepwell with a kind of flowering, i.e. Albizzia lebbek), *Khajjurīvāpī* (stepwell having a date palm tree), *Nimbavāpī* (stepwell with Nimb tree *Yamalvāpī* (Vav with couple tree), etc. These names of different trees indicate their commercial utility as well as their role in preventing soil erosion.[154]

The names of trees and their existence could also be known from the *Maitrika* plates. Generally while describing the boundaries of a donated land or field we come to know such references as of *Vaṭa* (banyan), *ankolla* (kind of medicinal plant, *alangium* lamarcki), *shāk* or *sāg* (teak tree), *khadir* or *kher* (a kind of tree, *acacia catechu*), *badari* or *boraḍī* (jujube tree), *shālmani* or *shīmaḷo* (silk cotton tree or *bombax mulbericum)* etc.[155]

Generally, along with the agricultural activity, the domestication of animal is associated. Over and above the domestication of war, serviceable animals was also in vogue. In the *Brhat Samhita,* its author Varāhamihira describes the good and evil characteristics of cow, bullock, dog, tortoise, cock, goat, horse, elephant, etc. The list of 64 arts also included the art of battle of *mesha* (ram), *kukkur* (cock), *lāvari* (a kind of bird), etc., and also the art of speaking to *shuka* and *sārikā* (parrot and the Indian jay). Such information also suggests that the domestication of animals and birds help the growth of agriculture as they help the corn from any sort of damage.

Mention of elephant, lion, horse, bull, deer, etc., is found in the *Prasastipādas* as they occur in Maitrika plates. They also include the lion, elephant, cow, camel, boar, cock, black bee, etc. Of these some animals were also useful for hunting, some for food and some for agricultural purposes.[156]

To save the crops from the thieves the fields were guarded by dogs. Made of straws the scarecrows were used to prevent the destruction of crops by the birds. This preventive act is still used in Gujarat. During the time of autumn harvest, the fields were guarded by farmers and their wives.[157]

During the Chalukyas regime, the grain was given on loan, as indicated by the term *Vṛuddhidhānya-ksharani* referred to in the *LP.* As noted in it, one man received

20 maunds of wheat on loan, and on the expiry of the harvesting season on the fifteenth day of the bright half of the month of *Jyeshtha*; it was decided to return 25 maunds of the borrowed wheat. Besides, there was a mention that the wheat must be of good quality and debtor should personally hand over the same through his own cart as decided. There is also a clarification in the document that such borrowed grains are returned after paying the state's share. If the debtor failed to return the wheat on fixed time, he was liable to pay in cash the sum of the market value of the equivalent of the borrowed grains. There was also a provision of two guarantees and they equally were considered responsible for the said loan when the debtor registered insolvency. This document bears the signatures of five witnesses.[158]

From this description, the following points require elucidation:

(1) That the grain was borrowed on loan with interest.
(2) That the loan be repaid in kind and with the same grain of good quality.
(3) That the remittance of the state-share must be made before the repayment of the loan.
(4) That the guarantees were too responsible to return the loan.
(5) That in case the debtor did not return the borrowed grain in time, he had to pay in cash the price equivalent to the borrowed grains but to the then market price.
(6) That the state should not be deprived of grain in any circumstances.

CONCLUSION

Finally and most importantly from the agricultural point of view; it is all very well to sum up that this region of India, being on the southern side of the tropic of cancer, has been characterized by less rain and hence manmade reservoirs are the main source for the growth of agricultural economy in Gujarat. Therefore, right from the Lothal and Dhoḷāvīrā cultures to the present day the unique feature of the cultural contour of Gujarat is reservoir-borne culture and more so agriculture.

NOTES AND REFERENCES

1. For more information see M.R. Majumdar, 1965, *Cultural History of Gujarat,* Popular Prakashan, Bombay, preface (henceforth CHG.) It is interesting to quote a stanza from Narmadashanker, a well-known Gujarati poet and social activist, who has poetically, described the boundaries of Gujarat thus:

   Uttarmāñ Aṁbā Mātā, Puravmāṁ Kālī Mātā;
   Chhe Dakshiṇa Dishāmāñ Karant Rakshā Kānteshvar Mahādev;
   Chee Somnāth Ne Dwārkesh E Pashchim Kerā Deva:
   Chee Sahāymāñ Sākshāt, Jai Jai Garvi Gujarat.

   (i.e. in the north resides *Ambāmātā* and in the east *Mātā Kāli* lives; *Kānteshvar* Mahadev is in the south and *Somnāth* and *Dwārkā* are located in the west—these deities protect the boundaries of Gujarat).

2. For more discussion on the origin of the word Gujarat see: A.K. Majumdar, 1956, *Chaulukyas of Gujarat,* Bharatiya Vidya Bhavan, Bombay, pp. 17ff (Henceforth *CGM*), *Epigraphia Indica,* Vol. II, p. 42. The first known reference to the name of this region appears in *Āburās* written in 1233 AD. The first known use of the word *Gurjar Deśa* is found in Kshemendra's Sanskrit work *Auchiyavi Aucityavičāračarčā* (997 AD).
3. For understanding the social and cultural tradition of Gujarat see: Rasesh Jamindar, 'Cultural Contours of Gujarat and its contribution in the making of National Heritage', Bulletin of the *Chunilal Gandhi Vidyabhavan, Surat,* 1985, No. 28, pp. 34–49.
4. For detailed history of Gujarat, see *CHG* and H.D. Sankalia, *Archaeology of Gujarat,* Bombay, 1941 (Henceforth AGS).
5. See *Ibid.* and also Rasesh Jamindar as in Fn. 3.
6. See *CHG.,* p. 52 and AGS. p. 3–4.
7. R.N. Mehta and Rasesh Jamindar, 'Geomorphology in relation to the development of Ahmadabad, *VISVAMBHARA* (Probings in Orientology), ed. by A.M. Shastri and others, New Delhi, 1995, pp. 41–42.
8. *AGS,* p. 3–4.
9. *CHG,* p. 261.
10. Samuel Beal, *Buddhist Records of the Western World, Indian Antiquary,* Vol. 2, 1884, p. 260. He further writes that the soil is rich and fertile and produces abundant harvests. Shrubs and trees are numerous and flourishing. Flowers and fruits are available with great quantities. The soil is suitable for winter wheat.
11. Gujarat does not have, then and now, four seasons. It seems that author's information is not correct.
12. Abdullah Wassaf, *Tazjryatu-i. Amsar was Tajriyatu-i Asar*', in Elliot and Dowson, *History of India as Told by its own Historian,* Vol. III, p. 31 (Henceforth ED).
13. But this is not the case with other areas. For other aspects of history and culture we have had good many sources at our disposal.
14. That of Yuan Chwang, Wassaf, Marco Polo, etc.
15. *Deshināmamālā, Lekhapaddhati, Dvyāsraya Mahākāvya* to name just few (Henceforth *DN.* and *LP.*).
16. The main data for Maitraka's history is their copperplates. It gives the importance of land in human life, significance of *bhūmidāna,* measurement of the land, ownership of it, etc.; for it deals with land and land-donation only.
17. *Aṅgavijjā* is a *Prakrut* work by an unknown author. This is an excellent work on prognostication, a science of divination through physical signs and symbols. It deals with many subject and things existing under the moon and sun. This includes valuable list of articles of food and drink over and above many other things. It was written sometime during the latter part of the Kshatrapas regime, i.e. fourth century AD (see Rasesh Jamindar, 'The Date and Place of *Aṅgavijjā*', *Prachya Pratibha,* Vol. 4, No. 2, Bhopal, pp. 101–106).
18. The lake, named as *Sudarśan,* was originally constructed in the time Chandragupta Maurya especially for the advancement of the agricultural activity. It was improved

with strengthening of the dam and had canals under Aśoka, who had his edicts carved on the rock for the first time, situated on the bank of the reservoir. Unfortunately, we do not have any inscription of Chandragupta in this regard. Four hundred years later, in AD 150, when the fresh repairs were found necessary, due to the unusual floods in the adjoining rivers in the month of *Mārgshīrsha,* Rudradaman, availed himself of the opportunity to give a brief history of the lake, which was otherwise would not have been possible to know and more so recounts his achievements on the same rock. In 457 AD, the lake-dam burst again. It was repaired by the local governor of the Gupta King Skandagupta who also used the same rock to record the event in verse, (for details see Rasesh Jamindar, *Itihas Samshodhan,* Ahmadabad, 1975, Ch. 17) (Henceforth *JIS*).

19. Its importance lay mainly in the circumstances that she possessed splendid harbours to which people flocked from all quarters of the globe with their merchandise.
20. Noted in *CHG.,* p. 53.
21. *Ratnamālā,* 1903 p. 288.
22. See a Gujarati verse

    Balihārī Tahārī Bājarī, Jenoñ Lomboñ Pān, Ghodi Pāṅkho
    Āviyo, Radhon Thayon Juvān.

    (Bravo for your *Bājarī,* which hath long leaves, and by eating which horse sprouted wings and old men became young).
23. This particular verse does not apply to the whole of Surashtra area; but to some areas, which were very interior and hard to reach there.
24. *CHG.,* p. 182.
25. See *Arthasāstra,* Book XII, 5.

    Kāmbojsurashtrakshatriya Shreṇyādayo
    *Vārtāshastropajeevinah*II
26. *El,* Vol. 8, p. 78 ff; *KGJ.,* 292.
27. *KGJ,* p. 65 and 72.
28. *Ibid.,* 292–293.
29. G.V. Acharya, 1933, *Gujarātnā Aitihāsik Lekho,* Vol. I, No. 16, Forbes Gujarati Sabha, Bombay, (Henceforth GAL).
30. Harprasad G. Shastri, 1955, *Maitrakakālīn Gujarat,* Vol. 2, Gujarat Vidyasabha, Ahmadabad, p. 529 (Henceforth SMG).
31. R.N. Saletore, 1943, *Life in the Gupta Age,* Bombay, p. 330 (Henceforth LGA).
32. *SMG,* II, p. 529.
33. *CGM,* p. 245.
34. *Ibid.,* p. 246.
35. Rasesh Jamindar, 1975, *Kshatrapakālnuṅ Gujarat,* Dept. of History and Culture, Gujarat Vidyapith, Ahmadabad, pp. 121–127 and 293 (Henceforth *JKG*).
36. *IA.,* V III, p. 141, *Periplus,* para 41.
37. R.N. Mehta, *Excavation at Nagarā,* Vadodara, 1970, p. 153.
38. Rasonānantarm vāyoho palāṇḍuhu paramauṣadham I

    Sākshādiva sthitam yatra śakādhipati jeevitam II

Yasyopayogen śakāṅganānām lāvaṇya sārādivi nirmitānām I
kapolkāntyā vijitah śaśāṅka rasātalam Iachchati nirvideva (निर्विदेय) II
(*Uttaratantra*, ch. 49)

39. Bāhlīkāhā pahlavāśreenāhā śūlīkā yavanah śakāhā
Māṅsagodhūm mādhvīkāśastra vaiśvānarochitāhā
(*Chikitsāsthān*, ch. 30, p. 316)
40. *Aṅgavijjā*, ed. by Muni Punyavijayji, Varanasi, 1957. (English Introduction was written by Motichandraji and that in Hindi by V.S. Agrawal.) Intr. p. 39 and 42 and *JIS.*, Ch. 16; *KGJ.*, p. 263–265.
41. *SMG.*, p. 616ff.
42. M.R. Majumdar, 1960, *Chronology of Gujarat*, M.S. University of Baroda, Vadodara, p. 261.
43. Kakkasuri, 1985, *Nābhinandanjirṇoddhārprabandha*, Ahmadabad, V.S.
44. Ramlal C. Modi, 1942, *Sanskrit Dvyāshraya Kāvyamān Madhyakālin Gujarātnī Sāmājik Sthiti*, Ahmadabad, p. 80.
45. B.J. Sandesara and R.N. Mehta, 1959, *Varṇaksammuchchaya*, Vol. II, M.S. University of Baroda, Vadodara, p. 128.
46. 1, 52; III, 50.
47. *Dvyāshraya*, III, 45; ND., V. 7.
48. Wassaf, III, p. 31.
49. *Dvyāshraya*, III, 3.
50. *ED*, I, p. 85.
51. *CGM.*, p. 257.
52. *Ibid.*
53. *DN*, 1, 21.
54. *CGM.*, p. 258.
55. *Ibid.*, p. 259.
56. *Ibid.*, p. 79.
57. *Ibid.*, pp. 79–80.
58. *Ibid.*, pp. 86–87.
59. Tr. by. A. Ricci, ed. by Sir Denison Ross, 1931, p. 332.
60. *DN*, 1, 117; Hemchandra, *Trishashtishalākāpurusha*, (ed. by Muni Punyavijayaji) Bhavnagar, 1950, III, 30 and 156.
61. Ratnamanirao Bhimarao, 1929, *Ahmādabad: Gujaratanuñ Pāṭanagar*, Gujarat Vidyasabha, Ahmadabad, p. 452.
62. *DN*, 1, 89,
63. *GAL*, Vol. II, No. 145.
64. *Ibid.*, Vol. 3, No. 222. The leaves of betel plant are dressed with other ingredients such as *sopārī*, *kātho* (catechu), lime, coriander seeds and eaten.
65. *Ai-ne-Akabari*, Vol. II, 2nd Ed., p. 257.
66. *DN*, VI, p. 32.
67. *CHG*, p. 256.
68. *GAL*, I, No. 72.

69. K.J. Virji, 1955, *Ancient History of Surashtra,* Konkan Institute of Arts and Sciences, Bombay, p. 246 (Henceforth *AHS*).
70. *GAL,* I, No. 72, Sukthankar, *CII,* XV, p. 256; *AHS.,* p. 246.
71. *CII,* Vol. 3, p. 170, fn.4.
72. *SMG,* Vol. II, p. 526.
73. *Ibid.,* p. 616.
74. *AHS,* p. 247.
75. Supra fn. 70.
76. Supra fn. 74.
77. Kane, *History of Hindu Dharmashastras,* Vol. III, p. 145. Bhandarkar Oriental Research Institute, Poona, 1946. Here, *danda* is a measure of length equal to 4 cubits and one cubit is equal to 18 to 22 inches. Hence, one *Danda* is equal to 72 to 78 inches.
78. *SMG.,* p. 526.
79. *Ibid.*
80. *Ibid.,* p. 280.
81. Monier William's Dictionary.
82. *LGA.,* p. 359.
83. *GAL.,* III, No. 243.
84. Book II, *Adhyaya* 6, and Verses 3 and 10.
85. Book II, *Adhyaya* 24, Verses 22 to 25. This chapter 24 is named as *Seetādhyaksha.*
86. R. Dikshitar, *Maurya Polity,* p. 224 ff; AGS., p. 195.
87. *Bombay Gazetteer,* Vol. I, part I, p. 81. (BG).
88. *SMG* 515–16. The former V.C. of the B.H.U. Acharya Anandshanker Dhruva had as his family name.
89. *BG.* I, I, p. 82.
90. J.F. Fleet, CII, Vol. 3, p. 96 U.N. Ghoshal, 1929, *Contributions to the History of Hindu Revenue System,* Calcutta, p. 210 (Henceforth HRS.)
91. *LGA.,* p. 348. It is a combination of *upari* and *kar. Upari* means above the right of proprietorship.
92. *AHS.,* p., 244; *SMG,* p. 527; *HRS.,* p. 210; *LGA.,* pp. 348–50.
93. *AHS.,* p. 244.
94. R. Dikshitar, 1929, *Hindu Administrative Institutions,* Madras, p. 163, (*HAI*).
95. *AHS.,* 244.
96. *Manusmriti,* VII, 5, VIII, 5; *Arthasastra,* II, VI, 58.
97. *HAL,* p. 180.
98. *CII.,* 111, p. 170.
99. *SMC.,* pp. 528–29.
100. *CHG.,* p. 242.
101. *Ibid.*
102. This is an authoritative institution with wide power for governing the administration of a town or a city or a village. The president of the *Pañchakula* was entitled to impose new taxes after consultation with the businessmen of the city, to extend sorts of donation. His importance was also recognized in the temple organization.

The term *Pañchakula* has been referred in inscription (see. *GAL.*, III, No. 223).
103. *CHG.*, pp. 242–43.
104. *IA.*, XIII, p. 69.
105. Supra fn. 103.
106. *Ibid.*
107. *Ibid.*, p. 243–244.
108. *EI.*, XV, p. 137.
109. *CHG* p. 244.
110. *Ibid.*, p. 245.
111. *Dvyāshraya,* III, p. 18.
112. *HRS.*, p. 244, 42, 65.
113. *Manusmriti,* VII, 127–28; *Arthasāstra,* 1, 13; II, 15.
114. *History of Rashtrakutas,* p. 215.
115. *CHG.*, p. 252.
116. Merutungacharya, *Prabandhachintamani,* (ed. by Muni Jinvijayji) 53.
117. *CHG.*, p. 252.
118. *Ibid.*, p. 253.
119. *Ibid.*
120. *JIS.*, Ch. 17.
121. *Ibid.*; *JKG.*, Appendix 6.
122. *Ibid.*
123. R.N. Mehta, 'Sudarshan Lake', *JOI,* Vol. 18, pp. 20 ff.
124. *Gujarātno Rājkīya Ane Sāṅskrutic Itihās,* Vol. 4, B.J. Institute of Learning and Research Ahmedabad, 1976, p. 420 (ed. By R.C. Parikh and H.G. Shastri).
125. अक ष्यायां भूमौ पशुभ्यो: विधीतानि प्रयच्देत। II, Ch. 20.
126. *SMG.* Vol. II, p. 524.
127. *CII,* III, p. 138 and *Arthasāstra,* II, 2.
128. *EI.*, Vol. 2, p. 349.
129. *CII,* II, p. 39; *EL,* IV, No. 8, p. 75; *LGA;* p. 335.
130. *LGA.*, p. 336–337.
131. The *Dānaśāsanas* of the Maitrakas refer to the position.
132. Generally, the *bhūt* was taken to mean the produce of the soil and *Pratyāya* means dues. The *vāt* denotes the wind-borne produce. It is more probable that *vūt* may indicate a sort of an octroi duty as the root *vūt* also means to wish to invoke to gain (*LGA.*, p. 352). Nevertheless, the exact nature of both *bhūt* and *vūt* is not explicable. Anyway they are other kinds of revenue.
133. *Ibid.*
134. *Dhānya* denotes the dues paid in kind, while *hirṇya* refers the dues paid in cash and *adeya* means taxes. *Hirṇya* here suggests a sort of currency. The *dhānya* and *hirṇya* both were obviously the crops growing on the soil and the gold preserved underneath it respectively.
135. *Ibid.*
136. The *bhāga* and *bhog* represented two land taxes due to the state. The *bhāga*

evidently means the land tax paid in the form of a fixed portion of the produce. Generally the portion due to the state was 1/6, and hence the king was known as शद भाग भग (*HAI.*, 163) *Manusmriti* (VIII, 130) prescribes the portion to very from 1/6 to 1/12. The *bhog* seems to have been a petty tax payable in the form of daily presents in kind such as fruits, vegetables, flowers or such objects which could be enjoyed without the right to alienate them.

137. *HAL,* p. 180; *LGA*; p. 352.
138. The term *utpadyamanvishti* means the right to exact forced labour whenever the occasion arises (Fleet, *CII,* p. 170). *Dashaparadha* was the right to fine culprits for any of the 10 offences enumerated therein.
139. The grantee of a religious gift of land was entitled to receive a number of varies dues (described just here before) as the proprietor of the land.
140. *SMG.*, p. 531. In the terminology of the *Dharmashastra* the donor was called *dātā,* the donee *pratigrahitā,* and things to be given in *dāna* as *deya.*
दाता प्रतिग्रहीता च श्रर्द्धा देय च धर्मयुक्त।
देशकालौ च दानानामडान्येतान्निषडविदु:।। (Hemadri, *Danakhada,* p. 14)
141. *JKG.*, 121–127; JIS. p.; *CHG* p. 61–62.
142. *CHG.*, p. 54.
143. *Ibid.*
144. Sarvanandasuri, *Jagaḍucharita,* Sarga 6, verses, 77–88.
145. *CHG.*, p. 63.
146. *Ibid.,* p. 64.
147. V.D. Krishnaswamy, 'Stone Age in India', *Ancient India,* Vol. III, 1948.
148. *CHG.*, p. 65.
149. *Ibid.,* pp. 65–66.
150. *Ibid.*, p. 66.
151. *SMG.*, pp. 616–617.
152. *Vasistha Dharma Sutra,* 29, 19; *Anuśāsanparva,* 62, 2.
153. *SMG.*, p. 524; *GAL;* III, Nos. 160, 170, 202, 217.
154. *SMG.*, p. 616.
155. *Ibid.*, p. 225.
166. *Ibid.*, p. 618.
167. *Dvyashraya,* III, 5; XV, 49; XVIII, 30.
158. *LP.*, p. 21.

CHAPTER 44

Agricultural Knowledge System in Ancient India and its Relevance in Sustainable Development

Anil Kumar Pandey, Nilam Pandey and V. K. Dubey

INTRODUCTION

The origin of crop cultivation and animal domestication is the distinguishing feature of the so-called Neolithic revolution. This term is used to describe the change from a hunting and food-gathering economy and a flaked-stone tool technology to an economy based on farming and a technology that included polished stone tools, pottery and weaving. The Neolithic revolution has been considered as important as the industrial revolution of modern age.

Generally, the Mesolithic age coincides with the warming trend that followed the retreat of the last glaciations some 10,000 years ago. Mesolithic adaptations ultimately led to the domestication of plants and animals. Traditionally, the beginning of food production marks the end of the Mesolithic age; so its length varies in different parts of the world. Therefore, the beginning of agriculture and domestication of animals varies in time and duration, accordingly.[1]

The domestication of plants and animals allowed people to produce more food on a given area of land and, thus, support a larger population.[2] Permanent settlements were formed near the fields and people began to hold title to pieces of land and call them their own. Conflicts began over ownership of land, property, resources and territory, and some groups within a farming society came to have higher prestige and more property and benefits than other groups.[3]

The food production may have begun over 10,000 years ago in the Middle East. Not only did agriculture begin fairly recently but it also spread very rapidly, compared to the role of change in the Paleolithic and Mesolithic ages. Between about 1,000 to 2000 years ago agriculture had replaced hunting and gathering as the mainstay of life almost everywhere.

ORIGIN OF AGRICULTURE IN INDIA

The origin of agriculture in the Indian subcontinent is supposed to have taken place simultaneously with the Middle East, but there is a difference of opinion in this regard. Some archaeologists and historians opine that in the Indian subcontinent, agricultural activities began around 7000 BC in the Neolithic period. The only remains of agricultural settlements of that period are found in this region at Mehrgarh, i.e. presently situated in Baluchistan of Pakistan. Unfortunately, it is the bias and predisposition of western thinkers to stretch the origin of Indian agriculture some 3000 years later than the Middle East.

The history of the beginning and development of agriculture and animal husbandry is correlated and unalienable with the human civilization. As per the changing situation and climatic conditions when the sustenance of life was not possible merely on the food gathering and hunting, then the increasing necessities of human life lead the process of the new beginning. There should be no confusion that the aforesaid condition of the Middle East and the Indian subcontinent was more or less the same at that time. If we compare the climatic condition, ecological situation, implements of hunting and holy book '*Zend Avesta* and *Ṛgveda*', it would be obviously clear that the evolution of agricultural practices in India simultaneously took place with the Middle East about 10,000 BC. The *Zend Avesta* of the Middle East and the Indian *Ṛgveda*, both are epics of the same period and deal with the same thought about agricultural practices.

THEORIES OF BEGINNING OF AGRICULTURE

Actually, the Neolithic people showed immense courage, wisdom and patience in the development of agriculture and animal husbandry and led a continuous process of invention, modification and adaptation. When they first began to cultivate wild plants and rear wild animals, they caused a kind of evolution, a change in gene frequencies. Perhaps the most potent force causing changes in genetic makeup is selection. In the case of domestication of crops and wild animals, humans were the selective agent. The people, knowingly or unknowingly, promoted the useful attributes through selection and introduction of good species of crops and animals. The natural process of self mutation also accorded the major role in the enhancement of qualitative and quantitative characters of human utility. As a result, the frequency of genes responsible for these traits increased. Plants and animals whose traits were an advantage in the wild but a liability under domestication, however, may have been less well protected by humans or purposely prevented from reproducing.

Human beings may have produced some of these changes unknowingly. Merely changing the environments of plants and animals by moving them to different areas or by protecting them from weeds, drought or predators permit more mutants and variants to survive and reproduce. Deliberate saving of seeds for future crops may have been a major step in domestication, since it probably led to planting in new environments and to further changes in the selective pressures acting on the genetic composition of the crops.[4] The soil of a new environment could differ in terms of moisture and nutrients. Such changed conditions might favour traits not previously selected for.

Cultivation of wild species of crops and rearing of wild animals in the primary stage was extremely difficult. Wild plants and animals are adapted to survival and reproduction in their natural environment. Often they have traits that make it difficult for human beings to use them effectively. According to archaeologist Kant Flannery (1965), wild grains posed three major problems:[5]

1. The brittle *rachis* of wild wheat and barley breaks easily and helps scatter the seeds.
2. Wild grains have tough, inedible glumes or husks, which hold the kernels tightly, despite vigorous threshing.
3. Wild grains grow in scattered patches on hillsides, not in concentrated stands suitable for easy reaping.

According to one expert (Helback–1959),[6] the earliest farmers overcame the problem of a fragile *rachis* by selectively reaping grains with tough *rachis*. Only deliberate sowing of tough *rachis* plants could lead to their domestication. The problem of grain being difficult to harvest was overcome through mutant form of barley better adapted to the drier spring weather of the level Mesopotamian plain. This mutation resulted in the barley with six fertile kernel rows instead of only two. The harvesting of barley, therefore, became both easier and more productive when it was planted in dense stands in lowland fields.

Likewise crops, the domestication of wild animals was much difficult. Early domesticators had to overcome several problems in keeping wild animals. They were either aggressive or good at escaping to survive in the wild. Wild cattle, for example, were large and powerful and had long horns for protection.[7] Hence, mutation and human directed changes in the frequency of genes controlling valued traits resulted in domesticated species. As they adapt to new human-dominated environments, they became dependent upon human intervention for survival.

The earliest means of subsistence of the people all over world was the fruits and roots of the trees and hunting. Why did people start sowing wild plants and go to the trouble of rearing wild animals? It is not easy to answer this question so we must imagine the thoughts and motives of Neolithic humans and climatological conditions of that period. Archaeologists have developed a number of hypotheses explaining the kinds of pressures that may have motivated the earliest farmers.

CHILDE'S OASIS MODEL

The most distinguished British anthropologist of the early twentieth century, V. Gordon Childe (1892–1957)[8] spent much of his life studying the Neolithic people of the near east. In his *New Light on the Most Ancient East* (1952), he published a theory linking the drying trend that presumably affected much of the world as the glaciers withdrew at the close of the Pleistocene period to the domestication of plants and animals.

According to Childe, as the Mesopotamian climate grew drier, already dry grasslands turned into deserts, dotted here and there with oasis. People, animals and plants were concentrated in these areas. This climatic crisis forced humans to domesticate plants and animals, since killing them would have left the oasis dwellers with no sources of food.

BRAIDWOOD'S NUCLEAR ZONE THEORY

Braidwood[9] believes that when people had acquired an indepth knowledge of the environment they lived in, they were ready to begin food production. They were culturally receptive to domestication. The other prerequisite for food production was an area rich in animals and plants that could be domesticated. In such an area, called a nuclear zone, people worked out the techniques of domestication. Braidwood believes that the ability to experiment has long been part of human nature.

POPULATION MODELS

Over the past twenty years, several theories have suggested that population growth beyond the land's ability to support the human biomass is responsible for the development and spread of agriculture. In other words, when there were not enough preferred food, people were forced to experiment with and eat secondary (and probably less desirable food).

Hence, it is clear that there is no consensus on the question of 'why agriculture began?' Most archaeologists would agree, however, that some combination of population pressures, ecological change and population movement was responsible for the origin of agriculture.

AGRICULTURAL SYSTEMS IN ANCIENT INDIA

Since the ancient period, agriculture and allied activities have provided livelihood for most of the population. Although the Indus valley civilization, according to available facts, is supposed to be an urban civilization, yet the agriculture was the mainstay of life and provided sound base to their economy. The archaeological remains of large granaries of Harappa and Mohenjo-daro prove it that agricultural activities systematically developed fully.[10] Unfortunately, our knowledge regarding this civilization is totally based on the archaeological excavations and the script is still unread.

The Harappan civilization glorifies the ancient agricultural system. Steven A. Weber (1991)[11] mentioned in his *Plants and Harappan Subsistence* that Harappan people cultivated several cereals, pulses, oilseeds, fibre crop, fruits and vegetables. They know crops of different seasons like *kharif* and *rabi.* In cereals, they cultivated wheat, barley, rice, *jowar, bajra* (pearl millet), *ragi* (finger millet), foxtail (Italian millet), *kedo, sawan* (*echinocholoo frumentaceae*); in pulses lentil, chick pea (Bengal gram), field pea, green gram (*mung*), black gram (*urd,* grass pea, horse pea and oilseed crops); linseed, mustard, sesame, and in fibre crops cotton. Fruits and vegetables had a minor interest and they cultivated date, jujube, grape, melon, etc.

It is interesting to discuss that Harappan people cultivated four species of wheat, like *Triticum aestivum, T. spharococcum, T. compactum* and *T. Vulgare.* They were also well versed in irrigation, water management through check dams, contouring, bunding, benching, terracing and levelling. They were aware of the application of irrigation water to crops as per to their water requirements and check the rainwater from very beginning in catchment areas to prevent soil erosion.[12]

The number of plants actually cultivated by the Harappans is significantly different from the total number of plants used by them. Like today, wheat and barley are the most common crop remains recovered from Harappan sites and, consequently, the definition of Harappan culture has tended to incorporate a wheat and barley subsistence base. In fact, it is not surprising that past cultivation practices were somewhat variable, because of the geographical differences represented in the Harappan areas.[13] Therefore, while wheat and barley did play an important role in Harappan culture, there may have been pulses like, pea, gram and lentil, oilseeds; mustard and sesame and fibre crops, cotton, sunhemp, flax, etc., major crops as well.

There are two main growing seasons in Indian subcontinent. The *rabi,* or winter season, extends from October to March and the *kharif* or summer season extends from June to October, wheat and barley do the best in a *rabi*-growing season, while many of the millets performed well in the *kharif* season. This may help explain the apparent preferences for millets over wheat and barley in archaeological sites in areas like Gujarat where the *kharif*-growing season is so popular today (Possehl, 1986).[14] The seasonal variations of rainfall and its distribution decided the fate of cropping patterns.

AGRICULTURE IN VEDIC AND LATER VEDIC PERIOD

The Aryans were an active pastoral people and agriculture was the backbone of economy at that time. Actually, the Vedic civilization was an obviously rural civilization. The process of cultivation of crops was almost the same, as it exists today in most of the Indian villages, particularly where there are small holdings. The land was tilled and cut into furrows in which the seeds were sown. The grain was harvested by human labour with sickles, threshed, shifted and winnowed. Irrigation of the fields through was carried out through various sources and by different techniques.

The Vedic literature is the main source of information regarding agricultural activities practiced at that time. In Vedic literature, *Vedas* are considered as primary source. They include the *Ṛgveda Saṃhitā*, the *Yajurveda Saṃhitā*, the *Sāmaveda Saṃhitā* and the *Atharvaveda Saṃhitā.* In spite of these *Vedas,* there are several other works of Vedic period, which provided important informations such as *Brāhamaṇa* (like: *Śatapatha Brāhamaṇa, Gopatha Brāhmaṇa,* etc.), *Upaniṣads, Saṃhitā, Purāṇas* and commentaries on *Vedas* like *Sāyaṇa's Bhāṣya,* etc.

The modern term for agriculture in Sanskrit literature is *kṛṣi* from *kṛṣ,* meaning to draw, to drag, to pull and thus to draw or make furrows to plough.[15] The word has the Vedic origin. Its occurrence is as follows in the Vedic literature:

Ṛgveda	*kṛṣatu* IV-57-4
	Kṛṣims X-34-13

	Kṛṣasva X-34-13
	Kṛṣan X-117-7
Yajurveda	*Kṛṣiḥ* IV-10
	Kṛṣantu XII-69
	kṛṣiḥ XIV-19 and XVIII-9
	kṛṣṭapachyaḥ XVIII-14
Atharvaveda	*kṛṣatu* III-17-16
	kṛṣim III-12-4, VIII-13-11 and 12, X-5-34 and X-6-12
	kṛṣi samsitaḥ I X-5-34
	kṛṣate XII-2-36.

The agricultural practices and animal husbandry adopted new dimensions of development in this period. Hunting with a bow and arrow played considerable part in the life of Aryan people. The practice of ploughing was certainly becoming an essential part of field preparation and use of cowdung as manure reveals that they knew soil fertility management. In the *Ṛgveda*, the word *Kṛṣ*[16], shows the use of plough for making furrows no doubt it means, ploughing and word *aśmachakra*[17] clearly indicated that the wheel was made of stone for lifting water.

In the *Ṛgveda*, giving serious attention to agriculture, it is, advised that 'play not with dice, pursue agriculture, delight in wealth acquired by you.[18] It is further mentioned that 'the ploughshare furrowing of the field provides food, (4-57-4). In the fourth *Maṇḍala* of the *Ṛgveda* for the overall welfare of agriculture it is elucidated; 'may the oxen (draw) happily, the men (labour) happily, the furrow and plough happily may the traces bind happily wield the good happily.'[19] In the *Śatapatha Brāhamaṇa*, it is enumerated that 'rejoice plants bearing abundant flowers and fruits, triumphing together over diseases like victorious horses' and further mentioned, 'plants! thus, I hail you, the divine mothers of mankind (*S.Br.*VII-2-4-28).

During the later Vedic period, great progress was made in agriculture and pastoral pursuits. The plough used for ploughing was large and heavy with as many as twenty-four oxen being harnessed to one. It had a sharp point and a smoothed handle. In addition to irrigation which has already been mentioned in *Ṛgveda*, the use of manure particularly composts (*desi khad*) became popular. Cereals like *yava* (barley), wheat, *jowar, bajra,* (*panicum miliacaum*) pulses, gram and pea and oilseeds such as sesame and mustard were cultivated on a large scale. The season of the different crops are briefly summed up in the *Taittirīya Saṃhitā.*

Since time immemorial, the Indian society gave much respect to trees, animals, flora and fauna. Most probably worshipping of some of these took place during Indus civilization. In the Vedic era, it was purified and *yajña* were organized for the well being. Trees like pipal, banyan (*vata*), mango, *sriphal, amala,* jujube, date palm, etc., had an important place. The reference of *kalpavṛkṣa* as the main source of livelihood of the people is a very common affair in the Brahmanical and Jaina traditions. The Buddhist tradition refers to *vanalata* and some sort of roots as the earliest means of subsistence.[20]

Agriculture in the Mauryan period remained mainly based upon the village community system of small holdings. Village life, apart from the exertions of the tax collectors, was quite happy. The rural economy at this stage was based chiefly on a system of village communities of landowners or what is known in Europe as peasant proprietorship. The author of *Arthaśāstra*, Kautilya, mentioned in detail the land distribution and ownership, tax structure, cropping system and cropping patterns, collection and storage of seeds, methods and time of irrigation etc. The arable ground of the *grāma* lay without the clustered dwellings, since a wall apparently enclosed these or stockage with gates, i.e. *grāmadvāra*.[21]

The history of Indian agriculture came to a major breakthrough after the introduction of iron. Archaeological evidences, as it stands today indicate that the unpainted black and red ware people at Noh introduced iron into Indo-Gangetic plains. But very soon the technology was mastered and popularized by the Painted Grey Ware (PGW) people who colonized northern India between 1100–800 BC.[22]

The hard alluvial soil and the deeply forested areas of the Gangetic plains posed ecological handicaps before people. Probably the broken pieces of socketed axes from Hastinapur and simple type axe from Atranjikhera, Jakhera and Noh were used by the PGW people for cutting the stumps of the burnt down trees of the dense forests, facilitating reclamation of the land for agricultural purposes.

The use of iron in various agricultural implements became conventional. The ploughshare from Alamgirpur[23] and those from Jakhera[24] must have been very helpful in breaking the hard alluvial soil of the Gangetic plains and making the fields fit for cultivation. A complete sickle (without handle) from Proto-PGW levels at Jakhera (Period II-A) and a smaller one from Atranjikhera (Period III) must have facilitated the agricultural operations at that time. Thus, in all likelihood, the newly acquainted iron technology was harnessed by the Plc people for overcoming the ecological handicaps and ushering in the era of large scale settlements and agricultural production.

The cultivation of rice (*Oryza sativa*) has been proved by excavations at Hastinapur and Noh. It may be noted that barley and rice were the staple food of the Gangetic plain during the Copper-Bronze age. Evidence from Atranjikhera clearly establishes that besides rice and barley, the PGW people started cultivating wheat (*triticum compactum*).[25] In the same context, referring to rice from Atranjikhera, Chawdhuri mentions that it was found in heaps and was mostly without husk. They further held that from these facts, one would be led to presume that production of cereal was then not only enough to meet the requirements of the entire community but there was surplus production.

The preparation of land before sowing of crops became more popular. About the cultivation of wheat, Watt[26] observes that during the rains in June and July the land was ploughed two or three times and smoothened. In northern parts of India this practice is still prevalent and better known as *chaumasa*. If the rain has been plentiful and ground remains moist, seed is sown in October–November and December. The ground is then again ploughed and promoted and the seedbed prepared for *rabi* crops.

The irrigation practices developed fully in this period. If there was subsequent rain, the fields were irrigated from wells six or seven times; if there was no rain nine or ten times. It meant that cultivation of wheat requires the fields to be ploughed at least four or five times. Moreover for this winter crop, artificial means of irrigation were also necessary in spite of the rains. It implies that iron tools were not only useful in breaking the hard crust of soil but also for procuring water for irrigation by topping artificial sources of water such as digging of wells, tanks and canals. This was more easily to be achieved with the help of iron tools. *Kachcha* wells are attested at Jakhera and Atranjikhera.[27]

Coming to the agricultural production of grains, it may be noted that barley (*yava*) and probably rice (*dhāna/dhānya*), etc., occur in the *Ṛgveda* and later Vedic texts. In the later *Vedic* texts the grains attested are *upavaka* (barley), *godhūma* (wheat), rice (*vrihi*), *plasuka* (fast-growing rice), *śāli* (rice), *khalakula* (kulattha—a kind of pulse), *mudga* and *kumasa* (beans), *till* (sesame), *khalva* (moong), *urvam* and *urvaruka* (cucumbers), *garamut* (wild beans), *karakandhu, kuvala* and *badara* (jujube) and other unidentifiable grains such as *anu, amba, govidhuka or gavadhuka, namba, priyaṅgu, masusya* and *sasya.*

Thus, on the basis of the above discussion, combined testimony of archaeological and literary sources it may be concluded that before the introduction of iron in the Gangetic valley agriculture was extensive without proper ploughing and management and that irrigation facility made it more and more intensive. The introduction of the iron tools leading to the development of multifarious agricultural activities produced some new crops. Wheat and rice were the most important crops grown. The agricultural development led to the rapid development of industrial life and transformation of society through subdivision of occupations.

CONCEPT AND CRISIS OF SUSTAINABLE DEVELOPMENT

The concept of development and its dimensions has become the matter of debate at global level between so-called developed and developing countries. The word development basically means, unfolding, revealing, and opening up something, which is latent. When applied to human beings, it therefore, means unfolding their potential powers. In the socio-economic development means a process of transition from a primitive or traditional stage to a developed and modern stage.

Whatever the geographic location, culture and historical stage of development of a society, there are at least three basic elements of development which are universally considered as constituting the true meaning of development. These are as follows[28]:

1. Life sustenance, which includes food, clothes, shelter, health care and security;
2. Self-respect or a sense of growth; and
3. Freedom, which includes both freedom in the political or ideological sense and freedom from social servitude.

Schumpeter defined development as only such changes in economic life as are not forced upon it from without, but arise by its own initiative from within.[29] Carry Jacobes observed that development is not a progranme executed by government but a social movement of the entire population, initiated, guided and led by government. Development calls for information, education, training, organization and administration at the widest possible level.

According to Duby Seers, development means creating conditions for the realization of human potential. Conclusively development as an innovative process leading to the structural transformation of social system.[30]

The concept of development and its wide dimensions were spelled out in the 1960s in the preamble of the international development strategy for the second United Nations Development Decade, as the ultimate purpose of development is to provide increasing opportunities to all people for a better life. It is essential to bring about a more equitable distribution of income and wealth for promoting both social justice and efficiency of production to raise substantially the level of employment, to achieve a greater degree of income security and to expand and improve facilities for education, health, nutrition, housing and social welfare and to safeguard the environment.

Unfortunately, the western concept of development does not consider the aforesaid dimensions of development. Its unilateral and fragmented approach is continuously confined to the purview of merely economic aspect. The national income, gross domestic product, per capita income and consumption level cannot determine the level of equitable distribution, internal prosperity, peace and pleasure of the society in any country. In the absence of equitable and balanced distribution of income and economic resources there is too much difference; between poverty and prosperity in the so-called developed countries. The unexpected negligence of environmental, social and cultural aspects of development resulted in the problems of pollution, breakdown of family institution, social disintegration and degradation of moral values, rituals and traditions of the society.

In the 55th *Indian Science Congress* (1968) held at Varanasi, renowned agricultural scientist Dr. M.S. Swaminathan presided over the agricultural section. He pointed out the emerging problems of development, such as the cultivable land may be changed into deserts, if we will be cultivating land continuously without properly preserving the structure of soil and its fertility. The soil may become saline and alkaline if we will be irrigating land without proper water management. The proportion of cancer and other likely diseases may increase due to over and unbalanced use of pesticides, fungicides, herbicides and other parts of edible portion. Water, the most valuable natural resource, is still abundant due to natural farming practiced since the very beginning. This resource may be decreased rapidly if unscientific use of water continues. The substitution of location specific innumerable many local species from few high yielding varieties may provide the opportunity for several pests and diseases which may become epidemic like havoc for cultivated crops.

The first UNO conference on Human Environment (1972) held at Stockholm in Sweden emphasized the need to create global awareness regarding all attempts of development should be correlated and interlinked with the basic elements of ecosystem. It was also highlighted that until poverty exists in large section of the society, it is impossible to develop cordial relationship with nature. After two decades 'Earth Summit' held in June 1992 at Rio de Janeiro in Brazil, observed that no significant achievements has been made on environment issues in any country. In this summit for amelioration of environmental problems time bound international programme, i.e. 'Agenda 21st century' has been adopted to preserve bio-diversity and conserve ecology.[31]

SUSTAINABLE AGRICULTURAL DEVELOPMENTS

According to the consultative Group on International Agricultural Research (*CGIAR*), 'sustainable agriculture is the successful management of resources to satisfy the quality of environment and conserving natural resources'.[32]

Land, water, climate, flora and fauna are the basic natural resources for agricultural development, which are subject to various kinds of deteriorating influences. It is imperative to develop strategies for conservation and improvement of resources because agricultural development can't subsist on the deteriorating natural resource base. The concept of sustainable resource management implies that the needs of the present can be met without compromising the ability of the resources to meet the needs of the future.[33]

The sustainable development is simply the process of development, which improves the quality of human life while living within the carrying capacity of supporting ecosystems. In fact, the alternative pathways of food chain insure the stability of ecosystem. As many pathways are available in the food chain, the stability of the system proportionately becomes sustainable. Hence, for sustaining the production, the process and restoration of natural process balanced ecosystem is inevitable. At the initial level of production pyramid, the number of producers is maximum and it is obviously clear from the biodiversity point of view that the conservation of biodiversity, specially flora and fauna is the conservation of production system. It was the reason behind Vedic literature emphasising affinity with all the components of nature.

Actually, the food chain established the inter-relationship within living being of the whole community. But the modern concept of development provided the place of human being merely consumer in the centre of ecosystem. The negligence of other factors of ecosystem led the imbalance in process of development. The dependence for energy use of renewable sources of energy, like coal, petroleum products and other inorganic substances gave birth to the problems of pollution loss of bio-diversity and several other socio-ecological problems.

The economic principle, 'law of diminishing returns', can be applied on agriculture very soon. So, in the long run, it will be very difficult to maintain the higher productivity in the field of agricultural production. It is imperative to sustain the developmental process through sustainable development. The indigenous knowledge has been gaining importance in this regard all over the world day by day, we cannot ignore our traditional agricultural knowledge system and peoples wisdom which is still prevalent in some parts of our country from the *desi* plough to *kumhar's* earthen pot, stone wheel.

The famous environmentalist, Sunderlal Bahuguna, rightly pointed out that forest destruction brought climatic changes, recession of glaciers and scarcity of water, erosion of topsoils and finally, the poverty of the people. The effort to modernize the Himalayan region has resulted in both ecological disaster and economic ruin. It has been difficult for the policy makers to understand that ecology and economy are interrelated. This has happened all over the world and it is only as a result of this that 'the income of the wealthiest 20 per cent of the population is 150 times greater than that of the poorest 20 per cent'.[34]

The knowledge of ancient agricultural practices in ecological farming, if practiced with relevant, compatible and sound preparation, increases the yield while reducing the cost of cultivation. Thus, it has dual advantages on the one side being sustainable without endangering the environment and on the other side being highly cost effective as there are negligible inputs that are bought into markets. Generally, it is a misconception that ecological farming gives poor yields and is an expensive practice. In several countries of the world it has been proved that ecological agriculture to be economically rewarding, intellectually challenging and environmentally safe.[35]

Energy calculations of ecological farming show that it has a distinct advantage over the modern farming system. While the modern system of farming expanded nearly 31,000 M.J. ecological farming method accounted for only 23,400 M.J. High-energy inputs such as fertilizer, plant protection chemicals, herbicides and petroleum products like high speed diesel are substituted with biological inputs and low energy human and cattle power. The output/input ratio in the case of modern farming is 3.76 while in the case of ecological farming it is more than 4.95. The efficiency of ecological farming can be best realized only through energy calculations.[35]

The gravity of the environment degradation, arising from faulty practices, has compelled to focus the scholar's attention on ecologically sound, viable and sustainable farming systems. The increasing trend in the use of agro-chemicals in farming practices are not only destroying the ecological balance but decreasing soil fertility and creating human hazards through chemical deposition, i.e. biomagnification in the food products. It has been observed since seventies that use of pesticides specially *B.H.C.* (*benzyme hexa chloride*) and *D.D.T.* (*dichloro-diphenyl trichloroethane*) and few others leaves residual effects for a long duration in food products and their use should be discarded. Use of several pesticides including *B.H.C.* and *D.D.T.* has been banned in most of the countries, while in India unfortunately these are still used at a large scale.

It is estimated that worldwide 22,000 people die every year due to immediate or slow pesticide poisoning. Pesticides not only kill unwanted pests, they also harm human beings at all stages, during production, application of pesticides, farmers and workers can get affected by inhaling and touching toxic pesticides. In India, as early as 1972, *the Indian Journal of Public Health*, declared the most serious threat to health is the uncontrolled use of pesticides for agronomic practices.[36] Hence, it is obviously clear that negative impact of insecticides and fungicides and other agro-chemicals on the environment and public health has been well recognized and assessed and its residues in the food chain have endangered the whole life sustaining systems.

In India, the present situation of soil fertility level and ground water table has became the matter of concern. The areas where the Green Revolution has been observed in sixties it has become the very common feature of media reporting that ground water table is rapidly decreasing and posing threat to drinking water availability as well as irrigation water (*Navbharat Times*, July 11, 1993, p. 10). Two-third area of Punjab and Haryana have turned grey and dark regarding the groundwater table. It is alarming feature that grey area blocks will become very soon dark area. Likewise, due to overuse of agrochemicals, the organic matter content in the soil is also decreasing. The organic matter in the soil system functions as blood in the human body. Decreasing trend in the soil organic matter not only affects the soil fertility and retention capacity but also affects the soil structure in long run.

Environmental factors, ecosystems, land, and bioinhabitants constitute the natural wealth. Land, the primary natural resource includes soil, water bodies, vegetation landscape and microclimate. India's total arable land area is 329 million hectares with reporting area of 305 mha. Agricultural sectors, consisting of net cultivated area, current fallows and cultivable wastes about 180 million hectares. An ecological sector constitutes nearly 103 million hectares comprising forests, miscellaneous trees, crops and groves, grazing and barren lands. Non-agricultural sectors, including lands under non-agricultural uses, are about 22 million hectares. In the agricultural operations the net sown area of 142 mha has remained constant over the last three decades.

The area under forests (not necessarily the forested area) has been varying since 1972. In the overall assessment during the last three decades, the area under the ecological sector, especially miscellaneous trees and groves, grazing lands and barren lands has decreased significantly. The non-expanding agricultural area and decreasing ecological lands have contributed to an unexpected rise in area under the non-agricultural sector.[37] However, it is the concept of sustainable development as adopted by the Earth Summit 1992 which ought to be the cardinal point of any shift in the land use pattern.

Renowned agricultural scientist Dr. R.S. Paroda pointed out that in the 21st century, ecological and environmental considerations would become important challenges owing to the damage now being inflicted to our land, water, flora, fauna and even the atmosphere. Demand for food grains for human consumption will continue to grow in view of our increasing human population. The emerging opportunities in the world market will further push this demand on a higher side. Further, the availability of agricultural land will be restricted and exploitation of water resources for agricultural use will reach its limit; sustainability will be a major concern for all technological change, will emerge as a major determinant of sustainable agricultural growth.[38]

The United States government task force in 1979 comprising scientists and economists concluded 'some farmers actually experienced no reduction at all in yields when they gave up the use of chemicals. And those who did not lose some production still made more money because they did not have to pay for expensive chemicals.' In another study conducted by the centre for the study of Biological systems, University of Washington at St. Louis, two groups of farms with similar soil and environmental conditions, with one using chemicals and other without it, were evaluated for five years. The study concluded: 'a five-year average shows that the organic farms yielded in dollars per acre exactly the same returns. In terms of yield, the organic farms were down about 10 per cent. The reason why the economics came out is that the savings in chemicals made up for the difference'.[39]

It would be better to examine the emerging barriers to crop sustainability. Punjab has often been hailed as the country's granary. The land, which once produced a rich golden harvest, is now beginning to collapse under its own artificial burden of intensive cultivation. The warning bells have been sounding for quite some time.

Intensive monoculture of wheat and rice has already exhausted the nutrient reservoir of the soil. The indiscriminate marketing of chemical fertilizers, without the accompanying doses of organic manures, has drastically reduced the soil fertility, with the organic content of soil hovering around a pathetically low of 0.1 per cent. Punjab soils are now solely dependent upon chemical fertilizers. As a consequence, it is only a matter of time before Punjab too becomes like Sind, turning into manmade desert.

ANCIENT AGRICULTURAL KNOWLEDGE SYSTEM AND SUSTAINABLE DEVELOPMENT

The basic motive of ancient agricultural knowledge system (*AANKS*) was ecologically sustainable and economically feasible and biologically viable. The age-old established rituals, traditions and harmonious relationship with nature always conserve the degradation and preserve the diversity of the ecosystem. The agricultural practices prevalent in ancient period were based on the life-long experiences of ecological conditions compatible to human needs.

Unfortunately, it has been dismissed as belonging to ancient period and unscientific. However, in recent times, in the purview of environmental pollution and ecological degradation and uncertainty in the process of production, its importance in the development of appropriate technologies, which can sustain the overall system ecologically, is being recognized at the global level. Many of these traditional agro-practices are still practiced in several parts of our country. It indicates that traditional knowledge and wisdom of agricultural activities has scientific rationality and practical utility.

The ancient agricultural knowledge system can be discussed in the purview of different aspects and dimensions of sustainable development as follows:

1. Eco-farming and natural balance
2. Conservation of biodiversity
3. Compatible agronomic practices
4. Integrated farming system
5. Soil fertility management, irrigation and water management
6. Management of insect pests and diseases
7. Post-harvest management
8. Sustainable food security.

ECO-FARMING AND NATURAL BALANCE

The ancient agricultural knowledge and wisdom of ecological farming is not as simple as one may presume. It was a highly knowledge-intensive, labour-oriented and complex system integrating several organic recycling processes. A high degree of motivation, a firm commitment and conviction, and innovative ability made successful organic farmers. Ecological farming is a self-reliant method and has to be an integrated system. Since it is a knowledge and intensive practice which is necessary to one has to keep pace with the dynamics of nature to increase the biological productivity of the soil.[40]

The farmers of ancient period have broad dimension of ecological farming. We cannot challenge their wisdom even today. Their holistic approach of development of all (not only human beings) restored the natural balance. It may be seen in the *Ṛgveda*[41] and the other Vedic literature that the ploughshare furrowing the field provides food for the ploughman; a man travelling along a road acquires wealth for his master by his movements.

Similarly, there are some relevant and very significant passages in the *Yajurveda* also:

> Wise through desire of bliss with gods, the skilful bind the traces fast and lay the yokes on either side.[42]
>
> Lay on the yokes and fasten well the traces; formed is the furrow sow the seed within it.
>
> Through song may we find hearing fraught with plenty's near to the ripened grain approach the sickle.[43]
>
> Happily let the shares turn up the plough land happily go to the ploughers with the oxen.
>
> *Suna* and *sira* pleased with our oblation cause ye our plants to bear abundant fruitage.[44]

These passages clarify the whole system of agriculture and speaks about yokes, traces, furrow, seeds sowing, ripening of the grain, harvesting of the crops by sickle, plough and share. Further, it is mentioned in the *Yajurveda* which provides the knowledge of inter-dependency of different factors of ecosystems. The sacrifices (*yajña*) were organized and offerings were provided for the peace and prosperity of all living and non-living being like;

> Approved by Viśvadevas and by Maruts, blamed be the furrow with sweet, flavoured fatness, succulent teeming with thy milky treasure, turn hitherward to us with milk; O furrow.[45]
>
> May ploughing and husbandry and my superiority and my prominence prosper by sacrifice.[46]
>
> May my fire and my water and my creepers and my plants with culture ripened fruit (*kṛṣṭapachyaḥ*) and my plants with fruit ripened without culture (*akṛṣṭapachyaḥ*)... be produced by sacrifice.[47]

In the *Taittirīya Saṃhitā*, we have the different passages of great significance in this regard that is also based on the Rgvedic and Yajurvedic verses;

> With prosperity may our ploughs cleave the ground. With prosperity may the ploughers go round the yokes; Prosperity may *parjanya* (give with honey and milk and do ye; O *sura* and *sira* accord prosperity to us.[48]

It is obviously clear that agricultural practices prevalent in the ancient period were absolutely systematic, environmentally compatible and correlated with the ecological conditions. Forestry, agriculture and animal husbandry were interdependent. The life style and needs of the people at that time were fully compatible with the ecosystem. Going into deep the concept of sustainable development lies in our traditions and we must know three major elements of the system.

(a) Self-sustained villages

The village system of ancient period was not only self-sustained in their day-to-day activities but also self-reliant in their different needs. The author of the *Arthaśāstra*, Kautilya,[49] has provided sufficient knowledge of that period. According to him the rural economy of ancient period was self-sustained, a higher level of coordination and harmonization was achieved by the society.

In the ancient period, villages were settled in the centre place of greeneries. On the surrounding of these villages, dense garden of fruits and flowers were situated. In Sanskrit literature, this natural beauty is called *vana-vaibhava.*[50] In the centre of these gardens, well-prepared ponds were made under which the rainwater were conserved. There was different use of stored water like, drinking, bathing, irrigation, etc. In the summer season, it becomes the boon for birds and animals. These gardens and forests provided appropriate harbour for living being.

Herbs and trees surrounded most of the agricultural fields and that is why the ecological balance was naturally maintained. These forest provided not only useful wood for agricultural purposes but also provided fuel, fodder and shelter. The total supply of fuel ensured from forest and the total by-product of animals were used for *desi khad* (manures). The level of organic matter in the soil was ensured by adequate supply of manures. Due to a high level of organic matter content in the soil the fertility level and retention capacity was naturally maintained at higher level.

The local availability of seed, manure, irrigation water and implements were ensured. The different aspects of the life miraculously harmonized and fulfilled in the periphery of village. A high degree of labour division was ensured and experts in the different fields being used for special task such as sowing, implements manufacturing and repairing, deciding appropriate time for sowing irrigation and harvesting. These experts were also a part of village community. Therefore, the inter-dependency of economic activities provided better harmonious relationship in the society.

(b) The gomātā cow

The cow has got great importance from very beginning of civilization in Hindu religion. Rural life was deeply correlated and dependent on cow for milk, manures and its bullocks were used for ploughing and other activities. In the Vedic literature, the cow is treated as a most useful holy animal. It is mentioned clearly in the *Ṛgveda* that cow is a source of prosperity. Cow is treated as *aghanya* (1-164-40) that cow is not to be killed and for this offence there was strict penalty and provision of penitence. The Aryan culture was essentially the culture of cow and horse. They stressed on the animals belonging to the family of cow like *ṛsabha* (special bull), *babhru-gau* (brown ox), *pṛsna-*

gau (pie-bold bullock), *vasa* (sterile cow) *sitiprastha-gau* (white backed bullock), *vasa-prsui* (spatted sterile cow), *śyama-gau* (dark-grey bull), *dhenu* (milch cow) *syeta anandvan* (reddish-white-draught bullock), *anandvan* (draught bullock), *Yama-gau* (twin bullocks), *dvirūpa-gau* (bicoloured bullock), *napunsaka-gau* (castrated bull) *pasthavat* (ox), *uksa* (bull).[51] Bovine animals in the *Śatapatha* are known as *gavyaḥpaśavaḥ*.[52]

In the *Ṛgveda* and other Vedic literature, at several places, the importance of cows is described. But in the *Ṛgveda*, the 19th *Sūkta* of the 10th *Manḍala* is totally dedicated to cow. Some important verses are:

> Let them return to us again under this herdsman let them feed. Do thou, O Agni, keep them have and let the wealth we have remain (3) I call upon their herdsman him who knows well their coming high.
>
> Their parting and their home return and watcheth their approach and rest. I offer you on every side butter and milk and strengthening food. May all the holy deities pour down on us a flood of wealth (7) These verses represent the importance of cows during the period. It does not mean that other domesticated animals were treated as secondary, but due to overall usefulness cows were regarded as pious and multipurpose, useful animal.

The word *karīshinim* were frequently used in the Vedic literature means cow dung. According to Satwalekar,[53] the different words used in the Vedas like *karīshinim, śakṛt, śakan, śak, śāk* were used for cowdung. Eminent translator 'Sayan' has supported the *karīshinim* as dried cowdung.[54] Now it has been clear that there are medicinal property in the cow-dung and urine. In the ancient period, *pañchgavya*, were prepared from cow's milk, curd, ghee, urine and dung. It is used for several purposes. The practising farmer of Maharashtra, N.D. Pandhari Pande has distinguished cow's dung from the dung of other animals and prepared several medicinal formulations.[55]

Cowdung contains phenol, ammonia, methanol, indol, formalin and bacteriophages. These gives medicinal properties to the cowdung, Indian villagers still use cow dung as a wall coating. According to Russian scientists, this product absorbs radiation. Cowdung powder smoke has a beneficial effect on respiratory diseases. According to Italian scientists, the smell of fresh cowdung cures malarial fever and the pathogenic bacteria vanishes. Cowdung can be used as a disinfectant also. The urine of cow also has medicinal value.[56]

(c) Lifeline the Ganga

Since the time immemorial, the River Ganga has achieved holy recognition, achieved great importance and treated as deity mother. The river system formed by the Ganga river drains, a large part of northern and central India. The plains of the Rivers Ganga and Yamuna are still one of the most important fertile agricultural fields in the world. The river Ganga is merely not a simple river it represents the importance of river system for the availability of water in the whole country. Ganga is the representative of all the rivers flowing continuously for the welfare of overall living system on the earth, like rivers provided very fundamental base for living beings as the veins of blood in the human body.[57]

In the VIIth Maṇḍala of *Ṛgveda,* there are illustrations of seven rivers—Sindhu, Sarasvati, etc. Panini has described several large and small rivers used for irrigation purposes at that time.[58] In the *Milindapañho*[59] references of 500 rivers of the Himalayan origin is found, among these the zone important rivers are the Ganga, Yamuna, Asravati, Sarayu (Ghaghara), Sindhu, Mahi, Sarasvati, Vibhavati, Vilmasa and Kandabhaga, etc. In *Viṣṇu, Purāṇ*[60] it is mentioned that Balaram, brother of Shri Krishna, diverted the stream of the Yamuna towards Brindavan for irrigation purposes.

Concluding the above discussions, it has become obviously clear that from ancient period onwards people have been providing great respect to the cows and rivers. Their needs were always compatible with ecological conditions and life style fully immense and harmoniously related with natural surroundings. Agricultural practices being developed taking into account the holistic approach regarding human needs and ecological conditions.

CONSERVATION OF BIODIVERSITY

The Indian subcontinent, is regarded as a major centre of origin in the world. With a wide range of climatic and physical conditions from the torrid to the Arctic, India has a rich and varied vegetation in the world. It is estimated that there are about 45,000 species of plants including shrubs in the country.[61] The vascular flora, which forms the conspicuous vegetation over comprises 15,000 species, of these 35 per cent is endemic and has so far not been reported anywhere in the world. The total plant wealth of the country include not only the useful large flowered plants, including flowering shrubs but also a large number of non-flowering plants like ferns, liverworts, algae and fungi. India has a great variety of fauna numbering a little over 75,000 known species. Of these, insects constitute about 40,000 mollusc, a little over 5,000, mammals 372, birds, 1228, reptiles 428, amphibians 204 and fisheries 2546.[62] India has about 2 per cent of the world area but it contains nearly 18 per cent of the biodiversity of the world.[63]

India can be divided into eight distinct floristic regions, namely, the western Himalayas, the eastern Himalayas, Assam, the Indus plain, the Ganga plain, the Deccan, Malabar and the Andamans. Hence, it is obviously clear that India also has wide climatic diversity from the Himalayas to the Deccan and from Assam to the western Ghats. This climatic regional variation enables us to have such a large number of genetic diversity in the world. It should be very clear that at the global level, the so-called developing countries have the minimum number of biodiversity. But, unfortunately, the benefits are being exploited by the rich and so-called developed countries.

The eminent environmentalist Prof. T.N. Khoshoo[64] has rightly pointed out that the indigenous people living in the third world were responsible for enhancing the bio-diversity of the globe though they were not conversant with the basic principles enunciated by Mendel and Darwin, traditional wisdom has developed over thousands of years enabled the indigenous population in the so-called backward countries to not only increase the diversity of the species but also to domesticate the large number of plants and animals.

In the Earth Summit (1992) of Rio de Janeiro, it has been observed that major threat to biodiversity conservation has emerged from industrialization of developed countries. In a report of an institution monitoring global environment the dangerous situation of bio-diversity conservation has emerged like—on an average, the extinction rate is 140 species of plants and animals per day, and about 17 lakh hectare forest land is decreasing per day. The six time highest recorded temperature out of seven from middle of nineteenth century are after 1980.[65] The unexpected rate of ozone layer depletion since eightees posing major threat against existence of life on this planet.

Since very beginning of civilisation in the Indian subcontinent, conservation of biodiversity were ensured through religious sentiments. As we already know, different type of trees and animals are associated with one or more god or goddess, since time immemorial we are worshipping some trees and animals. Few trees like pipal and banyan are treated as the shelter of gods. Evidence reveals that in the Indus valley civilization, pipal tree was most important consecrated tree. On the coins and seals of this civilization it can be seen with the pictures of cow, bull, sheep, deer, parrot, peacock, etc.[66] It reveals simply the affinity of people towards plants and animals.

In the *Ṛgveda*,[67] it is clearly mentioned that in the summer season, extremely heated and dried soils requires water. The sun provided rainfall and through this water mother earth and father sun come together and establish seeds of life. Hence in the rainy season, plants and animals flourished. In the 10th Manḍala, of this holy epic it is enumerated that Vedic people knew 107 places of medicinal plants.[68] In the later Vedic era, for the welfare of plants and animals, *yajña* (sacrifices) were organized.[69] In these *yajña*, people offering *āhutis* of *havya* for prosperity and increase in the production and productivity of crops. *Atharvaveda* also provides useful information regarding biodiversity conservation. In the sixth *Kāṇḍa*, it is enumerated that pipal tree is *deva sadan-ashvattho deva sadanaḥ*. *Yajuraveda* deals it like: *Aeshvattho, O nishadanam parṇe O vasatishtrita*' (12–79). In the fourth *Kāṇḍa* of *Atharvaveda*,[70] it is enumerated that where pipal, banyan trees peacock and, *chunamani*, tulsi and *kakyachi* plants were present, there no disease will attack on the people.

In the ancient, period digging of ponds and wells and planting trees were treated as holy work.[71] In the *Śatapatha Brāhamaṇa*,[72] it is clearly indicated that trees are the shelter of gods. According to this, the shelter of Brahma is the *palash* tree. In *Taittirīya Saṃhitā*,[73] Pipal were treated as shelter of Prajapati, i.e. god of living being. In the *Padma Purāṇa* it is mentioned that there is life system present in the plants.[74]

In the *Arthaśāstra*, Kautilya has mentioned useful ways and means for the conservation and protection of the biodiversity. He pointed out that uncultivable and barren lands should be used for biological gardens. According to this, in the peripheries of two *krosas*[75] (nearly 6 kms), deer parks should be developed under which useful fruit plants, herbs and other plants should be planted. He further revealed that parallel to this deer park another deer park should be established and in this park, animals should be brought from outside and far-flung areas. There was also provision for separate forest for wood purposes. The management of these forests rests on the president of forests and inhabitants. It was a very good example of participatory forest management through local residents at that time which we have not adopted even today.[76]

In *Mahābhārata*, it is assumed that there are life systems in the trees and that is why it was the provision at that time that he who instigate fire in the forest was liable for hard penalty and penitence. Shukra[77] clarifies that the burden and responsibility of forestation is on the king. As we know that Lord Krishna was closely associated with *kadamba* tree of Brindavan. Lord Rama resides in Panchvati with Lakshman and Sita. Gautama Buddha, in his whole life, was closely related to plants. He was born under an Asoka tree, received enlightment under a pipal tree, preached his new gospel in mango groves and under shady banyans and died in a sal grove. It reminds us the relationship of human life with plants and obviously, religion interlinks humans with ecosystems.

Actually, in the Hindu religion, the dimensions of biodiversity conservation is very broad. Through restoration of relationship of god with the nature, welfare of all living being restored in the whole system like in the water, earth and space. Lord Shiva is called 'Pashupati' (god of animals) because he is supposed to be universal conservator of animals. He resides on 'Kailash' mountain for conserving hill bio-diversity. Likewise lord Vishnu resides in the water for the protection of *jalchar* (water flora and fauna).

The *Śrimad Bhāgavata*[78] reveals the fact that once upon a time, a forest area caught fire suddenly and simultaneously wind velocity increased. The animals of forest were in the great danger of their life. Lord Krishna absorbed the fire for protecting animals, plants and other flora and fauna. The concept of *ahiṃsā parmo dharmaḥ* (non-violence is great religion) helps in the conservation of biodiversity. It is scientifically important and mentioned in the *Hemdro Brahmāṇḍa of Purāṇa*[79] that pipal, banyan, ficus (pakar), gular and mango are known as *pañcha pallava*. As we know, mango is the king of fruit trees and all this rest, for it restores the ecological balance. Scientific researches have also proved that these are very important for environment. In the *Agni Purāṇa*, it is mentioned that one tree is supposed to be equal to 10 sons.[80]

In fact, it is a well-established fact that diversity is the characteristics of nature and the basis of ecological stability. Diverse ecosystem gives rise to diverse life forms and to diverse cultures and this provides the foundation of sustainability. The coevolution of cultures, life forms and habitats has conserved the biological diversity on this planet. Communities everywhere in the world have developed knowledge and found ways to derive livelihoods from the bounties of nature's diversity, in wild and domesticated forms. In ancient period people knew thoroughly all these things and adopted compatible agro-practices for their respective eco-system.

COMPATIBLE AGRONOMIC PRACTICES

The agronomic practices followed in the ancient period were a revolutionary change, moving from subsistence agriculture at the initial stage to self-sufficiency and food producing economy at a further stage. Certainly, it took inventory, wisdom, patience of long duration and continuous effort to develop systematic practices of crop production. In this complex process they learnt adequately from nature by trial and error. At the initial lével, it was very tough task to interfere in the natural process and that is why their techniques were very much compatible towards respective ecosystem.

The agronomic practices include different activities which are interdependent on each other in the process of crop-production. These are as follows:

(a) Knowledge of meteorology
(b) Field management
(c) Selection of crops and sowing
(d) Soil fertility management
(e) Inter-culture and weed management
(f) Irrigation and water management
(g) Harvesting and storage.

(a) Knowledge of meteorology

The successful cultivation of crops depends on the good meteorological conditions even today. The absolute knowledge of monsoon and process of rainfall is enumerated in the *Ṛgveda.* In the very beginning of the first *Maṇḍala,* it is described that; truly the bright fierce and vigorous vital forces cherished by the cosmic power send down the rain on the inner realm like clouds raining without the aid of wind upon the desert.[81] Again, it is mentioned that 'the smooth gliding waters of the rain, the solar rays clothing the waters with a dark cloud ascend to heaven. They come down again from the dwelling of the rain and immediately the earth is moistened with water'.[82] In the next verse, it is highlighted that the sun is the germ of water.

In the *Bṛhataśaṁhitā* of Varahmihira, it is explained that 'the prediction of an astrologer who observes attentively day and night the symptoms of pregnancy of the clouds will never be falsified.[83] The symptoms of pregnancy are to be detected when the 'moon passes through the asterism *pūrvāṣāḍha* beginning from the first day of the bright fortnight of the *Mārgaśirṣa* month.'[84] It is again scientifically enumerated that the 'foetus formed during the moon's stay in a particular constellation will be released 195 (solar) days then after when the moon passes through the same constellation', according to the laws of her revolution.[85]

Therefore, it has been obviously clear that the knowledge of meteorology in ancient period was no doubt very perfect. They knew the different crops growing season, the different months and the appropriate time of sowing for different crops. On the basis of some other literature we may point out here that they knew when to start ploughing, sowing, irrigation, harvesting and storage. Their vision regarding meteorology was very broad and observation regarding each and every happening of nature was very sharp.[86] This science of meteorology was developed in ancient India by sages like Garga, Parasara, Kasyapa, Rsiputra and Siddhasena, the renowned figures of Indian astrology.[87] We can take advantage of this knowledge even today in sustaining the agricultural production.

(b) Field management

In agronomic practices, the field management is an important aspect. When to plough and how much to plough can only be decided by a well-experienced person. Miraculously, the concept of 'Xero-tillage' is mentioned in the *Vedas*. In the *Yajurveda*,[88] it is clearly mentioned that through *kṛṣṭapachyaḥ* (with cultivation) and *akṛṣṭapachyaḥ* (without cultivation), we can take up crop production. We can see even today, in some parts of northern India, few crops like lathyrus, lentil, etc., are being grown by the farmers without cultivating the land.

Eight years of research and experiments of the Punjab Agricultural University, Ludhiana have recently shown that tilling the land is not essential for the germination and growth of crops like wheat and maize they have also found that tilling is required only for weed control.[89] Lambrick pointed out that wheat and barley, the principal foodgrains of the Indus, are cultivated without ploughing and manuring. It does not mean that in the ancient period they were ignorant of advantages of the ploughing. The evidences have proved that in the Harappan period at Kalibangan field preparation were done through furrowing. Tilling operation was done as per the requirement of particular crops and the condition of the soil.[90]

(c) Selection of crops and its management

As we have discussed since Indus-valley civilization we are cultivating a number of crops including coarse grains to pulses and oilseeds. Unfortunately, due to specialization and monoculture of wheat and rice after the so-called green revolution, the cultivation of a number of coarse grains has been replaced by a few profitable crops. Hence, it is a grim situation and the total number of crops grown in ancient period is continuously decreasing day-to-day. It reduces the crops diversity which leads the problem of pests and diseases attack.

The wisdom of selection of crops for specific soil condition is highly appreciable. For drought and rained condition they selected tolerant varieties of crops and crops of hardy nature. According to Bakshi and Ranga (1974) cultivation of barley was the most profitable crop for saline and alkaline condition.[91] The scientific studies are tolerant to adverse soil conditions low soil fertility with little precipitation. According to Singh (1946), the wheat variety *triticum spharococcum* particularity is highly resistant to drought.[92]

The managerial skills of ancient farmers are praiseworthy. They practised mixed cropping avoiding the risk of the vagaries of nature. The evidence of mixed cropping of black gram and sesame is found in the *Bhāṣhya* of Patanjali.[93] Another example of mixed cropping of five crops is enumerated in the *Milindapaño*.[94] At present in northern India, we can observe the prevalence of mixed cropping of nearly 8 to 10 crops simultaneously, i.e. known as *ashadi kheti*. There are several advantages of mixed cropping:

i. The absorption of solar light per unit area is high due to multi-storey cropping.
ii. Less weed problem due to dense crop canopy.
iii. Less effect of insect pest and diseases due to diversity.
iv. Full utilization of nutrients due to different level of root sizes
v. Fixation of atmospheric nitrogen due to symbiotic effects in pulses.

The farmers of the ancient period followed the practice of crops rotation. It is mentioned in the *Ṛgveda*[95] that matured crops were harvested separately on different times. There are several crops mentioned in the *Yajurveda.*[96] It is enumerated in the *Taittirīya Saṁhitā* that two crops were grown in a rotation.[97]

(d) *Seed-sowing and treatment*

The successful cultivation of crops depends on the viability of seeds, their good health, proper sowing time and good soil moisture conditions, collection of seeds depends on the expertise and selection of the good crops. The healthy plants were selected before ripening and their care and maintenance were ensured properly. Harvesting and threshing were performed separately and in the appropriate time after performing religious rituals seeds were stored for next season.

The importance of good seed for better crop has been emphasized by 'Manu' that, 'as good seed alone sown in good land is successfully raised'.[98] The seed sown in poor land is within that very (land) destroyed.[99] In the *Ṛgveda* too, it is mentioned that seed should be sown only in the properly prepared seed bed.[100]

The treatment of seed was another important aspect. Different methods were adopted for the seed treatment like.[101]

i. The solar ray's treatment
ii. The full moon treatment
iii. The wind velocity treatment
iv. Water treatment, specially hot water treatment
v. The treatment from different substances like *agnihotra,* cow's urine, etc.

In the *Arthaśāstra*, it is properly mentioned that the seeds of paddy lay open for 7 days in the day and night. For treatment of sugarcane, it has been advised to paste honey, ghee in mixture with cowdung on the cutting edge.[102] There were several other methods traditionally used in that period and some of them are still used today.

The sowing techniques prevailed in ancient period was an expert's work. Only a few selected persons were deputed for this task. There was a definite quantity of seed for different crops and it depends on the moisture content of the soil, seeds health and sowing time. Sowing was done either in furrows or through broadcasting. In the *Aṣṭādhyāyi*, the term *bigakaroti* is mentioned, it means the sowing of seeds in the furrows.[103] According to this method, one person makes furrows with the help of plough and bullocks and other person following him, pours seeds in the furrows.[104] The quantity of seeds decided by some pots with which measurement capacity was predetermined.

The time of sowing was decided by an expert on the basis of *lagna* and *muhūrta,* i.e. called *śubha-muhūrta* (appropriate time). The sowing time was ultimately decided on taking into account the species and nature of crops. Panini has mentioned the best time of sowing of black gram is full moon day *(Pūraṇamāsī)*.[105] In the *Śatapatha Brāhamaṇa,* it is mentioned that beginning of sowing was done after planting '*kuśa*' (*saccharam sp.*) grass.[106] It is also quite interesting and pioneering that they knew of transplanting, especially in the case of paddy.[107]

INTER-CULTURE AND WEED MANAGEMENT

Inter-culture was one of the basic practice in crop cultivation and followed since the very beginning as a weed control method. Indirectly, it helped in the moisture conservation, best utilization of nutrients by the major crops. Inter-culture was done with the help of spade or other small implements. The synonym of inter-culture is secondary village. It has also been used for breaking of the upper layer, i.e. the crust of the soil formed by rains after sowing of seeds.

It is very interesting aspect from the biodiversity conservation point of view that traditional agriculture does not recognize the concept of weed. All plants have their uses; some plants have more than one use. Daniel Querol writes about a farm in Mexico which had over 214 'weeds'. The farmer, however, had a specific use for all these plants. In Sanskrit literature, it is clearly mentioned that there is no plant in nature that will be called useless.

However, it is mentioned in the *Ṛgveda*[108] that there are three types of crop weed competition as follows:

i. Competition for light
ii. Competition for moisture
iii. Competition for plant nutrients

The management of weed was done accordingly through mixed cropping—by increasing the canopy of main crops, through inter-culture and sowing of smothering crops. The other agronomic practices are basically the part of other portion and that is why those will be discussed later.

INTEGRATED FARMING SYSTEM

At present, it has been recognized that production process of agriculture must have a long-term sustainability of way of sustenance of natural resources economic viability and social acceptability of production systems and protection of environment. Here, the meaning of integration is to integrate agricultural activities, forestry, horticulture and animal husbandry simultaneously as a diversified system which restores the cycle of development in a continuous circle. Rained agriculture is one of the most neglected part of the modern development, but taking into account the sowing of coarse grains in these areas in ancient period, we can restore and enhance the production process and total food grain production.

The following table shows the inter-dependency and correlation of different factors of agricultural production system. By adding chemical fertilizers and agro-chemicals, we have disturbed the chain of natural fertility conservation providing adequate amount of organic matters. By-products of crop cultivation, forestry and animals can be successfully used by each other and cycle is naturally maintained. From Vedic period onwards, this cycle has been maintained and that is why agricultural ecology was self-sustained. The biofixation of nitrogen through pulse crops and sesbania is still quite useful for natural fertility maintenance.

Symbiotic Relationship and Interdependence of Land-Draught Animal-Man

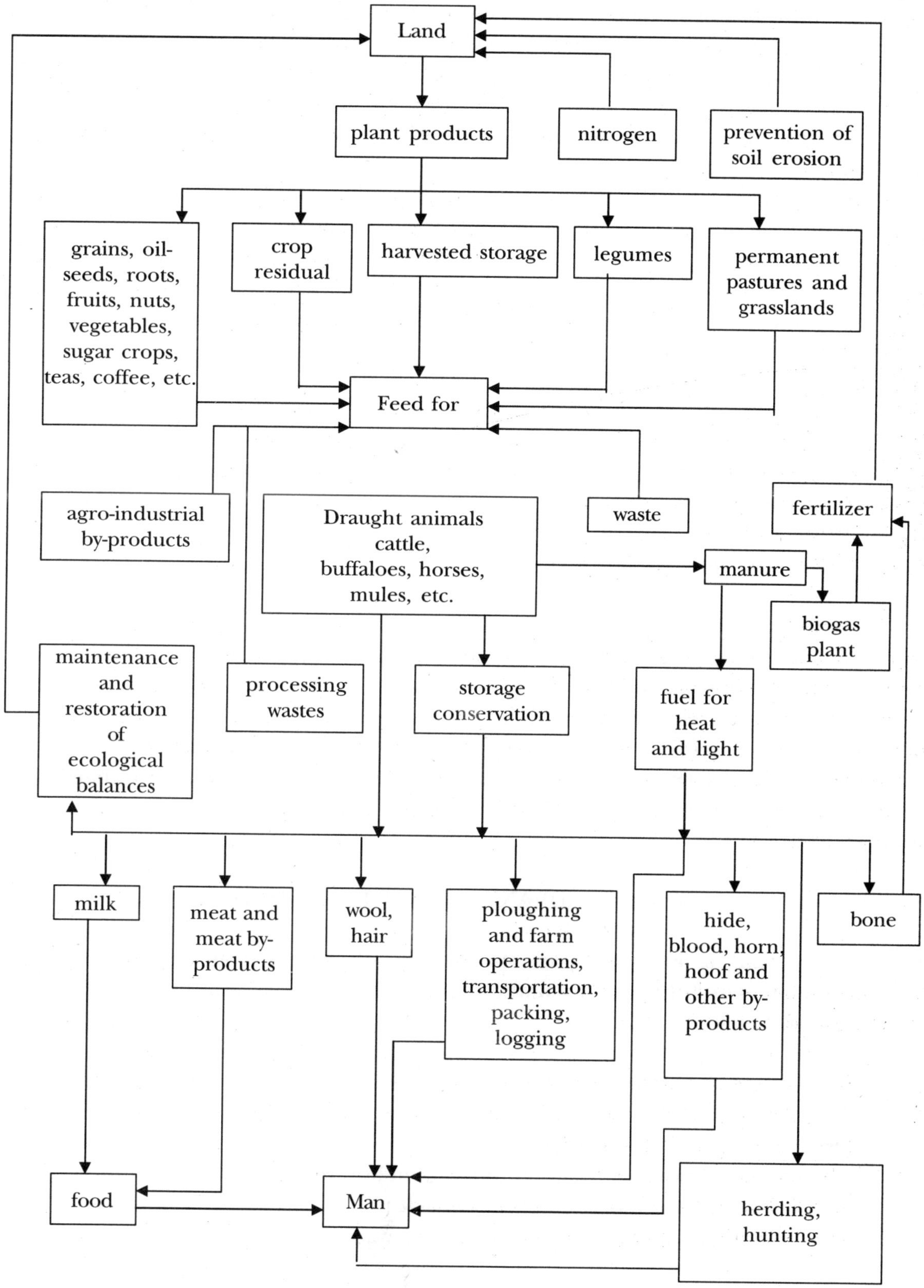

SOIL FERTILITY MANAGEMENT

Great attention has been paid by the farmers since the very beginning of agricultural practices. The available evidences have proved that Harappan farmers knew the advantages of soil fertility and that is why they used crystal gypsum. It is provided double advantages—on one side it was helpful providing double advantages on other side it was helpful in soil moisture conservation by farming mulch on the top of the soil and on the other side enhances the nutrients absorption.[109]

In the *Ṛgveda*, the cultivable land is mentioned as *urvara.*[110] The word *kariṣinim*, i.e. cowdung is frequently mentioned in the *Vedas, Śatapatha Brāhamaṇa* and in other Sanskrit literature. It seems to be used in reference to application of cowdung to enhance fertility. Actually, at that time, there was no chemicalization except few natural minerals and they used manure and composts commonly known as *desi khad.* They used *desi khad* in different forms as given below:

1. Manures prepared from dung and urine of animals
2. Compost prepared from litter and by-products of fodder
3. Compost prepared from by-products of kitchen in the form of ash
4. By-products of farm produce
5. In the form of green manure
6. The fertile silt deposits of ponds

In the later phase the forms of providing additional nutrients to soil system were changed. The use of green manure has been significantly increased and "basically the crops ground for this purpose were smothered in the same field and it is called in situ green manuring. In the *Atharvaveda*, it is mentioned that for the purpose of green manure the upper portion of *arjun (terminallea arjun)* and sesame were used.[111] In the *Bṛhatsaṃhitā*, it is clearly mentioned that at first step before sowing of the main crop, sesame should be sown and when the crop reaches at the blooming stage, it should be smothered in the same field.[112]

SPECIFIC ADVANTAGES OF DESI KHAD

It is more relevant to know what are the specific advantages from *desi khad.* The *desi khad* represents here both manure and composts. There are several specific characters as mentioned below:

1. It is complete manure which provides nearly all essential elements to plants.
2. The use of this increases the organic matter content, particularly at the human level, the most vital part of soil.
3. The increase in organic matter content ameliorates the physical condition of soil and enhances nutrients retention capacity.
4. The functional activities of microorganism accelerated.

5. In the *desi khad* there are more number of fungi and bacteria presented (bacterio phases) provides immunity to plants.[113]

METHOD OF PREPARATION

Actually, since time immemorial, manure from dung and urine and compost from farm by produce being prepared in the same pit and, therefore, it is very difficult to distinguish between the two. In the later phase, a few examples are also available suggesting that compost were made separately. The fementation pit always prepared far from the habitat and most probably at any corner of the field.

Parashara has elucidated certain rituals which have been performed before lifting the *khad.*[114] In the month of the *Mārgaśīrṣa* (January–February), on the appropriate day and *nakṣatra* (constellations), with the help of spade during sunshine, the heap of manure should be dug. After this in the month of the *phālguna*[115] (February–March) it should be dispersed and mixed with soil in the field. It is further mentioned that without providing manure, adequate crop production was impossible.[116]

METHOD OF APPLICATION

There are several methods practised in the ancient period for providing nutrients to the plants as enumerated below:

(1) Diluting nutrients with irrigation water[117]
(2) Through fumigation of nutrients containing substances[118]
(3) By pasting the nutrients[119]
(4) Treatment of seeds and application of nutrients at the time of sowing
(5) Through injection of nutrients.[120]

MANAGEMENT OF FERTILITY

The management of soil and its fertility is as old as the history of agriculture. The entire 12th *Kāṇḍa* of *Atharvaveda* is devoted to the mother earth and it is known as *Pṛthvi Sūkta.* Some important and appropriate verses are being discussed here:

> In the very first mantra it is enumerated that, 'truth', material prosperity, justice, military, strength, efficiency, hard work, knowledge, mutual regard, unity and charity sustain a state. May this motherland of us grant us ample scope for advancement (12-1-1).

In the next verses, we can assess the wisdom and broader concept of Aryan's regarding management of soil and its fertility as 'all bearing store of treasurers advancer of glory, gold-breasted, labourer of all that month, the sustainers of mankind, and their well-wisher the king ferocious like fire, the establisher over it of the rule of a sovereign may this motherland give us great possessions' (12.1-6). Again in the next shloka it is mentioned that 'may our vast motherland, always protected with ceaseless cane by the learned kings who are free from idleness, may she pour out for us full, lovely knowledge, may she bedew us with a flood of splendour' (12-l-7).

The concept of management reaches its climax in these verses: 'O motherland, auspicious be thy woodlands, auspicious be thy hills and snow-clad mountains, unslain, unwounded, unsubdued, I rule over the Earth on earth, the nourisher of all, cultivated by the peasants, full of various kinds of cereals and plants inhabitant by men of different complexions on the firm vast earth, well guarded by the king against danger' (12-1-11). Further, 'O mother land, whatever act of justice is thine. Whatever act of welfare for the military men is, thine, whatever invigorating articles than growest out of thy body, set us in their midst purify us. I am the son of earth, and earth is my mother cloud is my father, may it nourish us' (12-1-12).

The concept of sustainability can be seen in this mantra: 'let what I dig from thee, O Earth! rapidly spring and grow again. O Purifier! let me not pierce through they vital or they hear' 12-1-35. The vision of ancient people regarding sustainability was, in fact, vast and the concept was very broad. They know that each and every factor of ecosystem is interdependent and correlated complementarily. They practised several methods for soil and fertility management as follows:

(i) Classification of land on the basis of land capability,
(ii) Classification of soil according to different crops suitability,
(iii) Use of land as per their capability classification,
(iv) Use of crop rotation, including one or more pulse crop,
(v) Use of composts and manures for enhancing fertility,
(vi) Biological utilization of atmospheric nitrogen, and
(vii) Plant protection through biological means.,

IRRIGATION AND WATER MANAGEMENT

Since time immemorial, farmers have been using advance techniques of irrigation and water management. The Harappan fields were probably located near permanent or seasonal water sources where water control would be accomplished by either digging a trench between the river and the fields by natural overbank flooding or by use of store or earthen embankments for water division. Use of walled fields or we can say the high bunding was very common and helpful in controlling and harvesting rainwater as well as to check the wandering animals.[121]

According to Agrawal and Mckean (1983),[122] the farmers of Harappan civilization frequently used contouring, bunding, terracing and benching for water management, *gabarbands* (dams) and canals were commonly built to harness, utilize or slow on water and soil movements. It consisted of stone walls, raised earthen banks, brick walls or any combination of three. Raikes (1965) opine that artificially constructed dams served a variety of functions, terracing or diverting water in order to retain and distribute silt from flood waters and thereby add to the fertility of the agricultural fields.[123]

In the first *Maṇḍala* of *Ṛgveda,* it is clearly mentioned that 'Maruts, with their vigorous strength pursue dark deep forces upon high and cleave as under the mountains that obstructs the path of water'.[124] In another verse it is elucidated that, 'just as the expert engineer constructs canals of water through the winding channels to the place where water is scarce similarity the vital principles bring rejuvenation to the thirsty soils and restore the production of foodgrains'.[125]

In the *Ṛgveda,* it is indicated that there are four types of waters:[126]

(i) *divya jala* (rainwater)
(ii) *prasavya jala* (water of waterfall)
(iii) *khanitrima jala* (water of a well or a pond)
(iv) *svayañja jala* (self-source spring water)

In the *Yajurveda* and the *Atharvaveda* it is also mentioned that there are several types of water depending on the nature of source. In the *Yajurveda* it is elucidated that there are eleven types of *Jala.*[127]

IRRIGATION TECHNIQUES

The irrigation techniques used by the farmers of ancient period were not so complex and water wasting. They irrigated crops, as and when crops showed water requirements. The decision was taken on the basis of crops' appearance, field conditions and experiences. It was decided by the experts who divided fields in small segments, each segment was completely bonded, that is why less quality of water needed. There were five main techniques used at that time as given below:

(i) **Contour method**: This method is mostly used in the hill areas or sloppey soils. Rain water harvested in the catchment area and it was preserved for the future use. Contour bunds were used for checking irrigation water flowing as runoff and it was constructed before rainy season and on contour line.[128]
(ii) **Flood method**: This method was applied in such condition where adequate availability of water was ensured. In this method, the whole field was flooded with water.[129]
(iii) **Furrow method**: In this method, furrows were made for irrigation. It is mentioned in the *Śatapatha Brāhamaṇa*[130] that furrows made by plough were irrigated by mixture of milk and ghee. Though it was actually the rituals performed before irrigation, yet it indicated the furrow irrigation. Still today it is used in the vegetable farming.
(iv) **Ring method**: It was mostly used in gardening. It is elucidated in the *Agni Purāṇa*[131] that for irrigating trees ring methods should be followed. The requirement of water in this method is very less.
(v) **Sprinkling method**: This method was used specially in the flowering plants. It is still today the best water utilisation technique and used in different fashion like drip irrigation, sprinkling and in most advance phase as microirrigation.

WATER MANAGEMENT

In almost all traditions of mankind, the water occupy a special place. In the *Vedas,* it is elucidated clearly that the 'water is life'. In the seventh *Maṇḍala* of *Ṛgveda,* it is mentioned that 'water provides three facilities: food, drinking water and medicinal plants'.[132] The ninth *Sūkta* of 10th *Maṇḍala* of this epic were dedicated to *Apa* (Water) some *mantras* are as follows:

> These waters to be use for drink,
> Divine are they for aid and joy
> May they import to us health and strength(X-9-4)

In the *Śatapatha Brāhamaṇa,* it is mentioned that 'the essence of plant life is water' (III-6-1-7) The waters are the foundation of all this universe (XII-5-2-14) and from the waters is this universe produced (VI-8-2-3). Water is the elixir of immortality. Of this universe it is truth that waters were made first. Hence when the water flows then everything whatsoever exists is produced.

It is miraculous to note here that the farmers of ancient period were well-versed in water management techniques they knew water plant relationship. In the fifth *Maṇḍala* of *Ṛgveda,*[133] it is mentioned that the water of rainfall is just like semen of clouds and entering in the plants provide ability for flowering and fruiting. Several methods which were used for effective managements of water are as follows:

(i) Irrigation according to water requirements;
(ii) frequent irrigation for effective utilization of water;
(iii) use of engineering techniques and construction to check the loss of water through seepage and percolation;
(iv) irrigation in small segments of fields to check the over flooding of water;
(v) to check the rainfall in catchment area and water harvested at high-lands facilitate irrigation and check soil erosion;
(vi) use of organic matter sufficiently enhances the water retention capacity of soil; and
(vii) *in situ* conservation of moisture using mulch, inter-culture, etc.

MANAGEMENT OF INSECT, PESTS AND DISEASES

The management of insect, pests and diseases are an important aspect of cultivation of crops. In the ancient period, any type of insect, pest and disease attack were supposed to be nature provided and religious rituals were performed as precautionary measures, and preventive action. It is mentioned in the *Ṛgveda* that 'fire with flame, the sun with rays and wind with wave control diseases'.[134]

In the *Atharvaveda,* effective technique were enumerated for insect, pest and disease control. In the 2nd Kanda, it is mentioned that solar rays kill the pathogen of diseases. 'Let the sun at sunrise and sunset, destroy with his beams the worms that live on the earth'.[135] In other verse, it is enumerated: 'O disease-producing worms, I destroy ye as a violent flesh-eater destroys his bird of prey, as a fowl eats up all grains one by one as kindled fire extinguishes everything all at one stroke. I bray and bruise the worms to pieces with the Vedic knowledge of god'.[136]

Most probably, in the ancient period, loss caused by insect, pest and diseases was below the economic injury level. This is why it was not of major concern for them. There are two main reasons namely:

1. Greater tolerance limit of crops; and
2. Crops diversity due to mixed cropping.

Actually, the tolerance limit of any crops provides immunity to plants to resist against any untoward attack of insect, pests and diseases. Most of the crops cultivated in the ancient period have broad genetic base and due to mixed cropping the diversity of crops maintained at higher level. These factors on one side increased tolerance in crops and on the other side, impeded the virulence of pathogen and that is why attack of insect, pests and diseases never took the epidemic form.[137]

In fact, the dimensions of insect, pest and disease control of ancient India was very broad. On the one hand, they tried to restore and maintain the ecological balance and on the other hand, they cautiously check the population of insect, pest and diseases below economic injury level. They have clear concept that disturbed ecological balance is responsible for attack of insect, pest and diseases. In *Veda*, Asvina devas supposed to be the destroyer of pest and diseases. To control this, they consider three aspects as given below:

(i) Conservation of ecological balance;
(ii) population dynamics of insect pest and diseases; and
(iii) to keep population dynamics below economic injury level.

TECHNIQUES OF EFFECTIVE CONTROL

Techniques used in the ancient period for controlling insect pest and diseases were behavioural and economically feasible for each. In this regard, in the *Ṛgveda*, it is enumerated that expert person decides the following characteristics and control measures as follows:

(i) exact taxonomy
(ii) morphology of insect
(iii) lifecycle and seasonal history
(iv) natural history

Taking into consideration the above characteristics, the following methods were used successfully as given below:

(a) Agronomic practices

(i) Summer ploughing
(ii) Deep ploughing
(iii) Change in the time of sowing
(iv) Use of crop rotation
(v) Resting of field in rainy season (*chaumāsa*)
(vi) Mixed cropping
(vii) Appropriate techniques of water management
(viii) Use of organic manure
(ix) In tea culture and weed control
(x) use of trap crop

(b) Mechanical methods

(i) Hand picking
(ii) Tree bending
(iii) Fencing
(iv) Flooding and trenching
(v) Use of light traps

(c) Physical methods

(i) Performing *yajna*
(ii) Fire and fumes
(iii) High velocity sound
(iv) Rapid flowing water
(v) Using medicinal plants

(d) Autocidal techniques to kill or infertile the male insect.
(e) Elimination of female insects and their habitats.
(f) Use of ash (*bhasma*) of *Agnihotra.*
(g) Use of bio-pesticides of 'neem', *chrysanthemum,* etc.

All these techniques continue to be effective even today and can be used successfully to sustain the agricultural production process. Use of biopesticides in different forms are increasing every day. Several other remedial and curative methods have been developed with manufacture of medicinal plants or their produce or produce of animals.

POST-HARVEST MANAGEMENT

Merely successful cultivation of crops is not sufficient; it is also vital to harvest the crop at appropriate maturity and on appropriate time. As we know, in our country, a huge quantity of foodgrains are destroyed before reaching the hands of farmers. This loss can be minimized by taking decision appropriately in right time about when to harvest.

There are several other factors which affect the profitability of farmers—poor storage facilities, loss in storage, lack of marketing facilities, long chain of intermediaries in agricultural marketing, lacunae of agricultural price policy, poor transportation and faulty marketing and bargaining. In this process poor farmers have sufferered the most. In ancient period, the system of marketing was either group or community, so there was no question of suffering from anamoly of market.

SUSTAINABLE FOOD SECURITY

The concept of food security has gradually evolved in recent decades. It has been fully observed that merely production of high quantity of foodgrains is not sufficient, but equal emphasis on equitable distribution became inevitable for restoring sustainability in food security. The distance between resourceful and resourceless farmers is increasing continuously. Hence, there has an been emerging trend of fear for future availability among poor and marginal farmers. The corporation of farming system and marginalisation of poor farmers never restore food security.

A widely accepted definition of food security now is the one proposed by the Science Academics summit held at Chennai in July 1996 is; 'that every individual has the physical, economic, social and environmental access to a balanced diet that includes the necessary macro- and micro-nutrients, safe drinking water, sanitation, environmental hygiene, primary health care and education so as to lead a healthy and productive life.' The food originates from efficient and environmentally benign production technologies that conserve and enhance the natural resource base of crops, animal husbandry, forestry inland and moving fisheries.[138]

It we look at the aforesaid discussion in the perspective of food security of the ancient Indian systems for restoring it, we will find more saturated and purified system. In the *Tattirīya Saṁhitā*, it is elucidated that food is Brahman, for certainly all beings here are indeed born from food having been born they remain alive by food.[139] In another verse it is mentioned that one should not blame food. That should be observed as a pious rule. 'Life indeed is food'.[140] It is further enumerated that 'one should not reject food. That should be observed a pious rule.[141] And for ensuring proper availability, it is narrated that 'one should produce abundant food'.[142]

Since ancient period onwards, the food security of the society was successfully ensured by community approach. It is mentioned in the *Ṛgveda* that man is not merely an individual. He belongs to a concept of society. The hymns of the *Veda* inspire the entire society for concordance, love, humanity and well-knit affectionate discipline, which arises out of mutual respect, fondness and reverence and also mutual understanding. In the last 10th *Maṇḍala*, it is mentioned that, 'meet together, talk together, let your mind apprehend alike, in the like manner as the ancient people of wisdom concurring accepted their rewards in the social selfless sacrifices'. 'It should be very clear that each and every individual has right to get food. It would be better to provide food to hungry person or nation on cheaper price rather than destroying like America and some European countries for obtaining higher price and maintaining market rate. It may be beneficial for developed countries but it is not in the favour of humanity at all.

According to M.S. Swaminathan research foundation for ensuring food security, science must work in partnership with farmers to create a new agriculture and in this endeavour the efforts would aim towards:

(a) transformation of most marginalised farmers through the blending of traditional and frontier technologies in socially equitable, economically viable and environmentally sustainable eco-technologies; and
(b) production of more food from a diminishing resource base requiring new agricultural technologies and management systems focussing research on neglected crops such as minor millets, grain, legumes, pulses and oilseeds.

In fact that time has come to reconsider the farming system according to the specific ecological conditions and basic human needs. Transition from chemical and machinery-intensive specialized farming to knowledge, labour-intensive ecologically accountable technologies has became inevitable and need of the hour for food and nutrition security. The Vedic concept of traditional multiple income-earning opportunities like; crop-livestock, forestry and fisheries integration and restoration of soil health care water harvesting management and the conservation of bio-diversity may be successfully used as a complementary to modern agriculture for sustainable development.

Last but not the least, we can not minimize the importance of traditional knowledge and wisdom of ancient farmers. They had developed such technologies which were suitable to their needs and compatible to their ecosystem. Once again, the Vedic lifestyle and dietary patterns have became inevitable for ensuring sustainability of system. Globalization of dietary pattern has done immense loss to biodegradation. As we know, this process led to specialization in agriculture. The cultivation of minor millets and coarse grains have been marginalised by the monoculture of wheat and rice and some other cash crops. But it should be very clear that change in cropping pattern has disturbed the agro-ecosystem. The cost of this shifting has been paid by the people as health maintenance cost, because the loss of staple food of specific location has compelled the people to use globalize fast foods.

SUMMARY AND CONCLUSION

The development of agricultural practices and wisdom in the ancient period was the result of the process of long-term change, the dimension of which was very broad and nature was balanced and sustainable. Due to the integrated approach of contemporary farmers towards farming, animal husbandry and forestry in the process of production and consumption, a natural harmonious relationship was maintained. According to the ecological conditions and as per the human needs, there has been a continuous changing process going on for the alteration, improvement and purification of agricultural knowledge system, which is still flowing continuously as a traditional indigenous knowledge at the present time in some of the areas like, hilly tracks, tribal areas, etc.

The importance of this experience oriented and time tested traditional agricultural knowledge is still relevant at the present time. It has become inevitable to adopt the ancient approach for preserving environmental balance and for the protection of bio-diversity. The non-arable barren land, which is abundant, can be brought under production system through land-capability classification and integrated use of this land accordingly. The soil moisture can be preserved insitu and utilized for crop cultivation through proper water management techniques. The systematic development of check dams, contouring, benching, terracing of water management of ancient period are still relevant. The fertility and quality of the soil can also be enhanced through increasing organic matters content of soil and effective control of soil erosion.

The social-cultural dimensions of the society cannot be ignored the order to achieve the goal of sustainable development. The social harmony in the society can be re-established through creating awareness of social responsibility and engagement of all the classes of society in the different aspects of agricultural production system.

NOTES AND REFERENCES

1. Miller, E.S. and C.A. Weits, 1931, *Introduction to Anthropology*, Englewood Cliffs, New Jersey, pp. 228–235.
2. Clarke, D.L. 1977, Spacial Information in Archaeology, *In Spacial Archaeology*, Academic Press, New York.
3. Miller, E.S. and C.A. Weits, *op. cit.*, pp. 230–245.
4. Patterson, T.C. 1973, *America's Past; A New World Archaeology*, Glenview III, Scatt Foresman.
5. Flannery, K.V., 1965, 'The Ecology of early food production in Mesopatamia' *Science*-147.
6. Helback, H. 1959, 'Domestication of Food Plants' in *Old World Science*–1959–130.
7. Isaac, E. 1971, 'On the Domestication of Cattle'. In S. Struever (ed.), *Prehistoric Agriculture—Garden City*, New York, Natural History Press.
8. Childe, G.V. 1951, *Man Makes Himself*, New York, New American Library.
9. R.J. Braidwood, (1967), *Prehistoric Men* (7th ed.), Glenview-III Scott, Foresman.
10. Ghosh, A. 1989, *An Encyclopaedia of Indian Archaeology*, New Delhi, pp. 161–62.
11. Steven, A. Weber, 1991, 'Plants and Harappan subsistence an example of stability and change from Rojdi', *American Institute of Indian Studies*, New Delhi, pp. 21–32.
12. Mckean, M.B. 1983, *The Palynology of Balakot, a Pre-Harappan and Harappan Age Site* in *Bela Pakistan*, Ref. Weber Steven A., pp. 27–30.
13. Weber, Steven, A. 1991, *op. cit.*, pp. 28–32.
14. Possehl, G.L. 1986, *Studies in the Archaeology of India and Pakistan* (ed.), New Delhi, pp. 237–256.

15. Swami Satya Prakash Sarasvati, 1988, '*The Critical and Cultural Study of the Satapatha Brahmanam*', pp. 621–669.
16. *Ṛgveda,* 4-57-4, 10-34-13, 10-117-7 and *Śabdakalpadruma* and *Vāchaspatyam* etc.)
    शब्द कल्पद्रुम: क षि: स्त्री० (क ष विलेखने। "सर्व्व धातुथ्य इक") वेश्य व त्तिविशेष:। "त्वथ्यापत्तं क षिफलमिति भ्रूविकारानभिज्ञे:"
    वाचस्पत्यम्: क षि (स्त्री०) क ष इक। शस्योत्पादनार्थ भूमिकर्षणरूपे वैश्यव त्तिभेदे।
17. *Ṛgveda,* 10-101-7.- द्रोणा हावमतमश्मचक्रमंसत्राकोश सिन्चता न पाणाम्"।
18. *Ṛgveda,* 10-34-13.- अक्षैर्मादीव्य: क षिमित्क षस्व वित्ते रमस्व बहुमन्यमान:।
19. *Ṛgveda,* 4-57-4.- शुनं वाहा: शनं नर: शुनं क षतु लांगलम्।
    शुनं वरत्रा बध्यन्ता शुनमषट्रामुदिङ्गय।।
20. Chattopadhyay, Brajadulal, 1987, *Essays in Ancient Indian Economic History* (ed.).
21. *Jat* I-239, II-76, 135, III-9, IV-370.
22. Lal, B.B. 1981, Excavations at Hastinapur, *Ancient India* No. 10-11 and *PGW Culture; A History of the Civilization of Central Asia.*
23. Banerjee, N.R., *The Iron Age in India,* Pt. I.
24. Sahi, M.D.N., New Light on PGW People as revealed from the Excavation at Jakhera, *Man and Environment,* Vol. II, p. 104.
25. Chowdhury, K.A., 1987, *et al.*, 'Ancient Agriculture and Forestry in Northern India', pp. 63–66, ref. in *Essays in Ancient Indian Economic History* of Brajadulal Chattopadhyaya, *op. cit.*
26. Watt, G., *Dictionary of Economic Product of India,* Vol. VI, Pt. IV, p. 125.
27. Sahi, M.D.N., *op. cit.*, p. 104.
28. Singh, K., 1991, *Rural Development Management,* India's experience; *Modern Management Review,* New Delhi, 13.
29. Schumpeter, J.A., (1934): *The Theory of Economic Development.*
30. Duley, Seers, 1972, What are the Trying to Measure, *Journal of Development Studies.*
31. 'The RIO Declaration an Environment and Development': *The U.N. Conference and Environment and Development,* June 3–14, 1992.
32. Katyal, J. C., (1997) Sustainability, Relevance of Integrated Concept: *The Hindu Survey of Indian Agriculture* p. 25.
33. Paroda, R.S., 1996, *Farm Research: New Paradigms; The Hindu Survey of Indian Agriculture,* pp. 17–27.
34. Bahuguna, Sunderlal (1996), The Himalayan Threat: *Development is Wanton; The Hindustan Times,* Sept. 29, p. 13.
35. Venkataramani, G., 1995, Ecological Farming: A Viable option for the future; *The Hindu Survey of Agriculture,* pp. 23–31.
36. A Close look at pesticides: *Bulletin of Voluntary Health Association of India.*
37. Paroda, R.S., (1998), *Yojna* Vol. 42, No. 7, July 1998 , pp. 40–41.
38. Sharma, Devinder, 1997, 'Chemical Fertilizers hit. Sustainability'; *The Hindustan Times,* Dec. 15, 1997, p. 16.
39. Sharma, Devinder, (1997): *op. cit.*, p. 16.
40. Venkataramani, G., (1995): *op. cit.*, p. 25.
41. *Ṛgveda* 10-117-7.- क षान्नित् फाल अषितं क णोति यन्नथ्वान मपव ड़क्ते चरित्रै:।
    वदन् ब्रह्मा वदतो वनीयान् प णन्न पिरप णन्तमाभिष्यात्।।
42. *Yajurveda*-12-67. सीरा युश्जन्ति मवयो युगा वितन्वते प थक् धीरा देवेषु सुम्नया।।

43. *Yajurveda*-12-68

44. *Yajurveda*-12-69- शुन सुफाला विक षन्तु भूमि शुनं की नाशाडकाभियन्तु वाहै:।
शुनासीरा हविषा तोशमाना सुपिधलाडओषधी: कर्लनास्मे।।

45. *Yajurveda*-12-70- घ तेन सीता मधुना समज्यतां विश्वेदेवेरनुमता मरूद्भि:।
उर्जस्वती पयसा पिन्यमाना स्मान्त्सीते पयसाम्या व व त्स्व।।

46. *Yajurveda* 18-9 उर्क च मे सुन ता च में पयश्च में रसश्च मे घ तं च मे मधु च मे सग्धिश्च मे सपीतिश्च में क षिश्च में द ष्टिश्च में जैत्र च मडओदमिदं च में यज्ञेन कल्पंताम्।

47. *Yajurveda* 18-14- अग्निश्च मडआपश्च मे निरूधश्च मडओषधयश्च मे ... यज्ञेन कल्पताम्।।

48. *Taittirīya Saṁhitā*- 4-2-6.
शुनं न: फाला वितुदन्तु भूमि शुनं कीनाशा अभियन्तु वाहान्।
शुनं पर्जन्यो मधुन: पयोभि: शुनानीरा शुनमस्मासु धत्तम्।।

49. *Kautilya's Arth.* 2-171 and 2-40-24.

50. Prime, Ranchor, (1992), *Hinduism and Ecology-Seeds of Truth,* Delhi, p. 19.

51. Swami Sarasvati, S.P., (1988), *op.cit.*, pp. 660-661.

52. *Satapatha Br.* 13-3-2-3.

53. सातवलेकर, एस०डी० *वेदों में क षि विद्या,* प १९-२०

54. Gopal, Lallanji, 1980, *Aspects of History of Agriculture in Ancient India,* pp. 90–91.

55. Pande, N.D., Pandhari, 1996, Practising Farmers of Pusad Maharashtra delivered, lecture in the Deptt. of Extension Education, Instt. of Agril. Sciences, BHU, Varanasi on 23rd September.

56. Deshpandey, Madan, *et al., Process of Purification of the atmosphere,* Published by Institute of Studies in Vedic Sciences Shivpuri Akkalkot, February, 1990.
b- M.S. Parkhe, "*Vedic Solution for Present day Problems*", Vaidika Samshadhan Mandal, Pune, 1982. (Ref. by S.B. Singh in Rashtrayoga, 1993, pp. 760–62.

57. *Ṛgveda* 7-36-7 आयत् साके यशसो वावशाना: सरस्वती सप्तमी सिंधुमाता।
या: सुष्वयन्त सुदुधा: सुधारा अभिस्वेन वयसा पीप्याना:।।

58. Asthadhyayi of Panini.

59. *Questions of King Milinda* 4-1-36

60. हरिवंश विष्णुपर्व अध्याय 192.

61. *Conservation of Biological Deversity in India: An Approach* (1994), Ministry of Environment and Forest, Government of India.

62. Negi, S.S., 1994, *India's Forests, Forestry and Wild Life,* New Delhi, p. 360, Annex.-I.

63. Khoshoo, T.N., 1998, 'Preservation of India's Bio-diversity', *The Hindustan Times,* June 28, New Delhi.

64. Khoshoo, T.N., 1998, *op. cit.*, "Lecture on the award receiving Ceremony of Madhu Bhasin Memorial Award" at Guru Jambheshwar University, Hissar.

65. Earth Summit, 1992, A Report published in the *Rojgar Samachar,* 6–12th June, 1992, New Delhi, p. 4.

66. Ghosh, A., 1989, *op. cit.*, p. 40–70.

67. *Ṛgveda* 1-164-8.
माता पितरम त पोषमेव आ वभाज धीत्यग्रे मनसा संहिजग्मे।
सा बीभत्सुर्गर्भरंसानिविद्धा नमस्वन्त इदुपवाकमीणु:।।

68. *Ṛgveda.* 10-97-1 या ओषधी: पूर्वजाता देवेश्यस्ति युग्रं पुरा।
69. *Yaj. Veda,* 8-9- उर्क् च मे सून ता च मे पयश्च मे रसश्च मे ... यज्ञेन कल्पताम्।
18-12- व्रीहयंच मे यवाश्च मे माषाश्च मे तिलाश्च मे गुग्दाश्च मे खल्वाश्च मे प्रियंगवश्च मे व्वश्च मे श्यामाकाश्च मे नीवाराश्च मे गोधूमाश्च मे मसूराश्च मे यज्ञेन कल्पताम्।।
70. *Atharva Veda,* 4-37-5.
71. *Matsya Pūrāna,* 59-1 to 20 shlokas.
72. *Śatapath Brāhmaṇa.* 5-2-4-18. ब्रह्मवै पालाश:
73. *Tatt. S.* 3-8-12-2. पजापतिर्देविभ्यो निकायत सो श्वत्थे।
74. *Padma P.* तत्र विल्वे स्थित: शम्भुरश्वत्थे हरिश्व्यय:।
75. *Arth. S.* 2-18-2. सर्वातिथिम गं प्रत्यन्ते चान्यन्म गवनं भूमिवशेन वा निवेशयत्।
76. *Maha. Bh.* (Shanti) - 35-7.
77. *Shukra* 4-4-59. ग्रामे ग्राम्यान वने वन्यान् व क्षात् संरोपयेन्न प:।
78. *Srimad Bhāgvata* 19-7
79. हेमाद्रौ ब्रहमाण्डपुराण - अश्वत्थो दुम्बरप्लक्ष-यूतन्यग्रोध पल्लवा: पन्चभंगा इतिख्याता सर्वकर्मसु शोभना:। पंन्चभंगा पंचपल्लवा:।
80. अग्निपुराण - दशकूप समावापी दशवापी समोह द:। दशह द सम:पुत्रो दश पुत्रो समोद्रुम:।
81. *Ṛgveda* 1-38-7 - सत्यं त्वेषा अमवन्तो धन्वचिदा रूद्रियास:। मिहं क ण्वन्त्यवाताम्।।
82. *Ṛgveda* 1-164-40.
83. *Bṛhat Saṁhitā* 21-4.
84. *Ibid.,* 21–6.
85. *Ibid.,* 21–7.
86. *Śānkhāyana Gṛhya Sūtram* 13th Khanda.
87. *Bṛhat Samhitā of Varahmihira*
88. *Yaj. Veda* 28-14. अग्निश्च मडआपश्च मे विरूधश्च मडओषधयश्च मे क ष्टपच्याश्च मे ड क ष्टपच्याश्च ... पज्ञेन कल्पताम्।
89. Lal, B.B., 1971, Perhaps the Earliest Ploughed field so far Excavated, anywhere in the World, *Puratattva* 4: 1–3.
90. Lambrick, H. T. (1970), The Indus Flood Plain and the Indus Civilization, Geographical *Journal* 133: 484–94.
91. Bakshi, J.S. and R.S.Ranga, 1974, *Barley, In Evolutionary Studies in World Crops,* pp. 47-52 Combridge
92. Singh, S.S., 1986, Principles of Agronomy, Allahabad.
93. भाष्य 2-3-13 तिलै सहभाषान् वपति। तिलैर्मिश्रीक त्य भाषा उप्यन्ते।।
94. म्वेश्वन्स आफ किंग मिलिन्द: यू०उ० प० 90–105
95. *Ṛgveda* 10-131-3.
96. *Yaj. Veda*
97. *Tatt.* S. 5-1-7-3
98. *Manu Smrit.* 10-7. सुबीजं चैव सुक्षेत्रे जातंम सम्पद्यते यथा
99. Manu S. 10-70.
100. *Ṛgveda* 1-101-3.
101. *Ṛgveda* 1-115-3, 8-18-9
102. *Arth.* 2-24-1 to 10.
103. अष्टाध्यायी 5-4-48
104. अग्निहोत्री प्रभुदयाल: पतंजलिकालीन भारत, प 254

105. अष्टाध्यायी 4-3-45: आवश्ययुज्जा वञ।
106. अष्टाध्यायी 5-4-48:
107. शत०ब्रा० 7-2-2-9: अथ दर्भस्त म्वमुवदधाति ...।
108. स्थानांग 4-355
109. *Ṛgveda* 10-68-1.
110. Mittre, Vishnu, and R. Savithri, 1992, *Food Economy of the Harappans Civilisation*, Ed. G.L. Posshel, New Delhi, pp. 205–222.
111. ऋग० 2-21-1 - विश्वजिते धनजिते स्वर्जिते सताजिते न जिते उर्वराजिते।
112. *Ath. Veda* 2-8-3 & 2-8-4.
113. *Br. S.* 55-2. म दी भू: सर्वव क्षाणां हिता तस्यां तिलान् वयेत्।
पुष्पितांस्ताश्च म द्वीयात् कर्मेतत्यप्रथमं भुव:।।
114. Tom Young, (1996), *Ecological Science Ltd.,* An Independent Scientific Research Organization, London (B.B.C. News Service, Sept. 7, 1996).
115. वही - १११ - ततो वपनकाले तु कुर्यात् सार विमोचनम्।
विना सारेण यद्धान्यं वर्धतिफलवर्जितम्।
116. विष्णुधर्मोत्तर पुराण 2-30-32, अग्निपुराण 247-31
117. व क्षायुर्वेद - श्लोक 128. 159
118. वही - श्लोक 136
119. काशिका सूत्र 24-2 और पाराशर ग हसूत्र 3-1-6
120. M.B. Mckean, (1983): *op.cit.*
121. D.P. Agrawal, (1988): The Harappan Legacy: Break and continuity Paper presented in the *Lecture series, the Indus Civilization*, University of Pennsylvania, Sept.-Nov., 1988.
122. Steven, A. Weber, (1991): *op.cit.* (referred Raikes–1965).
123. *Ṛgveda,* 1-85-11.
124. *Ṛgveda,* 1-95-10.
125. *Ṛgveda,* 7-49-2.
126. *Yaj. Veda,* 22-25
127. *Ṛgveda,* 1-85-10 to 1-85-11.
128. *Regveda,* 1-95-10.
129. *Sat. Br,* 8-2-2-10 (Ref. Max Muller: *Sacred Book of the East,* p. 328, Vol. XLI).
130. अग्निपुराण २८२-८
131. *Ṛgveda,* 7-101-2.
132. *Ṛgveda,* 5-83-1.
133. *Ṛgveda,*1-9-2. एमेन स जतासुते मन्दिमिन्द्रायमन्दिने। चक्रिंकं विश्वनि चक्रये।
134. *Ath. Veda,* 2-32-1.
135. *Ath. Veda,* 2-32-3.
136. Vandana Shiva (1991), *The Violence of the Green Revolution: Third World Agriculture Ecology and Politics,* Goa, pp. 65–90.
137. M.S. Swaminathan, (1997), Pathway to rural prosperity; *The Hindu: Survey of Indian Agriculture,* pp. 9–11.
138. *Tatt. Up,* 3-2.
139. *Tatt. Up,* 3-7.
140. *Tatt. Up,* 3-8.
141. *Tatt. Up,* 3-9
142. *Ṛgveda,* 10-191-2. संगच्छध्वंसंवध्वं सं वो मनांति जानताम्।
देवाभागं यथापूर्वे सज्जानाना उपासते।।

CHAPTER 45

History of Agriculture as Reflected in the Art of India

Pushpa Tiwari

INTRODUCTION

The history of agriculture constitutes the most important chapter in the history of human civilization in general and that of Indian civilization in particular, in so far, as India is considered to be one of the original zones where agriculture of rice and other related crops was first experimented, developed and refined.[1] Before proceeding to the analysis of history of agriculture as reflected in the art of India, it is imperative to define the meaning of agriculture and art. As the objective of this chapter is to scrutinize art-evidences pertaining to agriculture, it becomes necessary to first, deal with the meaning of art-evidences.

Art is a knowledge by which *things* are created, i.e. art remains in the mind of the artist in the form of intellect, skill and expertise. Things created by this skill are called *artifacts*. Therefore, art historians' domain of study relates to the study of artifacts. The philosophy of art deals with the epistemology of art. Generally, material content discovered in the archaeological excavations and explorations are described as *artifacts*. In this sense, tools, equipments, structural evidences, goods used in everyday life—all are treated as artifacts. This practice and usage of the term *artifacts* by the archaeologists broadens the scope of this chapter. Everything created by human intellect and skill involves the play of artifice. Hence, manmade things are *artificial* in contrast to natural objects. Artificer is a technical term used for artists/artisans. Thus, I will also focus on the *artifacts* in the sense this term is used by the archaeologists besides examining well defined art-evidences *vis-à-vis* history of agriculture.

MEANING OF AGRICULTURE

The definition and meaning of agriculture also requires attention to remain focused on the desired objective of the chapter. We can approach it from two points of view: firstly, what the current usage of the term *agriculture* means and secondly, what is the traditional Indian world view on the subject. As per the meaning given in New Webster's

Dictionary[2], the term agriculture and its related derivatives have come to English language via Old French, Latin and Greek. In Latin, roots *ager* meaning field and *culture* meaning cultivation from the Latin equivalent *agriculture.* Agriculture is direct English adoption of the Latin term, meaning the cultivation of ground/land; the raising of crops and feeding of cattle or other livestock; husbandry; tillage; farming; agronomy. Agronomy is derived from Greek roots *'agros'*, i.e. field and *'nomos'*, i.e. law. It means agriculture, management of farms, science of crop production.[3] Likewise, *'Agro-techny'* is another related term which has come to the English language via Greek *'agros'* (field) and 'techne' (art/skill). It means the branch of agricultural science that deals with the conversion of raw farm products into manufactured commodities, as in dairying, canning, etc. Horticulture and husbandry are two cognate terms meaning the science and art of cultivating flowers, herbs, shrubs, fruits, garden vegetables and agricultural cultivation, production of crops, the breeding of animals for food, household management of agricultural products, respectively.[4] The New Encyclopaedia Britannica[5] defines the technology of agriculture as the application of techniques to control the growth and harvesting of animal and vegetable products. It lists following areas of specialized agricultural knowledge:

1. Horticulture (ornamental vegetation)
2. Irrigation and drainage
3. Pest control
4. Weed control
5. Livestock and poultry farming
6. Animal breeding
7. Plant breeding
8. Tobacco production
9. Natural Fibres
10. Vegetable and vegetable farming
11. Fruits and fruit farming
12. Cereals and other starch products, etc.

AGRICULTURE AND HORTICULTURE

It will also be useful to take note of the legal definitions of the terms, 'agriculture and horticulture', as they reflect the codified and widely accepted meanings of these terms as prevalent in day-to-day practice of socio-legal and economic dimensions of agriculture. The U.S. Department of Labor, in its code of Federal Regulations Pertaining to USA, gives the general meaning of 'agriculture or horticultural commodities' as follows: "Section 3(f) of the Act defines "agriculture" as the "production, cultivation, growing, and harvesting" of "agricultural or horticultural commodities". In general, within the meaning of the Act, *"Agricultural or horticultural commodities"* refer to commodities resulting from the application of agricultural or horticultural techniques. Insofar as the term refers to products of the soil, it means commodities that are planted and cultivated by man. Among such commodities are: grains, forage crops, fruits, vegetables, nuts, sugar crops, fibre crops, tobacco and nursery products. Thus, employees engaged in growing wheat, corn, hay, onions, carrots, sugarcane, seed or any other agricultural or horticultural commodity are engaged in *agriculture.* In addition to such products of the soil, the term also includes domesticated animals and some of their products such as milk, wool, eggs and honey.[6]

It will be of great help to compare this definition with Indian legal system:

> 'Agriculture' and 'Agricultural purpose' include making land fit for cultivation, cultivation of land, improvement of land [including development of sources of irrigation], raising and harvesting of crops, horticulture, forestry, cattle breeding, animal husbandry, dairy farming, piggery, poultry farming, seed farming, pisciculture, apiculture, sericulture, and such other activities, as are generally carried on by persons engaged in any of the aforementioned activities, including marketing of agricultural products, their storage and transport and the acquisition of implements and machinery in connection with any such activity.[7]

A perusal of important court judgements of different states in India makes it clear that agriculture, in law, means the art or science of cultivating the ground and raising or harvesting crops and 'an agriculturist is a person who actively takes part in the process of cultivation either by tilling or actually directing and supervising on the spot the process of agriculture.'[8]

The Indian legal system makes it clear that 'any operation on land to make it yield fruits or crops may be called agriculture. Application of human skill is a must. Spontaneous growth like grass, forest, trees, is not an agriculture.'[9]

Thus, a comparative legal definition of agriculture makes it clear that the art or science of cultivation and related activities can be safely called agriculture which is distinguished from natural growth by the application of human skill and management.

The above legal definition gives much leeway to broaden the scope of agricultural history as reflected in Indian art. Philosophy of language and historiographic debates now-a-days are much concerned with the problem of locating *meaning* of the *'text'* in the synchronic/diachronic framework of history writing. If language is the medium of human understanding, then one has to take into account the structural and post-structural ongoing debates on understanding of language, consciousness and meaning. Language is an artificial construct situated in the collective cultural consciousness from where it constructs and deconstructs meaning. Collective cultural consciousness is not a fixed commodity having all time acceptable concrete shape and form. It is always in a flux, changing in response to the needs of society/community along the trajectory of timeline. The term 'agricultural history' is the reference point from which we have to proceed to art-evidence. But the question is: is there any fixed meaning of agriculture? There can be three alternatives to the answer of this question. Firstly, we have to accept the contemporary meaning of the term both for understanding contemporary and/ or past agricultural history. Secondly, we can borrow the universal principles of language and construct universal meaning of the term agriculture and superimpose it on the past while writing agricultural history of that time. Thirdly, we can adopt a comparative model by contrasting the modern meaning of agriculture with the traditional meaning as contained in ancient texts and linguistic evidences. All the three alternatives have their advantages and disadvantages. They can be well taken care of if we take ourselves as the reference point, i.e., always keeping in mind that we are approaching the past equipped with present biases/prejudices and linguistic capabilities; hence, any historical endeavour will require a cautious and sensitive approach to avoid generalities and over simplification of the given text and context. Here, this question acquires more relevance as we are proposing to construct agricultural history through the lens of art. In the portrayal of myriad narratives/motifs/symbols, we have to isolate and identify those evidences which directly or indirectly tell us about agricultural activities. This presupposes the knowledge of agricultural definitions and meanings prevalent in the past, which in itself is a difficult task. To cull out agricultural activities depictions in Indian art, a broad spectrum definition of agriculture will be taken into account which is compatible both with modern and ancient meanings of the term.

AGRICULTURE IN ANCIENT INDIA

An overview of the technical terms for agriculture, as contained in ancient textual tradition, requires an independent analysis. Here it would suffice to say that Ṛgvedic terms *Ara* (अर) and *Kṛṣ* (कष्) connote two different techniques of agriculture: hoeing and ploughing, respectively. Husbandry and related activities are described as an essential part of agriculture. Kautilya[10] in his Arthaśāstra states: आन्वीक्षकी त्रायी वार्ता दण्डनीतिश्चेति विद्या, i.e. Ānvīksakī (knowledge of Sānkhya, Yoga and Lokāyata), Trayi (Study of *Rk, Yajur, Sāmavedas*), Vārtā and Dandanītī are included in Vidyā, i.e., domains of specialized knowledge. Kautilya[11] further explains Vārtā as 'क षिपाशुपाल्ये वाणिज्या च वार्ता' (1.3.1) i.e., agriculture; husbandry and commerce are specific domains of *vārtā*. Knowledge of *vārtā* is said to bestow prosperity in the form of agricultural produce (धान्य), animals (पशु), gold, copper, etc. (हिरण्यकुप्य) and manpower to be used as servants/labourers[12] (विष्टि) (1.3.1). Surprisingly, this definition is not far from that used in modern Labour Laws of India and other countries where, agricultural/horticultural commodities, domesticated animals, employees engaged therein, seeds, wild commodities, agricultural marketing, etc., are described under the broad definition of agriculture. The fifth-sixth century AD Sanskrit lexicon *Amarkośa*[13] has classified agriculture, cattle rearing, dairy products, food processing, conservation and preservation related trade and commerce activities under *Vaiśyavarga* (2.9.1 – 111). It seems in agreement with *Arthaśāstra* in its definition of *Vārtā* as 'क षि: पाशुपाल्यं वाणिज्यं चेतिव त्तय:' (agriculture, cattle rearing and commerce are livelihood or profession, i.e. Vārtā/vritti) (2.9.1–2).[14] The *Brihat Saṃhitā* is a veritable storehouse of information pertaining to astrology and its impact on the important aspects of society and human life. Its chapters, titled Sasyajātakādhyāya[15] (सस्यजातकाध्याय:), Dravyaniśchayādhyāya[16] (द्रव्यनिश्चयाध्याय:), and Arghakāndādhyāya[17] (सस्यजातकाध्याय:), are devoted to agriculture, horticulture and their produce, cattles and trade and commerce of agricultural and horticultural products along with precious and semi-precious stones while describing the impact of different planets and stars in different constellations. Point to be noted is that herein also, agricultural activities are intimately associated with horticulture, husbandry and commerce (*Brihat Saṃhitā* chapters 40, 41, 42).

We can sum up that agriculture does not mean only crop harvesting of different cereals. Horticulture, fruits, vegetables, herbs, cattle rearing, dairy products and management of grains, fruits and dairy products all come within the purview of agricultural activities. This conclusion will help to isolate agricultural activities as portrayed in Indian art.

AGRICULTURE IN INDIAN ART

Indian art has two levels of reference: spiritual/symbolic and material/mundane sphere. They are not mutually exclusive in the Indian world view. Dichotomy of sacred and profane or spiritual and sensual are modern notions. Bulk of the art-evidences are religious in nature, but their subject-matter reflects both spiritual and material spheres. It is upto the onlooker how he/she approaches to interpret it. Therefore, it is not surprising that one can find mundane activities depicted in religious art. Depiction of agricultural activities are no exception.

We can divide Indian art-evidences depicting the agricultural scenes in two broad groups: (i) Direct depictions where agricultural scenes such as ploughing, husking, chaffing and winnowing, etc., are depicted. (ii) Indirect depictions such as cooking, eating scenes, picnicing, orchards, groves, drinking, eating vessels, cattle, fireplaces, hearths, etc., which throw significant light on the existence of agriculture.

Chronologically, we can divide art-evidences into following groups: (1) Prehistoric/Protohistoric evidences which include pottery sherds, stone tools specific to agricultural activities, hearths, cattle bone, pottery paintings, engravings, relief sculpture such as seals/sealings, etc., and sculptural evidences; (2) Historic evidences which can be divided into (A) pre-Mauryan, (B) Mauryan-Sunga, (C) Kushana-Gupta, and (D) Early medieval. Historic art-evidences are both sculptural and architectural in nature. It is important to note that stupa art and cave paintings are more important in so far as they retain a narrative art style which has more scope for depicting agricultural activities, whereas Gupta and post-Gupta/Early medieval art is more of an iconographic nature which does not leave much scope for narratives and detailed scenes of mundane life. But there are symbols, motifs, attributes which indirectly throw significant light on agricultural history.

EARLIEST FORMS OF AGRICULTURE: ART/ARCHAEOLOGY

As has been stated earlier, agriculture does not mean only crop harvesting, but also includes horticulture and cattle breeding. Therefore, generally speaking, the beginning of agriculture in its primitive archaic form belongs to that phase of human history described in archaeology as Mesolithic period. Hunting, gathering, foraging were the basis of subsistence. Domestication of animals along with hutments, collection of wild seeds of grains, handmade pottery with cord-impression, geometric designs and most importantly creation of rock paintings depicting agricultural activities—all these are intimately associated with the Mesolithic period. Evidence of microliths used to construct sickles and other tools used in crop harvesting throw significant light on the existence of agricultural activities.[18] Bagor (Rajasthan), Adamgarh (Hoshangabad), Chopanimando (Allahabad, U.P.), Damdama (Pratpagarh, U.P.), are some important excavated sites of Mesolithic cultures which throw significant light on the settlement pattern, hunting-foraging economy, domestication of animals, gathering of wild seeds of wheat, barley, etc., processing of wild seeds through quern-muller and floors of huts, indicating a settled life. Remains of domesticated animals like cow, bull, etc., along with quern-muller and storage pits and sherds of handmade pottery—all indicate the foreplay of the beginning of agriculture and settled life in the Mesolithic period.[19] Mesolithic rock-paintings also reflect the hunting-foraging-gathering way of life. Scene of agricultural activities are not depicted in Mesolithic age paintings. But some evidences of gathering wild grains and their processing are portrayed in Mesolithic art (Fig. 1). These art evidences provide a glimpse of Mesolithic hunter-gatherers in a stage where they have domesticated cattle, sheep and goat, and acquired the knowledge to process plant food, including wild seeds.

Fig. 1: Mesolithic Rock Paintings

AGRICULTURE IN THE NEOLITHIC PHASE

Neolithic phase constitutes the real beginning of agriculture in the true sense of the term. The earliest evidence of agricultural life based on wheat, barley, cattle, sheep and goat in the Indian subcontinent comes from Mehrgarh in Baluchistan dating approximately 7000 BC. In fact, Neolithic cultures coincide with the growth of well-planned villages, cultivation of both wheat, barley and rice along with other artifacts of great importance such as painted wheel-made pottery sherds.

Neolithic agriculture is now-a-days being analyzed and debated by several archaeologists from two important angles: (1) Wheat, barley cultivation concentrated mainly in the region of Baluchistan to Haryana and Gujarat; (2) Rice cultivation concentrated mainly in mid and lower Ganga valley and now the region of Orissa is also included in this earliest rice-growing belt. Both these points of view are also concerned with the analysis of outside impact/influence/association/diffusion particularly west Asia-Iran *vis-á-vis* wheat, barley cultivation and China, Southeast Asia *vis-á-vis* rice cultivation. Koldihwa has produced the earliest date of rice cultivation in India, which comes to 6000 BC.[20]

Neolithic phase art-evidences are very important for reconstructing earliest phase of agricultural history in India. Art mirrors the image of society both at physical and mental levels. This is true as far as the reflection of agriculture is concerned in Neolithic art. Neolithic art includes these specific group of art-evidences: (1) Pottery sherds, (2) Archaeological artifacts such as quern-muller, etc., and (3) Rock paintings, generally assigned to the second phase of rock paintings, i.e., Neolithic/Chalcolithic stage.

In the rock paintings of Mirzapur, the scenes belonging to Neolithic/Chalcolithic phase are painted over Ist phase (Mesolithic) scenes and often below the IIIrd phase (historical). They illustrate a settled society portraying activities like load-bearing, cattle breeding, group dances along with humped bulls.[21] There are several such groups of rock paintings across the length and breadth of India, particularly in the Vindhyan ranges of central India where one can find the glimpses of early stage of settled life and related activities. Unfortunately, it is very difficult to fix the chronological sequence of these rock paintings, as different layers are superimposed over each other. The broad division of these paintings into Mesolithic, Neolithic/Chalcolithic and Historic phases is not satisfactory. Particularly, the transitory phases sandwiched between mesolithic and chalcolithic layers desire much attention and careful perusal to isolate Neolithic from those of Chalcolithic paintings.

Continuous sequence of occupation in the Kachi/Bolan region from the Neolithic settlement of Mehrgarh, beginning around 7000 BC to the end of Iron Age deposits of Period III at Pirak has been discovered through scientific archaeological excavations.[22] Mehrgarh is considered as an art centre due to the discovery of several types of artifacts such as clay figurines, pottery sherds, glazed material objects, etc. At Mehrgarh, Periods IA, IB and II belong to Mesolithic culture. From these periods, steatite ornaments[23], crude clay figurines with schematic representations of a trunk and legs on both sides of bulky hips suggesting a sitting position have come to light.[24] These Neolithic art-evidences suggest the existence of a complex, well-settled society based on agricultural economy. Pirak and Kachi/Bolan regions suggest the existence of pre-Harappan culture going back to Neolithic period which, in due course of time, were absorbed in the Harappan culture. But now, archaeologists have begun questioning the classical model of uniform Harappan culture empire. Archaeological studies throw significant light on the existence, continuity and incorporation of original local traditions in several regions of Harappan culture. Kachi/Bolan and Pirak regions betray this tendency.[25]

AGRICULTURE IN THE CHALCOLITHIC PHASE

Chalcolithic cultures have produced a rich data of art-evidences relating to agricultural history. Mostly, they constitute pottery sherds and architectural remains shedding light on the complex structure of social organisation including urban way of life. Chalcolithic horizon of India is very vast. Now, most archaeologists are in favour of dividing Chalcolithic horizon in 3 distinct phases: (1) Pre-Harappan, (2) Harappan, and (3) cultures beyond Harappan distribution zone. Among them, the most important phase from the point of view of history of agriculture is the pre-Harappan phase concentrated in the northwestern part of the Indian subcontinent from Baluchistan to Haryana and Gujarat. This phase reveals the growth of villages and subsistence economy based on agriculture and related local and regional trade of raw material. The excavated sites have yielded large pottery sherds reflecting the distinctive regional styles. These pottery sherds throw significant light on the early history of agriculture. They help us in two ways—firstly, their shapes and size speak volumes about agricultural activities and secondly, they help us in reconstructing the agricultural history through the content of paintings on their outer surface. In the regions of Baluchistan and Sindh (Pakistan) and in Rajasthan, Haryana, Gujarat (India), several sites have been systematically excavated, throwing significant light on Pre-Harappan cultures. The following cultural zones have been proposed on the basis of archaeological findings:

1. Amri-Nal Culture
2. Quetta Culture
3. Kulli Culture
4. Zhob Culture
5. Kot Diji/Sothi Cultures
6. Banawali and other sites in Punjab/Haryana

These cultures evolved out of agricultural practices and developed into a complex pattern of settled life, culminating in the growth of villages. A wide range of pottery has come to light. Prominent forms are: storage jars, perforated cylindrical jars, beakers, bowls, shallow dishes, dish on stand, cups, pedestalled cups, plates, jugs, etc.[26] These vessels in themselves are important evidences of agricultural and other related activities such as storing, processing, cooking, and preserving various raw material availed through agriculture/horticulture. As artifacts, they speak volumes about the culinary art and aesthetics of their times. Cereals, grains, vegetables, fruits, herbs and other vegetal raw material were handled and processed with considerable skill and knowledge to create wide variety of foods and drinks (beverages, dairy products) which required vessels of various shapes and sizes—a fact also corroborated by archaeological excavations and methods.

More relevant from the point of the theme of this chapter is the painted subject matter of pre-Harappan pottery. Almost all the sites of pre-Harappan cultures have yielded a painted ware tradition with strong regional variations and styles. Two distinctive traditions are discernible from the analysis: (1) Buff ware having paintings in black colour; and (2) Red ware with black coloured paintings. Buff tradition has many shades and nuances such as pinkish white and yellowish white. A critical analysis of these evidences throws considerable light on the contemporary practice of agriculture and related activities. We were able to cull out the following facts relating to agriculture from the art of pottery painting of pre-Harappan horizon:

1. Amri-Nal ceramics contain some motifs which display loops and suspended lines suggesting tassels. Its variant is seen in the ceramic tradition of Harappan culture and that belonging to West Asian Chalcolithic horizon particularly from Samarra and Tell Halaf[27] (Fig. 2).

Fig. 2: Pre Harappan Pottery

2. Quetta ceramic tradition displays many motifs whose parallels can be found both in West Asia and Harappan cultures. They are simple in form and suggest highly stylized and abstract nature of representation relating to mundane life and objective reality. For example, the so-called opposed triangle motif is suggestive of both a geometric pattern or an abstract notion of some objectified reality, and also of stylized form of an animal or axe. Either way, it shows a relationship with agricultural way of life. Ethno-archaeology and anthropological studies show the prevalence of these motifs in agricultural communities across time and space. Opposed triangles could also suggest conical baskets or other such objects used in agricultural activities. Quetta ware also displays tree motif in a highly stylized manner. These tree motifs reveal close association of trees with agriculture as they provide wood both for fuel and equipments/huts/houses. It is a fact to be noted that tree worship is a common feature in agricultural societies across the world. Hence, the depiction of trees could also be taken as reflecting the agricultural/horticultural way of life[28] (Fig. 3).

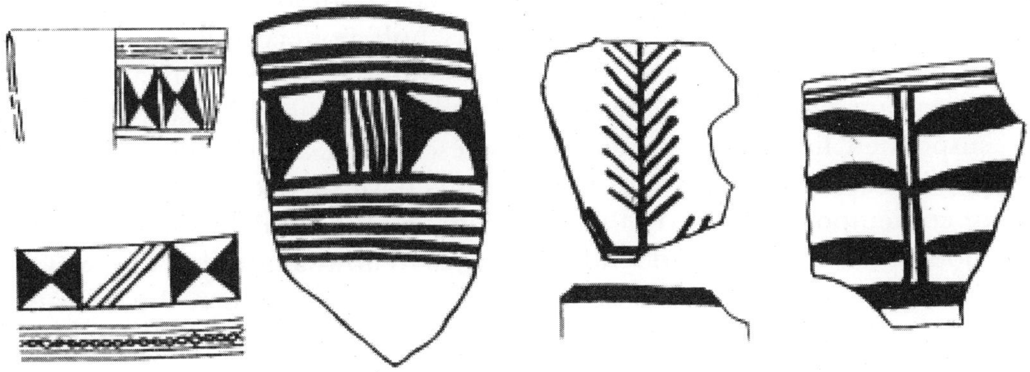

Fig. 3: Quetta ceramic tradition

3. Kulli or Kulli-Mehi culture has produced a distinct ceramic tradition having close parallels both in West Asiatic cultures and the Harappan culture. The pottery is typically buff or pinkish and designs are painted generally in black with occasional pale red or whitish slip also used. Instead of simple geometric and stylized forms, the Kulli ceramic tradition displays stereotype naturalistic scenes having a narrative quality. They consist of elongated humped cattle (probably a cow and a bull) depicted in a setting of stylized trees and minute stylized horned animals looking like goats. Various other symbols such as triangles, circles, W-and S-shaped designs, dotted eyelike forms, etc., produce a sense of overcrowding in the background. The whole composition is narrative in character, suggestive of religio-magical beliefs and practices of the agricultural society of Kulli. The elongated horned cattle, stylized forms of goats, pipal tree, stand like 'cult object', comb like symbol—they all betray parallels of Harappan pottery. Importance of cattle in agricultural society does not need any elaboration. The bull and cow are worshipped for their magical powers. Kulli was in close contact with Mesopotamia. A pot found at Susa depicts an almost similar scene with slightly different mannerism which suggests the local adaptation of Kulli ware designs probably by the artists from the Kulli itself[29] (Fig. 4).

Fig. 4: Kulli Ceramic Tradition

4. Zhob valley culture is famous for its red ware tradition, particularly for its more distinctive ceramic tradition described as 'bull' pottery by Ross and as Sur Jangal painted variant 2 by Fairservise at Rana Ghundai and Sur Jangal, respectively.[30] Important motifs are highly stylized abstraction like animal frieze, loop with suspended lines, angularised shapes resembling abstraction of human form, parallel lines, lozenges, etc. These motifs have survived in the living tradition of contemporary folk art of various regions of the Indian subcontinent. These motifs are so stylized and lend themselves to abstraction that one may not find any direct correlation with agricultural activities or culture. But the suspended loops and lines, lozenges are such motifs which could easily symbolize the multitude usages of crops—husks, corn, dried plants seeds, etc., which are still used for various religious and decorative purposes in folk art. Some pottery also display the symbol of snake and bird, possessing some mythological meaning. These motifs are correlated with the cult of the mother goddess also. Snake worship is a common feature in agricultural societies. Snakes protect crops by destroying rodents and mice. They also symbolise water and longevity of life. Birds are also important as they eat insects which are injurious to crops. The later version of Indian mythology pertaining to *Garuda* (Eagle) and snake portrays *Garuda* as the destroyer of snakes. Snakes are useful only when they could be controlled. Hence, this version of snake and bird motif as depicted in Zhob pottery is suggestive of the fragile balance of ecology experienced by the Zhob agricultural community[31] (Fig. 5).

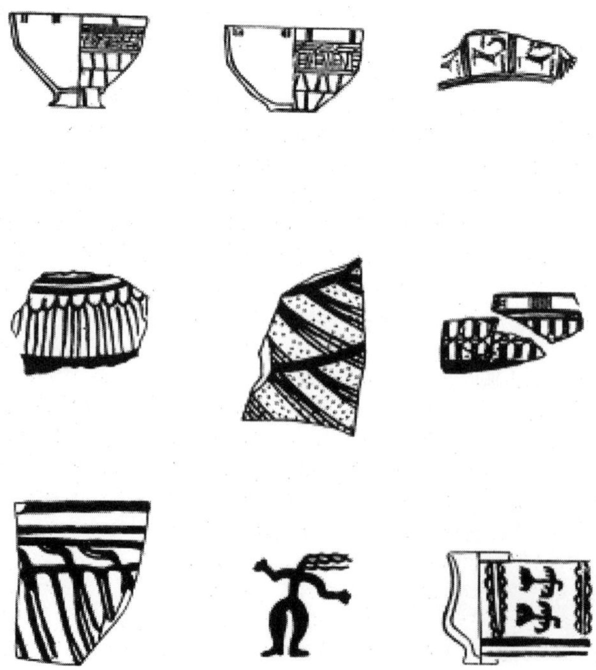

Fig. 5: Zhob Ceramic Tradition

5. Kot Diji and Sothi cultures have yielded a rich pottery assemblage from Cholistan, Sindh, and Rajasthan. Among painted designs, prominence is given to pipal leaves, fish scales and to the unique horned deity like animal with floral pattern[32] (Fig. 6). Scholars have identified two structural phases of Kot Diji/Sothi cultures at Kalibangan. Parts of a cultivated field have been excavated near the site. It has north-south furrow marks spaced at 1.9 m and the east-west marks at 30 cm (Fig. 7).[33] This evidence is very important for the history of agriculture as well as for the relevance of pottery art in the reconstruction of the agricultural history of ancient India.

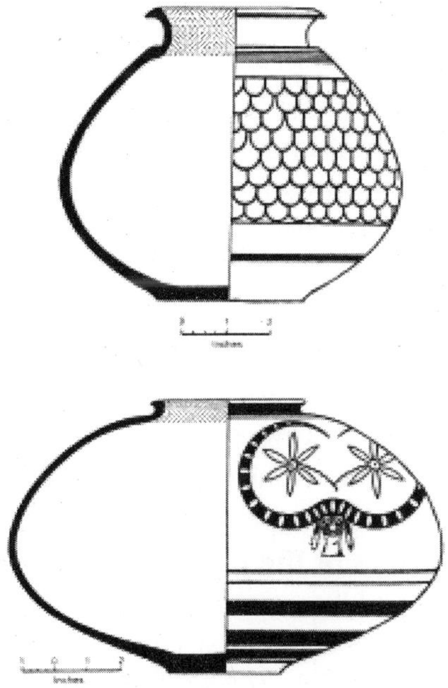

Fig. 6: Sothi/Kot Diji Ceramic Tradition

Fig. 7: Kalibangan (KLB2) Furrow Marks of Ploughed Land

6. Banawali and other places in Punjab and Haryana have yielded data throwing light on the beginning of agriculture and settled life in this region. Banawali pottery is similar to Kalibangan I type and the sherd shows the black painting of a canopied cart with spoke wheels.[34] Toy bullock carts have been discovered from various sites including Mohenjo-daro.[35] Pottery painting from Banawali is of greater importance as it portrays the canopy on the cart to protect goods from rain, sun, etc. Carts were important for local and regional trade of agricultural produces such as vegetables, grains, oils, etc. Plastered storage pits have also been unearthed in the courtyards of mud-brick houses.

AGRICULTURE AND THE HARAPPAN CIVILIZATION

The most important agriculture-related art evidences come from the horizon of Harappan civilization [Indus/Indus-Sarasvati are other names given to it]. This civilization was a complex urban civilization, but its economy was based both on trade, commerce, industry and agriculture. The discovery of granaries from Harappa, Mohenjo-daro, etc., reflects on the importance of agriculture in this urban civilization. We could divide art evidences in two distinct categories: (A) architectural remains, and (B) sculptural and pottery paintings.

(A) Architectural Evidences:

In this category, we shall describe those structures which directly or indirectly tell us about the history of agriculture. Two such structures have come to light from Mahenjo-daro and Harappa. They are described as granaries, i.e., huge storehouses or *godowns* where grains were stored for their use both in trade and monetary system. A warehouse-like structure has also been unearthed at Lothal. Existence of these granaries suggests the large-scale systematic agricultural industry. The largest granary has been discovered at Harappa[36] in the shape of a mud platfrom which is 52 by 42 metres in size and 1.2 metres high. On this huge mud platfrom two identical granary blocks were constructed each of 17 by 6 metres dimension, placed 7 metres apart and with 3 metre thick walls each block had six chambers with corridor like spaces between them. These chambers were approached by a short flight of steps and had opening only on the outside. Each chamber was divided into four storage spaces by full length walls. The eco-friendly scientific architecture displays ingenuity and a well-planned layout. The floor of the granary block rests on huge, large sleepers spaced apart so that air could circulate in the void below and enter the storage spaces of the chamber for proper aeration through small triangular vents. This design permitted free air circulation through the grain to avoid their destruction through pests and insects. The sheer size of the granary suggests some politico-economic mechanism of management, as also the fact that agriculture occupied a place of central importance in the economy of Harappan civilization (Fig. 8).

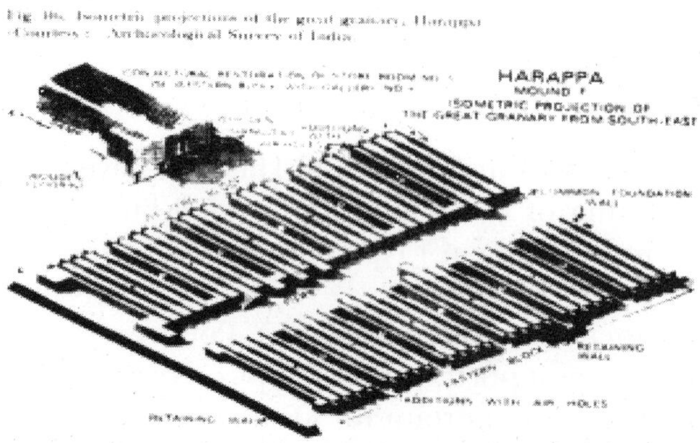

Fig. 8: Isometric Projections of the Great Granary, Harappa
(*Courtesy:* Archaeological Survey of India)

At Mohenjo-daro[37] the granary was constructed on a massive brick platform with steep sloping walls. On this platform bases of 27 mud-brick storage blocks arranged crosswise to facilitate air-circulation were found in the excavation. Built integrally with it in the north was a loading platform made of burnt bricks. It is believed that these storage blocks have had wooden superstructures filled with grain. The high-level granary was constructed to protect it from floods and other natural and man-made calamities. The construction of passages between the storage blocks is important from the point of aeration or free air circulation.[38]

At Lothal[39], a structure resembling a warehouse has come to light. It faces the dock to facilitate the passage of goods, particularly grains. The warehouse stands on a mud-brick podium which is 49 metres by 41 metres long and wide respectively and 4.1 metre high. This podium has an outer platform also. There are 12 solid blocks of mud brick on the podium separated by 1 metre wide criss-cross passage for free air circulation. These mud-brick blocks are 3.65 metre square and 0.91 metre high, and covered with wooden superstructures which were probably destroyed in the fire. Sixty-five terrocotta sealings with impressions of reed, woven fibre, matting and even twisted cords have been found in this complex, which suggests that 'packages of goods were examined and stored' in this complex (Fig 9).

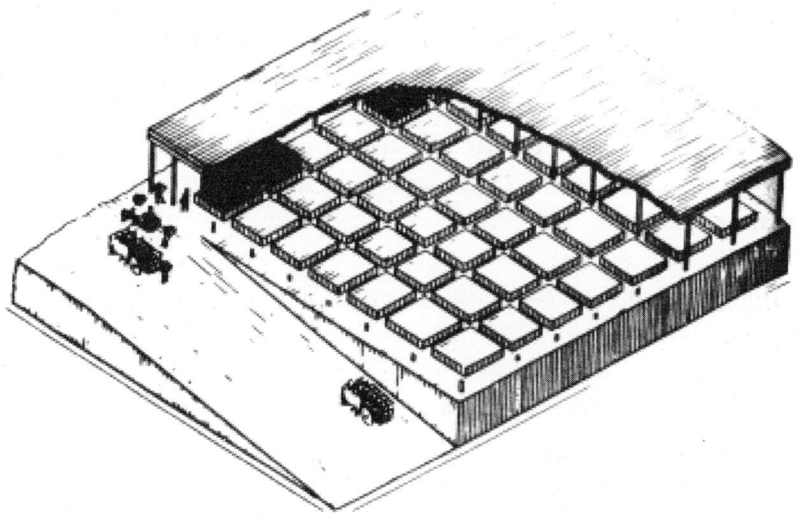

Fig. 9: Lothal Warehouse

Thus, the art of architecture, as revealed in the excavations of Mohenjo-daro, Harappa and Lothal, throws significant light on the history of agriculture relating to Indus-Hakra or Sindhu-Sarasvati culture of ancient India. Recently, Steven A. Weber[40] has analyzed the archaeobotanical data collected in the recent excavations (1993) through systematic flotation system to explain the Harappan subsistence pattern. Archaeobotanical data included pollen, seeds, charcoal, plant impressions and phytoliths.

The samples belong to two general periods: Period 3 B (2400–2350 BC) and Period 3 C (2180–2034 BC). He has derived the relative abundance value from the three principal means of measuring the archaeobotanical record: percentage of a given plant remains, its ubiquity and its density.[41] He has constructed a three-tiered hierarchical model of plant use which is like an inverted pyramid as given below:

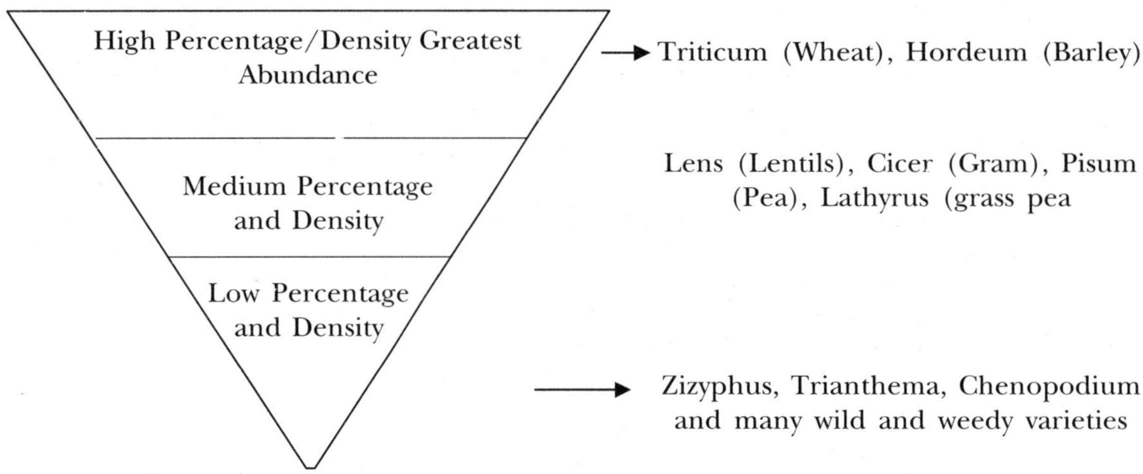

THREE-TIERED PLANT USE MODEL OF WINTER MONOCROPPING

This archaeobotanical analysis corroborates the interpretation of architectural evidences of Harappa, Mohenjo-daro, Lothal, etc., relating to granaries. This inverted pyramid model can be compared with the chart given below shedding light on the site-wise mapping of plant remains:

Site-wise Distribution of Crop and Plant Remains at Some Harappan Sites[42]
(After Chakrabarti)

Mohenjo-daro	Wheat, barley
Chanhudaro	Wheat, barley, mustard
Harappa	Wheat, barley, peas (*Pisum arvense*), sesame (*Sesamum indicum*), rice, and millets
Banawali	Basically unpublished, but wheat reported
Kalibangan	Wheat, barley, chickpea (*Cicer arietinum*), pea (*Pisum arvense*)
Rohira (district Sangrur, Indian *Punjab)*	Early Harappan context—modern *babul* (Acacia), *kareel* (*Capparis aphylla*), *jhau* (*Tamarix dioica*), teak (*Tectona grandis*), toon (*Cedrella toona*), khirani (*Manilkara hexandra*), and deodar (*Cedrus deodara*) woods; modern henna or *mehendi* (*Lawsonia inermis*) and grapevine (*Vinus vitifera*); mature Harappan context—in addition to *babul*

	and *jhau* trees and grape, *khejri* (*Prosopis spicigera*) and *sheesham* (*Dalbergia* sp.) woods, *ber* (*Zizyphus jujube*) and *harsingar/shephalika* flower
Mahorana (district Sangrur, Indian Punjab)	In addition to some of the things found at Rohira, an important find is hyacinth bean (*Dolichos lablab*), which needs 'frequent irrigation'
Hulas (Saharanpur district, UP Doab)	Wheat, barley, wild *(Oryza rupifogon)* and cultivated rice (*Oryza sativa*) as impressions on potsherds, sorghum and ragi millets, cotton, castor, almond, walnut and a variety of pulses, including *Dolichos biflorus, Pisum arvense, Pisum sativum, Lathyrus sativas, Vigan radiatus,* and *Vigna mungo*
Lothal	*Babul, siris* (*Albizzia* sp.), teak (*Tectona grandis*), *haldu* (*Adina cordifolia*) and *Rohini* (*Soymida febrifuga*) wood; rice (identified with certainty in the Lothal report), and possibly, millet and sesame
Surkotada	Mainly Italian millet (*Setaria Italica*) and ragi millet (*Eleusing coracana)* in addition to a large variety of wild grass and plant species
Shikarpur (Kutchh)	Wheat, ragi millet and Italian millet in addition to woods inclusive of silk-cotton (*Salmalia malabarica*) and sal (*Shorea robusta*)
Rojdi	A variety of millets inclusive of *ragi,* Italian and sorghum millets and a variety of wild grasses and plants, some of which can be used as fodder
Kuntasi	Wheat, barley, a variety of millets including Panicum, Italian and Kodo millets, and Job's tears, the seeds of which were probaby used for beads
Shortughai	Barley, wheat, Panicum millet, lentil, pea, almond, pistachio, grapes and linseed.

Some archaeologists have suggested the idea of Mohenjo-daro being the principal pilgrimage centre/sacred city of this civilization on the basis of cosmology, as reflected in the eco-social environment of city planning, particularly the citadel section where the great bath and granary were built. According to them the agricultural products had been collected at Mohenjo-daro for religious offerings as well as for trading.[43] They have suggested the religious nature of the citadel area in general and that of granary and the great bath in particular. In their view, though the function of the granary appears to have been the preservation of agricultural produce, the main function seems to have been a symbolic display of authority and power. This reflects the religio-social dimension of agriculture in the complex urban structure of this civilization. In the recent excavations conducted in 1994 and 1995 field seasons, Harappa has disclosed various architectural remains pertaining to Periods 2 and 3 A, B and C (Early Harappan and Harappan). They include circular hearths, storage pot set into the floor, semicircular brick bins, etc. Semicircular bins were probably used for setting a grinding stone. A cloth draped over the bricks could have been used to collect flour or other food stuffs that may have been processed with such a grinding, stone.[44] As the biggest granary has been discovered from this site, existence of such food processing area could add new meaning to this structure.

(B) Sculpture and Pottery Painting

Harappan civilization has yielded a wide range of sculptural artifacts, including terracotta, metal figurines, seals and sealings and pottery sherds along with paintings. They throw significant light on the socio-economic and religio-cultural spheres of this complex civilization. The history of agriculture is no exception. A terracotta toy plough (Fig. 10) has been discovered from the Banawali (Hissar in Haryana) relating to mature phase of Harappan civilization.[45] Explorations/excavations conducted in the Cholistan (Pakistan) region have also yielded broken fragments of teracotta ploughs. Pre-Harappan horizon of Kalibangan has disclosed the ploughed fields dating back to before 2800 BC. Beautiful clay models of the plough have also been discovered at the site of Mohenjo-daro.[46] Preponderance of terracotta figurines of bulls suggest its importance in the rural-uban setting of Indus civilization. The bull was useful both in agriculture and trade. Clay models of bullock carts (Fig. 11a, b, c) also suggest the centrality of bull in the economic sphere of this civilization. One of the most depicted figures on seals

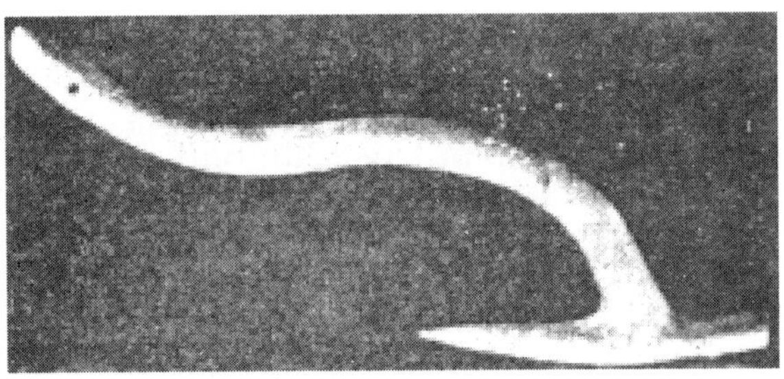

Fig. 10: Terracotta Toy Plough, Banawali

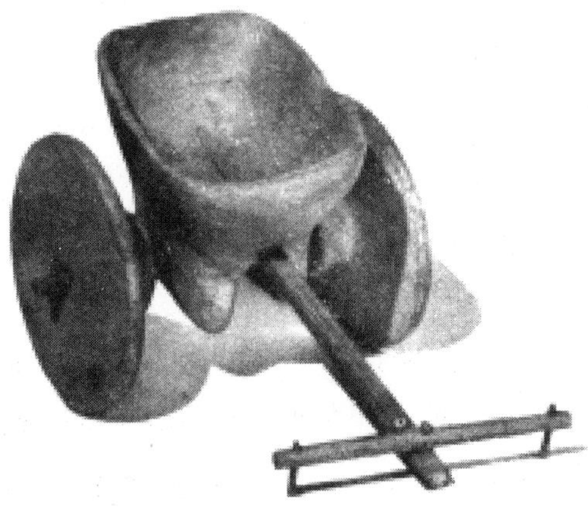

Fig. 11a: Clay Models of Bullock Carts

and sealings is also bull—both short-horned and Brahmani bull types which suggests its importance in the socio-religious sphere of this civilization.[47] The naturalistic rendering of lines, suggestive of muscular tension and movement of these creatures, is accepted as the evidence of highly skilled technique and aesthetics of Indus art.

Fig. 11b

Fig. 11c

Terracotta bull figurines recovered from Mohenjo-daro and Kalibangan are considered as excellent examples of Indus Plastic art. Some seals also portray buffalo figures in a naturalistic style. Buffaloes were also used in agricultural activities (Fig. 12) besides for dairy products.[48]

Seals representing plant forms are rare. Depiction of tree is seen on number of seals. Reference may be made to the depiction of Pipal tree in association with some mythological narratives. One such seal depicts pipal tree issuing from a jarlike object held by to unicorn like mythical creatures' conjoint heads[49] (Fig. 13). The whole motif

Fig. 12: Buffaloe-cart-Bronze Diamabad

Fig. 13: Harappan Seal

betrays strong resemblance with later Buddhist motif of 'Triratna' variety. Another seal portrays an upside down female figure with a plant issuing from her womb.[50] Some seals depict elaborate ritualistic scenes where a deity is depicted between the branches of Pipal tree, faced by a supplicating figure. In the foreground, a row of seven standing figures with plumed headdresses has been depicted which suggests the religious ritual being performed in front of the deity.[51] These depictions are indicative of the existence of a tree cult associated with fertility. Sociologists and cultural anthropologists tell us about the existence of fertility cult ritual in agricultural societies. Therefore, these cultic/ritualistic depictions indirectly throw some light on the agricultural society of Indus period. Pipal trees were necessarily planted in villages. More direct evidences on horticulture may be considered as those depictions which portray trees surrounded by a railing or platform.[52]

Harappan painted pottery reveals both the distinctiveness of artistic tradition and fundamental affinity with the pre-Harappan wares of Baluchistan and Sindh and also with West Asian ethos. Pottery sherds belong to the categories of the black-on-red, red, gray, buff and, black-and-red pottery traditions. Only 10% of the total pottery material belongs to the painted varieties in all the major sites. There are mainly black paintings on red surface, chocolate/purple black on buff, and white/creamy on the black-and-red surface. Designs are generally geometric but there are naturalistic to conventionalised animal and plant representations as well with occasional examples of human figures.[53]

The checkerboard with its variations is a common motif on pottery sherds. An interesting example of this motif is represented in a complex pattern on a sherd from Harappa.[54] It has narrative quality with human and animal figures (Fig. 14). Mackay has suggested that checkerboard patterns have been derived from basketry.[55] Baskets of various shapes and size are an essential part of agricultural societies. Another typical design is described as the comb motif, which is an abstracted or conventionalized form of loop and suspended thread motif seen on pre-Harappan wares. An intersecting cricle motif is an excellent geometric design used on jars and pots.[56] Another design is like block printing motif of seed or rosette like form.

Fig. 14: Pottery Sherd from Harappa

Harappan plant motifs are naturalistic and constitute a purely local development within the Indus valley. Pipal tree, pipal leaves, spiky tree with a shield-like shape, lanceolate leaf along with roseattes, peacocks, fish scale patterns, seven-petalled flower, wavy lines, etc., are prominently represented on the painted pottery of this culture.[57] Several abstracted or conventional forms would have been representational but their context of meaning is lost to us. Sigma and chevron designs were very common which may have some representational/symbolic meaning attached to them (Fig. 15). A sherd from Mohenjodaro portrays highly conventionalised animal figures amidst plant motifs. Another sherd from the same site has conventionalised figure of an animal, tree and banana plant like leaves (Fig. 16).

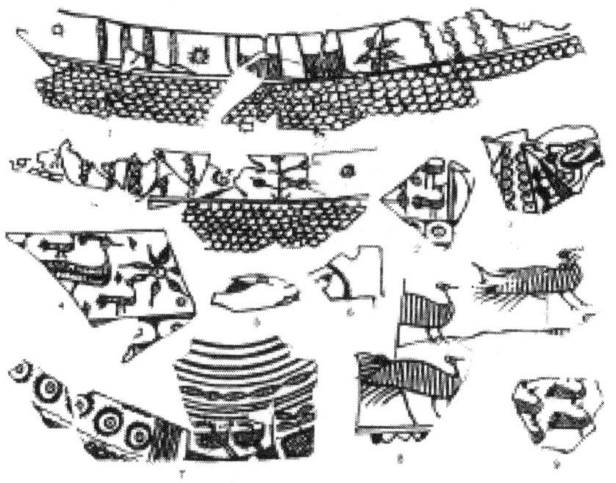

Fig. 15: Harappa Ceramic Tradition

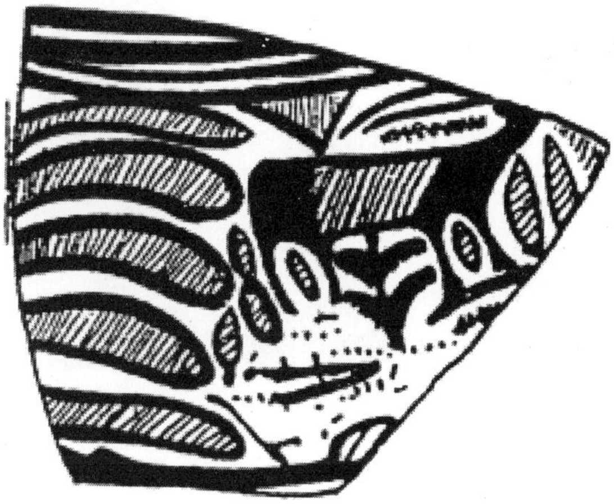

Fig. 16: Mohenjo-daro Pottery Sherd

From the above discussion of art-evidences it appears that flora and fauna constituted an integral part of human life in Harappan civilization. One cannot talk about agriculture solely as crop raising enterprise. Husbandry, horticulture and processing of their produce were essential parts of agriculture as there exists an intricate web of interrelatedness between agriculture and these abovementioned activities. There is no doubt in the existence of full-fledged agriculture in Indus society. Big storage jars, variety of vessels, copper frying pans, saddle querns and cylindrical rollers, grinding stones, platform for pounding grains they all shed light on the food processing, cooking and storage habits of this civilization. Archaeobotanical evidences and figurative art both corroborate the existence of highly developed agricultural activities in Harappan civilization (Fig. 17). Saddle-quern, hearths, a terracotta figurine from Mohenjo-daro kneading flour and carbonized grains from various sites throw valuable light on the history of agriculture (Fig. 18). Thus on the basis of archaeological artifacts and other art evidences one can safely infer the existence of systematic and complex agricultural activities in Harappan times, including both agriculture, horticulture and animal husbandry.

Between the protohistoric Harappan horizon and the beginnings of early historic India there exists a complex web of cultural pattern across the length and breadth of the Indian subcontinent, which, in the lack of any other suitable definition, is generally described as "Neolithic-Chalcolithic/Iron-bearing Cultures beyond the Harappan Distribution Zone".[58] The above archaeological term is used to map out the emergence and consolidation of early village farming communities of non-Harappan India to explain the foundations of early historic India. It is important to take note of following facts in relation with the history of agriculture:

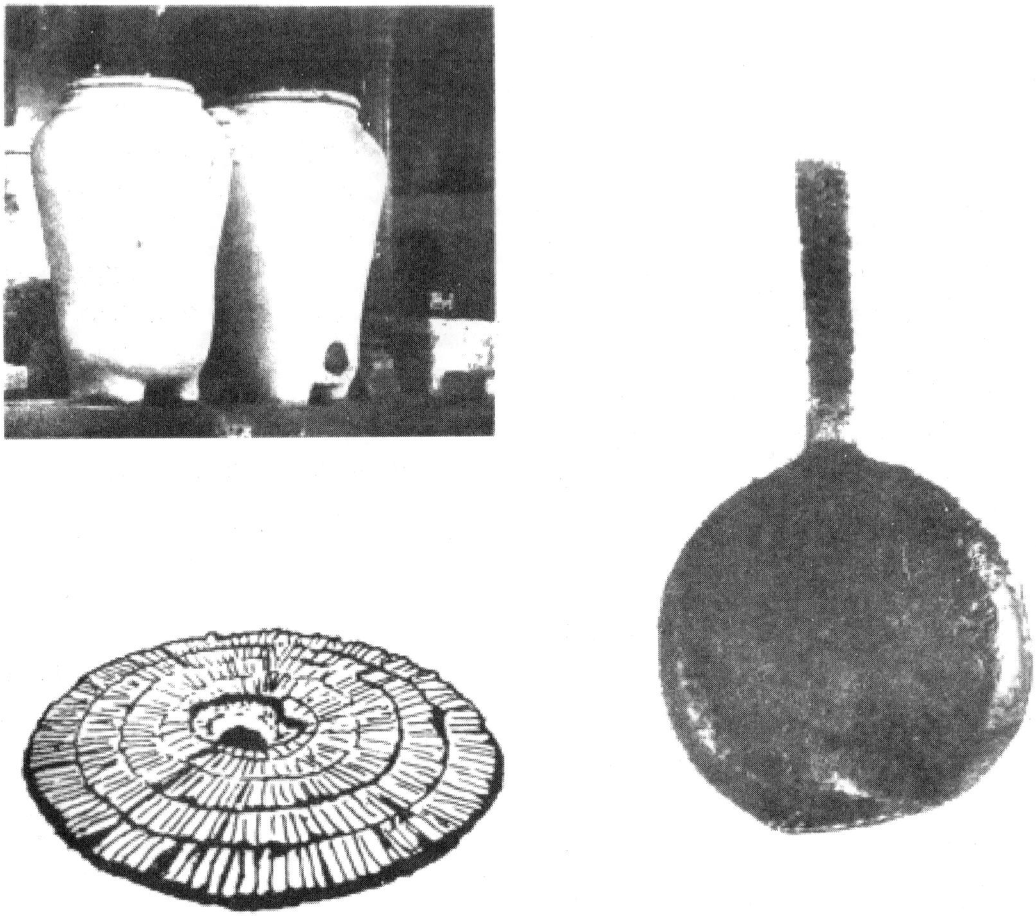

Fig. 17: Harappan Storage Jars, Bin, Copper-pan

Fig. 18: Harappan Saddle-quern and Terracotta Figurine Kneading Dough

(i) Interaction of Harappans with the cultures of this vast area both directly and indirectly.

(ii) Archaeological evidences of Domesticated rice from Koldihawa (Belan Valley, UP), indicating the emergence of rice cultivation in mainland paralleling the wheat-barley-cattle-sheep-goat strand of food-production of the Baluchistan uplands.

(iii) Complex archaeo-cultural sequence showing overlapping, late Harappans' and OCP horizon along with intricate pattern of black-and-red ware level succeeded by PGW (Painted Gray Ware) phase in Indo-Gangetic divide including Ganga-Yamuna Doab.[59]

(iv) Emergence of agriculture based on two crops in trans Sarayu plain during the first half of the second millennium BC as proved from the archaeobotanical remains of this period from Imlidih, Narhan and Sohgaura.[60]

(v) Iron technology (1400–1300 BC), fortified walls, embankments, ditches—all these archaeological facts indicate the foundations of early historic India. In Ganga Plain, the next phase is marked by NBP (Northern Black Polished Ware) and with this emerges the early historical period generally described as the period of 'second urbanisation' or 'pre-Mauryan' phase of politico-cultural dimension.

As the theme of the article relates to art evidences throwing light on history of agriculture, therefore, only artifacts will be discussed. The most prolific evidences come from PGW phase in the form of various vessels such as bowls, dishes, basins with black geometrical and sun/floral designs[61] (Fig. 19). Beads of ivory and semi-precious stones,

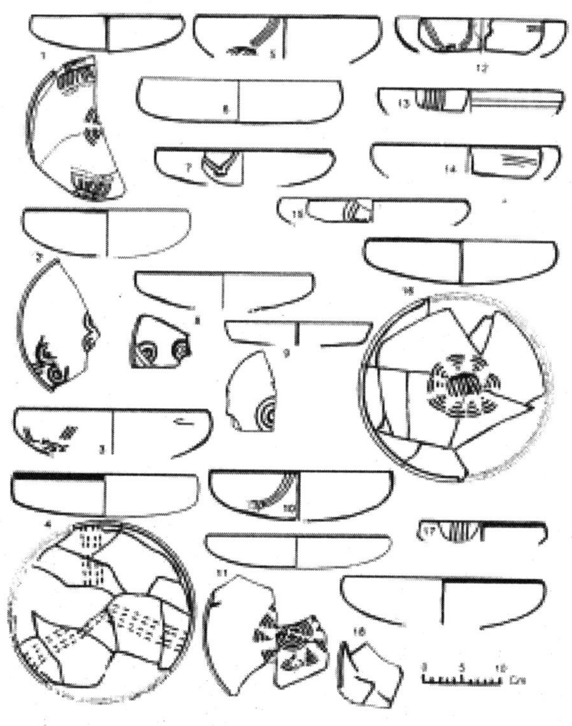

Fig. 19: Painted Gray Ware Pottery Forms and Patterns

lapis lazuli, ritualistic firepits, terracotta figurines, underground storage pits, overground clay storage bins, existence of moats, water channels, wells indicative of local irrigation systems for double cropping village-farming economy—all these characteristic features suggest the growth and development of agriculture in the post-Harappan phase of India's archaeological history (1500 BC onward) culminating at the first chapter of early historic India, i.e. pre-Mauryan phase. Archaeological and art remains definitely belonging to pre-Mauryan date are few and far between. Examples of terracotta of the mother goddess type generally classified as pre-Mauryan on stylistic grounds are important from two points of view. Firstly, they provide a link between Indus Valley figurines of the Mother goddess and those belonging to Sunga-Satavahana sculptures of Yakshini and/ or Mother godesses. For example, a pre-Mauryan terracotta figurine recovered from Mathura and presently in Boston Museum of fine Arts[62] is closely related—both iconographically and stylistically—to the earlier tradition of Mother goddess found not only in India but all over the near east. The bold display of the features of fecundity such as exaggerated and broad pelvis and heavy breasts are meant to suggest her fertility cult association, which in itself, is one of the most important characteristic features of agricultural society all across the world. Secondly, they symbolise the fertility cult tradition of agricultural society.

A number of tumuli or stupa-like solid earthen mounds have been discovered and excavated at Lauriya Nandangarah in Bihar. These mounds are described as the burial sites having religio-political connection of pre-Mauryan funerary beliefs and practices echo of which may be found in Ṛgveda itself.[63] From these mounds, two gold *repousse* figures have been recovered in a systematic excavation.[64] These small figures portray a nude Mother goddess with emphatic display of an enormously broad pelvis and heavy breasts, suggestive of fertility attributes. These small statuettes, along with pre-Mauryan terracottas, suggest the ritualistic and symbolic nature of art representing the cult of Mother goddess/fertility rites prevalent in agricultural society. The earth and its fertility are worshipped and ritually enhanced through religio-magical practices to protect the environment and agricultural production. Thus, these Mother goddess figures could be described as art-evidences throwing significant light on agricultural rituals and practices.

AGRICULTURE AND MAURYAN ART

Mauryan art seems to be a turning point in the long history of ancient Indian art. Within it, several new forms and transformations start appearing. The most revolutionary feature of Mauryan art is the use of Chunar sandstone in various forms as well as mirror-like polish on the surface. Another noteworthy feature is its royal court patronage having a direct impact on the form and technique of Mauryan art. Even folk art idiom, running parallel to court art, took stone as its medium. There does not occur any direct depiction of agricultural scenes in Mauryan art. But the continuity of the earlier mother goddess tradition in terracotta, along with Yaksha-Yakshini life-size stone sculptures, suggest the existence of fertility cult in the society. The earth was propitiated as mother goddess in order to yield rich crops and prosperity. Another feature reflecting the religio-magical practices prevalent in agricultural society is the evidence generally described as ring stones/disc stones recovered from several sites stretching from Taxila to Patna. In all, sixty-four of these are known till date. These carved ring stones are exquisitely carved on the obverse in jewel-like workmanship. These carved ring stones are possibly the earliest examples in which human, plant and animal life co-mingle within a framework of geometric patterns.[65] Nude figures of the Mother goddess are depicted amidst lotus flowers, geometric patterns, palm trees, running frieze of real and mythical animals such as winged creatures, stags, alligators, lion, elephants, horse, goose, lizards, etc., and in some cases, accompanied by nude males or males wearing kilt-like garments. Marshall was of the view that these ring stones were directly related to the earlier tradition of yoni worship in Harappan civilization.[66] Colonel Gordon and Banerjea have also endorsed Marshall's view.[67] Banerjea also suggested their association with cult objects like Sri-Yantra/ Vishnupattas/Ayagpattas of the Saktas, Vaishnava and Jainas respectively.[68] Pramod Chandra and Moti Chandra have drawn the attention to the West Asiatic, particularly Bobylonian parallels of the nude Mother goddess in context of these ring stones.[69] Most scholars are in agreement on the identification of the nude female deity (i) as Mother goddess symbolizing Earth or Goddess Prithvi; (ii) as a cult object of fertility rites performed in association to agricultural activities and birth process; (iii) as the great Mother Goddess symbolizing death and rebirth; and (iv) as heavenly female patronizing deity of prosperity and abundance associated with rain/water symbology and its life-giving, life-nourishing quality. Anthropological surveys have proved the continuity of such cult objects in modern village communities of India and other countries. Hence, we could conclusively take these ringstones (Fig. 20) suggestive of agricultural religio-magical beliefs and practices of Mauryan phase of Indian History, though some of these ring stones may belong to second or first centuries BC.

Fig. 20: Ringstones

Animal figures found on Asokan pillars have multivariate meanings and symbology attached to them. On the one hand, they are related to the Vedic Cosmology and Solar cult, on the other, they are direct descendent of prehistoric tradition of animal worship such as discovered in the pre-Harappan, Harappan and other Neolithic-Chalcolithic cultures of the Indian subcontinent. The Rampurwa Bull betrays strong Indian link, both iconographically and stylistically to Harappan Bulls depicted on seals and sealings (Fig. 21). From prehistoric times, the bull was intimately associated with fertility cult, concept of

Fig. 21: Asokan Pillar Bull Capital, Rampurva

virility and reproductive power all over the world. In Indian village life till date Bulls are treated with religio magical reverence and awe, particularly, in Saiva sects. Therefore, depiction of bulls on Asokan pillar capitals—as topmost animal figure as well as on the border of abacus—may be described as art-evidence directly related to agricultural practices and beliefs.

Lifesize statues of Yaksha and Yakshini have been discovered from several historical sites such as Mathura, Patna, Paways, etc.[70] (Fig. 22). They belong to folk religion and beliefs being formed over the time and consolidated legacies of successive waves of human cultures ever since the first phase of prehistoric beginnings of community life. They symbolize worldly power, prosperity and well-being. Health-wealth syndrome of popular beliefs were expressed through these Yaksha-Yakshini figures. They have

Fig. 22: Mauryan-Sunga Yaksha-Yakshini figures

multilinear meanings such as solar and water symbology, supernatural power and semi-divine nature associated with them. They were popular both in rural/agricultural and urban/money-economy related communities.

AGRICULTURE AND POST-MAURYAN/SUNGA ART

We come to concrete art-evidences relating to history of agriculture in India with the art tradition of post-Mauryan or Sunga period. This phase is generally described as the emergence of the first classical idiom of ancient Indian art which gave expression to popular Buddhist religion. Monumental stupa art emerged in a pan-Indian fashion across the length and breadth of the Indian subcontinent, including Ceylon. Important centres in northern India are Bharahut, Sanchi, Bodhgaya, Kaushambi, Mathura and, in southern part of the country, Amaravati and Nagarjunakonda were two most important centres, besides several others.

The collective art-evidences of these centres have portrayed in a narrative form several religio-social customs, beliefs along with mundane activities of general mass of humanity located in rural and urban settings. Besides these narratives, the realistic portrayal of rich flora and fauna sheds a significant light on the history of agriculture as depicted in ancient Indian art. We could divide art-evidences relating to agriculture in three distinct groups: (i) direct evidences depicting ploughing, harvesting, groves, orchards, fruits, animal herds, cattle; (ii) household activities and vessels relating to agricultural processing, dehusking, pounding and cooking; and (iii) scenes of merriment and enjoyment depicting various cooked foods and drinks.

Direct evidences relating to agriculture, horticulture and animal husbandry are largely depicted on the railings and gateways of Bharahut, Sanchi, Bodhgaya and earliest phase of Amaravati. Rajaram Hegde has reproduced a line drawing of plough on the basis of a Bodhgaya panel.[71] The panel clearly depicts a farmer engaged in ploughing with the help of Indian variety of plough pulled by two harnessed bulls. The farmer is holding a stick in his left hand while with the right hand he is controlling the plough (Fig. 23). Three parts of this type of plough are clearly visible in the panel: (i) A bent stock with elongated upper end for stilt which is lower than the tiller to

Fig. 23: Bodhgaya Panel Sculpture Depicting Ploughing of a Field

facilitate the process of pressing it while ploughing the field; (ii) The draught-beam fastened to the stock by inserting it into the hole in the stock; and (iii) the yoke tied to the draught beam and kept on the neck of the draught-bulls. Another panel from Bodhgaya itself portrays a farmer carrying a bent plough-stock without the draught beam.[72] This suggests that stock and draught-beam were joined together while tilling the land. They were detachable separate parts.

The above Bodhgaya panel, depicting the Indian plough, is very important from two points of view in relation with the history of agriculture as reflected in ancient Indian art. Firstly, after Harappan toy model this illustration seems to represent the early Indian plough used in second-first centuries BC. Secondly, this illustration shows the continuity of form and technique of plough and ploughing from protohistoric times till date as it provides a visual link documented in art between prehistoric times and second-first century BC on the one hand, and on the other, between early historic times and the present-day method of tilling the land. *Jātakās* and *Aṣṭādhyāyī* of Panini shed significant light on the technical names of various parts of plough. For example, they describe plough as *hala, sira* and *langala.* Bent stock was called *potra,* and the plough share at the tip was called *phala* or *Kusi.* These literary sources often describe *aya-phala,* i.e., plough-share made of iron. Draughtbeam was known as *isa* and yokes were called *yugas* or *dhuras. Yotra* or *Yoktra* was the rope used for fastening draught-bulls to the yoke.[73] Other than plough, various other tools and agricultural implements are represented in Sunga art. Spades, hammers, chisels, saws, pincers are used in different contexts of Jātaka stories depicted on railings. Bharhut sculptures contain many such agricultural tools which were used in rural economy for making different devices from wood and stone.[74]

Sanchi gateways depict a realistic mundane life of contemporary society which throws much light on agriculture, horticulture and husbandry practiced in those times. There are some elaborate scenes which portray the rural economy based on cultivation and husbandry. On the eastern gate of Sanchi at its south pillar-inner face-upper panel, there is a depiction of village life with all its nuances.[75] (Fig. 24). In the foreground

Fig. 24: Sanchi Sculpture Showing a Village Life

are shown animal herds, particularly buffaloes, bulls, goats and sheep amidst creepers and lotus flowers, indicating a pond. This scene is described as the village of Uruvila where the Buddha broke his fast. An empty platform under a *Chatra* or umbrella suggests the Buddha's association. The depiction of cattle is so realistic that it makes the whole ethos of village life and animal husbandry come alive before the eyes. Cattle played an important role in village economy both in cultivation and dairy farming.

At the west gate of Sanchi, there is another scene depicting buffaloes muddying a pool. In the background, small circular huts are depicted along with some archers while representing a *Jātaka* story.[76] Importance of buffaloes in the village life gets confirmed by this artistic portrayal. Fragment of a *torana* architrave discovered from Kaushambi district and presently kept in Allahabad Museum has a figure of Gajalakshmi standing amidst lotus creepers accompained by a majestic bull on her proper left.[77] The figure of bull betrays a strong resemblance with the bull carved on Rampurva bull capital of Asoka. This architrave suggests the religious background of bull associated with the popular folk beliefs of agricultural society. Besides these two Sanchi scenes, there are numerous representations of bulls and cattle in Bharahut and Sanchi sculptures. Amravati and Nagarjunakonda sculptures also depict bulls resembling those portrayed in Sanchi and Bharhut reliefs.[78] A beautiful bull capital from Manhai in Kaushambi district has been recovered. The majestic bull is shown wearing a jewelled garland suggesting its religious status.[79]

Thus, we may conclude that animal husbandry not only constituted an important part of agricultural society but was also responsible for introducing animal worship, fertility cults and several other religio-magical beliefs and practices absorbed in folk religion. Other than bulls and buffaloes, we find depictions of horse, camel, rat, squirrel, elephant, antelope, monkey, peacock, parrot, swan, fish, dog, crow, cock, stags, etc., in the sculptures of Sunga period.[80] These animals were an integral part of village life.

Flora constitutes a major part of Sunga sculptures. Many varieties of trees and flowers are depicted to represent the place of nature in human society and their mutual interdependence. Śālbhanjikā of the gateways at Sanchi are depicted as holding the branch of mango tree with bunches of mangoes hanging from them.[81] The rich variety of fauna depicted on Bharahut, Sanchi, Amaravati sculptures suggests the important role played by horticulture in the economy of village life. Tree worship was a prehistoric tradition. Harappan seals and sealings represent the religious background of the Pipal tree. Buddhism is a religion where different kinds of trees and their groves are associated with the great events of the Buddha's life. Bharahut Vedica sculptures contain depiction of specific trees associated with Manuṣi Buddhas or Purva-Buddhas. For example, Pipal is associated with Gautam a Buddha, Vatavriksha with Kashyapa, Udumbar with Kanakamuni, Patali with Vipasmin, Salavriksha with Vishvabhu, Sirisa with Krakucchanda Buddha.[82] This shows the cult of tree worship which was duly integrated with the popular practices of Buddhism so as to create a bridge between folk religion and Buddhist sects. From our point of view this evidence is important in that it reveals horticultural ethos of contemporary society. As the scenes are depicted in rural/urban settings, therefore, we can conclude that trees played an important role in the socio-religious and economic spheres of contemporary life. Trees were planted both for ornamentation and utilitarian purposes such as gaining flowers and fruits for medicinal and culinary delights. Jetavana (Fig. 25) and Ghoshitaram monasteries were attached with mango groves as Bharahut and Sanchi depictions bear witness to it.[83] Śālbhanjikā/Dohada motif (Fig. 26) is suggestive of fertility cult associated with female/earth as the symbol of reproductive power. In the magico-sexual rites, women are considered to increase the fertility of the soil and vegetation. They are involved in various such rites to bring rain and to enhance agricultural fruitfulness.

Fig. 25: Bharhut Jetavana Scene

Fig. 26: Sanchi Śālabhanjikā Motif Showing Mango Fruits

Anthropological studies suggest that in agricultural societies, cultivation of the soil is connected with the child-bearing function of women. Hence, Śālabhanjikā-Dohada motif and several other rituals were performed to enhance the power of fertility both in women and trees. The notion that a woman fertilizes trees and trees increase and protect female reproductive power has been deeply rooted in India's ancient tradition.[84] Bharahut and Sanchi reliefs have numerous depictions showing varieties of fruits. Among them, banana, mango, grapes, custard apple and jackfruits are noteworthy.[85] In Bharahut, custard apple is shown issuing from the wish-fulfilling creeper along with realistic leaves of the tree.[86] In one of the *Jataka* scenes on Bharahut railing, there is a beautiful depiction of a banana plant with clear portrayal of shaft, leaves and bunches of furits.[87] Both Bharahut and Sanchi sculptures contain realistic depiction of grape vines (Fig. 27) along with bunches of grapes.[88] At Sanchi on North Gate lower architrave, and on Western Gate garden scenes are depicted with banana plantation.[89] Ripe mangoes and jackfruits are shown issuing from Kalapalata or creeper scrolls.[90] Besides these depictions of groves, trees, plants and fruits,

Fig. 27: Bharhut Grape-Vines

there are certain scenes which portray the process of housekeeping, processing of grains, cooking and eating, etc. At Sanchi, one of the scenes on Torana depicts the natural surrounding of a village life (Fig. 24) along with household chores shedding light on agriculture. In the foreground, buffaloes, bulls, goats and sheep are depicted amidst foliage and creepers. In the middle, an empty seat with an umbrella occupies the place of honour. Keeping with the works relating to agriculture and husbandry, a person is shown carrying a staff on his right shoulder and bundle of rope in his left hand. In the background on the right side multistoreyed buildings are depicted. On the left side, two huts are shown. In front of the hut women are busy with domestic chores such as grinding either herbs or spices, pounding grain, probably rice in a mortar and winnowing or dehusking paddy with a *chaaj* (winnowing tray). A chaaj with its sides rounded, saddle and quern also find place in this realistic domestic scene. The saddle is four-legged and oblong, while the quern is cylindrical. A tall tripod-like object is depicted on which a woman is pounding something or rolling the cake. It is noteworthy that the lady is rolling the cake/chapati in standing position. Shape of the mortar is like an hour glass and seems to be made of stone. The pestle is like a heavy staff and the lady is shown standing while pounding or dehusking the paddy.[91] Baskets full of fruits and grains, various types of vessels and pots such as jugs, wine pot, spouted vessels, bowls, deep bowls with heaped food, laddles, jars, pots hanged on a pole with net of ropes, pots hanging from roof in a net, bullock carts — all shed significant light on eating habits relating to agriculture and horticulture/ dairy farming.[92] Thus one finds numerous scenes depicting mundane spheres of daily life such as hunting, eating cooking (Fig. 28), and other acts relating to agriculture, horticulture and dairy farming. Bharahut and Sanchi sculptures are veritable storehouse of such scenes informing us about domestic life located in rural-urban settings. K. Krishna Murthy[93] has

Fig. 28: Bharhut Jatak Scene Depicting Food and Fish

given a detailed account of bowls, basins, plates, vases, spouted vessels, goblet jars, water bottle, cups, ladles, pots, storage jars, baskets, large and small trays, hearths, fire tongs, bundle of sticks, mortar and pestle, legged quern and muller, winnowing basket, pegs, fans, axe and many such items used in daily life of rural-urban societies. Hunting scenes suggest the practice of meat eating, which is related to the agricultural way of life.

The terracotta art of this period is also important from the point of agricultural history. For example, the terracotta figurines described as Sinivali[94] shows corns attached to her head-dress. She is, no doubt, a goddess worshipped to gain fertility of soil and prosperity of crops. Another Sunga terracotta figure discovered from Kaushambi is unique

Fig. 29: Sunga Terracotta from Kaushambi Showing Picnic on a Cart

in that it portrays a picnic party moving in a chariot (Fig. 29). A huge platter or thal of cooked food is depicted with clear suggestion of rise, sweet balls, round cakes, etc.[95]

The famous Panchachuda figure is typical of Sunga period terracotta art. They have been recovered from various sites in mid and lower Ganga Valley (Fig. 30). The figurine discovered from Kaushambi portrays a divine female figure with beautiful smiling face. Her bejewelled dress and ornaments and majestic head-dress suggest her religio-magical nature. Five auspicious symbols like arrowhead, flag, goad, trident-like objects, axe, mirror and comb like symbols are attached to her head-dress on her right. Till date, objects such as mirror, trident, flag, etc., are used in popular folk cults of Indian culture. This figurine also suggests her fertility cult association and her agricultural background.[96] Likewise, a Sunga period Balaram statue has been recovered from Mathura. He is shown holding a club/pestle in his right hand and a plough-like implement in his left hand and shoulder.[97] Thus, we may conclude that art-evidences relating to Sunga-Satavahan period are very important from the point of shedding significant light on the history of agriculture.

Fig. 30: Panchachuda and Terracotta Figurines from Kaushambi and Other Sites

AGRICULTURE AND THE KUSHANA ART

Chronologically, the next phase of Indian art belongs to Kushana-Gupta art idiom. This phase has few of its own characteristic features. Stylistically Kushana art carries the legacies of preceding art idiom of Sunga-Satavahana phase. But, it also possesses time-bound and time-specific natural transformations of previous art forms without any abrupt change or discontinuity, as well as, anticipates future forms by paving the way for gentle process of assimilation, integration and innovation. Now narratives take a backseat and the human form occupies the central position with the result of full integration of nature with the human form itself. Stupa architecture continued but instead of narratives of *Jātaka* stories and events of the Buddha's life now lifesize statues of Yakshini figures in various postures were chosen as the theme of art-expression. Another major change came in the form of human Buddha images both in standing and sitting postures. The Kushana age was an age of free intermingling of outside art forms and cosmopolitan world view. Gandhar art bears testimony to this fact. Mainland art centres like Sarnath, Kaushambi, Mathura also absorbed these foreign traits and with their inherent art skills produced Indianised masterpieces relating to Buddhism, Jainism, Hinduism and other folk cults.

Archaeological artifacts discovered from excavations at different sites throw much light on agricultural implements. For example, from the excavations carried at the Bhir mound, Taxila, several iron agricultural implements have come to light. They are hoes, hoes with chisel-like blades, spuds with broad blades used like *khurpa,* true spade, weeding-forks, sickles and many such forms.[98]

Bhir and Sirkap mounds have produced in the excavation pivot stones, querns, mullers, pestles, mortars, grinding mills, etc.[99] Iron sickles and iron trowels of Kushana period are also kept in the Sanchi Museum.[100] Several excavated sites have revealed brick-wells, which were used for irrigation.[101]

Other than these artifacts there are some depictions in Gandhara-Mathura art of Kushana period which throw direct light on the history of agriculture. For example, in a Gandhara relief, which is now kept in Lahore Museum, Pakistan, a farmer is shown wearing an Indian *dhoti* above the knees engaged in the art of ploughing a field (Fig. 31). The plough is very heavy and resembles a kind of plough still used in the Solan area of Himachal Pradesh.[102] One can compare this Gandhara relief with that of Bodhagaya panel and bring out the tradition of continuity and regional variations.[103]

Fig. 31: Gandhara Relief Showing Ploughing of a Field

Almost identical scene of a farmer ploughing a field has been reproduced by Som Prakash Verma on the basis of a Bodhisattva statue from Sahri-Bahlol, kept in Peshawar Museum. On the pedestal, a farmer is shown pressing his right hand on the stilt of the plough which is yoked to the necks of two bulls.[104] Mathura railing pillars are another source of information throwing light on the mundane life of contemporary society. These railing pillars portray lifesize statues of beautiful charming ladies in pleasant postures and different moods of joyfulness and playful activities such as playing with a ball/a parrot, engaged in toiletries, playing a Veena like instrument, carrying swords, displaying an intoxicated state of being, standing under Asoka/Sala trees in Salabhanjika/Dohada posture, etc. Salabhanjika figures are important in that they suggest fertility cult and its wide popularity both in rural and urban communities. Originally, they were associated with agricultural societies as a magico-religious motif of increasing soil fertility and vegetative power of trees. Gradually, they were assimilated in all the three major religions, i.e., Buddhism, Jainism and Hinduism as semi-divine beings capable of bestowing untold riches and prosperity to their worshippers. These Mathura railing pillars bear testimony to the aesthetics and popular beliefs of the composite culture of Kushan India.[105] Among them, two types of images are more relevant to the theme of this chapter. Particularly important is the majestic lady standing with a wine jar holding in her left hand and a bunch of grapes in her right hand (Fig. 32). Above her, one couple is depicted in an inebriated mood trying to reach for the wine jar from above the railing. Indirectly this image gives information of grapes and wine, probably made of grapes.[106] Horticulture practices included grape vines and their processing to gain wine and other beverages. The double-sided bowl support from Palikhera, Mathura shows the pot-bellied Kubera being served wine, specifically grape wine, indicated by the grapes held by the figure on the left by a person dressed in Hellenistic dress. Import of Greek wine and ladies is mentioned in various historical sources both classical and Indian. This scene also suggests the impact of Dionysic rites in contemporary society.[107]

Fig. 32: Mathura Railing Pillar Lady Holding a Wine-jar and Grapes

Drinking wine became a pastime activity among the elites, as is suggested by the so-called famous scene of the inebriated woman supported by her husband.[108] Yaksha-Yakshini sculptures along with Naga images suggest the popularity of tree

Fig. 33: Mathura Railing Pillar-Purna-ghata Sri Lakshmi/Prithivi

worship along with serpent worship.[109] Another group of images are described either as Sri-Lakshmi or as Nourishing Prithivi. One railing pillar, kept in National Museum, New Delhi, shows a beautiful lady standing on Purna-Ghata from which lotus flowers and creepers are issuing in a stylized foliated form. On the back two beautiful peacocks are depicted to suggest natural ethos and beauty (Fig. 33). The lady is holding her right breast in her left hand in a motherly posture of feeding a baby.[110] Another railing pillar shows a lady standing under a Asoka tree, holding her right breast with her hand.[111] Both these images combine three elements relating to fertility cult: tree worship, Mother Goddess cult and water worship in the form of Purna-Ghata or a vase. They symbolize the reproductive power of a woman, the vegetative power of trees and fertility of the earth. All these traits are the legacies of agricultural society.

Gupta art has very few narratives. But iconography suggests prevalence of agricultural customs and beliefs such as tree, serpent and bull worship. At Kaushambi, a Naga worship scene depicts a lady carrying a heaped dish of food.[112] Divine images of Kaushambi school are depicted holding fruits in their hands, particularly custard apple or pomegranate variety.[113] Art historians who have tried to give materialistic interpretation of the Gupta art under the framework of Marxist philosophy have tried to situate the Gupta art in feudal milieu where process of deurbanization had paved the way for rural and agrarian way of life. In this set up land and cultivation occupied more importance than urban trade and commerce. Popularity of Varaha cult rescuing the goddess earth from the ocean of chaos is interpreted as symbolizing the feudal agrarian set up where more and more land was brought under cultivation. The famous Varaha mythology of Udayagiri cave 5 thus bears testimony to the expanding agricultural milieu in the Gupta period.[114]

Ajanta paintings contain some interesting scenes relating to cooking, hunting and other domestic activities. In Cave XVII, there is a scene of royal kitchen where several cooking pots, vessels of food and other implements are depicted.[115] In the same cave donation hall of King Sibi is portrayed where servants are shown bringing food and gifts in vessels and pots.[116] There is a realistic depiction of a village life where a farmer is portrayed along with his thatched hut, herd of cattle, banana plants, bunches of fruit/corn suspended from the peg.[117] In fact, Ajanta paintings are a veritable storehouse on the life of various strata of society. Variety of flora and fauna and scenic beauty depicted in these paintings throw significant light on the activities relating to cattle breeding, husbandry, horticulture, domestication of animals, eating and drinking habits, food and fruit, etc. Cave I has a beautiful scene (Fig. 34) of attendants carrying pitchers.[118] Cave XVII, Visvantara Jataka story has a realistic depiction of areca-nut tree behind Prince Visvantara.[119] Cave II contains scenes of devotees bringing offerings, a child carrying a hen,

Fig. 34: Ajanta Painting

beautiful ceilings with lotus scrolls, geese, cherub bringing a plate of offering, two men drinking together from a shallow bowl, and beautiful water lilies.[120] They all tell us about the complexities of life involving both agrarian and urban vocations of village and city life.

AGRICULTURE IN EARLY MEDIEVAL ART

Early medieval period is famous for its temple art and associated sculptures. This phase of art contains numerous scenes shedding light on horticulture, fruits, dairy-farming, irrigation techniques and ploughing the field. From Bhubaneshwar, Orissa, in the Lingaraja temple, a scene has been discovered where Yashoda is shown churning butter (Fig. 35). She is standing holding the cords to rotate the churner to bring out butter in the pot. An infant Krishna is depicted eating the butter from the pot.[121] This scene suggests the practice of cattle breeding and dairy farming for gaining milk, butter and other dairy products. Another important art-evidence relating to irrigation technique

Fig. 35: Yashoda churning Butter

has come from Mandor, Rajasthan, belonging to eleventh century AD. The panel on the north side of topmost terrace of shrine depicts a Persian wheel being turned. A wheel with a chain of terracotta buckets/jars is shown mounted on the wheel (Fig. 36). Persian wheel was moved through a gearing mechanism which, according to Irfan Habib, is absent in this panel. But Randhawa and others believe it to be the profile depiction; hence, the gearing mechanism is not shown. The controversy regarding it being a manually worked wheel or a true mechanically moved Persian wheel is too deep to be solved in the absence of earlier art-evidences. But this panel clearly shows the use of wheel carrying a chain of pots for irrigating the land—a practice still prevalent in some parts of India.[122] Randhawa has also reproduced an interesting illustration from Borobuddur, Java (eighth century AD). This panel depicts a ploughman tilling the land with his plough with the help of two bulls.[123] The plough resembles the indigenous variety of Indian plough being used from the earliest times. At Mahabalipuram, in one of the panels of Krishan Mandapa, Krishna is portrayed as milking a cow. The calf and a lady are shown standing in front of the cow.[124]

Fig. 36: Persian Wheel Irrigation

Most of the temple sculptures portray a variety of fruits, trees, flowers and animals which could be taken as shedding light on horticulture and husbandry. A Siva temple at Durga, M.P. belonging to thirteenth century AD has a beautiful panel depicting Matrikas with different fruit trees.[125] A panel of Lakshmi-Narayana riding their divine *Vahan Garuda* shows a beautiful tree in a highly stylized scroll form, from which bunch of mango fruits are suspended.[126] This panel is kept in the National Museum and has come from Halebid, Hassan district, Karnataka (twelfth century AD). At Amritpur (Chikmaglur, Karnataka), from the famous twelfth century AD Amriteshvara temple, a scene from Krishna's life, is also depicted. The scene refers to the episode where Yashoda tied Krishna with a cord to a mortar which was attached to the tree. A highly stylized tree is portrayed at the centre of the panel. The mortar is of an hour-glass shape.[127] The scene is suggestive of a typical village life. In one of the intricate panels of Hoysaleshwara temple, Halebid, Karnataka (twelfth century AD), the episode of Govardhana hill is portrayed with all the nuances of a village life (Fig. 37). In the foreground, cattle—bulls and cows—are depicted on either side of the majestic central figure of Lord Krishna. In the middle, villagers are depicted along with their implements used in daily life. The hill is depicted with thick forests, creepers, snakes, birds and other creatures. The whole scene is alive with the dramatic movement and excitement of the divine event.[128]

Fig. 37: Govardhana Hill, Halebid

Fruits and trees are depicted in various contexts on temple art. The Kadamba tree, with its globular fruits is often depicted in Krishna legends in the context of *Vastraharana* episodes. Brihadeshwara temple, Thanjavur (ninth–tenth AD) has a beautiful depiction of the *Vastraharana* episode where Krishna is sitting on the Kadamba tree and the *gopis* are standing below, pleading for the return of their clothes.[129]

Plantain or banana plants were cultivated in India from very early times. Religious rituals and Śāstric tradition describe it as highly auspicious plant, particularly, for marriage ceremonies. Banana fruit is considered to be a symbol of fertility and good luck. Its plant is associated with Vishnu and Lakshmi. There is a beautiful depiction of banana plantation on Airavateshwar temple at Darasuram, Tamil Nadu (twelfth century AD). The panel is divided in three horizontal rows. In the lowermost row, holy men are depicted sitting under banana plants in deep meditation. In the central row, they are standing in different Yogic postures. The upper panel has the depiction of a royal figure along with attendants carrying flywhisks and umbrellas.[130] The Jambukeshwar temple at Tiruchirapalli has a highly stylized form of a banana plant depicted with a bunch of banana fruits.[131] At Pattadakal, Karnataka, in one of the panels, Shakuntala and Dushyanta are depicted meeting under a banana plant in the forest.[132]

The most interesting depiction from the point of agriculture seems to be the depiction of corn or maize. The origin of the plant is rooted in controversy. Seven thousand years-old fossilised maize pollen have been discovered from Mexico. The Asiatic origin of Maize is linked with the regions of Assam, Meghalaya, Manipur, etc.[133] Various Hindu and Jaina deities on Karnataka temples are shown carrying a corn cob in their hands. The famous Chenna Keshava temple, Belur, Karnataka has many such depictions.[134] On the Lakshmi Narasimha temple, Nuggehalli, Karnataka, the eight-armed dancing Vishnu in his female form of Mohini is shown as holding a corn cob in one of its left hands. In the Trikuta Basti Mukhamandapa at Shravanabelagola, Karnataka (twelfth century AD), there is a unique depiction of Ambika Kushmandini, sitting on a lotus seat under a canopy of mangoes, holding a corn cob in her left hand.[135] All these temples belong to twelfth-fourteenth century AD. Thus these art-evidences throw significant light on the agriculture of Maize crop.

According to Gupta 'except in Khajuraho temples and Sanchi Stupas the author has not seen coconut motif on the temples of the north'.[136] In the ninth century. Nageshvaram temple, Kumbakonam, Tamil Nadu, the vase of plenty has been shown with a coconut fruit covering its mouth. Jambukeshwar temple, Tiruchirapalli (seventeenth century AD) has a beautiful depiction of Vishwamitra besides two coconut trees having large pinnate leaves and globular fruits.[137] On Sri Viswa Brahma temple, Alampur, Andhra Pradesh (eighth–ninth century AD), two females are portrayed flanked by two coconut trees.[138] All these depictions of coconut in south Indian temples suggest the important economic and religious value of this fruit in the horticultural life of this region. Pomegranate fruit is not sculpted frequently in Indian art. A statue of Tara, Sarnath is shown holding a pomegranate fruit in her left hand.[139] In Venkatachalapati temple of Vishnu, at Krishnapuram, Tirunelveli (sixteenth century AD), an *Apsara* has been shown holding a pomegranate fruit in her left hand. She has been depicted in a dancing posture of Bharatanatyam.[140] Brinjal motif is found on Saiva temples only. A single fruit of brinjal is sculpted on the pillar of Nataraja temple, Chidambaram, Tamil Nadu. In the seventeenth century Jambukeshwar temple, Tiruchirapalli, entire brinjal plant has been sculpted along with flowers, fruits, leaves.[141]

Sugarcane depictions are found only on Saiva temple of Tamil Nadu and Karnataka regions. Occasionally, Vaishnava temples also depict it. For example, on Lakshmi-Narasimha temple, Nuggehalli, Karnataka, Kamadeva is sculpted with his bow of sugarcane.[142] In the Jambukeshwar temple, Tiruchirapalli, Kamadeva is depicted carrying a Sugarcane bow and five arrows, riding his mount, the parrot.[143] On Airavateshwar temple, Darasuram (twelfth century AD), plantation of sugarcane is sculpted in a realistic manner. On the same temple Kamadeva is shown with his bow of sugarcane.[144] Betel vine is depicted on Meenakshi temple, Madura and Sri Lakshmi Narasimha temple, Namakkal, Tamil Nadu as a decorative design.[145] Temple sculptures are a veritable storehouse of flora and fauna. Numerous depictions of Asoka, Sala, Kadamba, Champa, Vata, Palasha trees, lotus flowers, scrolls, creepers, etc., along with deer, elephants, bulls, cows, sheep, lion, tiger, horse, buffaloes, goat, geese, parrots, peacocks, hen, dove, etc., create the entire ecology of ancient times where agriculture, horticulture, dairy farming, husbandry, forestry founded the backbone of society. Lord Ganesha depicted with modaka (Fig. 38) or his favourite sweet indirectly throws light on agriculture and related food habits.[146]

There is an interesting sculpture of oilseed processing in the twelfth century AD Darasuram temple in Tamil Nadu (Fig. 39). There are three terms in Indian languages for oilpress—*ghani, kolhu* and *ckekku*—all derivatives from early Sanskrit words

Fig. 38: Lord Ganesha Holding Modaka-bowl

Fig. 39: Oilseed-press, Darasuram Temple

for crushing devices.[147] The panel depicted on Darasuram temple shows a device whose loadbeam is drawn by bulls. A pestle rotating in a mortar is attached with the loadbeam.[148] Panini has described this device as *thaila-peshana-yantra.*[149] A sugarcane crushing Kolhu has been represented (Fig. 40) on a Jaina manuscript of AD. 1540.[150] Achaya has reproduced a scene from Chandigarh Museum where Krishna and Balarama are shown

Fig. 40: Sugarcane Crushing Scene, Jaina Manuscript

eating (Fig. 41) with their friends in a picnic using cups and plates made of Palasha leaves.[151] On a eighteenth century AD. Damodar temple at Kayanpur, Howrah, women are shown husking rice (Fig. 42) and using the *Dhenki* device.[152] From Ajanta Cave 1, we find an interesting scene of women ginning with cotton and scutching with roller-and-board.[153] In Bharat Kala Bhawan there is an illustration from Mirgavat (sixteenth century AD), where a peasant is depicted irrigating the land.[154] A scene from *Tarikh-i-Alfi* (sixteenth century AD), kept in British Museum, London shows a farmer using *Pata* to smoothen the soil with the help of two bulls.[155] In Bharat Kala Bhawan, Varanasi, there is a beautiful country scene from the sixteenth-century *Anwar-i-Suhaili.* Waterman is taking water from the well, cows and buffaloes are seen nearby, across the hill and water creek, a farmer is ploughing his field with the help of plough and two bulls, another farmer is standing carrying a spade on his shoulder, fisherman and bird-trappers are engaged in their works (Fig. 43). The whole scene is vibrating with village life.[156] An eighteenth-century painting kept in the National Museum, New Delhi, depicts people moving from Gokul to Vrindavan (Fig. 44). Among cattle, archers, bullock carts womenfolk are depicted carrying spinning wheel and cotton press. Baskets,

Fig. 41: Krishna-Balarama Eating Scene, Chandigarh Museum

Fig. 42: Women Dehusking Paddy, Damodar Temple, Howrah

Fig. 43: Country Scene Depicted in *'Anwar-i-Suhaili'*

Fig. 44: Gokul to Vrindavana—Painting, National Museum, New Delhi

churners, pitchers, spade are shown as routine items. This art-evidence shows the agricultural, horticultural and animal husbandry, dairy farming as the major activities of a village life.[157]

Art-evidences, thus, throw significant light on the history of agriculture, horticulture, hunting, fishing, husbandry, cooking and other related domestic chores.

NOTES AND REFERENCES

1. Sharma, G.R. *et al.*, 1980, *Beginnings of Agriculture*, Allahabad: Abinash Prakashan.
2. *New Webster's Dictionary*, 1986, Deluxe Encyclopaedic Edition, p. 22.
3. N. Web. Dic. *op. cit.*, p. 22.
4. N. Web. Dic. *op. cit.*, pp. 464, 468.
5. *The New Encyclopaedia Britannica*, 1977, Vol. I, 15th Ed; Encyclopaedia Britannica, Inc. p. 347.
6. U.S. Department of Labor in the 21st Century-www.dol.gov 29CFR 780. 112–116.
7. The UP Agricultural Credit Act, 1973, 2(a).
8. Chandrawati V. State of UP AIR 1961 All. 183(FB).

9. AIR 1954, Assam, 113.
10. *Kautiliyan Arthashastram* (Ed.) Gairola, Vachaspati), 1991, Chowkhamba Vidyabhavan, Varanasi, 4th Edition.
11. *Ibid.*
12. *Ibid.*
13. *Amarkośa of Amarasimha* (Ed. Satyadeva Mishra), 1972, University of Malaya, Malaysia.
14. *Ibid.*
15. *Brihat Samhita* of Varāhamihira (Ed.) Sri A. Jha, 1993, Chowkhamba Vidyabhavan, Varanasi.
16. *Ibid.*
17. *Ibid.*
18. Chakrabarti, D.K., 1999, *India: An Archaeological History*, Oxford University Press, New Delhi, p. 93.
19. Chakrabarti, *op. cit.*, pp. 91–116.
20. Chakrabarti, *op. cit.*, pp. 117–150; Pandey *op. cit.*, 327–361.
21. Tiwari, Rakesh, 1990, *Rock Paintings of Mirzapur*, Lucknow: U.P. State Archaeological Organisation, p. 42.
22. Jarrige, J.-F. 1995, "From Nausharo to Pirak: Continuity and change in the Kachi/Bolan Region from the 3rd to the 2nd Millennium BC" in *South Asian Archaeology*, vol. I, (Eds) R. and B. Allchins, 1997, New Delhi, Oxford & IBH Publishing Co. Pvt. Ltd. pp. 11–32, Fig. 1.
23. De Saizieu, Blanche Barthelemy and Bouquillon, Anne, "Evolution of Glazed Materials from the Chalcolithic to the Indus Period Based on the Date of Mehrgarh and Nausharo" in *South Asian Archaeology*, 1995, Eds. R. and B. Allchins, 1997, New Delhi, Oxford & IBH Publishing Co. Pvt. Ltd. pp. 63–76.
24. Jarrige, C. 1995, "The Figurines from Nausharo Period I and their further Development" in *South Asian Archaeology*, *op. cit.*, pp. 33–43.
25. Jarrige, J.F., *op. cit.*, p. 31.
26. Chakrabarti, D.K., *op. cit*, 1999, pp. 117–147.
27. Gajjar, Irene N. 1971, *Ancient Indian Art and the West*, D.B. Taraporevala Sons & Co. Pvt. Ltd., Bombay, p. 3.
28. *Ibid.*, pp. 4–7.
29. *Ibid.*, p. 9.
30. *Ibid.*, pp. 10–12.
31. *Ibid.*, p. 11.
32. Chakrabarti, Dilip K., 1999, *India—An Archaeological History*, Oxford University Press, New Delhi, pp. 142–144.
33. Randhawa, M.S., 1980, *A History of Agriculture in India*, Vol. I, ICAR, New Delhi, p. 170, fig. 80.
34. Chakrabarti, *op. cit.*, p. 146.
35. Randhawa, *op. cit.*, p. 172, fig. 83.
36. Chakrabarti, *op. cit.*, p. 168–169.
37. *Ibid.*, p. 166.

38. *Ibid.*, p. 166.
39. *Ibid.*, pp. 172–173, Figs. 18, 19.
40. Weber, Steven A., 1995, "Harappa Archaeobotany: A Model for Subsistence" in *South Asian Archaeology, op. cit.*, pp. 115–117.
41. Weber, S.A., 1991, *Plants and Harappan Subsistence*, Westview Press, Boulder.
42. Chakrabarti, *op. cit.*, pp. 190–191.
43. Kondo, R., Ichikawa, A., Morioka, T., 1995, "Taking a Bath in Mohenjo-daro" in *South Asian Archaeology, op. cit.*, pp. 127–137.
44. Meadow, Richard H., Kenoyer, J. Mark, 1995, "Excavations at Harappa 1994–1995: New Perspectives on the Indus Script, Craft Activities, and City Organization" in *South Asian Archaeology, op. cit.*, pp. 139–172.
45. Pandey, J.N., 2000, *Puratattva Vimarsha*, Prachya Vidya Samsthana, Allahabad, 7th Ed., p. 203, fig. 24.
46. Randhawa, M.S., 1980, *A History of Agriculture in India*, Vol I, Indian Council of Agricultural Research, New Delhi, p. 169.
47. Chakrabarti, *op. cit.*, p. 181.
48. Gajjar *op. cit.*, p. 36.
49. *Ibid.*, p. 38, fig. 136.
50. *Ibid.*, p. 31.
51. *Ibid.*
52. *Ibid.*
53. *Ibid.*, p. 17.
54. *Ibid.*, p. 18.
55. *Ibid.*
56. *Ibid.*, p. 20.
57. Chakrabarti, *op. cit.*, p. 184.
58. *Ibid.*, p. 205.
59. *Ibid.*, pp. 247–260.
60. *Ibid.*, pp. 247–246.
61. *Ibid.*, p. 257, fig. 41.
62. Rowland, Benjamin, 1970, *The Art and Architecture of India: Buddhist, Hindu, Jaina*, First Paperback Edition, Penguin Books, Maryland, USA, pp. 46–47, fig. 14.
63. *Ibid.*, pp. 45–46.
64. *Ibid.*, p. 46.
65. Gupta, S.P., 1980, *The Roots of Indian Art*, B.R. Pub. Corporation, Delhi, p. 53.
66. *Ibid.*, p. 67.
67. *Ibid.*, pp. 67–68.
68. *Ibid.*
69. *Ibid.*, pp. 66–67.
70. Huntington, Susan L., 1993, *The Art of Ancient India*, Weatherhill, New York, (1985), Second Edn., pp. 53–54, 59–60.
71. Hegde, R., 2002, *Sunga Art—Cultural Reflections*, Sharda Pub. House, Delhi, p. 152, fig. 45.

72. Hegde, *op. cit.,* p. 152.
73. *Ibid.,* p 151.
74. *Ibid.,* figs 40, 41, 42, 43, 44, *op. cit.,* pp. 151–152.
75. Acharya, K.T., 1998, *Indian Food,* Oxford University Press, Paperback Edition, New Delhi, p. 100.
76. Randhawa, *op. cit.,* p. 334, fig. 167.
77. *Ibid.,* p. 369, fig. 174–Top.
78. Sivaramamurti, C., 1956, Amaravati Scuptures in the Madras Govt. Museum, pl. IV, Fig. 5; Murthy, Krishna, 1977, *Nagarjunakonda—A Cultural Study,* Delhi, p. 245.
79. Ray, Niharranjan, 1975, *Maurya and Post-Maurya Art,* ICHR, New Delhi, Fig. 14.
80. Murthy, K. Krishna, 1983, *Material Culture of Sanchi,* Sundeep Prakashan, Delhi, pp. 226–236.
81. Rowland, Benjamin, *op. cit.,* pp. 97–98, fig. 43.
82. Agrawal, V.S., 1977, Bhartiya Kala, Prithivi Prakashan, Varanasi, (1966), IInd Edn., pp. 150–151; Mishra, R.N., 1978, *Bharatiya Murtikala Ka Itihasa,* Grantha Shilpi, 2002 Edn., p. 86.
83. Huntington, Susan L. 1985, *The Art of Ancient India,* Weatherhill, New York, Second Printing, 1993, fig. 5.16, p. 71.
84. Mall, Bansi Lal, 2000, *Trees in Indian Art: Mythology and Folklore,* Aryan Books International, New Delhi, p. 66.
85. Murthy, K. Krishna, *Material Culture of Sanchi,* Sundeep Prakashan, Delhi, 1983, p. 225; Marshall & Foucher, *The Monuments of Sanchi,* Calcutta, n.d., 55.3 (Vol. II), 23, 29, 30, 51, 52, 56, 58, 63, 65, 101 (Vol. II), 19, 27, 34 (Vol. II), 18, 23, 37 (Vol. II), 82 (Vol. III).
86. Gupta, Shakti M., 1996, *Plants in Indian Temple Art,* B.R. Publishing Corporation, Delhi, p. 20, plate 13.
87. *Ibid.,* 115, plate 145.
88. *Ibid.,* p. 172, plates 220, 222.
89. Srivastava, A.L., 1983, *Life in Sanchi Sculpture,* Abhinav Publications, plate XII, LV; Fergusson, James, 1971, Reprint Indological Book House, Delhi (Originally India Museum, London, 1868), Plate XXXVII, fig. 2, Plate XXXII, figs 1 and 2.
90. Gupta, *op. cit.,* p. 101, plate 132, p. 61, plate 70.
91. Dhavalikar, M.K., 1997, Sanchi, A Cultural Study, Deccan College Post Graduate & Research Institute, Pune, pp. 50–52, plate II and VI; Randhawa, *op. cit.,* p. 375, fig. 176 (Vol. I).
92. Hegde, *op. cit.,* pp 111–115, 122, 157, 165, figs 1–25, 48, p. 157– figs 1–3, p. 165– figs 24, 27, 29; Dhavalikar, op. cit., pp. 50–70; Srivastava, A. L., *op. cit.,* pp. 70–74, figs 28.1–12, fig 29.1–14, fig 30.1–10; Mathur, V.K., *Art & Culture Under the Sungas,* 1996, Delhi, figs 12, 13, 15, 17, 20.
93. Murthy, K. Krishna, *op. cit.,* pp. 132–144.
94. Sharma, G.R., 1980, *History To Prehistory,* Deptt. of Ancient History Culture and Archaeology, University of Allahabad, p. 47 extreme left fig.

95. *Ibid.*, p. 45, right side top fig.
96. Agrawal, V.S., *Bharatiya Kala*, p. 325, fig. 503; Huntington, *op. cit.*, pp. 88–89, fig 5.39.
97. Agrawal, *op. cit.*, p. 243, fig. 370.
98. Randhawa, *op. cit.*, p. 393, fig 185 (Vol. I).
99. *Ibid.*, p. 394, fig. 186.
100. *Ibid.*, 396. fig. 189.
101. *Ibid.*, p. 401.
102. *Ibid.*, p. 401, figs. 187–188.
103. Hegde, *op. cit.*, p. 152, fig. 45.
104. Verma, Som Prakash, 1994, India at work in Sculpture and Painting, Souvenir For IHC, 55th Session, Aligarh Muslim University, p. 1, fig. 2.
105. Huntington, *op. cit.*, pp. 156–158, figs. 8.35–38; Agrawal, V.S., *op. cit.*, pp. 230–234, figs. 343–356; Rowland, Benjamin, op. cit., p. 157, figs. 100–101.
106. Bussagli, Mario, Sivaramamurti, C., year *5000 years of the Art of India*, Harry N. Abrams Incorporated Publishers, New York, n.d., p. 98, fig. 106.
107. Huntington, S.L., *op. cit.*, p. 159, fig. 8.39; Agrawal, V. S., op. cit., p. 255, fig. 390.
108. Agrawal, V.S., *op. cit.*, p. 256, fig. 391.
109. Chakraberti, Kanchan, 1981, *Society, Religions and Art of the Kushana India*, K.P. Bagchi & Company, Calcutta, pp. 90–100.
110. Agrawal, V.S., *op. cit.*, p. 236, fig. 362; Bussagli and Sivaramamurti, *op. cit.*, p. 100, fig. 110; Huntington, S.L. p. 158, fig. 8.38.
111. Agrawal, V.S., *op. cit.*, p. 236, fig. 361.
112. Tripathi, Aruna, 2003, *The Buddhist Art of Kausambi*, D.K. Printworld (P) Ltd., New Delhi, p. 219, fig. 309.
113. Sharma, G.R., *op. cit.*, p. 39 – right fig, p. 40 – left fig, p. 55 – right fig.
114. Huntington, *op. cit.*, p. 192, fig. 10.8.
115. Sanyal, Narayan, 1984, *Immortal Ajanta*, Calcutta, p. 44, No. 19.
116. *Ibid.*, p. 42, No. 78, 49.
117. *Ibid.*, p. 50, 51, Cave XVII.
118. Verma, S.P., *op. cit.*, plate II.
119. Behl, Benoy K., 1998, *The Ajanta Caves*, Thames and Hundson, pp. 162–163.
120. Behl, *op. cit.*, pp. 126–143.
121. Verma, Som Prakash, *op. cit.*, plate IV.
122. Randhawa, *op. cit.*, p. 478, fig. 202.
123. *Ibid.*, p. 454, fig. 203.
124. *Ibid.*, p. 449, fig. 197.
125. Mall, Bansi Lal, *op. cit.*, plate 50.
126. *Ibid.*, plate 48.
127. *Ibid.*, plate 47.
128. Randhawa, *op. cit.*, p. 451, fig. 199.
129. Gupta, Shakti, M., 1996, *Plants in Indian Temple Art*, B.R. Publishing Corporation, Delhi, p. 22, plate 15.
130. Gupta, *op. cit.*, p. 111, plate 138.

131. *Ibid.*, p. 111, plate 139.
132. *Ibid.*, p. 111, plate 140.
133. *Ibid.*, p. 176.
134. *Ibid.*
135. *Ibid.*
136. *Ibi*d., p. 56.
137. *Ibid.*, p. 56, plate 63.
138. *Ibid.*, p. 57 plate 65.
139. *Ibid.*, p. 144, plate 179.
140. *Ibid.*, p. 144.
141. *Ibid.*, p. 165, plates 208, 209.
142. *Ibid.*, p. 148.
143. Ibid., p. 148, plate 184.
144. *Ibid.*, p. 151, plate 187, 188.
145. *Ibid.*, p. 141, plates 174, 175.
146. Achaya, K.T., *op. cit.*, p. 69.
147. *Ibid.*, p. 111.
148. *Ibid.*
149. *Ibid.*, p. 112.
150. *Ibid.*, p. 113.
151. *Ibid.*, p. 63.
152. Verma, S.P., *op. cit.*, plate XXIV.
153. *Ibid.*, plate, VIII.
154. *Ibid.*, fig. 6.
155. *Ibid.*, fig. 8.
156. *Ibid.*, plate IX, X.
157. *Ibid.*, plate XV.

CHAPTER 46

Therapeutic Value of Agricultural Produce and the Science of Nutrition in the *Purāṇas*

(Bio-chemical assessment and social significance of agricultural produce in the Purāṇas)

Susmita Pande

Ancient Indian knowledge system had been based upon an integrated view of health, medicine and diet. Religion, science and economic life interacted upon each other in holistic manner.[1] The *Purāṇas* provide a good example of this trend. The *Purāṇas* give evidence of the fact that serious research was being done in all the spheres, which had connection with the agricultural produce, e.g. the field of medicine (*Āyurveda*), science of agriculture (*Vṛkśāyurveda*), veterinary science (*aśvāyurveda* and *gajāyurveda,* etc.), cosmetology and the science of perfumes (*gandhayukti*) and also in the religious and esoteric practices.

Āyurveda (the science of life) aims at ensuring a healthy body along with a healthy mind. Here, the science of nutrition is not segregated from the science of medicine. Similarly, the effect of the food products on the body as well as the mind is evaluated. The property of an agricultural produce is determined by the influence of soil, season and in accordance with its morphology. Similarly, its therapeutic value varies according to its modes of cooking.

The modern science of nutrition analyzes the nutritional value of food products by the analysis of their nutrients, e.g. proteins, carbohydrates, minerals and vitamins. But the ancient science of nutrition analyzed the different properties of the agricultural products according to a detailed study of taxonomy also in accordance with their individual properties as potencies or *guṇas.* They correct the various degrees of humoral imbalance and lead to a healthy mind and body. However, an attempt is also made to tally the properties of various agricultural produce recognized in the ancient texts with the modern researches of the science of nutrition. But merely the analysis of nutrients in the modern science of nutrition does not give an idea of the wide range of the properties learnt by the ancient seers by their experience. The rich ancient heritage needs continuous research.

The *Viṣṇu Purāṇa* has classified the entire vegetation into three major categories —*gramauṣadhis, vanauṣadhis* and *yajñauṣadhis.*[2] The third category also includes the produce of the first two categories. The *Gramauṣadhi* includes the cultivated produce, mainly the common edible cereals. They are *vrīhi* (*dhāna* = paddy or whole rice grain, *oryza sativa*), *yava* (barley = *Horeum Vulgare*), *godhūma* (wheat = *Triticum Aestivam*), *caṇaka* (*cicre* = *arietinum*), *tila* (*Sesamum* = *indicum, sesame*). *Priyangu* has been interpreted variously as *priyāka,* perfumed cherry, *Aglaiya roxburghiana,*[3] Italian millet *Setaria italica* or *kanguni.*[4] *Callicarpa macrophylla vahl*[5] and *Prunas mahaleb.* Here, *priyangu* should be identified with *setaria italica* or *kanguni* as cereals are being enumerated here. The *Bhāva Prakāśa* refers to it under *kṣudradhanyam.* It is clear that *priyāngu* is also a synonym for *kanguha* or *kanguni-stṛyam kangu pṛyangu dve kṛṣṇārakta sitā tathā (Bhāva Prakāśa, nighantu. Dhanyavargaha* 75 Commentary of Shastri and Vaishya, p. 656, Chowkhamba) *callicarpa microphylla vahl* and *prunus mahaleb* are flowering trees and plants, respectively. *Aglaiya Roxburghiana* is also a flowering tree whose are popularly known as *priyangu.*[5a] Hence, the interpretation of *priyāngu* should be according to the context. *Hyudar'sa* (*jvar, sorghum vulgare*), *korūṣa* (*Kodo-paspalum scrobiculatum*), *satinaka* (small peas, *pisum sativam*), *māṣa* (urad, *phaseolus mungo*), *mudga* (*mūnga – phaseolus, aureus*), *masūraśca* (*masūra–lens esculenta*), *niṣpāva* (field bean *Dolichos lablab*), *kulattha* (*kulathi, dolichos biflorus,* horse gram), *ādhakya (arhar, cajanus cajan)* are some other vegetations.

The *yagñic auṣadhis* include *vrīhi* (*dhāna*), *yava* (barley), *māṣa* (urad), *godhūma* (wheat), *chaṇavah* (gram), *tila* (sesame), *priyangu* (*kangni*), *kulattha,* (*dolichas biflorus*) *śyāmāka,* (*śāvān, Panicum suma trense* or little millet), *Panicum frumenticeum* by P. Ray (*Suśruta Saṃhitā a scientific synopsis*) and as *Echinochloa frumentacia* by commentators of the *Bhāva Prakāśa* (Shastri and Rupalalji Vaish), *nīvāra* (Bengal wild rice *Hygroryza aristata*), *jartila (vanatila)*, *gavedhu* (job'stears *coix lacryma jobi*), *venuyava* (bamboo), *markata* which can also be *markati. Markati* is either *karanja* or *kevanch* according to *Amarakoṣa. Kevanca* is *mukuna prurita.* It is also called *kapti pachcha,* cowitch or cowhage. It has also been translated as *makoe* in the translation of the *Viṣṇu Purāṇa* (Hindi, Gīta Press, Gorakhpur). *Mokoe* is actually garden nightshade or *solanum nigrum* in the *Bhāva Prakāśa* (Ed. by Shastri and Vaisha).

In the above context, the word *auṣadhi* is not used in the modern context of medicine. The *Kāśikā* mentions four classes of plants—*vanaspati, vṛkṣa, oṣadhi* and *vīrudha.* Plants bearing fruits are called vanaspati; those having both visible flowers and fruits are called *Vṛkṣa.* Those that die after their fruits ripen are *auṣadhis* and the weak plants and shrubs are called *vīrudha.*[6] The classification of *sthāvara auṣadhis* by Suśruta is similar to the above classification,[7] although it is clear from the *Suśruta saṃhitā* that the word *auṣadhi* is used in a dual sense, viz., grains as well as medicine.[8] Śankara, in his commentary on the *Svetāśvatara Upaniṣad,* says that *oṣadhis* are small plants like paddy, while *vanaspatis* are big ones like the *pipal* tree.[9] The *Mitākṣarā* interprets the term *oṣadhi* as that which dries up when its fruits ripen like paddy, etc.[10] The term *oṣadhi* has also been described as that which contains heat or energy.[11] The *Bṛhatsaṃhitā*[12], the *Vāyu Purāṇa*[13], and the *Bhāgavat Purāṇa*[14] interpret the word *oṣadhi* as grain as well as medicinal plant. The *Viṣṇu Purāṇa* classifies the herb of food value and medicinal value as *phalini* and *mūlini,* respectively, on the basis of the use of their parts.[15]

The origin of the plants in the *Purāṇas* has been described from the milking of the earth, which, indicates that the plants received nutrition derived from the soil.[16]

If we consider the classification of the *Viṣṇu Purāṇa* the *gramauṣadhi* includes the cultivated cereals. The *Vanauṣadhi* list not only includes the forest produce but also trees, in parks and gardens as well as agriculturally cultivated fruits, vegetables and flowers. The *Amarakośa*[17,18] includes the forest and their produce as well as gardens, parks and cultivated flowers, fruits and vegetables, in the section *Vanauṣadhivarga.* It classifies the grains under the section *Vaiśya varga.* It defines *oṣadhi* which remain or dry up; often their fruits ripen. But the Indian tradition of nutrition and medicine includes both these types of agricultural produces.

In the *Agni Purāṇa,*[19] the chapter 279—*Siddhauṣadhāni* deals with the therapeutic value of the cereals and chapter 283—*nānāroga haraṇauṣadhāni* deals mostly with the specific forest produce presenting a complete encyclopaedia of the science of nutrition. The *Agni Purāṇa* in chapter 279[20] quotes Suśruta in prescribing the various diet charts for invalids. The tradition was further carried by the later *Ayurveda and Nighantu* works like *Aṣṭānga hṛdaya* of Vāgabhatta (most probably of the Gupta age), *Dhanvantari Nighantu, Rājanighantu* of Narahari (most probably of fifteenth century AD), *Bhāva Prakāśa* of Bhava Misra (of sixteenth century AD), *Kaiyadeva Nighantu* and many others.

Many of the diet charts of *Purāṇas* like *Agni Purāṇa* and *Garuḍa Purāṇa* are similar and carry the tradition of Suśruta. The various diet charts are in the form of recipes in combination with plants, fruits and herbs and also with non-vegetarian diet e.g., meat of animals.

Common cereals play an important role in these diets and their properties vary along with their combination with each other and different herbs.

Suśruta had classified the common cereals into three groups: *dhānyavarga, Kudhānyavarga, Śamidhanyavarga*[21]. (I. 46) The *dhānyavarga* included the *śāli varga* (paddy group), which included three types of rice *śali varga* (17 species), *śaśtika* (11 species) and *Vrīhi* (nine species). Many varieties have been identified by scholars like *rakta śāli* with *lalmati* (cultivated in Saharanpur district), *kalama* and *pramoda* with *kumuda* of Mumbai, *dīrgha sūka* with *hansarāja,* a variety of *basmati,* or *mahāśali* rice was grown in Magadha.[22] *Śāli* variety is easily digested, cooling, antidysenteric, *pitta* reducing. Red *śāli* is the best variety. It is diuretic, strength giving, febrifuge, antithirst, good for the heart.[23] *Śāli* grows in the *hemanta* season and is transplanted from one place to another. *Śaṣtic* ripens in the summer season. It ripens in 60 days and is not transplanted. *Vrīhi* ripens in the rainy season. The properties of these varieties vary with the different types of soil and season.

The *Purāṇas* mention the above different varieties of rice in different diet charts according to their properties learnt from the tradition of Suśruta.

Kudhānyavargha included 16 varieties—*Kordūṣaha, śyāmāka, nīvāra, śāntanu, varāka, uddālaka, Madhulikā, nandimukhi, Kuruvinda, gavedhuka, sara, varuka, todaparni, mukundaka and venuyava.*

The third group is of *śamidhānya varga* or legumes. It includes the varieties *vaidala* (11 species of pulses), *māṣa, kulattha, tila, yava, godhūma, śimbā, kusumbha, atasi* and *siddhārthaka.*

It is interesting that the *Garuḍa Purāṇa* enumerates these cereals (chapter 1.169) in the same order as does Suśruta.

The energy giving and easily digestive quality of rice was realized by Suśruta and also by the mediēval acaryas of Āyurveda like Bhava Misra of the *Bhava Prakāśa* and the *Nighantu* writers. The combinations of different varieties of rice is recommended so that they supplement the different nutrients. The modern analysis of rice reveals that apart from energy giving properties, it also has a rare variety of protein. The different forms of rice have 6–8 gm of protein as per 100 gm of edible portion. Parboiled and hand-pounded rice has 8.5 gm, and roasted and or *khīla* or puffed rice (often mentioned as *lājah* in the *Purāṇas* has 7.5 gm of protein. Its low fibre content (0.2, 0.3 gm) and more of moisture content 12–14 gm makes it a good remedy for the treatment of diarrhoea and dysentery.[24]

Suśruta recommends boiled rice with slightly seasoned meat soups, and soups prepared from *mudga* (*mūnga*).

Chapter 279 of the *Agni Purāṇa* gives the details of some popular diet charts containing common agricultural products.

To remove the weakness of a person suffering from fever, a gruel of roasted puffed rice (*lājā*), starch of rice mixed[25] with ginger powder is recommended.

The ayurvedic texts and the *Purāṇas* give due importance to ginger (*zingiber officinale*) as carminative. It was used for dropsy, alalgia, cephalalgia, asthma, cough, colic, diarrhoea, and cholera.[26]

The modern science of nutrition attributes to ginger protein (2.3 gm) minerals 1.2 gm, Carbohydrates (12.3 gm), calcium (20 gm), phosphorus (60 gm), iron (2.6 gm) the value is as per 100 gm of the edible portion.

For a convalescent patient, a diet of old *śastic* rice, *nīvāra,* red rice, *śāli* rice and *promodaka* variety of rice should be given.[27] Here, the *Purāṇa* accepts the rule of Suśruta in recommending old rice only. The nutrients of different varieties give a balanced diet to the patient. Suśruta also recommends a combination of *śāli* and *śastic* rice with honey. Sesame kalka (paste), mūnga *kalka* (paste) and *yūṣa* (thick soup) of *mūnga* is also given.[28]

Another diet chart prescribed by the *Agni Purāṇa* for a patient with fever includes preparations of barley, starch and *sattu.*[29]

In case of fever, another diet chart includes *mūnga (phaseolus aureus)*, *masoor* (*lens esculenta*), *canā* (*cicer aereticum*), *kulattha* (*dolichos biflorus*), *moth* (*phaseolos aconiti folius*), *arhar* (*cajanus cajan*), *lavāka* (common quail, *perdix chinensis*), *karkotaka* (*momardica mixta*) of the *katol* or bitter variety, *patola* (*parwar, tricho santhes dioica*), *sufalā* (is translated as *kunduru* by Tarinish Jha in the edition of the *Agni Purāṇa* by Hindi Sahitya Sammelan, Prayag). *Kunduru* or *coccinea indica* is called *pilu parne* and *bimbi* in the *Bhāva Prakāśa* but not *sufalā*), *nimba* (*neema* or *herbacia*), *parapata* (*citta pāpra*) and pomegranate. *Citta pāpra* has been identified with different plants in different parts of the country; hence, its Latin synonym varies according to the plants. The various plants that are identified with *parpata* are *olden landia corymbosa* or *oldenlandia herbacia, Fumaria indica, Polycarpia corymbosa, Justicia procumbens, Rungia repens, Justicia procumbens and Rungia parviflora, Perestraphe bicalyculata, Glossocardia linearifolia, Mallugo stricta.* The nutrients in per 100 gm of edible portion of *mūnga* are as follows: a high protein (24.5 gm per 100 gm of edible portion) and iron (8.5 gm) low fibre (0.8 in the lentil variety although in whole *mūnga* it is 4.1). Hence, it is wholesome and light, *Masoor* also has 25.1 gm protein, 4.8 iron and 0.7 fibre *chana* has 20.8 gm protein and 9.1 gm iron, *kulattha* has 22 gm, proteins and 8.9 gm iron. *Moth* has 23.6 gm protein and 9.5 gm iron and 4.5 gm, fibal content, *arhar* has 22.5 gm protein.

It is interesting to note that Suśruta had distinguished between *raktapitta* and *raktātisara.*[30] *Raktapitta* was internal bleeding in the intestines, liver or spleen which oozes out from both the ends of alimentary canal. Rectal bleeding indicates a difficult to cure disease. *Raktātisāra* is blood dysentery. The *Agni Purāṇa* is aware of the distinction between the two gastro intestinal diseases and mentions them separately. Both the texts classify *raktapitta* into two categories—one, when the blood oozes from the lower part of the alimentary canal the other when it oozes out from the upper part of the body like mouth, nose, eyes and ears. The mechanical treatment involves emiting from the upper part and purgative for the clearance from the lower part in the latter.[31] Suśruta recommends *Śāli* and Saṣtic rice with the decoction of lotus with milk (in case of *pitta-*dominated *raktapitta*).[32] It also recommends sugarcane pieces soaked in water and kept open under the sky at night and strained in the morning and mixed with blue lotus and honey.[33] It recommends six other medicines, including *yaṣti madhu* (*mulaithi* or *glycyrrhizza glabra*), rice soup, honey as the main ingredient[34]. The *Agni Purāṇa* recommends a diet of *sattu,* wheat, *laajā* (roasted puffed rice), *yava* (barley), *śāli,* (*oryza sativa*), *masoor,* (*lens esculenta*), *kuṣtha* (kūtha), gram, (*cicer aeretinum*), *mūnga* (*phaseolus aureus*) and wheat (*triticum aestivum*). Whole wheat flour contains 12.1 gm of protein, and 11.5 gm of iron per hundred grams of edible protein.[35] The juice of *Adusā* (*Adhatoda vasika*) cooked in clarified butter, cooled and mixed with honey is also considered a good diet.[36]

In diarrhoea (*atisāra*), old *śāli* rice is recommended. Constipating cereals should be given along with the decoction of *lodhra* (*symplecos racemosa*) bark,[37] *takra* or *matthā* or butter milk with *citraka* (white leadwort, *plumbago zeylanica*) is administered (*Agni Purāṇa* 279.24).

For flatulence and stomach ailments, one should protect from *vāyu. Atibalā* (Indian mallow or *abutilon indicum,* it is different from *mahābalā, sida rhombofolia*) with milk and *vāstūka* or *bathua* (*chinopodium album*) seasoned with ghee should be given. Wheat and *śāli* rice, both appetizing and health giving, should also be administered[38] in case of diarrhoea. Suśruta also recommends, soup of herbs like *balā* (*sida cordifolia*) along with bilva (*Aegle marmolos), paṣni parni* (*uaria logopoides*), lotus, coriander and ginger.[39] The names of four types of *balā-balā, mahābalā, atibalā* and *nāgabalā* occur in the texts. They have been identified with *sida cordifoliza, sida rhombifolia* (if identified with *sahadevi,* it is *veronica cinerea*) and *abutilion indicum,* respectively (6.40.95). *Bilva* (bela *Aegle marmelos)* and *yaṣtimadhu,* (*mulaithi, glycyrrhiza glabra*) also from important constituents for medicines in diarrhoea. Suśruta also recommends the paste of *tila, munga* and soup of *munga.*[40]

Leprosy was enumerated as one of the eight great diseases (*mahārogas*) which are, *vātavyadhi* (27 types of *vātavyādhi* have been described, including, paralysis), *prameha* poly urea (diabetes), *kuṣṭha* (leprosy), *arśa* (polyps and piles), *bhagandara* (fistula), *aśmari* (urinary calculi), *mūdhagarbha* (still birth) and *udara (udara* is of nine types and includes abdominal enlargements, abdominal dropsy, peritonitis, ascites, etc.).

The various diets for leprosy are given in the *Agni Purāṇa,* e.g. wheat, rice, *mūnga, āmlā* (*emblic myrobalan*), *khaera khadirodaka* (catechu tree—*acacia catechu;* a decoction of its bark is useful), *harada* (*chebulic myrobalan*), *pancakola* (pipplai—Indian piper longum), *peeparamūla* (root of long pepper), *cavya* (wild pepper or piper—*brachy stachyum*), *citraka* (rosy flowered leadwort or *plumbago indica*), *Sonth* (dry ginger—*zinziber officinale*), meat of animal, *nimba* (margosa), three types of *āmlas,* leaves of parwal, juice of lime with seeds, black cumin, dry radish and *saindhav* or rock salt and a decoction of *khadira* should be given.[41] Another diet chart consists of a soup of *masoor* (*lens esculenta*) and *mūnga* (*phaseolus aureus*), old *śāli rice (oryza sativa*), *nimba* (margosa) and *parpata (pittapāprā*) along with the soup of animals.[42]

The external applications of *Vāya Vidanga* (*embilia ribes*), black pepper, *nāgaramotha* (nut grass, muṣtām, *Cyprus rotundes*), *kuṣtha* (*Saussrea lappa* or *Sausurea auriculates*), *lodhra* (*symplecos racemosa*), *Suvarcikā* (*hurhur*), *mainsila* and vaca (*acorus calames*) or sweet flag plant) are used with cow urine.[43]

Agni Purāṇa chapter 285 writes of recipes made from concoctions of herbs to cure leprosy. The medicinal herbs were ground finely and cooked in water till a quarter of water remained. This was mixed with some more herbs cooked with ghee. The medicinal ghee was of various types. One such ghee was called *pancatikta,* which consisted of herbs like bark of margosa leaves of parwar (*trichosanthes dioica*), *Vyāghri* (*Solanam zanthocarpum*), *guduci* (*giloi, amḍta* or *Tenospora cordifolia*) and *vāsaka* (Malabar nut, tree *adūsa, adhatoda vasika*). The above mentioned herbs were cooked in 16 *seers* of water. When it remained one-fourth, it was cooled with the paste of *triphalā* and one *seer* of ghee. This ghee was *pancatikta* ghee given to the leprosy patient to eat.[44] Many other varieties of such *ghritas* are mentioned in detail in ch. 283 and 285 of the *Agni Purāṇa* as a cure for leprosy, boils, small pox, etc.[45]

In case of herpes (*Visarpa*), old wheat and barley, śāli rice, should be given with *mūnga,* (*phaseolus aureus*), *arhar* (*cajanus cajan*), lentils, rock salt, ghee and dry ginger. *Āmalaka (embilica officinalis, embolic myrobalan, phyllanthus emblica*), *ber* (*zizyphus*) and rock salt mixed with seasame, *munacca* (raisins) and animal soup should be given.[46]

Modern science of medicine also prescribes a diet of whole cereals or pulses rich in vitamin B group for herpes. For *śoth* or swelling, old barley, wheat and śāli rice with dry radish and *kulattha* (*dolichos biflorus*) with animal soup, jaggery and *harada* (*treminalis chebula* or *chebulic myrobalon*) are prescribed. Jaggery and dry ginger are also considered to be good.[47]

For a patient with breathing trouble or bronchitis, *pua* or doughnut of *kulattha (dolichos biflorus*), *munga* (*phaseolus aureus*), dry *ber* (Indian jujube—*zizyphus mauritiana, zizyphus sativa, zizyphus nummularia*), dry radish and meat should be soaked in curd and mixed with pomegranate juice, juice of *mātulinga* (a type of lime or lemon), honey, *kṣaudra* (*munacca,* dried black currants) and *vyoṣa* (dried ginger, pepper and long pepper). *Kulattha* dal seasoned with *daśamūla* and other herbs is also prescribed.[48]

Kulattha was well known as, antipyretic or febrifuge and was used in bronchitis and asthma. Caraka says *Kulatthā grahiṇāha kāsahikkā Śvāsārśasā hitah.* Suśruta also calls it *kaphamārutaghṇaha, śvāsāpahaha* and *kaphasyahantā.*[49]

A patient of rheumatism is prescribed a diet of old barley, wheat and *śāli* rice along with animal soup. *Mūnga* dal, along with *āmlā*, dates, munacca and ber is also given.[50] Honey, clarified butter (*ghee*), butter milk, margosa, *parpata* (*pitta pāparā olden landia herbatea*) are also given.[51]

A patient suffering from piles should be given various preparations of barley with meat soup. He is also to be given leafy vegetable, *sauvarchala* and *śati śākam* (*Sauvarcalam Śati. Śauvarchal* has been translated as *hurhur*).[52] According to Varier, the Sanskrit equivalent for *hurhur* is *Paśugandhā.*[53] Sauvarchal has also been translated as vegetable juices of emblic myrobalan[54] (amla). Rupalalji Vaishya (Commentary on *Bhāva Prakāśa,* p. 464) says that types of *suvarchala* are recognized—*gynandropis pentaphylla* which is the white variety and *cleome viscose* the yellow variety. There is one another purple variety or *cleome monophylla.* It is also recommended in breathing trouble, fever, polyuria, vaginitis, anaemia and meals as an anti thelmintic. The white variety removes flatulence, *kapha* and *vāta* (Vaish *op cit.*, p. 464). Varier quotes passages of *nighantus* giving the properties of *suvarchchala* under sunflower or *Helianthus annus.*[54a] *Śati* is *kachoor* (*curcuma zedoaria*). The powder of *harada* (*pathyā haritaki, chibulic myrobalan*) should be given with buttermilk.[55] Decoction of nutgrass, (*nāgaramothā, muṣtā* of two varieties—*cyperus rotundas* and *cyperus scariosus*) is to be administered frequently. A paste of *citraka* (rosy flowered leawort, *raktacitrā, plumbago indica*) with turmeric is to be applied externally.[56] *Yava* (barley), *śāli* (rice), *vāstukam* (*bathua*), and *āmlā* are also prescribed. The *Bhāva Prakāśa* recommends *Nāgarmotha* in all stomach and intestinal and urinary disorders. It is also used after delivery for the contraction of uterus and for proper lactation (commentary on *Bhāva Prakāśa* also for piles, leprosy, cough and for worms (*Bhāva Prasāśa nighantu* karpurādivarga, 95-96).

In retention of urine, cucumber (*Trapuṣa*) and *Irvaru* (*cucumis utilissimus* or *kakadi*) should be given. Wheat, milk along with sugarcane and clarified butter are also prescribed. *Manda* (thick creamy preparations) and wine should also be given.[57]

Wine has a diuretive effect.

In case of vomiting, puffed rice (*lājā*), *sattu,* honey, black currants (*ksaudram*), meat and *parūṣakam* (*fālsā, grewia asiatica*) or their juices are to be given to prevent parching. A patient is given milk diluted with equal portions in water. He is also given balls of jaggery mixed with nutgrass to chew.[58]

In *Urustambha* (swelling, numbness and lack of coordination in the thighs), many preparations of barley, *puas* (a type of doughnut) of dry radish, vegetables of *patola* (*paraval trichosanthes dioica*) and *vetra* (common rattan *calamus rotang*) are prescribed.[59] *Verta* sprouts are sweet, diuretic and thermogenic and are useful in a vitiated condition of *vāta* and *kapha.*[60]

In headaches, *snigdha* or lubricating and hot food should be administered. *Amla* and *ājya* (clarified butter) should be applied.[61] The diet for a patient of diabetes is *apūpa* or *pua* and other preparations of barley. *Mūnga, kulathi,* old *śāli* rice, green and bitter and dry vegetables like (*sesame*), *śigru* (drumstick), *vibhītaki* (*bahera, belleric myrobalan* are also prescribed).[62]

For a patient with tuberculosis, a year-old barley, wheat and *munga* should be given. Meat soup is also prescribed.[63]

The *Garuda Purāṇa* in 1.169 models the science of nutrition exactly on the lines of *Suśruta saṃhitā* (1.46). The names of the two chapters are also similar. *Suśruta Saṃhitā* calls the chapter *anna pānavidhi,* meaning dietetics and *anupānavidhi* in the *Garuḍa Purāṇa* also means supporting medication or dietetics.

Suśruta has assigned separate chapters for the classification of agricultural produce according to their therapeutic value in *drvyasangrahāṇiyam adhyaya* 1.38 and in *annapana vidhi* 1.46. In the former, the different agricultural produce are classified according to their common properties and in the latter the nutritive value of each of the different produce is discussed separately. The *Garuḍa Purāṇa* summarizes the ch. 1.46 of *Suśruta saṃhitā* in 1.169. The classification of Suśruta in 1.38 is based on the classification of plant according to their therapeutic values. The different classes are named after the leading or prominent plant in each. They are divided into 35 *vargas* or classes. The *Garuḍa Purāṇa* also classifies the edible products in 1.173 according to their common properties but names them in accordance with their prominent taste in each. They are divided into groups like *sweet group, acid group,* etc. It also enumerates the demerits of an excessive intake of these products. The nomenclature used in the *Garuḍa Purāṇa* for therapeutic nutritional products is *dravya* in 1.173. These are used as preventive diets to ward off illnesses and are also used as curative medicines. In chapter 1.202 (edited by Ramtej Pandey) or 1.204 (edited by J.L. Shastri), the various forest produce are called medicines or *auṣadhi. Nāmānyetāni ca hare vanyānām bheṣajām tathā and atha nāmaāni vakṣyāmi auṣadhi nāma samāsataha* (in the first *śloka* of the chapter 1.202 of the *Garuḍa Purāṇa* ed. By Ramtej Pandey, Chowkhamba).

Chapter 1.170 (*Garuḍa Purāṇa*) onwards lists in detail all the medicines in various permutation combinations and also in combination with different cereals. They are either taken internally or applied externally.

The Chapter 169 of the *Garuḍa Purāṇa* prescribes various cereals and plants as part of therapeutic nutrition. These are as following:

The red species of *śāli,* rice dispels excess of three *doṣas.* It removes thirst and fat[64] *rakta śāli tridoṣaghnam tṛṣṇāmedonivārakam.*

Mahaśali is a good aphrodisiac. *Kalama* variety dispels *kapha* and *pitta.* The *gaurṣaṣtika* dispels three *doṣas* and is heavy and śīta (cooling in potency).

These properties or potencies were later accepted by the Nighantu writers also. *Rāja nighantu* (by Narahari, Most probably in fifteenth century AD) says *rakta saliha sumadhuro laghuha snigdho balāvaha. Kaiyadeva Nighantu* also says that *rakta śāli* is aphrodisiac, diuretic, removes breathing troubles, acts as tonic and removes three *doṣas* the modern ayurvedists also claim its use in diarrhoea, consumption and colonopathy.[65]

Mahāśali is also called in aphrodisiac in later *Nighantus* like *Dhanvantri Nighantu.* Ksirasvāmi considers it older than *Amarakośa* but modern scholars regard it as belonging to approximately twelfth century AD.[66] *Dhanvantari Nighantu* emumerates the properties of different types of rice exactly in the manner of *Garuḍa Purāṇa.* It also says that *gauraṣaṣtika* is *śīta* and heavy and dispels three *doṣas* and *kalama* variety dispels *pitta* and *kapha.*[67]

Garuḍa Purāṇa further says that *śyāmāka* (*echinochola frumentacia*) is absorbant, parchifying and productive of *vāyu* principle. It dispels *kapha* and *pitta. Dhanvantari nighantu* and *Rajanighantu* both repeat the same properties.[68] *Priyangu* (*Setaria italica here*), *Nīvāra* (Bengal wild rice *Hygroryza aristata*), *Kordūṣa* (*paspalum scorbiculatum*) also have the same properties. Varier also quotes the *Bhava Prakāśa* for the various properties of *priyangu*. The flowers and fruits are cooling, anodyne, digestive and are useful in rheumatoid arthritis and dyspepsia and flatulence.[69] But here it may be identified with *Setaria Italica* or *kanguni* since its name occurs in the list of *dhanya* and *kudhānya* groups.

Yava (barley), taken frequently, generates *vāyu,* is cooling (*sīta*) in potency and dispels *kapha* and *pitta.* The properties of *yava* are repeated in later *Nighantus.* The *Dhanvantari nighantu* says *Vṛṣyaha sthaviryakaromūtra meda pitta kaphanjayet.* Because of its cooling properties, it is used in urocystiti, uretherites, gastropathy, ulcers and burns. It is an important part in the diet of invalids.[70] *Godhūma* (wheat) is an aphrodisiac and cooling in potency. It dispels *vāyu.* It is sweet and heavy.[71] The modern ayurvedists consider the grains as sweet, refrigerant, emollient, laxative, rejuvenating, aphrodisiac and tonic. The grains are useful in anorexia, flatulence, boils, pruritis, ulcers, constipation and general debility. The properties are corroborated in the *Dhanvantari Nighantu* and the *Bhāva Prakaśa.*[72]

Mudga (*munga, phaseolus aureus*) dispels excess of *kapha, pitta* and blood. It is astringent, sweet and light.[73] The later *nighantu* writers like *Bhava Prakāśa* mention five varieties of *mūnga* but say that the green variety is the best.[74]

The *Garuda Purāṇa* says that *māṣa* or black gram is strengthening, aphrodisiac and heavy. It dispels *pitta* and *kapha.*[75] But at another place, it says that *māṣa* removes *vāyu* and *pitta* and destroys the glossiness of skin.[76]

It is interesting to note that the *Garuḍa Purāṇa* says *rajamāṣa* (*vigna catjang* or *lobhia*) is not an aphrodisiac and this is accepted by modern ayurvedists also.[77] Further, the *Garuḍa Purāṇa* says that it dispels *kapha* and *pitta* and removes disorders due to *vāyu Kulattha* (*dolichos biflorus*), dispels *kapha* and *vāyu* and removes disorders of dyspnea, hiccough and enlarged spleen.[78] Suśruta also enumerates these preparations and says that it is beneficient in tumours and asthma.

Śukarśmarigulmaniṣūdanasya
Sāngrāhikaha pīnakāsahantā[79]

The *Āṣṭāngahṛdaya* agrees with this. *Bhāva Prakaśa* also points to its antithelmentic properties and the modern ayurvedists use it to cure cardiopathy.[80]

Makuṣtaka (*moth, phaseolus aconitifolius*) dispels fever due to *rakta pitta.* It is cooling in potency and astringent.[81] It is interesting to note that Suśruta says that it encourages worms; *Makuṣtakaha Kṛmi karāha.*[82]

It is very strange that gram or (*caṇaka*) is supposed to reduce the virile power in *Suśruta saṃhitā* and in *Garuḍa Purāṇa* too, although both say that *caṇaka* generates *vāyu,* dispels *kapha* and *pitta* and is styptic.[83] Suśruta saya *canakāh punsatvanāśanaha* and *Garuḍa Purāṇa* says—

Punstvāsḍkkaphapittaghnascanako Vālalaha smṛtaha.[84]

But the later *nighantus* and modern ayurvedists do not think so. *Kaiyādeva nighantu* says it is *Viṣtambhi punstvakārakaha.* Modern ayurvedists have analyzed its properties as a purgative, abortifacient and it is useful in anorexia, dyspepsia and bronchitis. The seeds are aphrodisiac, expectorant and cooling and are also useful in hyperdipsia, burning sensation spleno hepatomegaly[85] leprosy, pharyngopathy and skin diseases. Regarding its effect on virility, the diverse views are reconciled in the *Bhāva Prakāśa.* The *Bhāva Prakāśa* gives its different properties according to its different modes of cooking. It attributes the reduction of virility to soaked gram only. *Ārdroatikomalo rucyha pitta śukraharo himaha.*[86]

Masoor (*lens culinaris* or *lens esculenta*) is sweet, cooling in potency, easily digested (*sangrāhi*) and it dispels *kapha* and *pitta. Masooro madhuraha śītaha sangrāhi kapha pittahā*[87] Suśruta[88] says —*Vipāke madhurāha proktā masoorā baddhavarcasah. Bhāva Prakāśa*[89] attributes the same qualities to it as the *Garuḍa Purāṇa—masooro madhuraha pāke sangrāhi śītalo laghuha kapha pittastrajit.* The modern ayurvedists also use it in ophthalmopathy, diarrhoea, dysentery, tumours cardiopathy, indolent ulcers, skin deseases and anaemia.[90]

Masoor has a high protein content—25 gm as per 100 gm of edible protein, and low fibre 0.7 gm and 343 calories. Hence, it is supposed to be good for diarrhoea.[91]

Ādhaki (arhar—*cajanus cajan* or *cajanus indicus*) dispels *kapha* and *pitta* and is conducive to the generation of semen *ādhaki kapha pittaghni śukralā ca tathā smṛtā*[92]. Although the *Suśruta Saṃhitā* recommends the use of the oil of a *Tuvarak* found near the western seas for diabetes, urinary diseases and leprosy (*Suśruta Saṃhitā* 4.13.7–10), it is doubtful whether this can be identified with *Ādhaki* (which is also known as *tuvaraka*) as *ādhaki* is cultivated throughout India. However, the authors of the *Suśruta Saṃhitā scientific synopsis* have identified this with *cajanus indicus* or *ādhaki.*[93] The identification of Varier of *Tuvarakah* with *Hydnocarpus laurifolia* may be accepted here 93 (a).

Like the *Garuḍa Purāṇa,* the *Bhāva Prakāśa* of Bhava Misra also says that *ādhaki* removes *pitta* and *kapha.*[94]

Indian Medicinal Plants of Arya Vaidyasala recommends the leaves of *adhaki* (*cajanus cajan*) as a diuretic, laxative, cooling, anti-inflammatory and anodyne. They are useful in oral ulcers, odontalgia, gingivitis, strangury and inflammation. The seeds are useful in vitiated conditions of *pitta,* intestinal worms, oral, ulcers, tumours, bronchitis, cough, vomiting, haemorrphoids, fever and cardiac diseases.[95]

Kalāya (*pea, pisum sativum*) has the properties of *masoor* but also the property of inducing flatulence.[96]

Tila (sesame) is alkaline, sweet, oily, strengthening, heat producing and *pitta* generating.[97]

The nutritive property of common vegetables and fruits is also mentioned in the *Guruḍa Purāṇa* which are similar to those is *Suśruta Saṃhitā.*

The *Garuḍa Purāṇa* consider tandylīya and *pālankya* as antitoxic. *Tanduliya* is *chaulai* or *amaranthus spinoisus* and *pālankya* is identified with spinach or *spinacea oleracca* by most of the scholars. Varier, however, identifies it with *beeta vulgaris* or *chaukandar.*[97a]

Mūlaka mentioned in 1.169.16 (*Garuḍa Purāṇa*) can be easily identified with radish *raphanus sativus.* It is said that if cooked in steam it destroys *vāyu* and *kapha.* Varier distinguishes *mūlaka* and *mulakyā* and says that *mūlaka* is identified with a *kandaviṣa* which was poisonous in Suśruta and, hence, cannot be radish. He identified radish with *mūlikyā* (*raphanus sativus*).[98] But because of its properties and its identification in other Sanskrit texts with radish, *mūlaka* should be identified with radish only.[99] The *Bhāva Prakāśa* also identifies *mūlaka* with radish.[100]

Vārtāka (Brinjal or *solanum melangona*), *patolaka* (Parwar, *Trichosanthes dioica*) and *kāravellaka* (Karela, *momordica charantia*) are said to destroy all *dośas*. When cooked and eaten, they are said to be good for the heart and throat.[101]

Kūṣmāṇḍa (*cucurbita pepo* or *petha*) is diuretic, relishing and destroys all *doṣas*. It is highly efficacious in *kuṣṭa* or leprosy, urinary diseases, fever, dyspnea and asthma, as well as disorders due to the derangement of *pitta* and *kapha*.[102]

Alābu (*Lauki*, bottle gourd, *Lagenaria Siceraria* or *langenaria vulgaris*) destroy *pitta* and produces *vāyu*. *Trapuṣa* (cucumber, *cucumis sativus*) and *Irvaruka* (either *Kakadi* or *snake cucumber* or *musk melon* or *cucumis melo*) generate *vāta* and *kapha* and ward off *pitta*.[103]

Since the *Bhāva Prakāśa* gives the synonym *karkatah* (*kakadi*) to Ervāruka (not *Irvaruka* as mentioned in the *Purāṇa*), it is possible that *Irvāruka* is *kharbooja* (*musk melon, cucumis melo*).

Varieties of lemon (*jambīra, mātulinga* and *bījapūraka*) destroy *vāyu*.[104] *Bīja pūraka* are used for enlarged spleen, dyspnea, asthma and diseases of *vāyu* and *kapha*.[105] Pomegranate (*punica granatum*) is astringent and antiflatulent. *Nagarangaphala* (orange, *citrus reticulata* or *citrus aurentium*) is heavy.[106]

Āmlā (*emblica myrobalan, phyllanthus emblica*) and *hāritaki* (*harada, chebulic myrobalan*) are profusely used as medicines. *Āmlā* is said to be sweet, relishing and aphrodisiac but produces acidity if taken in greater quantity.[107] But *haritaki*[108] induces taste for food, is holy and nectar like. It is also said to be laxative and the destroyer of *kapha* and *tridoṣas*.[109]

The fruit of *Tintidi* (Tamarind, *Tamarindus Indica*) is laxative and the destroyer of *kapha* and *vāyu*.[110]

Mango generates *vāyu* and produces flesh, semen and heightens colour of the skin. *Jāmuna* (*Syzygium cumini* or *Eugenia jambolana*) destroys *vāyu, kapha* and *pitta*, is astringent but produces *viṣtambha* form of indigestion and is griping.[111]

Badara (*Ber, zizphus jujuba*) destroys *vāyu* and *pitta*.[112] In *ślokas* 11 and 12 of 1.169 of the *Garuḍa Purāṇa, bilva* (*Aegle marmelos*) destroys *kapha* and *pitta*, removes germs and is light and appetizing. *Bilvā kaphapittaghnāha krmighnā laghudīpikā*) but in 1.169.27 (*Garuḍa Purāṇa*) it is said that *bilva* generates *vāyu* and causes *Viṣtambha*. The two statements can be reconciled by referring to the *Bhāva Prakāśa*. According to the *Bhāva Prakāśa*, the raw fruit has properties of destroying *vāta* and *kapha* and it makes the intestines healthy. But the ripe fruit is sweet, aromatic and causes *Vistambhi*.[113] Varier, in *Indian Medicinal Plants*, says that the ripe fruits are astringent, sweet, aromatic, cooling, febrifuge, laxative and tonic, and are good for the heart and brain in dyspepsia.[114] But the differences in the properties of raw and ripe fruits given by the *nighantu* writers are quoted by Varier from the original passages, which distinguish between the properties of raw and ripe fruits of *bela or bilva*.[115]

Drakṣā (grapes), *madhūka* (*mahua—Madhuca longifolia* or *Bassia longifolia*) and dates (*khajoor—Phoenix sylvestris* and *chuhārā—Phoenix dactylifera*) and *kumkum* (*kesara, crocus sativus*) dispel disorders of blood and *vāyu*. Ripe *māgadhi* is sweet and destroys *pitta* and disorders due to dysprea.[116] *Māgadhi* refers both to *juhi* (*Jasminum auriculatum*) as well as *pippali* (*piper longum* or *pippali* or *peepari*). Here, *pippali* is indicated.

Ginger is relishing, appetizing, aphrodisiac and *kapha* and *vāyu* destroying. So is pepper. Asphoetida dispels enlarged spleen, pain, constipation and disorders of *vāyu* and *kapha*. Coriander and cumin seeds also destroy *vāyu* and *kapha*.[117]

Ch. 173 of the *Garuḍa Purāṇa* groups the various food produces in the following classes: *madhurādi* or the sweet class. This includes *śāli, ṣaṣtika, godhūma,* lotus, grapes, dates, *priyālaka*. (*Buchanania latifolia*), *madhūka, kūṣmānda,* etc. These are curative of epileptic fits and burning sensation. But if taken in excess, they produce worms, disorders or *kapha,* dypnea, aphonia, tumours, goitre, etc.[118]

The second group is *amla* or acidic. It includes *dādima* (pomegranate), *āmla* (*embolic myrobalan*), *āmra* (mango), *mātulinga* (lemon), *badar* (*zizyphus*), *Tintidi* (tamarind), etc. Although they are great digestants with ginger (*sunthi*), but if taken in excess, cause looseness of the limbs, burning sensation, coarseneess of voice, throat, mouth and *heart*. They also make wounds and ulcers suppurate.[119]

The next group is composed of salts. Although they are laxative and digestants, their excess leads to itching, tumours, *rakta vāta, rakta pitta,* etc.[120]

The fourth group is the pungent group (*katuka*). It includes *vyoṣa trikatu*— (*pipal, sontha, marich* pepper), *śigru* (drumstick or *sahjana*), *mūlaka* (radish), *devadaru cedrus deodara* or Himalayan cedar *Kuṣta* (Indian costus—*Saussurea lappa* or *saussarea auriculata*), *laśuna* (garlic), etc. Although they are anti *kapha* (anti expectorant) and subdue itching, remove corpulancy, lassitude and are anti thelmintic, when taken in excess, they produce vertigo, burning sensation, etc.[121]

The fifth or the bitter group (*tikta*) includes *haridrā* (haldi or turmeric), *Jāti* (nutmeg or *jayā phala*), *mandūka parni* (Indian pennywort—*Hydro cotyl asiatica*), *vartāku* (brinjal), *nima* (neema, margosa), *kāravellaka* (*karela,* bitter gourd), *punarnavā,* (Hogweed, Horse purslene, *Boerhavia diffusa*), *Tṛvṛt* (*Tribhandi, Tur peth* plant, *Ipomaea turpethium*), etc. Although these purify the ducts, remove fever and thirst, prevent fainting and reduce itching sensation, their excess intake dries up faeces, urine and mucous sensation.[122]

The next group is the astringent (*kaṣāya*) variety. This includes *triphala* (*harada, baherā, āmlā*), *śallaki* (*Indian olibanum, boswellia serrata*), *Jambu* (*jāmun*), *vata* (banian), *tinduki* (false mangosteen or *dyospyros embryopteris*), *śāla* (*shorea robusta*) *āmrataka* (Indian Hogplum, *spondivs mangifera* wild), *Pālanka* (*spinach*), *mudga* (*mūnga*), etc. They act as an absorbent, and dry up mucous sensations, produce granulations in ulcers but if taken in excess give chest pain, parch the mouth, cause fever, distention of the abdomen and paralysis of the mandible.[123]

Another group is *vātānta* and *kapha pitta hara* variety, which severally and in various combinations, removes excess of *vāyu pitta* and *kapha*. *Haridra* (turmeric), *Kuṣta* (*Kūtha, saussurea lappa* or *saussurea auricalata*), *punarnavā* (Hogweed, *Horse purslene Boerhaavia diffusa*), *yava* (barley) *kola* (*zizyphus, bera*), *kulattha* (*dolichus biflorus*) and *daśsamūla* (combination of ten herbs—five minor group of *pañcamūla* and five major groups of *pañcamūla*). The first includes *kanista panchmūla* which are as following—*trikantaka* (gokṣura, *tribulus terrestris*), *vṛhati* (*kākamāci,* black nightshade *solanum indicum*), *prithakparni* (painted leaved uraria, *uraria lagopoides*) and *vidārigandha* (*dīrghamūla śāla* parni, desmodium gangeticum).

Mahat pancamula includes the following: *bilva* (Bela, *aegle marmelos*), *agnimantha* (interpreted as premna *integrifolia* in *Suśruta Saṃhitā*). A Scientific Synopsis.

But the modern commentators of the *Bhāva Prakāśa* identify it with *Clerodendrum phlomidis arani*), *tuntuka* (Atrideva Vidyalankar and Bhaskar Govinda Ghanekar (commentatory of the *Suśruta Samhitā*) identify it with *śyonaka* or *colosanthes indica, patalā* (trumpet flower tree, *stercospermum suaveolens*), *kaśmari* (white teak, *gmelina arborea*).[124]

Next is *ślesma hara* variety or the *mucous kapha* subduing drugs. The variety includes *śatāvari* (Indian asparagus), *Vidāri* (*batatas peniculata* or *convolvus peniculata*), *bālaka* (fragement mallow plant or *pavonia odorata*), *uśīra* (*Khasa andropagan muricatum*), *candana* (sandalwood, *santalum album*), *dūrvā* (*cynodon dactylon*), *pippali* (long pepper plant or *piper longum, chavica roxburghie*), *badari* (*zizyphy jujuba*), *utpala* lotus, *Nelumbium speciosum willd), patolaka* (*Trichosanthes dioica*), *vyoṣa* (*peepal, sonth,* pepper or *kalimirch* (piper longum), zinziber officinalis and piper nigrum).[125]

Then comes the oil based group, a patient of *pitta* should take *ghee* alone; the patient of *vāta* should take it with salt. In a patient of excess of *kapha,* it should be administered with *vyoṣa* and alkalies.[126]

Various herbs and plants have acquired different therapeutic values according to different ways of cooking and also according to their various combinations.

Peya (gruel with plenty of water) is light, diuretic and restores deranged *vāyu,* e.g., peya with buttermilk, jaggery, *vyoṣa* pomegranate, honey and *pippali* dispels cough, bronchitis and diarrhoea.

Kṛṣara (*Khichadi*) destroys *vāyu.*[127]

Supa—Soup (*sūpa*) of watery nature, it is an emolient. Taken lukewarm, it is relishing and light. Soups with bulbous roots, radish and fruits well cooked with *ghee* or oil are heavy and rejuvenating. In a lukewarm state, it is light. Vegetables cooked in steam and well seasoned after squeezing out the water are wholesome.[128]

A *yūṣa* (soup which is thicker than sūpa) made of *dādima* and *amālaka* generates gastric fires and despels *vāyu* and *pitta.* A *yūṣa* prepared with radish dispels cough, breathing disorders and cold and cough.[129]

A *yūṣa* made with *yava* (barley), *kola* (ber) and *kulatha* (dolichos biflorus) dispels *vūya* and is beneficial to the voice. This prepared with *mudga* and *āmalaka,* is astringent and destroys *kapha* and *pitta.*[130]

Curd with jaggery dispels *vāyu, saktu* (flour of roasted grains) is sharp and generates vayu. *Saskuli* fried in ghee is an aphrodisiac and generates gastric fire.[131] Food stuffs with cooked meat are rejuvenating. *Piṣṭikas* (*pitthi* or ground paste of cereals) are heavy. Things fried in oil impair the eyesight. Those cooked in water or steam are difficult to digest.[132]

Hot *mandakas* (gruel) are wholesome, cooling in potency and also heavy.

Cereals and plants were also used externally in surgery related medicines and also as beauty aids. *Haridrā* was used for cleaning and sterilizing the interior of ulcers. *Godhūma* (wheat) was used as poultice; *yava* heals blisters if applied over them frequently.[133] Fried and powdered *tila* mixed with butter from buffalo's milk and *bhallāta* (*aruṣkara,* marking nut tree, *semicarpus anacardium*) applied as an errhine removes cardiac colic and when applied as plaster heals ulcers.[134]

Powdered roasted grains (*sattu*) mixed with goat milk and clarified butter can be applied to the feet to relieve them of aching sensation.[135]

A sword cut heals when the juice of *āmra* (Indian hogplum *spondius mangifera wild*) is applied. If spread with clarified butter, it does not turn into an ulcer.[136]

Several recipes of beauty aids are given in the *Purāṇas* which comprise mostly cereals and plants. A few examples can be given here: *Haridra* (turmeric) cooked in steam with rice, root of white mustard and seeds of *mātulinga* (a type of lemon) applied over the body for seven days result in a good complexion.[137]

A compound of *haritāla* (*orpiment*), *śankha* and ashes of plantain leaves is a good depilatory.[138] A paste of *lavaṇa, haritāla,* fruits of *tumbini* with juice of *lakṣa* (lac tree, *cocus lacca*) is also a fine depilatory.[139]

Atasi (linseed tree, *lenium usitatissimum*), *māṣa* (black gram, *phaseolus mungo* or *phaseolus radiatus*), *godhūma* (wheat, *Triticum, aestivum*), ground into a paste with *ghee* and powdered *pippali* when applied over the body makes it shine like that of Kandarpa.[140] Because of the styptic and purifying qualities of *godhūma,* it was also used as a poultice.

Paste of *pippali* and *haridra* with cow urine inserted into the rectum removes piles.[141]

A paste of *yava* (barley) with gingelly oil is a remedy for burns and scalds.[142]

The external uses of plants were popular in the field of obstetrics and gynaecology too. It is said that by applying the paste of leaves of *badari* (ber zizyphus) a rupture of vagina can be cured. A paste of *lodhra tilvak* (*symplicos racemosa*) and *tumbi phala* results in firm vaginal muscles.[143]

For sterile women, *aśvagandhā* (winter cherry, *withania somnifera*) taken with *ghee* and half an *ādhaka* of milk enables conception.[144] Suśruta also recommends *aśvagandhā* in promoting growth of normal tissues after surgery, for quick healing of wounds, in swellings and as an emetic.[145] Decoction of *daśamūla* quells all post-parturient pains. Powdered *śāli* rice taken with milk increases breast milk.[146] Similarly, a soup of *mudga* (*mūnga*) also purifies breast milk.[147]

Cereals were used profusely in the *Vṛkśāyurveda.* Chapter 282 of the *Agni Purāṇa* prescribes many recipes of different cereals for promoting the growth of plants and trees.

Some of the preparations presented in the *Purāṇas* are given below:

For trees bearing weak fruits, a mixture of *kulthi, urad, mūnga, til* and *yava* should be mixed with water.[148]

Similarly, *til* and powdered barley was mixed with the excreta of sheep and goat and ghee was smeared on this. This decoction was mixed in water and this water was then poured in the roots of fruit-bearing trees.[149]

Agriculture also promoted the science of astrology and *vāstu.* Various rules were prescribed for planting trees according to different constellations and periods. It is said that *plakṣa* (yellow fig tree *Ficus infectoria*) should be planted in the north, *Vata* (banyan tree) in the east, mango in the south and *aśvattha* (*peepul*) in the west is auspicious.[150]

Trees with thorns should be planted in the south. The moon and *brāhmaṇas* should be worshipped before planting trees. Rohini and Revati, constellations and *abhijit muhurta,* were preferred for planting of trees.[151] The time of planting of trees in the beginning of rainy season in the evening and morning and in winter it is the late afternoon. During the full onset of rainy season, the trees should be planted at night. The trees should be planted at a distance of 20 hands from each other. If not possible, 16 hands and 12 hands are the measured distances.[152]

The cereals were used in religious rituals too. With the growth of *Vaiṣṇavism,* animal sacrifices were replaced by herbs and cereals which were already used in Vedic *yajnas.* The *Mahābhāratā* (Santiparva) had recommended the use of *bijauṣadhis* in yajña *bījairyajñyeṣu yaṣṭavyamiti* (Santi Parva 322.19). A *Vaiṣṇava* devotee also offers to the god the same food that he partakes of everyday.

The various rituals in the tantric rites also required the use of cereals and herbs like *yava, śāli, śyāmāka, tila,* etc.[153]

The various deities ruling over children of various ages were pacified by cereals like *tila, māsa,* and *kulathi.*

The *Purāṇas* were not only religious texts. They included knowledge of all spheres of life as religion was not divorced from life. Hence, they incorporated the passages from texts of agriculture, medicine, *āyurveda* and science and went on to acquire an encyclopaedic character. The object of this was to spread knowledge among the masses who would not have read the above treatises of science, medicine, agriculture, etc. Thus, the *Purāṇas* are a valuable source of information on the medical properties of agricultural produces. Much of their information are corroborated by modern science of nutrition, medicine and the Āyurveda.

NOTES AND REFERENCES

1. V.S. Pathak *Smarta Religious tradition,* Susmita Pande, Smarta Religious tradition and the grants of Chandelas—A study in the contemporary socio-religious value in *Approaches in Art and Archaeology of Madhya Pradesh.*
2. *Viṣṇu Purāṇa* I. 6.21–28.
3. *Suśruta Saṃhitā (a scientific synopsis)* P. Roy, H.N. Gupta, M. Roy, p. 199.
4. C. Gopalan, B.V. Ramashastri and S.C. Balasubramanyam, *Nutritive value of Indian Foods.* p. 151.
5. Varier, *Indian Medicinal Plants,* 1.334.

5a. Shastri and Vaishya, *Bhāva Prakāśa,* pp. 248–251.

6. *Kāśikā* 8–4–6 *Phalivanaspatirgyeo vṛkṣah puṣpaphalopagā Oṣadhyah phala pakāntāh latāgulmaśca vīrudhah.*
7. Suśruta, 1.1.28–34. The *sthāvara auṣadhis* are plants and herbs. *Jangama* are birds and animals, *parthiva* metals and stones, etc., and *kāla kṛta* and sunlight, moonlight, shade, seasons, rains, etc.
8. *Ibid.,* 1.1.28–29.
9. *Oṣadhayaśca kṣudrā sthāvarā vanaspatayobharantaḥa.*
10. *Mitaksarā – oṣadhyah phalapakantaḥa śāliprabṛtayaḥ.*
11. *Kāśika* 6.2.42.
    Osodhīyate asyāmiti aṣadhiha.

12. *Bṛhat Saṃhitā* 19.12.
13. *Vāyu Purāṇa,* 53.19.
14. *Bhāgavata Purāṇa,* 1.8.40.
15. *Viṣṇu Purāṇa,* 1.5.50 *Oṣadhyah phalamūlinyo romebhyastasya jagyire.*
16. *Vāyu Purāṇa,* 44.187.
17. *Amarakośa,* II.4.
18. *Amarakośa,* II.4.6 *Oṣadhyah phalapākāntā.*
19. *Agni Purāṇa,* 279.3–4.
20. *Ibid.,* 279.6.
21. Suśruta, 1.46.
22. Huen Tsang was given *mahāśali* during his stay in Nalanda. *Atri deva vidya lalkara Ayurveda ka Brihat Itihāsa,* p. 447.
23. *Suśruta Saṃhitā,* I.46.6.
24. The values of different nutrients are taken from *Nutritive Value of Indian Foods,* Gopalan, B.V. Ramashastry and S.C. Balasubramanyam. The data is based on the researches carried out in the National Institute of Nutrition, Hyderabad.
25. *Agni Purāṇa,* 279. 3–4.
26. P.K. Varier, V.P.K. Nambiar and C. Ramankutty—*Indian Medicinal Plants* (Arya Vaidyasala) Kottakal, Vol. 5, p. 431.
27. *Agni Purāṇa,* 279.6.
28. *Suśruta Saṃhitā, Uttaratantra,* 40.95 and also *Uttaratantra* 4.115.
29. *Agni Purāṇa,* 279.6.
30. *Suśruta Saṃhitā, Uttaratantra,* VI, 45.
31. *Agni Purāṇa,* 279.8, also Suśruta VI 115.12.
32. *Suśruta Saṃhitā,* VI. 45.16.
33. *Ibid.,* VI. 45, 21–22.
34. *Ibid.,* VI. 45, 24–27.
35. *Nutritive Value of Indian Foods op cit.,* p. 62.
36. *Agni Purāṇā,* 279.9.
37. *Ibid.,* 279.10.
38. *Ibid.,* 279.11–12.
39. *Suśruta Saṃhitā,* 6.40.91.
40. *Ibid.,* VI. 6.40.115.
41. *Agni Purāṇa,* 179, 13–14.
42. *Ibid.,* 279.15.
43. *Ibid.,* 279.16.
44. *Ibid.,* 285.21–23.
45. *Ibid.,* 279.
46. *Ibid.,* 279.36–37.
47. *Ibid.,* 279.123–24.
48. *Ibid.,* 279.20–22.
49. *Indian Medicinal Plants* of P.S. Varier (Arya Vaidyasala, Kottakal) Vol. 3, p. 360.
50. *Agni Purāṇa,* 279.25.

51. *Ibid.*, 279.26.
52. Tarinish Jha (*Agni Purāṇa,* Hindi Sahitya Sammelan, Prayag) translates *Sauvarchal* as *hurhur.* Also Bhāva Prakāśa with the commentary of Brahma Shankar Mishra and Rupalalji Vaisya, p. 464.
53. *Indian Medicinal Plants, op. cit.* Vol. 2, p. 116.
54. P. Ray, H.W. Gupta, Meera Roy *Suśruta Samhitā* (*A Scientific Synopsis*) p. 233. One other meaning is soda neutralized with acid. *Ibid.* p. 23.
54a. Varier, *op. cit.*, Vol 3, p. 127.
55. *Agni Purāṇa,* 279.30.
56. *Agni Purāṇa,* 279.31.
57. *Ibid.*, 279.32.
58. *Ibid.*, 279.33–34.
59. *Ibid.*, 279.35.
60. Varier, *Indian Medicinal Plants, op. cit.*, Vol. 1, pp. 330–332. It quotes *kaiyādevanighantū Meham valāsam ca karoti vātam Vetrāgramuktam ruci kṛdviśeṣat.*
61. *Agni Purāṇa,* 279.43.
62. *Ibid.*, 279.17–18.
63. *Ibid.*, 279.19.
64. *Garuḍa Purāṇa,* 1.169.1.
65. Varier, *Indian Medicinal Plants* Vol. 4.193.
66. Atrideva Vidya Alankar *Ayurveda ka Brhat Itihasa,* p. 424.
67. Varier, *Indian Medicinal Plants,* Vol. 4, p. 196.
68. *Ibid.*, Vol. IV, p. 211.
69. *Garuaḍa Purāṇa,* 1.16.9.3 and Varier, *op.cit.*, Vol. 1.334–336.
70. Varier, *op. cit.*, Vol. 3, p. 175.
71. *Garuaḍa Purāṇa,* I, 169.4.
72. Varier, *op. cit.*, Vol 5, p. 335.
73. *Garuḍa Purāṇa,* 1.169.5.
74. Varier, *op. cit.*, Vol. 5, pp. 374–377.
75. *Garuḍa Purāṇa,* 1.169.5.
76. *Ibid.*, 1.169.21.
77. Varier, *op. cit.*, Vol. 5 p. 382.
78. *Garuda Purāṇa,* 1.169.6.
79. *Suśruta Samhitā,* 1.46.37.
80. *Indian Medicinal Plants, op. cit.*, Vol. 3, 358–359.
81. *Garuda Purāṇa,* 1.169.7.
82. *Suśruta Samhitā,* 1.46.30.
83. *Ibid.*, 1.169.31.
84. *Garuda Purāṇa,* 1.169.7.
85. Varier, *Indian Medicinal Plants, op. cit.*, Vol.2, p. 70.
86. *Bhava Prakāśa* ed. by Brahama Shankar Shastri and Rooplalji Vaishya Chaukhamha Varanasi) Vol. 1 *Bhāva Prakāśa Nighantu Dhānya Varga* pp, 648–49.
87. *Garuda Purāṇa,* 1.169.8.

88. *Suśruta Saṃhitā,* 1.46.29.
89. *Bhāva Prakāśa. op. cit.,* p. 647.
90. Varier, *op. cit.,* Vol. 3 p. 309.
91. Gopalan, Shastri, Balasubramanyam, *Nutritive Value of Indian Foods,* p. 63.
92. *Garuḍa Purāṇa,* 1.169.9.
93. Ray, Gupta, Ray *Suśruta Saṃhitā—A Scientific Synopsis,* p. 218.
93a. Varier, *op. cit.,* Vol. 3, p. 185.
94. *Bhāva Prakāśa, op. cit., Bhavaprakasa nighantu Dhānya Varga,* 52.
95. Varier, *Indian Medicinal Plants,* Vol. 1, p. 326.
96. *Garuḍa Purāṇa,* 1.169.8.
97. *Ibid.,* 1.169.10.
97a. Varier, *op. cit.,* Vol 1, p. 265.
98. Varier, *op. cit.,* Vol 5, p. 148.
99. Ray, Gupta, Ray *op. cit.,* also identify *Mūlaka* of Suïruta as radish p. 190.
100. *Bhāva Prakāśai op. cit., Bhāva Prakāśa nighantu, śākavarga* 99–103.
101. *Garuḍa Purāṇa,* 1.169.17 Varier *India Medicinal Plants* wrongly identifies it with *chicinda.* Actually, this is *parwar* of north India, which is agriculturally cultivated. This can be distinguished from its wild variety.
102. *Garuḍa Purāṇa,* I.169.18.
103. *Ibid.,* 1.169.19 Varier has identified *Irvāruka* as snake guard *op. cit.,* Vol. 2 p. 227 *Bhāva Prakāśa* gives the synonym *Ervaru to karkataḥ* or *kakadi. Bhāva Prakāśa* — Chowkhamba Ed. By Brahma Shankar Shastri and Roopalalji Vaish.
104. *Ibid.,* 1.169.20 for *jambira* and 21 for *matulinga.*
105. *Ibid.,* 1.169.23.
106. *Ibid.,* 1.169.20.
107. *Ibid.,* 1.169.22.
108. *Ibid.,* 1.169.22.
109. *Ibid.,* 1.169.24.
110. *Ibid.,* 1.169.24.
111. *Ibid.,* 1.169.26.
112. *Ibid.,* 1.169.27.
113. *Bhāva Prakāśa, op. cit.,* p. 275.
114. Varier, *op. cit.,* Vol. 1, p. 62.
115. *Ibid.,* p. 63–64.
116. *Garuḍa Purāṇa,* 1.169.29.
117. *Ibid.,* 1.169.30–32.
118. *Ibid.,* 1.173.1–6.
119. *Ibid.,* 1.173.6–9.
120. *Ibid.,* 1.173.10.
121. *Ibid.,* 1.173.12–13.
122. *Ibid.,* 1.173.14–18.
123. *Ibid.,* 1.173.19–20.
124. *Ibid.,* 1.173.21–22.

125. *Ibid.,* 1.173.23–25.
126. *Ibid.,* 1.173.26–29.
127. *Ibid.,* 1.169.55–56.
128. *Ibid.,* 1.169.57–58.
129. *Ibid.,* 1.169.59.
130. *Ibid.,* 1.169.60.
131. *Ibid.,* 1.169.61.
132. *Ibid.,* 1.169.62–63.
133. *Ibid.,* 1.177.60.
134. *Ibid.,* 1.177.61.
135. *Ibid.,* 1.177.56.
136. *Ibid.,* 1.177.63.
137. *Ibid.,* 1.177.69.
138. *Ibid.,* 1.181.7.
139. *Ibid.,* 1.181.8.
140. *Ibid.,* 1.182.4.
141. *Ibid.,* 1.183.
142. *Ibid.,* 1.183.7.
143. *Ibid.,* 1.172.4.
144. *Ibid.,* 1.172.8.
145. Ray, gupta, Ray *op. cit.* pp. 145–146.
146. *Garuḍa Purāṇa,* 1.172.12.
147. *Ibid.,* 1.172.13.
148. *Agni Purāṇa,* 282.10.
149. *Ibid.,* 282.11.
150. *Ibid.,* 282.1.
151. *Ibid.,* 282.2–4.
152. *Ibid.,* 282.7–9.
153. *Ibid.,* 282.14–18.

CHAPTER 47

Technique and Process of Agriculture in Early Medieval India (c. AD 700–1200)

Lallanji Gopal

The responsibility of the king to provide for irrigational facilities in his kingdom is traditional and very old in origin. The failure of rains was often ascribed to some sin or fault on the part of the king. Medhātithi[1] says that it is the nature of the rainy season that there should be rain, and yet, on account of the faulty action—either of the king or of the kingdom itself—there is some times drought. It was as a result of this notion that the kings undertook the excavations of wells, tanks and canals. The *Aparājitapṛcchā*[2] advised that crops should be protected to maintain the kingdom and water reservoirs should be provided for the crops. It was the recognition of the practical importance of irrigational work which led Lakṣmīdhara, the Gāhaḍavāla minister, to defy conventional notions and class such works of public utility separately under *dāna* and not under *iṣṭāpūrta* and ascribe them a high spiritual efficacy. It was a change of real significance. The classification had been followed by subsequent writers on the subject.[3]

IRRIGATION WORKS: DAMS, RESERVOIRS TANKS, WELLS, ETC.

The most remarkable irrigational project implying ingenious engineering skill comes from Kashmir in the reign of King Avantivarman. Under him, the Minister Suyya dammed the River Vitastā (Jhelum) to save Kashmir from devastating floods of the Mahāpadma Lake. Suyya deepened the bed of the Vitastā at its two ends, cleaned the river bed at its bottom after constructing temporary stone dams at all the threatened points and built protective stone embankments for seven *yojanas* along the river bank. Thus, he was able to shift the junction of the Vitastā and the Sindhu from its old location to its existing position. On the land raised from the water, he founded many villages protected by circular dykes and constructed extensive projects. Kalhāṇa notices the prosperity of Kashmir resulting from this work. He showers unstinted praise on the great engineer for this work of immense technological skill.[4] Another Kashmiri king, Lalitāditya Muktāpīḍa reclaimed many hitherto water-logged areas by making an arrangement at Cakradhara to conduct the water of the Vitastā and, by constructing a series of water wheels, distributing it to various villages.[5] The King Harṣa excavated the big Pampā Lake. It has been identified by Aurel Stein with the modern Pamba Sar.[6]

It is to be noted that the King Kalhaṇa of the Naḍḍula branch of the Cāhamānas and Ajayasiṃha, a son of his feudatory, constructed several wells.[7] A sister of Pūrṇapālā, a Paramāra ruler of Abu, also claims to have constructed tanks.[8] Bhoja, the Paramāra king also buit many reservoirs.[9]

The Calukyas of Gujarat have many important irrigational works to their credit. It would appear that even the first king of the dynasty, Mūlarāja I paid due attention to irrigation. Śrīdhara, in his *praśasti,* claims that one of his ancestors was appointed by Mūlarāja I to dig vāpīs, wells and tanks.[10] From the *Prabandhacintāmaṇi,* it is learnt that Udayamatī, the queen of Bhīma I excavated a very large reservoir at Pattana.[11] According to popular tradition, the stepwell known as Rānī-ki-vav also owes its construction to this queen.[12] Meruttuṅga credits her son, King Karṇa with the excavation of a large tank called Karṇasāgara.[13] Mayaṇallādevī, the wife of King Karṇa, is generally regarded as having excavated the still-existing Mansar Lake at Viramgam.[14] The Sahasraliṅga Lake was built by Siddharāja.[15] Archaeology testifies to the supercor engineering skill employed in this irrigational project. The lake received water from the river Saraswatī, with which it was connected by a 300 ft. long channel. Excavations have also revealed stone sluices through which water was conducted to the lake.[16] The *Prabandhacintāmaṇi* informs us that Tejapāla, the minister of King Viradhavala, also had to his credit the construction of many tanks.[17]

The extant records show that under the Kalacuris, many feudatories and ministers excavated tanks and wells. Thus Malayasiṃa, a feudatory, is reported to have dug a tank in AD 1192.[18] Vallāladevaka, another feudatory, also excavated a water channel.[19] There are several references to the construction of tanks and wells for irrigation by kings, queens, feudatories, ministers and aristocracies in different regions. References abound to the digging of tanks and wells in the Raipur-Bilaspur area by the Tummāna Kalacuri feudatory, Brahmadeva[20] and ministers Puruṣottama[21] and Gaṅgādhara.[22] In the Berar region, Gamiyaya, a minister of King Hemādrideva, excavated a tank and a well.[23]

A number of tanks existing even now bear eloquent testimony to the concern of the Candella kings towards irrigational facilities. Rahilya Sāgar, Kirat Sāgar (about 1 km in circumference), Madanasāgar (2½ km in circuit), Vijayasāgar and Kalyānsāgar tanks have been ascribed, respectively, to Rahila, Kīrtivarman, Madanavarman, Vijayapāla and Kalyāṇadevī. A tank in Ajayagadh is said to have been excavated by Paramardi.[24] In the Khajuraho inscription dated 1011 of the Vikrama era, there is reference to the construction of embankments to divert the course of a river.[25]

Srīharṣa refers to the charitable provisions for wells, etc., made by the hero of the *Naiṣadhīvacarita.*[26] Paṇḍita Dāmodara, who was associated with the Gahadavala court, refers in his Uktivyaktiprana not only to the cleaning of wells[27] and digging of tanks[28] but also to a state official named Surapala supervising the digging of a tank, his head shaded by an umbrella.[29] This reference is in line with what we have for the reign of King Mūlarāja I of the Calukya dynasty, indicative that the Calukya kings were not alone in commissioning officers to dig wells and tanks in their kingdoms. Lakṣmīdhara, the minister, lauding the merit of constructing a special type of reservoir called *dvārībandha*, observes that the merit which accrues from the Vedic sacrifices of the cow, man and the horse, is excelled by that accruing from the constructing of such a dam.[30] The term *Dvārībandha*, means the damming of a mountain spring and thus forming a high level reservoir useful for irrigation purposes. Where the bund is a long one, sluice gates are to be provided for draining of the excess water.[31] Laksmidhara suggests that flights of steps may be built into it, and gardens and temples near the bund may be provided. K.V.R. Aiyangar[32] opines that the description seems to be based on an irrigation dam actually existing at the time.

Gaṅgadharā, a minister of the King Rudramān, is said to have excavated a tank n the Gaya district of Bihar.[33] The *Rāmapālacarita*[34] has references to the reservoirs constructed by the King Rāmapāla. In the Pavanaduta, there is a reference to an embankment in the memory of the King Vallālasena. The Bhuvanesvar inscription testifies to the excavation of a tank by Bhaṭṭa Bhavadevā, the minister of Harivarman and refers to the advantages resulting from it.[35]

In addition to the construction of irrigational works by the State, there must have been many other irrigational works commissioned by private individuals. Digging of wells and tanks has been propagated in India as a charitable work of merit. There are many records testifying to the actual hold of this ideology on the Indian masses. Rich persons such as merchants built tanks and wells to reap the merit accruing from them. If we make allowance for cases in which the patrons of such work did not care to have their memory recorded on lasting material, and also for many more records of which have been lost or not recovered so far, we may form an idea of the extent of charitable works of this type. In the mathematical text *Līlāvatī*, there is an entire section devoted to the problem of measuring an excavation (*khāta-vyavahāra*).[36] It must have been included in the book because of its practical importance for the excavation of reservoirs.[37] Hemacandra mentions *uḍḍo* as the term for a man who digs wells, etc.[38] The *Upavana Vinoda* has a separate section for the examination of the soil where wells can be dug.[39] In the *Agni Purāṇa*, there is a list of asterisms auspicious for the excavation of tanks.[40]

That there were tanks and wells in the villages follows also from the land-grant of the period, which, in enumerating the boundaries of fields or villages granted, often mention canals, tanks, wells and embankments.[41]

Recognizing the importance of irrigational works, the *Agnī Purāna* says that for forts and fortified cities, that land is to be preferred which does not depend upon rains but has at its disposal water (in the form of reservoirs, etc.).[42] The *Aparājitaprcchā* says that famines do not visit the land where crops grow even with a little rain; that crops grow quickly if people have reserves of water with them or have built reservoirs; and that he who has collected water has achieved all his ends.[43] Medhātithi explains the term *sasyapradā* (fertile) as productive of rich harvests and not dependent entirely upon rains.[44]

The *Mānasollāsa* enumerates nine kinds of water according to the source.[45] The *Aparājitaprcchā* mentions rivers, streams, wells, *vāpīs*, tanks, river-dams, machine-wells and canals as the usual means of irrigation.[46] Of the different means of irrigation in the plains of the northern India, rivers naturally occupy the foremost place. This is implied by the references in the dictionaries of two kinds of countries, those living on river water and those depending on rains.[47] The Arab geographers Al Istakhri[48] and Ibn Haukal[49] notice how seed is sown in Sind after the River Mihran, having inundated the land, had subsided. The *Tilakamañjarī* also mentions about the fertile banks of a certain River Sarasvati whose water was canalised to many neighbouring villages.[50] In the *Agni Purāṇa*, we read of arms of rivers made to run into tanks.[51]

The *Mānasollāsa* classifies reservoirs into *kūpa* (well), *vāpikā*, *puṣkariṇī*, *dirghika* and *tadaga*.[52] Elsewhere, it refers to a misra tank *tādāka* as one which receives fresh water channels every year and is bound up with embankments.[53] It obviously refers to the tanks, so popular and useful in the plateau of the Deccan, in which rainwater is stored from year to year. The classification in the *Aparājitaprcchā* is quite elaborate when compared to that of the *Mānasollāsa*. It refers to ten categories of wells, *śrīmukha*, *vijaya*, *prānta*, *dundubhi*, *manohara*, *cūdāmaṇi*, *digbhadra*, *jaya*, *nanda* and *śaṅkara*, which varied from four to thirteen hands in length, each being one hand long than the one preceding it. It refers to small well as *kūpikās* and mentions two types of them, respectively, three and two hands long. It further refers to four classes of vapis and four classes of *kuṇḍas*. According to it, there are six types of tanks—*sara*, *mahāsara*, *bhadraka*, *subhadra*, *parigha* and *yugmaparigha* further classified as *jyeṣṭha*, *madhyama* and *kaniṣṭha*, according to the length and width of their fences (*pāli*).[54] Even if we make allowance for the academic emphasis on minute classification, we can utilize the list to form an estimate of the reservoirs. The dictionaries contain terms for a reservoir, a deep water reservoir, a pond, a small pond, a large pond, a natural pond, an artificial pond, a small tank, a large oblong tank, a dug spring, a natural spring, and a well.[55] The *Upamitibhavaprapañcākathā* mentions by name only three reservoirs, *dīrghikā*, *guñjālikā* and *yantravāpikā*.[56] The *Deśīnāmamālā* refers to the term *uttūho* for a steep well and *ulittaṃ* for a well in a high place.[57] Some idea may be formed about the working of tanks and reservoirs from the problem in mathematical texts of the period. The *Līlāvatī*, for example, refers to the regulation of fountains (*nirjharāh*) to fill a cistern (*vāpi*).[58] The *Gaṇitasāra-saṅgraha* enunciates the rules to determine the time taken to fill a *vāpī* when any number of optionally chose channels (*praṇālikās*) are together allowed to fill it.[59] The latter text also refers to water flowing out of a *vāpī* through a pump (*yantra*).[60] We can form an idea of how the wells worked from the dictionaries of the period which give terms for the cover or facing of a well, a small post near the well, the rope and bucket, water channel and a drain.[61]

Arghatta Hemachandra explains *āgattī*, *ukkā*, *ukkamdī*, *ukkaṃtī*, *dhemkā*, *nelicchī* and *vaḍḍhia* alike as *kūpatulā*[62] or a contrivance for raising water from a well in which a horizontal beam—from one end of which a bucket hangs—see-saws on a vertical post. It is not certain whether they stood for slightly differing but similar appliances. The *Aparājitaprchhā*[63] speaks of strong machines fitted to wells. In the dictionaries, *araghattaka* (usually branded as Persian wheel) is the name given to a wheel or machine for raising water from a well.[64] The revolving movements of the buckets in the wheel of *araghaṭṭa* have often been referred to in literary, mostly philosophical works.[65] The *arghaṭa* or *arghaṭa* occurs in land grants, sometimes as an object of gift[66] and sometimes in connection with the enumeration of boundaries.[67] But it does not appear likely that the *arghaṭa* was within the reach of a common cultivator with ordinary means. *Arghaṭa* as irrigational devices must have been quite valuable objects to have been donated as gifts. Significantly enough, they appear to have been owned mostly by kings, feudatories,[68] ministers and, in some cases, by rich merchants. The *Upamitibhavaprapañcākathā*, which gives elaborate similes connected with irrigation with the help of an *arghaṭṭa*, mentions the landlord together with his cultivators, workers and servants,[69] suggesting thereby that it was only a rich agriculturist who could possess this expensive irrigational mechanism.

From *Medhātithi*, we learn about the way in which cultivators used irrigational works. In this connection, he explains the term *yantra* for the building of embankments to regulate the flow of water.[70] He also refers to water being drawn from the well or the tank and preserved in a cistern and similar small reservoirs.[71] Elsewhere, he refers to cutting the embankments and taking the water to irrigate the field and to obstructing the water-courses by closing the source from which the water flows.[72] A story in the *Prabandhacintāmani* relates how workers were filling water from one field to another.[73] A stock example in the Vedanta school of philosophy speaks of the water of a reservoir issuing from sluices and flowing through a small channel (*kulyā*) on entering the fields, assuming rectangular and other shapes.[74]

The *Samarāṅgaṇasūtradhāra* refers to four water machines (*vāri-yantra*) to bring water down (*pāta*); to raise it first and then to bring it down (*ucchrāyasamapāta*); to bring it down and then to raise it (*pāta-samocchrāya*); and to raise to (*ucchraya*).[75] No doubt, mechanical contrivances in ponds are often referred to in literary works,[76] but it is highly improbable that these devices were regularly utilized for the irrigation of agricultural crops. The *Samarāṅgaṇasūtradhāra* refers to their use for recreation[77] and practical purposes and requires them to be installed in *vāpīs*, wells and *dīrghikas*.[78] It is highly improbable that these devices were utilized by the ordinary cultivator. Even in the case of rich noble or merchant, it is not likely that they were used in agriculture; at best they utilized these marvels in his gardens.[79]

OWNERSHIP OF IRRIGATIONAL WATER

In early medieval period, we find that with the growth in the claims and powers of the State, there was a distinct attempt on the part of the ruling powers to establish their ownership over irrigational water. The list of rights and privileges which the land grants transfer to the donee and which, therefore, existed earlier in the donor king or feudatory, include the ownership of water (*sajala*).[80] The authority quoted by Bhattasvamin also distinctly observes that those well versed in the sastras view the king as the owner of the land and water.[81] But in ancient India, the state never thought of deriving any income from its ownership of irrigation works. It was regarded as a pious duty on the part of the king to provide such works of public utility.[82] However, many demands on the purse of the feudal state led in—in its desire to tap new sources of revenue—to utilize its ownership of irrigational works as a separate source of income. We find the Gahadavalas imposing of *jalakara*,[83] for the first time in Indian history. It is likely that, pressed by the Muslims inroads, they resorted to all possible sources for increasing revenues.

But this is not to suggest that all the wells, tanks and ponds etc., in the village were owned by the state. It is obvious that the man who had a well dug on his own land was its natural owner.[84] In the same manner, there are references to show that machine wells called *arghaṭṭas* were owned by private individuals.[85]

It appears that the claim of the state to the ownership of water really amounted to its rights over natural ponds, lakes or rivers or streams situated in or passing through the village. *Medhātithi* would suggest that for all practical purposes, this right vested in the village as a whole which enjoyed it on behalf of the king. An example of *saṃsthā* (laws) given by him is 'this water should not be given for purposes of irrigation of crops to such and such a village until we have secured such and such a benefit from it in return'.[86] Elsewhere, referring to the compacts formed by the villagers, he says that if the inhabitants of a village decide to oppose those of a neighbouring village by diverting the water course and to be of one mind in case of a riot or legal suit, any one who is a party to the decision would be punished on this backing out of it on some inducement offered by the headman of the other village.[87]

PRIMARY OF RAIN

It is clear that all this irrigational activity going on in the country could little affect the ultimate dependence of Indian agriculture on the rains. Pandit Damodara expresses the helplessness of the common cultivator at many places in his *Uktivyaktiprakaraṇa.* He says that the clouds, by their timely rains, make the world new, as it were,[88] that fields yield crops only when the nature is kind,[89] and that if god produces rains then there will be grains.[90] *Medhātithi,* echoing the same sentiment, says that if god were to give rain, many persons would take to cultivation.[91] The arrival of rains was a source of pleasure to everybody and people looked to it with eagerness.[92] The *Aparājitapṛcchā* also shudders to think of the fate of the world if there are no timely rains or there is drought or only a little rain.[93] The picture of the Kali Age painted by the *Bṛhannāradīya* is that of people threatened with drought and with their eyes drawn towards the sky.[94] The testimony of Abū Zaid supports this position when he describes Indian rains or Fasara (=Skt. *varṣā*) during the three months of summer as 'the life of the Indians; were they to fail, they would be reduced to the deepest want. For their fields sown with rice, are watered only by rains, and are fecundated thereby.'

These sources indicate that the coming of the rains was always a subject of anxiety lest they should be inadequate as regarding the time, place and quantity. No doubt, there are references to rivers swelling from the rains[95] but the main fear was its paucity or absence, specially when required. In his helplessness, the cultivator could but pray the heavens for rain. Of course, the Vedic sacrifices had ceased to be very popular with the masses; even then, when threatened with drought, people must have performed sacrifices. Medhatithi refers to the sacrifices like the *kārīrī* performed by men desiring rain.[96] The *Agni Purāṇa* mentions that by casting libations into the fire with the *mantra Apām garbham,* etc., a man is sure to bring down a good shower of rain, while by performing a homa ceremony with the *mantra* of *Apaḥ piba*, etc., a man is sure to inundate the earth with a continuous downpour.[97]

FORECASTING OF WEATHER AND RAINFALL

There was an attempt to study the natural phenomena accompanying rains and create a somewhat systematic knowledge out of it. Parāśara, in his *Kṛṣiparāśara*, referring to the importance of rain for cultivation and of the latter for human existence, observes that one should first of all carefully acquire knowledge about rains.[98] An old text, still unpublished, variously called the *Mayūracitraka*, *Meghamālā* or Ratnamala ascribed to Narada, contains 'indications of coming rains, famine or plenty, etc., from, the appearance of the atmosphere'.[99] The *Śakunajñāna* of the *Śārṅgadharapaddhati* has a subsection on rains.[100]

The technique of forecasting weather involved the study of astronomy and depended on the experience and observation of natural phenomena accumulated over centuries. The *Kṛṣiparāśara* shows how these two approaches were utilized to predict monthly, yearly and immediate rainfall. In the first instance, the text describes the rainfall in any year according to the planets who happen to be its king, minister, superintendents of water and crops.[101] Later in the text, we read about rainfall or drought in any year as determined by the movement and position of planets.[102] It classifies clouds into four categories: *āvartta*, *saṃvartta*, *puṣkara* and drona and mentions annual rainfall accordingly, as one of them is supreme in the year.[103] Thereafter, it gives directions for predicting rainfall on the basis of observations made in different parts of the year. By dividing 30 days of *Pauṣa* into 12 equal parts of 2½ days each, corresponding to the 12 months in order, one can measure the rainfall in the 12 months by observing the course of the wind in their corresponding parts of the *Pauṣa* month as ascertained by the flag attached to a post.[104] Another device is to divide each day of Pausa into 12 parts of 5 *daṇḍas* each, and to observe the rainfall throughout the month; rainfall in the earlier or later half of those parts will indicate rainfall by day or by night, respectively, in the corresponding month.[105] Then follow indications of rainfall in the rainy season as they may be gathered from the climatic conditions, wind, lightning, rain, fog, clouds, etc., in the months of Pauṣa, Māgha and Phālguna.[106] In Caitra, the rainfall on the first day of the bright half of the moon is taken to foretell the annual rainfall according to the weekday on which it happens to occur. The position of the *nakṣatra* Citrā in Caitra also suggests the amount of rainfall.[107] By putting into a river a stick with certain marks on it in the month of Vaiśākha, one can ascertain the amount of rainfall and flood-water.[108] A shower in the months of Jyaiṣṭha and Srāvaṇa in certain *nakṣatras* is said to indicate the nature of the rainy season.[109] The direction of the wind during the full moon in Āṣāḍha also forebodes the weather in the rainy season. Rain on the ninth day of the bright half of Āṣāḍha also foretells the rainy season.[110] In Śrāvaṇa, rain in certain *nakṣatraṣ* indicates the nature of the rainy season.[111] In the month of Bhādra also rains are predicted on the basis of the *nakṣatras*.[112] Then the *Kṛṣiparāśara* mentions the behaviour on the part of living beings as indicating immediate rainfall such as a water-spout near or in the water, the rising of ants from their holes with their eggs, the sudden croaking of frogs, cats, ichneumons, serpents, other beings living in holes and moths flying around in an excited condition, boys making bridges with sand peacocks dancing, pain in the body of men afflicted with wound or rheumatism, snakes climbing to the tops of the trees, acquatic birds drying their wings in the sun, and the noise of crickets.[113]

It is not to be supposed that all this knowledge was only stored in books without being put to practical use. The very fact that books on the subject were composed in Sanskrit suggest their usefulness to the class which knew the language and was also integrated in the practical aspect of agriculture. The *Kṛṣiparāśara* merely collects the fund of knowledge which cultivators at that time possessed. Such practical knowledge was transmitted from generation to generation.[114] The popular aphorisms ascribed to Khanā indicate that such knowledge was actually possessed by the cultivators who remembered it and quoted it in the form of pithy sentences. It indicates how the weather was foretold on the basis of natural indications such as cloud, winds, rainbows and rain on particular days and months.[115] Some of these sayings of Khana would indicate that the knowledge of weather forecasting on the basis of astrological calculations or the position of planets was not altogether denied to the common cultivator.[116] That these aphorisms were meant for the guidance of the cultivators would appear from some of these sayings in which the cultivators are directly addressed or a housewife addresses her father-in-law, worrying about rains and cultivation.[117]

No doubt such knowledge must have been of great help to a cultivator. He could time the ploughing, sowing, etc., accordingly. It must have been useful in arranging for other sources to irrigate the fields. The knowledge of the amount of rainfall may have helped in the selection of crops for different regions. As early as the *Arthaśāstra*[118], we have a reference to the use of a bowl with its mouth as wide as an *aratni* (24 *aṅgulas*) as a rain-gauge. The reference indicates the way in which this branch of knowledge was developing, but in the absence of other references to rain-gauges, we cannot trace the evolution and perfection attained in the construction of such an instrument. The *Kṛṣiparāśara* refers to a crude way of determining the rainfall. It defines *ādhaka* as a vast expanse of water a hundred *yojanas* wide and thirty yojanas deep.[119] Likewise, the *Meghamālā* also states that a continuous rain for seven nights together was called *droṇa*.[120] The *Kṛṣiparāśara*[121] does not give us any idea of the comparative amount of total rainfall in different parts of the country. It simply says that of the total rainfall received by the sea, mountains and earth, respectively, is the ratio of 10, 6 and 4.

SOIL

It is well known that the most important thing in agriculture is the soil. It was the realization that the fields near rivers are more fertile which must have led people to attach importance to them.[122] The knowledge of the qualities of the soil is implied when Medhātithi[123] explains *ūsara* as that part of the land where, on account of the defects of the soil, seeds do not spring. The advance in the technical knowledge about the properties of soil is best indicated by the use in the dictionaries of different terms for an ordinary field, a fertile field suitable for every crop, a field unfit for cultivation, fallow land, an area with saline soil, desert, firm ground, clay, excellent soil, an area green with young grass and one abounding in reeds.[124] In the land grants, there are references to different types of land such as *sāra* and *ūsara*,[125] *kṣetra* and *khila*[126], etc., which indicate that these terms, coined to indicate the fertility and other features of the soil, were not merely academic but were of much practical utility in connection with the cultivation. It appears that the characteristics of the soils in the different parts of the country were also understood. The dictionaries have terms for countries with black or yellowish soil.[127] It was also realized that the properties of the soil made certain fields eminently suited for certain crops. Thus, there are terms for fields named after the crops grown in them, e.g., those of *mudga, kodrava, yava, vrīhi, śāli, ṣaṣti, tila.*[128] In the sayings of *Khanā*, there are specific details about the type of soil required for certain crops. 'You worthy cultivator, your aim will be fulfilled if you grow *pāṭol* (*Trochosanthes dioica*) in sandy alluvial soil.' 'The sandy soil for that of jute.' 'If yow grow arum on the bank of a river it will grow to the height of three cubits, i.e., it will fourish well.'[129]

THE ROLE OF MANURE

In agriculture, the knowledge and use of manure to maintain the fertility of soil is very essential. There seems to have been some recognition that however rich in chemical contents land may be, in due course of time, it becomes unproductive[130] because every crop takes away from it certain elements. That the Indian cultivator was using manure is supported by the testimony of the *Uktivyaktiprakaraṇa.*[131] The *Kṛṣiparāsara* recognizes the importance of manure for crops and says that without manure, the paddy simply grows up, but does not yield any fruit.[132] It describes the way in which cowdung, the chief form of manure in India was formed and used. It says 'In the month of Māgha, the heap of cowdung should be worshipped faithfully and on an auspicious day, the dung should be dried in the sun, and made into small balls. In the month of *Phalguna*, these balls are to be placed into pits dug in every field and at time time of sowing, the manure is to be spread over the field.'[133] Without suggesting that the passage implies knowledge of component elements in which modern science analyses cow-dung manure, it may be said to indicate an awareness of its fertilizing property and an appreciation of the way in which this property could be preserved and augmented. R. Gangopadhyay analyses this method of preparing dung-manure to show that the injunction not to disturb the dung-heap reduces to a minimum the loss of nitrogen, the chief fertilizing element; that of drying dung into balls results in reducing active ammonia which may be injurious to the plants, while that of placing the dung balls in pits increases the humus that contributes to the fertility of the soil.[134]

It has generally been observed that the modern Indian farmer does not make a proper and full use of manure. We have to see how far does the present practice represent the conditions in the early medieval period. No doubt, cowdung was used as a fuel.[135] It was recognised that dried cowdung increases fire[136] and so, dried dung-cakes were often collected in large numbers.[137] The *Deśīnāmamālā* refers to many words for a fire of dry cowdung: *kouā*,[138] *pouā*,[139] *phuṃphuā*[140] and *mummuro.*[141] In the *Aucityavicāracarcā*, a fire of dry cowdung is said to be as pleasant as the anger of a newly wedded wife[142] and there are several other definite references to its use.[143] But there is no evidence to suggest that cowdung was used as fuel to such an extent as to render it unavailable for use as manure. At that time when huge forests surrounded the villages, there was no dearth of wood as fuel as there is today; thus the cultivator must have found it advisable and profitable to use the cowdung for manuring the fields in addition to its utilization for fuel. It is only in modern times, when forests have been cleared, that the farmer has resorted to such a large use of cowdung for fuel. Bāṇa suggests that the cowdung and rubbish formed the usual manures. In one of his graphic pictures, he describes how a cultivator carried rubbish and cowdung in carts drawn by oxen to fields that had become unfertile.[144]

Some of the manures used in the period are referred to in the sayings of Khanā.[145] 'Those things (e.g., rotten cowdung) which injure man cure the plants.' 'If some water, in which a fish has been washed, is poured at the root of a gourd plant, then the plant will surely be benefited from it.' 'The land which contains rotten paddy as manure is fit for the rearing of chilies.' 'Betel-nut plants require cowdung as manure for their growth.' 'Pieces of rotten straw or chips of wood should be used as manure at the roots of *arums*.'

THE ROLE OF SEEDS

In agriculture, much depends upon the quality of seeds sown. The *Kṛṣiparāsara* mentions that if the seeds are unproductive, the efforts of other factors in cultivation become futile. The seeds are at the root of the crops; hence one should pay attention to the seeds.[146] Medhātithi mentions sowing of unripe seed as one of the faults on the part of the farmer which may hamper the crop.[147] Elsewhere he says that the corn, left over after consumption, when sown next year, does not reproduce.[148] The *Prabandhacintāmaṇi* of Merutunga at one place says that seeds which have been burnt do not sprout.[149] The commentary of Nīlakaṇṭha explains the *pulākā* grains mentioned in the *Mahābhārata* as grains rendered incapable of growing through the excessive heat of the soil.[150] Likewise, Nīlakaṇṭha observes that *kākayava* is the term for seeds separated from their outer skin and thus rendered unproductive.[151] The *Kṛṣiparāśara* has a detailed section on how to collect and preserve seeds.[152] It advises that all kinds of seeds should be collected in the sun and exposed to dew at night. The seeds are to be kept in small bundles (*puṭikās*). Mixed seeds (presumably seeds of more than one species sown together) result in bad crops and seeds of the same yield a rich harvest; hence one should collect with care seeds of the same class. After placing the seeds in it, the bundle should be closely tied up, and the seeds should be purged of grass particles, otherwise the crop will be full of grass. The *Kṛṣiparāśara* then requires the seeds to be kept away from impure associations. These suggestions, by attributing a degree of sanctity to the seeds, emphasized their importance in agriculture. It is thus said that the seeds should not be kept on an anthill, in a cowshed, a room where a woman has given brith, or a house inhabited by a barren women. Seeds should never be allowed to come in contact with the remnants of food, a woman in her menses, a barren woman, a pregnant woman or a woman who has just delivered a child. These may appear superstition but may have some scientific basis, at least from the point of view of hygienic atmosphere. Then follow some practical suggestions for preserving the seeds. A farmer should not even unknowingly place ghee, oil, whey, a lamp or salt over the seeds, and should avoid using seeds affected by lamps, fire, smoke and rain, and also those placed in pits. One should never sow seeds which have already germinated or sow mixed seeds together.

SOWING

Sowing had grown to be a technical and specialized procedure demanding careful attention. To bring home the importance of the process, the *Kṛṣiparāśara* treats it as a veritable ritual.[153] The proprietor of the field should himself sow, three handfuls of seeds moistened with cold water with a pure and concentrated mind and meditating upon Indra. He should invoke the goddess of earth (*vasundharā*) for the rich growth of crops, seasonal rainfall and the welfare of the cultivators. He should then feed the cultivators with ghee and other delicacies. These details would appear to be supported by the *Deśīnāmamālā* when it uses the term *maṅgalasajjaṃ* for a field ready for sowing seeds.[154]

Sowing has to be done on scheduled dates. *Medhātithi* implies that untimely sowing affects the yield of the crop.[155] The *Kṛṣiparāśara*, thinking in terms of paddy cultivation, observes that for the sowing of seeds *Vaiśākha* is the best, Jyaiṣṭha middling, Āṣāḍha bad, and *Śrāvaṇa* worst. It also mentions those *nakṣatras* which are auspicious for sowing and those which are less so. Saturdays and Sundays are to be avoided and certain *tithīs* and special periods are not auspicious for sowing.[156] We can never know how much of this advice was mere superstition and how much was the result of actual experience in the climatic conditions of the period and region of the text concerned. Khanā seems to be full of directions in this point.[157] 'By sowing paddy in the month of *Āṣāḍha* one gets a full harvest, in *Śrāvaṇa* only leaves and no fruits, in *Bhādra* only husks and in *Āśvina nothing.*' 'Sow paddy to your heart's content throughout the whole of *Śrāvaṇa* and the first twelve days of *Bhādra.*' 'During the last four days of *Bhādra* and the first four days of *Bhādra* and the first four days of *Āśvina* sow *kalāi* (*Phaseolus var. radiatus*) as much as you can.' 'One should sow peas after the first 19 days of *Āśvina* and within the first 19 days of *Kārtika.*' 'Khanā directs, good cultivator, sow mustard towards the end of Autumn.' 'The cultivator who does not plant, either in the month of Bhādra or *Āśvina* and whiles away his time and then too late in the month of *Kārtika* and *Agrahāyana* sets grown-plants in the field, is fated to see them perish from an attack of mildew and his barns empty.' 'Sow turmeric in *Vaiśākha* and *Jyaiṣṭha....*' 'This is the direction of Varāha's son Mihira: Excepting *Caitra* and *Vaisākha*, plant brinjal very cheerily throughout the other ten months...' 'Plant betel in Śrāvaṇa and produce will be too much to be chewed up even by Rāvaṇa.' '*Pātols* will doubly grow if planted in *Phālguna.*'

A cultivator was expected to know the requirements of his field as regards seeds.[158] This probably includes knowledge not only of the seeds likely to grow well in a particular field[159] but also the amount of seeds required for it. In the *Vaijayantī*, we find terms for naming the fields which require *droṇa āḍhaka, khāri*, etc., measures of seeds[160] to be sown in them. This knowledge would appear to have been quite common in this period. Many land grants of the period denote field in terms of its seed capacity.[161]

The cultivator was also required to know how seeds of particular types were to be sown, thickly or sparsely.[162] Generally, it appears that the seeds were scattered in the field by hand.[163] It was after the field had been ploughed that seeds were generally sown,[164] but sometimes the field was ploughed after sowing.[165] However, Medhātithi refers to seeds being sown with the help of the plough.[166] This probably refers to some such device by which a cultivator used to sow the seeds in the line of tilling through a cone attached to the pole of the plough.

ROTATION OF CROPS

It has to be discussed now whether the cultivators made use of the system of rotation of crops or resorted to the practice of leaving the fields fallow after one crop. As early as the times of the *Taittirīya Saṃhitā*, there are references to two seasonal crops[167] with the possibility of a third.[168] Khanā would suggest that a system of rotation was known, especially when she observes, 'My worthy cultivator, plant radish towards the end of the third season of the year, sow mustard towards the end of autumn, and if you mind to take money, sow maize in the following month of *Caitra*.'[169] But the system was not much in vogue if we believe in the testimony of the *Yuktikalpataru*.[170] It says that the land loses fertility from cultivation year after year; when one plot of land loses its fertility, cultivation should be done elsewhere. The word *khila*, which stands for a fallow land,[171] often occurs in the inscriptions. It would, therefore, follow that often plots of land were left to recover their fertility. However, we should also avoid the other extreme of suggesting that the rotation of crops was little utilized. In the inscriptions of the period, there are references to fields called *śāradyagraiṣmikakṣetrādi, graiṣmikakṣetrādi* and *śāradyakṣetrādi*.[172] It is evident from these terms that there were fields which had two crops in a year in *Śarada* and *Grīṣma*, obviously implying a practical application of the principal of rotation. But the reference to fields which had only one crop either in *Grīṣma* or *Śarada* would suggest that as in modern times, the cultivators, followed the system of rotation or keeping fallows according to the wealth of properties that their fields possessed, and also as their own resources permitted and their needs required.

AGRICULTURAL IMPLEMENTS

The cultivation requires suitable implements and tools. The *Kṛṣiparāśara*, emphasizing this need, says that they should be firm, otherwise the cultivator faces difficulty at each step.[173] In proportion to the importance of the plough, the *Kṛṣiparāśara* devotes considerable space to the different parts of the plough and their measurements. These are:[174] (1) *yuga* (yoke) which is described as extending upto the ears to the oxen;[175] (2) *aḍḍacalla* or the pins of the yoke where the bullocks are tied which measure 12 *aṅgulas* (9 inches); (3) *yotra* or yoke-tie which is to be 4 *hastas* (6 feet); (4) *rajju* or string 5 cubits (*karas*) long (7 feet 6 inches); (5) *īṣā* or the pole of the plough to be 5 *hastas* (7 feet 6 inches) long; (6) *niryola* or the rod of the plough exclusive of the pole and the share and measuring 1 *hastas* (2 feet 3 inches); (7) *śaula* or an extra piece of wood that tightly fixes the *niryola* to the pole and measures an *aratni* (cubit);[176] (8) *ābadha* or a rod of iron which prevents the *niryola* from getting out of the pole of circular in shape (*maṇḍalākāraḥ*) and is 54 *aṅgulas* (3 feet 4 inches) in length; (9) *niryolapāśikā* or the plates that fix the ploughshare to the *niryola*[177] and measure 12 *aṅgulas* (9 inches); (10) *phāla* or ploughshare which is 1 *hasta* and 5 *aṅgulas* (6 inches) and resembled an *arka* (*Calotropis gigantea*) leaf in shape; and (11) *halasthāṇu* (the handle of a plough) or a piece of wood fixed to the *niryola* at the end opposite of which the ploughshare is fixed, which the cultivator holds while ploughing and which measures 5 *vitastis*[178] (3 feet 9 inches). The *Kṛṣiparāśara* requires the goad (*paccanī* or *paccanikāl*; in local dialect in west Bengal, it is called *pāñcanbārī*) to be of strong bamboo with an iron top and measuring 12 or *muṣṭis* (fists) in length. The terms for some of the parts such as the ploughshare, yoke-tie, pin of a yoke, pole and the goad are to be found in the dictionaries of the period.[179] It may be that distinctive terms for other parts were still in the process of evolution or else the dictionaries confine themselves to the words in common use. These dictionaries employ the terms *śamyā*[180] in place of the *aḍḍacalla* of the *Kṛṣiparāśara*. Likewise, the *Deśīnāmamālā*[181] has *javaṇaṃ* for the *halasthāṇu* of the *Kṛṣiparāśara*. These differences may sugest regional variations in terminology.[182]

The *Kṛṣiparāśara* also mentions two other implements which appear to have been used as harrows. Of these, *viddhaka* is said to have 21 spikes (*śalyas*) and *madikā* (=*mai* of Bengali usage) is described as measuring 9 hands (13 feet 6 inches) in length. In the dictionaries, we find terms for some other agricultural implements whose connection with agriculture is indicated even by their position in the texts.[183] Thus, we have *matya* for a roller to level the sown field; *koṭīśa* for a spade or hoe; *godāraṇa* or *kuṇḍāla* for a scraper or shovel of which the main part was known as *abhriḥ* or *kṣūḥ* and *dātra, lavitra* or *asida* for a sickle of which the handle was called *vaṇṭaka*. The digging hoe (*kuddāla*) is also referred to.[184]

PLOUGHING

Ploughing being the first process in cultivation, naturally receives careful attention from a cultivator. It is full of rituals. The *Kṛṣiparāśara*[185] lays down the performance of the ceremony called *hala-prasāraṇa* before the commencement of ploughing and observes that he who starts cultivation without performing *hala-prasāraṇa* does it in vain. The ceremony consists in invoking a number of deities and worshipping them with offerings and also in whetting the ploughshares and besmearing them with honey and butter and ghee both sides of the faces of the oxen. The *Agni Purāṇa* gives details for worshipping the sun and the gods of the element and observes that these ceremonies are to be performed before transplanting or mowing down the rows of paddy plants, the unfastening or the ploughshare or removing the threshing plank.[186]

From the *Kṛtyakalpataru* of Lakṣmīdhara, it is known that the need for thorough ploughing was realized.[187] In the dictionaries of the period, we find terms for fields cultivated twice, thrice or even greater number of times.[188] The *Kṛṣiparāśara*[189] requires the furrows to be continuous and without break and to be one, three or five in number.

By the number of the furrows, the text probably means the number of times the field is to be ploughed. Khanā[190] has many practical tips for ploughing. 'One should plough the soil sixteen times for radish, half the number of times for cotton, half of that for paddy and none for betel.' 'One should cultivate the soil for radish, making it as soft as cotton and for sugarcane plough it to the dust.'

The *Kṛṣiparāśara* refers to the nature of soil and its suitability for ploughing at different periods of the year.[191] Having paddy cultivation in mind, Parāsāra regards Hemanta as the best season and the rainy season as the worst. The *Kṛṣiparāśara* requires actual ploughing to be commenced on an auspicious occasion. It gives a long list of *nakṣatras*, days, *tithis* and *rāṣis* (zodiac signs) auspicious for this purpose and goes on to add another list of days, *tithis* and *rāśis* which are inauspicious, noting their evil effects on the farmer and his bulls and crops and even on other people.[192] Such superstitious ideas appear to have had quite a hold on the minds of the cultivators; Khanā, likewise, requires the cultivator to begin the ploughing from the east and avoid the days of the full moon and the new moon for the commencement of ploughing.[193] The *Kṛṣiparāśara*[194] mentions a number of superstitions concerning the effects suggested by some portents in the course of ploughing, e.g., the raising of a tortoise by the plough when it is driven across the field,[195] the breaking of the ploughshare, plough, pole, yoke, *śaula*, or yoke-tie, exhausting of the cultivator or falling down, sudden running, bellowing, the licking of their noses or the passing of dung or urine by the oxen drawing the plough.

PLOUGH-OXEN

The utility of good plough depends upon the oxen that draw it.[196] The *Kṛṣiparāśara* gives much space to the proper upkeep of cows and bulls.[197] The *Bṛhat-Parāśara* repeatedly emphasizes the role of oxen. It observes that the world lives on the crops produced by oxen. They produce the grain, crush them and carry them; they are really objects of worship; the merit of keeping one ox is ten times that spoken of by the sages for keeping a cow; they are to be cared for properly; he who yokes them without care goes to hell.[198] The *Kṛṣiparāśara* also requires cultivation to be carried without cruelty to the oxen. If one earns a four-fold crop by oppressing the oxen, he is reduced to the condition of a pauper by their sighs.[199] The ideal of a humane treatment of the oxen[200] was at the base of the suggestion to increase the number of oxen yoked to a plough. Thus, the *Kṛṣiparāśara* remarks that a plough should have eight oxen yoked to it; one who yokes six is just a businessman; those who employ four are cruel, while those employing two are but beef-eaters.[201] A verse in the *Atrisaṃhitā* reads: plough-oxen are to be employed for one-quarter of the day when they are strong and two per plough, if the number of the oxen yoked to a plough is four they are to be engaged for half the day, if the number is six, for three-quarters of the day, and if the number is eight, for the whole day.[202] These considerations of compassion and pity imply the realization that the larger the number of the oxen, the better is the ploughing. It is, however, not clear how far this knowledge led to the practice of deep ploughing by using more oxen. It is not unlikely that in a few cases more than two oxen were yoked to a plough[203] but the modern practice would suggest that the six or eight oxen in the texts referred to above were to be used by rotation or relay.

The *Kṛṣiparāśara* favour black or red or black-and-red oxen and discards those having wide waists, or their tails and ears cut off or all-white ones. He realizes the practical need of keeping the oxen, like the cultivator, free from diseases.[204] The *Bṛhat-Parāśara* says that one should not yoke a bull with defective limbs, or one diseased, lifeless, weak, hungry, thirsty or tired. The bull should be able-bodied, free from disease, well-fed, hearty, not impotent or insolent but full of life and vigour. Bulls are to be yoked only for half of the day and after that they should be bathed.[205] After ploughing, their horns are to be smeared with oil or ghee and then they should be sent to graze. At night, they should be given fodder.[206] The *Kṛṣiparāśara* advises strength-giving fodder for the oxen and observes that if they are allowed to graze in the morning and evening, they always thrive.[207] It enumerates the *nakṣatras* and days auspicious and otherwise for a cow's *yātrā* and *praveśa*, probably referring to its going out to the field and entering the cowshed for the first time.[208]

The *Kṛṣiparāśara* appears to be full of concern for the welfare of cows and bulls. It provides for a fairly spacious cow-stall in a clean and hygienic condition. Its interest in the welfare of cattle is reflected in some of the superstitions it records, including those on the auspicious periods for constructing cow-stalls.[209] It describes certain festivities, which, when performed on the *tithi* called *laguḍapratipat*, are supposed to render cows free from diseases for one year.[210] On this occasion, the cows were marked with heated iron and their tails, hairs and ears were slightly cropped. The *Bṛhat-Parāśara* requires that the nose of the bull should be pierced with a wooden plug twelve fingers long.[211]

AGRICULTURAL PROCESSES

The *Kṛṣiparāśara* and other texts of the period inform us about the details of the different processes of the cultivation. The *Kṛṣiparāśara* requires that after the sowing of seeds, the field should be levelled with a harrow; otherwise the growth of the plants is not even.[212]

The *Kṛṣiparāśara* mentions that if a sown field is not hoed, the crops cannot grow in abundance, nor yield a good harvest. It adds that if hoeing is done in the month of Śrāvaṇa or Bhādra, the harvest is doubled, even if grass grows again, and that if another hoeing is done is the month of Āśvina, corn grows plentifully.[213] Medhātithi also refers to the weeder cleverly preserving the corn and plucking out the weeds though the two grow together and are in close touch with one another.[214]

Animals, birds and insects have to be checked from devouring or destroying the crops.[215] The *Bṛhat-Parāśara* advises that fences should be erected which animals cannot cross.[216] The *Prabandhacintāmaṇi* refers to the watchmen collecting the branches of throny shrub,[217] obviously for fencing the fields. Sometimes a scarecrow of grass was set up. The *Deśīnāmamāla* uses the expression *avaḍao,*[218] *jharaṃko* and *jharaṃto*[219] for these grass-men. The *Subhāṣitaratnakoṣa* gives a graphic picture of a herd of pigeons somehow overcoming their fear of the scarecrow and eating the seed in a corner of the field.[220] To protect the field, the watchmen had often to live in the fields,[221] keeping awake at night.[222] A verse in the *Subhāṣitaratnakoṣa*[223] refers to the platforms raised on the boundaries of the fields from which the watchmen scared away wild boars.[224] the noise made by the watchmen to scare away animals and birds is called *hiṃdolayaṃ, hilloḍaṇaṃ* and *hiḍolaṇayaṃ*[225] in the *Deśīnāmamāla.* According to Hemacandra, the expression *hiṃḍoloṃ* is also used to signify a contrivance to scare away birds, etc., from fields.[226] Even in modern times, Indian cultivators strike a horizontal hollow bamboo pole with a vertical one by pulling it with a rope while lying in their hut or in the mdst of their work in other parts of the field. The use of the term *hiṃḍolaṃ*, meaning swing suggests that Hemacandra had in his mind some such device.

Indian cultivators, renowned for their superstitious nature, are found resorting to many such practices for protecting their crops. Thus, in the *Śārṅgadharapaddhati,* it is said that if sand in a pot empowered with mantras is thrown in the field, then insects, cranes, parrots, boars, deer, rats and hares keep away.[227] The *Kṛṣiparāśara*[228] has an interesting *mantra* which, if written on the leaf of *Ketakī* and fastened in the northeast corner of the field, is said to protect the crop from diseases, insects and animals. The incantation requests Rāma to direct Hanumān to drive away with his tail all the insects, birds and animals—of which many are named—which destroy the crops. The *Kṛṣiparāśara*[229] further requires that on the occasion of the *Kārttikasaṃkrānti,* a cultivator should plant a leafy reed at the northeast corner of the field and worship it with suitable offerings to ensure a uniform and luxuriant growth of corps.

The *Agni Purāṇa* refers to auspicious occasions for harvesting.[230] Khanā gives many practical tips in connection with reaping.[231] 'Corn ripens within 20 days after the first appearance of the ear, and one should cut and thresh the corn in 10 days more.' 'The corn ripens 30 days after the first appearance of the spike, 20 days after the first appearance of flowers and 12 days after the ears bend like a horse's head...' Before the actual commencement of reaping, the *Kṛṣiparāśara*[232] requires the performance of the ceremony called *muṣṭigrahana* when the cultivator, after worshipping the plants, cuts off two and a half handfuls of plants in the northeast corner of the field and returns home with the plants on his head. Non-observance of the rite is said to create difficulties for the cultivator at every step and lead to the loss of the crops.

The expression used in the *Kṛṣiparāśara* for threshing (*mardayitvā* = *maṇāi* in Hindi and *malan* in Bengali) refers to the process of separating grain from straw by making the oxen tread on the corn. In the dictionaries of the period, there are terms for threshing floors. The term *methi* stands for the post of the threshing floor, or round which cattle turn to thresh out the grains.[233] In the *Rāmacarita,* we read about the threshing floor where the reaped crops were spread out and threshed by bullocks which went round and round over them.[234] The *Kṛṣiparāśara* attaches great significance to the threshing post and describes in details the rites accompanying the fixing up the post. It enumerates trees which are auspicious and inauspicious for making the post and also notes the days on which the rite is to be performed.[235] Threshing was sometimes done by human feet,[236] probably when the crops were not abundant or when farmer could not afford to have it done with the help of cattle.

To separate the grains from the husk, the corn was first pestled and then winnowed.[237] The *Vaijaynatī* has distinct terms for unhusked rice, husk, eye of corn, dust of clean rice, clean rice water and rice powder.[238] In the *Deśīnāmamālā,* we find many words for a pestle and a mortar.[239] It explains *kaṃcī* as the name for the metal ring at the end of a pestle.[240] Some other utensils used in this connection were the sieve[241] and the grinding stone.[242] The room for rice cleaning and pounding was called *kṣaṇā* or *buśālī.*[243]

The *Kṛṣiparāśara* requires that after all this has been done, the grains are to be measured and kept in the granary.[244] It then mentions the *nakṣatras* and days which are auspicious for storing the grains in the granary.[245] It again refers to two *mantras* which are to be written and kept in the granary to augment prosperity.[246] The *Agni Purāṇa* also advises that the *mantras* written on a leaf should be placed amidst the heap of grains.[247] The *Vaijayantī* mentions sacks, boxes and baskets in connection with the granary.[248]

To recapitulate and conclude, the early medieval period witnessed remarkable growth in the processes of cultivation, development of improved implements, extensive irrigational works, a knowledge system involving manuring system, and forecasting weather and rains based upon astronomy along with experience and observations. For the first time, we have evidence of technical texts on agriculture such as the *Kṛṣi Parāśāra*, though we did have lot of references to agriculture in other early medieval texts such as *Medhātithi on Manu* the *Aparājitapṛcchā*, the *Mānasollāsa*, the *Agni Purāṇa*, the *Brahat-Parāsara*, the *upamitabtavaprapancakatha* the *Samrangansutradhāra*, aphorism of *Khānā*, etc. This era also witnessed the emergence of the ownership of water by the state due to paucity of funds in some circles.

NOTES AND REFERENCES

1. On *Manu*, I, 30.
2. *Aparājitaprccha,* p. 186, v. *9-Rājyārthaṃ dhāryate śasyam śasyārtham tu jalāśayaḥ.*
3. K.V.R. Aiyangar is his Introduction (p. 114) to the *Dānakāṇḍa* of the *Kṛtyakalpataru.*
4. *Rājataranginī,* V., 84–121.
5. *Ibid.,* IV, 191.
6. *Ibid.,* IV, 940.
7. Ray, H.C., *Dynastic History of Northern India* (*D.H.N.I.*), Vol. II, p.1119.
8. *Epigraphia Indica,* Vol. IX, p. 15.
9. *Indian Antiquairy* (*I.A.*), Vol. XVII, pp. 350–52.
10. *Epigraphia Indica,* Vol. II, 438. Cf. *P.H.I.C.* VII, p. 645 for the grant of an *arghatta* by King Bhima II.
11. p. 54, II.3–4.
12. Majumdar, A.K., *Chaulukyas of Gujarat,* pp. 389–390. Cf. Burgess and Cousens, *Architectural Antiquities of Northern Gujarat,* p. 112 for a description if the *vav* of Vyad.
13. *Prabandhacintāmaṇi* (Tawney), p. 78. Cf. Forbes, *Rās Mālā-I,* 104.
14. *Archaeological Survey of Western India* (*A.S.W.I.*), VIII, ii, 91.
15. *Saraswatī Purāṇa,* XVI, 212; *Dvyāśrayakāvya,* XV, 120–21; cf. *P.H.I.C.,* 1939, p. 479 for other references.

16. A.S. Garde, *Archaeology in Baroda*, 1934–47, p. 8. See *Annual Reports of Baroda State*, 1934–35, p. 18 for an epigraphic reference to the water of the river Sarastvati having been canalized to fill up the lake.
17. p. 99, I. 27.
18. *E.I.*, XIX, 298 f.
19. *I.A.*, XVIII, p. 213.
20. *E.I.*, XXVI, 262.
21. *Ibid.*, XXVII, 283.
22. *Ibid.*, XXI, 164.
23. *Ibid.*, XXI, p.132.
24. *I.A.*, XXXVII, 132; *A.S.R.*, II, 439f referred to by N.S. Bose, *History of the Candellas*, p. 147.
25. *E.I.*, I, 122, v. 26.
26. *E.I.*, IV, 310.
27. p. 46, I.14.
28. p. 49, I.22.
29. p. 21, II, 15–16.
30. *Kṛtyakalpataru Dānakāṇḍa*, p. 296.
31. *Ibid.*, p. 292.
32. *Ibid.*, introduction, pp. 116–17.
33. *E.I.*, II, 338.
34. III, 42, p. 108.
35. *Inscriptions of Bengal*, III, p. 40.
36. pp. 93–95, nos. 217–223.
37. p. 94, no. 222.
38. *Deśīnāmamālā*, I, 85.
39. Pp. 77–81, vv. 94–146—*Kūpārtham bhūmiparīkṣā.*
40. CCLXXXII, 5–6
41. See, for example, Mitra, *Early Rulers of Khajuraho*, p. 180—*nālā, puskarṇī, bhiti;* Barua, *Cultural History of Assam*, p. 70–gaṅgī.
42. CCXII, 2—*Adevamātṛko bhaktajalo deśah paraśyate.*
43. p. 188, vv. 32–33, 40.
44. On Manu, VII, 212—*Bahusasyādevamātṛkā ca.*
45. II, p. 136, v. 1605—*divyāntariksaṃ nādeyṃ nairjharaṃ sārasam jalam. Bhaumaṃ caundam ca tāḍākam-audbhidaṃ navamaṃ smūṛtam.*
46. p. 188, VV. 34–36.
47. *Abhidhānaratnamālā*, II, 161; *Vaijayantī*, p. 38, 11.92–93.
48. Elliot and Dowson, I, 30.
49. *Ibid.*, I. 40, Cf. Al Idrisi (*ibid*, I, 81) about Tūbarān.
50. p. 181. Cf. *Aparājitaprcchā*, p. 188, v. 37-*Setubandhabhavā nadyāh śasyobhavatoddhavāḥ.*
51. CCI. XXII, 4-*Praveśayen-nadivāhān puṣkarināṃ tu kārayet.*
52. I, p. 8, vv. 71–72.
53. II, p. 137, v. 16–18.

54. pp. 183–85, vv. 2–36.
55. *Abhidhānaratnamālā*, vv. 675–677; *Vaijayantī*, p. 154, II, 9–15.
56. p. 56.
57. pp. 94, 89.
58. p. 39, no. 95.
59. VII, 33–3.
60. IV, 28–30.
61. *Abridhānaratnamālā* (*Halāyudhakova*), Ed. Jayaśaṅkara Jośī, vv. 684–85; *Vaijayantī* pp. 154–55, 11, 16–18, 40–42.
62. I, 63, 87; IV, 17, 44; VII, 36.
63. p. 214—*Kūpaśca dṛḍhayantrair-alaṅkṛtaḥ.*
64. *Abridhānaratnamālā*, v. 685; *Vaijatī*, p. 155, 1, 42.
65. *Upamitibhavaprapañcākathā*, pp. 66, 418, 723; *Prabandhacintāmaṇi*, p. 40, 1.24, 11.56—*Tavaṃ kiṃ na paśyasi ghatīr—jalayantracakre riktā bhavanti bharitāḥ punar-eva riktāḥ. Pañcatantra*, IV (Hertel), pp. 231–244.
66. *P.I.H.C.*, VII, 643; *E.I.*, XIV, p. 186, 1.26.
67. *E.I.*, XI, pp. 28f.
68. The Partabgarh Ins. (*E.I.*, XIVM 182 ff) refers to a field of the Provincial Governor Mādhava which was irrigated by a Persian wheel, *P.I.H.C.*, VII, 643—grant of an *arghaṭṭa* by Bhima II.
69. pp. 984–6.
70. On *Manu* XI, 62-*Yantāṇi setubandhādīni jala-pravādaniya-mārthāni.*
71. *Ibid.*, XI, 162—*Uddhṛtodakasyernādisthasya.*
72. *Ibid.*, III, 153.
73. p. 56, 11, 14–15.
74. *Vedāntaparibhāla*, p. 23.
75. I, pp. 178–79, vv. 109–114, Cf. *ibid.*, p. 175, vv. 75–76 for a device to convey water into a field and also to pour it out from it.
76. e.g., *Karpūramaññjarī*, III, 20; IV, 13; *Kuṭṭanīmata*. V. 684.
77. I, p. 178, v. 109—*Krīḍārthaṃ kārya-siddhyai ca.*
78. I, p. 179, v. 114.
79. Cf. *ibid.*, I, p. 178, v. 110—*Vāṭikādi-prayojanam.*
80. *E.I.*, XIV, 186, I. 24; XV, 295 ff; XXVIII, 327–8; *C.I.I.*, IV, 324–31.
81. On *Artha*, II, 24; *J.B.O.R.S.*, XII, p. 138-Rājā bhumer-patih *dṛṣtaḥ śāstrajñair-udakasya ca.*
82. *Digh*, I, 135; *Rāmāyaṇa*, II, 100, 48; *Mbh.*, II, 5, 78; V, 61, 17; V, 147, 26; *Artha*, IV, 3; *Raghu*, XVI, 2.
83. *E.I.*, IV, no. 11(O); VIII, no. 14 (D); X, no. 23; XIII, no. 20.
84. Cf. *Agni Purāṇa*, CCLVII, 8.
85. Cf. *Upamitibhavaprapañcākathā*, pp. 984–86.
86. On *Manu*, I, 21.
87. *Ibid.*, VIII, 219.

88. p. 22, I. 18—*Parjanyaḥ kcrute viśvaṃ kābhih punar-navam?—Vṛṣṭibhih.*
89. p. 35, I. 29—*Sukāle annaṃ niṣphāyate.*
90. p. 9, 11. 15–16—*Yadi devo vṛṣṭim akariṣyat, tadānnam abhaviṣyat.*
91. On *Manu*, II, 45—*Devaśced varṣed bahavaḥ kṛṣim kuryuḥ*. Cf. *Bṛhat-Parāśara*(Jivananda, *Vidyasagara, Dharmaśāstra-saṅgraha* II, p. 109)—*Deva-parjanya-bhū-sīra-yogāt kṛṣih, prajāyate.*
92. *Kathākośa*, pp. 3, 11.
93. p. 187, v. 28. Cf *Kathākośa*, p. 162—how shall the world manage to subsist if the moon shall withdraw her digit, if the giver tribute, the cloud rain?
94. XXXVIII, 80–81.
95. *Uktivyaktiprakarāṇa*, p. 40, 1. 14; *Kathākośa*, p. 3.
96. On *Manu*, II, 10 (Vol. I, p. 71).
97. CCLX, 50–51.
98. v. 10.
99. *I.A.*, XLVII, 262; SEE *J.O.I.*, IX, p. 408, f.n. 2 for four MSS. Of the *Meghamālā*.
100. vv. 2372–2377.
101. vv. 12–22.
102. vv.71–78.
103. vv. 23–25.
104. vv. 30–31.
105. vv. 32–33.
106. vv. 34–43.
107. vv. 44–47.
108. vv. 48–56.
109. vv. 57–58.
110. vv. 59–61.
111. vv. 62–64.
112. p. 17-found only in the Vaṅgavāsī edition of the *Kṛṣisaṅgraha* or *Kṛṣiparāśara.*
113. vv. 65–70.
114. Cf. Medhātithi II, 6 (Vol. I, p. 63)—*Pipīlikāṇḍāsañcāreṇa hi bhaviṣyantīṃ vṛṣṭimanuminmate; Raj.*, VIII, 722 refers to the cows looking above, serpents ascending trees, and ants moving with their eggs in their mouths as foretelling an immediate rainfall. Also, Khanā quoted by T.C. Dasgupta, *Aspects of Bengali Society*, pp. 230 ff, no. (f).
115. T.C. Dasgupta, *Aspects of Bengali Society*, pp. 230–236.
116. Cf. *ibid.*, pp. 225–26, nos. (b) and (d).
117. Cf. *ibid.*, pp. 230 ff, no. (d); G.P. Majumdar *Vanaspati*, pp. 211–12.
118. II, 5. F. 3.
119. vv. 26–28.
120. See *J.O.I.*, IX, p. 418.
121. v. 29 The *Agni Purāṇa*, CXXX, 1–11 mentions the asterisms belonging to the belt of fire under whose influence the food grains stand perched in the fields and

rainfall of the year is abnormally below the average as affecting the seven divisions of the Uttrāpatha, viz., Saindhava, Yāmuna, Gurjara, Bhoja, Bāhlika, Jālamdhara and Kāśmira. Likewise, it observes that the asterisms of the belt of wind affect the countries of Dāhala, Kāmarūpa, Kalinga, Kośala, Ayodhyā, Avantī, Koṅkaṇa and Andhraka.

122. Cf. *Kavīndravacanasamuccya*, no. 210—*Prāyaḥ stana-taṭībhūmiḥ prakāma-phala-dāyinī yasyām-agre karaṃ dattvā yojyate nakha-lāṅgalam.*
123. On *Manu*, II, 112.
124. *Abhidhānaratnamālā*, 158–159, *Vaijayantī*, p. 124, 11. 34–36.
125. Cf. Mitra, *Early Rulers of Khajuraho*, p. 165.
126. Cf. Choudhury, *Hist. of Assam*, p. 298.
127. *Abhidhānaratnamālā*, p. 160; *Vaijayntī*, p. 38, 1–89.
128. *Vaijayantī*, p. 124, 11. 37–41.
129. Majumdar, G.P., *Vanaspati*, pp. 212–13.
130. Cf. *Deśīnāmamālā*, VI, 63 for consumed land—*Polaccā kheṭitābhūmiḥ.*
131. p. 39, I. 17—*Hālikaḥ kṣetraṃ pasayati.*
132. v. 111—*Vinā sāreṇa yaddānyaṃ varddhate phala-varjitam.*
133. v. 109–111.
134. *Agriculture and agriculturists in ancient India*, pp. 58–59.
135. *Vaijayantī*, p. 10, 1. 40; p. 231, 1. 35.
136. *Vaijayantī*, p. 91, 1. 193.
137. *Ibid.*, p. 187, 1. 25.
138. II, 48.
139. VI, 61.
140. VI, 84.
141. VI, 147.
142. p. 6.
143. In a verse from the *Subhāṣitaratnakoṣa* (v. 302), we read of the womenfolk in the hamlet enjoying the fire of dried cowdung (cf. *ibid.*, v. 307). A very clear proof of cowdung being used as fuel is furnished by Bāṇa—a little earlier than the period of our study—who describes the houses of his kindred as filled with heaps of cowdung and fuel. *Harṣacarita*, p. 33—*Indhana-gomaya-piṇḍa–kūṭa–saṅkaṭāni.*
144. *Ibid.*, p. 202—*Purāṇa-pāṃsūtkira-karīṣa-kūta-vāhinīnāṃ... Śakaṭa-śreṇīāṃ sampātaiḥ saṃpādyamāna-durbalorvī-virūkṣa-kṣetrasaṃskāram.*
145. Dasgupta, T.C., *Aspects of Bengali Society*, pp. 236-39.
146. vv. 166–167.
147. On *Manu*—VIII, 243–*Nidānam-ayogya-bīja-vāpaḥ.*
148. On *Manu*, IX, 37—*Tadapi hi punarupabhuktaśeṣam-upyamānam-aparasminvatsare... nānuvartate.*
149. p. 82, I. 21—*Dagdhānāmiva bījānāṃ punarjanma na vidyate.*
150. On XII, 181, 7.
151. On II, 77, 13.
152. vv. 168–181.

153. vv. 177–181.
154. VI, 126.
155. *On Manu,* VIII, 243—*Akāle vāpanaṃ.*
156. vv. 168–76. Cf. *Agni Purāṇa,* CXXI, 49.
157. Majumdar, G.P., *Vanaspati,* pp. 214–5, Cf. *I.H.Q.*, VII, 23–24.
158. *Kṛtyakalpatru, Gārhasthyakāṇḍa,* p. 256—*Svakṣetrabīja guṇasya vettā syāt.*
159. Medhātithi on *Manu,* IX, 330-*Idaṃ bījamasmin kṣetra prarohatīdaṃ na.*
160. p. 124, 11. 42–43.
161. *E.I.*, no. 5 (I and II)—*Vrīhipiṭakavāpaṃ kṣetram* and *Vrīhidaśaprathavāpaṃ kṣetram; E.I.*, no. 16 (A)—*Dvādaśakhaṇḍīka-kodravabīja-saṃsthānaṃ kṣetram; E.I.*, XVIII, no. 6—*Viṃśati-khaṇḍikāvrīthi-biji-parimāṇam kṣetram; E.I.*, X, no. 11—*Vāpagatya koraḍe sārddha-droṇa-sapta-parikalitā.*
162. Medhātithi on Manu IX, 330—*Idaṃ bījaṃ vistṛtamupyat idaṃ saṃhatamupyata itthenāmuptiṃ viṃdyāt.* Cf. *Kṛtyakalpataru, Gārhasthyakāṇḍa,* p. 256—*'Bijānāmuptivit' samyak sasyotpatyanukūlaghanavirala-bījavapanavettā.*
163. Cf. *Vāsavadattā* (Tr. by Gray), pp. 100, 120.
164. Cf. Medhātithi on *Manu,* IX, 36—*Vapanakāle upapādite kṛṣṭasamīkarṇādinā saṃskṛte.*
165. *Vaijayantī,* p. 124, 1, 46.
166. Medhātithi On *Manu,* II, 112—*Śubhaṃ śreṣṭham vrīhyādikaṃ lāṅgalā dinopyate.*
167. *T.S.*, v, 1, 7, 3; *Gobhila* G.S., 1, 4, 29; 29 *Artha,* II, 24; *Mbh.*, XII, 100, 10–11; *Bṛhatsaṃhitā,* XXXIX, 13–14; Megasthenes (McCrindle), pp. 32, 55.
168. *T.S.*, VII, 2, 10, 2; *Artha,* IX, 1; Pāṇini, IV, 3, 44–46.
169. *I.H.Q.*, VII, p. 24.
170. p. 6—*Tathā varṣeṣu karṣanād bhūguṇākṣayah Ekasyāṃ guṇahīnāyāṃ kṛṣimanyatra kārayet.*
171. *Abhidhānaratnamālā,* v. 158; *Vaijayantī,* p. 24, 1.36.
172. *E.I.*, XX, 13, Dabok inscription dated AD 813.
173. vv. 119–120. Cf. Devala q. in the *Jyotiṣatattva* (*Smṛtitattva,* I, p. 687)—*Halādibhirdṛḍhaihḥ kṣemaṃ kudṛḍhairna śubhaiṃ vadet. Bṛhat-Parāśara*(Jivananda, II, p. 107), likewise, after giving the measurements of a plough, observes that it should not be in any way less than what has been mentioned.
174. vv. 111–20.—Cf. the descriptions of *Mānasāra,* V, 56–77 and *Bṛhat-Parāśara* (Jivananda, II, p. 107), which differ in the terms used and the measurements of the parts. Either these were earlier or were not the product of a man like the author of the *Kṛṣiparāśara* who had an intimate knowledge of agricultural details.
175. *Karṇasamānakam.* It may also mean 'of the shape of an ear'.
176. The distance between the elbow and the tip of the little finger.
177. Monier-Williams explains it is 'a strap of leather on a plough'.
178. Monier-Williams equates one *vitasti* with 12 *aṅgulas.*
179. *Abhidhānaratnamālā,* vv. 575, 577; *Vaijayantī,* p. 125, 11. 53–56, 58.
180. *Abhidhānaratnamālā,* v, 575; *Vaijayantī,* p. 125, 1–55.
181. III, 41.
182. Cf. T.C. Dasgupta, *Aspects of Bengali Society,* p. 230 for names of agricultural implements in early Bengali literature.

183. *Vaijayantī*, p. 125, 11. 56–60; *Abhidhānaratnamālā*, vv. 576, 577.
184. *Prabandhacintāmaṇi*, p. 108, 11. 1–2; *ibid.*, p. 109, 1. 16. Cf. Medhātithi on *Manu*, IX, 293.
185. vv. 121–56. *Halaprasāraṇaṃ naiva kṛtvā yaḥ karṣaṇaṃ caret. Kevalaṃ baladarpeṇa sa karoti kṛṣiṃ vṛthā*—v. 153. *Bṛhat-Parāśara*, Ch. III (Jivananda, II, pp. 107–108) describes the ceremony in detail under the name *halasya yojanam.*
186. CCLIX, 43–47.
187. *Gārhasthyakāṇḍa*, p. 195—*'Dvihalakṛte' punaḥ karṣaṇāya. Ibid.*, p. 182—*Taddhastāttrir-upahatāni svakṛṣtāni.*
188. *Vaijayantī*, p. 124, 11. 43–45; *Abhidhānaratnamala.* v. 576.
189. vv. 142–43.
190. Majumdar, G.P., *op.cit.*, p. 213.
191. vv. 154–55.
192. vv. 121–30. Cf *Agni Purāṇa*, CXXI, 46–48.
193. Majumdar, G.P., *op.cit.*, p. 213.
194. vv. 144–51.
195. Cf. *Samarāṅgaṇasūtradhāra*, I, p. 33, vv. 63–64—significance of wood, brick, stone or bone coming out of earth during ploughing.
196. Cf. *Bṛhat–Parāśara*, Ch, III (Jivasanda, II, p. 102)—*Kugavairna kṛṣiṃ kuryāt.*
197. vv. 84–108. Cf. *Bṛhat–Parāśara*, (Jivananda, II) Ch. III.
198. Ch. III (Jivananda, II, pp. 110, 105).
199. v. 85.
200. Cf. *Kṛtyakalpataru*, *'Gārhasthyakāṇḍa*, p. 192.
201. v. 96. Cf. *Parāśarasaṃhitā*, II, 3–4.
202. vv. 218–219.
203. Cf. *Abhidhānaratnamāla*, v. 283.
204. vv. 134, 141.
205. Ch. III (Jivananda, II, pp. 101 f, 110). Cf. *Parāśara-saṃāhitā*, IV, 2.
206. *Ibid.*, p. 110.
207. v. 86.
208. vv. 105–8, Cf. *Agni Purāṇa*, CXXI, 44–45.
209. vv. 87–95.
210. vv. 99–104.
211. Ch. III (Jivananda, II, p. 106).
212. v. 182. Medhātithi on *Manu*, IX, 36 suggest that after ploughing and before sowing, the field was prepared by levelling it. *Kāle varṣādau vapanakāla upapādite kṛsta-samī-karaṇādiva saṃskṛte.*
213. vv. 189–92.
214. On *Manu*, VII, 110.
215. Cf. *Uktivyaktiprakraṇa*, p. 46, 1.30.
216. Ch. III (Jivananda, II, p. 112).
217. p. 77, 11. 19–20.
218. I, 20.

219. III, 55.
220. V. 264.
221. *Deśīnāmamālā*, VI, 26.
222. *Ibid.*, III, 32. In the *Vaijayantī*, p. 161, 1. 54, the term for a field hut is *gargarī*.
223. V. 285.
224. VIII, 69.
225. VIII, 76.
226. *Deśīnāmamālā*, VIII, 69.
227. vv. 3117–3018.
228. pp. 48f. Cf. *J.B.O.R.S.*, III, p. 562 for a similar practice in vogue in Bihar with the difference that here the notice is addressed directly to the insects.
229. vv. 198–205.
230. CXXI, 50 Cf. Medhātithi on *Manu*, VIII, 243.
231. Majumdar, G.P., *Vanaspati*, pp. 215–16.
232. vv. 206–213.
233. *Abhidhānarotnamālā*, v. 578; *Vaijayantī*, p. 125, 1.61.
234. Ed. R.C. Majumdar, R.G., Basak and N.G. Banerji, *Kavipraśasti*, 13.
235. vv. 214–220.
236. *Deśīnāmamālā*, VI, 40; III, 37; cf. *ibid.*, VI, 34.
237. *Vaijananti*, p. 164, 11. 130–131, Cf. *Deśīnāmamālā*, I, 103; *Vaijayantī*, p. 193, 1.42.
238. *Ibid.*, p. 164, 11. 132–135.
239. I, 26, 32, 74; II, 56; III, 11; V, 23; VI, 15; VII, 94.
240. II, I.
241. *Vaijayantī*, p. 164, 1. 129.
242. *Deśīnāmamālā*, III, 10.
243. *Vaijayantī*, p. 161, 1.51.
244. v. 237.
245. vv. 241–42, Cf. *Agni purāṇa*, CXXI, 51.
246. p. 60.
247. CXXI, 52–54.
248. p. 164, 11. 127–28.

APPENDIX–1

Contributions of Asian Agri-History Foundation

Asian Agri-History Foundation, Secunderabad—a non-profit trust was established by a group of agricultural scientists led by Professor Y.L. Nene in 1994. The major objectives of the foundation are: (i) to disseminate information on the history of agriculture in the South and South-east Asia region and (ii) to stimulate interest in research on the history of agriculture in Asia with a special focus on South and South-east Asia.[1] The Foundation expects to learn from the traditional wisdom and indigenous time-tested technologies. This may provide clues for (i) understanding how farmers adjusted to changing environment in the past, and (ii) developing appropriate technologies leading to prosperous sustainable agriculture.[2] Its activities include (i) publication of quarterly international journal—*Asian Agri-History* in English, (ii) publication of translations of old manuscripts/ papers into English and Hindi, (iii) encouragement and support to agricultural research in universities and other institutions, (iv) organization of seminars, conferences and lectures for interaction among scholars, (v) establishment of a database Library of books/ papers on history of Asian agriculture. So far it has published *Asian-Agri-History Journal* regularly since March 1997. It has published translations of following manuscripts:

1. Surapala's *Vrkhscayurveda* (Bodleian manuscript U.K.) (1000 AD) (The Science of Plant Life) with text in Sanskrit, translation in English by Dr. Naleni Sadhale and commentary by Drs. Y.L. Nene, K.L. Mehra and S.M. Virmani, 1996.
2. *Krishi Parashara* (Agriculture by Parashara) with text in Sanskrit and translation in English by Dr. Nalini Sadhale and comments by Drs. H.V. Balkundi and Y.L. Nene, 1999.
3. *Nuskha Dar Fanni-Falahat* (The Art of Agriculture) (17th century) of Dara Shikoh with text in Persian and translation by Dr. Rajzea Akbar and comments by Drs. K.L. Mehra, J.S. Kanwar, K.L. Chadha and Y.L. Nene, 2000.
4. *Kashyapiyakrishisukti* (A Treatise on Agriculture by Kashyapa) (800 AD) (Adyar Library manuscript) with text in Sanskrit and translation into English by Dr. S.M. Ayachit with two commentaries by Dr. Nalini Sadhale, and the other by Dr. Y.L. Nene (2002).
5. Vishvavallabha (Dear to the world: The Science of Plant Life) by Chakrapani Mishra (1577 AD) (Rajasthan Prachya Vidya Pratisthan, Jodhpur manuscript) with text in Sanskrit, translation into English by Dr. Nalini Sadhale and comments by Dr. N. Sadhale and Dr. Y.L. Nene, 2004.

The Trust has provided financial assistance to researchers in universities/institutions for Vedic and Indian concepts of agriculture. The Trust has organized many seminars/ conferences/summer schools including summer school on Ancient and Medieval History of Indian Agriculture and its relevance to sustainable agriculture in the 21st century (1999) supported by ICAR, (its proceeding have been published): II organized a seminar on Vedic Agriculture (2001), National Conference on Agricultural Heritage of India, Udaipur (2002), National Seminar on Cow in Agricultural and Human Health supported by CSIR, New Delhi and PI Industries, Udaipur, 2003; International Conference on Agricultural Heritage of Asia (2004) in collaboration with National Academy of Agricultural Research Management and Acharya N.G. Ranga Agricultural University. It was supported by ICAR and some industries. It has organized an International Conference in December 2005. The Trust has been successful to a large extent in disseminating knowledge of traditional Asian agriculture and its historical development and some industries have adopted the knowledge in their products. It has a very relevant focus in the sense that ancient wisdom and modern technologies have to be harmonized as far as possible for sustainable development for food security and prosperity of India.

REFERENCES

1. For details see *Asian Agri-History Foundation: Activities and Achievements during the Food Decade* (1994–2004), V.C. Agarwal, S.L. Choudhary and S.P.S. Bansal, 2004.
2. www.agri-history.org.

APPENDIX–2

Contributions of Dr. M.S. Swaminathan and His Research Foundation

Dr. M.S. Swaminathan (b. 1925), an agricultural scientist, a plant geneticist, an environmentalist, an ecologist, an administrator, an educator, an advisor to governments around the world, a social worker, a philosopher and a visionary has innumerable contributions to the development of agriculture.[1] His dream is to rid the world of hunger and poverty and let mankind live in harmony with nature. He is former Director-General, Indian Council of Agricultural Research and Secretary to the Government of India, Department of Agricultural Research and Education (1972–79), Principal Secretary to the Government of India, Ministry of Agriculture and Irrigation (1979–80), Member, Planning Commission (1980–82), Director-General, International Rice Research Institute, Philippines (1982–88), Chairman, M.S. Swaminathan Research Foundation (1989 onwards) and UNESCO Chair on Eco-Technology (1994-onwards). He is recipient of Honorary D.Sc. degree from many universities in India and abroad. He has held prestigious honorary positions in International Committees/organizations and also in bodies devoted to Nature Conservation and Sustainable Development. He has received awards from scientific bodies of India and abroad. He has been awarded Padma Shri (1967), Padma Bhushan (1972) and Padma Vibhushan (1989). He played an important role in ushering in the Green Revolution (I) (Wheat) in India along with his team under Dr. Norman E. Borlaug. The use of a crossbreed wheat seed partly Japanese and partly Mexican developed by Dr. Borlaug which was modified to amber colour by scientific treatment to suit Indian taste led to increase in wheat production in late 1960s due to the team of Dr. Borlaug in which Dr. Swaminathan played significant role. Dr. Swaminathan's career is a legendary story of continuous achievements and contributions.

M.S. Swaminathan Research Foundation, an autonomous non-profit trust was started in 1988 at New Delhi but it was shifted to Chennai as a registered society in Tamil Nadu in 1990. The major scientific contributions of Dr. M.S. Swaminathan and his Research Foundation are as follows:

His contributions to education and extension include:

(i) the introduction of the concept of 'Techniracy'—concept of imparting training in the latest technical skill entirely through work experience.

(ii) fostered the establishment of a chain of Krishi Vigyan Kendras (Farm Science Centres) by ICAR.

(iii) initiated the lab to land programme and whole village operational research projects under ICAR.

The M.S. Swaminathan Research Foundation has a history of progress in research, institutional development, organization of conferences, seminars, dialogues and publications, partnership. It is not possible to do justice to the immense contributors of Dr. Swaminathan and his veritable dynamic Research Foundations in an appendix or even an article. For details in nutshell some select material mentioned below may be useful.[2] The Foundation's final objective of all programmes and partnerships is to promote an ever green revolution in agriculture and to create more income and livelihood opportunities[3] for rural families in an economically, and socially (gender equality) sustainable manner.

REFERENCES

1. For his biography see Gita Gopal Krishnan, *M.S. Swaminathan, one man's quest for a hunger-free world,* Chennai, 2002.
2. See M.S. Swaminathan, *Rio de Janerio to Johannesburg,* East West Books (Madras) Pvt. Ltd., Chennai, 2002.
3. See *M.S. Swaminathan Research Foundation,* 2003–2004, Chennai, *Towards a Knowledge Revolution in Rural India,* Chennai, 2003.
4. *A social vision for Science: The History of M.S. Swaminathan Research Foundation* (1990–2000), Chennai, 2000. *Towards an Era of Eco-Jobs and Eco-Entrepreneurship: Challenges and Opportunities,* Chennai, 2002, *Eco-villages,* Chennai, 1991.
5. World Wide Web at http://www.mssrp.org.

INDEX

* Part 4	Cultural Foundations of Mathematics: The Nature of Mathematical Proof and the Transmission of the Calculus from India to Europe in the 16th c. CE	C.K. Raju
* Part 5	Development of Modern Indian Thought and the Social Sciences	S. Bhattacharya (ed.)
* Part 6	Aspects of India's International Relations: South Asia and the World	Jayant Kumar Ray (ed.)
* Part 7	Political Ideas in Modern India: Thematic Explorations	V.R. Mehta (ed.)

VOLUME XI Consciousness, Science, Society, Value and Yoga (CONSSAVY)

* Part 1	Philosophical Consciousness and Scientific Knowledge: Conceptual Linkages and Civilizational Background	D.P. Chattopadhyaya (ed.)
* Part 2	Self, Society and Science: Theoretical and Historical Perspectives	D.P. Chattopadhyaya (ed.)
* Part 3	Consciousness, Indian Psychology and Yoga	Kireet Joshi & Matthijs Cornelissen (eds.)
* Part 4	The Enworlded Subjectivity: Its Three Worlds and Beyond	R. Balasubramanian (ed.)

VOLUME XII Levels of Reality (CONSSAVY)

† Part 1	A Conceptual-Analytic Study of Indian Philosophy of Morals	Rajendra Prasad (eds.)
† Part 2	A Historical-Developmental Study of Indian Philosophy of Morals	Rajendra Prasad (ed.)
* Part 3	Vedic Vision of Consciousness and Reality	S.P. Singh
† Part 4	Modern Atomism in Europe and India	J. Pasupathy (ed.)
† Part 5	Materialism and Immaterialism in India and Europe	Partha Ghose (ed.)
† Part 6	Life and Organicism	N.S. Rangaswamy (ed.)

VOLUME XIII Theories of Natural and Life Sciences (CONSSAVY)

† Part 1	India in the World of Physics: Then and Now	A.N. Mitra (ed.)
† Part 2	From Physiology and Chemistry to Biochemistry	D.P. Burma (ed.)
† Part 3	Psychology and Psychoanalysis	Girishwar Mishra (ed.)
† Part 4	Cognitive Sciences: Brain, Psychology and Computer	Shobini L. Rao (ed.)
† Part 5	Foundations of Sciences	B.V. Sreekantan (ed.)
† Part 6	History of Science and Philosophy of Science	P.K. Sengupta (ed.)
† Part 7	History of Science and Philosophy of Science	P.K. Sengupta (ed.)
† Part 8	Science in India	J.V. Narlihar (ed.)

VOLUME XIV Natural and Cultural Sciences (CONSSAVY)

† Part 1	Relations between Nature and Culture	Roddam Narasimha (ed.)
† Part 2	Social Sciences: Communication, Anthropology and Sociology	Yogendra Singh (ed.)
† Part 3	Economics and Ethics	Ashok Guha (ed.)
† Part 4	Different Types of History	Bharati Ray (ed.)
† Part 5	History of India's Polity, Governance and Constitutional Culture	Subhash C. Kashyap

VOLUME XV Science, Technology and Philosophy (CONSSAVY)

* Part 1	Science, Technology, Imperialism and War	J.B. Dasgupta (ed.)
† Part 2	Science and the Public	Ashok Jain (ed.)
† Part 3	Science, Literature and Aesthetics	Amiya Dev (ed.)
† Part 4	Science, Colonialism and Nationalism c. 1820 to c. 1940	Uma Das Gupta (ed.)
† Part 5	Old Wisdom and the New Horizon	Manoj Kumar Pal

VOLUME XVI Yoga (CONSSAVY)

† Part 1	Cognitve Anomalies, Consciousness and Yoga	K. Ramakrishna Rao
† Part 2	History of Yoga in India	S.P. Singh (ed.)
† Part 3	Varieties of Yogic Experience	Manoj Das (ed.)
† Part 4	Synthesis of Yoga	Kireet Joshi (ed.)
‡ Part 5	Yoga, Evolution and Mutation of Human Species	(Under Plan)

Monographs

* 1	Science, Philosophy and Culture in Historical Perspective	D.P. Chattopadhyaya & Ravinder Kumar (eds.)
* 2	Some Aspects of India's Philosophical & Scientific Heritage	D.P. Chattopadhyaya & Ravinder Kumar (eds.)
* 3	Mathematics, Astronomy and Biology in Indian Tradition: Some Conceptual Preliminaries	D.P. Chattopadhyaya & Ravinder Kumar (eds.)
* 4	Language, Logic and Science in India: Some Conceptual and Historical Perspectives	D.P. Chattopadhyaya & Ravinder Kumar (ed.)
* 5	Primal Spirituality of the Vedas: Its Renewal and Renaissance	R. Balasubramanian
* 6	Interdisciplinary Studies in Science, Technology, Philosophy and Culture	D.P. Chattopadhyaya
* 7	Ancient Yoga and Modern Science	T.R. Anantharaman
* 8	Prolegomena to Any Future Historiography of Cultures & Civilizations	Daya Krishna
* 9	Science and Spirituality: A Quantum Integration	Amit Goswami & Maggie Goswami
* 10	On Rational Historiography	V. Shekhawat
* 11	Kauṭilīya Arthaśāstra Revisited	Surendra Nath Mittal
* 12	Ways of Understanding the Human Past	D.P. Chattopadhyaya
* 13	The Architecture of Knowledge	Subhash Kak
* 14	Karnataka Music as Aesthetic Form	R. Sathyanarayana
* 15	Philosophy of Science: Some Perspectives from Indian Philosophical Traditions	Sundar Sarukkai

---

* Already Published
† In the Process of Publication
‡ Under Plan